Fractional Differential Equations, Inclusions and Inequalities with Applications

Fractional Differential Equations, Inclusions and Inequalities with Applications

Editor

Sotiris K. Ntouyas

MDPI • Basel • Beijing • Wuhan • Barcelona • Belgrade • Manchester • Tokyo • Cluj • Tianjin

Editor
Sotiris K. Ntouyas
University of Ioannina
Greece

Editorial Office
MDPI
St. Alban-Anlage 66
4052 Basel, Switzerland

This is a reprint of articles from the Special Issue published online in the open access journal *Mathematics* (ISSN 2227-7390) (available at: https://www.mdpi.com/journal/mathematics/special_issues/Fractional_Differential_Equations_Inclusions_Inequalities_Applications).

For citation purposes, cite each article independently as indicated on the article page online and as indicated below:

LastName, A.A.; LastName, B.B.; LastName, C.C. Article Title. *Journal Name* **Year**, *Article Number*, Page Range.

ISBN 978-3-03943-218-9 (Hbk)
ISBN 978-3-03943-219-6 (PDF)

Contents

About the Editor

Sotiris K. Ntouya is Professor Emeritus in the Department of Mathematics of the University of Ioannina, Greece. He received his B.S. and Ph.D. from the University of Ioannina in 1972 and 1980, respectively. His research interests include initial and boundary value problems for differential equations (ordinary, functional, with deviating arguments, neutral, partial, integrodifferential, inclusions, impulsive, fuzzy, stochastic, fractional), inequalities, asymptotic behavior and controllability. He has contributed to more than 640 papers that have been published in refereed journals. He is the co-author of the books *Impulsive differential equations and inclusions, Controllability for semilinear functional differential equations and inclusions, Quantum Calculus: New Concepts, Impulsive IVPs and BVPs, Inequalities and Hadamard-Type Fractional Differential Equations, and Inclusions and Inequalities*. He is a member of the Editorial Boards of 21 international journals and a reviewer for many international journals. He appears in the 2018 list of Highly Cited Researchers, published by Clarivate Analytics.

Preface to "Fractional Differential Equations, Inclusions and Inequalities with Applications "

In the past forty years, fractional calculus and its applications have gained significant importance, mainly because it has become a powerful tool with more accurate and successful results in modeling several complex phenomena in various fields of science and engineering. Fractional derivatives and integrals provide a much better tool for the description of memory and hereditary properties of various materials and processes than integer derivatives.

It is our great pleasure to publish this book. This selection of 33 papers focuses on recent developments in the area of fractional differential equations, inclusions, and inequalities. All contents were peer-reviewed by multiple referees and published as papers in the Special Issue "Fractional Differential Equations, Inclusions and Inequalities with Applications" in the journal *Mathematics*. They provide new and interesting results in different branches of fractional differential equations so that the readers will be able to obtain the latest developments in the fields of fractional differential equations and inequalities. We would like to thank the editors for their kind support on the publication of this Special Issue. We also wish to express our appreciation to the authors of all articles in this Special Issue for their excellent contributions as well as the reviewers for their work on analyzing the manuscripts.

Sotiris K. Ntouyas
Editor

mathematics

MDPI

Article

Nonlocal Fractional Evolution Inclusions of Order $\alpha \in (1,2)$

Jia Wei He [1], Yong Liang [2], Bashir Ahmad [3] and Yong Zhou [1,3,*]

[1] Faculty of Mathematics and Computational Science, Xiangtan University, Xiangtan 411105, China; hjw.haoye@163.com
[2] Faculty of Information Technology, Macau University of Science and Technology, Macau 999078, China; yliang@must.edu.mo
[3] Nonlinear Analysis and Applied Mathematics (NAAM) Research Group, Faculty of Science, King Abdulaziz University, Jeddah 21589, Saudi Arabia; bashirahmad_qau@yahoo.com
* Correspondence: yzhou@xtu.edu.cn

Received: 26 January 2019; Accepted: 14 February 2019; Published: 24 February 2019

Abstract: This paper studies the existence of mild solutions and the compactness of a set of mild solutions to a nonlocal problem of fractional evolution inclusions of order $\alpha \in (1,2)$. The main tools of our study include the concepts of fractional calculus, multivalued analysis, the cosine family, method of measure of noncompactness, and fixed-point theorem. As an application, we apply the obtained results to a control problem.

Keywords: fractional evolution inclusions; mild solutions; condensing multivalued map

MSC: 26A33; 34G25; 47D09

---

1. Introduction

In the past several decades, there has been a significant development in the theory and applications for fractional evolution equations and inclusions; for example, see the monographs by Miller and Ross [1], Podlubny [2], Kilbas et al. [3], Zhou [4], and the recent papers [5,6]. More recently, time-fractional diffusion and wave equations have been attracting the widespread attention of many fields of science and engineering [7,8]. The interest in the study of these topics arises from the fact that fractional diffusion equations $\alpha \in (0,1)$ or fractional wave equations $\alpha \in (1,2)$ can capture some nonlocal aspects of phenomena or systems. Examples of these phenomena include porous media, memory effects, anomalous diffusion, viscoelastic media, and so on. The papers [9–11] cover many of these applications.

By virtue of semigroup theory and the operator theoretical method, some fractional diffusion and wave equations can be abstracted as fractional evolution equations. Bajlekova [12] exploited the concept of the fractional resolvent solution operator to investigate the associated fractional abstract Cauchy problem. A number of papers [13–17] and the references therein were inspired by this concept, and the topic of the existence of mild solutions to fractional abstract equations of order $\alpha \in (1,2)$ was also studied. For further discussion in [18], the authors considered the controllability results for fractional evolution equations of order $\alpha \in (1,2)$ by applying the concepts of Mainardi's Wright function (a probability density function) and strongly continuous cosine families.

The study of fractional evolution inclusions of order $\alpha \in (0,1)$ also gained significant importance (see, e.g., [19,20]). However, the study of fractional evolution inclusions of order $\alpha \in (1,2)$ supplemented with nonlocal conditions is yet to be initiated. We need to point out that the work spaces are of finite dimension if the strongly continuous cosine families are compact (see, e.g., [21,22]). Motivated by this fact and the above-mentioned works and relying on the known material, we aim to develop a suitable definition for mild solutions of fractional evolution equations in terms of Mainardi's Wright function. For this purpose, we consider the following nonlocal problem of fractional evolution inclusions without further assumptions regarding the compactness of the cosine families or the associated sine families.

$$\begin{cases} {}^{C}D_t^{\alpha}x(t) \in Ax(t) + F(t, x(t)), & t \in J = [0, a],\ a > 0, \\ x(0) + g(x) = x_0,\ x'(0) = x_1, \end{cases} \tag{1}$$

where ${}^{C}D_t^{\alpha}$ is a Caputo fractional derivative of order $1 < \alpha < 2$; A is the infinitesimal generator of a strongly continuous cosine family $\{C(t)\}_{t\geq 0}$ of uniformly bounded linear operators in a Banach space X; $F : [0, a] \times X \to X$ is a multivalued map; g is a given appropriate function; and x_0, x_1 are elements of space X.

Here, we emphasize that the present work is also motivated by an inclusion of the following partial differential model:

$$\begin{cases} \partial_t^{\alpha}u(t,z) \in \partial_z^2 u(t,z) + F(t, z, u(t,z)), & z \in [0,\pi],\ t \in [0,a], \\ u(t,0) = u(t,\pi) = 0, & t \in [0,a], \\ u(0,z) + g(u) = u_0(z),\ u'(0,z) = u_1(z), & z \in [0,\pi], \end{cases}$$

where ∂_t^{α} is a Caputo fractional partial derivative. This model includes a class of fractional wave equations that have a memory effect and are not observed in integer-order differential equations; further, this class of equations indicates the coexistence of finite wave speed and absence of a wavefront (see, e.g., [9]). It is interesting that for the case of $\alpha = 2$, the above fractional partial differential inclusion reduces to a second-order differential inclusion involving one-dimensional wave equations with nonlocal initial-boundary conditions. For the case of $\alpha = 1$ or $\alpha \in (0,1)$ with the initial value $u_1(z)$ vanished, the model contains the classical diffusion equations or fractional diffusion equations. In addition, these types of equations can be handled by the method of semigroup theory (see, e.g., [20]) but not cosine families.

The rest of this paper is organized as follows. In Section 2, we recall some preliminary concepts related to our study. In Section 3, we establish an existence result for mild solutions of Equation (1) and discuss the compactness of the set of mild solutions. In Section 4, we show the utility of the obtained work by applying it to a control problem.

2. Preliminaries

Let X be a Banach space with the norm $\|\cdot\|$. Denote by $\mathcal{L}(X)$ the space of all bounded linear operators from X to X equipped with the norm $\|\cdot\|_{\mathcal{L}(X)}$. Let $C(J, X)$ denote the space of all continuous functions from J into X equipped with the usual sup-norm $\|x\|_C = \sup_{t\in J}\|x(t)\|$, where $J = [0, a], a > 0$. A measurable function $f : J \to X$ is Bochner integrable if $\|f\|$ is Lebesgue integrable. Let $L^p(J, X)$ $(p \geq 1)$ be the Banach space of measurable functions (defined in the sense of Bochner integral) endowed with the norm

$$\|f\|_p = \left(\int_J \|f(t)\|^p dt\right)^{\frac{1}{p}}.$$

Definition 1. *The fractional integral with the lower limit zero for a function $u : [0, \infty) \to X$ is given by*

$$I_{0+}^{\alpha}u(t) = \frac{1}{\Gamma(\alpha)}\int_0^t (t-s)^{\alpha-1}u(s)ds, \quad t > 0,\ \alpha \in \mathbb{R}_+,$$

provided the right side is point-wise defined on $[0, \infty)$, where $\Gamma(\cdot)$ is the gamma function.

Definition 2. *The Riemann–Liouville derivative with the lower limit zero for a function* $u : [0, \infty) \to X$ *is defined by*

$$^{L}D_{0+}^{\alpha}u(t) = \frac{1}{\Gamma(n-\alpha)}\frac{d^n}{dt^n}\int_0^t (t-s)^{n-\alpha-1}u(s)ds, \quad t>0,\ n-1<\alpha<n,\ \alpha \in \mathbb{R}_+.$$

Definition 3. *The Caputo derivative with the lower limit zero for a function u is defined by*

$$^{C}D_{0+}^{\alpha}u(t) = {}^{L}D_{0+}^{\alpha}\Big(u(t) - \sum_{k=0}^{n-1}\frac{u^{(k)}(0)}{k!}t^k\Big), \quad t>0,\ n-1<\alpha<n, \alpha \in \mathbb{R}_+.$$

Definition 4. [23] *A family of bounded linear operators* $\{C(t)\}_{t\in\mathbb{R}}$ *mapping the Banach space X into itself is called a strongly continuous cosine family if and only if* $C(0) = I$, $C(s+t) + C(s-t) = 2C(s)C(t)$ *for all* $s, t \in \mathbb{R}$, *and the map* $t \mapsto C(t)x$ *is strongly continuous for each* $x \in X$.

Let $\{S(t)\}_{t\in\mathbb{R}}$ denote the strongly continuous sine families associated with the strongly continuous cosine families $\{C(t)\}_{t\in\mathbb{R}}$, where

$$S(t)x = \int_0^t C(s)x ds, \quad x \in X,\ t \in \mathbb{R}. \tag{2}$$

In addition, an operator A is said to be an infinitesimal generator of cosine families $\{C(t)\}_{t\in\mathbb{R}}$ if

$$Ax = \frac{d^2}{dt^2}C(t)x\Big|_{t=0}, \quad \text{for all } x \in \mathcal{D}(A),$$

where the domain of A is given by $\mathcal{D}(A) = \{x \in X : C(t)x \in C^2(\mathbb{R}, X)\}$.

A multivalued map G is called upper semicontinuous (u.s.c.) on X if, for each $x_* \in X$, the set $G(x_*)$ is a nonempty subset of X, and for every open set $B \subseteq X$ such that $G(x_*) \subset B$, there exists a neighborhood V of x_* with the property that $G(V(x_*)) \subset B$. G is convex-valued if $G(x)$ is convex for all $x \in X$. G is closed if its graph $\Gamma_G = \{(x, y) \in X \times X : y \in G(x)\}$ is a closed subset of the space $X \times X$. The map G is bounded if $G(B)$ is bounded in X for every bounded set $B \subseteq X$. We say that G is completely continuous if $G(B)$ is relatively compact for every bounded subset B of X. Furthermore, if G is completely continuous with nonempty values, then G is u.s.c. if and only if G has a closed graph. If there exists an element $x \in X$ such that $x \in G(x)$, then G has a fixed point.

Let B be a subset of X. Then, we define

$$\begin{aligned}
&\mathcal{P}(X) = \{B \subseteq X : B \text{ is nonempty}\}, \quad \mathcal{P}_{cv}(X) = \{B \in \mathcal{P}(X) : B \text{ is convex}\},\\
&\mathcal{P}_{cl}(X) = \{B \in \mathcal{P}(X) : B \text{ is closed}\}, \quad \mathcal{P}_{bd}(X) = \{B \in \mathcal{P}(X) : B \text{ is bounded}\},\\
&\mathcal{P}_{cp}(X) = \{B \in \mathcal{P}(X) : B \text{ is compact}\}, \quad \mathcal{P}_{cl,cv}(X) = \mathcal{P}_{cl}(X) \cap \mathcal{P}_{cv}(X).
\end{aligned}$$

In addition, let $co(B)$ be the convex hull of a subset B, and let $\overline{co}(B)$ be the closed convex hull in X. A multivalued map $G : J \to \mathcal{P}_{cl}(X)$ is said to be measurable if, for each $x \in X$, the function $Z : J \to X$ defined by $Z(t) = d(x, G(t)) = \inf\{\|x - z\| : z \in G(t)\}$ is Lebesgue measurable. Let $G : J \to \mathcal{P}(X)$. A single-valued map $f : J \to X$ is called a selection of G if $f(t) \in G(t)$ for every $t \in J$.

Definition 5. *A multivalued map* $F : J \times X \to \mathcal{P}(X)$ *is called* L^1*-Carathéodory if*

(i) *the map* $t \mapsto F(t, x)$ *is measurable for each* $x \in X$;
(ii) *the map* $u \mapsto F(t, x)$ *is upper semicontinuous on X for almost all* $t \in J$;
(iii) *for each positive real number r, there exists* $h_r \in L^1(J, \mathbb{R}_+)$ *such that*

$$\|F(t, x)\|_{\mathcal{P}(X)} = \sup\{\|v\| : v(t) \in F(t, x)\} \le h_r(t), \ for\ \|x\| \le r,\ for\ a.e.\ t \in J.$$

For every $\Omega \in \mathcal{P}(X)$, the Hausdorff measure of noncompactness (MNC) is defined by

$$\chi(\Omega) = \inf\{\varepsilon > 0 : \ \Omega \text{ has a finite } \varepsilon\text{-net}\},$$

and the Kuratowski MNC is defined by

$$\tau(\Omega) = \inf\{d > 0 : \ \Omega \subset \bigcup_{j=1}^{n} M_j \text{ and } \mathrm{diam}(M_j) \leq d\},$$

where the diameter of M_j is given by $\mathrm{diam}(M_j) = \sup\{\|x - y\| : x, y \in M_j\}$, $j = 1, \ldots, n$. The Hausdorff and Kuratowski MNCs are connected by the relations:

$$\chi(\Omega) \leq \tau(\Omega) \leq 2\chi(\Omega).$$

A measure of noncompactness χ (or τ) is called: monotone if $\Omega_1, \Omega_2 \in \mathcal{P}(X)$ with $\Omega_1 \subseteq \Omega_2$ implies $\chi(\Omega_1) \leq \chi(\Omega_2)$; nonsingular if $\chi(\{c\} \cup \Omega) = \chi(\Omega)$ for every $c \in X$, $\Omega \in \mathcal{P}(X)$; regular if $\chi(\Omega) = 0$ is equivalent to the relative compactness of Ω.

We now introduce the MNC ν as follows: for a bounded set $D \subset C(J, X)$, we define

$$\nu(D) = \max_{D \in \Theta(D)} \left(\sup_{t \in J} \chi(D(t)), \ \mathrm{mod}_C(D) \right),$$

where $\Theta(D)$ is the collection of all denumerable subsets of D and $\mathrm{mod}_C(D)$ is the modulus of equicontinuity of the set of functions D that have the following form

$$\mathrm{mod}_C(D) = \lim_{\delta \to 0} \sup_{x \in D} \max_{|t_2 - t_1| < \delta} \|x(t_2) - x(t_1)\|.$$

It is known that the MNC ν is monotone, nonsingular, and regular. For more details on the MNC, we refer to [24,25].

Lemma 1. ([24]). *Let $W \subset X$ be bounded. Then, for each $\varepsilon > 0$, there exists a sequence $\{x_n\}_{n=1}^{\infty} \subset W$ such that*

$$\chi(W) \leq 2\chi\left(\{x_n\}_{n=1}^{\infty}\right) + \varepsilon.$$

Lemma 2. ([26]). *Let χ_C be the Hausdorff MNC on $C(J, X)$, and let $W(t) = \{x(t) : x \in W\}$. If $W \subset C(J, X)$ is bounded, then for every $t \in J$,*

$$\chi(W(t)) \leq \chi_C(W).$$

Furthermore, if W is equicontinuous, then the map $t \mapsto \chi(W(t))$ is continuous on J and

$$\chi_C(W) = \sup_{t \in J} \chi(W(t)).$$

Lemma 3. ([26]). *Let $\{x_n\}_{n=1}^{\infty}$ be a sequence of Bochner integrable functions from J into X. If there exists a function $\rho(\cdot) \in L^1(J, \mathbb{R}_+)$ satisfying $\|x_n(t)\| \leq \rho(t)$ for almost all $t \in J$ and for every $n \geq 1$, then the function $\psi(t) = \chi(\{x_n(t)\}_{n=1}^{\infty}) \in L^1(J, \mathbb{R}_+)$ satisfies*

$$\chi\left(\left\{\int_0^t x_n(s)ds : \ n \geq 1\right\}\right) \leq 2\int_0^t \psi(s)ds.$$

Lemma 4. ([27, Lemma 4]). *Let $\{f_n\}_{n=1}^{\infty} \subset L^p(J, X)$ $(p \geq 1)$ be an integrable bounded sequence satisfying*

$$\chi(\{f_n\}_{n=1}^{\infty}) \leq \gamma(t), \quad \text{a.e., } t \in J,$$

where $\gamma(\cdot) \in L^1(J, \mathbb{R}_+)$. *Then, for each* $\epsilon > 0$, *there exists a compact* $K_\epsilon \subseteq X$, *a measurable set* $J_\epsilon \subset J$ *with measure less than* ϵ, *and a sequence of functions* $\{g_n^\epsilon\}_{n=1}^\infty \subset L^p(J, X)$ *such that* $\{g_n^\epsilon(t)\}_{n=1}^\infty \subseteq K_\epsilon$, *for* $t \in J$, *and*

$$\|f_n(t) - g_n^\epsilon(t)\| < 2\gamma(t) + \epsilon, \quad for\ each\ n \geq 1\ and\ for\ every\ t \in J - J_\epsilon.$$

Lemma 5. ([28]) *Let* χ *be the Hausdorff MNC on* X. *If* $\{W_n\}_{n=1}^\infty \subset X$ *is a nonempty decreasing closed sequence and* $\lim_{n\to\infty} \chi(W_n) = 0$, *then* $\bigcap_{n=1}^\infty W_n$ *is nonempty and compact.*

Definition 6. *Let* D *be a subset of a Banach space* X. *A multivalued function* $F : D \to \mathcal{P}(X)$ *is said to be* ν*-condensing if* $\nu(F(\Omega)) \ngeq \nu(\Omega)$ *for every bounded and not relatively compact set* $\Omega \subseteq D$.

Lemma 6. ([25, Corollary 3.3.1]). *Let* Ω *be a convex closed subset of a Banach space* X *and* ν *be a nonsingular MNC defined on subsets of* Ω. *If* $F : \Omega \to P_{cv,cp}(\Omega)$ *is a closed* ν*-condensing multivalued map, then* F *has a fixed point.*

Lemma 7. ([25, Proposition 3.5.1]). *Let* Ω *be a closed subset of a Banach space* X *and* $F : \Omega \to P_{cp}(X)$ *be a closed multivalued function that is* ν*-condensing on every bounded subset of* Ω, *where* ν *is a monotone MNC in* X. *If the set of fixed points of* F *is bounded, then it is compact.*

Throughout this paper, we suppose that A is the infinitesimal generator of a strongly continuous cosine family of uniformly bounded linear operators $\{C(t)\}_{t\geq 0}$ in a Banach space X: that is, there exists $M \geq 1$ such that $\|C(t)\|_{\mathcal{L}(X)} \leq M$ for $t \geq 0$. In the sequel, we always set $q = \frac{\alpha}{2}$ for $\alpha \in (1, 2)$.

As argued in [18], we define a mild solution of Equation (1) as follows.

Definition 7. *A function* $x \in C(J, X)$ *is said to be a mild solution of Equation* (1) *if* $x(0) + g(x) = x_0$, $x'(0) = x_1$ *and there exists* $f \in L^1(J, X)$ *such that* $f(t) \in F(t, x(t))$ *on a.e.* $t \in J$ *and*

$$x(t) = C_q(t)(x_0 - g(x)) + K_q(t)x_1 + \int_0^t (t-s)^{q-1} P_q(t-s) f(s) ds,$$

where

$$C_q(t) = \int_0^\infty M_q(\theta) C(t^q \theta) d\theta, \quad K_q(t) = \int_0^t C_q(s) ds, \quad P_q(t) = \int_0^\infty q\theta M_q(\theta) S(t^q \theta) d\theta,$$

$$M_q(\theta) = \frac{1}{q}\theta^{-1-\frac{1}{q}} \xi_q(\theta^{-\frac{1}{q}}), \; \xi_q(\theta) = \frac{1}{\pi}\sum_{n=1}^\infty (-1)^{n-1}\theta^{-nq-1}\frac{\Gamma(nq+1)}{n!}\sin(n\pi q), \theta \in (0, \infty),$$

and $M_q(\cdot)$ *is the Mainardi's Wright-type function defined on* $(0, \infty)$ *such that*

$$M_q(\theta) \geq 0\ for\ \theta \in (0, \infty)\ and\ \int_0^\infty M_q(\theta) d\theta = 1.$$

Remark 1. *In considering the case of* $\alpha \in (0, 1)$, *we know from the references that there is a similar representation of mild solutions if the initial value* $x_1 = 0$ *for the case of* $\alpha \in (1, 2)$. *However, the biggest difference is that the operator* A *(typically the Laplacian operator) generates a* C_0*-semigroup, and one can use the method of semigroup theory to obtain some well-known results for the case of* $\alpha \in (0, 1)$ *instead of cosine families. Further, if* α *tends to 1, the method of semigroup theory can be also used to deal with first-order evolution problems; if* α *tends to 2, we can directly solve an evolution problem by using the concept of cosine families. Thus, the studied evolution problem in Equation* (1) *is more different from the case of* $\alpha \in (0, 1]$, *and it is valuable to consider the existence of Equation* (1).

Remark 2. *The setting $q = \alpha/2$ for $\alpha \in (1,2)$ is derived from the constraint of the Laplace transform of Mainardi's Wright-type function and the resolvent of cosine families (see [18]). This reflects the fact that the probability density function is closely related to the mild solutions of the corresponding evolution problems.*

Lemma 8. ([18]) *The operators $C_q(t)$, $K_q(t)$, and $P_q(t)$ (appearing in Definition 7) have the following properties:*

(i) *For any $t \geq 0$, the operators $C_q(t)$, $K_q(t)$, and $P_q(t)$ are linear operators;*
(ii) *For any fixed $t \geq 0$ and for any $x \in X$, the following estimates hold:*

$$\|C_q(t)x\| \leq M\|x\|, \quad \|K_q(t)x\| \leq M\|x\|t, \quad \|P_q(t)x\| \leq \frac{M}{\Gamma(2q)}\|x\|t^q;$$

(iii) *$\{C_q(t), t \geq 0\}$, $\{K_q(t), t \geq 0\}$, and $\{t^{q-1}P_q(t), t \geq 0\}$ are strongly continuous.*

Lemma 9. ([29]) *Let X be a separable metric space and let $G : \Omega \to \mathcal{P}_{cl}(X)$ be a multivalued map with nonempty closed images. Then, G is measurable if and only if there exist measurable single-valued maps $g_n : \Omega \to X$ such that*

$$G(\omega) = \overline{\bigcup\{g_n(\omega),\ n \geq 1\}}, \quad \text{for every } \omega \in \Omega.$$

Lemma 10. ([30, Theorem 8.2.10]) *Let $(\Omega, \mathcal{A}, \mu)$ be a complete σ-finite measurable space, and let X, Y be two complete separable metric spaces. If $F : \Omega \to \mathcal{P}(X)$ is a measurable multivalued map with nonempty closed images and $G : \Omega \times X \mapsto Y$ is a Carathéodory map (that is, for every $x \in X$, the multivalued map $\omega \mapsto G(\omega, x)$ is measurable, and for every $\omega \in \Omega$, the multivalued map $x \mapsto G(\omega, x)$ is continuous), then for every measurable map $h : \Omega \mapsto Y$ satisfying $h(\omega) \in G(\omega, F(\omega))$ for almost all $\omega \in \Omega$, there exists a measurable selection $f(\omega) \in F(\omega)$ such that $h(\omega) = G(\omega, f(\omega))$ for almost all $\omega \in \Omega$.*

3. Main Results

We need to state the following hypotheses for the forthcoming analysis.

Hypothesis 1. *The operator A is the infinitesimal generator of a uniformly bounded cosine family $\{C(t)\}_{t \geq 0}$ in X.*

Hypothesis 2. *The multivalued map $F : J \times X \to \mathcal{P}_{cl,cv}(X)$ is an L^1-Carathéodory multivalued map satisfying the following conditions:*

(i) *For every $t \in J$, the map $F(t, \cdot) : X \to \mathcal{P}_{cl,cv}(X)$ is u.s.c.;*
(ii) *For each $x \in X$, the map $F(\cdot, x) : J \to \mathcal{P}_{cl,cv}(X)$ is measurable and the set*

$$S_{F,x} = \{f \in L^1(J, X) : \ f(t) \in F(t, x(t)) \text{ for a.e. } t \in J\}$$

is nonempty.

Hypothesis 3. *There exists a function $k_f(\cdot) \in L^1(J, \mathbb{R}_+)$ such that*

$$\|F(t,x)\| = \sup\{\|f\| : \ f \in F(t,x)\} \leq k_f(t)(1 + \|x\|), \text{ for a.a. } t \in J \text{ and all } x \in X.$$

Hypothesis 4. *There exists a function $\beta(\cdot) \in L^1(J, \mathbb{R}_+)$ such that $\chi(F(t, D)) \leq \beta(t)\chi(D)$ for every bounded subset $D \subset C(J, X)$.*

Hypothesis 5. *$g : C(J, X) \to X$ is a continuous and compact function, and there exist constants N_{g1}, N_{g2} such that $\|g(x)\| \leq N_{g1}\|x\|_C + N_{g2}$ for $x \in C(J, X)$.*

Remark 3. *If X is a finite dimension Banach space, then for each $x \in C(J,X)$, $S_{F,x} \neq \emptyset$ (see, e.g., Lasota and Opial [31]). If X is an infinite dimension Banach space and $x \in C(J,X)$, it follows from Hu and Papageorgiou [32] that $S_{F,x} \neq \emptyset$ if and only if the function $\varsigma : J \to \mathbb{R}_+$ given by $\varsigma(t) := \inf\{\|v\| : v \in F(t,x)\}$ belongs to $L^1(J,\mathbb{R}_+)$.*

Lemma 11. ([31]). *Let X be a Banach space, let $F : J \times X \to \mathcal{P}_{cp,cv}(X)$ be a L^1-Carathéodory multivalued map with $S_{F,x} \neq \emptyset$ (see (H2)), and let Ψ be a linear continuous operator from $L^1(J,X)$ to $C(J,X)$. Then,*

$$\Psi \circ S_F : C(J,X) \to \mathcal{P}_{cp,cv}(C(J,X)), \quad x \mapsto (\Psi \circ S_F)(x) := \Psi(S_{F,x}),$$

is a closed graph operator in $C(J,X) \times C(J,X)$.

Theorem 1. *Assume that* (H1)–(H5) *are satisfied. Then, Equation (1) has at least one mild solution provided that $\|k_f\|_1 < (1 - MN_{g1})M^{-1}a^{1-2q}\Gamma(2q)$ and $\|\beta\|_1 < (8M)^{-1}a^{1-2q}\Gamma(2q)$.*

Proof. By (H2), we can define a multivalued map $\mathscr{P} : C(J,X) \to \mathcal{P}(C(J,X))$ as follows: for $x \in C(J,X)$, $\mathscr{P}(x)$ is the set of all functions $y \in \mathscr{P}(x)$ satisfying

$$y(t) = C_q(t)(x_0 - g(x)) + K_q(t)x_1 + \int_0^t (t-s)^{q-1}P_q(t-s)f(s)ds, \quad t \in J,$$

where $f \in S_{F,x}$. It will be verified in several steps, claims and parts that the operator $\mathscr{P}$ has fixed points that correspond to mild solutions of Equation (1).

Step 1. $\mathscr{P}$ maps a bounded closed convex set into a bounded closed convex set.

By the hypothesis of function $k_f(\cdot)$ in (H3), there exists $r > 0$ such that

$$M\|x_0\| + MN_{g1}r + MN_{g2} + Ma\|x_1\| + \frac{Ma^{2q-1}}{\Gamma(2q)}\|k_f\|_1 + \frac{Ma^{2q-1}}{\Gamma(2q)}\|k_f\|_1 r \leq r. \tag{3}$$

Furthermore, we introduce $W_0 = \{x \in C(J,X) : \|x\|_C \leq r\}$ and observe that W_0 is a nonempty bounded closed and convex subset of $C(J,X)$. Let $x \in W_0$ and $y \in \mathscr{P}(x)$, then, there exists $f \in S_{F,x}$ such that for each $t \in J$ and for any $x \in W_0$, we have

$$y(t) = C_q(t)(x_0 - g(x)) + K_q(t)x_1 + \int_0^t (t-s)^{q-1}P_q(t-s)f(s)ds.$$

By (H3) and (H4), we have

$$\begin{aligned}
\|y(t)\| \leq& \|C_q(t)\|_{\mathcal{L}(X)}\|x_0 - g(x)\| + \|K_q(t)\|_{\mathcal{L}(X)}\|x_1\| + \int_0^t (t-s)^{q-1}\|P_q(t-s)f(s)\|ds \\
\leq& M\|x_0\| + M\|g(x)\| + Mt\|x_1\| + \frac{M}{\Gamma(2q)}\int_0^t (t-s)^{2q-1}k_f(s)(1 + |x(s)|)ds \\
\leq& M\|x_0\| + MN_{g1}\|x\|_C + MN_{g2} + Mt\|x_1\| + \frac{Mt^{2q-1}}{\Gamma(2q)}\|k_f\|_1 + \frac{Mt^{2q-1}}{\Gamma(2q)}\|k_f\|_1\|x\|_C \\
\leq& M\|x_0\| + MN_{g1}r + MN_{g2} + Ma\|x_1\| + \frac{Ma^{2q-1}}{\Gamma(2q)}\|k_f\|_1 + \frac{Ma^{2q-1}}{\Gamma(2q)}\|k_f\|_1 r \\
\leq& r.
\end{aligned}$$

Therefore, $\|y\|_C \leq r$, which implies that $\mathscr{P}(W_0) \subseteq W_0$.

Define $W_1 = \overline{co}\mathscr{P}(W_0)$. Clearly, $W_1 \subset C(J, X)$ is a nonempty bounded closed and convex set. Repeating the arguments employed in the previous step, for any $x \in W_1$, $y \in \mathscr{P}(x)$, it follows that there exists $f \in S_{F,x}$ such that for each $t \in J$ and for any $x \in W_1$,

$$y(t) = C_q(t)(x_0 - g(x)) + K_q(t)x_1 + \int_0^t (t-s)^{q-1}P_q(t-s)f(s)ds.$$

By (H3) and (H4), together with Lemma 8 (ii), we have

$$\begin{aligned}\|y(t)\| \leq& \|C_q(t)(x_0 - g(x))\| + \|K_q(t)x_1\| + \int_0^t (t-s)^{q-1}\|P_q(t-s)f(s)\|ds \\ \leq& M\|x_0\| + MN_{g1}r + MN_{g2} + Ma\|x_1\| + \frac{Ma^{2q-1}}{\Gamma(2q)}\|k_f\|_1 + \frac{Ma^{2q-1}}{\Gamma(2q)}\|k_f\|_1 r \leq r,\end{aligned}$$

which implies that $\mathscr{P}(W_1) \subseteq W_1$ and $W_1 \subset W_0$.

Next, for every $n \geq 1$, we define $W_{n+1} = \overline{co}\mathscr{P}(W_n)$. From the above proof, it is easy to see that W_n is a nonempty bounded closed and convex subset of $C(J, X)$. Furthermore, $W_2 = \overline{co}\mathscr{P}(W_1) \subset W_1$. By induction, we know that the sequence $\{W_n\}_{n=1}^{\infty}$ is a decreasing sequence of nonempty bounded closed and convex subsets of $C(J, X)$. Furthermore, we set $W = \bigcap_{n=1}^{\infty} W_n$ and note that W is bounded closed and convex since W_n is bounded closed and convex for every $n \geq 1$.

Now, we establish that $\mathscr{P}(W) \subseteq W$. Indeed, $\mathscr{P}(W) \subseteq \mathscr{P}(W_n) \subseteq \overline{co}\mathscr{P}(W_n) = W_{n+1}$ for every $n \geq 1$. Therefore, $\mathscr{P}(W) \subseteq \bigcap_{n=2}^{\infty} W_n$. On the other hand, $W_n \subset W_1$ for every $n \geq 1$. Hence,

$$\mathscr{P}(W) \subseteq \bigcap_{n=2}^{\infty} W_n = \bigcap_{n=1}^{\infty} W_n = W.$$

Step 2. The multivalued map $\mathscr{P}$ is ν-condensing.

Let $\mathbb{B} \subseteq W$ be such that

$$\nu(\mathbb{B}) \leq \nu(\mathscr{P}(\mathbb{B})). \tag{4}$$

We show below that $\mathbb{B}$ is a relatively compact set; that is, $\nu(\mathbb{B}) = 0$.

Let $\sigma(\mathbb{B}) = \sup_{t\in J} \chi(\mathbb{B}(t))$, and let $\nu(\mathscr{P}(\mathbb{B}))$ be achieved on a sequence $\{y_n\}_{n=1}^{\infty} \subset \mathscr{P}(\mathbb{B})$; that is,

$$\nu(\{y_n\}_{n=1}^{\infty}) = \max\left(\sigma(\{y_n\}_{n=1}^{\infty}),\ \mathrm{mod}_C(\{y_n\}_{n=1}^{\infty})\right).$$

Then,

$$y_n(t) = C_q(t)(x_0 - g(x_n)) + K_q(t)x_1 + \int_0^t (t-s)^{q-1}P_q(t-s)f_n(s)ds, \quad t \in J,$$

where $\{x_n\}_{n=1}^{\infty} \subset \mathbb{B}$ and $f_n \in Sel_{F,x_n}$ for every $n \geq 1$.

Since g is compact, the set $\{g(x_n) : n \geq 1\}$ is relatively compact and $C_q(t)$, $K_q(t)$ are strongly continuous for $t \geq 0$. Hence, for every $t \in J$, we have

$$\nu(\{C_q(t)(x_0 - g(x_n)) + K_q(t)x_1,\ n \geq 1\}) = 0.$$

Therefore, it is enough to estimate that

$$\nu\left(\left\{\int_0^t (t-s)^{q-1}P_q(t-s)f_n(s)ds,\ n \geq 1\right\}\right) = 0.$$

Claim I. $\sigma(\{y_n\}_{n=1}^{\infty}) = 0$.

For any $t \in J$, using (H4), Lemma 3, and Lemma 8 (ii), we have

$$\begin{aligned}\sigma(\{y_n\}_{n=1}^{\infty}) =& \sup_{t\in J}\chi(\{y_n(t)\}_{n=1}^{\infty}) \leq 2\sup_{t\in J}\int_0^t (t-s)^{q-1}\chi\left(\{P_q(t-s)f_n(s)\}_{n=1}^{\infty}\right) ds \\ \leq& \sup_{t\in J}\frac{2M}{\Gamma(2q)}\int_0^t (t-s)^{2q-1}\beta(s)\chi\left(\{x_n(s)\}_{n=1}^{\infty}\right) ds \\ \leq& \sup_{t\in J}\frac{2M}{\Gamma(2q)}\int_0^t (t-s)^{2q-1}\beta(s)ds\sigma(\{x_n\}_{n=1}^{\infty}) \\ \leq& \frac{2Ma^{2q-1}}{\Gamma(2q)}\int_0^a \beta(s)ds\sigma(\{x_n\}_{n=1}^{\infty}) < \frac{1}{4}\sigma(\{x_n\}_{n=1}^{\infty}).\end{aligned}$$

On the other hand, Equation (4) implies that $\sigma(\{y_n\}_{n=1}^{\infty}) \geq \sigma(\{x_n\}_{n=1}^{\infty})$. In consequence, we have $\sigma(\{y_n\}_{n=1}^{\infty}) = 0$.

Claim II. $\text{mod}_C(\{y_n\}_{n=1}^{\infty}) = 0$; that is, the set $\mathbb{B}$ is equicontinuous.

Let

$$\widetilde{y}_n(\cdot) = \int_0^{\cdot} (\cdot - s)^{q-1}P_q(t-s)f_n(s)ds.$$

Therefore, it remains to be verified that $\text{mod}_C(\{\widetilde{y}_n\}_{n=1}^{\infty}) = 0$. Then, for any $t_1, t_2 \in J$ with $t_1 < t_2$, we have

$$\begin{aligned}\|\widetilde{y}_n(t_2) - \widetilde{y}_n(t_1)\| \leq& \int_{t_1}^{t_2} \|(t_2-s)^{q-1}P_q(t_2-s)f_n(s)\|ds \\ &+ \int_0^{t_1} \|((t_2-s)^{q-1}P_q(t_2-s) - (t_1-s)^{q-1}P_q(t_2-s))f_n(s)\|ds \\ =& I_1 + I_2.\end{aligned}$$

According to Lemma 8 (ii), we get

$$\begin{aligned}I_1 \leq& \frac{M}{\Gamma(2q)}\int_{t_1}^{t_2}(t_2-s)^{2q-1}k_f(s)(1+\|x_n(s)\|)ds \\ \leq& \frac{M}{\Gamma(2q)}(t_2-t_1)^{2q-1}\int_{t_1}^{t_2} k_f(s)ds(1+\|x_n\|_C) \to 0, \quad \text{as } t_2 \to t_1.\end{aligned}$$

Let $T_q(t) = t^{q-1}P_q(t)$ for $t \in J$. Then, we know from Lemma 8 (iii) that $T_q(t)$ is a strongly continuous operator. For I_2, taking $\varepsilon > 0$ to be small enough, we obtain

$$\begin{aligned}I_2 \leq& \int_0^{t_1-\varepsilon} \|(T_q(t_2-s) - T_q(t_1-s))f_n(s)\|ds + \int_{t_1-\varepsilon}^{t_1} \|(T_q(t_2-s) - T_q(t_1-s))f_n(s)\|ds \\ \leq& \int_0^{t_1} k_f(s)(1+\|x_n(s)\|)ds \sup_{s\in[0,t_1-\varepsilon]} \|T_q(t_2-s) - T_q(t_1-s)\|_{\mathcal{L}(X)} \\ &+ \left(\frac{M\varepsilon^{2q-1}}{\Gamma(2q)} + \frac{M(t_2-t_1+\varepsilon)^{2q-1}}{\Gamma(2q)}\right)\int_{t_1-\varepsilon}^{t_1} k_f(s)(1+\|x_n(s)\|)ds \\ \leq& \|k_f\|_1(1+\|x_n\|_C) \sup_{s\in[0,t_1-\varepsilon]} \|T_q(t_2-s) - T_q(t_1-s)\|_{\mathcal{L}(X)} \\ &+ \left(\frac{M\varepsilon^{2q-1}}{\Gamma(2q)} + \frac{M(t_2-t_1+\varepsilon)^{2q-1}}{\Gamma(2q)}\right)(1+\|x_n\|_C)\int_{t_1-\varepsilon}^{t_1} k_f(s)ds \\ \to& 0, \quad \text{as } t_2 \to t_1,\ \varepsilon \to 0.\end{aligned}$$

Consequently, we have

$$\text{mod}_C\left(\left\{\int_0^t (t-s)^{q-1}P_q(t-s)f_n(s)ds,\ n\geq 1\right\}\right)=0.$$

As a conclusion, it follows that $\text{mod}_C(\{y_n\}_{n=1}^{\infty})=0$. Hence, the multivalued map $\mathscr{P}$ is ν-condensing.

Step 3. The multimap $\mathscr{P}(x)$ is convex and compact for each $x\in W$.

Part I. $\mathscr{P}(x)$ has convex values for each $x\in W$.

In fact, if y_1, y_2 belong to $\mathscr{P}(x)$ for each $x\in W$, then there exist $f_1, f_2\in S_{F,x}$ such that for each $t\in J$, we have

$$y_i(t)=C_q(t)(x_0-g(x))+K_q(t)x_1+\int_0^t (t-s)^{q-1}P_q(t-s)f_i(s)ds,\quad i=1,2.$$

Let $\theta\in[0,1]$. Then, for each $t\in J$, we get

$$\begin{aligned}(\theta y_1+(1-\theta)y_2)(t)=&C_q(t)(x_0-g(x))+K_q(t)x_1\\&+\int_0^t (t-s)^{q-1}P_q(t-s)(\theta f_1+(1-\theta)f_2)(s)ds.\end{aligned}$$

As F has convex values by the definition of $S_{F,x}$, we deduce that $\theta f_1(s)+(1-\theta)f_2(s)\in S_{F,x}$. Thus, $\theta y_1+(1-\theta)y_2\in\mathscr{P}(x)$.

Part II. $\mathscr{P}$ has compact values. In view of the foregoing facts, it is enough to show that W is nonempty and compact in $C(J,X)$: that is, by Lemma 5, we need to show that

$$\lim_{n\to\infty}\nu(W_n)=0. \tag{5}$$

As in Step 2, we can show that $\text{mod}_C(W_n)=0$; that is, W_n is equicontinuous. Hence, it remains to be shown that $\sigma(W_n)=0$. By Lemma 1, for each $\varepsilon>0$, there exists a sequence $\{y_k\}_{k=1}^{\infty}$ in $\mathscr{P}(W_{n-1})$ such that

$$\sigma(W_n)=\sigma(\mathscr{P}(W_n))\leq 2\sigma(\{y_k\}_{k=1}^{\infty})+\varepsilon.$$

Therefore, by Lemma 2 and the nonsingularity of σ, it follows that

$$\sigma(W_n)\leq 2\sigma(\{y_k\}_{k=1}^{\infty})+\varepsilon=2\sup_{t\in J}\chi(\{y_k(t)\}_{k=1}^{\infty})+\varepsilon. \tag{6}$$

Since $y_k\in\mathscr{P}(W_{n-1})$ $(k\geq 1)$, there exists $x_k\in W_{n-1}$ such that $y_k\in\mathscr{P}(x_k)$. Hence, from the compactness of g and the strong continuity of $C_q(t)$ and $K_q(t)$ for $t\in J$, there exists $f_k\in S_{F,x_k}$ such that for every $t\in J$,

$$\begin{aligned}\chi(\{y_k(t)\}_{k=1}^{\infty})\leq&\chi(\{C_q(t)(x_0-g(\{x_k\}_{k=1}^{\infty}))+K_q(t)x_1\})\\&+\chi\left(\left\{\int_0^t (t-s)^{q-1}P_q(t-s)f_k(s)ds:\ k\geq 1\right\}\right)\\=&\chi\left(\left\{\int_0^t (t-s)^{q-1}P_q(t-s)f_k(s)ds:\ k\geq 1\right\}\right).\end{aligned}$$

By (H5) and Lemma 1, for a.e. $t\in J$, we have

$$\chi(\{f_k(t)\}_{k=1}^{\infty})\leq\chi(F(t,\{x_k(t)\}_{k=1}^{\infty}))\leq\beta(t)\chi(\{x_k(t)\}_{k=1}^{\infty})\leq\beta(t)\sigma(W_{n-1}):=\gamma(t).$$

On the other hand, by (H3), for almost all $t\in J$, $\|f_k(t)\|\leq k_f(t)(1+r)$ for every $k\geq 1$. Hence, $f_k\in L^1(J,X)$, $k\geq 1$. Note that $\gamma(\cdot)\in L^1(J,\mathbb{R}_+)$ from (H4). It follows from Lemma 4 that there exists a compact $K_\epsilon \subset X$, a measurable set $J_\epsilon \subset J$ with measure less than ϵ, and a sequence of functions $\{g_k^\epsilon\} \subset L^1(J,X)$ such that $\{g_k^\epsilon(s)\}_{k=1}^\infty \subseteq K_\epsilon$ for all $s \in J$, and

$$\|f_k(s) - g_k^\epsilon(s)\| < 2\gamma(s) + \epsilon, \quad \text{for every } k \geq 1 \text{ and every } s \in J'_\epsilon = J - J_\epsilon.$$

Then, using Minkowski's inequality and the property of the MNC, we obtain

$$\begin{aligned}
&\chi\left(\left\{\int_{J'_\epsilon}(t-s)^{q-1}P_q(t-s)(f_k(s)-g_k^\epsilon(s))ds : k \geq 1\right\}\right)\\
\leq&\frac{2M}{\Gamma(2q)}\int_{J'_\epsilon}(t-s)^{2q-1}\chi(\{(f_k(s)-g_k^\epsilon(s)) : k \geq 1\})ds\\
\leq&\frac{2M}{\Gamma(2q)}\int_{J'_\epsilon}(t-s)^{2q-1}\sup_{k\geq 1}\|f_k(s)-g_k^\epsilon(s)\|ds\\
\leq&\frac{2Ma^{2q-1}}{\Gamma(2q)}\int_{J'_\epsilon}(2\gamma(s)+\epsilon)ds\\
\leq&\frac{4Ma^{2q-1}}{\Gamma(2q)}\|\gamma\|_1+\frac{2Ma^{2q-1}}{\Gamma(2q)}\epsilon\\
\leq&\frac{4Ma^{2q-1}}{\Gamma(2q)}\sigma(W_{n-1})\|\beta\|_1+\frac{2Ma^{2q-1}}{\Gamma(2q)}\epsilon, \qquad (7)
\end{aligned}$$

and

$$\begin{aligned}
\chi\left(\left\{\int_{J_\epsilon}(t-s)^{q-1}P_q(t-s)f_k(s)ds : k \geq 1\right\}\right) \leq&\frac{2M}{\Gamma(2q)}\int_{J_\epsilon}(t-s)^{2q-1}\chi\left(\{f_k(s)\}_{k=1}^\infty\right)ds\\
\leq&\frac{2M}{\Gamma(2q)}\int_{J_\epsilon}(t-s)^{2q-1}\sup_{k\geq 1}\|f_k(s)\|ds\\
\leq&\frac{Ma^{2q-1}}{\Gamma(2q)}(1+r)\int_{J_\epsilon}k_f(s)ds. \qquad (8)
\end{aligned}$$

Using Equations (7) and (8), we have

$$\begin{aligned}
&\chi\left(\left\{\int_0^t(t-s)^{q-1}P_q(t-s)f_k(s)ds : k \geq 1\right\}\right) \leq \chi\left(\left\{\int_{J'_\epsilon}(t-s)^{q-1}P_q(t-s)f_k(s)ds : k \geq 1\right\}\right)\\
&\quad+\chi\left(\left\{\int_{J_\epsilon}(t-s)^{q-1}P_q(t-s)f_k(s)ds : k \geq 1\right\}\right)\\
\leq&\chi\left(\left\{\int_{J'_\epsilon}(t-s)^{q-1}P_q(t-s)(f_k(s)-g_k^\epsilon(s))ds : k \geq 1\right\}\right)\\
&\quad+\chi\left(\left\{\int_{J'_\epsilon}(t-s)^{q-1}P_q(t-s)g_k^\epsilon(s)ds : k \geq 1\right\}\right)\\
&\quad+\chi\left(\left\{\int_{J_\epsilon}(t-s)^{q-1}P_q(t-s)f_k(s)ds : k \geq 1\right\}\right)\\
\leq&\frac{4Ma^{2q-1}}{\Gamma(2q)}\sigma(W_{n-1})\|\beta\|_1+\frac{2Ma^{2q-1}}{\Gamma(2q)}\epsilon+\frac{Ma^{2q-1}}{\Gamma(2q)}(1+r)\int_{J_\epsilon}k_f(s)ds.
\end{aligned}$$

As ϵ is arbitrary, for all $t \in J$, we get

$$\chi\left(\left\{\int_0^t(t-s)^{q-1}P_q(t-s)f_k(s)ds\right\}\right) \leq \frac{4Ma^{2q-1}}{\Gamma(2q)}\|\beta\|_1\sigma(W_{n-1}).$$

Therefore, for each $t \in J$, we have

$$\chi(\{y_k(t)\}_{k=1}^{\infty}) \leq \frac{4Ma^{2q-1}}{\Gamma(2q)}\|\beta\|_1\sigma(W_{n-1}).$$

By the above inequality, together with Equation (6) and the arbitrary nature of ε, we can deduce that

$$\sigma(W_n) \leq \frac{8Ma^{2q-1}}{\Gamma(2q)}\|\beta\|_1\sigma(W_{n-1}).$$

Then, by induction, we find that

$$0 \leq \sigma(W_n) \leq \left(\frac{8Ma^{2q-1}}{\Gamma(2q)}\|\beta\|_1\right)^n \sigma(W_0), \quad \text{for all } n \geq 1.$$

Since this inequality is true for every $n \geq 1$, passing on to the limit $n \to \infty$ and by (H4), we obtain Equation (5). Hence, $W = \bigcap_{n=1}^{\infty} W_n$ is a nonempty compact set of X, and $\mathscr{P}$ has compact values in W.

Step 4. The values of $\mathscr{P}$ are closed.

Let $x_n, x_* \in W$ with $x_n \to x_*$ as $n \to \infty$, $y_n \in \mathscr{P}(x_n)$, and $y_n \to y_*$ as $n \to \infty$. We show that $y_* \in \mathscr{P}(x_*)$. Indeed, $y_n \in \mathscr{P}(x_n)$ means that there exists $f_n \in S_{F,x_n}$ such that

$$y_n(t) = C_q(t)(x_0 - g(x)) + K_q(t)x_1 + \int_0^t (t-s)^{q-1}P_q(t-s)f_n(s)ds.$$

Next, we must show that there exists $f_* \in S_{F,x_*}$ such that

$$y_*(t) = C_q(t)(x_0 - g(x)) + K_q(t)x_1 + \int_0^t (t-s)^{q-1}P_q(t-s)f_*(s)ds.$$

Since $x_n \to x_*$ and $y_n \in \mathscr{P}(x_n)$, we deduce that

$$\|(y_n(t) - C_q(t)x_0 + C_q(t)g(x_n) - K_q(t)x_1) - (y_*(t) - C_q(t)x_0 + C_q(t)g(x_*) - K_q(t)x_1)\| \to 0,$$

as $n \to \infty$.

Now, we consider the linear continuous operator

$$\mathscr{F} : L^1(J,X) \to C(J,X), \quad f \mapsto (\mathscr{F}f)(t) = \int_0^t (t-s)^{q-1}P_q(t-s)f(s)ds.$$

From Step 3 and Lemma 11, it follows that $\mathscr{F} \circ S_F$ is a closed graph operator. Furthermore, in view of the definition of $\mathscr{F}$, we have

$$(y_n(t) - C_q(t)x_0 + C_q(t)g(x_n) - K_q(t)x_1) \in \mathscr{F}(S_{F,x_n}).$$

In view of the fact that $x_n \to x_*$ as $n \to \infty$, the repeated application of Lemma 11 yields

$$y_*(t) - C_q(t)x_0 + C_q(t)g(x_*) - K_q(t)x_1 = \int_0^t (t-s)^{q-1}P_q(t-s)f(s)ds$$

for some $f \in S_{F,x_*}$. Thus, $\mathscr{P}$ is a closed multivalued map.

Therefore, as an implication of Steps 1–5, we deduce that $\mathscr{P} : W \to \mathcal{P}(W)$ is closed and ν-condensing with nonempty convex compact values. Thus, all the hypotheses of Lemma 6 are satisfied. Hence, there exists at least one fixed point $x \in W$ such that $x \in \mathscr{P}(x)$, which corresponds to a mild solution of Equation (1). □

Theorem 2. *Suppose that all the assumptions of Theorem 1 are satisfied. Then, the set of mild solutions of Equation (1) is compact in* $C(J,X)$.

Proof. Note that the set of mild solutions is nonempty by Theorem 1. Indeed, letting $r>0$, defined by Equation (3), we can get a mild solution in W_0. Now, we show that an arbitrary number of mild solutions of Equation (1) belongs to W_0. Let x be a mild solution of Equation (1). Then,

$$x(t)=C_q(t)(x_0-g(x))+K_q(t)x_1+\int_0^t(t-s)^{q-1}P_q(t-s)f(s)ds,$$

where $f\in S_{F,x}=\{f\in L^1(J,X):\ f(t)\in F(t,x(t)),\ \text{for a.e. } t\in J\}$. Using an argument similar to the one used in Step 1 of the proof of Theorem 1, we have

$$\begin{aligned}\|x\|_C=&\sup_{t\in J}\|x(t)\|\\ \le&\sup_{t\in J}\|C_q(t)(x_0-g(x))\|+\sup_{t\in J}\|K_q(t)x_1\|+\sup_{t\in J}\int_0^t(t-s)^{q-1}\|P_q(t-s)f(s)\|ds\\ \le&M\|x_0\|+MN_{g1}r+MN_{g2}+Ma\|x_1\|+\frac{Ma^{2q-1}}{\Gamma(2q)}\|k_f\|_1+\frac{Ma^{2q-1}}{\Gamma(2q)}\|k_f\|_1r\le r.\end{aligned}$$

This shows that the mild solutions of Equation (1) are bounded. Thus, the conclusion follows from Lemma 7. The proof is completed. □

4. An Application

Let $\Omega\subset\mathbb{R}^N$ $(N=1,2,3)$ be an open bounded set and $X=U=L^2(\Omega)$. Let us consider the following fractional partial differential equations with the constrained control u and a finite multi-point discrete mean condition:

$$\begin{cases}\partial_t^{\alpha}y(t,z)=\Delta y(t,z)+G(t,z,y(t,z),u(t,z)), & t\in[0,1],\ z\in\Omega,\ u\in U,\\ y(t,z)=0, \quad t\in[0,1], z\in\partial\Omega\\ y(0,z)-\sum_{i=0}^{n}\int_{\Omega}m(\xi,z)y(t_i,\xi)d\xi=0,\ y'(0,z)=0,\ z\in\Omega,\end{cases}\tag{9}$$

where ∂_t^{α} is the Caputo fractional partial derivative of order $\alpha\in(1,2)$, $0\le t_0<t_1<\cdots<t_n\le1$, $m(\xi,z):\Omega\times\Omega\to X$ is an L^2-Lebesgue integrable function, and $G:[0,1]\times\Omega\times X\times U\to X$ is a single-valued continuous measurable function.

We define $x(t)=y(t,\cdot)$, that is, $x(t)(z)=y(t,z)$, $t\in J$, $z\in\Omega$, here $J=[0,1]$. The set of the constraint functions $U:J\to\mathcal{P}_{cl,cv}(X)$ is a measurable multivalued map. If $u\in U$, then it means that $u(t)\in U(t,x(t))$, for a.e. $t\in J$. The function $f:J\times X\times U$ is given by $f(t,x(t),u(t))(z)=G(t,z,y(t,z),u(t,z))$. Equation (9) is solved if we show that there exists a control function u such that Equation (9) admits a mild solution. Let the multivalued map be given by

$$F(t,x(t))=\{f(t,x(t),u(t)),\quad u\in U\}.\tag{10}$$

Then, the set of mild solutions of the control problem in Equation (9), with the right-hand side given by Equation (10), coincides with the set of mild solutions of Equation (1).

Let A be the Laplace operator with Dirichlet boundary conditions defined by $A=\Delta$ with

$$\mathcal{D}(A)=\{v\in L^2(\Omega):\ v\in H_0^1(\Omega)\cap H^2(\Omega)\}.$$

Let $\{-\lambda_k, \phi_k\}_{k=1}^{\infty}$ be the eigensystem of the operator A. Then, $0 < \lambda_1 \leq \lambda_2 \leq \cdots, \lambda_k \to \infty$ as $k \to \infty$, and $\{\phi_k\}_{k=1}^{\infty}$ forms an orthonormal basis of X. Furthermore,

$$Ax = -\sum_{k=1}^{\infty} \lambda_k (x, \phi_k)\phi_k, \quad x \in \mathcal{D}(A),$$

where $(\cdot,\cdot)$ is the inner product in X. It is known that the operator A generates a strongly continuous uniformly bounded cosine family (see, e.g., [9]), which, in this case, is defined by

$$C(t)x = \sum_{k=1}^{\infty} \cos(\sqrt{\lambda_k} t)(x, \phi_k)\phi_k, \quad x \in X,$$

and then $\|C(t)\|_{\mathcal{L}(X)} \leq 1$ for every $t \geq 0$. Hence, (H1) holds.

Taking $\alpha = \frac{3}{2}$, we have $q = \frac{3}{4}$. Let $g : C(J,X) \to X$ be given by $g(x)(z) = \sum_{i=0}^{n} K_g x(t_i)(z)$ with $K_g v(z) = \int_{\Omega} m(\xi, z)v(\xi)d\xi$ for $v \in X$, $z \in \Omega$ (noting that $K_g : X \to X$ is completely continuous). Thus, the assumption in (H5) holds true. With the choice of operator A, Equation (9) can be reformulated in X as the following nonlocal control problem:

$$\begin{cases} {}^C D_t^{\alpha} x(t) = Ax(t) + f(t, x(t), u(t)), & t \in J,\ u \in U, \\ x(0) = g(x),\ x'(0) = 0. \end{cases} \tag{11}$$

Next, the results obtained in Section 4 apply to the following problem of fractional evolution inclusions:

$$\begin{cases} {}^C D_t^{\alpha} x(t) \in Ax(t) + F(t, x(t)), & t \in J, \\ x(0) = g(x),\ x'(0) = 0. \end{cases} \tag{12}$$

Theorem 3. *Assume that the following conditions hold:*

Hypothesis 6. *$U : J \to \mathcal{P}_{cl,cv}(X)$ is a measurable multivalued map.*

Hypothesis 7. *The function $f : J \times X \times X \to X$ is L^1-Carathéodory, linear in the third argument, and there exists a function $k_f(\cdot) \in L^1(J, \mathbb{R}_+)$ satisfying $\|k_f\|_1 < \sqrt{\pi}(1 - n\|m\|)/2$ such that $\|f(t,x,y)\| \leq k_f(t)(1 + \|x\|)$ for almost all $t \in J$ and all $x \in X$.*

Hypothesis 8. *There exists a function $\beta(\cdot) \in L^1(J, \mathbb{R}_+)$ satisfying $\|\beta\|_1 < \sqrt{\pi}/16$ such that*

$$\chi(f(t, D, U(t, D))) \leq \beta(t)\chi(D),$$

for every bounded subset $D \subset C(J,X)$.

Then, the control problem in Equation (9) has at least one mild solution. In addition, the set of mild solutions is compact.

Proof. From (H6) and (H7), the map $t \mapsto F(t,\cdot)$ is obviously a measurable multivalued map, and then $F(\cdot,\cdot) \in \mathcal{P}_{cv,cl}(X)$. Now, we show that the selection set of F is not empty. Since U is a measurable multivalued map, it follows by Lemma 9 that there exists a sequence of measurable selections $\{u_n\}_{n=1}^{\infty} \subset U$ such that

$$U(t) = \overline{\bigcup \{u_n(t),\ n \geq 1\}} \quad \text{for every } t \in J.$$

Let $v_n(t) = f(t, x(t), u_n(t))$ for $n \geq 1$ and $t \in J$. In view of the continuity of f, v_n is thus measurable. Hence, $\overline{\{v_n(t),\ n \geq 1\}} \subseteq F(t, x(t))$. Conversely, if $f(t, x(t), u(t)) \in F(t, x(t))$ for any $u \in U$, then there exists a subsequence in U which will be still defined by $\{u_n\}_{n=1}^{\infty}$ such that $u_n \to u$ as $n \to \infty$. It follows from the continuity of f that $f(t,x(t),u_n(t)) \to f(t,x(t),u(t))$ as $n \to \infty$. Hence, $f(t,x(t),u(t)) \in \overline{\{v_n(t),\ n \geq 1\}}$. This means that

$$F(t,x(t)) = \overline{\bigcup\{v_n(t),\ n \geq 1\}},$$

Consequently, from Lemma 9, $F(\cdot,x)$ is measurable.

Next, we show that the map $x \mapsto F(\cdot,x)$ is an u.s.c. multivalued map by means of contradiction. Firstly, we suppose that F is not u.s.c. at some point $x_0 \in \Omega$. Then, there exists an open neighborhood $W \subseteq X$ such that $F(t,x_0) \subset W$, and for every open neighborhood $V \subseteq \Omega$ of x_0, there exists $x_1 \in V$ such that $F(t,x_1) \not\subset W$. Let

$$V_n = \left\{x \in \Omega,\ \|x - x_0\| < \frac{1}{n},\ n = 1,2,\ldots\right\}.$$

Clearly, V_n is a open neighborhood of x_0. Then, for each $n \geq 1$, there exist $x_n \in V_n$, $v_n \in F(t,x_n)$, and $u_n \in U$ such that $v_n = f(t,x_n,u_n)$ and $v_n \notin W$. Moreover, as $\{u_n\}_{n=1}^{\infty} \subset U$, we set $u_n \to u$ as $n \to \infty$ for some $u \in U$. By the continuity of f, owing to $x_n \to x_0$ as $n \to \infty$, we have $v_n \to v$ as $n \to \infty$, where $v = f(t,x_0,u)$, which implies that $v \in F(t,x_0) \subset W$. This contradicts that $v_n \notin W$ for each $n \geq 1$. Thus, our supposition is false.

In addition, according to the condition in (H7), we find that F is an L^1-Carathéodory multivalued map. Hence, (H2) and (H3) are satisfied. On the other hand, the hypothesis (H8) corresponds to (H4). Thus, all of the hypotheses of Theorem 1 are satisfied. Hence, Equation (12) has at least one mild solution. Furthermore, the set of mild solutions of Equation (12) is compact by Theorem 2.

Finally, we show that the mild solutions of Equation (12) do coincide with the mild solutions of the control problem in Equation (11). Let x be a solution of Equation (12). Then, there exists a single-valued selection

$$\phi \in S_{F,x} = \{\phi \in L^1(J,X),\ \phi(t) \in F(t,x(t)),\ \text{a.e. } t \in J\}, \tag{13}$$

such that

$${}^{C}D_t^{\alpha}x(t) = Ax(t) + \phi(t),\ \text{a.e. } t \in J, \text{ and } x(0) = g(x),\ x'(0) = 0.$$

Now, we introduce a map $\Psi(t,u) = f(t,x(t),u(t))$ and note that it is Carathéodory. Moreover, let the equality in Equation (10) be satisfied. Then, for a.a. $t \in J$ and for every $\phi(t) \in \{f(t,x(t),u(t)), u \in U\} := \Psi(t,U(t))$, we deduce by Lemma 10 that there exists a measurable selection $u(t) \in U(t)$ such that $\phi(t) = \Psi(t,u(t)) = f(t,x(t),u(t))$ for a.a. $t \in J$. Thus, the mild solution satisfies the control problem in Equation (11).

On the other hand, let x satisfy the control problem in Equation (11). Then, x is obviously a mild solution of Equation (12), and the proof is completed. □

5. Conclusions

In the current paper, we study a class of fractional evolution inclusions with nonlocal initial conditions. We obtain the sufficient conditions for ensuring the existence of mild solutions and the compactness for set of mild solutions. We can see that the probability density function is closely related to the mild solutions of the corresponding evolution inclusion problems, which enrich the knowledge of the fractional calculus. Moreover, an illustrative example is provided to demonstrate the applicability of the proposed problem.

On the other hand, many evolution inclusion problems are focused on a finite interval. This is because the solutions of some physical models may blow up, or we can gain a clearer understanding of the state of a physical system in finite time. If the time goes to infinity, it urges us to extend the concept of mild solutions such as Equation (1) in $[0,\infty)$ and, furthermore, to find the existence of global mild solutions. However, the technique for an infinite interval is more complex, and this topic may be a future work. In addition, our future works also include the topological properties of solution sets (including R_δ, acyclicity, connectedness, compactness, and contractibility) for fractional evolution inclusions of order $\alpha \in (1,2)$.

Author Contributions: methodology, J.W.H. and Y.Z.; writing-original draft preparation; J.W.H. and Y.Z.; writing-review and editing, Y.L. and B.A.

Funding: This research was supported by the National Natural Science Foundation of China (no. 11671339).

Conflicts of Interest: The authors declare no conflict of interest.

References

1. Miller, K.S.; Ross, B. *An Introduction to the Fractional Calculus and Differential Equations*; John Wiley: New York, NY, USA, 1993.
2. Podlubny, I. *Fractional Differential Equations*; Academic Press: San Diego, CA, USA, 1999.
3. A. A. Kilbas, H. M. Srivastava, J. J. Trujillo, Theory and applications of fractional differential equations. In *North-Holland Mathematics Studies*; Elsevier Science B.V.: Amsterdam, The Netherlands, 2006; Volume 204.
4. Zhou, Y. *Basic Theory of Fractional Differential Equations*; World Scientific: Singapore, 2014.
5. Zhou, Y. Attractivity for fractional differential equations in Banach space. *Appl. Math. Lett.* **2018**, *75*, 1–6. [CrossRef]
6. Zhou, Y. Attractivity for fractional evolution equations with almost sectorial operators. *Fract. Calc. Appl. Anal.* **2018**, *21*, 786–800. [CrossRef]
7. Mainardi, F. *Fractional Calculus and Waves in Linear Viscoelasticity, An Introduction to Mathematical Models*; Imperial College Press: London, UK, 2010.
8. Tarasov, V.E. *Fractional Dynamics*; Springer: Berlin/Heidelberg, Germany, 2010.
9. Bazhlekova, E.; Bazhlekov, I. Subordination approach to multi-term time-fractional diffusion-wave equations. *J. Comput. Appl. Math.* **2018**, *339*, 179–192. [CrossRef]
10. Kian, Y.; Yamamoto, M. On existence and uniqueness of solutions for semilinear fractional wave equations. *Fract. Calc. Appl. Anal.* **2017**, *20*, 117–138. [CrossRef]
11. Zhou, Y.; Shangerganesh, L.; Manimaran, J.; Debbouche, A. A class of time-fractional reaction-diffusion equation with nonlocal boundary condition. *Math. Meth. Appl. Sci.* **2018**, *41*, 2987–2999. [CrossRef]
12. Bajlekova, E. Fractional Evolution Equations in Banach Spaces. Ph.D Thesis, Eindhoven University of Technology, Eindhoven, The Netherlands, 2001.
13. Fan, Z. Characterization of compactness for resolvents and its applications. *Appl. Math. Comput.* **2014**, *232*, 60–67. [CrossRef]
14. Li, K.; Peng, J.; Jia, J. Cauchy problems for fractional differential equations with Riemann-Liouville fractional derivatives. *J. Funct. Anal.* **2012**, *263*, 476–510. [CrossRef]
15. Li, Y.; Sun, H.; Feng, Z. Fractional abstract Cauchy problem with order $\alpha \in (1,2)$. *Dyn. Part. Differ. Eq.* **2016**, *13*, 155–177. [CrossRef]
16. Shu, X.B.; Wang, Q.Q. The existence and uniqueness of mild solutions for fractional differential equations with nonlocal conditions of order $1 < \alpha < 2$. *Comput. Math. Appl.* **2012**, *64*, 2100–2110.
17. Yan, Z. Approximate controllability of fractional neutral integro-differential inclusions with state-dependent delay in Hilbert spaces. *IMA J. Math. Control Inform.* **2013**, *30*, 443–462. [CrossRef]
18. Zhou, Y.; He, J.W. New results on controllability of fractional evolution systems with order $\alpha \in (1,2)$. *Evol. Eq. Control Theory* **2019**, in press.
19. Zhou, Y.; Peng, L.; Ahmad, B.; Alsaedi, A. Topological properties of solution sets of fractional stochastic evolution inclusions. *Adv. Differ. Equ.* **2017**, *2017*, 90. [CrossRef]
20. Zhou, Y.; Vijayakumar, V.; Murugesu, R. Controllability for fractional evolution inclusions without compactness. *Evol. Equ. Control Theory* **2015**, *4*, 507–524. [CrossRef]
21. Chalishajar, D.N. Controllability of second order impulsive neutral functional differential inclusions with infinite delay. *J. Optim. Theory Appl.* **2012**, *154*, 672–684. [CrossRef]
22. Hu, J.; Liu, X. Existence results of second-order impulsive neutral functional integrodifferential inclusions with unbounded delay in Banach spaces. *Math. Comput. Model.* **2009**, *49*, 516–526. [CrossRef]

23. Travis, C.C.; Webb, G.F. Cosine families and abstract nonlinear second order differential equations. *Acta Math. Hungar.* **1978**, *32*, 75–96. [CrossRef]
24. Bothe, D. Multivalued perturbations of m-accretive differential inclusions. *Israel J. Math.* **1998**, *108*, 109–138. [CrossRef]
25. Kamenskii, M.; Obukhowskii, V.; Zecca, P. *Condensing Multivalued Maps and Semilinear Differential Inclusions in Banach Spaces; De Gruyter Series in Nonlinear Analysis and Applications, 7*; Walter de Gruyter & Co.: Berlin, Germany, 2001.
26. Monch, H. Boundary value problems for nonlinear ordinary differential equations of second order in Banach spaces. *Nonlinear Anal.* **1980**, *49*, 985–999. [CrossRef]
27. Bader, R.; Kamenskii, M.; Obukhowskii, V. On some classes of operator inclusions with lower semicontinuous nonlinearities. *Topol. Methods Nonlinear Anal.* **2001**, *17*, 143–156. [CrossRef]
28. Cardinali, T.; Rubbioni, P. Impulsive mild solutions for semilinear differential inclusions with nonlocal conditions in Banach spaces. *Nonlinear Anal.* **2012**, *75*, 871–879. [CrossRef]
29. Castaing, C.; Valadier, M. *Convex Analysis and Measurable Multifunctions*; Lecture Notes in Mathematics; Springer: Berlin, Germany; New York, NY, USA, 1977.
30. Aubin, J.P.; Frankowska, H. *Set-Valued Analysis*; Birkhauser: Boston, MA, USA, 1990.
31. Lasota, A.; Opial, Z. An application of the Kakutani-Ky-Fan theorem in the theory of ordinary differential equations. *Bull. Acad. Pol. Sci. Ser. Sci. Math. Astronom. Phys.* **1965**, *13*, 781–786.
32. Hu, S.; Papageorgiou, N. *Handbook of Multivalued Analysis, Volume I: Theory*; Kluwer: Dordrecht, The Netherlands, 1997.

mathematics

MDPI

Article

On Ulam Stability and Multiplicity Results to a Nonlinear Coupled System with Integral Boundary Conditions

Kamal Shah [1], Poom Kumam [2,*] and Inam Ullah [1]

[1] Department of Mathematics, University of Malakand, Chakdara Dir(L), Khyber Pakhtunkhwa 18800, Pakistan; kamalshah408@gmail.com (K.S.); inamullahkhan829@gmail.com (I.U.)

[2] KMUTT Fixed Point Research Laboratory, KMUTT-Fixed Point Theory and Applications Research Group, Theoretical and Computational Science Center (TaCS), Science Laboratory Building, Faculty of Science, King Mongkut's University of Technology Thonburi (KMUTT), 126 Pracha-Uthit Road, Bang Mod, Thrung Khru, Bangkok 10140, Thailand

* Correspondence: poom.kum@kmutt.ac.th or poom.kumam@mail.kmutt.ac.th

Received: 1 February 2019; Accepted: 17 February 2019; Published: 27 February 2019

Abstract: This manuscript is devoted to establishing existence theory of solutions to a nonlinear coupled system of fractional order differential equations (FODEs) under integral boundary conditions (IBCs). For uniqueness and existence we use the Perov-type fixed point theorem. Further, to investigate multiplicity results of the concerned problem, we utilize Krasnoselskii's fixed-point theorems of cone type and its various forms. Stability analysis is an important aspect of existence theory as well as required during numerical simulations and optimization of FODEs. Therefore by using techniques of functional analysis, we establish conditions for Hyers-Ulam (HU) stability results for the solution of the proposed problem. The whole analysis is justified by providing suitable examples to illustrate our established results.

Keywords: arbitrary order differential equations; multiple positive solution; Perov-type fixed point theorem; HU stability

1. Introduction

Fractional order differential equations (FODEs) emerge in the scientific demonstration of numerous frameworks and different fields of science such as physics, chemistry , economics, polymer rheology, aerodynamics, electrodynamics of complicated medium, blood flow phenomena, biophysics, etc. (see [1–5]). Recently, many authors have studied FODEs from different aspects, one is the numerical and scientific techniques for finding solutions and the other is the theoretical perspective of uniqueness and existence of solutions. The interest of the researchers in the investigation of FODEs lies in the incontrovertible fact that fractional-order models (FOM) are found to be highly realistic and practical, compared to the integer order models. Because there are additional degrees of opportunity in the FOM, in consequence, the subject of FODEs is gaining more attention from researchers. Another facet of research, which has been completely studied for integer order differential equations is devoted to uniqueness and existence of solutions to boundary value problems (BVPs). The mentioned aspect has been very well studied for FODEs, we refer the readers [6–10]. Uniqueness and existence results of solutions to multi-point BVPs have been studied via classical fixed point theorems such as the Schauder fixed point theorem and the Banach contraction principle, see [11–17].

FODEs under integral boundary conditions (IBCs) have been investigated very well because these type of equations are increasingly used in fluid-mechanics and dynamical problems. Jankowski [18] studied the ordinary differential equation under IBCs given by

$$\begin{cases} y'(\vartheta) = F(\vartheta, y(\vartheta)), & \vartheta \in [0, T], \quad T > 0, \\ y(\vartheta)|_{\vartheta=0} = \delta \int_0^T y(s)ds + d_0, & d_0 \in R, \end{cases}$$

where $F \in C([0, T] \times R, R)$ and $\delta = 1$ or -1. He developed a sufficient condition for iterative approximate solutions to the above problem.

Nanware and Dhaigude [19] have investigated the aforementioned BVP under the IBCs for FODE as given by

$$\begin{cases} D^{\sigma}_{+0}y(\vartheta) = F(\vartheta, y(\vartheta)), & \vartheta \in [0, T], \quad T > 0, \\ y(\vartheta)|_{\vartheta=0} = \delta \int_0^T y(s)ds + d_0, & d_0 \in R, \end{cases}$$

where $0 < \sigma \leq 1, \delta$ is 1 or -1 and $F \in C([0, T] \times R, R)$, D^{σ}_{+0} is Riemann-Liouville fractional derivative of order σ is defined in (2). The aforementioned author also studied the iterative approximate solution to the above FODEs.

In the same line Cabada and Wang [20] studied the following problem under IBCs as

$$\begin{cases} {}^{C}D^{\sigma}_{+0}y(\vartheta) + \varphi(\vartheta, y(\vartheta)) = 0; & \vartheta \in (0, 1), \\ y(0) = y''(0) = 0, \quad y(1) = \delta \int_0^1 y(s)ds, \end{cases}$$

where $\sigma \in (2, 3]$, $\delta \in (0, 2)$ and $y : [0, 1] \times [0, \infty] \to [0, \infty]$ are the continuous functions. Also we remark that ${}^{C}D^{\sigma}_{+0}$ stands for Caputo's fractional derivative.

Inspired from the aforementioned work, in this article we investigate a system of nonlinear FODEs with IBCs as

$$\begin{cases} D^{\sigma}_{+0}y(\vartheta) + \varphi(\vartheta, y(\vartheta), z(\vartheta)) = 0; & \vartheta \in (0, 1); \quad m - 1 < \sigma \leq m, \\ D^{æ}_{+0}z(\vartheta) + \chi(\vartheta, y(\vartheta), z(\vartheta)) = 0; & \vartheta \in (0, 1); \quad m - 1 < æ \leq m, \\ y(0) = y'(0) = y''(0) = \cdots = y^{(m-2)}(0) = 0, & y(1) = \delta \int_0^1 y(s)ds, \\ z(0) = z'(0) = z''(0) = 0 \cdots = z^{(m-2)}(0) = 0, & z(1) = \varrho \int_0^1 z(s)ds, \end{cases} \tag{1}$$

such that $m \geq 3$, $\delta, \varrho \in (0, 2)$, the functions $\varphi, \chi : [0, 1] \times [0, \infty] \times [0, \infty] \to [0, \infty]$ are continuous functions and $D^{\sigma}_{+0}, D^{æ}_{+0}$ stand for Riemann-Liouville fractional derivatives is defined in (2). We claim that such a system of FODEs are very rarely considered for stability as well as multiplicity results. Our analysis is devoted to the existence theory of a solution, multiplicity results and stability analysis of the suggested problem.

During the last few decades another part of research, which has been considered for FODEs and got much attention from the researchers is stability analysis. Numerous forms of stabilities have been studied in literature which are Mittag-Leffer stability, exponential stability, Lyapunov stability etc., we refer [21–23].

The Ulam stability was first presented by Ulam in 1940 and then brilliantly explained by Hyers in 1941. For more information about HU stability, we refer [24,25]. The HU stability results were generalized and extended by many researchers for FODEs under IBCs. In 1978, Jung studied the said stability for ODEs. Oblaz, Benchohra, etc., have studied the said stability for FODEs but their investigation was limited to initial value problems, we refer to [26–28]. To the best of our information and knowledge, the HU stability has been very rarely studied for coupled system of FODEs under

IBCs. Therefore in this article we investigate HU stability to the considered problem. Here we remark that we also provide some necessary results for nonexistence of solution. Finally a series of examples are provided to support our analysis.

2. Axillary Results

In the current section, we review some fundamental definitions and useful results of functional analysis, fractional calculus and fixed point theory (see reference [1,2,8,29–32]). Here, first of all, we define the Banach space which is utilized throughout in this article.

Let us define $E = \{y(\vartheta)|y \in C[0,1]\}$ with the norm $||y|| = \max_{\vartheta\in[0,1]} |y(\vartheta)|$. We define the norm for the product space as $||(y,z)|| = ||y|| + ||z||$. Obviously $(E \times E, ||\cdot||)$ is a Banach space. Let $K = [\theta, 1-\theta]$ for each $\theta \in (0,1)$, then, we define the cone $C \subset E \times E$ by

$$C = \{(y,z) \in E \times E : \min_{\vartheta\in K}[y(\vartheta) + z(\vartheta)] \geq \lambda||(y,z)||\}.$$
$$C_r = \{(y,z) \in C : \|(y,z)\| \leq r\},\ \partial C_r = \{(y,z) \in K : \|(y,z)\| = r\}.$$

As in [31], we define positive solution as follows.

Definition 1. *A pair of functions $(y,z) \in E \times E$ is called a positive solution of problem (1) under the given IBCs if $D^{\sigma}_{+0}y, D^{\rho}_{+0}z \in L^1[0,1]$ with $(y,z) > (0,0)$ on $(0,1] \times (0,1]$, where the functions y, z satisfy the IBCs given in (1) respectively, for all $\vartheta \in [0,1]$.*

Definition 2. *The Riemann-Liouville fractional derivative of order $\sigma > 0$ of a continuous function $y : (0,\infty) \to R$ is defined as*

$$D^{\sigma}_{+0}y(\vartheta) = \frac{1}{\Gamma(m-\sigma)}\left(\frac{d}{d\vartheta}\right)^m \int_0^{\vartheta} (\vartheta - s)^{m-\sigma-1}y(s)ds, \tag{2}$$

where $m = [\sigma] + 1$ and $[\sigma]$ denotes the integer part of σ.

Definition 3. *The Riemann-Liouville fractions of integration of order $\sigma > 0$ of a continuous function $y : (0,\infty) \to R$ is defined by*

$$I^{\sigma}_{+0}y(\vartheta) = \frac{1}{\Gamma(\sigma)} \int_0^{\vartheta} (\vartheta - s)^{\sigma-1}y(s)ds, \tag{3}$$

where the integral is point-wise defined on $(0,\infty)$.

Lemma 1. *Let $\sigma > o$, then the FODE*

$$D^{\sigma}_{+0}y(\vartheta) = 0 \tag{4}$$

has a solution given by

$$y(\vartheta) = \sum_{i=1}^{m} \frac{y^i(0)}{i!}\vartheta^{-i}. \tag{5}$$

Lemma 2. *Let $\sigma > 0$. Then we have*

$$I^{\sigma}_{+0}[D^{\sigma}_{+0}y(\vartheta)] = y(\vartheta) - \sum_{i=0}^{m} \frac{y^i(0)}{i!}\vartheta^{-i}. \tag{6}$$

Lemma 3. *[2] Let $\sigma > o$ and $\vartheta \in C(0,1) \cap L(0,1)$, then the FODE*

$$D^{\sigma}_{+0}y(\vartheta) = h(\vartheta)$$

has a solution given by

$$y(\vartheta) = c_1\vartheta^{\sigma-1} + c_2\vartheta^{\sigma-2} + \cdots + c_m\vartheta^{\sigma-m} + I^{\sigma}_{+0}h(\vartheta),$$

where $c_i \in R$ *for* $i = 0, 1, 2, \ldots, m$ *and* $m = [\sigma] + 1$.

Definition 4. *[32,33] On the Banach space* E *defined afore, the mapping* $d : E \times E \to R^n$ *is called a generalized metric on* E *if* $\forall$ x, y, *and* $y, z \in E$ *with* $y \neq x, z \neq y, z \neq y$, *then the following hold*

$(A1)$ $d(y,z) = 0 \Leftrightarrow y = z,\ \forall\, y, z \in E$
$(A2)$ $d(y,z) = d(z,y),\ \forall\, y, z \in E$
$(A3)$ $d(x,y) = d(x,z) + d(z,y) + d(y,y), \forall\, x, y, y, z, \in E.$

Further the pair (E, d) *is called a generalized metric space.*

Definition 5. *[32,33] Let* $M = \{M_{m,m} \in R_+^{m\times m}\}$, *for any matrix* $\mathbf{B} \in M$ *the spectral radius is defined by* $æ(\mathbf{B}) = \sup\{|\widehat{\lambda}_i|, i = 1, 2, ..., m\}$, *where* $\widehat{\lambda}_i$, *for* $i = 1, 2, ..., m$ *are the eigenvalues of the matrix* $\mathbf{B}$ *and the matrix will converge to zero if* $æ(\mathbf{B}) < 1$.

Lemma 4. *[32,33] A complete generalized metric space* (M, d), *with operator* $B : M \to M$ *such that there* $\exists$ *a matrix* $\mathbf{B} \in M$ *with*

$$d(By, Bz) \leq \mathbf{B}d(y,z), \text{for all } y, z \in M,$$

if $æ(\mathbf{B}) < 1$, *then* B *has a fixed point in* M.

Lemma 5. *[32,33] Consider a Banach space* E *with cone* $C \subseteq E$ *and* $y \subset C$ *is relatively open set with* $0 \in y$ *and* $B : \overline{y} \to y$ *be a completely continuous mapping. Then one of the following hold*

$(A1)$ *The mapping* B *has a fixed point in* $\overline{y}$
$(A2)$ *There exist* $y \in \partial y$ *and* $\eta \in (0,1)$ *with* $y = \eta By$.

Lemma 6. *[33,34] Consider a cone* C *in the Banach space* E *and if* $\mathfrak{A}_1$ *and* $\mathfrak{A}_2$ *be two bounded open sets in* E, *such that* $0 \in \mathfrak{A}_1 \subset \overline{\mathfrak{A}}_1 \subset \mathfrak{A}_2$. *Let* $B : C \cap (\overline{\mathfrak{A}}_2 \setminus \mathfrak{A}_1) \to C$ *be completely continuous operator and one of the following satisfied:*

(1) $\|By\| \leq \|y\|\ \ \forall\, y \in C \cap \partial\mathfrak{A}_1; \|B\| \geqslant \|y\|, \forall\, y \in C \cap \partial\mathfrak{A}_2$
(2) $\|By\| \geqslant \|y\|\ \ \forall\, y \in C \cap \partial\mathfrak{A}_1; \|B\| \leq \|y\|, \forall\, y \in C \cap \partial\mathfrak{A}_2$

Then B has at least one fixed point in $C \cap (\overline{\mathfrak{A}}_2 \setminus \mathfrak{A}_1)$.

3. Existence of at Least One Solution

Lemma 7. *Let* $h \in C[0,1]$, *then the BVP*

$$\begin{cases} D^{\sigma}_{+0}y(\vartheta) + h(\vartheta) = 0; & \vartheta \in (0,1); \quad m-1 < \sigma \leq m, \\ y(0) = y'(0) = y''(0) = \cdots = y^{(m-2)}(0) = 0, & y(1) = \delta\displaystyle\int_0^1 y(s)ds,, \end{cases} \tag{7}$$

where $\delta \in (0,2)$, has the following unique solution

$$y(\vartheta) = \int_0^1 H_{\sigma}(\vartheta, s)h(s)ds,$$

where H_σ is the Green's function given by

$$H_\sigma(\vartheta, s) = \begin{cases} \dfrac{\vartheta^{\sigma-1}(1-s)^{\sigma-1}(\sigma-\delta+\delta s) - (\sigma-\delta)(\vartheta-s)^{\sigma-1}}{(\sigma-\delta)\Gamma(\delta)}, & 0 \le s \le \vartheta \le 1, \\ \dfrac{\vartheta^{\sigma-1}(1-s)^{\sigma-1}(\sigma-\delta+\delta s)}{(\sigma-\delta)\Gamma(\delta)}, & 0 \le \vartheta \le s \le 1. \end{cases} \quad (8)$$

Proof. Thanks to Lemma 3 for (7), one has

$$y(\vartheta) = -I^{\sigma}_{+0}h(\vartheta) + c_1\vartheta^{\sigma-1} + c_2\vartheta^{\sigma-2} + \cdots + c_m\vartheta^{\sigma-m}. \quad (9)$$

By using initial condition $y(0) = y'(0) = y''(0) = \cdots = y^{(m-2)}(0) = 0$, we get $c_2 = c_3 = \cdots = c_m = 0$. Therefore (9) implies that

$$y(\vartheta) = c_1\vartheta^{\sigma-1} - I^{\sigma}_{+0}h(\vartheta). \quad (10)$$

By using boundary condition $y(1) = \delta\int_0^1 y(s)ds$ in (10), we get

$$c_1 = \int_0^1 \frac{(\vartheta-s)^{\sigma-1}}{\Gamma(\sigma)}h(s)ds + \delta\int_0^1 y(s)ds.$$

Hence we have the following solution to (1)

$$y(\vartheta) = -\int_0^\vartheta \frac{(\vartheta-s)^{\sigma-1}}{\Gamma(\sigma)}h(s)ds + \vartheta^{\sigma-1}\int_0^1 \frac{(1-s)^{\sigma-1}}{\Gamma(\sigma)}h(s)ds + \delta\vartheta^{\sigma-1}\int_0^1 y(s)ds. \quad (11)$$

Let $B = \int_0^1 y(s)ds$, then from Equation (11), we have

$$B = -\int_0^1\int_0^\vartheta \frac{(\vartheta-s)^{\sigma-1}}{\Gamma(\sigma)}h(s)ds + \int_0^1\int_0^1 \frac{\vartheta^{\sigma-1}(\vartheta-s)^{\sigma-1}}{\Gamma(\sigma)}h(s)ds + \int_0^1 \delta B\vartheta^{\sigma-1}ds$$

$$B = -\int_0^1 \frac{(1-s)^{\sigma}}{\sigma\Gamma(\sigma)}h(s)ds + \int_0^1 \frac{(1-s)^{\sigma-1}}{\sigma\Gamma(\sigma)}h(s)ds + \frac{1}{\sigma}\delta B \quad (12)$$

implies Equation (12), so we get

$$B = -\frac{1}{\sigma-\delta}\int_0^1 \frac{(1-s)^{\sigma}}{\Gamma(\sigma)}h(s)ds + \frac{1}{\sigma-\delta}\int_0^1 \frac{(1-s)^{\sigma-1}}{\Gamma(\sigma)}h(s)ds.$$

Replacing this valve in (11), we get

$$\begin{aligned} y(\vartheta) &= -\int_0^t \frac{(\vartheta-s)^{\sigma-1}}{\Gamma(\sigma)}h(s)ds + \vartheta^{\sigma-1}\int_0^1 \frac{(1-s)^{\sigma-1}}{\Gamma(\sigma)}h(s)ds - \frac{\delta}{\sigma-\delta}\int_0^1 \frac{\vartheta^{\sigma-1}(1-s)^{\sigma}}{\sigma\Gamma(\sigma)}h(s)ds \\ &+ \frac{\delta}{\sigma-\delta}\int_0^1 \frac{\vartheta^{\sigma-1}(1-s)^{\sigma-1}}{\Gamma(\sigma)}h(s)ds. \\ &= -\int_0^\vartheta \frac{(\vartheta-s)^{\sigma-1}}{\Gamma(\sigma)}h(s)ds + \int_0^1 \frac{\vartheta^{\sigma-1}(1-s)^{\sigma-1}(\sigma-\delta+\delta s)}{(\sigma-\delta)\Gamma(\sigma)}h(s)ds \\ &= \int_0^\vartheta \frac{\vartheta^{\sigma-1}(1-s)^{\sigma-1}(\sigma-\delta+\delta s) - (\sigma-\delta)(\vartheta-s)^{\sigma-1}}{(\sigma-\delta)\Gamma(\sigma)}h(s)ds \\ &+ \int_\vartheta^1 \frac{\vartheta^{\sigma-1}(1-s)^{\sigma-1}(\sigma-\delta+\delta s)}{(\sigma-\delta)\Gamma(\sigma)}h(s)ds \\ &= \int_0^1 H_\sigma(\vartheta, s)h(s)ds, \end{aligned}$$

where $H_\sigma(\vartheta, s)$ is the Green's function of BVP (7). Similarly we can obtain $z(\vartheta) = \int_0^1 H_æ(\vartheta, s)h(s)ds$, where $H_æ(\vartheta, s)$ is the Green's function for the second equation of the system (1) and is given by

$$H_æ(\vartheta, s) = \begin{cases} \dfrac{\vartheta^{æ-1}(1-s)^{æ-1}(æ-\varrho+\varrho s) - (æ-\varrho)(\vartheta-s)^{æ-1}}{(æ-\varrho)\Gamma(æ)}, & 0 \le s \le \vartheta \le 1, \\ \dfrac{\vartheta^{æ-1}(1-s)^{æ-1}(æ-\varrho+\varrho s)}{(æ-\varrho)\Gamma(æ)}, & 0 \le \vartheta \le s \le 1. \end{cases} \tag{13}$$

□

Lemma 8. *Let* $H(\vartheta, s) = (H_\sigma(\vartheta, s),\ H_æ(\vartheta, s))$ *be the Green's function of* (1) *defined in Equations* (8) *and* (13). *This* $H(\vartheta, s)$ *has the given properties*

(F_1) $H(\vartheta, s)$ *is continuous function on the unit square* $\forall\ (\vartheta, s) \in [0,1] \times [0,1]$
(F_2) $H(\vartheta, s) \ge 0\ \forall\ \vartheta, s \in [0,1]$ *and* $H(\vartheta, s) > 0\ \forall\ \vartheta, s \in (0,1)$
(F_3) $\max\limits_{0 \le \vartheta \le 1} H(\vartheta, s) = H(1, s), \forall s \in [0,1]$
(F_4) $\min\limits_{\vartheta \in [\theta, 1-\theta]} H(\vartheta, s) \ge \lambda(s) H(1, s)$ *for each* $\theta,\ s \in (0,1)$,

where $\lambda = \min\{\lambda_\sigma = \theta^{\sigma-1}, \lambda_æ = \theta^{æ-1}\}$.

Now according to Lemma 7, we can write system (1) as follows

$$\begin{cases} y(\vartheta) = \int\limits_0^1 H_\sigma(\vartheta, s)\varphi(s, y(s), z(s))ds, \\ z(\vartheta) = \int\limits_0^1 H_æ(\vartheta, s)\chi(s, y(s), z(s))ds. \end{cases} \tag{14}$$

Let $B : E \times E \to E \times E$ be the operator defined as

$$\begin{aligned} B(y,z)(\vartheta) &= \left(\int\limits_0^1 H_\sigma(\vartheta, s)\varphi(s, y(s), z(s))ds,\ \int\limits_0^1 H_æ(\vartheta, s)\chi(s, y(s), z(s))ds \right) \\ &= \Big(B_1(y,z)(\vartheta), B_2(y,z)(\vartheta) \Big). \end{aligned} \tag{15}$$

Then the fixed point of operator B coincides with the solution of the coupled system (1).

Theorem 1. *Consider that* $u, v : [0,1] \times [0,\infty) \times [0,\infty) \to [0,\infty)$ *are continuous. Then* $B(C) \subset C$ *and* $B : C \to C$ *is completely continuous, where* B *is defined in* (15).

Proof. To prove that $B(C) \subset C$, let $(y,z) \in C$, then by Lemma 8, we have $B(y,z) \in C$ and from (F_4) and $\forall\ \vartheta \in K$, we obtain

$$B_1(y(\vartheta), z(\vartheta)) = \int\limits_0^1 H_\sigma(\vartheta, s)\varphi(s, y(s), z(s))ds \ge \lambda_\sigma \int\limits_0^1 H_\sigma(1, s)\varphi(s, y(s)), z(s)ds. \tag{16}$$

Also from (F_3), we obtain

$$B_1(y(\vartheta), z(\vartheta)) = \int\limits_0^1 H_\sigma(\vartheta, s)\varphi(s, y(s), z(s))ds \le \int\limits_0^1 H_\sigma(1, s)\varphi(s, y(s)), z(s)ds. \tag{17}$$

Thus from (16) and (17), we have

$$B_1(y(\vartheta), z(\vartheta)) \geq \lambda \|B_1(y,z)\|, \text{ for all } \vartheta \in K.$$

Similarly, one can write that

$$B_2(y(\vartheta), z(\vartheta)) \geq \lambda \|B_2(y,z)\|, \text{ for all } \vartheta \in K.$$

Thus

$$B_1(y(\vartheta), z(\vartheta)) + B_2(y(\vartheta), z(\vartheta)) \geq \lambda \|B(y,z)\|, \text{ for all } \vartheta \in K,$$
$$\min_{\vartheta \in K}[B_1(y(\vartheta), z(\vartheta)) + B_2(y(\vartheta), z(\vartheta))] \geq \lambda \|B(y,z)\|.$$

Hence we have $B(y,z) \in C \Rightarrow B(C) \subset C$. Next, like the proof of Theorem 1 of [35], and applying the Arzelà-Ascoli's theorem, it can be easily proven that $B : C \to C$ is completely continuous □

Theorem 2. *Consider that φ and χ are continuous on $[0,1] \times [0,\infty) \times (0,\infty) \to [0,\infty)$, and there exist $f_i(\vartheta), H_i(\vartheta), (i = 1,2) : (0,1) \to [0,\infty)$ that satisfy*

(A_1) $|\varphi(\vartheta, y, z) - \varphi(\vartheta, \bar{y}, \bar{z})| \leq u_1(\vartheta)|y - \bar{y}| + v_1(\vartheta)|z - \bar{z}|$, *for* $\vartheta \in (0,1)$ *and* $y, z, \bar{y}, \bar{z} \geq 0$
(A_2) $|\chi(\vartheta, y, z) - \chi(\vartheta, \bar{y}, \bar{z})| \leq u_2(\vartheta)|y - \bar{y}| + v_2(\vartheta)|z - \bar{z}|$, *for* $\vartheta \in (0,1)$ *and* $y, z, \bar{y}, \bar{z} \geq 0$
(A_3) $æ(\mathbf{B}) < 1$, *where* $\mathbf{B} \in \{M_{2,2} \in R_+^{2\times 2}\}$ *is a matrix given by*

$$\begin{bmatrix} \int_0^1 H_\sigma(1,s)u_1(s)ds & \int_0^1 H_\sigma(1,s)v_1(s)ds \\ \int_0^1 H_æ(1,s)u_2(s)ds & \int_0^1 H_æ(1,s)v_2(s)ds \end{bmatrix}.$$

Then the system (1) has a unique positive solution $(y,z) \in C$.

Proof. Let us define a generalized metric $d : E^2 \times E^2 \to R^2$ by

$$d((y,z),(\bar{y},\bar{z})) = \begin{pmatrix} \|y - \bar{y}\| \\ \|z - \bar{z}\| \end{pmatrix}, \text{ for all } (y,z),(\bar{y},\bar{z}) \in E \times E.$$

Obviously $(E \times E, d)$ is a generalized complete metric space. Then for any $(y,z),(\bar{y},\bar{z}) \in E \times E$ and using property (F_3) we get

$$\begin{aligned} |B_1(y,z)(\vartheta) - B_1(\bar{y},\bar{z})(\vartheta)| &\leq \max_{\vartheta \in [0,1]} \int_0^1 |H_\sigma(\vartheta,s)|[|\varphi(s,y(s),z(s)) - \varphi(s,\bar{y}(s),\bar{z}(s))|]ds \\ &\leq \int_0^1 H_\sigma(1,s)[u_1(s)\|y - \bar{y}\| + v_1(s)\|z - \bar{z}\|]ds \\ &\leq \int_0^1 u_1(s)H_\sigma(1,s)ds\|y - \bar{y}\| + \int_0^1 v_1(s)H_\sigma(1,s)ds\|z - \bar{z}\|. \end{aligned}$$

Similarly we can show that

$$|B_2(y,z) - B_2(\bar{y},\bar{z})| \leq \int_0^1 u_2(s)H_æ(1,s)ds\|y - \bar{y}\| + \int_0^1 v_2(s)H_æ(1,s)ds\|z - \bar{z}\|.$$

Thus we have

$$|B(y,z) - B(\bar{y},\bar{z})| \leq \mathbf{B}d((y,z),(\bar{y},\bar{z})), \ \forall (y,z),(\bar{y},\bar{z}) \in E \times E,$$

where

$$\mathbf{B} = \begin{bmatrix} \int_0^1 H_\sigma(1,s)u_1(s)ds & \int_0^1 H_\sigma(1,s)v_1(s)ds \\ \int_0^1 H_æ(1,s)u_2(s)ds & \int_0^1 H_æ(1,s)v_2(s)ds \end{bmatrix}.$$

As $æ(\mathbf{B}) < 1$, in the light of Lemma 4, system (1) has a unique positive solution. □

Theorem 3. *Consider that φ and χ are continuous on $[0,1] \times [0,\infty) \times (0,\infty) \to [0,\infty)$ and there exist $a_i, b_i, c_i (i = 1,2) : (0,1) \to [0,\infty)$ satisfying:*

(A_4) $\varphi(\vartheta, y(\vartheta), z(\vartheta)) \leq a_1(\vartheta) + b_1(\vartheta)y(\vartheta) + c_1(\vartheta)z(\vartheta),\ \vartheta \in (0,1), y, z \geq 0$

(A_5) $\chi(\vartheta, y(\vartheta), z(\vartheta)) \leq a_2(\vartheta) + b_2(\vartheta)y(\vartheta) + c_2(\vartheta)z(\vartheta),\ \vartheta \in (0,1), y, z \geq 0$

(A_6) $\Lambda_1 = \int_0^1 H_\sigma(1,s)a_1(s)ds < \infty,\ \Delta_1 = \int_0^1 H_\sigma(1,s)[b_1(s) + c_1(s)]ds < \frac{1}{2}$

(A_7) $\Lambda_2 = \int_0^1 H_æ(1,s)a_2(s)ds < \infty,\ \Delta_2 = \int_0^1 H_æ(s,s)[b_2(s) + c_2(s)]ds < \frac{1}{2}.$

Then the system (1) has at least one positive solution in

$$\Big\{(y,z) \in C : \|(y,z)\| \leq r\Big\},\ \textit{where}\ \max\left\{\frac{\Lambda_1}{\frac{1}{2} - \Delta_1}, \frac{\Lambda_2}{\frac{1}{2} - \Delta_2}\right\} < r.$$

Proof. Define $\Omega = \Big\{(y,z) \in C : \|(y,z)\| < r\Big\}$ with $\max\left\{\frac{\Lambda_1}{\frac{1}{2} - \Delta_1}, \frac{\Lambda_2}{\frac{1}{2} - \Delta_2}\right\} < r$.

According to the Theorem 1, the operator $B : \overline{\Omega} \to C$ is completely continuous. Let $(y,z) \in \Omega$, such that $\|(y,z)\| < r$. Then, we have

$$\begin{aligned} \|B_1(y,z)\| &= \max_{\vartheta \in [0,1]} \left| \int_0^1 H_\sigma(\vartheta,s)\varphi(s, y(s), z(s)) \right| ds \\ &\leq \left(\int_0^1 H_\sigma(1,s)a_1(s)ds + \int_0^1 H_\sigma(1,s)b_1(s)|y(s)|ds + \int_0^1 H_\sigma(1,s)c_1(s)|z(s)|ds \right) \\ &\leq \int_0^1 H_\sigma(1,s)a_1(s)ds + r\left[\int_0^1 H_\sigma(1,s)[b_1(s) + c_1(s)]ds \right] \\ &\leq \Lambda_1 + r\Delta_1 < \frac{r}{2}, \end{aligned}$$

Similarly, $\|B_2(y,z)\| < \frac{r}{2}$, thus $\|B(y,z)\| < r$. Therefore, thanks to Lemma 5, we have $B(y,z) \in \overline{\Omega}$, thus $B : \overline{\Omega} \to \overline{\Omega}$. Let there exist $\varsigma \in (0,1)$ and $(y,z) \in \partial\Omega$ such that $(y,z) = \varsigma B(y,z)$. Then in the light of assumptions $(A_4), (A_5)$ and by (F_4) of Lemma 8, we get $\forall\ \vartheta \in [0,1]$

$$\begin{aligned} |z(\vartheta)| &\leq \varsigma \int_0^1 H_\sigma(\vartheta,s)|\varphi(s, y(s), z(s))|ds \\ &\leq \varsigma\left(\int_0^1 H_\sigma(1,s)a_1(s)ds + \int_0^1 H_\sigma(1,s)b_1(s)y(s)ds + \int_0^1 H_\sigma(1,s)c_1(s)z(s)ds \right) \\ &\leq \zeta\Big(\Delta_1 + r\Lambda_1\Big) \\ &< \varsigma\frac{r}{2} \end{aligned}$$

which implies that $\|y\| < \varsigma\frac{r}{2}$. Similarly, it can be proved that $\|z\| < \varsigma\frac{r}{2}$. From which, we have $\|(y,z)\| < \varsigma r$, with $\varsigma \in (0,1)$ which is a contradiction that $(y,z) \in \partial\Omega$ as $r = \|(y,z)\|$. Thus, according to Lemma 5, B has at least one fixed point $(y,z) \in \overline{\Omega}$. □

Next the following assumptions and notations will be used:

(C_1) $\varphi, \chi : [0,1] \times [0,\infty) \times [0,\infty) \to [0,\infty)$ are continuous and $\varphi(\vartheta,0,0) = \chi(\vartheta,0,0) = 0$ uniformly with respect to ϑ on $[0,1]$

(C_2) $H_\sigma(1,s), H_{æ}(1,s)$ defined in Lemma 8 satisfy

$$0 < \int_0^1 H_\sigma(1,s)ds < \infty,\ 0 < \int_0^1 H_{æ}(1,s)ds < \infty$$

(C_3) Let these limits hold

$$\begin{aligned}
\varphi^{\alpha} &= \lim_{(y,z)\to(\alpha,\alpha)} \sup_{\vartheta\in[0,1]} \frac{\varphi(\vartheta,y,z)}{y+z},\ \chi^{\alpha} = \lim_{(y,z)\to(\alpha,\alpha)} \sup_{\vartheta\in[0,1]} \frac{\chi(\vartheta,y,z)}{y+z}, \\
\varphi_{\alpha} &= \lim_{(y,z)\to(\alpha,\alpha)} \in f_{\vartheta\in[0,1]} \frac{\varphi(\vartheta,y,z)}{y+z},\ \chi_{æ} = \lim_{(y,z)\to(\alpha,\alpha)} \in f_{\vartheta\in[0,1]} \frac{\chi(\vartheta,y,z)}{y+z},\ \text{where } \alpha \in \{0,\infty\}.
\end{aligned} \tag{18}$$

$$\alpha_\sigma = \max_{\vartheta\in[0,1]} \int_0^1 H_\sigma(\vartheta,s)ds,\ \alpha_{æ} = \max_{\vartheta\in[0,1]} \int_0^1 H_{æ}(\vartheta,s)ds. \tag{19}$$

Theorem 4. *If the assumptions $(C_1) - (C_2)$ hold and one of the following conditions is also satisfied:*

(D_1) $\varphi_0\left(\lambda^2 \int_\theta^{1-\theta} H_\sigma(1,s)ds\right) > 1,\ \varphi^{\infty}\alpha_\sigma < 1$ *and* $\chi_0\left(\lambda^2 \int_\theta^{1-\theta} H_{æ}(1,s)ds\right) > 1,\ \chi^{\infty}\alpha_{æ} < 1.$
Moreover, $\varphi_0 = \chi_0 = \infty$ *and* $\varphi^{\infty} = \chi^{\infty} = 0$

(D_2) *There exist two constants η_1, η_2 with $0 < \eta_1 \le \eta_2$ such that $\varphi(\vartheta,\cdot,\cdot)$ and $\chi(\vartheta,\cdot,\cdot)$ are nondecreasing on $[0,\eta_2]$ $\forall\ \vartheta \in [0,1]$,*

$$\varphi(\vartheta,\lambda_\sigma\eta_1,\lambda_{æ}\eta_1) \ge \frac{\eta_1}{2}\left(\lambda_\sigma \int_\theta^{1-\theta} H_\sigma(1,s)ds\right)^{-1},$$

$$\chi(\vartheta,\lambda_\sigma\eta_1,\lambda_{æ}\eta_1) \ge \frac{\eta_1}{2}\left(\lambda_{æ} \int_\theta^{1-\theta} H_{æ}(1,s)ds\right)^{-1}$$

$$\text{and } \varphi(\vartheta,\eta_2,\eta_2) \le \frac{\eta_2}{2\alpha_\sigma},\ \chi(\vartheta,\eta_2,\eta_2) \le \frac{\eta_2}{2\alpha_{æ}},\ \textit{for all } \vartheta \in [0,1],$$

where $\lambda, H_\sigma(1,s), H_{æ}(1,s)$ defined in Lemma 8 and $\varphi_0, \chi_0, \varphi^{\infty}, \chi^{\infty}, \alpha_\sigma, \sigma_\alpha$ defined in Equations (18) and (19). Then the coupled system (1) has at least one positive solution.

Proof. B as defined in (15) is completely continuous.

Case I. Let the condition (D_1) hold. Taking $\varphi_0\left(\lambda_\sigma^2 \int_\theta^{1-\theta} H_\sigma(1,s)ds\right) > 1$, then there exists a constant $\kappa_1 > 0$ such that

$$\varphi(\vartheta,y,z) \ge (\varphi_0 - r_1)(y(\vartheta) + z(\vartheta)),\ \chi(\vartheta,y,z) \ge (\chi_0 - r_2)(y(\vartheta) + z(\vartheta)),\ \text{for all } \vartheta \in [0,1], y,z \in [0,\kappa_1],$$

where $r_1 > 0$, and satisfies the conditions

$$(\varphi_0 - r_1)\frac{\lambda_\sigma^2}{2}\int\limits_\theta^{1-\theta} H_\sigma(1,s)ds \geq 1, \ \ (\chi_0 - r_1)\frac{\lambda_{æ}^2}{2}\int\limits_\theta^{1-\theta} H_{æ}(1,s)ds \geq 1.$$

So for $\vartheta \in [0,1], (y,z) \in \partial C_{\kappa_1}$, we have

$$\begin{aligned} B_1(y,z)(\vartheta) &= \int\limits_0^1 H_\sigma(\vartheta,s)\varphi(s,y(s),z(s))ds \geq \lambda_\sigma \int\limits_0^1 H_\sigma(1,s)\varphi(s,y(s),z(s))ds \\ &\geq (\varphi_0 - r_1)\frac{\lambda_\sigma^2}{2}\int\limits_0^1 H_\sigma(1,s)ds\|(y,z)\| \geq \frac{\|(y,z)\|}{2}. \end{aligned}$$

Analogously

$$\begin{aligned} B_2(y,z)(\vartheta) &= \int\limits_0^1 H_{æ}(\vartheta,s)\chi(s,y(s),z(s))ds \geq \lambda_{æ} \int\limits_0^1 H_{æ}(1,s)\varphi(s,y(s),z(s))ds \\ &\geq (\chi_0 - r_2)\frac{\lambda_{æ}^2}{2}\int\limits_0^1 H_{æ}(1,s)ds\|(y,z)\| \geq \frac{\|(y,z)\|}{2}. \end{aligned}$$

Therefore, we have

$$\|B(y,z)\| \geq \|B_1(y,z)\| + \|B_2(y,z)\| \geq \|(y,z)\|. \tag{20}$$

Also for $\varphi^\infty\alpha_\sigma < 1$ and $\chi^\infty\alpha_{æ} < 1$, there exists a constant say $\bar{\kappa_2} > 0$ such that $\varphi(\vartheta,y,z) \leq (\varphi^\infty + r_2)(y+z), \chi(\vartheta,y,z) \leq (\chi^\infty + r_2)(y+z)$, for $\vartheta \in [0,1], y,z \in (\bar{\kappa_2},\infty)$, where $r_2 > 0$ satisfies the conditions $\alpha_\sigma(\varphi^\infty + r_2) \leq 1$, $\alpha_{æ}(\chi^\infty + r_2) \leq 1$. Let $J = \max_{\vartheta\in[0,1],y,z\in[0,\bar{\kappa_2}]} \varphi(\vartheta,y,z), L = \max_{\vartheta\in[0,1],y,z\in[0,\bar{\kappa_2}]} \chi(\vartheta,y,z)$, then $\varphi(\vartheta,y,z) \leq J + (\varphi^\infty + r_2)(y,z), \chi(\vartheta,y,z) \leq L + (\chi^\infty + r_2)(y,z)$. Now setting $\max\{\kappa_1,\bar{\kappa_2}, J\alpha_\sigma(1-\alpha_\sigma(\varphi^\infty + r_2))^{-1}\} \leq \frac{\kappa_2}{2}, \max\{\kappa_1,\bar{\kappa_2}, L\alpha_{æ}(1-\alpha_{æ}(\chi^\infty + r_2))^{-1}\} \leq \frac{\kappa_2}{2}$.

So for any $\vartheta \in [0,1], (y,z) \in \partial C_{\kappa_2}$, we obtain

$$\begin{aligned} B_1(y,z)(\vartheta) &= \int\limits_0^1 H_\sigma(\vartheta,s)\varphi(s,y(s),z(s))ds \leq \lambda_\sigma \int\limits_0^1 H_\sigma(1,s)\varphi(s,y(s),z(s))ds \\ &\leq \int\limits_0^1 H_\sigma(1,s)(J + (\varphi^\infty + r_2)[u(s)+z(s)]ds \\ &\leq J\int_0^1 H_\sigma(1,s)ds + (\varphi^\infty + r_2)\int\limits_0^1 H_\sigma(1,s)ds\|(y,z)\| \\ &< \frac{\kappa_2}{2} - \alpha_\sigma(\varphi^\infty + r_2)\frac{\kappa_2}{2} + (\varphi^\infty + r_2)\alpha_\sigma\|(y,z)\| < \frac{\kappa_2}{2}. \end{aligned}$$

Similarly $B_2(y,z)(\vartheta) < \frac{\kappa_2}{2}$, as $(y,z) \in \partial C_{\kappa_2}$, thus we have

$$\|B(y,z)\| < \|(y,z)\|. \tag{21}$$

Case II. If assumptions in (D_2) hold, then in light of the definition of C for $(y,z) \in \partial C_{\eta_1}$, we have $\|(y,z)\| = \eta_1$, for $\vartheta \in K$. Then from (D_2), we have

$$B_1(y,z)(\vartheta) = \int_0^1 H_\sigma(\vartheta,s)\varphi(s,y(s),z(s))ds \geq \lambda_\sigma \int_\theta^{1-\theta} H_\sigma(1,s)\varphi(s,y(s),z(s))ds$$
$$\geq \left(\lambda_\sigma \int_\theta^{1-\theta} H_\sigma(1,s)ds\right)\frac{\eta_1}{2}\left(\lambda_\sigma \int_\theta^{1-\theta} H_\sigma(1,s)ds\right)^{-1} = \frac{\eta_1}{2}.$$

Similarly it can also be obtained that $B_2(y,z)(\vartheta) \geq \frac{\eta_1}{2}$, for $(y,z) \in \partial C_{\eta_1}$, and we get

$$\|B(y,z)\| = \|B_1(y,z)\| + \|B_2(y,z)\| \geq \|(y,z)\|. \tag{22}$$

Also for $(y,z) \in \partial C_{\eta_2}$, we get that $\|(y,z)\| = \eta_2$ for $\vartheta \in [0,1]$. Then from (D_2), one can get

$$B_1(y,z)(\vartheta) = \int_0^1 H_\sigma(\vartheta,s)\varphi(s,y(s),z(s))ds \leq \int_0^1 H_\sigma(1,s)\varphi(s,y(s),z(s))ds$$
$$\leq \frac{\eta_2}{2\alpha_\sigma}\int_0^1 H_\sigma(1,s)ds = \frac{\eta_2}{2}.$$

Similarly, it can also obtained that $B_2(y,z)(\vartheta) \leq \frac{\eta_2}{2}$, $(y,z) \in \partial C_{\eta_2}$. Hence, we have

$$\|B(y,z)\| = \|B_1(y,z)\| + \|B_2(y,z)\| \leq \|(y,z)\|. \tag{23}$$

Now according to the application of Lemma 6 to (20) and (21) or (22) and (23) implies that B has a fixed point $(y_1,z_1) \in \overline{C}_{\kappa,\eta}$ or $(y_1,z_1) \in \bar{C}_{\kappa_i,\eta_i}(i = 1,2)$ such that $y_1(\vartheta) \geq \lambda_\sigma\|y_1\| > 0$ and $z_1(\vartheta) \geq \lambda_{æ}\|z_1\| > 0, \vartheta \in [0,1]$. From which it follows that the coupled system (1) has at least one positive solution. □

Theorem 5. *Under the conditions* $(C_1) - (C_3)$ *and if the following assumptions hold*

(D_3) *If* $\varphi^0\alpha_\sigma < 1; \varphi^\infty\left(\lambda_\sigma^2 \int_\theta^{1-\theta} H_\sigma(1,s)ds\right) > 1$ *and* $\chi^0\alpha_{æ} < 1; \chi^\infty\left(\lambda_{æ}^2 \int_\theta^{1-\theta} H_\sigma(1,s)ds\right) > 1,$

then the coupled system (1) *has at least one positive solution. Further, if* $\varphi^0 = \chi^0 = 0$ *and* $\varphi^\infty = \chi^\infty = \infty$, *where* $\lambda, H_\sigma(1,s), H_{æ}(1,s)$ *defined in Lemma* 8 *and* $\varphi_0, \chi_0, \varphi^\infty, \chi^\infty, \alpha_\sigma, \sigma_\alpha$ *defined in Equations* (18) *and* (19), *then the the considered system* (1) *has at least one positive solution.*

Proof. Proof can be obtained as proof of Theorem 4. □

4. Existence of More Than One Solutions

Theorem 6. *Consider that* $(C_1) - (C_3)$ *hold and the following conditions are satisfied:*

(D_4) *If* $\varphi_0\left(\lambda_\sigma^2 \int_\theta^{1-\theta} H_\sigma(1,s)ds\right) > 1, \varphi_\infty\left(\lambda_\sigma^2 \int_\theta^{1-\theta} H_\sigma(1,s)ds\right) > 1$ *and*

$\chi_0\left(\lambda_{æ}^2 \int_\theta^{1-\theta} H_{æ}(1,s)ds\right) > 1, \chi_\infty\left(\lambda_{æ}^2 \int_\theta^{1-\theta} H_\sigma(1,s)ds\right) > 1.$

Moreover, $\varphi_0 = \chi_0 = \varphi_\infty = \chi_\infty = \infty$ *also hold:*

(D_5) *there exists* $a > 0$ *such that*
$\max_{\vartheta\in[0,1],(y,z)\in\partial C_a} \varphi(\vartheta,y,z) < \frac{a}{2\alpha_\sigma}$ *and* $\max_{\vartheta\in[0,1],(y,z)\in\partial C_a} \chi(\vartheta,y,z) < \frac{a}{2\alpha_{æ}}$.

Then the coupled system (1) *has at least two positive solutions* $(y,z),(\bar{y},\bar{z})$ *such that*

$$0 < \|(y,z)\| < a < \|(\bar{y},\bar{z})\|. \tag{24}$$

Where $\lambda, H_\sigma(1,s), H_æ(1,s)$ are defined in Lemma 8 and $\varphi_0,\chi_0,\varphi^\infty,\chi^\infty,\alpha_\sigma,\sigma_\alpha$ defined in Equations (18) and (19)

Proof. Let (D_4) hold. Select κ,η such that $0 < \kappa < \mu < \eta$. Now if $\varphi_0\left(\lambda_\sigma^2 \int\limits_\theta^{1-\theta} H_\sigma(1,s)ds\right) > 1$ and $\chi_0\left(\lambda_æ^2 \int\limits_\theta^{1-\theta} H_æ(1,s)ds\right) > 1$, then like the proof of Theorem 4, we have

$$\|B(y,z)\| \geq |(y,z)\|, \text{ for } (y,z) \in \partial C_\kappa. \tag{25}$$

Now, if $\varphi_\infty\left(\lambda_\sigma^2 \int\limits_\theta^{1-\theta} H_\sigma(1,s)ds\right) > 1$ and $\chi_\infty\left(\lambda_æ^2 \int\limits_\theta^{1-\theta} H_æ(1,s)ds\right) > 1$, then like the proof of Theorem 4, we have

$$\|B(y,z)\| \geq \|(y,z)\|, \text{ for } (y,z) \in \partial C_\eta. \tag{26}$$

Also from (D_5), $(y,z) \in \partial C_\mu$, we get

$$\begin{aligned} B_1(y,z)(\vartheta) &= \int_0^1 H_\sigma(\vartheta,s)\varphi(s,y(s),z(s))ds \\ &\leq \int_0^1 H_\sigma(1,s)\varphi(s,y(s),z(s))ds < \frac{\mu}{2\alpha_\sigma}\int_0^1 H_\sigma(1,s)ds = \frac{\mu}{2}. \end{aligned}$$

Similarly, we have $B_1(y,z)(\vartheta) < \frac{\mu}{2}$ as $(y,z) \in \partial C_\mu$. Hence, we have

$$\|B(y,z)\| < |(y,z)\|, \text{ for } (y,z) \in \partial C_\mu. \tag{27}$$

Now according to Lemma 6 for (25) and (27), we have gives that B has a fixed point $(y,z) \in \partial\overline{C}_{\kappa,\mu}$ and a fixed point in $(\bar{y},\bar{z}) \in \partial\overline{C}_{\mu,\eta}$. Therefore system (1) has at least two positive solutions $(y,z),(\bar{y},\bar{z})$ such that $\|(y,z)\| \neq \mu$ and $\|(\bar{y},\bar{z})\| \neq \mu$. Thus the relation (24) holds. □

Theorem 7. *Consider that* $(C_1)-(C_3)$ *hold together with the given conditions*

(D_6) $\alpha_\sigma\varphi_0 < 1$ *and* $\varphi_\infty\alpha_\sigma < 1$; $\alpha_æ\chi_0 < 1$, *and* $\chi_\infty\alpha_æ < 1$
(D_7) *there exist* $\mu > 0$ *such that*

$$\max_{\vartheta\in K,(y,z)\in\partial C_\mu} \varphi(\vartheta,y,z) > \frac{\mu}{2}\left(\lambda_\sigma^2\int\limits_\theta^{1-\theta} H_\sigma(1,s)ds\right)^{-1},$$

$$\max_{\vartheta\in K,(y,z)\in\partial C_\mu} \chi(\vartheta,y,z) > \frac{\mu}{2}\left(\lambda_æ^2\int\limits_\theta^{1-\theta} H_æ(1,s)ds\right)^{-1},$$

such that

$$0 < \|(y,z)\| < \mu < \|(\bar{y},\bar{z})\|,$$

where $\lambda, H_\sigma(1,s), H_æ(1,s)$ *defined in Lemma* 8 *and* $\varphi_0,\chi_0,\varphi^\infty,\chi^\infty,\alpha_\sigma,\sigma_\alpha$ *defined in Equations* (18) *and* (19). *Thus the system* (1) *has at least two positive solutions.*

Proof. We left the proof out, as it similar to the proof of Theorem 6. □

In same line for multiple solutions we give the following results.

Theorem 8. *Let* $(C_1)-(C_3)$ *hold. If there exist* $2m$ *positive numbers* $\mathbf{u}_L, \hat{\mathbf{u}}_L,\ L=1,2\ldots m$ *with* $\mathbf{u}_1 < \lambda_\sigma \hat{\mathbf{u}}_1 < \hat{\mathbf{u}}_1 < \mathbf{u}_2 < \lambda_\sigma \hat{\mathbf{u}}_2 < \hat{\mathbf{u}}_2 \ldots \mathbf{u}_m < \lambda_\sigma \hat{\mathbf{u}}_m < \hat{\mathbf{u}}_m$ *and* $\mathbf{u}_1 < \lambda_{æ} \hat{\mathbf{u}}_1 < \hat{\mathbf{u}}_1 < \mathbf{u}_2 < \lambda_{æ} \hat{\mathbf{u}}_2 < \hat{\mathbf{u}}_2 \ldots \mathbf{u}_m < \lambda_{æ} \hat{\mathbf{u}}_m < \hat{\mathbf{u}}_m$*, such that*

(D_8) $\varphi(\vartheta, y(\vartheta), z(\vartheta)) \geq \mathbf{u}_L \left(\lambda_\sigma \int\limits_0^1 H_\sigma(1,s)ds\right)^{-1}$, *for* $(\vartheta, y, z) \in [0,1] \times [\lambda_\sigma \mathbf{u}_L, \mathbf{u}_L] \times [\lambda_{æ} \mathbf{u}_L, \mathbf{u}_L]$, *and*

$$\varphi(\vartheta, y(\vartheta), z(\vartheta)) \leq \alpha_\sigma^{-1} \hat{\mathbf{u}}_L, \text{for } (\vartheta, y, z) \in [0,1] \times [\lambda_\sigma \hat{\mathbf{u}}_L, \hat{\mathbf{u}}_L] \times [\lambda_{æ} \mathbf{u}_L, \mathbf{u}_L], L = 1,2\ldots m,$$

(D_9) $\chi(\vartheta, y(\vartheta), z(\vartheta)) \geq \mathbf{u}_L \left(\lambda_{æ} \int\limits_0^1 H_{æ}(1,s)ds\right)^{-1}$, *for* $(\vartheta, y, z) \in [0,1] \times [\lambda_{æ} \mathbf{u}_L, \mathbf{u}_L] \times [\lambda_\sigma \mathbf{u}_L, \mathbf{u}_L]$, *and*

$$\chi(\vartheta, y(\vartheta), z(\vartheta)) \leq \alpha_{æ}^{-1} \hat{\mathbf{u}}_L, \text{for } (\vartheta, y, z) \in [0,1] \times [\lambda_\sigma \mathbf{u}_L, \mathbf{u}_L] \times [\lambda_{æ} \hat{\mathbf{u}}_L, \hat{\mathbf{u}}_L],\ L = 1,2\ldots m.$$

where $\lambda, H_\sigma(1,s), H_{æ}(1,s)$ *defined in Lemma 8.*

Then the coupled system (1) has at least m*-positive solutions* (y_L, z_L)*, satisfying*

$$\mathbf{u}_L \leq \|(y_L, z_L)\| \leq \hat{\mathbf{u}}_L,\ L = 1,2\ldots m.$$

Theorem 9. *Suppose that* $(C_1)-(C_3)$ *holds. If there exist* $2m$ *positive numbers* $\mathbf{u}_L, \hat{\mathbf{u}}_L,\ L=1,2\ldots m$*, with* $\mathbf{u}_1 < \hat{\mathbf{u}}_1 < \mathbf{u}_2 < \hat{\mathbf{u}}_2 \ldots < \mathbf{u}_m < \hat{\mathbf{u}}_m$ *such that*

(D_{10}) φ *and* χ *are non-decreasing on* $[0, \hat{\mathbf{u}}_m]$ $\forall\, \vartheta \in [0,1]$;

(D_{11}) $\varphi(\vartheta, y(\vartheta), z(\vartheta)) \geq \mathbf{u}_L \left(\lambda_\sigma \int\limits_\theta^{1-\theta} H_\sigma(1,s)ds\right)^{-1}$, $\varphi(\vartheta, y(\vartheta), z(\vartheta)) \leq \dfrac{\hat{\mathbf{u}}_L}{\alpha_\sigma}$, $L = 1,2\ldots m$,

$\chi(\vartheta, y(\vartheta), z(\vartheta)) \geq \mathbf{u}_L \left(\lambda_{æ} \int\limits_\theta^{1-\theta} H_{æ}(1,s)ds\right)^{-1}$, $\chi(\vartheta, y(\vartheta), z(\vartheta)) \leq \dfrac{\hat{\mathbf{u}}_L}{\alpha_{æ}}$, $L = 1,2\ldots m$.

Hence we conclude that there exist at least m *positive solutions* (y_L, z_L)*, corresponding to coupled system (1) which satisfy*

$$\mathbf{u}_L \leq \|(y_L, z_L)\| \leq \hat{\mathbf{u}}_L,\ L = 1,2\ldots m.$$

5. Hyers-Ulam Stability

Definition 6. *[30] Let* $B_1, B_2 : E \times E \to E \times E$ *be the two operators. Then the system of operator equations*

$$\begin{cases} y(\vartheta) = B_1(y,z)(\vartheta) \\ z(\vartheta) = B_2(y,z)(\vartheta) \end{cases} \tag{28}$$

is called the HU stability if we can find $J_i(i=1,2,3,4) > 0$*, with* $æ_i(i=1,2) > 0$ *and for each solution* $(y^*, z^*) \in E \times E$ *of the inequalities given by*

$$\begin{cases} ||y^* - \varphi(y^*, z^*)||_{E\times E} \leq æ_1, \\ ||z^* - \chi(y^*, z^*)||_{E\times E} \leq æ_2, \end{cases} \tag{29}$$

there exists a solution $(\overline{y}, \overline{z}) \in E \times E$ *of system (28) such that*

$$\begin{cases} ||y^* - \overline{y}||_{E\times E} \leq k_1 æ_1 + k_2 æ_2, \\ ||z^* - \overline{z}||_{E\times E} \leq k_3 æ_1 + k_4 æ_2, \end{cases} \tag{30}$$

Theorem 10. [30] *Let* $B_1, B_2 : E \times E \to E \times E$ *be the two operators such that*

$$\begin{cases} \|B_1(y,z) - B_1(y^*,z^*)\|_{E\times E} \le k_1\|y - y^*\|_{E\times E}ds + k_2\|z - z^*\|_{E\times E}ds, \\ \|B_2(y,z) - B_2(y^*,z^*)\|_{E\times E} \le k_3\|y - y^*\|_{E\times E}ds + k_4\|z - z^*\|_{E\times E}ds, \\ \textit{for all} \quad (y,z),(y^*,z^*) \in E \times E, \end{cases} \tag{31}$$

and if the matrix

$$\mathbf{B} = \begin{bmatrix} k_1 & k_2 \\ k_3 & k_3 \end{bmatrix}$$

converges to zero, then the fixed points corresponding to operator system (28) *are HU-stable. Further, the given condition holds* (M_{11}) *under the continuity of* $\varphi_i, i = 1,2$*, there exist* $f_i, H_i \in C(0,1), i = 1,2$ *and* $(y,z),(\overline{y},\overline{z})$ *such that*

$$|\varphi_i(\vartheta,y,z) - \varphi_i(\vartheta,\overline{y},\overline{z})| \le f_i(\vartheta)|y - \overline{y}| + H_i(\vartheta)|z - \overline{z}|, i = 1,2.$$

In this section, we study HU stability for the solutions of our proposed system.

Theorem 11. *Suppose that the assumption* (M_{11}) *along with condition that matrix*

$$\mathbf{B} = \begin{bmatrix} \int_0^1 H_\sigma(1,s)u_1(s)ds & \int_0^1 H_\sigma(1,s)v_1(s)ds \\ \int_0^1 H_æ(1,s)u_2(s)ds & \int_0^1 H_æ(1,s)v_2(s)ds \end{bmatrix}.$$

is converging to zero. Then, the solutions of (1) *are HU-stable.*

Proof. Thanks to Theorem 2, we have

$$\begin{cases} \|B_1(y,z) - B_1(y^*,z^*)\|_{E\times E} \le \int_0^1 H_\sigma(1,s)u_1(s)\|y - y^*\|_{E\times E}ds + \int_0^1 H_\sigma(1,s)v_1(s)\|z - z^*\|_{E\times E}ds, \\ \|B_2(y,z) - B_2(y^*,z^*)\|_{E\times E} \le \int_0^1 H_æ(1,s)u_2(s)\|y - y^*\|_{E\times E}ds + \int_0^1 H_æ(1,s)v_2(s)\|z - z^*\|_{E\times E}ds. \end{cases}$$

From which we get

$$\begin{cases} \|B_1(y,z) - B_1(y^*,z^*)\|_{E\times E} \le \left[\int_0^1 H_\sigma(1,s)u_1(s)ds\right] \|y - y^*\|_{E\times E} + \left[\int_0^1 H_\sigma(1,s)v_1(s)ds\right] \|z - z^*\|_{E\times E}, \\ \|B_2(y,z) - B_2(y^*,z^*)\|_{E\times E} \le \left[\int_0^1 H_æ(1,s)u_2(s)ds\right] \|y - y^*\|_{E\times E} + \left[\int_0^1 H_æ(1,s)v_2(s)ds\right] \|z - z^*\|_{E\times E}. \end{cases} \tag{32}$$

Analogously one has

$$\|P(y,z) - P(y^*,z^*)\|_{E\times E} \le \mathbf{B}\|(y,z) - (y^*,z^*)\|_{E\times E}, \tag{33}$$

such that

$$\mathbf{B} = \begin{bmatrix} \int_0^1 H_\sigma(1,s)u_1(s)ds & \int_0^1 H_\sigma(1,s)v_1(s)ds \\ \int_0^1 H_æ(1,s)u_2(s)ds & \int_0^1 H_æ(1,s)v_2(s)ds \end{bmatrix}.$$

Hence, we get the required results. □

6. Example

To verify the aforesaid established analysis we provide some test problems here in the given sequel.

Example 1. *Take the system of given BVPs with IBCs as*

$$\begin{cases} D^{\frac{7}{2}}_{+0}y(\vartheta) + \dfrac{\vartheta+1}{4}[\Gamma(\frac{5}{2})|y(\vartheta)| + \cos|z(\vartheta)|] = 0,\ \vartheta \in [0,1], y, z \geq 0 \\ D^{\frac{7}{2}}_{+0}z(\vartheta) + \dfrac{\vartheta^2+1}{4}[\sin|y(\vartheta)| + |z(\vartheta)|] = 0,\ \vartheta \in [0,1], y, z \geq 0 \\ y(0) = y'(0) = y''(0) = 0 \quad y(1) = \dfrac{1}{2}\displaystyle\int_0^1 y(s)ds, \\ z(0) = z'(0) = z''(0) = 0 \quad z(1) = \dfrac{1}{3}\displaystyle\int_0^1 z(s)ds. \end{cases} \tag{34}$$

Since $\varphi(\vartheta, y(\vartheta), z(\vartheta)) = \dfrac{\vartheta+1}{4}[\Gamma(\frac{5}{2})|y(\vartheta)| + \cos|z(\vartheta)|]$, $\chi(\vartheta, y(\vartheta), z(\vartheta)) = \dfrac{\vartheta^2+1}{4}[\sin|y(\vartheta)| + |z(\vartheta)|]$.

Also as $m = [3.5] + 1 = 4, \delta = \frac{1}{2}$ *and* $\varrho = \frac{1}{3}$.

Then

$$|\varphi(\vartheta, y_2, z_2) - \varphi(\vartheta, y_1, z_1)| \leq \Gamma(\frac{5}{2})\frac{\vartheta+1}{4}|y_2 - y_1| + \frac{\vartheta+1}{4}|z_2 - z_1|,$$
$$|\chi(\vartheta, y_2, z_2) - \chi(\vartheta, y_1, z_1)| \leq \frac{\vartheta^2+1}{4}|y_2 - y_1| + \frac{\vartheta^2+1}{4}|z_2 - z_1|.$$

where $u_1(\vartheta) = \frac{\vartheta+1}{4}\Gamma(\frac{5}{2}), v_1(\vartheta) = \frac{\vartheta+1}{4}$, $u_2(\vartheta) = v_2(\vartheta) = \dfrac{\vartheta^2+1}{4}$, *so one can get*

$$\mathbf{B} = \begin{bmatrix} \int_0^1 H_\sigma(1,s)u_1(s)ds & \int_0^1 H_\sigma(1,s)v_1(s)ds \\ \int_0^1 H_{æ}(1,s)u_2(s)ds & \int_0^1 H_{æ}(1,s)v_2(s)ds \end{bmatrix} = \begin{bmatrix} \frac{8}{11} & \frac{8}{165\sqrt{\pi}} \\ \frac{496}{11583\sqrt{\pi}} & \frac{496}{11583\sqrt{\pi}} \end{bmatrix}.$$

$$det(\mathbf{B} - \widehat{\lambda}I) = \begin{bmatrix} \frac{8}{11} - \widehat{\lambda} & \frac{8}{165\sqrt{\pi}} \\ \frac{496}{11583\sqrt{\pi}} & \frac{496}{11583\sqrt{\pi}} - \widehat{\lambda} \end{bmatrix}.$$

We get $\widehat{\lambda}_1 = 0.728$ *and* $\widehat{\lambda}_2 = 0.024$ *since* $æ(\mathbf{B}) = \sup\{|\widehat{\lambda}_i|, i = 1,2\} = 0.728 < 1$. *Therefore due to Theorem 2, BVPs (34) has a unique positive solution given by*

$$\begin{cases} y(\vartheta) = \displaystyle\int_0^1 H_{\frac{7}{2}}(\vartheta, s)\frac{s+1}{4}[\Gamma(\frac{5}{2})|y(s)| + \cos|z(s)|]ds, \\ z(\vartheta) = \displaystyle\int_0^1 H_{\frac{7}{2}}(\vartheta, s)\frac{s^2+1}{4}[\sin|y(s)| + |z(s)|]ds, \end{cases} \tag{35}$$

where $H_{\frac{7}{2}}(\vartheta, s)$ *and* $H_{\frac{7}{2}}(\vartheta, s)$ *are the Green's functions given by*

$$H_{\frac{7}{2}}(\vartheta, s) = \begin{cases} \dfrac{\vartheta^{\frac{5}{2}}(1-s)^{\frac{5}{2}}(3+\frac{1}{2}s) - 3(\vartheta - s)^{\frac{5}{2}}}{3\Gamma(\frac{7}{2})}, & 0 \leq s \leq \vartheta \leq 1, \\ \dfrac{\vartheta^{\frac{5}{2}}(1-s)^{\frac{5}{2}}(3+\frac{1}{2}s)}{3\Gamma(\frac{7}{2})}, & 0 \leq \vartheta \leq s \leq 1. \end{cases}$$

$$H_{\frac{7}{2}}(\vartheta,s)=\begin{cases}\dfrac{\vartheta^{\frac{5}{2}}(1-s)^{\frac{5}{2}}(\frac{19}{6}+\frac{1}{3}s)-\frac{19}{6}(\vartheta-s)^{\frac{5}{2}}}{\frac{19}{6}\Gamma(\frac{7}{2})}, & 0\le s\le\vartheta\le 1,\\ \dfrac{\vartheta^{\frac{5}{2}}(1-s)^{\frac{5}{2}}(\frac{19}{6}+\frac{1}{3}s)}{\frac{19}{6}\Gamma(\frac{7}{2})}, & 0\le\vartheta\le s\le 1.\end{cases}$$

Further, by the use of Theorem 11, the solution is HU-stable.

Example 2. *Taking a system of FODEs with IBCs as*

$$\begin{cases}D^{\frac{10}{3}}_{+0}y(\vartheta)+a(\vartheta)\sqrt{y(\vartheta)+z(\vartheta)}=0,\ D^{\frac{7}{2}}_{+0}z(\vartheta)+b(\vartheta)\sqrt[3]{y(\vartheta)+z(\vartheta)}=0,\ \vartheta\in(0,1),\\ y(0)=y'(0)=y''(0)=y'''(0)=0\quad y(1)=\int_0^1 y(s)ds,\\ z(0)=z'(0)=z''(0)=z'''(0)=0\quad z(1)=\int_0^1 z(s)ds.\end{cases}\tag{36}$$

From the given system one has

$$\varphi(\vartheta,y,z)=a(\vartheta)\sqrt{y(\vartheta)+z(\vartheta)}$$

and

$$\chi(\vartheta,y,z)=b(\vartheta)\sqrt[3]{y(\vartheta)+z(\vartheta)},\ m=4,\ \delta=\varrho=1.$$

Also $a,\ b:[0,1]\to[0,\infty)$ are continuous. Now $\varphi^0=\lim_{(y,z)\to 0}\frac{\varphi(\vartheta,y,z)}{y+z}=\infty$, similarly $\chi^0=\infty$.

Obviously we compute $\varphi^\infty=0=\chi^\infty$. Hence due to Theorem 4, system (36) has at least one positive solution.

Example 3. *Taking another test problem with IBCs as*

$$\begin{cases}D^{\frac{9}{2}}_{+0}y(\vartheta)+(1-\vartheta^2)[y(\vartheta)+z(\vartheta)]^2=0,\ D^{\frac{14}{3}}_{+0}z(\vartheta)+[y(\vartheta)+z(\vartheta)]^3=0,\ \vartheta\in(0,1),\\ y(0)=y'(0)=y''(0)=y'''(0)=y''''(0)=0,\quad y(1)=\frac{3}{2}\int_0^1 y(s)ds,\\ z(0)=z'(0)=z''(0)=z'''(0)=z''''(0)=0,\quad z(1)=\frac{3}{2}\int_0^1 z(s)ds.\end{cases}\tag{37}$$

From the considered problem (37), one has $\delta=\varrho=\frac{3}{2}$, as $m=5$. It is easy to see that $\varphi^0=\chi^0=0$ and $\varphi^\infty=\chi^\infty=\infty$. Therefore thanks to Theorem 5, the given system (37) has a positive solution.

Example 4. *Further we take another system of FODEs with IBCs as*

$$\begin{cases}D^{\frac{11}{2}}_{+0}y(\vartheta)+\dfrac{(1+\vartheta^2)[u^2(\vartheta)+z(\vartheta)]}{(4\vartheta^2+4)\alpha_\sigma}=0,\ \vartheta\in(0,1),\\ D^{\frac{16}{3}}_{+0}z(\vartheta)+\dfrac{(\vartheta^3+1)[y(\vartheta)+v^2(\vartheta)]}{(4\vartheta^3+4)\alpha_{æ}}=0,\ \vartheta\in(0,1),\\ y(0)=y'(0)=y''(0)=y'''(0)=y''''(0)=0,\quad y(1)=\frac{3}{2}\int_0^1 y(s)ds,\\ z(0)=z'(0)=z''(0)=z'''(0)=z''''(0)=0,\quad z(1)=\frac{3}{2}\int_0^1 z(s)ds.\end{cases}\tag{38}$$

where $\delta=\varrho=\frac{3}{2}$ and $m=6$. It is easy to obtain $\varphi_0=\chi_0=\infty$ and $\varphi_\infty=\chi_\infty=\infty$.

Further $\forall\ (\vartheta, y, z) \in [0,1] \times [0,1] \times [0,1]$, *we have*

$$\varphi(\vartheta, y, z) \leq \frac{(\vartheta^2+1)2}{4(\vartheta^2+1)\alpha_\sigma} = \frac{\alpha_\sigma^{-1}}{2},\ \chi(\vartheta, y, z) \leq \frac{(\vartheta^3+1)2}{4(\vartheta^3+1)\alpha_{æ}} = \frac{\alpha_{æ}^{-1}}{2}.$$

Hence all the conditions of Theorem 6 hold. Thanks to Theorem 6, the given system (38) has at least two positive solutions (y_1, z_1) *and* (y_2, z_2) *which satisfy*

$$0 < \|(y_1, z_1)\| < 1 < \|(y_2, z_2)\|.$$

7. Non-Existence of Positive Solution

Here some conditions are developed under which the coupled system (1) with given IBCs has no solution.

Theorem 12. *Consider that* $(C_1) - (C_3)$ *hold and* $\varphi(\vartheta, y, z) < \frac{\|(y,z)\|}{2\alpha_\sigma}$ *and* $\chi(\vartheta, y, z) < \frac{\|(y,z)\|}{2\alpha_{æ}}$ *for all* $\vartheta \in [0,1],\ y > 0,\ z > 0$, *then there is no positive solution for BVPs (1).*

Proof. Consider (y, z) to be the positive solution of BVPs (1). Then, $(y, z) \in C$ for $0 < \vartheta < 1$ and

$$\begin{aligned}
\|(y,z)\| &= \|y\| + \|z\| \\
&= \max_{\vartheta\in[0,1]} |y(\vartheta)| + \max_{\vartheta\in[0,1]} |z(\vartheta)| \\
&\leq \max_{\vartheta\in[0,1]} \int_0^1 H_\sigma(\vartheta, s)|\varphi(s, y(s), z(s))|ds + \max_{\vartheta\in[0,1]} \int_0^1 H_{æ}(\vartheta, s)|\chi(s, y(s), z(s)|ds \\
&< \int_0^1 H_\sigma(1, s)\frac{\|(y,z)\|}{2\alpha_{æ}}ds + \int_0^1 H_{æ}(1, s)\frac{\|(y,z)\|}{2\alpha_{æ}}ds \\
\Rightarrow \|(y,z)\| &< \|(y,z)\|,
\end{aligned}$$

which is contradiction. Hence the considered system (1) has no solution. □

Theorem 13. *Let the hypothesis* $(C_1) - (C_3)$ *hold along with the conditions*

$$\begin{aligned}
\varphi(\vartheta, y(\vartheta), z(\vartheta)) &> \frac{\|(y,z)\|}{2}\left(\lambda_\alpha^2 \int_\theta^{1-\theta} H_\sigma(1, s)ds\right)^{-1}, \\
\chi(\vartheta, y(\vartheta), z(\vartheta)) &> \frac{\|(y,z)\|}{2}\left(\lambda_\beta^2 \int_\theta^{1-\theta} H_{æ}(1, s)ds\right)^{-1}, \text{ for all } \vartheta \in [0,1],\ y > 0 \text{and } z > 0.
\end{aligned}$$

Then there does not exist positive solution to BVPs (1).

To demonstrate the results of Theorems 12 and 13 respectively, we give the following example.

Example 5. *Taking the given system of FODEs with given IBCs as*

$$\begin{cases} D^{\frac{5}{2}}_{+0}y(\vartheta) = 5 - 4\left(y + z + \frac{\pi}{3}\right)^{\frac{-5}{2}}, \ \vartheta \in [0,1], \\ D^{\frac{5}{2}}_{+0}z(\vartheta) = \left(30 + \frac{30}{\sqrt{y+z}}\right)^{\frac{-3}{2}} + \frac{1}{50}, \ \vartheta \in [0,1], \\ y(0) = y'(0) = y''(0) = 0, \ y(1) = \frac{1}{2}\int_0^1 y(\vartheta)d\vartheta, \\ z(0) = z'(0) = z''(0) = 0, \ z(1) = \frac{1}{2}\int_0^1 z(\vartheta)d\vartheta. \end{cases} \tag{39}$$

Also as $(C_1) - (C_3)$ *hold, where* $m = [2.5] + 1 = 3$ *and* $\delta = \varrho = \frac{1}{2}$. *We calculate*

$$\begin{aligned} \varphi^0 = 5 - \left(\frac{3}{\pi}\right)^{\frac{5}{2}}, \chi^0 &= \frac{1}{50}, \ \varphi^\infty = 5, \ \chi^\infty = 51, \\ \left(5 - \left(\frac{3}{\pi}\right)^{\frac{5}{2}}\right)\|(y,z)\| &< \varphi(\vartheta, y(\vartheta), z(\vartheta)) < 5\|(y,z)\|, \\ \frac{1}{50}\|(y,z)\| &< \chi(\vartheta, y(\vartheta), z(\vartheta)) < 51\|(y,z)\|. \end{aligned}$$

Therefore we have

$$5 - \left(\frac{3}{\pi}\right)^{\frac{5}{2}}\|(y,z)\| < \varphi(\vartheta, y(\vartheta), z(\vartheta)) < 5\|(y,z)\| \text{ and } \chi(\vartheta, y(\vartheta), z(\vartheta)) < 5\|(y,z)\| < \frac{\|(y,z)\|}{\alpha_\sigma},$$

where $\alpha_\sigma \approx 0.32239$ *and* $\alpha_{æ} \approx 0.32239$.

Case I: *Now*

$$\varphi(\vartheta, y(\vartheta), z(\vartheta)) < \frac{\|(y,z)\|}{\alpha_\sigma} \approx 1.1413\|(y,z)\|$$

yields that

$$\varphi(\vartheta, y(\vartheta), z(\vartheta)) < 5\|(y,z)\| \approx 3.1018\|(y,z)\|$$

and

$$\chi(\vartheta, y(\vartheta), z(\vartheta)) < 51\|(y,z)\| \approx 3.1018\|(y,z)\|.$$

Hence under the condition of Theorem 12, there is no solution corresponding to problem (39).

Case II: *Also*

$$\varphi(\vartheta, y(\vartheta), z(\vartheta)) > \left(5 - \left(\frac{3}{\pi}\right)^{\frac{5}{2}}\right)\|(y,z)\| > \|(y,z)\|\left(\lambda_\alpha^2 \int_{\frac{1}{100}}^{\frac{99}{100}} H_\alpha(1,s)ds\right)^{-1} \approx 0.615\|(y,z)\|$$

and

$$\chi(\vartheta, y(\vartheta), z(\vartheta)) > \frac{1}{50}\|(y,z)\| > \|(y,z)\|\left(\lambda_{æ}^2 \int_{\frac{1}{100}}^{\frac{99}{100}} H_{æ}(1,s)ds\right)^{-1} \approx 0.615\|(y,z)\|.$$

Hence under the condition of Theorem 13, there is no solution corresponding to coupled system (39).

8. Conclusions

In the above research work we have successfully investigated a coupled system of nonlinear FODEs with IBCs for multiplicity results. Further, the aforesaid investigation has been strengthened by developing some conditions under which the solutions of the proposed system are HU-stable. Further some results which demonstrate the conditions of nonexistence of solutions have been established. The whole results have been verified by considering some examples where needed.

Author Contributions: All authors have equal contribution in this paper.

Funding: This research work has been supported financially by KMUTT Fixed Point Research Laboratory, KMUTT-Fixed Point Theory and Applications Research Group, Faculty of Science, King Mongkut's University of Technology Thonburi (KMUTT), 126 Pracha-Uthit Road, Bang Mod, Thrung Khru, Bangkok 10140, Thailand.

Acknowledgments: All authors are very thankful to the referees for their useful comments and suggestions.

Conflicts of Interest: The authors declare no conflict of interest.

References

1. Miller, K.S.; Ross, B. *An Introduction to the Fractional Calculus and Fractional Differential Equations*; Wiley: New York, NY, USA, 1993.
2. Kilbas, A.A.; Srivastava, H.M.; Trujillo, J.J. Theory and Applications of Fractional Differential Equations. In *North-Holland Mathematics Studies*; Elsevier: Amsterdam, The Nederlands, 2006; Volume 204.
3. Podlubny, I. *Fractional Differential Equations, Mathematics in Science and Engineering*; Academic Press: New York, NY, USA, 1999.
4. Hilfer, R. *Applications of Fractional Calculus in Physics*; World Scientific: Singapore, 2000.
5. Lakshmikantham, V.; Vatsala, A.S. Basic theory of fractional differential equations. *Nonlinear Anal.* **2008**, *69*, 2677–2682. [CrossRef]
6. Benchohra, M.; Graef, J.R.; Hamani, S. Existence results for boundary value problems with nonlinear fractional differential equations. *Appl. Anal.* **2008**, *87*, 851–863. [CrossRef]
7. Shah, K.; Khan, R.A. Study of solution to a toppled system of fractional differential equations with integral boundary conditions. *Int. J. Appl. Comput. Math.* **2017**, *3*, 2369–2388. [CrossRef]
8. Ahmad, B.; Nieto, J.J. Existence results for a coupled system of nonlinear fractional differential equations with three-point boundary conditions. *Comput. Math. Appl.* **2009**, *58*, 1838–1843. [CrossRef]
9. Agarwal, R.P.; Benchohra, M.; Hamani, S. A survey on existence results for boundary value problems of nonlinear fractional differential equations and inclusions. *Acta Appl. Math.* **2010**, *109*, 973–1033. [CrossRef]
10. Li, C.F.; Luo, X.N.; Zhou, Y. Existence of positive solutions of the boundary value problem for nonlinear fractional differential equations. *Comput. Math. Appl.* **2010**, *59*, 1363–1375. [CrossRef]
11. Rehman, M.; Khan, R.A. Existence and uniqueness of solutions for multi-point boundary value problems for fractional differential equations. *Appl. Math. Lett.* **2010**, *23*, 1038–1044. [CrossRef]
12. Zhong, W.; Lin, W. Nonlocal and multiple-point boundary value problem for fractional differential equations. *Comput. Math. Appl.* **2010**, *59*, 1345–1351. [CrossRef]
13. El-Shahed, M.; Shammakh, W.M. Existence of positive solutions for m-point boundary value problem for nonlinear fractional differential equation. In *Abstract and Applied Analysis*; Hindawi: Cairo, Egypt, 2011; Volume 2011, p. 986575.
14. Khan, R.A.; Shah, K. Existence and uniqueness of solutions to fractional order multi point boundary value problems. *Commun. Appl. Anal.* **2015**, *19*, 515–526.
15. Khan, R.A. Three-point boundary value problems for higher order nonlinear fractional differential equations. *J. Appl. Math. Inform.* **2013**, *31*, 221–228. [CrossRef]
16. EL-Sayed, A.M.A.; Taher, E.O.B. Positive solutions for a nonlocal multi-point boundary-value problem of fractional and second order. *Electr. J. Differ. Equ.* **2013**, *64*, 1–8.
17. Yang, A.; Ge, W. Positive solutions of multi-point boundary value problems of nonlinear fractional differential equation at resonance. *J. Korea Soc. Math. Educ. Ser. B Pure Appl. Math.* **2009**, *16*, 181–193.
18. Jankowski, T. Differential equations with integral boundary condition. *J. Comput. Appl. Math.* **2002**, *147*, 2274–2280. [CrossRef]

19. Nanware, J.A.; Dhaigude, D.B. Existence and uniqueness of solution of differential equations of fractional order with integral boundary conditions. *J. Nonlinear Sci. Appl.* **2014**, *7*, 246–254. [CrossRef]
20. Cabada, A.; Wang, G. Positive solution of nonlinear fractional differential equations with integral boundary value conditions. *J. Math. Anal. Appl.* **2012**, *389*, 403–411. [CrossRef]
21. Deng, W. Smoothness and stability of the solutions for nonlinear fractional differential equations. *Nonlinear Anal. TMA* **2010**, *72*, 1768–1777. [CrossRef]
22. Li, Y.; Chen, Y.; Podlubny, I. Mittag-Leffler stability of fractional order nonlinear dynamic systems. *Automatica* **2009**, *45*, 1965–1969. [CrossRef]
23. Li, Y.; Chen, Y.; Podlubny, I. Stability of fractional-order nonlinear dynamic systems: Lyapunov direct method and generalized Mittag-Leffler stability. *Comput. Math. Appl.* **2010**, *59*, 1810–1821. [CrossRef]
24. Ulam, S.M. *Problems in Modern Mathematics*; John Wiley and Sons: New York, NY, USA, 1940; Chapter 6.
25. Hyers, D.H. On the stability of linear functional equation. *Proc. Natl. Acad. Sci. USA* **1941**, *27*, 222–224. [CrossRef] [PubMed]
26. Jung, S.M. On the Hyers-Ulam stability of the functional equations that have the quadratic property. *J. Math. Anal. Appl.* **1998**, *222*, 126–137. [CrossRef]
27. Jung, S.M. Hyers-Ulam stability of linear differential equations of first order. *Appl. Math. Lett.* **2006**, *19*, 854–858. [CrossRef]
28. Obloza, M. Hyers stability of the linear differential equation. *Rocznik Nauk. Dydakt. Prace Mat.* **1993**, *13*, 259–270.
29. Krasnoselskii, M.A. *Positive Solutions of Operator Equations*; Noordhoff: Groningen, The Netherlands, 1964.
30. Urs C. Coupled fixed point theorem and applications to periodic boundary value problem. *Miskolic Math. Notes* **2013**, *14*, 323–333. [CrossRef]
31. Stanêk, S. The existence of positive solutions of singular fractional boundary value problems. *Comput. Math. Appl.* **2011**, *62*, 1379–1388. [CrossRef]
32. Shah, K.; Khan, R.A. Multiple positive solutions to a coupled systems of nonlinear fractional differential equations. *SpringerPlus* **2016**, *5*, 1–20. [CrossRef] [PubMed]
33. Agarwal, R.; Meehan, M.; Regan, D.O. *Fixed Points Theory and Applications*; Cambridge University Press: Cambridge, UK, 2004.
34. Guo, D.; Lakshmikantham, V. *Nonlinear Problems in Abstract Cones*; Academic Press: New York, NY, USA, 1988.
35. Shah, K.; Khan, R.A. Existence and uniqueness of positive solutions to a coupled system of nonlinear fractional order differential equations with anti periodic boundary conditions. *Differ. Equ. Appl.* **2015**, *7*, 245–262. [CrossRef]

mathematics

MDPI

Article

Existence and Stability Results for a Fractional Order Differential Equation with Non-Conjugate Riemann-Stieltjes Integro-Multipoint Boundary Conditions

Bashir Ahmad [1], Ymnah Alruwaily [1], Ahmed Alsaedi [1,*] and Sotiris K. Ntouyas [1,2]

[1] Nonlinear Analysis and Applied Mathematics (NAAM)-Research Group, Department of Mathematics, Faculty of Science, King Abdulaziz University, P.O. Box 80203, Jeddah 21589, Saudi Arabia; bashirahmad_qau@yahoo.com (B.A.); ymnah@ju.edu.sa (Y.A.); sntouyas@uoi.gr (S.K.N.)

[2] Department of Mathematics, University of Ioannina, Ioannina 45110, Greece

* Correspondence: aalsaedi@hotmail.com

Received: 20 February 2019; Accepted: 5 March 2019; Published: 11 March 2019

Abstract: We discuss the existence and uniqueness of solutions for a Caputo-type fractional order boundary value problem equipped with non-conjugate Riemann-Stieltjes integro-multipoint boundary conditions on an arbitrary domain. Modern tools of functional analysis are applied to obtain the main results. Examples are constructed for the illustration of the derived results. We also investigate different kinds of Ulam stability, such as Ulam-Hyers stability, generalized Ulam-Hyers stability, and Ulam-Hyers-Rassias stability for the problem at hand.

Keywords: Caputo fractional derivative; nonlocal; integro-multipoint boundary conditions; existence; uniqueness; Ulam-Hyers stability

MSC: 34A08; 34B10; 34B15

1. Introduction

Fractional calculus played a pivotal role in improving the mathematical modeling of many real-world problems. The extensive application of fractional order (differential and integral) operators indeed reflects the popularity of this branch of mathematical analysis. In contrast to the integer order operators, such operators are nonlocal in nature and do have the capacity to trace the history of the phenomenon under investigation. A detailed account of the use of fractional calculus tools can be found in several scientific disciplines such as, chaos and fractional dynamics [1], evolution in honeycomb lattice via fractional Schrödinger equation [2], financial economics [3], ecology [4], bio-engineering [5], etc. For theoretical development and further application of the topic, see the texts [6–9].

During the past two decades, the study of fractional order boundary value problems has been one of the hot topics of scientific research. Several researchers contributed to the development of this class of problems by producing a huge number of articles, special issues, monographs, etc. Now the literature on the topic contains a variety of existence and uniqueness results, and analytic and numerical methods of solutions for these problems. In particular, there has been shown a great interest in the formulation and investigation of fractional order boundary value problems involving non-classical (nonlocal and integral) boundary conditions. The nonlocal boundary conditions are found to be of great utility in modeling the changes happening within the domain of the given scientific phenomena, while the concept of integral boundary conditions is applied to model the physical problems, such as blood flow problems on arbitrary structures and ill-posed backward problems. For some recent works on fractional order differential equations involving Riemann-Liouville, Caputo, and Hadamard type fractional derivatives, equipped with classical, nonlocal, and integral boundary conditions, we refer the reader to a series of papers [10–28] and the references cited therein.

In this paper, we study the existence of solutions for a nonlinear Liouville-Caputo-type fractional differential equation on an arbitrary domain:

$$^{c}D^{q}x(t) = f(t, x(t)),\ 3 < q \leq 4,\ t \in [a, b], \tag{1}$$

supplemented with non-conjugate Riemann-Stieltjes integro-multipoint boundary conditions of the form:

$$x(a) = \sum_{i=1}^{n-2} \alpha_i x(\eta_i) + \int_a^b x(s)dA(s),\ x'(a) = 0,\ x(b) = 0,\ x'(b) = 0, \tag{2}$$

where $^{c}D^{q}$ denotes the Caputo fractional derivative of order q, $a < \eta_1 < \eta_2 < \cdots < \eta_{n-2} < b$, $f : [a, b] \times \mathbb{R} \longrightarrow \mathbb{R}$ is a given continuous function, A is a function of bounded variation, and $\alpha_i \in \mathbb{R}$, $i = 1, 2, \cdots, n-2$.

The main emphasis in the present work is to introduce non-conjugate Riemann-Stieltjes integro-multipoint boundary conditions and develop the existence theory for a Caputo-type fractional order boundary value problem equipped with these conditions on an arbitrary domain. Conjugate conditions on the body/fluid interface provide continuity of the thermal fields by specifying the equalities of temperatures and heat fluxes of a body and a flow at the vicinity of interface. The results obtained in this paper may have potential applications in diffraction-free and self-healing optoelectronic devices. Moreover, propagation properties for fractional Schrödinger equation similar to our results are well known theoretically [29].

The rest of the paper is organized as follows. An auxiliary result related to the linear variant of the problems (1) and (2), which plays a key role in the forthcoming analysis, is presented in Section 2. Some basic ideas of fractional calculus are also given in this section. In Section 3, we obtain some existence results for the given problem, while Section 4 contains a uniqueness result for the problem at hand. Ulam stability of different kinds for the problem (1) and (2) is studied in Section 5.

2. Preliminary Material

We begin this section with some basic definitions of fractional calculus [6]. Later we prove an auxiliary lemma, which plays a key role in defining a fixed-point problem associated with the given problem.

Definition 1. *Let g be a locally integrable real-valued function on $-\infty \leq a < t < b \leq +\infty$. The Riemann-Liouville fractional integral I_a^p of order $p \in \mathbb{R}$ $(p > 0)$ for the function g is defined as*

$$I_a^p g(t) = (g * K_p)(t) = \frac{1}{\Gamma(p)} \int_a^t (t-s)^{p-1} g(s) ds,$$

where $K_p(t) = \frac{t^{p-1}}{\Gamma(p)}$, Γ denotes the Euler gamma function.

Definition 2. *The Caputo derivative of fractional order p for an $(m-1)$-times absolutely continuous function $g : [a, \infty) \longrightarrow \mathbb{R}$ is defined as*

$$^{c}D^{p}g(t) = \frac{1}{\Gamma(m-p)} \int_a^t (t-\bar{t})^{m-p-1} g^{(m)}(\bar{t}) d\bar{t},\ m-1 < p \leq m,\ m = [p]+1,$$

where $[p]$ denotes the integer part of the real number p.

Lemma 1. [6] *The general solution of the fractional differential equation* ${}^cD^q x(t) = 0,\ m-1 < q < m,\ t \in [a,b]$ *is*

$$x(t) = \omega_0 + \omega_1(t-a) + \omega_2(t-a)^2 + \ldots + \omega_{m-1}(t-a)^{m-1},$$

where $\omega_i \in \mathbb{R},\ i = 0,1,\ldots,m-1$. *Furthermore,*

$$I^q\, {}^cD^q x(t) = x(t) + \sum_{i=0}^{m-1} \omega_i (t-a)^i.$$

Lemma 2. *Let*

$$\gamma_1 = \frac{-A_1(b-a)^2}{3} + \frac{2A_3}{3(b-a)} - A_2 \neq 0. \tag{3}$$

For $\widehat{f} \in C([a,b],\mathbb{R})$, *the unique solution of the linear equation*

$${}^cD^q x(t) = \hat{f}(t),\ 3 < q \leq 4,\ t \in [a,b], \tag{4}$$

supplemented with the boundary conditions (2) *is given by*

$$\begin{aligned} x(t) &= \int_a^t \frac{(t-s)^{q-1}}{\Gamma(q)} \widehat{f}(s)ds - g_1(t) \int_a^b \frac{(b-s)^{q-1}}{\Gamma(q)} \widehat{f}(s)ds - g_2(t) \int_a^b \frac{(b-s)^{q-2}}{\Gamma(q-1)} \widehat{f}(s)ds \\ &+ g_3(t) \int_a^b \Big(\int_a^s \frac{(s-u)^{q-1}}{\Gamma(q)} \widehat{f}(u)du \Big) dA(s), \end{aligned} \tag{5}$$

where

$$\begin{aligned} g_1(t) &= \lambda_1 - \frac{(t-a)^2 A_1}{\gamma_1} + (t-a)^3\lambda_4,\ g_2(t) = \lambda_2 + \frac{(t-a)^2\gamma_2}{\gamma_1} + (t-a)^3\lambda_5, \\ g_3(t) &= \lambda_3 + \frac{(t-a)^2}{\gamma_1} + (t-a)^3\lambda_6, \end{aligned} \tag{6}$$

$$\gamma_2 = \frac{A_3 + (b-a)^3 A_1}{3(b-a)^2}, \tag{7}$$

$$\begin{aligned} \lambda_1 &= 1 + \frac{(b-a)^2 A_1}{3\gamma_1},\ \lambda_2 = \frac{-(b-a)}{3} - \frac{(b-a)^2\gamma_2}{3\gamma_1},\ \lambda_3 = -\frac{(b-a)^2}{3\gamma_1}, \\ \lambda_4 &= \frac{2A_1}{3(b-a)\gamma_1},\ \lambda_5 = \frac{\gamma_1 - 2(b-a)\gamma_2}{3(b-a)^2\gamma_1},\ \lambda_6 = \frac{-2}{3(b-a)\gamma_1}, \end{aligned} \tag{8}$$

$$\begin{aligned} A_1 &= 1 - \sum_{i=1}^{n-2} \alpha_i - \int_a^b dA(s),\ A_2 = \sum_{i=1}^{n-2} \alpha_i(\eta_i - a)^2 + \int_a^b (s-a)^2 dA(s), \\ A_3 &= \sum_{i=1}^{n-2} \alpha_i(\eta_i - a)^3 + \int_a^b (s-a)^3 dA(s). \end{aligned} \tag{9}$$

Proof. Applying the integral operator I^q to both sides of (4) and using Lemma 1, we get

$$x(t) = \int_a^t \frac{(t-s)^{q-1}}{\Gamma(q)} \hat{f}(s)ds + c_0 + c_1(t-a) + c_2(t-a)^2 + c_3(t-a)^3, \tag{10}$$

where $c_i \in \mathbb{R},\ i = 0,1,2,3$ are unknown arbitrary constants. Differentiating (10) with respect to t, we have

$$x'(t) = \int_a^t \frac{(t-s)^{q-2}}{\Gamma(q-1)} \hat{f}(s)ds + c_1 + 2c_2(t-a) + 3c_3(t-a)^2. \tag{11}$$

Using the boundary conditions (2) in (10) and (11), we obtain $c_1 = 0$ and

$$c_0 + (b-a)^2 c_2 + (b-a)^3 c_3 = I_1, \tag{12}$$

$$2(b-a)c_2 + 3(b-a)^2 c_3 = I_2, \tag{13}$$

$$A_1 c_0 - A_2 c_2 - A_3 c_3 = I_3, \tag{14}$$

where A_i $(i = 1, 2, 3)$ are given by (9) and

$$\begin{aligned} I_1 &= -\int_a^b \frac{(b-s)^{q-1}}{\Gamma(q)} \widehat{f}(s)ds, \;\; I_2 = -\int_a^b \frac{(b-s)^{q-2}}{\Gamma(q-1)} \widehat{f}(s)ds, \\ I_3 &= \int_a^b \Big(\int_a^s \frac{(s-u)^{q-1}}{\Gamma(q)} \widehat{f}(u)du \Big) dA(s). \end{aligned} \tag{15}$$

Solving (12) and (13), for c_0 and c_3 in terms of c_2, we get

$$c_0 = I_1 - \frac{(b-a)}{3} I_2 - \frac{(b-a)^2}{3} c_2, \tag{16}$$

$$c_3 = \frac{1}{3(b-a)^2} I_2 - \frac{2}{3(b-a)} c_2. \tag{17}$$

Substituting (16) and (17) in (14) yields

$$c_2 = \frac{\gamma_2}{\gamma_1} I_2 - \frac{A_1}{\gamma_1} I_1 + \frac{1}{\gamma_1} I_3, \;\; \gamma_1 \neq 0, \tag{18}$$

where γ_1 and γ_2 are defined by (3) and (7) respectively. Using (18) in (16) and (17), we find that

$$c_0 = \lambda_1 I_1 + \lambda_2 I_2 + \lambda_3 I_3,$$

$$c_3 = \lambda_4 I_1 + \lambda_5 I_2 + \lambda_6 I_3.$$

Inserting the values of c_0, c_1, c_2 and c_3 in (10) together with notations (6), we obtain the solution (5). The converse of the lemma can be proved by direct computation. □

3. Existence Results

Let $\mathcal{E} = C([a,b], \mathbb{R})$ denote the Banach space of all continuous functions from $[a,b] \longrightarrow \mathbb{R}$ equipped with the sup-norm $\|x\| = \sup\{|x(t)|, t \in [a,b]\}$. For computational convenience, we introduce

$$\Lambda = \left\{ \frac{(b-a)^q}{\Gamma(q+1)} + \bar{g_1} \frac{(b-a)^q}{\Gamma(q+1)} + \bar{g_2} \frac{(b-a)^{q-1}}{\Gamma(q)} + \bar{g_3} \int_a^b \frac{(s-a)^q}{\Gamma(q+1)} dA(s) \right\}, \tag{19}$$

where $\bar{g_1} = \sup_{t \in [a,b]} |g_1(t)|, \bar{g_2} = \sup_{t \in [a,b]} |g_2(t)|, \bar{g_3} = \sup_{t \in [a,b]} |g_3(t)|$. By Lemma 2, we transform the problems (1) and (2) into an equivalent fixed-point problem as

$$x = \mathcal{J} x, \tag{20}$$

where $\mathcal{J} : \mathcal{E} \longrightarrow \mathcal{E}$ is defined by

$$\begin{aligned} (\mathcal{J}x)(t) &= \int_a^t \frac{(t-s)^{q-1}}{\Gamma(q)} f(s, x(s))ds - g_1(t) \int_a^b \frac{(b-s)^{q-1}}{\Gamma(q)} f(s, x(s))ds \\ &- g_2(t) \int_a^b \frac{(b-s)^{q-2}}{\Gamma(q-1)} f(s, x(s))ds + g_3(t) \int_a^b \Big(\int_a^s \frac{(s-u)^{q-1}}{\Gamma(q)} f(u, x(u))du \Big) dA(s), \end{aligned} \tag{21}$$

where $g_1(t), g_2(t)$ and $g_3(t)$ are given by (6).

Evidently, the existence of fixed points of the operator $\mathcal{J}$ will imply the existence of solutions for the problems (1) and (2).

Now, the platform is set to present our main results. The following known fixed-point theorem [30] will be used in the proof of our first result.

Theorem 1. *Let X be a Banach space. Assume that $\mathcal{G} : X \longrightarrow X$ is a completely continuous operator and the set $\mathcal{P} = \{x \in X | x = \beta \mathcal{G} x,\ 0 < \beta < 1\}$ is bounded. Then $\mathcal{G}$ has a fixed point in X.*

Theorem 2. *Suppose that there exists $\varrho \in C([a,b], \mathbb{R}^+)$ such that $|f(t, x(t))| \leq \varrho(t)$, $\forall t \in [a,b]$, $x \in \mathcal{E}$, with $\sup_{t\in[a,b]} |\varrho(t)| = \|\varrho\|$. Then the problems (1) and (2) has at least one solution on $[a,b]$.*

Proof. Observe that continuity of the operator $\mathcal{J}$ follows from that of f. Let $\Phi \subset \mathcal{E}$ be bounded. Then, $\forall x \in \Phi$ together with the given assumption $|f(t, x(t))| \leq \varrho(t)$, we get

$$\begin{aligned}
|(\mathcal{J}x)| &\leq \sup_{t\in[a,b]} \left\{ \int_a^t \frac{(t-s)^{q-1}}{\Gamma(q)} |f(s,x(s))| ds + |g_1(t)| \int_a^b \frac{(b-s)^{q-1}}{\Gamma(q)} |f(s,x(s))| ds \right. \\
&\quad + |g_2(t)| \int_a^b \frac{(b-s)^{q-2}}{\Gamma(q-1)} |f(s,x(s))| ds \\
&\quad \left. + |g_3(t)| \int_a^b \Big(\int_a^s \frac{(s-u)^{q-1}}{\Gamma(q)} |f(u,x(u))| du \Big) dA(s) \right\} \\
&\leq \|\varrho\| \left[\frac{(b-a)^q}{\Gamma(q+1)} + \bar{g_1} \frac{(b-a)^q}{\Gamma(q+1)} + \bar{g_2} \frac{(b-a)^{q-1}}{\Gamma(q)} + \bar{g_3} \int_a^b \frac{(s-a)^q}{\Gamma(q+1)} dA(s) \right] \\
&= \|\varrho\| \Lambda = M_1,
\end{aligned}$$

which shows that $\mathcal{J}$ is bounded. Next, for $a < t_1 < t_2 < b$, we have

$$\begin{aligned}
& |(\mathcal{J}x)(t_2) - (\mathcal{J}x)(t_1)| \\
\leq\ & \int_a^{t_1} \frac{|(t_2-s)^{q-1} - (t_1-s)^{q-1}|}{\Gamma(q)} |f(s,x(s))| ds \\
& + \int_{t_1}^{t_2} \frac{|(t_2-s)^{q-1}|}{\Gamma(q)} |f(s,x(s))| ds + |g_1(t_2) - g_1(t_1)| \int_a^b \frac{|(b-s)^{q-1}|}{\Gamma(q)} |f(s,x(s))| ds \\
& + |g_2(t_2) - g_2(t_1)| \int_a^b \frac{|(b-s)^{q-2}|}{\Gamma(q-1)} |f(s,x(s))| ds \\
& + |g_3(t_2) - g_3(t_1)| \int_a^b \Big(\int_a^s \frac{|(s-u)^{q-1}|}{\Gamma(q)} |f(u,x(u))| du \Big) dA(s) \\
\leq\ & \|\varrho\| \left[\frac{|(t_2-a)^q - (t_1-a)^q| + 2(t_2-t_1)^q}{\Gamma(q+1)} + \frac{|g_1(t_2) - g_1(t_1)|\, |(b-a)^q|}{\Gamma(q+1)} \right. \\
& \left. + \frac{|g_2(t_2) - g_2(t_1)|\, |(b-a)^{q-1}|}{\Gamma(q)} + \frac{|g_3(t_2) - g_3(t_1)|}{\Gamma(q+1)} \int_a^b \Big(\int_a^s (s-a)^q dA(s) \Big) \right],
\end{aligned}$$

which tends to zero as $t_2 \longrightarrow t_1$ independent of x. Thus, $\mathcal{J}$ is equicontinuous on Φ. Hence, by Arzelá-Ascoli theorem, $\mathcal{J}$ is relatively compact on Φ. Therefore, $\mathcal{J}(\Phi)$ is a relatively compact subset of $\mathcal{E}$.

Now we consider a set $\mathcal{P} = \{x \in \mathcal{E} | x = \beta \mathcal{J} x,\ 0 < \beta < 1\}$, and show that the set $\mathcal{P}$ is bounded. Let $x \in \mathcal{P}$, then $x = \beta \mathcal{J} x$, $0 < \beta < 1$. For any $t \in [a,b]$, we have

$$|x(t)| = \beta |(\mathcal{J}x)(t)|$$

$$
\begin{aligned}
&\leq \sup_{t\in[a,b]}\left\{\int_a^t \frac{(t-s)^{q-1}}{\Gamma(q)}|f(s,x(s))|ds+|g_1(t)|\int_a^b \frac{(b-s)^{q-1}}{\Gamma(q)}|f(s,x(s))|ds\right.\\
&\quad +|g_2(t)|\int_a^b \frac{(b-s)^{q-2}}{\Gamma(q-1)}|f(s,x(s))|ds\\
&\quad \left.+|g_3(t)|\int_a^b\Big(\int_a^s \frac{(s-u)^{q-1}}{\Gamma(q)}|f(u,x(u))|du\Big)dA(s)\right\}\\
&\leq \|\varrho\|\left[\frac{(b-a)^q}{\Gamma(q+1)}+\bar{g_1}\frac{(b-a)^q}{\Gamma(q+1)}+\bar{g_2}\frac{(b-a)^{q-1}}{\Gamma(q)}+\bar{g_3}\int_a^b\frac{(s-a)^q}{\Gamma(q+1)}dA(s)\right]\\
&= \|\varrho\|\Lambda,
\end{aligned}
$$

where Λ given by (19). Thus, $\|x\| \leq \|\varrho\|\Lambda$ for any $t \in [a,b]$. Therefore, the set $\mathcal{P}$ is bounded. In consequence, the conclusion of Theorem 1 applies and that the operator $\mathcal{J}$ has at least one fixed point. Thus, there exists at least one solution for the problems (1) and (2) on $[a,b]$. □

Example 1. *Consider the fractional boundary value problem*

$$
\begin{cases}
{}^cD^{\frac{11}{3}}x(t) = \dfrac{6\,e^{-x^2}}{\sqrt{t^4+24}}+\dfrac{\cos t}{t^2+1}\Big(\dfrac{|x|^3}{1+|x|^3}\Big)+t^3+6,\ t\in[1,2],\\
x(1) = \displaystyle\sum_{i=1}^{4}\alpha_i x(\eta_i)+\int_1^2 x(s)dA(s),\ x'(1)=0,\ x(2)=0,\ x'(2)=0,
\end{cases}
\tag{22}
$$

where $q = 11/3,\ a = 1,\ b = 2,\ \alpha_1 = -1/2,\ \alpha_2 = -1/6,\ \alpha_3 = 1/6,\ \alpha_4 = 2,\ \eta_1 = 6/5,\ \eta_2 = 7/5,\ \eta_3 = 8/5,\ \eta_4 = 9/5$ *and* $f(t,x) = \dfrac{6\,e^{-x^2}}{\sqrt{t^4+24}}+\dfrac{\cos t}{t^2+1}\Big(\dfrac{|x|^3}{1+|x|^3}\Big)+t^3+6.$

Clearly, $|f(t,x)| \leq \frac{6}{\sqrt{t^4+24}}+\frac{\cos t}{t^2+1}+t^3+6 = \varrho(t) > 0$. *Therefore, there exists at least one solution for the problem (22) on* $[1,2]$ *by the conclusion of Theorem 2.*

Our next existence result is based on the following fixed-point theorem [30].

Theorem 3. *Let* Ω *be an open bounded subset of a Banach space* X *with* $0 \in \Omega$ *and the operator* $\mathcal{F} : \overline{\Omega} \longrightarrow X$ *is completely continuous satisfying* $\|\mathcal{F}x\| \leq \|x\|,\ \forall\, x \in \partial\Omega$. *Then the operator* $\mathcal{F}$ *has a fixed point in* $\overline{\Omega}$.

Theorem 4. *Let* $|f(t,x)| \leq \xi|x|$ *for* $0 < |x| < \tau$, *where* τ *and* ξ *are positive constants. Then the problems (1) and (2) has at least one solution for small values of* ξ.

Proof. Let us choose ξ such that

$$
\Lambda\xi < 1, \tag{23}
$$

where Λ given by (19). Define $B_{r_2} = \{x \in \mathcal{E}; \|x\| \leq r_2\}$ and take $x \in \mathcal{E}$ such that $\|x\| = r_2$, that is, $x \in \partial B_{r_2}$. As we argued in Theorem 2, it can be shown that $\mathcal{J}$ is completely continuous and

$$
\begin{aligned}
|(\mathcal{J}x)(t)| &\leq \sup_{t\in[a,b]}\left\{\int_a^t \frac{(t-s)^{q-1}}{\Gamma(q)}|f(s,x(s))|ds+|g_1(t)|\int_a^b \frac{(b-s)^{q-1}}{\Gamma(q)}|f(s,x(s))|ds\right.\\
&\quad +|g_2(t)|\int_a^b \frac{(b-s)^{q-2}}{\Gamma(q-1)}|f(s,x(s))|ds\\
&\quad \left.+|g_3(t)|\int_a^b\Big(\int_a^s \frac{(s-u)^{q-1}}{\Gamma(q)}|f(u,x(u))|du\Big)dA(s)\right\}
\end{aligned}
$$

$$
\begin{aligned}
&\leq \xi\|x\|\left[\frac{(b-a)^q}{\Gamma(q+1)}+\bar{g_1}\frac{(b-a)^q}{\Gamma(q+1)}+\bar{g_2}\frac{(b-a)^{q-1}}{\Gamma(q)}+\bar{g_3}\int_a^b\frac{(s-a)^q}{\Gamma(q+1)}dA(s)\right]\\
&= \xi\Lambda\|x\|.
\end{aligned}
$$

Using (23) and the norm $\|x\| = \sup_{t\in[a,b]} |x(t)|$, we get $\|\mathcal{J}x\| \leq \|x\|$, $x \in \partial B_{r_2}$. Therefore, conclusion of Theorem 3 applies and hence the problems (1) and (2) has a solution on $[a,b]$. □

Example 2. *Consider the fractional boundary value problem.*

$$
\begin{cases}
{}^cD^{\frac{27}{7}}x(t) = \dfrac{2|x|}{\sqrt{100+t^2}}\left(1+\dfrac{|x|}{1+|x|}\right),\ t\in[0,1],\\
x(0) = \displaystyle\sum_{i=1}^{4}\alpha_i x(\eta_i)+\int_0^1 x(s)dA(s),\ x'(0)=0,\ x(1)=0,\ x'(1)=0,
\end{cases}
\tag{24}
$$

where $q = 27/7$, $a = 0$, $b = 1$, $\alpha_1 = -2$, $\alpha_2 = -1/6$, $\alpha_3 = 1/6$, $\alpha_4 = 15/4$, $\eta_1 = 1/7$, $\eta_2 = 2/7$, $\eta_3 = 3/7$, $\eta_4 = 4/7$, *and* $f(t,x(t)) = \dfrac{2|x|}{\sqrt{100+t^2}}\left(1+\dfrac{|x|}{1+|x|}\right)$. *Let us take* $A(s) = \frac{s^2}{2}$. *Using the given data, we have that* $A_1 \approx -1.25$, $A_2 \approx 1.45068$, $A_3 \approx 0.903110$, $\gamma_1 \approx -0.43194$, $\gamma_2 \approx -0.115630$, $\lambda_1 \approx 1.96464$, $\lambda_2 \approx -0.422566$, $\lambda_3 \approx 0.771712$, $\lambda_4 \approx 1.92928$, $\lambda_5 \approx 0.154867$, $\lambda_6 \approx 1.54342$, $\bar{g_1} \approx 1.96464$, $\bar{g_2} \approx 0.422566$, $\bar{g_3} \approx 0.771712$, $\Lambda \approx 0.243646$, *where* Λ *is given by (19). Clearly the hypothesis of Theorem 4 is satisfied with* $\xi = \frac{2}{5}$. *Also* $\xi\Lambda \approx 0.097458 < 1$. *Therefore, the problem (24) has at least one solution on* $[0,1]$.

In the next result, we apply a fixed-point theorem due to Krasnoselskii [31] to establish the existence of solutions for the problems (1) and (2).

Theorem 5. *(Krasnoselskii [31]) Let M be a closed, convex, bounded and nonempty subset of a Banach space X and let* $\mathcal{F}_1, \mathcal{F}_2$ *be the operators defined from M to X such that: (i)* $\mathcal{F}_1x + \mathcal{F}_2y \in M$ *wherever* $x, y \in M$; *(ii)* $\mathcal{F}_1$ *is compact and continuous; (iii)* $\mathcal{F}_2$ *is a contraction. Then there exists* $z \in M$ *such that* $z = \mathcal{F}_1z + \mathcal{F}_2z$.

Theorem 6. *Assume that* $f : [a,b] \times \mathbb{R} \longrightarrow \mathbb{R}$ *is a continuous function such that the following conditions hold:*

(H_1) $|f(t,x) - f(t,y)| \leq L|x-y|$, $L > 0$, $\forall\, t \in [a,b]$, $x, y \in \mathbb{R}$;
(H_2) $|f(t,x)| \leq \mu(t)$, $\forall\, (t,x) \in [a,b]\times\mathbb{R}$, $\mu \in C([a,b],\mathbb{R}^+)$.

Then the problems (1) and (2) has at least one solution on $[a,b]$ *if*

$$
\left(\Lambda - \frac{(b-a)^q}{\Gamma(q+1)}\right)L < 1,
\tag{25}
$$

where Λ *is defined by (19).*

Proof. Consider a closed ball $B_r = \{x \in \mathcal{E} : \|x\| \leq r\}$ with $r \geq \Lambda\|\mu\|$, $\sup_{t\in[a,b]}|\mu(t)| = \|\mu\|$, and Λ is given by (19). Define operators $\mathcal{J}_1$ and $\mathcal{J}_2$ on B_r as

$$
\begin{aligned}
(\mathcal{J}_1x)(t) &= \int_a^t \frac{(t-s)^{q-1}}{\Gamma(q)}f(s,x(s))ds,\\
(\mathcal{J}_2x)(t) &= -g_1(t)\int_a^b \frac{(t-s)^{q-1}}{\Gamma(q)}f(s,x(s))ds - g_2(t)\int_a^b\frac{(t-s)^{q-2}}{\Gamma(q-1)}f(s,x(s))ds\\
&\quad +g_3(t)\int_a^b\left(\int_a^s\frac{(s-u)^{q-1}}{\Gamma(q)}f(u,x(u))du\right)dA(s).
\end{aligned}
$$

Please note that $\mathcal{J} = \mathcal{J}_1 + \mathcal{J}_2$. For $x, y \in B_r$, we have

$$\begin{aligned}\|\mathcal{J}_1 x + \mathcal{J}_2 y\| &\le \sup_{t\in[a,b]}\Big\{\int_a^t \frac{(t-s)^{q-1}}{\Gamma(q)}|f(s,x(s))|ds + |g_1(t)|\int_a^b \frac{(t-s)^{q-1}}{\Gamma(q)}|f(s,y(s))|ds \\ &+|(g_2(t)|\int_a^b \frac{(t-s)^{q-2}}{\Gamma(q-1)}|f(s,y(s))|ds \\ &+|g_3(t)|\int_a^b\Big(\int_a^s \frac{(s-u)^{q-1}}{\Gamma(q)}|f(u,y(u)|)du\Big)dA(s)\Big\} \\ &\le \|\mu\|\sup_{t\in[a,b]}\Big\{\int_a^t \frac{(t-s)^{q-1}}{\Gamma(q)}ds + |g_1(t)|\int_a^b \frac{(t-s)^{q-1}}{\Gamma(q)}ds \\ &+|g_2(t)|\int_a^b \frac{(t-s)^{q-2}}{\Gamma(q-1)}ds + |g_3(t)|\int_a^b\Big(\int_a^s \frac{(s-u)^{q-1}}{\Gamma(q)}du\Big)dA(s)\Big\} \\ &\le \|\mu\|\Lambda \le r,\end{aligned}$$

where we have used (19). Thus, $\mathcal{J}_1 x + \mathcal{J}_2 y \in B_r$. Next we show that $\mathcal{J}_2$ is a contraction. For $x, y \in B_r$, we have

$$\begin{aligned}\|\mathcal{J}_2 x - \mathcal{J}_2 y\| &= \sup_{t\in[a,b]}\Big|(\mathcal{J}_2 x)(t) - (\mathcal{J}_2 y)(t)\Big| \\ &\le \sup_{t\in[a,b]}\Big\{|g_1(t)|\int_a^b \frac{(b-s)^{q-1}}{\Gamma(q)}|f(s,x(s)) - f(s,y(s))|ds \\ &+|g_2(t)|\int_a^b \frac{(b-s)^{q-2}}{\Gamma(q-1)}|f(s,x(s)) - f(s,y(s))|ds \\ &+|g_3(t)|\int_a^b\Big(\int_a^s \frac{(s-u)^{q-1}}{\Gamma(q)}|f(u,x(u)) - f(u,y(u))|du\Big)dA(s)\Big\} \\ &\le L\|x-y\|\Big[\bar{g}_1\frac{(b-a)^q}{\Gamma(q+1)} + \bar{g}_2\frac{(b-a)^{q-1}}{\Gamma(q)} + \bar{g}_3\int_a^b \frac{(s-a)^q}{\Gamma(q+1)}dA(s)\Big] \\ &= L\Big(\Lambda - \frac{(b-a)^q}{\Gamma(q+1)}\Big)\|x-y\|,\end{aligned}$$

which shows that $\mathcal{J}_2$ is a contraction by the condition (25). Continuity of f implies that the operator $\mathcal{J}_1$ is continuous. Also, $\mathcal{J}_1$ is uniformly bounded on B_r as

$$\|\mathcal{J}_1 x\| \le \frac{(b-a)^q}{\Gamma(q+1)}\|\mu\|.$$

Next, we establish that the operator $\mathcal{J}_1$ is compact. Setting $\mathcal{S} = [a,b] \times B_r$, we define $\sup_{(t,x)\in\mathcal{S}}|f(t,x)| = M_r$. For $a < t_2 < t_1 < b$, we get

$$\begin{aligned}|(\mathcal{J}_1 x)(t_1) - (\mathcal{J}_1 x)(t_2)| &= \Big|\int_a^{t_2}\frac{(t_1-s)^{q-1} - (t_2-s)^{q-1}}{\Gamma(q)}f(s,x(s))ds \\ &+\int_{t_2}^{t_1}\frac{(t_1-s)^{q-1}}{\Gamma(q)}f(s,x(s))ds\Big| \\ &\le \frac{M_r}{\Gamma(q+1)}\Big[|(t_1-a)^q - (t_2-a)^q| + 2(t_1-t_2)^q\Big] \\ &\longrightarrow 0 \text{ when } t_1 - t_2 \longrightarrow 0,\end{aligned}$$

independent of x. Thus, $\mathcal{J}_1$ is equicontinuous on B_r. Hence, by Arzelá-Ascoli theorem, $\mathcal{J}_1$ is compact on B_r. Therefore, the conclusion of Theorem 5 applies to the problems (1) and (2). $\square$

Remark 1. *By interchanging the role of the operators $\mathcal{J}_1$ and $\mathcal{J}_2$ in Theorem 6, the condition (25) becomes:*

$$\frac{L(b-a)^q}{\Gamma(q+1)} < 1.$$

Example 3. *Consider the fractional differential equation:*

$$^cD^{\frac{11}{3}}x(t) = \frac{\delta}{\sqrt{t^2+8}}\, tan^{-1}x + e^{2t},\ t \in [1,2]. \tag{26}$$

subject to the boundary conditions of Example 1.

Let us take $A(s) = \frac{(s-1)^2}{2} + \frac{1}{2}$. Using the given data (from Example 1), it is found that $A_1 = -1$, $A_2 \approx 1.543333$, $A_3 \approx 1.245333$, $\gamma_1 \approx -0.379778$, $\gamma_2 \approx 0.081778$, $\lambda_1 \approx 1.877706$, $\lambda_2 \approx -0.261556$, $\lambda_3 \approx 0.877706$, $\lambda_4 \approx 1.755413$, $\lambda_5 \approx 0.476887$, $\lambda_6 \approx 1.755413$, $\bar{g}_1 \approx 1.877706$, $\bar{g}_2 \approx 0.264808$, $\bar{g}_3 \approx 0.877706$, $\Lambda \approx 0.272140$ (Λ is given by (19) and $\Lambda - \frac{(b-a)^2}{\Gamma(q+1)} \approx 0.204166$.

Clearly the hypotheses of Theorem 6 are satisfied with $L = \delta/3$ and $\mu(t) = \frac{\delta\pi}{2\sqrt{t^2+8}} + e^{2t}$. Also $L\Big(\Lambda - \frac{(b-a)^2}{\Gamma(q+1)}\Big) < 1$ for $\delta < 14.693960$. Therefore, there exists at least one solution for the problem (26) on $[1,2]$.

4. Uniqueness of Solution

Here, we prove the uniqueness of solutions for the problems (1) and (2).

Theorem 7. *Assume that $f : [a,b] \times \mathbb{R} \longrightarrow \mathbb{R}$ is a continuous function satisfying the condition (H_1). Then the problems (1) and (2) has a unique solution on $[a,b]$ if*

$$L\Lambda < 1, \tag{27}$$

where Λ is given by (19).

Proof. Setting $\sup_{t\in[a,b]} |f(t,0)| = N < \infty$, and selecting

$$r_1 \geq N\Lambda(1 - L\Lambda)^{-1},$$

we define $B_{r_1} = \{x \in \mathcal{E} : \|x\| \leq r_1\}$, and show that $\mathcal{J}B_{r_1} \subset B_{r_1}$, where the operator $\mathcal{J}$ is defined by (21). For $x \in B_{r_1}$,

$$\begin{aligned} |f(t,x(t))| &= |f(t,x(t)) - f(t,0) + f(t,0)| \leq |f(t,x(t)) - f(t,0)| + |f(t,0)| \\ &\leq L|x(t)| + N \leq L\|x\| + N \leq Lr_1 + N. \end{aligned}$$

Then,

$$\begin{aligned} \|\mathcal{J}x\| \leq \sup_{t\in[a,b]} \Bigg\{ & \int_a^t \frac{(t-s)^{q-1}}{\Gamma(q)}|f(s,x(s))|ds + |g_1(t)| \int_a^b \frac{(b-s)^{q-1}}{\Gamma(q)}|f(s,x(s))|ds \\ & +|g_2(t)| \int_a^b \frac{(b-s)^{q-2}}{\Gamma(q-1)}|f(s,x(s))|ds \\ & +|g_3(t)| \int_a^b \Big(\int_a^s \frac{(s-u)^{q-1}}{\Gamma(q)}|f(u,x(u))|du \Big) dA(s) \Bigg\} \end{aligned}$$

$$
\begin{aligned}
&\leq (L\|x\|+N)\left[\frac{(b-a)^q}{\Gamma(q+1)}+\bar{g_1}\frac{(b-a)^q}{\Gamma(q+1)}+\bar{g_2}\frac{(b-a)^{q-1}}{\Gamma(q)}+\bar{g_3}\int_a^b\frac{(s-a)^q}{\Gamma(q+1)}dA(s)\right]\\
&\leq (Lr_1+N)\Lambda\leq r_1.
\end{aligned}
$$

This shows that $\mathcal{J}x\in B_{r_1}$ for any $x\in B_{r_1}$. Therefore, $\mathcal{J}B_{r_1}\subset B_{r_1}$. Now, we show that $\mathcal{J}$ is a contraction. For $x,y\in\mathcal{E}$ and $t\in[a,b]$, we obtain

$$
\begin{aligned}
\|(\mathcal{J}x)-(\mathcal{J}y)\| &= \sup_{t\in[a,b]}\Big|(\mathcal{J}x)(t)-(\mathcal{J}y)(t)\Big|\\
&\leq \sup_{t\in[a,b]}\Bigg\{\int_a^t\frac{(t-s)^{q-1}}{\Gamma(q)}\Big|f(s,x(s))-f(s,y(s))\Big|ds\\
&\quad+|g_1(t)|\int_a^b\frac{(b-s)^{q-1}}{\Gamma(q)}\Big|f(s,x(s))-f(s,y(s))\Big|ds\\
&\quad+|g_2(t)|\int_a^b\frac{(b-s)^{q-2}}{\Gamma(q-1)}\Big|f(s,x(s))-f(s,y(s))\Big|ds\\
&\quad+|g_3(t)|\int_a^b\Big(\int_a^s\frac{(s-u)^{q-1}}{\Gamma(q)}\Big|f(u,x(u))-f(u,y(u))\Big|du\Big)dA(s)\Bigg\}\\
&\leq L\|x-y\|\left[\frac{(b-a)^q}{\Gamma(q+1)}+\bar{g_1}\frac{(b-a)^q}{\Gamma(q+1)}+\bar{g_2}\frac{(b-a)^{q-1}}{\Gamma(q)}+\bar{g_3}\int_a^b\frac{(s-a)^q}{\Gamma(q+1)}dA(s)\right]\\
&= L\Lambda\|x-y\|.
\end{aligned}
$$

By the condition (27), we deduce from the above inequality that $\mathcal{J}$ is a contraction. Thus, by the conclusion of Banach fixed-point theorem, the problems (1) and (2) has a unique solution on $[a,b]$. □

Example 4. *Let us take the problem considered in Example 1, and note that $L\Lambda<1$ for $\delta<11.023738$. Clearly the hypothesis of Theorem 7 is satisfied. Hence it follows by the conclusion of Theorem 7 that the problem (22) has a unique solution on $[1,2]$.*

5. Ulam Stability

In this section, we discuss the Ulam stability for the problems (1) and (2) by means of integral representation of its solution given by

$$
\begin{aligned}
y(t) &= \int_a^t\frac{(t-s)^{q-1}}{\Gamma(q)}f(s,y(s))ds-g_1(t)\int_a^b\frac{(b-s)^{q-1}}{\Gamma(q)}f(s,y(s))ds\\
&\quad-g_2(t)\int_a^b\frac{(b-s)^{q-2}}{\Gamma(q-1)}f(s,y(s))ds\\
&\quad+g_3(t)\int_a^b\Big(\int_a^s\frac{(s-u)^{q-1}}{\Gamma(q)}f(u,y(u))du\Big)dA(s). \qquad (28)
\end{aligned}
$$

Here $y\in C([a,b],\mathbb{R})$ possesses a fractional derivative of order $3<q\leq 4$ and $f:[a,b]\times\mathbb{R}\longrightarrow\mathbb{R}$ is a continuous function. Then the nonlinear operator $\mathcal{Q}:C([a,b],\mathbb{R})\longrightarrow C([a,b],\mathbb{R})$ defined by

$$\mathcal{Q}y(t)={}^cD^qy(t)-f(t,y(t))$$

is continuous.

Definition 3. *For each $\epsilon>0$ and for each solution y of (1) and (2) such that*

$$\|\mathcal{Q}y\|\leq\epsilon, \qquad (29)$$

the problems (1) and (2) is said to be Ulam-Hyers stable if we can find a positive real number ν *and a solution* $x \in C([a,b],\mathbb{R})$ *of (1) and (2) satisfying the inequality:*

$$\|x-y\| \leq \nu\,\epsilon_*,$$

where ϵ_* *is a positive real number depending on* ϵ.

Definition 4. *Let there exists* $\kappa \in C(\mathbb{R}^+,\mathbb{R}^+)$ *such that for each solution* y *of (1) and (2), we can find a solution* $x \in C([a,b],\mathbb{R})$ *of (1) and (2) such that*

$$|x(t)-y(t)| \leq \kappa(\epsilon),\ t \in [a,b].$$

Then the problems (1) and (2) is said to be generalized Ulam-Hyers stable

Definition 5. *For each* $\epsilon > 0$ *and for each solution* y *of (1) and (2), the problems (1) and (2) is called Ulam-Hyers-Rassias stable with respect to* $\sigma \in C([a,b],\mathbb{R}^+)$ *if*

$$|\mathcal{Q}y(t)| \leq \epsilon\sigma(t),\ t \in [a,b], \tag{30}$$

and there exist a real number $\nu > 0$ *and a solution* $x \in C([a,b],\mathbb{R})$ *of (1) and (2) such that*

$$|x(t)-y(t)| \leq \nu\epsilon_*\sigma(t),\ t \in [a,b],$$

where ϵ_* *is a positive real number depending on* ϵ.

Theorem 8. *Let the assumption* (H_1) *hold with* $L\Lambda < 1$, *where* Λ *is defined by (19). Then the problems (1) and (2) is both Ulam-Hyers and generalized Ulam-Hyers stable.*

Proof. Let $x \in C([a,b],\mathbb{R})$ be a solution of (1) and (2) satisfying (21) by Theorem 7. Let y be any solution satisfying (29). Then by Lemma 2, y satisfies the integral equation (28). Furthermore, the equivalence in Lemma 2 implies the equivalence between the operators $\mathcal{Q}$ and $\mathcal{J}-I$ (where I is identity operator) for every solution $y \in C([a,b],\mathbb{R})$ of (1) and (2) satisfying (27). Therefore, we deduce by the fixed-point property of the operator $\mathcal{J}$ (given by (21)) and (29) that

$$\begin{aligned}|y(t)-x(t)| &= |y(t)-\mathcal{J}y(t)+\mathcal{J}y(t)-\mathcal{J}x(t)|\\ &\leq |\mathcal{J}x(t)-\mathcal{J}y(t)|+|\mathcal{J}y(t)-y(t)|\\ &\leq L\Lambda\|x-y\|+\epsilon,\end{aligned}$$

where $\epsilon > 0$ and $L\Lambda < 1$. In consequence, we get

$$\|x-y\| \leq \frac{\epsilon}{1-L\Lambda}.$$

Fixing $\epsilon_* = \frac{\epsilon}{1-L\Lambda}$ and $\nu = 1$, we obtain the Ulam-Hyers stability condition. In addition, the generalized Ulam-Hyers stability follows by taking $\kappa(\epsilon) = \frac{\epsilon}{1-L\Lambda}$. □

Theorem 9. *Assume that* (H_1) *holds with* $L < \Lambda^{-1}$ *(where* Λ *is defined by (19)), and there exists a function* $\sigma \in C([a,b],\mathbb{R}^+)$ *satisfying the condition (30). Then the problems (1) and (2) is Ulam-Hyers-Rassias stable with respect to* σ.

Proof. Following the arguments employed in the proof of Theorem 8, we have

$$\|x-y\| \leq \epsilon_*\sigma(t),$$

where $\epsilon_* = \frac{\epsilon}{1-L\Lambda}$. This completes the proof. □

Example 5. *Consider the following fractional differential equation*

$$^{c}D^{\frac{27}{7}}x(t) = \frac{3}{t^2+16}\Big(\cos x + \frac{x}{1+x}\Big),\ t \in [0,1], \tag{31}$$

subject to the same data and the boundary conditions given in Example 1 with $f(t,x(t)) = \frac{3}{t^2+16}\Big(\cos x + \frac{x}{1+x}\Big)$. *Obviously* $|f(t,x) - f(t,y)| \leq \frac{3}{8}\|x-y\|$, *so,* $L = 3/8$ *and* $L\Lambda \approx 0.102053 < 1$. *Then the problem (31) is Ulam-Hyers stable, and generalized Ulam-Hyers stable. In addition, if there exists a continuous and positive function* $\sigma = e^{1+t^2} + 5$ *satisfying the condition (30), then the problem (31) is Ulam-Hyers-Rassias stable with the given value of* $f(t,x)$.

6. Conclusions

We have obtained several existence results for a new class of Caputo-type fractional differential equations of order $q \in (3,4]$, supplemented with non-conjugate Riemann-Stieltjes integro-multipoint boundary conditions on an arbitrary domain by imposing different kinds of conditions on the nonlinear function involved in the problem. The existence results, relying on different fixed-point theorems, are presented in Section 3. The uniqueness of solution for the given problem is studied in Section 4 with the aid of Banach fixed-point theorem. Section 5 is concerned with different kinds of Ulam stability for the problem at hand. Some new results follow as special cases of the ones presented in this paper. For example, taking $A(s) = s$, our results correspond to the ones for non-conjugate integro-multipoint boundary conditions of the form: $x(a) = \sum_{i=1}^{n-2}\alpha_i x(\eta_i) + \int_a^b x(s)ds,\ x'(a) = 0,\ x(b) = 0,\ x'(b) = 0$ and the value of Λ given by (19) takes the following form in this situation:

$$\Lambda = \Big\{\frac{(b-a)^q}{\Gamma(q+1)} + \bar{g_1}\frac{(b-a)^q}{\Gamma(q+1)} + \bar{g_2}\frac{(b-a)^{q-1}}{\Gamma(q)} + \bar{g_3}\frac{(b-a)^{q+1}}{\Gamma(q+2)}\Big\}.$$

Letting $A(s) = 0$ in our results, we get the ones for non-conjugate multipoint boundary conditions of the form: $x(a) = \sum_{i=1}^{n-2}\alpha_i x(\eta_i),\ x'(a) = 0,\ x(b) = 0,\ x'(b) = 0$. In case we take $\alpha_i = 0$ for all $i = 1,\dots,n-2$, our results reduce to the ones with non-conjugate Riemann-Stieltjes boundary conditions. By fixing $A(s) = 0$ and $\alpha_i = 0$ for all $i = 1,\dots,n-2$, our results correspond to a boundary value problem of fractional order $q \in (3,4]$ with conjugate boundary conditions.

Author Contributions: Formal Analysis, B.A., Y.A., S.K.N. and A.A.

Funding: This project was funded by the Deanship of Scientific Research (DSR) at King Abdulaziz University, Jeddah, Saudi Arabia, under grant no. (RG-25-130-38).

Acknowledgments: The authors acknowledge with thanks DSR technical and financial support. The authors also thank the reviewers for their constructive remarks on our work.

Conflicts of Interest: The authors declare no conflict of interest.

References

1. Zaslavsky, G.M. *Hamiltonian Chaos and Fractional Dynamics*; Oxford University Press: Oxford, UK, 2005.
2. Zhang, D.; Zhang, Y.; Zhang, Z.; Ahmed, N.; Zhang, Y.; Li, F.; Belic, M.R.; Xiao, M. Unveiling the link between fractional Schrödinger equation and light propagation in honeycomb lattice. *Ann. Phys.* **2017**, *529*, 1700149. [CrossRef]
3. Fallahgoul, H.A.; Focardi, S.M.; Fabozzi, F.J. *Fractional Calculus and Fractional Processes with Applications to Financial Economics. Theory and Application*; Elsevier/Academic Press: London, UK, 2017.
4. Javidi, M.; Ahmad, B. Dynamic analysis of time fractional order phytoplankton-toxic phytoplankton-zooplankton system. *Ecol. Model.* **2015**, *318*, 8–18. [CrossRef]
5. Magin, R.L. *Fractional Calculus in Bioengineering*; Begell House Publishers: Danbury, CT, USA, 2006.

6. Kilbas, A.A.; Srivastava, H.M.; Trujillo, J.J. *Theory and Applications of Fractional Differential Equations*; North-Holland Mathematics Studies, 204; Elsevier Science B.V.: Amsterdam, The Netherlands, 2006.
7. Lakshimikantham, V.; Leela, S.; Devi, J.V. *Theory of Fractional Dynamic Systems*; Cambridge Academic Publishers: Cambridge, UK, 2009.
8. Diethelm, K. *The Analysis of Fractional Differential Equations. An Application—Oriented Exposition Using Differential Operators of Caputo Type*; Lecture Notes in Mathematics 2004; Springer: Berlin, Germany, 2010.
9. Ahmad, B.; Alsaedi, A.; Ntouyas, S.K.; Tariboon, J. *Hadamard-Type Fractional Differential Equations, Inclusions and Inequalities*; Springer: Cham, Switzerland, 2017.
10. Ahmad, B.; Nieto, J.J. Riemann-Liouville fractional integro-differential equations with fractional nonlocal integral boundary conditions. *Bound. Value Probl.* **2011**, *2011*, 36. [CrossRef]
11. Liang, S.; Zhang, J. Existence of multiple positive solutions for m-point fractional boundary value problems on an infinite interval. *Math. Comput. Model.* **2011**, *54*, 1334–1346. [CrossRef]
12. Bai, Z.B.; Sun, W. Existence and multiplicity of positive solutions for singular fractional boundary value problems. *Comput. Math. Appl.* **2012**, *63*, 1369–1381. [CrossRef]
13. Agarwal, R.P.; O'Regan, D.; Stanek, S. Positive solutions for mixed problems of singular fractional differential equations. *Math. Nachr.* **2012**, *285*, 27–41. [CrossRef]
14. Ahmad, B.; Ntouyas, S.K.; Alsaedi, A. A study of nonlinear fractional differential equations of arbitrary order with Riemann-Liouville type multistrip boundary conditions. *Math. Probl. Eng.* **2013**, *2013*, 320415. [CrossRef]
15. Ahmad, B.; Ntouyas, S.K. Existence results for higher order fractional differential inclusions with multi-strip fractional integral boundary conditions. *Electron. J. Qual. Theory Differ. Equ.* **2013**, *2013*, 20. [CrossRef]
16. O'Regan, D.; Stanek, S. Fractional boundary value problems with singularities in space variables. *Nonlinear Dyn.* **2013**, *71*, 641–652. [CrossRef]
17. Zhai, C.; Xu, L. Properties of positive solutions to a class of four-point boundary value problem of Caputo fractional differential equations with a parameter. *Commun. Nonlinear Sci. Numer. Simul.* **2014**, *19*, 2820–2827. [CrossRef]
18. Graef, J.R.; Kong, L.; Wang, M. Existence and uniqueness of solutions for a fractional boundary value problem on a graph. *Fract. Calc. Appl. Anal.* **2014**, *17*, 499–510. [CrossRef]
19. Wang, G.; Liu, S.; Zhang, L. Eigenvalue problem for nonlinear fractional differential equations with integral boundary conditions. *Abstr. Appl. Anal.* **2014**, *2014*, 916260. [CrossRef]
20. Henderson, J.; Kosmatov, N. Eigenvalue comparison for fractional boundary value problems with the Caputo derivative. *Fract. Calc. Appl. Anal.* **2014**, *17*, 872–880. [CrossRef]
21. Zhang, L.; Ahmad, B.; Wang, G. Successive iterations for positive extremal solutions of nonlinear fractional differential equations on a half line. *Bull. Aust. Math. Soc.* **2015**, *91*, 116–128. [CrossRef]
22. Henderson, J.; Luca, R. Nonexistence of positive solutions for a system of coupled fractional boundary value problems. *Bound. Value Probl.* **2015**, *2015*, 138. [CrossRef]
23. Ntouyas, S.K.; Etemad, S. On the existence of solutions for fractional differential inclusions with sum and integral boundary conditions. *Appl. Math. Comput.* **2015**, *266*, 235–243. [CrossRef]
24. Mei, Z.D.; Peng, J.G.; Gao, J.H. Existence and uniqueness of solutions for nonlinear general fractional differential equations in Banach spaces. *Indag. Math.* **2015**, *26*, 669–678. [CrossRef]
25. Ahmad, B.; Ntouyas, S.K. Existence results for fractional differential inclusions with Erdelyi-Kober fractional integral conditions. *Analele Universitatii Ovidius Constanta-Seria Matematica* **2017**, *25*, 5–24. [CrossRef]
26. Srivastava, H.M. Remarks on some families of fractional-order differential equations. *Integr. Transf. Spec. Funct.* **2017**, *28*, 560–564. [CrossRef]
27. Wang, G.; Pei, K.; Agarwal, R.P.; Zhang, L.; Ahmad, B. Nonlocal Hadamard fractional boundary value problem with Hadamard integral and discrete boundary conditions on a half-line. *J. Comput. Appl. Math.* **2018**, *343*, 230–239. [CrossRef]
28. Ahmad, B.; Luca, R. Existence of solutions for sequential fractional integro-differential equations and inclusions with nonlocal boundary conditions. *Appl. Math. Comput.* **2018**, *339*, 516–534. [CrossRef]
29. Zhang, Y.; Wang, R.; Zhong, H.; Zhang, J.; Belic, M.R.; Zhang, Y. Resonant mode conversions and Rabi oscillations in a fractional Schrödinger equation. *Opt. Express* **2017**, *25*, 32401. [CrossRef]

30. Sun, J.X. *Nonlinear Functional Analysis and Its Application*; Science Press: Bejing, China, 2018.
31. Krasnoselskii, M.A. Two remarks on the method of successive approximations. *Uspekhi Matematicheskikh Nauk* **1955**, *10*, 123–127.

mathematics

MDPI

Article

A Coupled System of Fractional Difference Equations with Nonlocal Fractional Sum Boundary Conditions on the Discrete Half-Line

Jarunee Soontharanon [1], Saowaluck Chasreechai [1,*] and Thanin Sitthiwirattham [2,*]

[1] Department of Mathematics, Faculty of Applied Science, King Mongkut's University of Technology North Bangkok, Bangkok 10800, Thailand; jarunee.s@sci.kmutnb.ac.th

[2] Mathematics Department, Faculty of Science and Technology, Suan Dusit University, Bangkok 10700, Thailand

* Correspondence: saowaluck.c@sci.kmutnb.ac.th (S.C.); thanin_sit@dusit.ac.th (T.S.)

Received: 4 February 2019; Accepted: 8 March 2019; Published: 12 March 2019

Abstract: In this article, we propose a coupled system of fractional difference equations with nonlocal fractional sum boundary conditions on the discrete half-line and study its existence result by using Schauder's fixed point theorem. An example is provided to illustrate the results.

Keywords: existence; coupled system of fractional difference equations; fractional sum; discrete half-line

MSC: 39A05; 39A12

1. Introduction

Recently, many mathematicians and researchers have extensively studied fractional difference calculus since this subject can be used for describing many problems of real-world phenomena such as mechanical, control systems, flow in porous media, and electrical networks (see [1,2] and the references therein). The basic definitions and properties of fractional difference calculus are given in the book [3]. The applications and developments of the theory can be found in [4–47] and the references cited therein. For example, Ferreira [20] studied the fractional difference equation of order less than one. Goodrich [22] presented the fractional difference equation of order $1 < \alpha \leq 2$ with a constant boundary condition. Chen et al. [28] proposed the initial value problem of order less than one. Chen and Zhou [29] studied the antiperiodic boundary value problem of order $1 < \alpha \leq 2$. Sitthiwirattham et al. [38] initiated the study of the fractional sum boundary value problem of order $1 < \alpha \leq 2$. Sitthiwirattham [40] proposed the sequential fractional difference equation with the fractional sum boundary condition. We observe that these research works are fractional problems containing only one equation.

The study of coupled systems of fractional differential equations is an important topic in this area (see [48–53] and the references cited therein), and a recent example of the application of systems of fractional difference equations is [54].

For the boundary value problems for systems of discrete fractional equations, there are some studies in this area (see [55–60] and the references cited therein).

Pan et al. [55] proposed the system of discrete fractional difference equations as given by:

$$\begin{aligned} -\Delta^{\nu} y_1(t) &= f(y_1(t+\nu), y_2(t+\mu-1)), \\ -\Delta^{\mu} y_2(t) &= g(y_1(t+\nu), y_2(t+\mu-1)), \end{aligned} \tag{1}$$

for $t \in \mathbb{N}_{0,b+1} := \{0, 1, 2, ..., b+1\}$, with the difference boundary conditions:

$$y_1(\nu - 2) = \Delta y_1(\nu + b) = 0,$$
$$y_2(\mu - 2) = \Delta y_2(\mu + b) = 0, \tag{2}$$

where $b \in \mathbb{N}_0 := \mathbb{N} \cup \{0\}$; $1 < \mu, \nu \leq 2$; $0 < \beta \leq 1$; and $f, g : \mathbb{R} \times \mathbb{R} \to \mathbb{R}$ are continuous functions. Δ^ν and Δ^μ are fractional difference operator of order ν and μ, respectively.

In 2015, Goodrich [58] discussed the coupled system of discrete fractional difference equations:

$$\begin{aligned} -\Delta^{-\nu}x(t) &= \lambda_1 f(t + \nu - 1, y(t + \mu - 1)), \ t \in \mathbb{N}_{0,b+1}, \\ -\Delta^{-\mu}y(t) &= \lambda_2 g(t + \mu - 1, y(t + \nu - 1)), \end{aligned} \tag{3}$$

with the nonlinearities satisfying no growth conditions:

$$\begin{aligned} x(\nu - 2) &= H_1\left(\textstyle\sum_{i=1}^n a_i y(\xi_i)\right), & x(\nu + b + 1) &= 0, \\ y(\mu - 2) &= H_2\left(\textstyle\sum_{j=1}^m b_j x(\zeta_i)\right), & y(\mu + b + 1) &= 0, \end{aligned} \tag{4}$$

where $1 < \nu \leq 2$; $1 < \mu \leq 2$; $\lambda_1, \lambda_2 > 0$; $\{a_i\}_{i=1}^n, \{b_j\}_{j=1}^m \subseteq (0, \infty)$; and $H_1, H_2 : [0, \infty) \to [0, \infty)$ are continuous functions.

In this paper, we considered the coupled system of fractional difference equations:

$$\begin{cases} \Delta^{\alpha_1} u_1(t) = F_1(t + \alpha_1 - 1, t + \alpha_2 - 1, \Delta^{\beta_1} u_1(t + \alpha_1 - \beta_1), u_2(t + \alpha_2 - 1)), \\ \Delta^{\alpha_2} u_2(t) = F_2(t + \alpha_1 - 1, t + \alpha_2 - 1, \Delta^{\beta_2} u_2(t + \alpha_2 - \beta_1), u_1(t + \alpha_1 - 1)), \end{cases} \tag{5}$$

for $t \in \mathbb{N}_0$, subject to the nonlocal fractional sum boundary conditions on the discrete half-line $\mathbb{N}_0$:

$$\begin{cases} u_1(\alpha_1 - 2) = \phi_1(u_1, u_2), \\ u_2(\alpha_2 - 2) = \phi_2(u_1, u_2), \\ \lim_{t\to\infty} u_1(t + \alpha_1 - 2) = \lambda_2 \Delta^{-\theta_2} g_2(\eta_2 + \theta_2) u_2(\eta_2 + \theta_2), \\ \lim_{t\to\infty} u_2(t + \alpha_2 - 2) = \lambda_1 \Delta^{-\theta_1} g_1(\eta_1 + \theta_1) u_1(\eta_1 + \theta_1). \end{cases} \tag{6}$$

For $i = 1, 2$, $\alpha_i \in (1, 2]$; $\nu_i, \gamma_i, \theta_i \in (0, 1]$; $\beta_i \in (\alpha_i - 1, \alpha_i)$; $\lambda_1, \lambda_2 > 0$, and $\eta_i \in \mathbb{N}_{\alpha_i - 1, T + \alpha_i - 1}$ are given constants; $F_i \in C\left(\mathbb{N}_{\alpha_1 - 2} \times \mathbb{N}_{\alpha_2 - 2} \times \mathbb{R}^2, \mathbb{R}\right)$ and $g_i \in C\left(\mathbb{N}_{\alpha_i - 2, T + \alpha_i}, \mathbb{R}^+\right)$ are given functions; $\phi_i(u_1, u_2)$ are given functionals; and $\Delta^{-\theta_i}$ are fractional sums of order θ_i.

The goal of this study is to show the existence of solutions of the governing problems (5) and (6). The paper is structured as follows. Some definitions and basic lemmas are recalled in Section 2. In Section 3, we prove the existence of solutions of the boundary value problem (5) by employing Schauder's fixed point theorem. Finally, we present an example to illustrate our result in the last section.

2. Preliminaries

In what follows, the notation, definitions, and lemmas used in the main results are given.

Definition 1. *The generalized falling function is defined by* $t^{\underline{\alpha}} := \dfrac{\Gamma(t+1)}{\Gamma(t+1-\alpha)}$, *for any t and α for which the right-hand side is defined. If $t + 1 - \alpha$ is a pole of the Gamma function and $t + 1$ is not a pole, then $t^{\underline{\alpha}} = 0$.*

Lemma 1. [4] *Assume the falling factorial functions are well defined. If $t \leq r$, then $t^{\underline{\alpha}} \leq r^{\underline{\alpha}}$ for any $\alpha > 0$.*

Definition 2. *For $\alpha > 0$ and f defined on $\mathbb{N}_a$, the α-order fractional sum of f is defined by:*

$$\Delta^{-\alpha} f(t) := \frac{1}{\Gamma(\alpha)} \sum_{s=a}^{t-\alpha} (t-\sigma(s))^{\underline{\alpha-1}} f(s),$$

where $t \in \mathbb{N}_{a+\alpha}$ and $\sigma(s) = s+1$.

Definition 3. *For $\alpha > 0$ and f defined on $\mathbb{N}_a$, the α-order Riemann–Liouville fractional difference of f is defined by:*

$$\Delta^{\alpha} f(t) := \Delta^N \Delta^{-(N-\alpha)} f(t) = \frac{1}{\Gamma(-\alpha)} \sum_{s=a}^{t+\alpha} (t-\sigma(s))^{\underline{-\alpha-1}} f(s),$$

where $t \in \mathbb{N}_{a+N-\alpha}$ and $N \in \mathbb{N}$ are chosen so that $0 \le N-1 < \alpha \le N$.

Lemma 2. *[4] Let $0 \le N-1 < \alpha \le N$. Then,*

$$\Delta^{-\alpha}\Delta^{\alpha} y(t) = y(t) + C_1 t^{\underline{\alpha-1}} + C_2 t^{\underline{\alpha-2}} + \ldots + C_N t^{\underline{\alpha-N}},$$

for some $C_i \in \mathbb{R}$, with $1 \le i \le N$.

The following lemma deals with the linear variant of the boundary value problems (5) and (6) and gives a representation of the solution.

Lemma 3. *Let $\alpha_i \in (1,2]$, $\theta_i \in (0,1]$, $\lambda_1, \lambda_2 > 0$ and $\eta_i \in \mathbb{N}_{\alpha_i-1,T+\alpha_i-1}$ be given constants, $k_i \in C(\mathbb{N}_{\alpha_i-2}, \mathbb{R})$ and $g_i \in C(\mathbb{N}_{\alpha_i-2,T+\alpha_i}, \mathbb{R}^+)$ given functions, and $\phi_i(u_1,u_2)$ given functionals. For each $i,j \in \{1,2\}$ and $i \neq j$, then the problems:*

$$\Delta^{\alpha_i} u_i(t) = k_i(t+\alpha_i-1), \quad t \in \mathbb{N}_0, \tag{7}$$

$$u_i(\alpha_i-2) = \phi_i(u_1,u_2), \tag{8}$$

$$\lim_{t\to\infty} u_i(t+\alpha_i) = \lambda_j \Delta^{-\theta_j} g_j(\eta_j+\theta_j) u_j(\eta_j+\theta_j). \tag{9}$$

have the unique solutions:

$$\begin{aligned} u_1(t_1) &= t_1^{\underline{\alpha_1-1}} \left\{ \frac{\lambda_1}{\Lambda\Gamma(\theta_1)} \sum_{s=\alpha_1-2}^{\eta_1} (\eta_1+\theta_1-\sigma(s))^{\underline{\theta_1-1}} g_1(s) s^{\underline{\alpha_1-1}} \mathcal{P}(k_1,k_2) \right. \\ &\quad \left. - \frac{\lambda_2}{\Lambda\Gamma(\theta_2)} \sum_{s=\alpha_2-2}^{\eta_2} (\eta_2+\theta_2-\sigma(s))^{\underline{\theta_2-1}} g_2(s) s^{\underline{\alpha_2-1}} \mathcal{Q}(k_1,k_2) \right\} \\ &\quad + \frac{t_1^{\underline{\alpha_1-2}} \phi_1(u_1,u_2)}{\Gamma(\alpha_1)} + \frac{1}{\Gamma(\alpha_1)} \sum_{s=0}^{t_1-\alpha_1} (t_1-\sigma(s))^{\underline{\alpha_1-1}} k_1(s+\alpha_1-1), \quad t_1 \in \mathbb{N}_{\alpha_1-2}, \end{aligned} \tag{10}$$

$$\begin{aligned} u_2(t_2) &= t_2^{\underline{\alpha_2-1}} \left\{ \frac{\lim_{t_2\to\infty} t_2^{\underline{\alpha_2-1}}}{\Lambda} \mathcal{P}(k_1,k_2) - \frac{\lim_{t_1\to\infty} t_1^{\underline{\alpha_1-1}}}{\Lambda} \mathcal{Q}(k_1,k_2) \right\} \\ &\quad + \frac{t_2^{\underline{\alpha_2-2}} \phi_2(u_1,u_2)}{\Gamma(\alpha_2)} + \frac{1}{\Gamma(\alpha_2)} \sum_{s=0}^{t_2-\alpha_2} (t_2-\sigma(s))^{\underline{\alpha_2-1}} k_2(s+\alpha_2-1), \quad t_2 \in \mathbb{N}_{\alpha_2-2}, \end{aligned} \tag{11}$$

provided that both $u_1(t_1), u_2(t_2)$ are uniformly bounded on $\mathbb{N}_{\alpha_1-2}$ and $\mathbb{N}_{\alpha_2-2}$, respectively, and:

$$\Lambda = \frac{\lambda_2 \lim_{t_2\to\infty} t_2^{\underline{\alpha_2-1}}}{\Gamma(\alpha_2)} \sum_{s=\alpha_2-1}^{\eta_2} (\eta_2+\theta_2-\sigma(s))^{\underline{\theta_2-1}} g_2(s)\, s^{\underline{\alpha_2-1}} \tag{12}$$

$$-\frac{\lambda_1 \lim_{t_1\to\infty} t_1^{\underline{\alpha_1-1}}}{\Gamma(\alpha_1)} \sum_{s=\alpha_1-1}^{\eta_1} (\eta_1+\theta_1-\sigma(s))^{\underline{\theta_1-1}} g_1(s)\, s^{\underline{\alpha_1-1}}, \quad t_i \in \mathbb{N}_{\alpha_i-2},$$

$$\begin{aligned}\mathcal{P}(k_1,k_2) &= \frac{\lim_{t_1\to\infty} t_1^{\underline{\alpha_1-2}} \phi_1(u_1,u_2)}{\Gamma(\alpha_1)} - \frac{\lambda_2\phi_2(u_1,u_2)}{\Gamma(\alpha_2)\Gamma(\theta_2)} \sum_{s=\alpha_2-2}^{\eta_2} (\eta_2+\theta_2-\sigma(s))^{\underline{\theta_2-1}} g_2(s)\, s^{\underline{\alpha_2-2}} \\ &+ \frac{1}{\Gamma(\alpha_1)} \lim_{t_1\to\infty} \sum_{s=0}^{t_1-\alpha_1} (t_1-\sigma(s))^{\underline{\alpha_1-1}} k_1(s+\alpha_1-1) - \frac{\lambda_2}{\Gamma(\alpha_2)\Gamma(\theta_2)} \times \\ &\sum_{\xi=\alpha_2}^{\eta_2} \sum_{s=0}^{\xi-\alpha_2} (\eta_2+\theta_2-\sigma(\xi))^{\underline{\theta_2-1}} (\xi-\sigma(s))^{\underline{\alpha_2-1}} g_2(s+\alpha_2-1) k_2(s+\alpha_2-1),\end{aligned} \tag{13}$$

$$\begin{aligned}\mathcal{Q}(k_1,k_2) &= \frac{\lim_{t_2\to\infty} t_2^{\underline{\alpha_2-2}} \phi_2(u_1,u_2)}{\Gamma(\alpha_2)} - \frac{\lambda_1\phi_1(u_1,u_2)}{\Gamma(\alpha_1)\Gamma(\theta_1)} \sum_{s=\alpha_1-2}^{\eta_1} (\eta_1+\theta_1-\sigma(s))^{\underline{\theta_1-1}} g_1(s)\, s^{\underline{\alpha_1-2}} \\ &+ \frac{1}{\Gamma(\alpha_2)} \lim_{t_2\to\infty} \sum_{s=0}^{t_2-\alpha_2} (t_2-\sigma(s))^{\underline{\alpha_2-1}} k_2(s+\alpha_2-1) - \frac{\lambda_1}{\Gamma(\alpha_1)\Gamma(\theta_1)} \times \\ &\sum_{\xi=\alpha_1}^{\eta_1} \sum_{s=0}^{\xi-\alpha_1} (\eta_1+\theta_1-\sigma(\xi))^{\underline{\theta_1-1}} (\xi-\sigma(s))^{\underline{\alpha_1-1}} g_1(s+\alpha_1-1) k_1(s+\alpha_1-1).\end{aligned} \tag{14}$$

Proof. For each $i,j \in \{1,2\}$ and $i \neq j$, using Lemma 2 and the fractional sum of order $\alpha \in (1,2]$ for (7), we obtain:

$$u_i(t_i) = C_{1i} t_i^{\underline{\alpha_i-1}} + C_{2i} t_i^{\underline{\alpha_i-2}} + \frac{1}{\Gamma(\alpha_i)} \sum_{s=0}^{t_i-\alpha_i} (t_i-\sigma(s))^{\underline{\alpha_i-1}} k_i(s+\alpha_i-1), \tag{15}$$

for $t_i \in \mathbb{N}_{\alpha_i-2}$.

By using the boundary condition (8), we find that:

$$C_{2i} = \frac{\phi_i(u_1,u_2)}{\Gamma(\alpha_i)}. \tag{16}$$

Then, for $t_i \in \mathbb{N}_{\alpha_i-2}$, we have:

$$\begin{aligned} u_i(t_i) &= C_{1i} t_i^{\underline{\alpha_i-1}} + \frac{\phi_i(u_1,u_2)}{\Gamma(\alpha_i)} t_i^{\underline{\alpha_i-2}} \\ &+ \frac{1}{\Gamma(\alpha_i)} \sum_{s=0}^{t_i-\alpha_i} (t_i-\sigma(s))^{\underline{\alpha_i-1}} k_i(s+\alpha_i-1). \end{aligned} \tag{17}$$

Taking the fractional sum of order $0 < \theta_i \leq 1$ for (17), we obtain:

$$\begin{aligned} &\Delta^{-\theta_i} u_i(t_i) \\ &= \frac{C_{1i}}{\Gamma(\theta_i)} \sum_{s=\alpha_i-2}^{t_i} (t_i+\theta-\sigma(s))^{\underline{\theta_i-1}} g_i(s)\, s^{\underline{\alpha_i-1}} + \frac{\phi_i(u_1,u_2)}{\Gamma(\alpha_i)} \times \\ &\sum_{s=\alpha_i-2}^{t_i} (t_i+\theta_i-\sigma(s))^{\underline{\theta_i-1}} g_i(s)\, s^{\underline{\alpha_i-2}} + \frac{1}{\Gamma(\theta_i)\Gamma(\alpha_i)} \sum_{\xi=\alpha_i}^{t_i} \sum_{s=0}^{\xi-\alpha_i} (t_i+\theta_i-\sigma(\xi))^{\underline{\theta_i-1}} \times \\ &(\xi-\sigma(s))^{\underline{\alpha_i-1}} g_i(s+\alpha_i-1)\, k_i(s+\alpha_i-1), \end{aligned} \tag{18}$$

for $t_i \in \mathbb{N}_{\alpha_i-2}$.

Employing the boundary condition (9), this implies that:

$$
\begin{aligned}
& C_{11} \lim_{t_1\to\infty} t_1^{\underline{\alpha_1-1}} + \frac{\phi_1(u_1,u_2)}{\Gamma(\alpha_1)} \lim_{t_1\to\infty} t_1^{\underline{\alpha_1-2}} + \frac{\lim_{t_1\to\infty}}{\Gamma(\alpha_1)} \sum_{s=0}^{t_1-\alpha_1} (t_i-\sigma(s))^{\underline{\alpha_1-1}} k_1(s+\alpha_1-1) \\
&= \frac{\lambda_2 C_{12}}{\Gamma(\theta_2)} \sum_{s=\alpha_2-1}^{\eta_2} (\eta_2+\theta_2-\sigma(s))^{\underline{\theta_2-1}} g_2(s)\, s^{\underline{\alpha_2-1}} \\
&+ \frac{\lambda_2\phi_2(u_1,u_2)}{\Gamma(\alpha_2)\Gamma(\theta_2)} \sum_{s=\alpha_2-2}^{\eta_2} (\eta_2+\theta_2-\sigma(s))^{\underline{\theta_2-1}} g_2(s)\, s^{\underline{\alpha_2-2}} \\
&+ \frac{\lambda_2}{\Gamma(\alpha_2)\Gamma(\theta_2)} \sum_{\xi=\alpha_2}^{\eta_2} \sum_{s=0}^{\xi-\alpha_2} (\eta_2+\theta_2-\sigma(\xi))^{\underline{\theta_2-1}} (\xi-\sigma(s))^{\underline{\alpha_2-1}} g_2(s+\alpha_2-1)\, k_2(s+\alpha_2-1),
\end{aligned} \tag{19}
$$

and:

$$
\begin{aligned}
& C_{12} \lim_{t_2\to\infty} t_2^{\underline{\alpha_2-1}} + \frac{\phi_2(u_1,u_2)}{\Gamma(\alpha_2)} \lim_{t_2\to\infty} t_2^{\underline{\alpha_2-2}} + \frac{1}{\Gamma(\alpha_2)} \lim_{t_2\to\infty} \sum_{s=0}^{t_2-\alpha_2} (t_2-\sigma(s))^{\underline{\alpha_2-1}} k_2(s+\alpha_2-1) \\
&= \frac{\lambda_1 C_{11}}{\Gamma(\theta_1)} \sum_{s=\alpha_1-1}^{\eta_1} (\eta_1+\theta_1-\sigma(s))^{\underline{\theta_1-1}} g_1(s)\, s^{\underline{\alpha_1-1}} \\
&+ \frac{\lambda_1\phi_1(u_1,u_2)}{\Gamma(\alpha_1)\Gamma(\theta_1)} \sum_{s=\alpha_1-2}^{\eta_1} (\eta_1+\theta_1-\sigma(s))^{\underline{\theta_1-1}} g_1(s)\, s^{\underline{\alpha_1-2}} \\
&+ \frac{\lambda_1}{\Gamma(\alpha_1)\Gamma(\theta_1)} \sum_{\xi=\alpha_1}^{\eta_1} \sum_{s=0}^{\xi-\alpha_1} (\eta_1+\theta_1-\sigma(\xi))^{\underline{\theta_1-1}} (\xi-\sigma(s))^{\underline{\alpha_1-1}} g_1(s+\alpha_1-1)\, k_1(s+\alpha_1-1).
\end{aligned} \tag{20}
$$

After solving the system of Equations (19) and (20), we obtain:

$$
\begin{aligned}
C_{11} &= \frac{\lambda_1}{\Lambda\Gamma(\theta_1)} \sum_{s=\alpha_1-2}^{\eta_1} (\eta_1+\theta_1-\sigma(s))^{\underline{\theta_1-1}} g_1(s) s^{\underline{\alpha_1-1}} \mathcal{P}(k_1,k_2) \\
&\quad - \frac{\lambda_2}{\Lambda\Gamma(\theta_2)} \sum_{s=\alpha_2-2}^{\eta_2} (\eta_2+\theta_2-\sigma(s))^{\underline{\theta_2-1}} g_2(s) s^{\underline{\alpha_2-1}} \mathcal{Q}(k_1,k_2),
\end{aligned} \tag{21}
$$

and:

$$
C_{12} = \frac{\lim_{t_2\to\infty} t_2^{\underline{\alpha_2-1}}}{\Lambda} \mathcal{P}(k_1,k_2) - \frac{\lim_{t_1\to\infty} t_1^{\underline{\alpha_1-1}}}{\Lambda} \mathcal{Q}(k_1,k_2), \tag{22}
$$

where $\Lambda, \mathcal{P}(k_1,k_2)$ and $\mathcal{Q}(k_1,k_2)$ are defined as (12)–(14), respectively. □

The following lemma deals with the solutions $u_i(t_i)$, $i = 1,2$ of the problems (7)–(9), and $\Delta^{\beta_i} u_i(t_i-\beta_i+1)$ are uniformly bounded on $\mathbb{N}_{\alpha_i-2}$, $\beta_i \in (\alpha_i-1,\alpha_i)$.

Lemma 4. *For each $i,j \in \{1,2\}$ and $i \neq j$, let $k_i \in C(\mathbb{N}_{\alpha_i-2},\mathbb{R})$ and $g_i \in C(\mathbb{N}_{\alpha_i-2},\mathbb{R}^+)$ be given functions, $\phi_i(u_1,u_2)$ be given functionals, $\rho_i > \max\{\beta_i-\alpha_i\}$, $\beta_i \in (\alpha_i-1,\alpha_i)$, and $0 < g_i \leq g_i(s_i) \leq G_i$, for each $s_i \in \mathbb{N}_{\alpha_i-2,T+\alpha_i}$.*

The solution $u_i(t_i)$ of the problems (7)–(9) and $\Delta^{\beta_i} u_i(t_i-\beta_i+1)$ are uniformly bounded on $\mathbb{N}_{\alpha_i-2}$, if and only if $u_i(t_i)$ and $\Delta^{\beta_i} u_i(t_i-\beta_i+1)$ satisfy the following properties:

(A_1) *There exist constants $M_1, N_1, m_1, n_1 > 0$ such that, for u_1 and $\Delta^{\beta_1} u_1$,*

$$
|k_i(t_i)| \leq M_1\, e^{-m_1(2t_1+t_2)},
$$

$$|\phi_i(u_1,u_2)| \le N_1(t_1+\rho_1)^{\underline{\rho_1}} e^{-n_1(t_1+1)}.$$

(A_2) *There exist constants* $M_2, N_2, m_2, n_2 > 0$ *such that, for* u_2 *and* $\Delta^{\beta_2}u_2$,

$$|k_i(t_i)| \le M_2\,(t_2+\rho_2)^{\underline{\rho_2}} e^{-m_2(t_1+2t_2)},$$
$$|\phi_i(u_1,u_2)| \le N_2(t_1+\rho_1)^{\underline{\rho_1}} \left[(t_2+\rho_2)^{\underline{\rho_2}}\right]^2 e^{-n_2(t_2+1)}.$$

(A_3) *There exist constants* $\Omega_i > 0,\ i = 1,2$ *such that,*

$$\left\{\frac{(t_1-\alpha_1+1)^{\underline{2-\alpha_1}}(t_2-\alpha_2+1)^{\underline{2-\alpha_2}}}{1+(t_1+\rho_1)^{\underline{\rho_1}}(t_2+\rho_2)^{\underline{\rho_2}}}\right\}|u_i(t_i)| < \Omega_i,$$
$$\left\{\frac{(t_1-\alpha_1+1)^{\underline{2-\alpha_1}}(t_2-\alpha_2+1)^{\underline{2-\alpha_2}}}{1+(t_1+\rho_1)^{\underline{\rho_1}}(t_2+\rho_2)^{\underline{\rho_2}}}\right\}|\Delta^{\beta_i}u_i(t_i)| < \Omega_i.$$

Proof. Firstly, taking the fractional difference of order $\alpha_i - 1 < \beta_i < \alpha_i,\ i = 1,2$ for (10) and (11), we obtain:

$$\begin{aligned}
&\Delta^{\beta_1}u_1(t_1)\\
&= \frac{1}{\Gamma(-\beta_1)}\sum_{s=\alpha_1-1}^{t_1+1}(t_1-\beta_1+1-\sigma(s))^{\underline{-\beta_1-1}}s^{\underline{\alpha_1-1}} \times \left\{\frac{\lambda_1}{\Lambda\Gamma(\theta_1)}\times\right.\\
&\sum_{s=\alpha_1-2}^{\eta_1}(\eta_1+\theta_1-\sigma(s))^{\underline{\theta_1-1}}g_1(s)s^{\underline{\alpha_1-1}}\mathcal{P}(k_1,k_2) - \frac{\lambda_2}{\Lambda\Gamma(\theta_2)}\times\\
&\left.\sum_{s=\alpha_2-2}^{\eta_2}(\eta_2+\theta_2-\sigma(s))^{\underline{\theta_2-1}}g_2(s)s^{\underline{\alpha_2-1}}\mathcal{Q}(k_1,k_2)\right\} \qquad (23)\\
&+\frac{\phi_1(u_1,u_2)}{\Gamma(-\beta_1)\Gamma(\alpha_1)}\sum_{s=\alpha_1-2}^{t_1+1}(t_1-\beta_1+1-\sigma(s))^{\underline{-\beta_1-1}}s^{\underline{\alpha_1-2}}\\
&+\frac{1}{\Gamma(-\beta_1)\Gamma(\alpha_1)}\sum_{\xi=\alpha_1}^{t_1+1}\sum_{s=0}^{\xi-\alpha_1}(t_1-\beta_1+1-\sigma(s))^{\underline{-\beta_1-1}}(\xi-\sigma(s))^{\underline{\alpha_1-1}}k_1(s+\alpha_1-1),
\end{aligned}$$

and:

$$\begin{aligned}
&\Delta^{\beta_2}u_2(t_2)\\
&= \frac{1}{\Gamma(-\beta_2)}\sum_{s=\alpha_2-1}^{t_2+1}(t_2-\beta_2+1-\sigma(s))^{\underline{-\beta_2-1}}s^{\underline{\alpha_2-1}} \times \left\{\frac{\lim_{t_2\to\infty}t_2^{\underline{\alpha_2-1}}}{\Lambda}\mathcal{P}(k_1,k_2)\right. \qquad (24)\\
&\left.-\frac{\lim_{t_1\to\infty}t_1^{\underline{\alpha_1-1}}}{\Lambda}\mathcal{Q}(k_1,k_2)\right\} + \frac{\phi_2(u_1,u_2)}{\Gamma(-\beta_2)\Gamma(\alpha_2)}\sum_{s=\alpha_2-2}^{t_2+1}(t_2-\beta_1+1-\sigma(s))^{\underline{-\beta_2-1}}s^{\underline{\alpha_2-2}}\\
&+\frac{1}{\Gamma(-\beta_2)\Gamma(\alpha_2)}\sum_{\xi=\alpha_2}^{t_2+1}\sum_{s=0}^{\xi-\alpha_2}(t_2-\beta_2+1-\sigma(s))^{\underline{-\beta_2-1}}(\xi-\sigma(s))^{\underline{\alpha_2-1}}k_2(s+\alpha_2-1).
\end{aligned}$$

If $u_i(t_i)$ and $\Delta^{\beta_i}u_i(t_i-\beta_i+1)$ are uniformly bounded on $\mathbb{N}_{\alpha_i-2}$, we have:

$$|\Lambda| \le \left|\frac{\lambda_2 \lim_{t_2\to\infty}t_2^{\underline{\alpha_2-1}}}{\Gamma(\alpha_2)}\sum_{s=\alpha_2-1}^{\eta_2}(\eta_2+\theta_2-\sigma(s))^{\underline{\theta_2-1}}g_2(s)\,s^{\underline{\alpha_2-1}}\right.$$

$$-\frac{\lambda_1 \lim_{t_1\to\infty} t_1^{\underline{\alpha_1-1}}}{\Gamma(\alpha_1)} \sum_{s=\alpha_1-1}^{\eta_1} (\eta_1+\theta_1-\sigma(s))^{\underline{\theta_1-1}} g_1(s)\, s^{\underline{\alpha_1-1}} \Bigg| \tag{25}$$
$$\leq \max\left\{ \left| \lim_{t_2\to\infty} t_2^{\underline{\alpha_2-1}} (\eta_2+\theta_2-\alpha_2)^{\underline{\theta_2-1}} G_2\lambda_2\mathcal{A}_2 \right|, \left| \lim_{t_1\to\infty} t_1^{\underline{\alpha_1-1}} (\eta_1+\theta_1-\alpha_1)^{\underline{\theta_1-1}} G_1\lambda_1\mathcal{A}_1 \right| \right\}.$$

Furthermore, considering $u_1(t_i)$ and $\Delta^{\beta_1} u_i(t_i)$, we obtain:

$$|k_i(t_i)| < \begin{cases} M_1\, e^{1-(t_1+t_2)} & ,\text{for } u_1(t_1) \\ M_1\, e^{-(2t_1+t_2)} & ,\text{for } \Delta^{\beta_1} u_1(t_1-\beta_1+1) \\ M_2\,(t_2+\rho_2)^{\underline{\rho_2}}\, e^{-(t_1+t_2)} & ,\text{for } u_2(t_2) \\ M_2\,(t_2+\rho_2)^{\underline{\rho_2}}\, e^{-(t_1+2t_2)} & ,\text{for } \Delta^{\beta_2} u_2(t_2-\beta_2+1) \end{cases} \tag{26}$$

and:

$$|\phi_i(t_1,t_2)| < \begin{cases} N_1\,(t_1+\rho_1+1)^{\underline{\rho_1}} & ,\text{for } u_1(t_1) \\ N_1\,(t_1+\rho_1)^{\underline{\rho_1}}\, e^{-(t_1+1)} & ,\text{for } \Delta^{\beta_1} u_1(t_1-\beta_1+1) \\ N_2(t_1+\rho_1)^{\underline{\rho_1}} \left[(t_2+\rho_2+1)^{\underline{\rho_2}}\right]^2 & ,\text{for } u_2(t_2) \\ N_2(t_1+\rho_1)^{\underline{\rho_1}} \left[(t_2+\rho_2)^{\underline{\rho_2}}\right]^2 e^{-(t_2+1)} & ,\text{for } \Delta^{\beta_2} u_2(t_2-\beta_2+1) \end{cases} \tag{27}$$

where:

$$M_1 = \min\left\{ \frac{\lambda_1 G_1(\eta_1-\alpha_1+\theta_1)^{\underline{\theta_1-1}}\mathcal{A}_1}{\lambda_2 g_2\Gamma(\alpha_2)\mathcal{A}_2}, \frac{G_1(\eta_1-\alpha_1+\theta_1)^{\underline{\theta_1-1}}}{g_1}, \Gamma(\alpha_1), \frac{G_1(\eta_1-\alpha_1+\theta_1)^{\underline{\theta_1-1}}}{\lambda_2 g_1 g_2\Gamma(\alpha_1)\mathcal{C}_2}, \frac{G_1(\eta_1-\alpha_1+\theta_1)^{\underline{\theta_1-1}}\mathcal{A}_1}{\lambda_2 g_1 g_2\Gamma(\alpha_2)\mathcal{A}_2\mathcal{C}_1} \right\}, \tag{28}$$

$$M_2 = \min\left\{ \Gamma(\alpha_2), \lambda_2 G_2(\eta_2-\alpha_2+\theta_2)^{\underline{\theta_2-1}}\mathcal{A}_2, G_1(\eta_1-\alpha_1+\theta_1)^{\underline{\theta_1-1}}\mathcal{A}_1, \frac{G_1(\eta_1-\alpha_1+\theta_1)^{\underline{\theta_1-1}}\mathcal{A}_1}{g_1\mathcal{C}_1}, \frac{G_2(\eta_2-\alpha_2+\theta_2)^{\underline{\theta_2-1}}\mathcal{A}_2}{g_2\mathcal{C}_2} \right\}, \tag{29}$$

$$N_1 = \min\left\{ \Gamma(\theta_1), \Gamma(\theta_2), \Gamma(\alpha_1), \frac{\Gamma(\theta_1)}{\lambda_2 g_2\Gamma(\alpha_1)\mathcal{B}_2}, \frac{G_1(\eta_1-\alpha_1+\theta_1)^{\underline{\theta_1-1}}\mathcal{A}_1}{\lambda_2 g_1 g_2\Gamma(\alpha_1-1)\mathcal{B}_1\mathcal{A}_2} \right\}, \tag{30}$$

$$N_2 = \min\left\{ \Gamma(\alpha_2), \lambda_2 G_2\Gamma(\alpha_1)(\eta_2-\alpha_2+\theta_2)^{\underline{\theta_2-1}}\mathcal{A}_2, \lambda_1 G_1\Gamma(\alpha_2)(\eta_1-\alpha_1+\theta_1)^{\underline{\theta_1-1}}\mathcal{A}_1, \frac{G_2(\eta_2-\alpha_2+\theta_2)^{\underline{\theta_2-1}}\mathcal{A}_2}{g_2\mathcal{B}_2}, \frac{G_1(\eta_1-\alpha_1+\theta_1)^{\underline{\theta_1-1}}\mathcal{A}_1}{g_1\mathcal{B}_1} \right\}. \tag{31}$$

Consequently, the conditions $(A1)$ and $(A2)$ hold.

We next show that the condition $(A3)$ holds. By using the conditions $(A1)$ and $(A2)$, we obtain:

$$|u_i(t_i)| \leq t_1^{\underline{\alpha_1}} t_2^{\underline{\alpha_2}}\,\Omega_i < t_1^{\underline{\alpha_1+\rho_1-1}} t_2^{\underline{\alpha_2+\rho_2-1}}\,\Omega_i < \left[\frac{1+t_1^{\underline{\rho_1+2}} t_2^{\underline{\rho_2+2}}}{(t_1-\alpha_1+1)^{\underline{2-\alpha_1}}(t_2-\alpha_2+1)^{\underline{2-\alpha_2}}} \right]\Omega_i$$

and:

$$\left|\Delta^{\beta_i} u_i(t_i - \beta_i + 1)\right| \le (t_i - \alpha_i - \beta_i + 1)^{\underline{-\beta_i}} t_j^{\underline{\alpha_j}} \Omega_i < t_i^{\underline{\alpha_i}} t_j^{\underline{\alpha_j}} \Omega_i$$
$$< \left[\frac{1 + t_1^{\underline{\rho_1+2}} t_2^{\underline{\rho_2+2}}}{(t_1 - \alpha_1 + 1)^{\underline{2-\alpha_1}} (t_2 - \alpha_2 + 1)^{\underline{2-\alpha_2}}} \right] \Omega_i, \; i \neq j = 1, 2$$

where:

$$\Omega_1 = \max \left\{ \frac{N_1}{\Gamma(\theta_1)} + M_1 + \frac{\lambda_2 G_2 \Gamma(\alpha_1)(\eta_2 - \alpha_2 + \theta_2)^{\underline{\theta_2 - 1}}}{\Gamma(\theta_2)} \times \right.$$
$$\left(N_2 \mathcal{B}_2 + \frac{M_2 \mathcal{C}_2}{\lambda_1 G_1 (\eta_1 - \alpha_1 + \theta_1)^{\underline{\theta_1 - 1}} \mathcal{A}_1} \right),$$
$$\frac{N_2}{\Gamma(\theta_2)} + M_2 + \frac{\lambda_1 G_1 \Gamma(\alpha_2)(\eta_1 - \alpha_1 + \theta_1)^{\underline{\theta_1 - 1}}}{\Gamma(\theta_1)} \times$$
$$\left. \left(N_1 \mathcal{B}_1 + \frac{M_1 \mathcal{C}_1}{\lambda_2 G_2 (\eta_2 - \alpha_2 + \theta_2)^{\underline{\theta_2 - 1}} \mathcal{A}_2} \right) \right\}$$
$$+ \frac{1}{\Gamma(\alpha_1)} (M_1 + N_1), \tag{32}$$

$$\Omega_2 = \max \left\{ \frac{1}{G_2 (\eta_2 - \alpha_2 + \theta_2)^{\underline{\theta_2 - 1}} \mathcal{A}_2} \left[\frac{M_1 + N_1}{\Gamma(\alpha_1) \lambda_2} + \frac{1}{\Gamma(\theta_2) \mathcal{A}_2} \left(N_2 \mathcal{B}_2 + M_2 \mathcal{C}_2 \right) \right], \right.$$
$$\left. \frac{1}{G_1 (\eta_1 - \alpha_1 + \theta_1)^{\underline{\theta_1 - 1}} \mathcal{A}_1} \left[\frac{M_2 + N_2}{\Gamma(\alpha_2) \lambda_1} + \frac{1}{\Gamma(\theta_1) \mathcal{A}_1} \left(N_1 \mathcal{B}_1 + M_1 \mathcal{C}_1 \right) \right] \right\}$$
$$+ \frac{1}{\Gamma(\alpha_2)} (M_2 + N_2), \tag{33}$$

with

$$\mathcal{A}_i = {}_2F_1(\alpha_i, \alpha_i - \eta_i - 1; \alpha_i - \eta_i - \theta_i; 1) \tag{34}$$
$$\mathcal{B}_i = {}_2F_1(\alpha_i - 1, \alpha_i - \eta_i - 1; \alpha_i - \eta_i - \theta_i - 1; 1) \tag{35}$$
$$\mathcal{C}_i = {}_2F_1(\alpha_i + 1, \alpha_i - \eta_i; \alpha_i - \eta_i - \theta_i + 1; 1). \tag{36}$$

Therefore, the condition $(A3)$ holds.

Finally, if the conditions $(A1)$–$(A3)$ hold, it is clear that $u_i(t_i)$ and $\Delta^{\beta_i} u_i(t_i - \beta_i + 1)$ are uniformly bounded on $\mathbb{N}_{\alpha_i - 2}$. Our proof is complete. □

We next provide the following theorems used for proving the existence result for the problems (5) and (6).

Theorem 1. *(Arzelá–Ascoli theorem [61])*
A set of functions in $C[a, b]$ with the sup norm is relatively compact if and only if it is uniformly bounded and equicontinuous on $[a, b]$.

Theorem 2. *[61] If a set is closed and relatively compact, then it is compact.*

Theorem 3. *(Schauder's fixed point theorem [61])*
If S is a convex compact subset of a normed space, every continuous mapping of S into itself has a fixed point.

3. Main Result

In this section, we aim to establish the existence result for the problems (5) and (6). To accomplish this, we let $\mathcal{C}_i = C(\mathbb{N}_{\alpha_i-2}, \mathbb{R})$ be a Banach space of all functions on $\mathbb{N}_{\alpha_i-2}$, for each $i, j \in \{1,2\}$ and $i \neq j$. Obviously, the product spaces:

$$\mathcal{U}_i = \Big\{(u_1, u_2) \in \mathcal{C}_1 \times \mathcal{C}_2 : \ \Delta^{\beta_i} u_i(t_i - \beta_i + 1) \in \mathcal{C}_i \text{ and } \chi|u_j(t_j)|, \\ \chi|\Delta^{\beta_i} u_i(t_i - \beta_i + 1)| \ \text{ are bounded on } \mathbb{N}_{\alpha_j-2}, \mathbb{N}_{\alpha_i-2}, \text{respectively,}\Big\}$$

is also the Banach space endowed with the norm defined by:

$$\|(u_1, u_2)\|_{\mathcal{U}_i} = \|\Delta^{\beta_i} u_i\|_{\mathcal{C}_i} + \|u_j\|_{\mathcal{C}_j},$$

where:

$$\|\Delta^{\beta_i} u_i\|_{\mathcal{C}_i} = \max_{t_i \in \mathbb{N}_{\alpha_i-2}} \chi\left|\Delta^{\beta_i} u_i(t_i - \beta_i + 1, t_j)\right| \ \text{ and } \ \|u_j\|_{\mathcal{C}_j} = \max_{t_j \in \mathbb{N}_{\alpha_j-2}} \chi|u_j(t_i, t_j)|,$$

with for $\rho_i > \max\{\beta_i - \alpha_i\}$ and $\beta_i \in (\alpha_i - 1, \alpha_i)$,

$$\chi = \frac{(t_1 - \alpha_1 + 1)^{\underline{2-\alpha_1}} (t_2 - \alpha_2 + 1)^{\underline{2-\alpha_2}}}{1 + t_1^{\underline{\rho_1+2}} t_2^{\underline{\rho_2+2}}}. \tag{37}$$

Let $\mathcal{U} = \mathcal{U}_1 \cap \mathcal{U}_2$; clearly, the space $\Big(\mathcal{U}, \|(u_1, u_2)\|_{\mathcal{U}}\Big)$ is the Banach space with the norm:

$$\|(u_1, u_2)\|_{\mathcal{U}} = \max\left\{\|(u_1, u_2)\|_{\mathcal{U}_1}, \ \|(u_1, u_2)\|_{\mathcal{U}_2}\right\}.$$

Next, we define the operator $\mathcal{F} : \mathcal{U} \to \mathcal{U}$ by:

$$(\mathcal{F}(u_1, u_2))(t_1, t_2) \ = \ \Big((\mathcal{F}_1(u_1, u_2))(t_1, t_2), (\mathcal{F}_2(u_1, u_2))(t_1, t_2)\Big), \tag{38}$$

and:

$$\begin{aligned}(\mathcal{F}_1(u_1, u_2))(t_1, t_2) = {} & \frac{t_1^{\underline{\alpha_1-1}}}{\Lambda}\left\{\frac{\lambda_1}{\Gamma(\theta_1)} \sum_{s=\alpha_1-2}^{\eta_1} (\eta_1 + \theta_1 - \sigma(s))^{\underline{\theta_1-1}} g_1(s) s^{\underline{\alpha_1-1}} \mathcal{P}(F_1, F_2)\right. \\ & \left. - \frac{\lambda_2}{\Gamma(\theta_2)} \sum_{s=\alpha_2-2}^{\eta_2} (\eta_2 + \theta_2 - \sigma(s))^{\underline{\theta_2-1}} g_2(s) s^{\underline{\alpha_2-1}} \mathcal{Q}(F_1, F_2)\right\} \\ & + \frac{t_1^{\underline{\alpha_1-2}} \phi_1(u_1, u_2)}{\Gamma(\alpha_1)} + \frac{1}{\Gamma(\alpha_1)} \sum_{s=\alpha_1-1}^{t_1-1} (t_1 + \alpha_1 - 1 - \sigma(s))^{\underline{\alpha_1-1}} \times \\ & F_1\big(s, t_2, \Delta^{\beta_1} u_1(s - \beta_1 + 1), u_2(t_2)\big), \ \ t_i \in \mathbb{N}_{\alpha_i-2},\end{aligned} \tag{39}$$

$$\begin{aligned}(\mathcal{F}_2(u_1, u_2))(t_1, t_2) = {} & \frac{t_2^{\underline{\alpha_2-1}}}{\Lambda}\left\{\lim_{t_2 \to \infty} t_2^{\underline{\alpha_2-1}} \mathcal{P}(F_1, F_2) - \lim_{t_1 \to \infty} t_1^{\underline{\alpha_1-1}} \mathcal{Q}(F_1, F_2)\right\} \\ & + \frac{t_2^{\underline{\alpha_2-2}} \phi_2(u_1, u_2)}{\Gamma(\alpha_2)} + \frac{1}{\Gamma(\alpha_2)} \sum_{s=\alpha_2-1}^{t_2-1} (t_2 + \alpha_2 - 1 - \sigma(s))^{\underline{\alpha_2-1}} \times \\ & F_2\big(t_1, s, u_1(t_1), \Delta^{\beta_2} u_2(s - \beta_2 + 1)\big), \ \ t_i \in \mathbb{N}_{\alpha_i-2},\end{aligned} \tag{40}$$

where Λ is defined as (12), and:

$$\mathcal{P}(F_1,F_2) = \frac{\lim\limits_{t_1\to\infty} t_1^{\underline{\alpha_1-2}}\phi_1(u_1,u_2)}{\Gamma(\alpha_1)} - \frac{\lambda_2\phi_2(u_1,u_2)}{\Gamma(\alpha_2)\Gamma(\theta_2)}\sum_{s=\alpha_2-2}^{\eta_2}(\eta_2+\theta_2-\sigma(s))^{\underline{\theta_2-1}}g_2(s)\,s^{\underline{\alpha_2-2}}$$
$$+\frac{1}{\Gamma(\alpha_1)}\lim_{t_1\to\infty}\sum_{s=\alpha_1-1}^{t_1-1}(t_1+\alpha_1-1-\sigma(s))^{\underline{\alpha_1-1}}F_1(s,t_2,\Delta^{\beta_1}u_1(s-\beta_1+1),u_2(t_2))$$
$$-\frac{\lambda_2}{\Gamma(\alpha_2)\Gamma(\theta_2)}\sum_{\xi=\alpha_2}^{\eta_2}\sum_{s=\alpha_2-1}^{\xi-1}(\eta_2+\theta_2-\sigma(\xi))^{\underline{\theta_2-1}}(\xi+\alpha_2-1-\sigma(s))^{\underline{\alpha_2-1}}\times$$
$$g_2(s)F_2(t_1,s,u_1(t_1),\Delta^{\beta_2}u_2(s-\beta_2+1)), \tag{41}$$

$$\mathcal{Q}(F_1,F_2) = \frac{\lim\limits_{t_2\to\infty} t_2^{\underline{\alpha_2-2}}\phi_2(u_1,u_2)}{\Gamma(\alpha_2)} - \frac{\lambda_1\phi_1(u_1,u_2)}{\Gamma(\alpha_1)\Gamma(\theta_1)}\sum_{s=\alpha_1-2}^{\eta_1}(\eta_1+\theta_1-\sigma(s))^{\underline{\theta_1-1}}g_1(s)\,s^{\underline{\alpha_1-2}}$$
$$+\frac{1}{\Gamma(\alpha_2)}\lim_{t_2\to\infty}\sum_{s=\alpha_2-1}^{t_2-1}(t_2+\alpha_2-1-\sigma(s))^{\underline{\alpha_2-1}}F_2(t_1,s,u_1(t_1),\Delta^{\beta_2}u_2(s-\beta_2+1))$$
$$-\frac{\lambda_1}{\Gamma(\alpha_1)\Gamma(\theta_1)}\sum_{\xi=\alpha_1}^{\eta_1}\sum_{s=\alpha_1-1}^{\xi-1}(\eta_1+\theta_1-\sigma(\xi))^{\underline{\theta_1-1}}(\xi+\alpha_1-1-\sigma(s))^{\underline{\alpha_1-1}}\times$$
$$g_1(s)F_1(s,t_2,\Delta^{\beta_1}u_1(s-\beta_1+1),u_2(t_2)). \tag{42}$$

We next make the following assumptions:

(H_1) There exist positive numbers ${}_{ip}\rho_2 \in (-1,\rho_2)$ and $M_{ip}, m_{ip} > 0$ ($i = 1,2$ and $p = 1,2,3$) such that, for each $t_i \in \mathbb{N}_{\alpha_i-2}$ and $v_i \in \mathbb{R}$,

$$\left|F_i\left(t_1,t_2,\frac{1}{\chi}v_1,\frac{1}{\chi}v_2\right) - M_{i1}\,(t_2+{}_{i1}\rho_2)^{{}_{i1}\rho_2}\,e^{-m_{i1}(t_1+t_2)}\right|$$
$$\leq M_{i2}\,(t_2+{}_{i2}\rho_2)^{{}_{i2}\rho_2}\,e^{-m_{i2}(t_1+t_2)}\,|v_1| + M_{i3}\,(t_2+{}_{i3}\rho_2)^{{}_{i3}\rho_2}\,e^{-m_{i3}(t_1+t_2)}\,|v_2|.$$

(H_2) There exist positive numbers ${}_{ip}\tilde{\rho}_i \in (-1,\rho_i)$ and $N_{ip}, n_{ip} > 0$ ($i = 1,2$ and $p = 1,2,3$) such that, for $v_i \in \mathcal{C}_i$,

$$\left|\phi_i\left(\frac{1}{\chi}v_1,\frac{1}{\chi}v_2\right) - N_{i1}\,(t_1+{}_{i1}\tilde{\rho}_1)^{{}_{i1}\tilde{\rho}_1}\left[(t_2+{}_{i1}\tilde{\rho}_2)^{{}_{i1}\tilde{\rho}_2}\right]^2 e^{-n_{i1}(t_1+t_2)}\right|$$
$$\leq N_{i2}\,(t_1+{}_{i2}\tilde{\rho}_1)^{{}_{i2}\tilde{\rho}_1}\left[(t_2+{}_{i2}\tilde{\rho}_2)^{{}_{i2}\tilde{\rho}_2}\right]^2 e^{-n_{i2}(t_1+t_2)}\,\|v_1\| + N_{i3}\,(t_1+{}_{i3}\tilde{\rho}_1)^{{}_{i3}\tilde{\rho}_1}\left[(t_2+{}_{i3}\tilde{\rho}_2)^{{}_{i3}\tilde{\rho}_2}\right]^2 e^{-n_{i3}(t_1+t_2)}\,\|v_2\|.$$

(H_3) $g_i \leq g_i(\eta_i)$ for all $\eta_i \in \mathbb{N}_{\alpha_i-1,T+\alpha_i-1}$.

Lemma 5. *Suppose that (H_1)–(H_3) hold. Then, the fixed point of $\mathcal{F}$ coincides with the solution of the problems (5) and (6), and $\mathcal{F} : \mathcal{U} \to \mathcal{U}$ is completely continuous.*

Proof. Let $(u_1,u_2) \in \mathcal{U}$, for each $i,j \in \{1,2\}$ and $i \neq j$. By the above assumptions (H_1) and (H_2), it follows that:

$$\left|F_i\left(t_1,t_2,\Delta^{\beta_i}u_i(t_i-\beta_i+1),u_j(t_j)\right)\right|$$

$$
\begin{aligned}
&= \left| F_i \left(t_1, t_2, \frac{1}{\chi} \left[\chi \Delta^{\beta_i} u_i(t_i - \beta_i + 1) \right], \frac{1}{\chi} \left[\chi u_j(t_j) \right] \right) \right| \\
&\leq \; M_{i1} \left(t_2 + {}_{i1}\rho_2 \right)^{i1\rho_2} e^{-m_{i1}(t_1+t_2)} + M_{i2} \left(t_2 + {}_{i2}\rho_2 \right)^{i2\rho_2} e^{-m_{i2}(t_1+t_2)} \, \| \Delta^{\beta_i} u_i \|_{C_i} \\
&\quad + M_{i3} \left(t_2 + {}_{i3}\rho_2 \right)^{i3\rho_2} e^{-m_{i3}(t_1+t_2)} \, \| u_j \|_{C_j},
\end{aligned} \tag{43}
$$

and
$$
\begin{aligned}
|\phi_i(u_1, u_2)| &= \left| \phi_i \left(\frac{1}{\chi} [\chi u_1], \frac{1}{\chi} [\chi u_2] \right) \right| \\
&\leq \; N_{i1} \left(t_1 + {}_{i1}\tilde{\rho}_1 \right)^{i1\tilde{\rho}_1} \left[\left(t_2 + {}_{i1}\tilde{\rho}_2 \right)^{i1\tilde{\rho}_2} \right]^2 e^{-n_{i1}(t_1+t_2)} \\
&\quad + N_{i2} \left(t_1 + {}_{i2}\tilde{\rho}_1 \right)^{i2\tilde{\rho}_1} \left[\left(t_2 + {}_{i2}\tilde{\rho}_2 \right)^{i2\tilde{\rho}_2} \right]^2 e^{-n_{i2}(t_1+t_2)} \, \| u_1 \|_{C_1} \\
&\quad + N_{i3} \left(t_1 + {}_{i3}\tilde{\rho}_1 \right)^{i3\tilde{\rho}_1} \left[\left(t_2 + {}_{i3}\tilde{\rho}_2 \right)^{i3\tilde{\rho}_2} \right]^2 e^{-n_{i3}(t_1+t_2)} \, \| u_2 \|_{C_2}.
\end{aligned} \tag{44}
$$

The rest of the proof follows from Lemmas 3 and 4. This implies that the fixed point of $\mathcal{F}$ coincides with the solution of the problems (5) and (6).

To show that $\mathcal{F}$ is completely continuous, we organize the proof as the following four steps.

Step I. $\mathcal{F}$ is well defined and maps bounded sets into bounded sets.

Let $B_R = \{ (u_1, u_2) \in \mathcal{U} : \| (u_1, u_2) \|_{\mathcal{U}} \leq R \}$, then for $(u_1, u_2) \in \mathcal{U}$:

$$
\begin{aligned}
R \; &\geq \; \max \left\{ \| (u_1, u_2) \|_{\mathcal{U}_1}, \| (u_1, u_2) \|_{\mathcal{U}_2} \right\} \\
&= \; \max \Bigg\{ \chi \left[\left| \Delta^{\beta_1} u_1(t_1 - \beta_1 + 1) \right| + |u_2(t_2)| \right], \\
&\qquad \chi \left[|u_1(t_1)| + \left| \Delta^{\beta_2} u_2(t_2 - \beta_2 + 1) \right| \right] \Bigg\}.
\end{aligned} \tag{45}
$$

By the definition of $\mathcal{F}$, we get $\mathcal{F}_i(u_1, u_2), \Delta^{\beta_i} \mathcal{F}_i(u_1, u_2) \in \mathcal{U}$. Therefore, (43) and (44) imply that:

$$
\begin{aligned}
&\left| F_i(t_1, t_2, \Delta^{\beta_i} u_i(t_i - \beta_i + 1), u_j(t_j)) \right| \\
&\quad = \left| F_i \left(t_1, t_2, \frac{1}{\chi} \left[\chi \Delta^{\beta_i} u_i(t_i - \beta_i + 1) \right], \frac{1}{\chi} \left[\chi u_j(t_j) \right] \right) \right| \\
&\quad \leq M_{i1} \left(t_2 + {}_{i1}\rho_2 \right)^{i1\rho_2} e^{-m_{i1}(t_1+t_2)} + R M_{i2} \left(t_2 + {}_{i2}\rho_2 \right)^{i2\rho_2} \times \\
&\qquad e^{-m_{i2}(t_1+t_2)} + R M_{i3} \left(t_2 + {}_{i3}\rho_2 \right)^{i3\rho_2} e^{-m_{i3}(t_1+t_2)},
\end{aligned} \tag{46}
$$

and
$$
\begin{aligned}
|\phi_i(u_1, u_2)| &= \left| \phi_i \left(\frac{1}{\chi} [\chi u_1], \frac{1}{\chi} [\chi u_2] \right) \right| \\
&\leq N_{i1} \left(t_1 + {}_{i1}\tilde{\rho}_1 \right)^{i1\tilde{\rho}_1} \left[\left(t_2 + {}_{i1}\tilde{\rho}_2 \right)^{i1\tilde{\rho}_2} \right]^2 e^{-n_{i1}(t_1+t_2)} \\
&\quad + R N_{i2} \left(t_1 + {}_{i2}\tilde{\rho}_1 \right)^{i2\tilde{\rho}_1} \left[\left(t_2 + {}_{i2}\tilde{\rho}_2 \right)^{i2\tilde{\rho}_2} \right]^2 e^{-n_{i2}(t_1+t_2)} \\
&\quad + R N_{i3} \left(t_1 + {}_{i3}\tilde{\rho}_1 \right)^{i3\tilde{\rho}_1} \left[\left(t_2 + {}_{i3}\tilde{\rho}_2 \right)^{i3\tilde{\rho}_2} \right]^2 e^{-n_{i3}(t_1+t_2)}.
\end{aligned} \tag{47}
$$

Let

$$
\begin{aligned}
\tilde{\Omega}_1 =: \; & (N_{11} + N_{12}R + N_{13}R)\tilde{\Omega}_{11} + (N_{21} + N_{22}R + N_{23}R)\tilde{\Omega}_{12} \\
& + (M_{11} + M_{12}R + M_{13}R)\tilde{\Omega}_{13} + (M_{21} + M_{22}R + M_{23}R)\tilde{\Omega}_{14} \\
\tilde{\Omega}_2 =: \; & (N_{11} + N_{12}R + N_{13}R)\tilde{\Omega}_{21} + (N_{21} + N_{22}R + N_{23}R)\tilde{\Omega}_{22} \\
& + (M_{11} + M_{12}R + M_{13}R)\tilde{\Omega}_{23} + (M_{21} + M_{22}R + M_{23}R)\tilde{\Omega}_{24}
\end{aligned}
$$

where:

$$\begin{aligned}
\tilde{\Omega}_{11} &= \left[\max\left\{\frac{1}{\Gamma(\theta_1)}, \frac{\lambda_1 G_1 \Gamma(\alpha_2)}{\Gamma(\theta_1)}(\eta_1-\alpha_1+\theta_1)^{\underline{\theta_1-1}}\mathcal{B}_1\right\}+\frac{1}{\Gamma(\alpha_1)}\right],\\
\tilde{\Omega}_{12} &= \max\left\{\frac{1}{\Gamma(\theta_2)}, \frac{\lambda_2 G_2 \Gamma(\alpha_1)}{\Gamma(\theta_2)}(\eta_2-\alpha_2+\theta_2)^{\underline{\theta_2-1}}\mathcal{B}_2\right\},\\
\tilde{\Omega}_{13} &= \left[\max\left\{1, \frac{\lambda_1 G_1 \Gamma(\alpha_2)}{\Gamma(\theta_1)}(\eta_1-\alpha_1+\theta_1-1)^{\underline{\theta_1-1}}\mathcal{C}_1\right\}+\frac{1}{\Gamma(\alpha_1)}\right],\\
\tilde{\Omega}_{14} &= \max\left\{1, \frac{\lambda_2 G_2 \Gamma(\alpha_1)}{\Gamma(\theta_2)}(\eta_2-\alpha_2+\theta_2-1)^{\underline{\theta_2-1}}\mathcal{C}_2\right\},\\
\tilde{\Omega}_{21} &= \max\left\{\frac{1}{G_2\lambda_2\Gamma(\alpha_1)(\eta_2-\alpha_2+\theta_2)^{\underline{\theta_2-1}}\mathcal{A}_2}, \frac{\mathcal{B}_1}{\Gamma(\theta_1)\mathcal{A}_1}\right\},\\
\tilde{\Omega}_{22} &= \left[\max\left\{\frac{1}{G_1\lambda_1\Gamma(\alpha_2)(\eta_1-\alpha_1+\theta_1)^{\underline{\theta_1-1}}\mathcal{A}_1}, \frac{\mathcal{B}_2}{\Gamma(\theta_2)\mathcal{A}_2}\right\}+\frac{1}{\Gamma(\alpha_2)}\right],\\
\tilde{\Omega}_{23} &= \max\left\{\frac{1}{G_2\lambda_2\Gamma(\alpha_1)(\eta_2-\alpha_2+\theta_2)^{\underline{\theta_2-1}}\mathcal{A}_2}, \frac{\mathcal{C}_1}{\Gamma(\theta_1)\mathcal{A}_1}\right\},\\
\tilde{\Omega}_{24} &= \left[\max\left\{\frac{1}{G_1\lambda_1\Gamma(\alpha_2)(\eta_1-\alpha_1+\theta_1)^{\underline{\theta_1-1}}\mathcal{A}_1}, \frac{\mathcal{C}_2}{\Gamma(\theta_2)\mathcal{A}_2}\right\}+\frac{1}{\Gamma(\alpha_2)}\right].
\end{aligned}$$

Hence, we obtain:

$$\begin{aligned}
&\chi\left|\left(\mathcal{F}_1(u_1,u_2)\right)(t_1,t_2)\right|\\
&\le (N_{11}+N_{12}R+N_{13}R)\left[\max\left\{\frac{1}{\Gamma(\theta_1)}, \frac{\lambda_1 G_1 \Gamma(\alpha_2)}{\Gamma(\theta_1)}(\eta_1-\alpha_1+\theta_1)^{\underline{\theta_1-1}}\mathcal{B}_1\right\}+\frac{1}{\Gamma(\alpha_1)}\right]\\
&\quad+(N_{21}+N_{22}R+N_{23}R)\max\left\{\frac{1}{\Gamma(\theta_2)}, \frac{\lambda_2 G_2 \Gamma(\alpha_1)}{\Gamma(\theta_2)}(\eta_2-\alpha_2+\theta_2)^{\underline{\theta_2-1}}\mathcal{B}_2\right\}\\
&\quad+(M_{11}+M_{12}R+M_{13}R)\left[\max\left\{1, \frac{\lambda_1 G_1 \Gamma(\alpha_2)}{\Gamma(\theta_1)}(\eta_1-\alpha_1+\theta_1-1)^{\underline{\theta_1-1}}\mathcal{C}_1\right\}+\frac{1}{\Gamma(\alpha_1)}\right]\\
&\quad+(M_{21}+M_{22}R+M_{23}R)\max\left\{1, \frac{\lambda_2 G_2 \Gamma(\alpha_1)}{\Gamma(\theta_2)}(\eta_2-\alpha_2+\theta_2-1)^{\underline{\theta_2-1}}\mathcal{C}_2\right\}\\
&= \tilde{\Omega}_1,
\end{aligned} \tag{48}$$

and:

$$\begin{aligned}
&\chi\left|\left(\mathcal{F}_2(u_1,u_2)\right)(t_1,t_2)\right|\\
&\le (N_{11}+N_{12}R+N_{13}R)\max\left\{\frac{1}{G_2\lambda_2\Gamma(\alpha_1)(\eta_2-\alpha_2+\theta_2)^{\underline{\theta_2-1}}\mathcal{A}_2}, \frac{\mathcal{B}_1}{\Gamma(\theta_1)\mathcal{A}_1}\right\}\\
&\quad+(N_{21}+N_{22}R+N_{23}R)\times\\
&\quad\left[\max\left\{\frac{1}{G_1\lambda_1\Gamma(\alpha_2)(\eta_1-\alpha_1+\theta_1)^{\underline{\theta_1-1}}\mathcal{A}_1}, \frac{\mathcal{B}_2}{\Gamma(\theta_2)\mathcal{A}_2}\right\}+\frac{1}{\Gamma(\alpha_2)}\right]\\
&\quad+(M_{11}+M_{12}R+M_{13}R)\max\left\{\frac{1}{G_2\lambda_2\Gamma(\alpha_1)(\eta_2-\alpha_2+\theta_2)^{\underline{\theta_2-1}}\mathcal{A}_2}, \frac{\mathcal{C}_1}{\Gamma(\theta_1)\mathcal{A}_1}\right\}\\
&\quad+(M_{21}+M_{22}R+M_{23}R)\times\\
&\quad\left[\max\left\{\frac{1}{G_1\lambda_1\Gamma(\alpha_2)(\eta_1-\alpha_1+\theta_1)^{\underline{\theta_1-1}}\mathcal{A}_1}, \frac{\mathcal{C}_2}{\Gamma(\theta_2)\mathcal{A}_2}\right\}+\frac{1}{\Gamma(\alpha_2)}\right]
\end{aligned}$$

$$= \tilde{\Omega}_2. \tag{49}$$

Similarly, we have:

$$\chi\left|\Delta^{\beta_1}\left(\mathcal{F}_1(u_1,u_2)\right)(t_1-\beta_1+1,t_2)\right| < \tilde{\Omega}_1, \tag{50}$$
$$\chi\left|\Delta^{\beta_2}\left(\mathcal{F}_2(u_1,u_2)\right)(t_1,t_2-\beta_2+1)\right| < \tilde{\Omega}_2. \tag{51}$$

Therefore, $\mathcal{F}_i(u_1,u_2) \in \mathcal{U}$. This implies that $\mathcal{F}:\mathcal{U} \longrightarrow \mathcal{U}$ is well defined.

Furthermore, we obtain:

$$\begin{aligned}\|\mathcal{F}(u_1,u_2)\|_{\mathcal{U}_i} = \max\Big\{&\chi\left|\Delta^{\beta_i}\left(\mathcal{F}_i(u_1,u_2)\right)(t_i-\beta_i+1,t_j)\right| \\ &+\chi\left|\left(\mathcal{F}_j(u_1,u_2)\right)(t_i,t_j)\right| \text{ for } i,j\in\{1,2\},\ i\neq j\Big\}.\end{aligned} \tag{52}$$

Hence,

$$\|\mathcal{F}(u_1,u_2)\|_{\mathcal{U}} = \max\left\{\|\mathcal{F}(u_1,u_2)\|_{\mathcal{U}_1}, \|\mathcal{F}(u_1,u_2)\|_{\mathcal{U}_2}\right\} < \tilde{\Omega}_1+\tilde{\Omega}_2. \tag{53}$$

Thus, $\mathcal{F}$ maps bounded sets into bounded sets.

Step II. $\mathcal{F}$ is continuous.

Let $\epsilon > 0$ be given. Since F_i and ϕ_i are continuous, then F_i and ϕ_i are uniformly continuous. Therefore, there exists $\delta = \min\{\delta_i,\hat{\delta}_i\} > 0$ such that, for each $t_i \in \mathcal{N}_{\alpha_i-2}$, $u_i,v_i \in \mathcal{C}_i$ with $\max\left\{\chi|\Delta^{\beta_i}u_i(t_i-\beta_i+1) - \Delta^{\beta_i}v_i(t_i)| + \chi|u_j(t_i)-v_j(t_j)|\right\} < \delta_i$,

$$\begin{aligned}&\left|F_i(t_1,t_2,\Delta^{\beta_i}u_i(t_i-\beta_i+1),u_j(t_j)) - F_i\left(t_1,t_2,\Delta^{\beta_i}v_i(t_i-\beta_i+1),v_j(t_j)\right)\right| \\ =\ &\Bigg|F_i\left(t_1,t_2,\frac{1}{\chi}\left[\chi\Delta^{\beta_i}u_i(t_i-\beta_i+1)\right],\frac{1}{\chi}\left[\chi u_j(t_j)\right]\right) \\ &-F_i\left(t_1,t_2,\frac{1}{\chi}\left[\chi\Delta^{\beta_i}v_i(t_i-\beta_i+1)\right],\frac{1}{\chi}\left[\chi v_j(t_j)\right]\right)\Bigg| \\ <\ &2M_{i1}+2M_{i2}R+2M_{i3}R < \frac{\epsilon}{4\Omega_i}.\end{aligned} \tag{54}$$

For each $u_i,v_i \in \mathcal{C}_i$ with $|u_i-v_i| < \hat{\delta}_i$,

$$\begin{aligned}|\phi_i(u_1,u_2)-\phi_i(v_1,v_2)| &= \left|\phi_i\left(\frac{1}{\chi}[\chi u_1],\frac{1}{\chi}[\chi u_2]\right)-\phi_i\left(\frac{1}{\chi}[\chi v_1],\frac{1}{\chi}[\chi v_2]\right)\right| \\ &< 2N_{i1}+2N_{i2}R+2N_{i3}R < \frac{\epsilon}{4\Omega_i}.\end{aligned} \tag{55}$$

Similar to Step I, we obtain:

$$\chi\left|\left(\mathcal{F}_i(u_1,u_2)\right)-\left(\mathcal{F}_i(v_1,v_2)\right)\right| < 2\tilde{\Omega}_i < \tfrac{\epsilon}{2}$$

and $\quad \chi\left|\Delta^{\beta_i}\left(\mathcal{F}_i(u_1,u_2)\right)-\Delta^{\beta_i}\left(\mathcal{F}_i(v_1,v_2)\right)\right| < 2\tilde{\Omega}_i < \tfrac{\epsilon}{2}.$

Thus, we have:

$$\begin{aligned}&\|\mathcal{F}_i(u_1,u_2)-\mathcal{F}_i(v_1,v_2)\|_{\mathcal{U}_i} \\ =\ &\|\Delta^{\beta_i}\mathcal{F}_i(u_1,u_2)-\Delta^{\beta_i}\mathcal{F}_i(v_1,v_2)\|_{\mathcal{C}_i}+\|\mathcal{F}_j(u_1,u_2)-\mathcal{F}_j(v_1,v_2)\|_{\mathcal{C}_j} \\ <\ &2(\tilde{\Omega}_1+\tilde{\Omega}_2) < \epsilon.\end{aligned} \tag{56}$$

This means that each $\mathcal{F}_i$, $i = 1,2$ is continuous. This shows $\mathcal{F}$ is continuous.

In order to prove that $\mathcal{F}$ maps bounded sets of $\mathcal{U} \subset \mathcal{C}_1 \times \mathcal{C}_2$ to relatively compact sets of $\mathcal{U} \subset \mathcal{C}_1 \times \mathcal{C}_2$, it suffices to show that both $\mathcal{F}_1$ and $\mathcal{F}_2$ map bounded sets to relatively compact sets. Let $\Theta_i \subset \mathcal{C}_i$, $i = 1,2$ be bounded sets and $\Theta_1 \times \Theta_2 \subset \mathcal{U}$. Recall that Θ_i are relatively compact if:

- both Θ_i are bounded,
- both $\chi\Theta_i$ are equicontinuous on any closed subintervals of $\mathbb{N}_{\alpha_i-2}$,
- both $\chi\Theta_i$ are equiconvergent as $t_i \to \infty$.

It has been shown from in Step I that both $\mathcal{F}_i$ are uniformly bounded. Now, we show that $\mathcal{F}_i$ maps bounded sets into equicontinuous sets of $\mathcal{U}$.

Step III. Both $\mathcal{F}_i : \Theta_1 \times \Theta_2 \to \mathcal{U}$ are equicontinuous on $\big([a_1, b_1] \cap \mathbb{N}_{\alpha_1-2}\big) \times \big([a_2, b_2] \cap \mathbb{N}_{\alpha_2-2}\big) := \mathcal{D}$.

For any $\epsilon > 0$, there exists $\delta > 0$ such that, for each $t_{i1}, t_{i2} \in \mathbb{N}_{\alpha_i-2} \cap [a_i, b_i]$,

$$\left|(t_{11}+\rho_1)^{\underline{\rho_1}}(t_{21}+\rho_2)^{\underline{\rho_2}} - (t_{12}+\rho_1)^{\underline{\rho_1}}(t_{22}+\rho_2)^{\underline{\rho_2}}\right| \leq \frac{\epsilon}{2\max\{\tilde{\Omega}_1^* + \tilde{\Omega}_2^*\}} = \delta, \tag{57}$$

where:

$$\begin{aligned}
\tilde{\Omega}_1^* &= \Big([N_{11} + N_{12}R + N_{13}R] + [M_{11} + M_{12}R + M_{13}R]\Big)\left[\frac{1}{\Gamma(\theta_1)} + \frac{1}{\Gamma(\alpha_1)}\right] \\
&\quad + \Big([N_{11} + N_{12}R + N_{13}R]\mathcal{B}_1 + [M_{11} + M_{12}R + M_{13}R]\mathcal{C}_1\Big) \times \\
&\quad \frac{\lambda_1 G_1 \Gamma(\alpha_2)(\eta_1 - \alpha_1 + \theta_1)^{\underline{\theta_1-1}}}{\Gamma(\theta_1)\Gamma(\theta_2)} \\
&\quad + \Big([N_{21} + N_{22}R + N_{23}R] + [M_{21} + M_{22}R + M_{23}R]\Big)\frac{1}{\Gamma(\theta_2)} \\
&\quad + \Big([N_{21} + N_{22}R + N_{23}R]\mathcal{B}_2 + [M_{21} + M_{22}R + M_{23}R]\mathcal{C}_2\Big) \times \\
&\quad \frac{\lambda_2 G_2 \Gamma(\alpha_1)(\eta_2 - \alpha_2 + \theta_2)^{\underline{\theta_2-1}}}{\Gamma(\theta_1)\Gamma(\theta_2)},
\end{aligned} \tag{58}$$

$$\begin{aligned}
\tilde{\Omega}_2^* &= [N_{11} + N_{12}R + N_{13}R]\left[\frac{1}{\lambda_2 G_2(\eta_2 - \alpha_2 + \theta_2)^{\underline{\theta_2-1}}\mathcal{A}_2} + \frac{1}{\Gamma(\alpha_2)}\right] \\
&\quad + [N_{21} + N_{22}R + N_{23}R]\frac{1}{\lambda_1 G_1(\eta_1 - \alpha_1 + \theta_1)^{\underline{\theta_1-1}}\mathcal{A}_1} \\
&\quad + [M_{11} + M_{12}R + M_{13}R]\frac{\mathcal{B}_1}{\Gamma(\theta_1)\mathcal{A}_1} + [M_{21} + M_{22}R + M_{23}R]\frac{\mathcal{B}_2}{\Gamma(\theta_2)\mathcal{A}_2}.
\end{aligned} \tag{59}$$

Hence, for each $t_{i1}, t_{i2} \in \mathbb{N}_{\alpha_i-2} \cap [a_i, b_i]$, and $u_i \in \Theta_i$, we have:

$$\begin{aligned}
&\left|\chi(\mathcal{F}_1 u_1)(t_{11}, t_{21}) - \chi(\mathcal{F}_1 u_1)(t_{12}, t_{22})\right| \\
\leq\ & \chi\left|t_{11}^{\underline{\alpha_1}}\left\{\left[\frac{|\phi_1(u_1,u_2)|}{\Gamma(\theta_1)} - \frac{|\phi_2(u_1,u_2)|}{\Gamma(\theta_2)} + \frac{|\phi_1(u_1,u_2)|}{\Gamma(\alpha_1)}\right] + \frac{1}{\Gamma(\theta_1)\Gamma(\theta_2)} \times\right.\right. \\
&\left[\frac{\lambda_1\Gamma(\alpha_2)\mathcal{B}_1|\phi_1(u_1,u_2)|}{t_{21}^{\underline{\alpha_2-1}}\Gamma(\alpha_1)} + \frac{\lambda_2\Gamma(\alpha_1)\mathcal{B}_2|\phi_2(u_1,u_2)|}{t_{11}^{\underline{\alpha_1-1}}\Gamma(\alpha_2)}\right] + \frac{1}{\Gamma(\theta_1)\Gamma(\theta_2)} \times \\
&+ \left[\frac{\lambda_1\Gamma(\alpha_2)\mathcal{B}_1|F_1(t_1,t_2,\Delta^{\beta_1}u_1,u_2)|}{t_{22}^{\underline{\alpha_2-1}}\Gamma(\alpha_1)} + \frac{\lambda_2\Gamma(\alpha_1)\mathcal{B}_2|F_2(t_1,t_2,u_1,\Delta^{\beta_2}u_2)|}{t_{12}^{\underline{\alpha_1-1}}\Gamma(\alpha_2)}\right] \\
&\left.\left[\frac{|F_1(t_1,t_2,\Delta^{\beta_1}u_1,u_2)|}{\Gamma(\theta_1)} - \frac{|F_2(t_1,t_2,u_1,\Delta^{\beta_2}u_2)|}{\Gamma(\theta_2)} + \frac{|F_1(t_1,t_2,\Delta^{\beta_1}u_1,u_2)|}{\Gamma(\alpha_1)}\right]\right\}
\end{aligned}$$

$$
\begin{aligned}
& -t_{12}^{\alpha_1}\Bigg\{\left[\frac{|\phi_1(u_1,u_2)|}{\Gamma(\theta_1)}-\frac{|\phi_2(u_1,u_2)|}{\Gamma(\theta_2)}+\frac{|\phi_1(u_1,u_2)|}{\Gamma(\alpha_1)}\right]+\frac{1}{\Gamma(\theta_1)\Gamma(\theta_2)}\times \\
& \left[\frac{\lambda_1\Gamma(\alpha_2)\mathcal{B}_1|\phi_1(u_1,u_2)|}{t_{21}^{\alpha_2-1}\Gamma(\alpha_1)}+\frac{\lambda_2\Gamma(\alpha_1)\mathcal{B}_2|\phi_2(u_1,u_2)|}{t_{11}^{\alpha_1-1}\Gamma(\alpha_2)}\right]+\frac{1}{\Gamma(\theta_1)\Gamma(\theta_2)}\times \\
& +\left[\frac{\lambda_1\Gamma(\alpha_2)\mathcal{B}_1|F_1(t_1,t_2,\Delta^{\beta_1}u_1,u_2)|}{t_{22}^{\alpha_2-1}\Gamma(\alpha_1)}+\frac{\lambda_2\Gamma(\alpha_1)\mathcal{B}_2|F_2(t_1,t_2,u_1,\Delta^{\beta_2}u_2)|}{t_{12}^{\alpha_1-1}\Gamma(\alpha_2)}\right] \\
& \left[\frac{|F_1(t_1,t_2,\Delta^{\beta_1}u_1,u_2)|}{\Gamma(\theta_1)}-\frac{|F_2(t_1,t_2,u_1,\Delta^{\beta_2}u_2)|}{\Gamma(\theta_2)}+\frac{|F_1(t_1,t_2,\Delta^{\beta_1}u_1,u_2)|}{\Gamma(\alpha_1)}\right]\Bigg\}\Bigg| \\
< \; & \left|(t_{11}+\rho_1)^{\underline{\rho_1}}(t_{21}+\rho_2)^{\underline{\rho_2}}-(t_{12}+\rho_1)^{\underline{\rho_1}}(t_{22}+\rho_2)^{\underline{\rho_2}}\right|\tilde{\Omega}_1^* \\
< \; & \frac{\epsilon}{2},
\end{aligned} \tag{60}
$$

and:

$$
\begin{aligned}
& \left|\chi(\mathcal{F}_2u_2)(t_{11},t_{21})-\chi(\mathcal{F}_2u_2)(t_{12},t_{22})\right| \\
\leq \; & \chi\Bigg|t_{21}^{\alpha_2}\Bigg\{\left[\frac{t_{11}^{\alpha_1-1}|\phi_1(u_1,u_2)|}{\lambda_2G_2(\eta_2-\alpha_2+\theta_2)^{\underline{\theta_2-1}}\mathcal{A}_2}-\frac{t_{21}^{\alpha_2-1}|\phi_2(u_1,u_2)|}{\lambda_1G_1(\eta_1-\alpha_1+\theta_1)^{\underline{\theta_1-1}}\mathcal{A}_1}+\frac{|\phi_2(u_1,u_2)|}{\Gamma(\alpha_2)}\right] \\
& +t_{21}^{\alpha_2-1}\left[\frac{\mathcal{B}_1|\phi_1(u_1,u_2)|}{\Gamma(\theta_1)\mathcal{A}_1}-\frac{\mathcal{B}_2|\phi_2(u_1,u_2)|}{\Gamma(\theta_2)\mathcal{A}_2}\right]+\left[\frac{t_{11}^{\alpha_1-1}|F_1(t_1,t_2,\Delta^{\beta_1}u_1,u_2)|}{\lambda_2G_2(\eta_2-\alpha_2+\theta_2)^{\underline{\theta_2-1}}\mathcal{A}_2}\right. \\
& \left.-\frac{t_{21}^{\alpha_2-1}|F_2(t_1,t_2,u_1,\Delta^{\beta_2}u_2)|}{\lambda_1G_1(\eta_1-\alpha_1+\theta_1)^{\underline{\theta_1-1}}\mathcal{A}_1}+\frac{|F_2(t_1,t_2,u_1,\Delta^{\beta_2}u_2)|}{\Gamma(\alpha_2)}\right] \\
& +t_{21}^{\alpha_2-1}\left[\frac{t_{11}^{\alpha_1}\mathcal{C}_1|F_1(t_1,t_2,\Delta^{\beta_1}u_1,u_2)|}{\Gamma(\theta_1)\mathcal{A}_1}-\frac{t_{21}^{\alpha_2}\mathcal{C}_2|F_2(t_1,t_2,u_1,\Delta^{\beta_2}u_2)|}{\Gamma(\theta_2)\mathcal{A}_2}\right]\Bigg\} \\
& -t_{22}^{\alpha_2}\Bigg\{\left[\frac{t_{11}^{\alpha_1-1}|\phi_1(u_1,u_2)|}{\lambda_2G_2(\eta_2-\alpha_2+\theta_2)^{\underline{\theta_2-1}}\mathcal{A}_2}-\frac{t_{21}^{\alpha_2-1}|\phi_2(u_1,u_2)|}{\lambda_1G_1(\eta_1-\alpha_1+\theta_1)^{\underline{\theta_1-1}}\mathcal{A}_1}+\frac{|\phi_2(u_1,u_2)|}{\Gamma(\alpha_2)}\right] \\
& +t_{21}^{\alpha_2-1}\left[\frac{\mathcal{B}_1|\phi_1(u_1,u_2)|}{\Gamma(\theta_1)\mathcal{A}_1}-\frac{\mathcal{B}_2|\phi_2(u_1,u_2)|}{\Gamma(\theta_2)\mathcal{A}_2}\right]+\left[\frac{t_{11}^{\alpha_1-1}|F_1(t_1,t_2,\Delta^{\beta_1}u_1,u_2)|}{\lambda_2G_2(\eta_2-\alpha_2+\theta_2)^{\underline{\theta_2-1}}\mathcal{A}_2}\right. \\
& \left.-\frac{t_{21}^{\alpha_2-1}|F_2(t_1,t_2,u_1,\Delta^{\beta_2}u_2)|}{\lambda_1G_1(\eta_1-\alpha_1+\theta_1)^{\underline{\theta_1-1}}\mathcal{A}_1}+\frac{|F_2(t_1,t_2,u_1,\Delta^{\beta_2}u_2)|}{\Gamma(\alpha_2)}\right] \\
& +t_{21}^{\alpha_2-1}\left[\frac{t_{11}^{\alpha_1}\mathcal{C}_1|F_1(t_1,t_2,\Delta^{\beta_1}u_1,u_2)|}{\Gamma(\theta_1)\mathcal{A}_1}-\frac{t_{21}^{\alpha_2}\mathcal{C}_2|F_2(t_1,t_2,u_1,\Delta^{\beta_2}u_2)|}{\Gamma(\theta_2)\mathcal{A}_2}\right]\Bigg\} \\
< \; & \left|(t_{11}+\rho_1)^{\underline{\rho_1}}(t_{21}+\rho_2)^{\underline{\rho_2}}-(t_{12}+\rho_1)^{\underline{\rho_1}}(t_{22}+\rho_2)^{\underline{\rho_2}}\right|\tilde{\Omega}_2^* \\
< \; & \frac{\epsilon}{2}.
\end{aligned} \tag{61}
$$

Similarly, for each $i,j\in\{1,2\}$ and $j\neq i$, we obtain:

$$
\left|\Delta^{\beta_i}\left(\mathcal{F}_i(u_1,u_2)\right)(t_{i1}-\beta_i+1,t_{j1})-\Delta^{\beta_i}\left(\mathcal{F}_i(u_1,u_2)\right)(t_{i2}-\beta_i+1,t_{j2})\right|<\frac{\epsilon}{2}. \tag{62}
$$

Hence:

$$
\begin{aligned}
& \|\mathcal{F}_i(u_1,u_2)(t_{11},t_{21})-\mathcal{F}_i(u_1,u_2)(t_{12},t_{22})\|_{\mathcal{U}_i} \\
= \; & \|\Delta^{\beta_i}\mathcal{F}_i(u_1,u_2)(t_{i1}-\beta_i+1,t_{j1})-\Delta^{\beta_i}\mathcal{F}_i(u_1,u_2)(t_{i2}-\beta_i+1,t_{j2})\|_{\mathcal{C}_i}
\end{aligned}
$$

$$
\begin{aligned}
&+ \|\mathcal{F}_j(u_1,u_2)(t_{11},t_{21}) - \mathcal{F}_j(u_1,u_2)(t_{12},t_{22})\|_{C_j} \\
&< \frac{\epsilon}{2} + \frac{\epsilon}{2} = \epsilon.
\end{aligned} \tag{63}
$$

This implies that both $\mathcal{F}_1$ and $\mathcal{F}_2$ are equicontinuous on $\mathcal{D}$, which shows that $\mathcal{F}$ is equicontinuous on $\mathcal{D}$. Therefore, by the Arzelá–Ascoli theorem and Theorem 2, we can conclude that $\mathcal{F}$ is completely continuous.

Step IV. Both $\mathcal{F}_i : \Theta_1 \times \Theta_2 \to \mathcal{U}$ are equiconvergent as $t_1, t_2 \to \infty$.

By the assumption $(H1) - (H2)$, we obtain:

$$
\begin{aligned}
\chi\left| \left(\mathcal{F}_1(u_1,u_2)\right)(t_1,t_2)\right| &< \frac{\tilde{\Omega}_1^*}{t_1^{\alpha_1} t_2^{\alpha_2}} \to 0 \ \text{ uniformly in } \Theta_1 \times \Theta_2 \text{ as } t_1, t_2 \to \infty, \\
\chi\left| \left(\mathcal{F}_2(u_1,u_2)\right)(t_1,t_2)\right| &< \frac{\tilde{\Omega}_2^*}{t_1^{\alpha_1} t_2^{\alpha_2}} \to 0 \ \text{ uniformly in } \Theta_1 \times \Theta_2 \text{ as } t_1, t_2 \to \infty,
\end{aligned}
$$

where $\tilde{\Omega}_1^*, \tilde{\Omega}_2^*$ are defined as (58) and (59).

Furthermore, we have:

$$
\begin{aligned}
\chi\left|\Delta^{\beta_1}\left(\mathcal{F}_1(u_1,u_2)\right)(t_1-\beta_1+1,t_2)\right| &< \frac{\tilde{\Omega}_1^*}{t_1^{\alpha_1} t_2^{\alpha_2}} \to 0 \\
&\quad \text{uniformly in } \Theta_1 \times \Theta_2 \text{ as } t_1, t_2 \to \infty, \\
\chi\left|\Delta^{\beta_2}\left(\mathcal{F}_2(u_1,u_2)\right)(t_1,t_2-\beta_2+1)\right| &< \frac{\tilde{\Omega}_2^*}{t_1^{\alpha_1} t_2^{\alpha_2}} \to 0 \\
&\quad \text{uniformly in } \Theta_1 \times \Theta_2 \text{ as } t_1, t_2 \to \infty.
\end{aligned}
$$

Hence, both $\mathcal{F}_i$ are equiconvergent as $t_1, t_2 \to \infty$.

Consequently, from Step I–Step IV, we conclude that $\mathcal{F}$ is completely continuous. This complete the proof. □

Finally, we present the main result of the article. For the sake of convenience, we set:

$$
\begin{aligned}
\Psi_1 &= \left(N_{12}+N_{13}\right)\tilde{\Omega}_{11} + \left(N_{22}+N_{23}\right)\tilde{\Omega}_{12} + \left(M_{12}+M_{13}\right)\tilde{\Omega}_{13} \\
&\quad + \left(M_{22}+M_{23}\right)\tilde{\Omega}_{14},
\end{aligned} \tag{64}
$$

and (65)

$$
\begin{aligned}
\Psi_2 &= \left(N_{12}+N_{13}\right)\tilde{\Omega}_{21} + \left(N_{22}+N_{23}\right)\tilde{\Omega}_{22} + \left(M_{12}+M_{13}\right)\tilde{\Omega}_{23} \\
&\quad + \left(M_{22}+M_{23}\right)\tilde{\Omega}_{24},
\end{aligned} \tag{66}
$$

where $\tilde{\Omega}_{1p}, \tilde{\Omega}_{2p},\ p = 1,2,3,4$ are defined as (48) and (49).

Theorem 4. *Suppose that (H_1)–(H_2) hold. Then, the problems (5) and (6) has at least one solution if:*

$$
\Psi_1 + \Psi_2 < 1. \tag{67}
$$

Proof. Under the Banach space $\mathcal{U}$ equipped with the norm $\|\cdot\|_{\mathcal{U}}$, we let:

$$
\begin{aligned}
&\omega_1(t_1,t_2) \\
&= \frac{t_1^{\underline{\alpha_1-1}}}{\Lambda}\left\{\frac{\lambda_1 G_1 \Gamma(\alpha_1)(\eta_1-\alpha_1+\theta_1)^{\underline{\theta_1-1}}\mathcal{A}_1}{\Gamma(\theta_1)}\left[(t_1+\rho_1)^{\underline{\rho_1}} e^{-n_1(t_1+1)} \times\right.\right.
\end{aligned}
$$

$$\left(\frac{N_{11}t_1^{\underline{\alpha_1-2}}}{\Gamma(\alpha_1)} - \frac{N_{21}\lambda_2 G_2(\eta_2-\alpha_2+\theta_2)^{\underline{\theta_2-1}}\mathcal{B}_2}{\Gamma(\theta_2)}\right) + \frac{M_{11}e^{-m_1 t_2}}{\Gamma(\alpha_1)}\times$$

$$\sum_{s=0}^{t_1-\alpha_1}(t_1-\sigma(s))^{\underline{\alpha_1-1}}e^{-2m_1 s} - \frac{M_{21}G_2(\eta_2-\alpha_2+\theta_2-1)^{\underline{\theta_2-1}}e^{-m_1(2t_1+t_2)}}{\Gamma(\theta_2)}\Bigg]$$

$$+\frac{\lambda_2 G_2\Gamma(\alpha_2)(\eta_2-\alpha_2+\theta_2)^{\underline{\theta_2-1}}\mathcal{A}_2}{\Gamma(\theta_2)}\Bigg[(t_1+\rho_1)^{\underline{\rho_1}}e^{-n_1(t_1+1)}\times$$

$$\left(\frac{N_{21}t_2^{\underline{\alpha_2-2}}}{\Gamma(\alpha_2)} - \frac{N_{11}\lambda_1 G_1(\eta_1-\alpha_1+\theta_1)^{\underline{\theta_1-1}}\mathcal{B}_1}{\Gamma(\theta_1)}\right) + \frac{M_{21}e^{-2m_1 t_1}}{\Gamma(\alpha_2)}\times$$

$$\sum_{s=0}^{t_2-\alpha_2}(t_2-\sigma(s))^{\underline{\alpha_2-1}}e^{-m_1 s} - \frac{M_{11}G_1(\eta_1-\alpha_1+\theta_1-1)^{\underline{\theta_1-1}}e^{-m_1(2t_1+t_2)}}{\Gamma(\theta_1)}\Bigg]\Bigg\}$$

$$+\frac{N_{11}}{\Gamma(\alpha_1)}(t_1+\rho_1)^{\underline{\rho_1}}e^{-n_1(t_1+1)} + \frac{M_{11}e^{-m_1 t_2}}{\Gamma(\alpha_1)}\sum_{s=0}^{t_1-\alpha_1}(t_1-\sigma(s))^{\underline{\alpha_1-1}}e^{-2m_1 s}, \tag{68}$$

and

$$\omega_2(t_1,t_2)$$

$$=\frac{t_2^{\underline{\alpha_2-1}}}{\Lambda}\Bigg\{t_2^{\underline{\alpha_2-1}}\Bigg[(t_1+\rho_1)^{\underline{\rho_1}}\left[(t_2+\rho_2)^{\underline{\rho_2}}\right]^2 e^{-n_2(t_2+1)}\Bigg(\frac{N_{12}t_1^{\underline{\alpha_1-2}}}{\Gamma(\alpha_1)}$$

$$-\frac{N_{22}\lambda_2 G_2(\eta_2-\alpha_2+\theta_2)^{\underline{\theta_2-1}}\mathcal{B}_2}{\Gamma(\theta_2)}\Bigg) + \frac{M_{12}e^{-m_2 t_2}}{\Gamma(\alpha_1)}(t_2+\rho_2)^{\underline{\rho_2}}\sum_{s=0}^{t_1-\alpha_1}(t_1-\sigma(s))^{\underline{\alpha_1-1}}\times$$

$$e^{-2m_2 s} - \frac{M_{22}G_2(\eta_2-\alpha_2+\theta_2-1)^{\underline{\theta_2-1}}e^{-m_2(t_1+2t_2)}(t_2+\rho_2)^{\underline{\rho_2}}}{\Gamma(\theta_2)}\Bigg] + t_1^{\underline{\alpha_1-1}}\times$$

$$\Bigg[(t_1+\rho_1)^{\underline{\rho_1}}\left[(t_2+\rho_2)^{\underline{\rho_2}}\right]^2 e^{-n_2(t_2+1)}\left(\frac{N_{22}t_2^{\underline{\alpha_2-2}}}{\Gamma(\alpha_2)} - \frac{N_{12}\lambda_1 G_1(\eta_1-\alpha_1+\theta_1)^{\underline{\theta_1-1}}\mathcal{B}_1}{\Gamma(\theta_1)}\right)$$

$$+\frac{M_{22}e^{-m_2 t_1}}{\Gamma(\alpha_2)}\sum_{s=0}^{t_2-\alpha_2}(t_2-\sigma(s))^{\underline{\alpha_2-1}}e^{-2m_2 s}(t_2+\rho_2)^{\underline{\rho_2}}$$

$$-\frac{M_{22}G_1(\eta_1-\alpha_1+\theta_1-1)^{\underline{\theta_1-1}}e^{-m_2(t_2+2t_2)}}{\Gamma(\theta_1)}\Bigg] + \frac{N_{22}}{\Gamma(\alpha_2)}\left[(t_2+\rho_2)^{\underline{\rho_2}}\right]^2 e^{-n_2(t_2+1)}t_2^{\underline{\alpha_2-2}}\Bigg\}$$

$$+\frac{M_{22}e^{-m_2 t_1}}{\Gamma(\alpha_2)}\sum_{s=0}^{t_2-\alpha_2}(t_2-\sigma(s))^{\underline{\alpha_2-1}}e^{-2m_2 s}(t_2+\rho_2)^{\underline{\rho_2}}. \tag{69}$$

It is clear that $(\omega_1,\omega_2)\in\mathcal{U}$. For $\ell>0$, we define:

$$\Xi_\ell = \{(u_1,u_2)\in\mathcal{U} \,:\, \|(u_1,u_2)-(\omega_1,\omega_2)\|_{\mathcal{U}} \le \ell\}. \tag{70}$$

For $(u_1,u_2)\in\Xi_\ell$, we have:

$$\|(u_1,u_2)\|_{\mathcal{U}} \le \|(u_1,u_2)-(\omega_1,\omega_2)\|_{\mathcal{U}} + \|(\omega_1,\omega_2)\|_{\mathcal{U}} \le \ell + \|(\omega_1,\omega_2)\|_{\mathcal{U}}, \tag{71}$$

$$\|(u_1,u_2)\|_{\mathcal{U}} = \max\left\{\|(u_1,u_2)\|_{\mathcal{U}_1}, \|(u_1,u_2)\|_{\mathcal{U}_2}\right\} \le \ell + \|(\omega_1,\omega_2)\|_{\mathcal{U}}. \tag{72}$$

Using the conditions (H_1)–(H_2), together with the procedure employed in Lemma 5, we have:

$$\left|F_i\left(t_1,t_2,\Delta^{\beta_i}u_i,u_j\right) - M_{i1}\left(t_2+{}_{i1}\rho_2\right)^{\underline{i1\rho_2}}e^{-m_{i1}(t_1+t_2)}\right|$$

$$\le\Bigg\{M_{i2}\left(t_2+{}_{i2}\rho_2\right)^{\underline{i2\rho_2}}e^{-m_{i2}(t_1+t_2)}$$

$$+ M_{i3}\left(t_2 + {}_{i3}\rho_2\right)^{\underline{i3\rho_2}} e^{-m_{i3}(t_1+t_2)}\Big\} \|(u_1,u_2)\|_{\mathcal{U}}, \tag{73}$$

and:

$$\begin{aligned}&\left|\phi_i(u_1,u_2) - N_{i1}\left(t_1 + {}_{i1}\tilde{\rho}_1\right)^{\underline{i1\tilde{\rho}_1}}\left[\left(t_2 + {}_{i1}\tilde{\rho}_2\right)^{\underline{i1\tilde{\rho}_2}}\right]^2 e^{-n_{i1}(t_1+t_2)}\right| \\ =&\Big\{N_{i2}\left(t_1 + {}_{i2}\tilde{\rho}_1\right)^{\underline{i2\tilde{\rho}_1}}\left[\left(t_2 + {}_{i2}\tilde{\rho}_2\right)^{\underline{i2\tilde{\rho}_2}}\right]^2 e^{-n_{i2}(t_1+t_2)} \\ &+ N_{i3}\left(t_1 + {}_{i3}\tilde{\rho}_1\right)^{\underline{i3\tilde{\rho}_1}}\left[\left(t_2 + {}_{i3}\tilde{\rho}_2\right)^{\underline{i3\tilde{\rho}_2}}\right]^2 e^{-n_{i3}(t_1+t_2)}\Big\} \|(u_1,u_2)\|_{\mathcal{U}}.\end{aligned} \tag{74}$$

Therefore, we obtain:

$$\chi\left|\left(\mathcal{F}_i(u_i,u_j)\right)(t_1,t_2) - \omega_i(t_1,t_2)\right| \le \left(\ell + \|(\omega_1,\omega_2)\|_{\mathcal{U}}\right)\Psi_i. \tag{75}$$

Furthermore, we have:

$$\begin{aligned}&\chi\left|\Delta^{\beta_i}\left(\mathcal{F}_i(u_i,u_j)\right)(t_i-\beta_i+1,t_j) - \Delta^{\beta_i}\omega_i(t_i-\beta_i+1,t_j)\right| \\ \le&\left(\ell + \|(\omega_1,\omega_2)\|_{\mathcal{U}}\right)\Psi_i.\end{aligned} \tag{76}$$

Hence, it follows that:

$$\left\|\left(\mathcal{F}_i(u_1,u_2)\right) - \omega_i\right\|_{\mathcal{U}_i} \le \left(\ell + \|(\omega_1,\omega_2)\|_{\mathcal{U}}\right)2\Psi_i. \tag{77}$$

Therefore,

$$\left\|\left(\mathcal{F}(u_1,u_2)\right) - (\omega_1,\omega_2)\right\|_{\mathcal{U}} \le \left(\ell + \|(\omega_1,\omega_2)\|_{\mathcal{U}}\right)2\max\{\Psi_1,\Psi_2\}. \tag{78}$$

Choosing:

$$\ell \ge \frac{\|(\omega_1,\omega_2)\|_{\mathcal{U}}\,2\max\{\Psi_1,\Psi_2\}}{1-2\max\{\Psi_1,\Psi_2\}}, \tag{79}$$

and for $(u_1,u_2) \in \Xi_\ell$, we consequently obtain:

$$\left\|\left(\mathcal{F}(u_1,u_2)\right) - (\omega_1,\omega_2)\right\|_{\mathcal{U}} \le \ell. \tag{80}$$

From the Schauder fixed point theorem, this implies that $\mathcal{F}$ has a fixed point $(u_1,u_2) \in \Xi_\ell$, which is a bounded solution of the problems (5) and (6). The proof is complete. □

4. Example

In order to illustrate our result, we consider the following fractional sum boundary value problem:

$$\begin{aligned}\Delta^{\frac{3}{2}}u_1(t) =& \frac{2}{3}\left(t+\frac{4}{3}\right)e^{-(12t+5)} + \frac{\left(t+\frac{5}{3}\right)^{\frac{4}{3}} e^{-\left(25t+\frac{31}{3}\right)} u_2\left(t+\frac{1}{3}\right)}{4000\left(t+\frac{301}{3}\right)^2\left(1+\cos^2 u_2\pi\right)} \\ &+ \frac{\left(t+\frac{11}{6}\right)^{\frac{3}{2}} e^{-\left[(12t+5)+\left(t+\frac{1}{2}\right)\pi\right]}\Delta^{\frac{1}{3}}u_1\left(t+\frac{4}{3}\right)}{5000e + 10\cos^2\left(t+\frac{1}{2}\right)\pi}, \quad t \in \mathbb{N}_0\end{aligned}$$

$$\begin{aligned}
\Delta^{\frac{4}{3}}u_2(t) &= \frac{1}{2}\left(t+\frac{4}{3}\right)e^{-(12t+5)}+\frac{\left(t+\frac{5}{3}\right)^{\frac{4}{3}}e^{-\left(25t+\frac{21}{3}\right)}u_1\left(t+\frac{1}{2}\right)}{1000\left(e^{\left(t+\frac{1}{2}\right)}+10\right)^2}\\
&\quad+\frac{\left(t+\frac{11}{6}\right)^{\frac{3}{2}}e^{-(12t+5)}\arctan\left(\cos^2\left(t+\frac{1}{3}\right)\pi\right)\Delta^{\frac{3}{4}}u_2\left(t_2+\frac{7}{12}\right)}{1000\pi\left(t+\frac{10}{3}\right)^2},\quad t\in\mathbb{N}_0\\
u_1\Big(-\frac{1}{2}\Big) &= \phi_1(u_1,u_2)=\frac{|u_1|}{2000e^3}\cos^2|\pi u_1|+\frac{|u_2||u_2^2+2|^{\frac{1-|u_2^2+2|}{}}}{4000\pi^2(u_2^2+e)},\\
u_2\Big(-\frac{2}{3}\Big) &= \phi_2(u_1,u_2)=\frac{|u_2|}{5000e^2}\sin^2|\pi u_2|+\frac{|u_1||u_1^2+3|^{\frac{1-|u_1^2+3|}{}}}{2000\pi(u_1^2+\pi)},\\
\lim_{t\longrightarrow\infty}u_1\Big(t-\frac{1}{2}\Big) &= \frac{1}{2}\Delta^{-\frac{1}{4}}\Big(12e+\cos(4)\Big)^2u_2(4)\\
\lim_{t\longrightarrow\infty}u_2\Big(t-\frac{2}{3}\Big) &= \frac{3}{4}\Delta^{-\frac{2}{3}}\left(10e-\sin\Big(\frac{15}{4}\Big)\right)^3u_1\Big(\frac{15}{4}\Big).
\end{aligned}\tag{81}$$

Here, $\alpha_1=\frac{3}{2}$, $\alpha_2=\frac{4}{3}$, $\beta_1=\frac{1}{3}$, $\beta_2=\frac{3}{4}$, $\gamma_1=\frac{3}{4}$, $\gamma_2=\frac{5}{6}$, $\theta_1=\frac{1}{4}$, $\theta_2=\frac{2}{3}$, $\eta_1=\frac{7}{2}$, $\eta_2=\frac{10}{3}$, $\lambda_1=\frac{1}{2}$, $\lambda_2=\frac{3}{4}$, $T=4$, $g_1(t_1)=(10e-\sin t_1)^3$, $g_2(t_2)=(12e+\cos t_2)^2$, and:

$$\begin{aligned}
F_1\left(t_1,t_2,\Delta^{\frac{1}{3}}u_1\left(t_1+\tfrac{2}{3}\right),u_2(t_2)\right) &= \tfrac{2}{3}(t_2+1)e^{-6(t_1+t_2)}+\frac{\left(t_2+\frac{4}{3}\right)^{\frac{4}{3}}e^{-[12(t_1+t_2)+t_2]}u_2(t_2)}{4000(t_2+10)^2(1+\cos^2u_2\pi)}\\
&\quad+\frac{\left(t_2+\frac{3}{2}\right)^{\frac{3}{2}}e^{-[6(t_1+t_2)+t_1\pi]}\Delta^{\frac{1}{3}}u_1\left(t_1+\frac{2}{3}\right)}{5000e+10\cos^2t_1\pi}\\
F_2\left(t_1,t_2,\Delta^{\frac{3}{4}}u_2\left(t_2+\tfrac{1}{4}\right),u_1(t_1)\right) &= \tfrac{1}{2}(t_2+1)e^{-6(t_1+t_2)}+\frac{\left(t_2+\frac{4}{3}\right)^{\frac{4}{3}}e^{-[12(t_1+t_2)+t_1]}u_1(t_1)}{1000(e^{t_1}+10)^2}\\
&\quad+\frac{\left(t_2+\frac{3}{2}\right)^{\frac{3}{2}}e^{-6(t_1+t_2)}\arctan\left(\cos^2t_2\pi\right)\Delta^{\frac{3}{4}}u_2\left(t_2+\frac{1}{4}\right)}{1000\pi(t_2+3)^2}.
\end{aligned}$$

Choose $\rho_1=1$, $\rho_2=2$, ${}_{i1}\rho_1={}_{i1}\tilde{\rho}_1=\frac{1}{2}$, ${}_{i2}\rho_1={}_{i2}\tilde{\rho}_1=\frac{2}{3}$, ${}_{i3}\rho_1={}_{i3}\tilde{\rho}_1=\frac{3}{4}$ ${}_{i1}\rho_2=1$, ${}_{i2}\rho_2=\frac{3}{2}$, ${}_{i3}\rho_2=\frac{4}{3}$, $m_{i1}=n_{i1}=6$, $m_{i2}=n_{i2}=6$, $m_{i3}=n_{i3}=12$, where $\rho_i>\max\{\beta_1-\alpha_1,\beta_2-\alpha_2\}$, ${}_{ip}\rho_1\in(-1,1)$, ${}_{ip}\rho_2\in(-1,2)$ for $i=1,2$ and $p=1,2,3$.

Let $t_1\in\mathbb{N}_{-\frac{1}{2},\frac{11}{2}}$, $t_2\in\mathbb{N}_{-\frac{2}{3},\frac{16}{3}}$ and $\chi=\frac{(t_1-\frac{1}{2})^{1/2}(t_2-\frac{1}{3})^{2/3}}{1+t_1^{\frac{3}{2}}t_2^{\frac{4}{3}}}$. Since:

$$\begin{aligned}
&\left|F_1\left(t_1,t_2,\frac{1}{\chi}\Delta^{\frac{1}{3}}u_1,\frac{1}{\chi}u_2\right)-\frac{2}{3}(t_2+1)e^{-6(t_1+t_2)}\right|\\
\le\ &\frac{1}{361000}\left(t_2+\frac{4}{3}\right)^{\frac{4}{3}}e^{-12(t_1+t_2)}|u_2|+\frac{1}{5000e^{10}+10}\left(t_2+\frac{3}{2}\right)^{\frac{3}{2}}e^{-6(t_1+t_2)}\left|\Delta^{\frac{1}{3}}u_1\right|,\\
&\left|F_2\left(t_1,t_2,\frac{1}{\chi}\Delta^{\frac{3}{4}}u_2,\frac{1}{\chi}u_1\right)-\frac{1}{2}(t_2+1)e^{-6(t_1+t_2)}\right|\\
\le\ &\frac{1}{121000}\left(t_2+\frac{4}{3}\right)^{\frac{4}{3}}e^{-12(t_1+t_2)}|u_2|+\frac{9}{196000}\left(t_2+\frac{3}{2}\right)^{\frac{3}{2}}e^{-6(t_1+t_2)}\left|\Delta^{\frac{3}{4}}u_2\right|,
\end{aligned}$$

we find that (H_1) holds with $M_{11}=0.666$, $M_{12}=9.080$, $M_{13}=2.770$ and $M_{21}=0.500$, $M_{22}=0.000046$, $M_{23}=0.0000083$.

Furthermore, we obtain:

$$\left|\phi_1\left(\frac{1}{\chi}u_1,\frac{1}{\chi}u_2\right)-\frac{2}{5}\left(t_1+\frac{1}{2}\right)^{\frac{1}{2}}\left(t_2+\frac{1}{2}\right)e^{-6(t_1+t_2)}\right|$$

$$\leq \frac{1}{2000e^3}\left(t_1+\frac{2}{3}\right)^{\frac{2}{3}}\left(t_2+\frac{2}{3}\right)^{\frac{4}{3}}e^{-6(t_1+t_2)}\|u_1\|+\frac{1}{4000\pi^2}\left(t_1+\frac{3}{4}\right)^{\frac{3}{4}}\left(t_2+\frac{3}{4}\right)^{\frac{3}{2}}\times e^{-3(t_1+t_2)}\|u_2\|,$$

$$\left|\phi_2\left(\frac{1}{\chi}u_1,\frac{1}{\chi}u_2\right)-\frac{5}{6}\left(t_1+\frac{1}{2}\right)^{\frac{1}{2}}\left(t_2+\frac{1}{2}\right)e^{-6(t_1+t_2)}\right|$$

$$\leq \frac{1}{2000\pi}\left(t_1+\frac{2}{3}\right)^{\frac{2}{3}}\left(t_2+\frac{2}{3}\right)^{\frac{4}{3}}e^{-6(t_1+t_2)}\|u_1\|+\frac{1}{500e^2}\left(t_1+\frac{3}{4}\right)^{\frac{3}{4}}\left(t_2+\frac{3}{4}\right)^{\frac{3}{2}}\times e^{-3(t_1+t_2)}\|u_2\|,$$

and $(10e-1)^3 < g_1(t_1) < (10e+1)^3$ and $(12e-1)^2 < g_2(t_2) < (12e+1)^2$.

Thus, $(H_2),(H3)$ hold with $N_{11}=0.4$, $N_{12}=0.0000249$, $N_{13}=0.0000253$, $N_{21}=0.833$, $N_{22}=0.000159$, $N_{23}=0.000271$, $g_1=17949.37$, $g_2=999.79$, $G_1=22384.80$, and $G_2=1130.26$.

Finally, we find that:

$$\tilde{\Omega}_{11}=2313.238,\ \tilde{\Omega}_{12}=319.647,\ \tilde{\Omega}_{13}=27053.522,\ \tilde{\Omega}_{14}=2597.063,$$

$$\tilde{\Omega}_{21}=0.0212,\ \tilde{\Omega}_{22}=1.1403,\ \tilde{\Omega}_{23}=0.0198,\ \tilde{\Omega}_{24}=1.5093.$$

Therefore, we have:

$$\Psi_1+\Psi_2=0.00057+0.4692=0.4698<1.$$

Hence, by Theorem 4, this boundary value problem has at least one solution. □

Author Contributions: These authors contributed equally to this work.

Funding: This research was funded by King Mongkut's University of Technology, North Bangkok, Contract No. KMUTNB-ART-60-38.

Acknowledgments: The last author of this research was supported by Suan Dusit University.

Conflicts of Interest: The authors declare no conflicts of interest regarding the publication of this paper.

References

1. Wu, G.C.; Baleanu, D. Discrete fractional logistic map and its chaos. *Nonlinear Dyn.* **2014**, *75*, 283–287. [CrossRef]
2. Wu, G.C.; Baleanu, D. Chaos synchronization of the discrete fractional logistic map. *Signal Process.* **2014**, *102*, 96–99. [CrossRef]
3. Goodrich, C.S.; Peterson, A.C. *Discrete Fractional Calculus*; Springer: New York, NY, USA, 2015.
4. Atici, F.M.; Eloe, P.W. A transform method in discrete fractional calculus. *Int. J. Differ. Equ.* **2007**, *2*, 165–176.
5. Atici, F.M.; Eloe, P.W. Initial value problems in discrete fractional calculus. *Proc. Am. Math. Soc.* **2009**, *137*, 981–989. [CrossRef]
6. Atici, F.M.; Eloe, P.W. Two-point boundary value problems for finite fractional difference equations. *J. Differ. Equ. Appl.* **2011**, *17*, 445–456. [CrossRef]
7. Abdeljawad, T. On Riemann and Caputo fractional differences. *Comput. Math. Appl.* **2011**, *62*, 1602–1611. [CrossRef]
8. Abdeljawad, T. Dual identities in fractional difference calculus within Riemann. *Adv. Differ. Equ.* **2013**, *2013*, 36. [CrossRef]
9. Abdeljawad, T. On delta and nabla Caputo fractional differences and dual identities. *Discret. Dyn. Nat. Soc.* **2013**, *2013*, 406910. [CrossRef]
10. Abdeljawad, T.; Baleanu, D. Fractional differences and integration by parts. *J. Comput. Anal. Appl.* **2011**, *13*, 574–582.
11. Holm, M. Sum and difference compositions in discrete fractional calculus. *Cubo* **2011**, *13*, 153–184. [CrossRef]
12. Anastassiou, G. Foundations of nabla fractional calculus on time scales and inequalities. *Comput. Math. Appl.* **2010**, *59*, 3750–3762. [CrossRef]

13. Jia, B.; Erbe, L.; Peterson, A. Two monotonicity results for nabla and delta fractional differences. *Arch. Math.* **2015**, *104*, 589–597. [CrossRef]
14. Jia, B.; Erbe, L.; Peterson, A. Convexity for nabla and delta fractional differences. *J. Differ. Equ. Appl.* **2015**, *21*, 360–373.
15. Cermák, J.; Kisela, T.; Nechvátal, L. Stability and asymptotic properties of a linear fractional difference equation. *Adv. Differ. Equ.* **2012**, *2012*, 122. [CrossRef]
16. Jarad, F.; Abdeljawad, T.; Baleanu, D.; Biçen, K. On the stability of some discrete fractional nonautonomous systems. *Abstr. Appl. Anal.* **2012**, *2012*, 476581. [CrossRef]
17. Mozyrska, D.; Wyrwas, M. The Z-transform method and delta type fractional difference operators. *Discret. Dyn. Nat. Soc.* **2015**, *2015*, 12. [CrossRef]
18. Mozyrska, D.; Wyrwas, M. Explicit criteria for stability of fractional *h*-difference two-dimensional systems. *Int. J. Dyn. Control* **2017**, *5*, 4–9. [CrossRef]
19. Ferreira, R.A.C.; Torres, D.F.M. Fractional *h*-difference equations arising from the calculus of variations. *Appl. Anal. Discret. Math.* **2011**, *5*, 110–121. [CrossRef]
20. Ferreira, R.A.C. Existence and uniqueness of solution to some discrete fractional boundary value problems of order less than one. *J. Differ. Equ. Appl.* **2013**, *19*, 712–718. [CrossRef]
21. Ferreira, R.A.C.; Goodrich, C.S. Positive solution for a discrete fractional periodic boundary value problem. *Dyn. Contin. Discret. Impuls. Syst. Ser. A Math. Anal.* **2012**, *19*, 545–557.
22. Goodrich, C.S. Existence and uniqueness of solutions to a fractional difference equation with nonlocal conditions. *Comput. Math. Appl.* **2011**, *61*, 191–202. [CrossRef]
23. Goodrich, C.S. On a discrete fractional three-point boundary value problem. *J. Differ. Equ. Appl.* **2012**, *18*, 397–415. [CrossRef]
24. Goodrich, C.S. A convexity result for fractional differences. *Appl. Math. Lett.* **2014**, *35*, 58–62. [CrossRef]
25. Goodrich, C.S. The relationship between sequential fractional difference and convexity. *Appl. Anal. Discret. Math.* **2016**, *10*, 345–365. [CrossRef]
26. Dahal, R.; Goodrich, C.S. A monotonicity result for discrete fractional difference operators. *Arch. Math.* **2014**, *102*, 293–299. [CrossRef]
27. Erbe, L.; Goodrich, C.S.; Jia, B.; Peterson, A. Survey of the qualitative properties of fractional difference operators: Monotonicity, convexity, and asymptotic behavior of solutions. *Adv. Differ. Equ.* **2016**, *2016*, 43. [CrossRef]
28. Chen, F.; Luo, X.; Zhou, Y. Existence results for nonlinear fractional difference equation. *Adv. Differ. Equ.* **2011**, *2011*, 713201. [CrossRef]
29. Chen, F.; Zhou, Y. Existence and Ulam stability of solutions for discrete fractional boundary value problem. *Discret. Dyn. Nat. Soc.* **2013**, *2013*, 459161. [CrossRef]
30. Chen, Y.; Tang, X. Thee difference between a class of discrete fractional and integer order boundary value problems. *Commun. Nonlinear Sci.* **2014**, *19*, 4057–4067. [CrossRef]
31. Lv, W. Solvability for discrete fractional boundary value problems with a p-laplacian operator. *Discret. Dyn. Nat. Soc.* **2013**, *2013*, 679290. [CrossRef]
32. Lv, W. Solvability for a discrete fractional three-point boundary value problem at resonance. *Abstr. Appl. Anal.* **2014**, *2014*, 601092. [CrossRef]
33. Lv, W.; Feng, J. Nonlinear discrete fractional mixed type sum-difference equation boundary value problems in Banach spaces. *Adv. Differ. Equ.* **2014**, *2014*, 184. [CrossRef]
34. Chen, H.Q.; Cui, Y.Q.; Zhao, X.L. Multiple solutions to fractional difference boundary value problems. *Abstr. Appl. Anal.* **2014**, *2014*, 879380. [CrossRef]
35. Chen, H.Q.; Jin, Z.; Kang, S.G. Existence of positive solutions for Caputo fractional difference equation. *Adv. Differ. Equ.* **2015**, *2015*, 44. [CrossRef]
36. Kang, S.G.; Li, Y.; Chen, H.Q. Positive solutions to boundary value problems of fractional difference equations with nonlocal conditions. *Adv. Differ. Equ.* **2014**, *2014*, 7. [CrossRef]
37. Dong, W.; Xu, J.; Regan, D.O. Solutions for a fractional difference boundary value problem. *Adv. Differ. Equ.* **2013**, *2013*, 319. [CrossRef]
38. Sitthiwirattham, T.; Tariboon, J.; Ntouyas, S.K. Existence Results for fractional difference equations with three-point fractional sum boundary conditions. *Discret. Dyn. Nat. Soc.* **2013**, *2013*, 104276. [CrossRef]
39. Sitthiwirattham, T.; Tariboon, J.; Ntouyas, S.K. Boundary value problems for fractional difference equations with three-point fractional sum boundary conditions. *Adv. Differ. Equ.* **2013**, *2013*, 296. [CrossRef]

40. Sitthiwirattham, T. Existence and uniqueness of solutions of sequential nonlinear fractional difference equations with three-point fractional sum boundary conditions. *Math. Methods Appl. Sci.* **2015**, *38*, 2809–2815. [CrossRef]
41. Sitthiwirattham, T. Boundary value problem for p-Laplacian Caputo fractional difference equations with fractional sum boundary conditions. *Math. Methods Appl. Sci.* **2016**, *39*, 1522–1534. [CrossRef]
42. Chasreechai, S.; Kiataramkul, C.; Sitthiwirattham, T. On nonlinear fractional sum-difference equations via fractional sum boundary conditions involving different orders. *Math. Probl. Eng.* **2015**, *2015*, 519072. [CrossRef]
43. Reunsumrit, J.; Sitthiwirattham, T. Positive solutions of three-point fractional sum boundary value problem for Caputo fractional difference equations via an argument with a shift. *Positivity* **2016**, *20*, 861–876. [CrossRef]
44. Reunsumrit, J.; Sitthiwirattham, T. On positive solutions to fractional sum boundary value problems for nonlinear fractional difference equations. *Math. Methods Appl. Sci.* **2016**, *39*, 2737–2751. [CrossRef]
45. Soontharanon, J.; Jasthitikulchai, N.; Sitthiwirattham, T. Nonlocal Fractional Sum Boundary Value Problems for Mixed Types of Riemann-Liouville and Caputo Fractional Difference Equations. *Dyn. Syst. Appl.* **2016**, *25*, 409–414.
46. Laoprasittichok, S.; Sitthiwirattham, T. On a Fractional Difference-Sum Boundary Value Problems for Fractional Difference Equations Involving Sequential Fractional Differences via Different Orders. *J. Comput. Anal. Appl.* **2017**, *23*, 1097–1111.
47. Kaewwisetkul, B.; Sitthiwirattham, T. On Nonlocal Fractional Sum-Difference Boundary Value Problems for Caputo Fractional Functional Difference Equations with Delay. *Adv. Differ. Equ.* **2017**, *2017*, 219. [CrossRef]
48. Ahmad, B.; Ntouyas, S.K.; Alsaedi, A. On a coupled system of fractional differential equations with coupled nonlocal and integral boundary conditions. *Chaos Soliton Fractals* **2016**, *83*, 234–241. [CrossRef]
49. Ahmad, B.; Ntouyas, S.K. Existence results for a coupled system of Caputo type sequential fractional differential equations with nonlocal integral boundary conditions. *Appl. Math. Comput.* **2015**, *266*, 615–622. [CrossRef]
50. Henderson, J.; Luca, R. Nonexistence of positive solutions for a system of coupled fractional boundary value problems. *Bound. Value Probl.* **2015**, *2015*, 138. [CrossRef]
51. Henderson, J.; Luca, R.; Tudorache, A. On a system of fractional differential equations with coupled integral boundary conditions. *Fract. Calc. Appl. Anal.* **2015**, *18*, 361–386. [CrossRef]
52. Wang, J.R.; Zhang, Y. Analysis of fractional order differential coupled systems. *Math. Methods Appl. Sci.* **2015**, *38*, 3322–3338. [CrossRef]
53. Su, X. Boundary value problem for a coupled system of nonlinear fractional differential equations. *Appl. Math. Lett.* **2009**, *22*, 64–69. [CrossRef]
54. Ascione, G.; Leonenko, N.; Pirozzi, E. Fractional Queues with Catastrophes and Their Transient Behaviour. *Mathematics* **2018**, *6*, 159. [CrossRef]
55. Pan, Y.; Han, Z.; Sun, S.; Zhao, Y. The Existence of Solutions to a System of Discrete Fractional Boundary Value Problems. *Abstr. Appl. Anal.* **2012**, *2012*, 707631. [CrossRef]
56. Goodrich, C.S. Existence of a positive solution to a system of discrete fractional boundary value problems. *Appl. Math. Comput.* **2011**, *217*, 4740–4753. [CrossRef]
57. Dahal, R.; Duncan, D.; Goodrich, C.S. Systems of semipositone discrete fractional boundary value problems. *J. Differ. Equ. Appl.* **2014**, *20*, 473–491. [CrossRef]
58. Goodrich, C.S. Systems of discrete fractional boundary value problems with nonlinearities satisfying no growth conditions. *J. Differ. Equ. Appl.* **2015**, *21*, 437–453. [CrossRef]
59. Goodrich, C.S. Coupled systems of boundary value problems with nonlocal boundary conditions. *Appl. Math. Lett.* **2015**, *41*,17–22. [CrossRef]
60. Kunnawuttipreechachan, E.; Promsakon, C.; Sitthiwirattham, T. Nonlocal fractional sum boundary value problems for a coupled system of fractional sum-difference equations. *Dyn. Syst. Appl.* **2019**, *28*, 73–92. [CrossRef]
61. Griffel, D.H. *Applied Functional Analysis*; Ellis Horwood Publishers: Chichester, UK, 1981.

mathematics

MDPI

Article

Continuous Dependence of Solutions of Integer and Fractional Order Non-Instantaneous Impulsive Equations with Random Impulsive and Junction Points

Yu Chen [1] and JinRong Wang [1,2,*]

[1] Department of Mathematics, Guizhou University, Guiyang 550025, China; cyrainie@126.com
[2] School of Mathematical Sciences, Qufu Normal University, Qufu 273165, China
* Correspondence: jrwang@gzu.edu.cn

Received: 19 February 2019; Accepted: 28 March 2019; Published: 4 April 2019

Abstract: This paper gives continuous dependence results for solutions of integer and fractional order, non-instantaneous impulsive differential equations with random impulse and junction points. The notion of the continuous dependence of solutions of these equations on the initial point is introduced. We prove some sufficient conditions that ensure the solutions to perturbed problems have a continuous dependence. Finally, we use numerical examples to demonstrate the obtained theoretical results.

Keywords: non-instantaneous impulsive equations; random impulsive and junction points; continuous dependence

MSC: 34A37; 34A08

---

1. Introduction

Impulsive differential equations (IDEs) are applied in many fields, such as mechanical engineering, biology, and medical science. Generally speaking, there are two classes of impulsive equations. One is composed of instantaneous IDEs, for which the duration of the impulsive perturbation is very short compared to the entire evolution process, see for example, References [1,2]. The other class is composed of non-instantaneous IDEs, for which the impulsive action starts at a fixed point, and remains active over a period of time that may be related to the previous state.

Non-instantaneous IDEs were introduced in Reference [3] and address the shortcomings of instantaneous IDEs, which do not seem to describe some of the dynamics of evolution in pharmacotherapy. Wang and Fečkan [4] corrected non-instantaneous impulsive equations in Reference [3] and proposed new and generalized non-instantaneous IDEs by considering the impact from the previous system state. Wang [5] used the notion of a non-instantaneous impulsive operator to represent the solutions of linear problems, which are simplified from the model in Reference [4]. The existence and stability of solutions and control problems for these non-instantaneous IDEs, as well as inclusions have been studied in References [6–25]. Meanwhile, fractional differential equations provide an alternative model and are gaining much importance and attention. The qualitative theory of fractional differential equations was studied extensively in the literature; see References [26–33] and the references therein.

Recently, Dishlieva [34] studied a class of instantaneous IDEs with random impulsive effects and established sufficient conditions to ensure continuous dependence of the solutions. Motivated by

Reference [34], we investigate the continuous dependence of the solutions of the following first-order nonlinear differential equations with random non-instantaneous impulsive effects:

$$\begin{cases} x'(t) = \mathbf{f}(t, x(t)), \ t \in (s_i, t_{i+1}], \ i \in \mathbf{N} := \{0, 1, 2, \cdots\}, \\ x(t_i^+) = \mathbf{h_i}(t_i, x(t_i^-)), \ i \in \mathbf{N}^+ := \{1, 2, \cdots\}, \\ x(t) = \mathbf{h_i}(t, x(t_i^-)), \ t \in (t_i, s_i], \ i \in \mathbf{N}^+, \\ x(0) = x_0, \end{cases} \tag{1}$$

and also of the fractional-order random non-instantaneous IDEs:

$$\begin{cases} {}^C\mathbf{D}^{\alpha}_{s_i,t} x(t) = \mathbf{f}(t, x(t)), \ t \in (s_i, t_{i+1}], \ i \in \mathbf{N}, \ \alpha \in (0,1), \\ x(t_i^+) = \mathbf{h_i}(t_i, x(t_i^-)), \ i \in \mathbf{N}^+, \\ x(t) = \mathbf{h_i}(t, x(t_i^-)), \ t \in (t_i, s_i], \ i \in \mathbf{N}^+, \\ x(0) = x_0, \end{cases} \tag{2}$$

where ${}^C\mathbf{D}^{\alpha}_{s_i,t}$ denotes the classical Caputo fractional derivative of order α, by changing the lower limit s_i, as in Reference [35]. The random impulse and junction points, t_i and s_i, respectively, satisfy $t_0 = s_0 = 0 < t_1 < s_1 < t_2 < \cdots < s_i < t_{i+1} < \cdots, t_i \to \infty$. The symbol $x(t_i^+)$ and $x(t_i^-)$ represent the right and left limits of $x(t)$ at $t = t_i$, respectively. In addition, we set $x(t_i^-) = x(t_i)$. The function $\mathbf{f} : [0, \infty) \times \mathbf{R} \to \mathbf{R}$ is continuous, and the function $\mathbf{h_i} : [t_i, s_i] \times \mathbf{R} \to \mathbf{R}$ is continuous for all $i \in \mathbf{N}^+$. The piecewise continuous solutions of Equations (1) and (2) have been represented in Reference [14] [Equations (5) and (7), therein].

We also introduce the following related original and perturbed equations without impulses:

$$\begin{cases} X'(t) = \mathbf{f}(t, X(t)), \ t \in [s_i, t_{i+1}], \ i \in \mathbf{N}, \\ X(s_i) = x_{s_i}, \end{cases} \tag{3}$$

$$\begin{cases} X'(t) = \mathbf{f}(t, X(t)), \ t \in [s_i, t_{i+1}], \ i \in \mathbf{N}, \\ X(s_i) = \widetilde{x}_{s_i}, \end{cases} \tag{4}$$

$$\begin{cases} {}^C\mathbf{D}^{\alpha}_{s_i,t} X(t) = \mathbf{f}(t, X(t)), \ t \in [s_i, t_{i+1}], \ i \in \mathbf{N}, \ \alpha \in (0,1), \\ X(s_i) = x_{s_i}, \end{cases} \tag{5}$$

and:

$$\begin{cases} {}^C\mathbf{D}^{\alpha}_{s_i,t} X(t) = \mathbf{f}(t, X(t)), \ t \in [s_i, t_{i+1}], \ i \in \mathbf{N}, \ \alpha \in (0,1), \\ X(s_i) = \widetilde{x}_{s_i}. \end{cases} \tag{6}$$

Denote any solution of Equations (1) or (2) by $x(\cdot; 0, x_0) \in PC(\mathbf{J}, \mathbf{R}) := \{x : \mathbf{J} \to \mathbf{R} : x \in \mathbf{C}((t_k, t_{k+1}], \mathbf{R}), \ k = 0, 1, \cdots$ and there exists $x(t_k^+)$ and $x(t_k^-), k = 1, 2, \cdots$ with $x(t_k^-) = x(t_k)\}$, where $\mathbf{J} = [0, \infty)$ and $\mathbf{C}((t_k, t_{k+1}], \mathbf{R})$ denotes the space of all continuous functions from $(t_k, t_{k+1}]$ into $\mathbf{R}$. Additionally, denote any solution of Equations (3) or (5) by $X(\cdot; s_i, x_{s_i}) \in \mathbf{C}([s_i, t_{i+1}], \mathbf{R})$. For the interval $(t_i, s_i]$, we denote its solutions by $X(\cdot; t_i, x(t_i^+))$. Then, the following relationship is valid:

$$x(t; 0, x_0) = \begin{cases} X(t; 0, x_0), \ t \in [0, t_1]; \\ X(t; t_1, x(t_1^+)), \ t \in (t_1, s_1]; \\ X(t; s_1, x_{s_1}), \ t \in (s_1, t_2]; \\ \cdots\cdots \end{cases} \tag{7}$$

The main objective of this article is to present the continuous dependence of solutions with respect to the initial condition when random impulse and junction points are incorporated in Equations (1) and (2). We will take notice of the fact that the location and the number of the impulse points and junction points are not determined in a finite time interval, and so, we can assume that the impulse and junction points are random.

The main contributions of this paper are two folds. We extend the concept and results in Reference [34] to random non-instantaneous impulsive cases by imposing different conditions on the nonlinear term. We also extend the continuous dependence of solutions of first-order non-instantaneous impulsive equations to fractional-order non-instantaneous impulsive equations.

The rest of this paper is organized as follows. Section 2 gives the relevant definitions and notions for the continuous dependence of solutions and contains the main results. Section 3 gives two examples to demonstrate the application of our results. In Section 4, conclusions are drawn.

2. Main Results

Based on Reference [12] (Definition 2.1, therein) and Reference [34] (Definition 2, therein), we give the following definitions for the continuous dependence of solutions.

Definition 1. *The solution $x(\cdot;0,x_0) \in PC(\mathbf{J},\mathbf{R})$ of Equations (1) or (2) depends continuously on the initial point $(0,x_0)$, if for any $\varepsilon > 0$, $\mathbf{T} > 0$, there exists a $\delta = \delta(\varepsilon,\mathbf{T}) > 0$, such that for any $(0,\widetilde{x}_0) \in [0,\mathbf{T}] \times \mathbf{R}$, and $|\widetilde{x}_0 - x_0| < \delta$, then:*

$$|x(t;0,\widetilde{x}_0) - x(t;0,x_0)| < \varepsilon,\ t \in [0,t_1] \bigcup (t_i,s_i] \bigcup (s_i,t_{i+1}],\ i \in \mathbf{N}^+.$$

Definition 2. *The solution $X(\cdot;s_i,x_{s_i}) \in \mathbf{C}([s_i,t_{i+1}],\mathbf{R})$ of Equations (3) and (5) depends uniformly and continuously on the initial point (s_i,x_{s_i}), if for any $\varepsilon > 0$, $\mathbf{T} > 0$, there exists a $\delta = \delta(\varepsilon,\mathbf{T}) > 0$, such that for any $(s_i,\widetilde{x}_{s_i}) \in [0,\mathbf{T}] \times \mathbf{R}$, and $|\widetilde{x}_{s_i} - x_{s_i}| < \delta$, then:*

$$|X(t;s_i,\widetilde{x}_{s_i}) - X(t;s_i,x_{s_i})| < \varepsilon,\ t \in [s_i,t_{i+1}],\ i \in \mathbf{N}^+.$$

We introduce the following assumptions for further discussion:

$[H_1]$ The function $\mathbf{f}: \mathbf{J} \times \mathbf{R} \to \mathbf{R}$ is continuous and $\mathbf{h_i} \in \mathbf{C}([t_i,s_i] \times \mathbf{R},\mathbf{R}), i \in \mathbf{N}^+$.

$[H_2]$ There exists an $\mathbf{L_f} > 0$, such that $|\mathbf{f}(t,x) - \mathbf{f}(t,y)| \leq \mathbf{L_f}|x-y|$, for each $t \in [s_i,t_{i+1}]$, $i \in \mathbf{N}$, and for all $x,y \in \mathbf{R}$.

$[H_3]$ There exists a positive constant $\mathbf{L_{h_i}}$, $i \in \mathbf{N}^+$, such that $|\mathbf{h_i}(t,x) - \mathbf{h_i}(t,y)| \leq \mathbf{L_{h_i}}|x-y|$, for each $t \in [t_i,s_i]$, $i \in \mathbf{N}^+$, and for all $x,y \in \mathbf{R}$.

$[H_4]$ The solutions of Equations (3) or (5) depend uniformly and continuously on the initial point.

$[H_5]$ The functions $\mathbf{h_i}(t,x), i \in \mathbf{N}^+$, are uniformly bounded, i.e., for any $i \in \mathbf{N}^+$, there exists an $\mathbf{M} > 0$, for any $x \in \mathbf{R}$, $t \in \mathbf{J}$, such that $|\mathbf{h_i}(t,x)| \leq \mathbf{M}$.

Theorem 1. *Assume that $[H_1] - [H_3]$ are satisfied. Then the solution of Equation (1) depends continuously on the initial point $(0,x_0)$ at the random impulse and junction points.*

Proof. Let ε and $\mathbf{T}$ be two arbitrary positive constants and $\mathbf{\Omega} = \{t_1,s_1,t_2,s_2,\cdots\}$ be an arbitrary set of impulse and junction points. Note that if $t_i \to \infty$ as $i \to \infty$, then for any selection of the set of impulse points and junction points, there exists $k \in \mathbf{N}$ such that $s_k < \mathbf{T} \leq t_{k+1}$, i.e., there exists at most k impulse points and at most k junction points belonging to the interval $[0,\mathbf{T}]$. Without loss of generality, we assume that $\mathbf{T} = t_{k+1}$.

We divide the proof into several cases.

Case 1. For the interval $(s_k,t_{k+1}]$, the solutions of Equations (3) and (4) are given in Reference [14]:

$$X(t;s_k,x_{s_k}) = x_{s_k} + \int_{s_k}^{t} \mathbf{f}(s,X(s;s_k,x_{s_k}))ds, \tag{8}$$

and:

$$X(t;s_k,\widetilde{x}_{s_k}) = \widetilde{x}_{s_k} + \int_{s_k}^{t} \mathbf{f}(s,\widetilde{X}(s;s_k,x_{s_k}))ds. \tag{9}$$

Assume that there exists a $\delta_{k,k+1} = \delta_{k,k+1}(\varepsilon, \mathbf{T})$, where $0 < \delta_{k,k+1} < \varepsilon$ and $|\widetilde{x}_{s_k} - x_{s_k}| < \delta_{k,k+1}$, linking Equations (8) and (9), we get:

$$\begin{aligned} & |X(t; s_k, \widetilde{x}_{s_k}) - X(t; s_k, x_{s_k})| \\ \leq \ & |\widetilde{x}_{s_k} - x_{s_k}| + \mathbf{L_f} \int_{s_k}^{t} |X(s; s_k, \widetilde{x}_{s_k}) - X(s; s_k, x_{s_k})| ds \\ \leq \ & \delta_{k,k+1} + \mathbf{L_f} \int_{s_k}^{t} |X(s; s_k, \widetilde{x}_{s_k}) - X(s; s_k, x_{s_k})| ds. \end{aligned} \tag{10}$$

From Reference [36] (Theorem 1.1, therein), we get:

$$|X(t; s_k, \widetilde{x}_{s_k}) - X(t; s_k, x_{s_k})| \leq \delta_{k,k+1} e^{\mathbf{L_f}(t_{k+1} - s_k)}.$$

Since $0 < \delta_{k,k+1} < \varepsilon$, we have:

$$|X(t; s_k, \widetilde{x}_{s_k}) - X(t; s_k, x_{s_k})| < \varepsilon, \ t \in (s_k, t_{k+1}].$$

Case 2. For the interval $(t_k, s_k]$, we have the following expression of the solutions, respectively:

$$X(t; t_k, x(t_k^+)) = \mathbf{h_k}(t, x(t_k^-)) \quad \text{and} \quad X(t; t_k, \widetilde{x}(t_k)) = \mathbf{h_k}(t, \widetilde{x}(t_k^-)).$$

Assume there exists $\delta_{k,k} < \delta_{k,k+1}$, where $|\widetilde{x}(t_k^+) - x(t_k^+)| < \delta_{kk}$, then we have:

$$|X(t; t_k, \widetilde{x}(t_k^+)) - X(t; t_k, x(t_k^+))| \leq \mathbf{L_{h_k}} |\widetilde{x}(t_k^-) - x(t_k^-)|. \tag{11}$$

For $(s_{k-1}, t_k]$, similar to Equation (10), we assume there exists a $\delta_{k-1,k} < \frac{\delta_{k,k}}{\mathbf{L_{h_k}} e^{\mathbf{L_f}(t_k - s_{k-1})}}$. Then we obtain:

$$|X(t; s_{k-1}, \widetilde{x}_{s_{k-1}}) - X(t; s_{k-1}, x_{s_{k-1}})| \leq \delta_{k-1,k-1} e^{\mathbf{L_f}(t_k - s_{k-1})} < \frac{\delta_{k,k}}{\mathbf{L_{h_k}}} < \varepsilon, \ t \in (s_{k-1}, t_k]. \tag{12}$$

For $t = t_k$, we get:

$$|X(t_k; s_{k-1}, \widetilde{x}_{s_{k-2}}) - X(t_k; s_{k-1}, x_{s_{k-1}})| < \frac{\delta_{k,k}}{\mathbf{L_{h_k}}},$$

and:

$$|\mathbf{h_k}(t_k, \widetilde{x}(t_k^-)) - \mathbf{h_k}(t_k, x(t_k^-))| < \mathbf{L_{h_k}} \frac{\delta_{k,k}}{\mathbf{L_{h_k}}} < \delta_{k,k}.$$

Therefore, Equation (11) becomes:

$$|X(t; t_k, \widetilde{x}(t_k^+)) - X(t; t_k, x(t_k^+))| < \delta_{k,k} < \delta_{k,k+1} < \varepsilon, \ t \in (t_k, s_k]. \tag{13}$$

From the procedure of Equations (12) and (13), we finally establish the following facts:

For the interval $(t_1, s_1]$, if there is a $\delta_{11} < \delta_{12}$, where $|\widetilde{x}(t_1^+) - x(t_1^+)| < \delta_{11}$, then:

$$|X(t; t_1, \widetilde{x}(t_1^+)) - X(t; t_1, x(t_1^+))| \leq \mathbf{L_{h_1}} |\widetilde{x}(t_1^-) - x(t_1^-)|. \tag{14}$$

For the interval $[0, t_1]$, we assume there is a $\delta_{01} < \frac{\delta_{11}}{\mathbf{L_{h_1}} e^{\mathbf{L_f} t_1}}$, where $|\widetilde{x}_0 - x_0| < \delta_{01}$. Thus, we obtain:

$$|X(t; 0, \widetilde{x}_0) - X(t; 0, x_0)| \leq \delta_{01} e^{\mathbf{L_f} t_1} < \frac{\delta_{11}}{\mathbf{L_{h_1}}} < \varepsilon, \quad t \in [0, t_1]. \tag{15}$$

For $t = t_1$, we get:

$$|X(t_1; 0, \widetilde{x}_0) - X(t_1; 0, x_0)| < \frac{\delta_{11}}{\mathbf{L_{h_1}}},$$

and hence, we have:

$$|h_1(t_1, x(t_1^-)) - h_1(t_1, x(t_1^-))| < \mathbf{L_{h_1}} \frac{\delta_{11}}{\mathbf{L_{h_1}}} < \delta_{11}. \tag{16}$$

Thus, Equation (14) becomes:

$$|X(t; t_1, \widetilde{x}(t_1^+)) - X(t; t_1, x(t_1^+))| < \delta_{11} < \delta_{12} < \varepsilon, \quad t \in (t_1, s_1]. \tag{17}$$

Next, we notice that:

$$\begin{aligned} \delta_{01} &= \delta_{01}(\varepsilon, \mathbf{T}, \delta_{11}), \\ &\vdots \\ \delta_{k-1,k} &= \delta_{k-1,k}(\varepsilon, \mathbf{T}, \delta_{kk}), \\ \delta_{kk} &= \delta_{kk}(\varepsilon, \mathbf{T}, \delta_{k,k+1}), \\ \delta_{k,k+1} &= \delta_{k,k+1}(\varepsilon, \mathbf{T}). \end{aligned}$$

Therefore, we have $\delta_{01} = \delta_{01}(\varepsilon, \mathbf{T})$. Considering Equations (7) and (15), we obtain, for any $\varepsilon > 0$, there exists a $\delta_{01} > 0$, where $|\widetilde{x}_0 - x_0| < \delta_{01}$, thus:

$$|x(t; 0, \widetilde{x}_0) - x(t; 0, x_0)| < \varepsilon, \ t \in [0, t_1], \tag{18}$$

and from Equation (16) we get:

$$|\mathbf{h_1}(t_1, \widetilde{x}(t_1^-)) - \mathbf{h_1}(t_1, x(t_1^-))| = |\widetilde{x}(t_1^+) - x(t_1^+)| < \delta_{11}.$$

Taking into account Equations (17) and (7), we find that:

$$|x(t; 0, \widetilde{x}_0) - x(t; 0, x_0)| < \varepsilon, \ t \in (t_1, s_1], \tag{19}$$

and:

$$|\mathbf{h_1}(s_1, \widetilde{x}(t_1^-)) - \mathbf{h_1}(s_1, x(t_1^-))| = |\widetilde{x}_{s_1} - x_{s_1}| < \delta_{12}.$$

According to Equation (7) again, we obtain:

$$|x(t; 0, \widetilde{x}_0) - x(t; 0, x_0)| < \varepsilon, \ t \in (s_1, t_2].$$

Similarly, we achieve the following conclusion:

$$|x(t; 0, \widetilde{x}_0) - x(t; 0, x_0)| < \varepsilon, \ t \in (t_k, s_k], \tag{20}$$

and:

$$|\mathbf{h_k}(s_k, \widetilde{x}(t_k^-)) - \mathbf{h_k}(s_k, x(t_k^-))| = |\widetilde{x}_{s_k} - x_{s_k}| < \delta_{k,k+1}.$$

Then, it is further determined:

$$|x(t; 0, \widetilde{x}_0) - x(t; 0, x_0)| < \varepsilon, \ t \in (s_k, t_{k+1}]. \tag{21}$$

By Equations (18)–(21), we get that for any $\varepsilon > 0$ and $\widetilde{x}_0, x_0 \in \mathbf{R}$, there exists a $\delta > 0$, where $|\widetilde{x}_0 - x_0| < \delta$, such that $|x(t; 0, \widetilde{x}_0) - x(t; 0, x_0)| < \varepsilon$ for $t \in [0, t_1] \bigcup (t_i, s_i] \bigcup (s_i, t_{i+1}], i = 1, 2, \cdots, k$, where $\delta = \delta_{01}(\varepsilon, \mathbf{T})$. By Definition 1, the proof is completed. □

Theorem 2. *Suppose that $[H_1]-[H_3]$ are satisfied. Then the solution of Equation (2) depends continuously on the initial point $(0, x_0)$ for random impulse and junction points.*

Proof. Let ε and $\mathbf{T}$ be two arbitrary positive constants like in Theorem 1. We divide the proof into two cases.

Case 1. For the interval $(s_k, t_{k+1}]$, the representation of the solutions of Equations (5) and (6) are given by [14]:

$$X(t; s_k, x_{s_k}) = x_{s_k} + \frac{1}{\Gamma(\alpha)}\int_{s_k}^{t}(t-s)^{\alpha-1}\mathbf{f}(s, X(s; s_k, x_{s_k}))ds, \tag{22}$$

and:

$$X(t; s_k, \tilde{x}_{s_k}) = \tilde{x}_{s_k} + \frac{1}{\Gamma(\alpha)}\int_{s_k}^{t}(t-s)^{\alpha-1}\mathbf{f}(s, X(s; s_k, \tilde{x}_{s_k}))ds. \tag{23}$$

Assume that there exists a $\delta_{k,k+1} = \delta_{k,k+1}(\varepsilon, \mathbf{T}),\ 0 < \delta_{k,k+1} < \varepsilon$, where $|\tilde{x}_{s_k} - x_{s_k}| < \delta_{k,k+1}$, then from Equations (22) and (23), we have:

$$\begin{aligned}
& |X(t; s_k, \tilde{x}_{s_k}) - X(t; s_k, x_{s_k})| \\
\leq\ & |\tilde{x}_{s_k} - x_{s_k}| + \frac{\mathbf{L_f}}{\Gamma(\alpha)}\int_{s_k}^{t}(t-s)^{\alpha-1}|X(s; s_k, \tilde{x}_{s_k}) - X(s; s_k, x_{s_k})|ds \\
\leq\ & \delta_{k,k+1} + \frac{\mathbf{L_f}}{\Gamma(\alpha)}\int_{s_k}^{t}(t-s)^{\alpha-1}|X(s; s_k, \tilde{x}_{s_k}) - X(s; s_k, x_{s_k})|ds.
\end{aligned}$$

Using Reference [37] (Corollary 2, therein), we get:

$$|X(t; s_k, \tilde{x}_{s_k}) - X(t; s_k, x_{s_k})| \leq \delta_{k,k+1}\mathbf{E}_\alpha(\mathbf{L_f}(t_{k+1} - s_k)^\alpha),$$

where $\mathbf{E}_\alpha$ is the standard Mittag–Leffler function [35] defined as:

$$\mathbf{E}_\alpha(z) = \sum_{k=0}^{\infty}\frac{z^k}{\Gamma(k\alpha+1)}, \quad z \in \mathbb{C}.$$

Owing to $0 < \delta_{k,k+1} < \varepsilon$, one can get:

$$|X(t; s_k, \tilde{x}_{s_k}) - X(t; s_k, x_{s_k})| < \varepsilon,\ t \in (s_k, t_{k+1}].$$

Case 2. From the interval $(t_k, s_k]$, the expression of the solutions are given by:

$$X(t; t_k, x(t_k^+)) = \mathbf{h_k}(t, x(t_k^-)) \quad \text{and} \quad X(t; t_k, \tilde{x}(t_k)) = \mathbf{h_k}(t, \tilde{x}(t_k^-)).$$

Assume that there exists a $\delta_{kk} < \delta_{k,k+1}$, where $|\tilde{x}(t_k^+) - x(t_k^+)| < \delta_{kk}$, then we have:

$$|X(t; t_k, \tilde{x}(t_k^+)) - X(t; t_k, x(t_k^+))| \leq \mathbf{L_{h_k}}|\tilde{x}(t_k^-) - x(t_k^-)|. \tag{24}$$

Similar to the above procedure, for the interval $(s_{k-1}, t_k]$, one has:

$$|X(t; s_{k-1}, \tilde{x}_{s_{k-1}}) - X(t; s_{k-1}, x_{s_{k-1}})| \leq \delta_{k-1,k}\mathbf{E}_\alpha(L_f(t_k - s_{k-1})^\alpha).$$

Since $\delta_{k-1,k} < \frac{\delta_{kk}}{\mathbf{L_{h_k}}\mathbf{E}_\alpha(\mathbf{L_f}(t_k - s_{k-1})^\alpha)}$, one can obtain:

$$|X(t; s_{k-1}, \tilde{x}_{s_{k-1}}) - X(t; s_{k-1}, x_{s_{k-1}})| < \frac{\delta_{kk}}{\mathbf{L_{h_k}}} < \varepsilon,\ t \in (s_{k-1}, t_k].$$

For $t = t_k$, we get:

$$|X(t_k; s_{k-1}, \tilde{x}_{s_{k-1}}) - X(t_k; s_{k-1}, x_{s_{k-1}})| < \frac{\delta_{kk}}{\mathbf{L_{h_k}}},$$

and then:

$$|g_k(t_k, x(t_k^-)) - g_k(t_k, x(t_k^-))| < \mathbf{L_{h_k}} \frac{\delta_{kk}}{\mathbf{L_{h_k}}} < \delta_{kk}.$$

Therefore, (24) becomes:

$$|X(t; t_k, \tilde{x}(t_k^+)) - X(t; t_k, x(t_k^+))| < \delta_{kk} < \varepsilon.$$

From the above, one can deduce that for the interval $(t_1, s_1]$, there exists a $\delta_{11} < \delta_{12}$, and if $|\tilde{x}(t_1^-) - x(t_1^-)| < \delta_{11}$, then:

$$|X(t; t_1, x(t_1^+)) - X(t; t_1, x(t_1^+))| \leq \mathbf{L_{h_1}} |\tilde{x}(t_1^-) - x(t_1^-)|.$$

For $[0, t_1]$, there is $\delta_{01} < \frac{\delta_{11}}{\mathbf{L_{h_1}} \mathbf{E}_\alpha(\mathbf{L_f} t_1^\alpha)}$, if $|\tilde{x}_0 - x_0| < \delta_{01}$, then:

$$|X(t; 0, \tilde{x}_0) - X(t; 0, x_0)| \leq \delta_{01} \mathbf{E}_\alpha(\mathbf{L_f} t_1^\alpha) < \frac{\delta_{11}}{\mathbf{L_{h_1}}} < \varepsilon,\ t \in [0, t_1].$$

Thus, for $t = t_1$, we have:

$$|X(t_1; 0, \tilde{x}_0) - X(t_1; 0, x_0)| < \frac{\delta_{11}}{\mathbf{L_{h_1}}},$$

and then:

$$|\mathbf{h_1}(t_1, \tilde{x}(t_1)) - \mathbf{h_1}(t_1, x(t_1))| < \delta_{11}.$$

and:

$$|X(t; t_1, x(t_1^+)) - X(t; t_1, x(t_1^+))| < \delta_{11} < \delta_{12} < \varepsilon,\ t \in (t_1, s_1].$$

Similar to Theorem 1, we reach the conclusion. □

Theorem 3. *Assume that $[H_1]$, $[H_4]$, and $[H_5]$ are satisfied. The solution of Equation (1) depends continuously on the initial point $(0, x_0)$ at the random impulse and junction points provided that $2\mathbf{M} \leq |\tilde{x}_{s_i} - x_{s_i}|$.*

Proof. We divide proof into several steps.

Step 1. According to $[H_4]$, assume there is a $\delta_{k,k+1} = \delta_{k,k+1}(\varepsilon, \mathbf{T})$, where $0 < \delta_{k,k+1} < \varepsilon$, if $|\tilde{x}_{s_k} - x_{s_k}| < \delta_{k,k+1}$, then:

$$|X(t; s_k, \tilde{x}_{s_k}) - X(t; s_k, x_{s_k})| < \varepsilon,\ t \in (s_k, t_{k+1}].$$

Step 2. For the interval $(t_k, s_k]$, according to $[H_5]$, we obtain:

$$\begin{aligned} |X(t; t_k, \tilde{x}(t_k^+)) - X(t; t_k, x(t_k^+))| &= |\mathbf{h_k}(t, \tilde{x}(t_k^-)) - \mathbf{h_k}(t, x(t_k^-))| \\ &\leq 2\mathbf{M} \leq |\tilde{x}_{s_k} - x_{s_k}| < \delta_{k,k+1} < \varepsilon,\ t \in (t_k, s_k]. \end{aligned}$$

For $t = s_k$, we get:

$$|\mathbf{h_k}(s_k, \tilde{x}(t_k^-)) - \mathbf{h_k}(s_k, x(t_k^-))| < \delta_{k,k+1}. \tag{25}$$

Step 3. In this step, we check the continuity of the solutions in the interval $(s_{k-1}, t_k]$. From Equation (25), we assume there exists a $\delta_{k-1,k}$ depending on $\varepsilon, \mathbf{T}, \delta_{k,k+1}$. For brevity, we denote

$\delta_{k-1,k} = \delta_{k-1,k}(\varepsilon, \mathbf{T}, \delta_{k,k+1})$, where $0 < \delta_{k-1,k} < \varepsilon$. If $|\widetilde{x}_{s_{k-1}} - x_{s_{k-1}}| < \delta_{k-1,k}$, then by $[H_4]$ we deduce that:

$$|X(t; s_{k-1}, \widetilde{x}_{s_{k-1}}) - X(t; s_{k-1}, x_{s_{k-1}})| < \varepsilon,\ t \in (s_{k-1}, t_k].$$

Step 4. For $(t_{k-1}, s_{k-1}]$, we get:

$$\begin{aligned} |X(t; t_{k-1}, \widetilde{x}(t_{k-1}^{+})) - X(t; t_{k-1}, x(t_{k-1}^{+}))| &= |\mathbf{h_{k-1}}(t, \widetilde{x}(t_{k-1}^{-})) - \mathbf{h_{k-1}}(t, x(t_{k-1}^{-}))| \\ &\leq \mathbf{2M} \leq |\widetilde{x}_{s_{k-1}} - x_{s_{k-1}}| < \delta_{k-1,k} < \varepsilon,\ t \in (t_{k-1}, s_{k-1}]. \end{aligned}$$

For $t = s_{k-1}$, we have:

$$|\mathbf{h_{k-1}}(s_{k-1}, \widetilde{x}(t_{k-1}^{-})) - \mathbf{h_{k-1}}(s_{k-1}, x(t_{k-1}^{-}))| < \delta_{k-1,k}.$$

Similar to the above steps, we have the following general results.

Step $2k$. For the interval $(t_1, s_1]$, it follows that:

$$|X(t; t_1, \widetilde{x}(t_1^{+})) - X(t; t_1, x(t_1^{+}))| = |\mathbf{h_1}(t, \widetilde{x}(t_1^{-})) - \mathbf{h_1}(t, x(t_1^{-}))| < \delta_{12} < \varepsilon,\ t \in (t_1, s_1].$$

For $t = s_1$, we have:

$$|\mathbf{h_1}(s_1, \widetilde{x}(t_1^{-})) - \mathbf{h_1}(s_1, x(t_1^{-}))| < \delta_{12}. \tag{26}$$

Step $2k + 1$. For $[0, t_1]$, assume there exists a $\delta_{01} = \delta_{01}(\varepsilon, \mathbf{T}, \delta_{12})$, where $0 < \delta_{01} < \varepsilon$, if $|\widetilde{x}_0 - x_0| < \varepsilon$, then:

$$|X(t; 0, \widetilde{x}_0) - X(t; 0, x_0)| < \varepsilon,\ t \in [0, t_1].$$

Finally, we notice that:

$$\begin{aligned} \delta_{01} &= \delta_{01}(\varepsilon, \mathbf{T}, \delta_{12}), \\ &\vdots \\ \delta_{k-1,k} &= \delta_{k-1,k}(\varepsilon, \mathbf{T}, \delta_{k,k+1}), \\ \delta_{k,k+1} &= \delta_{k,k+1}(\varepsilon, \mathbf{T}). \end{aligned}$$

Therefore, we have $\delta_{01} = \delta_{01}(\varepsilon, \mathbf{T})$.

Now we apply the above results to the impulse case.

Step $1'$. Based on Equation (7) and Step 2k + 1, we obtain for any $\varepsilon > 0$, there exists a $\delta_{01} > 0$, such that if $|\widetilde{x}_0 - x_0| < \delta_{01}$, then:

$$|x(t; 0, \widetilde{x}_0) - x(t; 0, x_0)| < \varepsilon,\ t \in [0, t_1].$$

Step $2'$. Using Equation (7) and taking Step 2k into account, we find that:

$$|x(t; 0, \widetilde{x}_0) - x(t; 0, x_0)| < \varepsilon,\ t \in (t_1, s_1].$$

From Equation (26), we have:

$$|\mathbf{h_1}(s_1, \widetilde{x}(t_1^{-})) - \mathbf{h_1}(s_1, x(t_1^{-}))| = |\widetilde{x}_{s_1} - x_{s_1}| < \delta_{12}.$$

Step $3'$. According to Equation (7) and Step 2k − 1, it follows that:

$$|x(t; 0, \widetilde{x}_0) - x(t; 0, x_0)| < \varepsilon,\ t \in (s_1, t_2].$$

Step 4′. Again by Equation (7) and considering Step 2k − 2, we have:

$$|x(t;0,\tilde{x}_0) - x(t;0,x_0)| < \varepsilon,\ t \in (t_2, s_2],$$

and:

$$|\mathbf{h_2}(s_2, \tilde{x}(t_2^-)) - \mathbf{h_2}(s_2, x(t_2^-))| = |\tilde{x}_{s_2} - x_{s_2}| < \delta_{23}.$$

Similarly, we arrive at:

Step $(2k)'$. $|x(t;0,\tilde{x}_0) - x(t;0,x_0)| < \varepsilon,\ t \in (t_k, s_k]$.

Thus, from Equation (25), we get:

$$|\mathbf{h_k}(s_k, \tilde{x}(t_k^-)) - \mathbf{h_k}(s_k, x(t_k^-))| = |\tilde{x}_{s_k} - x_{s_k}| < \delta_{k,k+1}.$$

Step $(2k+1)'$. $|x(t;0,\tilde{x}_0) - x(t;0,x_0)| < \varepsilon,\ t \in (s_k, t_{k+1}]$.

From Step 1′ to Step $(2k+1)'$, we get that for any $\varepsilon > 0$, $\tilde{x}_0, x_0 \in \mathbf{R}$, there exists a $\delta > 0$, such that if $|\tilde{x}_0 - x_0| < \delta$, then $|x(t;0,\tilde{x}_0) - x(t;0,x_0)| < \varepsilon$, for $t \in [0,t_1] \bigcup (t_i, s_i] \bigcup (s_i, t_{i+1}], i = 1, 2, \cdots, k$, and $\delta = \delta_{01}(\varepsilon, \mathbf{T})$. □

By repeating the same proof procedure in Theorem 3, we have the result:

Remark 1. *Assume that $[H_1]$, $[H_4]$, and $[H_5]$ are satisfied. Then the solution of Equation (2) depends continuously on the initial point $(0, x_0)$ at the random impulse and junction points, provided that $2\mathbf{M} \leq |\tilde{x}_{s_i} - x_{s_i}|$.*

3. Numerical Examples

Let ε and $\mathbf{T}$ be two arbitrary positive constants.

Example 1. *Consider:*

$$\begin{cases} x'(t) = ax(t),\ t \in (s_i, t_{i+1}],\ i \in \mathbf{N},\ a > 0, \\ x(t_i^+) = \frac{\varepsilon}{2e^{at}} \cos\sqrt{t_i|x(t_i^-)|},\ i \in \mathbf{N}^+, \\ x(t) = \frac{\varepsilon}{2e^{at}} \cos\sqrt{t|x(t_i^-)|},\ t \in (t_i, s_i],\ i \in \mathbf{N}^+, \\ x(0) = x_0. \end{cases} \tag{27}$$

Then the solution of Equation (27) can be analytically determined, namely:

$$x(t) = \begin{cases} e^{at}x_0,\ \textit{for } t \in (s_0, t_1], \\ \frac{\varepsilon}{2e^{at}} \cos\sqrt{t|x(t_1^-)|},\ \textit{for } t \in (t_1, s_1], \\ \frac{\varepsilon}{2e^{as_1}} \cos\sqrt{s_1|x(t_1^-)|}e^{a(t-s_1)},\ \textit{for } t \in (s_1, t_2], \\ \vdots \\ \frac{\varepsilon}{2e^{at}} \cos\sqrt{t|x(t_{m-1}^-)|},\ \textit{for } t \in (t_{m-1}, s_m], \\ \frac{\varepsilon}{2e^{as_m}} (\cos\sqrt{s_m|x(t_{m-1}^-)}|e^{a(t-s_m)},\ \textit{for } t \in (s_m, t_{m+1}], \\ \vdots \end{cases} \tag{28}$$

Let $\mathbf{f}(t,x) = ax$ and $\mathbf{h_i}(t,x) = \cos\sqrt{t|x|}$. Note that $\mathbf{h_i} \in \mathbf{C}([t_i, s_i] \times \mathbf{R}, \mathbf{R}), i \in \mathbf{N}^+$. For any $x, y \in \mathbf{R}$, $|\mathbf{f}(t,x) - \mathbf{f}(t,y)| \leq a|x-y|$ and $|\mathbf{h_i}(t,x) - \mathbf{h_i}(t,y)| \leq \frac{\varepsilon}{2ea}|x-y|$. Set $\mathbf{L_f} = a$ and $\mathbf{L_{h_i}} = \frac{\varepsilon}{2ea}$. So, $[H_1]$–$[H_3]$ all hold. Therefore, all the assumptions in Theorem 1 are satisfied.

From Definition 2, choosing $\delta = \frac{\varepsilon}{e^{a\mathbf{T}}}$, we find that the solution of Equation (27) without impulses satisfies uniform and continuous dependence on the initial point. Choosing $2\mathbf{M} \leq \frac{\varepsilon}{e^{a\mathbf{T}}}$, then $\mathbf{h}_i(t,x)$ are uniformly bounded. Thus, the conditions $[H_1]$, $[H_4]$, and $[H_5]$ hold, and all the assumptions of Theorem 3 are satisfied.

The solutions of Equation (27) and the corresponding perturbation problem (with $x_0 = 1, \tilde{x_0} = 1.3, \mathbf{T} = 1, a = \frac{1}{2}, \varepsilon = \frac{1}{2}$*) are shown in Figure 1.*

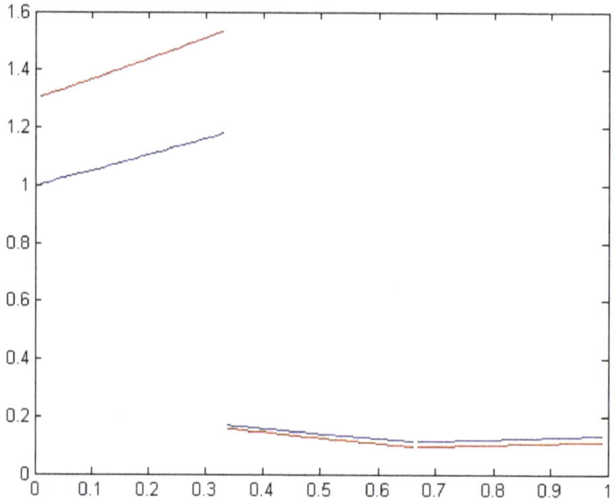

Figure 1. The blue line denotes the solution of Equation (27) and the red line denotes the corresponding perturbation problem.

Example 2. *Consider:*

$$\begin{cases} {}^{C}\mathbf{D}^{\frac{1}{2}}_{s_i,t}x(t) = ax(t),\ t \in (s_i, t_{i+1}],\ i \in \mathbf{N},\ a > 0, \\ x(t_i^+) = \frac{\varepsilon}{2\mathbf{E}_{\frac{1}{2}}(at^{\frac{1}{2}})}(\cos\sqrt{t_i|x(t_i^-)|}),\ i \in \mathbf{N}^+, \\ x(t) = \frac{\varepsilon}{2\mathbf{E}_{\frac{1}{2}}(at^{\frac{1}{2}})}(\cos\sqrt{t|x(t_i)^-|}),\ t \in (t_i, s_i],\ i \in \mathbf{N}^+, \\ x(0) = x_0. \end{cases} \tag{29}$$

Like Equation (27), let $\mathbf{f}(t,x) = ax$*, and* $\mathbf{h_i}(t,x) = \frac{\varepsilon}{2\mathbf{E}_{\frac{1}{2}}(at^{\frac{1}{2}})}(\cos\sqrt{t|x|})$*. Choosing* $\mathbf{L_f} = a$*,* $\mathbf{L_{h_i}} = \frac{\varepsilon}{2a^2}$*, therefore,* $[H_1] - [H_3]$ *hold. Then, one can obtain the solution of Equation (29), namely:*

$$x(t) = \begin{cases} x_0\mathbf{E}_{\frac{1}{2}}(at^{\frac{1}{2}}),\ \mathit{for}\ t \in (s_0, t_1], \\ \frac{\varepsilon}{2\mathbf{E}_{\frac{1}{2}}(at^{\frac{1}{2}})}\cos\sqrt{t|x(t_1^-)|},\ \mathit{for}\ t \in (t_1, s_1], \\ \frac{\varepsilon}{2\mathbf{E}_{\frac{1}{2}}(as_1^{\frac{1}{2}})}\cos\sqrt{s_1|x(t_1^-)|}E_{\frac{1}{2}}(a(t-s_1)^{\frac{1}{2}}),\ \mathit{for}\ t \in (s_1, t_2], \\ \vdots \\ \frac{\varepsilon}{2\mathbf{E}_{\frac{1}{2}}(at^{\frac{1}{2}})}\cos\sqrt{t|x(t_{m-1}^-)|},\ \mathit{for}\ t \in (t_{m-1}, s_m], \\ \frac{\varepsilon}{2\mathbf{E}_{\frac{1}{2}}(as_m^{\frac{1}{2}})}\cos\sqrt{s_m|x(t_{m-1}^-)|}E_{\frac{1}{2}}(a(t-s_m)^{\frac{1}{2}}),\ \mathit{for}\ t \in (s_m, t_{m+1}], \\ \vdots \end{cases} \tag{30}$$

From Definition 2, choosing $\delta = \frac{\varepsilon}{\mathbf{E}_{\frac{1}{2}}(a\mathbf{T}^{\frac{1}{2}})}$*, we get that the solution of Equation (29) without impulses satisfies uniform and continuous dependence on the initial point. Choosing* $2\mathbf{M} \leq \frac{\varepsilon}{\mathbf{E}_{\frac{1}{2}}(a\mathbf{T}^{\frac{1}{2}})}$*, then the* $\mathbf{h}_i(t,x)$ *are uniformly bounded. Thus, the conditions* $[H_1], [H_4]$*, and* $[H_5]$ *hold, and all the assumptions in Remark 1 are satisfied.*

The solutions of Equation (29) and the corresponding perturbation problem (with $x_0 = 1, \widetilde{x_0} = 1.3, \mathbf{T} = \frac{3}{2}, a = \frac{1}{5}, \varepsilon = \frac{1}{2}$*) are shown in Figure 2.*

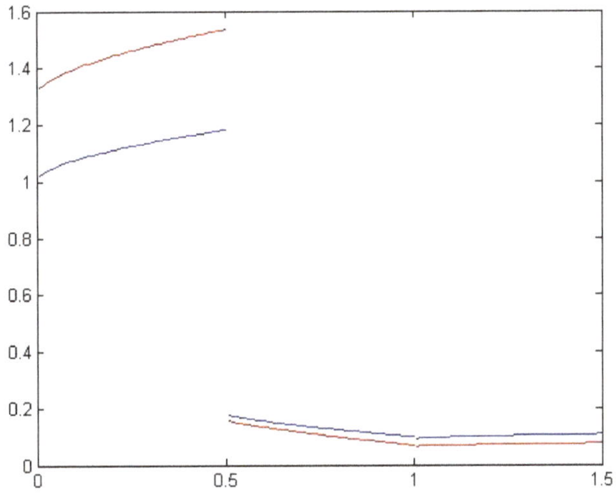

Figure 2. The blue line denotes the solution of Equation (29) and the red line denotes the corresponding perturbation problem.

Remark 2. *Equations (27) and (29) are called non-instantaneous impulsive logistic models, which are motivated from the instantaneous impulsive logistic equations. For more details of the models, one can refer to Reference [12] (Section 4, therein).*

4. Conclusions

In this paper, we presented the continuous dependence of the solutions to first order non-instantaneous IDEs with random impulse and junction points. Then, we extended the results to study the same problem for fractional order cases. The backward checking approach [34] (from the last subinterval to the first subinterval) is extended to differential and algebra equations and is used to prove the main results. The approach is different from Reference [13].

Author Contributions: The contributions of both authors (Y.C. and J.W.) are equal. All the main results and numerical examples were developed together.

Funding: This work was partially supported by the National Natural Science Foundation of China (Grant no. 11661016), Training Object of High Level and Innovative Talents of Guizhou Province ((2016)4006), Science and Technology Program of Guizhou Province ([2017]5788-10), and Major Research Project of Innovative Group in Guizhou Education Department ([2018]012).

Acknowledgments: The authors are grateful to the referees for their careful reading of the manuscript and their valuable comments. We also thank the editor.

Conflicts of Interest: The authors declare no conflict of interest.

References

1. Chalishajar, D.N.; Malar, K.; Karthikeyan, K. Approximate controllability of abstract impulsive fractional neutral evolution equations with infinite delay in Banach spaces. *Electron. J. Differ. Equ.* **2013**, *2013*, 1–21.
2. Chalishajar, D.N.; Karthikeyan, K.; Anguraj, A. Existence results for impulsive perturbed partial neutral functional differential equations in Frechet spaces. *Dyn. Contin. Discret. Impuls. Syst. Ser. Math. Anal.* **2015**, *22*, 25–45.
3. Hernández, E.; O'Regan, D. On a new class of abstract impulsive differential equations. *Proc. Am. Math. Soc.* **2013**, *141*, 1641–1649. [CrossRef]
4. Wang, J.; Fečkan, M. A general class of impulsive evolution equations. *Topol. Meth. Nonlinear Anal.* **2015**, *46*, 915–934. [CrossRef]
5. Wang, J. Stability of noninstantaneous impulsive evolution equations. *Appl. Math. Lett.* **2017**, *73*, 157–162. [CrossRef]
6. Chen, P.; Li, Y.; Yang, H. Perturbation method for nonlocal impulsive evolution equations. *Nonlinear Anal. Hybrid Syst.* **2013**, *8*, 22–30. [CrossRef]
7. Bai, L.; Nieto, J.J. Variational approach to differential equations with not instantaneous impulses. *Appl. Math. Lett.* **2017**, *73*, 44–48. [CrossRef]
8. Pierri, M.; O'Regan, D.; Rolnik, V. Existence of solutions for semi-linear abstract differential equations with not instantaneous impulses. *Appl. Math. Comput.* **2013**, *219*, 6743–6749. [CrossRef]
9. Pierri, M.; Henríquez, H.R.; Prokczyk, A. Global solutions for abstract differential equations with non-instantaneous impulses. *Mediterr. J. Math.* **2016**, *34*, 1685–1708. [CrossRef]
10. Hernández, E.; Pierri, M.; O'Regan, D. On abstract differential equations with non instantaneous impulses. *Topol. Methods Nonlinear Anal.* **2015**, *46*, 1067–1085.
11. Yang, D.; Wang, J. Non-instantaneous impulsive fractional-order implicit differential equations with random effects. *Stoch. Anal. Appl.* **2017**, *35*, 719–741. [CrossRef]
12. Yang, D.; Wang, J.; O'Regan, D. A class of nonlinear non-instantaneous impulsive differential equations involving parameters and fractional order. *Appl. Math. Comput.* **2018**, *321*, 654–671. [CrossRef]
13. Yang, D.; Wang, J.; O'Regan, D. Asymptotic properties of the solutions of nonlinear non-instantaneous impulsive differential equations. *J. Frankl. Inst.* **2017**, *354*, 6978–7011. [CrossRef]
14. Yang, D.; Wang, J.; O'Regan, D. On the orbital Hausdorff dependence of differential equations with non-instantaneous impulses. *C. R. Acad. Sci. Paris Ser. I* **2018**, *356*, 150–171. [CrossRef]
15. Abbas, S.; Benchohra, M. Uniqueness and Ulam stabilities results for partial fractional differential equations with not instantaneous impulses. *Appl. Math. Comput.* **2015**, *257*, 190–198. [CrossRef]
16. Muslim, M.; Kumar, A.; Fečkan, M. Existence, uniqueness and stability of solutions to second order nonlinear differential equations with non-instantaneous impulses. *J. King Saud Univ.* **2018**, *30*, 204–213. [CrossRef]
17. Colao, V.; Muglia, L.; Xu, H.K. An existence result for a new class of impulsive functional differential equations with delay. *J. Math. Anal. Appl.* **2016**, *441*, 668–683. [CrossRef]
18. Wang, J.; Fečkan, M. *Non-Instantaneous Impulsive Differential Equations Basic Theory and Computation*; IOP Publishing: Bristol, UK, 2018.
19. Wang, J.; Zhou, Y.; Lin, Z. On a new class of impulsive fractional differential equations. *Appl. Math. Comput.* **2014**, *242*, 649–657. [CrossRef]
20. Wang, J.; Fečkan, M.; Tian, Y. Stability analysis for a general class of non-instantaneous impulsive differential equations. *Mediterr. J. Math.* **2017**, *14*, 1–21. [CrossRef]
21. Wang, J.; Ibrahim, A.G.; O'Regan, D. Topological structure of the solution set for fractional non-instantaneous impulsive evolution inclusions. *J. Fixed Point Theory Appl.* **2018**, *20*, 1–25. [CrossRef]
22. Wang, J.; Ibrahim, A.G.; O'Regan, D.; Zhou, Y. Controllability for noninstantaneous impulsive semilinear functional differential inclusions without compactness. *Indag. Math.* **2018**, *29*, 1362–1392. [CrossRef]
23. Wang, J.; Ibrahim, A.G.; O'Regan, D. Hilfer type fractional differential switched inclusions with noninstantaneous impulsive and nonlocal conditions. *Nonlinear Anal. Model. Contr.* **2018**, *23*, 921–941. [CrossRef]
24. Wang, J.; Ibrahim, A.G.; O'Regan, D. Nonempty and compactness of solution set for fractional semilinear evolution inclusions with non-instantaneous impulses. *Electr. J. Differ. Equ.* **2019**, *2019*, 1–17.

25. Liu, S.; Wang, J.; Shen, D.; O'Regan, D. Iterative learning control for parabolic partial differential inclusions with noninstantaneous impulses. *Appl. Math. Comput.* **2019**, *350*, 48–59. [CrossRef]
26. Luo, D.; Wang, J.; Shen, D. Learning formation control for fractional-order multi-agent systems. *Math. Meth. Appl. Sci.* **2018**, *41*, 5003–5014. [CrossRef]
27. Zhang, J.; Wang, J. Numerical analysis for a class of Navier–Stokes equations with time fractional derivatives. *Appl. Math. Comput.* **2018**, *336*, 481–489.
28. Zhu, B.; Liu, L.; Wu, Y. Local and global existence of mild solutions for a class of nonlinear fractional reaction-diffusion equation with delay. *Appl. Math. Lett.* **2016**, *61*, 73–79. [CrossRef]
29. Ren, L.; Wang, J.; Fečkan, M. Asymptotically periodic solutions for Caputo type fractional evolution equations. *Fract. Calc. Appl. Anal.* **2018**, *21*, 1294–1312. [CrossRef]
30. Wang, X.; Wang, J.; Shen, D.; Zhou, Y. Convergence analysis for iterative learning control of conformable fractional differential equations. *Math. Meth. Appl. Sci.* **2018**, *41*, 8315–8328. [CrossRef]
31. Wang, Y.; Liu, L.; Zhang, X.; Wu, Y. Positive solutions of a fractional semipositone differential system arising from the study of HIV infection models. *Appl. Math. Comput.* **2015**, *258*, 312–324.
32. Zhang, X.; Mao, C.; Liu, L.; Wu, Y. Exact iterative solution for an abstract fractional dynamic system model for bioprocess. *Qual. Theory Dyn. Syst.* **2017**, *16*, 205–222. [CrossRef]
33. Zhang, X.; Liu, L.; Wu, B.Y. Wiwatanapataphee, Nontrivial solutions for a fractional advection dispersion equation in anomalous diffusion. *Appl. Math. Lett.* **2017**, *66*, 1–8. [CrossRef]
34. Dishlieva, K. On the qualitative theory of differential equaitons with random impulsive moments. *Int. J. Sci. Tech. Manag.* **2015**, *4*, 172–180.
35. Kilbas, A.A.; Srivastava, H.M.; Trujillo, J.J. *Theory and Applications of Fractional Differential Equations*; Elsevier: Amsterdam, The Netherlands, 2006.
36. Bainov, D.; Simeonov, P. *Integral Inequalities and Applications*; Kluwer Academic Publishers: Dordrecht, The Netherlands,1992.
37. Ye, H.; Gao, J.; Ding, Y. A generalized Gronwall inequality and its application to a fractional differential equation. *J. Math. Anal. Appl.* **2007**, *328*, 1075–1081. [CrossRef]

mathematics

MDPI

Article

Hyers–Ulam Stability and Existence of Solutions for Differential Equations with Caputo–Fabrizio Fractional Derivative

Kui Liu [1], Michal Fečkan [2,3], D. O'Regan [4] and JinRong Wang [1,5,*]

1 Department of Mathematics, Guizhou University, Guiyang 550025, China; liuk180916@163.com
2 Department of Mathematical Analysis and Numerical Mathematics, Faculty of Mathematics, Physics and Informatics, Comenius University in Bratislava, Mlynská dolina, 842 48 Bratislava, Slovakia; Michal.Feckan@fmph.uniba.sk
3 Mathematical Institute, Slovak Academy of Sciences, Štefánikova 49, 814 73 Bratislava, Slovakia
4 School of Mathematics, Statistics and Applied Mathematics, National University of Ireland, H91 TK33 Galway, Ireland; donal.oregan@nuigalway.ie
5 School of Mathematical Sciences, Qufu Normal University, Qufu 273165, China
* Correspondence: jrwang@gzu.edu.cn

Received: 14 March 2019; Accepted: 1 April 2019; Published: 5 April 2019

Abstract: In this paper, the Hyers–Ulam stability of linear Caputo–Fabrizio fractional differential equation is established using the Laplace transform method. We also derive a generalized Hyers–Ulam stability result via the Gronwall inequality. In addition, we establish existence and uniqueness of solutions for nonlinear Caputo–Fabrizio fractional differential equations using the generalized Banach fixed point theorem and Schaefer's fixed point theorem. Finally, two examples are given to illustrate our main results.

Keywords: Caputo–Fabrizio fractional differential equations; Hyers–Ulam stability

MSC: 34A08; 34D20

1. Introduction

Fractional differential operators describe mechanical and physical processes with historical memory and spatial global correlation and for the basic theory—see [1–3]. Results on existence, stability and controllability for differential equations with Caputo, Riemann–Liouville and Hilfer type fractional derivatives can be found, for example, in [4–19]. Caputo and Fabrizio [20] introduced a new nonlocal derivative without a singular kernel and Atangana and Nieto [21] studied the numerical approximation of this new fractional derivative and established a modified resistance loop capacitance (RLC) circuit model. Losada and Nieto [22] presented a fractional integral corresponding to the Caputo–Fabrizio fractional derivative and introduced Caputo–Fabrizio fractional differential equations and established existence and uniqueness results. Baleanu et al. [23] extended the study to Caputo–Fabrizio fractional integro-differential equations and obtained the approximate solution. Franc and Goufo [24] established a new Korteweg–de Vries–Burgers equation involving the Caputo–Fabrizio fractional derivative with no singular kernel and presented existence and uniqueness results and also gave numerical approximations.

Hyers–Ulam stability is a concept that provides an approximate solution for the exact solution in a simple form for differential equations. A Laplace transform method is applied to show the Hyers–Ulam stability for integer order differential equations in [25,26] and Wang and Li [27] adopted the idea and applied a Laplace transform method to show the Hyers–Ulam stability for fractional order differential equations involving Caputo derivatives. There are many papers on differential equations involving fractional derivatives–see, for example, [28–36]. However, there are only a few papers on the Hyers–Ulam stability for differential equations with the Caputo–Fabrizio fractional derivative. In [37], Wang et al. offered the Ulam stability for the fractional differential equations with the Caputo derivative.

First, we recall the well-known Caputo fractional derivative [2] of order β, given by

$$(\mathbb{D}^{\beta}y)(x) = \frac{1}{\Gamma(1-\beta)}\int_a^x \frac{\dot{f}(s)}{(x-s)^{\beta}}ds,\ 0<\beta<1,$$

where $f \in C^1(a,b), b > a$. By changing the kernel $(x-s)^{-\beta}$ with the function $\exp(-\frac{\beta}{1-\beta}(x-s))$ and $\frac{1}{\Gamma(1-\beta)}$ by $\frac{1}{\sqrt{2\pi(1-\alpha^2)}}$, we obtain the new definition of fractional derivative without a singular kernel $({}^{CF}\mathbb{D}^{\alpha}y)(x)$—see Definition 1 for details.

In this paper, we study Hyers–Ulam stability and existence and uniqueness of solutions for the following Caputo–Fabrizio fractional derivative equations:

$$({}^{CF}\mathbb{D}^{\alpha}y)(x) - \lambda({}^{CF}\mathbb{D}^{\beta}y)(x) = u(x),\ x \in [0,T],\ 0<\alpha,\beta<1, \tag{1}$$

and

$$({}^{CF}\mathbb{D}^{\alpha}y)(x) = f(x,y(x)),\ x \in [0,T],\ 0<\alpha<1, \tag{2}$$

where $({}^{CF}\mathbb{D}^{\gamma}y)(\cdot)$ denotes the Caputo–Fabrizio derivative for y with the order $0 < \gamma < 1$ (see Definition 1), $\lambda \in \mathbb{R}$, $u : [0,T] \to \mathbb{R}$ and $f : [0,T] \times \mathbb{R} \to \mathbb{R}$ will be specified later.

The main contributions are as follows: we obtain a simple result to check whether the approximate solution is near the exact solution for linear Equation (1), which implies Hyers–Ulam stability and generalized Hyers–Ulam stability on the finite time interval. In addition, we present a condition to derive existence and uniqueness of solutions for nonlinear Equation (2) using the generalized Banach fixed point theorem (this improves the result in (Theorem 1, [22])). In addition, we establish sufficient conditions to guarantee the existence of solutions for nonlinear Equation (2) using Schaefer's fixed point theorem. Based on the existence and uniqueness result, we prove the Hyers–Ulam stability of (2) via the Gronwall inequality.

2. Preliminaries

Let $C(I,\mathbb{R})$ be the Banach space of all continuous functions from I into $\mathbb{R}$ with the norm $\|y\|_C := \sup\{|y(x)| : x \in I\}$.

Definition 1 (see [22]). *Let $0<\alpha<1$, $h \in C^1[0,b)$ and $b>0$. The Caputo–Fabrizio fractional derivative for a function h of order α is defined by*

$${}^{CF}\mathbb{D}^{\alpha}h(\tau) = \frac{(2-\alpha)M(\alpha)}{2(1-\alpha)}\int_0^{\tau}\exp(-\frac{\alpha}{1-\alpha}(\tau-x))h'(x)dx,\ \tau\geq 0,$$

where $M(\alpha)$ is a normalization constant depending on α. Note that $({}^{CF}\mathbb{D}^{\alpha})(h) = 0$ if and only if h is a constant function.

Definition 2 (see Definition 1, [22]). *Let $0<\alpha<1$. The Caputo–Fabrizio fractional integral for a function h of order α is defined by*

$${}^{CF}I^{\alpha}h(\tau) = \frac{2(1-\alpha)}{(2-\alpha)M(\alpha)}h(\tau) + \frac{2\alpha}{(2-\alpha)M(\alpha)}\int_0^{\tau}h(x)dx,\ \tau \geq 0.$$

Theorem 1 (see [20,22]). *Let $\alpha \in (0,1)$. Then,*

$$\mathcal{L}[{}^{CF}\mathbb{D}^{\alpha}h(\tau)](s) = \frac{(2-\alpha)M(\alpha)}{2(s+\alpha(1-s))}(s\mathcal{L}[h(\tau)](s) - h(0)),\ s > 0.$$

Motivated by (Definition 2.3, [37]), we introduce the following definition.

Definition 3. *Let $0 < \alpha, \beta < 1$ and $u : [0,T] \to \mathbb{R}$ be a continuous function. Then, (1) is Hyers–Ulam stable if there exists $K > 0$ and $\epsilon > 0$ such that, for each solution $y \in C([0,T],\mathbb{R})$ of (1),*

$$|{}^{CF}\mathbb{D}^{\alpha}y(x) - \lambda {}^{CF}\mathbb{D}^{\beta}y(x) - u(x)| \leq \epsilon,\ \forall x \in [0,T], \tag{3}$$

and there exists a solution $z \in C([0,T),\mathbb{R})$ of (2) with

$$|y(x) - z(x)| \leq K\epsilon,\ \forall x \in [0,T].$$

Definition 4. *Let $0 < \alpha, \beta < 1$, $u : [0,T] \to \mathbb{R}$ be a continuous function and $G : [0,T] \to \mathbb{R}_+$ be continuous functions. Then, (1) is generalized Hyers–Ulam–Rassias stable with respect to G if there exists a constant $c_{f,G} > 0$ such that for each solution $y \in C([0,T],\mathbb{R})$ of (1),*

$$|{}^{CF}\mathbb{D}^{\alpha}y(x) - \lambda {}^{CF}\mathbb{D}^{\beta}y(x) - u(x))| \leq G(x),\ \forall x \in [0,T], \tag{4}$$

and there exists a solution $z \in C([0,T],\mathbb{R})$ of (2) with

$$|y(x) - z(x)| \leq c_{f,G}G(x),\ \forall x \in [0,T].$$

Definition 5. *Let $f : [0,T] \times \mathbb{R} \to \mathbb{R}$ be a continuous function. Then, (2) is Hyers–Ulam stable if there exists $K > 0$ and $\epsilon > 0$ such that for each solution $y \in C([0,T],\mathbb{R})$ of (2),*

$$|{}^{CF}\mathbb{D}^{\alpha}y(x) - f(x,y(x))| \leq \epsilon,\ \forall x \in [0,T], \tag{5}$$

and there exists a solution $z \in C([0,T),\mathbb{R})$ of (2) with

$$|y(x) - z(x)| \leq K\epsilon,\ \forall x \in [0,T].$$

Definition 6. *Let $f : [0,T] \times \mathbb{R} \to \mathbb{R}$ and $G : [0,T] \to \mathbb{R}_+$ be continuous functions. Then, (2) is generalized Hyers–Ulam–Rassias stable with respect to G if there exists a constant $c_{f,G} > 0$ such that, for each solution $y \in C([0,T],\mathbb{R})$ of (2),*

$$|{}^{CF}\mathbb{D}^{\alpha}y(x) - f(x,y(x))| \leq G(x),\ \forall x \in [0,T], \tag{6}$$

and there exists a solution $z \in C([0,T],\mathbb{R})$ of (2) with

$$|y(x) - z(x)| \leq c_{f,G}G(x),\ \forall x \in [0,T].$$

3. Stability Results for the Linear Equation

In this section, we study Hyers–Ulam and generalized Hyers–Ulam-Rassias stability of (1).

Theorem 2. *Let $0 < \beta, \alpha < 1$, $\lambda \in \mathbb{R}$, and $u(x)$ be a given real function on $[0,T]$. If a function $y : [0,T] \to \mathbb{R}$ satisfies the inequality*

$$|({}^{CF}\mathbb{D}^{\alpha}y)(x) - \lambda({}^{CF}\mathbb{D}^{\beta}y)(x) - u(x)| \leq \varepsilon \tag{7}$$

for each $x \in [0, T]$ *and* $\varepsilon > 0$, *then there exists a solution* $y_a : [0, T] \to \mathbb{R}$ *of* (1) *such that*

$$|y(x) - y_a(x)| \le 2|\frac{C}{A}|\varepsilon + 2\left|\frac{AD - BC}{A^2} - \frac{\alpha\beta}{B}\right| \max\{1, \exp(-\frac{B}{A}T)\}x\varepsilon + 2|\frac{\alpha\beta}{B}|x\varepsilon, \tag{8}$$

where

$$\begin{cases} A = (1-\beta)(2-\alpha)M(\alpha) - \lambda(2-\beta)M(\beta)(1-\alpha), \\ B = (2-\alpha)M(\alpha)\beta - \lambda(2-\beta)M(\beta)\alpha, \\ C = (1-\beta)(1-\alpha), \\ D = \alpha + \beta - 2\alpha\beta. \end{cases} \tag{9}$$

Proof. Let

$$F(x) = ({}^{CF}\mathbb{D}^{\alpha}y)(x) - \lambda({}^{CF}\mathbb{D}^{\beta}y)(x) - u(x),\ x \in [0, T]. \tag{10}$$

Taking the Laplace transform of (10) via Theorem 1, and we have

$$\begin{aligned} \mathcal{L}\{F(x)\}(s) &= \mathcal{L}\{({}^{CF}\mathbb{D}^{\alpha}y)(x) - \lambda({}^{CF}\mathbb{D}^{\beta}y)(x) - u(x)\}(s) \\ &= \mathcal{L}\{({}^{CF}\mathbb{D}^{\alpha}y)(x)\}(s) - \lambda\mathcal{L}\{({}^{CF}\mathbb{D}^{\beta}y)(x)\}(s) - \mathcal{L}\{u(x)\}(s) \\ &= [\frac{(2-\alpha)M(\alpha)}{2(s+\alpha(1-s))} - \lambda\frac{(2-\beta)M(\beta)}{2(s+\beta(1-s))}]s\mathcal{L}\{y(x)\}(s) \\ &\quad +[-\frac{(2-\alpha)M(\alpha)}{2(s+\alpha(1-s))} + \lambda\frac{(2-\beta)M(\beta)}{2(s+\beta(1-s))}]y(0) - \mathcal{L}\{u(x)\}(s), \end{aligned} \tag{11}$$

where $\mathcal{L}\{F\}$ denotes the Laplace transform of the function F. From (11), one has

$$\begin{aligned} &\mathcal{L}\{y(x)\}(s) \\ = &\ \frac{1}{s}y(0) + \frac{1}{s}\frac{2(s+\alpha(1-s))(s+\beta(1-s))}{(2-\alpha)M(\alpha)(s+\beta(1-s)) - \lambda(2-\beta)M(\beta)(s+\alpha(1-s))} \\ &\times\Big(\mathcal{L}\{u(x)\}(s) + \mathcal{L}\{F(x)\}(s)\Big) \\ = &\ \frac{1}{s}y(0) + 2\Big(\frac{C}{A} + \frac{AD-BC}{A^2}\frac{1}{s+\frac{B}{A}} + \frac{\alpha\beta}{B}\frac{1}{s} - \frac{\alpha\beta}{B}\frac{1}{s+\frac{B}{A}}\Big)(\mathcal{L}\{u(x)\}(s) + \mathcal{L}\{F(x)\}(s)), \end{aligned} \tag{12}$$

where A, B, C, D are defined as in (9). Set

$$y_a(x) = y(0) + 2\frac{C}{A}u(x) + 2\Big(\frac{AD-BC}{A^2} - \frac{\alpha\beta}{B}\Big)\int_0^x \exp(-\frac{B}{A}t)u(x-t)dt + 2\frac{\alpha\beta}{B}\int_0^x u(x-t)dt. \tag{13}$$

Taking the Laplace transform of (13), one has

$$\begin{aligned} &\mathcal{L}\{y_a(x)\}(s) \\ = &\ \frac{1}{s}y(0) + 2\frac{C}{A}\mathcal{L}\{u(x)\}(s) + 2\Big(\frac{AD-BC}{A^2} - \frac{\alpha\beta}{B}\Big)\frac{1}{s+\frac{B}{A}}\mathcal{L}\{u(x)\}(s) + 2\frac{\alpha\beta}{B}\frac{1}{s}\mathcal{L}\{u(x)\}(s) \\ = &\ \frac{1}{s}y(0) + 2\Big(\frac{C}{A} + \frac{AD-BC}{A^2}\frac{1}{s+\frac{B}{A}} + \frac{\alpha\beta}{B}\frac{1}{s} - \frac{\alpha\beta}{B}\frac{1}{s+\frac{B}{A}}\Big)\mathcal{L}\{u(x)\}(s). \end{aligned} \tag{14}$$

Note that

$$\begin{aligned} &\mathcal{L}\{({}^{CF}\mathbb{D}^{\alpha}y_a)(x) - \lambda({}^{CF}\mathbb{D}^{\beta}y_a)(x)\}(s) \\ = &\ \frac{(2-\alpha)M(\alpha)(s+\beta(1-s)) - \lambda(2-\beta)M(\beta)(s+\alpha(1-s))}{2(s+\alpha(1-s))(s+\beta(1-s))}(s\mathcal{L}\{y_a(x)\}(s) - y(0)). \end{aligned} \tag{15}$$

Substituting (14) into (15), we obtain

$$\mathcal{L}\{({}^{CF}\mathbb{D}^{\alpha}y_a)(x) - \lambda({}^{CF}\mathbb{D}^{\beta}y_a)(x)\}(s) = \mathcal{L}\{u(x)\},$$

which yields that $y_a(x)$ is a solution of Equation (1) since $\mathcal{L}$ is one-to-one. From (12) and (14), we have

$$\mathcal{L}\{y(x) - y_a(x)\}(s) = 2\left(\frac{C}{A} + \frac{AD-BC}{A^2}\frac{1}{s+\frac{B}{A}} + \frac{\alpha\beta}{B}\frac{1}{s} - \frac{\alpha\beta}{B}\frac{1}{s+\frac{B}{A}}\right)\mathcal{L}\{F(x)\}.$$

This implies that

$$y(x) - y_a(x) = 2\frac{C}{A}F(x) + 2(\frac{AD-BC}{A^2} - \frac{\alpha\beta}{B})(\exp(-\frac{B}{A}x) * F(x) + 2\frac{\alpha\beta}{B}(1 * F(x)),$$

so

$$\begin{aligned}
&|y(x) - y_a(x)| \\
= &\left|2\frac{C}{A}F(x) + 2(\frac{AD-BC}{A^2} - \frac{\alpha\beta}{B})(\exp(-\frac{B}{A}x) * F(x) + 2\frac{\alpha\beta}{B}(1 * F(x))\right| \\
\leq\; & 2|\frac{C}{A}F(x)| + 2\left|\frac{AD-BC}{A^2} - \frac{\alpha\beta}{B}\right||\exp(-\frac{B}{A}x) * F(x)| + 2|\frac{\alpha\beta}{B}||1 * F(x)| \\
\leq\; & 2|\frac{C}{A}||F(x)| + 2\left|\frac{AD-BC}{A^2} - \frac{\alpha\beta}{B}\right|\int_0^x |\exp(-\frac{B}{A}t)||F(x-t)|dt + 2|\frac{\alpha\beta}{B}|\int_0^x |F(x-t)|dt) \\
\leq\; & 2|\frac{C}{A}||F(x)| + 2\left|\frac{AD-BC}{A^2} - \frac{\alpha\beta}{B}\right|\varepsilon\int_0^x \max\{1, \exp(-\frac{B}{A}(T))\}dt + 2|\frac{\alpha\beta}{B}|\varepsilon\int_0^x 1dt) \\
\leq\; & 2|\frac{C}{A}|\varepsilon + 2\left|\frac{AD-BC}{A^2} - \frac{\alpha\beta}{B}\right|x\max\{1, \exp(-\frac{B}{A}T)\}\varepsilon + 2|\frac{\alpha\beta}{B}|x\varepsilon.
\end{aligned}$$

The proof is complete. □

Remark 1. *If $T < \infty$, then (1) is Hyers–Ulam stable with the constant*

$$K = 2|\frac{C}{A}| + 2|\frac{AD-BC}{A^2} - \frac{\alpha\beta}{B}|\max\{1, \exp(-\frac{B}{A}T)\}T + 2|\frac{\alpha\beta}{B}|T.$$

Remark 2. *Let $0 < \beta, \alpha < 1$, $\lambda \in \mathbb{R}$, and $u(x)$ be a given real function on $[0, T]$. If a function $y : [0, T] \to \mathbb{R}$ satisfies the inequality*

$$|({}^{CF}\mathbb{D}^{\alpha}y)(x) - \lambda({}^{CF}\mathbb{D}^{\beta}y)(x) - u(x)| \leq G(x), \tag{16}$$

this implies that

$$|F(x)| \leq G(x)$$

for each $x \in [0, T]$ and some function $G(x) > 0$, where F is defined in (10).

From Theorem 2, then there exists a solution $y_a : [0, T] \to \mathbb{R}$ of (1) such that

$$y(x) - y_a(x) = 2\frac{C}{A}F(x) + 2(\frac{AD-BC}{A^2} - \frac{\alpha\beta}{B})(\exp(-\frac{B}{A}x) * F(x) + 2\frac{\alpha\beta}{B}(1 * F(x)),$$

and

$$
\begin{aligned}
& |y(x) - y_a(x)| \\
\leq\ & 2|\frac{C}{A}F(x)| + 2\left|\frac{AD-BC}{A^2} - \frac{\alpha\beta}{B}\right||\exp(-\frac{B}{A}x) * F(x)| + 2|\frac{\alpha\beta}{B}||1 * F(x)| \\
\leq\ & 2|\frac{C}{A}||F(x)| + 2\left|\frac{AD-BC}{A^2} - \frac{\alpha\beta}{B}\right|\max\{1, \exp(-\frac{B}{A}T)\}|\int_0^x F(x-t)dt| + 2|\frac{\alpha\beta}{B}||\int_0^x F(x-t)dt| \\
\leq\ & 2|\frac{C}{A}||F(x)| + 2\left|\frac{AD-BC}{A^2} - \frac{\alpha\beta}{B}\right|\max\{1, \exp(-\frac{B}{A}T)\}|F(x)| + 2|\frac{\alpha\beta}{B}||F(x)| \\
\leq\ & 2\left[|\frac{C}{A}| + \left|\frac{AD-BC}{A^2} - \frac{\alpha\beta}{B}\right|\max\{1, \exp(-\frac{B}{A}(T))\} + |\frac{\alpha\beta}{B}|\right]G(x)
\end{aligned}
$$

provided that

$$\int_0^x F(t)dt \leq F(x)$$

for any $x \in [0, T]$, *where* F *is defined in* (10) *and* A, B, C, D *are defined as in* (9). *Thus,* (2) *is generalized Hyers–Ulam stable with respect to* G *on* $[0, T]$.

4. Existence and Stability Results for the Nonlinear Equation

We introduce the following conditions:

$[A1] : f : [0, T] \times \mathbb{R} \to \mathbb{R}$ is continuous.

$[A2]$: There exists a $k_f > 0$ such that

$$|f(x, y) - f(x, g)| \leq k_f|y - g|, \ \forall y, g \in \mathbb{R}, x \in [0, T].$$

$[A3]$: There exists a constant $L > 0$ such that

$$|f(x, y)| \leq L(1 + |y|)$$

for each $x \in [0, T]$ and all $y \in \mathbb{R}$.

Let $a_\alpha = \frac{2(1-\alpha)}{2-\alpha}M(\alpha)$, $b_\alpha = \frac{2\alpha}{2-\alpha}M(\alpha)$, $y(0) = y_0$ and $C_0 = -a_\alpha f(0, y_0) + y_0$.

Theorem 3. *Let* $0 < \alpha < 1$. *Assume that* $[A1]$ *and* $[A2]$ *hold. If* $a_\alpha k_f < 1$, *then* (2) *with* $y(0) = y_0$ *has a unique solution.*

Proof. Consider $P : C([0, T], \mathbb{R}) \to C([0, T], \mathbb{R})$ as follows:

$$(Py)(x) = C_0 + a_\alpha f(x, y(x)) + b_\alpha \int_0^x f(s, y(s))ds. \tag{17}$$

Note P is well defined because of $[A1]$. For all $y_1, y_2 \in C([0, T], \mathbb{R})$ and all $x \in [0, T]$, using $[A2]$, we have

$$
\begin{aligned}
& |(Py_1)(x) - (Py_2)(x)| \\
\leq\ & a_\alpha|f(x, y_1(x)) - f(x, y_2(x))| + b_\alpha \int_0^x |f(s, y_1(s)) - f(x, y_2(x))|ds \\
\leq\ & a_\alpha k_f|y_1(x) - y_2(x)| + b_\alpha \int_0^x k_f|y_1(s) - y_2(s)|ds \\
=\ & a_\alpha k_f\|y_1 - y_2\|_C + b_\alpha k_f x\|y_1 - y_2\|_C.
\end{aligned}
$$

Denote $C_n^i = \frac{n!}{(n-i)!i!}$. Next,

$$\begin{aligned}
&|(P^2y_1)(x) - (P^2y_2)(x)| \\
\leq\ & a_\alpha |f(x,(Py_1)(x)) - f(x,(Py_2)(x))| + b_\alpha \int_0^x |f(s,(Py_1)(s)) - f(x,(Py_2)(x))| ds \\
\leq\ & a_\alpha k_f |Py_1(x) - Py_2(x)| + b_\alpha \int_0^x k_f |Py_1(s) - Py_2(s)| ds \\
\leq\ & a_\alpha k_f (a_\alpha k_f \|y_1 - y_2\|_C + b_\alpha k_f x \|y_1 - y_2\|_C) \\
& + b_\alpha k_f \int_0^x (a_\alpha k_f \|y_1 - y_2\|_C + b_\alpha k_f x \|y_1 - y_2\|_C) ds \\
\leq\ & \left((k_f a_\alpha)^2 + 2 k_f a_\alpha (k_f b_\alpha x) + \frac{(k_f b_\alpha x)^2}{2!} \right) \|y_1 - y_2\|_C \\
=\ & \sum_{i=0}^{2} \frac{C_2^i (k_f a_\alpha)^{2-i} (k_f b_\alpha x)^i}{i!} \|y_1 - y_2\|_C.
\end{aligned}$$

For any $m \in \mathbb{N}^+$, suppose the following inequality hold

$$|(P^m y_1)(x) - (P^m y_2)(x)| \leq \sum_{i=0}^{m} \frac{C_m^i (k_f a_\alpha)^{m-i} (k_f b_\alpha x)^i}{i!} \|y_1 - y_2\|_C.$$

Then,

$$\begin{aligned}
&|(P^{m+1}y_1)(x) - (P^{m+1}y_2)(x)| \\
\leq\ & a_\alpha |f(x,(P^m y_1)(x)) - f(x,(P^m y_2)(x))| + b_\alpha \int_0^x |f(x,(P^m y_1)(s)) - f(x,(P^m y_2)(s))| ds \\
\leq\ & \left(k_f a_\alpha \sum_{i=0}^{m} \frac{C_m^i (k_f a_\alpha)^{m-i} (k_f b_\alpha x)^i}{i!} + k_f b_\alpha \int_0^x \sum_{i=0}^{m} \frac{C_m^i (k_f a_\alpha)^{m-i} (k_f b_\alpha s)^i}{i!} ds \right) \|y_1 - y_2\|_C \\
=\ & \sum_{i=0}^{m+1} \frac{C_{m+1}^i (k_f a_\alpha)^{m+1-i} (k_f b_\alpha x)^i}{i!} \|y_1 - y_2\|_C \\
\leq\ & S(m) \|y_1 - y_2\|_C,
\end{aligned}$$

where $S(m) := \sum_{i=0}^{m+1} \frac{C_{m+1}^i (k_f a_\alpha)^{m+1-i} (k_f b_\alpha T)^i}{i!}$. Thus, for any $m \in \mathbb{N}^+$,

$$\|P^{m+1}y_1 - P^{m+1}y_2\|_C \leq S(m) \|y_1 - y_2\|_C.$$

From the condition $k_f a_\alpha < 1$ via (Theorem 2.9, [38]), one has $S(m) \to 0$ as $m \to \infty$. This implies that for any large enough $m \in \mathbb{N}^+$, $S(m) < 1$. Thus, P^m is a contraction mapping. As a result, P has a fixed point. Thus, (2) with $y(0) = y_0$ has a unique solution. This proof is complete. □

Remark 3. *In (Theorem 1, [22]), an existence and uniqueness result for (2) with $y(0) = y_0$ is established by imposing a uniformly Lipschitz condition and applying Banach's fixed point theorem with the condition $a_\alpha k_f + b_\alpha T k_f < 1$, where k_f denotes the Lipschitz constant. Here, we use the generalized Banach fixed point theorem and we weaken the condition $a_\alpha k_f + b_\alpha T k_f < 1$ in (Theorem 1, [22]) to $a_\alpha k_f < 1$.*

Next, we show that the existence of solutions for (2) via Schaefer's fixed point theorem.

Theorem 4. *Assume that $[A1]$ and $[A3]$ hold. If $a_\alpha L < 1$, then (2) with $y(0) = y_0$ has at least one solution.*

Proof. Consider P as in (17). We divide our proof into several steps.

Step 1. P is continuous.

Let y_n be a sequence such that $y_n \to y$ in $C([0,T],\mathbb{R})$. For all $x \in [0,T]$, we get

$$\begin{aligned}
|Py_n(x) - Py(x)| &= |a_\alpha f(x,y_n(x)) + b_\alpha \int_0^x f(s,y_n(s))ds - a_\alpha f(x,y(x)) - b_\alpha \int_0^x f(s,y(s))ds| \\
&\leq a_\alpha |f(x,y_n(x)) - f(x,y(x))| + b_\alpha | \int_0^x f(s,y_n(s))ds - \int_0^x f(s,y(s))ds| \\
&\leq a_\alpha |f(x,y_n(x)) - f(x,y(x))| + b_\alpha \int_0^x |f(s,y_n(s)) - f(s,y(s))|ds. \\
&\leq (a_\alpha + b_\alpha T)\|f(\cdot,y_n) - f(\cdot,y)\|_C.
\end{aligned}$$

This shows that P is continuous since $\|fy_n - fy\|_C \to 0$ when $n \to \infty$.

Step 2. P maps bounded sets into bounded sets of $C([0,T],\mathbb{R})$.

Indeed, we prove that for all $r > 0$, there exists a $k > 0$ such that for every $y \in B_r = \{y \in C([0,T],\mathbb{R}) : \|y\|_C \leq r\}$, we have $\|Py\|_C \leq k$. In fact, for any $x \in [0,T]$, from $[A3]$, we have

$$\begin{aligned}
|Py(x)| &\leq |C_0| + a_\alpha |f(x,y(x))| + b_\alpha \int_0^x |f(s,y(s))|ds \\
&\leq |C_0| + a_\alpha L(1+|y|) + b_\alpha L \int_0^x (1+|y(s)|)ds \\
&\leq |C_0| + a_\alpha L(1+\|y\|_C) + b_\alpha TL|(1+\|y\|_C) \\
&\leq |C_0| + a_\alpha L(1+r) + b_\alpha TL(1+r) \\
&= |C_0| + (a_\alpha + b_\alpha T)L(1+r),
\end{aligned}$$

which implies that

$$\|Py\| \leq |C_0| + (a_\alpha + b_\alpha T)L(1+r) := k.$$

Step 3. P maps bounded sets into equicontinuous sets in $C([0,T],\mathbb{R})$.

Let $x_1, x_2 \in [0,T]$, with $0 \leq x_1 < x_2 \leq T, y \in B_r$. From $[A3]$, we have

$$\begin{aligned}
&|Py(x_1) - Py(x_2)| \\
= \;& |a_\alpha f(x_1,y(x_1)) + b_\alpha \int_0^{x_1} f(s,y(s))ds - a_\alpha f(x_2,y(x_2)) - b_\alpha \int_0^{x_2} f(s,y(s))ds| \\
\leq \;& a_\alpha |f(x_1,y(x_1)) - f(x_2,y(x_2))| + b_\alpha | \int_0^{x_1} f(s,y(s))ds - \int_0^{x_2} f(s,y(s))ds| \\
\leq \;& a_\alpha |f(x_1,y(x_1)) - f(x_1,y(x_2))| + a_\alpha |f(x_1,y(x_2)) - f(x_2,y(x_2))| + b_\alpha | \int_{x_1}^{x_2} f(s,y(s))ds| \\
\leq \;& a_\alpha |f(x_1,y(x_1)) - f(x_1,y(x_2))| + a_\alpha |f(x_1,y(x_2)) - f(x_2,y(x_2))| + b_\alpha L(1+r)(x_2 - x_1).
\end{aligned}$$

Then, as x_1 approaches x_2, the right-hand side of the above inequality tends to zero (because of $[A1]$) as $x_1 \to x_2$. Thus, P is equicontinuous.

We can conclude that P is completely continuous from Step 1–Step 3 with the Arzela–Ascoli theorem.

Step 4. A priori bounds.

Now, we show that the set $E(P) = \{y \in C([0,T],\mathbb{R}) : y = \lambda Py \text{ for some } \lambda \in (0,1)\}$ is bounded.

Let $y \in E(P)$. Then, $y = \lambda Py$ for some $\lambda \in (0,1)$. For each $x \in [0,T]$, we have

$$
\begin{aligned}
|y(x)| &\leq |C_0| + a_\alpha |f(x,y(x))| + b_\alpha \int_0^x |f(s,y(s))| ds \\
&\leq |C_0| + a_\alpha L(1+|y(x)|) + b_\alpha L \int_0^x (1+|y(s)|) ds \\
&\leq K + a_\alpha L |y(x)| + b_\alpha L \int_0^x |y(s)| ds \quad (K = |C_0| + a_\alpha L + b_\alpha LT).
\end{aligned}
$$

Using the condition $1 - a_\alpha L > 0$, one has

$$
|y(x)| \leq \frac{K}{1 - a_\alpha L} + \frac{b_\alpha L}{1 - a_\alpha L} \int_0^x |y(s)| ds,
$$

and Gronwall's inequality yields

$$
|y(x)| \leq \frac{K}{1 - a_\alpha L} \exp\left(\frac{b_\alpha LT}{1 - a_\alpha L}\right) < \infty.
$$

Then, the set $E(P)$ is bounded.

Schaefer's fixed point theorem guarantees that P has a fixed point, which is a solution of (2). The proof is finished. □

In the following, we consider (2) and (6) to discuss the generalized Ulam–Hyers–Rassias stability. We need the following condition.

$[A4]$: Let $G \in C([0,T], \mathbb{R}_+)$ be an increasing function and there exists $\lambda_G > 0$ such that

$$
\int_0^x G(s) ds \leq \lambda_G G(x), \ \forall x \in [0,T].
$$

Theorem 5. *Assumptions $[A1]$, $[A2]$ and $[A4]$ hold. If $a_\alpha k_f < 1$, then (2) is generalized Ulam–Hyers–Rassias stable with respect to G on $[0,T] (T < \infty)$.*

Proof. Let $g \in C([0,T], \mathbb{R})$ be a solution of (6). From Theorem 3,

$$
\begin{cases} {}^{CF}\mathbb{D}^\alpha y(x) = f(x,y(x)), \ 0 < \alpha < 1, \ t \in [0,T), \\ y(0) = C_0, \end{cases} \tag{18}
$$

has the unique solution

$$
y(x) = C_0 + a_\alpha f(x,y(x)) + b_\alpha \int_0^x f(s,y(s)) ds, \ x \in [0,T].
$$

From (6), we have

$$
\begin{aligned}
|g(x) - C_0 - a_\alpha f(x,g(x)) - b_\alpha \int_0^x f(s,g(s)) ds| &\leq a_\alpha G(x) + b_\alpha \int_0^x G(s) ds \\
&\leq (a_\alpha + b_\alpha \lambda_G) G(x), \ x \in [0,T].
\end{aligned}
$$

Thus,

$$\begin{aligned}
& |g(x)-y(x)| \\
\leq\ & \left|g(x)-C_0-a_\alpha f(x,y(x))-b_\alpha\int_0^x f(s,y(s))ds\right| \\
\leq\ & \left|g(x)-C_0-a_\alpha f(x,g(x))-b_\alpha\int_0^x f(s,g(s))ds\right. \\
& \left.+a_\alpha f(x,y(x))+b_\alpha\int_a^x f(s,y(s))ds-a_\alpha f(x,y(x))-b_\alpha\int_0^x f(s,y(s))ds\right| \\
\leq\ & \left|g(x)-C_0-a_\alpha f(x,g(x))-b_\alpha\int_0^x f(s,g(s))ds\right| \\
& +a_\alpha|f(x,y(x))-f(x,g(x))|+b_\alpha\int_0^x |f(s,y(s))-f(s,g(s))|ds \\
\leq\ & (a_\alpha+b_\alpha\lambda_G)G(x)+a_\alpha k_f|y(x)-g(x)|+b_\alpha k_f\int_0^x |y(s)-g(s)|ds.
\end{aligned}$$

Note that $a_\alpha k_f < 1$, and so,

$$|y(x)-g(x)| \leq \frac{(a_\alpha+b_\alpha\lambda_G)G(x)}{1-a_\alpha k_f}+\frac{b_\alpha k_f}{1-a_\alpha k_f}\int_0^x |y(s)-g(s)|ds.$$

From Gronwall's inequality, we have

$$|y(x)-g(x)| \leq \left[\frac{(a_\alpha+b_\alpha\lambda_G)}{1-a_\alpha k_f}\exp(x)\right]G(x),\ x\in[0,T]. \tag{19}$$

Set $K^* = \frac{a_\alpha+b_\alpha\lambda_G}{1-a_\alpha k_f}\exp(T)$. Note that one has

$$|y(x)-g(x)| \leq K^*G(x),\ x\in[0,T].$$

From Definition 6, (2) is generalized Ulam–Hyers–Rassias stable with respect to G on $[0,T]$. The proof is complete. □

5. Examples

In this section, two examples are given to illustrate our main results.

For convenience in calculating, we suppose that $M(\cdot)$ in Definition 2 is the roots of the following equation:

$$\frac{2(1-\cdot)}{(2-\cdot)M(\cdot)}+\frac{2\cdot}{(2-\cdot)M(\cdot)}=1.$$

Then, one can derive an explicit formula $M(\alpha)=\frac{2}{2-\alpha}$ and $M(\beta)=\frac{2}{2-\beta}$ (see (p. 89, [22])).

Example 1. *Consider*

$$({}^{CF}\mathbb{D}^{\frac{1}{2}}y)(x)-\frac{1}{3}({}^{CF}\mathbb{D}^{\frac{2}{3}}y)(x)=\frac{2}{3}e^x+\frac{1}{3}e^{-2x}-\frac{2}{3},\ x\in[0,T]. \tag{20}$$

Set $\alpha=\frac{1}{2}$, $\beta=\frac{2}{3}$, $u(x)=\frac{2}{3}e^x+\frac{1}{3}e^{-2x}-\frac{2}{3}$ *and* $\lambda=\frac{1}{3}$. *From (Definition 1, [22]),* $M(\frac{1}{2})=\frac{4}{3}$ *and* $M(\frac{2}{3})=\frac{3}{2}$.

Let $y_1(x) = e^x$, *and we have*

$$({}^{CF}\mathbb{D}^{\frac{1}{2}}y_1)(x) = 2\int_0^x e^{t-x}e^t dt = e^x - e^{-x},$$

$$({}^{CF}\mathbb{D}^{\frac{2}{3}}y_1)(x) = 3\int_0^x e^{-2(x-t)}e^t dt = e^x - e^{-2x}.$$

Choose $\varepsilon = \frac{2}{3}$. *Note* $y_1(x) = e^x$ *satisfies*

$$\begin{aligned}
&\left|({}^{CF}\mathbb{D}^{\frac{1}{2}}y_1)(x) - \frac{1}{3}({}^{CF}\mathbb{D}^{\frac{2}{3}}y_1)(x) - \frac{2}{3}e^x - \frac{1}{3}e^{-2x} + \frac{2}{3}\right| \\
= &\left|e^x - e^{-x} - \frac{1}{3}e^x + \frac{1}{3}e^{-2x} - \frac{2}{3}e^x - \frac{1}{3}e^{-2x} + \frac{2}{3}\right| \\
= &|\frac{2}{3} - e^{-x}| \leq \frac{2}{3}.
\end{aligned}$$

Note $y_1(0) = 1$ *and with the formulas of* A, B, C, D *in* (9) *and* (13), *we obtain an exact solution of Equation* (1) *as*

$$\begin{aligned}
y_a(x) &= y(0) + 2\frac{C}{A}u(x) + 2\left(\frac{AD-BC}{A^2} - \frac{\alpha\beta}{B}\right)\int_0^x \exp(-\frac{B}{A}t)u(x-t)dt \\
&\quad +2\frac{\alpha\beta}{B}\int_0^x u(x-t)dt \\
&= 1 + \frac{2}{3}e^x - \frac{1}{3}e^{-2x} - \frac{2}{3} - \frac{4}{9}\int_0^x e^{-3t}(e^{x-t} + \frac{e^{-2(x-t)}}{2} - 1)dt \\
&\quad +\frac{4}{9}\int_0^x (e^{x-t} + \frac{e^{-2(x-t)}}{2} - 1)dt \\
&= e^x + \frac{4}{27} + \frac{5}{27}e^{-3x} - \frac{2}{3}e^{-2x} - \frac{4}{9}x.
\end{aligned}$$

Clearly,

$$\begin{aligned}
|y_1(x) - y_a(x)| &= |e^x + \frac{4}{27} + \frac{5}{27}e^{-3x} - \frac{2}{3}e^{-2x} - \frac{4}{9}x - e^x| \\
&= |\frac{4}{27} + \frac{5}{27}e^{-3x} - \frac{2}{3}e^{-2x} - \frac{4}{9}x| \\
&\leq |\frac{4}{27} - \frac{4}{9}x| \\
&\leq \frac{2}{3} + \frac{8}{9}x = (1 + \frac{4}{3}x)\frac{2}{3}.
\end{aligned}$$

Note in Theorem 2 *(see Remark* 1*) that we have* $K = 2|\frac{C}{A}| + 2|\frac{AD-BC}{A^2} - \frac{\alpha\beta}{B}|\max\{1, \exp(-\frac{B}{A}T)\}T + 2|\frac{\alpha\beta}{B}|T = 1 + \frac{4}{3}T$ *and* $\varepsilon = \frac{2}{3}$. *Thus, Equation* (20) *is Hyers–Ulam stable when* $T < \infty$.

Example 2. *We consider the following fractional problem:*

$$({}^{CF}\mathbb{D}^{\frac{1}{3}}y)(x) = \frac{e^{-2x}}{1+e^x}\frac{|y|}{1+|y|}, \; x \in [0,2], \tag{21}$$

and the inequality

$$\left|({}^{CF}\mathbb{D}^{\frac{1}{3}}y)(x) - \frac{e^{-2x}}{1+e^x}\frac{|y|}{1+|y|}\right| \leq G(x), \; x \in [0,2]. \tag{22}$$

Set $\alpha = \frac{1}{3}$, $T = 2$ *and* $f(x,y) = \frac{e^{-2x}}{1+e^x}\frac{|y|}{1+|y|}$, $(x,y) \in [0,2] \times \mathbb{R}$. *Clearly,* $[A1]$ *holds. Then,* $M(\frac{1}{3}) = \frac{6}{5}$, $a_{\frac{1}{3}} = \frac{24}{25}$, $b_{\frac{1}{3}} = \frac{12}{25}$. *Let* $G(x) = e^x \in C([0,2],\mathbb{R})$ *and* $\int_0^x G(s)ds = \int_0^x e^s ds = e^x - 1 \leq e^x$. *Here,* $\lambda_G = 1 > 0$.

For any $x \in [0,2]$ *and* $y_1, y_2 \in \mathbb{R}$,

$$\begin{aligned} |f(x,y_1) - f(x,y_2)| &= \frac{e^{-2x}}{1+e^x}\left|\frac{|y_1|}{1+|y_1|} - \frac{|y_2|}{1+|y_2|}\right| \leq \frac{e^{-2x}|y_1 - y_2|}{(1+e^x)(1+|y_1|)(1+|y_2|)} \\ &\leq \frac{e^{-2x}}{(1+e^x)}|y_1 - y_2| \leq \frac{e^{-2x}}{2}|y_1 - y_2| \leq \frac{1}{2}|y_1 - y_2|. \end{aligned}$$

For all $x \in [0,2]$ *and* $y \in \mathbb{R}$,

$$|f(x,y)| = \frac{e^{-2x}}{1+e^x}\frac{|y|}{1+|y|} \leq \frac{e^{-2x}}{1+e^x}|y| \leq \frac{e^{-2x}}{2}|y| \leq \frac{1}{2}|y| \leq \frac{1}{2}(1+|y|).$$

Thus, $[A2]$ *and* $[A3]$ *hold.*

Set $L = \frac{1}{2} = k_f$. *Then* $a_\alpha k_f = \frac{24}{25} \times \frac{1}{2} = \frac{12}{25} < 1$. *From Theorem 3,* (21) *has an unique solution. Thus, all the assumptions in Theorem 4 are satisfied, so our results can be applied to (21).*

Let $g \in C([0,2],\mathbb{R})$ *be a solution of* (22). *We have*

$$\left|({}^{CF}\mathbb{D}^{\frac{1}{3}}g)(x) - f(x,g(x))\right| = \left|({}^{CF}\mathbb{D}^{\frac{1}{3}}g)(x) - \frac{e^{-2x}}{1+e^x}\frac{|g|}{1+|g|}\right| \leq G(x),\ x \in [0,2]. \tag{23}$$

From Theorem 3, we see (21) *with* $y(0) = C_0$ *has the unique solution*

$$\begin{aligned} y(x) &= C_0 + a_{\frac{1}{3}}f(x,y(x)) + b_{\frac{1}{3}}\int_0^x f(s,y(s))ds \\ &= C_0 + \frac{24}{25}\frac{e^{-2x}}{1+e^x}\frac{|y|}{1+|y|} + \frac{12}{25}\int_0^x \frac{e^{-2s}}{1+e^s}\frac{|y|}{1+|y|}ds. \end{aligned}$$

Applying the fractional integrating operator ${}^{CF}I^\alpha(\cdot)$ *on both sides of* (23), *we have*

$$\begin{aligned} &\left|g(x) - C_0 - a_{\frac{1}{3}}f(x,g(x)) - b_{\frac{1}{3}}\int_0^x f(s,g(s))ds\right| \\ \leq\ & a_{\frac{1}{3}}G(x) + b_{\frac{1}{3}}\int_0^x G(s)ds \\ \leq\ & (a_{\frac{1}{3}} + b_{\frac{1}{3}}\lambda_G)G(x),\ x \in [0,2]. \end{aligned}$$

In addition,

$$|y(x) - g(x)| \leq \left[\frac{(a_{\frac{1}{3}} + b_{\frac{1}{3}}\lambda_G)}{1 - a_{\frac{1}{3}}k_f}\exp(x)\right]G(x),\ x \in [0,2].$$

Set $K^* = \frac{a_{\frac{1}{3}} + b_{\frac{1}{3}}\lambda_G}{1 - a_{\frac{1}{3}}k_f}\exp(2) = \frac{\frac{24}{25} + \frac{12}{25}\times 1}{1 - \frac{24}{25}\times\frac{1}{2}}e^2 = \frac{36e^2}{13}$. *Note that one has*

$$|y(x) - g(x)| \leq K^* G(x),\ x \in [0,2].$$

6. Conclusions

By applying the well-known Gronwall inequality and fixed point theorems, we obtain the Hyers–Ulam stability of linear and semilinear Caputo–Fabrizio fractional differential equations. Existence and uniqueness theorems of solution are established. In a forthcoming work, we shall consider the impulsive Cauchy problem with Caputo–Fabrizio fractional derivative.

Author Contributions: The contributions of all authors (K.L., M.F., D.O. and J.W.) are equal. All the main results and examples were developed together.

Funding: This work is partially supported by the National Natural Science Foundation of China (11661016), the Training Object of High Level and Innovative Talents of Guizhou Province ((2016)4006), the Science and Technology Program of Guizhou Province ([2017]5788-10), the Major Research Project of Innovative Group in Guizhou Education Department ([2018]012), the Slovak Research and Development Agency under the contract No. APVV-14-0378, and the Slovak Grant Agency VEGA No. 2/0153/16 and No. 1/0078/17.

Acknowledgments: The authors thank the referees for their careful reading of the article and insightful comments.

Conflicts of Interest: The authors declare no conflict of interest.

References

1. Podlubny, I. *Fractional Differential Equations, Mathematics in Science and Engineering*; Academic Press: San Diego, CA, USA, 1999; Volume 198.
2. Kilbas, A.A.; Srivastava, H.M.; Trujillo, J.J. *Theory and Applications of Fractional Differential Equations*; Elsevier: Amsterdam, The Netherlands, 2006.
3. Tarasov, V.E. *Fractional Dynamics: Application of Fractional Calculuts to Dynamics of Particles, Fields and Media*; Springer: Berlin/Heidelberg, Germany, 2011.
4. Abbas, S.; Benchohra, M.; Darwish, M.A. New stability results for partial fractional differential inclusions with not instantaneous impulses. *Fract. Calc. Appl. Anal.* **2015**, *18*, 172–191. [CrossRef]
5. Li, M.; Wang, J. Exploring delayed Mittag-Leffler type matrix functions to study finite time stability of fractional delay differential equations. *Appl. Math. Comput.* **2018**, *324*, 254–265. [CrossRef]
6. Liu, S.; Wang, J.; Zhou, Y.; Fečkan, M. Iterative learning control with pulse compensation for fractional differential equations. *Math. Slov.* **2018**, *68*, 563–574. [CrossRef]
7. Wang, J.; Ibrahim, A.G.; O'Regan, D. Topological structure of the solution set for fractional non-instantaneous impulsive evolution inclusions. *J. Fixed Point Theory Appl.* **2018**, *20*, 59. [CrossRef]
8. Luo, D.; Wang, J.; Shen, D. Learning formation control for fractional-order multi-agent systems. *Math. Meth. Appl. Sci.* **2018**, *41*, 5003–5014. [CrossRef]
9. Peng, S.; Wang, J.; Yu, X. Stable manifolds for some fractional differential equations. *Nonlinear Anal. Model. Control* **2018**, *23*, 642–663. [CrossRef]
10. Chen, Y.; Wang, J. Continuous dependence of solutions of integer and fractional order non-instantaneous impulsive equations with random impulsive and junction points. *Mathematics* **2019**, *7*, 331. [CrossRef]
11. Zhang, J.; Wang, J. Numerical analysis for a class of Navier–Stokes equations with time fractional derivatives. *Appl. Math. Comput.* **2018**, *336*, 481–489
12. Zhu, B.; Liu, L.; Wu, Y. Local and global existence of mild solutions for a class of nonlinear fractional reaction-diffusion equation with delay. *Appl. Math. Lett.* **2016**, *61*, 73–79. [CrossRef]
13. Wang, Y.; Liu, L.; Wu, Y. Positive solutions for a nonlocal fractional differential equation. *Nonlinear Anal.* **2011**, *74*, 3599–3605. [CrossRef]
14. Zhang, X.; Liu, L.; Wu, Y. Existence results for multiple positive solutions of nonlinear higher order perturbed fractional differential equations with derivatives. *Appl. Math. Comput.* **2012**, *219*, 1420–1433. [CrossRef]
15. Wang, Y.; Liu, L.; Zhang, X.; Wu, Y. Positive solutions of a fractional semipositone differential system arising from the study of HIV infection models. *Appl. Math. Comput.* **2015**, *258*, 312–324.
16. Zhang, X.; Liu, L.; Wu, Y. Variational structure and multiple solutions for a fractional advection-dispersion equation. *Comput. Math. Appl.* **2014**, *68*, 1794–1805. [CrossRef]
17. Zhang, X.; Mao, C.; Liu, L.; Wu, Y. Exact iterative solution for an abstract fractional dynamic system model for bioprocess. *Qual. Theory Dyn. Syst.* **2017**, *16*, 205–222. [CrossRef]
18. Zhang, X.; Liu, L.; Wu, Y.; Wiwatanapataphee, B. Nontrivial solutions for a fractional advection dispersion equation in anomalous diffusion. *Appl. Math. Lett.* **2017**, *66*, 1–8. [CrossRef]
19. Jiang, J.; Liu, L.; Wu, Y. Multiple positive solutions of singular fractional differential system involving Stieltjes integral conditions. *Electron. J. Qual. Theory Differ. Equ.* **2012**, *43*, 1–18. [CrossRef]
20. Caputo, M.; Fabrizio, M. A new definition of fractional derivative without singular kernel. *Prog. Fract. Differ. Appl.* **2015**, *1*, 73–85.

21. Atangana, A.; Nieto, J.J. Numerical solution for the model of RLC circuit via the fractional derivative without singular kernel. *Adv. Mech. Eng.* **2015**, *7*, 1–7. [CrossRef]
22. Losada, J.; Nieto, J.J. Properties of a new fractional derivative without singular kernel. *Prog. Fract. Differ. Appl.* **2015**, *1*, 87–92.
23. Baleanu, D.; Mousalou, A.; Rezapour, S. On the existence of solutions for some infinite coefficient-symmetric Caputo–Fabrizio fractional integro-differential equations. *Bound. Value Prob.* **2017**, *2017*, 1–9. [CrossRef]
24. Franc, E.; Goufo, D. Application of the Caputo–Fabrizio fractional derivative without singular kernel to Korteweg–de Vries–Burgers equations. *Math. Model. Anal.* **2016**, *21*, 188–198.
25. Rezaei, H.; Jung, S.M.; Rassias, T.M. Laplace transform and Hyers–Ulam stability of linear differential equations. *J. Math. Anal. Appl.* **2013**, *403*, 244–251. [CrossRef]
26. Alqifiary, Q.H.; Jung, S.M. Laplace transform and generalized Hyers–Ulam stability of linear differential equations. *Electron. J. Diff. Equ.* **2014**, *2014*, 1–11.
27. Wang, J.; Li, X.Z. A uniform method to Ulam-Hyers stability for some Linear fractional equations. *Mediterr. J. Math.* **2016**, *13*, 625–635. [CrossRef]
28. Wang, J.; Zhang, Y. Ulam-Hyers-Mittag-Leffler stability of fractional-order delay differential equations. *Optimization* **2014**, *63*, 1181–1190. [CrossRef]
29. Capelas de Oliveira, E.; da C. Sousa, J.V. Ulam-Hyers-Rassias stability for a class of fractional integro-differential equations. *Result Math.* **2018**, *73*, 111. [CrossRef]
30. da C. Sousa, J.V.; Capelas de Oliveira, E. Ulam-Hyers stability of a nonlinear fractional Volterra integro-differential equation. *Appl. Math. Lett.* **2018**, *81*, 50–56.
31. da C. Sousa, J.V.; Kucche, K.D.; Capelas de Oliveira, E. Stability of ψ-Hilfer impulsive fractional differential equations. *Appl. Math. Lett.* **2018**, *88*, 73–80.
32. Wang, J.; Zhou, Y.; Fečkan, M. Nonlinear impulsive problems for fractional differential equations and Ulam stability. *Comput. Math. Appl.* **2012**, *64*, 3389–3405. [CrossRef]
33. da C. Sousa, J.V.; Capelas de Oliveira, E. On the Ulam-Hyers-Rassias stability for nonlinear fractional differential equations using the ψ-Hilfer operator. *J. Fixed Point Theory Appl.* **2018**, *20*, 5–21.
34. Shah, K.; Ali, A.; Bushnaq, S. Hyers–Ulam stability analysis to implicit Cauchy problem of fractional differential equations with impulsive conditions. *Math. Meth. Appl. Sci.* **2018**, *41*, 8329–8343. [CrossRef]
35. Ali, Z.; Zada, A.; Shah, K. Ulam stability to a toppled systems of nonlinear implicit fractional order boundary value problem. *Bound. Value Prob.* **2018**, *2018*, 175. [CrossRef]
36. Liu, K.; Wang, J.; O'Regan, D. Ulam-Hyers-Mittag-Leffler stability for ψ-Hilfer fractional-order delay differential equations. *Adv. Differ. Equ.* **2019**, *2019*, 50. [CrossRef]
37. Wang, J.; Lv, L.; Zhou, Y. Ulam stability and data depenaence for fractional differential equations with Caputo derivative. *Electron. J. Qual. Theory Differ. Equ.* **2011**, *63*, 1–10.
38. Wang, J.; Zhou, Y.; Fečkan, M. Abstract Cauchy problem for fractional differential equations. *Nonlinear Dyn.* **2013**, *71*, 685–700. [CrossRef]

mathematics

Article

Stability, Existence and Uniqueness of Boundary Value Problems for a Coupled System of Fractional Differential Equations

Nazim I Mahmudov * and Areen Al-Khateeb

Eastern Mediterranean University, Gazimagusa 99628, T.R. North Cyprus, Mersin 10, Turkey; khteb1987@live.com

* Correspondence: Nazim.mahmudov@emu.edu.tr; Tel.: +90-392-630-1227

Received: 20 February 2019; Accepted: 10 April 2019; Published: 16 April 2019

Abstract: The current article studies a coupled system of fractional differential equations with boundary conditions and proves the existence and uniqueness of solutions by applying Leray-Schauder's alternative and contraction mapping principle. Furthermore, the Hyers-Ulam stability of solutions is discussed and sufficient conditions for the stability are developed. Obtained results are supported by examples and illustrated in the last section.

Keywords: fractional derivative; fixed point theorem; fractional differential equation

1. Introduction

Fractional calculus is undoubtedly one of the very fast-growing fields of modern mathematics, due to its broad range of applications in various fields of science and its unique efficiency in modeling complex phenomena [1,2]. In particular, fractional differential equations with boundary conditions are widely employed to build complex mathematical models for numerous real-life problems such as blood flow problem, underground water flow, population dynamics, and bioengineering. As an example, consider the following equation that describes a thermostat model

$$-x'' = g(t)f(t,x), x(0) = 0, \beta x'(1) = x(\eta),$$

where $t \in (0,1), \eta \in (0,1]$ and β is a positive constant. Note that solutions of the above equation with the specified integral boundary conditions are in fact solutions of the one-dimensional heat equation describing a heated bar with a controller at point 1, which increases or reduces heat based on the temperature picked by a sensor at η. A few of the relevant studies on coupled systems of fractional differential equations with integral boundary conditions are briefly reviewed below and for further information on this topic, refer to References [3,4].

In Reference [5], Ntouyas and Obaid used Leray-Schauder's alternative and Banach's fixed-point theorem to prove the existence and uniqueness of solutions for the following coupled fractional differential equations with Riemann-Liouville integral boundary conditions:

$$\begin{cases} {}^cD^{\alpha}_{0+}u(t) = g(t,u(t),v(t)), t \in [0,1], \\ {}^cD^{\beta}_{0+}v(t) = g(t,u(t),v(t)), t \in [0,1], \\ u(0) = \gamma I^p u(\eta) = \gamma \int_0^{\eta} \frac{(\eta-s)^{p-1}}{\Gamma(p)} u(s)ds, 0 < \eta < 1, \\ v(0) = \delta I^q v(\zeta) = \delta \int_0^{\zeta} \frac{(\zeta-s)^{q-1}}{\Gamma(q)} v(s)ds, 0 < \zeta < 1. \end{cases}$$

Here, ${}^cD^{\alpha}_{0+}$ and ${}^cD^{\beta}_{0+}$ are Caputo fractional derivatives, $0 < \alpha, \beta \le 1$, $f, g \in C\left([0,1] \times \mathbb{R}^2, \mathbb{R}\right)$ and $p, q, \gamma, \delta \in \mathbb{R}$.

Similarly, Ahmed and Ntouyas [6] employed Banach fixed-point theorem and Leray-Schauder's alternative to prove the existence and uniqueness of solutions for the following coupled fractional differential system:

$$\begin{cases} {}^cD^q x(t) = f(t, x(t), y(t)), & t \in [0,1], \quad 1 < q \le 2, \\ {}^cD^p y(t) = g(t, x(t), y(t)), & t \in [0,1], \quad 1 < q \le 2, \end{cases}$$

supplemented with coupled and uncoupled slit-strips-type integral boundary conditions, respectively, given by

$$\begin{cases} x(0) = 0, \quad x(\zeta) = a\int_0^\eta y(s)ds + b\int_\xi^1 y(s)ds, & 0 < \eta < \zeta < \xi < 1, \\ y(0) = 0, \quad y(\zeta) = a\int_0^\eta x(s)ds + b\int_\xi^1 x(s)ds, & 0 < \eta < \zeta < \xi < 1, \end{cases}$$

and

$$\begin{cases} x(0) = 0, \quad x(\zeta) = a\int_0^\eta x(s)ds + b\int_\xi^1 x(s)ds, & 0 < \eta < \zeta < \xi < 1, \\ y(0) = 0, \quad y(\zeta) = a\int_0^\eta y(s)ds + b\int_\xi^1 y(s)ds, & 0 < \eta < \zeta < \xi < 1. \end{cases}$$

Furthermore, Alsulami et al. [7] investigated the following coupled system of fractional differential equations:

$$\begin{cases} {}^cD^\alpha x(t) = f(t, x(t), y(t)), t \in [0, T], 1 < \alpha \le 2, \\ {}^cD^\beta y(t) = g(t, x(t), y(t)), t \in [0, T], 1 < \beta \le 2, \end{cases}$$

subject to the following non-separated coupled boundary conditions:

$$\begin{cases} x(0) = \lambda_1 y(T), x'(0) = \lambda_2 y'(T), \\ y(0) = \mu_1 x(T), y'(0) = \mu_2 x'(T). \end{cases}$$

Note that ${}^cD^\alpha$ and ${}^cD^\beta$ denote Caputo fractional derivatives of order α and β. Moreover, λ_i, μ_i, $i = 1, 2$, are real constants with $\lambda_i\mu_i \ne 1$ and $f, g : [0, T] \times \mathbb{R} \times \mathbb{R} \to \mathbb{R}$ are appropriately chosen functions. For further details on this topic, refer to References [8–21].

The current paper studies the following coupled system of nonlinear fractional differential equations:

$$\begin{cases} {}^cD^\alpha x(t) = f(t, x(t), y(t)), & t \in [0, T], \quad 1 < \alpha \le 2, \\ \\ {}^cD^\beta y(t) = g(t, x(t), y(t)), & t \in [0, T], \quad 1 < \beta \le 2, \end{cases} \tag{1}$$

supplemented with boundary conditions of the form:

$$x(T) = \eta y'(\rho), \quad y(T) = \zeta x\prime(\mu), \quad x(0) = 0, \quad y(0) = 0, \rho, \mu \in [0, T] \tag{2}$$

Here, ${}^cD^k$ denotes Caputo fractional derivative of order k ($k = \alpha, \beta$); and $f, g \in C\big([0, T] \times \mathbb{R}^2, \mathbb{R}\big)$ are given continuous functions. Note that η, ζ are real constants such that $T^2 - \eta\zeta \ne 0$.

The rest of this paper is organized in the following manner: In Section 2, we briefly review some of the relevant definitions from fractional calculus and prove an auxiliary lemma that will be used later. Section 3 deals with proving the existence and uniqueness of solutions for the given problem, and Section 4 discusses the Hyers-Ulam stability of solutions and presents sufficient conditions for the stability. The paper concludes with supporting examples and obtained results.

2. Preliminaries

We begin this section by reviewing the definitions of fractional derivative and integral [1,2].

Definition 1. *The Riemann-Liouville fractional integral of order τ for a continuous function h is given by*

$$I^\tau h(s) = \frac{1}{\Gamma(\tau)}\int_0^s \frac{h(t)}{(s-t)^{1-\tau}}dt, \quad \tau > 0,$$

provided that the right-hand side is point-wise defined on $[0, \infty)$.

Definition 2. *The Caputo fractional derivatives of order* τ *for* $(h-1)$*—times absolutely continuous function* $g : [0, \infty) \to \mathbb{R}$ *is defined as*

$$^{c}D^{\tau} g(s) = \frac{1}{\Gamma(h-\tau)} \int_0^s (s-t)^{h-\tau-1} g^{(h)}(t) dt, \quad h-1 < \tau < h, \quad h = [\tau]+1,$$

where $[\tau]$ *is the integer part of real number* τ.

Here we prove the following auxiliary lemma that will be used in the next section.

Lemma 1. *Let* $u, v \in C([0,T], \mathbb{R})$ *then the unique solution for the problem*

$$\begin{cases} ^{c}D^{\alpha}x(t) = u(t), & t \in [0,T], \quad 1 < \alpha \leq 2, \\ ^{c}D^{\beta}y(t) = v(t), & t \in [0,T], \quad 1 < \beta \leq 2, \\ x(T) = \eta y'(\rho), \quad y(T) = \zeta x'(\mu), & x(0) = 0, \quad y(0) = 0, \rho, \mu \in [0,T] \end{cases} \tag{3}$$

is

$$\begin{aligned} x(t) = \; & \tfrac{t}{\Delta}\Big(\eta T \int_0^{\rho} \tfrac{(\rho-s)^{\beta-2}}{\Gamma(\beta-1)} v(s)ds - T \int_0^T \tfrac{(T-s)^{\alpha-1}}{\Gamma(\alpha)} u(s)ds + \eta\zeta \int_0^{\mu} \tfrac{(\mu-s)^{\alpha-2}}{\Gamma(\alpha-1)} u(s)ds - \eta \int_0^T \tfrac{(T-s)^{\beta-1}}{\Gamma(\beta)} v(s)ds\Big) \\ & + \int_0^t \tfrac{(t-s)^{\alpha-1}}{\Gamma(\alpha)} u(s)ds, \end{aligned} \tag{4}$$

and

$$\begin{aligned} y(t) = \; & \tfrac{t}{\Delta}\Big(\eta\zeta \int_0^{\rho} \tfrac{(\rho-s)^{\beta-2}}{\Gamma(\beta-1)} v(s)ds - \zeta \int_0^T \tfrac{(T-s)^{\alpha-1}}{\Gamma(\alpha)} u(s)ds + T\zeta \int_0^{\mu} \tfrac{(\mu-s)^{\alpha-2}}{\Gamma(\alpha-1)} u(s) - T \int_0^T \tfrac{(T-s)^{\beta-1}}{\Gamma(\beta)} v(s)ds\Big) \\ & + \int_0^t \tfrac{(t-s)^{\beta-1}}{\Gamma(\beta)} v(s)ds \end{aligned} \tag{5}$$

where $\Delta = T^2 - \eta\zeta \neq 0$.

Proof. General solutions of the fractional differential equations in (3) are known [6] as

$$\begin{aligned} x(t) &= at + b + \tfrac{1}{\Gamma(\alpha)} \int_0^t (t-s)^{\alpha-1} u(s)ds, \\ y(t) &= ct + d + \tfrac{1}{\Gamma(\beta)} \int_0^t (t-s)^{\beta-1} v(s)ds, \end{aligned} \tag{6}$$

where $a, b, c,$ and d are arbitrary constants.

Apply conditions $x(0) = 0$ and $y(0) = 0$, and we obtain $b = d = 0$.

Here

$$x\prime(t) = a + \frac{1}{\Gamma(\alpha-1)} \int_0^t (t-s)^{\alpha-2} u(s)ds,$$

$$y'(t) = c + \frac{1}{\Gamma(\beta-1)} \int_0^t (t-s)^{\beta-2} v(s)ds.$$

Considering boundary conditions

$$x(T) = \eta y'(\rho), \quad y(T) = \zeta x\prime(\mu)$$

we get

$$aT + \int_0^T \frac{(T-s)^{\alpha-1}}{\Gamma(\alpha)} u(s)ds = \eta c + \eta \int_0^{\rho} \frac{(\rho-s)^{\beta-2}}{\Gamma(\beta-1)} v(s)ds,$$

and

$$cT+\int_0^T \frac{(T-s)^{\beta-1}}{\Gamma(\beta)}v(s)ds = a\zeta+\zeta\int_0^{\mu}\frac{(\mu-s)^{\alpha-2}}{\Gamma(\alpha-1)}u(s)ds,$$

so

$$a=\frac{1}{T}\left(\eta c+\eta\int_0^{\rho}\frac{(\rho-s)^{\beta-2}}{\Gamma(\beta-1)}v(s)ds-\int_0^T\frac{(T-s)^{\alpha-1}}{\Gamma(\alpha)}u(s)ds\right),$$

$$c=\frac{1}{T}\left(a\zeta+\zeta\int_0^{\mu}\frac{(\mu-s)^{\alpha-2}}{\Gamma(\alpha-1)}u(s)ds-\int_0^T\frac{(T-s)^{\beta-1}}{\Gamma(\beta)}v(s)ds\right).$$

Hence, by substituting the value of a into c, we obtain the final result for these constants as

$$c=\frac{1}{T}\left(\frac{\zeta}{T}\left[\eta c+\eta\int_0^{\rho}\frac{(\rho-s)^{\beta-2}}{\Gamma(\beta-1)}v(s)ds-\int_0^T\frac{(T-s)^{\alpha-1}}{\Gamma(\alpha)}u(s)ds\right]+\zeta\int_0^{\mu}\frac{(\mu-s)^{\alpha-2}}{\Gamma(\alpha-1)}u(s)ds-\int_0^T\frac{(T-s)^{\beta-1}}{\Gamma(\beta)}v(s)ds\right),$$

$$c-\frac{\zeta\eta c}{T^2}=\frac{1}{T}\left(\frac{\zeta}{T}\left[\eta\int_0^{\rho}\frac{(\rho-s)^{\beta-2}}{\Gamma(\beta-1)}v(s)ds-\int_0^T\frac{(T-s)^{\alpha-1}}{\Gamma(\alpha)}u(s)ds\right]+\zeta\int_0^{\mu}\frac{(\mu-s)^{\alpha-2}}{\Gamma(\alpha-1)}u(s)ds-\int_0^T\frac{(T-s)^{\beta-1}}{\Gamma(\beta)}v(s)ds\right),$$

$$c\left(\frac{T^2-\zeta\eta}{T^2}\right)=\frac{1}{T}\left(\frac{\zeta}{T}\left[\eta\int_0^{\rho}\frac{(\rho-s)^{\beta-2}}{\Gamma(\beta-1)}v(s)ds-\int_0^T\frac{(T-s)^{\alpha-1}}{\Gamma(\alpha)}u(s)ds\right]+\zeta\int_0^{\mu}\frac{(\mu-s)^{\alpha-2}}{\Gamma(\alpha-1)}u(s)ds-\int_0^T\frac{(T-s)^{\beta-1}}{\Gamma(\beta)}v(s)ds\right),$$

$$c=\frac{T}{T^2-\zeta\eta}\left(\frac{\zeta}{T}\left[\eta\int_0^{\rho}\frac{(\rho-s)^{\beta-2}}{\Gamma(\beta-1)}v(s)ds-\int_0^T\frac{(T-s)^{\alpha-1}}{\Gamma(\alpha)}u(s)ds\right]+\zeta\int_0^{\mu}\frac{(\mu-s)^{\alpha-2}}{\Gamma(\alpha-1)}u(s)ds-\int_0^T\frac{(T-s)^{\beta-1}}{\Gamma(\beta)}v(s)ds\right),$$

$$c=\frac{1}{T^2-\zeta\eta}\left(\eta\zeta\int_0^{\rho}\frac{(\rho-s)^{\beta-2}}{\Gamma(\beta-1)}v(s)ds-\zeta\int_0^T\frac{(T-s)^{\alpha-1}}{\Gamma(\alpha)}u(s)ds+T\zeta\int_0^{\mu}\frac{(\mu-s)^{\alpha-2}}{\Gamma(\alpha-1)}u(s)ds-T\int_0^T\frac{(T-s)^{\beta-1}}{\Gamma(\beta)}v(s)ds\right)$$

$$c=\frac{1}{\Delta}\left(\eta\zeta\int_0^{\rho}\frac{(\rho-s)^{\beta-2}}{\Gamma(\beta-1)}v(s)ds-\zeta\int_0^T\frac{(T-s)^{\alpha-1}}{\Gamma(\alpha)}u(s)ds+T\zeta\int_0^{\mu}\frac{(\mu-s)^{\alpha-2}}{\Gamma(\alpha-1)}u(s)ds-T\int_0^T\frac{(T-s)^{\beta-1}}{\Gamma(\beta)}v(s)ds\right),$$

and

$$a=\frac{1}{\Delta}\left(\eta T\int_0^{\rho}\frac{(\rho-s)^{\beta-2}}{\Gamma(\beta-1)}v(s)ds-T\int_0^T\frac{(T-s)^{\alpha-1}}{\Gamma(\alpha)}u(s)ds+\eta\zeta\int_0^{\mu}\frac{(\mu-s)^{\alpha-2}}{\Gamma(\alpha-1)}u(s)ds-\eta\int_0^T\frac{(T-s)^{\beta-1}}{\Gamma(\beta)}v(s)ds\right),$$

Substituting the values of $a, b, c,$ and d in (6) and (7) we get (4) and (5). The converse follows by direct computation. This completes the proof. □

3. Existence and Uniqueness of Solutions

Consider the space $C([0,T],\mathbb{R})$ endowed with norm $\|x\| = \sup_{0\le t\le T}|x(t)|$. Consequently, the product space $C([0,T],\mathbb{R})\times C([0,T],\mathbb{R})$ is a Banach Space (endowed with $\|(x,y)\|=\|x\|+\|y\|$).

In view of Lemma 1, we define the operator $G: C([0,T],\mathbb{R})\times C([0,T],\mathbb{R})\to C([0,T],\mathbb{R})\times C([0,T],\mathbb{R})$ as:

$$G(x,y)(t)=(G_1(x,y)(t),G_2(x,y)(t)),$$

where

$$\begin{aligned}G_1(x,y)(t)=\frac{t}{\Delta}&\left(\eta T\int_0^{\rho}\frac{(\rho-s)^{\beta-2}}{\Gamma(\beta-1)}g(s,x(s),y(s))ds-T\int_0^T\frac{(T-s)^{\alpha-1}}{\Gamma(\alpha)}f(s,x(s),y(s))ds\right.\\&\left.+\eta\zeta\int_0^{\mu}\frac{(\mu-s)^{\alpha-2}}{\Gamma(\alpha-1)}f(s,x(s),y(s))ds-\eta\int_0^T\frac{(T-s)^{\beta-1}}{\Gamma(\beta)}g(s,x(s),y(s))ds\right)\\&+\int_0^t\frac{(t-s)^{\alpha-1}}{\Gamma(\alpha)}f(s,x(s),y(s))ds,\end{aligned}\quad(7)$$

and

$$G_2(x,y)(t) = \frac{t}{\Delta}\Big(\eta\zeta\int_0^{\rho}\frac{(\rho-s)^{\beta-2}}{\Gamma(\beta-1)}g(s,x(s),y(s))ds - \zeta\int_0^{T}\frac{(T-s)^{\alpha-1}}{\Gamma(\alpha)}f(s,x(s),y(s))ds$$
$$+T\zeta\int_0^{\mu}\frac{(\mu-s)^{\alpha-2}}{\Gamma(\alpha-1)}f(s,x(s),y(s))ds - T\int_0^{T}\frac{(T-s)^{\beta-1}}{\Gamma(\beta)}g(s,x(s),y(s))ds\Big) \tag{8}$$
$$+\int_0^{t}\frac{(t-s)^{\beta-1}}{\Gamma(\beta)}f(s,x(s),y(s))ds,$$

Here we establish the existence of the solutions for the boundary value problem (1) and (2) by using Banach's contraction mapping principle.

Theorem 1. *Assume $f,g : C([0,T]\times\mathbb{R}^2 \to \mathbb{R}$ are jointly continuous functions and there exist constants $\phi,\psi\in\mathbb{R}$, such that $\forall\, x_1,x_2,y_1,y_2\in\mathbb{R}, \forall t\in[0,T]$, we have*

$$\big|f(t,x_1,x_2) - f(t,y_1,y_2)\big| \le \phi\big(|x_2-x_1| + |y_2-y_1|\big),$$

$$\big|g(t,x_1,x_2) - f(t,y_1,y_2)\big| \le \psi\big(|x_2-x_1| + |y_2-y_1|\big),$$

where

$$\phi(Q_1+Q_3) + \psi(Q_2+Q_4) < 1,$$

then the BVP (1) and (2) has a unique solution on $[0,T]$. Here

$$\begin{aligned} Q_1 &= \frac{T}{|\Delta|}\left(\frac{T^{\alpha+1}}{\Gamma(\alpha+1)} + \frac{|\eta\zeta|\mu^{\alpha-1}}{\Gamma(\alpha)}\right) + \frac{T^{\alpha}}{\Gamma(\alpha+1)},\\ Q_2 &= \frac{T}{|\Delta|}\left(\frac{|\eta|T\rho^{\beta-1}}{\Gamma(\beta)} + \frac{|\eta|T^{\beta}}{\Gamma(\beta+1)}\right),\\ Q_3 &= \frac{T}{|\Delta|}\left(\frac{|\zeta|T^{\alpha}}{\Gamma(\alpha+1)} + \frac{T|\zeta|\mu^{\alpha-1}}{\Gamma(\alpha)}\right),\\ Q_4 &= \frac{T}{|\Delta|}\left(\frac{|\eta\zeta|\rho^{\beta-1}}{\Gamma(\beta)} + \frac{T^{\beta+1}}{\Gamma(\beta+1)}\right) + \frac{T^{\beta}}{\Gamma(\beta+1)}. \end{aligned} \tag{9}$$

Proof. Define $\sup_{0\le t\le T}\big|f(t,0,0)\big| = f_0 < \infty$, $\sup_{0\le t\le T}\big|g(t,0,0)\big| = g_0 < \infty$ and $\Omega_\varepsilon = \{(x,y)\in C([0,T],\mathbb{R})\times C([0,T],\mathbb{R}) : \|(x,y)\| \le \varepsilon\}$, and $\varepsilon > 0$, such that

$$\varepsilon \ge \frac{(Q_1+Q_3)f_0 + (Q_2+Q_4)g_0}{1-[\phi(Q_1+Q_3)+\psi(Q_2+Q_4)]}.$$

Firstly, we show that $G\Omega_\varepsilon \subseteq \Omega_\varepsilon$.
By our assumption, for $(x,y)\in\Omega_\varepsilon, t\in[0,T]$, we have

$$\begin{aligned} \big|f(t,x(t),y(t))\big| &\le \big|f(t,x(t),y(t)) - f(t,0,0)\big| + \big|f(t,0,0)\big|,\\ &\le \phi\big(|x(t)| + |y(t)|\big) + f_0 \le \phi(\|x\|+\|y\|) + f_0,\\ &\le \phi\varepsilon + f_0, \end{aligned}$$

and

$$\begin{aligned} \big|g(t,x(t),y(t))\big| &\le \psi\big(|x(t)| + |y(t)|\big) + g_0 \le \psi(\|x\|+\|y\|) + g_0,\\ &\le \psi\varepsilon + g_0, \end{aligned}$$

which lead to

$$\begin{aligned}\left|G_1(x,y)(t)\right| \le\ & \frac{T}{|\Delta|}\Big(|\eta| T\int_0^{\rho}\frac{(\rho-s)^{\beta-2}}{\Gamma(\beta-1)}ds(\psi(\|x\|+\|y\|)+g_0)\\ &+T\int_0^{T}\frac{(T-s)^{\alpha-1}}{\Gamma(\alpha)}ds(\phi(\|x\|+\|y\|)+f_0)\\ &+|\eta\zeta|\int_0^{\mu}\frac{(\mu-s)^{\alpha-2}}{\Gamma(\alpha-1)}ds(\phi(\|x\|+\|y\|)+f_0)\\ &+|\eta|\int_0^{T}\frac{(T-s)^{\beta-1}}{\Gamma(\beta)}ds(\psi(\|x\|+\|y\|)+g_0)\Big)\\ &+\sup_{0\le t\le T}\int_0^{t}\frac{(t-s)^{\alpha-1}}{\Gamma(\alpha)}ds(\phi(\|x\|+\|y\|)+f_0)\\ &\le(\phi(\|x\|+\|y\|)+f_0)\left[\frac{T}{|\Delta|}\left(\frac{T^{\alpha+1}}{\Gamma(\alpha+1)}+\frac{|\eta\zeta|\mu^{\alpha-1}}{\Gamma(\alpha)}\right)+\frac{T^{\alpha}}{\Gamma(\alpha+1)}\right]\\ &+(\psi(\|x\|+\|y\|)+g_0)\left[\frac{T}{|\Delta|}\left(\frac{|\eta|T\rho^{\beta-1}}{\Gamma(\beta)}+\frac{|\eta|T^{\beta}}{\Gamma(\beta+1)}\right)\right]\\ &\le(\phi(\|x\|+\|y\|)+f_0)Q_1+(\psi(\|x\|+\|y\|)+g_0)Q_2\\ &\le(\phi\varepsilon+f_0)Q_1+(\psi\varepsilon+g_0)Q_2.\end{aligned}$$

In a similar manner:

$$\left|G_2(x,y)(t)\right| \le (\phi(\|x\|+\|y\|)+f_0)Q_3+(\psi(\|x\|+\|y\|)+g_0)Q_4 \le (\phi\varepsilon+f_0)Q_3+(\psi\varepsilon+g_0)Q_4.$$

Hence,

$$\|G_1(x,y)\| \le (\phi\varepsilon+f_0)Q_1+(\psi\varepsilon+g_0)Q_2,$$

and

$$\|G_2(x,y)\| \le (\phi\varepsilon+f_0)Q_3+(\psi\varepsilon+g_0)Q_4.$$

Consequently,

$$\|G(x,y)\| \le (\phi\varepsilon+f_0)(Q_1+Q_3)+(\psi\varepsilon+g_0)(Q_2+Q_4) \le \varepsilon\ .$$

and we get $\|G(x,y)\| \le \varepsilon$ that is $G\Omega_\varepsilon \subseteq \Omega_\varepsilon$.

Now let $(x_1,y_1),(x_2,y_2) \in C([0,T],\mathbb{R}) \times C([0,T],\mathbb{R}), \forall t \in [0,T]$.

Then we have

$$\begin{aligned}\left|G_1(x_1,y_1)(t)-\right.&\left. G_1(x_2,y_2)(t)\right|\\ &\le \frac{T}{|\Delta|}\Big(|\eta| T\int_0^{\rho}\frac{(\rho-s)^{\beta-2}}{\Gamma(\beta-1)}ds\psi(\|x_2-x_1\|+\|y_2-y_1\|)\\ &+T\int_0^{T}\frac{(T-s)^{\alpha-1}}{\Gamma(\alpha)}ds\phi(\|x_2-x_1\|+\|y_2-y_1\|)\\ &+|\eta\zeta|\int_0^{\mu}\frac{(\mu-s)^{\alpha-2}}{\Gamma(\alpha-1)}ds\phi(\|x_2-x_1\|+\|y_2-y_1\|)\\ &+|\eta|\int_0^{T}\frac{(T-s)^{\beta-1}}{\Gamma(\beta)}ds\psi(\|x_2-x_1\|+\|y_2-y_1\|)\Big)\\ &+\sup_{0\le t\le T}\int_0^{t}\frac{(t-s)^{\alpha-1}}{\Gamma(\alpha)}ds\phi(\|x_2-x_1\|+\|y_2-y_1\|),\end{aligned}$$

$$\|G_1(x_1,y_1)-G_1(x_2,y_2)\| \le Q_1\phi(\|x_2-x_1\|+\|y_2-y_1\|)+Q_2\psi(\|x_2-x_1\|+\|y_2-y_1\|). \tag{10}$$

and likewise

$$\|G_2(x_1,y_1)-G_2(x_2,y_2)\| \le Q_3\phi(\|x_2-x_1\|+\|y_2-y_1\|)+Q_4\psi(\|x_2-x_1\|+\|y_2-y_1\|). \tag{11}$$

From (11) and (12) we have

$$\|G(x_1,y_1)-G(x_2,y_2)\| \le (\phi(Q_1+Q_3)+\psi(Q_2+Q_4))(\|x_2-x_1\|+\|y_2-y_1\|).$$

Since $\phi(Q_1 + Q_3) + \psi(Q_2 + Q_4) < 1$, therefore, the operator G is a contraction operator. Hence, by Banach's fixed-point theorem, the operator G has a unique fixed point, which is the unique solution of the BVP (1) and (2). This completes the proof. □

Next we will prove the existence of solutions by applying the Leray-Schauder alternative.

Lemma 2. *"(Leray-Schauder alternative [7], p. 4) Let $F : E \to E$ be a completely continuous operator (i.e., a map restricted to any bounded set in E is compact). Let $E(F) = \{x \in E : x = \lambda F(x)\ for\ some\ 0 < \lambda < 1\}$. Then either the set $E(F)$ is unbounded or F has at least one fixed point)".*

Theorem 2. *Assume $f, g : C([0,T] \times \mathbb{R}^2 \to \mathbb{R}$ are continuous functions and there exist $\theta_1, \theta_2, \lambda_1, \lambda_2 \geq 0$ where $\theta_1, \theta_2, \lambda_1, \lambda_2$ are real constants and $\theta_0, \lambda_0 > 0$ such that $\forall x_i, y_i \in \mathbb{R}$, $(i = 1, 2)$, we have*

$$\left|f(t, x_1, x_2)\right| \leq \theta_0 + \theta_1|x_1| + \theta_2|x_2|,$$

$$\left|g(t, x_1, x_2)\right| \leq \lambda_0 + \lambda_1|x_1| + \lambda_2|x_2|,$$

If

$$(Q_1 + Q_3)\theta_1 + (Q_2 + Q_4)\lambda_1 < 1,$$

and

$$(Q_1 + Q_3)\theta_2 + (Q_2 + Q_4)\lambda_2 < 1,$$

where $Q_i, i = 1, 2, 3, 4$ are defined in (10), then the problem (1) and (2) has at least one solution.

Proof. This proof will be presented in two steps.

Step 1: We will show that $G : C([0,T], \mathbb{R}) \times C([0,T], \mathbb{R}) \to C([0,T], \mathbb{R}) \times C([0,T], \mathbb{R})$ is completely continuous. The continuity of the operator G holds by the continuity of the functions f, g.

Let $B \subseteq C([0,T], \mathbb{R}) \times C([0,T], \mathbb{R})$ be bounded. Then there exists positive constants k_1, k_2 such that

$$\left|f(t, x(t), y(t))\right| \leq k_1, \quad \left|g(t, x(t), y(t))\right| \leq k_2, \quad \forall t \in [0,T].$$

Then $\forall (x, y) \in B$, and we have

$$\left|G_1(x, y)(t)\right| \leq Q_1 k_1 + Q_2 k_2,$$

which implies

$$\|G_1(x, y)\| \leq Q_1 k_1 + Q_2 k_2,$$

and similarly

$$\|G_2(x, y)\| \leq Q_3 k_1 + Q_4 k_2.$$

Thus, from the above inequalities, it follows that the operator G is uniformly bounded, since

$$\|G(x, y)\| \leq (Q_1 + Q_3)k_1 + (Q_2 + Q_4)k_2.$$

Next, we will show that operator G is equicontinuous. Let $\omega_1, \omega_2 \in [0,T]$ with $\omega_1 < \omega_2$. This yields

$$\begin{aligned}
\big|G_1(x,y)(\omega_2)- \quad & G_1(x,y)(\omega_1)\big| \\
& \leq \tfrac{\omega_2-\omega_1}{|\Delta|}\Big(\big|\eta\big|T\int_0^{\rho}\tfrac{(\rho-s)^{\beta-2}}{\Gamma(\beta-1)}\big|g(s,x(s),y(s))\big|ds \\
& +T\int_0^{T}\tfrac{(T-s)^{\alpha-1}}{\Gamma(\alpha)}\big|f(s,x(s),y(s))\big|ds+\big|\eta\zeta\big|\int_0^{\mu}\tfrac{(\mu-s)^{\alpha-2}}{\Gamma(\alpha-1)}\big|f(s,x(s),y(s))\big|ds \\
& +\big|\eta\big|\int_0^{T}\tfrac{(T-s)^{\beta-1}}{\Gamma(\beta)}\big|g(s,x(s),y(s))\big|ds\Big) \\
& +\Big|\int_0^{\omega_2}\tfrac{(\omega_2-s)^{\alpha-1}}{\Gamma(\alpha)}f(s,x(s),y(s))ds \\
& -\int_0^{\omega_1}\tfrac{(\omega_1-s)^{\alpha-1}}{\Gamma(\alpha)}f(s,x(s),y(s))ds\Big| \\
& \leq \tfrac{\omega_2-\omega_1}{|\Delta|}\Big(\big|\eta\big|Tk_2\int_0^{\rho}\tfrac{(\rho-s)^{\beta-2}}{\Gamma(\beta-1)}ds+Tk_1\int_0^{T}\tfrac{(T-s)^{\alpha-1}}{\Gamma(\alpha)}ds+\big|\eta\zeta\big|k_1\int_0^{\mu}\tfrac{(\mu-s)^{\alpha-2}}{\Gamma(\alpha-1)}ds \\
& +\big|\eta\big|k_2\int_0^{T}\tfrac{(T-s)^{\beta-1}}{\Gamma(\beta)}ds\Big)+\Big|\int_0^{\omega_1}\Big(\tfrac{(\omega_2-s)^{\alpha-1}}{\Gamma(\alpha)}-\tfrac{(\omega_1-s)^{\alpha-1}}{\Gamma(\alpha)}\Big)f(s,x(s),y(s))ds\Big| \\
& +\Big|\int_{\omega_1}^{\omega_2}\tfrac{(\omega_2-s)^{\alpha-1}}{\Gamma(\alpha)}f(s,x(s),y(s))ds\Big|, \\
& \leq \tfrac{\omega_2-\omega_1}{|\Delta|}\Big(\tfrac{k_2|\eta|T\rho^{\beta-1}}{\Gamma(\beta)}+\tfrac{k_1T^{\alpha+1}}{\Gamma(\alpha+1)}+\tfrac{k_1|\eta\zeta|\mu^{\alpha-1}}{\Gamma(\alpha)}+\tfrac{k_2|\eta|T^{\beta}}{\Gamma(\beta+1)}\Big) \\
& +\tfrac{k_1}{\Gamma(\alpha)}\Big(\int_0^{\omega_1}\big((\omega_2-s)^{\alpha-1}-(\omega_1-s)^{\alpha-1}\big)ds+\int_{\omega_1}^{\omega_2}(\omega_2-s)^{\alpha-1}ds\Big).
\end{aligned}$$

And we obtain

$$\begin{aligned}
\big|G_1(x,y)(\omega_2)-G_1(x,y)(\omega_1)\big| & \leq \tfrac{\omega_2-\omega_1}{|\Delta|}\Big(\tfrac{k_2|\eta|T\rho^{\beta-1}}{\Gamma(\beta)}+\tfrac{k_1T^{\alpha+1}}{\Gamma(\alpha+1)}+\tfrac{k_1|\eta\zeta|\mu^{\alpha-1}}{\Gamma(\alpha)}+\tfrac{k_2|\eta|T^{\beta}}{\Gamma(\beta+1)}\Big) \\
& +\tfrac{k_1}{\Gamma(\alpha+1)}[{\omega_2}^{\alpha}-{\omega_1}^{\alpha}].
\end{aligned}$$

Hence, we have $\|G_1(x,y)(\omega_2)-G_1(x,y)(\omega_1)\| \to 0$ independent of x and y as $\omega_2 \to \omega_1$. Furthermore, we obtain

$$\begin{aligned}
\big|G_2(x,y)(\omega_2)-G_2(x,y)(\omega_1)\big| & \leq \tfrac{\omega_2-\omega_1}{|\Delta|}\Big(\tfrac{k_2|\eta\zeta|\rho^{\beta-1}}{\Gamma(\beta)}+\tfrac{k_1|\zeta|T^{\alpha}}{\Gamma(\alpha+1)}+\tfrac{k_1T|\zeta|\mu^{\alpha-1}}{\Gamma(\alpha)}+\tfrac{k_2T^{\beta+1}}{\Gamma(\beta+1)}\Big) \\
& +\tfrac{k_2}{\Gamma(\beta+1)}\big[{\omega_2}^{\beta}-{\omega_1}^{\beta}\big],
\end{aligned}$$

which implies that $\|G_2(x,y)(\omega_2)-G_2(x,y)(\omega_1)\| \to 0$ independent of x and y as $\omega_2 \to \omega_1$.

Therefore, operator $G(x,y)$ is equicontinuous, and thus $G(x,y)$ is completely continuous.

Step 2: (Boundedness of operator)

Finally, we will show that $Z = \{(x,y) \in C([0,T],\mathbb{R}) \times C([0,T],\mathbb{R}) : (x,y) = hG(x,y), h \in [0,1]\}$ is bounded. Let $(x,y) \in \mathbb{R}$, with $(x,y) = hG(x,y)$ for any $t \in [0,T]$, we have

$$x(t) = hG_1(x,y)(t), \quad y(t) = hG_2(x,y)(t).$$

Then

$$\big|x(t)\big| \leq Q_1\big(\theta_0+\theta_1\big|x(t)\big|+\theta_2\big|y(t)\big|\big)+Q_2\big(\lambda_0+\lambda_1\big|x(t)\big|+\lambda_2\big|y(t)\big|\big),$$

and

$$\big|y(t)\big| \leq Q_3\big(\theta_0+\theta_1\big|x(t)\big|+\theta_2\big|y(t)\big|\big)+Q_4\big(\lambda_0+\lambda_1\big|x(t)\big|+\lambda_2\big|y(t)\big|\big).$$

Hence,

$$\|x\| \leq Q_1(\theta_0+\theta_1\|x\|+\theta_2\|y\|)+Q_2(\lambda_0+\lambda_1\|x\|+\lambda_2\|y\|),$$

and

$$\|y\| \leq Q_3(\theta_0+\theta_1\|x\|+\theta_2\|y\|)+Q_4(\lambda_0+\lambda_1\|x\|+\lambda_2\|y\|),$$

which implies

$$\begin{aligned}\|x\| + \|y\| &\le (Q_1 + Q_3)\theta_0 + (Q_2 + Q_4)\lambda_0 + ((Q_1 + Q_3)\theta_1 + (Q_2 + Q_4)\lambda_1)\|x\| \\ &+((Q_1 + Q_3)\theta_2 + (Q_2 + Q_4)\lambda_2)\|y\|.\end{aligned}$$

Therefore,

$$\|(x, y)\| \le \frac{(Q_1 + Q_3)\theta_0 + (Q_2 + Q_4)\lambda_0}{Q_0},$$

where $Q_0 = min\{1 - (Q_1 + Q_3)\theta_1 - (Q_2 + Q_4)\lambda_1, 1 - (Q_1 + Q_3)\theta_2 - (Q_2 + Q_4)\lambda_2\}$. This proves that Z is bounded and hence by Leray-Schauder alternative theorem, operator G has at least one fixed point. Therefore, the BVP (1) and (2) has at least one solution on $[0, T]$. This completes the proof. □

4. Hyers-Ulam Stability

In this section, we will discuss the Hyers-Ulam stability of the solutions for the BVP (1) and (2) by means of integral representation of its solution given by

$$x(t) = G_1(x, y)(t), y(t) = G_2(x, y)(t),$$

where G_1 and G_2 are defined by (8) and (9).

Define the following nonlinear operators $N_1, N_2 \in C([0, T], \mathbb{R}) \times C([0, T], \mathbb{R}) \to C([0, T], \mathbb{R})$;

$$\begin{aligned}{}^cD^\alpha x(t) - f(t, x(t), y(t)) &= N_1(x, y)(t), \quad t \in [0, T], \\ {}^cD^\beta y(t) - g(t, x(t), y(t)) &= N_2(x, y)(t), \quad t \in [0, T].\end{aligned}$$

For some $\varepsilon_1, \varepsilon_2 > 0$, we consider the following inequality:

$$N_1(x, y) \le \varepsilon_1, \quad N_2(x, y) \le \varepsilon_2. \tag{12}$$

Definition 3. ([8,9]). *The coupled system (1) and (2) is said to be Hyers-Ulam stable, if there exist $M_1, M_2 > 0$, such that for every solution $(x^*, y^*) \in C([0, T], \mathbb{R}) \times C([0, T], \mathbb{R})$ of the inequality (13), there exists a unique solution $(x, y) \in C([0, T], \mathbb{R}) \times C([0, T], \mathbb{R})$ of problems (1) and (2) with*

$$\|(x, y) - (x^*, y^*)\| \le M_1\varepsilon_1 + M_2\varepsilon_2.$$

Theorem 3. *Let the assumptions of Theorem 1 hold. Then the BVP (1) and (2) is Hyers-Ulam-stable.*

Proof. Let $(x, y) \in C([0, T], \mathbb{R}) \times C([0, T], \mathbb{R})$ be the solution of the problems (1) and (2) satisfying (8) and (9). Let (x^*, y^*) be any solution satisfying (13):

$$\begin{aligned}{}^cD^\alpha x^*(t) &= f(t, x^*(t), y^*(t)) + N_1(x^*, y^*)(t), \quad t \in [0, T], \\ {}^cD^\beta y^*(t) &= g(t, x^*(t), y^*(t)) + N_2(x^*, y^*)(t), \quad t \in [0, T].\end{aligned}$$

So

$$\begin{aligned}x^*(t) = \; & G_1(x^*, y^*)(t) \\ & + \tfrac{t}{\Delta}\Big(\eta T \int_0^\rho \tfrac{(\rho-s)^{\beta-2}}{\Gamma(\beta-1)} N_2(x^*, y^*)(s)ds - T \int_0^T \tfrac{(T-s)^{\alpha-1}}{\Gamma(\alpha)} N_1(x^*, y^*)(s)ds \\ & + \eta\zeta \int_0^\mu \tfrac{(\mu-s)^{\alpha-2}}{\Gamma(\alpha-1)} N_1(x^*, y^*)(s)ds - \eta \int_0^T \tfrac{(T-s)^{\beta-1}}{\Gamma(\beta)} N_2(x^*, y^*)(s)ds\Big) \\ & + \int_0^t \tfrac{(t-s)^{\alpha-1}}{\Gamma(\alpha)} N_1(x^*, y^*)(s)ds,\end{aligned}$$

It follows that

$$\begin{aligned}
\left|G_1(x^*,y^*)(t)- \right.&\left. x^*(t)\right| \\
&\le \frac{T}{|\Delta|}\Big(|\eta|T\int_0^\rho \frac{(\rho-s)^{\beta-2}}{\Gamma(\beta-1)}ds\varepsilon_2 + T\int_0^T \frac{(T-s)^{\alpha-1}}{\Gamma(\alpha)}ds\varepsilon_1 + |\eta\zeta|\int_0^\mu \frac{(\mu-s)^{\alpha-2}}{\Gamma(\alpha-1)}ds\varepsilon_1 \\
&+|\eta|\int_0^T \frac{(T-s)^{\beta-1}}{\Gamma(\beta)}ds\varepsilon_2\Big) + \int_0^T \frac{(T-s)^{\alpha-1}}{\Gamma(\alpha)}ds\varepsilon_1, \\
&\le \left[\frac{T}{|\Delta|}\Big(\frac{T^{\alpha+1}}{\Gamma(\alpha+1)} + \frac{|\eta\zeta|\mu^{\alpha-1}}{\Gamma(\alpha)}\Big) + \frac{T^\alpha}{\Gamma(\alpha+1)}\right]\varepsilon_1 + \frac{T}{|\Delta|}\Big(\frac{|\eta|T\rho^{\beta-1}}{\Gamma(\beta)} + \frac{|\eta|T^\beta}{\Gamma(\beta+1)}\Big)\varepsilon_2\,, \\
&\le Q_1\varepsilon_1 + Q_2\varepsilon_2.
\end{aligned}$$

Similarly,

$$\begin{aligned}
\left|G_1(x^*,y^*)(t)-x^*(t)\right| &\le \frac{T}{|\Delta|}\Big(\frac{|\zeta|T^\alpha}{\Gamma(\alpha+1)} + \frac{T|\zeta|\mu^{\alpha-1}}{\Gamma(\alpha)}\Big)\varepsilon_1 + \left[\frac{T}{|\Delta|}\Big(\frac{|\eta\zeta|\rho^{\beta-1}}{\Gamma(\beta)} + \frac{T^{\beta+1}}{\Gamma(\beta+1)}\Big) + \frac{T^\beta}{\Gamma(\beta+1)}\right], \\
&\le Q_3\varepsilon_1 + Q_4\varepsilon_2,
\end{aligned}$$

where $Q_i, i = 1,\ 2,\ 3,\ 4$ are defined in (10).

Therefore, we deduce by the fixed-point property of operator G, that is given by (8) and (9), which

$$\begin{aligned}
\left|x(t)-x^*(t)\right| &= \left|x(t) - G_1(x^*,y^*)(t) + G_1(x^*,y^*)(t) - x^*(t)\right| \\
&\le \left|G_1(x,y)(t) - G_1(x^*,y^*)(t)\right| + \left|G_1(x^*,y^*)(t) - x^*(t)\right| \\
&\le (Q_1\phi + Q_2\psi)(x,y) - (x^*,y^*) + Q_1\varepsilon_1 + Q_2\varepsilon_2,
\end{aligned} \tag{13}$$

and similarly

$$\begin{aligned}
\left|y(t)-y^*(t)\right| &= \left|y(t) - G_2(x^*,y^*)(t) + G_2(x^*,y^*)(t) - y^*(t)\right| \\
&\le \left|G_2(x,y)(t) - G_2(x^*,y^*)(t)\right| + \left|G_2(x^*,y^*)(t) - y^*(t)\right| \\
&\le (Q_3\phi + Q_4\psi)(x,y) - (x^*,y^*) + Q_3\varepsilon_1 + Q_4\varepsilon_2,
\end{aligned} \tag{14}$$

From (14) and (15) it follows that

$$\|(x,y)-(x^*,y^*)\| \le (Q_1\phi + Q_2\psi + Q_3\phi + Q_4\psi)\|(x,y)-(x^*,y^*)\| + (Q_1+Q_3)\varepsilon_1 + (Q_2+Q_4)\varepsilon_2,$$

$$\begin{aligned}
\|(x,y)-(x^*,y^*)\| &\le \frac{(Q_1+Q_3)\varepsilon_1 + (Q_2+Q_4)\varepsilon_2}{1-((Q_1+Q_3)\phi + (Q_2+Q_4)\psi)}, \\
&\le M_1\varepsilon_1 + M_2\varepsilon_2.
\end{aligned}$$

with

$$M_1 = \frac{(Q_1+Q_3)}{1-((Q_1+Q_3)\phi + (Q_2+Q_4)\psi)},$$

$$M_2 = \frac{(Q_2+Q_4)}{1-((Q_1+Q_3)\phi + (Q_2+Q_4)\psi)}.$$

Thus, sufficient conditions for the Hyers-Ulam stability of the solutions are obtained. □

5. Examples

Example 1. *Consider the following coupled system of fractional differential equations*

$$\begin{cases}
{}^cD^{\frac{3}{2}}x(t) = \frac{1}{6\pi\sqrt{81+t^2}}\Big(\frac{|x(t)|}{3+|x(t)|} + \frac{|y(t)|}{5+|x(t)|}\Big), \\
{}^cD^{\frac{7}{4}}y(t) = \frac{1}{12\pi\sqrt{64+t^2}}(sin(x(t)) + sin(y(t))), \\
x(1) = 2y'(1),\ \ y(1) = -x'(1/2),\ \ x(0) = 0,\ \ y(0) = 0,
\end{cases} \tag{15}$$

$$\alpha = \frac{3}{2}, \beta = \frac{7}{4}, T = 1, \eta = 2, \zeta = -1, \mu = \frac{1}{2}, \rho = 1.$$

Using the given data, we find that $\Delta = 3, Q_1 = 1.269, Q_2 = 1.1398, Q_3 = 0.5167, Q_4 = 1.554, \phi = \frac{1}{54\pi}, \psi = \frac{1}{48\pi}$.

It is clear that

$$f(t, x(t), y(t)) = \frac{1}{6\pi\sqrt{81+t^2}}\left(\frac{|x(t)|}{3+|x(t)|} + \frac{|y(t)|}{5+|x(t)|}\right),$$

and

$$g(t, x(t), y(t)) = \frac{1}{12\pi\sqrt{64+t^2}}(sin(x(t)) + sin(y(t))),$$

are jointly continuous functions and Lipschitz function with $\phi = \frac{1}{54\pi}, \psi = \frac{1}{48\pi}$. *Moreover,*

$$\frac{1}{54\pi}(1.269 + 0.5167) + \frac{1}{48\pi}(1.1398 + 1.554) = 0.0283 < 1.$$

Thus, all the conditions of Theorem 1 are satisfied, then problem (16) has a unique solution on $[0, 1]$, *which is Hyers-Ulam-stable.*

Example 2. *Consider the following system of fractional differential equation*

$$\begin{cases} {}^cD^{5/3}x(t) = \frac{1}{80+t^4} + \frac{|x(t)|}{120(1+y^2(t))} + \frac{1}{4\sqrt{2500+t^2}}e^{-3t}\cos(y(t)), \ t \in [0, 1] \\ {}^cD^{6/5}y(t) = \frac{1}{\sqrt{16+t^2}}\cos t + \frac{1}{150}e^{-3t}sin(y(t)) + \frac{1}{180}x(t), \ t \in [0, 1] \\ x(1) = -3y'(1/3), \ \ y(1) = x'(1), \ \ x(0) = 0, \ \ y(0) = 0, \end{cases} \tag{16}$$

$$\alpha = \frac{5}{3}, \beta = \frac{6}{5}, T = 1, \eta = -3, \zeta = 1, \mu = 1, \rho = 1/3.$$

Using the given data, we find that $\Delta = 3, Q_1 = 1.269, Q_2 = 1.1398, Q_3 = 0.5167, Q_4 = 1.554, \phi = \frac{1}{54\pi}, \psi = \frac{1}{48\pi}$.

It is clear that

$$|f(t, x, y)| \leq \frac{1}{80} + \frac{1}{120}|x| + \frac{1}{200}|y|,$$

$$|g(t, x, y)| \leq \frac{1}{4} + \frac{1}{180}|x| + \frac{1}{150}|y|.$$

Thus, $\theta_0 = \frac{1}{80}, \theta_1 = \frac{1}{120}, \theta_2 = \frac{1}{200}, \lambda_0 = \frac{1}{4}, \lambda_1 = \frac{1}{180}, \lambda_2 = \frac{1}{150}$.

Note that $(Q_1 + Q_3)\theta_1 + (Q_2 + Q_4)\lambda_1 = 0.0298 < 1$ *and* $(Q_1 + Q_3)\theta_2 + (Q_2 + Q_4)\lambda_2 = 0.0269 < 1$, *and hence by Theorem 2, problem (17) has at least one solution on* $[0, 1]$.

6. Conclusions

In this paper, the existence, uniqueness and the Hyers-Ulam stability of solutions for a coupled system of nonlinear fractional differential equations with boundary conditions were established and discussed.

Future studies may focus on different concepts of stability and existence results to a neutral time-delay system/inclusion, time-delay system/inclusion with finite delay.

Author Contributions: The authors have made the same contribution. All authors read and approved the final manuscript.

Funding: This research received no external funding.

Acknowledgments: The authors wish to thank the anonymous reviewers for their valuable comments and suggestions.

Conflicts of Interest: The authors declare no conflict of interest.

References

1. Podlubny, I. *Fractional Differential Equations*; Academic Press: San Diego, CA, USA, 1999.
2. Kilbas, A.A.; Srivastava, H.M.; Trujillo, J.J. *Theory and Applications of Fractional Differential Equations. North-Holland Mathematics Studies*; Elsevier: Amsterdam, The Netherlands, 2006; Volume 204.
3. Chalishajar, D.; Raja, D.S.; Karthikeyan, K.; Sundararajan, P. Existence results for nonautonomous impulsive fractional evolution equations. *Res. Nonlinear Anal.* **2018**, *1*, 133–147.
4. Chalishajar, D.; Kumar, A. Existence, uniqueness and Ulam's stability of solutions for a coupled system of fractional differential equations with integral boundary conditions. *Mathematics* **2018**, *6*, 96. [CrossRef]
5. Ntouyas, S.K.; Obaid, M. A coupled system of fractional differential equations with nonlocal integral boundary conditions. *Adv. Differ. Equ.* **2012**, *2012*, 130–139. [CrossRef]
6. Ahmad, B.; Ntouyas, S.K. A Coupled system of nonlocal fractional differential equations with coupled and uncoupled slit-strips-type integral boundary conditions. *J. Math. Sci.* **2017**, *226*, 175–196. [CrossRef]
7. Alsulami, H.H.; Ntouyas, S.K.; Agarwal, R.P.; Ahmad, B.; Alsaedi, A. A study of fractional-order coupled systems with a new concept of coupled non-separated boundary conditions. *Bound. Value Probl.* **2017**, *2017*, 68–74. [CrossRef]
8. Zhang, Y.; Bai, Z.; Feng, T. Existence results for a coupled system of nonlinear fractional three-point boundary value problems at resonance. *Comput. Math. Appl.* **2011**, *61*, 1032–1047. [CrossRef]
9. Granas, A.; Dugundji, J. *Fixed Point Theory*; Springer: New York, NY, USA, 2005.
10. Hyers, D.H. On the stability of the linear functional equation. *Proc. Nat. Acad. Sci. USA* **1941**, *27*, 222. [CrossRef] [PubMed]
11. Rus, I.A. Ulam stabilities of ordinary differential equations in a Banach space. *Carpathian J. Math.* **2010**, 103–107.
12. Cabada, A.; Wang, G. Positive solutions of nonlinear fractional differential equations with integral boundary value conditions. *J. Math. Anal. Appl.* **2012**, *389*, 403–411. [CrossRef]
13. Graef, J.R.; Kong, L.; Wang, M. Existence and uniqueness of solutions for a fractional boundary value problem on a graph. *Fract. Calc. Appl. Anal.* **2014**, *17*, 499–510. [CrossRef]
14. Ahmad, B.; Nieto, J.J. Existence results for a coupled system of nonlinear fractional differential equations with three-point boundary conditions. *Comput. Math. Appl.* **2009**, *58*, 1838–1843. [CrossRef]
15. Su, X. Boundary value problem for a coupled system of nonlinear fractional differential equations. *Appl. Math. Lett.* **2009**, *22*, 64–69. [CrossRef]
16. Wang, J.; Xiang, H.; Liu, Z. Positive solution to nonzero boundary values problem for a coupled system of nonlinear fractional differential equations. *Int. J. Differ. Equ.* **2010**, *10*, 12. [CrossRef]
17. Ahmad, B.; Ntouyas, S.K.; Alsaedi, A. On a coupled system of fractional differential equations with coupled nonlocal and integral boundary conditions. *Chaos Solitons Fractals* **2016**, *83*, 234–241. [CrossRef]
18. Zhai, C.; Xu, L. Properties of positive solutions to a class of four-point boundary value problem of Caputo fractional differential equations with a parameter. *Commun. Nonlinear Sci. Numer. Simul.* **2014**, *19*, 2820–2827. [CrossRef]
19. Ahmad, B.; Ntouyas, S.K. Existence results for a coupled system of Caputo type sequential fractional differential equations with nonlocal integral boundary conditions. *Appl. Math. Comput.* **2015**, *266*, 615–622. [CrossRef]
20. Tariboon, J.; Ntouyas, S.K.; Sudsutad, W. Coupled systems of Riemann-Liouville fractional differential equations with Hadamard fractional integral boundary conditions. *J. Nonlinear Sci. Appl.* **2016**, *9*, 295–308. [CrossRef]
21. Mahmudov, N.I.; Bawaneh, S.; Al-Khateeb, A. On a coupled system of fractional differential equations with four point integral boundary conditions. *Mathematics* **2019**, *7*, 279. [CrossRef]

mathematics

Article

On Separate Fractional Sum-Difference Equations with n-Point Fractional Sum-Difference Boundary Conditions via Arbitrary Different Fractional Orders

Saowaluck Chasreechai [1] **and Thanin Sitthiwiratham** [2,*]

[1] Department of Mathematics, Faculty of Applied Science, King Mongkut's University of Technology North Bangkok, Bangkok 10800, Thailand; saowaluck.c@sci.kmutnb.ac.th
[2] Mathematics Department, Faculty of Science and Technology, Suan Dusit University, Bangkok 10700, Thailand
* Correspondence: thanin_sit@dusit.ac.th

Received: 7 May 2019; Accepted: 20 May 2019; Published: 24 May 2019

Abstract: In this article, we study the existence and uniqueness results for a separate nonlinear Caputo fractional sum-difference equation with fractional difference boundary conditions by using the Banach contraction principle and the Schauder's fixed point theorem. Our problem contains two nonlinear functions involving fractional difference and fractional sum. Moreover, our problem contains different orders in $n+1$ fractional differences and $m+1$ fractional sums. Finally, we present an illustrative example.

Keywords: fractional sum-difference equations; boundary value problem; existence; uniqueness

JEL Classification: 39A05; 39A12

1. Introduction

Fractional calculus has recently been an attractive field to researchers because it is a powerful tool for explaining many engineering and scientific disciplines as the mathematical modeling of systems and processes which appear in nature, for example, ecology, biology, chemistry, physics, mechanics, networks, flow in porous media, electrical, control systems, viscoelasticity, mathematical biology, fitting of experimental data, and so forth. For example, Zhang et al. [1] proposed both analytical and numerical results from studying the propagation of optical beams in the fractional Schrödinger equation with a harmonic potential. In 2015, Zingales and Failla [2] solved the fractional-order heat conduction equation by using a pertinent finite element method. For Lazopoulos's [3] work, they defined the fractional curvature of plane curves, the fractional beam small deflection, the fractional curvature is approximate. In 2017, Sumelka and Voyiadjis [4] proposed a concept of short memory connected with the definition of damage parameter evolution in terms of fractional calculus for hyperelastic materials.

Basic definitions and properties of fractional difference calculus, appear in the book [5]. In particular, fractional calculus is a powerful tool for the processes which appear in nature, e.g., ecology, biology and other areas, one may see the papers [6–8] and the references therein. The interesting papers related to discrete fractional boundary value problems can be found in [9–29] and references cited therein. For previous works, Goodrich [10] considered the discrete fractional boundary value problem

$$\begin{cases} -\Delta^{\mu_1}\Delta^{\mu_2}\Delta^{\mu_3}y(t) = f(t+\mu_1+\mu_2+\mu_3-1, y(t+\mu_1+\mu_2+\mu_3-1)), \\ y(0) = 0 = y(b+2), \end{cases} \tag{1}$$

where $t \in \mathbb{N}_{2-\mu_1-\mu_2-\mu_3, b+2-\mu_1-\mu_2-\mu_3}$, $0 < \mu_1, \mu_2, \mu_3 < 1$, $1 < \mu_2 + \mu_3 < 2$, $1 < \mu_1 + \mu_2 + \mu_3 < 2$, $f : \mathbb{N}_0 \times \mathbb{R} \to [0, +\infty)$ is a continuous function, and Δ^{μ} is the Riemann-Liouville fractional difference operator of order μ. Existence of positive solutions are obtained by the use of the Krasnosel'skii fixed point theorem.

Weidong [12] examined the sequential fractional boundary value problem with a p-Laplacian

$$\begin{cases} \Delta_C^{\beta}[\phi_p(\Delta_C^{\alpha} x)](t) = f(t+\alpha+\beta-1, x(t+\alpha+\beta-1)), \quad t \in \mathbb{N}_{0,b}, \\ \Delta_C^{\beta} x(\beta-1) + \Delta_C^{\beta} x(\beta+b) = 0, \\ x(\alpha+\beta-2) + x(\alpha+\beta+b) = 0, \end{cases} \tag{2}$$

where $0 < \alpha, \beta \leq 1$, $1 < \alpha + \beta \leq 2$, $f : \mathbb{N}_{\alpha+\beta-1, \alpha+\beta+T-1} \times \mathbb{R} \to \mathbb{R}$ is a continuous function, ϕ_p is the p-Laplacian operator, and Δ_C^{β} is the Caputo fractional difference operator of order β. Existence and uniqueness of solutions are obtained by using the Schaefer's fixed point theorem.

Recently, Sitthiwirattham [19,20] investigated three-point fractional sum boundary value problems for sequential fractional difference equations of the forms

$$\begin{cases} \Delta_C^{\alpha}[\phi_p(\Delta_C^{\beta} x)](t) = f(t+\alpha+\beta-1, x(t+\alpha+\beta-1)), \\ \Delta_C^{\beta} x(\alpha-1) = 0, \quad x(\alpha+\beta+T) = \rho \Delta^{-\gamma} x(\eta+\gamma), \end{cases} \tag{3}$$

and

$$\begin{cases} \Delta_{\alpha}^{\alpha}(\Delta_{\alpha+\beta-1}^{\beta} + \lambda E_{\beta}) x(t) = f(t+\alpha+\beta-1, x(t+\alpha+\beta-1)), \\ x(\alpha+\beta-2) = 0, \quad x(\alpha+\beta+T) = \rho \Delta_{\alpha+\beta-1}^{-\gamma} x(\eta+\gamma), \end{cases} \tag{4}$$

where $t \in \mathbb{N}_{0,T}$, $0 < \alpha, \beta \leq 1$, $1 < \alpha + \beta \leq 2$, $0 < \gamma \leq 1$, $\eta \in \mathbb{N}_{\alpha+\beta-1, \alpha+\beta+T-1}$, ρ is a constant, $f : \mathbb{N}_{\alpha+\beta-2, \alpha+\beta+T} \times \mathbb{R} \to \mathbb{R}$ is a continuous function, $E_{\beta} x(t) = x(t+\beta-1)$ and ϕ_p is the p-Laplacian operator. Existence and uniqueness of solutions are obtained by using the Banach fixed point theorem and the Schaefer's fixed point theorem.

The results mentioned above are the motivation for this research. In this paper, we consider a separate nonlinear Caputo fractional sum-difference equation of the form

$$\begin{aligned} \left[\Delta_C^{\alpha} + (e+1)\,\Delta_C^{\alpha-1}\right] u(t) &= \lambda F\Big(t+\alpha-1, u(t+\alpha-1), (\Upsilon^{\vartheta} u)(t+\alpha-\vartheta)\Big) \\ &\quad + \mu H\Big(t+\alpha-1, u(t+\alpha-1), (\Psi^{\gamma} u)(t+\alpha+\gamma-1)\Big), \end{aligned} \tag{5}$$

with the fractional sum-difference boundary value conditions

$$\begin{aligned} u(\alpha-n) &= \Delta_C^{\beta_1} u(\alpha-n-\beta_1+2) = \Delta_C^{\beta_1+\beta_2} u(\alpha-n-\beta_1-\beta_2+4) = \ldots \\ &= \Delta_C^{\sum_{i=1}^{n-2}\beta_i} u\left(\alpha+n-4-\sum_{i=1}^{n-2}\beta_i\right) = 0, \\ u(T+\alpha) &= \tau \Delta^{-\sum_{i=1}^{m}\theta_i} g\left(\eta+\sum_{i=1}^{m}\theta_i\right) u\left(\eta+\sum_{i=1}^{m}\theta_i\right), \end{aligned} \tag{6}$$

where $t \in \mathbb{N}_{0,T} := \{0, 1, \ldots, T\}$, $\tau < \frac{\Gamma\left(\sum_{i=1}^{m}\theta_i\right) \sum_{s=\alpha-n}^{T+\alpha-n+1} e^{-s}}{\sum_{r=\alpha-n}^{\eta} \sum_{s=\alpha-n}^{r-n+1} \left(\eta+\sum_{i=1}^{m}\theta_i-\sigma(r)\right)^{\underline{\sum_{i=1}^{m}\theta_i-1}} e^{-s} g\left(\eta+\sum_{i=1}^{m}\theta_i\right)}$, $\alpha \in (n-1, n]$, $\beta_i, \theta_i, \gamma \in (0, 1]$, $m, n \in \mathbb{N}_4$, $m < n$, $T > n-3$, $\sum_{i=1}^{n-2}\beta_i \in (n-3, n-2]$, $\sum_{i=1}^{m}\theta_i \in (m-1, m]$ and $\lambda, \mu \in \mathbb{R}$ are given constants; $F \in C\big(\mathbb{N}_{\alpha-n, T+\alpha} \times \mathbb{R} \times \mathbb{R}, \mathbb{R}\big)$, $H \in C\big(\mathbb{N}_{\alpha-n, T+\alpha} \times \mathbb{R} \times \mathbb{R}, \mathbb{R}\big)$, $g \in C\big(\mathbb{N}_{\alpha-n, T+\alpha}, \mathbb{R}^{+}\big)$, and for $\varphi, \phi \in C\big(\mathbb{N}_{\alpha-n, T+\alpha} \times \mathbb{N}_{\alpha-n, T+\alpha}, [0, \infty)\big)$, we defined the operators

$$
\begin{aligned}
(\Upsilon^{\vartheta} u)(t-\vartheta+1) &:= [\Delta_C^{\vartheta} \phi\, u](t-\vartheta+1) \\
&= \frac{1}{\Gamma(1-\vartheta)} \sum_{s=\alpha-n+\vartheta-1}^{t+\vartheta-1} (t-\sigma(s))^{\underline{-\vartheta}} \phi(t, s-\vartheta+1) \Delta u(s-\vartheta+1),
\end{aligned}
$$

$$
\text{and } (\Psi^{\gamma} u)(t+\gamma) := [\Delta^{-\gamma} \varphi\, u](t+\gamma) = \frac{1}{\Gamma(\gamma)} \sum_{s=\alpha-n-\gamma}^{t-\gamma} (t-\sigma(s))^{\underline{\gamma-1}} \varphi(t, s+\gamma) u(s+\gamma).
$$

The plan of this paper is as follows. In Section 2 we recall some definitions and basic lemmas. We derive a representation for the solution of (5) by converting the problem to an equivalent summation equation. In Section 3, we prove existence results of the problem (5) by using the Banach contraction principle and the Schauder's theorem. Finally, an illustrative example is presented in Section 4.

2. Preliminaries

The notations, definitions, and lemmas which are used in the main results are as follows.

Definition 1. *We define the generalized falling function by* $t^{\underline{\alpha}} := \frac{\Gamma(t+1)}{\Gamma(t+1-\alpha)}$, *for any* t *and* α *for which the right-hand side is defined. If* $t+1-\alpha$ *is a pole of the Gamma function and* $t+1$ *is not a pole, then* $t^{\underline{\alpha}} = 0$.

Lemma 1 ([16]). *Assume the factorial functions are well defined. If* $t \leq r$, *then* $t^{\underline{\alpha}} \leq r^{\underline{\alpha}}$ *for any* $\alpha > 0$.

Definition 2. *For* $\alpha > 0$ *and* f *defined on* $\mathbb{N}_a := \{a, a+1, \ldots\}$, *the* α*-order fractional sum of* f *is defined by*

$$
\Delta^{-\alpha} f(t) := \frac{1}{\Gamma(\alpha)} \sum_{s=a}^{t-\alpha} (t-\sigma(s))^{\underline{\alpha-1}} f(s),
$$

where $t \in \mathbb{N}_{a+\alpha}$ *and* $\sigma(s) = s+1$.

Definition 3. *For* $\alpha > 0$ *and* f *defined on* $\mathbb{N}_a$, *the* α*-order Caputo fractional difference of* f *is defined by*

$$
\Delta_C^{\alpha} f(t) := \Delta^{-(N-\alpha)} \Delta^N f(t) = \frac{1}{\Gamma(N-\alpha)} \sum_{s=a}^{t-(N-\alpha)} (t-\sigma(s))^{\underline{N-\alpha-1}} \Delta^N f(s),
$$

where $t \in \mathbb{N}_{a+N-\alpha}$ *and* $N \in \mathbb{N}$ *is chosen so that* $0 \leq N-1 < \alpha < N$.

Lemma 2 ([14]). *Assume that* $\alpha > 0$ *and* $0 \leq N-1 < \alpha \leq N$. *Then*

$$
\Delta^{-\alpha} \Delta_C^{\alpha} y(t) = y(t) + C_0 + C_1 t^{\underline{1}} + C_2 t^{\underline{2}} + \ldots + C_{N-1} t^{\underline{N-1}},
$$

for some $C_i \in \mathbb{R}$, $0 \leq i \leq N-1$.

To investigate the solution of the boundary value problem (5) we need the following lemma involving a linear variant of the boundary value problem (5).

Lemma 3. *Let* $\tau < \frac{\Gamma\left(\sum_{i=1}^{m} \theta_i\right) \sum_{s=\alpha-n}^{T+\alpha-n+1} e^{-s}}{\sum_{r=\alpha-n}^{\eta} \sum_{s=\alpha-n}^{r-n+1} \left(\eta+\sum_{i=1}^{m} \theta_i - \sigma(r)\right)^{\underline{\sum_{i=1}^{m} \theta_i - 1}} e^{-s} g\left(\eta+\sum_{i=1}^{m} \theta_i\right)}$, $\alpha \in (n-1, n]$, $\beta_i, \theta_i, \gamma \in (0,1]$, $m, n \in \mathbb{N}_4$, $m < n$, $T > n-3$, $\sum_{i=1}^{n-2} \beta_i \in (n-3, n-2]$, $\sum_{i=1}^{m} \theta_i \in (m-1, m]$ *and* $h \in C(\mathbb{N}_{\alpha-n, T+\alpha} \times \mathbb{R}, \mathbb{R})$, $g \in C(\mathbb{N}_{\alpha-n, T+\alpha}, \mathbb{R}^+)$ *be given. Then the problem*

$$\left[\Delta_C^{\alpha} + (e-1)\,\Delta_C^{\alpha-1}\right] u(t) = h(t+\alpha-1), \quad t \in \mathbb{N}_{0,T} \tag{7}$$

$$\begin{aligned} u(\alpha-n) &= \Delta_C^{\beta_1} u\,(\alpha-n-\beta_1+2) \;=\; \Delta_C^{\beta_1+\beta_2} u\,(\alpha-n-\beta_1-\beta_2+4) \\ &= \ldots = \Delta_C^{\sum_{i=1}^{n-2}\beta_i} u\left(\alpha+n-4-\sum_{i=1}^{n-2}\beta_i\right) = 0, \end{aligned} \tag{8}$$

$$u(T+\alpha) = \tau\,\Delta^{-\sum_{i=1}^{m}\theta_i} g\left(\eta+\sum_{i=1}^{m}\theta_i\right) u\left(\eta+\sum_{i=1}^{m}\theta_i\right), \tag{9}$$

has the unique solution

$$\begin{aligned} u(t) &= \frac{\mathcal{O}[h]}{\Lambda}\sum_{s=\alpha-n}^{t-n+1} e^{-s} \\ &+\frac{1}{\Gamma(\alpha-n)}\sum_{s=\alpha-n}^{t-n+1}\sum_{v=\alpha-n}^{s-1}\sum_{\xi=0}^{v-\alpha+n} e^{v-s}(v-\sigma(\xi))^{\underline{\alpha-n-1}} h(\xi+\alpha-1), \end{aligned} \tag{10}$$

where the functional $\mathcal{O}[h]$ *and the constant* Λ *are defined by*

$$\begin{aligned} \mathcal{O}[h] &= \sum_{r=\alpha-n}^{\eta}\sum_{s=\alpha-n}^{r-n+1}\sum_{v=\alpha-n}^{s-1}\sum_{\xi=0}^{v-\alpha+n}\frac{\tau\,(\eta+\sum_{i=1}^{m}\theta_i-\sigma(r))^{\underline{\sum_{i=1}^{m}\theta_i-1}}}{\Gamma\,(\sum_{i=1}^{m}\theta_i)\,\Gamma(\alpha-n)}\times \\ &\quad e^{v-s}(v-\sigma(\xi))^{\underline{\alpha-n-1}} g\left(\eta+\sum_{i=1}^{m}\theta_i\right) h(\xi+\alpha-1) \\ &\quad -\frac{1}{\Gamma(\alpha-n)}\sum_{s=\alpha-n}^{T+\alpha-n+1}\sum_{v=\alpha-n}^{s-1}\sum_{\xi=0}^{v-\alpha+n} e^{v-s}(v-\sigma(x))^{\underline{\alpha-n-1}} h(x+\alpha-1), \end{aligned} \tag{11}$$

$$\begin{aligned} \Lambda &= \sum_{s=\alpha-n}^{T+\alpha-n+1} e^{-s} - \sum_{r=\alpha-n}^{\eta}\sum_{s=\alpha-n}^{r-n+1}\frac{\tau\,(\eta+\sum_{i=1}^{m}\theta_i-\sigma(r))^{\underline{\sum_{i=1}^{m}\theta_i-1}}}{\Gamma\,(\sum_{i=1}^{m}\theta_i)}\times \\ &\quad e^{-s} g\left(\eta+\sum_{i=1}^{m}\theta_i\right), \text{ respectively.} \end{aligned} \tag{12}$$

Proof. Using the fractional sum of order $\alpha : \Delta^{-\alpha}$ for (7), we obtain

$$\begin{aligned} u(t)+(e-1)\,\Delta^{-1}u(t) &= C_1 + C_2 t^{\underline{1}} + C_3 t^{\underline{2}} + \ldots + C_n t^{\underline{n-1}} \\ &+\frac{1}{\Gamma(\alpha)}\sum_{s=0}^{t-\alpha}(t-\sigma(s))^{\underline{\alpha-1}} h(s+\alpha-1), \quad t\in\mathbb{N}_{\alpha-n,T+\alpha}. \end{aligned} \tag{13}$$

For the forward difference of order $n : \Delta^n$ for (13), we have

$$\Delta^n u(t) + (e-1)\,\Delta^{n-1}u(t) = \frac{1}{\Gamma(\alpha-n)}\sum_{s=0}^{t-\alpha+n}(t-\sigma(s))^{\underline{\alpha-n-1}} h(s+\alpha-1).$$

Therefore,

$$\Delta\left[e^t\,\Delta^{n-1}u(t)\right] = \frac{e^t}{\Gamma(\alpha-n)}\sum_{s=0}^{t-\alpha+n}(t-\sigma(s))^{\underline{\alpha-n-1}} h(s+\alpha-1). \tag{14}$$

Taking the sum: Δ^{-1} to (14), we get

$$e^t\,\Delta^{n-1}u(t) = C_n + \frac{1}{\Gamma(\alpha-n)}\sum_{s=\alpha-n}^{t-1}\sum_{v=0}^{s-\alpha+n} e^s(s-\sigma(v))^{\underline{\alpha-n-1}}h(v+\alpha-1). \tag{15}$$

Next, taking the sum of order $n-1:\Delta^{-(n-1)}$ to (15), we obtain

$$\begin{aligned} u(t) &= C_1 + C_2t^{\underline{1}} + C_3t^{\underline{2}} + \dots + C_{n-1}t^{\underline{n-2}} + C_n\sum_{s=\alpha-n}^{t-n+1} e^{-s} \\ &+ \frac{1}{\Gamma(\alpha-n)}\sum_{s=\alpha-n}^{t-n+1}\sum_{v=\alpha-n}^{s-1}\sum_{x=0}^{v-\alpha+n} e^{v-s}(v-\sigma(x))^{\underline{\alpha-n-1}}h(x+\alpha-1),\ t\in\mathbb{N}_{\alpha-n,T+\alpha}. \end{aligned} \tag{16}$$

Using the Caputo fractional differences of order β_i for (16) where $i=1$ to $i=n-2$, we obtain

$$\begin{aligned} &\Delta_C^{\beta_1}u(t) \\ &= C_2\Delta_C^{\beta_1}t^{\underline{1}} + C_3\Delta_C^{\beta_1}t^{\underline{2}} + \dots + C_{n-1}\Delta_C^{\beta_1}t^{\underline{n-2}} \\ &+C_n\sum_{r=\alpha-n}^{t+\beta_1-1}\frac{(t-\sigma(r))^{\underline{-\beta_1}}}{\Gamma(1-\beta_1)}\Delta_r\left\{\sum_{s=\alpha-n}^{r-n+1}e^{-s}\right\} \\ &+\sum_{r=\alpha-n}^{t+\beta_1-1}\frac{(t-\sigma(r))^{\underline{-\beta_1}}}{\Gamma(1-\beta_1)}\Delta_r\left\{\sum_{s=\alpha-n}^{r-n+1}\sum_{v=\alpha-n}^{s-1}\sum_{x=0}^{v-\alpha+n}e^{v-s}\frac{(v-\sigma(x))^{\underline{\alpha-n-1}}}{\Gamma(\alpha-n)}h(x+\alpha-1)\right\}, \end{aligned} \tag{17}$$

for $t\in\mathbb{N}_{\alpha-n+1-\beta_1,T+\alpha+1-\beta_1}$.

$$\begin{aligned} &\Delta_C^{\beta_1+\beta_2}u(t) \\ &= C_3\Delta_C^{\beta_1+\beta_2}t^{\underline{2}} + \dots + C_{n-1}\Delta_C^{\beta_1+\beta_2}t^{\underline{n-2}} \\ &+C_n\sum_{r=\alpha-n}^{t+\beta_1+\beta_2-2}\frac{(t-\sigma(r))^{\underline{1-\beta_1-\beta_2}}}{\Gamma(2-\beta_1-\beta_2)}\Delta_r^2\left\{\sum_{s=\alpha-n}^{r-n+1}e^{-s}\right\} \\ &+\sum_{r=\alpha-n}^{t+\beta_1+\beta_2-2}\frac{(t-\sigma(r))^{\underline{1-\beta_1-\beta_2}}}{\Gamma(2-\beta_1-\beta_2)}\Delta_r^2\left\{\sum_{s=\alpha-n}^{r-n+1}\sum_{v=\alpha-n}^{s-1}\sum_{x=0}^{v-\alpha+n}e^{v-s}\frac{(v-\sigma(x))^{\underline{\alpha-n-1}}}{\Gamma(\alpha-n)}h(x+\alpha-1)\right\}, \end{aligned} \tag{18}$$

for $t\in\mathbb{N}_{\alpha-n+2-\beta_1-\beta_2,T+\alpha+2-\beta_1-\beta_2}$.

$$\vdots$$

$$\begin{aligned} &\Delta_C^{\sum_{i=1}^{n-2}\beta_i}u(t) \\ &= C_{n-1}\Delta_C^{\sum_{i=1}^{n-2}\beta_i}t^{\underline{n-2}} + C_n\sum_{r=\alpha-n}^{t-n+2+\sum_{i=1}^{n-2}\beta_i}\frac{(t-\sigma(r))^{\underline{n-3-\sum_{i=1}^{n-2}\beta_i}}}{\Gamma\left(n-2-\sum_{i=1}^{n-2}\beta_i\right)}\Delta_r^{n-2}\left\{\sum_{s=\alpha-n}^{r-n+1}e^{-s}\right\} \\ &+\sum_{r=\alpha-n}^{t-n+2+\sum_{i=1}^{n-2}\beta_i}\frac{(t-\sigma(r))^{\underline{n-3-\sum_{i=1}^{n-2}\beta_i}}}{\Gamma\left(n-2-\sum_{i=1}^{n-2}\beta_i\right)}\times \\ &\Delta_r^{n-2}\left\{\sum_{s=\alpha-n}^{r-n+1}\sum_{v=\alpha-n}^{s-1}\sum_{x=0}^{v-\alpha+n}e^{v-s}\frac{(v-\sigma(x))^{\underline{\alpha-n-1}}}{\Gamma(\alpha-n)}h(x+\alpha-1)\right\}, \end{aligned} \tag{19}$$

for $t \in \mathbb{N}_{\alpha-2-\sum_{i=1}^{n-2}\beta_i, T+\alpha+n-2-\sum_{i=1}^{n-2}\beta_i}$.

Employing the conditions of (8), we have the system of $n-1$ equations

$$\begin{aligned}
&(E_1)\quad C_1 + C_2(\alpha-n) + C_3(\alpha-n)^{\underline{2}} + \ldots + C_{n-1}(\alpha-n)^{\underline{n-2}} = 0,\\
&(E_2)\quad C_2\Delta_C^{\beta_1}(\alpha-n+2-\beta_1) + \ldots + C_{n-1}\Delta_C^{\beta_1}(\alpha-n+2-\beta_1)^{\underline{n-2}} = 0,\\
&(E_3)\quad C_3\Delta_C^{\beta_1+\beta_2}(\alpha-n+4-\beta_1-\beta_2)^{\underline{2}} + \ldots + C_{n-1}\Delta_C^{\beta_1+\beta_2}(\alpha-n+4-\beta_1-\beta_2)^{\underline{n-2}} = 0,\\
&\cdots\\
&(E_{n-1})\quad C_{n-1}\Delta_C^{\sum_{i=1}^{n-2}\beta_i}\left(\alpha+n-4-\sum_{i=1}^{n-2}\beta_i\right)^{\underline{n-2}} = 0.
\end{aligned} \tag{20}$$

Using the fractional sum of order $\sum_{i=1}^{m}\theta_i$ for (16), we have

$$\begin{aligned}
&\Delta_C^{-\sum_{i=1}^{m}\theta_i}u(t)\\
&= \sum_{r=\alpha-n}^{t-\sum_{i=1}^{m}\theta_i}\frac{(t-\sigma(r))^{\underline{\sum_{i=1}^{m}\theta_i-1}}}{\Gamma(\sum_{i=1}^{m}\theta_i)}\left[C_1 + C_2\Delta_C^{\beta_1}s^{\underline{1}} + C_3\Delta_C^{\beta_1}s^{\underline{2}} + \ldots + C_{n-1}\Delta_C^{\beta_1}s^{\underline{n-2}}\right]\\
&+C_n\sum_{r=\alpha-n}^{t-\sum_{i=1}^{m}\theta_i}\sum_{s=\alpha-n}^{r-n+1}\frac{(t-\sigma(r))^{\underline{\sum_{i=1}^{m}\theta_i-1}}}{\Gamma(\sum_{i=1}^{m}\theta_i)}e^{-s}\\
&+\sum_{r=\alpha-n}^{t-\sum_{i=1}^{m}\theta_i}\sum_{s=\alpha-n}^{r-n+1}\sum_{v=\alpha-n}^{s-1}\sum_{x=0}^{v-\alpha+n}\frac{(t-\sigma(r))^{\underline{\sum_{i=1}^{m}\theta_i-1}}}{\Gamma(\sum_{i=1}^{m}\theta_i)}\frac{(v-\sigma(x))^{\underline{\alpha-n-1}}}{\Gamma(\alpha-n)}e^{v-s}h(x+\alpha-1),
\end{aligned} \tag{21}$$

for $t \in \mathbb{N}_{\alpha-n+\sum_{i=1}^{m}\theta_i, T+\alpha+\sum_{i=1}^{m}\theta_i}$.

By substituting $t = \eta + \sum_{i=1}^{m}\theta_i$ into (21) and using the second condition of (9), we finally get

$$\begin{aligned}
(E_n)\quad & C_1\left\{1 - \sum_{r=\alpha-n}^{\eta}\frac{\tau(\eta+\sum_{i=1}^{m}\theta_i-\sigma(r))^{\underline{\sum_{i=1}^{m}\theta_i-1}}}{\Gamma(\sum_{i=1}^{m}\theta_i)}g\left(\eta+\sum_{i=1}^{m}\theta_i\right)\right\}\\
&+C_2\left\{(T+\alpha)^{\underline{1}} - \sum_{r=\alpha-n}^{\eta}\frac{\tau(\eta+\sum_{i=1}^{m}\theta_i-\sigma(r))^{\underline{\sum_{i=1}^{m}\theta_i-1}}s}{\Gamma(\sum_{i=1}^{m}\theta_i)}g\left(\eta+\sum_{i=1}^{m}\theta_i\right)\right\}\\
&+C_3\left\{(T+\alpha)^{\underline{2}} - \sum_{r=\alpha-n}^{\eta}\frac{\tau(\eta+\sum_{i=1}^{m}\theta_i-\sigma(r))^{\underline{\sum_{i=1}^{m}\theta_i-1}}s^{\underline{2}}}{\Gamma(\sum_{i=1}^{m}\theta_i)}g\left(\eta+\sum_{i=1}^{m}\theta_i\right)\right\}\\
&+\cdots\\
&+C_{n-1}\left\{(T+\alpha)^{\underline{n-2}} - \sum_{r=\alpha-n}^{\eta}\frac{\tau(\eta+\sum_{i=1}^{m}\theta_i-\sigma(r))^{\underline{\sum_{i=1}^{m}\theta_i-1}}s^{\underline{n-2}}}{\Gamma(\sum_{i=1}^{m}\theta_i)}g\left(\eta+\sum_{i=1}^{m}\theta_i\right)\right\}\\
&+C_n\left\{\sum_{s=\alpha-n}^{T+\alpha-n+1}e^{-s} - \sum_{r=\alpha-n}^{\eta}\sum_{s=\alpha-n}^{r-n+1}\frac{\tau(\eta+\sum_{i=1}^{m}\theta_i-\sigma(r))^{\underline{\sum_{i=1}^{m}\theta_i-1}}}{\Gamma(\sum_{i=1}^{m}\theta_i)}e^{-s}g\left(\eta+\sum_{i=1}^{m}\theta_i\right)\right\}\\
=&\sum_{r=\alpha-n}^{\eta}\sum_{s=\alpha-n}^{r-n+1}\sum_{v=\alpha-n}^{s-1}\sum_{x=0}^{v-\alpha+n}\frac{\tau(\eta+\sum_{i=1}^{m}\theta_i-\sigma(r))^{\underline{\sum_{i=1}^{m}\theta_i-1}}}{\Gamma(\sum_{i=1}^{m}\theta_i)}e^{v-s}\times\\
&\frac{(v-\sigma(x))^{\underline{\alpha-n-1}}}{\Gamma(\alpha-n)}g\left(\eta+\sum_{i=1}^{m}\theta_i\right)h(x+\alpha-1)\\
&-\sum_{s=\alpha-n}^{T+\alpha-n+1}\sum_{v=\alpha-n}^{s-1}\sum_{x=0}^{v-\alpha+n}e^{v-s}\frac{(v-\sigma(x))^{\underline{\alpha-n-1}}}{\Gamma(\alpha-n)}h(x+\alpha-1).
\end{aligned} \tag{22}$$

Solving the system of Equations $(E_1) - (E_n)$, we obtain

$$C_1 = C_2 = \dots = C_{n-1} = 0 \quad \text{and} \quad C_n = \frac{\mathcal{O}[h]}{\Lambda},$$

where $\mathcal{O}[h], \Lambda$ are defined by (11), (12), respectively. Substituting the constants $C_1 - C_n$ into (17), we obtain (10). This completes the proof. □

3. Main Results

The goal of this section is to show the existence results for the problem (5). To accomplish this, we denote $\mathcal{C} = C(\mathbb{N}_{\alpha-n,T+\alpha}, \mathbb{R})$, the Banach space of all functions u with the norm is defined by

$$\|u\|_{\mathcal{C}} = \|u\| + \|\Delta_C^{\vartheta} u\| + \|\Delta^{-\gamma} u\|,$$

where $\|u\| = \max_{t\in\mathbb{N}_{\alpha-n,T+\alpha}} |u(t)|$, $\|\Delta_C^{\vartheta} u\| = \max_{t\in\mathbb{N}_{t\in\mathbb{N}_{\alpha-n,T+\alpha}}} |\Delta_C^{\vartheta} x(t-\vartheta+1)|$ and $\|\Delta^{-\gamma} u\| = \max_{t\in\mathbb{N}_{t\in\mathbb{N}_{\alpha-n,T+\alpha}}} |\Delta^{-\gamma} x(t+\gamma)|$. In addition, we define the operator $\mathcal{F} : \mathcal{C} \to \mathcal{C}$ by

$$\begin{aligned}(\mathcal{F}u)(t) &= \frac{\mathcal{O}[F(u)+H(u)]}{\Lambda} \sum_{s=\alpha-n}^{t-n+1} e^{-s} + \frac{1}{\Gamma(\alpha-n)} \sum_{s=\alpha-n}^{t-n+1} \sum_{v=\alpha-n}^{s-1} \sum_{\xi=0}^{v-\alpha+n} e^{v-s} \times \\ &\quad (v-\sigma(\xi))^{\underline{\alpha-n-1}} \Big[\lambda F\Big(\xi+\alpha-1, u(\xi+\alpha-1), (\Upsilon^{\vartheta} u)(\xi+\alpha-\vartheta)\Big) \\ &\quad + \mu H\Big(\xi+\alpha-1, u(\xi+\alpha-1), (\Psi^{\gamma} u)(\xi+\alpha+\gamma-1)\Big)\Big], \end{aligned} \tag{23}$$

where Λ is defined by (12) and the functional $\mathcal{O}[F(u)+H(u)]$ is defined by

$$\begin{aligned} &\mathcal{O}[F(u)+H(u)] \\ &= \sum_{r=\alpha-n}^{\eta} \sum_{s=\alpha-n}^{r-n+1} \sum_{v=\alpha-n}^{s-1} \sum_{\xi=0}^{v-\alpha+n} \frac{\tau\left(\eta+\sum_{i=1}^m \theta_i - \sigma(r)\right)^{\underline{\sum_{i=1}^m \theta_i - 1}}}{\Gamma\left(\sum_{i=1}^m \theta_i\right)\Gamma(\alpha-n)} e^{v-s} g\left(\eta+\sum_{i=1}^m \theta_i\right) \times \\ &\quad (v-\sigma(\xi))^{\underline{\alpha-n-1}} \Big[\lambda F\Big(\xi+\alpha-1, u(\xi+\alpha-1), (\Upsilon^{\vartheta} u)(\xi+\alpha-\vartheta)\Big) \\ &\quad + \mu H\Big(\xi+\alpha-1, u(\xi+\alpha-1), (\Psi^{\gamma} u)(\xi+\alpha+\gamma-1)\Big)\Big] \\ &\quad - \frac{1}{\Gamma(\alpha-n)} \sum_{s=\alpha-n}^{T+\alpha-n+1} \sum_{v=\alpha-n}^{s-1} \sum_{\xi=0}^{v-\alpha+n} e^{v-s} (v-\sigma(x))^{\underline{\alpha-n-1}} \times \\ &\quad \Big[\lambda F\Big(\xi+\alpha-1, u(\xi+\alpha-1), (\Upsilon^{\vartheta} u)(\xi+\alpha-\vartheta)\Big) \\ &\quad + \mu H\Big(\xi+\alpha-1, u(\xi+\alpha-1), (\Psi^{\gamma} u)(\xi+\alpha+\gamma-1)\Big)\Big]. \end{aligned} \tag{24}$$

Clearly, the problem (5) has solutions if and only if the operator $\mathcal{F}$ has fixed points. The first show the existence and uniqueness of a solution to the problem (5) by using the Banach contraction principle.

Theorem 1. *Assume that $F, H : \mathbb{N}_{\alpha-n,T+\alpha} \times \mathbb{R} \times \mathbb{R} \to \mathbb{R}$ are continuous, $\varphi, \phi : \mathbb{N}_{\alpha-n,T+\alpha} \times \mathbb{N}_{\alpha-n,T+\alpha} \to [0,\infty)$ are continuous with $\varphi_0 = \max\{\varphi(t-1,s) : (t,s) \in \mathbb{N}_{\alpha-2,T+\alpha} \times \mathbb{N}_{\alpha-2,T+\alpha}\}$ and $\phi_0 = \max\{\phi(t-1,s) : (t,s) \in \mathbb{N}_{\alpha-2,T+\alpha} \times \mathbb{N}_{\alpha-2,T+\alpha}\}$. In addition, suppose that:*

(H_1) *there exist constants* $L_1, L_2 > 0$ *such that for each* $t \in \mathbb{N}_{\alpha-n,T+\alpha}$ *and* $u, v \in \mathcal{C}$

$$|F(t,u(t),(\Upsilon^{\vartheta}u)(t-\vartheta+1)) - F(t,v(t),(\Upsilon^{\vartheta}v)(t-\vartheta+1))| \leq L_1|u-v| + L_2|(\Upsilon^{\vartheta}u)-(\Upsilon^{\vartheta}v)|,$$

(H_2) *there exist constants* $\ell_1, \ell_2 > 0$ *such that for each* $t \in \mathbb{N}_{\alpha-n,T+\alpha}$ *and* $u, v \in \mathcal{C}$

$$|H(t,u(t),(\Psi^{\gamma}u)(t+\gamma)) - f(t,v(t),(\Psi^{\gamma}v)(t+\gamma))| \leq \ell_1|u-v| + \ell_2|(\Psi^{\gamma}u)-(\Psi^{\gamma}v)|,$$

(H_3) $0 < g(t) < K \neq \frac{(e^{T-n+3}-1)e^{\eta-2n+2}\Gamma(m+1)}{(\eta-\alpha+n+m)^{\underline{m}}\tau e^{T+\alpha-2n+2}(\eta-\alpha+n+m)^{\underline{m}}}$ *for each* $t \in \mathbb{N}_{\alpha-n,T+\alpha}$.

If
$$\chi := \left[\lambda\left(L_1 + L_2\frac{\varphi_0(T+n-\vartheta+1)^{\underline{1-\vartheta}}}{\Gamma(2-\vartheta)}\right) + \mu\left(\ell_1 + \ell_2\frac{\varphi_0(T+n+\gamma)^{\underline{\gamma}}}{\Gamma(\gamma+1)}\right)\right] \times (\Omega_1+\Omega_2+\Omega_3) < 1, \tag{25}$$

then the problem (5) has a unique solution on $\mathbb{N}_{\alpha-n,T+\alpha}$, *where*

$$\Omega_1 = \frac{\Theta e^{n-\alpha}(T+2)}{|\Lambda|} + \frac{e^{T-1}(T+\alpha-n+2)^{\underline{\alpha-n+2}}}{\Gamma(\alpha-n+3)}, \tag{26}$$

$$\Omega_2 = \left[\frac{\Theta e^{2n-\alpha-2}}{|\Lambda|} + \frac{e^{T-1}(T+\alpha-n+3)^{\underline{\alpha-n+2}}}{\Gamma(\alpha-n+3)}\right]\frac{(T+n-\vartheta+1)^{\underline{1-\vartheta}}}{\Gamma(2-\vartheta)}, \tag{27}$$

$$\Omega_3 = \left[\frac{\Theta e^{n-\alpha}(T+2)}{|\Lambda|} + \frac{e^{T-1}(T+\alpha-n+2)^{\underline{\alpha-n+2}}}{\Gamma(\alpha-n+3)}\right]\frac{(T+n+\gamma)^{\underline{\gamma}}}{\Gamma(\gamma+1)}, \tag{28}$$

$$\Theta = \frac{\tau K e^{\eta-\alpha-1}(\eta-\alpha+2)^{\underline{\alpha-n+2}}(\eta-\alpha+n+m)^{\underline{m}}}{\Gamma(m+1)\Gamma(\alpha-n+3)} - \frac{e^{T-1}(T+\alpha-n+2)^{\underline{\alpha-n+2}}}{\Gamma(\alpha-n+3)}. \tag{29}$$

Proof. We shall show that $\mathcal{F}$ is a contraction. For any $u, v \in \mathcal{C}$ and for each $t \in \mathbb{N}_{\alpha-n,T+\alpha}$, we have

$$\begin{aligned}
&\Big|\mathcal{O}[F(u)+H(u)] - \mathcal{O}[F(v)+H(v)]\Big| \\
&\leq \sum_{r=\alpha-n}^{\eta}\sum_{s=\alpha-n}^{r-n+1}\sum_{v=\alpha-n}^{s-1}\sum_{\xi=0}^{v-\alpha+n}\frac{\tau\left(\eta+\sum_{i=1}^{m}\theta_i-\sigma(r)\right)^{\underline{\sum_{i=1}^{m}\theta_i-1}}}{\Gamma\left(\sum_{i=1}^{m}\theta_i\right)\Gamma(\alpha-n)} e^{v-s}(v-\sigma(\xi))^{\underline{\alpha-n-1}}\times \\
&\quad\left[\lambda\Big(L_1|u-v| + L_2|(\Upsilon^{\vartheta}u)-(\Upsilon^{\vartheta}v)|\Big) + \mu\Big(\ell_1|u-v| + \ell_2|(\Psi^{\gamma}u)-(\Psi^{\gamma}v)|\Big)\right]\times \\
&\quad g\left(\eta+\sum_{i=1}^{m}\theta_i\right) - \frac{1}{\Gamma(\alpha-n)}\sum_{s=\alpha-n}^{T+\alpha-n+1}\sum_{v=\alpha-n}^{s-1}\sum_{\xi=0}^{v-\alpha+n} e^{v-s}(v-\sigma(x))^{\underline{\alpha-n-1}}\times \\
&\quad\left[\lambda\Big(L_1|u-v| + L_2|(\Upsilon^{\vartheta}u)-(\Upsilon^{\vartheta}v)|\Big) + \mu\Big(\ell_1|u-v| + \ell_2|(\Psi^{\gamma}u)-(\Psi^{\gamma}v)|\Big)\right] \\
&\leq \left[\lambda\left(L_1 + L_2\frac{\varphi_0(T+n-\vartheta+1)^{\underline{1-\vartheta}}}{\Gamma(2-\vartheta)}\right) + \mu\left(\ell_1 + \ell_2\frac{\varphi_0(T+n+\gamma)^{\underline{\gamma}}}{\Gamma(\gamma+1)}\right)\right]\Bigg|\,\|u-v\|_{\mathcal{C}}\times \\
&\quad\Bigg|\frac{\tau K e^{\eta-\alpha-1}(\eta-\alpha+2)^{\underline{\alpha-n+2}}(\eta-\alpha+n+m)^{\underline{m}}}{\Gamma(m+1)\Gamma(\alpha-n+3)} - \frac{e^{T-1}(T+\alpha-n+2)^{\underline{\alpha-n+2}}}{\Gamma(\alpha-n+3)} \\
&= \left[\lambda\left(L_1 + L_2\frac{\varphi_0(T+n-\vartheta+1)^{\underline{1-\vartheta}}}{\Gamma(2-\vartheta)}\right) + \mu\left(\ell_1 + \ell_2\frac{\varphi_0(T+n+\gamma)^{\underline{\gamma}}}{\Gamma(\gamma+1)}\right)\right]\|u-v\|_{\mathcal{C}}\Theta,
\end{aligned} \tag{30}$$

and

$$
\begin{aligned}
&\big|(\mathcal{F}u)(t) - (\mathcal{F}v)(t)\big| \\
&\leq \frac{1}{\Lambda}\Big|\mathcal{O}[F(u)+H(u)] - \mathcal{O}[F(v)+H(v)]\Big| \sum_{s=\alpha-n}^{t-n+1} e^{-s} \\
&\quad + \frac{1}{\Gamma(\alpha-n)} \sum_{s=\alpha-n}^{t-n+1} \sum_{v=\alpha-n}^{s-1} \sum_{\xi=0}^{v-\alpha+n} e^{v-s}(v-\sigma(\xi))^{\underline{\alpha-n-1}} \times \\
&\quad \Big[\lambda\Big(L_1|u-v| + L_2|(\Upsilon^{\vartheta}u) - (\Upsilon^{\vartheta}v)|\Big) + \mu\Big(\ell_1|u-v| + \ell_2|(\Psi^{\gamma}u) - (\Psi^{\gamma}v)|\Big)\Big] \\
&\leq \left[\lambda\left(L_1 + L_2\frac{\phi_0(T+n-\vartheta+1)^{\underline{1-\vartheta}}}{\Gamma(2-\vartheta)}\right) + \mu\left(\ell_1 + \ell_2\frac{\varphi_0(T+n+\gamma)^{\underline{\gamma}}}{\Gamma(\gamma+1)}\right)\right] \|u-v\|_{\mathcal{C}} \times \\
&\quad \left\{\frac{\Theta}{\Lambda}\sum_{s=\alpha-n}^{t-n+1} e^{-s} + \frac{1}{\Gamma(\alpha-n)} \sum_{s=\alpha-n}^{t-n+1} \sum_{v=\alpha-n}^{s-1} \sum_{\xi=0}^{v-\alpha+n} e^{v-s}(v-\sigma(\xi))^{\underline{\alpha-n-1}}\right\} \\
&\leq \left[\lambda\left(L_1 + L_2\frac{\phi_0(T+n-\vartheta+1)^{\underline{1-\vartheta}}}{\Gamma(2-\vartheta)}\right) + \mu\left(\ell_1 + \ell_2\frac{\varphi_0(T+n+\gamma)^{\underline{\gamma}}}{\Gamma(\gamma+1)}\right)\right] \|u-v\|_{\mathcal{C}} \times \\
&\quad \left(\frac{\Theta e^{n-\alpha}(T+2)}{|\Lambda|} + \frac{e^{T-1}(T+\alpha-n+2)^{\underline{\alpha-n+2}}}{\Gamma(\alpha-n+3)}\right) \\
&\leq \left[\lambda\left(L_1 + L_2\frac{\phi_0(T+n-\vartheta+1)^{\underline{1-\vartheta}}}{\Gamma(2-\vartheta)}\right) + \mu\left(\ell_1 + \ell_2\frac{\varphi_0(T+n+\gamma)^{\underline{\gamma}}}{\Gamma(\gamma+1)}\right)\right] \|u-v\|_{\mathcal{C}}\Omega_1.
\end{aligned} \tag{31}
$$

Next, we consider the following $(\Delta_C^{\vartheta}\mathcal{F}u)$ and $(\Delta^{\gamma}\mathcal{F}u)$ as

$$
\begin{aligned}
&(\Delta_C^{\vartheta}\mathcal{F}u)(t-\vartheta+1) \\
&\quad = \frac{\mathcal{O}[F(u)+H(u)]}{\Lambda\Gamma(1-\vartheta)} \sum_{s=\alpha-n}^{t} (t-\vartheta+1-\sigma(s))^{\underline{-\vartheta}} \Delta_s \sum_{v=\alpha-n}^{s-n+1} e^{-v} \\
&\quad + \frac{1}{\Gamma(\alpha-n)\Gamma(1-\vartheta)} \sum_{r=\alpha-n}^{t} (t-\vartheta+1-\sigma(r))^{\underline{-\vartheta}} \Delta_r \Bigg\{ \sum_{s=\alpha-n}^{r-n+1} \sum_{v=\alpha-n}^{s-1} \sum_{\xi=0}^{v-\alpha+n} e^{v-s} \times \\
&\quad (v-\sigma(\xi))^{\underline{\alpha-n-1}} \Big[\lambda F\Big(\xi+\alpha-1, u(\xi+\alpha-1), (\Upsilon^{\vartheta}u)(\xi+\alpha-\vartheta)\Big) \\
&\quad + \mu H\Big(\xi+\alpha-1, u(\xi+\alpha-1), (\Psi^{\gamma}u)(\xi+\alpha+\gamma-1)\Big)\Big]\Bigg\} \\
&\quad = \frac{\mathcal{O}[F(u)+H(u)]}{\Lambda\Gamma(1-\vartheta)} \sum_{s=\alpha-n}^{t} (t-\vartheta+1-\sigma(s))^{\underline{-\vartheta}} e^{n-s-2} \\
&\quad + \frac{1}{\Gamma(\alpha-n)\Gamma(1-\vartheta)} \sum_{r=\alpha-n}^{t} \sum_{s=\alpha-n}^{r-n+1} \sum_{v=\alpha-n}^{s-1} \sum_{\xi=0}^{v-\alpha+n} (t-\vartheta+1-\sigma(r))^{\underline{-\vartheta}} e^{v-s} \times \\
&\quad (v-\sigma(\xi))^{\underline{\alpha-n-1}} \Big[\lambda F\Big(\xi+\alpha-1, u(\xi+\alpha-1), (\Upsilon^{\vartheta}u)(\xi+\alpha-\vartheta)\Big) \\
&\quad + \mu H\Big(\xi+\alpha-1, u(\xi+\alpha-1), (\Psi^{\gamma}u)(\xi+\alpha+\gamma-1)\Big)\Big],
\end{aligned} \tag{32}
$$

$$(\Delta^{-\gamma}\mathcal{F}u)(t+\gamma) = \frac{\mathcal{O}[F(u)+H(u)]}{\Lambda\Gamma(\gamma)}\sum_{s=\alpha-n}^{t}\sum_{v=\alpha-n}^{s-n+1}(t+\gamma-\sigma(s))^{\underline{\gamma-1}}e^{-v}$$
$$+\frac{1}{\Gamma(\alpha-n)\Gamma(\gamma)}\sum_{r=\alpha-n}^{t}\sum_{s=\alpha-n}^{r-n+1}\sum_{v=\alpha-n}^{s-1}\sum_{\xi=0}^{v-\alpha+n}(t+\gamma-\sigma(r))^{\underline{\gamma-1}}e^{v-s}\times$$
$$(v-\sigma(\xi))^{\underline{\alpha-n-1}}\Big[\lambda F\Big(\xi+\alpha-1,u(\xi+\alpha-1),(\Upsilon^{\vartheta}u)(\xi+\alpha-\vartheta)\Big)$$
$$+\mu H\Big(\xi+\alpha-1,u(\xi+\alpha-1),(\Psi^{\gamma}u)(\xi+\alpha+\gamma-1)\Big)\Big]. \tag{33}$$

Similarly, we have

$$\begin{aligned}
&\left|(\Delta_C{}^{\vartheta}\mathcal{F}u)(t-\vartheta+1)-(\Delta_C^{\vartheta}\mathcal{F}v)(t-\vartheta+1)\right| \\
&\leq \left[\lambda\Big(L_1+L_2\frac{\phi_0(T+n-\vartheta+1)^{\underline{1-\vartheta}}}{\Gamma(2-\vartheta)}\Big)+\mu\Big(\ell_1+\ell_2\frac{\varphi_0(T+n+\gamma)^{\underline{\gamma}}}{\Gamma(\gamma+1)}\Big)\right]\|u-v\|_C\times \\
&\left[\frac{\Theta e^{2n-\alpha-2}}{|\Lambda|}+\frac{e^{T-1}(T+\alpha-n+3)^{\underline{\alpha-n+2}}}{\Gamma(\alpha-n+3)}\right]\frac{(T+n-\vartheta+1)^{\underline{1-\vartheta}}}{\Gamma(2-\vartheta)} \\
&\leq \left[\lambda\Big(L_1+L_2\frac{\phi_0(T+n-\vartheta+1)^{\underline{1-\vartheta}}}{\Gamma(2-\vartheta)}\Big)+\mu\Big(\ell_1+\ell_2\frac{\varphi_0(T+n+\gamma)^{\underline{\gamma}}}{\Gamma(\gamma+1)}\Big)\right]\|u-v\|_C\Omega_2,
\end{aligned}\tag{34}$$

and

$$\begin{aligned}
&\left|(\Delta^{-\gamma}\mathcal{F}u)(t+\gamma)-(\Delta^{-\gamma}\mathcal{F}v)(t+\gamma)\right| \\
&\leq \left[\lambda\Big(L_1+L_2\frac{\phi_0(T+n-\vartheta+1)^{\underline{1-\vartheta}}}{\Gamma(2-\vartheta)}\Big)+\mu\Big(\ell_1+\ell_2\frac{\varphi_0(T+n+\gamma)^{\underline{\gamma}}}{\Gamma(\gamma+1)}\Big)\right]\|u-v\|_C\times \\
&\left[\frac{\Theta e^{n-\alpha}(T+2)}{|\Lambda|}+\frac{e^{T-1}(T+\alpha-n+2)^{\underline{\alpha-n+2}}}{\Gamma(\alpha-n+3)}\right]\frac{(T+n+\gamma)^{\underline{\gamma}}}{\Gamma(\gamma+1)} \\
&\leq \left[\lambda\Big(L_1+L_2\frac{\phi_0(T+n-\vartheta+1)^{\underline{1-\vartheta}}}{\Gamma(2-\vartheta)}\Big)+\mu\Big(\ell_1+\ell_2\frac{\varphi_0(T+n+\gamma)^{\underline{\gamma}}}{\Gamma(\gamma+1)}\Big)\right]\|u-v\|_C\Omega_3.
\end{aligned}\tag{35}$$

Hence (31), (34) and (35) imply that

$$\begin{aligned}
\left\|(\mathcal{F}u)(t)-(\mathcal{F}v)(t)\right\|_C &\leq \left[\lambda\Big(L_1+L_2\frac{\phi_0(T+n-\vartheta+1)^{\underline{1-\vartheta}}}{\Gamma(2-\vartheta)}\Big)\right. \\
&\quad \left.+\mu\Big(\ell_1+\ell_2\frac{\varphi_0(T+n+\gamma)^{\underline{\gamma}}}{\Gamma(\gamma+1)}\Big)\right](\Omega_1+\Omega_2+\Omega_3)\|u-v\|_C \\
&= \chi\|u-v\|_C.
\end{aligned}\tag{36}$$

By (H_4), we have $\left\|(\mathcal{F}u)(t)-(\mathcal{F}v)(t)\right\|_C < \|u-v\|_C$.

Consequently, $\mathcal{F}$ is a contraction. Therefore, by the Banach fixed point theorem, we get that $\mathcal{F}$ has a fixed point which is a unique solution of the problem (5) on $t\in\mathbb{N}_{\alpha-n,T+\alpha}$. □

In the second result, we deduce the existence of at least one solution of (5) by the following, the Schauder's fixed point theorem.

Lemma 4 ([30]). *(Arzelá-Ascoli theorem) A set of function in $C[a,b]$ with the sup norm is relatively compact if and only it is uniformly bounded and equicontinuous on $[a,b]$.*

Lemma 5 ([30]). *If a set is closed and relatively compact then it is compact.*

Lemma 6 ([31]). *(Schauder fixed point theorem) Let (D, d) be a complete metric space, U be a closed convex subset of D, and $T : D \to D$ be the map such that the set $Tu : u \in U$ is relatively compact in D. Then the operator T has at least one fixed point $u^* \in U$: $Tu^* = u^*$.*

Theorem 2. *Assuming that $(H_1) - (H_3)$ hold, problem (5) has at least one solution on $\mathbb{N}_{\alpha-n,T+\alpha}$.*

Proof. We divide the proof into three steps as follows.
Step I. Verify $\mathcal{F}$ map bounded sets into bounded sets in $B_R = \{u \in \mathcal{C} : \|u\|_{\mathcal{C}} \leq R\}$. We consider $B_R = \{u \in C(\mathbb{N}_{\alpha-n,T+\alpha}) : \|u\|_{\mathcal{C}} \leq R\}$.

Let $\max\limits_{t\in\mathbb{N}_{\alpha-n,T+\alpha}} |F(t,0,0)| = M$, $\max\limits_{t\in\mathbb{N}_{\alpha-n,T+\alpha}} |H(t,0,0)| = N$ and choose a constant

$$R \geq \frac{(M+N)(\Omega_1+\Omega_2+\Omega_3)}{1-(\Omega_1+\Omega_2+\Omega_3)\left\{\lambda\left(L_1+L_2\frac{\phi_0(T+n-\vartheta+1)^{\underline{1-\vartheta}}}{\Gamma(2-\vartheta)}\right)+\mu\left(\ell_1+\ell_2\frac{\varphi_0(T+n+\gamma)^{\underline{\gamma}}}{\Gamma(\gamma+1)}\right)\right\}}. \tag{37}$$

Noting that

$$\begin{aligned}
|\mathcal{S}(t,u,0)| &= \Big|F(t+\alpha-1,u(t+\alpha-1),\Delta_C^{\vartheta}u(t+\alpha-\vartheta))-F(t+\alpha-1,0,0)\big| \\
&\quad +\big|F(t+\alpha-1,0,0)\Big|, \\
|\mathcal{T}(t,u,0)| &= \Big|H(t+\alpha-1,u(t+\alpha-1),\Delta^{-\gamma}u(t+\alpha+\gamma-1))-H(t+\alpha-1,0,0)\big| \\
&\quad +\big|H(t+\alpha-1,0,0)\Big|,
\end{aligned}$$

for each $u \in B_R$, we obtain

$$\begin{aligned}
&\Big|\mathcal{O}[F(u)+H(u)]\Big| \\
\leq & \sum_{r=\alpha-n}^{\eta}\sum_{s=\alpha-n}^{r-n+1}\sum_{v=\alpha-n}^{s-1}\sum_{\xi=0}^{v-\alpha+n}\frac{\tau\left(\eta+\sum_{i=1}^m\theta_i-\sigma(r)\right)^{\underline{\sum_{i=1}^m\theta_i-1}}}{\Gamma\left(\sum_{i=1}^m\theta_i\right)\Gamma(\alpha-n)}e^{v-s}(v-\sigma(\xi))^{\underline{\alpha-n-1}}\times \\
&\Big[|\mathcal{S}(\xi,u,0)|+\mu|\mathcal{T}(\xi,u,0)|\Big] \\
\leq & \sum_{r=\alpha-n}^{\eta}\sum_{s=\alpha-n}^{r-n+1}\sum_{v=\alpha-n}^{s-1}\sum_{\xi=0}^{v-\alpha+n}\frac{\tau\left(\eta+\sum_{i=1}^m\theta_i-\sigma(r)\right)^{\underline{\sum_{i=1}^m\theta_i-1}}}{\Gamma\left(\sum_{i=1}^m\theta_i\right)\Gamma(\alpha-n)}e^{v-s}(v-\sigma(\xi))^{\underline{\alpha-n-1}}\times \\
&\Big[\lambda\Big(L_1\|u\|+L_2\|\mathrm{Y}^{\vartheta}u\|+M\Big)+\mu\Big(\ell_1\|u\|+\ell_2\|\Psi^{\gamma}u\|+N\Big)\Big] \\
&-\frac{1}{\Gamma(\alpha-n)}\sum_{s=\alpha-n}^{T+\alpha-n+1}\sum_{v=\alpha-n}^{s-1}\sum_{\xi=0}^{v-\alpha+n}e^{v-s}(v-\sigma(x))^{\underline{\alpha-n-1}}\times \\
&\Big[\lambda\Big(L_1|u-v|+L_2|(\mathrm{Y}^{\vartheta}u)-(\mathrm{Y}^{\vartheta}v)|+M\Big)+\mu\Big(\ell_1|u-v|+\ell_2|(\Psi^{\gamma}u)-(\Psi^{\gamma}v)|+N\Big)\Big] \\
\leq & \Big\{\lambda\Big[\Big(L_1+L_2\frac{\phi_0(T+n-\vartheta+1)^{\underline{1-\vartheta}}}{\Gamma(2-\vartheta)}\Big)\|u\|_{\mathcal{C}}+M\Big] \\
&+\mu\Big[\Big(\ell_1+\ell_2\frac{\varphi_0(T+n+\gamma)^{\underline{\gamma}}}{\Gamma(\gamma+1)}\Big)\|u\|_{\mathcal{C}}+N\Big]\Big\}\times \\
&\left|\frac{\tau K e^{\eta-\alpha-1}(\eta-\alpha+2)^{\underline{\alpha-n+2}}(\eta-\alpha+n+m)^{\underline{m}}}{\Gamma(m+1)\Gamma(\alpha-n+3)}-\frac{e^{T-1}(T+\alpha-n+2)^{\underline{\alpha-n+2}}}{\Gamma(\alpha-n+3)}\right| \\
\leq & \Big\{\lambda\Big[\Big(L_1+L_2\frac{\phi_0(T+n-\vartheta+1)^{\underline{1-\vartheta}}}{\Gamma(2-\vartheta)}\Big)\|u\|_{\mathcal{C}}+M\Big] \\
&+\mu\Big[\Big(\ell_1+\ell_2\frac{\varphi_0(T+n+\gamma)^{\underline{\gamma}}}{\Gamma(\gamma+1)}\Big)\|u\|_{\mathcal{C}}+N\Big]\Big\}\Theta,
\end{aligned} \tag{38}$$

and

$$
\begin{aligned}
&\left|(\mathcal{F}u)(t)\right| \\
&\le \frac{1}{\Lambda}\left|\mathcal{O}[F(u)+H(u)]\right|\sum_{s=\alpha-n}^{t-n+1} e^{-s} \\
&+\frac{1}{\Gamma(\alpha-n)}\sum_{s=\alpha-n}^{t-n+1}\sum_{v=\alpha-n}^{s-1}\sum_{\xi=0}^{v-\alpha+n} e^{v-s}(v-\sigma(\xi))^{\underline{\alpha-n-1}}\Big[\lambda\left|\mathcal{S}(\xi,u,0)\right|+\mu\left|\mathcal{T}(\xi,u,0)\right|\Big] \\
&\le \Big[\lambda\Big(L_1\|u\|+L_2\|\Upsilon^{\vartheta}u\|+M\Big)+\mu\Big(\ell_1\|u\|+\ell_2\|\Psi^{\gamma}u\|+N\Big)\Big]\times \\
&\quad\left\{\frac{\Theta}{|\Lambda|}\sum_{s=\alpha-n}^{t-n+1} e^{-s}+\frac{1}{\Gamma(\alpha-n)}\sum_{s=\alpha-n}^{t-n+1}\sum_{v=\alpha-n}^{s-1}\sum_{\xi=0}^{v-\alpha+n} e^{v-s}(v-\sigma(\xi))^{\underline{\alpha-n-1}}\right\} \\
&\le \left\{\lambda\Big[\Big(L_1+L_2\frac{\varphi_0(T+n-\vartheta+1)^{\underline{1-\vartheta}}}{\Gamma(2-\vartheta)}\Big)\|u\|_{\mathcal{C}}+M\Big]\right. \\
&\quad\left.+\mu\Big[\Big(\ell_1+\ell_2\frac{\varphi_0(T+n+\gamma)^{\underline{\gamma}}}{\Gamma(\gamma+1)}\Big)\|u\|_{\mathcal{C}}+N\Big]\right\}\left(\frac{\Theta e^{n-\alpha}(T+2)}{|\Lambda|}+\frac{e^{T-1}(T+\alpha-n+2)^{\underline{\alpha-n+2}}}{\Gamma(\alpha-n+3)}\right) \\
&\le \left\{\lambda\Big[\Big(L_1+L_2\frac{\varphi_0(T+n-\vartheta+1)^{\underline{1-\vartheta}}}{\Gamma(2-\vartheta)}\Big)\|u\|_{\mathcal{C}}+M\Big]\right. \\
&\quad\left.+\mu\Big[\Big(\ell_1+\ell_2\frac{\varphi_0(T+n+\gamma)^{\underline{\gamma}}}{\Gamma(\gamma+1)}\Big)\|u\|_{\mathcal{C}}+N\Big]\right\}\Omega_1.
\end{aligned}
\tag{39}
$$

Furthermore, we have

$$
\begin{aligned}
\left|(\Delta_C^{\vartheta}\mathcal{F}u)(t-\vartheta+1)\right| \le &\left\{\lambda\Big[\Big(L_1+L_2\frac{\varphi_0(T+n-\vartheta+1)^{\underline{1-\vartheta}}}{\Gamma(2-\vartheta)}\Big)\|u\|_{\mathcal{C}}+M\Big]\right. \\
&\left.+\mu\Big[\Big(\ell_1+\ell_2\frac{\varphi_0(T+n+\gamma)^{\underline{\gamma}}}{\Gamma(\gamma+1)}\Big)\|u\|_{\mathcal{C}}+N\Big]\right\}\Omega_2,
\end{aligned}
\tag{40}
$$

and

$$
\begin{aligned}
\left|(\Delta^{-\gamma}\mathcal{F}u)(t+\gamma)\right| \le &\left\{\lambda\Big[\Big(L_1+L_2\frac{\varphi_0(T+n-\vartheta+1)^{\underline{1-\vartheta}}}{\Gamma(2-\vartheta)}\Big)\|u\|_{\mathcal{C}}+M\Big]\right. \\
&\left.+\mu\Big[\Big(\ell_1+\ell_2\frac{\varphi_0(T+n+\gamma)^{\underline{\gamma}}}{\Gamma(\gamma+1)}\Big)\|u\|_{\mathcal{C}}+N\Big]\right\}\Omega_3.
\end{aligned}
\tag{41}
$$

Hence (39)–(41) imply that

$$
\begin{aligned}
\|(\mathcal{F}u)(t)\|_{\mathcal{C}} \le &\left\{\lambda\Big[\Big(L_1+L_2\frac{\varphi_0(T+n-\vartheta+1)^{\underline{1-\vartheta}}}{\Gamma(2-\vartheta)}\Big)\|u\|_{\mathcal{C}}+M\Big]\right. \\
&\left.+\mu\Big[\Big(\ell_1+\ell_2\frac{\varphi_0(T+n+\gamma)^{\underline{\gamma}}}{\Gamma(\gamma+1)}\Big)\|u\|_{\mathcal{C}}+N\Big]\right\}(\Omega_1+\Omega_2+\Omega_3) \\
\le\ & R.
\end{aligned}
\tag{42}
$$

So, $\|\mathcal{F}u\|_{\mathcal{C}} \le R$. This implies that $\mathcal{F}$ is uniformly bounded.

Step II. Since F and H are continuous, the operator $\mathcal{F}$ is continuous on B_R.

Step III. Examine $\mathcal{F}$ is equicontinuous on B_R. For any $\epsilon > 0$, there exists a positive constant $\rho^* = \max\{\delta_1, \delta_2, \delta_3, \delta_4\}$ such that for $t_1, t_2 \in \mathbb{N}_{\alpha-n,T+\alpha}$

$$
\begin{aligned}
&\left|t_2 - t_1\right| < \frac{\epsilon\Lambda}{6e^{T-1}\Theta\left[\lambda\|F\|+\mu\|H\|\right]}, \qquad \text{whenever } |t_2 - t_1| < \delta_1, \\
&\left|(t_2 - n + 1)^{\underline{\alpha-n+2}} - (t_1 - n + 1)^{\underline{\alpha-n+2}}\right| < \frac{\epsilon\Gamma(\alpha-n+3)}{6e^{n-\alpha}\Theta\left[\lambda\|F\|+\mu\|H\|\right]}, \qquad \text{whenever } |t_2 - t_1| < \delta_2, \\
&\left|(t_2 - \alpha + n - \vartheta + 1)^{\underline{1-\vartheta}} - (t_1 - \alpha + n - \vartheta + 1)^{\underline{1-\vartheta}}\right| \\
&\quad < \frac{\epsilon}{3\left[\lambda\|F\| + \mu\|H\|\right]\left(\frac{\Theta e^{2n-\alpha-2}}{\Lambda\Gamma(2-\vartheta)} + \frac{e^{T-1}(T+\alpha-n+3)^{\underline{\alpha-n+2}}}{\Gamma(\alpha-n+3)\Gamma(2-\vartheta)}\right)} \\
&\quad < \frac{\epsilon\,\Gamma(1-\nu)\Gamma(\beta)\Gamma(\alpha)}{4\widetilde{\Theta}_R}, \qquad \text{whenever } |t_2 - t_1| < \delta_3, \\
&\left|(t_2 - \alpha + n + \gamma)^{\underline{\gamma}} - (t_1 - \alpha + n + \gamma)^{\underline{\gamma}}\right| < \frac{\epsilon}{3\left[\lambda\|F\| + \mu\|H\|\right]\left(\frac{\Theta e^{n-\alpha}(T+2)}{\Lambda\Gamma(\gamma+1)} + \frac{e^{T-1}(T+\alpha-n+2)^{\underline{\alpha-n+2}}}{\Gamma(\alpha-n+3)\Gamma(\gamma+1)}\right)}, \\
&\qquad \text{whenever } |t_2 - t_1| < \delta_4.
\end{aligned}
$$

Then we have

$$
\begin{aligned}
&\left|(\mathcal{F}x)(t_2) - (\mathcal{F}x)(t_1)\right| \\
&\le \frac{1}{|\Lambda|}\mathcal{O}[F(u) + H(u)]\left|\sum_{s=\alpha-n}^{t_2-n+1} e^{-s} - \sum_{s=\alpha-n}^{t_1-n+1} e^{-s}\right| + \frac{1}{\Gamma(\alpha-n)}\left|\sum_{s=\alpha-n}^{t_2-n+1}\sum_{v=\alpha-n}^{s-1}\sum_{\xi=0}^{v-\alpha+n} e^{v-s} \times\right. \\
&\quad (v - \sigma(\xi))^{\underline{\alpha-n-1}}\left[\lambda F\Big(\xi + \alpha - 1, u(\xi + \alpha - 1), (\Upsilon^{\vartheta}u)(\xi + \alpha - \vartheta)\Big)\right. \\
&\quad \left. + \mu H\Big(\xi + \alpha - 1, u(\xi + \alpha - 1), (\Psi^{\gamma}u)(\xi + \alpha + \gamma - 1)\Big)\right] - \sum_{s=\alpha-n}^{t_1-n+1}\sum_{v=\alpha-n}^{s-1}\sum_{\xi=0}^{v-\alpha+n} e^{v-s} \times \\
&\quad (v - \sigma(\xi))^{\underline{\alpha-n-1}}\left[\lambda F\Big(\xi + \alpha - 1, u(\xi + \alpha - 1), (\Upsilon^{\vartheta}u)(\xi + \alpha - \vartheta)\Big)\right. \\
&\quad \left.\left. + \mu H\Big(\xi + \alpha - 1, u(\xi + \alpha - 1), (\Psi^{\gamma}u)(\xi + \alpha + \gamma - 1)\Big)\right]\right| \\
&\le [\lambda\|F\| + \mu\|H\|]\left\{\frac{\Theta e^{n-\alpha}}{|\Lambda|}\left|t_2 - t_1\right| + \frac{e^{T-1}}{\Gamma(\alpha-n+3)}\left|(t_2 - n + 1)^{\underline{\alpha-n+2}} - (t_1 - n + 1)^{\underline{\alpha-n+2}}\right|\right\} \\
&< \frac{\epsilon}{6} + \frac{\epsilon}{6} = \frac{\epsilon}{3}.
\end{aligned} \tag{43}
$$

Furthermore, we have

$$
\begin{aligned}
&\left|(\Delta_C^{\vartheta}\mathcal{F}u)(t_2 - \vartheta + 1) - (\Delta_C^{\vartheta}\mathcal{F}u)(t_1 - \vartheta + 1)\right| \\
&\le [\lambda\|F\| + \mu\|H\|]\left\{\left[\frac{\Theta e^{2n-\alpha-2}}{|\Lambda|\Gamma(2-\vartheta)} + \frac{e^{T-1}(T + \alpha - n + 3)^{\underline{\alpha-n+2}}}{\Gamma(\alpha-n+3)\Gamma(2-\vartheta)}\right] \times\right. \\
&\quad \left.\left|(t_2 - \alpha + n - \vartheta + 1)^{\underline{1-\vartheta}} - (t_1 - \alpha + n - \vartheta + 1)^{\underline{1-\vartheta}}\right|\right\} \le \frac{\epsilon}{3},
\end{aligned} \tag{44}
$$

and

$$
\begin{aligned}
&\left|(\Delta^{-\gamma}\mathcal{F}u)(t_2+\gamma)-(\Delta^{-\gamma}\mathcal{F}u)(t_1+\gamma)\right| \\
&\leq\ [\lambda\|F\|+\mu\|H\|]\left\{\left[\frac{\Theta e^{n-\alpha}(T+2)}{|\Lambda|\Gamma(\gamma+1)}+\frac{e^{T-1}(T+\alpha-n+2)^{\underline{\alpha-n+2}}}{\Gamma(\alpha-n+3)\Gamma(\gamma+1)}\right]\times\right. \\
&\quad\left.\left|(t_2-\alpha+n+\gamma)^{\underline{\gamma}}-(t_1-\alpha+n+\gamma)^{\underline{\gamma}}\right|\right\}\ \leq\ \frac{\epsilon}{3}.
\end{aligned}
\tag{45}
$$

Hence

$$
\left\|(\mathcal{F}x)(t_2)-(\mathcal{F}x)(t_1)\right\|_{\mathcal{C}}\ <\ \frac{\epsilon}{3}+\frac{\epsilon}{3}+\frac{\epsilon}{3}\ =\ \epsilon. \tag{46}
$$

This implies that the set $\mathcal{F}(B_R)$ is an equicontinuous set. As a consequence of Steps I to III together with the Arzelá-Ascoli theorem, we find that $\mathcal{F}:\mathcal{C}\to\mathcal{C}$ is completely continuous. By Schauder fixed point theorem, we can conclude that problem (5) has at least one solution. The proof is completed. □

4. An Example

In order to study the existence of a solution to our problem, we obtain the conditions provided in Section 3. Since our designated problem is a theoretical problem, it is rare to find the application related to our results. However, for thorough explanation, we provide the following example to illustrate our results. Consider the following fractional difference boundary value problem

$$
\begin{aligned}
\left[\Delta_C^{\frac{9}{2}}+(e+1)\Delta_C^{\frac{7}{2}}\right]u(t) =& \frac{e^{-\cos^2\left(2\pi\left(t+\frac{7}{2}\right)+10\right)}}{100+e^{\sin^2\left(2\pi\left(t+\frac{7}{2}\right)\right)}}\cdot\frac{\left|u\left(t+\frac{7}{2}\right)\right|+\left|\Upsilon^{\frac{1}{3}}u\left(t+\frac{26}{6}\right)\right|}{\left[1+\left|u\left(t+\frac{7}{2}\right)\right|\right]} \\
&+\frac{\left(t+\frac{227}{2}\right)^{-2}\left|u\left(t+\frac{7}{2}\right)\right|+\left|\Psi^{\frac{4}{5}}u\left(t+\left(t+\frac{43}{10}\right)\right)\right|}{\left(t+\frac{27}{2}\right)^3\left[1+\left|u\left(t+\frac{7}{2}\right)\right|\right]}, \\
u\left(-\frac{1}{2}\right) =& D_C^{\frac{1}{4}}u\left(\frac{5}{4}\right)=D_C^{\frac{1}{2}}u\left(\frac{11}{4}\right)=D_C^{\frac{3}{4}}u\left(4\right)=0 \\
u\left(\frac{27}{2}\right) =& \frac{1}{e^5}\Delta^{-\frac{77}{60}}e^{-sin\left(\frac{587\pi}{60}\right)}u\left(\frac{587}{60}\right),
\end{aligned}
\tag{47}
$$

$$
\text{where}\quad (\Psi^{\frac{1}{3}}u)\left(t+\frac{26}{6}\right)=\sum_{s=-\frac{7}{6}}^{t-\frac{2}{3}}\frac{(t-\sigma(s))^{\underline{-\frac{1}{3}}}}{\Gamma\left(\frac{2}{3}\right)}\frac{e^{-(s+\frac{2}{3})}}{(t+50)^3}\Delta u\left(s+\frac{2}{3}\right),
$$

$$
(\Psi^{\frac{4}{5}}u)\left(t+\frac{4}{5}\right)=\sum_{s=-\frac{13}{10}}^{t-\frac{4}{5}}\frac{(t-\sigma(s))^{-\frac{1}{5}}}{\Gamma\left(\frac{4}{5}\right)}\frac{e^{-(s+\frac{4}{5})}}{(t+100)^2}u\left(s+\frac{4}{5}\right).
$$

Set $\alpha=\frac{9}{2}$, $n=5$, $\vartheta=\frac{1}{3}$, $\gamma=\frac{4}{5}$, $\beta_1=\frac{1}{4}$, $\beta_2=\frac{1}{2}$, $\beta_3=\frac{3}{4}$, $\lambda=e^{-10}$, $\mu=1$, $T=6$, $\eta=\frac{17}{2}$, $T=6$, $\tau=e^{-5}$, $m=4$, $\theta_1=\frac{1}{2}$, $\theta_2=\frac{1}{3}$, $\theta_3=\frac{1}{4}$, $\theta_4=\frac{1}{5}$, $g(t)=e^{\sin(\pi t)}$, $\phi(t,s-\vartheta+1)=\frac{e^{-(s+\frac{2}{3})}}{(t+50)^3}$ and $\varphi(t,s+\gamma)=\frac{e^{-s+\frac{4}{5}}}{(t+100)^2}$.

We can show that

$$
\begin{aligned}
\Theta=545.5721,\ |\Lambda|=202.553,\ \Omega_1=2189.264,\ \Omega_2=15715.32 \\
\Omega_3=17049.09\ \text{ and }\ \phi_0=\left(\tfrac{2}{99}\right)^3e^{-\frac{1}{6}},\ \varphi_0\leq\left(\tfrac{2}{199}\right)^2e^{-\frac{3}{10}}.
\end{aligned}
$$

Nothing that $(H_1)-(H_3)$ hold, for each $t\in\frac{1}{2}\mathbb{N}_{-\frac{1}{2},\frac{27}{2}}$, we obtain

$$
\left|F[t,u,\Upsilon^{\frac{1}{3}}u]-F[t,v,\Upsilon^{\frac{1}{3}}v]\right|\leq\tfrac{1}{101}|u-v|+\tfrac{1}{101}|\Upsilon^{\frac{1}{3}}u-\Upsilon^{\frac{1}{3}}v|,
$$

$$\left|F[t,u,\Psi^{\frac{4}{5}}u]-F[t,v,\Psi^{\frac{4}{5}}v]\right| \leq \left(\frac{2}{199}\right)^2\left(\frac{2}{19}\right)^3|u-v|+\left(\frac{2}{19}\right)^3|\Psi^{\frac{4}{5}}u-\Psi^{\frac{4}{5}}v|,$$
$$\text{and } \frac{1}{e}<g(t)<e,$$

so, $L_1 = L_2 = 0.0099$, $\ell_1 = 1.178\times 10^{-7}, \ell_2 = 0.0012$, $K = e$.

Finally, we find that

$$\chi = 0.0443 < 1.$$

Hence, by Theorem 1, the problem (47) has a unique solution on $\frac{1}{2}\mathbb{N}_{-\frac{1}{2},\frac{27}{2}}$.

5. Conclusions

We study the existence and unique results of the solution for a separate nonlinear Caputo fractional sum-difference equation with fractional sum-difference boundary conditions. Some conditions are obtained when Banach contraction principle is used as a tool. In addition, the conditions for the case of at least one solution are obtained by using the Schauder fixed point theorem.

Author Contributions: These authors contributed equally to this work.

Funding: This research was funded by King Mongkut's University of Technology North Bangkok. Contract no. KMUTNB-GOV-59-37.

Acknowledgments: The last author of this research was supported by Suan Dusit University.

Conflicts of Interest: The authors declare no conflict of interest regarding the publication of this paper.

References

1. Zhang, Y.; Liu, X.; Belić, M.R.; Zhong, W.; Zhang, Y.; Xiao, M. Propagation Dynamics of a Light Beam in a Fractional Schrödinger Equation. *Phys. Rev. Lett.* **2015**, *115*, 180403. [CrossRef] [PubMed]
2. Zingales, M.; Failla, G. The finite element method for fractional non-local thermal energy transfer in non-homogeneous rigid conductors. *Commun. Nonlinear Sci. Numer. Simul.* **2015**, *29*, 116–127. [CrossRef]
3. Lazopoulos, K.A.; Lazopoulos, A.K. On fractional bending of beams. *Arch. Appl. Mech.* **2016**, *86*, 1133–1145. [CrossRef]
4. Sumelka, W.; Voyiadjis, G.Z. A hyperelastic fractional damage material model with memory. *Int. J. Solids Struct.* **2017**, *124*, 151–160. [CrossRef]
5. Goodrich, C.S.; Peterson, A.C. *Discrete Fractional Calculus*; Springer: New York, NY, USA, 2015.
6. Wu, G.C.; Baleanu, D. Discrete fractional logistic map and its chaos. *Nonlinear Dyn.* **2014**, *75*, 283–287. [CrossRef]
7. Wu, G.C.; Baleanu, D. Chaos synchronization of the discrete fractional logistic map. *Signal Process.* **2014**, *102*, 96–99. [CrossRef]
8. Wu, G.C.; Baleanu, D.; Xie, H.P.; Chen, F.L. Chaos synchronization of fractional chaotic maps based on stability results. *Phys. A Stat. Mech. Its Appl.* **2016**, *460*, 374–383. [CrossRef]
9. Agarwal, R.P.; leanu, D.; Rezapour, S.; Salehi, S. The existence of solutions for some fractional finite difference equations via sum boundary conditions. *Adv. Differ. Equ.* **2014**, *2014*, 282. [CrossRef]
10. Goodrich, C.S. On discrete sequential fractional boundary value problems. *J. Math. Anal. Appl.* **2012**, *385*, 111–124. [CrossRef]
11. Goodrich, C.S. On a discrete fractional three-point boundary value problem. *J. Differ. Equ. Appl.* **2012**, *18*, 397–415. [CrossRef]
12. Weidong, L. Existence of solutions for discrete fractional boundary value problems witha p-Laplacian operator. *Adv. Differ. Equ.* **2012**, *2012*, 163.
13. Ferreira, R. Existence and uniqueness of solution to some discrete fractional boundary value problems of order less than one. *J. Differ. Equ. Appl.* **2013**, *19*, 712–718. [CrossRef]
14. Abdeljawad, T. On Riemann and Caputo fractional differences. *Comput. Math. Appl.* **2011**, *62*, 1602–1611. [CrossRef]

15. Atici, F.M.; Eloe, P.W. Two-point boundary value problems for finite fractional difference equations. *J. Differ. Equ. Appl.* **2011**, *17*, 445–456. [CrossRef]
16. Atici, F.M.; Eloe, P.W. A transform method in discrete fractional calculus. *Int. J. Differ. Equ.* **2007**, *2*, 165–176.
17. Sitthiwirattham, T.; Tariboon, J.; Ntouyas, S.K. Existence Results for fractional difference equations with three-point fractional sum boundary conditions. *Discret. Dyn. Nat. Soc.* **2013**, *2013*, 104276. [CrossRef]
18. Sitthiwirattham, T.; Tariboon, J.; Ntouyas, S.K. Boundary value problems for fractional difference equations with three-point fractional sum boundary conditions. *Adv. Differ. Equ.* **2013**, *2013*, *296*. [CrossRef]
19. Sitthiwirattham, T. Existence and uniqueness of solutions of sequential nonlinear fractional difference equations with three-point fractional sum boundary conditions. *Math. Method. Appl. Sci.* **2015**, *38*, 2809–2815. [CrossRef]
20. Sitthiwirattham, T. Boundary value problem for p-Laplacian Caputo fractional difference equations with fractional sum boundary conditions. *Math. Method Appl. Sci.* **2016**, *39*, 1522–1534. [CrossRef]
21. Chasreechai, S.; Kiataramkul, C.; Sitthiwirattham, T. On nonlinear fractional sum-difference equations via fractional sum boundary conditions involving different orders. *Math. Probl. Eng.* **2015**, *2015*, 519072. [CrossRef]
22. Reunsumrit, J.; Sitthiwirattham, T. Positive solutions of three-point fractional sum boundary value problem for Caputo fractional difference equations via an argument with a shift. *Positivity* **2016**, *20*, 761–1014. [CrossRef]
23. Reunsumrit, J.; Sitthiwirattham, T. On positive solutions to fractional sum boundary value problems for nonlinear fractional difference equations. *Math. Method. Appl. Sci.* **2016**, *39*, 2737–2751. [CrossRef]
24. Soontharanon, J.; Jasthitikulchai, N.; Sitthiwirattham, T. Nonlocal Fractional Sum Boundary Value Problems for Mixed Types of Riemann-Liouville and Caputo Fractional Difference Equations. *Dyn. Syst. Appl.* **2016**, *25*, 409–414.
25. Laoprasittichok, S.; Sitthiwirattham, T. On a Fractional Difference-Sum Boundary Value Problems for Fractional Difference Equations Involving Sequential Fractional Differences via Different Orders. *J. Comput. Anal. Appl.* **2017**, *23*, 1097–1111.
26. Kaewwisetkul, B.; Sitthiwirattham, T. On Nonlocal Fractional Sum-Difference Boundary Value Problems for Caputo Fractional Functional Difference Equations with Delay. *Adv. Differ. Equ.* **2017**, *2017*, 219. [CrossRef]
27. Reunsumrit, J.; Sitthiwirattham, T. A New Class of Four-Point Fractional Sum Boundary Value Problems for Nonlinear Sequential Fractional Difference Equations Involving Shift Operators. *Kragujevac J. Math.* **2018**, *42*, 371–387. [CrossRef]
28. Chasreechai, S.; Sitthiwirattham, T. Existence Results of Initial Value Problems for Hybrid Fractional Sum-Difference Equations. *Discret. Dyn. Nat. Soc.* **2018**, *2018*, 5268528. [CrossRef]
29. Chasreechai, S.; Sitthiwirattham, T. On Nonlocal Boundary Value Problems for Hybrid Fractional Sum-Difference Equations Involving Different Orders. *J. Nonlinear Funct. Anal.* **2018**. [CrossRef]
30. Griffel, D.H. *Applied Functional Analysis*; Ellis Horwood Publishers: Chichester, UK, 1981.
31. Guo, D.; Lakshmikantham, V. *Nonlinear Problems in Abstract Cone*; Academic Press: Orlando, FL, USA, 1988.

mathematics

MDPI

Article

Existence Result and Uniqueness for Some Fractional Problem

Guotao Wang [1,5,6,*], Abdeljabbar Ghanmi [2], Samah Horrigue [3] and Samar Madian [4]

1 School of Mathematics and Computer Science, Shanxi Normal University, Linfen, Shanxi 041004, China
2 Faculté des Sciences de Tunis, Université de Tunis El Manar, Tunis 1060, Tunisia; abdeljabbar.ghanmi@lamsin.rnu.tn
3 Department of Mathematics, Higher Institute of Applied Science and Technology, University of Monastir, Monastir 5000, Tunisia; samah.horrigue@fst.rnu.tn
4 Basic Sciences Department, Higher Institute for Engineering and Technology, New Damietta 34517, Egypt; samar_math@yahoo.com
5 College of Mathematics and System Science, Shandong University of Science and Technology, Qingdao, Shandong 266590, China
6 Nonlinear Analysis and Applied Mathematics (NAAM) Research Group, Department of Mathematics, Faculty of Science, King Abdulaziz University, Jeddah 21589, Saudi Arabia
* Correspondence: wgt2512@163.com

Received: 24 April 2019; Accepted: 23 May 2019; Published: 5 June 2019

Abstract: In this article, by the use of the lower and upper solutions method, we prove the existence of a positive solution for a Riemann–Liouville fractional boundary value problem. Furthermore, the uniqueness of the positive solution is given. To demonstrate the serviceability of the main results, some examples are presented.

Keywords: positive solution; green function; fractional differential equation; the method of lower and upper solutions

1. Introduction

The aim of this work is to study the existence and uniqueness of the positive solution for the following problem:

$$\begin{cases} D^{\alpha}_{0^+}k(t) = j(t,k(t)), \quad 1<\alpha\leq 2, \quad 0<t<1 \quad , \\ k(0)=0, \quad \beta k(1)-\gamma k(\eta)=0, \quad \eta\in[0,1] \quad , \end{cases} \tag{1}$$

where β, γ, and η are positive real numbers such that $\beta - 2\gamma\eta^{\alpha-1} > 0$, j is a nonnegative continuous function on $[0,1]\times[0,\infty)$, and $D^{\alpha}_{0^+}$ is the fractional derivative in the sense of Riemann–Liouville. This type of equation is important in many disciplines such as chemistry, aerodynamics, polymer rheology, etc.

Different techniques are used in such problems to obtain the existence of solutions, for example the variational method, the Adomian decomposition method, etc.; we refer the reader to [1–8] and references therein.

Existence results of nonlinear fractional problems are given by the use of fixed point theorems; see [9–15]. More precisely, the authors in [12] gave the existence of positive solutions for the following equation:

$$D^{\alpha}_{0^+}k(t) + j(t,k(t)) = 0, \quad 1<\alpha\leq 2, \quad 0<t<1,$$

according to some boundary conditions.

Using the theory of the fixed point index, the author in [11], presented the existence of the positive solution for the following system:

$$\begin{cases} D^{\alpha}_{0+}k(t) + j(t,k(t)) = 0, \quad 0 < t < 1, \quad 1 < \alpha \leq 2, \\ k(0) = 0, \quad \beta k(\eta) = k(1). \end{cases}$$

Recently, the upper solution method and lower solution method have been the aim of many papers; see for example the book [16] and the recent papers [17–30]. The main idea of this method is to study some modified problem and, then, give the existence results for the principal problem.

Motivated by the above works, in this article, we will present a new method to study the given problem, that is we combine the lower and upper solution method with the fixed point theorem method in order to prove the existence and uniqueness of the positive solution. Let us assume the following:

Hypothesis (H1). *The function j is nonnegative and continuous on $[0,1] \times [0,+\infty)$.*

Hypothesis (H2). *For each $t \in [0,1]$, the function $j(t,.)$ is bounded and increasing on $[0,+\infty)$.*

Hypothesis (H3). *There exists a function $a : [0,1] \to [0,\infty)$ such that the function j satisfies:*

$$|j(s,x) - j(s,y)| \leq a(s)|x-y|, \ \forall\, s \in [0,1], \ \forall\, x,y \geq 0. \tag{2}$$

The main theorems of this paper are summarized as follows.

Theorem 1. *Under Hypotheses (H_1)–(H_2). If $\beta - 2\gamma\eta^{\alpha-1} > 0$, then Equation (1) admits a positive solution.*

Theorem 2. *Under hypothesis (H_3), if $\beta - 2\gamma\eta^{\alpha-1} > 0$ and if:*

$$\int_0^1 [1 + \frac{\beta}{(\beta - \gamma\eta^{\alpha-1})}(1-\delta)^{\alpha-1}]\delta^{\alpha-1}a(\delta)d\delta < \Gamma(\alpha), \tag{3}$$

then Equation (1) admits a unique positive solution.

2. Preliminaries

In this section, we collect some basic results and notations that will be used in the forthcoming sections.

We denote by $L(0,1)$ the set of all integrable functions on $(0,1)$ and by $C(0,1)$ the set of functions that are continuous on $(0,1)$.

Lemma 1 ([31]). *Let $\alpha > 0$, $N = [\alpha] + 1$. Assume that the function k is in $C(0,1) \cap L(0,1)$. Then, the following equation:*

$$D^{\alpha}_{0+}k(t) = 0,$$

admits a unique solution. Moreover, this solution is given by:

$$k(t) = C_1 t^{\alpha-1} + C_2 t^{\alpha-2} + \ldots + C_N t^{\alpha-N},$$

for some $C_i \in \mathbb{R}$, where $i = 1,2,...,N$.

Lemma 2 ([31]). *Let $\alpha > 0$ and $N = [\alpha] + 1$. Assume that either k and $D^{\alpha}_{0+}k$ are in $C(0,1) \cap L(0,1)$. Then, there exists $C_i \in \mathbb{R}$, for $i = 1,2,...,N$, such that:*

$$I^{\alpha}_{0+}D^{\alpha}_{0+}k(t) = k(t) - C_1 t^{\alpha-1} - C_2 t^{\alpha-2} - \ldots - C_N t^{\alpha-N}.$$

Now, we give the Green function associated with the problem (1).

Theorem 3. *If* $1 < \alpha < 2$ *and* h *is continuous in* $[0,1]$, *then equation:*

$$D^{\alpha}_{0+}k(t) + h(t) = 0, \quad 0 < t < 1, \tag{4}$$

with the following conditions:

$$k(0) = 0, \quad \beta k(1) - \gamma k(\eta) = 0, \quad \eta \in [0,1], \tag{5}$$

admits a unique solution, which is given by:

$$k(t) = \int_0^1 G(t,\delta)h(\delta)d\delta,$$

with $G(t,\delta)$ *being the Green function defined by:*

$$\begin{aligned}&\Gamma(\alpha)G(t,\delta)\\ &= (t-\delta)^{\alpha-1}\chi_{[0,t]}(\delta) + \frac{t^{\alpha-1}}{\beta-\gamma\eta^{\alpha-1}}\left[\beta(1-\delta)^{\alpha-1} - \gamma(\eta-\delta)^{\alpha-1}\chi_{[0,\eta]}(\delta)\right],\end{aligned} \tag{6}$$

where χ_A *is the function defined by:*

$$\chi_A(x) = \begin{cases} 1 & \text{if } x \in A, \\ 0 & \text{if } x \notin A. \end{cases}$$

Proof. From Equation (4) and using Lemma 2, there exist two real numbers C_1 and C_2 such that:

$$k(t) = -I^{\alpha}_{0+}h(t) + C_1 t^{\alpha-1} + C_2 t^{\alpha-2}.$$

It follows that:

$$k(t) = \int_0^t \frac{(t-\delta)^{\alpha-1}}{\Gamma(\alpha)}h(\delta)d\delta + C_1 t^{\alpha-1} + C_2 t^{\alpha-2}. \tag{7}$$

Since $k(0) = 0$, then $C_2 = 0$.

On the other hand:

$$\begin{cases} k(1) = \int_0^1 \frac{(1-\delta)^{\alpha-1}}{\Gamma(\alpha)}h(\delta)d\delta + C_1, \\ k(\eta) = \int_0^\eta \frac{(\eta-\delta)^{\alpha-1}}{\Gamma(\alpha)}h(\delta)d\delta + C_1\eta^{\alpha-1}. \end{cases}$$

As $\beta k(1) - \gamma k(\eta) = 0$, then we have:

$$C_1 = \frac{\beta\int_0^1 \frac{(1-\delta)^{\alpha-1}}{\Gamma(\alpha)}h(\delta) - \gamma\int_0^\eta \frac{(\eta-\delta)^{\alpha-1}}{\Gamma(\alpha)}h(\delta)d\delta}{\beta-\gamma\eta^{\alpha-1}}. \tag{8}$$

By substituting the values of C_1 and C_2 into Equation (7), we get:

$$\begin{aligned}&k(t)\\ &= \int_0^t \frac{(t-\delta)^{\alpha-1}}{\Gamma(\alpha)}h(\delta)d\delta + \frac{1}{\beta-\gamma\eta^{\alpha-1}}\left[\beta\int_0^1 \frac{(1-\delta)^{\alpha-1}}{\Gamma(\alpha)}h(\delta)d\delta - \gamma\int_0^\eta \frac{(\eta-\delta)^{\alpha-1}}{\Gamma(\alpha)}h(\delta)d\delta\right]t^{\alpha-1}\\ &= \frac{1}{\Gamma(\alpha)}\int_0^1\left[(t-\delta)^{\alpha-1}\chi_{[0,t]}(\delta) + \frac{1}{\beta-\gamma\eta^{\alpha-1}}\left\{\beta(1-\delta)^{\alpha-1} - \gamma(\eta-\delta)^{\alpha-1}\chi_{[0,\eta]}(\delta)\right\}t^{\alpha-1}\right]h(\delta)d\delta.\end{aligned}$$

That is:

$$k(t) = \int_0^t G(t,\delta)h(\delta)d\delta.$$

It follows that for all real numbers $t, \delta \in [0,1]$, we have:

$$G(t,\delta) = \frac{(t-\delta)^{\alpha-1}}{\Gamma(\alpha)}\chi_{[0,t]}(\delta) + \frac{1}{\Gamma(\alpha)(\beta-\gamma\eta^{\alpha-1})}\left[\beta(1-\delta)^{\alpha-1} - \gamma(\eta-\delta)^{\alpha-1}\chi_{[0,\eta]}(\delta)\right]t^{\alpha-1}.$$

□

Proposition 1. *Let G be the function given by Equation (6), then we have the following properties:*

(i) *Put* $q(\delta) = \frac{2\beta-\gamma\eta^{\alpha-1}}{\Gamma(\alpha)(\beta-\gamma\eta^{\alpha-1})}(1-\delta)^{\alpha-1}$*, then we have:*

$$G(t,\delta) \le q(\delta), \forall\ t,\delta \in [0,1].$$

(ii) *Put* $p(t) = \frac{\beta}{\beta-\gamma\eta^{\alpha-1}}t^{\alpha-1}$*, then we obtain:*

$$G(t,\delta) \ge q(\delta)p(t),\ \forall\ t,\delta \in [0,1].$$

Proof. **(i)** Firstly, we remark that $G(1,\delta)$ is given by:

$$G(1,\delta) = \frac{1}{\Gamma(\alpha)}\begin{cases} (1-\delta)^{\alpha-1} + \frac{1}{\beta-\gamma\eta^{\alpha-1}}\left\{\beta(1-\delta)^{\alpha-1} - \gamma(\eta-\delta)^{\alpha-1}\right\} & , \text{if } 0 \le \delta \le \eta \\ (1-\delta)^{\alpha-1} + \frac{1}{\beta-\gamma\eta^{\alpha-1}}\beta(1-\delta)^{\alpha-1} & , \text{if } \eta \le \delta \le 1. \end{cases}$$

On the other hand, for all $\delta \in [0,1]$, the function $t \longmapsto G(t,\delta)$ is increasing, so, for any $t,\delta \in [0,1]$, we have:

$$G(t,\delta) \le G(1,\delta).$$

As $\gamma > 0$, it is easy to see that:

$$\begin{aligned}\Gamma(\alpha)G(t,\delta) &\le (1-\delta)^{\alpha-1} + \frac{1}{\beta-\gamma\eta^{\alpha-1}}\beta(1-\delta)^{\alpha-1} \\ &= \frac{2\beta-\gamma\eta^{\alpha-1}}{\beta-\gamma\eta^{\alpha-1}}(1-\delta)^{\alpha-1}, \quad \forall t,\delta \in [0,1].\end{aligned}$$

That is, if $t,\delta \in [0,1]$, then we have:

$$G(t,\delta) \le q(\delta).$$

(ii) If $0 \le \eta \le t < 1$, then using (6) and the fact that $\beta - 2\gamma\eta^{\alpha-1} > 0$, we obtain:

$$
\begin{aligned}
&G(t,\delta)\\
&= q(\delta)\begin{cases} \frac{\beta-\gamma\eta^{\alpha-1}}{2\beta-\gamma\eta^{\alpha-1}}\left(\frac{t-\delta}{1-\delta}\right)^{\alpha-1} + \frac{1}{2\beta-\gamma\eta^{\alpha-1}}\left\{\beta - \gamma(\frac{\eta-\delta}{1-\delta})^{\alpha-1}\right\}t^{\alpha-1}, & \text{if } 0 \le \delta \le \eta \\ \frac{\beta-\gamma\eta^{\alpha-1}}{2\beta-\gamma\eta^{\alpha-1}}\left(\frac{t-\delta}{1-\delta}\right)^{\alpha-1} + \frac{\beta}{2\beta-\gamma\eta^{\alpha-1}}t^{\alpha-1}, & \text{if } \eta \le \delta \le t \\ \frac{\beta}{2\beta-\gamma\eta^{\alpha-1}}t^{\alpha-1}, & \text{if } t \le \delta < 1 \end{cases}\\
&\ge q(\delta)\begin{cases} p(t) + \frac{\beta-\gamma\eta^{\alpha-1}}{2\beta-\gamma\eta^{\alpha-1}}\left(\frac{t-\delta}{1-\delta}\right)^{\alpha-1} - \frac{\gamma}{2\beta-\gamma\eta^{\alpha-1}}(\frac{\eta-\delta}{1-\delta})^{\alpha-1}t^{\alpha-1}, & \text{if } 0 \le \delta \le \eta \\ p(t) + \frac{\beta-\gamma\eta^{\alpha-1}}{2\beta-\gamma\eta^{\alpha-1}}\left(\frac{t-\delta}{1-\delta}\right)^{\alpha-1}, & \text{if } \eta \le \delta \le t \\ p(t), & \text{if } t \le \delta < 1, \end{cases}\\
&\ge q(\delta)\begin{cases} p(t) + \frac{(\beta-\gamma\eta^{\alpha-1})(t-\delta)^{\alpha-1} - \gamma(\eta-\delta)^{\alpha-1}t^{\alpha-1}}{2\beta-\gamma\eta^{\alpha-1}}, & \text{if } 0 \le \delta \le \eta \\ p(t), & \text{if } \eta \le \delta < 1, \end{cases}\\
&\ge q(\delta)\begin{cases} p(t) + \frac{1}{2\beta-\gamma\eta^{\alpha-1}}\left[(\beta-\gamma\eta^{\alpha-1})t^{\alpha-1} - \gamma\eta^{\alpha-1}t^{\alpha-1}\right], & \text{if } 0 \le \delta \le \eta \\ p(t), & \text{if } \eta \le \delta < 1, \end{cases}\\
&\ge q(\delta)\begin{cases} p(t) + \frac{1}{2\beta-\gamma\eta^{\alpha-1}}(\beta - 2\gamma\eta^{\alpha-1})t^{\alpha-1}, & \text{if } 0 \le \delta \le \eta \\ p(t), & \text{if } \eta \le \delta < 1, \end{cases}\\
&\ge p(t)q(\delta),
\end{aligned}
$$

where $p(t) = \frac{\beta}{2\beta-\gamma\eta^{\alpha-1}}t^{\alpha-1}$. The proof of Proposition 1 is now completed. □

3. Proof of the Main Results

This section is devoted to proving our main results. To this aim, we will apply the following lemma.

Lemma 3 (See [32]). *Let E be a semi-order Banach space and P be a cone in E. Let $D \subset P$ and a nondecreasing operator $T : D \to E$. Assume that the equation $x - T(x) = 0$ admits a lower solution $x_0 \in D$ and an upper solution $y_0 \in D$, with $x_0 \le y_0$. Assume that if $x_0 \le x \le y_0$, then $x \in D$. If one of the following statements holds:*

(i) *P is normal, and T is compact continuous.*
(ii) *P is regular, and T is continuous.*
(ii) *E is reflexive, P normal, and T continuous or weak continuous.*

Then, the equation

$$T(x) = x,$$

admits a maximum solution x^* and admits a minimum solution y^* such that $x_0 \le x^* \le y^* \le y_0$.

Note that a function v (resp. A function w) is called the lower solution (resp. upper solution) of operator T if:

$$v(t) \le Tv(t), (\text{ resp. } w(t) \ge Tw(t)).$$

Let $E = C[0,1]$, equipped with the supremum norm. Put $P = \{k \in E\, k(t) \ge 0; 0 \le t \le 1\}$, which is a cone in E. Define T on P by:

$$T(k)(t) = \int_0^1 G(t,\delta)j(\delta,k(\delta))d\delta,$$

so it is not difficult to see that k is a solution for Equation (1) if and only if $T(k) = k$.

Proof of Theorem 1. We divide the proof into four steps.

Step 1: We will prove that T maps P into itself and that it is completely continuous.

First, since G and j are nonnegative and continuous, it is easy to see that T maps P into itself and that it is continuous. Let Ω be a bounded subset of P, which is to say the existence of $M > 0$ with:

$$\|k\| \leq M, \forall\ k \in \Omega.$$

Put:

$$L = \max_{0 \leq t \leq 1, k \in \Omega} |h(t,k)|.$$

Then, for all k in Ω, we get:

$$|Tk(t)| \leq \int_0^1 G(t,\delta)|h(\delta,k(\delta))|d\delta \leq L \int_0^1 G(t,\delta)d\delta.$$

That is, $T(\Omega)$ is a bounded subset of P.
Now, for any $k \in \Omega$ and $0 \leq t_1 < t_2 \leq 1$, we have:

$$\begin{aligned}
&|Tk(t_2) - Tk(t_1)| = |\int_0^1 G(t_2,\delta)h(\delta,k(\delta))d\delta - \int_0^1 G(t_1,\delta)h(\delta,k(\delta))d\delta| \\
&= |\int_0^1 (G(t_2,\delta) - G(t_1,\delta))h(\delta,k(\delta))d\delta| \\
&\leq \int_0^1 |G(t_2,\delta) - G(t_1,\delta)||h(\delta,k(\delta))|d\delta \\
&\leq L \int_0^1 |G(t_2,\delta) - G(t_1,\delta)|d\delta \\
&\leq L\left(\int_0^{t_1} |G(t_2,\delta) - G(t_1,\delta)|d\delta + \int_{t_1}^{t_2} |G(t_2,\delta) - G(t_1,\delta)|d\delta + \int_{t_2}^1 |G(t_2,\delta) - G(t_1,\delta)|d\delta \right) \\
&\leq L\left(I_1 + I_2 + I_3 \right)
\end{aligned}$$

where:

$$\begin{aligned}
I_1 &= \int_0^{t_1} |G(t_2,\delta) - G(t_1,\delta)|d\delta \\
&\leq \int_0^{t_1} |(t_2-\delta)^{\alpha-1}\chi_{[0,t_2]}(\delta) - (t_1-\delta)^{\alpha-1}\chi_{[0,t_1]}(\delta)|d\delta \\
&+ \frac{(t_2^{\alpha-1} - t_1^{\alpha-1})}{\beta - \gamma\eta^{\alpha-1}} \int_0^{t_1} |\beta(1-\delta)^{\alpha-1} - \gamma(\eta-\delta)^{\alpha-1}\chi_{[0,\eta]}(\delta)|d\delta \\
&= \begin{cases} \frac{1}{\alpha}\left(t_2^\alpha - (t_2-t_1)^\alpha - t_1^\alpha\right) + \frac{(t_2^{\alpha-1}-t_1^{\alpha-1})}{\beta-\gamma\eta^{\alpha-1}}\left(\frac{\beta}{\alpha}(1-(1-t_1)^\alpha) + \frac{\gamma\eta^\alpha}{\alpha}\right), & \text{if } \eta < t_1 \\ \frac{t_2^\alpha - t_1^\alpha}{\alpha} + \frac{(t_2^{\alpha-1}-t_1^{\alpha-1})}{\beta-\gamma\eta^{\alpha-1}}\left(\frac{\beta}{\alpha}(1-(1-t_1)^\alpha) + \gamma\frac{\eta^\alpha - (\eta-t_1)^\alpha}{\alpha}\right), & \text{if } \eta > t_1 \end{cases}
\end{aligned}$$

$$
\begin{aligned}
I_2 &= \int_{t_1}^{t_2} |G(t_2,\delta) - G(t_1,\delta)|d\delta \\
&\leq \int_{t_1}^{t_2} |(t_2-\delta)^{\alpha-1}\chi_{[0,t_2]}(\delta) - (t_1-\delta)^{\alpha-1}\chi_{[0,t_1]}(\delta)|d\delta \\
&+ \frac{(t_2^{\alpha} - (t_2-t_1)^{\alpha})}{\beta - \gamma\eta^{\alpha-1}} \int_{t_1}^{t_2} |\beta(1-\delta)^{\alpha-1} - \gamma(\eta-\delta)^{\alpha-1}\chi_{[0,\eta]}(\delta)|d\delta \\
&= \begin{cases}
\frac{1}{\alpha}(t_2-t_1)^{\alpha} + \frac{(t_2^{\alpha-1}-t_1^{\alpha-1})}{\beta-\gamma\eta^{\alpha-1}}\frac{\beta}{\alpha}((1-t_1)^{\alpha} - (1-t_2)^{\alpha}), & \text{if } 0 \leq \eta < t_1 \\
\frac{t_2^{\alpha}-t_1^{\alpha}}{\alpha} + + \frac{(t_2^{\alpha-1}-t_1^{\alpha-1})}{\beta-\gamma\eta^{\alpha-1}}\left\{\frac{\beta}{\alpha}((1-t_1)^{\alpha} - (1-t_2)^{\alpha}) - \frac{\gamma}{\alpha}(t_1-\eta)^{\alpha}\right\}, & \text{if } t_1 \leq \eta \leq t_2 \\
\frac{t_2^{\alpha}-t_1^{\alpha}}{\alpha} + + \frac{(t_2^{\alpha-1}-t_1^{\alpha-1})}{\beta-\gamma\eta^{\alpha-1}}\left\{\frac{\beta}{\alpha}((1-t_1)^{\alpha} - (1-t_2)^{\alpha}) + \frac{\gamma}{\alpha}[(t_2-\eta)^{\alpha} - (t_1-\eta)^{\alpha}]\right\}, & \text{if } t_2 \leq \eta \leq 1
\end{cases},
\end{aligned}
$$

$$
\begin{aligned}
I_3 &= \int_{t_2}^{1} |G(t_2,\delta) - G(t_1,\delta)|d\delta \\
&\leq \int_{t_1}^{t_2} |(t_2-\delta)^{\alpha-1}\chi_{[0,t_2]}(\delta) - (t_1-\delta)^{\alpha-1}\chi_{[0,t_1]}(\delta)|d\delta \\
&\leq \int_{t_1}^{t_2} |(t_2-\delta)^{\alpha-1}\chi_{[0,t_2]}(\delta) - (t_1-\delta)^{\alpha-1}\chi_{[0,t_1]}(\delta)|d\delta \\
&+ \frac{(t_2^{\alpha} - (t_2-t_1)^{\alpha})}{\beta - \gamma\eta^{\alpha-1}} \int_{t_1}^{t_2} |\beta(1-\delta)^{\alpha-1} - \gamma(\eta-\delta)^{\alpha-1}\chi_{[0,\eta]}(\delta)|d\delta \\
&= \begin{cases}
\frac{t_2^{\alpha-1}-t_1^{\alpha-1}}{\beta-\gamma\eta^{\alpha-1}}\frac{\beta}{\alpha}(1-t_2)^{\alpha}, & \text{if } \eta < t_2 \\
\frac{t_2^{\alpha-1}-t_1^{\alpha-1}}{\beta-\gamma\eta^{\alpha-1}}\left\{\frac{\beta}{\alpha}(1-t_2)^{\alpha} - \frac{\gamma}{\alpha}(\eta-t_2)^{\alpha}\right\}, & \text{if } t_2 < 1.
\end{cases}
\end{aligned}
$$

Then, we obtain that:

$$
\begin{aligned}
|Tk(t_2) - Tk(t_1)| &\leq L(I_1 + I_2 + I_3) \\
&\leq L\left(\frac{t_2^{\alpha} - t_1^{\alpha}}{\alpha} + \frac{t_2^{\alpha-1} - t_1^{\alpha-1}}{\beta - \gamma\eta^{\alpha-1}}(\frac{\beta - \gamma\eta^{\alpha}}{\alpha})\right).
\end{aligned}
$$

Since t^{α} and $t^{\alpha-1}$ are uniformly continuous when $t \in [0,1]$ and $1 < \alpha < 2$, it is easy to prove that $T(\Omega)$ is equicontinuous. From the Arzela–Ascoli theorem (see [33], we deduce that $\overline{T(\Omega)}$ is a compact subset. That is, $T : P \to P$ is a completely continuous operator.

Step 2: T is an increasing operator.

Let $0 \leq t \leq 1$. Since the function $\delta \longmapsto j(t,\delta)$ is nondecreasing, then there exists $a > 0$, such that the function $[0,a] \ni \delta \longmapsto j(t,\delta)$ is strictly increasing. It follows that for $k_1 \leq k_2$, we have:

$$
Tk_1(t) = \int_0^1 G(t,\delta)h(\delta,k_1(\delta))d\delta \leq \int_0^1 G(t,\delta)h(\delta,k_2(\delta))d\delta = Tk_2(t).
$$

Step 3: For each $t \in [0,1]$ and from (H_2), there exists $M > 0$ with $0 < j(t,k(t)) < M$. It follows by applying Theorem 3 that equation:

$$
\begin{cases}
D_{0^+}^{\alpha} w(t) + M = 0, & 0 < t < 1,\, 1 < \alpha \leq 2, \\
w(0) = 0,\, \beta w(1) - \gamma w(\eta) = 0 & ,\, \eta \in [0,1].
\end{cases}
$$

has a solution w. Moreover, this solution satisfies:

$$w(t) = \int_0^1 G(t,\delta)Md\delta \geq \int_0^1 G(t,\delta)j(t,w(\delta))d\delta = Tw(t).$$

That is, the operator T admits w as an upper solution.

On the other hand, the operator T admits the zero function as a lower solution; moreover:

$$0 \leq w(t) \ \ \forall\, 0 \leq t \leq 1.$$

Step 4: Since P is a normal cone, Lemma 3 implies that T admits a fixed point $k \in \langle 0, w(t)\rangle$. Therefore, Equation (1) admits a positive solution. □

Proof of Theorem 2. To prove Theorem 2, we begin to prove that T has a fixed point. We remark that if for n large enough, T^n is a contraction operator, then T has a unique fixed point. Indeed, assume that for n large enough, T^n is a contraction operator, and fix $x \in E$. Since T is an increasing operator, which is uniformly bounded, then the sequence $\{T^m x\}_{m\in\mathbb{N}}$ is convergent, that is there is $p \in E$ such that $\lim\limits_{m\to\infty} T^m x = p$. Since T is continuous, we get:

$$Tp = T \lim_{m\to\infty} T^m p = \lim_{m\to\infty} T^{m+1} = p,$$

On the other hand, if p is a fixed point for the operator T, then it is also a fixed point for the operator T^n, so we obtain the uniqueness of p.

Now, let us prove that for n large enough, the operator T^n is a contraction. Let $k, v \in P$, then we have:

$$\begin{aligned}
|Tk(t) - Tv(t)| &= \int_0^1 G(t,\delta)|j(\delta,k(\delta)) - j(\delta,v(\delta))|d\delta \\
&\leq \int_0^1 G(t,\delta)a(\delta)|k(\delta) - v(\delta)|d\delta \\
&\leq \frac{\|k-v\|}{\Gamma(\alpha)} \int_0^1 [(t-\delta)^{\alpha-1} + \frac{t^{\alpha-1}\beta}{(\beta - \gamma\eta^{\alpha-1})}(1-\delta)^{\alpha-1}]a(\delta)d\delta \\
&\leq \frac{\|k-v\|t^{\alpha-1}}{\Gamma(\alpha)} \int_0^1 [1 + \frac{\beta}{(\beta - \gamma\eta^{\alpha-1})}(1-\delta)^{\alpha-1}]a(\delta)d\delta \\
&\leq \frac{\|k-v\|t^{\alpha-1}}{\Gamma(\alpha)} K,
\end{aligned}$$

where $K = \int_0^1 [1 + \frac{\beta}{(\beta-\gamma\eta^{\alpha-1})}(1-\delta)^{\alpha-1}]a(\delta)d\delta$.

Similarly, we have:

$$\begin{aligned}
|T^2k(t) - T^2v(t)| &= \int_0^1 G(t,\delta)|j(\delta,Tk(\delta)) - j(\delta,Tv(\delta))|d\delta \\
&\leq \int_0^1 G(t,\delta)a(\delta)|Tk(\delta) - Tv(\delta)|d\delta \\
&\leq \int_0^1 G(t,\delta)a(\delta)\frac{\|k-v\|\delta^{\alpha-1}}{\Gamma(\alpha)} K d\delta \\
&\leq \frac{\|k-v\|t^{\alpha-1}}{\Gamma(\alpha)^2} \int_0^1 [1 + \frac{\beta}{(\beta - \gamma\eta^{\alpha-1})}(1-\delta)^{\alpha-1}]\delta^{\alpha-1}a(\delta)d\delta \\
&\leq \frac{\|k-v\|t^{\alpha-1}}{\Gamma(\alpha)^2} KH,
\end{aligned}$$

where $H = \int_0^1 [1 + \frac{\beta}{(\beta - \gamma\eta^{\alpha-1})}(1-\delta)^{\alpha-1}]\delta^{\alpha-1}a(\delta)d\delta$.

By mathematical induction, it follows that:

$$|T^n k(t) - T^n v(t)| \leq \frac{\|k - v\| t^{\alpha-1}}{\Gamma(\alpha)^n} K H^{n-1}.$$

By using (3), we get:

$$\frac{H}{\Gamma(\alpha)} < 1.$$

Then, for n large enough, it follows that:

$$\frac{KH^{n-1}}{\Gamma(\alpha)^n} = \frac{K}{\Gamma(\alpha)}\left(\frac{H}{\Gamma(\alpha)}\right)^{n-1} < 1.$$

Hence, it holds that:

$$|T^n k(t) - T^n v(t)| < \|k - v\| t^{\alpha-1} \leq \|k - v\|,$$

and this completes the proof. □

4. Examples

In this section, some examples are presented in order to illustrate the usefulness of our main results.

Example 1. *Consider the system:*

$$\begin{cases} D^{\alpha}_{0^+} k(t) = \sqrt{t} e^{-k(t)}, & 0 < t < 1, \quad 1 < \alpha \leq 2, \\ k(0) = 0, \quad \beta k(1) - \gamma k(\eta) = 0, & \eta \in [0,1] \end{cases}, \tag{9}$$

where $\beta, \gamma > 0$, $\beta - 2\gamma\eta^{\alpha-1} > 0$.

Note that since for any $t \in [0,1]$, we have $0 < \sqrt{t}e^{-k(t)} < \sqrt{t}$, which implies that Conditions (H_1) and (H_2) hold. On the other hand, there is an equivalence between the solution of Problem (9) and the fixed point of the operator T given by:

$$Tk(t) = \int_0^1 G(t,\delta)\sqrt{\delta} e^{-k(\delta)} d\delta.$$

Take $w(t) = \int_0^1 G(t,\delta)\sqrt{\delta} d\delta$ and $v(t) \equiv 0$, then:

$$w(t) \geq \int_0^1 G(t,\delta)\sqrt{\delta} e^{-w(\delta)} d\delta = Tw(t),$$

which implies that the operator T admits the function w as an upper solution. Moreover, it is obvious that the zero function is a lower solution for T. Hence, from Theorem 1, we conclude that Problem (9) admits a positive solution.

Example 2. *In the second example, we study the following problem:*

$$\begin{cases} D^{3/2}_{0^+} k(t) = \frac{\sin t}{(1+t^2)} \arctan(1 + k(t)), & 0 < t < 1, \\ k(0) = 0, \quad \beta k(1) - \gamma k(\eta) = 0, \quad \eta \in [0,1] \end{cases}, \tag{10}$$

where $\beta, \gamma > 0$, $\beta - 2\gamma\sqrt{\eta} > 0$.

Note that $\arctan(1 + k(t)) < \frac{\pi}{2}$ for each $t \in [0,1]$, then $j(t, k(t)) = \frac{\sin t}{(1+t^2)} \arctan(1 + k(t))$ is an increasing and bounded function on k. Therefore, we can easily prove that Conditions (H_1) and (H_2) are

satisfied.
Take $w(t) = \int_0^1 G(t,\delta)\frac{\sin\delta}{(1+\delta^2)}d\delta$ *and* $v(t) \equiv 0$, *then:*

$$w(t) \geq \int_0^1 G(t,\delta)\sqrt{\delta}\frac{\sin\delta}{(1+\delta^2)}\arctan(1+k(\delta))d\delta = Tw(t),$$

which implies that T admits the function w as an upper solution and the zero function as a lower solution. Thus, from Theorem 1, we obtain a positive solution to Problem (9).

Example 3. *In this example, we take* $\beta = 1$, $\gamma = 0$, *and* $\eta \in [0,1]$, *and we consider the following problem:*

$$\begin{cases} D^{\frac{3}{2}}y(t) = \lambda y(t) + f(t), \\ y(0) = 0, \quad y(1) = 0, \end{cases} \tag{11}$$

where f is a nonnegative function. It is clear that we have:

$$j(t,x) = \lambda x + f(t),$$

and $\alpha = \frac{3}{2}$. *Moreover, for all* $t \in [0,1]$, *one has:*

$$|j(t,x_1) - j(t,x_2)| \leq \lambda|x_1 - x_2|, \text{ that is } a(t) = \lambda,$$

and:

$$\begin{aligned} \int_0^1 [1 + \frac{\beta}{(\beta - \gamma\eta^{\alpha-1})}(1-\delta)^{\alpha-1}]\delta^{\alpha-1}a(\delta)d\delta &= \lambda\int_0^1 \left(1 + (1-\delta)^{\alpha-1}\right)\delta^{\alpha-1}d\delta \\ &= \lambda(B(1,\alpha) + B(\alpha,\alpha)) \\ &= \lambda\Gamma(\alpha)\left(\frac{1}{\Gamma(\alpha+1)} + \frac{\Gamma(\alpha)}{\Gamma(2\alpha)}\right) \\ &= \lambda\Gamma(\frac{3}{2})\left(\frac{1}{\Gamma(\frac{5}{2})} + \frac{\Gamma(\frac{3}{2})}{\Gamma(3)}\right) < \Gamma(\frac{3}{2}), \end{aligned}$$

for all $0 < \lambda < \frac{12\sqrt{\pi}}{16+3\pi}$, *where* $B(.,.)$ *is the beta function.*

Finally, all conditions of Theorem 2 are satisfied. Therefore, for $0 < \lambda < \frac{12\sqrt{\pi}}{16+3\pi}$, *Problem (11) admits a unique solution.*

Author Contributions: All authors contributed equally to this work.

Funding: Professor Guotao Wang is supported by the National Natural Science Foundation of China (No.11501342), the Scientific and Technological Innovation Programs of Higher Education Institutions in Shanxi (Nos.201802068 and 201802069) and the NSF of Shanxi, China (No.201701D221007).

Conflicts of Interest: The authors declare no conflict of interest.

References

1. Ben Ali, K.; Ghanmi, A.; Kefi, K. Existence of solutions for fractional differential equations with dirichlet boundary conditions. *Electr. J. Differ. Equ.* **2016**, *2016*, 1–11.
2. Cabada, A.; Wang, G. Positive solutions of nonlinear fractional differential equations with integral boundary value conditions. *J. Math. Anal. Appl.* **2012**, *389*, 403–411. [CrossRef]
3. Jafari, H.; Gejji, V.D. Positive solutions of nonlinear fractional boundary value problems using adomian decomposition method. *Appl. Math. Comput.* **2006**, *180*, 700–706. [CrossRef]
4. Liang, S.H.; Zhang, J.H. Positive solutions for boundary value problems of nonlinear fractional differential equation. *Nonlinear Anal.* **2009**, *71*, 5545–5550. [CrossRef]

5. Miller, K.; Ross, B. *An Introduction to the Fractional Calculus and Fractional Differential Equations*; Wiley and Sons: New York, NY, USA, 1993.
6. Nieto, J.J. Maximum principles for fractional differential equations derived from Mittag-Leffler functions. *Appl. Math. Lett.* **2010**, *23*, 1248–1251. [CrossRef]
7. Podlubny, I. Fractional Differential Equations. In *Mathematics in Science and Engineering*; Academic Press: New York, NY, USA, 1999.
8. Wei, Z. Dong, W.; Che, J. Periodic boundary value problems for fractional differential equations involving a Riemann–Liouville fractional derivative. *Nonlinear Anal.* **2010**, *73*, 3232–3238. [CrossRef]
9. Ahmad, B.; Nieto, J.J. Existence results for nonlinear boundary value problems of fractional integro-differential equations with integral boundary conditions. *Bound Value Probl.* **2009**, *2009*, 708576.
10. Băleanu, D.; Octavian, G.M.; O'Regan, D. On a fractional differential equation with infinitely many solutions. *Adv. Diff. Equ.* **2012**, *2012*, 145. [CrossRef]
11. Bai, Z. On positive solutions of a nonlocal fractional boundary value problem. *Nonlinear Anal.* **2010**, *72*, 916–924. [CrossRef]
12. Bai, Z.; Lü, H.S. Positive solutions of boundary value problems of nonlinear fractional differential equation. *J. Math. Anal. Appl.* **2005**, *311*, 495–505. [CrossRef]
13. Ghanmi, A.; Kratou, M.; Saoudi, K. A Multiplicity Results for a Singular Problem Involving a Riemann–Liouville Fractional Derivative. *FILOMAT* **2018**, *32*, 653–669. [CrossRef]
14. Saoudi, K.; Agarwal, P.; Kumam, P.; Ghanmi, A.; Thounthong, P. The Nehari manifold for a boundary value problem involving Riemann–Liouville fractional derivative. *Adv. Diff. Equ.* **2018**, *2018*, 263. [CrossRef]
15. Ghanmi, A.; Horrigue, S. Existence Results for Nonlinear Boundary Value Problems. *FILOMAT* **2018**, *32*, 609–618. [CrossRef]
16. Lakshmikantham, V.; Leela, S.; Vasundhara, J. *Theory of Fractional Dynamic Systems*; Cambridge Academic Publishers: Cambridge, UK, 2009.
17. Wang, G.; Pei, K.; Agarwal, R.P.; Zhang, L.; Ahmad, B. Nonlocal Hadamard fractional boundary value problem with Hadamard integral and discrete boundary conditions on a half-line. *J. Comput. Appl. Math.* **2018**, *343*, 230–239. [CrossRef]
18. Wang, G.; Ren, X.; Bai, Z.; Hou, W. Radial symmetry of standing waves for nonlinear fractional Hardy-Schrödinger equation. *Appl. Math. Lett.* **2019**, *96*, 131–137. [CrossRef]
19. Pei, K.; Wang, G.; Sun, Y. Successive iterations and positive extremal solutions for a Hadamard type fractional integro-differential equations on infinite domain. *Appl. Math. Comput.* **2017**, *312*, 158–168. [CrossRef]
20. Bai, Z.; Zhang, S.; Sun, S.; Yin, C. Monotone iterative method for a class of fractional differential equations. *Electron. J. Differ. Equ.* **2016**, *2016*, 1–8.
21. Ding, Y.; Wei, Z.; Xu, J.; O'Regan, D. Extremal solutions for nonlinear fractional boundary value problems with p-Laplacian. *J. Comput. Appl. Math.* **2015**, *288*, 151–158. [CrossRef]
22. Wang, G. Twin iterative positive solutions of fractional q-difference Schrödinger equations. *Appl. Math. Lett.* **2018**, *76*, 103–109. [CrossRef]
23. Wang, G.; Sudsutad, W.; Zhang, L.; Tariboon, J. Monotone iterative technique for a nonlinear fractional q-difference equation of Caputo type. *Adv. Differ. Equ.* **2016** *211*. [CrossRef]
24. Hu, C.; Liu, B.; Xie, S. Monotone iterative solutions for nonlinear boundary value problems of fractional differential equation with deviating arguments. *Appl. Math. Comput.* **2013**, *222*, 72–81. [CrossRef]
25. Liu, Z.; Sun, J.; Szanto, I. Monotone iterative technique for Riemann–Liouville fractional integro- differential equations with advanced arguments. *Results Math.* **2013**, *63*, 1277–1287. [CrossRef]
26. Zhang, X.; Liu, L.; Wu, Y.; Lu, Y. The iterative solutions of nonlinear fractional differential equations. *Appl. Math. Comput.* **2013**, *219*, 4680–4691. [CrossRef]
27. Zhang, X.; Liu, L.; Wu, Y. The uniqueness of positive solution for a fractional order model of turbulent flow in a porous medium. *Appl. Math. Lett.* **2014**, *37*, 26–33. [CrossRef]
28. Zhang, L.; Ahmad, B.; Wang, G.; Agarwal, R.P. Nonlinear fractional integro-differential equations on unbounded domains in a Banach space. *J. Comput. Appl. Math.* **2013**, *249*, 51–56. [CrossRef]
29. Zhang, L.; Ahmad, B.; Wang, G. Successive iterations for positive extremal solutions of nonlinear fractional differential equations on a half-line. *Bull. Aust. Math. Soc.* **2015**, *91*, 116–128. [CrossRef]
30. Wang, G. Explicit iteration and unbounded solutions for fractional integral boundary value problem on an infinite interval. *Appl. Math. Lett.* **2015**, *47*, 1–7. [CrossRef]

31. Kilbas, A.A.; Srivastava, H.M.; Trujillo, J.J. Theory and Applications of Fractional Differential Equations. In *North-Holland Mathematics Studies*; Elsevier Science BV: Amsterdam, The Netherlands, 2006; Volume 204.
32. Zhong, C.; Fan, X.; Chen, W. *Nonlinear Functional Analysis and Its Application*; Lan Zhou University Press: Lanzhou, China, 1998.
33. Rudin, W. *Principles of Mathematical Analysis*; McGraw-Hill New York: 1976; ISBN 978-0-07-054235-8.

mathematics

MDPI

Article

Application of Fixed-Point Theory for a Nonlinear Fractional Three-Point Boundary-Value Problem

Ehsan Pourhadi [1,2], Reza Saadati [2] and Sotiris K. Ntouyas [3,4,*]

1 International Center for Mathematical Modelling in Physics and Cognitive Sciences, Department of Mathematics, Linnaeus University, SE-351 95 Växjö, Sweden; epourhadi@alumni.iust.ir
2 Department of Mathematics, Iran University of Science and Technology, Narmak, Tehran 16846-13114, Iran; rsaadati@eml.cc
3 Department of Mathematics, University of Ioannina, 451 10 Ioannina, Greece
4 Nonlinear Analysis and Applied Mathematics (NAAM)-Research Group, Department of Mathematics, Faculty of Science, King Abdulaziz University, P.O. Box 80203, Jeddah 21589, Saudi Arabia
* Correspondence: sntouyas@uoi.gr

Received: 10 February 2019; Accepted: 6 June 2019; Published: 10 June 2019

Abstract: Throughout this paper, via the Schauder fixed-point theorem, a generalization of Krasnoselskii's fixed-point theorem in a cone, as well as some inequalities relevant to Green's function, we study the existence of positive solutions of a nonlinear, fractional three-point boundary-value problem with a term of the first order derivative

$$({}_a^C D^\alpha x)(t) = f(t, x(t), x'(t)), \;\; a < t < b, \;\; 1 < \alpha < 2,$$

$$x(a) = 0, x(b) = \mu x(\eta), \;\; a < \eta < b, \;\; \mu > \lambda,$$

where $\lambda = \dfrac{b-a}{\eta - a}$ and ${}_a^C D^\alpha$ denotes the Caputo's fractional derivative, and $f : [a,b] \times \mathbb{R} \times \mathbb{R} \to \mathbb{R}$ is a continuous function satisfying the certain conditions.

Keywords: three-point boundary-value problem; Caputo's fractional derivative; Riemann-Liouville fractional integral; fixed-point theorems

---

1. Introduction

In the last decade, questions on positive solutions to two-point, three-point, and multi-point boundary value problems (BVPs) and integral boundary-value problems for nonlinear ordinary and fractional differential equations have attracted much interest. The investigation of three-point BVPs for nonlinear integer-order ordinary differential equations was initially begun by Gupta [1]. Since then, several authors have put their focus on the existence and multiplicity of solutions (or positive solutions) of three-point BVPs for nonlinear integer-order ordinary differential equations. Several papers are available in regard to the setting of integer orders of differential equations in the literature. In 2000, applying the fixed-point index theorems, the Leray-Schauder degree, and upper and lower solutions, Ma [2] studied a class of second-order three-point boundary value problems with a nonlinear term $f(x)$. In 2002, He and Ge [3], with the help of the Leggett-Williams fixed-point theorem [4], investigated the multiplicity of positive solutions of a problem with the nonlinear term $f(t,x)$ (see [5–15] and the references therein).

In recent years, multi-point boundary value problems have also been considered for fractional-order differential equations. For instance, employing the superlinearity and sublinearity, together with the well-known Guo-Lakshmikantham fixed-point theorem in cones, Ntouyas and Pourhadi [16] studied the existence of positive solutions to the boundary-value problem with a fractional order, $1 < \alpha < 2$. Furthermore, they investigated the convexity and concavity of the solutions with respect to the behavior of a given function as a coefficient of the subjected problem (see also [17–21]).

There were only a few papers available which focused on the existence of solutions for nonlinear fractional differential equations associated with three-point boundary conditions, which served as motivation for this work. The key idea of the current paper is that a term of the first-order derivative is involved in the subjected nonlinear problem, while most works (either fractional or ordinary differential equations) are done under the assumption that the first-order derivative is not involved explicitly in the nonlinear term.

In this paper, an analogy with a boundary-value problem for differential equations of integer orders via the Schauder fixed-point theorem, a generalized version of Krasnoselskii's fixed-point theorem in a cone [22], and also using the associated Green's function for the relevant problem, the existence of positive solutions for a fractional three-point boundary-value problem is investigated.

$$\begin{cases} ({}_a^C D^\alpha x)(t) = f(t, x(t), x'(t)), & a < t < b, \quad 1 < \alpha < 2, \\ x(a) = 0, \quad x(b) = \mu x(\eta), & a < \eta < b, \quad \mu > \lambda, \end{cases} \tag{1}$$

where $\lambda = \dfrac{b-a}{\eta - a}$ and ${}_a^C D^\alpha$ stands for the Caputo's fractional derivative, and $f : [a,b] \times \mathbb{R} \times \mathbb{R} \to \mathbb{R}$ is a continuous function which will be specified later on.

The organization of this paper is as follows. In Section 2, we recall some auxiliary facts and preliminaries. In Section 3, we first find the Green's function associated with (1), and then, using the inequalities related with this function and two well-known fixed-point theorems, we present our main results. An illustrative example is also given.

2. Preliminaries

This section is devoted to recall and gathering of some essential definitions and auxiliary facts in fractional calculus, as well as the results needed further on, which can be found in [23–25].

Definition 1. *Let $\alpha \geq 0$ and f be a real function defined in $[a,b]$. The Riemann-Liouville fractional integral of order α for a continuous function $f : (a, \infty) \to \mathbb{R}$ is defined by $({}_aI^0 f)(t) = f(t)$ and*

$$({}_aI^\alpha f)(t) = \frac{1}{\Gamma(\alpha)} \int_a^t (t-s)^{\alpha-1} f(s) ds, \quad \alpha > 0, \quad t \in [a,b],$$

where $\Gamma(\cdot)$ is the Gamma function.

Definition 2. *For a continuous function $f : (a, \infty) \to \mathbb{R}$, the Riemann-Liouville fractional derivative of fractional order $\alpha > 0$ is defined by*

$${}^{RL}D^\alpha_{a+} f(t) = \frac{1}{\Gamma(n-\alpha)} \left(\frac{d}{dt}\right)^n \int_a^t (t-s)^{n-\alpha-1} f(s) ds, \quad n = [\alpha] + 1,$$

where $[\alpha]$ denotes the integer part of the real number α.

For $\alpha < 0$ and the convenience of the reader, we use the denotation $D^\alpha y = I^{-\alpha} y$. Moreover, for $\beta \in [0, \alpha)$, it is valid that $D^\beta I^\alpha y = I^{\alpha - \beta} y$.

Definition 3. *Caputo's fractional derivative of order $\alpha \geq 0$ is given by $({}_a^C D^0 f)(t) = f(t)$, and $({}_a^C D^\alpha f)(t) = ({}_a I^{m-\alpha} D^m f)(t)$ for $\alpha > 0$, where m is the smallest integer greater or equal to α. Besides, it can be formulated by*

$$ {}^C D_{a+}^\alpha f(t) = \frac{1}{\Gamma(n-\alpha)} \int_a^t (t-s)^{n-\alpha-1} f^{(n)}(s) ds, \quad n = [\alpha] + 1, \quad \text{for } f \in AC^n([a,b]), $$

where $\alpha \notin \mathbb{N}_0$ and $AC^n([a,b])$ represents the space of all absolutely continuous functions having an absolutely continuous derivative up to $(n-1)$ (see also [23]).

In the sequel, the associated Green's function for the three-point BVP (1) is formulated by utilizing a crucial lemma derived by Zhang [26] as follows:

Lemma 1. *Let $\alpha > 0$; then, in $C(0,T) \cap L(0,T)$, the differential equation*

$$ {}^C D_{0+}^\alpha u(t) = 0 $$

has solutions $u(t) = c_0 + c_1 t + c_2 t^2 + \cdots + c_n t^{n-1}$, $c_i \in \mathbb{R}$, $i = 0, 1, \cdots, n$, $n = [\alpha] + 1$.

Furthermore, it has been proved that $I_{0+}^\alpha D_{0+}^\alpha u(t) = u(t) + c_0 + c_1 t + c_2 t^2 + \cdots + c_n t^{n-1}$ for some $c_i \in \mathbb{R}$, $i = 0, 1, \cdots, n$, $n = [\alpha] + 1$ (see Lemma 2.3 in [26]).

3. Main Results

In the following, we present a pivotal lemma which will play a crucial role in our next analysis and direct our attention to a variant of Problem (1).

Lemma 2. *Let $\Delta := \mu(\eta - a) - (b - a) > 0$. Then, $x \in C^1(I, \mathbb{R})$ is the solution of fractional three-point BVP (1) if, and only if x satisfies the integral equation*

$$ x(t) = \int_a^b G(t,s) f(s, x(s), x'(s)) ds, \qquad t \in I := [a,b] \tag{2} $$

where the Green's function $G(t,s) := G_1(t,s) + G_2(t,s)$ is given by

$$ G_1(t,s) = \begin{cases} \dfrac{(t-s)^{\alpha-1}}{\Gamma(\alpha)}, & a \leq s \leq t \leq b \\ 0, & a \leq t \leq s \leq b \end{cases} \tag{3} $$

$$ G_2(t,s) = \begin{cases} \dfrac{t-a}{\Delta\Gamma(\alpha)} \Big((b-s)^{\alpha-1} - \mu(\eta - s)^{\alpha-1} \Big), & a \leq s \leq \eta, \quad t \in I, \\ \dfrac{(b-s)^{\alpha-1}(t-a)}{\Delta\Gamma(\alpha)}, & \eta \leq s \leq b, \quad t \in I. \end{cases} \tag{4} $$

Moreover,

$$ \max_{t,s \in I} G(t,s) \leq \frac{\mu(b-a)^\alpha}{\lambda\Delta\Gamma(\alpha)}. \tag{5} $$

Proof. By employing the Riemann-Liouville fractional integral ${}_a I^\alpha$ for Equation (1), the imposed boundary conditions, and the knowledge received from the fractional calculus theory, we observe that $x \in C^1[a,b]$ is a solution of (1) if, and only if

$$ x(t) = c_0 + c_1(t-a) + \frac{1}{\Gamma(\alpha)} \int_a^t (t-s)^{\alpha-1} f(s, x(s), x'(s)) ds \tag{6} $$

for some real constants c_0 and c_1 (see Lemma 1). Since $x(a) = 0$, we immediately derive $c_0 = 0$. Now,

$$\begin{aligned} x(b) = \mu x(\eta) \;&\Leftrightarrow\; c_1(b-a) + \frac{1}{\Gamma(\alpha)}\int_a^b (b-s)^{\alpha-1} f(s,x(s),x'(s))ds \\ &= c_1\mu(\eta - a) + \frac{\mu}{\Gamma(\alpha)}\int_a^\eta (\eta-s)^{\alpha-1} f(s,x(s),x'(s))ds \\ &\Leftrightarrow\; c_1 = \frac{1}{\Delta\Gamma(\alpha)}\bigg(\int_a^b (b-s)^{\alpha-1} f(s,x(s),x'(s))ds \\ &\qquad - \mu\int_a^\eta (\eta-s)^{\alpha-1} f(s,x(s),x'(s))ds\bigg) \end{aligned}$$

which, together with (6), implies that

$$\begin{aligned} x(t) &= \frac{1}{\Gamma(\alpha)}\int_a^t (t-s)^{\alpha-1} f(s,x(s),x'(s))ds \\ &+ \frac{t-a}{\Delta\Gamma(\alpha)}\bigg(\int_a^b (b-s)^{\alpha-1} f(s,x(s),x'(s))ds - \mu\int_a^\eta (\eta-s)^{\alpha-1} f(s,x(s),x'(s))ds\bigg). \end{aligned} \tag{7}$$

This is also equivalent to

$$\begin{aligned} x(t) &= \frac{1}{\Gamma(\alpha)}\int_a^t (t-s)^{\alpha-1} f(s,x(s),x'(s))ds \\ &+ \frac{t-a}{\Delta\Gamma(\alpha)}\bigg(\int_a^\eta \Big((b-s)^{\alpha-1} - \mu(\eta-s)^{\alpha-1}\Big) f(s,x(s),x'(s))ds \\ &+ \int_\eta^b (b-s)^{\alpha-1} f(s,x(s),x'(s))ds\bigg). \end{aligned} \tag{8}$$

Now, (8) can be rewritten as follows:

$$x(t) = \int_a^b G(t,s) f(s,x(s),x'(s))ds, \qquad t \in I = [a,b],$$

where the associated Green's function $G(t,s) = G_1(t,s) + G_2(t,s)$ is defined by (3) and (4). Furthermore, for any $s \in I$,

$$\begin{aligned} \max_{t\in I} G(t,s) &= \frac{(b-s)^{\alpha-1}}{\Gamma(\alpha)} + \max_{t\in I}\left(\frac{(b-s)^{\alpha-1}(t-a)}{\Delta\cdot\Gamma(\alpha)}\right) \\ &= \frac{(b-s)^{\alpha-1}}{\Gamma(\alpha)}\left(1 + \frac{b-a}{\Delta}\right) \\ &= \frac{(b-s)^{\alpha-1}}{\Delta\Gamma(\alpha)}\Big(\mu(\eta-a)\Big) \\ &\le \frac{\mu(b-a)^\alpha}{\lambda\Delta\Gamma(\alpha)}. \end{aligned} \tag{9}$$

Therefore, the inequality (5) is proved. □

Throughout the remainder of this paper, we employ two well-known fixed-point results to study Equation (1).

3.1. Existence of Positive Solution with the Schauder Fixed-Point Principle

In the following, we investigate Equation (1) via the Schauder fixed-point theorem.

Theorem 1 (Schauder fixed-point Theorem, [27]). *Let U be a nonempty and convex subset of a normed space X. Let T be a continuous mapping of U into a compact set $K \subset U$. Then, T has a fixed point.*

In the sequel, we suppose the following condition:

(C_0) f satisfies Carathéodory-type conditions. That is, $f(\cdot, u, v)$ is measurable for the fixed u, v, and $f(t, \cdot, \cdot)$ is continuous for a.e. $t \in I$. Moreover, if $u \geq 0$, then $f(t, u, v) \geq 0$.

Under this condition, the equivalent representation for Equation (2) is given by

$$x(t) = [\mathcal{F}x](t), \qquad t \in I = [a, b],$$

where $\mathcal{F}$ is an operator defined by

$$[\mathcal{F}x](t) = \int_a^b G(t,s) f(s, x(s), x'(s)) ds, \qquad t \in I.$$

It is obvious to see that $x(t)$ is a solution to the problem (1) if it is a fixed point of the operator $\mathcal{F}$.

Theorem 2. *Suppose that f satisfies the condition (C_0) and the followings:*

(C_1) *There exists an L^1-function $\varphi : I \to \mathbb{R}^+$, such that*

$$|f(t, x(t), x'(t))| \leq \varphi(t) \Omega(\|x\|), \qquad x \in C^1(I, \mathbb{R}), \quad t \in I,$$

where $\Omega : I \to [0, \infty)$ is a non-decreasing continuous function and $\|\cdot\|$ denotes the supremum norm on I.

(C_2) *The point $\eta \in (a, b)$ is taken sufficiently close to a, such that*

$$\mu \int_a^\eta (\eta - s)^{\alpha-1} f(s, x(s), x'(s)) ds \leq \int_a^b (b - s)^{\alpha-1} f(s, x(s), x'(s)) ds, \quad \textit{for all } x \in C^1(I, \mathbb{R}).$$

Moreover, suppose that there exists a continuous function p defined on I satisfying the following inequality:

$$\frac{\mu (b-a)^\alpha}{\lambda \Delta \Gamma(\alpha)} \|\varphi\|_1 \Omega(\|p\|) \leq \|p\|. \tag{10}$$

Then, Equation (1) has at least one positive solution in $C^1(I, \mathbb{R})$, bounded above by $\|p\|$.

Proof. Let us define

$$S = \left\{ x \in C^1(I, \mathbb{R}) \;\middle|\; 0 \leq x(t) \leq \|p\| \text{ for } t \in I \right\},$$

where $\|\cdot\|_1$ denotes the L^1-norm on I, and p is a function satisfying the condition (C_2). Clearly, the set S is a non-empty, closed, bounded, and convex subset of $C^1(I, \mathbb{R})$. To establish that Equation (1) has a positive solution, it only suffices to show that the operator $\mathcal{F}$ has a fixed point in S. We first show that S is $\mathcal{F}$-invariant. Let $x(t)$ be a non-negative function; then, following condition (C_0), one finds that $f(t, x(t), x'(t))$ is non-negative too, and the right-hand side of (7), together with conditions (C_0), (C_2) and the fact that $\Delta > 0$ imply that $[\mathcal{F}x](t) \geq 0$.

On the other hand, using (2), (9), (10) and condition (C_1), one can see that

$$|[\mathcal{F}x](t)| \leq \|p\|, \qquad \text{for } t \in I.$$

Hence, $\mathcal{F}S \subset S$. Furthermore, to show the continuity of the operator $\mathcal{F} : S \to S$, we have

$$|[\mathcal{F}x_n](t) - [\mathcal{F}x](t)| \to 0 \quad \Longleftrightarrow \quad \int_a^b |G(t,s)| \cdot |f(s,x(s),x'(s)) - f(s,x_n(s),x_n'(s))| ds \to 0$$

for $x, x_n \in S \subset C^1(I,\mathbb{R})$.

Next, we show that $\mathcal{F}S$ is equicontinuous. Assume that $a \leq t_1 < t_2 \leq b$. Following the definition of $\mathcal{F}$ and the condition (C_1), we have

$$|[\mathcal{F}x](t_2) - [\mathcal{F}x](t_1)| \leq \|\varphi\|_1 \Omega(\|x\|) \cdot \max_{s \in I} |G(t_2,s) - G(t_1,s)|,$$

which tends to zero, as $t_1 \to t_2$. Consequently, we conclude that $\mathcal{F}S$ is equicontinuous. Furthermore, the equicontinuity of the set of functions $[\mathcal{F}S]' = \{y' : y = \mathcal{F}x, x \in S\}$ can also be shown. Indeed, suppose that $a \leq t_1 < t_2 \leq b$; then,

$$|[\mathcal{F}x]'(t_2) - [\mathcal{F}x]'(t_1)| \leq \|\varphi\|_1 \Omega(\|x\|) \cdot \max_{s \in I} |\frac{\partial}{\partial t} G(t_2,s) - \frac{\partial}{\partial t} G(t_1,s)| \to 0$$

whenever $t_1 \to t_2$. Therefore, we conclude that $[\mathcal{F}S]'$ is equicontinuous.

Besides, S is totally bounded (since every sequence in S has a Cauchy subsequence), so S is compact and $\mathcal{F}S$ is compact. Now, all the conditions of the Schauder fixed point are fulfilled; thus, the operator $\mathcal{F}$ as a self-map on S possesses a fixed point in this set, which yields that Equation (1) has a positive solution bounded above by $\|p\|$. □

3.2. Existence of Positive Solution via the Krasnoselskii Type Fixed-Point Theorem

In what follows, we recall a generalization of Krasnoselskii's fixed-point theorem of cone expansion and compression of a norm type. To do this, let us suppose $(X, \|\cdot\|)$ is a Banach space, and P is the cone in X. Assume that $\overline{\alpha}, \overline{\beta} : X \to \mathbb{R}^+$ are two continuous non-negative functionals that satisfy

$$\overline{\alpha}(rx) \leq |r|\overline{\alpha}(x), \quad \overline{\beta}(rx) \leq |r|\overline{\beta}(x), \quad \text{for } x \in X,\ r \in [0,1], \tag{11}$$

and

$$M_1 \max\{\overline{\alpha}(x), \overline{\beta}(x)\} \leq \|x\| \leq M_2 \max\{\overline{\alpha}(x), \overline{\beta}(x)\}, \quad \text{for } x \in X, \tag{12}$$

where M_1, M_2 are two positive constants.

The following lemma is understood as a special case of a result derived by Bai and Ge (see [22] Theorem 2.1).

Lemma 3. *Let $r_2 > r_1 > 0$, $L_2 > L_1 > 0$ be constants and*

$$\Omega_i = \{x \in X \mid \overline{\alpha}(x) < r_i,\ \overline{\beta}(x) < L_i\},\ i = 1,2$$

be two open subsets in X, such that $\theta \in \Omega_1 \subset \overline{\Omega}_1 \subset \Omega_2$. In addition, let

$$C_i = \{x \in X \mid \overline{\alpha}(x) = r_i,\ \overline{\beta}(x) \leq L_i\},\ i = 1,2;$$
$$D_i = \{x \in X \mid \overline{\alpha}(x) \leq r_i,\ \overline{\beta}(x) = L_i\},\ i = 1,2.$$

Assume $T : P \to P$ is a completely continuous operator satisfying

(S_1) $\overline{\alpha}(Tx) \leq r_1, x \in C_1 \cap P$; $\overline{\beta}(Tx) \leq L_1, x \in D_1 \cap P$;
$\overline{\alpha}(Tx) \geq r_2, x \in C_2 \cap P$; $\overline{\beta}(Tx) \geq L_2, x \in D_2 \cap P$;

or

(S_2) $\overline{\alpha}(Tx) \geq r_1, x \in C_1 \cap P$; $\overline{\beta}(Tx) \geq L_1, x \in D_1 \cap P$;
$\overline{\alpha}(Tx) \leq r_2, x \in C_2 \cap P$; $\overline{\beta}(Tx) \leq L_2, x \in D_2 \cap P$;

then, T has at least one fixed point in $(\overline{\Omega}_2 \setminus \Omega_1) \cap P$.

To apply the recent fixed-point theorem, let us consider the following settings.

Let X be a Banach space in $C^1(I, \mathbb{R})$, with

$$\|x\| = \max\{\max_{t\in I} |x(t)|, \max_{t\in I} |x'(t)|\}, \quad x \in X.$$

Define a cone P by

$$P = \left\{ x \in X \,\middle|\, x(t) \geq 0, \quad \text{for all } t \in I \right\},$$

and functionals

$$\overline{\alpha}(x) = \max_{t\in I} |x(t)|, \qquad \overline{\beta}(x) = \max_{t\in I} |x'(t)|, \quad \forall x \in X.$$

With the help of (11) and (12), $\overline{\alpha}$ and $\overline{\beta}$ are two continuous non-negative functionals, such that $\|x\| = \max\{\overline{\alpha}(x), \overline{\beta}(x)\}$. Let us consider the following notations:

$$L := \max_{t\in I} \left| \int_a^b G(t,s)ds \right|,$$

$$N := \left(\frac{(b-a)^{\alpha-1}}{\Gamma(\alpha)} + \frac{1}{\Delta\Gamma(\alpha+1)} \left[2(b-\xi)^{\alpha} - 2\mu(\eta-\xi)^{\alpha} + \mu(\eta-a)^{\alpha} - (b-a)^{\alpha} \right] \right),$$

where $\xi = \eta - \dfrac{b-\eta}{\mu^{\frac{1}{\alpha-1}} - 1} \in [a, \eta]$.

Accounting on condition (C_2), we get that the operator $\mathcal{F}$ (as defined before) transforms P into itself; moreover, a standard argument shows that it is completely continuous. In fact, $\mathcal{F}$ is continuous and maps any bounded subset of P into a relatively compact subset of P.

In the following result, we suppose that η is sufficiently close to a such that the Green function G is non-negative. For the possibility, we refer to Example (1).

Theorem 3. *Suppose there are four constants,* $k_2 > k_1 > 0$, $l_2 > l_1 > 0$, *such that* $\max\left\{\frac{k_1}{L}, \frac{l_1}{M}\right\} \leq \min\left\{\frac{k_2}{L}, \frac{l_2}{N}\right\}$, *and the following assumptions hold:*

(C_3) *There is an* L^1*-function* $\psi : I \to \mathbb{R}^+$ *which satisfies the following condition:*

$$\int_a^b (b-s)^{\alpha-1} f(s,x(s),x'(s))ds - \mu \int_a^{\eta} (\eta-s)^{\alpha-1} f(s,x(s),x'(s))ds \geq \int_a^b \psi(s) f(s,x(s),x'(s))ds,$$

for all $x \in C^1(I, \mathbb{R}^+)$.

(C_4) $f(t,u,v) \geq \max\left\{\frac{k_1}{L}, \frac{l_1}{M}\right\}$, *for* $(t,u,v) \in I \times [0,k_1] \times [-l_1, l_1]$;

(C_5) $f(t,u,v) \leq \min\left\{\frac{k_2}{L}, \frac{l_2}{N}\right\}$, *for* $(t,u,v) \in I \times [0,k_2] \times [-l_2, l_2]$, *where* $M = \dfrac{\|\psi\|_1}{\Delta\Gamma(\alpha)}$. *Then, problem* (1) *has at least one positive solution* $x(t)$, *such that*

$$k_1 \leq \max_{t\in I} x(t) \leq k_2 \qquad or \qquad l_1 \leq \max_{t\in I} |x'(t)| \leq l_2.$$

Proof. Let us take the following subsets of $X = C^1(I, \mathbb{R})$

$$\begin{aligned} \Omega_i &= \{x \in X \mid \overline{\alpha}(x) < k_i, \quad \overline{\beta}(x) < l_i\}, \quad i = 1,2; \\ \mathcal{P}_i &= \{x \in X \mid \overline{\alpha}(x) = k_i, \quad \overline{\beta}(x) \leq l_i\}, \quad i = 1,2; \\ \mathcal{Q}_i &= \{x \in X \mid \overline{\alpha}(x) \leq k_i, \quad \overline{\beta}(x) = l_i\}, \quad i = 1,2. \end{aligned}$$

For $x \in \mathcal{P}_1 \cap P$, by ($C_4$), there exists

$$\overline{\alpha}(\mathcal{F}x) = \max_{t\in I} \left| \int_a^b G(t,s) f(s, x(s), x'(s)) ds \right| \geq \frac{k_1}{L} \max_{t\in I} \left| \int_a^b G(t,s) ds \right| = k_1. \tag{13}$$

Since η is taken sufficiently close to a such that the Green function G is non-negative, the inequality (13) holds. Moreover, taking into account the continuity and properties of $\mathcal{F}$, we derive

$$\begin{aligned} (\mathcal{F}x)'(t) &= \frac{\alpha - 1}{\Gamma(\alpha)} \int_a^t (t-s)^{\alpha-2} f(s, x(s), x'(s)) ds \\ &\quad + \frac{1}{\Delta\Gamma(\alpha)} \left(\int_a^b (b-s)^{\alpha-1} f(s, x(s), x'(s)) ds - \mu \int_a^\eta (\eta - s)^{\alpha-1} f(s, x(s), x'(s)) ds \right), \\ (\mathcal{F}x)''(t) &= -\frac{(\alpha-1)(2-\alpha)}{\Gamma(\alpha)} \int_a^t (t-s)^{\alpha-3} f(s, x(s), x'(s)) ds \leq 0, \qquad t \in I. \end{aligned}$$

Therefore, $(\mathcal{F}x)(t)$ is concave on I, and so the absolute value of $(\mathcal{F}x)'$ takes its maximum only at the endpoints of I. That is,

$$\max_{t\in I} |(\mathcal{F}x)'(t)| = \max_{t\in I} \{|(\mathcal{F}x)'(a)|, |(\mathcal{F}x)'(b)|\} = |(\mathcal{F}x)'(b)|.$$

Therefore, for $x \in \mathcal{Q}_1 \cap P$, followed by ($C_3$) and ($C_4$), one can see that

$$\begin{aligned} \beta(\mathcal{F}x) &= \max_{t\in I} \{|(\mathcal{F}x)'(a)|, |(\mathcal{F}x)'(b)|\} \\ &\geq |(\mathcal{F}x)'(a)| \\ &= \frac{1}{\Delta\Gamma(\alpha)} \left(\int_a^b (b-s)^{\alpha-1} f(s, x(s), x'(s)) ds - \mu \int_a^\eta (\eta - s)^{\alpha-1} f(s, x(s), x'(s)) ds \right) \\ &\geq \frac{1}{\Delta\Gamma(\alpha)} \int_a^b \psi(s) f(s, x(s), x'(s)) ds \\ &\geq \frac{l_1}{M\Delta\Gamma(\alpha)} \|\psi\|_1 = l_1. \end{aligned}$$

Now, assuming $x \in \mathcal{P}_2 \cap P$, by ($C_5$), there is

$$\begin{aligned} \overline{\alpha}(\mathcal{F}x) &= \max_{t\in I} \left| \int_a^b G(t,s) f(s, x(s), x'(s)) ds \right| \\ &\leq \frac{k_2}{L} \max_{t\in I} \left| \int_a^b G(t,s) ds \right| = k_2. \end{aligned}$$

Finally, for $x \in \mathcal{Q}_2 \cap P$, by ($C_5$), one can find

$$\begin{aligned}
\beta(\mathcal{F}x) &= \max_{t\in I}\{|(\mathcal{F}x)'(a)|, |(\mathcal{F}x)'(b)|\} \\
&= |(\mathcal{F}x)'(b)| \\
&= \frac{\alpha-1}{\Gamma(\alpha)}\int_a^b (b-s)^{\alpha-2} f(s,x(s),x'(s))ds \\
&\quad + \frac{1}{\Delta\Gamma(\alpha)}\left(\int_a^\eta \left((b-s)^{\alpha-1} - \mu(\eta-s)^{\alpha-1}\right) f(s,x(s),x'(s))ds \right. \\
&\quad \left. + \int_\eta^b (b-s)^{\alpha-1} f(s,x(s),x'(s))ds\right) \\
&\leq \frac{l_2}{N}\left(\frac{(b-a)^{\alpha-1}}{\Gamma(\alpha)} + \frac{1}{\Delta\Gamma(\alpha)}\int_a^\eta \left|(b-s)^{\alpha-1} - \mu(\eta-s)^{\alpha-1}\right| ds + \frac{(b-\eta)^\alpha}{\Delta\Gamma(\alpha+1)}\right) \\
&= \frac{l_2}{N}\left(\frac{(b-a)^{\alpha-1}}{\Gamma(\alpha)} + \frac{1}{\Delta\Gamma(\alpha)}\left(\int_a^\xi \left[\mu(\eta-s)^{\alpha-1} - (b-s)^{\alpha-1}\right] ds \right.\right. \\
&\quad \left.\left. + \int_\xi^\eta \left[(b-s)^{\alpha-1} - \mu(\eta-s)^{\alpha-1}\right] ds\right) + \frac{(b-\eta)^\alpha}{\Delta\Gamma(\alpha+1)}\right) \\
&= \frac{l_2}{N}\left(\frac{(b-a)^{\alpha-1}}{\Gamma(\alpha)} + \frac{1}{\Delta\Gamma(\alpha+1)}\left[2(b-\xi)^\alpha - 2\mu(\eta-\xi)^\alpha + \mu(\eta-a)^\alpha - (b-a)^\alpha\right]\right) \\
&= l_2.
\end{aligned}$$

Now, all conditions of Lemma 3 are satisfied, and it implies that there exists $x \in (\overline{\Omega}_2 \setminus \Omega_1) \cap P$, such that $x = \mathcal{F}x$. That is, the problem (1) has at least one positive solution $x(t)$, such that

$$k_1 \leq \overline{\alpha}(x) \leq k_2 \quad \text{or} \quad l_1 \leq \overline{\beta}(x) \leq l_2.$$

In other words,

$$k_1 \leq \max_{t\in I} x(t) \leq k_2 \quad \text{or} \quad l_1 \leq \max_{t\in I} |x'(t)| \leq l_2,$$

which completes the proof. □

In the following, we illustrate the said result with an example.

Example 1. *Consider the boundary value problem:*

$$\begin{cases} ({}^C_aD^{\frac{3}{2}}x)(t) = f(t,x(t),x'(t)), \quad 0 < t < 1, \\ x(0) = 0, \quad x(1) = \mu x(\eta), \quad 0 < \eta < 1, \quad \mu\eta > 1, \end{cases} \tag{14}$$

where $\mu = a\eta^{-r}$ for some $a > 1$, $0 < r < 1.5$, and $f : [0,1] \times \mathbb{R} \times \mathbb{R} \to \mathbb{R}$ is given by

$$f(t,u,v) = \lambda_1 \sin^2 u + \lambda_2 \cos^2 v + \lambda_3 t + \lambda_4, \qquad t \in I = [0,1],\ u,v \in \mathbb{R}$$

such that $\lambda_i \geq 0$, $\lambda_4 > 0$, $i = 1,2,3$, are constant. Since f takes both supremum and infimum over domain D, let us set

$$\inf_D f(t,u,v) = M_1, \qquad \sup_D f(t,u,v) = M_2.$$

A direct computation shows that

$$L = \frac{4\mu\lambda}{3(\mu\lambda-1)\sqrt{\pi}}.$$

On the other hand, by considering η as being sufficiently close to 0 and $\psi(t) < \dfrac{2M_1}{3M_2}$, we see that

$$\int_0^1 \sqrt{1-s}f(s,x(s),x'(s))ds > \int_0^1 \psi(s)f(s,x(s),x'(s))ds.$$

Thus, condition (C_3) is satisfied. To give more detail, if $\eta \to 0$, then using the fact that $\mu = a\eta^{-r}$, together with Leibniz's rule, we see that

$$\begin{aligned}\lim_{\eta\to 0}\left|\mu\int_0^\eta \sqrt{\eta-s}f(s,x(s),x'(s))ds\right| &= \frac{a}{2r}\lim_{\eta\to 0}\frac{\left|\int_0^\eta \frac{1}{\sqrt{\eta-s}}f(s,x(s),x'(s))ds\right|}{\eta^{r-1}}\\ &\le \frac{a|M_2|}{2r}\lim_{\eta\to 0}\frac{\int_0^\eta \frac{1}{\sqrt{\eta-s}}ds}{\eta^{r-1}}\\ &= \frac{a|M_2|}{r}\lim_{\eta\to 0}\eta^{1.5-r} = 0,\end{aligned}$$

which shows that the second integral term in the left-hand side of the inequality in condition (C_3) vanishes for η sufficiently close to 0.

Furthermore, $M = \dfrac{2\|\psi\|_1}{(\mu\eta-1)\sqrt{\pi}} < L$ *and*

$$N = \frac{2}{\sqrt{\pi}} + \frac{L}{\mu\eta}\left(2(1-\xi)^{1.5} - 2\mu(\eta-\xi)^{1.5} + \mu\eta^{1.5} - 1\right) > L,$$

where $\xi = \dfrac{\eta\mu^2-1}{\mu^2-1} \in (0,1)$. Next, to check the conditions (C_4) and (C_5), choose $k_2 > l_2 > l_1 > k_1 > 0$, such that $l_1 = M \cdot M_1$ and $l_2 = N \cdot M_2$. Then, one can derive the followings:

$$f(t,u,v) \ge \max\left\{\frac{k_1}{L}, \frac{l_1}{M}\right\} = \frac{l_1}{M}, \qquad \text{for } (t,u,v) \in I \times [0,k_1] \times [-l_1,l_1];$$
$$f(t,u,v) \le \min\left\{\frac{k_2}{L}, \frac{l_2}{N}\right\} = \frac{l_2}{N}, \qquad \text{for } (t,u,v) \in I \times [0,k_2] \times [-l_2,l_2].$$

That is to say, all the assumptions of Theorem 3 are fulfilled, then problem (14) has at least one positive solution x, such that

$$k_1 \le \max_{t\in I} x(t) \le k_2 \qquad \text{or} \qquad M \cdot M_1 \le \max_{t\in I}|x'(t)| \le N \cdot M_2.$$

Author Contributions: Formal Analysis, E.P., R.S. and S.K.N.

Acknowledgments: The authors thank the reviewers for their useful remarks on our work.

Conflicts of Interest: The authors declare no conflict of interest.

References

1. Gupta, C.P. Solvability of a three-point nonlinear boundary value problem for a second order ordinary differential equation. *J. Math. Anal. Appl.* **1992**, *168*, 540–551. [CrossRef]
2. Ma, R. Multiplicity of positive solutions for second-order three-point boundary value problems. *Comput. Math. Appl.* **2000**, *40*, 193–204. [CrossRef]
3. He, X.; Ge, W. Triple solutions for second-order three-point boundary value problems. *J. Math. Anal. Appl.* **2002**, *268*, 256–265. [CrossRef]

4. Leggett, R.W.; Williams, L.R. Multiple positive fixed-points of nonlinear operators on ordered Banach spaces. *Indiana Univ. Math. J.* **1979**, *28*, 673–688. [CrossRef]
5. Bai, Z. Solvability for a class of fractional m-point boundary value problem at resonance. *Comput. Math. Appl.* **2011**, *62*, 1292–1302. [CrossRef]
6. Bai, Z.; Zhang, Y. Solvability of fractional three-point boundary value problems with nonlinear growth. *Appl. Math. Comput.* **2011**, *218*, 1719–1725. [CrossRef]
7. Cui, Y.; Sun, J. Positive solutions for second-order three-point boundary value problems in Banach spaces. *Acta Math. Sin.* **2011**, *4*, 743–751.
8. Guo, Y.; Ge, W. Positive solutions for three-point boundary value problems with dependence on the first order derivative. *J. Math. Anal. Appl.* **2004**, *290*, 291–301. [CrossRef]
9. Il'in, V.A.; Moiseev, E.I. Nonlocal boundary value problem of the first kind for a Sturm-Liouville operator in its differential and finite difference aspects. *Differ. Equ.* **1987**, *23*, 803–810.
10. Il'in, V.A.; Moiseev, E.I. Nonlocal boundary value problem of the second kind for a Sturm-Liouville operator. *Differ. Equ.* **1987**, *23*, 979–987.
11. Ji, D.; Bai, Z.; Ge, W. The existence of countably many positive solutions for singular multipoint boundary value problems. *Nonlinear Anal. Theory Methods Appl.* **2010**, *72*, 955–964. [CrossRef]
12. Li, H. Existence of nontrivial solutions for superlinear three-point boundary value problems. *Acta Math. Appl. Sin.* **2017**, *33*, 1043–1052. [CrossRef]
13. Ma, R. Positive solutions of nonlinear three-point boundary value problem. *Electron. J. Differ. Equations* **1999**, *34*, 1–8. [CrossRef]
14. Marano, S. A remark on a second order three-point boundary value problem. *J. Math. Anal. Appl.* **1994**, *183*, 518–522. [CrossRef]
15. Webb, J.R.L. Positive solutions of some three point boundary value problems via fixed-point index theory. *Nonlinear Anal. Theory Methods Appl.* **2001**, *47*, 4319–4332. [CrossRef]
16. Ntouyas, S.K.; Pourhadi, E. Positive solutions of nonlinear fractional three-point boundary-value problem. *Le Mat.* **2018**, *73*, 139–154.
17. Sudsutad, W.; Tariboon, J.; Ntouyas, S.K. Positive solutions for fractional differential equations with three-point multi-term fractional integral boundary conditions. *Adv. Differ. Equ.* **2014**, *2014*, 28.
18. Wang, G.; Ntouyas, S.K.; Zhang, L. Positive solutions of the three-point boundary value problem for fractional-order differential equations with an advanced argument. *Adv. Differ. Equ.* **2011**, *2011*, 2. [CrossRef]
19. Wang, G.; Zhang, L.; Ntouyas, S.K. Multiplicity of positive solutions for fractional order three-point boundary value problems. *Commun. Appl. Nonlinear Anal.* **2013**, *20*, 41–53.
20. Zhang, Y. Existence results for a coupled system of nonlinear fractional multi-point boundary value problems at resonance. *J. Inequal. Appl.* **2018**, *2018*, 198. [CrossRef]
21. Zou, Y.; Liu, L.; Cui, Y. The existence of solutions for four-point coupled boundary value problems of fractional differential equations at resonance. *Abstr. Appl. Anal.* **2014**, *2014*, 1–8. [CrossRef]
22. Bai, Z.; Ge, W. Existence of positive solutions to fourth-order quasilinear boundary value problems. *Acta Math. Sin.* **2006**, *22*, 1825–1830. [CrossRef]
23. Kilbas, A.A.; Srivastava, H.M.; Trujillo, J.J. *Theory and Applications of Fractional Differential Equations*; North-Holland Math. Stud.; Elsevier: Amsterdam, The Netherlands, 2006; Volume 204.
24. Podlubny, I. *Fractional Differential Equations*; Mathematics in Science and Engineering; Academic Press: New York, NY, USA; London, UK; Toronto, ON, Canada, 1999; Volume 198.
25. Samko, S.G.; Kilbas, A.A.; Marichev, O.I. *Fractional Integral and Derivatives (Theory and Applications)*; Gordon and Breach: Yverdon, Switzerland, 1993.
26. Zhang, S. Positive solutions for boundary-value problems of nonlinear fractional differential equations. *Electron. J. Differ. Equ.* **2006**, *2006*, 1–12. [CrossRef]
27. Smart, D.R. *Fixed-Point Theorems*; Cambridge University Press: London, UK; New York, NY, USA, 1974.

mathematics MDPI

Article

The Langevin Equation in Terms of Generalized Liouville–Caputo Derivatives with Nonlocal Boundary Conditions Involving a Generalized Fractional Integral

Bashir Ahmad [1], Madeaha Alghanmi [1], Ahmed Alsaedi [1] and Hari M. Srivastava [2,3,*] and Sotiris K. Ntouyas [1,4]

1 Nonlinear Analysis and Applied Mathematics (NAAM)-Research Group, Department of Mathematics, Faculty of Science, King Abdulaziz University, P.O. Box 80203, Jeddah 21589, Saudi Arabia; bashirahmad_qau@yahoo.com (B.A.); madeaha@hotmail.com (M.A.); aalsaedi@hotmail.com (A.A.); sntouyas@uoi.gr (S.K.N.)
2 Department of Mathematics and Statistics, University of Victoria, Victoria, BC V8W 3R4, Canada
3 Department of Medical Research, China Medical University Hospital, China Medical University, Taichung 40402, Taiwan
4 Department of Mathematics, University of Ioannina, 451 10 Ioannina, Greece
* Correspondence: harimsri@math.uvic.ca

Received: 25 April 2019; Accepted: 6 June 2019; Published: 11 June 2019

Abstract: In this paper, we establish sufficient conditions for the existence of solutions for a nonlinear Langevin equation based on Liouville-Caputo-type generalized fractional differential operators of different orders, supplemented with nonlocal boundary conditions involving a generalized integral operator. The modern techniques of functional analysis are employed to obtain the desired results. The paper concludes with illustrative examples.

Keywords: Langevin equation; generalized fractional integral; generalized Liouville–Caputo derivative; nonlocal boundary conditions; existence; fixed point

1. Introduction

The topic of fractional calculus has emerged as an interesting area of investigation in view of its widespread applications in social sciences, engineering and technical sciences. Mathematical models based on fractional order differential and integral operators are considered to be more realistic and practical than their integer-order counterparts as such models can reveal the history of the ongoing phenomena in systems and processes. This branch of mathematical analysis is now very developed and covers a wide range of interesting results, for instance [1–7].

The Langevin equation is an effective tool of mathematical physics, which can describe processes like anomalous diffusion in a descent manner. Examples of such processes include price index fluctuations [8], harmonic oscillators [9], etc. A generic Langevin equation for noise sources with correlations also plays a central role in the theory of critical dynamics [10]. The nature of the quantum noise can be understood better by means of a generalized Langevin equation [11]. The role of the Langevin equation in fractional systems, such as fractional reaction-diffusion systems [12,13], is very rich and beautiful. The fractional analogue (also known as the stochastic differential equation) of the usual Langevin equation is suggested for systems in which the separation between microscopic and macroscopic time scales is not observed; for example, see [8]. In [14], the author investigated moments, variances, position and velocity correlation for a Riemann-Liouville-type fractional Langevin equation in time and compared the results obtained with the ones derived for the same generalized Langevin equation involving the Liouville-Caputo fractional derivative. Some recent results on the Langevin equation with different boundary conditions can be found in the papers [15–20] and the references cited therein.

Motivated by the aforementioned work on the Langevin equation and its variants, in this paper, we introduce and study a new form of Langevin equation involving generalized Liouville-Caputo derivatives of different orders and solve it with nonlocal generalized fractional integral boundary conditions. In precise terms, we investigate the problem:

$$\begin{cases} {}^{\rho}_{c}D^{\alpha}_{a+}({}^{\rho}_{c}D^{\beta}_{a+} + \lambda)x(t) = f(t, x(t)), \ t \in J := [a, T], \ \lambda \in \mathbb{R}, \\ x(a) = 0, \ x(\eta) = 0, \ x(T) = \mu\, {}^{\rho}I^{\gamma}_{a+}x(\xi), \quad a < \eta < \xi < T, \end{cases} \tag{1}$$

where ${}^{\rho}_{c}D^{\alpha}_{a+}$, ${}^{\rho}_{c}D^{\beta}_{a+}$ denote the Liouville–Caputo-type generalized fractional differential operators of order $1 < \alpha \le 2$, $0 < \beta < 1$, $\rho > 0$, respectively, ${}^{\rho}I^{\gamma}_{a+}$ is the generalized fractional integral operator of order $\gamma > 0$ and $\rho > 0$, and $f : [a, T] \times \mathbb{R} \to \mathbb{R}$ is a given continuous function.

Here, we emphasize that the present work may have useful applications in fractional quantum mechanics and fractional statistical mechanics, in relation to further generalization of the Feynman and Weiner path integrals [21].

We compose the rest of the article as follows. Section 2 contains the basic concepts of generalized fractional calculus and an auxiliary lemma dealing with the linear variant of the given problem. In Section 3, we present the main results and illustrative examples.

2. Preliminaries

Definition 1 ([22]). *The generalized left-sided fractional integral of order $\beta > 0$ and $\rho > 0$ of $g \in X^p_c(a, b)$ for $-\infty < a < t < b < \infty$, is defined by:*

$$({}^{\rho}I^{\beta}_{a+}g)(t) = \frac{\rho^{1-\beta}}{\Gamma(\beta)} \int_a^t \frac{s^{\rho-1}}{(t^{\rho} - s^{\rho})^{1-\beta}} g(s)ds, \tag{2}$$

where $X^p_c(a, b)$ denotes the space of all complex-valued Lebesgue measurable functions ϕ on (a, b) equipped with the norm:

$$\|\phi\|_{X^p_c} = \Big(\int_a^b |x^c\phi(x)|^p \frac{dx}{x}\Big)^{1/p} < \infty, \ c \in \mathbb{R}, 1 \le p \le \infty.$$

Similarly, the right-sided fractional integral ${}^{\rho}I^{\beta}_{b-}g$ is defined by:

$$({}^{\rho}I^{\beta}_{b-}g)(t) = \frac{\rho^{1-\alpha}}{\Gamma(\beta)} \int_t^b \frac{s^{\rho-1}}{(s^{\rho} - t^{\rho})^{1-\beta}} g(s)ds. \tag{3}$$

Definition 2 ([23]). *For $\beta > 0$, $n = [\beta] + 1$, $\rho > 0$ and $0 \le a < x < b < \infty$, we define the generalized fractional derivatives in terms of the generalized fractional integrals (2) and (3) as:*

$$\begin{aligned} ({}^{\rho}D^{\beta}_{a+}g)(t) &= \Big(t^{1-\rho}\frac{d}{dt}\Big)^n ({}^{\rho}I^{n-\beta}_{a+}g)(t) \\ &= \frac{\rho^{\beta-n+1}}{\Gamma(n-\beta)} \Big(t^{1-\rho}\frac{d}{dt}\Big)^n \int_a^t \frac{s^{\rho-1}}{(t^{\rho} - s^{\rho})^{\beta-n+1}} g(s)ds, \end{aligned} \tag{4}$$

and:

$$\begin{aligned} ({}^{\rho}D^{\beta}_{b-}g)(t) &= \Big(-t^{1-\rho}\frac{d}{dt}\Big)^n ({}^{\rho}I^{n-\beta}_{b-}g)(t) \\ &= \frac{\rho^{\beta-n+1}}{\Gamma(n-\beta)} \Big(-t^{1-\rho}\frac{d}{dt}\Big)^n \int_t^b \frac{s^{\rho-1}}{(s^{\rho} - t^{\rho})^{\beta-n+1}} g(s)ds, \end{aligned} \tag{5}$$

if the integrals in the above expressions exist.

Definition 3 ([24]). *For $\beta > 0, n = [\beta] + 1$ and $g \in AC^n_\delta[a,b]$, the Liouville–Caputo-type generalized fractional derivatives ${}^{\rho}_{c}D^{\beta}_{a+}g$ and ${}^{\rho}_{c}D^{\beta}_{b-}g$ are respectively defined via (4) and (5) as follows:*

$$ {}^{\rho}_{c}D^{\beta}_{a+}g(x) = {}^{\rho}D^{\beta}_{a+}\Big[g(t) - \sum_{k=0}^{n-1} \frac{\delta^k g(a)}{k!}\Big(\frac{t^\rho - a^\rho}{\rho}\Big)^k\Big](x), \quad \delta = x^{1-\rho}\frac{d}{dx}, \tag{6} $$

$$ {}^{\rho}_{c}D^{\beta}_{b-}g(x) = {}^{\rho}D^{\beta}_{b-}\Big[g(t) - \sum_{k=0}^{n-1} \frac{(-1)^k\delta^k g(b)}{k!}\Big(\frac{b^\rho - t^\rho}{\rho}\Big)^k\Big](x), \quad \delta = x^{1-\rho}\frac{d}{dx}, \tag{7} $$

where $AC^n_\delta[a,b]$ denotes the class of all absolutely-continuous functions g possessing δ^{n-1}-derivative ($\delta^{n-1}g \in AC([a,b],\mathbb{R})$), equipped with the norm $\|g\|_{AC^n_\delta} = \sum_{k=0}^{n-1}\|\delta^k g\|_C$.

Remark 1 ([24]). *For $\alpha \geq 0$ and $g \in AC^n_\delta[a,b]$, the left and right generalized Liouville–Caputo derivatives of g are respectively defined by the expressions:*

$$ {}^{\rho}_{c}D^{\beta}_{a+}g(t) = \frac{1}{\Gamma(n-\beta)}\int_a^t \Big(\frac{t^\rho - s^\rho}{\rho}\Big)^{n-\beta-1}\frac{(\delta^n g)(s)ds}{s^{1-\rho}}, \tag{8} $$

$$ {}^{\rho}_{c}D^{\beta}_{b-}g(t) = \frac{1}{\Gamma(n-\beta)}\int_t^b \Big(\frac{s^\rho - t^\rho}{\rho}\Big)^{n-\alpha-1}\frac{(-1)^n(\delta^n g)(s)ds}{s^{1-\rho}}. \tag{9} $$

Lemma 1 ([24]). *Let $g \in AC^n_\delta[a,b]$ or $C^n_\delta[a,b]$ and $\beta \in \mathbb{R}$. Then:*

$$ {}^{\rho}I^{\beta}_{a+}{}^{\rho}_{c}D^{\beta}_{a+}g(x) = g(x) - \sum_{k=0}^{n-1}\frac{(\delta^k g)(a)}{k!}\Big(\frac{x^\rho - a^\rho}{\rho}\Big)^k, $$

$$ {}^{\rho}I^{\beta}_{b-}{}^{\rho}_{c}D^{\beta}_{b-}g(x) = g(x) - \sum_{k=0}^{n-1}\frac{(-1)^k(\delta^k g)(a)}{k!}\Big(\frac{b^\rho - x^\rho}{\rho}\Big)^k. $$

In particular, for $0 < \beta \leq 1$, we have:

$$ {}^{\rho}I^{\beta}_{a+}{}^{\rho}_{c}D^{\beta}_{a+}g(x) = g(x) - g(a), \quad {}^{\rho}I^{\beta}_{b-}{}^{\rho}_{c}D^{\beta}_{b-}g(x) = g(x) - g(b). $$

Definition 4. *A function $x \in C([a,T],\mathbb{R})$ is called a solution of (1) if x satisfies the equation ${}^{\rho}_{c}D^{\alpha}_{a+}({}^{\rho}_{c}D^{\beta}_{a+} + \lambda)x(t) = f(t,x(t))$ on $[a,T]$, and the conditions $x(a) = 0, x(\eta) = 0, x(T) = \mu^\rho I^\gamma_{a+}x(\xi)$.*

In the next lemma, we solve the linear variant of Problem (1).

Lemma 2. *Let $h \in C([a,T],\mathbb{R})$, $x \in AC^3_\delta(J)$ and:*

$$ \Omega = \Big[\frac{(T^\rho - a^\rho)^\beta(T^\rho - \eta^\rho)}{\rho^{\beta+1}\Gamma(\beta+2)} - \frac{\mu(\xi^\rho - a^\rho)^{\beta+\gamma}[(\beta+1)(\xi^\rho - \eta^\rho) - \gamma(\eta^\rho - a^\rho)]}{\rho^{\beta+\gamma+1}\Gamma(\beta+\gamma+2)(\beta+1)}\Big] \neq 0. \tag{10} $$

Then, the unique solution of linear problem:

$$ \begin{cases} {}^{\rho}_{c}D^{\alpha}_{a+}({}^{\rho}_{c}D^{\beta}_{a+} + \lambda)x(t) = h(t), \quad t \in J := [a,T], \\ x(a) = 0, \quad x(\eta) = 0, \quad x(T) = \mu^\rho I^\gamma_{a+}x(\xi), \quad a < \eta < \xi < T, \end{cases} \tag{11} $$

is given by:

$$\begin{aligned} x(t) &= {}^{\rho}I_{a+}^{\alpha+\beta}h(t) - \lambda^{\rho}I_{a+}^{\beta}x(t) + \frac{(t^{\rho}-a^{\rho})^{\beta}(\eta^{\rho}-t^{\rho})}{\rho^{\beta+1}\Gamma(\beta+2)\Omega}\Big\{{}^{\rho}I_{a+}^{\alpha+\beta}h(T) - \lambda^{\rho}I_{a+}^{\beta}x(T) \\ &\quad -\mu^{\rho}I_{a+}^{\alpha+\beta+\gamma}h(\xi) + \mu\lambda^{\rho}I_{a+}^{\beta+\gamma}x(\xi)\Big\} - \frac{(t^{\rho}-a^{\rho})^{\beta}}{\Omega(\eta^{\rho}-a^{\rho})^{\beta}}\Big(\frac{(T^{\rho}-a^{\rho})^{\beta}(T^{\rho}-t^{\rho})}{\rho^{\beta+1}\Gamma(\beta+2)} \\ &\quad -\frac{\mu(\xi^{\rho}-a^{\rho})^{\beta+\gamma}[(\beta+1)(\xi^{\rho}-t^{\rho})-\gamma(t^{\rho}-a^{\rho})]}{\rho^{\beta+\gamma+1}\Gamma(\beta+\gamma+2)(\beta+1)}\Big)\Big\{{}^{\rho}I_{a+}^{\alpha+\beta}h(\eta) - \lambda^{\rho}I_{a+}^{\beta}x(\eta)\Big\}. \end{aligned} \tag{12}$$

Proof. Applying ${}^{\rho}I_{a+}^{\alpha}$ on the fractional differential equation in (11) and using Lemma 1 yield:

$$({}_{c}^{\rho}D_{a+}^{\beta} + \lambda)x(t) = {}^{\rho}I_{a+}^{\alpha}h(t) + c_1 + c_2\frac{(t^{\rho}-a^{\rho})}{\rho}, \tag{13}$$

for some $c_1, c_2 \in \mathbb{R}$.

Applying ${}^{\rho}I_{a+}^{\beta}$ to both sides of Equation (13), the general solution of the Langevin equation in (11) is found to be:

$$x(t) = {}^{\rho}I_{a+}^{\alpha+\beta}h(t) - \lambda^{\rho}I_{a+}^{\beta}x(t) + c_1\frac{(t^{\rho}-a^{\rho})^{\beta}}{\rho^{\beta}\Gamma(\beta+1)} + c_2\frac{(t^{\rho}-a^{\rho})^{\beta+1}}{\rho^{\beta+1}\Gamma(\beta+2)} + c_3, \tag{14}$$

where $c_3 \in \mathbb{R}$.

Using the condition $x(a) = 0$ in (14), we find that $c_3 = 0$. Inserting the value of c_3 in (14) and then applying the operator ${}^{\rho}I_{a+}^{\gamma}$ on the resulting equation, we get:

$${}^{\rho}I_{a}^{\gamma}x(t) = {}^{\rho}I_{a+}^{\alpha+\beta+\gamma}h(t) - \lambda^{\rho}I_{a+}^{\beta+\gamma}x(t) + c_1\frac{(t^{\rho}-a^{\rho})^{\beta+\gamma}}{\rho^{\beta+\gamma}\Gamma(\beta+\gamma+1)} + c_2\frac{(t^{\rho}-a^{\rho})^{\beta+\gamma+1}}{\rho^{\beta+\gamma+1}\Gamma(\beta+\gamma+2)}. \tag{15}$$

Using the boundary conditions $x(\eta) = 0$ and $x(T) = \mu^{\rho}I_{a+}^{\gamma}x(\xi)$ together with (14) and (15) leads to a system of algebraic equations in c_1 and c_2, which, upon solving, yields:

$$\begin{aligned} c_1 &= \frac{\rho^{\beta}\Gamma(\beta+1)}{\Omega(\eta^{\rho}-a^{\rho})^{\beta}}\Big\{\frac{(\eta^{\rho}-a^{\rho})^{\beta+1}}{\rho^{\beta+1}\Gamma(\beta+2)}\Big({}^{\rho}I_{a+}^{\alpha+\beta}h(T) - \lambda^{\rho}I_{a+}^{\beta}x(T) - \mu^{\rho}I^{\alpha+\beta+\gamma}h(\xi) + \mu\lambda^{\rho}I^{\beta+\gamma}x(\xi)\Big) \\ &\quad -\Big(\frac{(T^{\rho}-a^{\rho})^{\beta+1}}{\rho^{\beta+1}\Gamma(\beta+2)} - \frac{\mu(\xi^{\rho}-a^{\rho})^{\beta+\gamma+1}}{\rho^{\beta+\gamma+1}\Gamma(\beta+\gamma+2)}\Big)\Big({}^{\rho}I_{a+}^{\alpha+\beta}h(\eta) - \lambda^{\rho}I_{a+}^{\beta}x(\eta)\Big)\Big\}, \\ c_2 &= -\frac{\rho^{\beta}\Gamma(\beta+1)}{\Omega(\eta^{\rho}-a^{\rho})^{\beta}}\Big\{\frac{(\eta^{\rho}-a^{\rho})^{\beta}}{\rho^{\beta}\Gamma(\beta+1)}\Big({}^{\rho}I_{a+}^{\alpha+\beta}h(T) - \lambda^{\rho}I_{a+}^{\beta}x(T) - \mu^{\rho}I^{\alpha+\beta+\gamma}h(\xi) + \mu\lambda^{\rho}I^{\beta+\gamma}x(\xi)\Big) \\ &\quad -\Big(\frac{(T^{\rho}-a^{\rho})^{\beta}}{\rho^{\beta}\Gamma(\beta+1)} - \frac{\mu(\xi^{\rho}-a^{\rho})^{\beta+\gamma}}{\rho^{\beta+\gamma}\Gamma(\beta+\gamma+1)}\Big)\Big({}^{\rho}I_{a+}^{\alpha+\beta}h(\eta) - \lambda^{\rho}I_{a+}^{\beta}x(\eta)\Big)\Big\}. \end{aligned}$$

Inserting the values of c_1, c_2 and c_3 in (13) yields the solution (12). The converse of the Lemma 2, can be obtained by direct computation. This finishes the proof. □

3. Existence and Uniqueness Results

In view of Lemma 2, we introduce an operator $\mathcal{F} : \mathcal{C} \to \mathcal{C}$ by:

$$\begin{aligned} \mathcal{F}(x)(t) &= {}^{\rho}I_{a+}^{\alpha+\beta}f(t,x(t)) - \lambda^{\rho}I_{a+}^{\beta}x(t) + \frac{(t^{\rho}-a^{\rho})^{\beta}(\eta^{\rho}-t^{\rho})}{\rho^{\beta+1}\Gamma(\beta+2)\Omega}\Big\{{}^{\rho}I_{a+}^{\alpha+\beta}f(T,x(T)) - \lambda^{\rho}I_{a+}^{\beta}x(T) \\ &\quad - \mu^{\rho}I_{a+}^{\alpha+\beta+\gamma}f(\xi,x(\xi)) + \mu\lambda^{\rho}I_{a+}^{\beta+\gamma}x(\xi)\Big\} - \frac{(t^{\rho}-a^{\rho})^{\beta}}{\Omega(\eta^{\rho}-a^{\rho})^{\beta}}\Big[\frac{(T^{\rho}-a^{\rho})^{\beta}(T^{\rho}-t^{\rho})}{\rho^{\beta+1}\Gamma(\beta+2)} \\ &\quad - \frac{\mu(\xi^{\rho}-a^{\rho})^{\beta+\gamma}[(\beta+1)(\xi^{\rho}-t^{\rho})-\gamma(t^{\rho}-a^{\rho})]}{\rho^{\beta+\gamma+1}\Gamma(\beta+\gamma+2)(\beta+1)}\Big]\Big\{{}^{\rho}I_{a+}^{\alpha+\beta}f(\eta,x(\eta)) - \lambda^{\rho}I_{a+}^{\beta}x(\eta)\Big\}. \end{aligned} \tag{16}$$

Here, $\mathcal{C}$ denotes the Banach space of all continuous functions from $[a,T]$ to $\mathbb{R}$ equipped with the norm $\|x\| = \sup_{t\in[a,T]} |x(t)|$.

For the sake of computational convenience, we set:

$$\begin{aligned}\Lambda_1 &= \frac{(T^\rho - a^\rho)^{\alpha+\beta}}{\rho^{\alpha+\beta}\Gamma(\alpha+\beta+1)}\Big[1+\frac{\zeta_1}{\rho^{\beta+1}\Gamma(\beta+2)|\Omega|}\Big] + \frac{|\mu|(\xi^\rho - a^\rho)^{\alpha+\beta+\gamma}\zeta_1}{\rho^{\alpha+2\beta+\gamma+1}\Gamma(\alpha+\beta+\gamma+1)\Gamma(\beta+2)|\Omega|}\\ &\quad+\frac{(\eta^\rho - a^\rho)^{\alpha}\zeta_2}{\rho^{\alpha+\beta}\Gamma(\alpha+\beta+1)|\Omega|},\end{aligned} \tag{17}$$

$$\begin{aligned}\Lambda_2 &= \frac{|\lambda|(T^\rho - a^\rho)^{\beta}}{\rho^{\beta}\Gamma(\beta+1)}\Big[1+\frac{\zeta_1}{\rho^{\beta+1}\Gamma(\beta+2)|\Omega|}\Big] + \frac{|\mu||\lambda|(\xi^\rho - a^\rho)^{\beta+\gamma}\zeta_1}{\rho^{2\beta+\gamma+1}\Gamma(\beta+\gamma+1)\Gamma(\beta+2)|\Omega|}\\ &\quad+\frac{|\lambda|\zeta_2}{\rho^{\beta}\Gamma(\beta+1)|\Omega|},\end{aligned} \tag{18}$$

where:

$$\zeta_1 := \max_{t\in[a,T]}\Big|(t^\rho - a^\rho)^\beta(\eta^\rho - t^\rho)\Big|, \tag{19}$$

$$\zeta_2 := \max_{t\in[a,T]}\Big|(t^\rho - a^\rho)^\beta\Big[\frac{(T^\rho - a^\rho)^\beta(T^\rho - t^\rho)}{\rho^{\beta+1}\Gamma(\beta+2)} - \frac{\mu(\xi^\rho - a^\rho)^{\beta+\gamma}[(\beta+1)(\xi^\rho - t^\rho) - \gamma(t^\rho - a^\rho)]}{\rho^{\beta+\gamma+1}\Gamma(\beta+\gamma+2)(\beta+1)}\Big]\Big|. \tag{20}$$

Now, we are in a position to present our main results. Our first existence result for the problem (1) is based on Krasnoselskii's fixed point theorem [25], which is stated below.

Lemma 3. *(Krasnoselskii's fixed point theorem) Let $\mathcal{S}$ be a closed convex and non-empty subset of a Banach space E. Let $\mathcal{G}_1, \mathcal{G}_2$ be the operators from $\mathcal{S}$ to E such that (a) $\mathcal{G}_1 x + \mathcal{G}_2 y \in \mathcal{S}$ whenever $u, v \in \mathcal{S}$; (b) $\mathcal{G}_1$ is compact and continuous; and (c) $\mathcal{G}_2$ is a contraction mapping. Then, there exists a fixed point $\omega \in \mathcal{S}$ such that $\omega = \mathcal{G}_1\omega + \mathcal{G}_2\omega$.*

Theorem 1. *Let $f : J \times \mathbb{R} \to \mathbb{R}$ be a continuous function such that the following condition holds:*

(A_1) There exists a continuous function $\phi \in C([a,T], \mathbb{R}^+)$ such that:

$$|f(t,u)| \le \phi(t), \quad \forall (t,u) \in J \times \mathbb{R}.$$

Then, the problem (1) has at least one solution on J, provided that:

$$\Lambda_2 < 1. \tag{21}$$

Proof. Introduce a closed ball $B_r = \{x \in \mathcal{C} : \|x\| \le r\}$, with $r > \frac{\|\phi\|\Lambda_1}{1-\Lambda_2}$, $\|\phi\| = \sup_{t\in[a,T]} |\phi(t)|$, where Λ_2 is given by (18). Then, we define operators $\mathcal{F}_1$ and $\mathcal{F}_2$ from B_r to $\mathcal{C}$ by:

$$\begin{aligned}\mathcal{F}_1(x)(t) &= {}^\rho I_{a+}^{\alpha+\beta} f(t,x(t)) + \frac{(t^\rho - a^\rho)^\beta(\eta^\rho - t^\rho)}{\rho^{\beta+1}\Gamma(\beta+2)\Omega}\Big\{{}^\rho I_{a+}^{\alpha+\beta} f(T,x(T)) - \mu\,{}^\rho I_{a+}^{\alpha+\beta+\gamma} f(\xi,x(\xi))\Big\}\\ &\quad-\frac{(t^\rho - a^\rho)^\beta}{\Omega(\eta^\rho - a^\rho)^\beta}\Big[\frac{(T^\rho - a^\rho)^\beta(T^\rho - t^\rho)}{\rho^{\beta+1}\Gamma(\beta+2)} - \frac{\mu(\xi^\rho - a^\rho)^{\beta+\gamma}[(\beta+1)(\xi^\rho - t^\rho) - \gamma(t^\rho - a^\rho)]}{\rho^{\beta+\gamma+1}\Gamma(\beta+\gamma+2)(\beta+1)}\Big]\\ &\quad\times{}^\rho I_{a+}^{\alpha+\beta} f(\eta, x(\eta)),\\ \mathcal{F}_2(x)(t) &= -\lambda\,{}^\rho I_{a+}^{\beta} x(t) - \frac{(t^\rho - a^\rho)^\beta(\eta^\rho - t^\rho)}{\rho^{\beta+1}\Gamma(\beta+2)\Omega}\Big\{\lambda\,{}^\rho I_{a+}^{\beta} x(T) - \mu\lambda\,{}^\rho I_{a+}^{\beta+\gamma} x(\xi)\Big\} + \frac{\lambda(t^\rho - a^\rho)^\beta}{\Omega(\eta^\rho - a^\rho)^\beta}\times\\ &\quad\times\Big[\frac{(T^\rho - a^\rho)^\beta(T^\rho - t^\rho)}{\rho^{\beta+1}\Gamma(\beta+2)} - \frac{\mu(\xi^\rho - a^\rho)^{\beta+\gamma}[(\beta+1)(\xi^\rho - t^\rho) - \gamma(t^\rho - a^\rho)]}{\rho^{\beta+\gamma+1}\Gamma(\beta+\gamma+2)(\beta+1)}\Big]{}^\rho I_{a+}^{\beta} x(\eta).\end{aligned}$$

Note that $\mathcal{F} = \mathcal{F}_1 + \mathcal{F}_2$ on B_r. For $x, y \in B_r$, we find that:

$$\begin{aligned}
&\|\mathcal{F}_1 x + \mathcal{F}_2 y\| \\
\leq\ & \sup_{t \in J} \Big\{ {}^{\rho}I_{a+}^{\alpha+\beta}|f(t,x(t))| + |\lambda|^{\rho}I_{a+}^{\beta}|y(t)| \\
& + \frac{(t^{\rho}-a^{\rho})^{\beta}|(\eta^{\rho}-t^{\rho})|}{\rho^{\beta+1}\Gamma(\beta+2)|\Omega|} \Big\{ {}^{\rho}I_{a+}^{\alpha+\beta}|f(T,x(T))| + |\lambda|^{\rho}I_{a+}^{\beta}|y(T)| \\
& + |\mu|^{\rho}I_{a+}^{\alpha+\beta+\gamma}|f(\xi,x(\xi))| + |\mu||\lambda|^{\rho}I_{a+}^{\beta+\gamma}|x(\xi)| \Big\} + \frac{(t^{\rho}-a^{\rho})^{\beta}}{|\Omega|(\eta^{\rho}-a^{\rho})^{\beta}} \Big| \frac{(T^{\rho}-a^{\rho})^{\beta}(T^{\rho}-t^{\rho})}{\rho^{\beta+1}\Gamma(\beta+2)} \\
& - \frac{\mu(\xi^{\rho}-a^{\rho})^{\beta+\gamma}[(\beta+1)(\xi^{\rho}-t^{\rho}) - \gamma(t^{\rho}-a^{\rho})]}{\rho^{\beta+\gamma+1}\Gamma(\beta+\gamma+2)(\beta+1)} \Big| \Big\{ {}^{\rho}I_{a+}^{\alpha+\beta}|f(\eta,x(\eta))| + |\lambda|^{\rho}I_{a+}^{\beta}|y(\eta)| \Big\} \\
\leq\ & \|\phi\| \Big\{ \frac{(T^{\rho}-a^{\rho})^{\alpha+\beta}}{\rho^{\alpha+\beta}\Gamma(\alpha+\beta+1)} \Big[1 + \frac{\zeta_1}{\rho^{\beta+1}\Gamma(\beta+2)|\Omega|} \Big] + \frac{|\mu|(\xi^{\rho}-a^{\rho})^{\alpha+\beta+\gamma}\zeta_1}{\rho^{\alpha+2\beta+\gamma+1}\Gamma(\alpha+\beta+\gamma+1)\Gamma(\beta+2)|\Omega|} \\
& + \frac{(\eta^{\rho}-a^{\rho})^{\alpha}\zeta_2}{\rho^{\alpha+\beta}\Gamma(\alpha+\beta+1)|\Omega|} \Big\} + \|x\| \Big\{ \frac{|\lambda|(T^{\rho}-a^{\rho})^{\beta}}{\rho^{\beta}\Gamma(\beta+1)} \Big[1 + \frac{\zeta_1}{\rho^{\beta+1}\Gamma(\beta+2)|\Omega|} \Big] \\
& + \frac{|\mu||\lambda|(\xi^{\rho}-a^{\rho})^{\beta+\gamma}\zeta_1}{\rho^{2\beta+\gamma+1}\Gamma(\beta+\gamma+1)\Gamma(\beta+2)|\Omega|} + \frac{|\lambda|\zeta_2}{\rho^{\beta}\Gamma(\beta+1)|\Omega|} \Big\} \\
\leq\ & \|\phi\|\Lambda_1 + r\Lambda_2 < r.
\end{aligned}$$

Thus, $\mathcal{F}_1 x + \mathcal{F}_2 y \in B_r$.
Next, it will be shown that $\mathcal{F}_2$ is a contraction. For that, let $x, y \in \mathcal{C}$. Then:

$$\begin{aligned}
&\|\mathcal{F}_2 x - \mathcal{F}_2 y\| \\
\leq\ & \sup_{t \in J} \Big\{ |\lambda|^{\rho}I_{a+}^{\beta}|x(t)-y(t)| + \frac{|(t^{\rho}-a^{\rho})^{\beta}(\eta^{\rho}-t^{\rho})|}{\rho^{\beta+1}\Gamma(\beta+2)|\Omega|} \times \\
& \times \Big\{ |\lambda|^{\rho}I_{a+}^{\beta}|x(T)-y(T)| + |\mu||\lambda|^{\rho}I_{a+}^{\beta+\gamma}|x(\xi)-y(\xi)| \Big\} \\
& + \frac{|\lambda||(t^{\rho}-a^{\rho})^{\beta}|}{|\Omega|(\eta^{\rho}-a^{\rho})^{\beta}} \Big| \frac{(T^{\rho}-a^{\rho})^{\beta}(T^{\rho}-t^{\rho})}{\rho^{\beta+1}\Gamma(\beta+2)} - \frac{\mu(\xi^{\rho}-a^{\rho})^{\beta+\gamma}[(\beta+1)(\xi^{\rho}-t^{\rho}) - \gamma(t^{\rho}-a^{\rho})]}{\rho^{\beta+\gamma+1}\Gamma(\beta+\gamma+2)(\beta+1)} \Big| \times \\
& \times {}^{\rho}I_{a+}^{\beta}|x(\eta)-y(\eta)| \Big\} \\
\leq\ & \Big\{ \frac{|\lambda|(T^{\rho}-a^{\rho})^{\beta}}{\rho^{\beta}\Gamma(\beta+1)} \Big[1 + \frac{\zeta_1}{\rho^{\beta+1}\Gamma(\beta+2)|\Omega|} \Big] + \frac{|\mu||\lambda|(\xi^{\rho}-a^{\rho})^{\beta+\gamma}\zeta_1}{\rho^{2\beta+\gamma+1}\Gamma(\beta+\gamma+1)\Gamma(\beta+2)|\Omega|} \\
& + \frac{|\lambda|\zeta_2}{\rho^{\beta}\Gamma(\beta+1)|\Omega|} \Big\} \|x-y\| \\
=\ & \Lambda_2 \|x-y\|,
\end{aligned}$$

which, by the condition (21), implies that $\mathcal{F}_2$ is a contraction. The continuity of the operator $\mathcal{F}_1$ follows from that of f. Furthermore, $\mathcal{F}_1$ is uniformly bounded on B_r as:

$$\|\mathcal{F}_1 x\| \leq \|\phi\|\Lambda_1.$$

Finally, we establish the compactness of the operator $\mathcal{F}_1$. Let us set $\sup_{(t,x)\in J\times B_r}|f(t,x)| = \bar{f} < \infty$. Then, for $t_1, t_2 \in J$, $t_1 < t_2$, we have:

$$|(\mathcal{F}_1 x)(t_2) - (\mathcal{F}_1 x)(t_1)|$$

$$
\begin{aligned}
= & \left|\frac{\rho^{1-(\alpha+\beta)}}{\Gamma(\alpha+\beta)}\Big[\int_0^{t_1} s^{\rho-1}[(t_2^\rho-s^\rho)^{\alpha+\beta-1}-(t_1^\rho-s^\rho)^{\alpha+\beta-1}]f(s,x(s))ds\right.\\
& +\int_{t_1}^{t_2} s^{\rho-1}(t_2^\rho-s^\rho)^{\alpha+\beta-1}f(s,x(s))ds\Big]\\
& +\Big[\frac{(t_2^\rho-a^\rho)^\beta(\eta^\rho-t_2^\rho)}{\rho^{\beta+1}\Gamma(\beta+2)\Omega}-\frac{(t_1^\rho-a^\rho)^\beta(\eta^\rho-t_1^\rho)}{\rho^{\beta+1}\Gamma(\beta+2)\Omega}\Big]\Big\{{}^\rho I_{a+}^{\alpha+\beta}f(T,x(T))-\mu{}^\rho I_{a+}^{\alpha+\beta+\gamma}f(\xi,x(\xi))\Big\}\\
& -\Big[\frac{(t_2^\rho-a^\rho)^\beta}{\Omega(\eta^\rho-a^\rho)^\beta}\Big[\frac{(T^\rho-a^\rho)^\beta(T^\rho-t_2^\rho)}{\rho^{\beta+1}\Gamma(\beta+2)}-\frac{\mu(\xi^\rho-a^\rho)^{\beta+\gamma}[(\beta+1)(\xi^\rho-t_2^\rho)-\gamma(t^\rho-a^\rho)]}{\rho^{\beta+\gamma+1}\Gamma(\beta+\gamma+2)(\beta+1)}\Big]\\
& -\frac{(t_1^\rho-a^\rho)^\beta}{\Omega(\eta^\rho-a^\rho)^\beta}\Big[\frac{(T^\rho-a^\rho)^\beta(T^\rho-t_1^\rho)}{\rho^{\beta+1}\Gamma(\beta+2)}-\frac{\mu(\xi^\rho-a^\rho)^{\beta+\gamma}[(\beta+1)(\xi^\rho-t_1^\rho)-\gamma(t_1^\rho-a^\rho)]}{\rho^{\beta+\gamma+1}\Gamma(\beta+\gamma+2)(\beta+1)}\Big]\Big]\times\\
& \left.\times{}^\rho I_{a+}^{\alpha+\beta}f(\eta,x(\eta))\right|\\
\leq & \frac{\bar{f}}{\rho^{\alpha+\beta}\Gamma(\alpha+\beta+1)}\Big\{|t_2^{\rho(\alpha+\beta)}-t_1^{\rho(\alpha+\beta)}|+2(t_2^\rho-t_1^\rho)^{\alpha+\beta}\Big\}\\
& +\Big|\frac{(t_2^\rho-a^\rho)^\beta(\eta^\rho-t_2^\rho)}{\rho^{\beta+1}\Gamma(\beta+2)\Omega}-\frac{(t_1^\rho-a^\rho)^\beta(\eta^\rho-t_1^\rho)}{\rho^{\beta+1}\Gamma(\beta+2)\Omega}\Big|\Big\{{}^\rho I_{a+}^{\alpha+\beta}|f(T,x(T))|+\mu{}^\rho I_{a+}^{\alpha+\beta+\gamma}|f(\xi,x(\xi))|\Big\}\\
& +\Big|\frac{(t_2^\rho-a^\rho)^\beta}{\Omega(\eta^\rho-a^\rho)^\beta}\Big[\frac{(T^\rho-a^\rho)^\beta(T^\rho-t_2^\rho)}{\rho^{\beta+1}\Gamma(\beta+2)}-\frac{\mu(\xi^\rho-a^\rho)^{\beta+\gamma}[(\beta+1)(\xi^\rho-t_2^\rho)-\gamma(t_2^\rho-a^\rho)]}{\rho^{\beta+\gamma+1}\Gamma(\beta+\gamma+2)(\beta+1)}\Big]\\
& -\frac{(t_1^\rho-a^\rho)^\beta}{\Omega(\eta^\rho-a^\rho)^\beta}\Big[\frac{(T^\rho-a^\rho)^\beta(T^\rho-t_1^\rho)}{\rho^{\beta+1}\Gamma(\beta+2)}-\frac{\mu(\xi^\rho-a^\rho)^{\beta+\gamma}[(\beta+1)(\xi^\rho-t_1^\rho)-\gamma(t_1^\rho-a^\rho)]}{\rho^{\beta+\gamma+1}\Gamma(\beta+\gamma+2)(\beta+1)}\Big]\Big|\times\\
& \times{}^\rho I_{a+}^{\alpha+\beta}|f(\eta,x(\eta))|,
\end{aligned}
$$

which tends to zero as $t_2 \to t_1$, independently of $x \in B_r$. Thus, $\mathcal{F}_1$ is equicontinuous. Therefore, $\mathcal{F}_1$ is relatively compact on B_r. As a consequence, we deduce by the the Arzelá–Ascoli theorem that $\mathcal{F}_1$ is compact on B_r. Thus, the hypothesis of Lemma 3 is satisfied. Therefore, the conclusion of Lemma 3 applies, and hence, there exists at least one solution for the problem (1) on J. □

In the next result, the uniqueness of solutions for the problem (1) is shown by means of the Banach contraction mapping principle.

Theorem 2. *Let $f : J \times \mathbb{R} \to \mathbb{R}$ be a continuous function satisfying the Lipschitz condition:*

(A_2)

$$|f(t,u)-f(t,v)| \leq L|u-v|, L>0, /, \quad \textit{for } t \in J \textit{ and every } u,v \in \mathbb{R}.$$

Then, there exists a unique solution for the problem (1) on $[a,T]$, provided that:

$$L\Lambda_1+\Lambda_2<1, \tag{22}$$

where Λ_1 and Λ_2 are respectively given by (17) and (18).

Proof. In the first step, we show that $\mathcal{F}B_{\bar{r}} \subset B_{\bar{r}}$, where $B_{\bar{r}} = \{x \in C([a,T],\mathbb{R}) : \|x\| \leq \bar{r}\}$, $M = \sup_{t\in[a,T]}|f(t,0)|$, $\bar{r} \geq \frac{\Lambda_1 M}{1-L\Lambda_1-\Lambda_2}$, and the operator $\mathcal{F} : C \to C$ is given by (16). For $x \in B_{\bar{r}}$, using (A_2), we get:

$$
\begin{aligned}
& |\mathcal{F}(x)(t)|\\
\leq & {}^\rho I_{a+}^{\alpha+\beta}[|f(t,x(t))-f(t,0)|+|f(t,0)|]+|\lambda|{}^\rho I_{a+}^{\beta}|x(t)|\\
& +\frac{(t^\rho-a^\rho)^\beta|(\eta^\rho-t^\rho)|}{\rho^{\beta+1}\Gamma(\beta+2)|\Omega|}\Big\{{}^\rho I_{a+}^{\alpha+\beta}[|f(T,x(T))-f(T,0)|+|f(T,0)|]+|\lambda|{}^\rho I_{a+}^{\beta}|x(T)|\\
& +|\mu|{}^\rho I_{a+}^{\alpha+\beta+\gamma}[|f(\xi,x(\xi))-f(\xi,0)|+|f(\xi,0)|]+|\mu||\lambda|{}^\rho I_{a+}^{\beta+\gamma}|x(\xi)|\Big\}
\end{aligned}
$$

$$
\begin{aligned}
& +\frac{(t^\rho - a^\rho)^\beta}{|\Omega|(\eta^\rho - a^\rho)^\beta}\left|\frac{(T^\rho - a^\rho)^\beta (T^\rho - t^\rho)}{\rho^{\beta+1}\Gamma(\beta+2)} - \frac{\mu(\xi^\rho - a^\rho)^{\beta+\gamma}[(\beta+1)(\xi^\rho - t^\rho) - \gamma(t^\rho - a^\rho)]}{\rho^{\beta+\gamma+1}\Gamma(\beta+\gamma+2)(\beta+1)}\right| \\
& \times\left\{{}^\rho I_{a+}^{\alpha+\beta}[|f(\eta,x(\eta)) - f(\eta,0)| + |f(\eta,0)|] + |\lambda|{}^\rho I_{a+}^{\beta}|x(\eta)|\right\} \\
\leq\ & (L\bar{r}+M)\Big(\frac{(T^\rho - a^\rho)^{\alpha+\beta}}{\rho^{\alpha+\beta}\Gamma(\alpha+\beta+1)}\Big[1+\frac{\zeta_1}{\rho^{\beta+1}\Gamma(\beta+2)|\Omega|}\Big] + \frac{|\mu|(\xi^\rho - a^\rho)^{\alpha+\beta+\gamma}\zeta_1}{\rho^{\alpha+2\beta+\gamma+1}\Gamma(\alpha+\beta+\gamma+1)\Gamma(\beta+2)|\Omega|} \\
& +\frac{(\eta^\rho - a^\rho)^\alpha \zeta_2}{\rho^{\alpha+\beta}\Gamma(\alpha+\beta+1)|\Omega|}\Big) \\
& +\bar{r}\Big(\frac{|\lambda|(T^\rho - a^\rho)^\beta}{\rho^\beta\Gamma(\beta+1)}\Big[1+\frac{\zeta_1}{\rho^{\beta+1}\Gamma(\beta+2)|\Omega|}\Big] + \frac{|\mu||\lambda|(\xi^\rho - a^\rho)^{\beta+\gamma}\zeta_1}{\rho^{2\beta+\gamma+1}\Gamma(\beta+\gamma+1)\Gamma(\beta+2)|\Omega|} + \frac{|\lambda|\zeta_2}{\rho^\beta\Gamma(\beta+1)|\Omega|}\Big) \\
=\ & (L\bar{r}+M)\Lambda_1 + \Lambda_2\bar{r} \leq \bar{r},
\end{aligned}
$$

which, on taking the norm for $t \in [a,T]$, implies that $\|\mathcal{F}(x)\| \leq \bar{r}$. Thus, the operator $\mathcal{F}$ maps $B_{\bar{r}}$ into itself. Now, we proceed to prove that the operator $\mathcal{F}$ is a contraction. For $x, y \in C([a,T],\mathbb{R})$ and $t \in [a,T]$, we have:

$$
\begin{aligned}
& |\mathcal{F}(x)(t) - \mathcal{F}(y)(t)| \\
\leq\ & {}^\rho I_{a+}^{\alpha+\beta}|f(t,x(t)) - f(t,y(t))| + |\lambda|{}^\rho I_{a+}^{\beta}|x(t) - y(t)| \\
& +\frac{(t^\rho - a^\rho)^\beta|(\eta^\rho - t^\rho)|}{\rho^{\beta+1}\Gamma(\beta+2)|\Omega|}\Big\{{}^\rho I_{a+}^{\alpha+\beta}|f(T,x(T)) - f(T,y(T))| + |\lambda|{}^\rho I_{a+}^{\beta}|x(T) - y(T)| \\
& +|\mu|{}^\rho I_{a+}^{\alpha+\beta+\gamma}|f(\xi,x(\xi)) - f(\xi,y(\xi))| + |\mu||\lambda|{}^\rho I_{a+}^{\beta+\gamma}|x(\xi) - y(\xi)|\Big\} \\
& +\frac{(t^\rho - a^\rho)^\beta}{|\Omega|(\eta^\rho - a^\rho)^\beta}\left|\frac{(T^\rho - a^\rho)^\beta (T^\rho - t^\rho)}{\rho^{\beta+1}\Gamma(\beta+2)} - \frac{\mu(\xi^\rho - a^\rho)^{\beta+\gamma}[(\beta+1)(\xi^\rho - t^\rho) - \gamma(t^\rho - a^\rho)]}{\rho^{\beta+\gamma+1}\Gamma(\beta+\gamma+2)(\beta+1)}\right| \times \\
& \times\left\{{}^\rho I_{a+}^{\alpha+\beta}|f(\eta,x(\eta)) - f(\eta,y(\eta))| + |\lambda|{}^\rho I_{a+}^{\beta}|x(\eta) - y(\eta)|\right\} \\
\leq\ & L\|x-y\|\Big(\frac{(T^\rho - a^\rho)^{\alpha+\beta}}{\rho^{\alpha+\beta}\Gamma(\alpha+\beta+1)}\Big[1+\frac{\zeta_1}{\rho^{\beta+1}\Gamma(\beta+2)|\Omega|}\Big] \\
& +\frac{|\mu|(\xi^\rho - a^\rho)^{\alpha+\beta+\gamma}\zeta_1}{\rho^{\alpha+2\beta+\gamma+1}\Gamma(\alpha+\beta+\gamma+1)\Gamma(\beta+2)|\Omega|} \\
& +\frac{(\eta^\rho - a^\rho)^\alpha \zeta_2}{\rho^{\alpha+\beta}\Gamma(\alpha+\beta+1)|\Omega|}\Big) + \|x-y\|\Big(\frac{|\lambda|(T^\rho - a^\rho)^\beta}{\rho^\beta\Gamma(\beta+1)}\Big[1+\frac{\zeta_1}{\rho^{\beta+1}\Gamma(\beta+2)|\Omega|}\Big] \\
& +\frac{|\mu||\lambda|(\xi^\rho - a^\rho)^{\beta+\gamma}\zeta_1}{\rho^{2\beta+\gamma+1}\Gamma(\beta+\gamma+1)\Gamma(\beta+2)|\Omega|} + \frac{|\lambda|\zeta_2}{\rho^\beta\Gamma(\beta+1)|\Omega|}\Big) \\
=\ & (L\Lambda_1 + \Lambda_2)\|x-y\|.
\end{aligned}
$$

Taking the norm of the above inequality for $t \in [a,T]$, we get:

$$\|\mathcal{F}(x) - \mathcal{F}(y)\| \leq (L\Lambda_1 + \Lambda_2)\|x-y\|,$$

which implies that the operator $\mathcal{F}$ is a contraction on account of the condition (22). Thus, we deduce by the Banach contraction mapping principle that the operator $\mathcal{F}$ has a unique fixed point. Hence, there exists a unique solution for the problem (1). The proof is complete. □

Example 1. *Let us consider the following boundary value problem:*

$$
\begin{cases}
{}^{1/3}_{c}D^{5/4}\left({}^{1/3}_{c}D^{1/4} + 1/5\right)x(t) = \dfrac{1}{\sqrt{400+t}}\Big(\dfrac{|x(t)|+2}{|x(t)|+1} + e^{-t}\Big), \quad t \in J := [1,2], \\
x(1) = 0, \quad x(3/2) = 0, \quad x(2) = 2/7\, {}^{1/3}I^{3/4}x(7/4).
\end{cases} \tag{23}
$$

Here, $\rho = 1/3, \alpha = 5/4, \beta = 1/4, \gamma = 3/4, \lambda = 1/5, \mu = 2/7, a = 1, \eta = 3/2, \xi = 7/4, T = 2$ and $f(t,x) = \frac{1}{\sqrt{400+t}}\left(\frac{|x|+2}{|x|+1} + e^{-t}\right)$. Using the given data, we find that $|\Omega| \approx 0.293634, \Lambda_1 \approx 1.336009, \Lambda_2 \approx 0.673563, \zeta_1 \approx 0.082260, \zeta_2 \approx 0.232036$, where $\Omega, \Lambda_1, \Lambda_2, \zeta_1$, and ζ_2 are given by (10), (17), (18), (19) and (20) respectively.

For illustrating Theorem 1, we show that all the conditions of Theorem 1 are satisfied. Clearly, $f(t,x)$ is continuous and satisfies the condition (A_1) with $\phi(t) = \frac{2+e^{-t}}{\sqrt{400+t}}$. Furthermore, $\Lambda_2 \approx 0.673563 < 1$. Thus, all the conditions of Theorem 1 are satisfied, and consequently, the problem (23) has at least one solution on $[1,2]$.

Furthermore, Theorem 2 is applicable to the problem (23) with $L = 1/20$ as $L\Lambda_1 + \Lambda_2 \approx 0.740363 < 1$. Thus, all the assumptions of Theorem 2 are satisfied. Therefore, the conclusion of Theorem 2 applies to the problem (23) on $[1,2]$.

4. Conclusions

We have introduced a new type of nonlinear Langevin equation in terms of Liouville-Caputo-type generalized fractional differential operators of different orders and solved it with nonlocal generalized integral boundary conditions. The existence result was obtained by applying the Krasnoselskii fixed point theorem without requiring the nonlinear function to be of the Lipschitz type, while the uniqueness of solutions for the given problem was based on a celebrated fixed point theorem due to Banach. Here, we remark that many known existence results, obtained by means of the Krasnoselskii fixed point theorem, demand the associated nonlinear function to satisfy the Lipschitz condition. Moreover, by fixing the parameters involved in the given problem, we can obtain some new results as special cases of the ones presented in this paper. For example, letting $\rho = 1$, $\mu = 0$, $a = 0$ and $T = 1$ in the results of Section 3, we get the ones derived in [15].

Author Contributions: Formal analysis, B.A., M.A., A.A., H.M.S. and S.K.N.

Funding: This project was funded by the Deanship of Scientific Research (DSR), King Abdulaziz University, Jeddah, Saudi Arabia, under Grant No. KEP-PhD-24-130-40. The authors acknowledge with thanks the DSR's technical and financial support. The authors also acknowledge the reviewers for their constructive remarks on our work.

Conflicts of Interest: The authors declare no conflict of interest.

References

1. Zaslavsky, G.M. *Hamiltonian Chaos and Fractional Dynamics*; Oxford University Press: Oxford, UK, 2005.
2. Magin, R.L. *Fractional Calculus in Bioengineering*; Begell House Publishers: Danbury, CT, USA, 2006.
3. Kilbas, A.A.; Srivastava, H.M.; Trujillo, J.J. *Theory and Applications of Fractional Differential Equations*; North-Holland Mathematics Studies, 204; Elsevier Science B.V.: Amsterdam, The Netherlands, 2006.
4. Diethelm, K. *The Analysis of Fractional Differential Equations. An Application-Oriented Exposition Using Differential Operators of Liouville-Caputo Type. Lecture Notes in Mathematics 2004*; Springer: Berlin, Germany, 2010.
5. Javidi, M.; Ahmad, B. Dynamic analysis of time fractional order phytoplankton-toxic phytoplankton-zooplankton system. *Ecol. Model.* **2015**, *318*, 8–18. [CrossRef]
6. Fallahgoul, H.A.; Focardi, S.M.; Fabozzi, F.J. *Fractional Calculus and Fractional Processes with Applications to Financial Economics. Theory and Application*; Elsevier/Academic Press: London, UK, 2017.
7. Ahmad, B.; Alsaedi, A.; Ntouyas, S.K.; Tariboon, J. *Hadamard-Type Fractional Differential Equations, Inclusions and Inequalities*; Springer: Cham, Switzerland, 2017.
8. West, B.J.; Picozzi, S. Fractional Langevin model of memory in financial time series. *Phys. Rev. E* **2002**, *65*, 037106. [CrossRef] [PubMed]
9. Vinales, A.D.; Desposito, M.A. Anomalous diffusion: Exact solution of the generalized Langevin equation for harmonically bounded particle. *Phys. Rev. E* **2006**, *73*, 016111. [CrossRef] [PubMed]
10. Hohenberg, P.C.; Halperin, B.I. Theory of dynamic critical phenomena. *Rev. Mod. Phys.* **1977**, *49*, 435–479. [CrossRef]

11. Metiu, H.; Schon, G. Description of Quantum noise by a Langevin equation. *Phys. Rev. Lett.* **1984**, *53*, 13. [CrossRef]
12. Datsko, B.; Gafiychuk, V. Complex nonlinear dynamics in subdiffusive activator–inhibitor systems. *Commun. Nonlinear Sci. Numer. Simul.* **2012**, *17*, 1673–1680. [CrossRef]
13. Datsko, B.; Gafiychuk, V. Complex spatio-temporal solutions in fractional reaction-diffusion systems near a bifurcation point. *Fract. Calc. Appl. Anal.* **2018**, *21*, 237–253. [CrossRef]
14. Fa, K.S. Fractional Langevin equation and Riemann–Liouville fractional derivative. *Eur. Phys. J. E* **2007**, *24*, 139–143.
15. Ahmad, B.; Nieto, J.J.; Alsaedi, A.; El-Shahed, M. A study of nonlinear Langevin equation involving two fractional orders in different intervals. *Nonlinear Anal. Real World Appl.* **2012**, *13*, 599–606. [CrossRef]
16. Wang, G.; Zhang, L.; Song, G. Boundary value problem of a nonlinear Langevin equation with two different fractional orders and impulses. *Fixed Point Theory Appl.* **2012**, *2012*, 200. [CrossRef]
17. Ahmad, B.; Ntouyas, S.K. New existence results for differential inclusions involving Langevin equation with two indices. *J. Nonlinear Convex Anal.* **2013**, *14*, 437–450.
18. Muensawat, T.; Ntouyas, S.K.; Tariboon, J. Systems of generalized Sturm-Liouville and Langevin fractional differential equations. *Adv. Differ. Equ.* **2017**, *2017*, 63. [CrossRef]
19. Fazli, H.; Nieto, J.J. Fractional Langevin equation with anti-periodic boundary conditions. *Chaos Solitons Fractals* **2018**, *114*, 332–337. [CrossRef]
20. Ahmad, B.; Alsaedi, A.; Salem, S. On a nonlocal integral boundary value problem of nonlinear Langevin equation with different fractional orders. *Adv. Differ. Equ.* **2019**, *2019*, 57. [CrossRef]
21. Laskin, N. Fractional quantum mechanics and Levy path integrals. *Phys. Lett. A* **2000**, *268*, 298–305. [CrossRef]
22. Katugampola, U.N. New Approach to a generalized fractional integral. *Appl. Math. Comput.* **2015**, *218*, 860–865. [CrossRef]
23. Katugampola, U.N. A new approach to generalized fractional derivatives. *Bull. Math. Anal. Appl.* **2014**, *6*, 1–15.
24. Jarad, F.; Abdeljawad, T.; Baleanu, D. On the generalized fractional derivatives and their Caputo modification. *J. Nonlinear Sci. Appl.* **2017**, *10*, 2607–2619. [CrossRef]
25. Krasnoselskii, M.A. Two remarks on the method of successive approximations. *Uspekhi Matematicheskikh Nauk* **1955**, *10*, 123–127.

mathematics

MDPI

Article

Existence of Solutions for Anti-Periodic Fractional Differential Inclusions Involving ψ-Riesz-Caputo Fractional Derivative

Dandan Yang * and Chuanzhi Bai

School of Mathematical Science, Huaiyin Normal University, Huaian 223300, Jiangsu, China
* Correspondence: ydd@hytc.edu.cn

Received: 9 June 2019; Accepted: 12 July 2019; Published: 15 July 2019

Abstract: In this paper, we investigate the existence of solutions for a class of anti-periodic fractional differential inclusions with ψ-Riesz-Caputo fractional derivative. A new definition of ψ-Riesz-Caputo fractional derivative of order α is proposed. By means of Contractive map theorem and nonlinear alternative for Kakutani maps, sufficient conditions for the existence of solutions to the fractional differential inclusions are given. We present two examples to illustrate our main results.

Keywords: fractional differential inclusions; ψ-Riesz-Caputo derivative; existence of solutions; fixed point theorem; anti-periodic boundary value problems

---

1. Introduction

Fractional order models, providing excellent description of memory and hereditary processes, are more adequate than integer order ones. Some recent contributions to fractional differential equations and inclusions have been carried out, see the monographs [1–8], and the references cited therein. The study of fractional differential equations or inclusions with anti-periodic boundary problems, which are applied in different fields, such as physics, chemical engineering, economics, populations dynamics and so on, have recently received considerable attention, see the references ([9,10]) and papers cited therein.There are several definitions of fractional differential derivatives and integrals, such like Caputo type, Rimann-Liouville type, Hadamard type and Erdelyi-Kober type and so on. In order to develop the fractional calculus, some different and special form of differential operators are chosen, for example, see [11–15] and the references therein. The α order ψ-Caputo fractional derivative was first introduced by Almeida in [3]. Some properties, like semigroup law, Taylor's Theorem, Fermat's Thorem, etc., were presented. This newly defined fractional derivative could model more accurately the process using differential kernels for the fractional operator. In 2018, Samet and Aydi in [16] considered the following fractional differential equation with anti-periodic boundary conditions:

$$\begin{cases} {}^cD^{\alpha,\psi}u(x) + f(x,u(x)) = 0, \quad a < x < b, \\ u(a) + u(b) = 0, u'(a) + u'(b) = 0 \end{cases} \tag{1}$$

where $(a,b) \in R^2, a < b, 1 < \alpha < 2, \psi \in C^2([a,b]), \psi'(x) > 0, x \in [a,b]$ ${}^cD^{\alpha,\psi}$ is the ψ-Caputo fractional derivative of order α, and $f : [a,b] \times R \to R$ is a given function. A Lyapunov-type inequality is established for problem (1). The authors also give some examples to illustrate the applications of their main results.

Very recently, Chen et al. in [10] studied the following anti-periodic boundary problem involving the Riesz-Caputo derivative

$$\begin{cases} {}_{0}^{RC}D_{T}^{\gamma}y(\tau) = g(\tau, y(\tau)), \quad \tau \in [0, T], 1 < \gamma \leq 2, \\ y(0) = -y(T), \quad y'(0) = -y'(T), \end{cases} \tag{2}$$

where ${}_{0}^{RC}D_{T}^{\gamma}$ is a Riesz-Caputo derivative, which can reflect both the past and the future nonlocal memory effects and $g : [0, T] \times R \to R$ is a continuous function with respect to τ and y. Some existence results of solutions are given based on the Lipschitz condition, the growth condition and the comparison condition. Most of the present work are concerned with fractional differential equations or inclusions involving Riemann-Liouville or Caputo fractional derivative, merely reflecting the past or future memory effect. Riesz derivative is a two-sided fractional operator, whose advantage is that it could reflect both the past and the future memory effects. We take anomalous diffusion problem for example. The fractional differential equation with the Riesz derivative is adopted to describe the anomalous diffusion problem, in which the Riesz derivative stands for the nonlocality and the dependence on path of the diffusion concentration. Some applications of Riesz derivative about anomalous diffusion, we refer the reader to [17,18]. Another typical example is stocks. According to the price trend of the past and future time, investors would buy or sell a stock at an agreed-on price within a period of time. This process depends on both past state and its development in the future, which is the characteristic of Riesz derivative. There are some other applications of this derivative, and we refer the reader to [10,19,20]. In 2009, Ahamad and Otero-Espinar [1] investigated the following fractional inclusions with anti-periodic boundary conditions

$$\begin{cases} {}^{c}D^{q}x(t) \in F(t, x(t)), \quad t \in [0, T], 1 < q \leq 2, \\ x(0) = -x(T), \quad x'(0) = -x'(T), \end{cases} \tag{3}$$

where ${}^{c}D^{q}x(t)$ is the standard Caputo derivative of order q, $F : [0, T] \times R \to \mathcal{P}(R)$ is a multivalued map, $\mathcal{P}(R)$ is the family of all subsets of R. Some sufficient conditions for the existence of solutions are given by means of Bohnenblust-Karlin fixed point theorem.

Inspired by the above-mentioned works, in this paper, we are concerned with the following anti-periodic fractional inclusions with ψ-Riesz-Caputo derivative:

$$\begin{cases} {}_{a}^{RC}D_{b}^{\alpha,\psi}u(x) \in F(x, u(x)), \quad a < x < b, \\ u(a) + u(b) = 0, u'(a) + u'(b) = 0, \end{cases} \tag{4}$$

where $(a, b) \in R^2, a < b, 1 < \alpha < 2, \psi \in C^2([a, b]), \psi'(x) > 0, x \in [a, b]$. ${}_{a}^{RC}D_{b}^{\alpha,\psi}$ is the ψ-Riesz-Caputo fractional derivative of order α, and $F : [a, b] \times R \to \mathcal{P}(R)$ is a multivalued map. Sufficient conditions for the existence of solutions are given in view of the fixed point theorems for multi-valued mapping. The aim of this paper is to develop the calculus of fractional derivatives. We shall combine the two definitions of Riesz-Caputo derivative and ψ-Caputo fractional derivative. Then we investigate the existence of solutions of anti-periodic inclusions (4). The rest of this paper is organized as follows. We first present some basic definitions of fractional calculus, ψ-Caputo derivative, Riesz-Caputo derivative and multi-valued maps, and then a new definition of ψ-Riesz-Caputo fractional derivative of order α is given. In Section 3, the main results on the existence of solutions for anti-periodic boundary value problem (4) are provided. We present two examples in order to illustrate our main results in last section. Our results generalize some published known results. There is no literature to research the fractional differential inclusions with ψ-Riesz-Caputo fractional derivative. If we take $F(x, u) = \{f(x, u)\}$, where $f : [a, b] \times R \to R$ is a given continuous function, then the problem (4) corresponds to the single-valued problem (1). If we take $a = 0, b = T, \psi(x) = x, F(x, u) = \{g(x, u)\}$, where $g : [0, T] \times R \to R$ is a given continuous function, then the problem (4) corresponds to the single-valued problem (2).

2. Preliminaries

In this section, we recall some notation, definitions and preliminaries about fractional calculus [6,7,21], ψ-Caputo fractional calculus [3–5,22,23], and Riesz or Riesz-Caputo fractional derivative [17–19].

Definition 1 ([6]). *The left Caputo fractional derivative order α $(1 < \alpha \leq 2)$ of a function $f \in C^2([a,b])$ is given by*

$$ {}_a^CD_x^\alpha f(x) = ({}_aI_x^{2-\alpha} f'')(x), a < x < b, $$

that is,

$$ {}_a^CD_x^\alpha f(x) = \frac{1}{\Gamma(2-\alpha)} \int_a^x (x-t)^{1-\alpha} f''(t)dt, a < x < b. $$

Similarly, the right Caputo fractional integral order α $(1 < \alpha \leq 2)$ of a function $f \in C^2([a,b])$ is given by

$$ {}_x^CD_b^\alpha f(x) = \frac{1}{\Gamma(2-\alpha)} \int_x^b (t-x)^{1-\alpha} f''(t)dt, a < x < b. $$

Definition 2 ([6]). *The fractional left, right and Riemann-Liouville integrals of order $\beta > 0$ are defined as*

$$ {}_aI_\tau^\beta g(\tau) = \frac{1}{\Gamma(\beta)} \int_a^\tau (\tau - s)^{\beta-1} g(s)ds, $$

$$ {}_\tau I_b^\beta g(\tau) = \frac{1}{\Gamma(\beta)} \int_\tau^b (s - \tau)^{\beta-1} g(s)ds, $$

$$ {}_aI_b^\beta g(\tau) = \frac{1}{\Gamma(\beta)} \int_a^b |s - \tau|^{\beta-1} g(s)ds. $$

Let $\psi \in C^2([a,b])$ be a given function such that

$$ \psi'(x) > 0, a \leq x \leq b. $$

Definition 3 ([3]). *The fractional left, right integral of order $\alpha > 0$ of a function $f \in C([a,b])$ with respect to ψ are defined by*

$$ ({}_aI_x^{\alpha,\psi}) f(x) = \frac{1}{\Gamma(\alpha)} \int_a^x \psi'(t)(\psi(x) - \psi(t))^{\alpha-1} f(t)dt, a \leq x \leq b, \tag{5} $$

$$ ({}_xI_b^{\alpha,\psi}) f(x) = \frac{1}{\Gamma(\alpha)} \int_x^b \psi'(t)(\psi(t) - \psi(x))^{\alpha-1} f(t)dt, a \leq x \leq b. \tag{6} $$

Definition 4 ([3]). *The left, right ψ-Caputo fractional derivative of order α $(1 < \alpha \leq 2)$ of a function $f \in C^2([a,b])$ are defined as*

$$ {}_a^CD_x^{\alpha,\psi} f(x) = \frac{1}{\Gamma(2-\alpha)} \int_a^x \psi'(t)(\psi(x) - \psi(t))^{1-\alpha} (\frac{1}{\psi'(t)} \frac{d}{dt})^2 f(t)dt, a < x < b, \tag{7} $$

$$ {}_x^CD_b^{\alpha,\psi} f(x) = \frac{1}{\Gamma(2-\alpha)} \int_x^b \psi'(t)(\psi(t) - \psi(x))^{1-\alpha} (\frac{1}{\psi'(t)} \frac{d}{dt})^2 f(t)dt, a < x < b. \tag{8} $$

Remark 1. *Consider $\psi(x) = x$, $\psi(x) = \ln x$, the Riemann-Liouville and Hadamard fractional operators are obtained.*

Inspired by the above definitions, we shall present a new definition of ψ-Riesz-Caputo fractional derivative of order α, which is a combination of ψ-Caputo fractional derivative and Riesz-Caputo fractional derivative.

Definition 5. *Let $f \in C^2([a,b])$. For $x \in [a,b]$, the ψ-Riesz-Caputo fractional derivative ${}_a^{RC}D_b^{\alpha,\psi} f(x)$ of order α $(1 < \alpha \leq 2)$ could be defined by*

$$ {}_a^{RC}D_b^{\alpha,\psi} f(x) = \frac{1}{2}\left({}_a^C D_x^{\alpha,\psi} + {}_x^C D_b^{\alpha,\psi}\right) f(x). \tag{9}$$

If we take $\psi(x) = x$, it follows from (7)–(9) that the classic Riesz-Caputo derivative fractional order α $(1 < \alpha \leq 2)$ of a function $f \in C^2([a,b])$ is given by

$$ {}_a^{RC}D_b^{\alpha} f(x) = \frac{1}{2}\left({}_a^C D_x^{\alpha} + {}_x^C D_b^{\alpha}\right) f(x), \tag{10}$$

which is defined as in [19]. For convenience, denote

$$P_{cl}(X) = \{Y \in \mathcal{P}(X) : Y \quad \text{is closed}\},$$

$$P_b(X) = \{Y \in \mathcal{P}(X) : Y \quad \text{is bounded}\},$$

$$P_{cp}(X) = \{Y \in \mathcal{P}(X) : Y \quad \text{is compact}\},$$

$$P_{cp,c}(X) = \{Y \in \mathcal{P}(X) : Y \quad \text{is convex and compact}\}.$$

The following are definitions and properties concerning multi-valued maps [24–28] which will be used in the remainder of this paper.

Definition 6 ([28]). *A multivalued map $G : X \to \mathcal{P}(X)$:*

(a) *denote the set $Gr(G) = \{(x,y) \in X \times Y, y \in G(x)\}$ as the graph of G,*

$$t \mapsto d(y, G(t)) = \inf\{|y - z| : z \in G(t)\}$$

is measurable.

(b) *if $G : X \to \mathcal{P}_{cl}(X)$ is called $\gamma-$Lipschitz if and only if there exists $\gamma > 0$ such that*

$$H_d(N(x), N(y)) \leq \gamma d(x,y), \text{for each } x, y \in X.$$

(c) *if $G : X \to \mathcal{P}_{cl}(X)$ is called contraction if and only if it is $\gamma-$Lipschitz with $\gamma < 1$.*

(d) *G is said to be measurable if for every $y \in R$, the function*

Definition 7 ([26]). *Assume that $F : J \times R \to \mathcal{P}(R)$ is a multivalued map with nonempty compact values. Denote a multivalued operator $\mathcal{F} : C(J \times R) \to \mathcal{P}(L^1(J,R)$ associated with F as*

$$\mathcal{F}(x) = \{w \in L^1(J,R) : w(t) \in F(t, x(t))\}$$

for a.e. $t \in J := [a,b]$ is a closed interval from a to b.

Definition 8 ([26]). *Assume that Y is a separable metric space and $N : Y \to \mathcal{P}(L^1(J,R))$ is a multivalued operator. If N is lower semi-continuous(l.s.c.) and has nonempty closed and decomposable values, we say N has a property (BC) .*

Definition 9 ([28]). *For each $u \in C(J,R)$, $t \in J = [a,b]$, denote the selection set of F as*

$$S_{F,y} := \{f \in L^1(J,R) : f(t) \in F(t, u(t)) \quad a.e. \quad t \in J\}.$$

Definition 10 ([28]). *Let $A, B \in \mathcal{P}_{cl}(X.)$ The Pompeiu-Hausdorff distance of A, B is defined by*

$$H_d(A,B) = \max\{\sup_{a\in A} d(a,B), \sup_{b \in B} d(A,b)\},$$

where $d(A,b) = \inf_{a\in A} d(a,b)$, $d(a,B) = \inf_{b\in B} d(a,b)$.

Property 1 ([24]). *Let G be a completely continuous multi-valued map with nonempty compact values, then T is u.s.c.* $\Longleftrightarrow$ *G has a closed graph.*

The following lemmas play important roles in the proof of our main results.

Lemma 1 ([28]). *(Nonlinear alternative for Kakutani maps). Assume that E is a Banach space, C is a closed convex subset of E, and U is an open subset of C with* $0 \in U$. *Let* $F : \bar{U} \to \mathcal{P}_{c,cv}(C)$ *be a upper semicontinuous compact map. Then either*

(i) *F has a fixed point in* $\bar{U}$, *or*

(ii) *there exist a* $u \in \partial U$ *and* $\lambda \in (0,1)$ *satisfying* $u \in \lambda F(u)$.

Lemma 2 ([29]). *Let* (X,d) *be a complete metric space. If* $N : X \to \mathcal{P}_{cl}(X)$ *is a contraction, then* $FixN \neq \emptyset$.

Lemma 3 ([30]). *Let X be a Banach space, and* $F : J \times X \to (P)(X)$ *be a* L^1-*Carathédory set-valued map with* $S_F \neq \emptyset$ *and let* $\Theta : L^1(J,X) \to C(J,X)$ *be a linear continuous mapping. Then the set-valued map* $\Gamma \circ S_F : C(J,X) \to \mathcal{P}(C(J,X))$ *defined by*

$$(\Theta \circ S_F)(u) : C(J \times X) \to \mathcal{P}_{cp,c}(C(J,X)),\ x \mapsto (\Theta \circ S_F)(u) = \Theta(S_{F,u})$$

is a closed graph operator in $C(J,X) \times C(J,X)$.

Lemma 4 ([20]). *Assume that Y is a separable metric space and* $N : Y \to \mathcal{P}(L^1(J,R))$ *is a multivalued operator with the property (BC). Then there exists a continuous single-valued function* $g : Y \to L^1(J,R)$ *satisfying* $g(x) \in N(x)$ *for every* $x \in Y$, *i.e., N has a continuous selection.*

From [6], we have

Lemma 5. *If* $1 < \beta \leq 2$ *and* $g \in C^2[a,b]$, *then*

$$ {}_aI^{\beta}_{\tau}{}^C_aD^{\beta}_{\tau}g(\tau) = g(\tau) - g(a) - g'(a)(\tau - a),$$
$$ {}_{\tau}I^{\beta}_{b}{}^C_{\tau}D^{\beta}_{b}g(\tau) = g(\tau) - g(b) + g'(b)(b - \tau).$$

From (10) and Lemma 2.1 in [15], for $u \in C^2[a,b]$, *and* $1 < \alpha \leq 2$, *we have that*

$$\begin{aligned} {}_aI^{\alpha}_{b}{}^C_aD^{\alpha}_{b}u(\tau) &= \frac{1}{2}\left({}_aI^{\alpha}_{\tau}{}^C_aD^{\alpha}_{\tau} + {}_{\tau}I^{\alpha}_{b}{}^C_{\tau}D^{\alpha}_{b}\right)u(\tau) \\ &= u(\tau) - \frac{1}{2}(u(a)+u(b)) - \frac{1}{2}u'(a)(\tau - a) + \frac{1}{2}u'(b)(b-\tau). \end{aligned} \tag{11}$$

By (11), similar to the proof of Lemma 2.2 in [10], we have the following lemma.

Lemma 6. *Assume that* $h \in C[a,b]$. *A function* $u \in C^2[a,b]$ *given by*

$$\begin{aligned} u(t) = &-\frac{b-a}{2\Gamma(\alpha-1)}\int_a^b (b-s)^{\alpha-2}h(s)ds \\ &+\frac{1}{\Gamma(\alpha)}\int_a^t (t-s)^{\alpha-1}h(s)ds + \frac{1}{\Gamma(\alpha)}\int_t^b (s-t)^{\alpha-1}h(s)ds, \end{aligned} \tag{12}$$

is a unique solution of the following anti-periodic boundary value problem

$$\begin{cases} ({}^{RC}_aD^{\alpha}_bu)(t) = h(t), \quad t \in (a,b), 1 < \alpha \leq 2, \\ u(a) + u(b) = 0, \quad u'(a) + u'(b) = 0. \end{cases} \tag{13}$$

As the same argument of Lemma 2.1 in [16], we can easily obtain the following result, which plays a very important role in proving the main results.

Lemma 7. *If $f, \psi \in C^2([a,b])$, and $\psi'(x) > 0$ for each $x \in [a,b]$, then*

$$({}_{a}^{RC}D_{x}^{\alpha,\psi} f)(\psi^{-1}(y)) = ({}_{\psi(a)}^{RC}D_{\psi(x)}^{\alpha}(f \circ \psi^{-1}))(y), \psi(a) < y < \psi(b), \tag{14}$$

and

$$({}_{x}^{RC}D_{b}^{\alpha,\psi} f)(\psi^{-1}(y)) = ({}_{\psi(x)}^{RC}D_{\psi(b)}^{\alpha}(f \circ \psi^{-1}))(y), \psi(a) < y < \psi(b). \tag{15}$$

Moreover, we have

$$({}_{a}^{RC}D_{b}^{\alpha,\psi} f)(\psi^{-1}(y)) = ({}_{\psi(a)}^{RC}D_{\psi(b)}^{\alpha}(f \circ \psi^{-1}))(y), \psi(a) < y < \psi(b). \tag{16}$$

Lemma 8. *If $f : [a,b] \times R \to R$, $\psi \in C^2[a,b]$ with $\psi'(x) > 0$, and $\psi'(a) = \psi'(b)$, then the problem*

$$\begin{cases} {}_{a}^{RC}D_{b}^{\alpha,\psi} u(x) = f(x, u(x)), \quad a < x < b, \\ u(a) + u(b) = 0, u'(a) + u'(b) = 0, \end{cases} \tag{17}$$

could be transformed into the following problem

$$\begin{cases} {}_{\psi(a)}^{RC}D_{\psi(b)}^{\alpha} v(y) = f(\psi^{-1}(y), v(y)), \quad \psi(a) < y < \psi(b), \\ v(\psi(a)) + v(\psi(b)) = 0, \quad v'(\psi(a)) + v'(\psi(b)) = 0. \end{cases} \tag{18}$$

A nontrivial solution to (18) is given by $\quad v(y) = -\frac{B-A}{2\Gamma(\alpha-1)} \int_A^B (B-s)^{\alpha-2} f(\psi^{-1}(s), v(s))ds$

$$+\frac{1}{\Gamma(\alpha)} \int_A^y (y-s)^{\alpha-1} f(\psi^{-1}(s), v(s))ds + \frac{1}{\Gamma(\alpha)} \int_y^B (s-y)^{\alpha-1} f(\psi^{-1}(s), v(s))ds, \tag{19}$$

where $A = \psi(a)$ and $B = \psi(b)$.

Proof. We introduce the function $v : [\psi(a), \psi(b)] \to R$, defined by

$$v(y) = u(\psi^{-1}(y)), \psi(a) \le y \le \psi(b).$$

In virtue of (16), one has

$${}_{\psi(a)}^{RC}D_{\psi(b)}^{\alpha} v(\psi^{-1}(y)) = {}_{a}^{RC}D_{b}^{\alpha,\psi} u(x), \quad \psi(a) < y < \psi(b). \tag{20}$$

By a chain rule, we have

$$v'(y) = \frac{1}{\psi'(\psi^{-1}(y))} u'(\psi^{-1}(y)), \psi(a) \le y \le \psi(b).$$

Thus, we have

$$v'(\psi(a)) = \frac{1}{\psi'(a)} u'(a), \text{and } v'(\psi(b)) = \frac{1}{\psi'(b)} u'(b).$$

From boundary condition (17) and condition $\psi'(a) = \psi'(b)$, we have that

$$v(\psi(a)) + v(\psi(b)) = 0, v'(\psi(a)) + v'(\psi(b)) = 0. \tag{21}$$

Therefore, the problem (17) could be transformed into problem (18). By virtue of Lemma 6, we obtain $v \in C^2[A,B]$ is a nontrivial solution to (18).

From Lemma 8, we can easily know that

$$u(x) = -\frac{\psi(b)-\psi(a)}{2\Gamma(\alpha-1)}\int_a^b (\psi(b)-\psi(t))^{\alpha-2}\psi'(t)f(t,u(t))dt$$
$$+\frac{1}{\Gamma(\alpha)}\int_a^x (\psi(x)-\psi(t))^{\alpha-1}\psi'(t)f(t,u(t))dt + \frac{1}{\Gamma(\alpha)}\int_x^b (\psi(t)-\psi(x))^{\alpha-1}\psi'(t)f(t,u(t))dt \quad (22)$$

is a unique solution of problem (17). □

3. Main Results

We pose the following hypotheses:

(H_1) $F : [a,b] \times R \to \mathcal{P}(R)$ is Carathéodory and it has nonempty compact and convex values;
(H_2) there exist a continuous nondecreasing function $q : [0,\infty) \to [0,\infty)$ and a function $p \in C([a,b], R^+)$ satisfying

$$\|F(t,u)\| := \sup\{|f| : f \in F(t,u)\} \leq p(t)q(\|u\|), \text{ for each } (t,x) \in [a,b] \times R.$$

(H_3) $1 < \alpha < 2$, $\psi \in C^2([a,b])$, $\psi'(x) > 0$, $x \in [a,b]$.
(H_4) $\psi'(a) = \psi'(b)$.
(H_5) $F : [a,b] \times R \to \mathcal{P}_{cp}(R)$ is such that, for every $u \in R$, $F(\cdot,u)$ is measurable.
(H_6) There exists $m \in L^1([a,b)], R^+)$ for almost all $t \in [a,b]$, such that

$$d_H(F(t,u), F(t,\bar{u})) \leq m(t)|u-\bar{u}|, \quad \forall u,\bar{u} \in R$$

with $d(0, F(t,0)) \leq m(t)$ for almost all $t \in [a,b]$.
(H_7) $F : [a,b] \times R \to \mathcal{P}(R)$ is a nonempty compact-valued multivalued map such that

(a) $(x,u) \mapsto F(x,u)$ is $\mathcal{L} \bigotimes \mathcal{B}$ is measurable.
(b) $u \mapsto F(x,u)$ is lower semicontinuous for each $x \in [a,b]$,

Now we are in the position to state our main results. The first theorem is dealing with the Carathéodory case.

Theorem 1. *Assume that (H_1)–(H_4) hold. Moreover, if there exists a constant $M > 0$, such that*

$$M\left[q(M)\left(\frac{\psi(b)-\psi(a)}{2\Gamma(\alpha-1)}\int_a^b (\psi(b)-\psi(s))^{\alpha-2}p(s)\psi'(s)ds\right.\right.$$
$$\left.\left.+\frac{2}{\Gamma(\alpha)}(\psi(b)-\psi(a))^{\alpha-1}\int_a^b p(s)\psi'(s)ds\right)\right]^{-1} > 1. \quad (23)$$

Then (4) has at least one solution on $[a,b]$.

Proof. The operator $T : C([a,b],R) \to \mathcal{P}(C[A,B],R)$ is defined as follows:

$$T(u) = \{h \in C([a,b],R) : h(t) = -\frac{\psi(b)-\psi(a)}{2\Gamma(\alpha-1)}\int_a^b (\psi(b)-\psi(s))^{\alpha-2}\psi'(s)f(s)ds$$
$$+\frac{1}{\Gamma(\alpha)}\int_a^t (\psi(t)-\psi(s))^{\alpha-1}\psi'(s)f(s)ds + \frac{1}{\Gamma(\alpha)}\int_t^b (\psi(s)-\psi(t))^{\alpha-1}\psi'(s)f(s)ds, \quad f \in S_{F,u}\}. (24)$$

We divide the proof into 5 parts, which shows that T satisfies all the conditions of Lemma 1.

Part (i). T maps the bounded sets into bounded sets of $C([a,b],R)$. Set $B_r = \{v \in C([a,b],R) : \|v\| \leq r, \ \ r > 0\}$, which is a bounded ball in $C([a,b],R)$, then for $h \in T(u)$, $u \in B_r$, there exists $f \in S_{F,u}$ such that

$$h(t) = -\frac{\psi(b)-\psi(a)}{2\Gamma(\alpha-1)}\int_a^b (\psi(b)-\psi(s))^{\alpha-2}\psi'(s)f(s)ds$$

$$+\frac{1}{\Gamma(\alpha)}\int_a^t(\psi(t)-\psi(s))^{\alpha-1}\psi'(s)f(s)ds+\frac{1}{\Gamma(\alpha)}\int_t^b(\psi(s)-\psi(t))^{\alpha-1}\psi'(s)f(s)ds. \quad (25)$$

Then

$$\begin{aligned}|h(t)| &\le \frac{\psi(b)-\psi(a)}{2\Gamma(\alpha-1)}\int_a^b(\psi(b)-\psi(s))^{\alpha-2}\psi'(s)|f(s)|ds\\&+\frac{1}{\Gamma(\alpha)}\int_a^t(\psi(t)-\psi(s))^{\alpha-1}\psi'(s)|f(s)|ds+\frac{1}{\Gamma(\alpha)}\int_t^b(\psi(s)-\psi(t))^{\alpha-1}\psi'(s)|f(s)|ds.\\&\le q(\|u\|)\left[\frac{\psi(b)-\psi(a)}{2\Gamma(\alpha-1)}\int_a^b(\psi(b)-\psi(s))^{\alpha-2}\psi'(s)p(s)ds\right.\\&\left.+\frac{1}{\Gamma(\alpha)}\int_a^t(\psi(t)-\psi(s))^{\alpha-1}\psi'(s)p(s)ds+\frac{1}{\Gamma(\alpha)}\int_t^b(\psi(s)-\psi(t))^{\alpha-1}\psi'(s)p(s)ds\right]\\&\le q(r)\left(\frac{\psi(b)-\psi(a)}{2\Gamma(\alpha-1)}\int_a^b(\psi(b)-\psi(s))^{\alpha-2}\psi'(s)p(s)ds\right.\\&\left.+\frac{2}{\Gamma(\alpha)}(\psi(b)-\psi(a))^{\alpha-1}\int_a^b\psi'(s)p(s)ds\right). \quad (26)\end{aligned}$$

Part (ii). T maps bounded set into equicontinuous sets. Let $u\in B_r$, $t_1,t_2\in[a,b]$, $t_1<t_2$, where B_r is a bounded set in $C([a,b],R)$, for $u\in T(u)$, we have

$$\begin{aligned}|h(t_2)-h(t_1)| &\le \frac{1}{\Gamma(\alpha)}\int_a^{t_1}[(\psi(t_2)-\psi(s))^{\alpha-1}-(\psi(t_1)-\psi(s))^{\alpha-1}]\psi'(s)|f(s)|ds\\&+\frac{1}{\Gamma(\alpha)}\int_{t_1}^{t_2}[(\psi(t_2)-\psi(s))^{\alpha-1}-(\psi(t_1)-\psi(s))^{\alpha-1}]\psi'(s)|f(s)|ds\\&+\frac{1}{\Gamma(\alpha)}\int_{t_2}^{b}[(\psi(t_2)-\psi(s))^{\alpha-1}-(\psi(t_1)-\psi(s))^{\alpha-1}]\psi'(s)|f(s)|ds\\&\le \frac{q(r)}{\Gamma(\alpha)}\int_a^{t_1}[(\psi(t_2)-\psi(s))^{\alpha-1}-(\psi(t_1)-\psi(s))^{\alpha-1}]\psi'(s)p(s)ds\\&+\frac{q(r)}{\Gamma(\alpha)}\int_{t_1}^{t_2}[(\psi(t_2)-\psi(s))^{\alpha-1}-(\psi(t_1)-\psi(s))^{\alpha-1}]\psi'(s)p(s)ds\\&+\frac{q(r)}{\Gamma(\alpha)}\int_{t_2}^{b}[(\psi(t_2)-\psi(s))^{\alpha-1}-(\psi(t_1)-\psi(s))^{\alpha-1}]\psi'(s)p(s)ds, \quad (27)\end{aligned}$$

independent of $u\in B_r$ as $t_1\to t_2$, the right side hand of above inequality tends to 0. According to the Ascoli-Arzelá Theorem, T is completely continuous.

Part (iii). T has a closed graph. Set $u_n\to u_*$, $h_n\in T(u_n)$ and $h_n\to h_*$. Then, we shall show that $h_*\in T(u_*)$. For $h_n\in T(u_n)$, there exist $f_n\in S_{F,u_n}$ such that

$$h_n(t)=-\frac{\psi(b)-\psi(a)}{2\Gamma(\alpha-1)}\int_a^b(\psi(b)-\psi(s))^{\alpha-2}\psi'(s)f_n(s)ds \quad (28)$$

$$+\frac{1}{\Gamma(\alpha)}\int_a^t(\psi(t)-\psi(s))^{\alpha-1}\psi'(s)f_n(s)ds+\frac{1}{\Gamma(\alpha)}\int_t^b(\psi(s)-\psi(t))^{\alpha-1}\psi'(s)f_n(s)ds.$$

Hence, it suffices to show that there exists $f_*\in S_{F,u_*}$, such that for each $t\in[a,b]$,

$$h_*(t)=-\frac{\psi(b)-\psi(a)}{2\Gamma(\alpha-1)}\int_a^b(\psi(b)-\psi(s))^{\alpha-2}\psi'(s)f_*(s)ds$$

$$+\frac{1}{\Gamma(\alpha)}\int_a^t(\psi(t)-\psi(s))^{\alpha-1}\psi'(s)f_*(s)ds+\frac{1}{\Gamma(\alpha)}\int_t^b(\psi(s)-\psi(t))^{\alpha-1}\psi'(s)f_*(s)ds. \quad (29)$$

Define the continuous linear the operator $\Phi : L^1([a,b],R) \to C([a,b],R)$:

$$f \mapsto \Phi(f)(t) = -\frac{\psi(b)-\psi(a)}{2\Gamma(\alpha-1)}\int_a^b(\psi(b)-\psi(s))^{\alpha-2}\psi'(s)f(s)ds$$

$$+\frac{1}{\Gamma(\alpha)}\int_a^t(\psi(t)-\psi(s))^{\alpha-1}\psi'(s)f(s)ds+\frac{1}{\Gamma(\alpha)}\int_t^b(\psi(s)-\psi(t))^{\alpha-1}\psi'(s)f(s)ds. \quad (30)$$

We have $\|h_n - h\| \to 0$, as $n \to \infty$. Thus, in light of Lemma 3, $\Phi \circ S_F$ is a closed graph operator. Furthermore, we have $h_n(t) \in \Phi(S_{F,u_n})$. By $u_n \to u_*$, we obtain

$$h_*(t) = -\frac{\psi(b)-\psi(a)}{2\Gamma(\alpha-1)}\int_a^b(\psi(b)-\psi(s))^{\alpha-2}\psi'(s)f_*(s)ds$$

$$+\frac{1}{\Gamma(\alpha)}\int_a^t(\psi(t)-\psi(s))^{\alpha-1}\psi'(s)f_*(s)ds+\frac{1}{\Gamma(\alpha)}\int_t^b(\psi(s)-\psi(t))^{\alpha-1}\psi'(s)f_*(s)ds, \quad (31)$$

for some $f_* \in S_{F,u_*}$.

Part (iv). T is convex for each $x \in C([a,b],R)$. Since $S_{F,u}$ is convex, it is obviously true.

Part (v). We show that there exists a open set $U \subset C([a,b],R)$, with $u \notin T(u)$ for any $\eta \in (0,1)$ and all $u \in \partial U$. Let $\eta \in (0,1)$, $u \in \eta T(u)$. Then for $t \in [a,b]$, there exists $f \in S_{F,u}$ such that

$$h(t) = -\frac{\psi(b)-\psi(a)}{2\Gamma(\alpha-1)}\int_a^b(\psi(b)-\psi(s))^{\alpha-2}\psi'(s)f(s)ds$$

$$+\frac{1}{\Gamma(\alpha)}\int_a^t(\psi(t)-\psi(s))^{\alpha-1}\psi'(s)f(s)ds+\frac{1}{\Gamma(\alpha)}\int_t^b(\psi(s)-\psi(t))^{\alpha-1}\psi'(s)f(s)ds. \quad (32)$$

A similar discussion as in part (i), we have

$$\|h\| \le q(\|u\|)\left(\frac{\psi(b)-\psi(a)}{2\Gamma(\alpha-1)}\int_a^b(\psi(b)-\psi(s))^{\alpha-2}\psi'(s)p(s)ds\right.$$

$$\left.+\frac{2}{\Gamma(\alpha)}(\psi(b)-\psi(a))^{\alpha-1}\int_a^b\psi'(s)p(s)ds\right). \quad (33)$$

Consequently, we have

$$\frac{\|u\|}{q(\|u\|)\left(\frac{\psi(b)-\psi(a)}{2\Gamma(\alpha-1)}\int_a^b(\psi(b)-\psi(s))^{\alpha-2}\psi'(s)p(s)ds+\frac{2}{\Gamma(\alpha)}(\psi(b)-\psi(a))^{\alpha-1}\int_a^b\psi'(s)p(s)ds\right)} \le 1. \quad (34)$$

By (23), there exists M such that $\|u\| \neq M$. Let

$$U = \{x \in C([a,b],R) : \|u\| < M\}.$$

It is clear that the operator $T : \bar{U} \to \mathcal{P}(C([a,b],R))$ is upper semicontinuous and completely continuous. If we choose U properly, for some $\eta \in (0,1)$, there is no $u \in \partial U$ such that $u \in \eta T(u)$. Thus, by means of Lemma 1, we can get the conclusion that thereexists a fixed point $u \in \bar{U}$, that is, it is a solution of problem (4). We complete the proof. □

We shall give the second theorem which is concerned with the Lipschitz case.

Theorem 2. *Suppose that the conditions* (H_3)–(H_6) *are satisfied. Moreover, if*

$$\gamma := \frac{\psi(b)-\psi(a)}{2\Gamma(\alpha-1)}\int_a^b(\psi(b)-\psi(s))^{\alpha-2}\psi'(s)m(s)ds+\frac{2}{\Gamma(\alpha)}(\psi(b)-\psi(a))^{\alpha-1}\int_a^b\psi'(s)m(s)ds<1 \quad (35)$$

then problem (4) has at least a solution on $[a,b]$.

Proof. By (22), we define the operator $T : C([a,b],R) \to \mathcal{P}(C[a,b],R)$ as follows:

$$T(u) = \{h \in C([a,b],R) : h(t) = -\frac{\psi(b)-\psi(a)}{2\Gamma(\alpha-1)}\int_a^b(\psi(b)-\psi(s))^{\alpha-2}\psi'(s)g(s)ds$$
$$+\frac{1}{\Gamma(\alpha)}\int_a^t(\psi(t)-\psi(s))^{\alpha-1}\psi'(s)g(s)ds$$
$$+\frac{1}{\Gamma(\alpha)}\int_t^b(\psi(s)-\psi(t))^{\alpha-1}\psi'(s)g(s)ds, \quad g \in S_{F,u}\}. \tag{36}$$

Obviously, the fixed point of T is the solution of (4). Our aim is to prove that the operator T satisfies all the conditions in Lemma 2. The proof will be given in two claims.

Claim 1. For each $h \in C([a,b],\mathbb{R})$ the operator T is closed. Let $\{h_n\}_{n\geq 0} \in T(u)$ be such that $h_n \to h(n \to \infty)$ in $C([a,b],R)$. Then $h \in C([a,b],\mathbb{R})$, and there exists $v_n \in S_{F,u}$ such that for each $t \in [a,b]$,

$$h_n(t) = -\frac{\psi(b)-\psi(a)}{2\Gamma(\alpha-1)}\int_a^b(\psi(b)-\psi(s))^{\alpha-2}\psi'(s)v_n(s)ds$$
$$+\frac{1}{\Gamma(\alpha)}\int_a^t(\psi(t)-\psi(s))^{\alpha-1}\psi'(s)v_n(s)ds+\frac{1}{\Gamma(\alpha)}\int_t^b(\psi(s)-\psi(t))^{\alpha-1}\psi'(s)v_n(s)ds. \tag{37}$$

For F has compact values, we get a subsequence v_n which converges to $v \in L^1([a,b],R)$. Thus, $v \in S_{F,u}$, and for each $t \in [a,b]$, one has

$$h_n(t) \to h(t) = -\frac{\psi(b)-\psi(a)}{2\Gamma(\alpha-1)}\int_a^b(\psi(b)-\psi(s))^{\alpha-2}\psi'(s)v(s)ds$$
$$+\frac{1}{\Gamma(\alpha)}\int_a^t(\psi(t)-\psi(s))^{\alpha-1}\psi'(s)v(s)ds+\frac{1}{\Gamma(\alpha)}\int_t^b(\psi(s)-\psi(t))^{\alpha-1}\psi'(s)v(s)ds. \tag{38}$$

Therefore, $h \in T(u)$.

Claim 2. We shall show that there exists $\gamma < 1$ such that

$$H_d(F(t,u),F(t,\bar{u})) \leq \gamma\|u-\bar{u}\|.$$

Let $u,\bar{u} \in C([a,b],R)$ and $h_1 \in T(u)$. There exists $v_1(t) \in F(t,u(t))$ such that for each $t \in [a,b]$,

$$h_1(t) = -\frac{\psi(b)-\psi(a)}{2\Gamma(\alpha-1)}\int_a^b(\psi(b)-\psi(s))^{\alpha-2}\psi'(s)v_1(s)ds$$
$$+\frac{1}{\Gamma(\alpha)}\int_a^t(\psi(t)-\psi(s))^{\alpha-1}\psi'(s)v_1(s)ds+\frac{1}{\Gamma(\alpha)}\int_t^b(\psi(s)-\psi(t))^{\alpha-1}\psi'(s)v_1(s)ds. \tag{39}$$

By (H_6), there exists $w \in F(t,\bar{u}(t))$ such that

$$|v_1(t)-w(t)| \leq m(t)|u(t)-\bar{u}(t)|, t \in [a,b].$$

$U : [a,b] \to \mathcal{P}(R)$ is defined as

$$U(t) := \{w \in R : |v_1(t)-w(t)| \leq m(t)|u(t)-\bar{u}(t)|\}.$$

The multivalued operator $U(t) \cap F(t,\bar{u}(t))$ is measurable, so there exits a measurable selection for $U(t) \cap F(t,\bar{u}(t))$. We denote this function as $v_2(t)$. For each $t \in [a,b]$, one has

$$|v_1(t)-v_2(t)| \leq m(t)|u(t)-\bar{u}(t)|.$$

Then, we define for each $t \in [a,b]$,

$$h_2(t) = -\frac{\psi(b)-\psi(a)}{2\Gamma(\alpha-1)}\int_a^b(\psi(b)-\psi(t))^{\alpha-2}\psi'(s)v_2(s)ds$$
$$+\frac{1}{\Gamma(\alpha)}\int_a^t(\psi(t)-\psi(s))^{\alpha-1}\psi'(s)v_2(s)ds+\frac{1}{\Gamma(\alpha)}\int_t^b(\psi(s)-\psi(t))^{\alpha-1}\psi'(s)v_2(s)ds, \quad (40)$$

it follows that

$$\begin{aligned}|h_1(t)-h_2(t)| &\le \frac{\psi(b)-\psi(a)}{2\Gamma(\alpha-1)}\int_a^b(\psi(b)-\psi(t))^{\alpha-2}\psi'(s)|v_1(s)-v_2(s)|ds \\
&+\frac{1}{\Gamma(\alpha)}\int_a^t(\psi(t)-\psi(s))^{\alpha-1}\psi'(s)|v_1(s)-v_2(s)|ds \\
&+\frac{1}{\Gamma(\alpha)}\int_t^b(\psi(s)-\psi(t))^{\alpha-1}\psi'(s)|v_1(s)-v_2(s)|ds \\
&\le \frac{\psi(b)-\psi(a)}{2\Gamma(\alpha-1)}\|u-\bar{u}\|\int_a^b(\psi(b)-\psi(t))^{\alpha-2}\psi'(s)m(s)ds \\
&+\frac{1}{\Gamma(\alpha)}\|u-\bar{u}\|\int_a^t(\psi(t)-\psi(s))^{\alpha-1}\psi'(s)m(s)ds \\
&+\frac{1}{\Gamma(\alpha)}\|u-\bar{u}\|\int_t^b(\psi(s)-\psi(t))^{\alpha-1}\psi'(s)m(s)ds \\
&\le \Bigg[\frac{\psi(b)-\psi(a)}{2\Gamma(\alpha-1)}\int_a^b(\psi(b)-\psi(t))^{\alpha-2}\psi'(s)m(s)ds \\
&+\frac{2}{\Gamma(\alpha)}(\psi(b)-\psi(a))^{\alpha-1}\int_a^b\psi'(s)m(s)ds\Bigg]\|u-\bar{u}\|. \\
&=\gamma\|u-\bar{u}\|. \end{aligned} \quad (41)$$

Therefore,

$$\|h_1-h_2\| \le \gamma\|u-\bar{u}\|.$$

Interchanging u and $\bar{u}$ yields

$$H_d(F(t,\bar{u}),F(t,u)) \le \gamma\|u-\bar{u}\|.$$

Thus, T is a contraction by $\gamma<1$. Since Lemma 2, we conclude that T admits a fixed point which is a solution to problem (4). □

The third theorem is about the lower semicontinuous case.

Theorem 3. *Assume that (H_1)–(H_4) hold, if (H_7) is also satisfied, then the anti-periodic boundary problem (4) has at least one solution on $[a,b]$.*

Proof. It is clear that F is of l.s.c. type as condition (H_7) is satisfied. By means of Lemma 4, there exists a continuous function $f: C(J,R)\to L^1(J,R)$ such that $f(u)\in\mathcal{F}(u)$ for all $u\in C(J,R)$.

Next, we shall consider the following problem

$$\begin{cases} ({}^{RC}_{\psi(a)}D^{\alpha}_{\psi(b)}u)(x)=f(u(x)), \quad \psi(a)<x<\psi(b), \\ u(\psi(a))+u(\psi(b))=0, \quad u'(\psi(a))+u'(\psi(b))=0, \end{cases} \quad (42)$$

Note that if $u \in C^2([a,b],R)$ is a solution to (42), then u is a solution to the problem (4). we define the operator $\bar{\mathcal{T}}$ as

$$\bar{\mathcal{T}}u(x) = -\frac{\psi(b)-\psi(a)}{2\Gamma(\alpha-1)}\int_a^b (\psi(b)-\psi(s))^{\alpha-2}\psi'(s)f(u(s))ds$$

$$+\frac{1}{\Gamma(\alpha)}\int_a^t (\psi(t)-\psi(s))^{\alpha-1}\psi'(s)f(u(s))ds + \frac{1}{\Gamma(\alpha)}\int_t^b (\psi(s)-\psi(t))^{\alpha-1}\psi'(s)f(u(s))ds. \quad (43)$$

We transform the problem (42) into a fixed point problem. Obviously, the operator $\bar{\mathcal{T}}$ is continuous and completely continuous. As the remainder of the proof is similar to that of Theorem 1, we omit it here. □

Remark 2. *If we take $F(x,u) = \{f(x,u)\}$, where $f:[a,b]\times R \to R$ is a given continuous function, then the problem (4) corresponds to the single-valued problem (1).*

Remark 3. *If we take $a = 0$, $b = T$, $\psi(x) = x$, $F(x,u) = \{g(x,u)\}$, where $g:[0,T]\times R \to R$ is a given continuous function, then the problem (4) corresponds to the single-valued problem (2).*

4. Applications

Example 1. *Consider the fractional differential inclusion involving ψ-Riesz-Caputo derivative with anti-periodic boundary value conditions*

$$\begin{cases} {}^{RC}_{-1}D_1^{\frac{3}{2},\psi}u(x) \in F(x,u(x)), \\ u(-1)+u(1)=0, u'(-1)+u'(1)=0, \end{cases} \quad (44)$$

where $\psi(x) = \sinh(x)$, $-1 \le x \le 1$. $\alpha = \frac{3}{2}$. Observe that $\psi \in C^2([-1,1])$, $\psi'(x) = \cosh(x) > 0$, $-1 \le x \le 1$. Moreover, we have

$$\psi'(-1) = \cosh(-1) = \cosh(1) = \psi'(1),$$

which implies condition (H_3)–(H_4) hold.

$$x \to F(x,u(x)) := \left[\frac{|u|^5}{|u|^5+3} + x^2 + 1, \frac{|u|}{|u|+1} + x^3 + 1\right], u \in R,$$

and

$$\|F(x,u)\| := \sup|v| : v \in F(x,u) \le 3 := p(x)q(\|u\|)\ u \in R.$$

Obviously, condition (H_1) is satisfied. And $p(x) = 1$, $q(\|u\|) = 3$, we can find a positive constant M such that

$$M\left[3\left(\frac{|\sinh(1)-\sinh(-1)|}{2\Gamma(1/2)}\int_{-1}^1 (\sinh(1)-\sinh(s))^{-1/2}\cosh(s))ds\right.\right.$$

$$\left.\left.+\frac{2(\sinh(1)-\sinh(-1))^{1/2}}{\Gamma(3/2)}\int_{-1}^1 \cosh(s)ds\right)\right]^{-1} > 1,$$

that is, $M > 30.486$. All the conditions in Theorem 1 are satisfied. Therefore, the fractional differential inclusion with anti-periodic boundary value conditions (44) has at least one solution.

Example 2. *Consider the fractional differential inclusion involving ψ-Riesz-Caputo derivative with anti-periodic boundary value conditions*

$$\begin{cases} {}^{RC}_{-1}D_1^{\frac{6}{5},\psi}u(x) \in F(x,u(x)), \\ u(-1)+u(1)=0, u'(-1)+u'(1)=0, \end{cases} \quad (45)$$

where $\psi(x) = \sin(x)$, $-1 \le x \le 1$. $\alpha = \frac{6}{5}$. *Observe that* $\psi \in C^2([-1,1])$, $\psi'(x) = \cos(x) > 0$, $-1 \le x \le 1$. *Moreover, we have*

$$\psi'(-1) = \cos(-1) = \cos(1) = \psi'(1),$$

which implies condition (H_3)–(H_4) *hold.*

$$x \to F(x, u(x)) := \left[0, \frac{|x|}{3}\frac{|u|}{|u|+1}\right], u \in R,$$

and

$$d_H(F(t,u), F(t,\bar{u})) \le \frac{|t|}{3}|u - \bar{u}|, u \in R.$$

we can find out that

$$\begin{aligned}\gamma &= \frac{\psi(1)-\psi(-1)}{2\Gamma(\frac{1}{5})}\int_{-1}^{1}(\psi(1)-\psi(s))^{-\frac{4}{5}}\psi'(s)\frac{|s|}{3}ds + \frac{2}{\Gamma(\frac{1}{5})}(\psi(1)-\psi(-1))^{\frac{1}{5}}\int_{-1}^{1}\psi'(s)\frac{|s|}{3}ds \\ &= \frac{\sin(1)-\sin(-1)}{2\Gamma(\frac{1}{5})}\int_{-1}^{1}(\sin(1)-\sin(s))^{-\frac{4}{5}}\cos(s)\frac{|s|}{3}ds \\ &+ \frac{2}{\Gamma(\frac{6}{5})}(\sin(1)-\sin(-1))^{\frac{1}{5}}\int_{-1}^{1}\cos(s)\frac{|s|}{3}ds \\ &< 0.967,\end{aligned}$$

that is, $\gamma < 1$. *All the conditions in Theorem 2 are satisfied. Therefore, the fractional differential inclusion with anti-periodic boundary value conditions (45) has at least one solution.*

5. Conclusions

Riesz derivative, which is different from one-sided fractional derivative, as the Caputo or Riemann-Liouville derivative, is a two-sided fractional operator. It is of great use due to its reflecting both the past and the future memory effects. We study the existence of solutions for a class of anti-periodic fractional differential inclusions with ψ-Riesz-Caputo fractional derivative in this paper. Firstly, combining ψ-Caputo derivative with Riesz-Caputo derivative, we give a new definition of ψ-Riesz-Caputo fractional derivative of order α. Then, in virtue of fixed-point theorems for multi-valued maps, some sufficient conditions for the existence of solutions to the fractional differential inclusions are presented. Last but not least, we present two examples to illustrate our main results.

Author Contributions: Both authors contributed equally and significantly in writing this paper. Both authors read and approved the final manuscript.

Funding: This work is supported by the Natural Science Foundation of China (11571136).

Acknowledgments: The authors thanks anonymous referees for their remarkable comments, suggestion, and ideas that help to improve this paper.

Conflicts of Interest: The authors declare that they have no competing interests.

References

1. Ahmad, B.; Otero-Espinar, V. Existence of solutions for fractional differential inclusions with antiperiodic boundary conditions. *Bound. Value Probl.* **2009**, *2009*, 625347. [CrossRef]
2. Ahmad, B.; Ntouyas, S.K.; Zhou, Y.; Alsaedi, A. A Study of Fractional Differential Equations and Inclusions with Nonlocal Erdelyi-Kober Type Integral Boundary Conditions. *Bull. Iran. Math. Soc.* **2018**, *44*, 1315–1328. [CrossRef]

3. Almeida, R. A Caputo fractional derivative of a function with respect to another function. *Commun. Nonlinear Sci. Numer. Simula* **2017**, *44*, 460–481. [CrossRef]
4. Almeida, R.; Malinowska, A.B.; Monteiro, M.T.T. Fractional differential equations with a Caputo derivative with respect to a kernel function and their applications. *Math. Meth. Appl. Sci.* **2018**, *41*, 336–352. [CrossRef]
5. Almeida, R.; Malinowska, A.B.; Odzijewicz, T. On systems of fractional differential equations with the ψ-Caputo derivative and their applications. *Math. Meth. Appl. Sci.* **2019**. [CrossRef]
6. Kilbas, A.A.; Srivastava, H.M.; Trujillo, J.J. *Theory and Applications of Fractional Differential Equations*; North-Holl and Mathematics Studies: Elsevier: Amsterdam, The Netherlands, 2006.
7. Podlubny, I. *Fractional Differential Equations*; Academic Press: San Diego, CA, USA, 1999.
8. Wang, J.R.; Ibrahim, A.G.; O'Regan, D. Topological structure of the solution set for fractional non-instantaneous impulsive evolution inclusions. *J. Fixed Point Theory Appl.* **2018**, *20*, 59. [CrossRef]
9. Ahmad, B.; Ntouyas, S.K.; Alsaedi, A. On fractional differential inclusions with anti-periodic type integral boundary conditions. *Bound. Value Probl.* **2013**, *2013*, 82. [CrossRef]
10. Chen, F.L.; Chen, A.P.; Wu, X. Anti-periodic boundary value problems with Riesz-Caputo derivative. *Adv. Differ. Equ.* **2019**, *2019*, 119. [CrossRef]
11. Ahmad, B.; Ntouyas, S.K. On Hadmard fractional integro-differential boundary value problems. *J. Appl. Math. Camput.* **2015**, *47*, 119–131. [CrossRef]
12. Baghani, H. Existence and uniqueness of solutions to fractional Langevin equations involving two fractional orders. *J. Fixed Point Theory Appl.* **2013**, *20*, 63. [CrossRef]
13. Baleanu, D.; Mousalou, A.; Rezapour, S. A new method for investigating approximate solutions of some fractional integro-differential equations involving the Caputo-Fabrizio derivative. *Adv. Differ. Equ.* **2017**, *2017*, 51. [CrossRef]
14. da C. Sousa, J.V.; Rodrigues, F.G.; de Oliveira, E.C. Stability of the fractional Volterra integro-differential equation by means of ψ-Hilfer operator. *Math. Meth. Appl. Sci.* **2019**, *42*, 3033–3043.
15. Vivek, D.; Kanagarajan, K.; Elsayed, E.M. Some existence and stability results for Hilfer-fractional implicit differential equations with nonlocal conditions. *Mediterr. J. Math.* **2018**, *15*, 15. [CrossRef]
16. Samet, B.; Aydi, H. Lyapunov-type inequalities for an anti-periodic fractional boundary value problem involving ψ-Caputo fractional derivative. *J. Inequal. Appl.* **2018**, *2018*, 286. [CrossRef] [PubMed]
17. Wu, G.; Baleanu, D.; Deng, Z.; Zeng, S. Lattice fractional diffusion equation in terms of a Riesz-Caputo difference. *Physica A* **2015**, *438*, 335–339. [CrossRef]
18. Yang, Q.; Liu, F.; Turner, I. Numerical methods for fractional partial differential equations with Riesz space fractional derivatives. *Appl. Math. Model.* **2013**, *34*, 200–218. [CrossRef]
19. Agrawal, O.P. Fractional variational calculus in terms of Riesz fractional derivatives. *J. Phys.* **2007**, *40*, 6287–6303. [CrossRef]
20. Chen, F.L.; Baleanu, D.; Wu, G.C. Existence results of fractional differential equations with Riesz-Caputo derivative. *Eur. Phys. J. Spec. Top.* **2017**, *226*, 3411–3425. [CrossRef]
21. Lakshmikantham, V.; Vatsala, A.S. Basic theory of fractional differential equations. *Nonlinear Anal.* **2008**, *69*, 2677–2682. [CrossRef]
22. Awadalla, M.; Yameni, Y.Y. Modeling exponential growth and exponential decay real phenomena by ψ-Caputo fractional derivative. *J. Adv. Math. Comput. Sci.* **2018**, *28*, 1–13. [CrossRef]
23. Luo, D.F.; Luo, Z.G. Existence and finite-time stability of solutions for a class of nonlinear fractional differential equations with time-varying delays and non-instantaneous impulses. *Adv. Dffer. Equ.* **2019**, *2019*, 155. [CrossRef]
24. Bohnenblust, H.F.; Karlin, S. On a theorem of Ville. In *Contributions to the Theory of Games, Vol. I*; Annals of Mathematics Studies; Princeton University Press: Princeton, NJ, USA, 1950; Volume 24, pp. 155–160.
25. Bressan, A.; Colombo, G. Extensions and selections of maps with decomposable values. *Studia Math.* **1988**, *90*, 69–86. [CrossRef]
26. Deimling, K. *Multivalued Differential Equations*; De Gruyter: Berlin, Germany, 1992.
27. Granas, A.; Dugundji, J. *Fixed Point Theory*; Springer: New York, NY, USA, 2005.
28. Papageorgiou; Shouchuan, H.; Nikolaos, S. *Handbook of Multivalued Analysis Theory I*; Kluwer: Dordrecht, The Netherlands, 1997.

29. Covitz, H.; Nadler, S.B., Jr. Multivalued contraction mappings in generalized metric spaces. *Israel J. Math.* **1970**, *8*, 5–11. [CrossRef]
30. Lasota, A.; Opial, Z. An application of the Kakutani-Ky Fan theorem in the theory of ordinary differential equations. *Bull. Acad. Pol. Sci. Ser. Sci. Math. Astronom. Phys.* **1965**, *13*, 781–786.

mathematics

MDPI

Article

Existence Theory for a Fractional q-Integro-Difference Equation with q-Integral Boundary Conditions of Different Orders

Sina Etemad [1], Sotiris K. Ntouyas [2,3] and Bashir Ahmad [3,*]

1 Department of Mathematics, Azarbaijan Shahid Madani University, Azarshahr, Tabriz, Iran
2 Department of Mathematics, University of Ioannina, 45110 Ioannina, Greece
3 Nonlinear Analysis and Applied Mathematics (NAAM)-Research Group, Department of Mathematics, Faculty of Science, King Abdulaziz University, P.O. Box 80203, Jeddah 21589, Saudi Arabia
* Correspondence: bashirahmad_qau@yahoo.com

Received: 18 June 2019; Accepted: 20 July 2019; Published: 24 July 2019

Abstract: In this paper, we study the existence of solutions for a new class of fractional q-integro-difference equations involving Riemann-Liouville q-derivatives and a q-integral of different orders, supplemented with boundary conditions containing q-integrals of different orders. The first existence result is obtained by means of Krasnoselskii's fixed point theorem, while the second one relies on a Leray-Schauder nonlinear alternative. The uniqueness result is derived via the Banach contraction mapping principle. Finally, illustrative examples are presented to show the validity of the obtained results. The paper concludes with some interesting observations.

Keywords: q-integro-difference equation; boundary value problem; existence; fixed point

1. Introduction and Preliminaries

Fractional calculus, dealing with differential and integral operators of arbitrary order, serves as a powerful modelling tool for many real-world phenomena. An interesting feature of such operators is their nonlocal nature that accounts for the history of the phenomena involved in the fractional models. Motivated by the extensive applications of fractional calculus, many researchers turned to the theoretical development of fractional-order initial and boundary value problems. Now, the literature on the topic contains many interesting and important results on the existence and uniqueness of solutions, and other properties of solutions for fractional-order problems. The available material includes different types of derivatives such as Riemann-Liouville, Caputo, Hadamard, etc. and a variety of boundary conditions. For some recent works on the topic, for instance, see [1–8] and the references therein.

Fractional q-difference equations (fractional analogue of q-difference equations) also received significant attention. One can find preliminary work on the topic in [9], while some interesting details about initial and boundary value problems of q-difference and fractional q-difference equations can be found in the book [10].

In 2012, Ahmad et al. [11] discussed the existence and uniqueness of solutions for the nonlocal boundary value problem of fractional q-difference equations:

$$\begin{cases} {}^cD_q^{\alpha}x(t) = f(t, x(t)), \quad 0 \le t \le 1,\ 1 < \alpha \le 2,\ 0 < q < 1, \\ \alpha_1 x(0) - \beta_1 D_q x(0) = \gamma_1 x(\eta_1), \qquad \alpha_2 x(1) - \beta_2 D_q x(1) = \gamma_2 x(\eta_2), \end{cases}$$

where $f \in C([0,1] \times \mathbb{R}, \mathbb{R})$, ${}^cD_q^{\alpha}$ is the fractional q-derivative of the Caputo type, and $\alpha_i, \beta_i, \gamma_i, \eta_i \in \mathbb{R}, i = 1, 2$.

In 2013, Zhou and Liu [12] applied Mönch's fixed point theorem together with the technique of measure of weak noncompactness to investigate the existence of solutions for the following fractional q-difference equation with boundary conditions:

$$\begin{cases} {}^cD_q^\alpha u(t) + f(t,u(t)) = 0, & 0 \le t \le 1,\ 0 < q < 1, \\ u(0) = (D_q^2 u)(0) = 0, & \gamma(D_q u)(1) + \beta(D_q^2 u)(1) = 0, \end{cases}$$

where $2 < \alpha \le 3$, $\gamma, \beta \ge 0$ and $f : [0,1] \times \mathbb{R} \to \mathbb{R}$ is a continuous function.

In 2014, Ahmad et al. [13] derived some existence results for a nonlinear fractional q-difference equation with four-point nonlocal integral boundary conditions given by

$$\begin{cases} {}^cD_q^\beta({}^cD_q^\gamma + \lambda)u(t) = f(t,u(t)), & 0 \le t \le 1,\ 0 < q < 1,\ \lambda \in \mathbb{R}, \\ u(0) = aI_q^{\alpha-1}u(\eta), \quad u(1) = bI_q^{\alpha-1}u(\sigma), & a, b \in \mathbb{R}, \end{cases}$$

where $0 < \beta, \gamma \le 1$, $0 < \eta, \sigma < 1$, $\alpha > 2$, $f : [0,1] \times \mathbb{R} \to \mathbb{R}$ is a continuous function and I_q^α denotes the Riemann-Liouville fractional q-integral of order α.

Later, Niyom et al. [14] studied the following boundary value problem containing Riemann-Liouville fractional derivatives of different orders:

$$\begin{cases} (\lambda D^\alpha + (1-\lambda)D^\beta)u(t) = f(t,u(t)), & t \in [0,T],\ 1 < \alpha, \beta < 2, \\ u(0) = 0, \quad \mu D^{\gamma_1}u(T) + (1-\mu)D^{\gamma_2}u(T) = \gamma_3, & 0 < \gamma_1, \gamma_2 < \alpha - \beta, \end{cases}$$

where D^ϕ is the ordinary Riemann-Liouville fractional derivative of order $\phi \in \{\alpha, \beta, \gamma_1, \gamma_2\}$ such that $0 < \lambda \le 1$, $0 \le \mu \le 1$, $\gamma_3 \in \mathbb{R}$ and $f \in C([0,T] \times \mathbb{R}, \mathbb{R})$ for $T > 0$.

Some recent results on fractional q-difference equations equipped with different kinds of boundary conditions can be found in the papers [15–25].

Now, we recall some important results on fractional q-integro-difference equations. In [26], the authors studied a nonlocal four-point boundary value problem of nonlinear fractional q-integro-difference equations given by

$$\begin{cases} {}^cD_q^\beta({}^cD_q^\gamma + \lambda)x(t) = pf(t,x(t)) + kI_q^\xi g(t,x(t)),\ 0 \le t \le 1,\ 0 < q < 1, \\ \alpha_1 x(0) - \beta_1\left(t^{(1-\gamma)}D_q x(0)\right)\Big|_{t=0} = \sigma_1 x(\eta_1),\ \alpha_2 x(1) + \beta_2 D_q x(1) = \sigma_2 x(\eta_2), \end{cases}$$

where ${}^cD_q^\beta$ and ${}^cD_q^\gamma$ denote the fractional q-derivative of the Caputo type, $0 < \beta, \gamma \le 1$, $I_q^\xi(.)$ represents a Riemann-Liouville fractional integral of order $\xi \in (0,1)$, $f, g : [0,1] \times \mathbb{R} \to \mathbb{R}$ are continuous functions, $\lambda \ne 0$ and $p, k, \alpha_i, \beta_i, \sigma_i \in \mathbb{R}$, $\eta_i \in (0,1)$, $i = 1,2$. For some recent works on boundary value problems of fractional q-integro-difference equations, for instance, see [27–31].

Motivated by aforementioned works, in this paper, we study the following nonlinear fractional q-integro-difference equation

$$(\lambda D_q^\alpha + (1-\lambda)D_q^\beta)u(t) = af(t,u(t)) + bI_q^\delta g(t,u(t)), \quad t \in [0,1], a, b \in \mathbb{R}^+, \tag{1}$$

supplemented with q-integral boundary conditions

$$u(0) = 0,\ \mu\int_0^1 \frac{(1-qs)^{(\gamma_1-1)}}{\Gamma_q(\gamma_1)}u(s)d_qs + (1-\mu)\int_0^1 \frac{(1-qs)^{(\gamma_2-1)}}{\Gamma_q(\gamma_2)}u(s)d_qs = 0,\ \gamma_1, \gamma_2 > 0, \tag{2}$$

where $0 < q < 1$, $1 < \alpha, \beta < 2$, $0 < \delta < 1$, $0 < \lambda \le 1$, $0 \le \mu \le 1$, $\alpha - \beta > 1$ and D_q^α denotes the Riemann-Liouville fractional q-derivative of order α and $f, g : [0,1] \times \mathbb{R} \to \mathbb{R}$ are continuous functions. Notice that Equation (1) contains q-derivatives of fractional orders α and β and a fractional q-integral of orders δ, while fractional q-integrals of orders γ_1 and γ_2 are involved in the boundary conditions (2).

We make use of Krasnoselskii's fixed point theorem and a Leray-Schauder nonlinear alternative to prove the existence results, while the uniqueness result is proved via Banach contraction mapping principle for the given problem.

Let us first recall some necessary concepts and definitions about q-fractional calculus and fixed point theory.

Let $0 < q < 1$ be an arbitrary real number. For every $a \in \mathbb{R}$, the $q-$number $[a]_q$ is defined by $[a]_q = \dfrac{1-q^a}{1-q}$ [9]. In addition, the q-shifted factorial of real number a is defined by $(a;q)_0 = 1$ and $(a;q)_n = \prod_{j=0}^{n-1}(1-aq^j)$ for $n \in \mathbb{N} \cup \{\infty\}$. For $a, b \in \mathbb{R}$, the q-analogue of the power function $(a-b)^n$ with $n \in \mathbb{N}_0 := \{0, 1, 2, \ldots\}$ is given by

$$(a-b)^{(0)} = 1, \quad (a-b)^{(n)} = \prod_{j=0}^{n-1}(a - bq^j).$$

In general, if α is real number, then $(a-b)^{(\alpha)} = a^\alpha \prod_{j=0}^{\infty} \dfrac{a-bq^j}{a-bq^{\alpha+j}}$ and $a^{(\alpha)} = a^\alpha$ when $b = 0$. If $\alpha > 0$ and $0 \le a \le b \le t$, then $(t-b)^{(\alpha)} \le (t-a)^{(\alpha)}$. The q-Gamma function $\Gamma_q(\alpha)$ is defined as

$$\Gamma_q(\alpha) = \frac{(1-q)^{(\alpha-1)}}{(1-q)^{\alpha-1}}, \quad \alpha \in \mathbb{R} \setminus \{0, -1, -2, \ldots\}$$

and satisfies the relation $\Gamma_q(\alpha+1) = [\alpha]_q \Gamma_q(\alpha)$ [9].

Let $\alpha \ge 0$ and $u : (0, \infty) \to \mathbb{R}$ be a continuous function. The Riemann-Liouville fractional q-integral for the function u of order α is defined by $(I_q^0 u)(t) = u(t)$ and

$$(I_q^\alpha u)(t) = \frac{1}{\Gamma_q(\alpha)} \int_0^t (t-qs)^{(\alpha-1)} u(s) d_q s, \quad \alpha > 0$$

for $t \in (0, \infty)$, provided that the right-hand side is pointwise defined on $(0, \infty)$ [9].

Recall that $I_q^\beta I_q^\alpha u(t) = I_q^{\beta+\alpha} u(t)$ for $\alpha, \beta \in \mathbb{R}^+$ [9] and

$$I_q^\alpha t^\beta = \frac{\Gamma_q(\beta+1)}{\Gamma_q(\alpha+\beta+1)} t^{\alpha+\beta}, \ \beta \in (-1, \infty), \ \alpha \ge 0, \ t > 0.$$

If $f \equiv 1$, then $I_q^\alpha 1(t) = \dfrac{1}{\Gamma_q(\alpha+1)} t^\alpha$ for all $t > 0$.

The Riemann-Liouville fractional q-derivative of order $\alpha > 0$ for a function $u : (0, \infty) \to \mathbb{R}$ is defined by [9]

$$D_q^\alpha u(t) = \frac{1}{\Gamma_q(n-\alpha)} \int_0^t \frac{u(s)}{(t-qs)^{\alpha-n+1}} d_q s, \quad n-1 < \alpha < n.$$

Next, we state some fixed point theorems related to our work.

Lemma 1. *Let M be a closed, bounded, convex and nonempty subset of a Banach space E. Let A and B be operators mapping M into E, such that*

(i) $Ax + By \in M$, where $x, y \in M$;
(ii) A is compact and continuous;
(iii) B is a contraction mapping.

Then, there exists $z \in M$ such that $z = Az + Bz$ (Krasnoselskii's fixed point theorem [32]).

Lemma 2. *Let $\mathcal{X}$ be a closed and convex subset of a Banach space $\mathcal{E}$ and let $\mathcal{Y}$ be an open subset of $\mathcal{X}$ with $0 \in \mathcal{Y}$. Then, a continuous compact map $\mathcal{H} : \overline{\mathcal{Y}} \to \mathcal{X}$ has a fixed point in $\overline{Y}$ or there is a $y \in \partial Y$ and $\sigma \in (0,1)$ such that $y = \sigma\mathcal{H}(y)$, where $\partial\mathcal{Y}$ is the boundary of $\mathcal{Y}$ in $\mathcal{X}$ (Nonlinear alternative for single-valued maps [33]).*

2. Main Results

Let $E = C([0,1], \mathbb{R})$ be the set of continuous functions defined on $[0,1]$. The set E is a Banach space with the following norm

$$\|u\|_E = \sup_{t\in[0,1]} |u(t)|, \quad u \in E.$$

Now, we prove the following lemma which characterizes the structure of solutions for boundary value problems (1) and (2).

Lemma 3. *Let $h \in C([0,1], \mathbb{R})$ and*

$$\Delta := \frac{\mu\Gamma_q(\alpha)}{\Gamma_q(\alpha+\gamma_1)} + \frac{(1-\mu)\Gamma_q(\alpha)}{\Gamma_q(\alpha+\gamma_2)} \neq 0. \tag{3}$$

The function u is a solution for the fractional q-difference boundary value problem

$$\begin{cases} (\lambda D_q^\alpha + (1-\lambda)D_q^\beta)u(t) = h(t), \quad t \in [0,1], \\ u(0) = 0, \ \mu \int_0^1 \frac{(1-qs)^{(\gamma_1-1)}}{\Gamma_q(\gamma_1)} u(s)d_qs + (1-\mu)\int_0^1 \frac{(1-qs)^{(\gamma_2-1)}}{\Gamma_q(\gamma_2)} u(s)d_qs = 0, \end{cases} \tag{4}$$

if and only if u is a solution for the fractional q-integral equation

$$\begin{aligned} u(t) \ = \ & \frac{(\lambda-1)}{\lambda\Gamma_q(\alpha-\beta)} \int_0^t (t-qs)^{(\alpha-\beta-1)} u(s)d_qs + \frac{1}{\lambda\Gamma_q(\alpha)} \int_0^t (t-qs)^{(\alpha-1)} h(s)d_qs \\ & + \frac{t^{\alpha-1}}{\Delta}\Big[- \frac{\mu(\lambda-1)}{\lambda\Gamma_q(\alpha-\beta+\gamma_1)} \int_0^1 (1-qs)^{(\alpha-\beta+\gamma_1-1)} u(s)d_qs \\ & - \frac{\mu}{\lambda\Gamma_q(\alpha+\gamma_1)} \int_0^1 (1-qs)^{(\alpha+\gamma_1-1)} h(s)d_qs \\ & - \frac{(1-\mu)(\lambda-1)}{\lambda\Gamma_q(\alpha-\beta+\gamma_2)} \int_0^1 (1-qs)^{(\alpha-\beta+\gamma_2-1)} u(s)d_qs \\ & - \frac{(1-\mu)}{\lambda\Gamma_q(\alpha+\gamma_2)} \int_0^1 (1-qs)^{(\alpha+\gamma_2-1)} h(s)d_qs\Big]. \end{aligned} \tag{5}$$

Proof. Let u be a solution of the q-fractional boundary value problem (4). Then, we have

$$D_q^\alpha u(t) = \frac{\lambda-1}{\lambda} D_q^\beta u(t) + \frac{1}{\lambda} h(t).$$

Taking the Riemann-Liouville fractional q-integral of order α to both sides of the above equation, we get

$$u(t) = \frac{\lambda-1}{\lambda} I_q^\alpha D_q^\beta u(t) + \frac{1}{\lambda} I_q^\alpha h(t) + c_1 t^{\alpha-1} + c_2 t^{\alpha-2},$$

where $c_1, c_2 \in \mathbb{R}$ are arbitrary constants. Since $1 < \alpha < 2$, it follows from the first boundary condition that $c_2 = 0$. Thus,

$$u(t) = \frac{\lambda-1}{\lambda\Gamma_q(\alpha-\beta)} \int_0^t (t-qs)^{(\alpha-\beta-1)} u(s)d_qs + \frac{1}{\lambda\Gamma_q(\alpha)} \int_0^t (t-qs)^{(\alpha-1)} h(s)d_qs + c_1 t^{\alpha-1}. \tag{6}$$

On the other hand, if $\sigma \in \{\gamma_1, \gamma_2\}$, then we have

$$\begin{aligned} I_q^{\sigma} u(t) &= \frac{\lambda - 1}{\lambda \Gamma_q(\alpha - \beta + \sigma)} \int_0^t (t - qs)^{(\alpha - \beta + \sigma - 1)} u(s) d_q s \\ &+ \frac{1}{\lambda \Gamma_q(\alpha + \sigma)} \int_0^t (t - qs)^{(\alpha + \sigma - 1)} h(s) d_q s + c_1 \frac{\Gamma_q(\alpha)}{\Gamma_q(\alpha + \sigma)} t^{\alpha + \sigma - 1}. \end{aligned}$$

Now, by using the second boundary value condition and substituting the values $\sigma \in \{\gamma_1, \gamma_2\}$ into the above expression, we obtain

$$\begin{aligned} &\frac{\mu(\lambda - 1)}{\lambda \Gamma_q(\alpha - \beta + \gamma_1)} \int_0^1 (1 - qs)^{(\alpha - \beta + \gamma_1 - 1)} u(s) d_q s \\ &+ \frac{\mu}{\lambda \Gamma_q(\alpha + \gamma_1)} \int_0^1 (1 - qs)^{(\alpha + \gamma_1 - 1)} h(s) d_q s + c_1 \frac{\mu \Gamma_q(\alpha)}{\Gamma_q(\alpha + \gamma_1)} \\ &+ \frac{(1 - \mu)(\lambda - 1)}{\lambda \Gamma_q(\alpha - \beta + \gamma_2)} \int_0^1 (1 - qs)^{(\alpha - \beta + \gamma_2 - 1)} u(s) d_q s \\ &+ \frac{(1 - \mu)}{\lambda \Gamma_q(\alpha + \gamma_2)} \int_0^1 (1 - qs)^{(\alpha + \gamma_2 - 1)} h(s) d_q s + c_1 \frac{(1 - \mu) \Gamma_q(\alpha)}{\Gamma_q(\alpha + \gamma_2)} = 0. \end{aligned}$$

Solving the above equation for c_1, we find that

$$\begin{aligned} c_1 &= \frac{1}{\Delta} \Big[- \frac{\mu(\lambda - 1)}{\lambda \Gamma_q(\alpha - \beta + \gamma_1)} \int_0^1 (1 - qs)^{(\alpha - \beta + \gamma_1 - 1)} u(s) d_q s \\ &- \frac{\mu}{\lambda \Gamma_q(\alpha + \gamma_1)} \int_0^1 (1 - qs)^{(\alpha + \gamma_1 - 1)} h(s) d_q s \\ &- \frac{(1 - \mu)(\lambda - 1)}{\lambda \Gamma_q(\alpha - \beta + \gamma_2)} \int_0^1 (1 - qs)^{(\alpha - \beta + \gamma_2 - 1)} u(s) d_q s \\ &- \frac{(1 - \mu)}{\lambda \Gamma_q(\alpha + \gamma_2)} \int_0^1 (1 - qs)^{(\alpha + \gamma_2 - 1)} h(s) d_q s \Big], \end{aligned}$$

where Δ is defined in (3).

Substituting the value of c_1 in (6), we get the solution (5). Conversely, it is clear that u is a solution for the fractional q-difference Equation (4) whenever u is a solution for the fractional q-integral Equation (5). This completes the proof. □

In relation to the problems (1) and (2), we introduce an operator $\mathcal{T} : E \to E$ by

$$\begin{aligned}
(\mathcal{T}u)(t) &= \frac{(\lambda-1)}{\lambda\Gamma_q(\alpha-\beta)}\int_0^t (t-qs)^{(\alpha-\beta-1)}u(s)d_qs + \frac{a}{\lambda\Gamma_q(\alpha)}\int_0^t (t-qs)^{(\alpha-1)}f(s,u(s))d_qs \\
&+ \frac{b}{\lambda\Gamma_q(\alpha+\delta)}\int_0^t (t-qs)^{(\alpha+\delta-1)}g(s,u(s))d_qs \\
&+ \frac{t^{\alpha-1}}{\Delta}\Big[-\frac{\mu(\lambda-1)}{\lambda\Gamma_q(\alpha-\beta+\gamma_1)}\int_0^1 (1-qs)^{(\alpha-\beta+\gamma_1-1)}u(s)d_qs \\
&- \frac{a\mu}{\lambda\Gamma_q(\alpha+\gamma_1)}\int_0^1 (1-qs)^{(\alpha+\gamma_1-1)}f(s,u(s))d_qs \\
&- \frac{b\mu}{\lambda\Gamma_q(\alpha+\delta+\gamma_1)}\int_0^1 (1-qs)^{(\alpha+\delta+\gamma_1-1)}g(s,u(s))d_qs \\
&- \frac{(1-\mu)(\lambda-1)}{\lambda\Gamma_q(\alpha-\beta+\gamma_2)}\int_0^1 (1-qs)^{(\alpha-\beta+\gamma_2-1)}u(s)d_qs \\
&- \frac{a(1-\mu)}{\lambda\Gamma_q(\alpha+\gamma_2)}\int_0^1 (1-qs)^{(\alpha+\gamma_2-1)}f(s,u(s))d_qs \\
&- \frac{b(1-\mu)}{\lambda\Gamma_q(\alpha+\delta+\gamma_2)}\int_0^1 (1-qs)^{(\alpha+\delta+\gamma_2-1)}g(s,u(s))d_qs\Big],
\end{aligned} \tag{7}$$

where $u \in E$ and $t \in [0,1]$. In the sequel, we set

$$\begin{aligned}
\Lambda_0 &:= \frac{|\lambda-1|}{\lambda\Gamma_q(\alpha-\beta+1)} + \frac{\mu|\lambda-1|}{\lambda|\Delta|\Gamma_q(\alpha-\beta+\gamma_1+1)} + \frac{(1-\mu)|\lambda-1|}{\lambda|\Delta|\Gamma_q(\alpha-\beta+\gamma_2+1)}, \\
\Lambda_1 &:= \frac{a}{\lambda\Gamma_q(\alpha+1)} + \frac{a\mu}{\lambda|\Delta|\Gamma_q(\alpha+\gamma_1+1)} + \frac{a(1-\mu)}{\lambda|\Delta|\Gamma_q(\alpha+\gamma_2+1)}, \\
\Lambda_2 &:= \frac{b}{\lambda\Gamma_q(\alpha+\delta+1)} + \frac{b\mu}{\lambda|\Delta|\Gamma_q(\alpha+\delta+\gamma_1+1)} + \frac{b(1-\mu)}{\lambda|\Delta|\Gamma_q(\alpha+\delta+\gamma_2+1)}.
\end{aligned} \tag{8}$$

Now, we are ready to present our main results. The first existence result is based on Krasnoselskii's fixed point theorem.

Theorem 1. *Suppose that $f, g : [0,1] \times \mathbb{R} \to \mathbb{R}$ are continuous functions satisfying the following conditions:*

(i) *there exists a positive constant L such that for each $u, v \in \mathbb{R}$,*

$$|f(t,u) - f(t,v)| \le L|u-v|, \quad t \in [0,1];$$

(ii) *For each $u \in \mathbb{R}$, there exists a continuous function m on $[0,1]$ such that*

$$|g(t,u)| \le m(t), \quad t \in [0,1].$$

If $\Lambda_0 + L\Lambda_1 < 1$, then the fractional q-integro-difference Equation (1) with q-integral boundary conditions (2) has at least one solution on $[0,1]$, where Λ_0 and Λ_1 are defined by (8).

Proof. Let $\|m\| = \sup_{t\in[0,1]} |m(t)|$. Define $B_r := \{u \in E : \|u\| \le r\}$ with

$$r \ge \frac{\|m\|\Lambda_2 + K\Lambda_1}{1-(\Lambda_0 + L\Lambda_1)}, \tag{9}$$

where $K := \sup_{t\in[0,1]} |f(t,0)|$ and Λ_1 and Λ_2 are given by (8). Clearly, B_r is a closed, bounded, convex and nonempty subset of Banach space E. We consider the operator $\mathcal{T} : E \to E$ as (7). By Lemma 3, it is obvious that the fixed point of $\mathcal{T}$ is the solution of problems (1) and (2). Now, for each $t \in [0,1]$, we define two operators from B_r to E as follows:

$$\begin{aligned}
\mathcal{T}_1 u(t) &= \frac{(\lambda-1)}{\lambda\Gamma_q(\alpha-\beta)}\int_0^t (t-qs)^{(\alpha-\beta-1)}u(s)d_qs + \frac{a}{\lambda\Gamma_q(\alpha)}\int_0^t (t-qs)^{(\alpha-1)}f(s,u(s))d_qs \\
&\quad + \frac{t^{\alpha-1}}{\Delta}\Big[-\frac{\mu(\lambda-1)}{\lambda\Gamma_q(\alpha-\beta+\gamma_1)}\int_0^1 (1-qs)^{(\alpha-\beta+\gamma_1-1)}u(s)d_qs \\
&\quad - \frac{a\mu}{\lambda\Gamma_q(\alpha+\gamma_1)}\int_0^1 (1-qs)^{(\alpha+\gamma_1-1)}f(s,u(s))d_qs \\
&\quad - \frac{(1-\mu)(\lambda-1)}{\lambda\Gamma_q(\alpha-\beta+\gamma_2)}\int_0^1 (1-qs)^{(\alpha-\beta+\gamma_2-1)}u(s)d_qs \\
&\quad - \frac{a(1-\mu)}{\lambda\Gamma_q(\alpha+\gamma_2)}\int_0^1 (1-qs)^{(\alpha+\gamma_2-1)}f(s,u(s))d_qs\Big]
\end{aligned}$$

and

$$\begin{aligned}
\mathcal{T}_2 u(t) &= \frac{b}{\lambda\Gamma_q(\alpha+\delta)}\int_0^t (t-qs)^{(\alpha+\delta-1)}g(s,u(s))d_qs \\
&\quad + \frac{t^{\alpha-1}}{\Delta}\Big[\frac{b\mu}{\lambda\Gamma_q(\alpha+\delta+\gamma_1)}\int_0^1 (1-qs)^{(\alpha+\delta+\gamma_1-1)}g(s,u(s))d_qs \\
&\quad - \frac{b(1-\mu)}{\lambda\Gamma_q(\alpha+\delta+\gamma_2)}\int_0^1 (1-qs)^{(\alpha+\delta+\gamma_2-1)}g(s,u(s))d_qs\Big].
\end{aligned}$$

By the condition (H_1), we have that $|f(t,u(t))| \le |f(t,u(t)) - f(t,0)| + |f(t,0)| \le L\|u\| + K \le Lr + K$ for any $u \in \mathbb{R}$ and $t \in [0,1]$. Thus, for any $u, v \in B_r$ and $t \in [0,1]$, it follows by means of (8) and (9) that

$$\begin{aligned}
|\mathcal{T}_1 u(t) + \mathcal{T}_2 v(t)| &\le \frac{|\lambda-1|}{\lambda\Gamma_q(\alpha-\beta+1)}\|u\| + \frac{a}{\lambda\Gamma_q(\alpha+1)}\Big(L\|u\|+K\Big) \\
&\quad + \frac{\mu|\lambda-1|}{\lambda|\Delta|\Gamma_q(\alpha-\beta+\gamma_1+1)}\|u\| + \frac{a\mu}{\lambda|\Delta|\Gamma_q(\alpha+\gamma_1+1)}\Big(L\|u\|+K\Big) \\
&\quad + \frac{(1-\mu)|\lambda-1|}{\lambda|\Delta|\Gamma_q(\alpha-\beta+\gamma_2+1)}\|u\| + \frac{a(1-\mu)}{\lambda|\Delta|\Gamma_q(\alpha+\gamma_2+1)}\Big(L\|u\|+K\Big) \\
&\quad + \frac{b}{\lambda\Gamma_q(\alpha+\delta+1)}\|m\| + \frac{b\mu}{\lambda|\Delta|\Gamma_q(\alpha+\delta+\gamma_1+1)}\|m\| \\
&\quad + \frac{b(1-\mu)}{\lambda|\Delta|\Gamma_q(\alpha+\delta+\gamma_2+1)}\|m\| \\
&\le (\Lambda_0 + L\Lambda_1)r + \Lambda_2\|m\| + \Lambda_1 K \le r,
\end{aligned}$$

which implies that $\|\mathcal{T}_1 u + \mathcal{T}_2 v\| \le r$ and so $\mathcal{T}_1 u + \mathcal{T}_2 v \in B_r$ for all $u, v \in B_r$.

Now, we prove that $\mathcal{T}_2$ is continuous. Let $\{u_n\}_{n\ge 1}$ be a sequence in B_r such that $u_n \to u$. Then, for each $t \in [0,1]$, we have

$$\begin{aligned}
|\mathcal{T}_2 u_n(t) - \mathcal{T}_2 u(t)| &\le \frac{b}{\lambda\Gamma_q(\alpha+\delta+1)}|g(s,u_n(s)) - g(s,u(s))| \\
&\quad + \frac{b\mu}{\lambda|\Delta|\Gamma_q(\alpha+\delta+\gamma_1+1)}|g(s,u_n(s)) - g(s,u(s))| \\
&\quad + \frac{b(1-\mu)}{\lambda|\Delta|\Gamma_q(\alpha+\delta+\gamma_2+1)}|g(s,u_n(s)) - g(s,u(s))|.
\end{aligned}$$

Since g is continuous, we get $\|\mathcal{T}_2 u_n - \mathcal{T}_2 u\| \to 0$ as $u_n \to u$. In consequence, it follows that the operator $\mathcal{T}_2$ is continuous on B_r.

In the next step, we show that the operator $\mathcal{T}_2$ is compact. Let us first show that $\mathcal{T}_2$ is uniformly bounded. For each $u \in B_r$ and $t \in [0,1]$, we have

$$\begin{aligned}
|\mathcal{T}_2 u(t)| &\leq \frac{bt^{\alpha+\delta}}{\lambda\Gamma_q(\alpha+\delta+1)}|g(s,u(s))| + \frac{b\mu t^{\alpha-1}}{\lambda|\Delta|\Gamma_q(\alpha+\delta+\gamma_1+1)}|g(s,u(s))| \\
&\quad + \frac{b(1-\mu)t^{\alpha-1}}{\lambda|\Delta|\Gamma_q(\alpha+\delta+\gamma_2+1)}|g(s,u(s))| \\
&\leq \|m\|\Big[\frac{b}{\lambda\Gamma_q(\alpha+\delta+1)} + \frac{b\mu}{\lambda|\Delta|\Gamma_q(\alpha+\delta+\gamma_1+1)} + \frac{b(1-\mu)}{\lambda|\Delta|\Gamma_q(\alpha+\delta+\gamma_2+1)}\Big] \\
&= \Lambda_2\|m\|,
\end{aligned}$$

which implies that $\|\mathcal{T}_2 u\| \leq \Lambda_2\|m\|$.

In order to establish the equicontinuity of the operator $\mathcal{T}_2$, we assume that $t_1, t_2 \in [0,1]$ such that $t_2 > t_1$. We will show that $\mathcal{T}_2$ maps bounded sets into equicontinuous sets. For each $u \in B_r$, we have

$$\begin{aligned}
|\mathcal{T}_2 u(t_2) - \mathcal{T}_2 u(t_1)| &\leq \frac{b}{\lambda\Gamma_q(\alpha+\delta)}\int_0^{t_1}[(t_2-qs)^{(\alpha+\delta-1)} - (t_1-qs)^{(\alpha+\delta-1)}]|g(s,u(s))|d_qs \\
&\quad + \frac{b}{\lambda\Gamma_q(\alpha+\delta)}\int_{t_1}^{t_2}(t_2-qs)^{(\alpha+\delta-1)}|g(s,u(s))|d_qs \\
&\quad + \frac{|t_2^{\alpha-1} - t_1^{\alpha-1}|}{|\Delta|}\Big[\frac{b\mu}{\lambda\Gamma_q(\alpha+\delta+\gamma_1)}\int_0^1(1-qs)^{(\alpha+\delta+\gamma_1-1)}|g(s,u(s))|d_qs \\
&\quad + \frac{b(1-\mu)}{\lambda\Gamma_q(\alpha+\delta+\gamma_2)}\int_0^1(1-qs)^{(\alpha+\delta+\gamma_2-1)}|g(s,u(s))|d_qs\Big] \\
&\leq \|m\|\Big[\frac{2b(t_2-t_1)^{\alpha+\delta} + b|t_2^{\alpha+\delta} - t_1^{\alpha+\delta}|}{\lambda\Gamma_q(\alpha+\delta+1)} + \frac{b\mu|t_2^{\alpha-1} - t_1^{\alpha-1}|}{\lambda|\Delta|\Gamma_q(\alpha+\delta+\gamma_1+1)} \\
&\quad + \frac{b(1-\mu)|t_2^{\alpha-1} - t_1^{\alpha-1}|}{\lambda|\Delta|\Gamma_q(\alpha+\delta+\gamma_2+1)}\Big].
\end{aligned}$$

Observe that the right-hand side of the above inequality is independent of $u \in B_r$ and tends to zero as $t_1 \to t_2$. This shows that $\mathcal{T}_2$ is equicontinuous. Therefore, the operator $\mathcal{T}_2$ is relatively compact on B_r and the Arzelá-Ascoli theorem implies that $\mathcal{T}_2$ is completely continuous and so $\mathcal{T}_2$ is compact operator on B_r.

Finally, we prove that the operator $\mathcal{T}_1$ is a contraction. For any $u, v \in B_r$ and $t \in [0,1]$, we obtain

$$\begin{aligned}
|\mathcal{T}_1 u(t) - \mathcal{T}_1 v(t)| &\leq \frac{|\lambda-1|}{\lambda\Gamma_q(\alpha-\beta+1)}|u(s)-v(s)| + \frac{a}{\lambda\Gamma_q(\alpha+1)}L|u(s)-v(s)| \\
&\quad + \frac{\mu|\lambda-1|}{\lambda|\Delta|\Gamma_q(\alpha-\beta+\gamma_1+1)}|u(s)-v(s)| + \frac{a\mu}{\lambda|\Delta|\Gamma_q(\alpha+\gamma_1+1)}L|u(s)-v(s)| \\
&\quad + \frac{(1-\mu)|\lambda-1|}{\lambda|\Delta|\Gamma_q(\alpha-\beta+\gamma_2+1)}|u(s)-v(s)| + \frac{a(1-\mu)}{\lambda|\Delta|\Gamma_q(\alpha+\gamma_2+1)}L|u(s)-v(s)| \\
&\leq (\Lambda_0 + L\Lambda_1)\|u-v\|.
\end{aligned}$$

Since $\Lambda_0 + L\Lambda_1 < 1$, $\mathcal{T}_1$ is a contraction. Thus, all the assumptions of Lemma 1 are satisfied. Therefore, the fractional q-integro-difference Equation (1) with q-integral boundary conditions (2) has at least one solution on $[0,1]$ and the proof is completed. □

In the following result, we prove the existence of solutions for the problem (1) and (2) by means of a Leray-Schauder nonlinear alternative.

Theorem 2. *Let $f : [0,1] \times \mathbb{R} \to \mathbb{R}$ be a continuous function satisfying the conditions:*

(H_3) *there exist continuous nondecreasing functions $\psi_1, \psi_2 : [0,\infty) \to (0,\infty)$ and functions $\phi_1, \phi_2 \in C([0,1], \mathbb{R}^+)$ such that $|f(t,u)| \leq \phi_1(t)\psi_1(|u|)$ and $|g(t,u)| \leq \phi_2(t)\psi_2(|u|)$ for each $(t,u) \in [0,1] \times \mathbb{R}$;*

(H_4) *there exists a constant $\Xi > 0$ such that*

$$\frac{(1-\Lambda_0)\Xi}{\Lambda_1\|\phi_1\|\psi_1(\Xi) + \Lambda_2\|\phi_2\|\psi_2(\Xi)} > 1, \ \ \Lambda_0 < 1,$$

where $\Lambda_0, \Lambda_1, \Lambda_2$ are defined by (8).

Then, the fractional q-integro-difference Equation (1) with q-integral boundary conditions (2) has at least one solution on $[0,1]$.

Proof. We verify the hypothesis of a Leray-Schauder nonlinear alternative (Lemma 2) in several steps. Let us first show that the operator $\mathcal{T}$, defined by (7), *maps bounded sets (balls) into bounded sets in E*. For a positive number R, let $B_R = \{u \in E : \|u\| \leq R\}$ be a bounded ball in E. Then, for $t \in [0,1]$, we have

$$\begin{aligned}
|\mathcal{T}u(t)| \leq\ & \frac{|\lambda-1|}{\lambda\Gamma_q(\alpha-\beta)}\int_0^t (t-qs)^{(\alpha-\beta-1)}\|u\|d_qs + \frac{a}{\lambda\Gamma_q(\alpha)}\int_0^t (t-qs)^{(\alpha-1)}\|\phi_1\|\psi_1(\|u\|)d_qs \\
& + \frac{b}{\lambda\Gamma_q(\alpha+\delta)}\int_0^t (t-qs)^{(\alpha+\delta-1)}\|\phi_2\|\psi_2(\|u\|)d_qs \\
& + \frac{t^{\alpha-1}}{|\Delta|}\Big[\frac{\mu|\lambda-1|}{\lambda\Gamma_q(\alpha-\beta+\gamma_1)}\int_0^1 (1-qs)^{(\alpha-\beta+\gamma_1-1)}\|u\|d_qs \\
& + \frac{a\mu}{\lambda\Gamma_q(\alpha+\gamma_1)}\int_0^1 (1-qs)^{(\alpha+\gamma_1-1)}\|\phi_1\|\psi_1(\|u\|)d_qs \\
& + \frac{b\mu}{\lambda\Gamma_q(\alpha+\delta+\gamma_1)}\int_0^1 (1-qs)^{(\alpha+\delta+\gamma_1-1)}\|\phi_2\|\psi_2(\|u\|)d_qs \\
& + \frac{(1-\mu)|\lambda-1|}{\lambda\Gamma_q(\alpha-\beta+\gamma_2)}\int_0^1 (1-qs)^{(\alpha-\beta+\gamma_2-1)}\|u\|d_qs \\
& + \frac{a(1-\mu)}{\lambda\Gamma_q(\alpha+\gamma_2)}\int_0^1 (1-qs)^{(\alpha+\gamma_2-1)}\|\phi_1\|\psi_1(\|u\|)d_qs \\
& + \frac{b(1-\mu)}{\lambda\Gamma_q(\alpha+\delta+\gamma_2)}\int_0^1 (1-qs)^{(\alpha+\delta+\gamma_2-1)}\|\phi_2\|\psi_2(\|u\|)d_qs\Big] \\
\leq\ & \lambda_0\|u\| + \Lambda_1\|\phi_1\|\psi_1(\|u\|) + \Lambda_2\|\phi_2\|\psi_1(\|u\|).
\end{aligned}$$

Therefore,

$$\|\mathcal{T}u\| \leq \lambda_0 R + \Lambda_1\|\phi_1\|\psi_1(R) + \Lambda_2\|\phi_2\|\psi_1(R).$$

Secondly, we show that $\mathcal{T}$ *maps bounded sets into equicontinuous sets of E*. Let $t_1, t_2 \in [0,1]$ with $t_1 < t_2$ and $u \in B_R$. Then, we have

$$\begin{aligned}
& |\mathcal{T}u(t_2) - \mathcal{T}u(t_1)| \\
\leq\ & \frac{|\lambda-1|R}{\lambda\Gamma_q(\alpha-\beta)}\left[\int_0^{t_1}[(t_2-qs)^{(\alpha-\beta-1)} - (t_1-qs)^{(\alpha-\beta-1)}]d_qs + \int_{t_1}^{t_2}(t_2-qs)^{(\alpha-\beta-1)}d_qs\right] \\
& + \frac{a\|\phi_1\|\psi_1(R)}{\lambda\Gamma_q(\alpha)}\left[\int_0^{t_1}[(t_2-qs)^{(\alpha-1)} - (t_1-qs)^{(\alpha-1)}]d_qs + \int_{t_1}^{t_2}(t_2-qs)^{(\alpha-1)}d_qs\right] \\
& + \frac{|t_2^{\alpha-1} - t_1^{\alpha-1}|}{|\Delta|}\Big[\frac{\mu|\lambda-1|R}{\lambda\Gamma_q(\alpha-\beta+\gamma_1+1)} + \frac{a\mu\|\phi_1\|\psi_1(R)}{\lambda\Gamma_q(\alpha+\gamma_1+1)}
\end{aligned}$$

$$
\begin{aligned}
&+\frac{(1-\mu)|\lambda-1|R}{\lambda\Gamma_q(\alpha-\beta+\gamma_2+1)}+\frac{a(1-\mu)\|\phi_1\|\psi_1(R)}{\lambda\Gamma_q(\alpha+\gamma_2+1)}\Bigg]\\
&+\frac{b\|\phi_2\psi_2(R)}{\lambda\Gamma_q(\alpha+\delta)}\left[\int_0^{t_1}[(t_2-qs)^{(\alpha+\delta-1)}-(t_1-qs)^{(\alpha+\delta-1)}]d_qs+\int_{t_1}^{t_2}(t_2-qs)^{(\alpha+\delta-1)}d_qs\right]\\
&+\frac{|t_2^{\alpha-1}-t_1^{\alpha-1}|}{|\Delta|}\left[\frac{b\mu\|\phi_2\psi_2(R)}{\lambda\Gamma_q(\alpha+\delta+\gamma_1+1)}+\frac{b(1-\mu)\|\phi_2\psi_2(R)}{\lambda\Gamma_q(\alpha+\delta+\gamma_2+1)}\right]\\
\leq\ &\frac{|\lambda-1|R}{\lambda\Gamma_q(\alpha-\beta+1)}[2(t_2-t_1)^{\alpha-\beta}+|t_2^{\alpha-\beta}-t_1^{\alpha-\beta}|]+\frac{a\|\phi_1\|\psi_1(R)}{\lambda\Gamma_q(\alpha+1)}[2(t_2-t_1)^{\alpha}+|t_2^{\alpha}-t_1^{\alpha}|]\\
&+\frac{|t_2^{\alpha-1}-t_1^{\alpha-1}|}{|\Delta|}\Bigg[\frac{\mu|\lambda-1|R}{\lambda\Gamma_q(\alpha-\beta+\gamma_1+1)}+\frac{a\mu\|\phi_1\|\psi_1(R)}{\lambda\Gamma_q(\alpha+\gamma_1+1)}\\
&+\frac{(1-\mu)|\lambda-1|R}{\lambda\Gamma_q(\alpha-\beta+\gamma_2+1)}+\frac{a(1-\mu)\|\phi_1\|\psi_1(R)}{\lambda\Gamma_q(\alpha+\gamma_2+1)}\Bigg]\\
&+\frac{b\|\phi_2\|\psi(R)}{\lambda\Gamma_q(\alpha+\delta+1)}[2(t_2-t_1)^{\alpha+\delta}+|t_2^{\alpha+\delta}-t_1^{\alpha+\delta}|]\\
&+\frac{b\|\phi_2\|\psi(R)\mu|t_2^{\alpha-1}-t_1^{\alpha-1}|}{\lambda|\Delta|\Gamma_q(\alpha+\delta+\gamma_1+1)}+\frac{b(1-\mu)\|\phi_2\|\psi(R)|t_2^{\alpha-1}-t_1^{\alpha-1}|}{\lambda|\Delta|\Gamma_q(\alpha+\delta+\gamma_2+1)}\Bigg]\\
&\longrightarrow 0 \text{ as } t_2\to t_1 \text{ independent of } u\in B_R.
\end{aligned}
$$

Thus, the Arzelá-Ascoli theorem applies and hence $\mathcal{T}:E\to E$ is completely continuous.

In the last step, we show that all solutions to the equation $u=\theta\mathcal{T}u$ are bounded for $\theta\in[0,1]$. For that, let u be a solution of $u=\theta\mathcal{T}u$ for $\theta\in[0,1]$. Then, for $t\in[0,1]$, we apply the strategy used in the first step to obtain

$$\|u\|\leq\Lambda_0\|u\|+\Lambda_1\|\phi_1\|\psi_1(\|u\|)+\Lambda_2\|\phi_2\|\psi_1(\|u\|).$$

Consequently, we have

$$\frac{(1-\Lambda_0)\|u\|}{\Lambda_1\|\phi_1\|\psi_1(\|u\|)+\Lambda_2\|\phi_2\|\psi_1(\|u\|)}\leq 1.$$

By the condition (H_4), we can find a positive number Ξ such that $\|u\|\neq\Xi$. Introduce a set

$$U=\{u\in E:\|u\|<\Xi\},\tag{10}$$

and observe that the operator $\mathcal{T}:\overline{U}\to E$ is continuous and completely continuous. With this choice of U, we cannot find $u\in\partial U$ satisfying the relation $u=\theta\mathcal{T}x$ for some $\theta\in(0,1)$. Therefore, it follows by a nonlinear alternative of the Leray-Schauder type (Lemma 2) that the operator $\mathcal{T}$ has a fixed point in $\overline{U}$. Thus, there exists a solution of problems (1) and (2) on $[0,1]$. The proof is complete. □

In our final result, the uniqueness of solutions for the given problem is shown with the aid of a Banach contraction mapping principle [34].

Theorem 3. *Let $f:[0,1]\times\mathbb{R}\to\mathbb{R}$ be a function satisfying the assumption (H_1). In addition, assume that the function $g:[0,1]\times\mathbb{R}\to\mathbb{R}$ satisfies the condition*

(H_5) *there exists a positive constant M such that, for each $u,v\in\mathbb{R}$,*

$$|g(t,u)-g(t,v)|\leq M|u-v|,\quad t\in[0,1].$$

Then, the fractional q-integro-difference Equation (1) with q-integral boundary conditions (2) has a unique solution on $[0,1]$, provided that $\Lambda_0 + L\Lambda_1 + M\Lambda_2 < 1$, where $\Lambda_0, \Lambda_1, \Lambda_2$ are defined by (8).

Proof. By a Banach contraction mapping principle, we will show that the operator $\mathcal{T} : E \to E$ defined by (7) has a unique fixed point which corresponds to the unique solution of problems (1) and (2). Setting $\sup_{t\in[0,1]} |f(t,0)| = K < \infty$ and $\sup_{t\in[0,1]} |g(t,0)| = N < \infty$ and selecting

$$r \geq \frac{N\Lambda_2 + K\Lambda_1}{1 - (\Lambda_0 + L\Lambda_1 + M\Lambda_2)},$$

we show that $\mathcal{T}B_r \subset B_r$, where $B_r = \{u \in E : \|u\| \leq r\}$. For any $u \in B_r$, following the arguments used in the proof of Theorem 1, one can obtain

$$\|\mathcal{T}u\| \leq (\Lambda_0 + L\Lambda_1 + M\Lambda_2)r + \Lambda_2 N + \Lambda_1 K < r,$$

which implies that $\mathcal{T}B_r \subset B_r$. For any $t \in [0,1]$ and any $u, v \in \mathbb{R}$, we obtain

$$\begin{aligned}
\|(\mathcal{T}u) - (\mathcal{T}v)\| \leq\ & \frac{|\lambda - 1|}{\lambda\Gamma_q(\alpha - \beta)} \int_0^t (t - qs)^{(\alpha-\beta-1)} |u(s) - v(s)| d_q s \\
& + \frac{a}{\lambda\Gamma_q(\alpha)} \int_0^t (t - qs)^{(\alpha-1)} |f(s,u(s)) - f(s,v(s))| d_q s \\
& + \frac{b}{\lambda\Gamma_q(\alpha+\delta)} \int_0^t (t - qs)^{(\alpha+\delta-1)} |g(s,u(s)) - g(s,v(s))| d_q s \\
& + \frac{t^{\alpha-1}}{|\Delta|} \Big[\frac{\mu(\lambda - 1)}{\lambda\Gamma_q(\alpha - \beta + \gamma_1)} \int_0^1 (1 - qs)^{(\alpha-\beta+\gamma_1-1)} |u(s) - v(s)| d_q s \\
& + \frac{a\mu}{\lambda\Gamma_q(\alpha+\gamma_1)} \int_0^1 (1 - qs)^{(\alpha+\gamma_1-1)} |f(s,u(s)) - f(s,v(s))| d_q s \\
& + \frac{b\mu}{\lambda\Gamma_q(\alpha+\delta+\gamma_1)} \int_0^1 (1 - qs)^{(\alpha+\delta+\gamma_1-1)} |g(s,u(s)) - g(s,v(s))| d_q s \\
& + \frac{(1-\mu)|\lambda - 1|}{\lambda\Gamma_q(\alpha - \beta + \gamma_2)} \int_0^1 (1 - qs)^{(\alpha-\beta+\gamma_2-1)} |u(s) - v(s)| d_q s \\
& + \frac{a(1-\mu)}{\lambda\Gamma_q(\alpha+\gamma_2)} \int_0^1 (1 - qs)^{(\alpha+\gamma_2-1)} |f(s,u(s)) - f(s,v(s))| d_q s \\
& + \frac{b(1-\mu)}{\lambda\Gamma_q(\alpha+\delta+\gamma_2)} \int_0^1 (1 - qs)^{(\alpha+\delta+\gamma_2-1)} |g(s,u(s)) - g(s,v(s))| d_q s \Big] \\
\leq\ & (\Lambda_0 + L\Lambda_1 + M\Lambda_2)\|u - v\|.
\end{aligned}$$

As $\Lambda_0 + L\Lambda_1 + M\Lambda_2 < 1$, therefore $\mathcal{T}$ is a contraction. Hence, we deduce by the conclusion of the Banach contraction mapping principle that the operator $\mathcal{T}$ has a unique fixed point, which is the unique solution of problems (1) and (2). The proof is completed. □

3. Examples

I. Illustration of Theorem 1

Example 1. *Consider the fractional q-integro-difference equation*

$$(0.9D_{0.5}^{1.5} + (1-0.9)D_{0.5}^{1.01})u(t) = 0.2\frac{|90t\sin(u(t))|}{100(|\sin(u(t))|+1)} + 0.3I_{0.5}^{0.35}\frac{\sin t|u^3(t)|}{(2+t)^3(1+|u^3(t)|)}, \quad (11)$$

subject to q-integral boundary conditions

$$u(0)=0,\ \ 0.1\int_0^1 \frac{(1-qs)^{(0.1-1)}}{\Gamma_q(0.1)}u(s)d_qs+(1-0.1)\int_0^1 \frac{(1-qs)^{(0.1-1)}}{\Gamma_q(0.1)}u(s)d_qs=0. \tag{12}$$

Here, $\alpha=1.5$, $q=0.5$, $\beta=1.01$, $a=0.2$, $b=0.3$, $\delta=0.35$, $\lambda=0.9$, $\mu=0.1$, $\gamma_1=\gamma_2=0.1$, $t\in[0,1]$ *and* $f,g:[0,1]\times\mathbb{R}\to\mathbb{R}$ *are*

$$f(t,u)=\frac{|90t\sin u|}{100(|\sin u|+1)},\qquad g(t,u)=\frac{\sin t|u^3|}{(2+t)^3(1+|u^3|)}.$$

For each $u,v\in\mathbb{R}$, *notice that* $|f(t,u)-f(t,v)|\le L|u-v|$ *with* $L=0.9>0$. *On the other hand, there exists a continuous function* $m(t)=\dfrac{1}{(2+t)^3}$ *on* $[0,1]$ *such that* $|g(t,u)|\le m(t)$ *for all* $u\in\mathbb{R}$. *In addition, we have* $\|m\|=\sup_{t\in[0,1]}m(t)=0.125$. *Using the given values, it is found that* $\Delta=0.9935$ *and* $\Lambda_0+L\Lambda_1=0.5567<1$. *Clearly, all the assumptions of Theorem 1 are satisfied. Therefore, the conclusion of Theorem 1 implies that the fractional q-integro-difference Equation (11) with q-integral boundary conditions (12) has at least one solution on* $[0,1]$.

II. Illustration of Theorem 2

Example 2. *We consider the fractional q-integro-difference equation*

$$(0.9D_{0.5}^{1.5}+(1-0.9)D_{0.5}^{1.01})u(t)=\tfrac{0.2}{\sqrt{256+t^2}}\Big(\sin u(t)+\tfrac{|u(t)|}{1+|u(t)|}\Big)+0.3I_{0.5}^{0.35}\tfrac{1}{4+t}\Big(\tfrac{1}{2}+\tfrac{|\arcsin u(t)|}{1+|\arcsin u(t)|}\Big) \tag{13}$$

supplemented with q-integral boundary conditions

$$u(0)=0,\ \ 0.1\int_0^1 \frac{(1-qs)^{(0.1-1)}}{\Gamma_q(0.1)}u(s)d_qs+(1-0.1)\int_0^1 \frac{(1-qs)^{(0.1-1)}}{\Gamma_q(0.1)}u(s)d_qs=0, \tag{14}$$

where $\alpha=1.5$, $q=0.5$, $\beta=1.01$, $a=0.2$, $b=0.3$, $\delta=0.35$, $\lambda=0.9$, $\mu=0.1$, $\gamma_1=\gamma_2=0.1$, $t\in[0,1]$ *and*

$$f(t,u(t))=\frac{1}{\sqrt{256+t^2}}\Big(\sin u(t)+\frac{|u(t)|}{1+|u(t)|}\Big),\ g(t,u(t))=\frac{1}{4+t}\Big(\frac{1}{2}+\frac{|\arcsin u(t)|}{1+|\arcsin u(t)|}\Big).$$

Obviously,

$$|f(t,u(t))|\le\frac{1}{\sqrt{256+t^2}}(1+\|u\|),\ |g(t,u(t))|\le\frac{1}{4+t}(1+\|u\|),$$

with $\phi_1(t)=\dfrac{1}{\sqrt{256+t^2}}$, $\phi_2(t)=\dfrac{1}{4+t}$ *and* $\psi_1(\|u\|)=\psi_2(\|u\|)=1+\|u\|$. *Note that* $\|\phi_1\|=\dfrac{1}{16}=0.0625$, $\|\phi_2\|=\dfrac{1}{4}=0.25$ *and* $\psi_1(\Xi)=\psi_2(\Xi)=1+\Xi$. *Using the given data, we find that* $\Delta=0.9935$, $\Lambda_0=0.2414<1$, $\Lambda_1=0.3504$, *and* $\Lambda_2=0.4685$. *Then, by condition* (H_4), *we get* $\Xi>0.22438$. *Thus, all the assumptions of Theorem 2 are satisfied. Therefore, by Theorem 2, problems (13) and (14) have at least one solution on* $[0,1]$.

III. Illustration of Theorem 3

Example 3. *Let us consider the fractional q-integro-difference equation*

$$(0.9D_{0.5}^{1.5}+(1-0.9)D_{0.5}^{1.01})u(t)=0.2\frac{0.9t|\sin(\frac{\pi}{2}(t))||u(t)|}{5+|u(t)|}+0.3I_{0.5}^{0.35}\frac{80|\cos(\pi t)||u(t)|}{100(7+|u(t)|)} \tag{15}$$

with q-integral boundary conditions

$$u(0) = 0, \;\; 0.1\int_0^1 \frac{(1-qs)^{(0.1-1)}}{\Gamma_q(0.1)}u(s)d_qs + (1-0.1)\int_0^1 \frac{(1-qs)^{(0.1-1)}}{\Gamma_q(0.1)}u(s)d_qs = 0, \tag{16}$$

where $\alpha = 1.5, q = 0.5, \beta = 1.01, a = 0.2, b = 0.3, \delta = 0.35, \lambda = 0.9, \mu = 0.1, \gamma_1 = \gamma_2 = 0.1, t \in [0,1]$ *and*

$$f(t,u(t)) = \frac{0.9t|\sin(\frac{\pi}{2}(t))||u(t)|}{5+|u(t)|}, \;\; g(t,u(t)) = \frac{80|\cos(\pi t)||u(t)|}{100(7+|u(t)|)}.$$

Then, $L = 9/10, M = 8/10$ *as*

$$|f(t,u(t)) - f(t,v(t))| \le \frac{9}{10}(|u(t)-v(t)|), \; |g(t,u(t)) - g(t,v(t))| \le \frac{8}{10}(|u(t)-v(t)|).$$

With the given data, it is found that $\Delta = 0.9935$, $\Lambda_0 + L\Lambda_1 + M\Lambda_2 = 0.9315 < 1$. *Clearly, the assumptions of Theorem 3 hold. Thus, by the conclusion of Theorem 3, problems (15) and (16) have a unique solution* $[0,1]$.

4. Conclusions

We have derived some new existence and uniqueness results for a nonlinear fractional q-integro-difference equation equipped with q-integral boundary conditions. The obtained results significantly contribute to the literature on boundary value problems of fractional q-integro-difference equations and yield several new results as special cases. Some of these results are listed below.

(a) By letting $\lambda = 1/2$ in the results of this paper, we obtain the ones for a nonlinear fractional q-integro-difference equation of the form:

$$\left(D_q^\alpha + D_q^\beta\right)u(t) = 2af(t,u(t)) + 2bI_q^\delta g(t,u(t)), \quad t \in [0,1], a,b \in \mathbb{R}^+.$$

(b) For $\mu = 1/2$, our results correspond to the following boundary conditions:

$$u(0) = 0, \quad \int_0^1 \Big[\frac{(1-qs)^{(\gamma_1-1)}}{\Gamma_q(\gamma_1)} + \frac{(1-qs)^{(\gamma_2-1)}}{\Gamma_q(\gamma_2)}\Big]u(s)d_qs = 0, \;\; \gamma_1, \gamma_2 > 0.$$

(c) Our results with $a = 0$ and $b = 0$ correspond to the ones with purely integral nonlinearity and purely non-integral nonlinearity, respectively.

Author Contributions: Formal Analysis, S.E., S.K.N. and B.A.

Funding: This research received no external funding.

Acknowledgments: The authors acknowledge the reviewers for their constructive remarks on our work.

Conflicts of Interest: The authors declare no conflict of interest.

References

1. Kilbas, A.A.; Srivastava, H.M.; Trujillo, J.J. *Theory and Applications of Fractional Differential Equations*; North-Holland Mathematics Studies; Elsevier: Amsterdam, The Netherlands, 2003; Volume 204.
2. Ahmad, B.; Sivasundaram, S. On four-point nonlocal boundary value problems of nonlinear integro-differential equations of fractional order. *Appl. Math. Comput.* **2010**, *217*, 480–487. [CrossRef]
3. Agarwal, R.P.; Ntouyas, S.K.; Ahmad, B.; Alhothuali, M. Existence of solutions for integro-differential equations of fractional order with nonlocal three-point fractional boundary conditions. *Adv. Differ. Equ.* **2013**, *2013*, 128. [CrossRef]
4. Baleanu, D.; Rezapour, S.; Etemad, S.; Alsaedi, A. On a time-fractional integro-differential equation via three-point boundary value conditions. *Math. Probl. Eng.* **2015**, *2015*, 785738. [CrossRef]

5. Ntouyas, S.K.; Etemad, S. On the existence of solutions for fractional differential inclusions with sum and integral boundary conditions. *Appl. Math. Comp.* **2015**, *266*, 235–243. [CrossRef]
6. Ahmad, B.; Alsaedi, A.; Ntouyas, S.K.; Tariboon, J. *Hadamard-Type Fractional Differential Equations, Inclusions and Inequalities*; Springer International Publishing AG: London, UK, 2017.
7. Ahmad, B.; Alghanmi, M.; Ntouyas, S.K.; Alsaedi, A. Fractional differential equations involving generalized derivative with Stieltjes and fractional integral boundary conditions. *Appl. Math. Lett.* **2018**, *84*, 111–117. [CrossRef]
8. Ahmad, B.; Alruwaily, Y.; Ntouyas, S.K.; Alsaedi, A. Existence and stability results for a fractional order differential equation with non-conjugate Riemann-Stieltjes integro-multipoint boundary conditions. *Mathematics* **2019**, *7*, 249. [CrossRef]
9. Annaby, M.H.; Mansour, Z.S. *q-Fractional Calculus and Equations*; Lecture Notes in Mathematics 2056; Springer: Berlin, Germany, 2012.
10. Ahmad, B.; Ntouyas, S.K.; Tariboon, J. *Quantum Calculus. New Concepts, Impulsive IVPs and BVPs, Inequalities. Trends in Abstract and Applied Analysis*; World Scientific Publishing Co. Pte. Ltd.: Hackensack, NJ, USA, 2016.
11. Ahmad, B.; Ntouyas, S.K.; Purnaras, I.K. Existence results for nonlocal boundary value problems of nonlinear fractional q-difference equations. *Adv. Differ. Equ.* **2012**, *2012*, 140. [CrossRef]
12. Zhou, W.X.; Liu, H.Z. Existence solutions for boundary value problem of nonlinear fractional q-difference equations. *Adv. Differ. Equ.* **2013**, *2013*, 113. [CrossRef]
13. Ahmad, B.; Nieto, J.J.; Alsaedi, A.; Al-Hutami, H. Boundary value problems of nonlinear fractional q-difference (integral) equations with two fractional orders and four-point nonlocal integral boundary conditions. *Filomat* **2014**, *28*, 1719–1736. [CrossRef]
14. Niyom, S.; Ntouyas, S.K.; Laoprasittichok, S.; Tariboon, J. Boundary value problems with four orders of Riemann-Liouville fractional derivatives. *Adv. Differ. Equ.* **2016**, *2016*, 165. [CrossRef]
15. Ferreira, R.A.C. Positive solutions for a class of boundary value problems with fractional q-differences. *Comput. Math. Appl.* **2011**, *61*, 367–373. [CrossRef]
16. Graef, J.R.; Kong, L. Positive solutions for a class of higher order boundary value problems with fractional q-derivatives. *Appl. Math. Comput.* **2012**, *218*, 9682–9689. [CrossRef]
17. Liang, S.; Zhang, J. Existence and uniqueness of positive solutions for three-point boundary value problem with fractional q-differences. *J. Appl. Math. Comput.* **2012**, *40*, 277–288. [CrossRef]
18. Almeida, R.; Martins, N. Existence results for fractional q-difference equations of order $\alpha \in [2,3]$ with three-point boundary conditions. *Commun. Nonlinear Sci. Numer. Simul.* **2014**, *19*, 1675–1685. [CrossRef]
19. Etemad, S.; Ettefagh, M.; Rezapour, S. On the existence of solutions for nonlinear fractional q-difference equations with q-integral boundary conditions. *J. Adv. Math. Stud.* **2015**, *8*, 265–285.
20. Alsaedi, A.; Ntouyas, S.K.; Ahmad, B. An existence theorem for fractional q-difference inclusions with nonlocal sub-strip type boundary conditions. *Sci. World J.* **2015**, *2015*, 424306. [CrossRef] [PubMed]
21. Ahmad, B.; Etemad, S.; Ettefagh, M.; Rezapour, S. On the existence of solutions for fractional q-difference inclusions with q-anti-periodic boundary conditions. *Bull. Math. Soc. Sci. Math. Roum.* **2016**, *59*, 119–134.
22. Ren, J.; Zhai, C. A fractional q-difference equation with integral boundary conditions and comparison theorem. *Int. J. Nonlinear Sci. Numer. Simul.* **2017**, *18*, 575–583. [CrossRef]
23. Zhai, C.B.; Ren, J. Positive and negative solutions of a boundary value problem for a fractional q-difference equation. *Adv. Differ. Equ.* **2017**, *2017*, 82. [CrossRef]
24. Abdeljawad, T.; Alzabut, J. On Riemann-Liouville fractional q-difference equations and their application to retarded logistic type model. *Math. Methods Appl. Sci.* **2018**, *41*, 8953–8962. [CrossRef]
25. Zhang, L.; Sun, S. Existence and uniqueness of solutions for mixed fractional q-difference boundary value problems. *Bound. Value Probl.* **2019**, *2019*, 100. [CrossRef]
26. Ahmad, B.; Nieto, J.J.; Alsaedi, A.; Al-Hutami, H. Existence of solutions for nonlinear fractional q-difference integral equations with two fractional orders and nonlocal four-point boundary conditions. *J. Frankl. Inst.* **2014**, *351*, 2890–2909. [CrossRef]
27. Ahmad, B.; Ntouyas, S.K.; Alsaedi, A.; Alsulami, H.H. Nonlinear q-fractional differential equations with nonlocal and sub-strip type boundary conditions. *Electron. J. Qual. Theory Differ. Equ.* **2014**, *2014*, 1–12. [CrossRef]

28. Ahmad, B.; Alsaedi, A.; Al-Hutami, H. A study of sequential fractional q-integro-difference equations with perturbed anti-periodic boundary conditions. In *Fractional Dynamics*; De Gruyter Open: Berlin, Germany, 2015; pp. 110–128.
29. Ahmad, B.; Ntouyas, S.K.; Tariboon, J.; Alsaedi, A.; Alsulami, H.H. Impulsive fractional q-integro-difference equations with separated boundary conditions. *Appl. Math. Comput.* **2016**, *281*, 199–213. [CrossRef]
30. Patanarapeelert, N.; Sriphanomwan, U.; Sitthiwirattham, T. On a class of sequential fractional q-integrodifference boundary value problems involving different numbers of q in derivatives and integrals. *Adv. Differ. Equ.* **2016**, *2016*, 148. [CrossRef]
31. Sitthiwirattham, T. On a fractional q-integral boundary value problems for fractional q-difference equations and fractional q-integro-difference equations involving different numbers of order q. *Bound. Value Probl.* **2016**, *2016*, 12. [CrossRef]
32. Krasnoselskii, M.A. Two remarks on the method of successive approximations. *Uspekhi Mat. Nauk* **1955**, *10*, 123–127.
33. Granas, A.; Dugundji, J. *Fixed Point Theory*; Springer: New York, NY, USA, 2003.
34. Deimling, K. *Nonlinear Functional Analysis*; Springer: New York, NY, USA, 1985.

mathematics

MDPI

Article

Ostrowski Type Inequalities Involving ψ-Hilfer Fractional Integrals

Yasemin Basci [1,*] and Dumitru Baleanu [2,3]

1 Department of Mathematics, Faculty of Arts and Sciences, Bolu Abant Izzet Baysal University, 14280 Golkoy, Bolu, Turkey
2 Department of Mathematics, Faculty of Arts and Sciences, Çankaya University, 06530 Ankara, Turkey
3 Institute of Space Sciences, 077125 Bucharest-Magurele, Romania
* Correspondence: basci_y@ibu.edu.tr

Received: 29 June 2019; Accepted: 16 August 2019; Published: 21 August 2019

Abstract: In this study we introduce several new Ostrowski-type inequalities for both left and right sided fractional integrals of a function g with respect to another function ψ. Our results generalized the ones presented previously by Farid. Furthermore, two illustrative examples are presented to support our results.

Keywords: fractional calculus; fractional integrals; Ostrowski type inequality

MSC: 26A24; 26A33; 26B15

1. Introduction and Preliminaries

Since 1695 [1–3], fractional calculus has been studied by many researchers from both theoretical and applied viewpoints [4,5]. Particularly, fractional calculus is used to generalize classical inequalities. Studies involving integral inequalities are important in several areas such as mathematics, physics, chemistry, biology, engineering and others [6–15]. We recall that there are many definitions of fractional operators, including Riemann–Liouville (RL), Hadamard, Liouville, Weyl (see [16–19]). From such fractional integrals, one can obtain generalizations of the inequalities: Hadamard, Hermite–Hadamard, Hardy, Opial, Gruss, and Montgomery, among others [20–32].

We mention that the following inequality was developed by Ostrowski [33]:

Theorem 1. *Let $g : I \to \mathbb{R}$ be a mapping differentiable in I° such that I is an interval in $\mathbb{R}$, I° is the interior of I and $a_1, b_1 \in I^\circ$, $a_1 < b_1$. If $|g'(\xi)| \le M$ for all $\xi \in [a_1, b_1]$, then the integral inequality holds*

$$\left| g(x) - \frac{1}{b_1 - a_1} \int_{a_1}^{b_1} g(\xi)\, d\xi \right| \le \left[\frac{1}{4} + \frac{\left(x - \frac{a_1 + b_1}{2} \right)^2}{(b_1 - a_1)^2} \right] (b_1 - a_1) M, \tag{1}$$

for all $x \in [a_1, b_1]$.

In the literature, the inequality (1) is called the Ostrowski inequality, see [34]. This inequality has a great importance while studying the error bounds of different numerical quadrature rules. In recent years, such inequalities have been generalized and developed by many researchers. Various authors obtained new Ostrowski-type inequalities for different fractional operators, see [16–19,35–47] and the references therein.

In 2009, Anastassiou et al. [20] obtained Montgomery identities for fractional integrals and a generalization for double fractional integrals. For fractional integrals they discussed both Ostrowski and Grüss inequalities. In 2010, Alomari and Darus [36] presented some Ostrowski-type inequalities for the class of convex (or concave) functions. In 2012, Set [41] obtained some new fractional Ostrowski-type inequalities. In the same year, Liu [40] established some Ostrowski-type inequalities involving RL fractional integrals for the h-convex function. His results are generalizations of [41,42]. He also provided new estimates on Ostrowski-type inequalities for fractional integrals.

In 2013, Yue [38] obtained Ostrowski inequalities for both fractional integrals and associated fractional integrals. In 2014, Aljinević [16] studied Montgomery identities for fractional integrals of a function f with respect to another function g. Also, he gave the Ostrowski inequality for fractional integrals for functions whose first derivatives belong to L_p spaces. In the same year, Yıldırım and Kırtay [46] established new generalizations for Ostrowski inequalities by using the generalized RL fractional integral.

Yıldız et al. [47] used the RL fractional integrals to obtain several new generalizations of Ostrowski type inequalities. Farid [35] found a new version of Ostrowski type inequalities in a very simple way for RL fractional integrals. He also derived some related results. Recently, Dragomir studied several generalizations of the Ostrowski type integral inequality involving RL fractional integrals of bounded variation: Hölder and Lipschitzian functions, see [17–19]. In 2018, Yaldız and Set [45] obtained some new Ostrowski type inequalities for generalized fractional integral operators.

Recently, Sousa and Oliveira [43] introduced the left and right sided fractional integrals and the so-called ψ-Hilfer fractional derivative with respect to another function. They studied Gronwall inequalities and the Cauchy-type problem by means of the ψ- Hilfer operator in [44]. Consequently, they opened a window for new applications.

The following definitions are special approaches for when the kernel is unknown, involving a function ψ. Let $\alpha_1 > 0$ and $I = [a_1, b_1]$ be a finite or infinite interval. Also, let the function g be integrable defined on I, and the function ψ be increasing and positive monotone on $(a_1, b_1]$, having a continuous derivative $\psi'(x)$ on (a_1, b_1).

The expressions of the left sided and right sided fractional integrals of a function g with respect to another function ψ can be seen [4,5], respectively:

$$I_{a_1^+}^{\alpha_1;\psi} g(x) := \frac{1}{\Gamma(\alpha_1)} \int_{a_1}^{x} \psi'(\xi) (\psi(x) - \psi(\xi))^{\alpha_1 - 1} g(\xi) d\xi \tag{2}$$

and

$$I_{b_1^-}^{\alpha_1;\psi} g(x) := \frac{1}{\Gamma(\alpha_1)} \int_{x}^{b_1} \psi'(\xi) (\psi(\xi) - \psi(x))^{\alpha_1 - 1} g(\xi) d\xi. \tag{3}$$

If we take $\psi(x) = x$ and $\psi(x) = \ln x$, then we obtain RL and Hadamard fractional integrals, respectively.

The organization of this manuscript is as follows. In Section 1, we give the introduction and preliminaries. Motivated by [35,43], several Ostrowski-type inequalities for the left sided and right sided fractional integrals of a function g with respect to another function ψ are established in Section 2. Illustrative examples are presented in Section 3 to support our results. Section 4 deals with our conclusions.

2. Main Results

Below, we will show several new Ostrowski-type inequalities for both left and right sided fractional integrals of a function g with respect to another function ψ.

Theorem 2. *Assume that the conditions of the Theorem 1 are satisfied. Also, suppose that the function $\psi \in C^1(I)$ is increasing and positive monotone, and $\psi'(x) \geq 1$ $(\forall x \in I)$. Let $I_{a_1^+}^{\alpha_1;\psi}$ and $I_{b_1^-}^{\beta_1;\psi}$ be defined as (2) and (3), respectively. Then the following inequality holds:*

$$\Big| \left((\psi(b_1) - \psi(x))^{\beta_1} + (\psi(x) - \psi(a_1))^{\alpha_1} \right) g(x)$$

$$-\left(\Gamma\left(\beta_1+1\right) I_{b_1^-}^{\beta_1;\psi} g\left(x\right)+\Gamma\left(\alpha_1+1\right) I_{a_1^+}^{\alpha_1;\psi} g\left(x\right)\right)\Bigg|$$

$$\leq M\left(\frac{\beta_1}{\beta_1+1}\left(\psi\left(b_1\right)-\psi\left(x\right)\right)^{\beta_1+1}+\frac{\alpha_1}{\alpha_1+1}\left(\psi\left(x\right)-\psi\left(a_1\right)\right)^{\alpha_1+1}\right), \tag{4}$$

where $\alpha_1, \beta_1 > 0$ *and* $x \in [a_1, b_1]$.

Proof of Theorem 2. Taking into account that ψ is an increasing and positive monotone function, for $\alpha_1 > 0$ and $\xi \in [a_1, x]$ we get

$$\left(\psi\left(x\right)-\psi\left(\xi\right)\right)^{\alpha_1} \leq \left(\psi\left(x\right)-\psi\left(a_1\right)\right)^{\alpha_1}. \tag{5}$$

Utilizing (5) and the given condition on g', we obtain

$$\int_{a_1}^{x}\left(M\psi'(\xi)-g'(\xi)\right)\left(\psi(x)-\psi(\xi)\right)^{\alpha_1}d\xi \leq \left(\psi(x)-\psi(a_1)\right)^{\alpha_1}\int_{a_1}^{x}\left(M\psi'(\xi)-g'(\xi)\right)d\xi$$

and

$$\int_{a_1}^{x}\left(M\psi'(\xi)+g'(\xi)\right)\left(\psi(x)-\psi(\xi)\right)^{\alpha_1}d\xi \leq \left(\psi(x)-\psi(a_1)\right)^{\alpha_1}\int_{a_1}^{x}\left(M\psi'(\xi)+g'(\xi)\right)d\xi.$$

If the above integrals are calculated, we obtain the following inequalities, respectively:

$$\left(\psi\left(x\right)-\psi\left(a_1\right)\right)^{\alpha_1} g(x)-\Gamma\left(\alpha_1+1\right) I_{a_1^+}^{\alpha_1;\psi} g\left(x\right) \leq \frac{M\alpha_1}{\alpha_1+1}\left(\psi\left(x\right)-\psi\left(a_1\right)\right)^{\alpha_1+1} \tag{6}$$

and

$$\Gamma\left(\alpha_1+1\right) I_{a_1^+}^{\alpha_1;\psi} g(x)-\left(\psi(x)-\psi(a_1)\right)^{\alpha_1} g(x) \leq \frac{M\alpha_1}{\alpha_1+1}\left(\psi(x)-\psi(a_1)\right)^{\alpha_1+1}. \tag{7}$$

By using (6) and (7), we report the following inequality:

$$\left|\left(\psi(x)-\psi(a_1)\right)^{\alpha_1} g(x)-\Gamma\left(\alpha_1+1\right) I_{a_1^+}^{\alpha_1;\psi} g(x)\right| \leq \frac{M\alpha_1}{\alpha_1+1}\left(\psi(x)-\psi(a_1)\right)^{\alpha_1+1}. \tag{8}$$

On the other hand, if ψ is an increasing and positive function, for $\xi \in [x, b_1]$ and $\beta_1 > 0$ we get

$$\left(\psi(\xi)-\psi(x)\right)^{\beta_1} \leq \left(\psi(b_1)-\psi(x)\right)^{\beta_1}. \tag{9}$$

By using (9) and the given condition on g', we conclude

$$\int_{x}^{b_1}\left(M\psi'(\xi)-g'(\xi)\right)\left(\psi(\xi)-\psi(x)\right)^{\beta_1}d\xi \leq \left(\psi(b_1)-\psi(x)\right)^{\beta_1}\int_{x}^{b_1}\left(M\psi'(\xi)-g'(\xi)\right)d\xi$$

and

$$\int_{x}^{b_1}\left(M\psi'(\xi)+g'(\xi)\right)\left(\psi(\xi)-\psi(x)\right)^{\beta_1}d\xi \leq \left(\psi(b_1)-\psi(x)\right)^{\beta_1}\int_{x}^{b_1}\left(M\psi'(\xi)+g'(\xi)\right)d\xi.$$

If the above integrals are calculated, we obtain the following inequalities, respectively:

$$\Gamma\left(\beta_1+1\right) I_{b_1^-}^{\beta_1;\psi} g(x)-\left(\psi(b_1)-\psi(x)\right)^{\beta_1} g(x) \leq \frac{M\beta_1}{\beta_1+1}\left(\psi(b_1)-\psi(x)\right)^{\beta_1+1} \tag{10}$$

and

$$\left(\psi\left(b_1\right)-\psi(x)\right)^{\beta_1} g(x)-\Gamma\left(\beta_1+1\right) I_{b_1^-}^{\beta_1;\psi} g(x) \leq \frac{M\beta_1}{\beta_1+1}\left(\psi(b_1)-\psi(x)\right)^{\beta_1+1}. \tag{11}$$

By using (10) and (11), the following inequality will appear:

$$\left|(\psi(b_1)-\psi(x))^{\beta_1} g(x)-\Gamma\left(\beta_1+1\right) I_{b_1^-}^{\beta_1;\psi} g(x)\right| \leq \frac{M\beta_1}{\beta_1+1}(\psi(b_1)-\psi(x))^{\beta_1+1}. \tag{12}$$

So, by utilizing (8) and (12), we obtain (4). □

Theorem 3. *Let $g: I \rightarrow \mathbb{R}$ be a mapping differentiable in I° (the interior of I) such that I is an interval in $\mathbb{R}$ and $a_1, b_1 \in I^\circ$, $a_1 < b_1$. Assume that the function $\psi \in C^1(I)$ is increasing and positive monotone, and $\psi'(x) \geq 1$ $(\forall x \in I)$. Also, let $I_{a_1^+}^{\alpha_1;\psi}$ and $I_{b_1^-}^{\beta_1;\psi}$ be defined as (2) and (3), respectively. If $m \leq g'(t) \leq M$ for $M \geq 0$, $m \leq 0$ and all $\xi \in [a_1, b_1]$, then the following inequalities hold:*

$$\begin{aligned}
&\left(\left(\psi(x)-\psi\left(a_1\right)\right)^{\alpha_1}-\left(\psi\left(b_1\right)-\psi\left(x\right)\right)^{\beta_1}\right) g\left(x\right) \\
&\qquad -\left(\Gamma\left(\alpha_1+1\right) I_{a_1^+}^{\alpha_1;\psi} g\left(x\right)-\Gamma\left(\beta_1+1\right) I_{b_1^-}^{\beta_1;\psi} g\left(x\right)\right) \\
&\leq M\left(\frac{\alpha_1}{\alpha_1+1}(\psi(x)-\psi(a_1))^{\alpha_1+1}+\frac{\beta_1}{\beta_1+1}(\psi(b_1)-\psi(x))^{\beta_1+1}\right)
\end{aligned} \tag{13}$$

and

$$\begin{aligned}
&\left(\left(\psi(b_1)-\psi\left(x\right)\right)^{\beta_1}-\left(\psi\left(x\right)-\psi\left(a_1\right)\right)^{\alpha_1}\right) g\left(x\right) \\
&\qquad +\left(\Gamma\left(\alpha_1+1\right) I_{a_1^+}^{\alpha_1;\psi} g(x)-\Gamma\left(\beta_1+1\right) I_{b_1^-}^{\beta_1;\psi} g\left(x\right)\right) \\
&\leq -m\left(\frac{\beta_1}{\beta_1+1}(\psi(b_1)-\psi(x))^{\beta_1+1}+\frac{\alpha_1}{\alpha_1+1}(\psi(x)-\psi(a_1))^{\alpha_1+1}\right),
\end{aligned} \tag{14}$$

where $\alpha_1, \beta_1 > 0$ and $x \in [a_1, b_1]$.

Proof of Theorem 3. Using the given comparing conditions on g', the proof is similar to one of Theorem 2. That is, from (5) and by using the given condition on g', we conclude

$$\int_{a_1}^{x}\left(M\psi'\left(\xi\right)-g'\left(\xi\right)\right)(\psi(x)-\psi(\xi))^{\alpha_1} d\xi \leq (\psi(x)-\psi(a_1))^{\alpha_1} \int_{a_1}^{x}\left(M\psi'\left(\xi\right)-g'\left(\xi\right)\right) d\xi$$

and

$$\int_{a_1}^{x}\left(g'(\xi)-m\psi'\left(\xi\right)\right)(\psi(x)-\psi(\xi))^{\alpha_1} d\xi \leq (\psi(x)-\psi(a_1))^{\alpha_1} \int_{a_1}^{x}\left(g'(\xi)-m\psi'\left(\xi\right)\right) d\xi.$$

If the above integrals are calculated, we obtain the following inequalities, namely:

$$(\psi(x)-\psi(a_1))^{\alpha_1} g(x)-\Gamma\left(\alpha_1+1\right) I_{a_1^+}^{\alpha_1;\psi} g(x) \leq \frac{M\alpha_1}{\alpha_1+1}(\psi(x)-\psi(a_1))^{\alpha_1+1} \tag{15}$$

and

$$\Gamma(\alpha_1+1) I_{a_1^+}^{\alpha_1;\psi} g(x)-(\psi(x)-\psi(a_1))^{\alpha_1} g(x) \leq -\frac{m\alpha_1}{\alpha_1+1}(\psi(x)-\psi(a_1))^{\alpha_1+1}. \tag{16}$$

On the other hand, by using (9) and the given condition on g', we have

$$\int_{x}^{b_1}\left(M\psi'(\xi)-g'(\xi)\right)(\psi(\xi)-\psi(x))^{\beta_1} d\xi \leq (\psi(b_1)-\psi(x))^{\beta_1} \int_{x}^{b_1}\left(M\psi'(\xi)-g'(\xi)\right) d\xi$$

and

$$\int_x^{b_1} (g'(\xi) - m\psi'(\xi)) (\psi(\xi) - \psi(x))^{\beta_1} d\xi \leq (\psi(b_1) - \psi(x))^{\beta_1} \int_x^{b_1} (g'(\xi) - m\psi'(\xi)) d\xi.$$

If the above integrals are calculated, we obtain the following inequalities, namely:

$$\Gamma(\beta_1 + 1) I_{b_1^-}^{\beta_1;\psi} g(x) - (\psi(b_1) - \psi(x))^{\beta_1} g(x) \leq \frac{M\beta_1}{\beta_1 + 1} (\psi(b_1) - \psi(x))^{\beta_1 + 1} \tag{17}$$

and

$$(\psi(b_1) - \psi(x))^{\beta_1} g(x) - \Gamma(\beta_1 + 1) I_{b_1^-}^{\beta_1;\psi} g(x) \leq -\frac{m\beta_1}{\beta_1 + 1} (\psi(b_1) - \psi(x))^{\beta_1 + 1}, \tag{18}$$

respectively. By using (15) and (17), we obtain (13). In addition, by using (16) and (18), we provide (14). □

Theorem 4. *Let $g : I \to \mathbb{R}$ be a mapping differentiable in I° (the interior of I) such that I is an interval in $\mathbb{R}$ and $a_1, b_1 \in I^\circ$, $a_1 < b_1$. Assume that the function $\psi \in C^1(I)$ is increasing and positive monotone, and $\psi'(x) \geq 1$ $(\forall x \in I)$. Also, let $I_{a_1^+}^{\alpha_1;\psi}$ and $I_{b_1^-}^{\beta_1;\psi}$ be defined as (2) and (3), respectively. If $m \leq g'(t) \leq M$ for $M \geq 0$, $m \leq 0$ and all $\xi \in [a_1, b_1]$, then the following inequalities hold:*

$$\begin{aligned} &\left((\psi(x) - \psi(a_1))^{\alpha_1} + (\psi(b_1) - \psi(x))^{\beta_1} \right) g(x) \\ &\quad - \left(\Gamma(\alpha_1 + 1) I_{a_1^+}^{\alpha_1;\psi} g(x) + \Gamma(\beta_1 + 1) I_{b_1^-}^{\beta_1;\psi} g(x) \right) \\ &\leq \frac{M\alpha_1}{\alpha_1 + 1} (\psi(x) - \psi(a_1))^{\alpha_1 + 1} - \frac{m\beta_1}{\beta_1 + 1} (\psi(b_1) - \psi(x))^{\beta_1 + 1} \end{aligned} \tag{19}$$

and

$$\begin{aligned} &-\left((\psi(b_1) - \psi(x))^{\beta_1} + (\psi(x) - \psi(a_1))^{\alpha_1} \right) g(x) \\ &\quad + \left(\Gamma(\alpha_1 + 1) I_{a_1^+}^{\alpha_1;\psi} g(x) + \Gamma(\beta_1 + 1) I_{b_1^-}^{\beta_1;\psi} g(x) \right) \\ &\leq \frac{M\beta_1}{\beta_1 + 1} (\psi(b_1) - \psi(x))^{\beta_1 + 1} - \frac{m\alpha_1}{\alpha_1 + 1} (\psi(x) - \psi(a_1))^{\alpha_1 + 1}, \end{aligned} \tag{20}$$

where $\alpha_1, \beta_1 > 0$ and $x \in [a_1, b_1]$.

Proof of Theorem 4. Proof is constructed in the same line as the proof of Theorem 3. By using (15) and (18), we obtain (19). In addition, from (16) and (17), we get (20). □

Theorem 5. *Suppose that the conditions of the Theorem 2 are satisfied. Also, assume that the function $\psi \in C^1(I)$ is increasing and positive monotone, and $\psi'(x) \geq 1$ $(\forall x \in I)$. Let $I_{a_1^+}^{\alpha_1;\psi}$ and $I_{b_1^-}^{\beta_1;\psi}$ be defined as (2) and (3), respectively. Then the following inequality holds:*

$$\begin{aligned} &\Big| \left((\psi(b_1) - \psi(x))^{\beta_1} g(b_1) + (\psi(x) - \psi(a_1))^{\alpha_1} g(a_1) \right) \\ &\quad - \left(\Gamma(\beta_1 + 1) I_{x^+}^{\beta_1;\psi} g(b_1) + \Gamma(\alpha_1 + 1) I_{x^-}^{\alpha_1;\psi} g(a_1) \right) \Big| \\ &\leq M \left(\frac{\beta_1}{\beta_1 + 1} (\psi(b_1) - \psi(x))^{\beta_1 + 1} + \frac{\alpha_1}{\alpha_1 + 1} (\psi(x) - \psi(a_1))^{\alpha_1 + 1} \right), \end{aligned} \tag{21}$$

where $\alpha_1, \beta_1 > 0$ and $x \in [a_1, b_1]$.

Proof of Theorem 5. Recalling ψ is an increasing and positive monotone function, for $\alpha_1 > 0$ and $\xi \in [a_1, x]$ we obtain

$$(\psi(\xi) - \psi(a_1))^{\alpha_1} \leq (\psi(x) - \psi(a_1))^{\alpha_1}. \tag{22}$$

By using (22) and the given condition on g', we have

$$\int_{a_1}^{x} (M\psi'(\xi) - g'(\xi))(\psi(\xi) - \psi(a_1))^{\alpha_1} d\xi \leq (\psi(x) - \psi(a_1))^{\alpha_1} \int_{a_1}^{x} (M\psi'(\xi) - g'(\xi))\, d\xi$$

and

$$\int_{a_1}^{x} (M\psi'(\xi) + g'(\xi))(\psi(\xi) - \psi(a_1))^{\alpha_1} d\xi \leq (\psi(x) - \psi(a_1))^{\alpha_1} \int_{a_1}^{x} (M\psi'(\xi) + g'(\xi))\, d\xi.$$

If the above integrals are calculated, we obtain the following inequalities, respectively:

$$\Gamma(\alpha_1 + 1) I_{x^-}^{\alpha_1;\psi} g(a_1) - (\psi(x) - \psi(a_1))^{\alpha_1} g(a_1) \leq \frac{M\alpha_1}{\alpha_1 + 1} (\psi(x) - \psi(a_1))^{\alpha_1 + 1} \tag{23}$$

and

$$(\psi(x) - \psi(a_1))^{\alpha_1} g(a_1) - \Gamma(\alpha_1 + 1) I_{x^-}^{\alpha_1;\psi} g(a_1) \leq \frac{M\alpha_1}{\alpha_1 + 1} (\psi(x) - \psi(a_1))^{\alpha_1 + 1}. \tag{24}$$

By utilizing (23) and (24), the following inequality holds:

$$\left| (\psi(x) - \psi(a_1))^{\alpha_1} g(a_1) - \Gamma(\alpha_1 + 1) I_{x^-}^{\alpha_1;\psi} g(a_1) \right|$$
$$\leq \frac{M\alpha_1}{\alpha_1 + 1} (\psi(x) - \psi(a_1))^{\alpha_1 + 1}. \tag{25}$$

Using the fact that ψ is an increasing and positive monotone function, for $\xi \in [x, b_1]$ and $\beta_1 > 0$ we get

$$(\psi(b_1) - \psi(\xi))^{\beta_1} \leq (\psi(b_1) - \psi(x))^{\beta_1}. \tag{26}$$

By using (26) and the given condition on g', we have

$$\int_{x}^{b_1} (M\psi'(\xi) - g'(\xi))(\psi(b_1) - \psi(\xi))^{\beta_1} d\xi \leq (\psi(b_1) - \psi(x))^{\beta_1} \int_{x}^{b_1} (M\psi'(\xi) - g'(\xi))\, d\xi$$

and

$$\int_{x}^{b_1} (M\psi'(\xi) + g'(\xi))(\psi(b_1) - \psi(\xi))^{\beta_1} d\xi \leq (\psi(b_1) - \psi(x))^{\beta_1} \int_{x}^{b_1} (M\psi'(\xi) + g'(\xi))\, d\xi.$$

If the above integrals are calculated, we obtain the following inequalities, respectively:

$$(\psi(b_1) - \psi(x))^{\beta_1} g(b_1) - \Gamma(\beta_1 + 1) I_{x^+}^{\beta_1;\psi} g(b_1) \leq \frac{M\beta_1}{\beta_1 + 1} (\psi(b_1) - \psi(x))^{\beta_1 + 1} \tag{27}$$

and

$$\Gamma(\beta_1 + 1) I_{x_1^+}^{\beta_1;\psi} g(b_1) - (\psi(b_1) - \psi(x))^{\beta_1} g(b_1) \leq \frac{M\beta_1}{\beta_1 + 1} (\psi(b_1) - \psi(x))^{\beta_1 + 1}. \tag{28}$$

By making use of (27) and (28), the following inequality holds:

$$|(\psi(b_1) - \psi(x))^{\beta_1} g(b_1) - \Gamma(\beta_1 + 1) I_{x^+}^{\beta_1;\psi} g(b_1)| \leq \frac{M\beta_1}{\beta_1 + 1} (\psi(b_1) - \psi(x))^{\beta_1 + 1}. \tag{29}$$

So, from (25) and (29), we obtain (21). □

Corollary 1. *If $\beta_1 = \alpha_1$ in Theorem 2, then the following fractional Ostrowski inequality holds:*

$$\begin{aligned}&\left|\left((\psi(b_1)-\psi(x))^{\alpha_1}+(\psi(x)-\psi(a_1))^{\alpha_1}\right)g(x)\right.\\&\qquad\left.-\Gamma(\alpha_1+1)\left(I_{b_1^-}^{\alpha_1;\psi}g(x)+I_{a_1^+}^{\alpha_1;\psi}g(x)\right)\right|\\&\le\frac{M\alpha_1}{\alpha_1+1}\left((\psi(b_1)-\psi(x))^{\alpha_1+1}+(\psi(x)-\psi(a_1))^{\alpha_1+1}\right),\end{aligned}$$

where $\alpha_1 > 0$ and $x \in [a_1, b_1]$.

Corollary 2. *If $\alpha_1 = \beta_1 = 1$ and $\psi(x) = x$, then we lead to the Ostrowski inequality (1).*

Corollary 3. *If $\alpha_1 = \beta_1$ in Theorem 5, then we obtain the following fractional Ostrowski inequality:*

$$\begin{aligned}&\left|\left((\psi(b_1)-\psi(x))^{\alpha_1}g(b_1)+(\psi(x)-\psi(a_1))^{\alpha_1}g(a_1)\right)\right.\\&\qquad\left.-\Gamma(\alpha_1+1)\left(I_{x^+}^{\alpha_1;\psi}g(b_1)+I_{x^-}^{\alpha_1;\psi}g(a_1)\right)\right|\\&\le\frac{M\alpha_1}{\alpha_1+1}\left((\psi(b_1)-\psi(x))^{\alpha_1+1}+(\psi(x)-\psi(a_1))^{\alpha_1+1}\right),\end{aligned}$$

where $\alpha_1 > 0$ and $x \in [a_1, b_1]$.

Remark 1. *If we take $\psi(x) = x$, then Theorem 2, Theorem 3, and Theorem 5 reduce to Theorem 1.2–Theorem 1.4 in Farid [35], respectively. But, in [35], $-m$ should be M in the first inequality in Theorem 1.3. Also, M should be $-m$ in the second inequality.*

Remark 2. *After following the steps of the proof of Theorem 2 with $\psi(x) = x$ and $\alpha_1 = \beta_1 = 1$, an alternative proof of the Ostrowski inequality is obtained (see [37]).*

3. Examples

In this section, we support our main results by presenting two examples.

Example 1. *Let $\alpha_1 = 0.5$, $\beta_1 = 2.2$, $\psi(x) = e^x$, $g(x) = \sin x$ and $[a_1, b_1] = [0, \pi]$. Then, we obtain $|g'(x)| = |\cos x| \le 1$, that is, $M = 1$. Also, $\psi(x) = e^x$ is an increasing continuous derivative and positive monotone function with $\psi'(x) = e^x \ge 1$ for all $x \in [0, \pi]$. Then, using Theorem 2, for $[0, \pi]$ we obtain the following Ostrowski type inequality:*

$$\begin{aligned}&\left|\left((e^\pi-e^x)^{2.2}+(e^x-1)^{0.5}\right)\sin x-\left(\Gamma(3.2)\,I_{\pi^-}^{2.2;\psi}\sin x+\Gamma(1.5)\,I_{0^+}^{0.5;\psi}\sin x\right)\right|\\&\qquad\le\frac{11}{16}(e^\pi-e^x)^{3.2}+\frac{1}{3}(e^x-1)^{1.5}.\end{aligned}$$

Example 2. *Let $\alpha_1 = 0.5$, $\beta_1 = 2.2$, $\psi(x) = 6\sqrt{x+2}$, $g(x) = (x-1)^2$ and $[a_1, b_1] = [0, 2]$. Then, we get $g'(x) = 2(x-1)$. Let $m = -2$ and $M = 2$. Also, $\psi(x) = 6\sqrt{x+2}$ is an increasing continuous derivative and positive monotone function with $\psi'(x) = \frac{3}{\sqrt{x+2}} \ge 1$ for all $x \in I = [0, 2]$.Then, using Theorem 3, for $x \in [0, 2]$ we obtain the following Ostrowski type inequality:*

$$\begin{aligned}&\left(\left(6\sqrt{x+2}-6\sqrt{2}\right)^{0.5}-\left(12-6\sqrt{x+2}\right)^{2.2}\right)(x-1)^2-\left(\Gamma(1.5)\,I_{0^+}^{0.5;\psi}(x-1)^2-\Gamma(3.2)\,I_{2^-}^{2.2;\psi}(x-1)^2\right)\\&\qquad\le 2\left(\frac{1}{3}\left(6\sqrt{x+2}-6\sqrt{2}\right)^{1.5}+\frac{11}{16}\left(12-6\sqrt{x+2}\right)^{3.2}\right)\end{aligned}$$

and

$$\left(\left(12-6\sqrt{x+2}\right)^{2.2}-\left(6\sqrt{x+2}-6\sqrt{2}\right)^{1.5}\right)(x-1)^2+\left(\Gamma(1.5)\,I_{0^+}^{0.5;\psi}(x-1)^2-\Gamma(3.2)\,I_{2^-}^{2.2;\psi}(x-1)^2\right)$$
$$\leq 2\left(\frac{11}{16}\left(6\sqrt{2}-6\sqrt{x+2}\right)^{3.2}+\frac{1}{3}\left(6\sqrt{x+2}-6\sqrt{2}\right)^{1.5}\right).$$

4. Conclusions

Studies involving integral inequalities play an important role in several areas of science and engineering. During recent years, such inequalities have been generalized and developed by many researchers. Ostrowski inequalities have a great importance while studying the error bounds of different numerical quadrature rules, for example the midpoint rule, Simpson's rule, the trapezoidal rule and other generalized Riemann types. In this paper, by generalizing the inequalities in [35], we proposed, within four theorems and their related corollaries, several new Ostrowski-type integral inequalities for the left and right sided fractional integrals of a function g with respect to another function ψ. Finally, we investigated in detail two examples to show the reported results.

Author Contributions: All authors contributed to each part of this work equally, and they all read and approved the final manuscript.

Funding: This research received no external funding.

Acknowledgments: The authors would like to thank to the referees for their useful comments and remarks.

Conflicts of Interest: The authors declare no conflicts of interest.

References

1. Leibniz, G.W. Letter from Hanover, Germany to G.F.A L'Hospital, September 30, 1695. In *Leibniz Mathematische Schriften*; Olms-Verlag: Hildesheim, Germany, 1962; pp. 301–302.
2. Leibniz, G.W. Letter from Hanover, Germany, to Johann Bernoulli, December 28, 1695. In *Leibniz Mathematische Schriften*; Olms-Verlag: Hildesheim, Germany, 1962; p. 226.
3. Leibniz, G.W. Letter from Hanover, Germany, to John Wallis, May 30, 1697. In *Leibniz Mathematische Schriften*; Olms-Verlag: Hildesheim, Germany, 1962; p. 25.
4. Kilbas, A.A.; Srivastava, H.M.; Trujillo, J.J. Theory and applications of fractional differential equations. In *North-Holland Mathematics Studies*; Elsevier: Amsterdam, The Netherlands, 2006; p. 204.
5. Samko, S.G.; Kilbas, A.A.; Marichev, O.I. *Fractional Integrals and Derivatives*; Gordon and Breach: Yverdon, Switzerland, 1993.
6. Atanackovic, T.M.; Pilipovic, S.; Stankovic, B.; Zorica, D. *Fractional Calculus with Applications in Mechanics: Vibrations and Diffusion Processes*; Wiley: London, UK; Hoboken, NJ, USA, 2014.
7. Bandle, C.; Gilányi, A.; Losonczi, L.; Páles, Z.; Plum, M. *Inequalities and Applications: Conference on Inequalities and Applications, Noszvaj (Hungary), September 2007*; Springer: Berlin/Heidelberg, Germany, 2008; p. 157.
8. Bainov, D.D.; Simeonov, P.S. *Integral Inequalities and Applications*; Springer: Berlin/Heidelberg, Germany, 2013; p. 57.
9. Corlay, S.; Lebovits, J.; Véhel, J.L. Multifractional stochastic volatility models. *Math. Finance* **2014**, *24*, 364–402. [CrossRef]
10. Herrmann, R. *Fractional Calculus: An Introduction for Physicists*; World Scientific Publishing Company: Singapore, 2011.
11. Magin, R.L.; Ingo, C; Colon-Perez, L.; Triplett, W.; Mareci, T.H. Characterization of anomalous diffusion in porous biological tissues using fractional order derivatives an, entropy. *Microporous Mesoporous Mater.* **2013**, *178*, 39–43. [CrossRef] [PubMed]
12. Magin, R.L. Fractional calculus models of complex dynamics in biological tissues. *Comput. Math. Appl.* **2010**, *59*, 1586–1593. [CrossRef]
13. Meral, F.; Royston, T.; Magin, R. Fractional calculus in viscoelasticity: An experimental study. *Commun. Nonlinear Sci. Numer. Simul.* **2010**, *15*, 939–945. [CrossRef]

14. Silva Costa, F.; Soares, J.C.S.; Gomez Plata, A.R.; Capelas de Oliveira, E. On the fractional Harry Dym equation. *Comput. Appl. Math.* **2018**, *37*, 2862–2876. [CrossRef]
15. Silva Costa, F.; Contharteze Grigoletto, E.; Vaz Jr, J.; Capelas de Oliveira, E. Slowing-down of neutrons: A fractional model. *Commun. Appl. Ind. Math.* **2015**, *6*. [CrossRef]
16. Aljinević, A.A. Montgomery identity and Ostrowski type inequalities for Riemann-Liouville fractional integral. *J. Math. Vol.* **2014**, 503195. [CrossRef]
17. Dragomir, S.S. Ostrowski type inequalities for Riemann-Liouville fractional integrals of bounded variation, Hölder and Lipschitzian functions. *RGMIA Res. Rep. Coll.* **2017**, *20*, 48.
18. Dragomir, S.S. Ostrowski type inequalities for generalized Riemann-Liouville fractional integrals of functions with bounded variation. *RGMIA Res. Rep. Coll.* **2017**, *20*, 58.
19. Dragomir, S.S. Further Ostrowski and Trapezoid type inequalities for Riemann-Liouville fractional integrals of functions with bounded variation. *RGMIA Res. Rep. Coll.* **2017**, *20*, 84.
20. Anastassiou, G.M.; Hooshmandasl, R.; Ghasemi, A.; Moftakharzahed, F. Montgomery identities for fractional integrals and related fractional inequalities. *J. Inequal. Pure Appl. Math.* **2009**, *10*, 97.
21. Chen, F. Extensions of the Hermite-Hadamard inequality for convex functions via fractional integrals. *J. Math. Inequal.* **2016**, *10*, 75–81. [CrossRef]
22. Chen, H.; Katugampola, U.N. Hermite-Hadamard and Hermite-Hadamard-Fejér type inequalities for generalized fractional integrals. *J. Math. Anal. Appl.* **2013**, *446*, 1274–1291. [CrossRef]
23. Iqbal, S.; Krulić, K.; Pečarić, J. Weighted Hardy-type inequalities for monotone convex functions with some applications. *Fract. Differ. Calc.* **2013**, *3*, 31–53. [CrossRef]
24. Iqbal, S.; Krulić, K.; Pečarić, J. On refined Hardy-type inequalities with fractional integrals and fractional derivatives. *Math. Slovaca* **2014**, *64*, 879–892. [CrossRef]
25. Iqbal S, Krulić K, Pečarić J. On a new class of Hardy-type inequalities with fractional integrals and fractional derivatives. *Rad Hazu. Matematičke Znanosti* **2014**, *18=519*, 91–106.
26. Iqbal, S.; Pečarić, J.; Samraiz, M.; Tomovski, Z. Hardy-type inequalities for generalized fractional integral operators. *Tbilisi Math. J.* **2017**, *10*, 75–90. [CrossRef]
27. Iqbal, S.; Pečarić, J.; Samraiz, M.; Tomovski, Z. On some Hardy-type inequalities for fractional calculus operators. *Banach J. Math. Anal.* **2017**, *11*, 438–457. [CrossRef]
28. Jain, R.; Pathan, M.A. On Weyl fractional integral operators. *Tamkang J. Math.* **2004**, *35*, 169–173. [CrossRef]
29. Nasibullin, R. Hardy-type inequalities for fractional integrals and derivatives of Riemann-Liouville. *Lobachevskii J. Math.* **2017**, *38*, 709–718. [CrossRef]
30. Sarıkaya, M.Z.; Budak, H. New inequalities of Opial type for conformable fractional integrals. *Turk. J. Math.* **2017**, *41*, 1164–1173. [CrossRef]
31. Set, E.; İscan, I.; Sarıkaya, M.Z.S.; Özdemir, M.E. On new inequalities of Hermite-Hadamard Fejér type for convex functions via fractional integrals. *Appl. Math. Comput.* **2015**, *259*, 875–881. [CrossRef]
32. Wang, J.; Li, X.; Fekan, M.; Zhou, Y. Hermite-Hadamard-type inequalities for Riemann-Liouville fractional integrals via two kinds of convexity. *Appl. Anal.* **2013**, *92*, 2241–2253. [CrossRef]
33. Ostrowski, A. Über die Absolutabweichung einer differentiierbaren Funktion von ihrem Integralmittelwert. *Comment. Math. Helv.* **1937**, *10*, 226–227. [CrossRef]
34. Mitrinovic, D.S.; Pecaric, J.; Fink, A.M. *Inequalities Involving Functions and Their Integrals and Derivatives*; Mathematics and its Applications (East European Series), *53*, Kluwer Academic Publishers: Dortrecht, The Netherlands, 1991.
35. Farid, G. Some new Ostrowski type inequalities via fractional integrals. *Int. J. Anal. Appl.* **2017**, *14*, 64–68.
36. Alomari, M.; Darus, M. Some Ostrowski type inequalities for convex functions with applications. *RGMIA* **2010**, *13*, 3.
37. Farid, G. Straightforward proofs of Ostrowski inequality and some related results. *Int. J. Anal.* **2016**, *2016*, 3918483. [CrossRef]
38. Yue, Hu. Ostrowski inequality for fractional integrals and related fractional inequalities. *Transylv. J. Math. Mech.* **2013**, *5*, 85–89.
39. Khan, M.A.; Begum, S.; Khurshid, Y.; Chu, Y.-M. Ostrowski type inequalities involving conformable fractional integral. *J. Inequal. Appl.* **2018**, *2018*, 70. [CrossRef]
40. Liu, W. Some Ostrowski type inequalities via Riemann-Liouville fractional integrals for h-convex functions. *J. Comput. Anal. Appl.* **2012**, *16*, 998–1004.

41. Set, E. New inequalities for Ostrowski type for mappings whose derivatives are s-convex in the second sense via fractional integrals. *Comput. Math. Appl.* **2012**, *63*, 1147–1154. [CrossRef]
42. Tunç, M. Ostrowski type inequalities via h-convex functions with applications for special means. *J. Inequal. Appl.* **2013**, *2013*, 326. [CrossRef]
43. Vanterler da C. Sousa, J.; Capelas de Oliveira, E. On the ψ-Hilfer fractional derivative. *Commun. Nonlinear Sci. Numer. Simul.*, **2018**, *60*, 72–91. [CrossRef]
44. Vanterler da C. Sousa, J.; Capelas de Oliveira, E. A Gronwall inequality and the Cauchy-type problem by means of ψ- Hilfer operator. *Differ. Equ. Appl.* **2019** *11*, 87–106. [CrossRef]
45. Yaldız, H.; Set, E. Some new Ostrowski type inequalities for generalized fractional integrals. *AIP Conf. Proc.* **2018**, *1991*, 020018. [CrossRef]
46. Yıldırım, H.; Kırtay, Z. Ostrowski inequality for generalized fractional integral and related inequalities. *Malaya J. Mat.* **2014**, *2*, 322–329.
47. Yıldız, Ç.; Özdemir, M.E.; Sarıkaya, M.Z. New generalizations of Ostrowski-Like type inequalities for fractional integral. *Kyungpook Math. J.* **2016**, *56*, 161–172. [CrossRef]

mathematics

MDPI

Article

Some Estimates for Generalized Riemann-Liouville Fractional Integrals of Exponentially Convex Functions and Their Applications

Saima Rashid [1], Thabet Abdeljawad [2,*], Fahd Jarad [3] and Muhammad Aslam Noor [1]

1 Department of Mathematics, COMSATS University Islamabad, Islamabad 45550, Pakistan
2 Department of Mathematics and General Sciences, Prince Sultan University, P.O. Box 66833, Riyadh 11586, Saudi Arabia
3 Department of Mathematics, Çankaya University, Etimesgut, 06790 Ankara, Turkey
* Correspondence: tabdeljawad@psu.edu.sa

Received: 29 July 2019; Accepted: 20 August 2019; Published: 2 September 2019

Abstract: In the present paper, we investigate some Hermite-Hadamard ($\mathcal{HH}$) inequalities related to generalized Riemann-Liouville fractional integral ($\mathcal{GRLFI}$) via exponentially convex functions. We also show the fundamental identity for $\mathcal{GRLFI}$ having the first order derivative of a given exponentially convex function. Monotonicity and exponentially convexity of functions are used with some traditional and forthright inequalities. In the application part, we give examples and new inequalities for the special means.

Keywords: convex function; exponentially convex function; fractional integrals; generalized Riemann-liouville fractional integrals

1. Introduction

Recently, several researchers have attracted the fractional calculus, see References [1–4]. The effect and motivation of this fractional calculus in both theoretical and applied science and engineering rose substantially. Numerous studies related to the discrete versions of this fractional calculus have been established, which benefit from countless applications in the theory of time scales, physics, different fields of engineering, chemistry and so forth (e.g, see References [4–32] and the references therein).

A few decades ago, a lot of new operator definitions were given and the properties and structures of these operators have been examined. Some of these operators are very close to classical operators in terms of their characteristics and definitions. It is known that the $\mathcal{GRLFI}$, which was introduced in reference [33], extends several well-known fractional integral operators (see Remark 1 below). Both the generalized Riemann-Liouville fractional derivative and the integral operator are useful in the study of transform theory, quantum theory and fractional intgerodifferential equations.

Almost every mathematician knows the importance of convexity theory in every field of mathematics, for example in nonlinear programming and optimization theory. By using the concept of convexity, several integral inequalities have been introduced such as Jensen, $\mathcal{HH}$ and Slater inequalities, and so forth. But the well-known one is the celebrated $\mathcal{HH}$ inequality.

Let $\mathcal{I} \subseteq \mathcal{R}$ be an interval and $\mathcal{U} : \mathcal{I} \rightarrow \mathcal{R}$ be a convex function. Then the double inequality

$$\mathcal{U}\Big(\frac{d_1+d_2}{2}\Big) \leq \frac{1}{d_2-d_1}\int_{d_1}^{d_2} \mathcal{U}(x)dx \leq \frac{\mathcal{U}(d_1)+\mathcal{U}(d_2)}{2}, \tag{1}$$

holds for all $d_1, d_2 \in \mathcal{I}$ with $d_1 < d_2$. It is easy to see that if $\mathcal{U}$ is concave on $\mathcal{I}$, one has the reverse of this inequality. This inequality provides bounds for the mean value of a convex function.

Recently, mathematicians have focused on obtaining new variants of the $\mathcal{HH}$ inequality by giving generalizations, improvements, refinements and extensions, see References [34–36].

Exponentially convex functions have emerged as a significant new class of convex functions, which have important applications in technology, data science and statistics. The main motivation of this paper depends on new inequalities that have been proved via $\mathcal{GRLFI}$ and applied for exponentially convex functions. This identity offers new upper bounds and estimations of $\mathcal{HH}$ type integral inequalities. Some particular cases have been discussed, which can be deduced from these consequences.

Recall the definition of an exponentially convex function, which is investigated by Dragomir and Gomm [34]:

Definition 1. *(See [34]) A positive real-valued function $\mathcal{U} : K \subseteq \mathcal{R} \longrightarrow (0,\infty)$ is said to be exponentially convex on K if the inequality*

$$e^{\mathcal{U}(\tau x+(1-\tau)y)} \leq \tau e^{\mathcal{U}(x)} + (1-\tau)e^{\mathcal{U}(y)}$$

holds for $x, y \in K$ and $\tau \in [0,1]$.

Exponentially convex functions are used to manipulate for statistical learning, sequential prediction and stochastic optimization, see References [37–39].

After the class of exponentially convex functions was introduced by Dragomir and Gomm [34], Alirezai and Mathar [37] have investigated the mathematical perspectives along with their fertile applications in statistics and information theory, see References [37,39]. Due to its significance, Pecaric and Jaksetic [40,41] used another kind of exponentially convex function introduced in Reference [42] and have provided some applications in Euler-Radau expansions and stolarsky means. Our intention is to use the exponentially convexity property of the functions as well as the absolute values of their derivatives in order to establish estimates for $\mathcal{GRLFI}$.

Definition 2. *([1–3]) Let (d_1, d_2) $(-\infty \leq d_1 < d_2 \leq \infty)$ be a finite or infinite real interval and $\rho > 0$. Let $\Psi(x)$ be an increasing and positive monotone function on $(d_1, d_2]$ with a continuous derivative on (d_1, d_2). Then the left and right-sided generalized Riemann-Liouville fractional integrals of a function $\mathcal{U}$ on $[d_1, d_2]$ are defined by*

$$I^{\rho,\Psi}_{d_1^+}\mathcal{U}(\tau) = \frac{1}{\Gamma(\rho)}\int_{d_1}^{\tau} \Psi'(x)(\Psi(\tau)-\Psi(x))^{\rho-1}\mathcal{U}(x)dx, \tag{2}$$

and

$$I^{\rho,\Psi}_{d_2^-}\mathcal{U}(\tau) = \frac{1}{\Gamma(\rho)}\int_{\tau}^{d_2} \Psi'(x)(\Psi(x)-\Psi(\tau))^{\rho-1}\mathcal{U}(x)dx, \tag{3}$$

respectively; with $\Gamma(.)$, the classical gamma function.

Remark 1. *Many known defined fractional integral operators are just special cases of (2) and (3).*

1. *Setting $\mathcal{U}(\tau) = \tau$, it turns into the both sided (RLI).*
2. *Setting $\mathcal{U}(\tau) = \log\tau$, the Hadamard fractional integrals are obtained [1,3].*
3. *Setting $\mathcal{U}(\tau) = \frac{\tau^\beta}{\beta}$, $\beta > 0$ it turns into the both sided Katugampola fractional integrals given in Reference [33].*
4. *Setting $\mathcal{U}(\tau) = \frac{(\tau-a)^\beta}{\beta}$, $\beta > 0$, the operators in reference [43] are obtained.*
5. *Setting $\mathcal{U}(\tau) = \frac{\tau^{\epsilon+\varsigma}}{\epsilon+\varsigma}$, it turns into the both sided generalized conformable fractional integrals defined by Khan et al. in reference [44].*

The principal objective of this paper is to use a new convex class and a new integral operator to obtain new versions of $\mathcal{HH}$-inequality that give bounds for the mean value of convexity. Also, we will establish some more general estimates and related modulus inequalities for $\mathcal{GRLFI}$ via exponentially convex functions. In addition, the accuracy of the results was tested with applications of special means and with some examples.

2. $\mathcal{HH}$ Inequality for $\mathcal{GRLFI}$

Theorem 1. *Let $\mathcal{U} : [d_1, d_2] \to \mathcal{R}$ be a positive function, for $0 \le d_1 < d_2$, and $e^{\mathcal{U}} \in L_1([d_1, d_2])$. Let $\Psi(x)$ is an increasing and positive monotone function on $(d_1, d_2]$, with continuous derivative $\Psi'(x)$ on (d_1, d_2). Let $\mathcal{U}$ is an exponentially convex function and $\rho \in (0, 1)$. Then*

$$e^{\mathcal{U}\left(\frac{d_1+d_2}{2}\right)} \le \frac{\Gamma(\rho+1)}{(d_2-d_1)^\rho}\left[I^{\rho,\Psi}_{d_1^+}(e^{\mathcal{U}} \circ \Psi)(\Psi^{-1}(d_2)) + I^{\rho,\Psi}_{d_2^-}(e^{\mathcal{U}} \circ \Psi)(\Psi^{-1}(d_1))\right] \le \frac{e^{\mathcal{U}(d_1)} + e^{\mathcal{U}(d_2)}}{2}. \tag{4}$$

Proof. Since $\mathcal{U}$ is an exponentially convex function for $[d_1, d_2]$, we have

$$e^{\mathcal{U}\left(\frac{u+v}{2}\right)} \le \frac{e^{\mathcal{U}(u)} + e^{\mathcal{U}(v)}}{2}. \tag{5}$$

Let $u = \tau d_1 + (1-\tau)d_2$ and $v = (1-\tau)d_1 + \tau d_2$, we get

$$2e^{\mathcal{U}\left(\frac{d_1+d_2}{2}\right)} \le e^{\mathcal{U}(\tau d_1+(1-\tau)d_2)} + e^{\mathcal{U}((1-\tau)d_1+\tau d_2)}. \tag{6}$$

Multiplying by $\tau^{\rho-1}$ on both sides of inequality (6) and then integrating w.r.t τ over $[0, 1]$, implies

$$\frac{2}{\rho}e^{\mathcal{U}\left(\frac{d_1+d_2}{2}\right)} \le \int_0^1 \tau^{\rho-1}e^{\mathcal{U}(\tau d_1+(1-\tau)d_2)}d\tau + \int_0^1 \tau^{\rho-1}e^{\mathcal{U}((1-\tau)d_1+\tau d_2)}d\tau. \tag{7}$$

Now consider,

$$\begin{aligned}
&\frac{\Gamma(\rho+1)}{2(d_2-d_1)^\rho}\left[I^{\rho,\Psi}_{d_1^+}(e^{\mathcal{U}} \circ \Psi)(\Psi^{-1}(d_2)) + I^{\rho,\Psi}_{d_2^-}(e^{\mathcal{U}} \circ \Psi)(\Psi^{-1}(d_1))\right] \\
&= \frac{\Gamma(\rho+1)}{2(d_2-d_1)^\rho}\frac{1}{\Gamma(\rho)}\left[\int_{\Psi^{-1}(d_1)}^{\Psi^{-1}(d_2)} \Psi'(z)(d_2-\Psi(z))^{\rho-1}(e^{\mathcal{U}} \circ \Psi)(z)dz \right. \\
&\quad \left. + \int_{\Psi^{-1}(d_1)}^{\Psi^{-1}(d_2)} \Psi'(z)(\Psi(z)-d_1)^{\rho-1}(e^{\mathcal{U}} \circ \Psi)(z)dz\right] \\
&= \frac{\rho}{2(d_2-d_1)^\rho}\left[\int_{\Psi^{-1}(d_1)}^{\Psi^{-1}(d_2)} \Psi'(z)(d_2-\Psi(z))^{\rho-1}e^{\mathcal{U}}(\Psi(z))dz \right. \\
&\quad \left. + \int_{\Psi^{-1}(d_1)}^{\Psi^{-1}(d_2)} \Psi'(z)(\Psi(z)-d_1)^{\rho-1}e^{\mathcal{U}}(\Psi(z))dz\right] \\
&= \frac{\rho}{2}\left[\int_0^1 \tau^{\rho-1}e^{\mathcal{U}(\tau d_1+(1-\tau)d_2)}d\tau + \int_0^1 \tau^{\rho-1}e^{\mathcal{U}((1-\tau)d_1+\tau d_2)}d\tau\right] \\
&\ge e^{\mathcal{U}\left(\frac{d_1+d_2}{2}\right)},
\end{aligned}$$

by using (7). Thus the first inequality of (4) is proved.

Next, we again use the exponential convexity of $\mathcal{U}$, that is,

$$e^{\mathcal{U}(\tau d_1+(1-\tau)d_2)}+e^{\mathcal{U}((1-\tau)d_1+\tau d_2)}\le e^{\mathcal{U}(d_1)}+e^{\mathcal{U}(d_2)}. \tag{8}$$

Multiplying by $\tau^{\rho-1}$ on both sides of inequality (8), and then integrating w.r.t τ over $[0,1]$, implies

$$\int_0^1 \tau^{\rho-1}e^{\mathcal{U}(\tau d_1+(1-\tau)d_2)}d\tau+\int_0^1 \tau^{\rho-1}e^{\mathcal{U}((1-\tau)d_1+\tau d_2)}d\tau\le\frac{e^{\mathcal{U}(d_1)}+e^{\mathcal{U}(d_2)}}{\rho}.$$

That is,

$$\frac{\Gamma(\rho+1)}{(d_2-d_1)^{\rho}}\left[I^{\rho,\Psi}_{d_1^+}(e^{\mathcal{U}}\circ\Psi)(\Psi^{-1}(d_2))+I^{\rho,\Psi}_{d_2^-}(e^{\mathcal{U}}\circ\Psi)(\Psi^{-1}(d_1))\right]\le\frac{e^{\mathcal{U}(d_1)}+e^{\mathcal{U}(d_2)}}{2}.$$

Hence the proof is completed. □

Our next result is the subsequent lemma, which is useful for our coming results.

Lemma 1. *Let $\mathcal{U}:[d_1,d_2]\to\mathcal{R}$ be a differentiable mapping, for $0\le d_1<d_2$, and $e^{\mathcal{U}}\in L_1([d_1,d_2])$. Let $\Psi(x)$ is an increasing and positive monotone function on $(d_1,d_2]$, with continuous derivative $\Psi'(x)$ on (d_1,d_2) and $\rho\in(0,1)$. Then*

$$\begin{aligned}&\frac{e^{\mathcal{U}(d_1)}+e^{\mathcal{U}(d_2)}}{2}-\frac{\Gamma(\rho+1)}{2(d_2-d_1)^{\rho}}\left[I^{\rho,\Psi}_{d_1^+}(e^{\mathcal{U}}\circ\Psi)(\Psi^{-1}(d_2))+I^{\rho,\Psi}_{d_2^-}(e^{\mathcal{U}}\circ\Psi)(\Psi^{-1}(d_1))\right]\\&=\frac{1}{2}\left[\int_{\Psi^{-1}(d_1)}^{\Psi^{-1}(d_2)}\Psi'(z)\Big(\frac{\Psi(z)-d_1}{d_2-d_1}\Big)^{\rho}(e^{\mathcal{U}}\circ\Psi)(z)\mathcal{U}'(\Psi(z))dz\right.\\&\left.-\int_{\Psi^{-1}(d_1)}^{\Psi^{-1}(d_2)}\Psi'(z)\Big(\frac{d_2-\Psi(z)}{d_2-d_1}\Big)^{\rho}(e^{\mathcal{U}}\circ\Psi)(z)\mathcal{U}'(\Psi(z))dz\right].\end{aligned} \tag{9}$$

Proof. Consider

$$\begin{aligned}\mathbb{I}&=\frac{d_2-d_1}{2}\int_0^1[(1-\tau)^{\rho}-\tau^{\rho}]e^{\mathcal{U}\left((1-\tau)d_1+\tau d_2\right)}\mathcal{U}'((1-\tau)d_1+\tau d_2)d\tau\\&=\frac{d_2-d_1}{2}\left[\int_0^1(1-\tau)^{\rho}e^{\mathcal{U}\left((1-\tau)d_1+\tau d_2\right)}\mathcal{U}'((1-\tau)d_1+\tau d_2)d\tau\right.\\&\left.-\int_0^1\tau^{\rho}e^{\mathcal{U}\left((1-\tau)d_1+\tau d_2\right)}\mathcal{U}'((1-\tau)d_1+\tau d_2)d\tau\right].\end{aligned}$$

By making a change of variable in the above equation $\Psi(z)=(1-\tau)d_1+\tau d_2$, we have

$$\begin{aligned}\mathbb{I} &= \frac{1}{2}\left[\int\limits_{\Psi^{-1}(d_1)}^{\Psi^{-1}(d_2)} \Psi'(z)\Big(\frac{d_2-\Psi(z)}{d_2-d_1}\Big)^{\rho}(e^{\mathcal{U}}\circ\Psi)(z)\mathcal{U}'(\Psi(z))dz\right.\\ &\quad\left.-\int\limits_{\Psi^{-1}(d_1)}^{\Psi^{-1}(d_2)} \Psi'(z)\Big(\frac{\Psi(z)-d_1}{d_2-d_1}\Big)^{\rho}(e^{\mathcal{U}}\circ\Psi)(z)\mathcal{U}'(\Psi(z))dz\right]\\ &= I_1+I_2. \end{aligned} \tag{10}$$

Now

$$\begin{aligned}&\frac{\Gamma(\rho+1)}{2(d_2-d_1)^{\rho}} I^{\rho,\Psi}_{d_1^+}(e^{\mathcal{U}}\circ\Psi)(\Psi^{-1}(d_2))\\ &= \frac{\rho}{2(d_2-d_1)^{\rho}}\int\limits_{\psi^{-1}(d_1)}^{\psi^{-1}(d_2)} \psi'(z)(d_2-\psi(z))^{\rho-1}(e^{\mathcal{U}}\circ\psi)(z)dz\\ &= -\frac{1}{2(d_2-d_1)^{\rho}}\int\limits_{\psi^{-1}(d_1)}^{\psi^{-1}(d_2)} (e^{\mathcal{U}}\circ\psi)d(d_2-\psi(z))^{\rho}\\ &= \frac{1}{2(d_2-d_1)^{\rho}}\left[(d_2-d_1)^{\rho}e^{\mathcal{U}(d_1)}+\int\limits_{\psi^{-1}(d_1)}^{\psi^{-1}(d_2)} \psi'(z)(d_2-\Psi(z))^{\rho}(e^{\mathcal{U}}\circ\psi)(z)\mathcal{U}'(\Psi(z))dz\right]\\ &= \frac{e^{\mathcal{U}(d_1)}}{2}+I_1\end{aligned}$$

and

$$\begin{aligned}&\frac{\Gamma(\rho+1)}{2(d_2-d_1)^{\rho}} I^{\rho,\Psi}_{d_2^-}(e^{\mathcal{U}}\circ\Psi)(\Psi^{-1}(d_1))\\ &= \frac{\rho}{2(d_2-d_1)^{\rho}}\int\limits_{\Psi^{-1}(d_1)}^{\psi^{-1}(d_2)} \psi'(z)(\Psi(z)-d_1)^{\rho-1}(e^{\mathcal{U}}\circ\Psi)(z)dz\\ &= \frac{1}{2(d_2-d_1)^{\rho}}\int\limits_{\Psi^{-1}(d_1)}^{\psi^{-1}(d_2)} (e^{\mathcal{U}}\circ\psi)d(\Psi(z)-d_1)^{\rho}\\ &= \frac{1}{2(d_2-d_1)^{\rho}}\left[(d_2-d_1)^{\rho}e^{\mathcal{U}(d_2)}-\int\limits_{\Psi^{-1}(d_1)}^{\Psi^{-1}(d_2)} \Psi'(z)(\Psi(z)-d_1)^{\rho}(e^{\mathcal{U}}\circ\psi)(z)\mathcal{U}'(\Psi(z))dz\right]\\ &= \frac{e^{\mathcal{U}(d_2)}}{2}-I_2.\end{aligned}$$

It follows that

$$\frac{e^{\mathcal{U}}(d_2)+e^{\mathcal{U}}(d_1)}{2}-I_1-I_2=\mathbb{I}.$$

This completes the proof. □

Theorem 2. *Let $\mathcal{U} : [d_1, d_2] \to \mathcal{R}$ be a differentiable mapping, for $0 \le d_1 < d_2$, and $e^{\mathcal{U}} \in L_1([d_1, d_2])$. Let $\Psi(x)$ be an increasing and positive monotone function on $(d_1, d_2]$, with continuous derivative $\Psi'(x)$ on (d_1, d_2) and $\rho \in (0,1)$. If $|\mathcal{U}'|^q$ is exponentially convex and $q \ge 1$, then*

$$\left|\frac{e^{\mathcal{U}(d_1)}+e^{\mathcal{U}(d_2)}}{2} - \frac{\Gamma(\rho+1)}{2(d_2-d_1)^\rho}\left[I^{\rho,\Psi}_{d_1^+}(e^{\mathcal{U}}\circ\Psi)(\Psi^{-1}(d_2)) + I^{\rho,\Psi}_{d_2^-}(e^{\mathcal{U}}\circ\Psi)(\Psi^{-1}(d_1))\right]\right|$$
$$\frac{d_2-d_1}{2}\left(\frac{2}{\rho+1}\left(1-\frac{1}{2^\rho}\right)\right)^{1-\frac{1}{q}}\left[\frac{1}{\rho+3}\left(1-\frac{1}{2^{\rho+2}}\right)\right]^{\frac{1}{q}}\left[|e^{\mathcal{U}(d_1)}\mathcal{U}'(d_1)|^q + |e^{\mathcal{U}(d_2)}\mathcal{U}'(d_2)|^q\right]^{\frac{1}{q}}. \tag{11}$$

Proof. First note that, for every $z \in \left(\Psi^{-1}(d_1)\Psi^{-1}(d_2)\right)$, we have $d_1 < \Psi(z) < d_2$. Let $\tau = \frac{d_2-\Psi(z)}{d_2-d_1}$, then we have $\Psi(z) = \tau d_1 + (1-\tau)d_2$. Applying Lemma 1, Hölder's inequality and exponentially convexity of $|\mathcal{U}'|$, we obtain

$$\left|\frac{e^{\mathcal{U}(d_1)}+e^{\mathcal{U}(d_2)}}{2} - \frac{\Gamma(\rho+1)}{2(d_2-d_1)^\rho}\left[I^{\rho,\Psi}_{d_1^+}(e^{\mathcal{U}}\circ\Psi)(\Psi^{-1}(d_2)) + I^{\rho,\Psi}_{d_2^-}(e^{\mathcal{U}}\circ\Psi)(\Psi^{-1}(d_1))\right]\right|$$
$$\le \frac{1}{2}\left[\int_{\Psi^{-1}(d_1)}^{\Psi^{-1}(d_2)}\left|\left(\frac{\Psi(z)-d_1}{d_2-d_1}\right)^\rho - \left(\frac{d_2-\Psi(z)}{d_2-d_1}\right)^\rho\right||(e^{\mathcal{U}}\circ\Psi)(z)|d(\mathcal{U}(\Psi(z)))\right]$$
$$= \frac{d_2-d_1}{2}\left(\int_0^1 |(1-\tau)^\rho - \tau^\rho|d\tau\right)^{1-\frac{1}{q}}\left(\int_0^1 |(1-\tau)^\rho - \tau^\rho||e^{\mathcal{U}(\tau d_1+(1-\tau)d_2)}\mathcal{U}'(\tau d_1 + (1-\tau)d_2)|^q d\tau\right)^{\frac{1}{q}}$$
$$\le \frac{d_2-d_1}{2}\left(\int_0^1 |(1-\tau)^\rho - \tau^\rho|d\tau\right)^{1-\frac{1}{q}}$$
$$\times\left(\int_0^1 |(1-\tau)^\rho - \tau^\rho||\{\tau|e^{\mathcal{U}(d_1)}|^q + (1-\tau)|e^{\mathcal{U}(d_2)}|^q\}\{\tau|\mathcal{U}'(d_1)|^q + (1-\tau)|\mathcal{U}'(d_2)|^q\}d\tau\right)^{\frac{1}{q}}$$
$$= \frac{d_2-d_1}{2}\left(\frac{2}{\rho+1}\left(1-\frac{1}{2^\rho}\right)\right)^{1-\frac{1}{q}}\left(\int_0^1 |(1-\tau)^\rho - \tau^\rho|\left[\tau^2|e^{\mathcal{U}(d_1)}\mathcal{U}'(d_1)|^q\right.\right.$$
$$\left.\left. +(1-\tau)^2|e^{\mathcal{U}(d_2)}\mathcal{U}'(d_2)|^q + \tau(1-\tau)\{|e^{\mathcal{U}(d_1)}\mathcal{U}'(d_2)|^q + |e^{\mathcal{U}(d_2)}\mathcal{U}'(d_1)|^q\}\right]d\tau\right)^{\frac{1}{q}}$$
$$= \frac{d_2-d_1}{2}\left(\frac{2}{\rho+1}\left(1-\frac{1}{2^\rho}\right)\right)^{1-\frac{1}{q}}$$
$$\times\left(\int_0^1 |(1-\tau)^\rho - \tau^\rho|\left[\tau^2|e^{\mathcal{U}(d_1)}\mathcal{U}'(d_1)|^q + (1-\tau)^2|e^{\mathcal{U}(d_2)}\mathcal{U}'(d_2)|^q + \tau(1-\tau)\Delta(d_1,d_2)\right]d\tau\right)^{\frac{1}{q}}$$
$$:= \frac{d_2-d_1}{2}\left(\frac{2}{\rho+1}\left(1-\frac{1}{2^\rho}\right)\right)^{1-\frac{1}{q}}(S_1+S_2). \tag{12}$$

where

$$S_1 := \int_0^{\frac{1}{2}} [(1-\tau)^\rho - \tau^\rho]\left[\tau^2|e^{\mathcal{U}(d_1)}\mathcal{U}'(d_1)|^q + (1-\tau)^2|e^{\mathcal{U}(d_2)}\mathcal{U}'(d_2)|^q + \tau(1-\tau)\Delta(d_1,d_2)\right]d\tau,$$
$$S_2 := \int_{\frac{1}{2}}^1 [\tau^\rho - (1-\tau)^\rho]\left[\tau^2|e^{\mathcal{U}(d_1)}\mathcal{U}'(d_1)|^q + (1-\tau)^2|e^{\mathcal{U}(d_2)}\mathcal{U}'(d_2)|^q + \tau(1-\tau)\Delta(d_1,d_2)\right]d\tau.$$

Now

$$\int_0^{\frac{1}{2}} \tau^2(1-\tau)^\rho d\tau = \int_{\frac{1}{2}}^1 \tau^\rho(1-\tau)^2 d\tau = \beta_{\frac{1}{2}}(3,\rho+1),$$

$$\int_0^{\frac{1}{2}} \tau^\rho(1-\tau)^2 d\tau = \int_{\frac{1}{2}}^1 \tau^2(1-\tau)^\rho d\tau = \beta_{\frac{1}{2}}(\rho+1,3),$$

$$\int_0^{\frac{1}{2}} \tau^{\rho+2} d\tau = \int_{\frac{1}{2}}^1 (1-\tau)^{\rho+2} d\tau = \frac{1}{(\rho+3)2^{\rho+3}},$$

$$\int_0^{\frac{1}{2}} (1-\tau)^{\rho+2} d\tau = \int_{\frac{1}{2}}^1 \tau^{\rho+2} d\tau = \frac{1}{\rho+3} - \frac{1}{(\rho+3)2^{\rho+3}},$$

$$\int_0^{\frac{1}{2}} \left[(1-\tau)^{\rho+1}\tau - \tau^{\rho+1}(1-\tau)\right] d\tau = \beta_{\frac{1}{2}}(\rho+2,2) - \beta_{\frac{1}{2}}(2,\rho+2) = 0,$$

$$\int_{\frac{1}{2}}^1 \left[\tau^{\rho+1}(1-\tau) - (1-\tau)^{\rho+1}\tau\right] d\tau = \beta_{\frac{1}{2}}(\rho+2,2) - \beta_{\frac{1}{2}}(2,\rho+2) = 0.$$

By substituting the above integral values in (12) and after some simplification, we get the required inequality (13). □

Corollary 1. *Letting $q = 1$, then under the assumption of Theorem 2, we have*

$$\begin{aligned} &\left|\frac{e^{\mathcal{U}(d_1)} + e^{\mathcal{U}(d_2)}}{2} - \frac{\Gamma(\rho+1)}{2(d_2-d_1)^\rho}\left[I_{d_1^+}^{\rho,\Psi}\left(e^{\mathcal{U}} \circ \Psi\right)\left(\Psi^{-1}(d_2)\right) + I_{d_2^-}^{\rho,\Psi}\left(e^{\mathcal{U}} \circ \Psi\right)\left(\Psi^{-1}(d_1)\right)\right]\right| \\ &\leq \frac{d_2-d_1}{2}\left[\frac{1}{\rho+3}\left(1-\frac{1}{2^{\rho+2}}\right)\right]\left[|e^{\mathcal{U}(d_1)}\mathcal{U}'(d_1)| + |e^{\mathcal{U}(d_2)}\mathcal{U}'(d_2)|\right]. \end{aligned} \tag{13}$$

Proof. Since Ψ is differentiable and strictly increasing function, we can write $(\Psi(x) - \Psi(\tau))^{\rho-1} \leq (\Psi(x) - \Psi(d_1))^{\rho-1}$, where as $x \in [d_1, d_2]$ and $\tau \in [d_1, x]$, $\rho \geq 1$, and $\Psi'(\tau) > 0$. Then, the subsequent inequality holds true

$$\Psi'(\tau)(\Psi(x) - \Psi(\tau))^{\rho-1} \leq \Psi'(\tau)(\Psi(x) - \Psi(d_1))^{\rho-1}. \tag{14}$$

By exponentially convexity of $\mathcal{U}$, we have

$$e^{\mathcal{U}(\tau)} \leq \frac{x-\tau}{x-d_1}e^{\mathcal{U}(d_1)} + \frac{\tau-d_1}{x-d_1}e^{\mathcal{U}(x)}. \tag{15}$$

From (14) and (15), one has

$$\begin{aligned} &\int_{d_1}^x \Psi'(\tau)(\Psi(x) - \Psi(\tau))^{\rho-1} e^{\mathcal{U}(\tau)} d\tau \\ \leq\ &\frac{(\Psi(x) - \Psi(d_1))^{\rho-1}}{x-d_1}\left[e^{\mathcal{U}(d_1)}\int_{d_1}^x (x-\tau)\Psi'(\tau)d\tau + e^{\mathcal{U}(x)}\int_{d_1}^x (\tau-d_1)\Psi'(\tau)d\tau\right]. \end{aligned}$$

By using (2) of Definition 2, we get

$$\Gamma(\rho) I^{\rho,\Psi}_{d_1^+} e^{\mathcal{U}(x)} \leq \frac{\left(\Psi(x)-\Psi(d_1)\right)^{\rho-1}}{x-d_1}\left[(x-d_1)\left[\Psi(x)e^{\mathcal{U}(x)} - e^{\mathcal{U}(d_1)}\Psi(d_1)\right] - \left(e^{\mathcal{U}(x)} - e^{\mathcal{U}(d_1)}\right)\int_{d_1}^{x}\Psi(\tau)d\tau\right]. \tag{16}$$

Now for $x \in [d_1, d_2]$, $\tau \in [x, d_2]$ and $\delta \geq 1$, the subsequent inequality holds true

$$\Psi'(\tau)\left(\Psi(\tau)-\Psi(x)\right)^{\delta-1} \leq \Psi'(\tau)\left(\Psi(d_2)-\Psi(x)\right)^{\delta-1}. \tag{17}$$

By exponentially convexity of $\mathcal{U}$, we have

$$e^{\mathcal{U}(\tau)} \leq \frac{\tau - x}{d_2 - x} e^{\mathcal{U}(d_2)} + \frac{d_2 - \tau}{d_2 - x} e^{\mathcal{U}(x)}. \tag{18}$$

Adopting the same procedure as we did for (14) and (15) , one can get from (17) and (18) the coming inequality

$$\Gamma(\delta) I^{\delta,\Psi}_{d_2^-} e^{\mathcal{U}(x)} \leq \frac{\left(\Psi(d_2)-\Psi(x)\right)^{\delta-1}}{d_2-x}\left[(d_2-x)\left[\Psi(d_2)e^{\mathcal{U}(d_2)} - e^{\mathcal{U}(x)}\Psi(x)\right] - \left(e^{\mathcal{U}(d_2)} - e^{\mathcal{U}(x)}\right)\int_{x}^{d_2}\Psi(\tau)d\tau\right]. \tag{19}$$

From inequalities (16) and (19), we get (13). Hence the proof is completed. □

Particular cases are stated as follows.

Corollary 2. *Choosing $\rho = \delta$ in (13), then we have a new inequality for $\mathcal{GRLFI}$;*

$$\begin{aligned}
&\Gamma(\rho) I^{\rho,\Psi}_{d_1^+} e^{\mathcal{U}(x)} + \Gamma(\rho) I^{\rho,\Psi}_{d_2^-} e^{\mathcal{U}(x)} \\
\leq\ & \frac{\left(\Psi(x)-\Psi(d_1)\right)^{\rho-1}}{x-d_1}\left[(x-d_1)\left[\Psi(x)e^{\mathcal{U}(x)} - e^{\mathcal{U}(d_1)}\Psi(d_1)\right] - \left(e^{\mathcal{U}(x)} - e^{\mathcal{U}(d_1)}\right)\int_{d_1}^{x}\Psi(\tau)d\tau\right] \\
&+ \frac{\left(\Psi(d_2)-\Psi(x)\right)^{\rho-1}}{d_2-x}\left[(d_2-x)\left[\Psi(d_2)e^{\mathcal{U}(d_2)} - e^{\mathcal{U}(x)}\Psi(x)\right] - \left(e^{\mathcal{U}(d_2)} - e^{\mathcal{U}(x)}\right)\int_{x}^{d_2}\Psi(\tau)d\tau\right].
\end{aligned}$$

Corollary 3. *Choosing $x = d_1$ and $x = d_2$ in (13), adding the resulting inequalities, then the conditions of Theorem 1 are satisfied, we have*

$$\begin{aligned}
\Gamma(\rho) I^{\rho,\Psi}_{d_1^+} e^{\mathcal{U}(d_2)} + \Gamma(\delta) I^{\delta,\Psi}_{d_2^-} e^{\mathcal{U}(d_1)} &\leq \left(\frac{\left(\Psi(d_2)-\Psi(d_1)\right)^{\rho-1} + \left(\Psi(d_2)-\Psi(d_1)\right)^{\delta-1}}{d_2-d_1}\right) \\
\times\left((d_2-d_1)\left[e^{\mathcal{U}(d_2)}\Psi(d_2) - e^{\mathcal{U}(d_1)}\Psi(d_1)\right] - \left[e^{\mathcal{U}(d_2)} - e^{\mathcal{U}(d_1)}\right]\int_{d_1}^{d_2}\Psi(\tau)d\tau\right)&.
\end{aligned} \tag{20}$$

Corollary 4. *If we take $\rho = \delta$ in (20), then we get the following inequality for $\mathcal{GRLFI}$*

$$\Gamma(\rho)\left[I_{d_1^+}^{\rho+1,\Psi}\Psi(d_2) + I_{d_2^-}^{\rho+1,\Psi}\Psi(d_1)\right] \leq \left(\frac{2(\Psi(d_2)-\Psi(d_1))^{\rho-1}}{d_2-d_1}\right)$$
$$\times\left((d_2-d_1)\left[e^{\mathcal{U}(d_2)}\Psi(d_2) - e^{\mathcal{U}(d_1)}\Psi(d_1)\right] - \left[e^{\mathcal{U}(d_2)} - e^{\mathcal{U}(d_1)}\right]\int_{d_1}^{d_2}\Psi(\tau)d\tau\right).$$

Theorem 3. *Let $\mathcal{U}, \Psi : [d_1, d_2] \longrightarrow \mathcal{R}$ be the functions such that $e^{\mathcal{U}}$ be differentiable function, Ψ is also differentiable and strictly increasing with $\Psi' \in L_1([d_1, d_2])$. Then for $\rho, \delta > 0$, $x \in [d_1, d_2]$ we have*

$$\begin{aligned}
&\left|\Gamma(\rho+1)I_{d_1^+}^{\rho,\Psi}e^{\mathcal{U}(x)} + \Gamma(\delta+1)I_{d_2^+}^{\delta,\Psi}e^{\mathcal{U}(x)} - (\Psi(x)-\Psi(d_1))^{\rho}e^{\mathcal{U}(d_1)} + (\Psi(d_2)-\Psi(x))^{\delta}e^{\mathcal{U}(d_2)}\right| \\
\leq\ & \frac{(\Psi(x)-\Psi(d_1))^{\rho}(x-d_1)|e^{\mathcal{U}(d_1)}\mathcal{U}'(d_1)| + (\Psi(d_2)-\Psi(x))^{\delta}(d_2-x)|e^{\mathcal{U}(d_2)}\mathcal{U}'(d_2)|}{3} \\
&+ \frac{\left\{(\Psi(d_2)-\Psi(x))^{\delta}(d_2-x) + (\Psi(x)-\Psi(d_1))^{\rho}(x-d_1)\right\}|e^{\mathcal{U}(x)}\mathcal{U}'(x)|}{3} \\
&+ \frac{(\Psi(x)-\Psi(d_1))^{\rho}(x-d_1)\Delta(d_1,x) + (\Psi(d_2)-\Psi(x))^{\delta}(d_2-x)\Delta(d_2,x)}{6},
\end{aligned} \tag{21}$$

where

$$\Delta(d_1, x) = |e^{\mathcal{U}(x)}\mathcal{U}'(d_1)| + |e^{\mathcal{U}(d_1)}\mathcal{U}'(x)|, \tag{22}$$

$$\Delta(d_2, x) = |e^{\mathcal{U}(x)}\mathcal{U}'(d_2)| + |e^{\mathcal{U}(d_2)}\mathcal{U}'(x)|. \tag{23}$$

respectively.

Proof. From the convexity of $|(e^{\mathcal{U}})'|$, we obtain

$$\begin{aligned}
|e^{\mathcal{U}(\tau)}\mathcal{U}'(\tau)| &\leq \left[\frac{x-\tau}{x-d_1}|e^{\mathcal{U}(d_1)}| + \frac{\tau-d_1}{x-d_1}|e^{\mathcal{U}(x}|\right]\left[\frac{x-\tau}{x-d_1}|\mathcal{U}'(d_1)| + \frac{\tau-d_1}{x-d_1}|\mathcal{U}'(x)|\right] \\
&\leq \left(\frac{x-\tau}{x-d_1}\right)^2|e^{\mathcal{U}(d_1)}\mathcal{U}'(d_1)| + \left(\frac{\tau-d_1}{x-d_1}\right)^2|e^{\mathcal{U}(x)}\mathcal{U}'(x)| \\
&\quad + \left(\frac{x-\tau}{x-d_1}\right)\left(\frac{\tau-d_1}{x-d_1}\right)\{|e^{\mathcal{U}(d_1)}\mathcal{U}'(x)| + |e^{\mathcal{U}(x)}\mathcal{U}'(d_1)|\} \\
&= \left(\frac{x-\tau}{x-d_1}\right)^2|e^{\mathcal{U}(d_1)}\mathcal{U}'(d_1)| + \left(\frac{\tau-d_1}{x-d_1}\right)^2|e^{\mathcal{U}(x)}\mathcal{U}'(x)| + \left(\frac{x-\tau}{x-d_1}\right)\left(\frac{\tau-d_1}{x-d_1}\right)\Delta(d_1,x).
\end{aligned} \tag{24}$$

From (24), we have

$$e^{\mathcal{U}(\tau)}\mathcal{U}'(\tau) \leq \left(\frac{x-\tau}{x-d_1}\right)^2|e^{\mathcal{U}(d_1)}\mathcal{U}'(d_1)| + \left(\frac{\tau-d_1}{x-d_1}\right)^2|e^{\mathcal{U}(x)}\mathcal{U}'(x)| + \left(\frac{x-\tau}{x-d_1}\right)\left(\frac{\tau-d_1}{x-d_1}\right)\Delta(d_1,x). \tag{25}$$

Since Ψ is a differentiable and strictly increasing function, we have the subsequent inequality

$$(\Psi(x) - \Psi(\tau))^{\rho} \leq (\Psi((x) - \Psi(d_1))^{\rho}, \tag{26}$$

where as $x \in [d_1, d_2]$ and $\tau \in [d_1, x]$, $\rho > 0$.

From (25) and (26), one has

$$
\begin{aligned}
& (\Psi(x)-\Psi(\tau))^{\rho} e^{\mathcal{U}(\tau)} \mathcal{U}'(\tau) \\
\leq\ & \frac{(\Psi(x)-\Psi(d_1))^{\rho}}{(x-d_1)^2}\Big[(x-\tau)^2|e^{\mathcal{U}(d_1)}\mathcal{U}'(d_1)|+(\tau-d_1)^2|e^{\mathcal{U}(x)}\mathcal{U}'(x)|+(x-\tau)(\tau-d_1)\Delta(d_1,x)\Big].
\end{aligned}
$$

Integrating over $[d_1, x]$, we have

$$
\begin{aligned}
& \int_{d_1}^{x} (\Psi(x)-\Psi(\tau))^{\rho} e^{\mathcal{U}(\tau)} \mathcal{U}'(\tau) d\tau \\
\leq\ & \frac{(\Psi(x)-\Psi(d_1))^{\rho}}{(x-d_1)^2}\Big[|e^{\mathcal{U}(d_1)}\mathcal{U}'(d_1)|\int_{d_1}^{x}(x-\tau)^2 d\tau+|e^{\mathcal{U}(x)}\mathcal{U}'(x)|\int_{d_1}^{x}(\tau-d_1)^2 d\tau+\Delta(d_1,x)\int_{d_1}^{x}(x-\tau)(\tau-d_1)d\tau\Big] \\
=\ & (\Psi(x)-\Psi(d_1))^{\rho}(x-d_1)\left[\frac{2\{|e^{\mathcal{U}(d_1)}\mathcal{U}'(d_1)|+|e^{\mathcal{U}(x)}\mathcal{U}'(x)|\}+\Delta(d_1,x)}{6}\right],
\end{aligned} \tag{27}
$$

and

$$
\begin{aligned}
& \int_{d_1}^{x} (\Psi(x)-\Psi(\tau))^{\rho} e^{\mathcal{U}(\tau)} \mathcal{U}'(\tau) d\tau = e^{\mathcal{U}(\tau)}(\Psi(x)-\Psi(\tau))^{\rho}|_{d_1}^{x} + \rho\int_{d_1}^{x}(\Psi(x)-\Psi(\tau))^{\rho-1}e^{\mathcal{U}(\tau)}\Psi'(\tau)d\tau \\
=\ & -e^{\mathcal{U}(d_1)}(\Psi(x)-\Psi(d_1))^{\rho}+\Gamma(\rho+1)I_{d_1^+}^{\rho,\Psi}e^{\mathcal{U}(x)}.
\end{aligned} \tag{28}
$$

Therefore (28) takes the form

$$
\begin{aligned}
& \Gamma(\rho+1)I_{d_1^+}^{\rho,\Psi}e^{\mathcal{U}(x)} - e^{\mathcal{U}(d_1)}(\Psi(x)-\Psi(d_1))^{\rho} \\
\leq\ & (\Psi(x)-\Psi(d_1))^{\rho}(x-d_1)\left[\frac{2\{|e^{\mathcal{U}(d_1)}\mathcal{U}'(d_1)|+|e^{\mathcal{U}(x)}\mathcal{U}'(x)|\}+\Delta(d_1,x)}{6}\right]
\end{aligned} \tag{29}
$$

Also from (24), one has

$$
e^{\mathcal{U}(\tau)}\mathcal{U}'(\tau) \geq -\left[\left(\frac{x-\tau}{x-d_1}\right)^2|e^{\mathcal{U}(d_1)}\mathcal{U}'(d_1)|+\left(\frac{\tau-d_1}{x-d_1}\right)^2|e^{\mathcal{U}(x)}\mathcal{U}'(x)|+\left(\frac{x-\tau}{x-d_1}\right)\left(\frac{\tau-d_1}{x-d_1}\right)\Delta(d_1,x)\right]. \tag{30}
$$

Following the same procedure as we did for (25), we also have

$$
\begin{aligned}
& e^{\mathcal{U}(d_1)}(\Psi(x)-\Psi(d_1))^{\rho} - \Gamma(\rho+1)I_{d_1^+}^{\rho,\Psi}e^{\mathcal{U}(x)} \\
\leq\ & (\Psi(x)-\Psi(d_1))^{\rho}(x-d_1)\left[\frac{2\{|e^{\mathcal{U}(d_1)}\mathcal{U}'(d_1)|+|e^{\mathcal{U}(x)}\mathcal{U}'(x)|\}+\Delta(d_1,x)}{6}\right]
\end{aligned} \tag{31}
$$

From (29) and (31), we get

$$
\begin{aligned}
& |\Gamma(\rho+1)I_{d_1^+}^{\rho,\Psi}e^{\mathcal{U}(x)} - e^{\mathcal{U}(d_1)}(\Psi(x)-\Psi(d_1))^{\rho}| \\
\leq\ & (\Psi(x)-\Psi(d_1))^{\rho}(x-d_1)\left[\frac{2\{|e^{\mathcal{U}(d_1)}\mathcal{U}'(d_1)|+|e^{\mathcal{U}(x)}\mathcal{U}'(x)|\}+\Delta(d_1,x)}{6}\right].
\end{aligned} \tag{32}
$$

By convexity of $|(e^{\mathcal{U}})'|$, we have

$$
\begin{aligned}
|e^{\mathcal{U}(\tau)}\mathcal{U}'(\tau)| \leq\ & \left(\frac{\tau-x}{d_2-x}\right)^2|e^{\mathcal{U}(d_2)}\mathcal{U}'(d_2)|+\left(\frac{d_2-\tau}{d_2-x}\right)^2|e^{\mathcal{U}(x)}\mathcal{U}'(x)| \\
& +\left(\frac{\tau-x}{d_2-x}\right)\left(\frac{d_2-\tau}{d_2-x}\right)\Delta(d_2,x).
\end{aligned} \tag{33}
$$

Now for $x \in [d_1, d_2]$ and $\tau \in [x, d_2]$ and $\delta > 0$, the following inequality holds true

$$\left(\Psi(\tau) - \Psi(x)\right)^{\delta} \leq \left(\Psi(d_2) - \Psi(x)\right)^{\delta}. \tag{34}$$

Following the same way as we have done for (25), (26) and (30) we can get from (33) and (34) the subsequent inequality

$$\begin{aligned} & |\Gamma(\delta+1) I^{\delta,\Psi}_{d_2^+} e^{\mathcal{U}(x)} - e^{\mathcal{U}(d_2)} \left(\Psi(d_2) - \Psi(x)\right)^{\delta}| \\ \leq \; & \left(\Psi(d_2) - \Psi(x)\right)^{\delta} (d_2 - x) \left[\frac{2\{|e^{\mathcal{U}(d_2)} \mathcal{U}'(d_2)| + |e^{\mathcal{U}(x)} \mathcal{U}'(x)|\} + \Delta(d_2, x)}{6} \right]. \end{aligned} \tag{35}$$

From inequalities (32) and (35) using triangular inequality, we get (21) which is desired. □

Particular cases are stated as follows.

Corollary 5. *Choosing $\rho = \delta$ in (21), then we have a new inequality for $\mathcal{GRLFI}$*

$$\begin{aligned} & \left|\Gamma(\rho+1)\left[I^{\rho,\Psi}_{d_1^+} e^{\mathcal{U}(x)} + I^{\rho,\Psi}_{d_2^+} e^{\mathcal{U}(x)}\right] - \left(\Psi(x) - \Psi(d_1)\right)^{\rho} e^{\mathcal{U}(d_1)} + \left(\Psi(d_2) - \Psi(x)\right)^{\rho} e^{\mathcal{U}(d_2)}\right| \\ \leq \; & \frac{\left(\Psi(x) - \Psi(d_1)\right)^{\rho} (x - d_1) |e^{\mathcal{U}(d_1)} \mathcal{U}'(d_1)| + \left(\Psi(d_2) - \Psi(x)\right)^{\rho} (d_2 - x) |e^{\mathcal{U}(d_2)} \mathcal{U}'(d_2)|}{3} \\ & + \frac{\left\{\left(\Psi(d_2) - \Psi(x)\right)^{\rho} (d_2 - x) + \left(\Psi(x) - \Psi(d_1)\right)^{\rho} (x - d_1)\right\} |e^{\mathcal{U}(x)} \mathcal{U}'(x)|}{3} \\ & + \frac{\left(\Psi(x) - \Psi(d_1)\right)^{\rho} (x - d_1) \Delta(d_1, x) + \left(\Psi(d_2) - \Psi(x)\right)^{\rho} (d_2 - x) \Delta(d_2, x)}{6}. \end{aligned}$$

To prove our next result we need the following Lemma.

Lemma 3. *Suppose that $\mathcal{U} : [d_1, d_2] \to \mathcal{R}$ is an exponentially convex function which is symmetric about $\frac{d_1+d_2}{2}$. Then we have*

$$e^{\mathcal{U}\left(\frac{d_1+d_2}{2}\right)} \leq e^{\mathcal{U}(x)}, \quad x \in [d_1, d_2]. \tag{36}$$

Proof. Write

$$\frac{d_1+d_2}{2} = \frac{1}{2}\left(\frac{x-d_1}{d_2-d_1} d_2 + \frac{d_2-x}{d_2-d_1} d_1\right) + \frac{1}{2}\left(\frac{x-d_1}{d_2-d_1} d_1 + \frac{d_2-x}{d_2-d_1} d_2\right).$$

Since $e^{\mathcal{U}}$ is convex, therefore we have

$$\begin{aligned} e^{\mathcal{U}\left(\frac{d_1+d_2}{2}\right)} & \leq \frac{1}{2} e^{\mathcal{U}\left(\frac{x-d_1}{d_2-d_1} d_2 + \frac{d_2-x}{d_2-d_1} d_1\right)} + \frac{1}{2} e^{\mathcal{U}\left(\frac{x-d_1}{d_2-d_1} d_1 + \frac{d_2-x}{d_2-d_1} d_2\right)} \\ & = \frac{1}{2}\left[e^{\mathcal{U}(x)} + e^{\mathcal{U}(d_1+d_2-x)}\right]. \end{aligned}$$

Also, $e^{\mathcal{U}}$ is symmetric about $\frac{d_1+d_2}{2}$, therefore we have $e^{\mathcal{U}(x)} = e^{\mathcal{U}(d_1+d_2-x)}$ and the inequality in (36) holds. □

Theorem 4. *Suppose that $\mathcal{U} : [d_1, d_2] \longrightarrow \mathcal{R}$ be an exponentially convex function such that $e^{\mathcal{U}}$ is positive convex and symmetric about $\frac{d_1+d_2}{2}$, Ψ is a differentiable and strictly increasing function having $\Psi' \in L_1([d_1, d_2])$. Then $\rho, \delta > 0$, we have*

$$e^{\mathcal{U}\left(\frac{d_1+d_2}{2}\right)}\left[\frac{(\Psi(d_2)-\Psi(d_1))^{\rho+1}}{\rho+1}+\frac{(\Psi(d_2)-\Psi(d_1))^{\delta+1}}{\delta+1}\right] \tag{37}$$

$$\leq \Gamma(\rho+1) I_{d_1^+}^{\rho+1,\Psi}\Psi(d_2)+\Gamma(\delta+1) I_{d_2^-}^{\delta+1,\Psi}\Psi(d_1) \leq \frac{(\Psi(d_2)-\Psi(d_1))^{\delta}+(\Psi(d_2)-\Psi(d_1))^{\rho}}{d_2-d_1}$$

$$\times\left[(d_2-d_1)[e^{\mathcal{U}(d_2)}\Psi(d_2)-e^{\mathcal{U}(d_1)}\Psi(d_1)]-[e^{\mathcal{U}(d_2)}-e^{\mathcal{U}(d_1)}]\int_{d_1}^{d_2}\Psi(\tau)d\tau\right].$$

Proof. Since Ψ is differentiable and strictly increasing function therefore $(\Psi(x)-\Psi(d_1))^{\delta} \leq (\Psi(d_2) - \Psi(d_1))^{\delta}$, where as $x \in [d_1, d_2]$, $\delta > 0$, and $\Psi'(x) > 0$. Hence, the following inequality holds true

$$\Psi'(x)(\Psi(x)-\Psi(d_1))^{\delta} \leq \Psi'(x)(\Psi(d_2)-\Psi(d_1))^{\delta}. \tag{38}$$

From the exponential convexity of $\mathcal{U}$, it can be obtained

$$e^{\mathcal{U}(x)} \leq \frac{x-d_1}{d_2-d_1}e^{\mathcal{U}(d_2)}+\frac{d_2-x}{d_2-d_1}e^{\mathcal{U}(d_1)}. \tag{39}$$

From (38) and (39), one can have

$$\int_{d_1}^{d_2} e^{\mathcal{U}(x)}\Psi'(x)(\Psi(x)-\Psi(d_1))^{\delta}dx \tag{40}$$

$$\leq \frac{(\Psi(d_2)-\Psi(d_1))^{\delta}}{d_2-d_1}\left[e^{\mathcal{U}(d_2)}\int_{d_1}^{d_2}(x-d_1)\Psi'(x)dx+e^{\mathcal{U}(d_1)}\int_{d_1}^{d_2}(d_2-x)\Psi'(x)dx\right].$$

By using (2) of Definition 2, we get

$$\Gamma(\delta+1) I_{d_2^-}^{\delta+1,\Psi} e^{\mathcal{U}(d_1)} \tag{41}$$

$$\leq \frac{(\Psi(d_2)-\Psi(d_1))^{\delta}}{d_2-d_1}\left[(d_2-d_1)[e^{\mathcal{U}(d_2)}\Psi(d_2)-e^{\mathcal{U}}(d_1)\Psi(d_1)]-[e^{\mathcal{U}(d_2)}-e^{\mathcal{U}(d_1)}]\int_{d_1}^{d_2}\Psi(x)dx\right].$$

Now for $x \in [d_1, d_2]$, $\tau \in [x, d_2]$ and $\rho > 0$, the following inequality holds true

$$\Psi'(x)(\Psi(d_2)-\Psi(x))^{\rho} \leq \Psi'(x)(\Psi(d_2)-\Psi(d_1))^{\rho}. \tag{42}$$

Adopting the same procedure as we did for (38) and (39) one can get from (40) and (42) the subsequent inequality

$$\Gamma(\rho+1) I_{d_1^+}^{\rho+1,\Psi} e^{\mathcal{U}(d_2)} \tag{43}$$

$$\leq \frac{(\Psi(d_2)-\Psi(d_1))^{\rho}}{d_2-d_1}\left[(d_2-d_1)[e^{\mathcal{U}(d_2)}\Psi(d_2)-e^{\mathcal{U}}(d_1)\Psi(d_1)]-[e^{\mathcal{U}(d_2)}-e^{\mathcal{U}(d_1)}]\int_{d_1}^{d_2}\Psi(x)dx\right].$$

From (41) and (43), we get

$$\Gamma(\delta+1)I_{d_2^-}^{\delta+1,\Psi}e^{\mathcal{U}(d_1)}+\Gamma(\rho+1)I_{d_1^+}^{\rho+1,\Psi}e^{\mathcal{U}(d_2)}\leq\frac{(\Psi(d_2)-\Psi(d_1))^{\delta}+(\Psi(d_2)-\Psi(d_1))^{\rho}}{d_2-d_1} \tag{44}$$
$$\times\left[(d_2-d_1)\left[e^{\mathcal{U}(d_2)}\Psi(b)-e^{\mathcal{U}(d_1)}\Psi(d_1)\right]-\left[e^{\mathcal{U}(d_2)}-e^{\mathcal{U}(d_1)}\right]\int_{d_1}^{d_2}\Psi(x)dx\right].$$

Using Lemma 3 and multiplying (36) with $(\Psi(x)-\Psi(d_1))^{\delta}\Psi'(x)$ and integrating over $[d_1,d_2]$, we get

$$e^{\mathcal{U}\left(\frac{d_1+d_2}{2}\right)}\int_{d_1}^{d_2}(\Psi(x)-\Psi(d_1))^{\delta}\Psi'(x)dx\leq\int_{d_1}^{d_2}e^{\mathcal{U}(x)}(\Psi(x)-\Psi(d_1))^{\delta}\Psi'(x)dx. \tag{45}$$

By using (2) of Definition 2 we get

$$e^{\mathcal{U}\left(\frac{d_1+d_2}{2}\right)}\left[\frac{(\Psi(d_2)-\Psi(d_1))^{\delta+1}}{\delta+1}\right]\leq\Gamma(\delta+1)I_{d_2^-}^{\delta,\Psi}e^{\mathcal{U}(d_1)}. \tag{46}$$

Similarly, using Lemma 3 and multiplying (36) with $(\Psi(d_2)-\Psi(x))^{\rho}\Psi'(x)$ and integrating over $[d_1,d_2]$, we get

$$e^{\mathcal{U}\left(\frac{d_1+d_2}{2}\right)}\left[\frac{(\Psi(d_2)-\Psi(d_1))^{\rho+1}}{\rho+1}\right]\leq\Gamma(\rho+1)I_{d_1^+}^{\rho+1,\Psi}e^{\mathcal{U}(d_2)}. \tag{47}$$

From (44) and (47), we get (37) which is the required result. □

Corollary 6. *Choosing $\rho=\delta$ in (37), then we have a new inequality for $\mathcal{GRLFI}$*

$$2\frac{(\Psi(d_2)-\Psi(d_1))^{\rho+1}}{\rho+1}e^{\mathcal{U}\left(\frac{d_1+d_2}{2}\right)}\leq\Gamma(\rho+1)\left[I_{d_1^+}^{\rho+1,\Psi}\Psi(d_2)+I_{d_2^-}^{\rho+1,\Psi}\Psi(d_1)\right]\left(\frac{2(\Psi(d_2)-\Psi(d_1))^{\rho-1}}{d_2-d_1}\right)$$
$$\times\left((d_2-d_1)\left[e^{\mathcal{U}(d_2)}\Psi(d_2)-e^{\mathcal{U}(d_1)}\Psi(d_1)\right]-\left[e^{\mathcal{U}(d_2)}-e^{\mathcal{U}(d_1)}\right]\int_{d_1}^{d_2}\Psi(\tau)d\tau\right).$$

3. Examples

Example 5. *Let $d_1=2, d_2=4, \rho=2, \delta=2, \mathcal{U}(x)=2\ln x, e^{\mathcal{U}(x)}=x^2, \Psi(x)=x^2$ and $x\in[2,4]$. Then all the assumptions in Theorem 1 are satisfied.*

Clearly,

$$\begin{aligned}\Gamma(\rho)I_{d_1^+}^{\rho,\Psi}e^{\mathcal{U}(x)} &= \int_{d_1}^{x}\Psi'(\tau)(\Psi(x)-\Psi(\tau))^{\rho-1}e^{\mathcal{U}(\tau)}d\tau\\ &= 2\int_{2}^{3}(9-\tau^2)\tau^3dt\approx 70.83,\end{aligned}$$

and

$$\Gamma(\delta) I_{d_2^-}^{\delta,\Psi} e^{\mathcal{U}(x)} = \int_x^{d_2} \Psi'(t)\big(\Psi(\tau)-\Psi(x)\big)^{\rho-1} e^{\mathcal{U}(\tau)} d\tau$$
$$= 2\int_3^4 (\tau^2-9)\tau^3 d\tau \approx 334.83.$$

Adding the above equations, we have the left-hand side term of (13)

$$\Gamma(\rho) I_{d_1^+}^{\rho,\Psi} e^{\mathcal{U}(x)} + \Gamma(\delta) I_{d_2^-}^{\delta,\Psi} e^{\mathcal{U}(x)} \approx 405.66. \tag{48}$$

On the other hand,

$$\frac{\big(\Psi(x)-\Psi(d_1)\big)^{\rho-1}}{x-d_1}\Bigg[(x-d_1)\big[\Psi(x)e^{\mathcal{U}(x)} - e^{\mathcal{U}(d_1)}\Psi(d_1)\big] - \big(e^{\mathcal{U}(x)} - e^{\mathcal{U}(d_1)}\big)\int_{d_1}^{x}\Psi(\tau)d\tau\Bigg]$$
$$+\frac{\big(\Psi(d_2)-\Psi(x)\big)^{\delta-1}}{d_2-x}\Bigg[(d_2-x)\big[\Psi(d_2)e^{\mathcal{U}(d_2)} - e^{\mathcal{U}(x)}\Psi(x)\big] - \big(e^{\mathcal{U}(d_2)} - e^{\mathcal{U}(x)}\big)\int_{x}^{d_2}\Psi(\tau)d\tau\Bigg]$$
$$= \frac{(9-4)}{3-2}\Bigg[(3-2)[9(9)-4(4)] - (9-4)\int_2^3 \tau^2 d\tau\Bigg]$$
$$+\frac{(16-9)}{4-3}\Bigg[(4-3)[16(16)-9(9)] - (16-9)\int_3^4 \tau^2 d\tau\Bigg]$$
$$= 166.675 + 620.83 \approx 785.50. \tag{49}$$

It is nice to see that the following implications hold in (48) and (49),

$$405.66 < 785.50.$$

Example 6. *Let* $d_1 = 2, d_2 = 4, \rho = 2, \delta = 2, \mathcal{U}(x) = 2\ln x, e^{\mathcal{U}(x)} = x^2, \Psi(x) = x^2$ *and* $x \in [2,4]$. *where* $\Delta(x,d_1)$ *and* $\Delta(x,d_2)$ *are given in* (22) *and* (23)*, respectively. Then all the assumptions in Theorem 3 are satisfied.*

Clearly,

$$\Gamma(\rho+1) I_{d_1^+}^{\rho,\Psi} e^{\mathcal{U}(x)} - \big(\Psi(x)-\Psi(d_1)\big)^{\rho} e^{\mathcal{U}(a)} = \int_{d_1}^{x} \big(\Psi(x)-\Psi(\tau)\big)^{\rho} e^{\mathcal{U}(\tau)}\mathcal{U}'(\tau)d\tau$$
$$= 2\int_2^3 \tau(9-\tau^2)^2 d\tau \approx 41.67$$

and

$$\Gamma(\delta+1) I_{d_2^-}^{\delta,\Psi} e^{\mathcal{U}(x)} - \big(\Psi(d_2)-\Psi(x)\big)^{\delta} e^{\mathcal{U}(d_2)} = \int_{x}^{d_2} \big(\Psi(\tau)-\Psi(x)\big)^{\delta} e^{\mathcal{U}(\tau)}\mathcal{U}'(\tau)d\tau$$
$$= 2\int_3^4 \tau(\tau^2-9)^2 d\tau \approx 114.33,$$

Adding the above equations, we get the left-hand term of (21)

$$|\Gamma(\rho+1)I_{d_1^+}^{\rho,\Psi}e^{\mathcal{U}(x)}-(\Psi(x)-\Psi(d_1))^{\rho}e^{\mathcal{U}(d_1)}+\Gamma(\delta+1)I_{d_2^-}^{\delta,\Psi}e^{\mathcal{U}(x)}-(\Psi(d_2)-\Psi(x))^{\delta}e^{\mathcal{U}(d_2)}|$$
$$=41.67+114.33\approx 156. \tag{50}$$

Next,

$$\frac{(\Psi(x)-\Psi(d_1))^{\rho}(x-d_1)|e^{\mathcal{U}(d_1)}\mathcal{U}'(d_1)|+(\Psi(d_2)-\Psi(x))^{\delta}(d_2-x)|e^{\mathcal{U}(d_2)}\mathcal{U}'(d_2)|}{3}$$
$$=\frac{(9-4)^2(3-2)|4(\frac{2}{2})|+(16-9)^2(4-3)|16(\frac{2}{4})|}{3}=164,$$

$$\frac{\{(\Psi(d_2)-\Psi(x))^{\delta}(d_2-x)+(\Psi(x)-\Psi(d_1))^{\rho}(x-d_1)\}|e^{\mathcal{U}(x)}\mathcal{U}'(x)|}{3}$$
$$=\frac{(16-9)^2(4-3)+(9-4)^2(3-2)}{3}|9(\frac{2}{3})|=148,$$

$$\frac{(\Psi(x)-\Psi(d_1))^{\rho}(x-d_1)\Delta(d_1,x)+(\Psi(d_2)-\Psi(x))^{\delta}(d_2-x)\Delta(d_2,x)}{6}$$
$$\approx 172.47.$$

Adding the above equations we get the right-hand side term of (21)

$$\frac{(\Psi(x)-\Psi(d_1))^{\rho}(x-d_1)|e^{\mathcal{U}(d_1)}\mathcal{U}'(d_1)|+(\Psi(d_2)-\Psi(x))^{\delta}(d_2-x)|e^{\mathcal{U}(d_2)}\mathcal{U}'(d_2)|}{3}$$
$$+\frac{\{(\Psi(d_2)-\Psi(x))^{\delta}(d_2-x)+(\Psi(x)-\Psi(d_1))^{\rho}(x-d_1)\}|e^{\mathcal{U}(x)}\mathcal{U}'(x)|}{3}$$
$$+\frac{(\Psi(x)-\Psi(d_1))^{\rho}(x-d_1)\Delta(d_1,x)+(\Psi(d_2)-\Psi(x))^{\delta}(d_2-x)\Delta(d_2,x)}{6}$$
$$=164+148+172.47\approx 484.47. \tag{51}$$

It is nice to see that the following implications hold (50) and (51),

$$156<484.47.$$

4. Applications

We consider the following special means for arbitrary real numbers $\mu,\nu,\mu\neq\nu$:

$$\begin{aligned}
\mathcal{A}(\mu,\nu) &= \frac{\mu+\nu}{2}, \quad \mu,\nu\in\mathcal{R},\\
\mathcal{H}(\mu,\nu) &= \frac{2\mu\nu}{\mu+\nu}, \quad \mu,\nu\in\mathcal{R}\setminus\{0\},\\
\mathcal{L}(\mu,\nu) &= \frac{\nu-\mu}{\ln|\nu|-\ln|\mu|}, \quad |\mu|\neq|\nu|,\mu\nu\neq 0,\\
\mathcal{L}_n(\mu,\nu) &= \Big[\frac{\nu^{n+1}-\mu^{n+1}}{(n+1)(\nu-\mu)}\Big], \quad n\in\mathcal{Z}\setminus\{-1,0\},\mu,\nu\in\mathcal{R},\mu\neq\nu.
\end{aligned}$$

Now using the results in Section 1, we have some applications to the special means of real numbers.

Proposition 6. *For* $a_1, a_2 \in \mathcal{R}^+,\ a_1 < a_2$, *then*

$$|\mathcal{A}(e^{a_1}, e^{a_2}) - \mathcal{L}(e^{a_1}, e^{a_2})| \leq \frac{a_2 - a_1}{2^{2-\frac{1}{q}}} \left(\frac{7}{32}\right)^{\frac{1}{q}} \left[|e^{a_1}|^q + |e^{a_2}|^q\right]^{\frac{1}{q}}.$$

Proof. Apply Theorem 2 with $e^{\mathcal{U}(x)} = e^x, \Psi(x) = x, \rho = 1$, and we obtain the desired result. □

Proposition 6. *For* $a_1, a_2 \in \mathcal{R}^+,\ a_1 < a_2$, *then*

$$|\mathcal{H}^{-1}(a_1, a_2) - \mathcal{L}^{-1}(a_1, a_2)| \leq \frac{a_2 - a_1}{2^{2-\frac{1}{q}}} \left(\frac{7}{32}\right)^{\frac{1}{q}} \left[|\frac{1}{a_1}|^{2q} + |\frac{1}{a_2}|^{2q}\right]^{\frac{1}{q}}.$$

Proof. Apply Theorem 2 with $e^{\mathcal{U}(x)} = \frac{1}{x}, \Psi(x) = x, \rho = 1$, and we obtain the desired result. □

Proposition 6. *Let* $a_1, a_2 \in \mathcal{R}^+, a_1 < a_2$, *then*

$$|\mathcal{A}(a_1^2, a_2^2) - \mathcal{L}_2^2(a_1, a_2)| \leq \frac{a_2 - a_1}{2^{1-\frac{1}{q}}} \left(\frac{7}{32}\right)^{\frac{1}{q}} \left[|a_1|^q + |a_2|^q\right]^{\frac{1}{q}}.$$

Proof. Apply Theorem 2 with $e^{\mathcal{U}(x)} = x^2, \Psi(x) = x, \rho = 1$, and we obtain the desired result. □

Proposition 6. *Let* $a_1, a_2 \in \mathcal{R}^+, a_1 < a_2$, *then*

$$|\mathcal{A}(a_1^n, a_2^n) - \mathcal{L}_n^n(a_1, a_2)| \leq \frac{n(a_2 - a_1)}{2^{2-\frac{1}{q}}} \left(\frac{7}{32}\right)^{\frac{1}{q}} \left[|a_1|^{(n-1)q} + |a_2|^{(n-1)q}\right]^{\frac{1}{q}}.$$

Proof. Apply Theorem 2 with $e^{\mathcal{U}(x)} = x^n, \Psi(x) = x, \rho = 1$, and we obtain the general result. □

5. Conclusions

In this article, we have investigated a few fractional integral inequalities for $\mathcal{GRLFI}$ via exponentially convexity. These inequalities have bounds of the sum of left-sided and right-sided fractional integrals and inequalities for the function, and their first derivative in absolute value is exponentially convex. Also, fractional inequalities of $\mathcal{HH}$ type for a symmetric and exponentially convex function are proved. These estimates, bounds and inequalities exist for all fractional operators are stated in Remark 1. The method followed to produce fractional inequalities is innovative and simple. It could be followed to broaden further consequences for other classes of functions related to exponentially convex functions, using convenient fractional integral operators.

Author Contributions: All authors contributed equally to the writing of this paper. All authors read and approved the final manuscript.

Funding: The second author would like to thank Prince Sultan University for funding this work through research, group Nonlinear Analysis Methods in Applied Mathematics (NAMAM) group number RG-DES-2017-01-17.

Acknowledgments: The authors are grateful to the reviewers for their valuable and constructive suggestions.

Conflicts of Interest: The authors declare no conflict of interest.

References

1. Samko, S.G.; Kilbas, A.A.; Marichev, O.I. *Fractional Integrals and Derivatives: Theory and Applications*; Gordon and Breach: Yverdon, Switzerland, 1993.
2. Podlubny, I. *Fractional Differential Equations*; Academic Press: San Diego, CA, USA, 1999.
3. Kilbas, A.; Srivastava, H.M.; Trujillo, J.J. *Theory and Application of Fractional Differential Equations*; North Holland Mathematics Studies: Berlin, Germany, 2006; Volume 204.

4. Miller, K.; Ross, B. *An Introduction to the Fractional Differential Equations*; John Wiley and Sons Inc.: New York, NY, USA, 1993.
5. Sonin, N.Y. On differentiation with arbitrary index. *Mosc. Matem. Sbornik.* **1869**, *6*, 1–38.
6. Letnikov, A.V. Theory of differentiation with an arbitray index. *Matem. Sbornik.* **1868**, *3*, 1–66. (In Russian)
7. Laurent, H. On the Calculation of Derivatives with Any Indices. 2004, Volume 3, pp. 240–252. Available online: http://www.numdam.org/item/NAM_1884_3_3__240_0/ (accessed on 25 July 2019). (In French)
8. Magin, R.L. Magin, R.L. Fractional Calculus in Bioengineering. *Crit. Rev. Biomed. Eng.* **2004**, *32*, 1–104. [CrossRef] [PubMed]
9. Hilfer, R. *Applications of Fractional Calculus in Physics*; Word Scientific: Singapore, 2000.
10. Srivastava, H.M.; Tomovski, Z. Fractional calculus with an integral operator containing generalized Mittag-Leffler function in the kernal. *Appl. Math. Comput.* **2009**, *211*, 198–210.
11. Tomovski, Z.; Hilfer, R.; Srivastava, H.M. Fractional and operational calculus with generalized fractional derivative operators and Mittag-Leffler function. *Integral Transforms Spec. Funct.* **2011**, *21*, 797–814. [CrossRef]
12. Belarbi, S.; Dahmani, Z. On some new fractional integral inequalities. *J. Inequal. Pure Appl. Math.* **2009**, *10*, 86.
13. Dahmani, Z. On Minkowski and Hermite-Hadamard integral inequalities via fractional integration. *Ann. Funct.* **2010**, *1*, 51–58. [CrossRef]
14. Dahmani, Z. New inequalities in fractional integrals. *Int. J. Nonlinear Sci.* **2010**, *9*, 493–497.
15. Dahmani, Z. The Riemann-Liouville operator to generate some new inequalities. *Int. J. Nonlinear Sci.* **2011**, *12*, 452–455.
16. Denton, Z.; Vatsala, A.S. Fractional integral inequalities and applications. *Comput. Math. Appl.* **2010**, *59*, 1087–1094. [CrossRef]
17. Anastassiou, G.A. *Fractional Differentiation Inequalities*; Springer: NewYork, NY, USA, 2009.
18. Abdeljawad, T. A Lyapunov type inequality for fractional operators with nonsingular Mittag-Leffler kernel. *J. Inequal. Appl.* **2017**, *2017*, 130. [CrossRef] [PubMed]
19. Abdeljawad, T.; Alzabut, J. The q-Fractional Analogue for Gronwall-Type Inequality. *J. Funct. Spaces Appl.* **2013**, *2013*, 543839. [CrossRef]
20. Adjabi, Y.; Jarad, F.; Abdeljawad, T. On generalized fractional operators and a Gronwall type inequality with applications. *Filomat* **2017**, *31*, 5457–5473. [CrossRef]
21. Jarad, F.; Abdeljawad, T.; Hammouch, Z. On a class of ordinary differential equations in the frame of Atangana-Baleanu fractional derivative. *Chaos Solitons Fractals* **2018**, *117*, 16–20. [CrossRef]
22. Abdeljawad, T.; Al-Mdallal, Q.M.; Hajji, M.A. Arbitrary order fractional difference operators with discrete exponential kernels and applications. *Discret. Dyn. Nat. Soc.* **2017**, *2017*, 4149320. [CrossRef]
23. Abdeljawad, T. Fractional operators with exponential kernels and a Lyapunov type inequality. *Adv. Differ. Equ.* **2017**, *2017*, 313. [CrossRef]
24. Alzabut, J.; Abdeljawad, T.; Jarad, F.; Sudsutad, W. A Gronwall inequality via the generalized proportional fractional derivative with applications. *J. Inequal. Appl.* **2019**, *2019*, 101. [CrossRef]
25. Khan, H.; Abdeljawad, T.; Tunç, C.; Alkhazzan, A.; Khan, A. Minkowski's inequality for the AB-fractional integral operator. *J. Inequal. Appl.* **2019**, *2019*, 96. [CrossRef]
26. Sarikaya, M.Z.; Set, E.; Yaldiz, H.; Başak, N. Hermite-Hadamard's inequalities for fractional integrals and related fractional inequalities, *Math. Comput. Model.* 2013, 57, 2403–2407. [CrossRef]
27. Chalishajar, D.; Ravichandran, C.; Dhanalakshmi, S.; Murugesu, R. Existence of fractional impulsive functional integro-differential equations in Banach spaces. *Appl. Syst. Innov.* **2019**, *2*, 18. [CrossRef]
28. Zhang, D.; Zhang, Y.; Zhang, Z.; Ahmed, N.; Zhang, Y.; Li, F.; Belic, M.R.; Xiao, M. On unveiling the link between fractional Schrödinger equation and light propogation in honeycomb lattice. *Ann. Phys.* **2017**, *529*, 1–6. [CrossRef]
29. Zhang, Y.; Zhang, H.; Belic, M.R.; Zhu, Y.; Zhang, W.; Zhang, Y.; Christodoulides, D.N.; Xiao, M. On PT symmetry in a fractional Schrödinger equation. *Laser Photonics Rev.* **2016**, *10*, 526. [CrossRef]
30. Zhang, Y.; Liu, X.; Belic, M.R.; Zhang, W.; Zhang, Y.; Xiao, M. On propogation dynamics of a light beam in a fractional Schrödinger equation. *Phys. Rev. Lett.* **2015**, *115*, 180403. [CrossRef]
31. Jarad, F.; Abdeljawad, T. Generalized fractional derivatives and Laplace transform. *Discrete Contin. Dyn. Syst.* **2018**. [CrossRef]
32. Ameen, R.; Jarad, F.; Abdeljawad, T. Ulam stability of fractional differential equations with a generalized Caputo derivative. *Filomat* **2018**, *32*, 5265–5274. [CrossRef]

33. Chen, H.; Katugampola, U.N. Hermite-Hadamard and Hermite-Hadamard-Fejér type inequalities for generalized fractional integrals. *J. Math. Anal. Appl.* **2017**, *446*, 1274–1291. [CrossRef]
34. Dragomir, S.S.; Gomm, I. Some Hermite-Hadamard type inequalities for functions whose exponentials are convex. *Stud. Univ. Babes-Bolyai Math.* **2015**, *60*, 527–534
35. Dragomir, S.S.; Agarwal, R.P. Two inequalities for differentiable mappings and applications to special means of real numbers and to trapezoidal formula. *Appl. Math. Lett.* **1998**, *11*, 91–95. [CrossRef]
36. Farid, G., Nazeer, W., Saleem, M. S., Mehmood. S., Kang, S. M.: Bounds of Riemann-Liouville fractional integrals in General form via convex functions and their applications. *Mathematics* **2018**, *6*, 248. [CrossRef]
37. Alirezaei, G.; Mathar, R. On exponentially concave functions and their impact in information theory. *J. Inform. Theory Appl.* **2018**, *9*, 265–274
38. Antczak, T. On (p, r)-invex sets and functions. *J. Math. Anal. Appl.* **2001**, *263*, 355–379. [CrossRef]
39. Pal, S.; Wong, T.K.L. On exponentially concave functions and a new information geometry.*Ann. Probab.* **2018**, *46*, 1070–1113. [CrossRef]
40. Pecaric, J.; Jaksetic, J. On exponential convexity, Euler-Radau expansions and stolarsky means, Rad Hrvat. *Matematicke Znanosti* **2013**, *17*, 81–94.
41. Jakestic, J.; Pecaric, J. Exponential convexity method. *J. Convex Anal.* **2012**, *20*, 181–187.
42. Bernstein, S.N. Sur les fonctions absolument monotones. *Acta. Math.* **1929**, *52*, 1–66. [CrossRef]
43. Jarad, F.; Ugurlu, E.; Abdeljawad, T.; Baleanu, D. On a new class of fractional operators. *Adv. Differ. Equ.* **2017**, *2017*, 247. [CrossRef]
44. Khan, T.U.; Khan, M.A. Generalized conformable fractional operators. *J. Comput. Appl. Math.* **2019**, *346*, 378–389. [CrossRef]

mathematics

MDPI

Article

Hermite–Hadamard-Type Inequalities for Convex Functions via the Fractional Integrals with Exponential Kernel

Xia Wu [1], JinRong Wang [2,3,*] and Jialu Zhang [1]

1 School of Mathematics and Finance, Xiangnan University, Chenzhou 411105, China; 230169714@seu.edu.cn (X.W.); zhangjl@xnu.edu.cn (J.Z.)
2 Department of Mathematics, Guizhou University, Guiyang 550025, China
3 School of Mathematical Sciences, Qufu Normal University, Qufu 273165, China
* Correspondence: jrwang@gzu.edu.cn

Received: 24 August 2019; Accepted: 9 September 2019; Published: 12 September 2019

Abstract: In this paper, we establish three fundamental integral identities by the first- and second-order derivatives for a given function via the fractional integrals with exponential kernel. With the help of these new fractional integral identities, we introduce a few interesting Hermite–Hadamard-type inequalities involving left-sided and right-sided fractional integrals with exponential kernels for convex functions. Finally, some applications to special means of real number are presented.

Keywords: convex functions, Hermite–Hadamard-type inequalities, fractional integrals, exponential kernel

MSC: 26A33

---

1. Introduction

Let $h : [a,b] \subset \Re \rightarrow \Re$ be a convex function. Then, h meets the following classic Hermite–Hadamard inequality (see [1])

$$h\left(\frac{a+b}{2}\right) \leq \frac{1}{b-a}\int_a^b h(s)ds \leq \frac{h(a)+h(b)}{2}. \tag{1}$$

If h is a concave function, the inequalities in (1) are presented in the negative direction. The Hermite–Hadamard inequality provides us the estimates for the integral average of a continuous convex function on a compact interval.

For the latest results on generalizing, improving, and extending this classical Hermite–Hadamard inequality, one can see [2–9] and the references therein.

In [10], Dragomir and Agarwal proved the following result connected with the right part of (1). In [11], Alomart also elicited the similar result for functions whose second derivatives absolute values are convex.

Lemma 1 (see [10], Theorem 2.2). *Assuming $h : [a,b] \subseteq \Re \rightarrow \Re$ is a differentiable function, $h' \in L[a,b]$ and $|h''|$ is convex on $[a,b]$. Then, the bellow inequality holds:*

$$\left|\frac{h(a)+h(b)}{2} - \frac{1}{b-a}\int_a^b h(s)ds\right| \leq \frac{(b-a)}{8}(|h'(a)|+|h'(b)|). \tag{2}$$

Lemma 2 (see [11], Theorem 3). *Assuming $h : [a,b] \subseteq \Re \to \Re$ is a twice differentiable function, $h'' \in L[a,b]$ and $|h''|$ is convex on $[a,b]$. Then, the following inequality holds*

$$\left|\frac{h(a)+h(b)}{2} - \frac{1}{b-a}\int_a^b h(s)ds\right| \leq \frac{(b-a)^2}{24}(|h''(a)| + |h''(b)|). \tag{3}$$

Now, fractional calculus has turned into an enchanting field of mathematics. Many extensive investigations have been carried out in this area. Due to the wide applications of Hermite–Hadamard inequalities and fractional integrals, many researchers have extended their research to Hermite–Hadamard inequalities involving fractional integrals rather than integer integrals, see [12–22]. Sarikaya et al. [12] have deduced an amusing inequality of Hermite–Hadamard-type involving fractional integrals in the place of ordinary integrals. This research fascinates many researchers to consider this respect. As a result, some new integral inequalities by the approach of fractional calculus have been obtained in the literature until now. In addition, Ahmad et al. [16] gave the new fractional integral operators with an exponential kernel and proved similar inequalities.

Definition 1 (see [16], Definition 2). *Let $h \in L[a,b]$. The fractional left-side integral $\mathcal{J}_{a^+}^{\alpha} h$ and right-side integral $\mathcal{J}_{b^-}^{\alpha} h$ of order $\alpha \in (0,1)$ are, respectively, defined by*

$$\mathcal{J}_{a^+}^{\alpha} h(x) = \frac{1}{\alpha}\int_a^x e^{-\frac{1-\alpha}{\alpha}(x-s)} h(s)ds, \ \ x > a, \tag{4}$$

and

$$\mathcal{J}_{b^-}^{\alpha} h(x) = \frac{1}{\alpha}\int_x^b e^{-\frac{1-\alpha}{\alpha}(x-s)} h(s)ds, \ \ x < b. \tag{5}$$

Lemma 3 (see [16], Theorem 1). *Let $h : [a,b] \to \Re$ be a positive convex function with $0 \leq a < b$ and $h \in L[a,b]$. The following inequality for fractional integrals (4) and (5) holds:*

$$h(\frac{a+b}{2}) \leq \frac{1-\alpha}{2(1-e^{-\rho})}[\mathcal{J}_{a^+}^{\alpha} h(b) + \mathcal{J}_{b^-}^{\alpha} h(a)] \leq \frac{h(a)+h(b)}{2}. \tag{6}$$

Remark 1. *In (6), note that*

$$\rho = \frac{1-\alpha}{\alpha}(b-a).$$

In addition, Ahmad et al. [16] derived the bound estimate of the difference between the mean value of the endpoints and the average of the fractional integrals.

Lemma 4 (see [16], Theorem 3). *Assuming $h : [a,b] \subseteq \Re \to \Re$ is differentiable, $h' \in L[a,b]$ and $|h'|$ are convex on $[a,b]$. Then, the following inequality holds:*

$$|Q_{mr}| \leq \frac{(b-a)}{2\rho}\tanh(\frac{\rho}{4})(|h'(a)| + |h'(b)|),$$

where

$$Q_{mr} := \frac{h(a)+h(b)}{2} - \frac{1-\alpha}{2(1-e^{-\rho})}[\mathcal{J}_{a^+}^{\alpha} h(b) + \mathcal{J}_{b^-}^{\alpha} h(a)]$$

denotes the bound estimate of the difference between the mean value of the endpoints and the average of the fractional integrals.

However, the bound for the left of the Hermite–Hadamard inequality (6) has not been studied. It will be interesting to find

$$|Q_{ml}| \leq \textbf{what?}$$

Here,

$$Q_{ml} := \frac{1-\alpha}{2\,(1-e^{-\rho})}[\mathcal{J}^{\alpha}_{a^+}h(b) + \mathcal{J}^{\alpha}_{b^-}h(a)] - h(\frac{a+b}{2})$$

denotes the bound estimate of the difference between the value of the midpoint and the average of the fractional integrals.

Furthermore, if $|h''|$ is convex, it is natural to study the right- and left-type Hermite–Hadamard inequality via the fractional integral with an exponential kernel similar to Lemma 2, i.e., we want to find the constants ρ_1 and ρ_2 satisfying the following inequities:

$$|Q_{mr}| \leq \rho_1 \cdot (|h''(a)| + |h''(b)|),$$

and

$$|Q_{ml}| \leq \rho_2 \cdot (|h''(a)| + |h''(b)|).$$

Motivated by [12,15,16], we will demonstrate three new fractional-type integral identities and set up their corresponding Hermite–Hadamard-type inequalities involving left-sided and right-sided fractional integrals for convex functions, respectively.

2. New Fractional Integral Identity and Hermite–Hadamard-Type Inequality for First Order Derivative

We firstly prove the following lemma in order to attest the following result.

Lemma 5. *Assuming $h : [a,b] \subseteq \Re \rightarrow \Re$ is a differentiable mapping and $h' \in L[a,b]$. Then, the following equality for the fractional integrals (4) and (5) holds:*

$$\begin{aligned} Q_{ml} &= \frac{b-a}{2}\int_0^1 kh'(sa+(1-s)b)ds \\ &\quad -\frac{b-a}{2(1-e^{-\rho})}\left[\int_0^1 e^{-\rho s}h'(sa+(1-s)b)ds - \int_0^1 e^{-\rho(1-s)}h'(sa+(1-s)b)ds\right], \end{aligned} \tag{7}$$

where

$$k = \begin{cases} 1, \; 0 \leq s < \frac{1}{2}, \\ -1, \; \frac{1}{2} \leq s \leq 1. \end{cases}$$

Proof. Define

$$\begin{aligned} V &:= \int_0^1 e^{-\rho s}h'(sa+(1-s)b)ds - \int_0^1 e^{-\rho(1-s)}h'(sa+(1-s)b)ds \\ &= V_1 - V_2, \end{aligned} \tag{8}$$

where

$$\begin{aligned} V_1 &= \int_0^1 e^{-\rho s}h'(sa+(1-s)b)ds, \\ V_2 &= \int_0^1 e^{-\rho(1-s)}h'(sa+(1-s)b)ds. \end{aligned}$$

Integrating by parts, one has

$$\begin{aligned}
V_1 &= \int_0^1 e^{-\rho s} h'(sa+(1-s)b)ds \\
&= \frac{1}{a-b}\int_0^1 e^{-\rho s} d\left(h(sa+(1-s)b)\right) \\
&= \frac{1}{a-b}\left[e^{-\rho s} f(sa+(1-s)b)\Big|_0^1 - \int_0^1 h(sa+(1-s)b) d\left(e^{-\rho s}\right)\right] \\
&= \frac{1}{a-b}\left[e^{-\rho} h(a) - h(b) + \rho \int_0^1 h(sa+(1-s)b) e^{-\rho s} ds\right] \\
&= \frac{e^{-\rho} h(a) - h(b)}{a-b} + \frac{\rho}{(a-b)^2}\int_b^a h(x) e^{-\rho \frac{x-b}{a-b}} dx \\
&= \frac{e^{-\rho} h(a) - h(b)}{a-b} - \frac{\rho}{(a-b)^2}\int_a^b e^{-\frac{1-\alpha}{\alpha}(b-x)} h(x) dx \\
&= \frac{e^{-\rho} h(a) - h(b)}{a-b} - \frac{\rho}{(a-b)^2}\alpha \mathcal{J}_{a^+}^{\alpha} h(b) \\
&= \frac{e^{-\rho} h(a) - h(b)}{a-b} - \frac{1-\alpha}{(b-a)}\mathcal{J}_{a^+}^{\alpha} h(b), \qquad (9)
\end{aligned}$$

and

$$\begin{aligned}
V_2 &= \int_0^1 e^{-\rho(1-s)} h'(sa+(1-s)b)ds \\
&= \frac{1}{a-b}\int_0^1 e^{-\rho(1-s)} d\left(h(sa+(1-s)b)\right) \\
&= \frac{1}{a-b}\left[e^{-\rho(1-s)} h(sa+(1-s)b)\Big|_0^1 - \int_0^1 h(sa+(1-s)b) d\left(e^{-\rho(1-s)}\right)\right] \\
&= \frac{1}{a-b}\left[h(a) - e^{-\rho} h(b) - \rho \int_0^1 h(sa+(1-s)b) e^{-\rho(1-s)} ds\right] \\
&= \frac{h(a) - e^{-\rho} h(b)}{a-b} - \frac{\rho}{(a-b)^2}\int_b^a h(x) e^{-\rho \frac{a-x}{a-b}} dx \\
&= \frac{h(a) - e^{-\rho} h(b)}{a-b} + \frac{\rho}{(a-b)^2}\int_a^b e^{-\frac{1-\alpha}{\alpha}(x-a)} h(x) dx \\
&= \frac{h(a) - e^{-\rho} h(b)}{a-b} + \frac{\rho}{(a-b)^2}\alpha \mathcal{J}_{b^-}^{\alpha} h(a) \\
&= \frac{h(a) - e^{-\rho} h(b)}{a-b} + \frac{1-\alpha}{(b-a)}\mathcal{J}_{b^-}^{\alpha} h(a). \qquad (10)
\end{aligned}$$

Substituting (9) and (10) into (8), we get that

$$\begin{aligned}
V &= V_1 - V_2 \\
&= \frac{\left(1-e^{-\rho}\right)\left(h(a)+h(b)\right)}{b-a} - \frac{1-\alpha}{(b-a)}\left[\mathcal{J}_{a^+}^{\alpha} h(b) + \mathcal{J}_{b^-}^{\alpha} h(a)\right]. \qquad (11)
\end{aligned}$$

Note

$$
\begin{aligned}
& \frac{b-a}{2}\int_0^1 kh'(sa+(1-s)b)dt \\
= & \frac{b-a}{2}\int_0^{\frac{1}{2}} h'(sa+(1-s)b)ds - \frac{b-a}{2}\int_{\frac{1}{2}}^1 h'(sa+(1-s)b)ds \\
= & \frac{f(b)-h\left(\frac{a+b}{2}\right)}{2} + \frac{h(a)-h\left(\frac{a+b}{2}\right)}{2} \\
= & \frac{h(a)+h(b)}{2} - h\left(\frac{a+b}{2}\right).
\end{aligned} \tag{12}
$$

Substituting (12) and (11) into the right-side of (7), we obtain the left of (7). This testifies the proof. □

Then, we can declare the first theorem including Hermite–Hadmard-type inequality.

Theorem 1. *If $h : [a,b] \subseteq \Re \to \Re$ is differentiable, $|h'|$ is convex on [a,b], and $h' \in L[a,b]$, then the following inequality about the fractional integrals (4) and (5) holds:*

$$
|Q_{ml}| \leq \frac{b-a}{2}\left[\frac{1}{2} - \frac{\tanh(\frac{\rho}{4})}{\rho}\right](|h'(a)|+|h'(b)|). \tag{13}
$$

Proof. Using Lemma 5, convexity of $|h'|$, and $e^{-\rho s} \geq e^{-\rho}$ and $e^{-\rho(1-s)} \geq e^{-\rho}$ for any $s \in [0,1]$, we obtain

$$
\begin{aligned}
|Q_{ml}| = & \left|\frac{b-a}{2}\int_0^1 kh'(sa+(1-s)b)ds \right.\\
& \left. -\frac{b-a}{2(1-e^{-\rho})}\left[\int_0^1 e^{-\rho s}h'(sa+(1-s)b)ds - \int_0^1 e^{-\rho(1-s)}h'(sa+(1-s)b)ds\right]\right| \\
= & \frac{b-a}{2(1-e^{-\rho})}\left|\int_0^{\frac{1}{2}}[(1-e^{-\rho})-e^{-\rho s}+e^{-\rho(1-s)}]h'(sa+(1-s)b)ds \right.\\
& \left. -\int_{\frac{1}{2}}^1[(1-e^{-\rho})-e^{-\rho(1-s)}+e^{-\rho s}]h'(sa+(1-s)b)ds\right| \\
\leq & \frac{b-a}{2(1-e^{-\rho})}[\int_0^{\frac{1}{2}}(1-e^{-\rho}-e^{-\rho s}+e^{-\rho(1-s)})(s|h'(a)|+(1-s)|h'(b)|)ds \\
& +\int_{\frac{1}{2}}^1(1-e^{-\rho}-e^{-\rho(1-s)}+e^{-\rho s})(s|h'(a)|+(1-s)|h'(b)|)ds] \\
= & \frac{b-a}{2(1-e^{-\rho})}[\int_0^{\frac{1}{2}}(1-e^{-\rho}-e^{-\rho s}+e^{-\rho(1-s)})(s|h'(a)|+(1-s)|h'(b)|)ds \\
& +\int_0^{\frac{1}{2}}(1-e^{-\rho}-e^{-\rho s}+e^{-\rho(1-s)})((1-s)|h'(a)|+s|h'(b)|)ds] \\
= & \frac{b-a}{2(1-e^{-\rho})}\int_0^{\frac{1}{2}}(1-e^{-\rho}-e^{-\rho s}+e^{-\rho(1-s)})(|h'(a)|+|h'(b)|)ds \\
= & \frac{b-a}{2(1-e^{-\rho})}\left[\frac{1-e^{-\rho}}{2} - \frac{1}{\rho}(1-e^{-\frac{\rho}{2}})^2\right](|h'(a)|+|h'(b)|) \\
= & \frac{b-a}{2}\left[\frac{1}{2} - \frac{\tanh(\frac{\rho}{4})}{\rho}\right](|h'(a)|+|h'(b)|).
\end{aligned}
$$

The proof is completed. □

3. New Fractional Integral Identity and Hermite–Hadamard-Type Inequality for Second Order Derivative

In [16] Lemma 4, Ahmad et al. gave the equality

$$Q_{mr} = \frac{(b-a)}{2(1-e^{-\rho})}\left[\int_0^1 e^{-\rho s}h'(sa+(1-s)b)ds - \int_0^1 e^{-\rho(1-s)}h'(sa+(1-s)b)ds\right]. \tag{14}$$

By (14), we will prove the Hermite–Hadamard-type inequality of the order derivatives via the fractional integrals with an exponential kernel for convex functions. Before we prove our main results in this section, we give the following lemmas.

Lemma 6. *Assuming $h : [a,b] \to \Re$ is a twice differentiable function. If $h'' \in L[a,b]$, then the following equality for fractional integrals holds:*

$$Q_{mr} = \frac{(b-a)^2}{2\rho(1-e^{-\rho})}\int_0^1 \left(1+e^{-\rho}-e^{-\rho s}-e^{-\rho(1-s)}\right)h''(sa+(1-s)b)ds. \tag{15}$$

Proof. By using equality (14), we note

$$\begin{aligned} K_1 &= \int_0^1 e^{-\rho s}h'(sa+(1-s)b)ds = -\frac{1}{\rho}\int_0^1 h'(sa+(1-s)b)d(e^{-\rho s}) \\ &= \frac{1}{\rho}\left[h'(b)-e^{-\rho}h'(a)+(a-b)\int_0^1 e^{-\rho s}h''(sa+(1-s)b)ds\right], \end{aligned} \tag{16}$$

and

$$\begin{aligned} K_2 &= \int_0^1 e^{-\rho(1-s)}h'(sa+(1-s)b)ds = \frac{1}{\rho}\int_0^1 h'(sa+(1-s)b)d(e^{-\rho(1-s)}) \\ &= \frac{1}{\rho}\left[h'(a)-e^{-\rho}h'(b)-(a-b)\int_0^1 e^{-\rho(1-s)}h''(sa+(1-s)b)ds\right]. \end{aligned} \tag{17}$$

Inserting the values of K_1 and K_2 in (14), we obtain

$$\begin{aligned} &\frac{h(a)+h(b)}{2} - \frac{1-\alpha}{2(1-e^{-\rho})}[\mathcal{J}_{a+}^{\alpha}h(b)+\mathcal{J}_{b-}^{\alpha}h(a)] \\ =& \frac{b-a}{2\rho(1-e^{-\rho})}\Big[(1+e^{-\rho})\left(h'(b)-h'(a)\right) \\ &-(b-a)\int_0^1\left(e^{-\rho s}+e^{-\rho(1-s)}\right)h''(sa+(1-s)b)ds\Big] \\ =& \frac{b-a}{2\rho(1-e^{-\rho})}\Big[(1+e^{-\rho})(b-a)\int_0^1 h''(sa+(1-s)b)ds \\ &-(b-a)\int_0^1\left(e^{-\rho s}+e^{-\rho(1-s)}\right)h''(sa+(1-s)b)ds\Big] \\ =& \frac{(b-a)^2}{2\rho(1-e^{-\rho})}\int_0^1\left(1+e^{-\rho}-e^{-\rho s}-e^{-\rho(1-s)}\right)h''(sa+(1-s)b)ds. \end{aligned} \tag{18}$$

This completes the proof. □

Lemma 7. *Assuming $h : [a,b] \to \Re$ is a twice differentiable function. If $h'' \in L[a,b]$, then the following equality for fractional integrals holds:*

$$Q_{ml} = \frac{(b-a)^2}{2}\int_0^1 m(s)h''(sa+(1-s)b)ds, \tag{19}$$

where

$$m(s)=\begin{cases} s-\frac{1+e^{-\rho}-e^{-\rho s}-e^{-\rho(1-s)}}{\rho(1-e^{-\rho})}, & 0\le s<\frac{1}{2},\\ (1-s)-\frac{1+e^{-\rho}-e^{-\rho s}-e^{-\rho(1-s)}}{\rho(1-e^{-\rho})}, & \frac{1}{2}\le s<1. \end{cases}$$

Proof. By using the proof of the Lemma 5, we can get

$$\begin{aligned} Q_{ml} &= \frac{b-a}{2}\int_0^1 kh'(sa+(1-s)b)ds \\ &-\frac{b-a}{2(1-e^{-\rho})}\left[\int_0^1 e^{-\rho s}h'(sa+(1-s)b)ds-\int_0^1 e^{-\rho(1-s)}h'(sa+(1-s)b)ds\right]. \end{aligned} \tag{20}$$

Thus,

$$\begin{aligned} &\frac{b-a}{2}\int_0^1 kh'(sa+(1-s)b)ds \\ =&\frac{b-a}{2}\int_0^{\frac{1}{2}} h'(sa+(1-s)b)ds-\frac{b-a}{2}\int_{\frac{1}{2}}^1 h'(sa+(1-s)b)ds \\ =&\frac{b-a}{2}\left[sh'(sa+(1-s)b)\Big|_0^{\frac{1}{2}}-(a-b)\int_0^{\frac{1}{2}} sh''(sa+(1-s)b)ds\right] \\ &-\frac{b-a}{2}\left[sh'(sa+(1-s)b)\Big|_{\frac{1}{2}}^1-(a-b)\int_{\frac{1}{2}}^1 sh''(sa+(1-s)b)ds\right] \\ =&\frac{b-a}{2}\left[\frac{1}{2}h'(\frac{a+b}{2})-(a-b)\int_0^{\frac{1}{2}} sh''(sa+(1-s)b)ds\right] \\ &-\frac{b-a}{2}\left[h'(a)-\frac{1}{2}h'(\frac{a+b}{2})-(a-b)\int_{\frac{1}{2}}^1 sh''(sa+(1-s)b)ds\right] \\ =&\frac{b-a}{2}\left[h'(\frac{a+b}{2})-h'(a)\right] \\ &+\frac{(b-a)^2}{2}\int_0^{\frac{1}{2}} sh''(sa+(1-s)b)ds-\frac{(b-a)^2}{2}\int_{\frac{1}{2}}^1 sh''(sa+(1-s)b)ds \\ =&\frac{(b-a)^2}{2}\int_{\frac{1}{2}}^1 h''(sa+(1-s)b)ds \\ &+\frac{(b-a)^2}{2}\int_0^{\frac{1}{2}} sh''(sa+(1-s)b)ds-\frac{(b-a)^2}{2}\int_{\frac{1}{2}}^1 sh''(sa+(1-s)b)ds \\ =&\frac{(b-a)^2}{2}\int_0^{\frac{1}{2}} sh''(sa+(1-s)b)ds+\frac{(b-a)^2}{2}\int_{\frac{1}{2}}^1 (1-s)h''(sa+(1-s)b)ds. \end{aligned} \tag{21}$$

Submitting (16), (17), and (21) to (20), we get (19). This completes the proof. □

Now, we can prove our Hermite–Hadamard-type inequalities by the second order derivatives.

Theorem 2. *Assuming $h:[a,b]\to\Re$ is a twice differentiable function. If $h''\in L[a,b]$ and $|h''|$ is convex on $[a,b]$, the following inequality for fractional integrals with exponential kernel holds:*

$$|Q_{mr}|\le\frac{(b-a)^2}{2\rho(1-e^{-\rho})}\left(\frac{1+e^{-\rho}}{2}-\frac{1-e^{-\rho}}{\rho}\right)(|h''(a)|+|h''(b)|). \tag{22}$$

Proof. Note that

$$
\begin{aligned}
& \int_0^1 \left(1+e^{-\rho}-e^{-\rho s}-e^{-\rho(1-s)}\right) sds \\
= \;& (1+e^{-\rho}) \int_0^1 sds - \int_0^1 se^{-\rho s}ds - \int_0^1 se^{-\rho(1-s)}ds \\
= \;& \frac{1+e^{-\rho}}{2} + \frac{1}{\rho}\left(e^{-\rho} - \frac{1-e^{-\rho}}{\rho}\right) - \frac{1}{\rho}\left(1 - \frac{1-e^{-\rho}}{\rho}\right) \\
= \;& \frac{1+e^{-\rho}}{2} - \frac{1-e^{-\rho}}{\rho},
\end{aligned} \tag{23}
$$

and

$$
\begin{aligned}
& \int_0^1 \left(1+e^{-\rho}-e^{-\rho s}-e^{-\rho(1-s)}\right)(1-s)ds \\
= \;& \int_0^1 \left(1+e^{-\rho}-e^{-\rho s}-e^{-\rho(1-s)}\right) ds - \int_0^1 s\left(1+e^{-\rho}-e^{-\rho s}-e^{-\rho(1-s)}\right) ds \\
= \;& (1+e^{-\rho})\Big|_0^1 + \frac{e^{-\rho s}}{\rho}\Big|_0^1 - \frac{e^{-\rho(1-s)}}{\rho}\Big|_0^1 - \left(\frac{1+e^{-\rho}}{2} - \frac{1-e^{-\rho}}{\rho}\right) \\
= \;& \frac{1+e^{-\rho}}{2} - \frac{1-e^{-\rho}}{\rho}.
\end{aligned} \tag{24}
$$

According to (15), (23), (24), and the convex of $|h''|$, we can get

$$
\begin{aligned}
|Q_{mr}| &= \left|\frac{(b-a)^2}{2\rho(1-e^{-\rho})} \int_0^1 \left(1+e^{-\rho}-e^{-\rho s}-e^{-\rho(1-s)}\right) h''(sa+(1-s)b)ds\right| \\
&\le \frac{(b-a)^2}{2\rho(1-e^{-\rho})} \int_0^1 \left|1+e^{-\rho}-e^{-\rho s}-e^{-\rho(1-s)}\right| \left|h''(sa+(1-s)b)\right| ds \\
&\le \frac{(b-a)^2}{2\rho(1-e^{-\rho})} \int_0^1 \left(1+e^{-\rho}-e^{-\rho s}-e^{-\rho(1-s)}\right) \left(s|h''(a)|+(1-s)|h''(b)|\right) ds \\
&= \frac{(b-a)^2}{2\rho(1-e^{-\rho})} \left(\frac{1+e^{-\rho}}{2} - \frac{1-e^{-\rho}}{\rho}\right) (|h''(a)|+|h''(b)|).
\end{aligned}
$$

The proof is finished. □

Remark 2. *$\alpha \to 1$ in (22) of Theorem 2, then $\rho = \frac{1-\alpha}{\alpha}(b-a) \to 0$, one obtains*

$$
\lim_{\rho\to 0} \frac{1-\alpha}{2(1-e^{-\rho})} = \lim_{x\to 1} \frac{1-\frac{b-a}{b-a-\ln x}}{2(1-x)} = \lim_{x\to 1} \frac{-\ln x}{2(1-x)(b-a-\ln x)} = \frac{1}{2(b-a)}, \tag{25}
$$

and

$$
\lim_{\rho\to 0} \frac{1}{\rho(1-e^{-\rho})}\left(\frac{1+e^{-\rho}}{2} - \frac{1-e^{-\rho}}{\rho}\right) = \lim_{\rho\to 0} \frac{\rho+\rho e^{-\rho}-2+2e^{-\rho}}{2\rho^2(1-e^{-\rho})} = \frac{1}{12}. \tag{26}
$$

So (22) is transformed to

$$
\left|\frac{h(a)+h(b)}{2} - \frac{1}{b-a}\int_a^b h(s)ds\right| \le \frac{(b-a)^2}{24}(|h''(a)|+|h''(b)|).
$$

This result coincides the conclusion in [11], Theorem 3.

Theorem 3. *Assuming $h : [a,b] \to \Re$ is a twice differentiable function. If $h'' \in L[a,b]$ and $|h''|$ is convex on $[a,b]$, then the following inequality for fractional integrals with exponential kernel holds:*

$$|Q_{ml}| \leq \frac{(b-a)^2}{2}\left[\frac{1}{8}+\frac{1+e^{-\rho}}{2\rho(1-e^{-\rho})}-\frac{1}{\rho^2}\right](|h''(a)|+|h''(b)|). \tag{27}$$

Proof. According to Lemma 7 and the convex of $|h''|$, we can get

$$\begin{aligned}
|Q_{ml}| &= \left|\frac{(b-a)^2}{2}\int_0^1 m(s)h''(sa+(1-s)b)ds\right| \leq \frac{(b-a)^2}{2}\int_0^1 |m(s)|\,|h''(sa+(1-s)b)|\,ds \\
&\leq \frac{(b-a)^2}{2}\int_0^{\frac{1}{2}}\left(s+\frac{1+e^{-\rho}-e^{-\rho s}-e^{-\rho(1-s)}}{\rho(1-e^{-\rho})}\right)(s|h''(a)|+(1-s)|h''(b)|)\,ds \\
&\quad+\frac{(b-a)^2}{2}\int_{\frac{1}{2}}^1\left(1-s+\frac{1+e^{-\rho}-e^{-\rho s}-e^{-\rho(1-s)}}{\rho(1-e^{-\rho})}\right)(s|h''(a)|+(1-s)|h''(b)|)\,ds \\
&= \frac{(b-a)^2}{2}\left[\int_0^{\frac{1}{2}}\left(s^2|h''(a)|+t(1-t)|h''(b)|\right)ds+\int_{\frac{1}{2}}^1\left(s(1-s)|h''(a)|ds+(1-s)^2|h''(b)|\right)ds\right. \\
&\quad\left.+\frac{1}{\rho(1-e^{-\rho})}\int_0^1\left(1+e^{-\rho}-e^{-\rho s}-e^{-\rho(1-s)}\right)(s|h''(a)|+(1-s)|h''(b)|)\,ds\right] \\
&= \frac{(b-a)^2}{2}\left[\frac{1}{24}|h''(a)|+\frac{1}{12}|h''(b)|+\frac{1}{12}|h''(a)|+\frac{1}{24}|h''(b)|\right. \\
&\quad\left.+\frac{1}{\rho(1-e^{-\rho})}\left(\frac{1+e^{-\rho}}{2}-\frac{1-e^{-\rho}}{\rho}\right)(|h''(a)|+|h''(b)|)\right] \\
&= \frac{(b-a)^2}{2}\left[\frac{1}{8}+\frac{1+e^{-\rho}}{2\rho(1-e^{-\rho})}-\frac{1}{\rho^2}\right](|h''(a)|+|h''(b)|).
\end{aligned}$$

This completes the proof. □

Remark 3. *Let $\alpha \to 1$ in (27), one has*

$$\left|\frac{1}{b-a}\int_a^b h(s)ds-h(\frac{a+b}{2})\right| \leq \frac{5(b-a)^2}{48}(|h''(a)|+|h''(b)|).$$

4. Application to Special Means

Think on the following particular means [23] for $\forall p,q \in R, p \neq q$ as follows:

$$(i)\quad H(p,q)=\frac{2}{\frac{1}{p}+\frac{1}{q}},\ \ p,q\in R\backslash\{0\};$$

$$(ii)\quad A(p,q)=\frac{p+q}{2},\ \ p,q\in R;$$

$$(iii)\quad L(p,q)=\frac{q-p}{\ln|p|-\ln|q|},\ \ |p|\neq|q|, pq\neq 0;$$

$$(iv)\quad L_m(p,q)=\left[\frac{q^{m+1}-p^{m+1}}{(m+1)(q-p)}\right]^{\frac{1}{m}},\ \ m\in Z\backslash\{-1,0\}, p,q\in R, p\neq q.$$

Next, making use of the acquired results in Section 3, we give some applications to particular means of real number.

Proposition 1. *Let $p,q \in \Re,\ p<q,\ pq>0$ and $m \in \mathbb{Z},\ |m| \geq 2$. Then,*

$$|L_m^m(p,q)-A^m(p,q)| \leq \frac{5}{24}(q-p)^2|m(m-1)|A(|p|^{m-2},|q|^{m-2}). \tag{28}$$

Proof. Applying Remark 3 for $h(x) = x^m$, we can get the conclusion immediately. □

The upper bound is smaller than the result of Proposion 3.1 in [5] when $|q-p| \leq 1$ and $|p|, |q| > 1$ obviously.

Proposition 2. *Let $p, q \in \Re$, $p < q$, $pq > 0$. Then,*

$$\left|L^{-1}(p,q) - A^{-1}(p,q)\right| \leq \frac{5}{12}(q-p)^2 A(|p|^{-3}, |q|^{-3}). \tag{29}$$

Proof. The inference follows from Remark 3 used for $h(x) = \frac{1}{x}$. □

Proposition 3. *Let $p, q \in \Re$, $p < q$, $pq > 0$ and $m \in \mathbb{Z}$, $|m| \geq 2$. Then, we have*

$$\left|L_m^m(q^{-1}, p^{-1}) - H^{-m}(q,p)\right| \leq \frac{5}{24}\left(\frac{1}{p} - \frac{1}{q}\right)^2 |m(m-1)| H^{-1}(|p|^{m-2}, |q|^{m-2}), \tag{30}$$

and

$$\left|L^{-1}(p,q) - H(p,q)\right| \leq \frac{5}{12}(q-p)^2 H^{-1}(|p|^3, |q|^3). \tag{31}$$

Proof. Doing he replacement $q^{-1} \to p, p^{-1} \to q$ in the inequalities (28) and (29), we can obtain the required inequalities (30) and (31), respectively. Here, we have observed $A^{-1}(p^{-1}, q^{-1}) = H(p,q) = 2/(\frac{1}{p} + \frac{1}{q}), q^{-1} < p^{-1}$. □

At last, we will present an application to a midpoint formula. In [23], let w be a division $p = s_0 < s_1 \cdots < s_{m-1} < s_m = q$ of the interval $[p, q]$ and inspect the quadrature formula

$$\int_p^q h(s)ds = T(h,w) + E(h,w), \tag{32}$$

where

$$T(h,w) = \sum_{i=0}^{m-1} h(\frac{s_i + s_{i+1}}{2})(s_{i+1} - s_i)$$

is the midpoint version and $E(h, w)$ refers to the approximation error. Here, we deduce the error estimate for the midpoint formula.

Proposition 4. *Let $h : [p,q] \to \Re$ be a twice differentiable mapping on (p,q) with $p < q$. If $h'' \in L[a,b]$ and $|h''|$ is convex on $[p,q]$, then in (32), for every division w of $[p,q]$, the following inequality holds:*

$$|E(f,w)| \leq \frac{5}{48}\sum_{i=0}^{m-1}(s_{i+1} - s_i)^3 \left(|h''(s_i)| + |h''(s_{i+1})|\right).$$

Proof. Applying Remark 3 on subinterval $[s_i, s_{i+1}] (i = 0, 1, \cdots, m-1)$ of the division w, we derive

$$\left|\int_{s_i}^{s_{i+1}} h(s)ds - h(\frac{s_i + s_{i+1}}{2})(s_i - s_{i+1})\right| \leq \frac{5}{48}(s_{i+1} - s_i)^3 \left(|h''(s_i)| + |h''(s_{i+1})|\right).$$

Summing over from 0 to $m-1$ and making use of the convexity of $|h''|$, we infer that

$$\left|\int_p^q h(s)ds - T(h,w)\right| = \left|\sum_{i=0}^{m-1}\left[\int_{s_i}^{s_{i+1}} h(s)ds - f(\frac{s_i+s_{i+1}}{2})(s_i - s_{i+1})\right]\right|$$
$$\leq \sum_{i-=0}^{m-1}\left|\int_{s_i}^{s_{i+1}} h(s)ds - h(\frac{s_i+s_{i+1}}{2})(s_i - s_{i+1})\right|$$
$$\leq \frac{5}{48}\sum_{i=0}^{m-1}(s_{i+1}-s_i)^3\left(|h''(s_i)| + |h''(s_{i+1})|\right).$$

The proof is completed. □

5. Conclusions

Based on the above interpretation, we acquire the bound estimates of the difference between the average of the fractional integrals with an exponential kernel and the mean values of the endpoints and the midpoint.

By comparing these bound estimates, we have obtained the following conclusions:

(i) With the first and second order derivatives of a given function, the Hermite–Hadamard-type inequalities involving left-sided and right-sided, the fractional integrals are different. The Hermite–Hadamard-type inequalities with the second order derivatives of a given function are more accurate.

(ii) With the same order derivatives of a given function, the Hermite–Hadamard-type inequalities involving different fractional integrals finally tend to be same when $\alpha \to 1$.

Author Contributions: Conceptualization and supervision, J.W.; Formal analysis and writing-original draft preparation, X.W., J.Z.; Writing-review and editing, X.W., J.W..

Funding: The authors thank the referees for their careful reading of the article and insightful comments. This work is partially supported by "Applied Mathematics" as a Key Construction Subject in the 12th Five-Year Plan of Hunan Province, Hunan Natural Science Foundation (2017J2241) and Hunan Social Science Foundation Subsidized Project (16YBA329).

Acknowledgments: The authors thank the referees for their careful reading of the article and insightful comments.

Conflicts of Interest: The authors declare no conflict of interest.

References

1. Mitrinović, D.S.; Lackovixcx, I.B. Hermite and convexity. *Aequ. Math.* **1985**, *28*, 229–232. [CrossRef]
2. Abramovich, S.; Barić, J.; Pexcxarixcx, J. Fejer and Hermite-Hadamard type inequalities for superquadratic functions. *J. Math. Anal. Appl.* **2008**, *344*, 1048–1056. [CrossRef]
3. Cal, J.; Carcamob, J.; Escauriaza, L. A general multidimensional Hermite-Hadamard type inequality. *J. Math. Anal. Appl.* **2009**, *356*, 659–663. [CrossRef]
4. Ödemir, M.E.; Avci, M.; Set, E. On some inequalities of Hermite-Hadamard type via m-convexity. *Appl. Math. Lett.* **2010**, *23*, 1065–1070. [CrossRef]
5. Xiao, Z.; Zhang, Z.; Wu, Y. On weighted Hermite-Hadamard inequalities. *Appl. Math. Comput.* **2011**, *218*, 1147–1152. [CrossRef]
6. Niculescu, C.P. The Hermite-Hadamard inequality for log-convex functions. *Nonlinear Anal. TMA* **2012**, *75*, 662–669. [CrossRef]
7. Dragomir, S.S. Inequalities of Hermite-Hadamard type for h-convex functions on linear spaces. *Proyecc. J. Math.* **2015**, *34*, 323–341. [CrossRef]
8. Muhammad, A.K.; Chu, Y.; Tahir, U.K.; Jamroz, K. Some new inequalities of Hermite-Hadamard type for s-convex functions with applications. *Open Math.* **2017**, *15*, 1414–1430.
9. Budak, H.; Sarikaya, M.Z. Journal of Mathematical Extension, Some new generalized Hermite-Hadamard inequalities for generalized convex functions and applications. *Math. Ext.* **2018**, *12*, 51–56.

10. Dragomir, S.S.; Agarwal, R.P. Two inequalities for differentiable mappings and applications to special means of real numbers and to trapezoidal formula. *Appl. Math. Lett.* **1998**, *11*, 91–95. [CrossRef]
11. Alomari, M.; Darus, M.; Dragomir, S.S. New inequalities of Hermite-Hadamard type for functions whose second derivates absolute values are quasiconvex. *RGMIA Res. Rep. Coll.* **2010**, *12*. [CrossRef]
12. Sarikaya, M.Z.; Set, E.; Yaldiz, H.; Başak, N. Hermite-Hadamard's inequalities for fractional integrals and related fractional inequalities. *Math. Comput. Modell.* **2013**, *57*, 2403–2407. [CrossRef]
13. İŞcan, I.; Wu, S. Hermite-Hadamard type inequalities for harmonically convex functions via fractional integrals. *Appl. Math. Comput.* **2014**, *238*, 237–244. [CrossRef]
14. Jleli, M.; O'Regan, M.; Samet, B. On Hermite-Hadamard type inequalities via generalized fractional integrals. *Turk. J. Math.* **2016**, *40*, 1221–1230. [CrossRef]
15. Noor, M.A.; Noor, K.I.; Awan, M.U.; Khan, S. Fractional Hermite-Hadamard inequalities for some new classes of Godunova-Levin functions. *Appl. Math. Inf. Sci.* **2014**, *8*, 2865–2872. [CrossRef]
16. Ahmad, B.; Alsaedi, A.; Kirane, M.; Torebek, B.T. Hermite-Hadamard, Hermite-Hadamard-Fejér, Dragomir-Agarwal and Pachpatte type inequalities for convex functions via new fractional integrals. *J. Comput. Appl. Math.* **2019**, *353*, 120–129. [CrossRef]
17. Wang, J.; Zhu, C.; Zhou, Y. New generalized Hermite-Hadamard type inequalities and applications to specialmeans. *J. Inequ. Appl.* **2013**, *325*, 1–15.
18. Set, E.; Çelik, B. Generaliaed Fractional Hermite-Hadamard type inequalities for m-Convex and (α, m)-Convex funciton. *Commun. Fac. Sci. Univ. Ank. Ser. Math. Stat.* **2018**, *67*, 333–344.
19. Chen, H.; Katugampola, U.N. Hermite-Hadamard and Hermite-Hadamard-Fejér type inequalities for generalized fractional integrals. *J. Math. Anal. Appl.* **2017**, *446*, 1274–1291. [CrossRef]
20. Set, E.; Noor, M.A.; Awan, M.U.; Gözpinar, A. Generalized Hermite-Hadamard type inequalities involving fractional integral operators. *J. Inequ. Appl.* **2017**, *169*, 1–10. [CrossRef]
21. Wang, J.; Li, X.; Zhou, Y. Hermite-Hadamard inequalities involving Riemann-Liouville fractional integrals via s-convex functions and applications to special means. *Filomat* **2016**, *30*, 1143–1150,. [CrossRef]
22. Liu, W.; Wen, W.; Park, J. Hermite-Hadamard type inequalities for MT-convex functions via classical integrals and fractional integrals. *J. Nonlinear Sci. Appl.* **2016**, *9*, 766–777. [CrossRef]
23. Pearce, C.E.M.; Pećarixcx, J. Inequalities for Differentiable Mappings with Application to Special Means and Quadrature Formula. *Appl. Math. Lett.* **2000**, *13*, 51–55. [CrossRef]

mathematics

MDPI

Article

Stability Results for a Coupled System of Impulsive Fractional Differential Equations

Akbar Zada [1], Shaheen Fatima [1], Zeeshan Ali [1], Jiafa Xu [2] and Yujun Cui [3,*]

[1] Department of Mathematics, University of Peshawar, Khyber Pakhtunkhwa 25000, Pakistan; akbarzada@uop.edu.pk (A.Z.); shaheenfatima072@gmail.com (S.F.); zeeshanmaths1@gmail.com (Z.A.)

[2] School of Mathematical Sciences, Chongqing Normal University, Chongqing 401331, China; 20150028@cqnu.edu.cn

[3] State Key Laboratory of Mining Disaster Prevention and Control Co-founded by Shandong Province and the Ministry of Science and Technology, Shandong University of Science and Technology, Qingdao 266590, China

* Correspondence: cyj720201@sdust.edu.cn

Received: 2 September 2019; Accepted: 2 October 2019; Published: 6 October 2019

Abstract: In this paper, we establish sufficient conditions for the existence, uniqueness and Ulam–Hyers stability of the solutions of a coupled system of nonlinear fractional impulsive differential equations. The existence and uniqueness results are carried out via Banach contraction principle and Schauder's fixed point theorem. The main theoretical results are well illustrated with the help of an example.

Keywords: caputo fractional derivative; coupled system; impulses; existence theory; stability theory

1. Introduction

Fractional differential equations (FDEs) provide an excellent tool for the description of memory and hereditary properties of different processes and materials. Thus, contrary to the classical derivative, the fractional derivative is nonlocal. Fractional calculus has played a very important role in enhancing the mathematical modeling of several phenomena occurring in engineering and scientific disciplines, such as blood flow systems, control theory, aerodynamics, the nonlinear oscillation of earthquake, the fluid-dynamic traffic model, polymer rheology, regular variation in thermodynamics, etc. FDEs are more accurate than the integer-order derivatives. Therefore, in the last few decades, fractional calculus has received great attention from researchers [1–13]. On the other hand, it is impossible to describe the complicated systems and processes with a single differential equation. Therefore, the coupled systems involving FDEs have also received incredible attention; consequently, many results are devoted to them [14–31].

It is well known that the effects of a pulse cannot be ignored in many processes and phenomena. For example, in biological systems such as heart beats, blood flows, mechanical systems with impact, population dynamical systems and so on. Thus, researchers used differential equations with impulses to describe the aforesaid kinds of phenomena. Therefore, many mathematicians studied impulsive FDEs with different boundary conditions; see [32–40] and references cited therein.

In fields such as numerical analysis, optimization theory, and nonlinear analysis, we mostly deal with the approximate solutions and hence we need to check how close these solutions are to the actual solutions of the related system. For this purpose, many approaches can be used, but the approach of Ulam–Hyers stability is a simple and easy one. The aforesaid stability was first initiated by Ulam in 1940 and then was confirmed by Hyers in 1941 [41,42]. That's why this stability is known as Ulam–Hyers stability. In 1978 [43], Rassias generalized the Ulam–Hyers stability by considering variables. Thereafter, mathematicians extended the work mentioned above to functional, differential, integrals and FDEs; for more information about the topic, the reader is recommended to [44–59].

Inspired from the above discussion, in this article, we study the existence, uniqueness and stability analysis of a coupled system of nonlinear FDEs with impulses of the form:

$$\begin{cases} {}^c\mathbf{D}^{\alpha}x(t) + h(t, {}^c\mathbf{D}^{a}x(t), {}^c\mathbf{D}^{b}y(t)) = 0, \ t \neq t_m, \ m = 1, 2\ldots, n, \\ {}^c\mathbf{D}^{\beta}y(t) + w(t, {}^c\mathbf{D}^{a}x(t), {}^c\mathbf{D}^{b}y(t)) = 0, \ t \neq t_m, \ m = 1, 2\ldots, n, \\ \Delta x \mid_{t=t_m} = M_{1m}(x(t_m)), \ \Delta x' \mid_{t=t_m} = N_{1m}(x(t_m)), \ \Delta x'' \mid_{t=t_m} = O_{1m}(x(t_m)), \\ \Delta y \mid_{t=t_m} = M_{2m}(y(t_m)), \ \Delta y' \mid_{t=t_m} = N_{2m}(y(t_m)), \ \Delta y'' \mid_{t=t_m} = O_{2m}(y(t_m)), \\ x(0) = x'(0) = 0, \ {}^c\mathbf{D}^{\varepsilon}x(\Omega) = x''(1), \\ y(0) = y'(0) = 0, \ {}^c\mathbf{D}^{\rho}y(\Phi) = y''(1), \end{cases} \tag{1}$$

where $t \in J = [0, 1]$, $2 < \alpha, \beta \leqslant 3$, $0 < a, b, \varepsilon, \Omega, \rho, \Phi < 1$. ${}^c\mathbf{D}$ stands for Caputo fractional derivative and $h, w : J \times \mathbb{R}^3 \to \mathbb{R}$ are continuous functions. $M_{1m}, M_{2m}, N_{1m}, N_{2m}, O_{1m}, O_{2m} \in C(\mathbb{R}, \mathbb{R})$ and t_m satisfied $0 = t_0 < t_1 < \cdots < t_n < t_{n+1} = 1$, $\Delta x \mid_{t=t_m} = x(t_m^+) - x(t_m^-)$, $\Delta x' \mid_{t=t_m} = x'(t_m^+) - x'(t_m^-)$, $\Delta x'' \mid_{t=t_m} = x''(t_m^+) - x''(t_m^-)$, $\Delta y \mid_{t=t_m} = y(t_m^+) - y(t_m^-)$, $\Delta y' \mid_{t=t_m} = y'(t_m^+) - y'(t_m^-)$, $\Delta y'' \mid_{t=t_m} = y''(t_m^+) - y''(t_m^-)$, $x(t_m^+), y(t_m^+)$, and $x(t_m^-), y(t_m^-)$ represent the right and left limits of $x(t), y(t)$, respectively, at $t = t_m$.

The remaining article is organized as follows: In Section 2, we give some definitions and lemmas related to fractional calculus. In Section 3, we establish our main results about the existence and uniqueness of solutions for the proposed system (1). In Section 4, we study the Ulam–Hyers stability. In Section 5, we provide an example to support our main results.

2. Background Materials

In this section, we give some basic definitions of fractional calculus that will be used throughout the article.

Definition 1. *(see [60]) If $x : (0, \infty) \to \mathbb{R}$ and $\alpha > 0$, then the Caputo fractional derivative of order α is defined as*

$$ {}^c\mathbf{D}^{\alpha}x(t) = \frac{1}{\Gamma(n-\alpha)} \int_0^t (t-s)^{n-\alpha-1} x^{(n)}(s)ds, \ n-1 < \alpha < n, \ n = [\alpha] + 1,$$

where $[\alpha]$ denotes the integer part of real number α, provided that the right side is pointwise defined on $(0, \infty)$.

Definition 2. *(see [60]) The Riemann–Liouville fractional integral of order $\alpha > 0$ for a function $x : (0, \infty) \to \mathbb{R}$ is defined as*

$$\mathbf{I}^{\alpha}x(t) = \frac{1}{\Gamma(\alpha)} \int_0^t (t-s)^{\alpha-1} x(s)ds, \quad t > 0,$$

provided that the right side is pointwise defined on $(0, \infty)$, where Γ is the Euler Gamma function.

Lemma 1. *(see [60]) The solution of the differential equations involving Caputo derivative ${}^c\mathbf{D}^{\alpha}x(t) = f(t)$, $t \in J$, has the form:*

$$\mathbf{I}^{\alpha c}\mathbf{D}^{\alpha}x(t) = \mathbf{I}^{\alpha}f(t) + e_0 + e_1 t + \cdots + e_{n-1}t^{n-1},$$

for some $e_i \in \mathbb{R}$, $i = 0, 1, \ldots, n-1$, $n = [\alpha] + 1$.

Lemma 2. *(see [60]) If $\alpha, \beta > 0$, $t \in J$, then, for $x(t)$, we have*

$$ {}^c\mathbf{D}^{\alpha}\mathbf{I}^{\alpha}x(t) = x(t), \quad \mathbf{I}^{\alpha}\mathbf{I}^{\beta}x(t) = \mathbf{I}^{\alpha+\beta}x(t).$$

Lemma 3. *(**Banach contraction principle**, see [59]) If X is real Banach space and $\mathbf{W} : X \longrightarrow X$ is a contraction mapping, then $\mathbf{W}$ has a unique fixed point in X.*

Theorem 1. *(**Schauder fixed point theorem**, see [59]) If ω is a closed bounded convex subset of a Banach space X and $\mathbf{W} : \omega \longrightarrow \omega$ is completely continuous, then $\mathbf{W}$ has at least one fixed point in ω.*

For the sake of convenience, we introduce the Banach space as follows:
Let $J = [0,1]$, $J' = J/\{t_1, t_2, \ldots, t_n\}$. Define the set by

$$X = PC(J) = \{x(t) : x(t), x'(t), x''(t), {}^c\mathbf{D}^a x(t), {}^c\mathbf{D}^b x(t) \in C(J'), x(t_m^+) \text{ and } x(t_m^-) \text{ exists and satisfying } x(t_m^-) = x(t_m), 1 \leqslant m \leqslant n\}.$$

It is easy to verify that X is a Banach space equipped with the norm:

$$\|x\|_0 = \max\{\sup_{t\in J}|x(t)|, \sup_{t\in J}|x'(t)|, \sup_{t\in J}|x''(t)|, \sup_{t\in J}|{}^c\mathbf{D}^a x(t)|, \sup_{t\in J}|{}^c\mathbf{D}^b x(t)|\}, \forall\, x(t) \in PC(J).$$

Similarly, we can define a set $Y = PC(J)$, which is a Banach space endowed with the defined norm:

$$\|y\|_0 = \max\{\sup_{t\in J}|y(t)|, \sup_{t\in J}|y'(t)|, \sup_{t\in J}|y''(t)|, \sup_{t\in J}|{}^c\mathbf{D}^a y(t)|, \sup_{t\in J}|{}^c\mathbf{D}^b y(t)|\}, \forall\, y(t) \in PC(J).$$

Furthermore, we define the Banach space $Y' = X \times Y$ with the norms $\|(x,y)\| = \|x\|_0 + \|y\|_0$ and $\|(x,y)\| = \max\{\|x\|_0, \|y\|_0\}$.

Definition 3. *A pair of functions $(x(t), y(t)) \in Y'$ is called a solution of (1) if $(x(t), y(t))$ satisfy all the equations and boundary value conditions of the system (1).*

Lemma 4. *Assume that $f \in C(J, \mathbb{R})$. A function $x \in PC(J)$ is a solution of the boundary value system*

$$\begin{cases} {}^c\mathbf{D}^\alpha x(t) + f(t) = 0,\ 2 < \alpha \leqslant 3, \\ \Delta x\,|_{t=tm} = M_{1m}(x(t_m)),\ m = 1,2,\ldots,n, \\ \Delta x'\,|_{t=tm} = N_{1m}(x(t_m)),\ m = 1,2,\ldots,n, \\ \Delta x''\,|_{t=tm} = O_{1m}(x(t_m)),\ m = 1,2,\ldots,n, \\ x(0) = x'(0) = 0,\ {}^c\mathbf{D}^\varepsilon x(\Omega) = x''(1),\ 0 < \varepsilon, \Omega < 1, \end{cases} \tag{2}$$

if and only if $x \in PC(J)$ is the solution of integral equation

$$x(t) = \begin{cases} -\frac{1}{\Gamma(\alpha)}\int_0^t (t-s) f(s)ds + e^* t^2,\ t \in [0, t_1], \\ -\frac{1}{\Gamma(\alpha)}\int_{t_m}^t (t-s)^{\alpha-1} f(s)ds - \frac{1}{\Gamma(\alpha)}\sum_{j=1}^m \int_{t_{j-1}}^{t_j} (t_j - s)^{\alpha-1} f(s)ds \\ -\frac{1}{\Gamma(\alpha-1)}\sum_{j=1}^m (t-t_j)\int_{t_{j-1}}^{t_j}(t_j-s)^{\alpha-2} f(s)ds - \frac{1}{2\Gamma(\alpha-2)}\sum_{j=1}^m (t-t_j)^2 \\ \times \int_{t_{j-1}}^{t_j} (t_j-s)^{\alpha-3} f(s)ds + \sum_{j=1}^m M_{1j}(x(t_j)) + \sum_{j=1}^m (t-t_j) N_{1j}(x(t_j)) \\ + \sum_{j=1}^m \frac{(t-t_j)^2}{2} O_{1j}(x(t_j)) + e^* t^2,\ t \in (t_m, t_{m+1}],\ 1 \leqslant m \leqslant n, \end{cases} \tag{3}$$

where $t \in (t_m, t_{m+1}], 1 \leqslant m \leqslant n$,

$$e^* = \frac{\Gamma(3-\varepsilon)}{2(\Gamma(3-\varepsilon)-\Omega^{2-\varepsilon})}\Bigg[-\frac{1}{\Gamma(\alpha-\varepsilon)}\int_{t_j}^{\Omega}(\Omega-s)^{\alpha-\varepsilon-1}f(s)ds - \frac{\Omega^{1-\varepsilon}}{\Gamma(\alpha-1)\Gamma(2-\varepsilon)}$$
$$\times\sum_{j=1}^{m}\int_{t_{j-1}}^{t_j}(t_j-s)^{\alpha-2}f(s)ds - \frac{\Omega^{2-\varepsilon}}{\Gamma(\alpha-2)\Gamma(3-\varepsilon)}\sum_{j=1}^{m}\int_{t_{j-1}}^{t_j}(t_j-s)^{\alpha-3}f(s)ds$$
$$+\frac{\Omega^{1-\varepsilon}}{\Gamma(\alpha-2)\Gamma(2-\varepsilon)}\sum_{j=1}^{m}t_j\int_{t_{j-1}}^{t_j}(t_j-s)^{\alpha-3}f(s)ds + \frac{\Omega^{1-\varepsilon}}{\Gamma(2-\varepsilon)}\sum_{j=1}^{m}N_{1j}(x(t_j))$$
$$+\frac{\Omega^{2-\varepsilon}}{\Gamma(3-\varepsilon)}\sum_{j=1}^{m}O_{1j}(x(t_j)) - \frac{\Omega^{1-\varepsilon}}{\Gamma(2-\varepsilon)}\sum_{j=1}^{m}t_jO_{1j}(x(t_j)) + \frac{1}{\Gamma(\alpha-2)}\int_{t_m}^{1}(1-s)^{\alpha-3}f(s)ds$$
$$+\frac{1}{\Gamma(\alpha-2)}\sum_{j=1}^{m}\int_{t_{j-1}}^{t_j}(t_j-s)^{\alpha-3}f(s)ds - \sum_{j=1}^{m}O_{1j}(x(t_j))\Bigg].$$

Proof. Applying Lemma 1, for some constants $e_0, e_1, e_2 \in \mathbb{R}$, we have

$$\begin{aligned} x(t) &= -\mathbf{I}^{\alpha}f(s)ds + e_0 + e_1t + e_2t^2 \\ &= -\frac{1}{\Gamma(\alpha)}\int_0^t (t-s)^{\alpha-1}f(s)ds + e_0 + e_1t + e_2t^2,\ t \in [0, t_1]. \end{aligned} \tag{4}$$

Then, we obtain

$$\begin{cases} x'(t) = -\dfrac{1}{\Gamma(\alpha-1)}\displaystyle\int_0^t (t-s)^{\alpha-2}f(s)ds + e_1 + 2e_2t, \\ x''(t) = -\dfrac{1}{\Gamma(\alpha-2)}\displaystyle\int_0^t (t-s)^{\alpha-3}f(s)ds + 2e_2. \end{cases} \tag{5}$$

When $t \in (t_1, t_2)$, we have

$$\begin{cases} x(t) = -\dfrac{1}{\Gamma(\alpha)}\displaystyle\int_{t_1}^t (t-s)^{\alpha-1}f(s)ds + e_3 + e_4(t-t_1) + e_5(t-t_1)^2, \\ x'(t) = -\dfrac{1}{\Gamma(\alpha-1)}\displaystyle\int_{t_1}^t (t-s)^{\alpha-2}f(s)ds + e_4 + 2e_5(t-t_1), \\ x''(t) = -\dfrac{1}{\Gamma(\alpha-2)}\displaystyle\int_{t_1}^t (t-s)^{\alpha-3}f(s)ds + 2e_5, \end{cases} \tag{6}$$

where e_3, e_4, e_5 are arbitrary constants, from (4)–(6), we can find

$$x(t_1^-) = -\frac{1}{\Gamma(\alpha)}\int_0^{t_1}(t_1-s)^{\alpha-1}f(s)ds + e_0 + e_1t_1 + e_2t_1^2,\ \ x(t_1^+) = e_3, \tag{7}$$

$$x'(t_1^-) = -\frac{1}{\Gamma(\alpha-1)}\int_0^{t_1}(t_1-s)^{\alpha-2}f(s)ds + e_1 + 2e_2t_1,\ \ x'(t_1^+) = e_4, \tag{8}$$

$$x''(t_1^-) = -\frac{1}{\Gamma(\alpha-2)}\int_0^{t_1}(t_1-s)^{\alpha-3}f(s)ds + 2e_2,\ \ x''(t_1^+) = 2e_5. \tag{9}$$

Furthermore, $\Delta x\ |_{t=t_1} = M_{11}(x(t_1))$, $\Delta x'\ |_{t=t_1} = N_{11}(x(t_1))$, $\Delta x''\ |_{t=t_1} = O_{11}(x(t_1))$, and (7)–(9) give us:

$$\begin{cases} e_3 = -\dfrac{1}{\Gamma(\alpha)}\displaystyle\int_0^{t_1}(t_1-s)^{\alpha-1}f(s)ds + e_0 + e_1t_1 + e_2t_1^2 + M_{11}(x(t_1)), \\ e_4 = -\dfrac{1}{\Gamma(\alpha-1)}\displaystyle\int_0^{t_1}(t_1-s)^{\alpha-2}f(s)ds + e_1 + 2e_2t_1 + N_{11}(x(t_1)), \\ e_5 = -\dfrac{1}{2\Gamma(\alpha-2)}\displaystyle\int_0^{t_1}(t_1-s)^{\alpha-3}f(s)ds + e_2 + \dfrac{1}{2}O_{11}(x(t_1)). \end{cases}$$

Plugging e_3, e_4, and e_5 into the first equation of (6) for $t \in (t_1, t_2]$, we have

$$\begin{aligned} x(t) = & -\frac{1}{\Gamma(\alpha)}\int_{t_1}^{t}(t-s)^{\alpha-1}f(s)ds - \frac{1}{\Gamma(\alpha)}\int_0^{t_1}(t_1-s)f(s)ds - \frac{(t-t_1)}{\Gamma(\alpha-1)}\int_0^{t_1}(t_1-s)^{\alpha-2}f(s)ds \\ & - \frac{(t-t_1)^2}{2\Gamma(\alpha-2)}\int_0^{t_1}(t_1-s)^{\alpha-3}f(s)ds + M_{11}(x(t_1)) + (t-t_1)N_{11}(x(t_1)) + \frac{(t-t_1)^2}{2}O_{11}(x(t_1)) \\ & + e_0 + e_1 t + e_2 t^2. \end{aligned}$$

Repeating the same process for $t \in (t_m, t_{m+1}]$ such that $(m = 1, 2, \ldots, n)$, then we can write

$$\begin{aligned} x(t) = & -\frac{1}{\Gamma(\alpha)}\int_{t_m}^{t}(t-s)^{\alpha-1}f(s)ds - \frac{1}{\Gamma(\alpha)}\sum_{j=1}^{m}\int_{t_{j-1}}^{t_j}(t_j-s)^{\alpha-1}f(s)ds - \frac{1}{\Gamma(\alpha-1)}\sum_{j=1}^{m}(t-t_j) \\ & \times \int_{t_{j-1}}^{t_j}(t_j-s)^{\alpha-2}f(s)ds - \frac{1}{2\Gamma(\alpha-2)}\sum_{j=1}^{m}(t-t_j)^2\int_{t_{j-1}}^{t_j}(t_j-s)^{\alpha-3}f(s)ds + \sum_{j=1}^{m}M_{1j}(x(t_j)) \\ & + \sum_{j=1}^{m}(t-t_j)N_{1j}(x(t_j)) + \sum_{j=1}^{m}\frac{(t-t_j)^2}{2}O_{1j}(x(t_j)) + e_0 + e_1 t + e_2 t^2. \end{aligned} \quad (10)$$

Furthermore, we have

$$\begin{aligned} x''(t) = & -\frac{1}{\Gamma(\alpha-2)}\int_{t_m}^{t}(t-s)^{\alpha-3}f(s)ds - \frac{1}{\Gamma(\alpha-2)}\sum_{j=1}^{m}\int_{t_{j-1}}^{t_j}(t_j-s)^{\alpha-3}f(s)ds + O_{1j}(x(t_j)) \\ & +2e_2. \end{aligned} \quad (11)$$

By utilizing conditions $x(0) = x'(0) = 0$ in (4), we get $e_0 = e_1 = 0$. In addition, it follows from (11) that

$$\begin{aligned} x''(1) = & -\frac{1}{\Gamma(\alpha-2)}\int_{t_m}^{1}(1-s)^{\alpha-3}f(s)ds - \frac{1}{\Gamma(\alpha-2)}\sum_{j=1}^{m}\int_{t_{j-1}}^{t_j}(t_j-s)^{\alpha-3}f(s)ds + O_{1j}(x(t_j)) \\ & +2e_2. \end{aligned} \quad (12)$$

In view of $j \in \{0, 1, \ldots, n\}$ such that $\Omega \in (t_j, t_{j+1}]$, we have

$$\begin{aligned} x(\Omega) = & -\frac{1}{\Gamma(\alpha)}\int_{t_j}^{\Omega}(\Omega-s)^{\alpha-1}f(s)ds - \frac{1}{\Gamma(\alpha)}\sum_{j=1}^{m}\int_{t_{j-1}}^{t_j}(t_j-s)^{\alpha-1}f(s)ds - \frac{1}{\Gamma(\alpha-1)}\sum_{j=1}^{m}(\Omega-t_j) \\ & \times \int_{t_{j-1}}^{t_j}(t_j-s)^{\alpha-2}f(s)ds - \frac{1}{2\Gamma(\alpha-2)}\sum_{j=1}^{m}(\Omega-t_j)^2\int_{t_{j-1}}^{t_j}(t_j-s)^{\alpha-3}f(s)ds \\ & + \sum_{j=1}^{m}M_{1j}(x(t_j)) + \sum_{j=1}^{m}(\Omega-t_j)N_{1j}(x(t_j)) + \frac{(\Omega-t_j)^2}{2}O_{1j}(x(t_j)) + e_2\Omega^2. \end{aligned} \quad (13)$$

By applying result (5), we get

$$\begin{aligned} {}^{c}D_{0^+}^{\varepsilon}x(\Omega) = & -\frac{1}{\Gamma(\alpha-\varepsilon)}\int_{t_j}^{\Omega}(\Omega-s)^{\alpha-\varepsilon-1}f(s)ds - \frac{\Omega^{1-\varepsilon}}{\Gamma(\alpha-1)\Gamma(2-\varepsilon)}\sum_{j=1}^{m}\int_{t_{j-1}}^{t_j}(t_j-s)^{\alpha-2}f(s)ds \\ & - \frac{\Omega^{2-\varepsilon}}{\Gamma(\alpha-2)\Gamma(3-\varepsilon)}\sum_{j=1}^{m}\int_{t_{j-1}}^{t_j}(t_j-s)^{\alpha-3}f(s)ds + \frac{\Omega^{1-\varepsilon}}{\Gamma(\alpha-2)\Gamma(2-\varepsilon)}\sum_{j=1}^{m}t_j \\ & \times \int_{t_{j-1}}^{t_j}(t_j-s)^{\alpha-3}f(s)ds + \frac{\Omega^{1-\varepsilon}}{\Gamma(2-\varepsilon)}\sum_{j=1}^{m}N_{1j}(x(t_j)) + \frac{\Omega^{2-\varepsilon}}{\Gamma(3-\varepsilon)}\sum_{j=1}^{m}O_{1j} \\ & - \frac{\Omega^{1-\varepsilon}}{\Gamma(2-\varepsilon)}\sum_{j=1}^{m}t_j O_{1j} + \frac{2\Omega^{2-\varepsilon}}{\Gamma(3-\varepsilon)}e_2. \end{aligned} \quad (14)$$

Since ${}^cD^{\varepsilon}_{0+}x(\Omega) = x''(1)$, thus (13) and (14) gives

$$e_2 = \frac{\Gamma(3-\varepsilon)}{2(\Gamma(3-\varepsilon)-\Omega^{2-\varepsilon})}\Bigg[-\frac{1}{\Gamma(\alpha-\varepsilon)}\int_{t_j}^{\Omega}(\Omega-s)^{\alpha-\varepsilon-1}f(s)ds - \frac{\Omega^{1-\varepsilon}}{\Gamma(\alpha-1)\Gamma(2-\varepsilon)}$$
$$\times\sum_{j=1}^{m}\int_{t_{j-1}}^{t_j}(t_j-s)^{\alpha-2}f(s)ds - \frac{\Omega^{2-\varepsilon}}{\Gamma(\alpha-2)\Gamma(3-\varepsilon)}\sum_{j=1}^{m}\int_{t_{j-1}}^{t_j}(t_j-s)^{\alpha-3}f(s)ds$$
$$+\frac{\Omega^{1-\varepsilon}}{\Gamma(\alpha-2)\Gamma(2-\varepsilon)}\sum_{j=1}^{m}t_j\int_{t_{j-1}}^{t_j}(t_j-s)^{\alpha-3}f(s)ds + \frac{\Omega^{1-\varepsilon}}{\Gamma(2-\varepsilon)}\sum_{j=1}^{m}N_{1j}(x(t_j))$$
$$+\frac{\Omega^{2-\varepsilon}}{\Gamma(3-\varepsilon)}\sum_{j=1}^{m}O_{1j} - \frac{\Omega^{1-\varepsilon}}{\Gamma(2-\varepsilon)}\sum_{j=1}^{m}t_jO_{1j} + \frac{1}{\Gamma(\alpha-2)}\int_{t_m}^{1}(1-s)^{\alpha-3}f(s)ds$$
$$+\frac{1}{\Gamma(\alpha-2)}\sum_{j=1}^{m}\int_{t_{j-1}}^{t_j}(t_j-s)^{\alpha-3}f(s)ds - \sum_{j=1}^{m}O_{1j}(x(t_j))\Bigg].$$

Plugging the values of e_0, e_1 and e_2 into (4) and (10), (3) can thus be obtained. Conversely, we consider that $x(t)$ is a solution of (3). Then, it is obvious that (3) satisfies (2). □

Similarly as in Lemma 4, we can prove the following:

Lemma 5. *Let $\vartheta \in C(J, \mathbb{R})$. A function $y \in PC(J)$ is the solution of*

$$\begin{cases} {}^c\mathbf{D}^{\beta}y(t) + \vartheta(t) = 0,\ 2 < \beta \leqslant 3, \\ \Delta y\mid_{t=t_m} = M_{2m}(y(t_m)),\ m = 1,2,\dots,n, \\ \Delta y'\mid_{t=t_m} = N_{2m}(y(t_m)),\ m = 1,2,\dots,n, \\ \Delta y''\mid_{t=t_m} = O_{2m}(y(t_m)),\ m = 1,2,\dots,n, \\ y(0) = y'(0) = 0,\ {}^c\mathbf{D}^{\rho}_{0+}y(\Phi) = y''(1),\ 0 < \rho, \Phi < 1, \end{cases} \tag{15}$$

if and only if $y \in PC(J)$ is the solution of the integral equation

$$y(t) = \begin{cases} -\frac{1}{\Gamma(\beta)}\int_0^t(t-s)^{\beta-1}\vartheta(s)ds + c^*t^2,\ t \in [0, t_1], \\ -\frac{1}{\Gamma(\beta)}\int_{t_m}^t(t-s)^{\beta-1}\vartheta(s)ds - \frac{1}{\Gamma(\beta)}\sum_{j=1}^m\int_{t_{j-1}}^{t_j}(t_j-s)^{\beta-1}\vartheta(s)ds \\ -\frac{1}{\Gamma(\beta-1)}\sum_{j=1}^m(t-t_j)\int_{t_{j-1}}^{t_j}(t_j-s)^{\beta-2}\vartheta(s)ds - \frac{1}{2\Gamma(\beta-2)}\sum_{j=1}^m(t-t_j)^2 \\ \times\int_{t_{j-1}}^{t_j}(t_j-s)^{\beta-3}\vartheta(s)ds + \sum_{j=1}^m M_{2j}(y(t_j)) + \sum_{j=1}^m(t-t_j)N_{2j}(y(t_j)) \\ +\sum_{j=1}^m\frac{(t-t_j)^2}{2}O_{2j}(y(t_j)) + c^*t^2,\ t \in (t_m, t_{m+1}],\ 1 \leqslant m \leqslant n, \end{cases} \tag{16}$$

where $c \in \{0, 1, \dots, n\}$ such that $\Phi \in (t_c, t_{c+1}]$, and

$$
\begin{aligned}
c^* =& \frac{\Gamma(3-\rho)}{2(\Gamma(3-\rho)-\Phi^{2-\rho})}\Big[-\frac{1}{\Gamma(\beta-\rho)}\int_{t_j}^{\Phi}(\Phi-s)^{\beta-\rho-1}\vartheta(s)ds-\frac{\Phi^{1-\rho}}{\Gamma(\beta-1)\Gamma(2-\rho)}\\
&\times\sum_{j=1}^{m}\int_{t_{j-1}}^{t_j}(t_j-s)^{\beta-2}\vartheta(s)ds-\frac{\Phi^{2-\rho}}{\Gamma(\beta-2)\Gamma(3-\rho)}\sum_{j=1}^{m}\int_{t_{j-1}}^{t_j}(t_j-s)^{\beta-3}\vartheta(s)ds\\
&+\frac{\Phi^{1-\rho}}{\Gamma(\beta-2)\Gamma(2-\rho)}\sum_{j=1}^{m}t_j\int_{t_{j-1}}^{t_j}(t_j-s)^{\beta-3}\vartheta(s)ds+\frac{\Phi^{1-\rho}}{\Gamma(2-\rho)}\sum_{j=1}^{m}N_{2j}(y(t_j))\\
&+\frac{\Phi^{2-\rho}}{\Gamma(3-\rho)}\sum_{j=1}^{m}O_{2j}(y(t_j))-\frac{\Phi^{1-\rho}}{\Gamma(2-\rho)}\sum_{j=1}^{m}t_jO_{2j}(y(t_j))+\frac{1}{\Gamma(\beta-2)}\int_{t_m}^{1}(1-s)^{\beta-3}\vartheta(s)ds\\
&+\frac{1}{\Gamma(\beta-2)}\sum_{j=1}^{m}\int_{t_{j-1}}^{t_j}(t_j-s)^{\beta-3}\vartheta(s)ds-\sum_{j=1}^{m}O_{2j}(y(t_j))\Big].
\end{aligned}
$$

3. Main Results

In this section, we use fixed point theorems to prove the existence of solutions to problem (1). According to Lemmas 4 and 5, we define operator $\mathbf{W}: Y' \longrightarrow Y'$ by

$$
\mathbf{W}(x,y)(t) = \big(\mathbf{W}_1(x,y)(t), \mathbf{W}_2(x,y)(t)\big)^T, \forall\ (x,y) \in Y',\ t \in [0,1], \tag{17}
$$

where

$$
\mathbf{W}_1(x,y)(t) = \begin{cases}
-\frac{1}{\Gamma(\alpha)}\int_0^t(t-s)h(s,{}^c\mathbf{D}^a x(s),{}^c\mathbf{D}^b y(s))ds + e^*t^2,\ t\in[0,t_1],\\
-\frac{1}{\Gamma(\alpha)}\int_{t_m}^t(t-s)^{\alpha-1}h(s,{}^c\mathbf{D}^a x(s),{}^c\mathbf{D}^b y(s))ds\\
-\frac{1}{\Gamma(\alpha)}\sum_{j=1}^m\int_{t_{j-1}}^{t_j}(t_j-s)^{\alpha-1}h(s,{}^c\mathbf{D}^a x(s),{}^c\mathbf{D}^b y(s))ds\\
-\frac{1}{\Gamma(\alpha-1)}\sum_{j=1}^m(t-t_j)\int_{t_{j-1}}^{t_j}(t_j-s)^{\alpha-2}h(s,{}^c\mathbf{D}^a x(s),{}^c\mathbf{D}^b y(s))ds\\
-\frac{1}{2\Gamma(\alpha-2)}\sum_{j=1}^m(t-t_j)^2\int_{t_{j-1}}^{t_j}(t_j-s)^{\alpha-3}h(s,{}^c\mathbf{D}^a x(s),{}^c\mathbf{D}^b y(s))ds\\
+\sum_{j=1}^m M_{1j}(x(t_j))+\sum_{j=1}^m(t-t_j)N_{1j}(x(t_j))+\sum_{j=1}^m\frac{(t-t_j)^2}{2}O_{1j}(x(t_j))+e^*t^2,\\
t\in(t_m,t_{m+1}],\ 1\leqslant m\leqslant n,
\end{cases}
$$

and

$$
\mathbf{W}_2(x,y)(t) = \begin{cases}
-\frac{1}{\Gamma(\beta)}\int_0^t(t-s)^{\beta-1}w(s,{}^c\mathbf{D}^a x(s),{}^c\mathbf{D}^b y(s))ds + c^*t^2,\ t\in[0,t_1],\\
-\frac{1}{\Gamma(\beta)}\int_{t_m}^t(t-s)^{\beta-1}w(s,{}^c\mathbf{D}^a x(s),{}^c\mathbf{D}^b y(s))ds\\
-\frac{1}{\Gamma(\beta)}\sum_{j=1}^m\int_{t_{j-1}}^{t_j}(t_j-s)^{\beta-1}w(s,{}^c\mathbf{D}^a x(s),{}^c\mathbf{D}^b y(s))ds\\
-\frac{1}{\Gamma(\beta-1)}\sum_{j=1}^m(t-t_j)\int_{t_{j-1}}^{t_j}(t_j-s)^{\beta-2}w(s,{}^c\mathbf{D}^a x(s),{}^c\mathbf{D}^b y(s))ds\\
-\frac{1}{2\Gamma(\beta-2)}\sum_{j=1}^m(t-t_j)^2\int_{t_{j-1}}^{t_j}(t_j-s)^{\beta-3}w(s,{}^c\mathbf{D}^a x(s),{}^c\mathbf{D}^b y(s))ds\\
+\sum_{j=1}^m M_{2j}(y(t_j))+\sum_{j=1}^m(t-t_j)N_{2j}(y(t_j))+\sum_{j=1}^m\frac{(t-t_j)^2}{2}O_{2j}(y(t_j))+c^*t^2,\\
t\in(t_m,t_{m+1}],\ 1\leqslant m\leqslant n,
\end{cases}
$$

with

$$
\begin{aligned}
e^* =& \mathcal{Z}_\alpha\Big[-\frac{1}{\Gamma(\alpha-\varepsilon)}\int_{t_j}^{\Omega}(\Omega-s)^{\alpha-\varepsilon-1}h(s,{}^c\mathbf{D}^a x(s),{}^c\mathbf{D}^b y(s))ds\\
&-\frac{\Omega^{1-\varepsilon}}{\Gamma(\alpha-1)\Gamma(2-\varepsilon)}\sum_{j=1}^{m}\int_{t_{j-1}}^{t_j}(t_j-s)^{\alpha-2}h(s,{}^c\mathbf{D}^a x(s),{}^c\mathbf{D}^b y(s))ds\\
&-\frac{\Omega^{2-\varepsilon}}{\Gamma(\alpha-2)\Gamma(3-\varepsilon)}\sum_{j=1}^{m}\int_{t_{j-1}}^{t_j}(t_j-s)^{\alpha-3}h(s,{}^c\mathbf{D}^a x(s),{}^c\mathbf{D}^b y(s))ds
\end{aligned}
$$

$$
\begin{aligned}
&+\frac{\Omega^{1-\varepsilon}}{\Gamma(\alpha-2)\Gamma(2-\varepsilon)}\sum_{j=1}^{m} t_j\int_{t_{j-1}}^{t_j}(t_j-s)^{\alpha-3}h(s,{}^c\mathbf{D}^a x(s),{}^c\mathbf{D}^b y(s))ds\\
&+\frac{\Omega^{1-\varepsilon}}{\Gamma(2-\varepsilon)}\sum_{j=1}^{m}N_{1j}(x(t_j))+\frac{\Omega^{2-\varepsilon}}{\Gamma(3-\varepsilon)}\sum_{j=1}^{m}O_{1j}(x(t_j))-\frac{\Omega^{1-\varepsilon}}{\Gamma(2-\varepsilon)}\sum_{j=1}^{m}t_jO_{1j}(x(t_j))\\
&+\frac{1}{\Gamma(\alpha-2)}\int_{t_m}^{1}(1-s)^{\alpha-3}h(s,{}^c\mathbf{D}^a x(s),{}^c\mathbf{D}^b y(s))ds\\
&+\frac{1}{\Gamma(\alpha-2)}\sum_{j=1}^{m}\int_{t_{j-1}}^{t_j}(t_j-s)^{\alpha-3}h(s,{}^c\mathbf{D}^a x(s),{}^c\mathbf{D}^b y(s))ds-\sum_{j=1}^{m}O_{1j}(x(t_j))\Bigg],\\
c^* =&\mathcal{Z}_\beta\Bigg[-\frac{1}{\Gamma(\beta-\rho)}\int_{t_j}^{\Phi}(\Phi-s)^{\beta-\rho-1}w(s,{}^c\mathbf{D}^a x(s),{}^c\mathbf{D}^b y(s))ds\\
&-\frac{\Phi^{1-\rho}}{\Gamma(\beta-1)\Gamma(2-\rho)}\sum_{j=1}^{m}\int_{t_{j-1}}^{t_j}(t_j-s)^{\beta-2}w(s,{}^c\mathbf{D}^a x(s),{}^c\mathbf{D}^b y(s))ds\\
&-\frac{\Phi^{2-\rho}}{\Gamma(\beta-2)\Gamma(3-\rho)}\sum_{j=1}^{m}\int_{t_{j-1}}^{t_j}(t_j-s)^{\beta-3}w(s,{}^c\mathbf{D}^a x(s),{}^c\mathbf{D}^b y(s))ds\\
&+\frac{\Phi^{1-\rho}}{\Gamma(\beta-2)\Gamma(2-\rho)}\sum_{j=1}^{m}t_j\int_{t_{j-1}}^{t_j}(t_j-s)^{\beta-3}w(s,{}^c\mathbf{D}^a x(s),{}^c\mathbf{D}^b y(s))ds)\\
&+\frac{\Phi^{1-\rho}}{\Gamma(2-\rho)}\sum_{j=1}^{m}N_{2j}(y(t_j)+\frac{\Phi^{2-\rho}}{\Gamma(3-\rho)}\sum_{j=1}^{m}O_{2j}(y(t_j))-\frac{\Phi^{1-\rho}}{\Gamma(2-\rho)}\sum_{j=1}^{m}t_jO_{2j}(y(t_j))\\
&+\frac{1}{\Gamma(\beta-2)}\int_{t_m}^{1}(1-s)^{\beta-3}w(s,{}^c\mathbf{D}^a x(s),{}^c\mathbf{D}^b y(s))ds\\
&+\frac{1}{\Gamma(\beta-2)}\sum_{j=1}^{m}\int_{t_{j-1}}^{t_j}(t_j-s)^{\beta-3}w(s,{}^c\mathbf{D}^a x(s),{}^c\mathbf{D}^b y(s))ds-\sum_{j=1}^{m}O_{2j}(y(t_j))\Bigg],
\end{aligned}
$$

where $\mathcal{Z}_\alpha = \frac{\Gamma(3-\varepsilon)}{2(\Gamma(3-\varepsilon)-\Omega^{2-\varepsilon})}$ and $\mathcal{Z}_\beta = \frac{\Gamma(3-\rho)}{2(\Gamma(3-\rho)-\Phi^{2-\rho})}$. Thus, solving problem (1) is equivalent to obtain a fixed point of the operator $\mathbf{W}$. Next, we have to prove the uniqueness of solutions of problem (1).

Theorem 2. *Let the following conditions* $(\mathbf{M}_1)-(\mathbf{M}_3)$ *hold, and then the boundary value problem* (1) *has a unique solution.*
$(\mathbf{M}_1)$: *For all* $t \in J$ *and* $x_j, y_j \in \mathbb{R}$ $(j = 1, 2)$ *there exists some positive constants* μ_j, μ_j' $(j = 1, 2)$ *such that*

$$
\begin{aligned}
\big|h(t,x_1,y_1)-h(t,x_2,y_2)\big| &\leqslant \mu_1|x_1-x_2|+\mu_2|y_1-y_2|,\\
\big|w(t,x_1,y_1)-w(t,x_2,y_2)\big| &\leqslant \mu_1'|x_1-x_2|+\mu_2'|y_1-y_2|.
\end{aligned}
$$

$(\mathbf{M}_2)$: *For all* $x, y \in \mathbb{R}$, *there exist some positive constants* $I_{jk}, \hat{I}_{jk}, \check{I}_{jk}(j = 1, 2;\ m = 1, 2, \ldots, n)$ *such that*

$$
\begin{aligned}
\big|M_{jm}(x)-M_{jm}(y)\big| &\leqslant I_{jm}|x-y|,\\
\big|N_{jm}(x)-N_{jm}(y)\big| &\leqslant \hat{I}_{jm}|x-y|,\\
\big|O_{jm}(x)-O_{jm}(y)\big| &\leqslant \check{I}_{jm}|x-y|.
\end{aligned}
$$

$(\mathbf{M}_3)$:

$$\Big[(\mu_1+\mu_2)\Big(\frac{2}{\Gamma(\alpha+1)}+\frac{1}{\Gamma(\alpha)}+\frac{1}{2\Gamma(\alpha-1)}+\frac{\mathcal{Z}_\alpha}{\Gamma(\alpha-\varepsilon+1)}+\frac{\mathcal{Z}_\alpha\Omega^{1-\varepsilon}}{\Gamma(\alpha)\Gamma(2-\varepsilon)}+\frac{\mathcal{Z}_\alpha\Omega^{2-\varepsilon}}{\Gamma(\alpha-1)\Gamma(3-\varepsilon)}$$
$$+\frac{\mathcal{Z}_\alpha\Omega^{1-\varepsilon}}{\Gamma(\alpha-1)\Gamma(2-\varepsilon)}+\frac{2\mathcal{Z}_\alpha}{\Gamma(\alpha-1)}\Big)+\sum_{j=1}^{m}I_{1j}+\sum_{j=1}^{m}\hat{I}_{1j}+\frac{\mathcal{Z}_\alpha\Omega^{1-\varepsilon}}{\Gamma(2-\varepsilon)}\sum_{j=1}^{m}\hat{I}_{1j}+\frac{\mathcal{Z}_\alpha\Omega^{2-\varepsilon}}{\Gamma(3-\varepsilon)}\sum_{j=1}^{m}\check{I}_{1j}$$
$$+\frac{\mathcal{Z}_\alpha\Omega^{1-\varepsilon}}{\Gamma(2-\varepsilon)}\sum_{j=1}^{m}\check{I}_{1j}+\mathcal{Z}_\alpha\sum_{j=1}^{m}\check{I}_{1j}+\frac{1}{2}\sum_{j=1}^{m}\check{I}_{1j}\Big]<1$$

and

$$\Big[(\mu_1'+\mu_2')\Big(\frac{2}{\Gamma(\beta+1)}+\frac{1}{\Gamma(\beta)}+\frac{1}{2\Gamma(\beta-1)}+\frac{\mathcal{Z}_\beta}{\Gamma(\beta-\rho+1)}+\frac{\mathcal{Z}_\beta\Phi^{1-\rho}}{\Gamma(\beta)\Gamma(2-\rho)}+\frac{\mathcal{Z}_\beta\Phi^{2-\rho}}{\Gamma(\beta-1)\Gamma(3-\rho)}$$
$$+\frac{\mathcal{Z}_\beta\Phi^{1-\rho}}{\Gamma(\beta-1)\Gamma(2-\rho)}+\frac{2\mathcal{Z}_\beta}{\Gamma(\beta-1)}\Big)+\sum_{j=1}^{m}I_{2j}+\sum_{j=1}^{m}\hat{I}_{2j}+\frac{\mathcal{Z}_\beta\Phi^{1-\rho}}{\Gamma(2-\rho)}\sum_{j=1}^{m}\hat{I}_{2j}+\frac{\mathcal{Z}_\beta\Phi^{2-\rho}}{\Gamma(3-\rho)}\sum_{j=1}^{m}\check{I}_{2j}$$
$$+\frac{\mathcal{Z}_\beta\Phi^{1-\rho}}{\Gamma(2-\rho)}\sum_{j=1}^{m}\check{I}_{2j}+\mathcal{Z}_\beta\sum_{j=1}^{m}\check{I}_{2j}+\frac{1}{2}\sum_{j=1}^{m}\check{I}_{2j}\Big]<1.$$

Proof. By using the Banach contraction principle, we can prove that $\mathbf{W}$, defined by (17), has a fixed point. Before proving the main result first, we will prove the contraction. When $t \in J$, from (17) and conditions $(\mathbf{M}_1)-(\mathbf{M}_2)$, for all $(x_1, y_1), (x_2, y_2) \in Y'$, we have

$$\begin{aligned}
&\big|\mathbf{W}_1(x_1,y_1)(t)-\mathbf{W}_1(x_2,y_2)(t)\big|\\
\leqslant&\frac{1}{\Gamma(\alpha)}\int_0^t(t-s)^{\alpha-1}\big|h(s,{}^c\mathbf{D}^a x_1(s),{}^c\mathbf{D}^b y_1(s))-h(s,{}^c\mathbf{D}^a x_2(s),{}^c\mathbf{D}^b y_2(s))\big|ds\\
&+\frac{\mathcal{Z}_\alpha t^2}{\Gamma(\alpha-\varepsilon)}\int_{t_j}^{\Omega}(\Omega-s)^{\alpha-\varepsilon-1}\big|h(s,{}^c\mathbf{D}^a x_1(s),{}^c\mathbf{D}^b y_1(s))-h(s,{}^c\mathbf{D}^a x_2(s),{}^c\mathbf{D}^b y_2(s))\big|ds\\
&+\frac{\mathcal{Z}_\alpha t^2\Omega^{1-\varepsilon}}{\Gamma(\alpha-1)\Gamma(2-\varepsilon)}\sum_{j=1}^{m}\int_{t_{j-1}}^{t_j}(t_j-s)^{\alpha-2}\big|h(s,{}^c\mathbf{D}^a x_1(s),{}^c\mathbf{D}^b y_1(s))-h(s,{}^c\mathbf{D}^a x_2(s),{}^c\mathbf{D}^b y_2(s))\big|ds\\
&+\frac{\mathcal{Z}_\alpha t^2\Omega^{2-\varepsilon}}{\Gamma(\alpha-2)\Gamma(3-\varepsilon)}\sum_{j=1}^{m}\int_{t_{j-1}}^{t_j}(t_j-s)^{\alpha-3}\big|h(s,{}^c\mathbf{D}^a x_1(s),{}^c\mathbf{D}^b y_1(s))-h(s,{}^c\mathbf{D}^a x_2(s),{}^c\mathbf{D}^b y_2(s))\big|ds\\
&+\frac{\mathcal{Z}_\alpha t^2\Omega^{1-\varepsilon}}{\Gamma(\alpha-2)\Gamma(2-\varepsilon)}\sum_{j=1}^{m}t_j\int_{t_{j-1}}^{t_j}(t_j-s)^{\alpha-3}\big|h(s,{}^c\mathbf{D}^a x_1(s),{}^c\mathbf{D}^b y_1(s))-h(s,{}^c\mathbf{D}^a x_2(s),{}^c\mathbf{D}^b y_2(s))\big|ds\\
&+\frac{\mathcal{Z}_\alpha t^2\Omega^{1-\varepsilon}}{\Gamma(2-\varepsilon)}\sum_{j=1}^{m}\big|N_{1j}(x_1(t_j))-N_{1j}(x_2(t_j))\big|+\frac{\mathcal{Z}_\alpha t^2\Omega^{2-\varepsilon}}{\Gamma(3-\varepsilon)}\sum_{j=1}^{m}\big|O_{1j}(x_1(t_j))-O_{1j}(x_2(t_j))\big|\\
&+\frac{\mathcal{Z}_\alpha t^2}{\Gamma(\alpha-2)}\int_{t_m}^{1}(1-s)^{\alpha-3}\big|h(s,{}^c\mathbf{D}^a x_1(s),{}^c\mathbf{D}^b y_1(s))-h(s,{}^c\mathbf{D}^a x_2(s),{}^c\mathbf{D}^b y_2(s))\big|ds\\
&+\frac{\mathcal{Z}_\alpha t^2}{\Gamma(\alpha-2)}\sum_{j=1}^{m}\int_{t_{j-1}}^{t_j}(t_j-s)^{\alpha-3}\big|h(s,{}^c\mathbf{D}^a x_1(s),{}^c\mathbf{D}^b y_1(s))-h(s,{}^c\mathbf{D}^a x_2(s),{}^c\mathbf{D}^b y_2(s))\big|ds\\
&+\frac{\mathcal{Z}_\alpha t^2\Omega^{1-\varepsilon}}{\Gamma(2-\varepsilon)}\sum_{j=1}^{m}|t_j|\big|O_{1j}(x_1(t_j))-O_{1j}(x_2(t_j))\big|+\mathcal{Z}_\alpha t^2\sum_{j=1}^{m}\big|O_{1j}(x_1(t_j))-O_{1j}(x_2(t_j))\big|,\\
\leqslant&\frac{1}{\Gamma(\alpha+1)}\big[\mu_1\big|{}^c\mathbf{D}^a x_1(s)-{}^c\mathbf{D}^a x_2(s)\big|+\mu_2\big|{}^c\mathbf{D}^b y_1(s)-{}^c\mathbf{D}^b y_2(s)\big|\big]
\end{aligned}$$

$$
\begin{aligned}
&+\frac{\mathcal{Z}_\alpha}{\Gamma(\alpha-\varepsilon+1)}\big[\mu_1|{}^c\mathbf{D}^a x_1(s)-{}^c\,\mathbf{D}^a x_2(s)|+\mu_2|{}^c\mathbf{D}^b y_1(s)-{}^c\,\mathbf{D}^b y_2(s)|\big]\\
&+\frac{\mathcal{Z}_\alpha\Omega^{1-\varepsilon}}{\Gamma(\alpha)\Gamma(2-\varepsilon)}\big[\mu_1|{}^c\mathbf{D}^a x_1(s)-{}^c\,\mathbf{D}^a x_2(s)|+\mu_2|{}^c\mathbf{D}^b y_1(s)-{}^c\,\mathbf{D}^b y_2(s)|\big]\\
&+\frac{\mathcal{Z}_\alpha\Omega^{2-\varepsilon}}{\Gamma(\alpha-1)\Gamma(3-\varepsilon)}\big[\mu_1|{}^c\mathbf{D}^a x_1(s)-{}^c\,\mathbf{D}^a x_2(s)|+\mu_2|{}^c\mathbf{D}^a y_1(s)-{}^c\,\mathbf{D}^a y_2(s)|\big]\\
&+\frac{\mathcal{Z}_\alpha\Omega^{1-\varepsilon}}{\Gamma(\alpha-1)\Gamma(2-\varepsilon)}\big[\mu_1|{}^c\mathbf{D}^a x_1(s)-{}^c\,\mathbf{D}^a x_2(s)|+\mu_2|{}^c\mathbf{D}^b y_1(s)-{}^c\,\mathbf{D}^b y_2(s)|\big]\\
&+\frac{\mathcal{Z}_\alpha\Omega^{1-\varepsilon}}{\Gamma(2-\varepsilon)}\sum_{j=1}^{m}|N_{1j}(x_1(t_j))-N_{1j}(x_2(t_j))|+\frac{\mathcal{Z}_\alpha\Omega^{2-\varepsilon}}{\Gamma(3-\varepsilon)}\sum_{j=1}^{m}|O_{1j}(x_1(t_j))-O_{1j}(x_2(t_j))|\\
&+\frac{\mathcal{Z}_\alpha}{\Gamma(\alpha-1)}\big[\mu_1|{}^c\mathbf{D}^a x_1(s)-{}^c\,\mathbf{D}^a x_2(s)|+\mu_2|{}^c\mathbf{D}^b y_1(s)-{}^c\,\mathbf{D}^b y_2(s)|\big]\\
&+\frac{\mathcal{Z}_\alpha}{\Gamma(\alpha-1)}\big[\mu_1|{}^c\mathbf{D}^a x_1(s)-{}^c\,\mathbf{D}^a x_2(s)|+\mu_2|{}^c\mathbf{D}^b y_1(s)-{}^c\,\mathbf{D}^b y_2(s)|\big]\\
&+\frac{\mathcal{Z}_\alpha\Omega^{1-\varepsilon}}{\Gamma(2-\varepsilon)}\sum_{j=1}^{m}t_j|O_{1j}(x_1(t_j))-O_{1j}(x_2(t_j))|+\mathcal{Z}_\alpha\sum_{j=1}^{m}|O_{1j}(x_1(t_j))-O_{1j}(x_2(t_j))|,\\
\leqslant&\frac{1}{\Gamma(\alpha+1)}[\mu_1\|x_1-x_2\|_0+\mu_2\|y_1-y_2\|_0]+\frac{\mathcal{Z}_\alpha}{\Gamma(\alpha-\varepsilon+1)}[\mu_1\|x_1-x_2\|_0+\mu_2\|y_1-y_2\|_0]\\
&+\frac{\mathcal{Z}_\alpha\Omega^{1-\varepsilon}}{\Gamma(\alpha)\Gamma(2-\varepsilon)}[\mu_1\|x_1-x_2\|_0+\mu_2\|y_1-y_2\|_0]+\frac{\mathcal{Z}_\alpha\Omega^{2-\varepsilon}}{\Gamma(\alpha-1)\Gamma(3-\varepsilon)}[\mu_1\|x_1-x_2\|_0+\mu_2\|y_1-y_2\|_0]\\
&+\tfrac{\mathcal{Z}_\alpha\Omega^{1-\varepsilon}}{\Gamma(\alpha-1)\Gamma(2-\varepsilon)}[\mu_1\|x_1-x_2\|_0+\mu_2\|y_1-y_2\|_0]+\tfrac{\mathcal{Z}_\alpha\Omega^{1-\varepsilon}}{\Gamma(2-\varepsilon)}\textstyle\sum_{j=1}^{m}\hat{I}_{1j}\|x_1-x_2\|_0+\tfrac{\mathcal{Z}_\alpha\Omega^{2-\varepsilon}}{\Gamma(3-\varepsilon)}\sum_{j=1}^{m}\check{I}_{1j} \qquad (18)\\
&\times\|x_1-x_2\|_0+\frac{\mathcal{Z}_\alpha}{\Gamma(\alpha-1)}[\mu_1\|x_1-x_2\|_0+\mu_2\|y_1-y_2\|_0]+\frac{\mathcal{Z}_\alpha}{\Gamma(\alpha-1)}[\mu_1\|x_1-x_2\|_0+\mu_2\|y_1-y_2\|_0]\\
&+\frac{\mathcal{Z}_\alpha\Omega^{1-\varepsilon}}{\Gamma(2-\varepsilon)}\sum_{j=1}^{m}\check{I}_{1j}\|x_1-x_2\|_0+\mathcal{Z}_\alpha\sum_{j=1}^{m}\check{I}_{1j}\|x_1-x_2\|_0,\\
\leqslant&\frac{1}{\Gamma(\alpha+1)}[\mu_1+\mu_2]\|(x_1-x_2,y_1-y_2)\|+\frac{\mathcal{Z}_\alpha}{\Gamma(\alpha-\varepsilon+1)}[\mu_1+\mu_2]\|(x_1-x_2,y_1-y_2)\|\\
&+\frac{\mathcal{Z}_\alpha\Omega^{1-\varepsilon}}{\Gamma(\alpha)\Gamma(2-\varepsilon)}[\mu_1+\mu_2]\|(x_1-x_2,y_1-y_2)\|+\frac{\mathcal{Z}_\alpha\Omega^{2-\varepsilon}}{\Gamma(\alpha-1)\Gamma(3-\varepsilon)}[\mu_1+\mu_2]\|(x_1-x_2,y_1-y_2)\|\\
&+\frac{\mathcal{Z}_\alpha\Omega^{1-\varepsilon}}{\Gamma(\alpha-1)\Gamma(2-\varepsilon)}[\mu_1+\mu_2]\|(x_1-x_2,y_1-y_2)\|+\frac{\mathcal{Z}_\alpha\Omega^{1-\varepsilon}}{\Gamma(2-\varepsilon)}\sum_{j=1}^{m}\hat{I}_{1j}\|(x_1-x_2,y_1-y_2)\|\\
&+\frac{\mathcal{Z}_\alpha\Omega^{2-\varepsilon}}{\Gamma(3-\varepsilon)}\sum_{j=1}^{m}\check{I}_{1j}\|(x_1-x_2,y_1-y_2)\|+\frac{\mathcal{Z}_\alpha}{\Gamma(\alpha-1)}[\mu_1+\mu_2]\|(x_1-x_2,y_1-y_2)\|\\
&+\frac{\mathcal{Z}_\alpha}{\Gamma(\alpha-1)}[\mu_1+\mu_2]\|(x_1-x_2,y_1-y_2)\|+\frac{\mathcal{Z}_\alpha\Omega^{1-\varepsilon}}{\Gamma(2-\varepsilon)}\sum_{j=1}^{m}\check{I}_{1j}\|(x_1-x_2,y_1-y_2)\|\\
&+\mathcal{Z}_\alpha\sum_{j=1}^{m}\check{I}_{1j}\|(x_1-x_2,y_1-y_2)\|,\\
\leqslant&\mathcal{Z}_\alpha\Bigg[(\mu_1+\mu_2)\bigg(\frac{1}{\mathcal{Z}_\alpha\Gamma(\alpha+1)}+\frac{1}{\Gamma(\alpha-\varepsilon+1)}+\frac{\Omega^{1-\varepsilon}}{\Gamma(\alpha)\Gamma(2-\varepsilon)}+\frac{\Omega^{2-\varepsilon}}{\Gamma(\alpha-1)\Gamma(3-\varepsilon)}+\frac{\Omega^{1-\varepsilon}}{\Gamma(\alpha-1)\Gamma(2-\varepsilon)}\\
&+\frac{2}{\Gamma(\alpha-1)}\bigg)+\frac{\Omega^{1-\varepsilon}}{\Gamma(2-\varepsilon)}\sum_{j=1}^{m}\hat{I}_{1j}+\frac{\Omega^{2-\varepsilon}}{\Gamma(3-\varepsilon)}\sum_{j=1}^{m}\check{I}_{1j}+\frac{\Omega^{1-\varepsilon}}{\Gamma(2-\varepsilon)}\sum_{j=1}^{m}\check{I}_{1j}+\sum_{j=1}^{m}\check{I}_{1j}\Bigg]\\
&\times\|(x_1-x_2,y_1-y_2)\|,\ t\in[0,t_1].
\end{aligned}
$$

When $t \in (t_m, t_{m+1}]$, then

$$\begin{aligned}
&\left|\mathbf{W}_1(x_1,y_1)(t)-\mathbf{W}_1(x_2,y_2)(t)\right| \\
&\leqslant \frac{1}{\Gamma(\alpha)}\int_{t_m}^{t}(t-s)^{\alpha-1}\left|h(s,{}^c\mathbf{D}^a x_1(s),{}^c\mathbf{D}^b y_1(s))-h(s,{}^c\mathbf{D}^a x_2(s),{}^c\mathbf{D}^b y_2(s))\right|ds \\
&+\frac{1}{\Gamma(\alpha)}\sum_{j=1}^{m}\int_{t_{j-1}}^{t_j}(t_j-s)^{\alpha-1}\left|h(s,{}^c\mathbf{D}^a x_1(s),{}^c\mathbf{D}^b y_1(s))-h(s,{}^c\mathbf{D}^a x_2(s),{}^c\mathbf{D}^b y_2(s))\right|ds \\
&+\frac{1}{\Gamma(\alpha-1)}\sum_{j=1}^{m}\left|(t-t_j)\right|\int_{t_{j-1}}^{t_j}(t_j-s)^{\alpha-2}\left|h(s,{}^c\mathbf{D}^a x_1(s),{}^c\mathbf{D}^b y_1(s))-h(s,{}^c\mathbf{D}^a x_2(s),{}^c\mathbf{D}^b y_2(s))\right|ds \\
&+\frac{1}{2\Gamma(\alpha-2)}\sum_{j=1}^{m}\left|(t-t_j)^2\right|\int_{t_{j-1}}^{t_j}(t_j-s)^{\alpha-3}\left|h(s,{}^c\mathbf{D}^a x_1(s),{}^c\mathbf{D}^b y_1(s))-h(s,{}^c\mathbf{D}^a x_2(s),{}^c\mathbf{D}^b y_2(s))\right|ds \\
&+\sum_{j=1}^{m}\left|M_{1j}(x_1(t_j))-M_{1j}(x_2(t_j))\right|+\sum_{j=1}^{m}\left|(t-t_j)\right|\left|N_{1j}(x_1(t_j))-N_{1j}(x_2(t_j))\right| \\
&+\frac{\mathcal{Z}_\alpha t^2}{\Gamma(\alpha-\varepsilon)}\int_{t_j}^{\Omega}(\Omega-s)^{\alpha-\varepsilon-1}\left|h(s,{}^c\mathbf{D}^a x_1(s),{}^c\mathbf{D}^b y_1(s))-h(s,{}^c\mathbf{D}^a x_2(s),{}^c\mathbf{D}^b y_2(s))\right|ds \\
&+\frac{\mathcal{Z}_\alpha t^2\Omega^{1-\varepsilon}}{\Gamma(\alpha-1)\Gamma(2-\varepsilon)}\sum_{j=1}^{m}\int_{t_{j-1}}^{t_j}(t_j-s)^{\alpha-2}\left|h(s,{}^c\mathbf{D}^a x_1(s),{}^c\mathbf{D}^b y_1(s))-h(s,{}^c\mathbf{D}^a x_2(s),{}^c\mathbf{D}^b y_2(s))\right|ds \quad (19)\\
&+\frac{\mathcal{Z}_\alpha t^2\Omega^{2-\varepsilon}}{\Gamma(\alpha-2)\Gamma(3-\varepsilon)}\sum_{j=1}^{m}\int_{t_{j-1}}^{t_j}(t_j-s)^{\alpha-3}\left|h(s,{}^c\mathbf{D}^a x_1(s),{}^c\mathbf{D}^b y_1(s))-h(s,{}^c\mathbf{D}^a x_2(s),{}^c\mathbf{D}^b y_2(s))\right|ds \\
&+\frac{\mathcal{Z}_\alpha t^2\Omega^{1-\varepsilon}}{\Gamma(\alpha-2)\Gamma(2-\varepsilon)}\sum_{j=1}^{m}|t_j|\int_{t_{j-1}}^{t_j}(t_j-s)^{\alpha-3}\left|h(s,{}^c\mathbf{D}^a x_1(s),{}^c\mathbf{D}^b y_1(s))-h(s,{}^c\mathbf{D}^a x_2(s),{}^c\mathbf{D}^b y_2(s))\right|ds \\
&+\frac{\mathcal{Z}_\alpha t^2\Omega^{1-\varepsilon}}{\Gamma(2-\varepsilon)}\sum_{j=1}^{m}\left|N_{1j}(x_1(t_j))-N_{1j}(x_2(t_j))\right|+\frac{\mathcal{Z}_\alpha t^2\Omega^{2-\varepsilon}}{\Gamma(3-\varepsilon)}\sum_{j=1}^{m}\left|O_{1j}(x_1(t_j))-O_{1j}(x_2(t_j))\right| \\
&+\frac{\mathcal{Z}_\alpha t^2}{\Gamma(\alpha-2)}\int_{t_m}^{1}(1-s)^{\alpha-3}\left|h(s,{}^c\mathbf{D}^a x_1(s),{}^c\mathbf{D}^b y_1(s))-h(s,{}^c\mathbf{D}^a x_2(s),{}^c\mathbf{D}^b y_2(s))\right|ds \\
&+\frac{\mathcal{Z}_\alpha t^2}{\Gamma(\alpha-2)}\sum_{j=1}^{m}\int_{t_{j-1}}^{t_j}(t_j-s)^{\alpha-3}\left|h(s,{}^c\mathbf{D}^a x_1(s),{}^c\mathbf{D}^b y_1(s))-h(s,{}^c\mathbf{D}^a x_2(s),{}^c\mathbf{D}^b y_2(s))\right|ds \\
&+\frac{\mathcal{Z}_\alpha t^2\Omega^{1-\varepsilon}}{\Gamma(2-\varepsilon)}\sum_{j=1}^{m}|t_j|\left|O_{1j}(x_1(t_j))-O_{1j}(x_2(t_j))\right|+\mathcal{Z}_\alpha t^2\sum_{j=1}^{m}\left|O_{1j}(x_1(t_j))-O_{1j}(x_2(t_j))\right| \\
&+\sum_{j=1}^{m}\frac{\left|(t-t_j)^2\right|}{2}\left|O_{1j}(x_1(t_j))-O_{1j}(x_2(t_j))\right|.
\end{aligned}$$

Utilizing $(\mathbf{M}_1)$ and $(\mathbf{M}_2)$ in (19) and taking the maximum, we get

$$\begin{aligned}
&\leqslant \frac{1}{\Gamma(\alpha+1)}\left[\mu_1\|x_1-x_2\|_0+\mu_2\|y_1-y_2\|_0\right]+\frac{1}{\Gamma(\alpha+1)}\left[\mu_1\|x_1-x_2\|_0+\mu_2\|y_1-y_2\|_0\right] \\
&+\frac{1}{\Gamma(\alpha)}\left[\mu_1\|x_1-x_2\|_0+\mu_2\|y_1-y_2\|_0\right]+\frac{1}{2\Gamma(\alpha-1)}\left[\mu_1\|x_1-x_2\|_0+\mu_2\|y_1-y_2\|_0\right] \\
&+\sum_{j=1}^{m}I_{1j}\|x_1-x_2\|_0+\sum_{j=1}^{m}\hat{I}_{1j}\|x_1-x_2\|_0+\frac{\mathcal{Z}_\alpha}{\Gamma(\alpha-\varepsilon+1)}\left[\mu_1\|x_1-x_2\|_0+\mu_2\|y_1-y_2\|_0\right] \\
&+\frac{\mathcal{Z}_\alpha\Omega^{1-\varepsilon}}{\Gamma(\alpha)\Gamma(2-\varepsilon)}\left[\mu_1\|x_1-x_2\|_0+\mu_2\|y_1-y_2\|_0\right]+\frac{\mathcal{Z}_\alpha\Omega^{2-\varepsilon}}{\Gamma(\alpha-1)\Gamma(3-\varepsilon)}\left[\mu_1\|x_1-x_2\|_0+\mu_2\|y_1-y_2\|_0\right]
\end{aligned}$$

$$+\frac{\mathcal{Z}_\alpha\Omega^{1-\varepsilon}}{\Gamma(\alpha-1)\Gamma(2-\varepsilon)}\left[\mu_1\|x_1-x_2\|_0+\mu_2\|y_1-y_2\|_0\right]+\frac{\mathcal{Z}_\alpha\Omega^{1-\varepsilon}}{\Gamma(2-\varepsilon)}\sum_{j=1}^{m}\hat{I}_{1j}\|x_1-x_2\|_0$$
$$+\frac{\mathcal{Z}_\alpha\Omega^{2-\varepsilon}}{\Gamma(3-\varepsilon)}\sum_{j=1}^{m}\check{I}_{1j}\|x_1-x_2\|_0+\frac{\mathcal{Z}_\alpha}{\Gamma(\alpha-1)}\left[\mu_1\|x_1-x_2\|_0+\mu_2\|y_1-y_2\|_0\right]$$
$$+\frac{\mathcal{Z}_\alpha}{\Gamma(\alpha-1)}\left[\mu_1\|x_1-x_2\|_0+\mu_2\|y_1-y_2\|_0\right]+\frac{\mathcal{Z}_\alpha\Omega^{1-\varepsilon}}{\Gamma(2-\varepsilon)}\sum_{j=1}^{m}\check{I}_{1j}\|x_1-x_2\|_0+\mathcal{Z}_\alpha\sum_{j=1}^{m}\check{I}_{1j}\|x_1-x_2\|_0$$
$$+\frac{1}{2}\sum_{j=1}^{m}\check{I}_{1j}\|x_1-x_2\|_0,$$
$$\leqslant\frac{1}{\Gamma(\alpha+1)}[\mu_1+\mu_2]\|(x_1-x_2,y_1-y_2)\|+\frac{1}{\Gamma(\alpha+1)}[\mu_1+\mu_2]\|(x_1-x_2,y_1-y_2)\|$$
$$+\frac{1}{\Gamma(\alpha)}[\mu_1+\mu_2]\|(x_1-x_2,y_1-y_2)\|+\frac{1}{2\Gamma(\alpha-1)}[\mu_1+\mu_2]\|(x_1-x_2,y_1-y_2)\|$$
$$+\textstyle\sum_{j=1}^{m}I_{1j}\|(x_1-x_2,y_1-y_2)\|+\sum_{j=1}^{m}\hat{I}_{1j}\|(x_1-x_2,y_1-y_2)\|+\frac{\mathcal{Z}_\alpha}{\Gamma(\alpha-\varepsilon+1)}[\mu_1+\mu_2]\|(x_1-x_2,y_1-y_2)\|$$
$$+\tfrac{\mathcal{Z}_\alpha\Omega^{1-\varepsilon}}{\Gamma(\alpha)\Gamma(2-\varepsilon)}[\mu_1+\mu_2]\|(x_1-x_2,y_1-y_2)\|+\tfrac{\mathcal{Z}_\alpha\Omega^{2-\varepsilon}}{\Gamma(\alpha-1)\Gamma(3-\varepsilon)}[\mu_1+\mu_2]\|(x_1-x_2,y_1-y_2)\| \tag{20}$$
$$+\frac{\mathcal{Z}_\alpha\Omega^{1-\varepsilon}}{\Gamma(\alpha-1)\Gamma(2-\varepsilon)}[\mu_1+\mu_2]\|(x_1-x_2,y_1-y_2)\|+\frac{\mathcal{Z}_\alpha\Omega^{1-\varepsilon}}{\Gamma(2-\varepsilon)}\sum_{j=1}^{m}\hat{I}_{1j}\|(x_1-x_2,y_1-y_2)\|$$
$$+\frac{\mathcal{Z}_\alpha\Omega^{2-\varepsilon}}{\Gamma(3-\varepsilon)}\sum_{j=1}^{m}\check{I}_{1j}\|(x_1-x_2,y_1-y_2)\|+\frac{\mathcal{Z}_\alpha}{\Gamma(\alpha-1)}[\mu_1+\mu_2]\|(x_1-x_2,y_1-y_2)\|$$
$$+\frac{\mathcal{Z}_\alpha}{\Gamma(\alpha-1)}[\mu_1+\mu_2]\|(x_1-x_2,y_1-y_2)\|+\frac{\mathcal{Z}_\alpha\Omega^{1-\varepsilon}}{\Gamma(2-\varepsilon)}\sum_{j=1}^{m}\check{I}_{1j}\|(x_1-x_2,y_1-y_2)\|$$
$$+\mathcal{Z}_\alpha\sum_{j=1}^{m}\check{I}_{1j}\|(x_1-x_2,y_1-y_2)\|+\frac{1}{2}\sum_{j=1}^{m}\check{I}_{1j}\|(x_1-x_2,y_1-y_2)\|$$
$$\leqslant\Bigg[(\mu_1+\mu_2)\bigg(\frac{2}{\Gamma(\alpha+1)}+\frac{1}{\Gamma(\alpha)}+\frac{1}{2\Gamma(\alpha-1)}+\frac{\mathcal{Z}_\alpha}{\Gamma(\alpha-\varepsilon+1)}+\frac{\mathcal{Z}_\alpha\Omega^{1-\varepsilon}}{\Gamma(\alpha)\Gamma(2-\varepsilon)}+\frac{\mathcal{Z}_\alpha\Omega^{2-\varepsilon}}{\Gamma(\alpha-1)\Gamma(3-\varepsilon)}$$
$$+\frac{\mathcal{Z}_\alpha\Omega^{1-\varepsilon}}{\Gamma(\alpha-1)\Gamma(2-\varepsilon)}+\frac{2\mathcal{Z}_\alpha}{\Gamma(\alpha-1)}\bigg)+\sum_{j=1}^{m}I_{1j}+\sum_{j=1}^{m}\hat{I}_{1j}+\frac{\mathcal{Z}_\alpha\Omega^{1-\varepsilon}}{\Gamma(2-\varepsilon)}\sum_{j=1}^{m}\hat{I}_{1j}+\frac{\mathcal{Z}_\alpha\Omega^{2-\varepsilon}}{\Gamma(3-\varepsilon)}\sum_{j=1}^{m}\check{I}_{1j}$$
$$+\frac{\mathcal{Z}_\alpha\Omega^{1-\varepsilon}}{\Gamma(2-\varepsilon)}\sum_{j=1}^{m}\check{I}_{1j}+\mathcal{Z}_\alpha\sum_{j=1}^{m}\check{I}_{1j}+\frac{1}{2}\sum_{j=1}^{m}\check{I}_{1j}\Bigg]\|(x_1-x_2,y_1-y_2)\|.$$

In the same fashion, we can obtain

$$\left|\mathbf{W}_2(x_1,y_1)(t)-\mathbf{W}_2(x_2,y_2)(t)\right|$$
$$\leqslant\ \mathcal{Z}_\beta\Bigg[(\mu_1'+\mu_2')\bigg(\tfrac{1}{\mathcal{Z}_\beta\Gamma(\beta+1)}+\tfrac{1}{\Gamma(\beta-\rho+1)}+\tfrac{\Phi^{1-\rho}}{\Gamma(\beta)\Gamma(2-\rho)}+\tfrac{\Phi^{2-\rho}}{\Gamma(\beta-1)\Gamma(3-\rho)}+\tfrac{\Phi^{1-\rho}}{\Gamma(\beta-1)\Gamma(2-\rho)}$$
$$+\tfrac{2}{\Gamma(\beta-1)}\bigg)+\tfrac{\Phi^{1-\rho}}{\Gamma(2-\rho)}\textstyle\sum_{j=1}^{m}\hat{I}_{2j}+\tfrac{\Phi^{2-\rho}}{\Gamma(3-\rho)}\sum_{j=1}^{m}\check{I}_{2j}+\tfrac{\Phi^{1-\rho}}{\Gamma(2-\rho)}\sum_{j=1}^{m}\check{I}_{2j}+\sum_{j=1}^{m}\check{I}_{2j}\Bigg] \tag{21}$$
$$\times\|(x_1-x_2,y_1-y_2)\|,\ t\in[0,t_1],$$

and

$$\begin{aligned}
&\left|\mathbf{W}_2(x_1,y_1)(t)-\mathbf{W}_2(x_2,y_2)(t)\right| \\
\leqslant\;& \Bigg[(\mu_1'+\mu_2')\left(\tfrac{2}{\Gamma(\beta+1)}+\tfrac{1}{\Gamma(\beta)}+\tfrac{1}{2\Gamma(\beta-1)}+\tfrac{\mathcal{Z}_\beta}{\Gamma(\beta-\rho+1)}+\tfrac{\mathcal{Z}_\beta\Phi^{1-\rho}}{\Gamma(\beta)\Gamma(2-\rho)}+\tfrac{\mathcal{Z}_\beta\Phi^{2-\rho}}{\Gamma(\beta-1)\Gamma(3-\rho)}\right. \\
&\left.+\tfrac{\mathcal{Z}_\beta\Phi^{1-\rho}}{\Gamma(\beta-1)\Gamma(2-\rho)}+\tfrac{2\mathcal{Z}_\beta}{\Gamma(\beta-1)}\right)+\textstyle\sum_{j=1}^m \mathrm{I}_{2j}+\sum_{j=1}^m \hat{\mathrm{I}}_{2j}+\tfrac{\mathcal{Z}_\beta\Phi^{1-\rho}}{\Gamma(2-\rho)}\sum_{j=1}^m \hat{\mathrm{I}}_{2j}+\tfrac{\mathcal{Z}_\beta\Phi^{2-\rho}}{\Gamma(3-\rho)}\sum_{j=1}^m \check{\mathrm{I}}_{2j} \\
&+\tfrac{\mathcal{Z}_\beta\Phi^{1-\rho}}{\Gamma(2-\rho)}\textstyle\sum_{j=1}^m \check{\mathrm{I}}_{2j}+\mathcal{Z}_\beta\sum_{j=1}^m \check{\mathrm{I}}_{2j}+\tfrac{1}{2}\sum_{j=1}^m \check{\mathrm{I}}_{2j}\Bigg]\|(x_1-x_2,y_1-y_2)\|,\ t\in(t_m,t_{m+1}].
\end{aligned} \tag{22}$$

Thus, from (18)–(22) and $(\mathbf{M}_3)$, we infer that $\mathbf{W}$ is a contraction mapping. According to Lemma 3, $\mathbf{W}$ has a fixed point $(x^*(t),y^*(t))\in Y'$, which is unique. Therefore, problem (1) has a unique solution $(x^*(t),y^*(t))$. □

Theorem 3. *Let* $(\mathbf{M}_1)-(\mathbf{M}_2)$, $(\mathbf{M}_4)$ *and for all* $t\in J$ *such that* $h(t,0,0)=w(t,0,0)=0$, $M_{ik}=N_{ik}=O_{ik}=0$, $(i=1,2;\ k=1,2,\dots,n)$ *hold. Then,* (1) *has at least one solution* $(x^*(t),y^*(t))$.

Proof. For the sake of simplicity, let us denote

$$\begin{aligned}
\varpi=&\Bigg[(\mu_1+\mu_2)\left(\frac{3}{\Gamma(\alpha+1)}+\frac{1}{\Gamma(\alpha)}+\frac{1}{2\Gamma(\alpha-1)}+\frac{\mathcal{Z}_\alpha}{\Gamma(\alpha-\varepsilon+1)}+\frac{\mathcal{Z}_\alpha\Omega^{1-\varepsilon}}{\Gamma(\alpha)\Gamma(2-\varepsilon)}+\frac{\mathcal{Z}_\alpha\Omega^{2-\varepsilon}}{\Gamma(\alpha-1)\Gamma(3-\varepsilon)}\right. \\
&\left.+\frac{\mathcal{Z}_\alpha\Omega^{1-\varepsilon}}{\Gamma(\alpha-1)\Gamma(2-\varepsilon)}+\frac{2\mathcal{Z}_\alpha}{\Gamma(\alpha-1)}\right)+\sum_{j=1}^m \mathrm{I}_{1j}+\sum_{j=1}^m \hat{\mathrm{I}}_{1j}+\frac{\mathcal{Z}_\alpha\Omega^{1-\varepsilon}}{\Gamma(2-\varepsilon)}\sum_{j=1}^m \hat{\mathrm{I}}_{1j}+\frac{\mathcal{Z}_\alpha\Omega^{2-\varepsilon}}{\Gamma(3-\varepsilon)}\sum_{j=1}^m \check{\mathrm{I}}_{1j} \\
&+\frac{\mathcal{Z}_\alpha\Omega^{1-\varepsilon}}{\Gamma(2-\varepsilon)}\sum_{j=1}^m \check{\mathrm{I}}_{1j}+\mathcal{Z}_\alpha\sum_{j=1}^m \check{\mathrm{I}}_{1j}+\frac{1}{2}\sum_{j=1}^m \check{\mathrm{I}}_{1j}\Bigg], \\
\xi=&\Bigg[(\mu_1'+\mu_2')\left(\frac{3}{\Gamma(\beta+1)}+\frac{1}{\Gamma(\beta)}+\frac{1}{2\Gamma(\beta-1)}+\frac{\mathcal{Z}_\beta}{\Gamma(\beta-\rho+1)}+\frac{\mathcal{Z}_\beta\Phi^{1-\rho}}{\Gamma(\beta)\Gamma(2-\rho)}+\frac{\mathcal{Z}_\beta\Phi^{2-\rho}}{\Gamma(\beta-1)\Gamma(3-\rho)}\right. \\
&\left.+\frac{\mathcal{Z}_\beta\Phi^{1-\rho}}{\Gamma(\beta-1)\Gamma(2-\rho)}+\frac{2\mathcal{Z}_\beta}{\Gamma(\beta-1)}\right)+\sum_{j=1}^m \mathrm{I}_{2j}+\sum_{j=1}^m \hat{\mathrm{I}}_{2j}+\frac{\mathcal{Z}_\beta\Phi^{1-\rho}}{\Gamma(2-\rho)}\sum_{j=1}^m \hat{\mathrm{I}}_{2j}+\frac{\mathcal{Z}_\beta\Phi^{2-\rho}}{\Gamma(3-\rho)}\sum_{j=1}^m \check{\mathrm{I}}_{2j} \\
&+\frac{\mathcal{Z}_\beta\Phi^{1-\rho}}{\Gamma(2-\rho)}\sum_{j=1}^m \check{\mathrm{I}}_{2j}+\mathcal{Z}_\beta\sum_{j=1}^m \check{\mathrm{I}}_{2j}+\frac{1}{2}\sum_{j=1}^m \check{\mathrm{I}}_{2j}\Bigg],
\end{aligned}$$

and $\mathcal{R}_\upsilon=\max\{(\frac{1}{\varpi}+1,\frac{1}{\xi}+1)\}$. Define the operator $\mathbf{W}$, as in (17), and a closed ball of Banach space Y' as follows:

$$\upsilon=\left\{(x,y)\in Y':\|(x,y)\|\leqslant\mathcal{R}_\upsilon\right\}. \tag{23}$$

Similar to (18)–(22), we easily show that $\mathbf{W}(\upsilon)\subset\upsilon$ by applying $(\mathbf{M}_4)$. $\mathbf{W}(\upsilon)\subset\upsilon$ indicates that $\mathbf{W}(\upsilon)$ is uniformly bounded in Y'. The continuity of the operator $\mathbf{W}$ is follows from the continuity of h,w,M_{im},N_{im} and O_{im}. Now, we need to prove that $\mathbf{W}:\upsilon\to\upsilon$ is equicontinuous. Let $(x,y)\in\upsilon$ and $\ell_1,\ell_2\in[0,1]$ with $\ell_1<\ell_2$. When $0\leqslant\ell_1<\ell_2\leqslant t_1$, similar to Equation (18), we have

$$\begin{aligned}
&\left|\mathbf{W}_1(x,y)(\ell_2)-\mathbf{W}_1(x,y)(\ell_1)\right| \\
=&\Bigg|\frac{1}{\Gamma(\alpha)}\int_0^{\ell_1}\left[(\ell_2-s)^{\alpha-1}-(\ell_1-s)^{\alpha-1}\right]h(s,{}^c\mathbf{D}^a x(s),{}^c\mathbf{D}^b y(s))ds+\frac{1}{\Gamma(\alpha)}\int_{\ell_1}^{\ell_2}(\ell_2-s)^{\alpha-1} \\
&\times h(s,{}^c\mathbf{D}^a x(s),{}^c\mathbf{D}^b y(s))ds-e^*(\ell_2-\ell_1)^2\Bigg|,
\end{aligned}$$

$$\leqslant \frac{1}{\Gamma(\alpha)} \int_0^{\ell_1} \left[(\ell_2 - s)^{\alpha-1} - (\ell_1 - s)^{\alpha-1}\right] \left|h(s, {}^c\mathbf{D}^a x(s), {}^c\mathbf{D}^b y(s))\right| ds + \frac{1}{\Gamma(\alpha)} \int_{\ell_1}^{\ell_2} (\ell_2 - s)^{\alpha-1}$$
$$\times \left|h(s, {}^c\mathbf{D}^a x(s), {}^c\mathbf{D}^b y(s))\right| ds + \left|e^*\right| (\ell_2 - \ell_1)^2,$$
$$\leqslant \frac{\mu_1 + \mu_2}{\Gamma(\alpha)} \|(x, y)\| \int_0^{\ell_1} \left[(\ell_2 - s)^{\alpha-1} - (\ell_1 - s)^{\alpha-1}\right] ds + \frac{\mu_1 + \mu_2}{\Gamma(\alpha+1)} \|(x, y)\| (\ell_2 - \ell_1)^\alpha + \left|e^*\right| (\ell_2 - \ell_1)^2,$$
$$\leqslant \frac{\mu_1 + \mu_2}{\Gamma(\alpha+1)} \|(x, y)\| \left[(\ell_2 - \ell_1)^\alpha - (\ell_2^\alpha - \ell_1^\alpha)\right] + \frac{\mu_1 + \mu_2}{\Gamma(\alpha+1)} \|(x, y)\| (\ell_2 - \ell_1)^\alpha + |e^*| (\ell_2 - \ell_1)^2, \tag{24}$$
$$\leqslant \mathcal{Z}_\alpha \Bigg[(\mu_1 + \mu_2) \left(\frac{2}{\mathcal{Z}_\alpha \Gamma(\alpha+1)} + \frac{1}{\Gamma(\alpha-\varepsilon+1)} + \frac{\Omega^{1-\varepsilon}}{\Gamma(\alpha)\Gamma(2-\varepsilon)} + \frac{\Omega^{2-\varepsilon}}{\Gamma(\alpha-1)\Gamma(3-\varepsilon)} + \frac{\Omega^{1-\varepsilon}}{\Gamma(\alpha-1)\Gamma(2-\varepsilon)}$$
$$+ \frac{2}{\Gamma(\alpha-1)} \right) + \frac{\Omega^{1-\varepsilon}}{\Gamma(2-\varepsilon)} \sum_{j=1}^m \hat{I}_{1j} + \frac{\Omega^{2-\varepsilon}}{\Gamma(3-\varepsilon)} \sum_{j=1}^m \check{I}_{1j} + \frac{\Omega^{1-\varepsilon}}{\Gamma(2-\varepsilon)} \sum_{j=1}^m \check{I}_{1j} + \sum_{j=1}^m \check{I}_{1j} \Bigg] \|x, y\| (\ell_2 - \ell_1)^2,$$
$$\leqslant \varpi \mathcal{R}_\upsilon (\ell_2 - \ell_1)^2.$$

In the same fashion, we obtain

$$\left|\mathbf{W}_2(x, y)(\ell_2) - \mathbf{W}_2(x, y)(\ell_1)\right| \leqslant \xi \mathcal{R}_\upsilon (\ell_2 - \ell_1)^2. \tag{25}$$

In addition, we obtain the same result when $t_m < \ell_1 < \ell_2 \leqslant t_{m+1}$, $1 \leqslant m \leqslant n$, similar to (20)

$$\left|\mathbf{W}_1(x, y)(\ell_2) - \mathbf{W}_1(x, y)(\ell_1)\right| \leqslant \varpi \mathcal{R}_\upsilon (\ell_2 - \ell_1)^2 \tag{26}$$

and

$$\left|\mathbf{W}_2(x, y)(\ell_2) - \mathbf{W}_2(x, y)(\ell_1)\right| \leqslant \xi \mathcal{R}_\upsilon (\ell_2 - \ell_1)^2. \tag{27}$$

Thus, it follows from (24)–(27) that, for any $\epsilon > 0$, there exists a positive constant $\sigma = \frac{\epsilon}{\mathcal{R}_\upsilon} \min\{\frac{1}{\varpi}, \frac{1}{\xi}\}$ independent of ℓ_1, ℓ_2 and (x, y) such that $\|\mathbf{W}(x, y)(\ell_2) - \mathbf{W}(x, y)(\ell_1)\| < \epsilon$, whenever $|\ell_2 - \ell_1| \leqslant \sigma$. Thereby, $\mathbf{W} : Y' \to Y'$ is equicontinuous. By the Arzela–Ascoli theorem, we know that $\mathbf{W} : Y' \to Y'$ is completely continuous. In view of Theorem 1, $\mathbf{W}$ has a unique fixed point $(x^*(t), y^*(t)) \in \overline{\upsilon}$, which is a solution of system (1). □

4. Ulam–Hyers Stability

In this section, we are interested in Ulam–Hyers stability and its types for the solution of (1).

Definition 4. [61] *Problem* (1) *is Ulam–Hyers stable if there exists a constant* $\mathbf{K}_{\alpha,\beta} = (\mathbf{K}_\alpha, \mathbf{K}_\beta) > 0$ *such that, for any* $\epsilon = (\epsilon_\alpha, \epsilon_\beta) > 0$, *and* $m = 1, 2, \ldots, n$, *there exists a solution* $(x, y) \in Y'$ *of:*

$$\begin{cases} \left|{}^c\mathbf{D}^\alpha x(t) - h(t, {}^c\mathbf{D}^a x(t), {}^c\mathbf{D}^b y(t))\right| \leqslant \epsilon_\alpha, \\ \left|\Delta x(t_m) - M_{1m}(x(t_m))\right| \leqslant \epsilon_\alpha, \\ \left|\Delta x'(t_m) - N_{1m}(x(t_m))\right| \leqslant \epsilon_\alpha, \\ \left|\Delta x''(t_m) - O_{1m}(x(t_m))\right| \leqslant \epsilon_\alpha, \\ \left|{}^c\mathbf{D}^\beta y(t) - w(t, {}^c\mathbf{D}^a x(t), {}^c\mathbf{D}^b y(t))\right| \leqslant \epsilon_\beta, \\ \left|\Delta y(t_m) - M_{2m}(y(t_m))\right| \leqslant \epsilon_\beta, \\ \left|\Delta y'(t_m) - N_{2m}(y(t_m))\right| \leqslant \epsilon_\beta, \\ \left|\Delta y''(t_m) - O_{2m}(y(t_m))\right| \leqslant \epsilon_\beta, \end{cases} \tag{28}$$

corresponding to a solution $(\zeta, \chi) \in Y'$ *of* (1) *such that*

$$\left|(x, y)(t) - (\zeta, \chi)(t)\right| \leqslant \mathbf{K}_{\alpha,\beta}\, \epsilon.$$

Definition 5. *[61] Problem (1) is generalized Ulam–Hyers stable if there exists a function $\Theta_{\alpha,\beta} \in C(\mathbb{R}_+, \mathbb{R}_+)$, $\Theta_{\alpha,\beta}(0) = 0$ for each $\epsilon > 0$, such that for every solution $(x, y) \in Y'$ of the inequality (28). there is a solution $(\zeta, \chi) \in Y'$ of (1) such that*

$$\big|(x,y)(t) - (\zeta,\chi)(t)\big| \leqslant \Theta_{\alpha,\beta}(\epsilon).$$

Definition 6. *[61] Problem (1) is Ulam–Hyers–Rassias stable with respect to $(\Psi_{\alpha,\beta}, \varphi_{\alpha,\beta})$, where $\Psi_{\alpha,\beta} = (\Psi_\alpha, \Psi_\beta) \in C(J, \mathbb{R})$ and $\varphi_{\alpha,\beta} = (\varphi_\alpha, \varphi_\beta) \in C(J, \mathbb{R})$, if, for every $\epsilon = (\epsilon_\alpha, \epsilon_\beta) > 0$, there exists a real number $\mathbf{K}_{\Psi,\varphi} > 0$, such that for $m = 1, 2, \ldots, n$ and for a solution $(x, y) \in Y'$ of:*

$$\begin{cases} \big|{}^c\mathbf{D}^\alpha x(t) - h(t, {}^c\mathbf{D}^a x(t), {}^c\mathbf{D}^b y(t))\big| \leqslant \Psi_\alpha(t), \\ \big|\Delta x(t_m) - M_{1m}(x(t_m))\big| \leqslant \varphi_\alpha, \\ \big|\Delta x'(t_m) - N_{1m}(x(t_m))\big| \leqslant \varphi_\alpha, \\ \big|\Delta x''(t_m) - O_{1m}(x(t_m))\big| \leqslant \varphi_\alpha, \\ \big|{}^c\mathbf{D}^\beta y(t) - w(t, {}^c\mathbf{D}^a x(t), {}^c\mathbf{D}^b y(t))\big| \leqslant \Psi_\beta(t), \\ \big|\Delta y(t_m) - M_{2m}(y(t_m))\big| \leqslant \varphi_\beta, \\ \big|\Delta y'(t_m) - N_{2m}(y(t_m))\big| \leqslant \varphi_\beta, \\ \big|\Delta y''(t_m) - O_{2m}(y(t_m))\big| \leqslant \varphi_\beta, \end{cases} \tag{29}$$

there is a solution $(\zeta, \chi) \in Y'$ of (1) such that

$$\big|(x,y)(t) - (\zeta,\chi)(t)\big| \leqslant \mathbf{K}_{\Psi,\varphi}\big(\Psi_{\alpha,\beta}(t) + \varphi_{\alpha,\beta}\big)\epsilon.$$

Definition 7. *[61] Problem (1) is generalized Ulam–Hyers–Rassias stable with respect to $(\Psi_{\alpha,\beta}, \varphi_{\alpha,\beta}) \in C(J, \mathbb{R})$, if there exists a real number $\mathbf{K}_{\Psi,\varphi} > 0$, such that for $m = 1, 2, \ldots, n$ and for every solution $(x, y) \in Y'$ of the following:*

$$\begin{cases} \big|{}^c\mathbf{D}^\alpha x(t) - h(t, {}^c\mathbf{D}^a x(t), {}^c\mathbf{D}^b y(t))\big| \leqslant \Psi_\alpha(t)\epsilon_\alpha, \\ \big|\Delta x(t_m) - M_{1m}(x(t_m))\big| \leqslant \varphi_\alpha\epsilon_\alpha, \\ \big|\Delta x'(t_m) - N_{1m}(x(t_m))\big| \leqslant \varphi_\alpha\epsilon_\alpha, \\ \big|\Delta x''(t_m) - O_{1m}(x(t_m))\big| \leqslant \varphi_\alpha\epsilon_\alpha, \\ \big|{}^c\mathbf{D}^\beta y(t) - w(t, {}^c\mathbf{D}^a x(t), {}^c\mathbf{D}^b y(t))\big| \leqslant \Psi_\beta(t)\epsilon_\beta, \\ \big|\Delta y(t_m) - M_{2m}(y(t_m))\big| \leqslant \varphi_\beta\epsilon_\beta, \\ \big|\Delta y'(t_m) - N_{2m}(y(t_m))\big| \leqslant \varphi_\beta\epsilon_\beta, \\ \big|\Delta y''(t_m) - O_{2m}(y(t_m))\big| \leqslant \varphi_\beta\epsilon_\beta, \end{cases} \tag{30}$$

there is a solution $(\zeta, \chi) \in Y'$ of (1) such that

$$\big|(x,y)(t) - (\zeta,\chi)(t)\big| \leqslant \mathbf{K}_{\Psi,\varphi}\big(\Psi_{\alpha,\beta}(t) + \varphi_{\alpha,\beta}\big).$$

Remark 1. *A function $(x, y) \in Y'$ is a solution of the inequality (28), if and only if there exist functions $F_h, \hat{F}_w \in Y'$ and a sequence $F_m, \hat{F}_m$, $m = 1, 2, \ldots, n$ depending on (x, y), such that*

- $|F_h(t)| \leqslant \epsilon_\alpha$, $|\hat{F}_w(t)| \leqslant \epsilon_\beta$, $|F_m| \leqslant \epsilon_\alpha$, $|\hat{F}_m| \leqslant \epsilon_\beta$, $t \in J_m$, $m = 1, \ldots, n$;
- ${}^cD^\alpha x(t) = -h(t, {}^cD^a x(t), {}^cD^b y(t)) + F_h(t)$;
- $\Delta x\,|_{t=t_m} = M_{1m}(x(t_m)) + F_m$;
- $\Delta x\,|_{t=t_m} = N_{1m}(x(t_m)) + F_m$;
- $\Delta x\,|_{t=t_m} = O_{1m}(x(t_m)) + F_m$;
- ${}^cD^\beta x(t) = -w(t, {}^cD^a x(t), {}^cD^b y(t)) + \hat{F}_w(t)$;
- $\Delta y\,|_{t=t_m} = M_{2m}(y(t_m)) + \hat{F}_m$;
- $\Delta y\,|_{t=t_m} = N_{2m}(y(t_m)) + \hat{F}_m$;
- $\Delta y\,|_{t=t_m} = O_{2m}(y(t_m)) + \hat{F}_m$.

Remark 2. *A function* $(x, y) \in Y'$ *is a solution of the inequality* (29), *if and only if there exist functions* $F_h, \hat{F}_w \in Y'$ *and a sequence* $F_m, \hat{F}_m$, $m = 1, 2, \ldots, n$ *depending on* (x, y), *such that*

- $|F_h(t)| \leqslant \Psi_\alpha$, $|\hat{F}_w(t)| \leqslant \Psi_\beta$, $|F_m| \leqslant \varphi_\alpha$, $|F_m| \leqslant \varphi_\beta$, $t \in J_m$, $m = 1, \ldots, n$;
- ${}^cD^\alpha x(t) = -h(t, {}^cD^a x(t), {}^cD^b y(t)) + F_h(t)$;
- $\Delta x \mid_{t=t_m} = M_{1m}(x(t_m)) + F_m$;
- $\Delta x \mid_{t=t_m} = N_{1m}(x(t_m)) + F_m$;
- $\Delta x \mid_{t=t_m} = O_{1m}(x(t_m)) + F_m$;
- ${}^cD^\beta x(t) = -w(t, {}^cD^a x(t), {}^cD^b y(t)) + \hat{F}_w(t)$;
- $\Delta y \mid_{t=t_m} = M_{2m}(y(t_m)) + \hat{F}_m$;
- $\Delta y \mid_{t=t_m} = N_{2m}(y(t_m)) + \hat{F}_m$;
- $\Delta y \mid_{t=t_m} = O_{2m}(y(t_m)) + \hat{F}_m$.

Similarly, one can easily state such a remark for the inequality (30).

Theorem 4. *If the assumptions* $(\mathbf{M}_1) - (\mathbf{M}_2)$ *hold with*

$$\Lambda_0 = 1 - \frac{\Lambda_2^*\Lambda_4^*}{(1-\Lambda_1^*)(1-\Lambda_3^*)} > 0, \tag{31}$$

then (1) *is Ulam–Hyers and generalized Ulam–Hyers stable.*

Proof. Let $(x, y) \in Y'$ be any solution of the inequality (28) and let $(\zeta, \chi) \in Y'$ be the unique solution of the following:

$$\begin{cases} {}^c\mathbf{D}^\alpha \zeta(t) + h(t, {}^c\mathbf{D}^a \zeta(t), {}^c\mathbf{D}^b \chi(t)) = 0, \ t \neq t_m, \ m = 1, 2, \ldots, n, \\ {}^c\mathbf{D}^\beta \chi(t) + w(t, {}^c\mathbf{D}^a \zeta(t), {}^c\mathbf{D}^b \chi(t)) = 0, \ t \neq t_m, \ m = 1, 2, \ldots, n, \\ \Delta\zeta \mid_{t=t_m} = M_{1m}(\zeta(t_m)), \ \Delta\zeta' \mid_{t=t_m} = N_{1m}(\zeta(t_m)), \ \Delta\zeta'' \mid_{t=t_m} = O_{1m}(\zeta(t_m)), \\ \Delta\chi \mid_{t=t_m} = M_{2m}(\chi(t_m)), \ \Delta\chi' \mid_{t=t_m} = N_{2m}(\chi(t_m)), \ \Delta\chi'' \mid_{t=t_m} = O_{2m}(\chi(t_m)), \\ \zeta(0) = \zeta'(0) = 0, \ {}^cD^\varepsilon \zeta(\Omega) = \zeta''(1), \\ \chi(0) = \chi'(0) = 0, \ {}^cD^\rho \chi(\Phi) = \chi''(1). \end{cases} \tag{32}$$

By Lemma 2.4, we have

$$\begin{cases} {}^c\mathbf{D}^\alpha x(t) + h(t, {}^c\mathbf{D}^a x(t), {}^c\mathbf{D}^b y(t)) = F_h(t), \ t \neq t_m, \ m = 1, 2 \ldots, n, \\ {}^c\mathbf{D}^\beta y(t) + w(t, {}^c\mathbf{D}^a x(t), {}^c\mathbf{D}^b y(t)) = F_w(t), \ t \neq t_m, \ m = 1, 2 \ldots, n, \\ \Delta x \mid_{t=t_m} = M_{1m}(x(t_m)) + F_m, \ \Delta x' \mid_{t=t_m} = N_{1m}(x(t_m)) + F_m, \ \Delta x'' \mid_{t=t_m} = O_{1m}(x(t_m)) + F_m, \\ \Delta y \mid_{t=t_m} = M_{2m}(y(t_m)) + \hat{F}_m, \ \Delta y' \mid_{t=t_m} = N_{2m}(y(t_m)) + \hat{F}_m, \ \Delta y'' \mid_{t=t_m} = O_{2m}(y(t_m)) + \hat{F}_m, \\ x(0) = x'(0) = 0, \ {}^c\mathbf{D}^\varepsilon x(\Omega) = x''(1), \\ y(0) = y'(0) = 0, \ {}^c\mathbf{D}^\rho y(\Phi) = y''(1). \end{cases} \tag{33}$$

Since (x, y) is a solution of the inequality (28) and $t \in J$; hence, by Remark 1, we obtain

$$x(t)=\begin{cases} -\frac{1}{\Gamma(\alpha)}\int_0^t (t-s)\left[h(s,{}^c\mathbf{D}^a x(s),{}^c\mathbf{D}^b y(s))+\mathcal{F}_h(s)\right]ds+e^*t^2, \ t\in[0,t_1],\\ -\frac{1}{\Gamma(\alpha)}\int_{t_m}^t (t-s)^{\alpha-1}\left[h(s,{}^c\mathbf{D}^a x(s),{}^c\mathbf{D}^b y(s))+\mathcal{F}_h(s)\right]ds\\ -\frac{1}{\Gamma(\alpha)}\sum_{j=1}^m\int_{t_{j-1}}^{t_j}(t_j-s)^{\alpha-1}\left[h(s,{}^c\mathbf{D}^a x(s),{}^c\mathbf{D}^b y(s))+\mathcal{F}_h(s)\right]ds\\ -\frac{1}{\Gamma(\alpha-1)}\sum_{j=1}^m(t-t_j)\int_{t_{j-1}}^{t_j}(t_j-s)^{\alpha-2}\left[h(s,{}^c\mathbf{D}^a x(s),{}^c\mathbf{D}^b y(s))+\mathcal{F}_h(s)\right]ds\\ -\frac{1}{2\Gamma(\alpha-2)}\sum_{j=1}^m(t-t_j)^2\int_{t_{j-1}}^{t_j}(t_j-s)^{\alpha-3}\left[h(s,{}^c\mathbf{D}^a x(s),{}^c\mathbf{D}^b y(s))+\mathcal{F}_h(s)\right]ds\\ +\sum_{j=1}^m\left[M_{1j}(x(t_j))+\mathcal{F}_m\right]+\sum_{j=1}^m(t-t_j)\left[N_{1j}(x(t_j))+\mathcal{F}_m\right]+\sum_{j=1}^m\frac{(t-t_j)^2}{2}\\ \times\left[O_{1j}(x(t_j))+\mathcal{F}_m\right]+e^*t^2, \ t\in(t_m,t_{m+1}], \ 1\leqslant m\leqslant n. \end{cases}$$

$$y(t)=\begin{cases} -\frac{1}{\Gamma(\beta)}\int_0^t (t-s)^{\beta-1}\left[w(s,{}^c\mathbf{D}^a x(s),{}^c\mathbf{D}^b y(s))+\mathcal{F}_w(s)\right]ds+c^*t^2, \ t\in[0,t_1],\\ -\frac{1}{\Gamma(\beta)}\int_{t_m}^t (t-s)^{\beta-1}\left[w(s,{}^c\mathbf{D}^a x(s),{}^c\mathbf{D}^b y(s))+\mathcal{F}_w(s)\right]ds\\ -\frac{1}{\Gamma(\beta)}\sum_{j=1}^m\int_{t_{j-1}}^{t_j}(t_j-s)^{\beta-1}w(s,{}^c\mathbf{D}^a x(s),{}^c\mathbf{D}^b y(s))ds\\ -\frac{1}{\Gamma(\beta-1)}\sum_{j=1}^m(t-t_j)\int_{t_{j-1}}^{t_j}(t_j-s)^{\beta-2}w(s,{}^c\mathbf{D}^a x(s),{}^c\mathbf{D}^b y(s))ds\\ -\frac{1}{2\Gamma(\beta-2)}\sum_{j=1}^m(t-t_j)^2\int_{t_{j-1}}^{t_j}(t_j-s)^{\beta-3}\left[w(s,{}^c\mathbf{D}^a x(s),{}^c\mathbf{D}^b y(s))+\mathcal{F}_w(s)\right]ds\\ +\sum_{j=1}^m\left[M_{2j}(y(t_j))+\mathcal{F}_m\right]+\sum_{j=1}^m(t-t_j)\left[M_{2j}(y(t_j))+\mathcal{F}_m\right]+\sum_{j=1}^m\frac{(t-t_j)^2}{2}\\ \times\left[O_{2j}(y(t_j))+\mathcal{F}_m\right]+c^*t^2, \ t\in(t_m,t_{m+1}], \ 1\leqslant m\leqslant n, \end{cases}$$

where

$$\begin{aligned} e^* =&\mathcal{Z}_\alpha\Bigg[-\frac{1}{\Gamma(\alpha-\varepsilon)}\int_{t_j}^{\Omega}(\Omega-s)^{\alpha-\varepsilon-1}\left[h(s,{}^c\mathbf{D}^a x(s),{}^c\mathbf{D}^b y(s))+\mathcal{F}_h(s)\right]ds\\ &-\frac{\Omega^{1-\varepsilon}}{\Gamma(\alpha-1)\Gamma(2-\varepsilon)}\sum_{j=1}^m\int_{t_{j-1}}^{t_j}(t_j-s)^{\alpha-2}\left[h(s,{}^c\mathbf{D}^a x(s),{}^c\mathbf{D}^b y(s))+\mathcal{F}_h(s)\right]ds\\ &-\frac{\Omega^{2-\varepsilon}}{\Gamma(\alpha-2)\Gamma(3-\varepsilon)}\sum_{j=1}^m\int_{t_{j-1}}^{t_j}(t_j-s)^{\alpha-3}\left[h(s,{}^c\mathbf{D}^a x(s),{}^c\mathbf{D}^b y(s))+\mathcal{F}_h(s)\right]ds\\ &+\frac{\Omega^{1-\varepsilon}}{\Gamma(\alpha-2)\Gamma(2-\varepsilon)}\sum_{j=1}^m t_j\int_{t_{j-1}}^{t_j}(t_j-s)^{\alpha-3}\left[h(s,{}^c\mathbf{D}^a x(s),{}^c\mathbf{D}^b y(s))+\mathcal{F}_h(s)\right]ds\\ &+\frac{\Omega^{1-\varepsilon}}{\Gamma(2-\varepsilon)}\sum_{j=1}^m\left[N_{1j}(x(t_j))+\mathcal{F}_m\right]+\frac{\Omega^{2-\varepsilon}}{\Gamma(3-\varepsilon)}\sum_{j=1}^m\left[O_{1j}(x(t_j))+\mathcal{F}_m\right]-\frac{\Omega^{1-\varepsilon}}{\Gamma(2-\varepsilon)}\sum_{j=1}^m t_j\\ &\times\left[O_{1j}(x(t_j))+\mathcal{F}_m\right]+\frac{1}{\Gamma(\alpha-2)}\int_{t_m}^1(1-s)^{\alpha-3}\left[h(s,{}^c\mathbf{D}^a x(s),{}^c\mathbf{D}^b y(s))+\mathcal{F}_h(s)\right]ds\\ &+\frac{1}{\Gamma(\alpha-2)}\sum_{j=1}^m\int_{t_{j-1}}^{t_j}(t_j-s)^{\alpha-3}\left[h(s,{}^c\mathbf{D}^a x(s),{}^c\mathbf{D}^b y(s))+\mathcal{F}_h(s)\right]ds-\sum_{j=1}^m\left[O_{1j}(x(t_j))+\mathcal{F}_m\right]\Bigg],\\ c^* =&\mathcal{Z}_\beta\Bigg[-\frac{1}{\Gamma(\beta-\rho)}\int_{t_j}^{\Phi}(\Phi-s)^{\beta-\rho-1}\left[w(s,{}^c\mathbf{D}^a x(s),{}^c\mathbf{D}^b y(s))+\mathcal{F}_w(s)\right]ds\\ &-\frac{\Phi^{1-\rho}}{\Gamma(\beta-1)\Gamma(2-\rho)}\sum_{j=1}^m\int_{t_{j-1}}^{t_j}(t_j-s)^{\beta-2}\left[w(s,{}^c\mathbf{D}^a x(s),{}^c\mathbf{D}^b y(s))+\mathcal{F}_w(s)\right]ds\\ &-\frac{\Phi^{2-\rho}}{\Gamma(\beta-2)\Gamma(3-\rho)}\sum_{j=1}^m\int_{t_{j-1}}^{t_j}(t_j-s)^{\beta-3}\left[w(s,{}^c\mathbf{D}^a x(s),{}^c\mathbf{D}^b y(s))+\mathcal{F}_w(s)\right]ds\\ &+\frac{\Phi^{1-\rho}}{\Gamma(\beta-2)\Gamma(2-\rho)}\sum_{j=1}^m t_j\int_{t_{j-1}}^{t_j}(t_j-s)^{\beta-3}\left[w(s,{}^c\mathbf{D}^a x(s),{}^c\mathbf{D}^b y(s))+\mathcal{F}_w(s)\right]ds \end{aligned}$$

$$+\frac{\Phi^{1-\rho}}{\Gamma(2-\rho)}\sum_{j=1}^{m}\left[N_{2j}(y(t_j))+\digamma_m\right]+\frac{\Phi^{2-\rho}}{\Gamma(3-\rho)}\sum_{j=1}^{m}\left[O_{2j}(y(t_j))+\digamma_m\right]-\frac{\Phi^{1-\rho}}{\Gamma(2-\rho)}\sum_{j=1}^{m}t_j$$
$$\times\left[O_{2j}(y(t_j))+\digamma_m\right]+\frac{1}{\Gamma(\beta-2)}\int_{t_m}^{1}(1-s)^{\beta-3}\left[w(s,{}^c\mathbf{D}^a x(s),{}^c\mathbf{D}^b y(s))+\digamma_w(s)\right]ds$$
$$+\frac{1}{\Gamma(\beta-2)}\sum_{j=1}^{m}\int_{t_{j-1}}^{t_j}(t_j-s)^{\beta-3}\left[w(s,{}^c\mathbf{D}^a x(s),{}^c\mathbf{D}^b y(s))+\digamma_w(s)\right]ds-\sum_{j=1}^{m}\left[O_{2j}(y(t_j))+\digamma_m\right]\Bigg].$$

For $t\in[0,t_1]$, we have

$$\begin{aligned}
x(t)=&\frac{1}{\Gamma(\alpha)}\int_0^t(t-s)^{\alpha-1}\left[h(s,{}^c\mathbf{D}^a x_1(s),{}^c\mathbf{D}^b y_1(s))+\digamma_h(s)\right]ds\\
&+\frac{\mathcal{Z}_\alpha t^2}{\Gamma(\alpha-\varepsilon)}\int_{t_j}^{\Omega}(\Omega-s)^{\alpha-\varepsilon-1}\left[h(s,{}^c\mathbf{D}^a x_1(s),{}^c\mathbf{D}^b y_1(s))+\digamma_h(s)\right]ds\\
&+\frac{\mathcal{Z}_\alpha t^2\Omega^{1-\varepsilon}}{\Gamma(\alpha-1)\Gamma(2-\varepsilon)}\sum_{j=1}^{m}\int_{t_{j-1}}^{t_j}(t_j-s)^{\alpha-2}\left[h(s,{}^c\mathbf{D}^a x_1(s),{}^c\mathbf{D}^b y_1(s))+\digamma_h(s)\right]ds\\
&+\frac{\mathcal{Z}_\alpha t^2\Omega^{2-\varepsilon}}{\Gamma(\alpha-2)\Gamma(3-\varepsilon)}\sum_{j=1}^{m}\int_{t_{j-1}}^{t_j}(t_j-s)^{\alpha-3}\left[h(s,{}^c\mathbf{D}^a x_1(s),{}^c\mathbf{D}^b y_1(s))+\digamma_h(s)\right]ds\\
&+\frac{\mathcal{Z}_\alpha t^2\Omega^{1-\varepsilon}}{\Gamma(\alpha-2)\Gamma(2-\varepsilon)}\sum_{j=1}^{m}t_j\int_{t_{j-1}}^{t_j}(t_j-s)^{\alpha-3}\left[h(s,{}^c\mathbf{D}^a x_1(s),{}^c\mathbf{D}^b y_1(s))+\digamma_h(s)\right]ds \qquad (34)\\
&+\frac{\mathcal{Z}_\alpha t^2\Omega^{1-\varepsilon}}{\Gamma(2-\varepsilon)}\sum_{j=1}^{m}\left[N_{1j}(x(t_j))+\digamma_m\right]+\frac{\mathcal{Z}_\alpha t^2\Omega^{2-\varepsilon}}{\Gamma(3-\varepsilon)}\sum_{j=1}^{m}\left[O_{1j}(x(t_j))+\digamma_m\right]\\
&+\frac{\mathcal{Z}_\alpha t^2}{\Gamma(\alpha-2)}\int_{t_m}^{1}(1-s)^{\alpha-3}\left[h(s,{}^c\mathbf{D}^a x_1(s),{}^c\mathbf{D}^b y_1(s))+\digamma_h(s)\right]ds\\
&+\frac{\mathcal{Z}_\alpha t^2}{\Gamma(\alpha-2)}\sum_{j=1}^{m}\int_{t_{j-1}}^{t_j}(t_j-s)^{\alpha-3}\left[h(s,{}^c\mathbf{D}^a x_1(s),{}^c\mathbf{D}^b y_1(s))+\digamma_h(s)\right]ds\\
&+\frac{\mathcal{Z}_\alpha t^2\Omega^{1-\varepsilon}}{\Gamma(2-\varepsilon)}\sum_{j=1}^{m}t_j\left[O_{1j}(x(t_j))+\digamma_m\right]+\mathcal{Z}_\alpha t^2\sum_{j=1}^{m}\left[O_{1j}(x(t_j))+\digamma_m\right].
\end{aligned}$$

For computational convenience, we use $s(t)$ for the sum of terms which are free of $\digamma$; then, (34) becomes

$$\begin{aligned}
&\left|x(t)-s_1(t)\right|\\
\leqslant&\frac{1}{\Gamma(\alpha)}\int_0^t(t-s)^{\alpha-1}\left|\digamma_h(s)\right|ds+\frac{\mathcal{Z}_\alpha}{\Gamma(\alpha-\varepsilon)}\int_{t_j}^{\Omega}(\Omega-s)^{\alpha-\varepsilon-1}\left|\digamma_h(s)\right|ds\\
&+\frac{\mathcal{Z}_\alpha\Omega^{1-\varepsilon}}{\Gamma(\alpha-1)\Gamma(2-\varepsilon)}\sum_{j=1}^{m}\int_{t_{j-1}}^{t_j}(t_j-s)^{\alpha-2}\left|\digamma_h(s)\right|ds+\frac{\mathcal{Z}_\alpha\Omega^{2-\varepsilon}}{\Gamma(\alpha-2)\Gamma(3-\varepsilon)}\sum_{j=1}^{m}\int_{t_{j-1}}^{t_j}(t_j-s)^{\alpha-3}\left|\digamma_h(s)\right|ds\\
&+\frac{\mathcal{Z}_\alpha\Omega^{1-\varepsilon}}{\Gamma(\alpha-2)\Gamma(2-\varepsilon)}\sum_{j=1}^{m}|t_j|\int_{t_{j-1}}^{t_j}(t_j-s)^{\alpha-3}\left|\digamma_h(s)\right|ds+\frac{\mathcal{Z}_\alpha\Omega^{1-\varepsilon}}{\Gamma(2-\varepsilon)}\sum_{j=1}^{m}\left|\digamma_m\right|+\frac{\mathcal{Z}_\alpha\Omega^{2-\varepsilon}}{\Gamma(3-\varepsilon)}\sum_{j=1}^{m}\left|\digamma_m\right|\\
&+\frac{\mathcal{Z}_\alpha}{\Gamma(\alpha-2)}\int_{t_m}^{1}(1-s)^{\alpha-3}\left|\digamma_h(s)\right|ds+\frac{\mathcal{Z}_\alpha}{\Gamma(\alpha-2)}\sum_{j=1}^{m}\int_{t_{j-1}}^{t_j}(t_j-s)^{\alpha-3}\left|\digamma_h(s)\right|ds\\
&+\frac{\mathcal{Z}_\alpha\Omega^{1-\varepsilon}}{\Gamma(2-\varepsilon)}\sum_{j=1}^{m}|t_j|\left|\digamma_m\right|+\mathcal{Z}_\alpha\sum_{j=1}^{m}\left|\digamma_m\right|.
\end{aligned}$$

By utilizing Remark 1, we get

$$\begin{aligned}
&\left|x(t)-s_1(t)\right| \\
&\leqslant\left[\frac{1}{\Gamma(\alpha+1)}+\frac{\mathcal{Z}_\alpha}{\Gamma(\alpha-\varepsilon+1)}+\frac{\mathcal{Z}_\alpha\Omega^{1-\varepsilon}}{\Gamma(\alpha)\Gamma(2-\varepsilon)}+\frac{\mathcal{Z}_\alpha\Omega^{2-\varepsilon}}{\Gamma(\alpha-1)\Gamma(3-\varepsilon)}+\frac{\mathcal{Z}_\alpha\Omega^{1-\varepsilon}}{\Gamma(\alpha-1)\Gamma(2-\varepsilon)}+\frac{2\mathcal{Z}_\alpha\Omega^{1-\varepsilon}}{\Gamma(2-\varepsilon)}\right. \\
&\left.+\frac{\mathcal{Z}_\alpha\Omega^{2-\varepsilon}}{\Gamma(3-\varepsilon)}+\frac{2\mathcal{Z}_\alpha}{\Gamma(\alpha-1)}+\mathcal{Z}_\alpha\right]\epsilon_\alpha.
\end{aligned} \tag{35}$$

Let

$$\begin{aligned}
\mathcal{Q}_1=&\frac{1}{\Gamma(\alpha+1)}+\frac{\mathcal{Z}_\alpha}{\Gamma(\alpha-\varepsilon+1)}+\frac{\mathcal{Z}_\alpha\Omega^{1-\varepsilon}}{\Gamma(\alpha)\Gamma(2-\varepsilon)}+\frac{\mathcal{Z}_\alpha\Omega^{2-\varepsilon}}{\Gamma(\alpha-1)\Gamma(3-\varepsilon)}+\frac{\mathcal{Z}_\alpha\Omega^{1-\varepsilon}}{\Gamma(\alpha-1)\Gamma(2-\varepsilon)}+\frac{2\mathcal{Z}_\alpha\Omega^{1-\varepsilon}}{\Gamma(2-\varepsilon)} \\
&+\frac{\mathcal{Z}_\alpha\Omega^{2-\varepsilon}}{\Gamma(3-\varepsilon)}+\frac{2\mathcal{Z}_\alpha}{\Gamma(\alpha-1)}+\mathcal{Z}_\alpha.
\end{aligned}$$

Thus, (35) becomes

$$\left|x(t)-s_1(t)\right|\leqslant\mathcal{Q}_1\epsilon_\alpha. \tag{36}$$

Let

$$\left|x(t)-\zeta(t)\right|=\left|x(t)-s_1(t)+s_1(t)-\zeta(t)\right|\leqslant\left|x(t)-s_1(t)\right|+\left|s_1(t)-\zeta(t)\right|. \tag{37}$$

Using (36) in (37), we have

$$\begin{aligned}
&\left|x(t)-\zeta(t)\right| \\
&\leqslant\mathcal{Q}_1\epsilon_\alpha+\frac{1}{\Gamma(\alpha)}\int_0^t(t-s)^{\alpha-1}\left|h(s,{}^c\mathbf{D}^a x(s),{}^c\mathbf{D}^b y(s))-h(s,{}^c\mathbf{D}^a\zeta(s),{}^c\mathbf{D}^b\chi(s))\right|ds \\
&+\frac{\mathcal{Z}_\alpha t^2}{\Gamma(\alpha-\varepsilon)}\int_{t_j}^{\Omega}(\Omega-s)^{\alpha-\varepsilon-1}\left|h(s,{}^c\mathbf{D}^a x(s),{}^c\mathbf{D}^b y(s))-h(s,{}^c\mathbf{D}^a\zeta(s),{}^c\mathbf{D}^b\chi(s))\right|ds \\
&+\frac{\mathcal{Z}_\alpha t^2\Omega^{1-\varepsilon}}{\Gamma(\alpha-1)\Gamma(2-\varepsilon)}\sum_{j=1}^{m}\int_{t_{j-1}}^{t_j}(t_j-s)^{\alpha-2}\left|h(s,{}^c\mathbf{D}^a x(s),{}^c\mathbf{D}^b y(s))-h(s,{}^c\mathbf{D}^a\zeta(s),{}^c\mathbf{D}^b\chi(s))\right|ds \\
&+\frac{\mathcal{Z}_\alpha t^2\Omega^{2-\varepsilon}}{\Gamma(\alpha-2)\Gamma(3-\varepsilon)}\sum_{j=1}^{m}\int_{t_{j-1}}^{t_j}(t_j-s)^{\alpha-3}\left|h(s,{}^c\mathbf{D}^a x(s),{}^c\mathbf{D}^b y(s))-h(s,{}^c\mathbf{D}^a\zeta(s),{}^c\mathbf{D}^b\chi(s))\right|ds \\
&+\frac{\mathcal{Z}_\alpha t^2\Omega^{1-\varepsilon}}{\Gamma(\alpha-2)\Gamma(2-\varepsilon)}\sum_{j=1}^{m}t_j\int_{t_{j-1}}^{t_j}(t_j-s)^{\alpha-3}\left|h(s,{}^c\mathbf{D}^a x(s),{}^c\mathbf{D}^b y(s))-h(s,{}^c\mathbf{D}^a\zeta(s),{}^c\mathbf{D}^b\chi(s))\right|ds \\
&+\frac{\mathcal{Z}_\alpha t^2\Omega^{1-\varepsilon}}{\Gamma(2-\varepsilon)}\sum_{j=1}^{m}\left|N_{1j}(x(t_j))-N_{1j}(\zeta(t_j))\right|+\frac{\mathcal{Z}_\alpha t^2\Omega^{2-\varepsilon}}{\Gamma(3-\varepsilon)}\sum_{j=1}^{m}\left|O_{1j}(x(t_j))-O_{1j}(\zeta(t_j))\right| \\
&+\frac{\mathcal{Z}_\alpha t^2}{\Gamma(\alpha-2)}\int_{t_m}^{1}(1-s)^{\alpha-3}\left|h(s,{}^c\mathbf{D}^a x(s),{}^c\mathbf{D}^b y(s))-h(s,{}^c\mathbf{D}^a\zeta(s),{}^c\mathbf{D}^b\chi(s))\right|ds \\
&+\frac{\mathcal{Z}_\alpha t^2}{\Gamma(\alpha-2)}\sum_{j=1}^{m}\int_{t_{j-1}}^{t_j}(t_j-s)^{\alpha-3}\left|h(s,{}^c\mathbf{D}^a x(s),{}^c\mathbf{D}^b y(s))-h(s,{}^c\mathbf{D}^a\zeta(s),{}^c\mathbf{D}^b\chi(s))\right|ds \\
&+\frac{\mathcal{Z}_\alpha t^2\Omega^{1-\varepsilon}}{\Gamma(2-\varepsilon)}\sum_{j=1}^{m}|t_j|\left|O_{1j}(x(t_j))-O_{1j}(\zeta(t_j))\right|+\mathcal{Z}_\alpha t^2\sum_{j=1}^{m}\left|O_{1j}(x(t_j))-O_{1j}(\zeta(t_j))\right|.
\end{aligned}$$

Utilizing $(\mathbf{M}_1)$ and $(\mathbf{M}_2)$, we get

$$\begin{aligned}
&\leqslant \mathfrak{Q}_1\epsilon_\alpha + \frac{1}{\Gamma(\alpha+1)}\left[\mu_1\|x-\zeta\|_0 + \mu_2\|y-\chi\|_0\right] + \frac{\mathcal{Z}_\alpha}{\Gamma(\alpha-\varepsilon+1)}\left[\mu_1\|x-\zeta\|_0 + \mu_2\|y-\chi\|_0\right] \\
&+ \frac{\mathcal{Z}_\alpha\Omega^{1-\varepsilon}}{\Gamma(\alpha)\Gamma(2-\varepsilon)}\left[\mu_1\|x-\zeta\|_0 + \mu_2\|y-\chi\|_0\right] + \frac{\mathcal{Z}_\alpha\Omega^{2-\varepsilon}}{\Gamma(\alpha-1)\Gamma(3-\varepsilon)}\left[\mu_1\|x-\zeta\|_0 + \mu_2\|y-\chi\|_0\right] \\
&+ \frac{\mathcal{Z}_\alpha\Omega^{1-\varepsilon}}{\Gamma(\alpha-1)\Gamma(2-\varepsilon)}\left[\mu_1\|x-\zeta\|_0 + \mu_2\|y-\chi\|_0\right] + \frac{\mathcal{Z}_\alpha\Omega^{1-\varepsilon}}{\Gamma(2-\varepsilon)}\sum_{j=1}^{m}\hat{I}_{1j}\|x-\zeta\|_0 \qquad (38)\\
&+ \frac{\mathcal{Z}_\alpha\Omega^{2-\varepsilon}}{\Gamma(3-\varepsilon)}\sum_{j=1}^{m}\check{I}_{1j}\|x-\zeta\|_0 + \frac{\mathcal{Z}_\alpha}{\Gamma(\alpha-1)}\left[\mu_1\|x-\zeta\|_0 + \mu_2\|y-\chi\|_0\right] + \frac{\mathcal{Z}_\alpha}{\Gamma(\alpha-1)} \\
&\times \left[\mu_1\|x-\zeta\|_0 + \mu_2\|y-\chi\|_0\right] + \frac{\mathcal{Z}_\alpha\Omega^{1-\varepsilon}}{\Gamma(2-\varepsilon)}\sum_{j=1}^{m}\check{I}_{1j}\|x-\zeta\|_0 + \mathcal{Z}_\alpha\sum_{j=1}^{m}\check{I}_{1j}\|x-\zeta\|_0.
\end{aligned}$$

After some calculation and rearrangement in (38), we get

$$\|x-\zeta\|_0 - \frac{\Lambda_2}{(1-\Lambda_1)}\|y-\chi\|_0 \leqslant \frac{\mathfrak{Q}_1\epsilon_\alpha}{(1-\Lambda_1)}, \qquad (39)$$

where

$$\begin{aligned}
\Lambda_1 =& \mathcal{Z}_\alpha\Bigg[\mu_1\Bigg(\frac{1}{\mathcal{Z}_\alpha\Gamma(\alpha+1)} + \frac{1}{\Gamma(\alpha-\varepsilon+1)} + \frac{\Omega^{1-\varepsilon}}{\Gamma(\alpha)\Gamma(2-\varepsilon)} + \frac{\Omega^{2-\varepsilon}}{\Gamma(\alpha-1)\Gamma(3-\varepsilon)} + \frac{\Omega^{1-\varepsilon}}{\Gamma(\alpha-1)\Gamma(2-\varepsilon)} \\
&+ \frac{2}{\Gamma(\alpha-1)}\Bigg) + \frac{\Omega^{1-\varepsilon}}{\Gamma(2-\varepsilon)}\sum_{j=1}^{m}\hat{I}_{1j} + \frac{\Omega^{2-\varepsilon}}{\Gamma(3-\varepsilon)}\sum_{j=1}^{m}\check{I}_{1j} + \frac{\Omega^{1-\varepsilon}}{\Gamma(2-\varepsilon)}\sum_{j=1}^{m}\check{I}_{1j} + \sum_{j=1}^{m}\check{I}_{1j}\Bigg], \\
\Lambda_2 =& \mathcal{Z}_\alpha\Bigg[\mu_2\Bigg(\frac{1}{\mathcal{Z}_\alpha\Gamma(\alpha+1)} + \frac{1}{\Gamma(\alpha-\varepsilon+1)} + \frac{\Omega^{1-\varepsilon}}{\Gamma(\alpha)\Gamma(2-\varepsilon)} + \frac{\Omega^{2-\varepsilon}}{\Gamma(\alpha-1)\Gamma(3-\varepsilon)} + \frac{\Omega^{1-\varepsilon}}{\Gamma(\alpha-1)\Gamma(2-\varepsilon)} \\
&+ \frac{2}{\Gamma(\alpha-1)}\Bigg)\Bigg].
\end{aligned}$$

In addition, for $t \in (t_m, t_{m+1}]$, we have

$$\begin{aligned}
x(t) =& \frac{1}{\Gamma(\alpha)}\int_{t_m}^{t}(t-s)^{\alpha-1}\left[h(s,{}^c\mathbf{D}^a x(s),{}^c\mathbf{D}^b y(s)) + \digamma_h(s)\right]ds \\
&+ \frac{1}{\Gamma(\alpha)}\sum_{j=1}^{m}\int_{t_{j-1}}^{t_j}(t_j-s)^{\alpha-1}\left[h(s,{}^c\mathbf{D}^a x(s),{}^c\mathbf{D}^b y(s)) + \digamma_h(s)\right]ds \\
&+ \frac{1}{\Gamma(\alpha-1)}\sum_{j=1}^{m}(t-t_j)\int_{t_{j-1}}^{t_j}(t_j-s)^{\alpha-2}\left[h(s,{}^c\mathbf{D}^a x(s),{}^c\mathbf{D}^b y(s)) + \digamma_h(s)\right]ds \\
&+ \frac{1}{2\Gamma(\alpha-2)}\sum_{j=1}^{m}(t-t_j)^2\int_{t_{j-1}}^{t_j}(t_j-s)^{\alpha-3}\left[h(s,{}^c\mathbf{D}^a x(s),{}^c\mathbf{D}^b y(s)) + \digamma_h(s)\right]ds \\
&+ \sum_{j=1}^{m}\left[M_{1j}x(t_j) + \digamma_m\right] + \sum_{j=1}^{m}(t-t_j)\left[N_{1j}(x(t_j)) + \digamma_m\right] \\
&+ \frac{\mathcal{Z}_\alpha t^2}{\Gamma(\alpha-\varepsilon)}\int_{t_j}^{\Omega}(\Omega-s)^{\alpha-\varepsilon-1}\left[h(s,{}^c\mathbf{D}^a x(s),{}^c\mathbf{D}^b y(s)) + \digamma_h(s)\right]ds \\
&+ \frac{\mathcal{Z}_\alpha t^2\Omega^{1-\varepsilon}}{\Gamma(\alpha-1)\Gamma(2-\varepsilon)}\sum_{j=1}^{m}\int_{t_{j-1}}^{t_j}(t_j-s)^{\alpha-2}\left[h(s,{}^c\mathbf{D}^a x(s),{}^c\mathbf{D}^b y(s)) + \digamma_h(s)\right]ds
\end{aligned}$$

$$
\begin{aligned}
&+\frac{\mathcal{Z}_\alpha t^2\Omega^{2-\varepsilon}}{\Gamma(\alpha-2)\Gamma(3-\varepsilon)}\sum_{j=1}^{m}\int_{t_{j-1}}^{t_j}(t_j-s)^{\alpha-3}\big[h(s,{}^c\mathbf{D}^a x(s),{}^c\mathbf{D}^b y(s))+\digamma_h(s)\big]ds\\
&+\frac{\mathcal{Z}_\alpha t^2\Omega^{1-\varepsilon}}{\Gamma(\alpha-2)\Gamma(2-\varepsilon)}\sum_{j=1}^{m}t_j\int_{t_{j-1}}^{t_j}(t_j-s)^{\alpha-3}\big[h(s,{}^c\mathbf{D}^a x(s),{}^c\mathbf{D}^b y(s))+\digamma_h(s)\big]ds\\
&+\frac{\mathcal{Z}_\alpha t^2\Omega^{1-\varepsilon}}{\Gamma(2-\varepsilon)}\sum_{j=1}^{m}\big[N_{1j}(x(t_j))+\digamma_m\big]+\frac{\mathcal{Z}_\alpha t^2\Omega^{2-\varepsilon}}{\Gamma(3-\varepsilon)}\sum_{j=1}^{m}\big[O_{1j}(x(t_j))+\digamma_m\big]\\
&+\frac{\mathcal{Z}_\alpha t^2}{\Gamma(\alpha-2)}\int_{t_m}^{1}(1-s)^{\alpha-3}\big[h(s,{}^c\mathbf{D}^a x(s),{}^c\mathbf{D}^b y(s))+\digamma_h(s)\big]ds\\
&+\frac{\mathcal{Z}_\alpha t^2}{\Gamma(\alpha-2)}\sum_{j=1}^{m}\int_{t_{j-1}}^{t_j}(t_j-s)^{\alpha-3}\big[h(s,{}^c\mathbf{D}^a x(s),{}^c\mathbf{D}^b y(s))+\digamma_h(s)\big]ds\\
&+\frac{\mathcal{Z}_\alpha t^2\Omega^{1-\varepsilon}}{\Gamma(2-\varepsilon)}\sum_{j=1}^{m}t_j\big[O_{1j}(x(t_j))+\digamma_m\big]+\mathcal{Z}_\alpha t^2\sum_{j=1}^{m}\big[O_{1j}(x(t_j))+\digamma_m\big]\\
&+\sum_{j=1}^{m}\frac{(t-t_j)^2}{2}\big[O_{1j}(x(t_j))+\digamma_m\big].
\end{aligned}
$$

For computational convenience, we use $s_1^*(t)$ for the sum of terms which are free of $\digamma$, so we have

$$
\begin{aligned}
&|x(t)-s_1^*(t)|\\
\leqslant&\frac{1}{\Gamma(\alpha)}\int_{t_m}^{t}(t-s)^{\alpha-1}|\digamma_h(s)|ds+\frac{1}{\Gamma(\alpha)}\sum_{j=1}^{m}\int_{t_{j-1}}^{t_j}(t_j-s)^{\alpha-1}|\digamma_h(s)|ds\\
&+\frac{1}{\Gamma(\alpha-1)}\sum_{j=1}^{m}|(t-t_j)|\int_{t_{j-1}}^{t_j}(t_j-s)^{\alpha-2}|\digamma_h(s)|ds+\frac{1}{2\Gamma(\alpha-2)}\sum_{j=1}^{m}|(t-t_j)^2|\int_{t_{j-1}}^{t_j}(t_j-s)^{\alpha-3}\\
&\times|\digamma_h(s)|ds+\sum_{j=1}^{m}|\digamma_m|+\sum_{j=1}^{m}|(t-t_j)||\digamma_m|+\frac{\mathcal{Z}_\alpha}{\Gamma(\alpha-\varepsilon)}\int_{t_j}^{\Omega}(\Omega-s)^{\alpha-\varepsilon-1}|\digamma_h(s)|ds\\
&+\frac{\mathcal{Z}_\alpha\Omega^{1-\varepsilon}}{\Gamma(\alpha-1)\Gamma(2-\varepsilon)}\sum_{j=1}^{m}\int_{t_{j-1}}^{t_j}(t_j-s)^{\alpha-2}|\digamma_h(s)|ds+\frac{\mathcal{Z}_\alpha\Omega^{2-\varepsilon}}{\Gamma(\alpha-2)\Gamma(3-\varepsilon)}\sum_{j=1}^{m}\int_{t_{j-1}}^{t_j}(t_j-s)^{\alpha-3}|\digamma_h(s)|ds\\
&+\frac{\mathcal{Z}_\alpha\Omega^{1-\varepsilon}}{\Gamma(\alpha-2)\Gamma(2-\varepsilon)}\sum_{j=1}^{m}t_j\int_{t_{j-1}}^{t_j}(t_j-s)^{\alpha-3}|\digamma_h(s)|ds+\frac{\mathcal{Z}_\alpha\Omega^{1-\varepsilon}}{\Gamma(2-\varepsilon)}\sum_{j=1}^{m}|\digamma_m|+\frac{\mathcal{Z}_\alpha\Omega^{2-\varepsilon}}{\Gamma(3-\varepsilon)}\sum_{j=1}^{m}|\digamma_m|\\
&+\frac{\mathcal{Z}_\alpha}{\Gamma(\alpha-2)}\int_{t_m}^{1}(1-s)^{\alpha-3}|\digamma_h(s)|ds+\frac{\mathcal{Z}_\alpha}{\Gamma(\alpha-2)}\sum_{j=1}^{m}\int_{t_{j-1}}^{t_j}(t_j-s)^{\alpha-3}|\digamma_h(s)|ds\\
&+\frac{\mathcal{Z}_\alpha\Omega^{1-\varepsilon}}{\Gamma(2-\varepsilon)}\sum_{j=1}^{m}|t_j||\digamma_m|+\mathcal{Z}_\alpha\sum_{j=1}^{m}|\digamma_m|+\sum_{j=1}^{m}\frac{|(t-t_j)^2|}{2}|\digamma_m|.
\end{aligned}
$$

By utilizing Remark 1, we get

$$
\begin{aligned}
&|x(t)-s_1^*(t)|\\
\leqslant&\Bigg(\frac{2}{\Gamma(\alpha+1)}+\frac{1}{\Gamma(\alpha)}+\frac{4\mathcal{Z}_\alpha+1}{2\Gamma(\alpha-1)}+\frac{3\mathcal{Z}_\alpha+4}{2}+\frac{\mathcal{Z}_\alpha}{\Gamma(\alpha-\varepsilon+1)}+\frac{\mathcal{Z}_\alpha\Omega^{1-\varepsilon}}{\Gamma(\alpha)\Gamma(2-\varepsilon)}+\frac{2\mathcal{Z}_\alpha\Omega^{1-\varepsilon}}{\Gamma(2-\varepsilon)}+\frac{\mathcal{Z}_\alpha\Omega^{2-\varepsilon}}{\Gamma(3-\varepsilon)}\\
&+\frac{\mathcal{Z}_\alpha\Omega^{2-\varepsilon}}{\Gamma(\alpha-1)\Gamma(3-\varepsilon)}+\frac{\mathcal{Z}_\alpha\Omega^{1-\varepsilon}}{\Gamma(\alpha-1)\Gamma(2-\varepsilon)}\Bigg)\epsilon_\alpha.
\end{aligned}
\tag{40}
$$

For computational convenience, let

$$\mathcal{Q}_\alpha = \frac{2}{\Gamma(\alpha+1)} + \frac{1}{\Gamma(\alpha)} + \frac{4\mathcal{Z}_\alpha+1}{2\Gamma(\alpha-1)} + \frac{3\mathcal{Z}_\alpha+4}{2} + \frac{\mathcal{Z}_\alpha}{\Gamma(\alpha-\varepsilon+1)} + \frac{\mathcal{Z}_\alpha\Omega^{1-\varepsilon}}{\Gamma(\alpha)\Gamma(2-\varepsilon)} + \frac{2\mathcal{Z}_\alpha\Omega^{1-\varepsilon}}{\Gamma(2-\varepsilon)} + \frac{\mathcal{Z}_\alpha\Omega^{2-\varepsilon}}{\Gamma(3-\varepsilon)}$$
$$+ \frac{\mathcal{Z}_\alpha\Omega^{2-\varepsilon}}{\Gamma(\alpha-1)\Gamma(3-\varepsilon)} + \frac{\mathcal{Z}_\alpha\Omega^{1-\varepsilon}}{\Gamma(\alpha-1)\Gamma(2-\varepsilon)}.$$

Thus, (40) becomes

$$\left|x(t) - s_1^*(t)\right| \leqslant \mathcal{Q}_\alpha \epsilon_\alpha. \tag{41}$$

Let

$$\left|x(t) - \zeta(t)\right| = \left|x(t) - s_1^*(t) + s_1^*(t) - \zeta(t)\right| \leqslant \left|x(t) - s_1^*(t)\right| + \left|s_1^*(t) - \zeta(t)\right|. \tag{42}$$

Using (41) in (42), we get

$$\begin{aligned}
&\left|x(t) - \zeta(t)\right| \\
\leqslant &\mathcal{Q}_\alpha \epsilon_\alpha + \frac{1}{\Gamma(\alpha)} \int_{t_m}^{t} (t-s)^{\alpha-1} \left|h(s, {}^c\mathbf{D}^a x(s), {}^c\mathbf{D}^b y(s)) - h(s, {}^c\mathbf{D}^a \zeta(s), {}^c\mathbf{D}^b \chi(s))\right| ds \\
&+ \frac{1}{\Gamma(\alpha)} \sum_{j=1}^{m} \int_{t_{j-1}}^{t_j} (t_j-s)^{\alpha-1} \left|h(s, {}^c\mathbf{D}^a x(s), {}^c\mathbf{D}^b y(s)) - h(s, {}^c\mathbf{D}^a \zeta(s), {}^c\mathbf{D}^b \chi(s))\right| ds \\
&+ \frac{1}{\Gamma(\alpha-1)} \sum_{j=1}^{m} \left|(t-t_j)\right| \int_{t_{j-1}}^{t_j} (t_j-s)^{\alpha-2} \left|h(s, {}^c\mathbf{D}^a x(s), {}^c\mathbf{D}^b y(s)) - h(s, {}^c\mathbf{D}^a \zeta(s), {}^c\mathbf{D}^b \chi(s))\right| ds \\
&+ \frac{1}{2\Gamma(\alpha-2)} \sum_{j=1}^{m} \left|(t-t_j)^2\right| \int_{t_{j-1}}^{t_j} (t_j-s)^{\alpha-3} \left|h(s, {}^c\mathbf{D}^a x(s), {}^c\mathbf{D}^b y(s)) - h(s, {}^c\mathbf{D}^a \zeta(s), {}^c\mathbf{D}^b \chi(s))\right| ds \\
&+ \sum_{j=1}^{m} \left|M_{1j}(x(t_j)) - M_{1j}(\zeta(t_j))\right| + \sum_{j=1}^{m} \left|(t-t_j)\right| \left|N_{1j}(x(t_j)) - N_{1j}(\zeta(t_j))\right| \\
&+ \frac{\mathcal{Z}_\alpha t^2}{\Gamma(\alpha-\varepsilon)} \int_{t_j}^{\Omega} (\Omega-s)^{\alpha-\varepsilon-1} \left|h(s, {}^c\mathbf{D}^a x(s), {}^c\mathbf{D}^b y(s)) - h(s, {}^c\mathbf{D}^a \zeta(s), {}^c\mathbf{D}^b \chi(s))\right| ds \\
&+ \frac{\mathcal{Z}_\alpha t^2 \Omega^{1-\varepsilon}}{\Gamma(\alpha-1)\Gamma(2-\varepsilon)} \sum_{j=1}^{m} \int_{t_{j-1}}^{t_j} (t_j-s)^{\alpha-2} \left|h(s, {}^c\mathbf{D}^a x(s), {}^c\mathbf{D}^b y(s)) - h(s, {}^c\mathbf{D}^a \zeta(s), {}^c\mathbf{D}^b \chi(s))\right| ds \\
&+ \frac{\mathcal{Z}_\alpha t^2 \Omega^{2-\varepsilon}}{\Gamma(\alpha-2)\Gamma(3-\varepsilon)} \sum_{j=1}^{m} \int_{t_{j-1}}^{t_j} (t_j-s)^{\alpha-3} \left|h(s, {}^c\mathbf{D}^a x(s), {}^c\mathbf{D}^b y(s)) - h(s, {}^c\mathbf{D}^a \zeta(s), {}^c\mathbf{D}^b \chi(s))\right| ds \\
&+ \frac{\mathcal{Z}_\alpha t^2 \Omega^{1-\varepsilon}}{\Gamma(\alpha-2)\Gamma(2-\varepsilon)} \sum_{j=1}^{m} \left|t_j\right| \int_{t_{j-1}}^{t_j} (t_j-s)^{\alpha-3} \left|h(s, {}^c\mathbf{D}^a x(s), {}^c\mathbf{D}^b y(s)) - h(s, {}^c\mathbf{D}^a \zeta(s), {}^c\mathbf{D}^b \chi(s))\right| ds \\
&+ \frac{\mathcal{Z}_\alpha t^2 \Omega^{1-\varepsilon}}{\Gamma(2-\varepsilon)} \sum_{j=1}^{m} \left|N_{1j}(x(t_j)) - N_{1j}(\zeta(t_j))\right| + \frac{\mathcal{Z}_\alpha t^2 \Omega^{2-\varepsilon}}{\Gamma(3-\varepsilon)} \sum_{j=1}^{m} \left|O_{1j}(x(t_j)) - O_{1j}(\zeta(t_j))\right| \\
&+ \frac{\mathcal{Z}_\alpha t^2}{\Gamma(\alpha-2)} \int_{t_m}^{1} (1-s)^{\alpha-3} \left|h(s, {}^c\mathbf{D}^a x(s), {}^c\mathbf{D}^b y(s)) - h(s, {}^c\mathbf{D}^a \zeta(s), {}^c\mathbf{D}^b \chi(s))\right| ds \\
&+ \frac{\mathcal{Z}_\alpha t^2}{\Gamma(\alpha-2)} \sum_{j=1}^{m} \int_{t_{j-1}}^{t_j} (t_j-s)^{\alpha-3} \left|h(s, {}^c\mathbf{D}^a x(s), {}^c\mathbf{D}^b y(s)) - h(s, {}^c\mathbf{D}^a \zeta(s), {}^c\mathbf{D}^b \chi(s))\right| ds
\end{aligned}$$

$$+\frac{\mathcal{Z}_\alpha t^2\Omega^{1-\varepsilon}}{\Gamma(2-\varepsilon)}\sum_{j=1}^{m}|t_j||O_{1j}(x(t_j))-O_{1j}(\zeta(t_j))|+\mathcal{Z}_\alpha t^2\sum_{j=1}^{m}|O_{1j}(x(t_j))-O_{1j}(\zeta(t_j))|$$
$$+\sum_{j=1}^{m}\frac{|(t-t_j)^2|}{2}|O_{1j}(x(t_j))-O_{1j}(\zeta(t_j))|.$$

Using $(\mathbf{M}_1)$ and $(\mathbf{M}_2)$, we have

$$\begin{aligned}
&\leqslant \mathcal{Q}_\alpha\epsilon_\alpha+\tfrac{1}{\Gamma(\alpha+1)}[\mu_1\|x-\zeta\|_0+\mu_2\|y-\chi\|_0]+\tfrac{1}{\Gamma(\alpha+1)}[\mu_1\|x-\zeta\|_0+\mu_2\|y-\chi\|_0]\\
&+\tfrac{1}{\Gamma(\alpha)}[\mu_1\|x-\zeta\|_0+\mu_2\|y-\chi\|_0]+\tfrac{1}{2\Gamma(\alpha-1)}[\mu_1\|x-\zeta\|_0+\mu_2\|y-\chi\|_0]\\
&+\textstyle\sum_{j=1}^{m}I_{1j}\|x-\zeta\|_0+\sum_{j=1}^{m}\hat{I}_{1j}\|x-\zeta\|_0+\tfrac{\mathcal{Z}_\alpha}{\Gamma(\alpha-\varepsilon+1)}[\mu_1\|x-\zeta\|_0+\mu_2\|y-\chi\|_0]\\
&+\tfrac{\mathcal{Z}_\alpha\Omega^{1-\varepsilon}}{\Gamma(\alpha)\Gamma(2-\varepsilon)}[\mu_1\|x-\zeta\|_0+\mu_2\|y-\chi\|_0]+\tfrac{\mathcal{Z}_\alpha\Omega^{2-\varepsilon}}{\Gamma(\alpha-1)\Gamma(3-\varepsilon)}[\mu_1\|x-\zeta\|_0+\mu_2\|y-\chi\|_0]\\
&+\tfrac{\mathcal{Z}_\alpha\Omega^{1-\varepsilon}}{\Gamma(\alpha-1)\Gamma(2-\varepsilon)}[\mu_1\|x-\zeta\|_0+\mu_2\|y-\chi\|_0]+\tfrac{\mathcal{Z}_\alpha\Omega^{1-\varepsilon}}{\Gamma(2-\varepsilon)}\textstyle\sum_{j=1}^{m}\hat{I}_{1j}\|x-\zeta\|_0\\
&+\tfrac{\mathcal{Z}_\alpha\Omega^{2-\varepsilon}}{\Gamma(3-\varepsilon)}\textstyle\sum_{j=1}^{m}\check{I}_{1j}\|x-\zeta\|_0+\tfrac{\mathcal{Z}_\alpha}{\Gamma(\alpha-1)}[\mu_1\|x-\zeta\|_0+\mu_2\|y-\chi\|_0]\\
&+\tfrac{\mathcal{Z}_\alpha}{\Gamma(\alpha-1)}[\mu_1\|x-\zeta\|_0+\mu_2\|y-\chi\|_0]+\tfrac{\mathcal{Z}_\alpha\Omega^{1-\varepsilon}}{\Gamma(2-\varepsilon)}\textstyle\sum_{j=1}^{m}\check{I}_{1j}\|x-\zeta\|_0+\mathcal{Z}_\alpha\sum_{j=1}^{m}\check{I}_{1j}\|x-\zeta\|_0\\
&+\tfrac{1}{2}\textstyle\sum_{j=1}^{m}\check{I}_{1j}\|x-\zeta\|_0.
\end{aligned}\tag{43}$$

After some calculation and rearrangement in (43), we get

$$\|x-\zeta\|_0-\frac{\Lambda_2^*}{(1-\Lambda_1^*)}\|y-\chi\|_0\leqslant\frac{\mathcal{Q}_\alpha\epsilon_\alpha}{(1-\Lambda_1^*)},\tag{44}$$

where

$$\begin{aligned}
\Lambda_1^*=&\Bigg[\mu_1\Bigg(\frac{2}{\Gamma(\alpha+1)}+\frac{1}{\Gamma(\alpha)}+\frac{1}{2\Gamma(\alpha-1)}+\frac{\mathcal{Z}_\alpha}{\Gamma(\alpha-\varepsilon+1)}+\frac{\mathcal{Z}_\alpha\Omega^{1-\varepsilon}}{\Gamma(\alpha)\Gamma(2-\varepsilon)}+\frac{\mathcal{Z}_\alpha\Omega^{2-\varepsilon}}{\Gamma(\alpha-1)\Gamma(3-\varepsilon)}\\
&+\frac{\mathcal{Z}_\alpha\Omega^{1-\varepsilon}}{\Gamma(\alpha-1)\Gamma(2-\varepsilon)}+\frac{2\mathcal{Z}_\alpha}{\Gamma(\alpha-1)}\Bigg)+\sum_{j=1}^{m}I_{1j}+\sum_{j=1}^{m}\hat{I}_{1j}+\frac{\mathcal{Z}_\alpha\Omega^{1-\varepsilon}}{\Gamma(2-\varepsilon)}\sum_{j=1}^{m}\hat{I}_{1j}+\frac{\mathcal{Z}_\alpha\Omega^{2-\varepsilon}}{\Gamma(3-\varepsilon)}\sum_{j=1}^{m}\check{I}_{1j}\\
&+\frac{\mathcal{Z}_\alpha\Omega^{1-\varepsilon}}{\Gamma(2-\varepsilon)}\sum_{j=1}^{m}\check{I}_{1j}+\mathcal{Z}_\alpha\sum_{j=1}^{m}\check{I}_{1j}+\frac{1}{2}\sum_{j=1}^{m}\check{I}_{1j}\Bigg],\\
\Lambda_2^*=&\mu_2\Bigg(\frac{2}{\Gamma(\alpha+1)}+\frac{1}{\Gamma(\alpha)}+\frac{1}{2\Gamma(\alpha-1)}+\frac{\mathcal{Z}_\alpha}{\Gamma(\alpha-\varepsilon+1)}+\frac{\mathcal{Z}_\alpha\Omega^{1-\varepsilon}}{\Gamma(\alpha)\Gamma(2-\varepsilon)}+\frac{\mathcal{Z}_\alpha\Omega^{2-\varepsilon}}{\Gamma(\alpha-1)\Gamma(3-\varepsilon)}\\
&+\frac{\mathcal{Z}_\alpha\Omega^{1-\varepsilon}}{\Gamma(\alpha-1)\Gamma(2-\varepsilon)}+\frac{2\mathcal{Z}_\alpha}{\Gamma(\alpha-1)}\Bigg).
\end{aligned}$$

On the similar fashion, for $t\in[0,t_1]$, and utilizing $(\mathbf{M}_1)-(\mathbf{M}_2)$, we can find

$$|y(t)-s_2(t)|\leqslant\mathcal{Q}_2\epsilon_\beta,\tag{45}$$

where $s_2(t)$ are those terms which are free of $\mathcal{F}$ and

$$\begin{aligned}
\mathcal{Q}_2=&\frac{1}{\Gamma(\beta+1)}+\frac{\mathcal{Z}_\beta}{\Gamma(\beta-\rho+1)}+\frac{\mathcal{Z}_\beta\Phi^{1-\rho}}{\Gamma(\beta)\Gamma(2-\rho)}+\frac{\mathcal{Z}_\beta\Phi^{2-\rho}}{\Gamma(\beta-1)\Gamma(3-\rho)}+\frac{\mathcal{Z}_\beta\Phi^{1-\rho}}{\Gamma(\beta-1)\Gamma(2-\rho)}+\frac{2\mathcal{Z}_\beta\Phi^{1-\rho}}{\Gamma(2-\rho)}\\
&+\frac{\mathcal{Z}_\beta\Phi^{2-\rho}}{\Gamma(3-\rho)}+\frac{2\mathcal{Z}_\beta}{\Gamma(\beta-1)}+\mathcal{Z}_\beta,
\end{aligned}$$

and

$$\|y-\chi\|_0 - \frac{\Lambda_4}{(1-\Lambda_3)}\|x-\zeta\|_0 \leqslant \frac{\mathcal{Q}_2\epsilon_\beta}{(1-\Lambda_3)}, \tag{46}$$

where

$$\begin{aligned}\Lambda_3 =& \mathcal{Z}_\beta\Bigg[\mu_1'\Bigg(\frac{1}{\mathcal{Z}_\beta\Gamma(\beta+1)} + \frac{1}{\Gamma(\beta-\rho+1)} + \frac{\Phi^{1-\rho}}{\Gamma(\beta)\Gamma(2-\rho)} + \frac{\Phi^{2-\rho}}{\Gamma(\beta-1)\Gamma(3-\rho)} + \frac{\Phi^{1-\rho}}{\Gamma(\beta-1)\Gamma(2-\rho)}\\ &+ \frac{2}{\Gamma(\beta-1)}\Bigg) + \frac{\Phi^{1-\rho}}{\Gamma(2-\rho)}\sum_{j=1}^{m}\hat{I}_{2j} + \frac{\Phi^{2-\rho}}{\Gamma(3-\rho)}\sum_{j=1}^{m}\check{I}_{2j} + \frac{\Phi^{1-\rho}}{\Gamma(2-\rho)}\sum_{j=1}^{m}\check{I}_{2j} + \sum_{j=1}^{m}\check{I}_{2j}\Bigg],\\ \Lambda_4 =& \mathcal{Z}_\beta\Bigg[\mu_2'\Bigg(\frac{1}{\mathcal{Z}_\beta\Gamma(\beta+1)} + \frac{1}{\Gamma(\beta-\rho+1)} + \frac{\Phi^{1-\rho}}{\Gamma(\beta)\Gamma(2-\rho)} + \frac{\Phi^{2-\rho}}{\Gamma(\beta-1)\Gamma(3-\rho)} + \frac{\Phi^{1-\rho}}{\Gamma(\beta-1)\Gamma(2-\rho)}\\ &+ \frac{2}{\Gamma(\beta-1)}\Bigg)\Bigg].\end{aligned}$$

In addition, for $t \in (t_m, t_{m+1}]$, $1 \leqslant m \leqslant n$, we can get

$$\left|y(t) - s_2^*(t)\right| \leqslant \mathcal{Q}_\beta\epsilon_\beta, \tag{47}$$

where $s_2^*(t)$ are those terms which are free of $\mathcal{F}$ and

$$\begin{aligned}\mathcal{Q}_\beta =& \frac{2}{\Gamma(\beta+1)} + \frac{1}{\Gamma(\beta)} + \frac{4\mathcal{Z}_\beta+1}{2\Gamma(\beta-1)} + \frac{3\mathcal{Z}_\beta+4}{2} + \frac{\mathcal{Z}_\beta}{\Gamma(\beta-\rho+1)} + \frac{\mathcal{Z}_\beta\Phi^{1-\rho}}{\Gamma(\beta)\Gamma(2-\rho)} + \frac{2\mathcal{Z}_\beta\Phi^{1-\rho}}{\Gamma(2-\rho)} + \frac{\mathcal{Z}_\beta\Phi^{2-\rho}}{\Gamma(3-\rho)}\\ &+ \frac{\mathcal{Z}_\beta\Phi^{2-\rho}}{\Gamma(\beta-1)\Gamma(3-\rho)} + \frac{\mathcal{Z}_\beta\Phi^{1-\rho}}{\Gamma(\beta-1)\Gamma(2-\rho)}.\end{aligned}$$

In addition,

$$\|y-\chi\|_0 - \frac{\Lambda_4^*}{(1-\Lambda_3^*)}\|x-\zeta\|_0 \leqslant \frac{\mathcal{Q}_\beta\epsilon_\beta}{(1-\Lambda_3^*)}, \tag{48}$$

where

$$\begin{aligned}\Lambda_3^* =& \Bigg[\mu_2'\Bigg(\frac{2}{\Gamma(\beta+1)} + \frac{1}{\Gamma(\beta)} + \frac{1}{2\Gamma(\beta-1)} + \frac{\mathcal{Z}_\beta}{\Gamma(\beta-\rho+1)} + \frac{\mathcal{Z}_\beta\Phi^{1-\rho}}{\Gamma(\beta)\Gamma(2-\rho)} + \frac{\mathcal{Z}_\beta\Phi^{2-\rho}}{\Gamma(\beta-1)\Gamma(3-\rho)}\\ &+ \frac{\mathcal{Z}_\beta\Phi^{1-\rho}}{\Gamma(\beta-1)\Gamma(2-\rho)} + \frac{2\mathcal{Z}_\beta}{\Gamma(\beta-1)}\Bigg) + \sum_{j=1}^{m}I_{2j} + \sum_{j=1}^{m}\hat{I}_{2j} + \frac{\mathcal{Z}_\beta\Phi^{1-\rho}}{\Gamma(2-\rho)}\sum_{j=1}^{m}\hat{I}_{2j} + \frac{\mathcal{Z}_\beta\Phi^{2-\rho}}{\Gamma(3-\rho)}\sum_{j=1}^{m}\check{I}_{2j}\\ &+ \frac{\mathcal{Z}_\beta\Phi^{1-\rho}}{\Gamma(2-\rho)}\sum_{j=1}^{m}\check{I}_{2j} + \mathcal{Z}_\beta\sum_{j=1}^{m}\check{I}_{2j} + \frac{1}{2}\sum_{j=1}^{m}\check{I}_{2j}\Bigg],\\ \Lambda_4^* =& \mu_1'\Bigg(\frac{2}{\Gamma(\beta+1)} + \frac{1}{\Gamma(\beta)} + \frac{1}{2\Gamma(\beta-1)} + \frac{\mathcal{Z}_\beta}{\Gamma(\beta-\rho+1)} + \frac{\mathcal{Z}_\beta\Phi^{1-\rho}}{\Gamma(\beta)\Gamma(2-\rho)} + \frac{\mathcal{Z}_\beta\Phi^{2-\rho}}{\Gamma(\beta-1)\Gamma(3-\rho)}\\ &+ \frac{\mathcal{Z}_\beta\Phi^{1-\rho}}{\Gamma(\beta-1)\Gamma(2-\rho)} + \frac{2\mathcal{Z}_\beta}{\Gamma(\beta-1)}\Bigg).\end{aligned}$$

The equivalent matrix of Equations (44) and (48) is given as:

$$\begin{bmatrix} 1 & -\frac{\Lambda_2^*}{(1-\Lambda_1^*)} \\ -\frac{\Lambda_4^*}{(1-\Lambda_3^*)} & 1 \end{bmatrix}\begin{bmatrix} \|x-\zeta\|_0 \\ \|y-\chi\|_0 \end{bmatrix} \leqslant \begin{bmatrix} \frac{\mathcal{Q}_\alpha\epsilon_\alpha}{(1-\Lambda_1^*)} \\ \frac{\mathcal{Q}_\beta\epsilon_\beta}{(1-\Lambda_3^*)} \end{bmatrix}.$$

Solving the above inequality, we get

$$\begin{bmatrix} \|x-\zeta\|_0 \\ \\ \|y-\chi\|_0 \end{bmatrix} \leqslant \begin{bmatrix} \frac{1}{\Lambda_0} & \frac{\Lambda_2^*}{\Lambda_0(1-\Lambda_1^*)} \\ \\ \frac{\Lambda_4^*}{\Lambda_0(1-\Lambda_3^*)} & \frac{1}{\Lambda_0} \end{bmatrix} \begin{bmatrix} \frac{\Omega_\alpha \epsilon_\alpha}{(1-\Lambda_1^*)} \\ \\ \frac{\Omega_\beta \epsilon_\beta}{(1-\Lambda_3^*)} \end{bmatrix},$$

where

$$\Lambda_0 = 1 - \frac{\Lambda_2^* \Lambda_4^*}{(1-\Lambda_1^*)(1-\Lambda_3^*)} > 0.$$

Further simplification of the above system gives

$$\|x-\zeta\|_0 \leqslant \frac{\Omega_\alpha \epsilon_\alpha}{\Lambda_0(1-\Lambda_1^*)} + \frac{\Lambda_2^* \Omega_\beta \epsilon_\beta}{\Lambda_0(1-\Lambda_1^*)(1-\Lambda_3^*)},$$
$$\|y-\chi\|_0 \leqslant \frac{\Omega_\beta \epsilon_\beta}{\Lambda_0(1-\Lambda_3^*)} + \frac{\Lambda_4^* \Omega_\alpha \epsilon_\alpha}{\Lambda_0(1-\Lambda_1^*)(1-\Lambda_3^*)},$$

from which we have

$$\|x-\zeta\|_0 + \|y-\chi\|_0 \leqslant \frac{\Omega_\alpha \epsilon_\alpha}{\Lambda_0(1-\Lambda_1^*)} + \frac{\Omega_\beta \epsilon_\beta}{\Lambda_0(1-\Lambda_3^*)} + \frac{\Lambda_4^* \Omega_\alpha \epsilon_\alpha}{\Lambda_0(1-\Lambda_1^*)(1-\Lambda_3^*)} + \frac{\Lambda_2^* \Omega_\beta \epsilon_\beta}{\Lambda_0(1-\Lambda_1^*)(1-\Lambda_3^*)}. \quad (49)$$

Let $\max\{\epsilon_\alpha, \epsilon_\beta\} = \epsilon$; then, from (49), we get

$$\|(x-\zeta, y-\chi)\| \leqslant \mathbf{K}_{\alpha,\beta}\,\epsilon,$$

which implies that

$$\|(x,y)-(\zeta,\chi)\| \leqslant \mathbf{K}_{\alpha,\beta}\,\epsilon, \quad (50)$$

where

$$\mathbf{K}_{\alpha,\beta} = \left[\frac{\Omega_\alpha}{\Lambda_0(1-\Lambda_1^*)} + \frac{\Omega_\beta}{\Lambda_0(1-\Lambda_3^*)} + \frac{\Lambda_4^* \Omega_\alpha}{\Lambda_0(1-\Lambda_1^*)(1-\Lambda_3^*)} + \frac{\Lambda_2^* \Omega_\beta}{\Lambda_0(1-\Lambda_1^*)(1-\Lambda_3^*)}\right].$$

Hence, problem (1) is Ulam–Hyers stable. Moreover, if we set $\Theta(\epsilon) = \mathbf{K}_{x,y}\epsilon$; $\Theta(0) = 0$ in (50), then problem (1) is generalized Ulam–Hyers stable. □

$(\mathbf{M}_5)$: Let $\Psi_\alpha, \Psi_\beta \in PC(J, \mathbb{R}^+)$ be nondecreasing functions; then, for $t \in J$, there are $\lambda_\alpha, \lambda_\beta > 0$ such that

$$\mathbf{I}^\alpha \Psi_\alpha(t) \leqslant \lambda_\alpha \Psi_\alpha(t), \;\; \mathbf{I}^{\alpha-1}\Psi_\alpha(t) \leqslant \lambda_\alpha \Psi_\alpha(t),$$
$$\mathbf{I}^{\alpha-2}\Psi_\alpha(t) \leqslant \lambda_\alpha \Psi_\alpha(t), \;\; \mathbf{I}^{\alpha-\varepsilon}\Psi_\alpha(t) \leqslant \lambda_\alpha \Psi_\alpha(t).$$

Similarly,

$$\mathbf{I}^\beta \Psi_\beta(t) \leqslant \lambda_\beta \Psi_\beta(t), \;\; \mathbf{I}^{\beta-1}\Psi_\beta(t) \leqslant \lambda_\beta \Psi_\beta(t),$$
$$\mathbf{I}^{\beta-2}\Psi_\beta(t) \leqslant \lambda_\beta \Psi_\beta(t), \;\; \mathbf{I}^{\beta-\rho}\Psi_\beta(t) \leqslant \lambda_\beta \Psi_\beta(t).$$

Theorem 5. *Assume that $(\mathbf{M}_1) - (\mathbf{M}_2)$ and $(\mathbf{M}_5)$ are satisfied; then, by Definition 6 and Definition 7, Problem (1) is Ulam–Hyers–Rassias stable with respect to $(\Psi_{\alpha,\beta}, \varphi_{\alpha,\beta})$, as well as generalized Ulam–Hyers–Rassias stable.*

5. Example

To substantiate the aforemention demonstrated theory, we supply the following problem:

$$\begin{cases} {}^{c}\mathbf{D}^{\alpha}x(t)+h(t,{}^{c}\mathbf{D}^{a}x(t),{}^{c}\mathbf{D}^{b}y(t))=0, t\neq t_m, \\ {}^{c}\mathbf{D}^{\beta}y(t)+w(t,{}^{c}\mathbf{D}^{a}x(t),{}^{c}\mathbf{D}^{b}y(t))=0, \\ \Delta x\mid_{t=t_m}=M_{1m}x(t_m),\ \Delta x'\mid_{t=t_m}=N_{1m}x(t_m),\ \Delta x''\mid_{t=t_m}=O_{1m}x(t_m), \\ \Delta y\mid_{t=t_m}=M_{2m}y(t_m),\ \Delta y'\mid_{t=t_m}=N_{2m}y(t_m),\ \Delta y''\mid_{t=t_m}=O_{2m}y(t_m), \\ x(t)='x(t)=0,\ {}^{c}\mathbf{D}^{\varepsilon}x(\Omega)=x''(1),\ y(0)=y'(0)=0,\ {}^{c}\mathbf{D}^{\rho}y(\Phi)=y''(1), \\ 0<\varepsilon,\Omega<1,\ 0<\rho,\Phi<1. \end{cases} \tag{51}$$

Take $J=[0,1]$, $\alpha=\frac{5}{2}$, $\beta=\sqrt{5}$, $\varepsilon=\frac{1}{2}$, $\Omega=\frac{1}{3}$, $a=\frac{1}{2}$, $b=\frac{1}{8}$, $t_1=\frac{1}{2}$, $\rho=\frac{1}{3}$, $\Phi=\frac{1}{5}$, $h(t,x,y)=e^t+\frac{|x|+|y|}{40(1+|x|+|y|)}$, $w(t,x,y)=\frac{\cos t+|x|+|y|}{40(1+|x|+|y|)}$, $M_{11}(x)=M_{21}(x)=\frac{|x|}{5+|x|}$, $N_{11}(x)=N_{21}(x)=\frac{|x|}{10+|x|}$, $O_{11}(x)=O_{21}(x)=\frac{|x|}{15+|x|}$. By direct computation, we have $\mu_1=\mu_2=\mu_1'=\mu_2'=\frac{1}{40}$, $I_{11}=I_{21}=\frac{1}{5}$, $\hat{I}_{11}=\hat{I}_{21}=\frac{1}{10}$, $\check{I}_{11}=\check{I}_{21}=\frac{1}{15}$,

$$\Bigg[(\mu_1+\mu_2)\bigg(\frac{2}{\Gamma(\alpha+1)}+\frac{1}{\Gamma(\alpha)}+\frac{1}{2\Gamma(\alpha-1)}+\frac{\mathcal{Z}_\alpha}{\Gamma(\alpha-\varepsilon+1)}+\frac{\mathcal{Z}_\alpha\Omega^{1-\varepsilon}}{\Gamma(\alpha)\Gamma(2-\varepsilon)}+\frac{\mathcal{Z}_\alpha\Omega^{2-\varepsilon}}{\Gamma(\alpha-1)\Gamma(3-\varepsilon)}$$
$$+\frac{\mathcal{Z}_\alpha\Omega^{1-\varepsilon}}{\Gamma(\alpha-1)\Gamma(2-\varepsilon)}+\frac{2\mathcal{Z}_\alpha}{\Gamma(\alpha-1)}\bigg)+\sum_{j=1}^{m}I_{1j}+\sum_{j=1}^{m}\hat{I}_{1j}+\frac{\mathcal{Z}_\alpha\Omega^{1-\varepsilon}}{\Gamma(2-\varepsilon)}\sum_{j=1}^{m}\hat{I}_{1j}+\frac{\mathcal{Z}_\alpha\Omega^{2-\varepsilon}}{\Gamma(3-\varepsilon)}\sum_{j=1}^{m}\check{I}_{1j}$$
$$+\frac{\mathcal{Z}_\alpha\Omega^{1-\varepsilon}}{\Gamma(2-\varepsilon)}\sum_{j=1}^{m}\check{I}_{1j}+\mathcal{Z}_\alpha\sum_{j=1}^{m}\check{I}_{1j}+\frac{1}{2}\sum_{j=1}^{m}\check{I}_{1j}\Bigg]\approx 0.680277<1.$$

Similarly,

$$\Bigg[(\mu_1'+\mu_2')\bigg(\frac{2}{\Gamma(\beta+1)}+\frac{1}{\Gamma(\beta)}+\frac{1}{2\Gamma(\beta-1)}+\frac{\mathcal{Z}_\beta}{\Gamma(\beta-\rho+1)}+\frac{\mathcal{Z}_\beta\Phi^{1-\rho}}{\Gamma(\beta)\Gamma(2-\rho)}+\frac{\mathcal{Z}_\beta\Phi^{2-\rho}}{\Gamma(\beta-1)\Gamma(3-\rho)}$$
$$+\frac{\mathcal{Z}_\beta\Phi^{1-\rho}}{\Gamma(\beta-1)\Gamma(2-\rho)}+\frac{2\mathcal{Z}_\beta}{\Gamma(\beta-1)}\bigg)+\sum_{j=1}^{m}I_{2j}+\sum_{j=1}^{m}\hat{I}_{2j}+\frac{\mathcal{Z}_\beta\Phi^{1-\rho}}{\Gamma(2-\rho)}\sum_{j=1}^{m}\hat{I}_{2j}+\frac{\mathcal{Z}_\beta\Phi^{2-\rho}}{\Gamma(3-\rho)}\sum_{j=1}^{m}\check{I}_{2j}$$
$$+\frac{\mathcal{Z}_\beta\Phi^{1-\rho}}{\Gamma(2-\rho)}\sum_{j=1}^{m}\check{I}_{2j}+\mathcal{Z}_\beta\sum_{j=1}^{m}\check{I}_{2j}+\frac{1}{2}\sum_{j=1}^{m}\check{I}_{2j}\Bigg]\approx 0.427420<1.$$

Thus, by Theorem 2, Problem (51) has a unique solution. Furthermore, all of the assumptions are satisfied, so the Problem (51) is Ulam–Hyers, generalized Ulam–Hyers, Ulam–Hyers–Rassias and generalized Ulam–Hyers–Rassias stable.

6. Conclusions

In the above study, we have successfully built up existence theory for the solutions of system (1). The required analysis has been developed with the help of the Banach contraction principle and Schauder fixed point theorem. We found that the fractional order coupled system is additionally complicated and challenging as compared to the single FDEs. We also concluded that, if we increase the order or boundary conditions, then the end result turns into extra accurate. Our results are new and fascinating. Our methods can be used to study the existence of solutions for the high order or multiple-point boundary value systems of a nonlinear coupled system of FDEs. Furthermore, we have presented different kinds of Ulam–Hyers stability results for the solution of the considered system (1). In addition, we have presented our main theoretical results with the help of an example. In the future, this concept can be extended to more applied and complicated problems of applied nature. The obtained results can be used in fields like numerical analysis and managerial sciences including business mathematics and economics, etc.

Author Contributions: S.F. and Z.A. contributed equally in writing this article; supervision, A.Z., J.X. and Y.C. All authors read and approved the final manuscript.

Funding: This work is supported by the Talent Project of Chongqing Normal University (Grant No. 02030307-0040), the National Natural Science Foundation of China (Grant No. 11601048, 11571207), the Natural Science Foundation of Chongqing Normal University (Grant No. 16XYY24), the Shandong Natural Science Foundation (ZR2018MA011), and the Tai'shan Scholar Engineering Construction Fund of Shandong Province of China.

Conflicts of Interest: The authors declare that they have no competing interests.

References

1. Brikaa, M. Existence results for a couple system of nonlinear fractional differential equation with three point boundary conditions. *J. Fract. Calc. Appl.* **2015**, *3*, 1–10.
2. Henderson, J.; Luca, R. Positive solutions for a system of fractional differential equations with coupled integral boundary conditions. *Appl. Math. Comput.* **2014**, *249*, 182–197. [CrossRef]
3. Kilbas, A.; Srivastava, H.; Trujillo, J. Theory and Applications of Fractional Differential Equations. In *North–Holland Mathematics Studies*; Elsevier: Amsterdam, The Netherlands, 2006; Volume 204.
4. Klafter, J.; Lim, S.C. *Fractional Dynamics in Physics*; Metzler, R., Ed.; World Scientific: Singapore, 2011.
5. Podlubny, I. *Fractional Differential Equations*; Academic Press: San Diego, CA, USA, 1999.
6. Yang, W. Positive solutions for a coupled system of nonlinear fractional differential equations with integral boundary conditions. *Comput. Math. Appl.* **2012**, *63*, 288–297. [CrossRef]
7. Zhang, K.; Fu, Z. Solutions for a class of Hadamard fractional boundary value problems with sign-changing nonlinearity. *J. Funct. Spaces* **2019**, *2019*, 9046472. [CrossRef]
8. Zhang, K.; Wang, J.; Ma, W. Solutions for integral boundary value problems of nonlinear Hadamard fractional differential equations. *J. Funct. Spaces* **2018**, *2018*, 2193234. [CrossRef]
9. Zou, Y.; He, G. On the uniqueness of solutions for a class of fractional differential equations. *Appl. Math. Lett.* **2017**, *74*, 68–73. [CrossRef]
10. Yue, Z.; Zou, Y. New uniqueness results for fractional differential equation with dependence on the first order derivative. *Adv. Differ. Equ.* **2019**, *2019*, 38. [CrossRef]
11. Fu, Z.; Bai, S.; O'Regan, D.; Xu, J. Nontrivial solutions for an integral boundary value problem involving Riemann–Liouville fractional derivatives. *J. Inequal. Appl.* **2019**, *2019*, 104. [CrossRef]
12. Zhang, K.; O'Regan, D.; Xu, J.; Fu, Z. Nontrivial solutions for a higher order nonlinear fractional boundary value problem involving Riemann–Liouville fractional derivatives. *J. Funct. Spaces* **2019**, *2019*, 2381530. [CrossRef]
13. Pu, R.; Zhang, X.; Cui, Y.; Li, P.; Wang, W. Positive solutions for singular semipositone fractional differential equation subject to multipoint boundary conditions. *J. Funct. Spaces* **2017**, *2017*, 5892616. [CrossRef]
14. Zhang, Y. Existence results for a coupled system of nonlinear fractional multi-point boundary value problems at resonance. *J. Inequal. Appl.* **2018**, *2018*, 198. [CrossRef] [PubMed]
15. Zhang, Y.; Bai, Z.; Feng, T. Existence results for a coupled system of nonlinear fractional three-point boundary value problems at resonance. *Comput. Math. Appl.* **2011**, *61*, 1032–1047. [CrossRef]
16. Qi, T.; Liu, Y.; Zou, Y. Existence result for a class of coupled fractional differential systems with integral boundary value conditions. *J. Nonlinear Sci. Appl.* **2017**, *10*, 4034–4045. [CrossRef]
17. Qi, T.; Liu, Y.; Cui, Y. Existence of solutions for a class of coupled fractional differential systems with nonlocal boundary conditions. *J. Funct. Spaces* **2017**, *2017*, 6703860. [CrossRef]
18. Xu, J.; Goodrich, C.S.; Cui, Y. Positive solutions for a system of first-order discrete fractional boundary value problems with semipositone nonlinearities. *Revista de la Real Academia de Ciencias Exactas Físicas y Naturales Serie A Matemáticas* **2019**, *113*, 1343–1358. [CrossRef]
19. Cheng, W.; Xu, J.; Cui, Y.; Ge, Q. Positive solutions for a class of fractional difference systems with coupled boundary conditions. *Adv. Differ. Equ.* **2019**, *2019*, 249. [CrossRef]
20. Cheng, W.; Xu, J.; Cui, Y. Positive solutions for a system of nonlinear semipositone fractional q-difference equations with q-integral boundary conditions. *J. Nonlinear Sci. Appl.* **2017**, *10*, 4430–4440. [CrossRef]

21. Wang, F.; Cui, Y. Positive solutions for an infinite system of fractional order boundary value problems. *Adv. Differ. Equ.* **2019**, *2019*, 169. [CrossRef]
22. Qiu, X.; Xu, J.; O'Regan, D.; Cui, Y. Positive solutions for a system of nonlinear semipositone boundary value problems with Riemann–Liouville fractional derivatives. *J. Funct. Spaces* **2018**, *2018*, 7351653. [CrossRef]
23. Hao, X.; Wang, H.; Liu, L.; Cui, Y. Positive solutions for a system of nonlinear fractional nonlocal boundary value problems with parameters and p-Laplacian operator. *Bound. Value Probl.* **2017**, *2017*, 182. [CrossRef]
24. Zhang, X.; Liu, L.; Zou, Y. Fixed-point theorems for systems of operator equations and their applications to the fractional differential equations. *J. Funct. Spaces* **2018**, *2018*, 7469868. [CrossRef]
25. Zhang, X.; Liu, L.; Wu, Y.; Zou, Y. Existence and uniqueness of solutions for systems of fractional differential equations with Riemann-Stieltjes integral boundary condition. *Adv. Differ. Equ.* **2018**, *2018*, 204. [CrossRef]
26. Li, H.; Zhang, J. Positive solutions for a system of fractional differential equations with two parameters. *J. Funct. Spaces* **2018**, *2018*, 1462505. [CrossRef]
27. Zhao, Y.; Hou, X.; Sun, Y.; Bai, Z. Solvability for some class of multi-order nonlinear fractional systems. *Adv. Differ. Equ.* **2019**, *2019*, 23. [CrossRef]
28. Zhao, Y.; Sun, Y.; Wang, Y.; Bai, Z. Asymptotical stabilization of the nonlinear upper triangular fractional-order systems. *Adv. Differ. Equ.* **2019**, *2019*, 157. [CrossRef]
29. Cui, Y. Multiplicity results for positive solutions to differential systems of singular coupled integral boundary value problems. *Math. Probl. Eng.* **2017**, *2017*, 3608352. [CrossRef]
30. Jiang, J.; O'Regan, D.; Xu, J.; Fu, Z. Positive solutions for a system of nonlinear Hadamard fractional differential equations involving coupled integral boundary conditions. *J. Inequal. Appl.* **2019**, *2019*, 204. [CrossRef]
31. Zhai, C.; Wang, W.; Li, H. A uniqueness method to a new Hadamard fractional differential system with four-point boundary conditions. *J. Inequal. Appl.* **2018**, *2018*, 207. [CrossRef]
32. Wang, J.; Fečkan, M.; Zhou, Y. A survey on impulsive fractional differential equations. *Fract. Calc. Appl. Anal.* **2016**, *19*, 806–831. [CrossRef]
33. Yukunthorn, W.; Ahmad, B.; Ntouyas, S.; Tariboon, J. On Caputo-Hadamard type fractional impulsive hybrid systems with nonlinear fractional integral conditions. *Nonlinear Anal. Hybrid Syst.* **2016**, *19*, 77–92. [CrossRef]
34. Zhao, K.; Liang, J. Solvability of triple-point integral boundary value problems for a class of impulsive fractional differential equations. *Adv. Differ. Equ.* **2017**, *2017*, 50. [CrossRef]
35. Fu, X.; Bao, X. Some existence results for nonlinear fractional differential equations with impulsive and fractional integral boundary conditions. *Adv. Differ. Equ.* **2014**, *2014*, 129. [CrossRef]
36. Zhao, K. Multiple positive solutions of integral BVPs for high-order nonlinear fractional differential equations with impulses and distributed delays. *Dyn. Syst.* **2015**, *30*, 208–223. [CrossRef]
37. Zhao, K. Impulsive boundary value problems for two classes of fractional differential equation with two different Caputo fractional derivatives. *Mediterr. J. Math.* **2016**, *13*, 1033–1050. [CrossRef]
38. Wang, Y.; Liu, Y.; Cui, Y. Infinitely many solutions for impulsive fractional boundary value problem with p-Laplacian. *Bound. Value Probl.* **2018**, *2018*, 94. [CrossRef]
39. Zuo, M.; Hao, X.; Liu, L.; Cui, Y. Existence results for impulsive fractional integro-differential equation of mixed type with constant coefficient and antiperiodic boundary conditions. *Bound. Value Probl.* **2017**, *2017*, 161. [CrossRef]
40. Bai, Z.; Dong, X.; Yin, C. Existence results for impulsive nonlinear fractional differential equation with mixed boundary conditions. *Bound. Value Probl.* **2016**, *2016*, 63. [CrossRef]
41. Ulam, S.M. *Problems in Modern Mathematics*; John Wiley and Sons: New York, NY, USA, 1940.
42. Ulam, S.M. *A Collection of Mathematical Problems*; Interscience: New York, NY, USA, 1960.
43. Rassias, T.M. On the stability of the linear mapping in Banach spaces. *Proc. Am. Math. Soc.* **1978**, *72*, 297–300. [CrossRef]
44. Ali, Z.; Zada, A.; Shah, K. Ulam stability to a toppled systems of nonlinear implicit fractional order boundary value problem. *Bound. Value Probl.* **2018**, *2018*, 175. [CrossRef]
45. Ali, Z.; Zada, A.; Shah, K. On Ulam's stability for a coupled systems of nonlinear implicit fractional differential equations. *Bull. Malays. Math. Sci. Soc.* **2019**, *42*, 2681–2699. [CrossRef]
46. Ahmad, N.; Ali, Z.; Shah, K.; Zada, A.; Rahman, G. Analysis of implicit type nonlinear dynamical problem of impulsive fractional differential equations. *Complexity* **2018**, *2018*, 6423974. [CrossRef]

47. Zada, A.; Shah, S.O.; Shah, R. Hyers–Ulam stability of non-autonomous systems in terms of boundedness of Cauchy problem. *Appl. Math. Comput.* **2015**, *271*, 512–518. [CrossRef]
48. Shah, S.O.; Zada, A. Existence, uniqueness and stability of solution to mixed integral dynamic systems with instantaneous and noninstantaneous impulses on time scales. *Appl. Math. Comput.* **2019**, *359*, 202–213. [CrossRef]
49. Zada, A.; Mashal, A. Stability analysis of nth order nonlinear impulsive differential equations in Quasi-Banach space. *Numer. Funct. Anal. Optim.* **2019**. [CrossRef]
50. Ali, A.; Rabiei, F.; Shah, K. On Ulam's type stability for a class of impulsive fractional differential equations with nonlinear integral boundary conditions. *J. Nonlinear Sci. Appl.* **2017**, *10*, 4760–4775. [CrossRef]
51. Riaz, U.; Zada, A.; Ali, Z.; Ahmad, M.; Xu, J.; Fu, Z. Analysis of nonlinear coupled systems of impulsive fractional differential equations with Hadamard derivatives. *Math. Probl. Eng.* **2019**, *2019*, 5093572. [CrossRef]
52. Riaz, U.; Zada, A.; Ali, Z.; Cui, Y.; Xu, J. Analysis of coupled systems of implicit impulsive fractional differential equations involving Hadamard derivatives. *Adv. Differ. Equ.* **2019**, *2019*, 226. [CrossRef]
53. Ali, Z.; Kumam, P.; Shah, K.; Zada, A. Investigation of Ulam stability results of a coupled system of nonlinear implicit fractional differential equations. *Mathematics* **2019**, *7*, 341. [CrossRef]
54. Wang, J.; Zada, A.; Waheed, H. Stability analysis of a coupled system of nonlinear implicit fractional anti-periodic boundary value problem. *Math. Meth. App. Sci.* **2019**. [CrossRef]
55. Shah, K.; Shah, L.; Ahmad, S.; Rassias, J.M.; Li, Y. Monotone iterative techniques together with Hyers-Ulam-Rassias stability. *Math. Meth. Appl. Sci.* **2019**, 1–18. [CrossRef]
56. Zada, A.; Ali, W.; Park, C. Ulam's type stability of higher order nonlinear delay differential equations via integral inequality of Grönwall–Bellman–Bihari's type. *Appl. Math. Comput.* **2019**, *350*, 60–65.
57. Jung, S.M. *Hyers–Ulam–Rassias Stability of Functional Equations in Nonlinear Analysis*; Springer: New York, NY, USA, 2011.
58. Kumam, P.; Ali, A.; Shah, K.; Khan, R.A. Existence results and Hyers–Ulam stability to a class of nonlinear arbitrary order differential equations. *J. Nonlinear Sci. Appl.* **2017**, *10*, 2986–2997. [CrossRef]
59. Guo, D.; Lakshmikantham, V. *Nonlinear Problems in Abstract Cone*; Academic Press: Orlando, FL, USA, 1988.
60. Miller, K.; Ross, B. *An Introduction to the Fractional Calculus and Fractional Differential Equations*; Wiley: New York, NY, USA, 1993.
61. Rus, I.A. Ulam stabilities of ordinary differential equations in a Banachspace. *Carpathian J. Math.* **2010**, *26*, 103–107.

mathematics

MDPI

Article

A Note on Double Conformable Laplace Transform Method and Singular One Dimensional Conformable Pseudohyperbolic Equations

Hassan Eltayeb [1,*], Said Mesloub [1], Yahya T. Abdalla [2] and Adem Kılıçman [3]

[1] Mathematics Department, College of Science, King Saud University, P.O. Box 2455, Riyadh 11451, Saudi Arabia; mesloub@ksu.edu.sa
[2] Department of Mathematics, College of Science, Sudan University of Science and Technology, P.O. Box 407, Khartoum 11111, Sudan; amynt2005@gmail.com
[3] Department of Mathematics and Institute for Mathematical Research, Universiti Putra Malaysia, Serdang 43400 UPM, Selangor, Malaysia; akilic@upm.edu.my
* Correspondence: hgadain@ksu.edu.sa

Received: 5 September 2019; Accepted: 9 October 2019; Published: 12 October 2019

Abstract: The purpose of this article is to obtain the exact and approximate numerical solutions of linear and nonlinear singular conformable pseudohyperbolic equations and conformable coupled pseudohyperbolic equations through the conformable double Laplace decomposition method. Further, the numerical examples were provided in order to demonstrate the efficiency, high accuracy, and the simplicity of present method.

Keywords: conformable derivative; conformable partial derivative; conformable double Laplace decomposition method; conformable Laplace transform; singular one dimensional coupled Burgers' equation

1. Introduction

In recent years, many mathematicians have been studying and discussing the linear and nonlinear fractional differential equations (FDEs) which arise in various fields of physical sciences, as well as in engineering. These types of equations play a significant role and also help to develop mathematical tools in order to understand fractional modelling.

However, there are many different methods to obtain exact and approximate solutions of these kinds of equations. In [1], the author point out a major flaw in the so-called conformable calculus. Recently, many researchers have also paid much attention to study the numerical and exact methods for finding the solution of conformable differential equations. In [2], the authors proposed so-called conformable derivatives. In [3], the conformable heat equation was studied. Similarly, in [4], the nonlinear conformable problems were also studied. The authors in [5] discussed the concepts underlying the formulation of operators capable of being interpreted as fractional derivatives or fractional integrals. In a very short period of time, many mathematicians became interested and provided mathematical models related to conformable derivatives, for the details we refer reader to see [6–9]. In [10,11], the conformable derivatives were applied to some problems in mechanics, and in [12] total frational derivative and directional fractional derivative of functions of several variables were studied.

In order to solve the conformable derivatives, the single Laplace transform method was first introduced and used in [13]. In [14], the idea was extended to the conformable double Laplace transform. In [15], the modified Laplace transform was applied to solve some ordinary differential equations in the frame work of a certain type generalized fractional derivatives. The authors in [16] applied the double Laplace decomposition method to solve singular linear and nonlinear one-dimensional pseudohyperbolic equations.

In this present research, the main objective is to solve linear and nonlinear singular pseudohyperbolic equations by using the conformable double Laplace transform decomposition method, which is a combination of the conformable double Laplace transformation and decomposition method.

2. Properties of Conformable Derivative and Conformable Double Laplace Transform

In this part, we present some background about the nature of the conformable Laplace transform. In the following example, we present the conformable partial derivatives of certain functions as follows.

Example 1. *Let $\mu, \nu \in (0,1]$ and $a, b, m, n, \lambda, \mu \in R$, then the conformable derivative follows*

$$\begin{aligned}
\frac{\partial^{\mu}}{\partial x^{\mu}}\left(au\left(x,t\right)+bv\left(x,t\right)\right) &= a\frac{\partial^{\mu}u}{\partial x^{\mu}}+b\frac{\partial^{\mu}u}{\partial x^{\mu}},\\
\frac{\partial^{\mu+\nu}}{\partial x^{\mu}\partial t^{\nu}}\left(x^{\mu}t^{\lambda}\right) &= \mu\lambda x^{\kappa-\mu}t^{\lambda-\nu},\\
\frac{\partial^{\mu}}{\partial x^{\mu}}\left(e^{\lambda\frac{x^{\mu}}{\mu}+\frac{\tau t^{\nu}}{\nu}}\right) &= \lambda e^{\lambda\frac{x^{\mu}}{\mu}+\frac{\tau t^{\nu}}{\nu}},\\
\frac{\partial^{\nu}}{\partial t^{\nu}}\left(e^{\lambda\frac{x^{\mu}}{\mu}+\frac{\tau t^{\nu}}{\nu}}\right) &= \tau e^{\lambda\frac{x^{\mu}}{\mu}+\frac{\tau t^{\nu}}{\nu}},\\
\frac{\partial^{\nu}}{\partial t^{\nu}}\left(\frac{x^{\mu}}{\mu}\right)^{n}\left(\frac{t^{\nu}}{\nu}\right)^{m} &= m\left(\frac{x^{\mu}}{\mu}\right)^{n}\left(\frac{t^{\nu}}{\nu}\right)^{m-1},\\
\frac{\partial^{\mu}}{\partial x^{\mu}}\left(\frac{x^{\mu}}{\mu}\right)^{n}\left(\frac{t^{\nu}}{\nu}\right) &= n\left(\frac{x^{\mu}}{\mu}\right)^{n-1}\left(\frac{t^{\nu}}{\nu}\right),\\
\frac{\partial^{\nu}}{\partial t^{\nu}}\left(\sin\left(\frac{x^{\mu}}{\mu}\right)\sin\left(\frac{t^{\nu}}{\nu}\right)\right) &= \sin\left(\frac{x^{\mu}}{\mu}\right)\cos\left(\frac{t^{\nu}}{\nu}\right),\\
\frac{\partial^{\mu}}{\partial x^{\mu}}\left(\sin a\left(\frac{x^{\mu}}{\mu}\right)\sin\left(\frac{t^{\nu}}{\nu}\right)\right) &= a\cos\left(\frac{x^{\mu}}{\mu}\right)\sin\left(\frac{t^{\nu}}{\nu}\right).
\end{aligned}$$

Next we recall the conformable single and double Laplace transforms, see [14,17], repectively.

Definition 1. *Let $f : [0,\infty) \to \mathbb{R}$ be a real valued function. The conformable single Laplace transform of f is defined by*

$$L_t^{\nu}\left(f\left(\frac{t^{\nu}}{\nu}\right)\right)=\int_0^{\infty}e^{-s\frac{t^{\nu}}{\nu}}f\left(\frac{t^{\nu}}{\nu}\right)t^{\nu-1}dt.$$

Similarly, if we let $u\left(\frac{x^{\mu}}{\mu},\frac{t^{\nu}}{\nu}\right)$ be a piecewise continuous function on $[0,\infty)\times[0,\infty)$ of exponential order and for some $a, b \in \mathbb{R}$,

$$\sup\left\{\frac{x^{\mu}}{\mu},\frac{t^{\nu}}{\nu}\right\}>0\text{ , and }\frac{\left|u\left(\frac{x^{\mu}}{\mu},\frac{t^{\nu}}{\nu}\right)\right|}{e^{a\frac{x^{\mu}}{\mu}+b\frac{t^{\nu}}{\nu}}}\leq 1.$$

Then the conformable double Laplace transform is defined by

$$L_x^{\mu}L_t^{\nu}\left(u\left(\frac{x^{\mu}}{\mu},\frac{t^{\nu}}{\nu}\right)\right)=F_{\mu,\nu}\left(p,s\right)=\int_0^{\infty}\int_0^{\infty}e^{-p\frac{x^{\mu}}{\mu}-s\frac{t^{\nu}}{\nu}}u\left(\frac{x^{\mu}}{\mu},\frac{t^{\nu}}{\nu}\right)t^{\nu-1}x^{\mu-1}dtdx \tag{1}$$

where $p,s \in \mathbb{C}$, $0 < \mu,\nu \leq 1$ and integrals by means of conformable integrals with respect to x and t, respectively.

Further, the first and second order partial derivatives of the conformable double Laplace transform with respect to $\frac{x^\mu}{\mu}$ are given by

$$L_x^\mu L_t^\nu\left[\frac{\partial^\mu u}{\partial x^\mu}\right] = pU_{\mu,\nu}(p,s) - U_\nu(0,s), \quad (2)$$

$$L_x^\mu L_t^\nu\left(\frac{\partial^{2\mu} u}{\partial x^{2\mu}}\right) = p^2U_{\mu,\nu}(p,s) - pU_\nu(0,s) - L_t^\nu\left(\frac{\partial^\mu}{\partial x^\mu}u\left(0,\frac{t^\nu}{\nu}\right)\right). \quad (3)$$

Similarly, with respect to $\frac{t^\nu}{\nu}$ they are given by

$$L_x^\mu L_t^\nu\left(\frac{\partial^\nu u}{\partial t^\nu}\right) = sU_{\mu,\nu}(p,s) - U_\mu(p,0), \quad (4)$$

$$L_x^\mu L_t^\nu\left(\frac{\partial^{2\nu} u}{\partial t^{2\nu}}\right) = s^2U_{\mu,\nu}(p,s) - sU_\mu(p,0) - L_x^\mu\left(\frac{\partial^\nu}{\partial t^\nu}u\left(\frac{x^\mu}{\mu},0\right)\right). \quad (5)$$

In the following examples we state some conformable Laplace transforms of certain functions which are useful in this to Examples 3, 4, and 5.

Example 2. *In this example we calculate the conformable double Laplace for certain functions*

1. $L_x^\mu\left[(\frac{x^\mu}{\mu})^2\right] = L_x\left[x^2\right] = \frac{2!}{p^3}$.
2. $L_x^\mu L_t^\nu\left[(\frac{x^\mu}{\mu})^3\sin(\tau\frac{t^\nu}{\nu})\right] = L_xL_t\left[(x)^3\sin(t)\right] = \frac{3!}{p^4}\frac{1}{s^2+1}$.
3. $L_x^\mu L_t^\nu\left[(\frac{x^\mu}{\mu})\cos(\frac{t^\nu}{\nu})\right] = L_xL_t\left[(x\cos(t)\right] = \frac{1}{p^2}\frac{s}{s^2+1}$.
4. $L_x^\mu L_t^\nu\left[4\left(\frac{t^\nu}{\nu}\right) - 4\sin\left(\frac{t^\nu}{\nu}\right)\right] = L_xL_t\left[(4t - 4\sin(t)\right] = \frac{4}{ps^2(s^2+1)}$.
5. $L_x^\mu L_t^\nu\left[(4 - 4\cos\left(\frac{t^\nu}{\nu}\right)\right] = L_xL_t\left[(4 - 4\cos(t)\right] = \frac{4s}{ps^2(s^2+1)}$.

The next result generalizes the conformable double Laplace transform, see [14].

Theorem 1. *Let* $0 < \mu,\nu \leq 1$ *and* $m,n \in \mathbb{N}$ *such that* $u\left(\frac{x^\mu}{\mu},\frac{t^\nu}{\nu}\right) \in C^l(\mathbb{R}^+\times\mathbb{R}^+)$, *and* $l = \max(m,n)$. *Further, we also let the conformable Laplace transforms of the functions be denoted by* $u\left(\frac{x^\mu}{\mu},\frac{t^\nu}{\nu}\right)$, $\frac{\partial^{m\mu}u}{\partial x^{m\mu}}$, and $\frac{\partial^{n\nu}u}{\partial t^{n\nu}}$. *Then*

$$L_x^\mu L_t^\nu\left(\frac{\partial^{m\mu}}{\partial x^{m\mu}}u\left(\frac{x^\mu}{\mu},\frac{t^\nu}{\nu}\right)\right) = p^mU_{\mu,\nu}(p,s) - p^{m-1}U_\nu(0,s) - \sum_{i=1}^{m-1}p^{m-1-i}L_t^\nu\left(\frac{\partial^{i\mu}}{\partial x^{i\mu}}u\left(0,\frac{t^\nu}{\nu}\right)\right)$$

$$L_x^\mu L_t^\nu\left(\frac{\partial^{n\nu}}{\partial t^{n\nu}}u\left(\frac{x^\mu}{\mu},\frac{t^\nu}{\nu}\right)\right) = s^nU_{\mu,\nu}(p,s) - s^{n-1}U_\mu(p,0) - \sum_{j=1}^{n-1}s^{n-1-j}L_x^\mu\left(\frac{\partial^{j\nu}}{\partial t^{j\nu}}u\left(\frac{x^\mu}{\mu},0\right)\right)$$

where

$$\frac{\partial^{m\mu}}{\partial x^{m\mu}}u\left(\frac{x^\mu}{\mu},\frac{t^\nu}{\nu}\right) \text{ and } \frac{\partial^{n\nu}}{\partial t^{n\nu}}u\left(\frac{x^\mu}{\mu},\frac{t^\nu}{\nu}\right)$$

denotes m,n *times conformable derivatives of function* $u(x,t)$ *respectively.*

Theorem 2. *If the conformable double Laplace transform of the conformable derivatives $\frac{\partial^{\nu} u}{\partial t^{\nu}}$ is given by Equation (4), then the double Laplace transforms of*

$$\left(\frac{x^{\mu}}{\mu}\right)^{n} \frac{\partial^{\nu}}{\partial t^{\nu}} f\left(\frac{x^{\mu}}{\mu}, \frac{t^{\nu}}{\nu}\right) \text{ and } \frac{x^{\mu}}{\mu} g\left(\frac{x^{\mu}}{\mu}, \frac{t^{\nu}}{\nu}\right)$$

are given by

$$(-1)^{n} \frac{d^{n}}{dp^{n}}\left(L_{x}^{\mu} L_{t}^{\nu}\left[f\left(\frac{x^{\mu}}{\mu}, \frac{t^{\nu}}{\nu}\right)\right]\right) = L_{x}^{\mu} L_{t}^{\nu}\left[\left(\frac{x^{\mu}}{\mu}\right)^{n} f\left(\frac{x^{\mu}}{\mu}, \frac{t^{\nu}}{\nu}\right)\right], \tag{6}$$

$$(-1)^{n} \frac{d^{n}}{dp^{n}}\left(L_{x}^{\mu} L_{t}^{\nu}\left[\frac{\partial^{\nu} f}{\partial t^{\nu}}\right]\right) = L_{x}^{\mu} L_{t}^{\nu}\left[\left(\frac{x^{\mu}}{\mu}\right)^{n} \frac{\partial^{\nu} f}{\partial t^{\nu}}\right] \tag{7}$$

and where $n = 1, 2, 3, \ldots$.

Proof. Using the definition of conformable double Laplace transform for Equation (6), we get

$$L_{x}^{\mu} L_{t}^{\nu}\left[f(\frac{x^{\mu}}{\mu}, \frac{t^{\nu}}{\nu})\right] = \int_{0}^{\infty} \int_{0}^{\infty} e^{-p\frac{x^{\mu}}{\mu} - s\frac{t^{\nu}}{\nu}} f\left(\frac{x^{\mu}}{\mu}, \frac{t^{\nu}}{\nu}\right) t^{\nu-1} x^{\mu-1} dt\, dx, \tag{8}$$

by taking the ^{n}th derivative with respect to p for both sides of Equation (8), we have

$$\begin{aligned}\frac{d^{n}}{dp^{n}}\left(L_{x}^{\mu} L_{t}^{\nu}\left[f(\frac{x^{\mu}}{\mu}, \frac{t^{\nu}}{\nu})\right]\right) &= \int_{0}^{\infty} \int_{0}^{\infty} \frac{d^{n}}{dp^{n}}\left(e^{-p\frac{x^{\mu}}{\mu} - s\frac{t^{\nu}}{\nu}} f\left(\frac{x^{\mu}}{\mu}, \frac{t^{\nu}}{\nu}\right)\right) t^{\nu-1} x^{\mu-1} dt dx \\ &= (-1)^{n} \int_{0}^{\infty} \int_{0}^{\infty} \left(\frac{x^{\mu}}{\mu}\right)^{n} e^{-p\frac{x^{\mu}}{\mu} - s\frac{t^{\nu}}{\nu}} t^{\nu-1} x^{\mu-1} f\left(\frac{x^{\mu}}{\mu}, \frac{t^{\nu}}{\nu}\right) dt\, dx \\ &= (-1)^{n} L_{x}^{\mu} L_{t}^{\nu}\left[\left(\frac{x^{\mu}}{\mu}\right)^{n} f\left(\frac{x^{\mu}}{\mu}, \frac{t^{\nu}}{\nu}\right)\right],\end{aligned}$$

and further we obtain

$$(-1)^{n} \frac{d^{n}}{dp^{n}}\left(L_{x}^{\mu} L_{t}^{\nu}\left[f\left(\frac{x^{\mu}}{\mu}, \frac{t^{\nu}}{\nu}\right)\right]\right) = L_{x}^{\mu} L_{t}^{\nu}\left[\left(\frac{x^{\mu}}{\mu}\right)^{n} f\left(\frac{x^{\mu}}{\mu}, \frac{t^{\nu}}{\nu}\right)\right].$$

Similarly, we can prove the Equation (7). □

3. Conformable Derivatives Double Laplace Transform Decomposition Method Applied to Singular Pseudohyperbolic Equation

The main aim of this section is to discuss the applicability of the conformable double Laplace transform decomposition method (CDLDM) for the linear and nonlinear singular pseudohyperbolic equation. The pseudo-hyperbolic equations arise, for example, in the description of the electron diffusion processes in a plate, and they also arise in hydrodynamics in the study of fluid motion with an alternating viscosity. In this study we define the conformable double Laplace transform of the function $u\left(\frac{x^{\mu}}{\mu}, \frac{t^{\nu}}{\nu}\right)$ by $U_{\mu,\nu}(p, s)$. To illustrate the idea of our method, let us suggest here two important problems.

The first problem:

Consider the linear pseudohyperbolic equations

$$\frac{\partial^{2\nu} u}{\partial t^{2\nu}} - \frac{\mu}{x^{\mu}} \frac{\partial^{\mu}}{\partial x^{\mu}}\left(\frac{x^{\mu}}{\mu} \frac{\partial^{\mu} u}{\partial x^{\mu}}\right) - \frac{\mu}{x^{\mu}} \frac{\partial^{\mu+\nu}}{\partial x^{\mu} \partial t^{\nu}}\left(\frac{x^{\mu}}{\mu} \frac{\partial^{\mu} u}{\partial x^{\mu}}\right) = f\left(\frac{x^{\mu}}{\mu}, \frac{t^{\nu}}{\nu}\right), \quad x, t > 0 \tag{9}$$

subject to the condition

$$u(x,0)=f_1\left(\frac{x^\mu}{\mu}\right),\quad \frac{\partial^\nu u(x,0)}{\partial t^\nu}=f_2\left(\frac{x^\mu}{\mu}\right), \tag{10}$$

where $f\left(\frac{x^\mu}{\mu},\frac{t^\nu}{\nu}\right)$, $f_1\left(\frac{x^\mu}{\mu}\right)$, and $f_2\left(\frac{x^\mu}{\mu}\right)$ are source term and initial conditions, respectively.

The method:

In order to obtain the solution of Equation (9) by using conformable double Laplace transform decomposition methods, we applying the following steps:

Step 1: Multiplying both sides of Equation (9) by the term $\frac{x^\mu}{\mu}$, we have

$$\frac{x^\mu}{\mu}\frac{\partial^{2\nu}u}{\partial t^{2\nu}}-\frac{\partial^\mu}{\partial x^\mu}\left(\frac{x^\mu}{\mu}\frac{\partial^\mu u}{\partial x^\mu}\right)-\frac{\partial^{\mu+\nu}}{\partial x^\mu\partial t^\nu}\left(\frac{x^\mu}{\mu}\frac{\partial^\mu u}{\partial x^\mu}\right)=\frac{x^\mu}{\mu}f\left(\frac{x^\mu}{\mu},\frac{t^\nu}{\nu}\right), \tag{11}$$

Step 2: Applying conformable double Laplace transform for Equation (11) we get

$$\begin{aligned} L_x^\mu L_t^\nu\left[\frac{x^\mu}{\mu}\frac{\partial^{2\nu}u}{\partial t^{2\nu}}\right] &= L_x^\mu L_t^\nu\left[\frac{\partial^\mu}{\partial x^\mu}\left(\frac{x^\mu}{\mu}\frac{\partial^\mu u}{\partial x^\mu}\right)+\frac{\partial^{\mu+\nu}}{\partial x^\mu\partial t^\nu}\left(\frac{x^\mu}{\mu}\frac{\partial^\mu u}{\partial x^\mu}\right)\right] \\ &\quad +L_x^\mu L_t^\nu\left[\frac{x^\mu}{\mu}f\left(\frac{x^\mu}{\mu},\frac{t^\nu}{\nu}\right)\right], \end{aligned} \tag{12}$$

Step 3: On using Equations (5)–(7), we obtain

$$-\frac{d}{dp}\left[s^2U_{\mu,\nu}(p,s)-sU_\mu(p,0)-\frac{\partial^\nu u(p,0)}{\partial t^\nu}\right]=L_x^\mu L_t^\nu[\Psi]-\frac{d}{dp}\left[L_x^\mu L_t^\nu\left(f\left(\frac{x^\mu}{\mu},\frac{t^\nu}{\nu}\right)\right)\right], \tag{13}$$

where the conformable Laplace transforms of $u\left(\frac{x^\mu}{\mu},0\right)$ and $\frac{\partial^\nu u\left(\frac{x^\mu}{\mu},0\right)}{\partial t^\nu}$ are denoted by

$$U_\mu(p,0)=F_1(p,0),\quad \frac{\partial^\nu u(p,0)}{\partial t^\nu}=F_2(p,0)$$

respectively, and

$$\Psi=\frac{\partial^\mu}{\partial x^\mu}\left(\frac{x^\mu}{\mu}\frac{\partial^\mu u}{\partial x^\mu}\right)+\frac{\partial^{\mu+\nu}}{\partial x^\mu\partial t^\nu}\left(\frac{x^\mu}{\mu}\frac{\partial^\mu u}{\partial x^\mu}\right),$$

using given initial condition Equation (13) becomes

$$\frac{d}{dp}\left[U_{\mu,\nu}(p,s)\right]=\frac{1}{s}\frac{d}{dp}F_1(p,0)+\frac{1}{s^2}\frac{d}{dp}F_2(p,0)-\frac{1}{s^2}L_x^\mu L_t^\nu[\Psi]+\frac{1}{s^2}\frac{d}{dp}F_{\mu,\nu}(p,s). \tag{14}$$

Step 4: By applying the integral for both sides of Equation (14), from 0 to p with respect to p, where p is transform of the variable $\frac{x^\mu}{\mu}$, we have

$$U_{\mu,\nu}(p,s)=\frac{F_1(p,0)}{s}+\frac{F_2(p,0)}{s^2}-\frac{1}{s^2}\int_0^p L_x^\mu L_t^\nu[\Psi]\,dp+\frac{1}{s^2}F_{\mu,\nu}(p,s), \tag{15}$$

where $F_{\mu,\nu}(p,s)$, $F_1(p,0)$, and $F_2(p,0)$ are conformable Laplace transforms of the functions $f\left(\frac{x^\mu}{\mu},\frac{t^\nu}{\nu}\right)$, $f_1\left(\frac{x^\mu}{\mu}\right)$, and $f_2\left(\frac{x^\mu}{\mu}\right)$, respectively.

Step 5: By taking the inverse conformable double Laplace transform for Equation (15), we can compute the solution $u\left(\frac{x^\mu}{\mu},\frac{t^\nu}{\nu}\right)$ as follows

$$u\left(\frac{x^\mu}{\mu},\frac{t^\nu}{\nu}\right)=f_1\left(\frac{x^\mu}{\mu}\right)+\frac{t^\nu}{\nu}f_2\left(\frac{x^\mu}{\mu}\right)+L_p^{-1}L_s^{-1}\left[\frac{1}{s^2}F_{\mu,\nu}(p,s)\right]-L_p^{-1}L_s^{-1}\left[\frac{1}{s^2}\int_0^p L_x^\mu L_t^\nu\left[\Psi\right]dp\right] \tag{16}$$

where $L_p^{-1}L_s^{-1}$ indicates the double inverse conformable derivatives double Laplace transform. Here, we assume that the double inverse Laplace transform with respect to p and s exists for each term in the right hand side of Equation (16).

Step 6: The conformable double Laplace transform decomposition method (CDLDM) defines the solutions $u(\frac{x^\mu}{\mu},\frac{t^\nu}{\nu})$ with the help of infinite series as:

$$u\left(\frac{x^\mu}{\mu},\frac{t^\nu}{\nu}\right)=\sum_{n=0}^{\infty}u_n\left(\frac{x^\mu}{\mu},\frac{t^\nu}{\nu}\right). \tag{17}$$

By substituting Equation (17) into Equation (16), we obtain

$$\begin{aligned}\sum_{n=0}^{\infty}u_n\left(\frac{x^\mu}{\mu},\frac{t^\nu}{\nu}\right) &= f_1\left(\frac{x^\mu}{\mu}\right)+\frac{t^\nu}{\nu}f_2\left(\frac{x^\mu}{\mu}\right)+L_p^{-1}L_s^{-1}\left[\frac{1}{s^2}F_{\mu,\nu}(p,s)\right]\\ &-L_p^{-1}L_s^{-1}\left[\frac{1}{s^2}\int_0^p L_x^\mu L_t^\nu\left[\frac{\partial^\mu}{\partial x^\mu}\left(\frac{x^\mu}{\mu}\frac{\partial^\mu}{\partial x^\mu}\left(\sum_{n=0}^{\infty}u_n\right)\right)\right]dp\right]\\ &-L_p^{-1}L_s^{-1}\left[\frac{1}{s^2}\int_0^p L_x^\mu L_t^\nu\left[\left(\frac{\partial^{\mu+\nu}}{\partial x^\mu\partial t^\nu}\left(\frac{x^\mu}{\mu}\frac{\partial^\mu}{\partial x^\mu}\sum_{n=0}^{\infty}u_n\right)\right)\right]dp\right].\end{aligned} \tag{18}$$

The zeroth component u_0, as suggested by Adomian method, is always identified by the given initial condition and the source term $L_p^{-1}L_s^{-1}\left[\frac{1}{s^2}F_{\mu,\nu}(p,s)\right]$, both of which are assumed to be known. Accordingly, we set

$$u_0\left(\frac{x^\mu}{\mu},\frac{t^\nu}{\nu}\right)=f_1\left(\frac{x^\mu}{\mu}\right)+\frac{t^\nu}{\nu}f_2\left(\frac{x^\mu}{\mu}\right)+L_p^{-1}L_s^{-1}\left[\frac{1}{s^2}F_{\mu,\nu}(p,s)\right].$$

The other components $u_{k+1}, k\geq 0$ are given by using the relation

$$\begin{aligned}u_{k+1}\left(\frac{x^\mu}{\mu},\frac{t^\nu}{\nu}\right) &= -L_p^{-1}L_s^{-1}\left[\frac{1}{s^2}\int_0^p L_x^\mu L_t^\nu\left[\frac{\partial^\mu}{\partial x^\mu}\left(\frac{x^\mu}{\mu}\frac{\partial^\mu u_k}{\partial x^\mu}\right)\right]dp\right]\\ &- L_p^{-1}L_s^{-1}\left[\frac{1}{s^2}\int_0^p L_x^\mu L_t^\nu\left[\frac{\partial^{\mu+\nu}}{\partial x^\mu\partial t^\nu}\left(\frac{x^\mu}{\mu}\frac{\partial^\mu u_k}{\partial x^\mu}\right)\right]dp\right],\end{aligned} \tag{19}$$

the first few components from the last recursive relation are, at $k=0$,

$$\begin{aligned}u_1\left(\frac{x^\mu}{\mu},\frac{t^\nu}{\nu}\right) &= -L_p^{-1}L_s^{-1}\left[\frac{1}{s^2}\int_0^p L_x^\mu L_t^\nu\left[\frac{\partial^\mu}{\partial x^\mu}\left(\frac{x^\mu}{\mu}\frac{\partial^\mu u_0}{\partial x^\mu}\right)\right]dp\right]\\ &- L_p^{-1}L_s^{-1}\left[\frac{1}{s^2}\int_0^p L_x^\mu L_t^\nu\left[\frac{\partial^{\mu+\nu}}{\partial x^\mu\partial t^\nu}\left(\frac{x^\mu}{\mu}\frac{\partial^\mu u_0}{\partial x^\mu}\right)\right]dp\right],\end{aligned}$$

at $k = 1$

$$\begin{aligned} u_2\left(\frac{x^\mu}{\mu},\frac{t^\nu}{\nu}\right) &= -L_p^{-1}L_s^{-1}\left[\frac{1}{s^2}\int_0^p L_x^\mu L_t^\nu\left[\frac{\partial^\mu}{\partial x^\mu}\left(\frac{x^\mu}{\mu}\frac{\partial^\mu u_1}{\partial x^\mu}\right)\right]dp\right] \\ &- L_p^{-1}L_s^{-1}\left[\frac{1}{s^2}\int_0^p L_x^\mu L_t^\nu\left[\frac{\partial^{\mu+\nu}}{\partial x^\mu \partial t^\nu}\left(\frac{x^\mu}{\mu}\frac{\partial^\mu u_1}{\partial x^\mu}\right)\right]dp\right], \end{aligned}$$

at $k = 2$

$$\begin{aligned} u_3\left(\frac{x^\mu}{\mu},\frac{t^\nu}{\nu}\right) &= -L_p^{-1}L_s^{-1}\left[\frac{1}{s^2}\int_0^p L_x^\mu L_t^\nu\left[\frac{\partial^\mu}{\partial x^\mu}\left(\frac{x^\mu}{\mu}\frac{\partial^\mu u_2}{\partial x^\mu}\right)\right]dp\right] \\ &- L_p^{-1}L_s^{-1}\left[\frac{1}{s^2}\int_0^p L_x^\mu L_t^\nu\left[\frac{\partial^{\mu+\nu}}{\partial x^\mu \partial t^\nu}\left(\frac{x^\mu}{\mu}\frac{\partial^\mu u_2}{\partial x^\mu}\right)\right]dp\right], \end{aligned}$$

etc. The important terms used in infinite series depend on the problems and may be three terms or four terms, etc.

In order to give a clear overview of this method, we present the following example:

Example 3. *Consider singular conformable derivatives in one dimensional pseudohyperbolic equations with the indicated initial condition*

$$\begin{aligned} \frac{\partial^{2\nu} u}{\partial t^{2\nu}} - \frac{\mu}{x^\mu}\frac{\partial^\mu}{\partial x^\mu}\left(\frac{x^\mu}{\mu}\frac{\partial^\mu u}{\partial x^\mu}\right) - \frac{\mu}{x^\mu}\frac{\partial^{\mu+\nu}}{\partial x^\mu \partial t^\nu}\left(\frac{x^\mu}{\mu}\frac{\partial^\mu u}{\partial x^\mu}\right) &= -\left(\frac{x^\mu}{\mu}\right)^2 \sin\left(\frac{t^\nu}{\nu}\right) \\ &-4\sin\left(\frac{t^\nu}{\nu}\right) - 4\cos\left(\frac{t^\nu}{\nu}\right), \end{aligned} \tag{20}$$

and

$$u\left(\frac{x^\mu}{\mu},0\right) = 0, \quad \frac{\partial u\left(\frac{x^\mu}{\mu},0\right)}{\partial t} = \left(\frac{x^\mu}{\mu}\right)^2. \tag{21}$$

By using the aforesaid method subject to the initial condition, we have

$$\begin{aligned} \frac{dU_{\mu,\nu}(p,s)}{dp} &= -\frac{1}{s^2}L_x^\mu L_t^\nu\left[\frac{\mu}{x^\mu}\frac{\partial^\mu}{\partial x^\mu}\left(\frac{x^\mu}{\mu}\frac{\partial^\mu u}{\partial x^\mu}\right) + \frac{\mu}{x^\mu}\frac{\partial^{\mu+\nu}}{\partial x^\mu \partial t^\nu}\left(\frac{x^\mu}{\mu}\frac{\partial^\mu u}{\partial x^\mu}\right)\right] \\ &+\frac{3!}{p^4 s^2(s^2+1)} + \frac{4}{p^2 s^2(s^2+1)} + \frac{4}{p^2 s(s^2+1)} - \frac{3!}{p^4 s^2}, \end{aligned} \tag{22}$$

taking the integral for Equation (22), from 0 to p with respect to p, we get

$$\begin{aligned} U_{\mu,\nu}(p,s) &= -\frac{1}{s^2}\int_0^p L_x^\mu L_t^\nu\left[\frac{\partial^\mu}{\partial x^\mu}\left(\frac{x^\mu}{\mu}\frac{\partial^\mu u}{\partial x^\mu}\right) + \frac{\partial^{\mu+\nu}}{\partial x^\mu \partial t^\nu}\left(\frac{x^\mu}{\mu}\frac{\partial^\mu u}{\partial x^\mu}\right)\right]dp \\ &-\frac{2}{p^3 s^2(s^2+1)} - \frac{4}{p s^2(s^2+1)} - \frac{4}{p s(s^2+1)} + \frac{2!}{p^3 s^2}. \end{aligned} \tag{23}$$

Employing the inverse conformable derivatives double Laplace transform to Equation (23), we get

$$\begin{aligned} u\left(\frac{x^\mu}{\mu},\frac{t^\nu}{\nu}\right) &= -L_p^{-1}L_s^{-1}\left[\frac{1}{s^2}\int_0^p L_x^\mu L_t^\nu\left[\frac{\partial^\mu}{\partial x^\mu}\left(\frac{x^\mu}{\mu}\frac{\partial^\mu u}{\partial x^\mu}\right) + \frac{\partial^{\mu+\nu}}{\partial x^\mu \partial t^\nu}\left(\frac{x^\mu}{\mu}\frac{\partial^\mu u}{\partial x^\mu}\right)\right]dp\right] \\ &+\left(\frac{x^\mu}{\mu}\right)^2 \sin\left(\frac{t^\nu}{\nu}\right) - 4\left(\frac{t^\nu}{\nu}\right) + 4\sin\left(\frac{t^\nu}{\nu}\right) + 4\cos\left(\frac{t^\nu}{\nu}\right) - 4, \end{aligned} \tag{24}$$

by substituting Equation (17) into Equation (24), we obtain:

$$\begin{aligned}
\sum_{n=0}^{\infty} u_n\left(\frac{x^\mu}{\mu},\frac{t^\nu}{\nu}\right) &= -L_p^{-1}L_s^{-1}\left[\frac{1}{s^2}\int_0^p L_x^\mu L_t^\nu\left[\frac{\partial^\mu}{\partial x^\mu}\left(\frac{x^\mu}{\mu}\frac{\partial^\mu}{\partial x^\mu}\sum_{n=0}^{\infty}u_n\right)\right]dp\right] \\
&\quad -L_p^{-1}L_s^{-1}\left[\frac{1}{s^2}\int_0^p L_x^\mu L_t^\nu\left[\frac{\partial^{\mu+\nu}}{\partial x^\mu \partial t^\nu}\left(\frac{x^\mu}{\mu}\frac{\partial^\mu}{\partial x^\mu}\sum_{n=0}^{\infty}u_n\right)\right]dp\right] \\
&\quad +\left(\frac{x^\mu}{\mu}\right)^2\sin\left(\frac{t^\nu}{\nu}\right)-4\left(\frac{t^\nu}{\nu}\right)+4\sin\left(\frac{t^\nu}{\nu}\right)+4\cos\left(\frac{t^\nu}{\nu}\right)-4.
\end{aligned}$$

By applying the conformable double Laplace transform decomposition method, we obtain

$$u_0=\left(\frac{x^\mu}{\mu}\right)^2\sin\left(\frac{t^\nu}{\nu}\right)-4\left(\frac{t^\nu}{\nu}\right)+4\sin\left(\frac{t^\nu}{\nu}\right)+4\cos\left(\frac{t^\nu}{\nu}\right)-4,$$

eventually, we have the general recursive relation, given by

$$\begin{aligned}
u_{k+1}\left(\frac{x^\mu}{\mu},\frac{t^\nu}{\nu}\right) &= -L_p^{-1}L_s^{-1}\left[\frac{1}{s^2}\int_0^p L_x^\mu L_t^\nu\left[\frac{\partial^\mu}{\partial x^\mu}\left(\frac{x^\mu}{\mu}\frac{\partial^\mu u_k}{\partial x^\mu}\right)\right]dp\right] \\
&\quad -L_p^{-1}L_s^{-1}\left[\frac{1}{s^2}\int_0^p L_x^\mu L_t^\nu\left[\frac{\partial^{\mu+\nu}}{\partial x^\mu \partial t^\nu}\left(\frac{x^\mu}{\mu}\frac{\partial^\mu u_k}{\partial x^\mu}\right)\right]dp\right],
\end{aligned}$$

where $k \geq 0$, therefore

$$\begin{aligned}
u_1 &= -L_p^{-1}L_s^{-1}\left[\frac{1}{s^2}\int_0^p L_x^\mu L_t^\nu\left[\frac{\partial^\mu}{\partial x^\mu}\left(\frac{x^\mu}{\mu}\frac{\partial^\mu}{\partial x^\mu}u_0\right)+\frac{\partial^{\mu+\nu}}{\partial x^\mu \partial t^\nu}\left(\frac{x^\mu}{\mu}\frac{\partial^\mu}{\partial x^\mu}u_0\right)\right]dp\right] \\
u_1 &= -L_p^{-1}L_s^{-1}\left[\frac{1}{s^2}\int_0^p L_x^\mu L_t^\nu\left[4\left(\frac{x^\mu}{\mu}\right)\sin\left(\frac{t^\nu}{\nu}\right)+4\left(\frac{x^\mu}{\mu}\right)\cos\left(\frac{t^\nu}{\nu}\right)\right]dp\right] \\
&= -L_p^{-1}L_s^{-1}\left[\frac{1}{s^2}\int_0^p\left[\frac{4}{p^2(s^2+1)}+\frac{4s}{p^2(s^2+1)}\right]dp\right], \\
u_1 &= L_p^{-1}L_s^{-1}\left[\frac{1}{s^2}\frac{4}{p(s^2+1)}+\frac{4s}{p(s^2+1)}\right]=L_s^{-1}\left[\frac{4}{s^2(s^2+1)}+\frac{4s}{s^2(s^2+1)}\right],
\end{aligned}$$

by using partial fractional and inverse Laplace transform with respect to s, we have

$$\begin{aligned}
u_1 &= L_s^{-1}\left[\frac{4}{s^2(s^2+1)}+\frac{4s}{s^2(s^2+1)}\right] \\
&= L_s^{-1}\left[\frac{4}{s^2}-\frac{4}{s^2+1}+\frac{4}{s}-\frac{4s}{s^2+1}\right] \\
&= 4\left(\frac{t^\nu}{\nu}\right)-4\sin\left(\frac{t^\nu}{\nu}\right)+4-4\cos\left(\frac{t^\nu}{\nu}\right),
\end{aligned}$$

and

$$\begin{aligned}
u_2 &= -L_p^{-1}L_s^{-1}\left[\frac{1}{s^2}\int_0^p L_x^\mu L_t^\nu\left[\frac{\partial^\mu}{\partial x^\mu}\left(\frac{x^\mu}{\mu}\frac{\partial^\mu}{\partial x^\mu}u_1\right)+\frac{\partial^{\mu+\nu}}{\partial x^\mu \partial t^\nu}\left(\frac{x^\mu}{\mu}\frac{\partial^\mu}{\partial x^\mu}u_1\right)\right]dp\right] \\
u_2 &= -L_p^{-1}L_s^{-1}\left[\frac{1}{s^2}\int_0^p L_x^\mu L_t^\nu\left[0+0\right]dp\right]=0.
\end{aligned}$$

In the view of the above equations, the series solution is given by

$$\begin{aligned}\sum_{n=0}^{\infty} u_n &= u_0+u_1+u_2+....\\ &= \left(\frac{x^{\mu}}{\mu}\right)^2 \sin\left(\frac{t^{\nu}}{\nu}\right)-4\left(\frac{t^{\nu}}{\nu}\right)+4\sin\left(\frac{t^{\nu}}{\nu}\right)+4\cos\left(\frac{t^{\nu}}{\nu}\right)-4\\ &\quad +4\left(\frac{t^{\nu}}{\nu}\right)-4\sin\left(\frac{t^{\nu}}{\nu}\right)+4-4\cos\left(\frac{t^{\nu}}{\nu}\right)+0+0+\dots\end{aligned}$$

Hence, the exact solution of Equation (20) is given by:

$$u\left(\frac{x^{\mu}}{\mu},\frac{t^{\nu}}{\nu}\right)=\left(\frac{x^{\mu}}{\mu}\right)^2 \sin\left(\frac{t^{\nu}}{\nu}\right).$$

Second problem: Consider the following general form of the nonlinear singular pseudohyperbolic equations in one dimension of the form:

$$\frac{\partial^{2\nu}u}{\partial t^{2\nu}}-\frac{\mu}{x^{\mu}}\frac{\partial^{\mu}}{\partial x^{\mu}}\left(\frac{x^{\mu}}{\mu}\frac{\partial^{\mu}u}{\partial x^{\mu}}\right)-\frac{\mu}{x^{\mu}}\frac{\partial^{\mu+\nu}}{\partial x^{\mu}\partial t^{\nu}}\left(\frac{x^{\mu}}{\mu}\frac{\partial^{\mu}u}{\partial x^{\mu}}\right)-a\left(\frac{x^{\mu}}{\mu}\right)u\frac{\partial^{\mu}u}{\partial x^{\mu}}+u^2=f\left(\frac{x^{\mu}}{\mu},\frac{t^{\nu}}{\nu}\right) \tag{25}$$

with initial condition

$$u\left(\frac{x^{\mu}}{\mu},0\right)=g_1\left(\frac{x^{\mu}}{\mu}\right),\quad \frac{\partial^{\nu}u\left(\frac{x^{\mu}}{\mu},0\right)}{\partial t^{\nu}}=g_2\left(\frac{x^{\mu}}{\mu}\right), \tag{26}$$

where the functions $a(\frac{x^{\mu}}{\mu})$ are arbitrary. In order to obtain the solution of Equation (25), we use the following steps:

First step: By multiplying Equation (25) by $\frac{x^{\mu}}{\mu}$ and taking conformable double Laplace transform, we have

$$\begin{aligned}L_x^{\mu}L_t^{\nu}\left[\frac{x^{\mu}}{\mu}\frac{\partial^{2\nu}u}{\partial t^{2\nu}}\right] &= L_x^{\mu}L_t^{\nu}\left[\frac{\partial^{\mu}}{\partial x^{\mu}}\left(\frac{x^{\mu}}{\mu}\frac{\partial^{\mu}u}{\partial x^{\mu}}\right)+\frac{\partial^{\mu+\nu}}{\partial x^{\mu}\partial t^{\nu}}\left(\frac{x^{\mu}}{\mu}\frac{\partial^{\mu}u}{\partial x^{\mu}}\right)\right]\\ &+L_x^{\mu}L_t^{\nu}\left[a(x)\frac{x^{\mu}}{\mu}\frac{x^{\mu}}{\mu}u\frac{\partial^{\mu}u}{\partial x^{\mu}}-\frac{x^{\mu}}{\mu}u^2\right]+L_x^{\mu}L_t^{\nu}\left[\frac{x^{\mu}}{\mu}f\left(\frac{x^{\mu}}{\mu},\frac{t^{\nu}}{\nu}\right)\right],\end{aligned} \tag{27}$$

where conformable Laplace transform of $u\left(\frac{x^{\mu}}{\mu},0\right)$ and $\frac{\partial^{\nu}u\left(\frac{x^{\mu}}{\mu},0\right)}{\partial t^{\nu}}$ are given by

$$U_{\mu}(p,0)=G_1(p,0),\quad \frac{\partial^{\nu}u(p,0)}{\partial t^{\nu}}=G_2(p,0). \tag{28}$$

Second step: Applying Equations (5)–(28) into Equation (27), one can get that

$$\frac{d}{dp}\left[U_{\mu,\nu}(p,s)\right]=\frac{1}{s}\frac{d}{dp}G_1(p,0)+\frac{1}{s^2}\frac{d}{dp}G_2(p,0)-\frac{1}{s^2}L_x^{\mu}L_t^{\nu}[\Phi]+\frac{1}{s^2}\frac{d}{dp}F_{\mu,\nu}(p,s), \tag{29}$$

where,

$$\Phi=\frac{\partial^{\mu}}{\partial x^{\mu}}\left(\frac{x^{\mu}}{\mu}\frac{\partial^{\mu}u}{\partial x^{\mu}}\right)+\frac{\partial^{\mu+\nu}}{\partial x^{\mu}\partial t^{\nu}}\left(\frac{x^{\mu}}{\mu}\frac{\partial^{\mu}u}{\partial x^{\mu}}\right)+a(x)\frac{x^{\mu}}{\mu}\frac{x^{\mu}}{\mu}u\frac{\partial^{\mu}u}{\partial x^{2\mu}}-\frac{x^{\mu}}{\mu}u^2.$$

Third step: By taking the integral for Equation (29), from 0 to p with respect to p, where p is a transform of $\frac{x^{\mu}}{\mu}$, we have

$$U_{\mu,\nu}(p,s) = \frac{G_1(p,0)}{s} + \frac{vG_2(p,0)}{s^2} - \frac{1}{s^2}\int_0^p L_x^{\mu} L_t^{\nu}[\Phi]\,dp + F_{\mu,\nu}(p,s). \tag{30}$$

Fourth step: Using CFDLDM, the solution can be written in the infinite series as in Equation (17). By using the inverse Laplace transformation to Equation (30), we obtain.

$$\begin{aligned} u\left(\frac{x^{\mu}}{\mu},\frac{t^{\nu}}{\nu}\right) &= g_1\left(\frac{x^{\mu}}{\mu}\right) + \frac{t^{\nu}}{\nu} g_2\left(\frac{x^{\mu}}{\mu}\right) + L_p^{-1}L_s^{-1}\left[\frac{1}{s^2}F_{\mu,\nu}(p,s)\right] \\ &\quad - L_p^{-1}L_s^{-1}\left[\frac{1}{s^2}\int_0^p L_x^{\mu}L_t^{\nu}[\Phi]\,dp\right] \end{aligned} \tag{31}$$

furthermore, the nonlinear terms $u\frac{\partial^{\mu}u}{\partial x^{\mu}}$ and u^2 can be defined by:

$$u^2 = N_1 = \sum_{n=0}^{\infty} A_n,\ u\frac{\partial^{\mu}u}{\partial x^{\mu}} = N_2 = \sum_{n=0}^{\infty} B_n. \tag{32}$$

We have a few terms of the Adomian polynomials for A_n and B_n that are denoted by

$$A_n = \frac{1}{n!}\left(\frac{d^n}{d\lambda^n}\left[N_1\sum_{i=0}^{\infty}(\lambda^n u_n)\right]\right)_{\lambda=0}, \tag{33}$$

and

$$B_n = \frac{1}{n!}\left(\frac{d^n}{d\lambda^n}\left[N_2\sum_{i=0}^{\infty}(\lambda^n u_n)\right]\right)_{\lambda=0}, \tag{34}$$

where $n = 0,1,2,\ldots$. By putting Equations (33)–(32) into Equation (31), we get

$$\begin{aligned} \sum_{n=0}^{\infty} u_n\left(\frac{x^{\mu}}{\mu},\frac{t^{\nu}}{\nu}\right) &= f_1\left(\frac{x^{\mu}}{\mu}\right) + \frac{t^{\nu}}{\nu} f_2\left(\frac{x^{\mu}}{\mu}\right) + L_p^{-1}L_s^{-1}\left[\frac{1}{s^2}F_{\mu,\nu}(p,s)\right] \\ &\quad - L_p^{-1}L_s^{-1}\left[\frac{1}{s^2}\int_0^p L_x^{\mu}L_t^{\nu}\left[\frac{\partial^{\mu}}{\partial x^{\mu}}\left(\frac{x^{\mu}}{\mu}\frac{\partial^{\mu}}{\partial x^{\mu}}\left(\sum_{n=0}^{\infty} u_n\right)\right)\right]dp\right] \\ &\quad - L_p^{-1}L_s^{-1}\left[\frac{1}{s^2}\int_0^p L_x^{\mu}L_t^{\nu}\left[\left(\frac{x^{\mu}}{\mu}\frac{\partial^{\mu+\nu}}{\partial x^{\mu}\partial t^{\nu}}\left(\sum_{n=0}^{\infty} u_n\left(\frac{x^{\mu}}{\mu},\frac{t^{\nu}}{\nu}\right)\right)\right)\right]dp\right] \\ &\quad - L_p^{-1}L_s^{-1}\left[\frac{1}{s^2}\int_0^p L_x^{\mu}L_t^{\nu}\left[\frac{x^{\mu}}{\mu}\left(\sum_{n=0}^{\infty} A_n\right)\right]dp\right] \\ &\quad + L_p^{-1}L_s^{-1}\left[\frac{1}{s^2}\int_0^p L_x^{\mu}L\left[a(x)\frac{x^{\mu}}{\mu}\sum_{n=0}^{\infty} B_n\right]dp\right], \end{aligned} \tag{35}$$

the few components of the Adomian polynomials of Equations (33) and (34) are given as follows

$$\begin{aligned} A_0 &= u_0^2, \\ A_1 &= 2u_0u_1, \\ A_2 &= 2u_0u_2 + u_1^2 \\ A_3 &= 2u_0u_3 + 2uu_2, \end{aligned} \tag{36}$$

and

$$\begin{aligned} B_0 &= u_0\frac{\partial^\mu u_0}{\partial x^\mu}, \\ B_1 &= u_0\frac{\partial^\mu u_1}{\partial x^\mu}+u_1\frac{\partial^\mu u_0}{\partial x^\mu}, \\ B_2 &= u_0\frac{\partial^\mu u_2}{\partial x^\mu}+u_1\frac{\partial^\mu u_1}{\partial x^\mu}+u_2\frac{\partial^\mu u_0}{\partial x^\mu}, \\ B_3 &= u_0\frac{\partial^\mu u_3}{\partial x^\mu}+u_1\frac{\partial^\mu u_2}{\partial x^\mu}+u_2\frac{\partial^\mu u_1}{\partial x^\mu}+u_3\frac{\partial^\mu u_0}{\partial x^\mu}. \end{aligned} \tag{37}$$

Hence, the zeroth component u_0 from Equation (35) is given by

$$u_0\left(\frac{x^\mu}{\mu},\frac{t^\nu}{\nu}\right)=f_1\left(\frac{x^\mu}{\mu}\right)+\frac{t^\nu}{\nu}f_2\left(\frac{x^\mu}{\mu}\right)+L_p^{-1}L_s^{-1}\left[\frac{1}{s^2}F_{\mu,\nu}(p,s)\right] \tag{38}$$

and

$$\begin{aligned} u_{k+1}\left(\frac{x^\mu}{\mu},\frac{t^\nu}{\nu}\right) &= -L_p^{-1}L_s^{-1}\left[\frac{1}{s^2}\int_0^p L_x^\mu L_t^\nu\left[\frac{\partial^\mu}{\partial x^\mu}\left(\frac{x^\mu}{\mu}\frac{\partial^\mu u_k}{\partial x^\mu}\right)\right]dp\right] \\ & -L_p^{-1}L_s^{-1}\left[\frac{1}{s^2}\int_0^p L_x^\mu L_t^\nu\left[\left(\frac{x^\mu}{\mu}\frac{\partial^{\mu+\nu}u_k}{\partial x^\mu \partial t^\nu}\right)\right]dp\right] \\ & +L_p^{-1}L_s^{-1}\left[\frac{1}{s^2}\int_0^p L_x^\mu L_t^\nu\left[a(x)\frac{x^\mu}{\mu}A_k-\frac{x^\mu}{\mu}B_k\right]dp\right], \end{aligned} \tag{39}$$

where $k \geq 0$.

Example 4. *Consider the nonlinear singular pseudohyperbolic equation in one dimensional is governed by*

$$\frac{\partial^{2\nu}u}{\partial t^{2\nu}}-\frac{\mu}{x^\mu}\frac{\partial^\mu}{\partial x^\mu}\left(\frac{x^\mu}{\mu}\frac{\partial^\mu u}{\partial x^\mu}\right)-\frac{\mu}{x^\mu}\frac{\partial^{\mu+\nu}}{\partial x^\mu \partial t^\nu}\left(\frac{x^\mu}{\mu}\frac{\partial^\mu u}{\partial x^\mu}\right)-\frac{1}{2}\frac{x^\mu}{\mu}u\frac{\partial^\mu u}{\partial x^\mu}+u^2=\left(\frac{x^\mu}{\mu}\right)^2 e^{-\frac{t^\nu}{\nu}}, \tag{40}$$

subject to the following initial conditions

$$u\left(\frac{x^\mu}{\mu},0\right)=\left(\frac{x^\mu}{\mu}\right)^2, \quad \frac{\partial^\nu u\left(\frac{x^\mu}{\mu},0\right)}{\partial t^\nu}=-\left(\frac{x^\mu}{\mu}\right)^2. \tag{41}$$

The conformable double Laplace transform decomposition method leads to the following scheme

$$u_0\left(\frac{x^\mu}{\mu},\frac{t^\nu}{\nu}\right)=\left(\frac{x^\mu}{\mu}\right)^2 e^{-\frac{t^\nu}{\nu}},$$

and

$$\begin{aligned} u_1\left(\frac{x^\mu}{\mu},\frac{t^\nu}{\nu}\right) &= -L_p^{-1}L_s^{-1}\left[\frac{1}{s^2}\int_0^p L_x^\mu L_t^\nu\left[\frac{\partial^\mu}{\partial x^\mu}\left(\frac{x^\mu}{\mu}\frac{\partial^\mu u_0}{\partial x^\mu}+\frac{x^\mu}{\mu}\frac{\partial^{\mu+\nu}u_0}{\partial x^\mu \partial t^\nu}\right)\right]dp\right] \\ & +L_p^{-1}L_s^{-1}\left[\frac{1}{s^2}\int_0^p L_x^\mu L_t^\nu\left[\frac{x^\mu}{\mu}u_0\frac{\partial^\mu u_0}{\partial x^\mu}-\frac{x^\mu}{\mu}u_0^2\right]dp\right] \\ &= -L_p^{-1}L_s^{-1}\left[\frac{1}{s^2}\int_0^p L_x^\mu L_t^\nu\left[4\left(\frac{x^\mu}{\mu}\right)e^{-\frac{t^\nu}{\nu}}-4\left(\frac{x^\mu}{\mu}\right)e^{-\frac{t^\nu}{\nu}}\right]dp\right] \\ & -L_p^{-1}L_s^{-1}\left[\frac{1}{s^2}\int_0^p L_x^\mu L_t^\nu\left[\left(\frac{x^\mu}{\mu}\right)^5 e^{-\frac{t^\nu}{\nu}}-\left(\frac{x^\mu}{\mu}\right)^5 e^{-\frac{t^\nu}{\nu}}\right]dp\right] \\ &= 0, \end{aligned}$$

proceeding in a similar manner, we have

$$u_2\left(\frac{x^\mu}{\mu},\frac{t^\nu}{\nu}\right)=0,\ u_3\left(\frac{x^\mu}{\mu},\frac{t^\nu}{\nu}\right)=0,\ \ u_4\left(\frac{x^\mu}{\mu},\frac{t^\nu}{\nu}\right)=0,\ldots$$

so that the solution $u\left(\frac{x^\mu}{\mu},\frac{t^\nu}{\nu}\right)$ *is given by*

$$\begin{aligned} u\left(\frac{x^\mu}{\mu},\frac{t^\nu}{\nu}\right) &= \sum_{n=0}^{\infty}u_n\left(\frac{x^\mu}{\mu},\frac{t^\nu}{\nu}\right)=u_0+u_1+u_2+\ldots \\ &= \left(\frac{x^\mu}{\mu}\right)^2 e^{-\frac{t^\nu}{\nu}} \end{aligned}$$

and hence the conformable solution is given by

$$u\left(\frac{x^\mu}{\mu},\frac{t^\nu}{\nu}\right)=\left(\frac{x^\mu}{\mu}\right)^2 e^{-\frac{t^\nu}{\nu}}. \tag{42}$$

By substituting $\mu=1$ *and* $\nu=1$ *into Equation (42), the solution becomes*

$$\psi(x,t)=x^2e^{-t}.$$

Conformable double Laplace transform method and Singular conformable coupled pseudohyperbolic equation.

In this section, conformable double Laplace decomposition method is considered for the one-dimensional conformable derivatives coupled pseudohyperbolic equation since the method is much simpler and more efficient in the study of linear equations.

The thrid problem: Let us consider the conformable derivatives coupled pseudohyperbolic equations

$$\begin{aligned} \frac{\partial^{2\nu}u}{\partial t^{2\nu}}-\frac{\mu}{x^\mu}\frac{\partial^\mu}{\partial x^\mu}\left(\frac{x^\mu}{\mu}\frac{\partial^\mu}{\partial x^\mu}u\right)-\frac{\mu}{x^\mu}\frac{\partial^{\mu+\nu}}{\partial x^\mu\partial t^\nu}\left(\frac{x^\mu}{\mu}\frac{\partial^\mu}{\partial x^\mu}u\right)+\zeta v &= f\left(\frac{x^\mu}{\mu},\frac{t^\nu}{\nu}\right) \\ \frac{\partial^{2\nu}v}{\partial t^{2\nu}}-\frac{\mu}{x^\mu}\frac{\partial^\mu}{\partial x^\mu}\left(\frac{x^\mu}{\mu}\frac{\partial^\mu}{\partial x^\mu}v\right)-\frac{\mu}{x^\mu}\frac{\partial^{\mu+\nu}}{\partial x^\mu\partial t^\nu}\left(\frac{x^\mu}{\mu}\frac{\partial^\mu}{\partial x^\mu}v\right)+\zeta u &= g\left(\frac{x^\mu}{\mu},\frac{t^\nu}{\nu}\right), \end{aligned} \tag{43}$$

subject to

$$u\left(\frac{x^\mu}{\mu},0\right)=f_1\left(\frac{x^\mu}{\mu}\right),\ \frac{\partial^\nu u\left(\frac{x^\mu}{\mu},0\right)}{\partial t^\nu}=f_2\left(\frac{x^\mu}{\mu}\right)\ \text{and}\ v\left(\frac{x^\mu}{\mu},0\right)=g_1(x),\ \frac{\partial^\nu u\left(\frac{x^\mu}{\mu},0\right)}{\partial t^\nu}=g_2\left(\frac{x^\mu}{\mu}\right) \tag{44}$$

where the linear terms $\frac{\mu}{x^\mu}\frac{\partial^\mu}{\partial x^\mu}\left(\frac{x^\mu}{\mu}\frac{\partial^\mu}{\partial x^\mu}\right)$ are the so-called conformable Bessel operators. Here, $f\left(\frac{x^\mu}{\mu},\frac{t^\nu}{\nu}\right)$, $g\left(\frac{x^\mu}{\mu},\frac{t^\nu}{\nu}\right)$, $f_1\left(\frac{x^\mu}{\mu}\right)$, $f_2\left(\frac{x^\mu}{\mu}\right)$, $g_1\left(\frac{x^\mu}{\mu}\right)$, and $g_2\left(\frac{x^\mu}{\mu}\right)$ are given functions, ζ is the coupling parameter. One can obtain the solution of Equation (43), by using the following steps.

(1): Multiply both sides of Equation (43) by $\frac{x^\mu}{\mu}$, we have

$$\begin{aligned} \frac{x^\mu}{\mu}\frac{\partial^{2\nu}u}{\partial t^{2\nu}}-\frac{\partial^\mu}{\partial x^\mu}\left(\frac{x^\mu}{\mu}\frac{\partial^\mu}{\partial x^\mu}u\right)-\frac{\partial^{\mu+\nu}}{\partial x^\mu\partial t^\nu}\left(\frac{x^\mu}{\mu}\frac{\partial^\mu}{\partial x^\mu}u\right)+\zeta\frac{x^\mu}{\mu}v &= \frac{x^\mu}{\mu}f\left(\frac{x^\mu}{\mu},\frac{t^\nu}{\nu}\right) \\ \frac{x^\mu}{\mu}\frac{\partial^{2\nu}v}{\partial t^{2\nu}}-\frac{\partial^\mu}{\partial x^\mu}\left(\frac{x^\mu}{\mu}\frac{\partial^\mu}{\partial x^\mu}v\right)-\frac{\partial^{\mu+\nu}}{\partial x^\mu\partial t^\nu}\left(\frac{x^\mu}{\mu}\frac{\partial^\mu}{\partial x^\mu}v\right)+\zeta\frac{x^\mu}{\mu}u &= \frac{x^\mu}{\mu}g\left(\frac{x^\mu}{\mu},\frac{t^\nu}{\nu}\right). \end{aligned} \tag{45}$$

(2): We apply conformable double Laplace transform on both sides of Equation (45) and single conformable Laplace transform for Equation (44), we get

$$\begin{aligned} L_x^{\mu}L_t^{\nu}\left[\frac{x^{\mu}}{\mu}\frac{\partial^{2\nu}u}{\partial t^{2\nu}}\right] &= L_x^{\mu}L_t^{\nu}\left[\frac{\partial^{\mu}}{\partial x^{\mu}}\left(\frac{x^{\mu}}{\mu}\frac{\partial^{\mu}}{\partial x^{\mu}}u\right)+\frac{\partial^{\mu+\nu}}{\partial x^{\mu}\partial t^{\nu}}\left(\frac{x^{\mu}}{\mu}\frac{\partial^{\mu}}{\partial x^{\mu}}u\right)-\zeta\frac{x^{\mu}}{\mu}v+\frac{x^{\mu}}{\mu}f\left(\frac{x^{\mu}}{\mu},\frac{t^{\nu}}{\nu}\right)\right], \\ L_x^{\mu}L_t^{\nu}\left[\frac{x^{\mu}}{\mu}\frac{\partial^{2\nu}v}{\partial t^{2\nu}}\right] &= L_x^{\mu}L_t^{\nu}\left[\frac{\partial^{\mu}}{\partial x^{\mu}}\left(\frac{x^{\mu}}{\mu}\frac{\partial^{\mu}}{\partial x^{\mu}}v\right)+\frac{\partial^{\mu+\nu}}{\partial x^{\mu}\partial t^{\nu}}\left(\frac{x^{\mu}}{\mu}\frac{\partial^{\mu}}{\partial x^{\mu}}v\right)-\zeta\frac{x^{\mu}}{\mu}u+\frac{x^{\mu}}{\mu}g\left(\frac{x^{\mu}}{\mu},\frac{t^{\nu}}{\nu}\right)\right], \end{aligned} \tag{46}$$

on using theorem 1 and theorem 2, we obtain

$$\begin{aligned} \frac{d}{dp}\left[U_{\mu,\nu}(p,s)\right] &= \frac{1}{s}\frac{d}{dp}F_1(p,0)+\frac{1}{s^2}\frac{d}{dp}F_2(p,0) \\ &\quad \frac{1}{s^2}L_x^{\mu}L_t^{\nu}\left[\frac{\partial^{\mu}}{\partial x^{\mu}}\left(\frac{x^{\mu}}{\mu}\frac{\partial^{\mu}}{\partial x^{\mu}}u\right)+\frac{\partial^{\mu+\nu}}{\partial x^{\mu}\partial t^{\nu}}\left(\frac{x^{\mu}}{\mu}\frac{\partial^{\mu}}{\partial x^{\mu}}u\right)-\zeta\frac{x^{\mu}}{\mu}v\right] \\ &\quad +\frac{1}{s^2}\frac{d}{dp}\left(L_x^{\mu}L_t^{\nu}\left[f\left(\frac{x^{\mu}}{\mu},\frac{t^{\nu}}{\nu}\right)\right]\right), \\ \frac{d}{dp}\left[U_{\mu,\nu}(p,s)\right] &= \frac{1}{s}\frac{d}{dp}G_1(p,0)+\frac{1}{s^2}\frac{d}{dp}G_2(p,0) \\ &\quad -\frac{1}{s^2}L_x^{\mu}L_t^{\nu}\left[\frac{\partial^{\mu}}{\partial x^{\mu}}\left(\frac{x^{\mu}}{\mu}\frac{\partial^{\mu}}{\partial x^{\mu}}v\right)+\frac{\partial^{\mu+\nu}}{\partial x^{\mu}\partial t^{\nu}}\left(\frac{x^{\mu}}{\mu}\frac{\partial^{\mu}}{\partial x^{\mu}}v\right)-\zeta\frac{x^{\mu}}{\mu}u\right] \\ &\quad +\frac{1}{s^2}\frac{d}{dp}\left(L_x^{\mu}L_t^{\nu}\left[g\left(\frac{x^{\mu}}{\mu},\frac{t^{\nu}}{\nu}\right)\right]\right). \end{aligned} \tag{47}$$

(3): By integrating both sides of Equation (47) from 0 to p with respect to p, we have

$$\begin{aligned} U_{\mu,\nu}(p,s) &= \frac{F_1(p,0)}{s}+\frac{F_2(p,0)}{s^2} \\ &\quad -\frac{1}{s^2}\int_0^p L_x^{\mu}L_t^{\nu}\left[\frac{\partial^{\mu}}{\partial x^{\mu}}\left(\frac{x^{\mu}}{\mu}\frac{\partial^{\mu}}{\partial x^{\mu}}u\right)+\frac{\partial^{\mu+\nu}}{\partial x^{\mu}\partial t^{\nu}}\left(\frac{x^{\mu}}{\mu}\frac{\partial^{\mu}}{\partial x^{\mu}}u\right)-\zeta\frac{x^{\mu}}{\mu}v\right]dp \\ &\quad +\frac{1}{s^2}\int_0^p\left(\frac{d}{dp}\left(L_x^{\mu}L_t^{\nu}\left[f\left(\frac{x^{\mu}}{\mu},\frac{t^{\nu}}{\nu}\right)\right]\right)\right)dp, \\ V_{\mu,\nu}(p,s) &= \frac{G_1(p,0)}{s}+\frac{G_2(p,0)}{s^2} \\ &\quad -\frac{1}{s^2}\int_0^p L_x^{\mu}L_t^{\nu}\left[\frac{\partial^{\mu}}{\partial x^{\mu}}\left(\frac{x^{\mu}}{\mu}\frac{\partial^{\mu}}{\partial x^{\mu}}v\right)+\frac{\partial^{\mu+\nu}}{\partial x^{\mu}\partial t^{\nu}}\left(\frac{x^{\mu}}{\mu}\frac{\partial^{\mu}}{\partial x^{\mu}}v\right)-\zeta\frac{x^{\mu}}{\mu}u\right]dp \\ &\quad +\frac{1}{s^2}\int_0^p\left(\frac{d}{dp}\left(L_x^{\mu}L_t^{\nu}\left[g\left(\frac{x^{\mu}}{\mu},\frac{t^{\nu}}{\nu}\right)\right]\right)\right)dp, \end{aligned} \tag{48}$$

where $F_1(p,0)$, $F_2(p,0)$, $G_1(p,0)$, and $G_2(p,0)$ are conformable Laplace transform of the functions $f_1\left(\frac{x^{\mu}}{\mu}\right)$, $f_2\left(\frac{x^{\mu}}{\mu}\right)$, $g_1\left(\frac{x^{\mu}}{\mu}\right)$, and $g_2\left(\frac{x^{\mu}}{\mu}\right)$, respectively. By applying double inverse Laplace transform for Equation (48), we have

$$\begin{aligned} u\left(\frac{x^{\mu}}{\mu},\frac{t^{\nu}}{\nu}\right) &= f_1\left(\frac{x^{\mu}}{\mu}\right)+\frac{t^{\nu}}{\nu}f_2\left(\frac{x^{\mu}}{\mu}\right)+L_p^{-1}L_s^{-1}\left[\frac{1}{s^2}\int_0^p\left(\frac{d}{dp}\left(L_x^{\mu}L_t^{\nu}\left[f\left(\frac{x^{\mu}}{\mu},\frac{t^{\nu}}{\nu}\right)\right]\right)\right)dp\right] \\ &\quad -L_p^{-1}L_s^{-1}\left[\frac{1}{s^2}\int_0^p L_x^{\mu}L_t^{\nu}\left[\frac{\partial^{\mu}}{\partial x^{\mu}}\left(\frac{x^{\mu}}{\mu}\frac{\partial^{\mu}}{\partial x^{\mu}}u\right)\right]dp\right] \\ &\quad -L_p^{-1}L_s^{-1}\left[\frac{1}{s^2}\int_0^p L_x^{\mu}L_t^{\nu}\left[\frac{\partial^{\mu+\nu}}{\partial x^{\mu}\partial t^{\nu}}\left(\frac{x^{\mu}}{\mu}\frac{\partial^{\mu}}{\partial x^{\mu}}u\right)-\zeta\frac{x^{\mu}}{\mu}v\right]dp\right] \end{aligned} \tag{49}$$

and

$$
\begin{aligned}
v\left(\frac{x^{\mu}}{\mu},\frac{t^{\nu}}{\nu}\right) &= g_1\left(\frac{x^{\mu}}{\mu}\right)+\frac{t^{\nu}}{\nu}g_2\left(\frac{x^{\mu}}{\mu}\right)+L_p^{-1}L_s^{-1}\left[\frac{1}{s^2}\int_0^p\left(\frac{d}{dp}\left(L_x^{\mu}L_t^{\nu}\left[g\left(\frac{x^{\mu}}{\mu},\frac{t^{\nu}}{\nu}\right)\right]\right)\right)dp\right] \\
&-L_p^{-1}L_s^{-1}\left[\frac{1}{s^2}\int_0^p L_x^{\mu}L_t^{\nu}\left[\frac{\partial^{\mu}}{\partial x^{\mu}}\left(\frac{x^{\mu}}{\mu}\frac{\partial^{\mu}}{\partial x^{\mu}}v\right)\right]dp\right] \\
&-L_p^{-1}L_s^{-1}\left[\frac{1}{s^2}\int_0^p L_x^{\mu}L_t^{\nu}\left[\frac{\partial^{\mu+\nu}}{\partial x^{\mu}\partial t^{\nu}}\left(\frac{x^{\mu}}{\mu}\frac{\partial^{\mu}}{\partial x^{\mu}}v\right)-\zeta\frac{x^{\mu}}{\mu}u\right]dp\right]. \qquad (50)
\end{aligned}
$$

The conformable double Laplace decomposition methods represent the solutions of Equation (43), by the infinite series

$$
u\left(\frac{x^{\mu}}{\mu},\frac{t^{\nu}}{\nu}\right)=\sum_{n=0}^{\infty}u_n\left(\frac{x^{\mu}}{\mu},\frac{t^{\nu}}{\nu}\right),\quad v\left(\frac{x^{\mu}}{\mu},\frac{t^{\nu}}{\nu}\right)=\sum_{n=0}^{\infty}v_n\left(\frac{x^{\mu}}{\mu},\frac{t^{\nu}}{\nu}\right). \qquad (51)
$$

By substituting Equation (51) into Equations (49) and (50), we get

$$
\begin{aligned}
\sum_{n=0}^{\infty}u_n\left(\frac{x^{\mu}}{\mu},\frac{t^{\nu}}{\nu}\right) &= f_1\left(\frac{x^{\mu}}{\mu}\right)+\frac{t^{\nu}}{\nu}f_2\left(\frac{x^{\mu}}{\mu}\right)+L_p^{-1}L_s^{-1}\left[\frac{1}{s^2}\int_0^p\left(\frac{d}{dp}\left(L_x^{\mu}L_t^{\nu}\left[f\left(\frac{x^{\mu}}{\mu},\frac{t^{\nu}}{\nu}\right)\right]\right)\right)dp\right] \\
&-L_p^{-1}L_s^{-1}\left[\frac{1}{s^2}\int_0^p\left(L_x^{\mu}L_t^{\nu}\left[\frac{\partial^{\mu}}{\partial x^{\mu}}\left(\frac{x^{\mu}}{\mu}\frac{\partial^{\mu}}{\partial x^{\mu}}\left(\sum_{n=0}^{\infty}u_n\right)\right)\right]\right)dp\right] \\
&-L_p^{-1}L_s^{-1}\left[\frac{1}{s^2}\int_0^p\left(L_x^{\mu}L_t^{\nu}\left[\frac{\partial^{\mu+\nu}}{\partial x^{\mu}\partial t^{\nu}}\left(\frac{x^{\mu}}{\mu}\frac{\partial^{\mu}}{\partial x^{\mu}}\left(\sum_{n=0}^{\infty}u_n\right)\right)\right]\right)dp\right] \\
&+L_p^{-1}L_s^{-1}\left[\frac{1}{s^2}L_xL_t\left[\zeta\frac{x^{\mu}}{\mu}\sum_{n=0}^{\infty}v_n\right]\right], \qquad (52)
\end{aligned}
$$

and

$$
\begin{aligned}
\sum_{n=0}^{\infty}v_n\left(\frac{x^{\mu}}{\mu},\frac{t^{\nu}}{\nu}\right) &= g_1\left(\frac{x^{\mu}}{\mu}\right)+\frac{t^{\nu}}{\nu}g_2\left(\frac{x^{\mu}}{\mu}\right)+L_p^{-1}L_s^{-1}\left[\frac{1}{s^2}\int_0^p\left(\frac{d}{dp}\left(L_x^{\mu}L_t^{\nu}\left[g\left(\frac{x^{\mu}}{\mu},\frac{t^{\nu}}{\nu}\right)\right]\right)\right)dp\right] \\
&-L_p^{-1}L_s^{-1}\left[\frac{1}{s^2}\int_0^p\left(L_x^{\mu}L_t^{\nu}\left[\frac{\partial^{\mu}}{\partial x^{\mu}}\left(\frac{x^{\mu}}{\mu}\frac{\partial^{\mu}}{\partial x^{\mu}}\left(\sum_{n=0}^{\infty}v_n\right)\right)\right]\right)dp\right] \\
&-L_p^{-1}L_s^{-1}\left[\frac{1}{s^2}\int_0^p\left(L_x^{\mu}L_t^{\nu}\left[\frac{\partial^{\mu+\nu}}{\partial x^{\mu}\partial t^{\nu}}\left(\frac{x^{\mu}}{\mu}\frac{\partial^{\mu}}{\partial x^{\mu}}\left(\sum_{n=0}^{\infty}v_n\right)\right)\right]\right)dp\right] \\
&+L_p^{-1}L_s^{-1}\left[\frac{1}{s^2}L_xL_t\left[\zeta\frac{x^{\mu}}{\mu}\sum_{n=0}^{\infty}u_n\right]\right]. \qquad (53)
\end{aligned}
$$

Our method suggests that the zeroth components u_0 and v_0 are identified by the initial conditions and from source terms as follows

$$
\begin{aligned}
u_0 &= f_1\left(\frac{x^{\mu}}{\mu}\right)+\frac{t^{\nu}}{\nu}f_2\left(\frac{x^{\mu}}{\mu}\right)+L_p^{-1}L_s^{-1}\left[\frac{1}{s}\int_0^p\left(\frac{d}{dp}\left(L_x^{\mu}L_t^{\nu}\left[f\left(\frac{x^{\mu}}{\mu},\frac{t^{\nu}}{\nu}\right)\right]\right)\right)dp\right], \\
v_0 &= g_1\left(\frac{x^{\mu}}{\mu}\right)+\frac{t^{\nu}}{\nu}g_2\left(\frac{x^{\mu}}{\mu}\right)+L_p^{-1}L_s^{-1}\left[\frac{1}{s}\int_0^p\left(\frac{d}{dp}\left(L_x^{\mu}L_t^{\nu}\left[g\left(\frac{x^{\mu}}{\mu},\frac{t^{\nu}}{\nu}\right)\right]\right)\right)dp\right]. \qquad (54)
\end{aligned}
$$

The remaining terms are given by

$$
\begin{aligned}
u_{k+1} = & -L_p^{-1}L_s^{-1}\left[\frac{1}{s^2}\int_0^p\left(L_x^{\mu}L_t^{\nu}\left[\frac{\partial^{\mu}}{\partial x^{\mu}}\left(\frac{x^{\mu}}{\mu}\frac{\partial^{\mu}}{\partial x^{\mu}}u_k\right)\right]\right)dp\right] \\
& -L_p^{-1}L_s^{-1}\left[\frac{1}{s^2}\int_0^p\left(L_x^{\mu}L_t^{\nu}\left[\frac{\partial^{\mu+\nu}}{\partial x^{\mu}\partial t^{\nu}}\left(\frac{x^{\mu}}{\mu}\frac{\partial^{\mu}}{\partial x^{\mu}}u_k\right)\right]\right)dp\right] \\
& +L_p^{-1}L_s^{-1}\left[\frac{1}{s^2}L_xL_t\left[\zeta\,\frac{x^{\mu}}{\mu}v_k\right]\right],
\end{aligned}
\tag{55}
$$

and

$$
\begin{aligned}
v_{k+1} = & -L_p^{-1}L_s^{-1}\left[\frac{1}{s^2}\int_0^p\left(L_x^{\mu}L_t^{\nu}\left[\frac{\partial^{\mu}}{\partial x^{\mu}}\left(\frac{x^{\mu}}{\mu}\frac{\partial^{\mu}}{\partial x^{\mu}}v_k\right)\right]\right)dp\right] \\
& -L_p^{-1}L_s^{-1}\left[\frac{1}{s^2}\int_0^p\left(L_x^{\mu}L_t^{\nu}\left[\frac{\partial^{\mu+\nu}}{\partial x^{\mu}\partial t^{\nu}}\left(\frac{x^{\mu}}{\mu}\frac{\partial^{\mu}}{\partial x^{\mu}}v_k\right)\right]\right)dp\right] \\
& +L_p^{-1}L_s^{-1}\left[\frac{1}{s^2}L_xL_t\left[\zeta\,\frac{x^{\mu}}{\mu}u_k\right]\right].
\end{aligned}
\tag{56}
$$

Here, we assume that the double inverse Laplace transform with respect to p and s exists for each term in the right hand side of the above equations.

To illustrate our method for solving the conformable derivatives coupled pseudohyperbolic equations, we will consider the following example:

Example 5. *Consider the following homogeneous form of a conformable derivatives coupled pseudohyperbolic equation*

$$
\begin{aligned}
\frac{\partial^{2\nu}v}{\partial t^{2\nu}}-\frac{\mu}{x^{\mu}}\frac{\partial^{\mu}}{\partial x^{\mu}}\left(\frac{x^{\mu}}{\mu}\frac{\partial^{\mu}}{\partial x^{\mu}}u\right)-\frac{\mu}{x^{\mu}}\frac{\partial^{\mu+\nu}}{\partial x^{\mu}\partial t^{\nu}}\left(\frac{x^{\mu}}{\mu}\frac{\partial^{\mu}}{\partial x^{\mu}}u\right)-v &= 0 \\
\frac{\partial^{2\nu}v}{\partial t^{2\nu}}-\frac{\mu}{x^{\mu}}\frac{\partial^{\mu}}{\partial x^{\mu}}\left(\frac{x^{\mu}}{\mu}\frac{\partial^{\mu}}{\partial x^{\mu}}v\right)-\frac{\mu}{x^{\mu}}\frac{\partial^{\mu+\nu}}{\partial x^{\mu}\partial t^{\nu}}\left(\frac{x^{\mu}}{\mu}\frac{\partial^{\mu}}{\partial x^{\mu}}v\right)-u &= 0,
\end{aligned}
\tag{57}
$$

with initial condition

$$
\begin{aligned}
u\left(\frac{x^{\mu}}{\mu},0\right) &= \left(\frac{x^{\mu}}{\mu}\right)^2, \quad \frac{\partial^{\nu}u\left(\frac{x^{\mu}}{\mu},0\right)}{\partial t^{\nu}}=-\left(\frac{x^{\mu}}{\mu}\right)^2 \\
v\left(\frac{x^{\mu}}{\mu},0\right) &= \left(\frac{x^{\mu}}{\mu}\right)^2, \quad \frac{\partial^{\nu}v\left(\frac{x^{\mu}}{\mu},0\right)}{\partial t^{\nu}}=-\left(\frac{x^{\mu}}{\mu}\right)^2.
\end{aligned}
\tag{58}
$$

By applying above method for Equations (57) and (58), we obtain

$$
\begin{aligned}
\sum_{n=0}^{\infty}u_n\left(\frac{x^{\mu}}{\mu},\frac{t^{\nu}}{\nu}\right)=\left(\frac{x^{\mu}}{\mu}\right)^2 & -\left(\frac{x^{\mu}}{\mu}\right)^2\frac{t^{\nu}}{\nu} \\
& -L_p^{-1}L_s^{-1}\left[\frac{1}{s^2}\int_0^p\left(L_x^{\mu}L_t^{\nu}\left[\frac{\partial^{\mu}}{\partial x^{\mu}}\left(\frac{x^{\mu}}{\mu}\frac{\partial^{\mu}}{\partial x^{\mu}}\left(\sum_{n=0}^{\infty}u_n\right)\right)\right]\right)dp\right] \\
& -L_p^{-1}L_s^{-1}\left[\frac{1}{s^2}\int_0^p\left(L_x^{\mu}L_t^{\nu}\left[\frac{\partial^{\mu+\nu}}{\partial x^{\mu}\partial t^{\nu}}\left(\frac{x^{\mu}}{\mu}\frac{\partial^{\mu}}{\partial x^{\mu}}\left(\sum_{n=0}^{\infty}u_n\right)\right)\right]\right)dp\right] \\
& -L_p^{-1}L_s^{-1}\left[\frac{1}{s^2}L_xL_t\left[\frac{x^{\mu}}{\mu}\sum_{n=0}^{\infty}v_n\right]\right],
\end{aligned}
\tag{59}
$$

and

$$\begin{aligned}
\sum_{n=0}^{\infty} v_n\left(\frac{x^{\mu}}{\mu},\frac{t^{\nu}}{\nu}\right) &= \left(\frac{x^{\mu}}{\mu}\right)^2-\left(\frac{x^{\mu}}{\mu}\right)^2\frac{t^{\nu}}{\nu} \\
&-L_p^{-1}L_s^{-1}\left[\frac{1}{s^2}\int_0^p\left(L_x^{\mu}L_t^{\nu}\left[\frac{\partial^{\mu}}{\partial x^{\mu}}\left(\frac{x^{\mu}}{\mu}\frac{\partial^{\mu}}{\partial x^{\mu}}\left(\sum_{n=0}^{\infty}v_n\right)\right)\right]\right)dp\right] \\
&-L_p^{-1}L_s^{-1}\left[\frac{1}{s^2}\int_0^p\left(L_x^{\mu}L_t^{\nu}\left[\frac{\partial^{\mu+\nu}}{\partial x^{\mu}\partial t^{\nu}}\left(\frac{x^{\mu}}{\mu}\frac{\partial^{\mu}}{\partial x^{\mu}}\left(\sum_{n=0}^{\infty}v_n\right)\right)\right]\right)dp\right] \\
&-L_p^{-1}L_s^{-1}\left[\frac{1}{s^2}L_xL_t\left[\zeta\,\frac{x^{\mu}}{\mu}\sum_{n=0}^{\infty}u_n\right]\right]. \qquad (60)
\end{aligned}$$

By applying equations Equations (54)–(56), we have

$$\begin{aligned}
u_0 &= \left(\frac{x^{\mu}}{\mu}\right)^2-\left(\frac{x^{\mu}}{\mu}\right)^2\frac{t^{\nu}}{\nu},\quad v_0=\left(\frac{x^{\mu}}{\mu}\right)^2-\left(\frac{x^{\mu}}{\mu}\right)^2\frac{t^{\nu}}{\nu} \\
u_1 &= -L_p^{-1}L_s^{-1}\left[\frac{1}{s^2}L_xL_t\left[\frac{\partial^{\mu}}{\partial x^{\mu}}\left(\frac{x^{\mu}}{\mu}\frac{\partial^{\mu}}{\partial x^{\mu}}u_0\right)+\frac{\partial^{\mu+\nu}}{\partial x^{\mu}\partial t^{\nu}}\left(\frac{x^{\mu}}{\mu}\frac{\partial^{\mu}}{\partial x^{\mu}}u_0\right)+\frac{x^{\mu}}{\mu}v_0\right]\right] \\
&= -\frac{2}{3}\left(\frac{t^{\nu}}{\nu}\right)^3+\frac{1}{2}\left(\frac{x^{\mu}}{\mu}\right)^2\left(\frac{t^{\nu}}{\nu}\right)^2-\frac{1}{6}\left(\frac{x^{\mu}}{\mu}\right)^2\left(\frac{t^{\nu}}{\nu}\right)^3, \\
v_1 &= -L_p^{-1}L_s^{-1}\left[\frac{1}{s^2}L_xL_t\left[\frac{\partial^{\mu}}{\partial x^{\mu}}\left(\frac{x^{\mu}}{\mu}\frac{\partial^{\mu}}{\partial x^{\mu}}v_0\right)+\frac{\partial^{\mu+\nu}}{\partial x^{\mu}\partial t^{\nu}}\left(\frac{x^{\mu}}{\mu}\frac{\partial^{\mu}}{\partial x^{\mu}}v_0\right)+\frac{x^{\mu}}{\mu}u_0\right]\right] \\
&= -\frac{2}{3}\left(\frac{t^{\nu}}{\nu}\right)^3+\frac{1}{2}\left(\frac{x^{\mu}}{\mu}\right)^2\left(\frac{t^{\nu}}{\nu}\right)^2-\frac{1}{6}\left(\frac{x^{\mu}}{\mu}\right)^2\left(\frac{t^{\nu}}{\nu}\right)^3, \\
u_2 &= -L_p^{-1}L_s^{-1}\left[\frac{1}{s^2}L_xL_t\left[\frac{\partial^{\mu}}{\partial x^{\mu}}\left(\frac{x^{\mu}}{\mu}\frac{\partial^{\mu}}{\partial x^{\mu}}u_1\right)+\frac{\partial^{\mu+\nu}}{\partial x^{\mu}\partial t^{\nu}}\left(\frac{x^{\mu}}{\mu}\frac{\partial^{\mu}}{\partial x^{\mu}}u_1\right)+\frac{x^{\mu}}{\mu}v_1\right]\right] \\
&= \frac{2}{3}\left(\frac{t^{\nu}}{\nu}\right)^3+\frac{1}{24}\left(\frac{x^{\mu}}{\mu}\right)^2\left(\frac{t^{\nu}}{\nu}\right)^4-\frac{1}{120}\left(\frac{x^{\mu}}{\mu}\right)^2\left(\frac{t^{\nu}}{\nu}\right)^5, \\
v_2 &= -L_p^{-1}L_s^{-1}\left[\frac{1}{s^2}L_xL_t\left[\frac{\partial^{\mu}}{\partial x^{\mu}}\left(\frac{x^{\mu}}{\mu}\frac{\partial^{\mu}}{\partial x^{\mu}}v_1\right)+\frac{\partial^{\mu+\nu}}{\partial x^{\mu}\partial t^{\nu}}\left(\frac{x^{\mu}}{\mu}\frac{\partial^{\mu}}{\partial x^{\mu}}v_1\right)+\frac{x^{\mu}}{\mu}u_1\right]\right] \\
&= \frac{2}{3}\left(\frac{t^{\nu}}{\nu}\right)^3+\frac{1}{24}\left(\frac{x^{\mu}}{\mu}\right)^2\left(\frac{t^{\nu}}{\nu}\right)^4-\frac{1}{120}\left(\frac{x^{\mu}}{\mu}\right)^2\left(\frac{t^{\nu}}{\nu}\right)^5,
\end{aligned}$$

and

$$\begin{aligned}
u_3 &= -L_p^{-1}L_s^{-1}\left[\frac{1}{s^2}L_xL_t\left[\frac{\partial^{\mu}}{\partial x^{\mu}}\left(\frac{x^{\mu}}{\mu}\frac{\partial^{\mu}}{\partial x^{\mu}}u_2\right)+\frac{\partial^{\mu+\nu}}{\partial x^{\mu}\partial t^{\nu}}\left(\frac{x^{\mu}}{\mu}\frac{\partial^{\mu}}{\partial x^{\mu}}u_2\right)+\frac{x^{\mu}}{\mu}v_2\right]\right] \\
&= \frac{1}{15}\left(\frac{t^{\nu}}{\nu}\right)^5-\frac{1}{1260}\left(\frac{t^{\nu}}{\nu}\right)^7+\frac{1}{720}\left(\frac{x^{\mu}}{\mu}\right)^2\left(\frac{t^{\nu}}{\nu}\right)^6-\frac{1}{5040}\left(\frac{x^{\mu}}{\mu}\right)^2\left(\frac{t^{\nu}}{\nu}\right)^7, \\
v_3 &= -L_p^{-1}L_s^{-1}\left[\frac{1}{s}L_xL_t\left[\frac{\partial^{\mu}}{\partial x^{\mu}}\left(\frac{x^{\mu}}{\mu}\frac{\partial^{\mu}}{\partial x^{\mu}}v_2\right)+\frac{\partial^{\mu+\nu}}{\partial x^{\mu}\partial t^{\nu}}\left(\frac{x^{\mu}}{\mu}\frac{\partial^{\mu}}{\partial x^{\mu}}v_2\right)+\frac{x^{\mu}}{\mu}u_2\right]\right] \\
&= \frac{1}{15}\left(\frac{t^{\nu}}{\nu}\right)^5-\frac{1}{1260}\left(\frac{t^{\nu}}{\nu}\right)^7+\frac{1}{720}\left(\frac{x^{\mu}}{\mu}\right)^2\left(\frac{t^{\nu}}{\nu}\right)^6-\frac{1}{5040}\left(\frac{x^{\mu}}{\mu}\right)^2\left(\frac{t^{\nu}}{\nu}\right)^7,
\end{aligned}$$

and so on for other components. Using Equation (51), our required solutions are given below

$$u\left(\frac{x^{\mu}}{\mu},\frac{t^{\nu}}{\nu}\right) = u_0+u_1+u_2+u_3+\ldots = \left(1-\frac{t^{\nu}}{\nu}+\frac{\left(\frac{t^{\nu}}{\nu}\right)^2}{2!}-\frac{\left(\frac{t^{\nu}}{\nu}\right)^3}{3!}+\frac{\left(\frac{t^{\nu}}{\nu}\right)^4}{4!}-\ldots\right)\left(\frac{x^{\mu}}{\mu}\right)^2$$

$$v\left(\frac{x^{\mu}}{\mu},\frac{t^{\nu}}{\nu}\right) = v_0+v_1+v_2+v_3+\ldots = \left(1-\frac{t^{\nu}}{\nu}+\frac{\left(\frac{t^{\nu}}{\nu}\right)^2}{2!}-\frac{\left(\frac{t^{\nu}}{\nu}\right)^3}{3!}+\frac{\left(\frac{t^{\nu}}{\nu}\right)^4}{4!}-\ldots\right)\left(\frac{x^{\mu}}{\mu}\right)^2$$

and hence the exact solution becomes

$$u\left(\frac{x^{\mu}}{\mu},\frac{t^{\nu}}{\nu}\right)=\left(\frac{x^{\mu}}{\mu}\right)^2 e^{-\frac{t^{\nu}}{\nu}},\quad v\left(\frac{x^{\mu}}{\mu},\frac{t^{\nu}}{\nu}\right)=\left(\frac{x^{\mu}}{\mu}\right)^2 e^{-\frac{t^{\nu}}{\nu}}.$$

By taking $\mu=1$ and $\nu=1$, the conformable solution becomes

$$u(x,t)=x^2e^{-t},\quad v(x,t)=x^2e^{-t}.$$

4. Numerical Result

In this section, we shall illustrate the accuracy and effciency of the double conformable Laplace transform method by numerical results of $u(x,t)$ for the exact solution when $\mu=\nu=1$, and approximate solutions when μ and ν taken different fractional values in Equations (20) and (40), which are depicted through Figures 1–4, respectively.

The three dimensional surface in Figure 1 shows the exact solution of Equation (20) in standard form of singular pseudohyperbolic equation at $\mu=\nu=1$. Figure 2 compares the approximate solutions of Equation (20) when $t=\frac{\pi}{2}$. In Figure 2a, the numerical solution at $0<\mu=\nu\leq 1$, in this case $u(x,t)$, increases hastily at fractional derivative decrease, Figure 2b shows the solution at $\mu=0.99$ and $\nu=0.95, 0.90, 0.85$ and we see $u(x,t)$ increasing regularly when ν decreases, and in Figure 2c we can observe $u(x,t)$ increasing slowly at $\mu=0.95, 0.90, 0.85$ and $\nu=0.99$ when μ decreases.

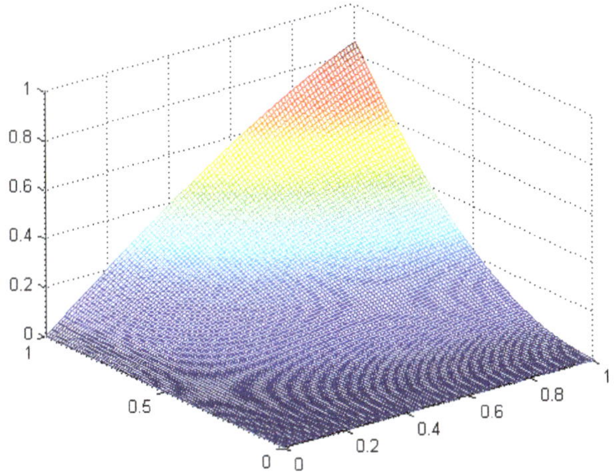

Figure 1. The Exact Solutions $u(x,t)$ for Equation (20) when $\mu=\nu=1$.

Similarly, the exact solution and approximate solution of Equation (40) were demonstrated in Figures 3 and 4 when $t = 1$. In the case $\nu = \mu = 1$, we get the exact solution of a singular pseudohyperbolic equation, as seen in Figure 3. Figure 4 shows the approximate solution of Equation (40) with different values of μ and ν. Figure 4a gives plots of the behavior of Equation (40) when $0 < \mu = \nu \leq 1$, in this case the function $u(x,t)$ increases quickly, and in Figure 4b we have obtained the solution for the values of $\mu = 0.99$ and different values of $0 < \nu \leq 1$, in this case the function $u(x,t)$ increases gradually, and Figure 4c gives the behavior of Equation (40) at $\nu = 0.99$ and different values of μ, in this case the function $u(x,t)$ increasing tardily.

It is clear from the solutions of Equations (20) and (40) that the conformable double Laplace decomposition method has good agreement with the exact solutions of the problems. The fractional-order solution of these two problems and exact solution of integer order problems are equal at $0 < \mu = \nu \leq 1$, in this case we have no error.

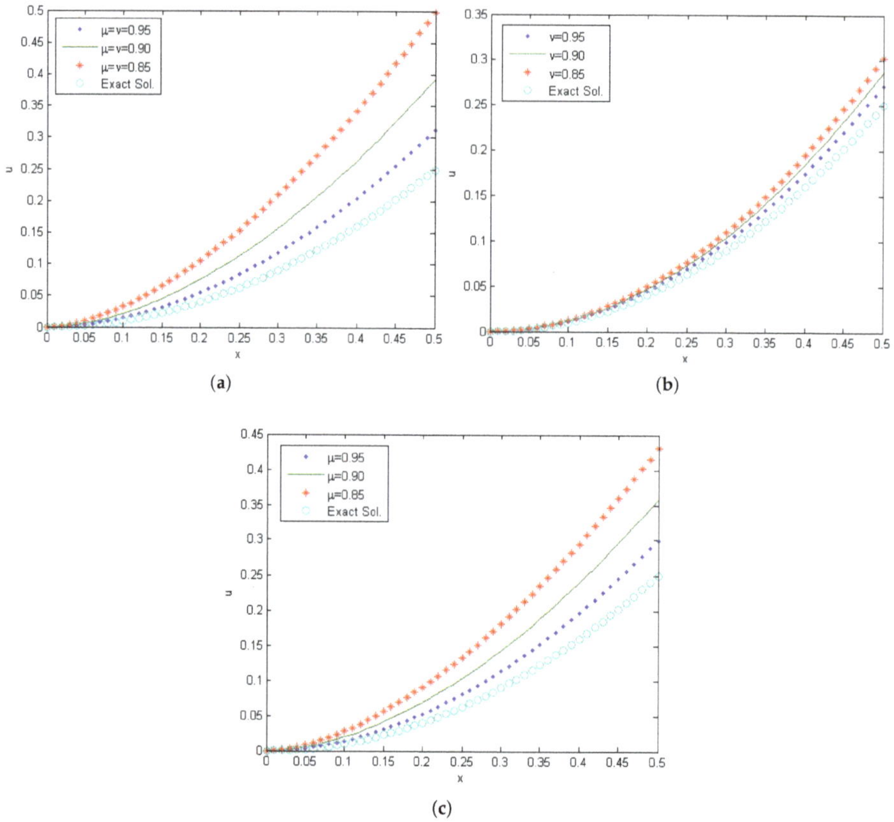

Figure 2. The solutions $u(x,t)$ for Equation (20) for different values of μ and ν when $t = \frac{\pi}{2}$. (**a**) Plot solutions $u(x,t)$ for Equation (20) at $\mu = \nu$. (**b**) Plot solutions $u(x,t)$ for Equation (20) when $\mu = 0.99$ and different values of ν. (**c**) Plot solutions $u(x,t)$ for Equation (20) for different values of μ at $\nu = 0.99$.

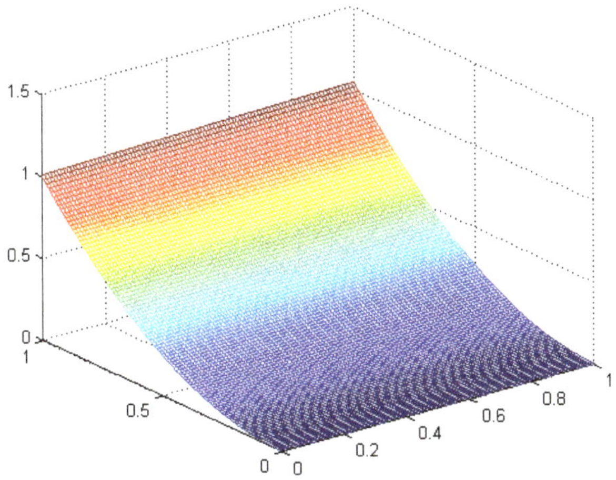

Figure 3. The Exact Solutions $u(x,t)$ for Equation (40) when $\mu = \nu = 1$.

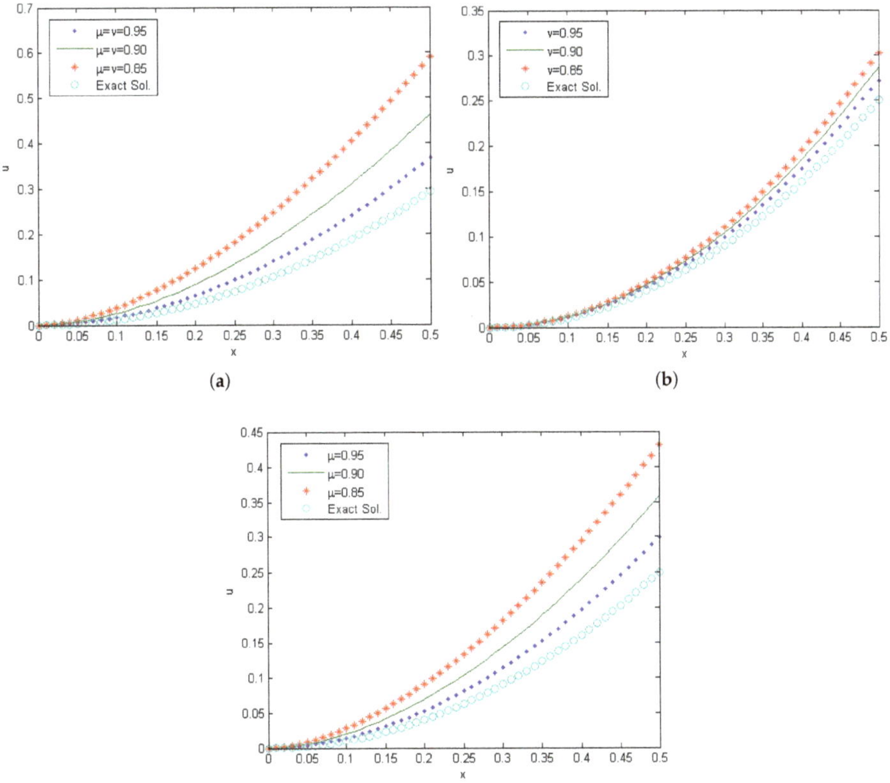

(a)

(b)

(c)

Figure 4. The solutions $u(x,t)$ for Equation (40) for different values of μ and ν when $t = 1$. (**a**) Plot solutions $u(x,t)$ for Equation (40) at $\mu = \nu$. (**b**) Plot solutions $u(x,t)$ for Equation (40) when $\mu = 0.99$ and different values of ν. (**c**) Plot solutions $u(x,t)$ for Equation (40) for different values of μ at $\nu = 0.99$.

5. Conclusions

In the present work we have studied singular linear and nonlinear pseudohyperbolic equations by employing the conformable double Laplace transform decomposition method (CDLDM), and we obtain analytic solutions when $\mu = \nu = 1$ and numerical solutions for different fractional values. Further, we also studied singular coupled pseudohyperbolic equations. It is clear that the solutions of Equations (20) and (40) were obtained as infinite series by using the conformable double Laplace decomposition method and they are in good agreement with the exact solutions of the problems. We have provided three different examples in order to demonstrate the efficiency, high accuracy, and the simplicity of the present method. Further, we plot the exact solutions, as well as the numerical solutions, in Figures 1–4, and we can easily see the efficieny of and agreement among the solutions.

Author Contributions: Conceptualization, H.E. and Y.T.; Data curation, H.E.; Formal analysis, H.E. and Y.T.; Methodology, H.E.; Supervision, H.E.; Validation, S.M.; Visualization, H.E. and A.K.; Writing, H.E. draft preparation, H.E.; Writing review and editing, A.K.

Funding: The authors would like to extend their sincere appreciation to the Deanship of Scientific Research at King Saud University for its funding this Research group No (RG-1440-030).

Conflicts of Interest: The authors declare no conflict of interest.

References

1. Abdelhakim, A.A. The flaw in the conformable calculus: It is conformable because it is not fractional. *Fract. Calcul. Appl. Anal.* **2019**, *22*, 242–254. [CrossRef]
2. Khalil, R.; Al Horani, M.; Yousef, A.; Sababheh, M. A new definition of fractional derivative. *J. Comput. Appl. Math.* **2014**, *264*, 65–70. [CrossRef]
3. Çenesiz, Y.; Baleanu, D.; Kurt, A.; Tasbozan, O. New exact solutions of burgers' type equations with conformable derivative. *Waves Random Complex Media* **2017**, *27*, 103–116. [CrossRef]
4. Korkmaz, A.; Hosseini, K. Exact solutions of a nonlinear conformable time-fractional parabolic equation with exponential nonlinearity using reliable methods. *Opt. Quantum Electron.* **2017**, *49*, 278. [CrossRef]
5. Ortigueira, M.D.; Machado, J.T. What is a fractional derivative? *J. Comput. Phys.* **2015**, *293*, 4–13. [CrossRef]
6. Aminikhah, H.; Sheikhani, A.R.; Rezazadeh, H. Sub-equation method for the fractional regularized long-wave equations with conformable fractional derivatives. *Sci. Iranica Trans. B Mech. Eng.* **2016**, *23*, 1048. [CrossRef]
7. Chung, W.S. Fractional newton mechanics with conformable fractional derivative. *J. Comput. Appl. Math.* **2015**, *290*, 150–158. [CrossRef]
8. Eslami, M.; Rezazadeh, H. The first integral method for wu–zhang system with conformable time-fractional derivative. *Calcolo* **2016**, *53*, 475–485. [CrossRef]
9. Ünal, E.; Gökdoğan, A. Solution of conformable fractional ordinary differential equations via differential transform method. *Optik* **2017**, *128*, 264–273. [CrossRef]
10. Rahimi, Z.; Rezazadeh, G.; Sumelka, W.; Yang, X. A study of critical point instability of micro and nano beams under a distributed variable-pressure force in the framework of the inhomogeneous non-linear nonlocal theory. *Arch. Mech.* **2017**, *69*, 413–433.
11. Rahimi, Z.; Sumelka, W.; Yang, X.-J. A new fractional nonlocal model and its application in free vibration of timoshenko and euler-bernoulli beams. *Eur. Phys. J. Plus* **2017**, *132*, 479. [CrossRef]
12. Tallafha, A.; Al Hihi, S. Total and directional fractional derivatives. *Int. J. Pure Appl. Math.* **2016**, *107*, 1037–1051. [CrossRef]
13. Hashemi, M. Invariant subspaces admitted by fractional differential equations with conformable derivatives. *Chaos Solitons Fractals* **2018**, *107*, 161–169. [CrossRef]
14. Özkan, O.; Kurt, A. On conformable double laplace transform. *Opt. Quantum Electron.* **2018**, *50*, 103. [CrossRef]
15. Jarad, F.; Abdeljawad, T. A modified laplace transform for certain generalized fractional operators. *Results Nonlinear Anal.* **2018**, *1*, 88–98.

16. Eltayeb, H.; Mesloub, S.; Kılıçman, A. Application of double laplace decomposition method to solve a singular one-dimensional pseudohyperbolic equation. *Adv. Mech. Eng.* **2017**, *9*, 1687814017716638. [CrossRef]
17. Eroğlu, B.; Avcı, D.; Özdemir, N. Optimal control problem for a conformable fractional heat conduction equation. *Acta Phys. Pol. A* **2017**, *132*, 658–662. [CrossRef]

mathematics

MDPI

Article

Existence and Iterative Method for Some Riemann Fractional Nonlinear Boundary Value Problems

Imed Bachar [1,*], Habib Mâagli [2,3] and Hassan Eltayeb [1]

1 Mathematics Department, College of Science, King Saud University, P.O. Box 2455, Riyadh 11451, Saudi Arabia; hgadain@ksu.edu.sa
2 Department of Mathematics, College of Sciences and Arts, King Abdulaziz University, Rabigh Campus, P.O. Box 344, Rabigh 21911, Saudi Arabia; abobaker@kau.edu.sa or habib.maagli@fst.rnu.tn
3 Analyse Harmonique et Théorie du Potentiel, LR10ES09 Modélisation Mathématique, Faculté des Sciences de Tunis, Université de Tunis El Manar, Tunis 2092, Tunisie
* Correspondence: abachar@ksu.edu.sa; Tel.: +966-114676518

Received: 3 September 2019; Accepted: 30 September 2019; Published: 13 October 2019

Abstract: In this paper, we prove the existence and uniqueness of solution for some Riemann–Liouville fractional nonlinear boundary value problems. The positivity of the solution and the monotony of iterations are also considered. Some examples are presented to illustrate the main results. Our results generalize those obtained by Wei et al (Existence and iterative method for some fourth order nonlinear boundary value problems. Appl. Math. Lett. 2019, 87, 101–107.) to the fractional setting.

Keywords: fractional differential equation; Green's function; existence and uniqueness of solution; positivity of solution; iterative method

1. Introduction

Forth-order boundary value problems, can be used to model the deformation of the elastic beam, which is considered to be one of the most used elements in structures such as bridges, buildings and aircraft (see, for instance, [1,2]).

In the literature problems of the form

$$u^{(4)}(x) = f(x, u(x), u''(x)), \ x \in (0,1), \tag{1}$$

subject to different types of boundary conditions have been extensively studied (see, for example, [1–11] and the references therein).

Under adequate conditions imposed on f and using different approach, the existence, uniqueness and qualitative properties of solutions have been considered.

In [1], Aftabizadeh considered Equation (1) together with the boundary conditions:

$$u(0) = u(1) = u''(0) = u''(1) = 0, \tag{2}$$

where $f : [0,1] \times \mathbb{R}^2 \to \mathbb{R}$ is continuous. Under adequate conditions imposed on f he proved that problem (1)–(2) has a unique solution. To do this, he transforms Equation (1) into a second-order integro-differential equation and apply the Schauder's fixed point theorem.

In [4], by using the method of lower and upper solutions for a fourth-order equation and some restrictive conditions on f, Bai established an existence result to problem (1)–(2).

In [7], Dang et al., to prove the existence and uniqueness of a solution of the problem (1)–(2), they reduced the problem to an operator equation for the right-hand side function and proved the contraction of the operator under some convenient conditions on f. The positivity of the solution and the monotony of iterations are also considered. This idea was also used by Dang and Qey for cantilever beam equation [12].

Recently, in [11], Wei et al. considered the following problem

$$\begin{cases} u^{(4)}(x) = f(x, u(x), u'(x)), & t \in (0,1), \\ u(0) = u'(0) = u'(1) = u''(1) = 0. \end{cases} \tag{3}$$

Observe that problem (3) cannot be reduced to two second-order problems. Nevertheless, following the idea developed in [7], they proved the existence and uniqueness of this problem.

Motivated by the mentioned works, in this paper, we generalize the results obtained in [11] to the fractional setting.

More precisely, we are concerned with the following problem

$$\begin{cases} D^{\beta}(D^{\alpha}u)(x) = f(x, u(x), D^{\alpha}u(x)), & x \in (0,1), \\ u(0) = D^{\alpha}u(0) = D^{\alpha}u(1) = (D^{\alpha}u)'(1) = 0, \end{cases} \tag{4}$$

where $0 < \alpha \leq 1, 2 < \beta \leq 3$, and $f : [0,1] \times \mathbb{R}^2 \to \mathbb{R}$ continuous function satisfying some adequate assumptions. Here D^{α} (resp. D^{β}) denotes the Riemann–Liouville fractional derivative of order α (resp. β).

It is worth mentioning that many authors studied fractional differential equations which were applied in many fields such as physics, mechanics, chemistry, and engineering; (see, for instance [13–32] and the references therein).

Following a different approach, they addressed the question of existence and uniqueness of positive continuous solution.

In [31], the authors considered the two-dimensional fractional Schrödinger equation (FSE) without potential

$$i\frac{\partial \psi}{\partial z} - \left(-\frac{\partial^2}{\partial x^2} - \frac{\partial^2}{\partial y^2}\right)^{\frac{\alpha}{2}} \psi = 0, \tag{5}$$

for the slowly varying envelope ψ of the optical field and $1 < \alpha \leq 2$.

They transformed Equation (5) into a Dirac–Weyl-like equation, which is used to establish a link with light propagation in the honeycomb lattice (HCL). They discovered a very similar behavior—the conical diffraction. This similarity in behavior is broken if an additional potential is brought into system.

Our paper is organized as follows. In Section 2, we establish some estimates on the Green's function and we prove appropriate inequalities on some integral operators involving the Green' function. In Section 3, under adequate conditions imposed on function f, we prove the existence and uniqueness of a solution of problem (4). Our approach is based on the Banach contraction principle. The positivity of the solution and the monotony of iterations are also considered. Some examples are given to illustrate our existence results.

Throughout this paper, we denote by $C([0,1])$ the set of continuous functions in $[0,1]$. We recall that the space $C([0,1])$ equipped with the uniform norm $\|u\| := \max_{x \in [0,1]} |u(x)|$ is a Banach space.

2. Preliminary Results

2.1. Fractional Calculus

We recall in this section some basic definitions on fractional calculus (see [33–36]).

Definition 1. *The Riemann–Liouville fractional integral of order $\gamma > 0$ for a measurable function $f : (0, \infty) \to \mathbb{R}$ is defined as*

$$I^{\gamma} f(x) = \frac{1}{\Gamma(\gamma)} \int_0^x (x-t)^{\gamma-1} f(t)\, dt, \quad x > 0,$$

provided that the right-hand side is pointwise defined on $(0,\infty)$. *Here* Γ *is the Euler Gamma function.*

Definition 2. *The Riemann–Liouville fractional derivative of order* $\gamma > 0$ *for a measurable function* $f : (0,\infty) \to \mathbb{R}$ *is defined as*

$$D^{\gamma} f(x) = \frac{1}{\Gamma(n-\gamma)} \left(\frac{d}{dx}\right)^{n} \int_0^x (x-t)^{n-\gamma-1} f(t)\, dt = \left(\frac{d}{dx}\right)^{n} I^{n-\gamma} f(x),$$

provided that the right-hand side is pointwise defined on $(0,\infty)$. *Here* $n = [\gamma]+1$, *where* $[\gamma]$ *denotes the integer part of* γ.

Please note that if $\gamma = m \in \mathbb{N}\backslash\{0\}$, *then we obtain the classical derivative of order* m.

Lemma 1. *Let* $\gamma > 0$ *and* $u \in C(0,1) \cap L^1(0,1)$. *Then we have*

(*i*) *For* $0 < \gamma < \delta$, $D^{\gamma} I^{\delta} u = I^{\delta-\gamma} u$ *and* $D^{\gamma} I^{\gamma} u = u$.

(*ii*) $D^{\gamma} u(x) = 0$ *if and only if* $u(x) = c_1 x^{\gamma-1} + c_2 x^{\gamma-2} + \ldots + c_n x^{\gamma-m}$,

where m *is the smallest integer greather than or equal to* γ *and* $c_i \in \mathbb{R}$ $(i = 1,\ldots,m)$ *are arbitrary constants.*

(*iii*) *Assume that* $D^{\gamma} u \in C(0,1) \cap L^1(0,1)$, *then*

$$I^{\gamma} D^{\gamma} u(x) = u(x) + c_1 x^{\gamma-1} + c_2 x^{\gamma-2} + \ldots + c_m x^{\gamma-m},$$

where m *is the smallest integer greather than or equal to* γ *and* $c_i \in \mathbb{R}$ $(i = 1,\ldots,m)$ *are arbitrary constants.*

Proof. For the convenience of the reader, we provide the proof of property (*ii*) which plays an important role in the rest of the paper.

The property is clear if $\gamma = m \in \mathbb{N}\backslash\{0\}$. Next we assume that $m-1 < \gamma < m$.

We claim that for $i = 1, 2, \ldots, m$,

$$D^{\gamma}(t^{\gamma-i})(x) = 0.$$

Indeed, by elementary calculus, we have

$$I^{m-\gamma}(t^{\gamma-i})(x) = \frac{1}{\Gamma(m-\gamma)} \int_0^x (x-t)^{m-\gamma-1} t^{\gamma-i} dt = \frac{\Gamma(\gamma+1-i)}{\Gamma(m-i+1)} x^{m-i}.$$

Hence

$$D^{\gamma}(t^{\gamma-i})(x) = \left(\frac{d}{dx}\right)^{m} \left(I^{m-\gamma}(t^{\gamma-i})\right)(x) = 0.$$

Therefore, if $u(x) = \sum_{i=1}^{m} c_i x^{\gamma-i}$, then $D^{\gamma} u(x) = 0$.

Conversely, assume that $D^{\gamma} u(x) = 0$.

From Definition 2, we obtain

$$I^{m-\gamma} u(x) = a_0 + a_1 x + \ldots + a_{m-1} x^{m-1},$$

where $a_i \in \mathbb{R}(i = 0, 1, \ldots, m-1)$ are arbitrary constants.

Using property (i), we deduce that

$$\begin{aligned} u(x) &= D^{m-\gamma}(I^{m-\gamma}u)(x) \\ &= \sum_{i=0}^{m-1} a_i D^{m-\gamma}(t^i)(x) \\ &= \sum_{i=0}^{m-1} a_i \frac{\Gamma(1+i)}{\Gamma(1+i-m+\gamma)} x^{i-m+\gamma} \\ &= \sum_{i=1}^{m} c_i x^{\gamma-i}, \end{aligned}$$

where $c_i \in \mathbb{R}(i = 1, ..., m)$ are arbitrary constants. □

2.2. Estimates on the Green's Function

Lemma 2. *Let $2 < \beta \leq 3$ and $\varphi \in C([0,1])$, then the boundary-value problem,*

$$\begin{cases} D^{\beta}v(x) = \varphi(x) \text{ in } (0,1), \\ v(0) = v(1) = v'(1) = 0, \end{cases} \tag{6}$$

has a unique solution

$$v(x) = \int_0^1 G_{\beta}(x,t)\,\varphi(t)\,dt, \tag{7}$$

where for $x, t \in [0,1]$,

$$\begin{aligned} G_{\beta}(x,t) &= \frac{1}{\Gamma(\beta)} \begin{cases} G(x,t), & \text{for } 0 \leq x \leq t \leq 1, \\ G(x,t) + (x-t)^{\beta-1}, & \text{for } 0 \leq t \leq x \leq 1, \end{cases} \\ &= \frac{1}{\Gamma(\beta)} (G(x,t) + (\max(x-t,0))^{\beta-1}), \end{aligned} \tag{8}$$

with

$$G(x,t) := x^{\beta-2}(1-t)^{\beta-2}\left[(\beta-1)(t-x) + (\beta-2)x(1-t)\right] \tag{9}$$

$$= (\beta-1)t(1-x)x^{\beta-2}(1-t)^{\beta-2} - x^{\beta-1}(1-t)^{\beta-1}. \tag{10}$$

$G_{\beta}(x,t)$ is called Green's function of boundary-value problem (6).

Proof. By means of Lemma 1, we can reduce equation $D^{\beta}v(x) = \varphi(x)$ to an equivalent integral equation

$$v(x) = c_1 x^{\beta-1} + c_2 x^{\beta-2} + c_3 x^{\beta-3} + I^{\beta}\varphi(x), \tag{11}$$

where $(c_1, c_2, c_3) \in \mathbb{R}^3$.

The boundary condition $v(0) = 0$ implies that $c_3 = 0$, while the condition $v(1) = 0$, gives

$$c_1 + c_2 + I^{\beta}\varphi(1) = 0. \tag{12}$$

On the other hand, since $v'(1) = 0$, we obtain

$$(\beta-1)c_1 + (\beta-2)c_2 + I^{\beta-1}\varphi(1) = 0.$$

Hence

$$c_1 = (\beta-2)\,I^{\beta}\varphi(1) - I^{\beta-1}\varphi(1) \text{ and } c_2 = I^{\beta-1}\varphi(1) - (\beta-1)\,I^{\beta}\varphi(1).$$

Therefore the unique solution of problem (6) is

$$\begin{aligned}
v(x) &= \frac{(\beta-2)}{\Gamma(\beta)}\int_0^1 x^{\beta-1}(1-t)^{\beta-1}\varphi(t)\,dt - \frac{1}{\Gamma(\beta-1)}\int_0^1 x^{\beta-1}(1-t)^{\beta-2}\varphi(t)\,dt \\
&\quad +(\frac{1}{\Gamma(\beta-1)}\int_0^1 x^{\beta-2}(1-t)^{\beta-2}\varphi(t)\,dt - \frac{(\beta-1)}{\Gamma(\beta)}\int_0^1 x^{\beta-2}(1-t)^{\beta-1}\varphi(t)\,dt \\
&\quad +\frac{1}{\Gamma(\beta)}\int_0^x (x-t)^{\beta-1}\varphi(t)\,dt \\
&= \frac{1}{\Gamma(\beta)}\int_0^1 x^{\beta-2}(1-t)^{\beta-2}((\beta-1)(t-x)+(\beta-2)x(1-t))\varphi(t)\,dt \\
&\quad +\frac{1}{\Gamma(\beta)}\int_0^x (x-t)^{\beta-1}\varphi(t)\,dt \\
&= \int_0^1 G_\beta(x,t)\,\varphi(t)\,dt.
\end{aligned}$$

The proof is completed. □

In the following, for some values of β we give the representation of the Green function $G_\beta(x,t)$ with the contours and the projections on some coordinate planes (see Figures 1–3). These details give an immediate idea of the behavior of these functions.

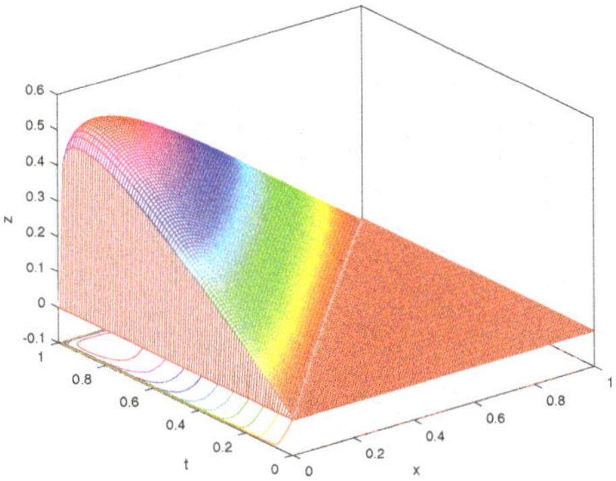

(**a**) $G_\beta(x,t)$ and contours.

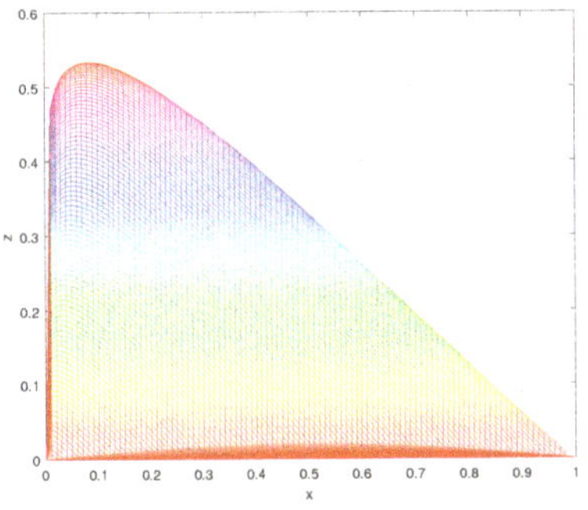

(**b**) Projection of graph of $G_\beta(x,t)$ on the plane xz.

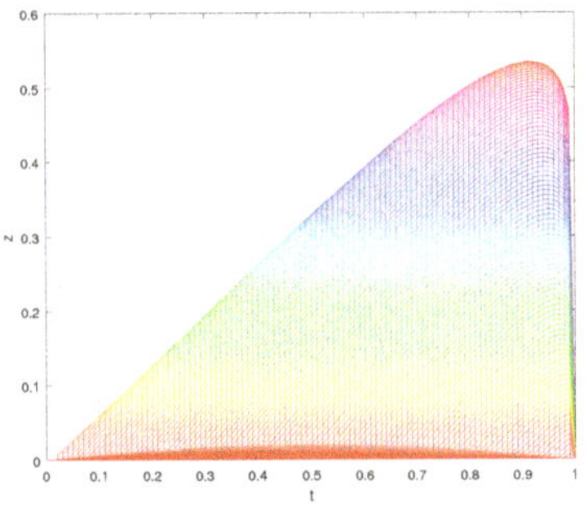

(**c**) Projection of graph of $G_\beta(x,t)$ on the plane tz.

Figure 1. The Green function for $\beta = 2.1$.

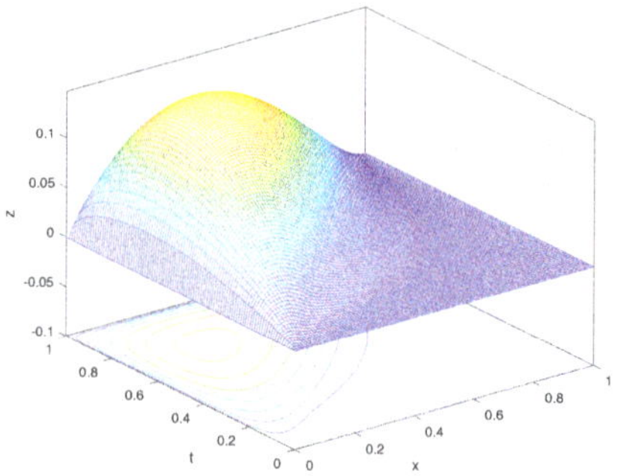

(**a**) $G_\beta(x,t)$ and contours.

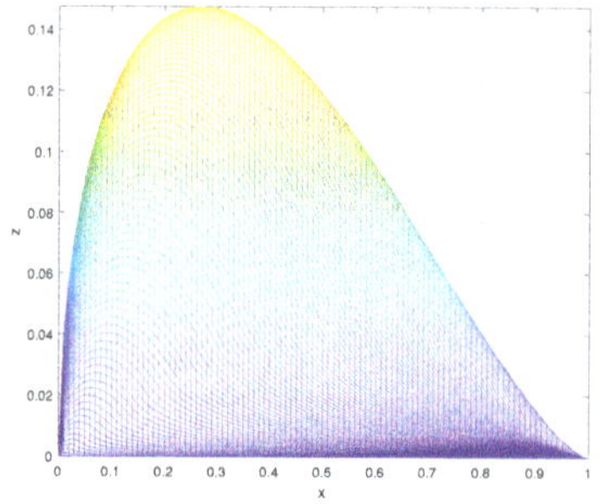

(**b**) Projection of graph of $G_\beta(x,t)$ on the plane xz.

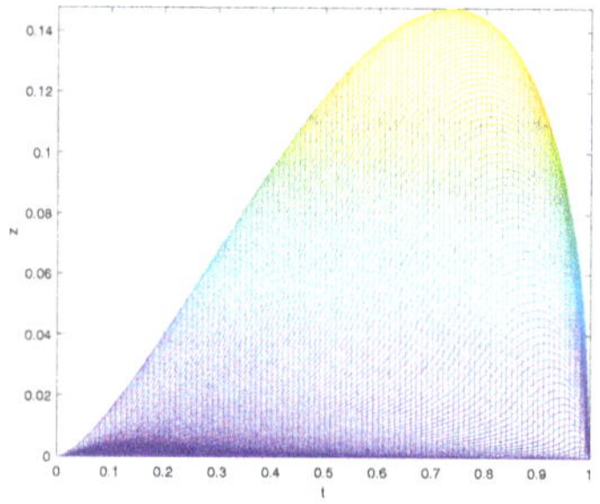

(**c**) Projection of graph of $G_\beta(x,t)$ on the plane tz.

Figure 2. The Green function for $\beta = 5/2$.

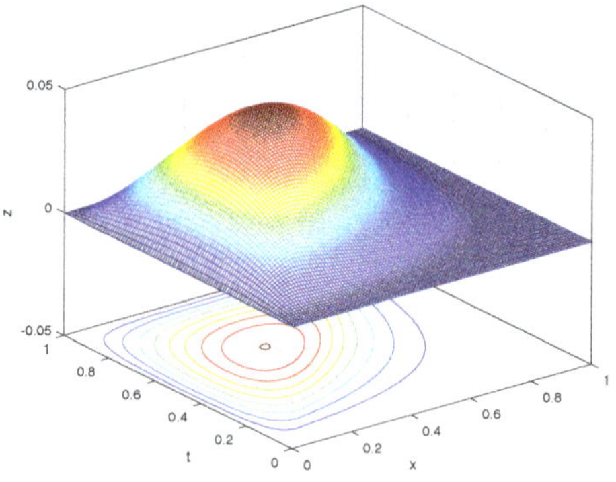

(**a**) $G_\beta(x,t)$ and contours.

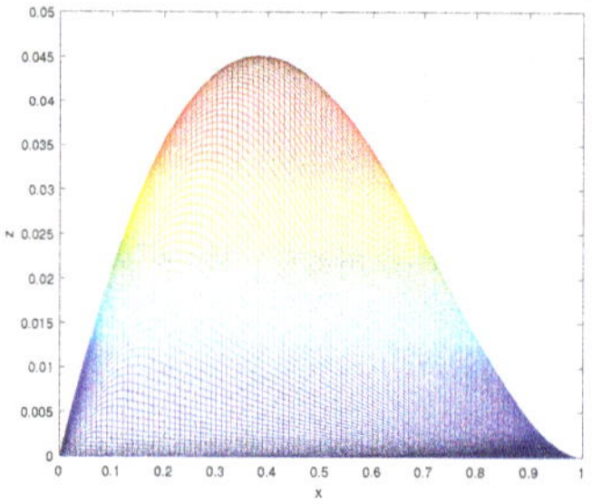

(**b**) Projection of graph of $G_\beta(x,t)$ on the plane xz.

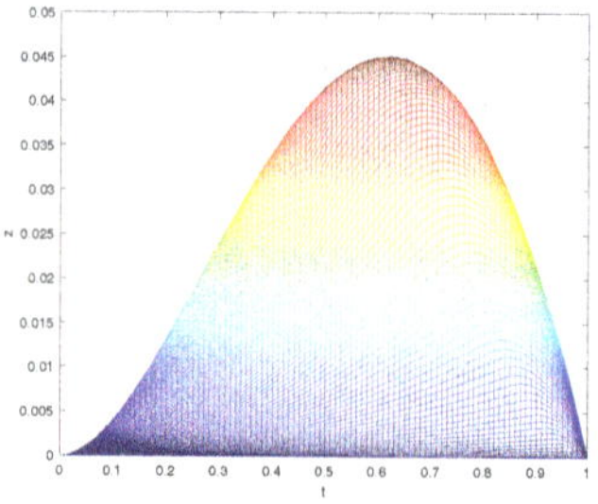

(**c**) Projection of graph of $G_\beta(x,t)$ on the plane tz.

Figure 3. The Green function for $\beta = 3$.

Proposition 1. *Let $2 < \beta \leq 3$. The Green function $G_\beta(x,t)$ satisfies the following properties.*

(i) $(x,t) \to G_\beta(x,t)$ *is continuous on* $[0,1] \times [0,1]$.

(ii) *For* $0 \leq x \leq t \leq 1$, *we have*

$$(\beta - 2)\, H(x,t) \leq \Gamma(\beta)\, G_\beta(x,t) \leq (\beta - 1)\, H(x,t),$$

where $H(x,t) := t\,(1-x)\, x^{\beta-2}\,(1-t)^{\beta-2}$.

(iii) *For* $0 \leq t \leq x \leq 1$, *we have*

$$(\beta - 2)\, \overline{H}(x,t) \leq 2\Gamma(\beta - 1)\, G_\beta(x,t) \leq \overline{H}(x,t),$$

where $\overline{H}(x,t) := t^2\,(1-x)^2\, x^{\beta-3}\,(1-t)^{\beta-3}$.

Proof. *(i)* It is clear.
(ii) Assume that $0 \leq x \leq t \leq 1$. From (8) and (9) we have

$$\begin{aligned}\Gamma(\beta)\, G_\beta(x,t) &= x^{\beta-2}\,(1-t)^{\beta-2}\,[(\beta-1)(t-x) + (\beta-2)\,x\,(1-t)] \\ &\leq (\beta-1)x^{\beta-2}\,(1-t)^{\beta-2}\,[(t-x) + x\,(1-t)] \\ &\leq (\beta-1)H(x,t).\end{aligned}$$

On the other hand, since $t - x \geq 0$, we get

$$\Gamma(\beta)\, G_\beta(x,t) \geq (\beta-2)H(x,t).$$

(iii) Now, assume that $0 \leq t \leq x \leq 1$.
Since

$$x^{\beta-1}\,(1-t)^{\beta-1} - (x-t)^{\beta-1} = (\beta-1)t(1-x)\int_0^1 (x - t + st(1-x))^{\beta-2} ds,$$

it follows from (8) and (10) that

$$G_\beta(x,t) = \frac{1}{\Gamma(\beta-1)} t(1-x)x^{\beta-2}\,(1-t)^{\beta-2}\int_0^1 (1 - (\frac{x-t+st(1-x)}{x(1-t)})^{\beta-2}) ds. \tag{13}$$

Now, using the fact that

$$(\beta-2)(1 - (\frac{x-t+st(1-x)}{x(1-t)})) \leq 1 - (\frac{x-t+st(1-x)}{x(1-t)})^{\beta-2} \leq 1 - (\frac{x-t+st(1-x)}{x(1-t)}), \tag{14}$$

we deduce from (13) that

$$\begin{aligned}\Gamma(\beta-1)\, G_\beta(x,t) &\leq t(1-x)x^{\beta-3}\,(1-t)^{\beta-3}\int_0^1 t(1-x)(1-s) ds \\ &\leq \frac{1}{2} t^2(1-x)^2 x^{\beta-3}\,(1-t)^{\beta-3}.\end{aligned}$$

Similarly, using again (13) and (14), we obtain

$$\Gamma(\beta-1)\, G_\beta(x,t) \geq \frac{(\beta-2)}{2} t^2(1-x)^2 x^{\beta-3}\,(1-t)^{\beta-3}.$$

□

Throughout this paper, for $2 < \beta \leq 3$ and $\varphi \in C([0,1])$, we denote by

$$G_\beta \varphi(x) = \int_0^1 G_\beta(x,t)\, \varphi(t)dt, \text{ for } x \in [0,1], \tag{15}$$

where $G_\beta(x,t)$ is given by (8).

Lemma 3. *Let $0 < \alpha \leq 1, 2 < \beta \leq 3$ and $\varphi \in C([0,1])$. Then the following assertions hold:*

$$\|G_\beta \varphi\| \leq K_\beta \|\varphi\| \text{ and } \|I^\alpha (G_\beta \varphi)\| \leq M_{\alpha,\beta} \|\varphi\|, \tag{16}$$

where

$$K_\beta := \frac{4}{\beta^2 \Gamma(\beta+1)} \left(\frac{\beta-2}{\beta}\right)^{\beta-2}, \tag{17}$$

and

$$M_{\alpha,\beta} := \frac{\omega^{\alpha+\beta-2}}{\beta^2 \Gamma(\alpha+\beta+1)} \left(\left(1 + \sqrt{\frac{(\alpha+\beta-1)(1-\alpha)}{\beta-1}}\right)^2 + \frac{\alpha(\alpha+\beta)}{\beta-1} \right), \tag{18}$$

with $\omega := \frac{\alpha+\beta-1}{\beta} - \frac{1}{\beta}\sqrt{\frac{(\alpha+\beta-1)(1-\alpha)}{\beta-1}}$.

Proof. Let $\varphi \in C([0,1])$. By (15), we have for $x \in [0,1]$

$$|G_\beta \varphi(x)| \leq \|\varphi\| \int_0^1 G_\beta(x,t)\, dt. \tag{19}$$

Using Lemma 2, we obtain

$$\begin{aligned}
\int_0^1 G_\beta(x,t)\, dt &= \frac{1}{\Gamma(\beta)} \int_0^1 G(x,t)\, dt + \frac{1}{\Gamma(\beta)} \int_0^x (x-t)^{\beta-1}\, dt \\
&= \frac{(\beta-1)}{\Gamma(\beta)} (1-x) x^{\beta-2} \int_0^1 t(1-t)^{\beta-2}\, dt \\
&\quad - \frac{1}{\Gamma(\beta)} x^{\beta-1} \int_0^1 (1-t)^{\beta-1}\, dt + \frac{1}{\Gamma(\beta)} \int_0^x (x-t)^{\beta-1}\, dt \\
&= \frac{1}{\Gamma(\beta+1)} ((1-x)x^{\beta-2} - x^{\beta-1} + x^\beta) \\
&= \frac{1}{\Gamma(\beta+1)} (1-x)^2 x^{\beta-2} := \theta(x).
\end{aligned} \tag{20}$$

By simple computation we obtain

$$\|\theta\| = \max_{x\in[0,1]} |\theta(x)| = \theta\left(\frac{\beta-2}{\beta}\right) = K_\beta. \tag{21}$$

Hence from (19) and (21), we get the first inequality in (16).

Now, using Definition 1 and (20), we obtain for $x \in [0,1]$

$$\begin{aligned} |I^{\alpha}(G_{\beta}\varphi)(x)| &\leq \frac{\|\varphi\|}{\Gamma(\alpha)\Gamma(\beta+1)}\int_0^x (x-t)^{\alpha-1}(1-t)^2 t^{\beta-2}dt \\ &= \frac{\|\varphi\|}{\Gamma(\alpha)\Gamma(\beta+1)}\int_0^x (x-t)^{\alpha-1}(t^{\beta}-2t^{\beta-1}+t^{\beta-2})dt \\ &= \frac{\|\varphi\|}{\Gamma(\beta+1)}\Big(\frac{\Gamma(\beta+1)}{\Gamma(\alpha+\beta+1)}x^{\alpha+\beta}-2\frac{\Gamma(\beta)}{\Gamma(\alpha+\beta)}x^{\alpha+\beta-1} \\ &\quad +\frac{\Gamma(\beta-1)}{\Gamma(\alpha+\beta-1)}x^{\alpha+\beta-2}\Big) \\ &= \frac{\|\varphi\|}{\Gamma(\alpha+\beta)}\psi(x), \end{aligned} \tag{22}$$

where
$$\psi(x) = \frac{1}{\alpha+\beta}x^{\alpha+\beta} - \frac{2}{\beta}x^{\alpha+\beta-1} + \frac{\alpha+\beta-1}{\beta(\beta-1)}x^{\alpha+\beta-2}.$$

Observe that

$$\begin{aligned} \psi'(x) &= x^{\alpha+\beta-3}\Big(x^2-2\frac{(\alpha+\beta-1)}{\beta}x+\frac{(\alpha+\beta-1)(\alpha+\beta-2)}{\beta(\beta-1)}\Big) \\ &= x^{\alpha+\beta-3}(x-\omega)(x-\overline{\omega}), \end{aligned}$$

where $\omega = \frac{(\alpha+\beta-1)}{\beta} - \frac{1}{\beta}\sqrt{\frac{(\alpha+\beta-1)(1-\alpha)}{\beta-1}}$ and $\overline{\omega} = \frac{(\alpha+\beta-1)}{\beta} + \frac{1}{\beta}\sqrt{\frac{(\alpha+\beta-1)(1-\alpha)}{\beta-1}}$.

Since $\omega \in (0,1]$ and $\overline{\omega} \geq 1$, it follows that $\psi'(x) \geq 0$ on $[0,\omega]$ and $\psi'(x) \leq 0$ on $[\omega,1]$.

Hence
$$\|\psi\| = \psi(\omega). \tag{23}$$

By combining (22) and (23), we obtain the second inequality in (16). □

3. Main Results

Let $0 < \alpha \leq 1$ and $2 < \beta \leq 3$. For each real number $M > 0$, denote by

$$\mathcal{D}_M = \{(x,u,v) \in \mathbb{R}^3 : \ 0 \leq x \leq 1, \ |u| \leq MM_{\alpha,\beta}, \ |v| \leq MK_{\beta}\},$$

where K_β and $M_{\alpha,\beta}$ are respectively given by (17) and (18).

By $B[O,M]$, we denote the closed ball centered at O with radius M in the space $C([0,1])$.

3.1. Existence and Uniqueness of a Solution

Theorem 1. *Let $f : [0,1] \times \mathbb{R}^2 \to \mathbb{R}$ be a continuous function and assume that there exist numbers $M, L_1, L_2 \geq 0$ such that*

(i) $|f(x,u,v)| \leq M$ *for any* $(x,u,v) \in \mathcal{D}_M$.

(ii) $|f(x,u_2,v_2) - f(x,u_1,v_1)| \leq L_1|u_2-u_1| + L_2|v_2-v_1|$,

for any $(x,u_i,v_i) \in \mathcal{D}_M$, $i = 1,2$.

(iii) $q := L_1M_{\alpha,\beta} + L_2K_\beta < 1$.

Then the boundary value problem (4) has a unique solution $u \in C([0,1])$ satisfying

$$\|u\| \leq MM_{\alpha,\beta} \text{ and } \|D^{\alpha}u\| \leq MK_{\beta}. \tag{24}$$

Proof. Consider the operator $T : C([0,1]) \to C([0,1])$ defined for $\varphi \in C([0,1])$ by

$$T\varphi(x) = f(x, I^{\alpha}(G_\beta\varphi)(x), G_\beta\varphi(x)), \ x \in [0,1], \tag{25}$$

where $G_\beta\varphi$ is defined by (15) and I^α is the Riemann–Liouville fractional integral operator given by Definition 1.

We shall investigate problem (4) via the operator equation (25).

Observe that if φ is a fixed point of the operator T, then by Lemma 1, (15) and Lemma 2,

$$u(x) := I^\alpha(G_\beta\varphi)(x), \tag{26}$$

is a solution of problem (4) and vice versa.

We claim that T is a contraction operator from $B[O,M]$ into itself.

First, we show that the operator T maps $B[O,M]$ into itself.

Indeed, since φ is continuous and by Proposition 1 (i) the Green' s function $G_\beta(x,t)$ is continuous on $[0,1]\times[0,1]$, it is not difficult to check that $T\varphi$ is continuous on $[0,1]$.

Now, for any $\varphi \in B[O,M]$, we have by Lemma 3

$$\|G_\beta\varphi\| \leq MK_\beta \text{ and } \|I^\alpha(G_\beta\varphi)\| \leq MM_{\alpha,\beta}. \tag{27}$$

Hence, for $x \in [0,1]$, we have $(x, I^\alpha(G_\beta\varphi)(x), G_\beta\varphi(x)) \in \mathcal{D}_M$. Therefore, from assumption (i), it follows that $\|T\varphi\| \leq M$. Therefore, the operator T maps $B[O,M]$ into itself.

Secondly, we prove that $T : B[O,M] \to B[O,M]$ is a contraction operator. Indeed, for any $\varphi_1, \varphi_2 \in B[O,M]$, by using assumption (ii) and Lemma 3, we obtain for $x \in [0,1]$,

$$\begin{aligned}
|T\varphi_2(x) - T\varphi_1(x)| &= |f(x, I^\alpha(G_\beta\varphi_2)(x), G_\beta\varphi_2(x)) - f(x, I^\alpha(G_\beta\varphi_1)(x), G_\beta\varphi_1(x))| \\
&\leq L_1 \|I^\alpha(G_\beta\varphi_2) - I^\alpha(G_\beta\varphi_1)\| + L_2 \|G_\beta\varphi_2 - G_\beta\varphi_1)\| \\
&= L_1 \|I^\alpha(G_\beta(\varphi_2 - \varphi_1))\| + L_2 \|G_\beta(\varphi_2 - \varphi_1)\| \\
&\leq L_1 M_{\alpha,\beta} \|\varphi_2 - \varphi_1\| + L_2 K_\beta \|\varphi_2 - \varphi_1\| \\
&= q \|\varphi_2 - \varphi_1\|,
\end{aligned}$$

where q is defined in assumption (iii).

Therefore, T is a contraction operator in $B[O,M]$. Hence, it has a unique fixed point φ in $B[O,M]$.

Therefore, problem (4) has a unique solution $u \in C([0,1])$ given by (26). The estimates (24) follow from Lemma 3 and the fact that $\|\varphi\| \leq M$.

The the proof is completed. □

Next, we present a particular case of Theorem 1. To this end, denote

$$\mathcal{D}_M^+ = \{(x,u,v) \in \mathbb{R}^3 : 0 \leq x \leq 1, 0 \leq u \leq MM_{\alpha,\beta}, 0 \leq v \leq MK_\beta\}.$$

Corollary 1. *Let $f : [0,1] \times \mathbb{R}^2 \to \mathbb{R}$ be a continuous function and assume that there exists numbers $M, L_1, L_2 \geq 0$ such that*

(i) $0 \leq f(x,u,v) \leq M$ *for any* $(x,u,v) \in \mathcal{D}_M^+$.

(ii) $|f(x,u_2,v_2) - f(x,u_1,v_1)| \leq L_1 |u_2 - u_1| + L_2 |v_2 - v_1|$,

for any $(x,u_i,v_i) \in \mathcal{D}_M^+$, $i = 1,2$.

(iii) $q := L_1 M_{\alpha,\beta} + L_2 K_\beta < 1$.

Then the boundary value problem (4) has a unique nonnegative solution $u \in C([0,1])$ satisfying

$$0 \leq u(x) \leq MM_{\alpha,\beta} \text{ and } 0 \leq D^\alpha u \leq MK_\beta. \tag{28}$$

3.2. Iterative Method and Examples

Consider the following iterative process.

$$\begin{cases} \text{Let } \varphi_0 \in B[O,M], \\ \varphi_{k+1}(x) := T\varphi_k(x) = f(x, I^\alpha(G_\beta\varphi_k)(x), G_\beta\varphi_k(x)), \text{ for } k = 0,1,...;\ x \in [0,1]. \end{cases} \tag{29}$$

Theorem 2. *Assume that hypotheses of Theorem 1 are satisfied. The sequence $(\varphi_k)_{k\geq 0}$ converges with the rate of geometric progression and we have*

$$\left\| I^\alpha(G_\beta\varphi_k) - u \right\| \leq M_{\alpha,\beta}\frac{q^k}{1-q}\left\| \varphi_1 - \varphi_0 \right\|, \tag{30}$$

where u is the exact solution of problem (4) and q is given in assumption (iii) in Theorem 1.

Proof. It is known by the Banach contracting mapping principle that the sequence $(\varphi_k)_{k\geq 0}$ converges with the rate of geometric progression and we have

$$\left\| \varphi_k - \varphi \right\| \leq \frac{q^k}{1-q}\left\| \varphi_1 - \varphi_0 \right\|, \tag{31}$$

where φ is the unique fixed point of the operator T in $B[O,M]$.

Using this fact and Lemma 3, we obtain

$$\begin{aligned} \left\| I^\alpha(G_\beta\varphi_k) - u \right\| &= \left\| I^\alpha(G_\beta\varphi_k) - I^\alpha(G_\beta\varphi) \right\| \\ &= \left\| I^\alpha(G_\beta(\varphi_k - \varphi)) \right\| \\ &\leq M_{\alpha,\beta}\left\| \varphi_k - \varphi \right\| \\ &\leq M_{\alpha,\beta}\frac{q^k}{1-q}\left\| \varphi_1 - \varphi_0 \right\|. \end{aligned}$$

The proof is completed. □

Proposition 2. (*Monotony*)*Assume that hypotheses of Theorem 1 are satisfied. In addition, we assume that the function $f(x,u,v)$ is nondecreasing in u and v for any $(x,u,v) \in \mathcal{D}_M$. Let $\varphi_0, \psi_0 \in B[O,M]$ be initial approximations such that $\varphi_0(x) \leq \psi_0(x)$, for all $x \in [0,1]$. Then*

(i) for all $k \in \mathbb{N}$ and $x \in [0,1]$,

$$I^\alpha(G_\beta\varphi_k)(x) \leq I^\alpha(G_\beta\psi_k)(x). \tag{32}$$

(ii) Suppose further that for all $(x,u,v) \in \mathcal{D}_M$

$$\varphi_0(x) \leq f(x,u,v) \leq \psi_0(x). \tag{33}$$

Then the sequences $(I^\alpha(G_\beta\varphi_k))_{k\geq 0}$ and $(I^\alpha(G_\beta\psi_k))_{k\geq 0}$ converge to the unique solution u of problem (4) and

$$I^\alpha(G_\beta\varphi_k) \leq I^\alpha(G_\beta\varphi_{k+1}) \leq u \leq I^\alpha(G_\beta\psi_{k+1}) \leq I^\alpha(G_\beta\psi_k). \tag{34}$$

In particular, if $\varphi_0 \geq 0$ (resp. $\psi_0 \leq 0$), then u is nonnegative (resp. nonpositive) solution.

Proof. (i) We claim that for all $k \in \mathbb{N}$, we have

$$\varphi_k(x) \leq \psi_k(x), \text{ on } [0,1]. \tag{35}$$

We proceed by induction. From hypothesis, the inequality is clear for $k = 0$. For a given $k \in \mathbb{N}$, assume that $\varphi_k(x) \leq \psi_k(x)$.

Since the Green function is nonnegative, we deduce from (15) and Definition 1 that

$$G_\beta \varphi_k \leq G_\beta \psi_k \text{ and } I^\alpha(G_\beta \varphi_k) \leq I^\alpha(G_\beta \psi_k).$$

Combining this fact and that the function $f(x, u, v)$ is nondecreasing in u and v, we obtain

$$\varphi_{k+1}(x) := f(x, I^\alpha(G_\beta \varphi_k)(x), G_\beta \varphi_k(x)) \leq f(x, I^\alpha(G_\beta \psi_k)(x), G_\beta \psi_k(x)) = \psi_{k+1}(x).$$

So our claim is proved.

Using (35), (15) and Definition 1 we get inequality in (32)

(ii) From Theorem 2, we know that the sequences $(I^\alpha(G_\beta \varphi_k))_{k \geq 0}$ and $(I^\alpha(G_\beta \psi_k))_{k \geq 0}$ converge to the unique solution u of problem (4).

We claim that the sequence $(\varphi_k)_{k \geq 0}$ is nondecreasing.

Indeed, since for $x \in [0, 1]$, we have $(x, I^\alpha(G_\beta \varphi_0)(x), G_\beta \varphi_0(x)) \in \mathcal{D}_M$, we deduce from (33) that

$$\varphi_0(x) \leq f(x, I^\alpha(G_\beta \varphi_0)(x), G_\beta \varphi_0(x)) = \varphi_1(x).$$

Assume that $\varphi_k(x) \leq \varphi_{k+1}(x)$. From (15), Definition 1 and the monotony of the function f, we deduce that

$$\varphi_{k+1}(x) = f(x, I^\alpha(G_\beta \varphi_k)(x), G_\beta \varphi_k(x)) \leq f(x, I^\alpha(G_\beta \varphi_{k+1})(x), G_\beta \varphi_{k+1}(x)) = \varphi_{k+2}(x).$$

Hence the sequence $(\varphi_k)_{k \geq 0}$ is nondecreasing.

Therefore, by using again (15) and Definition 1, it follows that the sequence $(I^\alpha(G_\beta \varphi_k))_{k \geq 0}$ is nondecreasing.

Since the sequence $(I^\alpha(G_\beta \varphi_k))_{k \geq 0}$ converges to u, we obtain

$$I^\alpha(G_\beta \varphi_k) \leq I^\alpha(G_\beta \varphi_{k+1}) \leq u$$

Similarly, we prove that the sequence $(I^\alpha(G_\beta \psi_k))_{k \geq 0}$ is nonincreasing and that

$$u \leq I^\alpha(G_\beta \psi_{k+1}) \leq I^\alpha(G_\beta \psi_k).$$

So inequalities in (34) are proved.

Finally, from (34), we have

$$I^\alpha(G_\beta \varphi_0) \leq u \leq I^\alpha(G_\beta \psi_0).$$

This implies that if $\varphi_0 \geq 0$ (resp. $\psi_0 \leq 0$), then u is nonnegative (resp. nonpositive) solution.

This completes the proof. $\square$

Example 1. *Consider the following boundary value problem:*

$$\begin{cases} D^{\frac{5}{2}}(D^{\frac{1}{2}}u)(x) = xu(x) + x^2(D^{\frac{1}{2}}u(x))^2 + 2x + 1, & x \in (0,1), \\ u(0) = D^{\frac{1}{2}}u(0) = D^{\frac{1}{2}}u(1) = (D^{\frac{1}{2}}u)'(1) = 0. \end{cases} \tag{36}$$

In this case $K_{\frac{5}{2}} = 8.6123 \times 10^{-2}$, $M_{\frac{1}{2},\frac{5}{2}} = 5.4279 \times 10^{-2}$ *and* $f(x, u, v) = xu + x^2v^2 + 2x + 1$. *So condition* (i) *in Theorem 1 will be satisfied if we choose* $M > 0$ *such that*

$$MM_{\frac{1}{2},\frac{5}{2}} + M^2K^2_{\frac{5}{2}} + 3 \leq M.$$

It is easy to verify that $M = 4$ *is an example of suitable choice.*

Since

$$f'_u = x \text{ and } f'_v = 2x^2 v,$$

it follows that for any $(x, u, v) \in \mathcal{D}_4 = \{(x, u, v), 0 \le x \le 1, |u| \le 4M_{\frac{1}{2},\frac{5}{2}}, |v| \le 4K_{\frac{5}{2}}\}$,

$$|f'_u| \le 1 \text{ and } |f'_v| \le 8K_{\frac{5}{2}} \le 1.$$

Hence, $L_1 = 1$ *and* $L_2 = 1$ *satisfy the condition (ii) in Theorem 1. Also, we have* $q := L_1 M_{\frac{1}{2},\frac{5}{2}} + L_2 K_{\frac{5}{2}} = M_{\frac{1}{2},\frac{5}{2}} + K_{\frac{5}{2}} < 1$.

Thus by Theorem 1, problem (36) has a unique solution, and the iterative method converges.

In Figure 4, we present the approximation of the unique solution of problem (36) with $u_k(x) := I^{\frac{1}{2}}(G_{\frac{5}{2}}\varphi_k)(x)$ *and* $\varphi_0(x) := 2x + 1$.

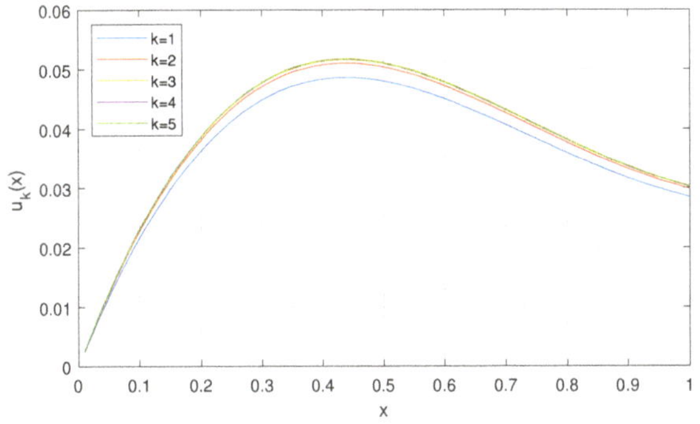

Figure 4. The approximation of the solution of problem (36).

Example 2. *Consider the following boundary value problem:*

$$\begin{cases} D^{\frac{8}{3}}(u')(x) = -3u^2(u'(x))^2 + 3u(x) + 4u'(x) + \sin(\pi x), \;\; x \in (0,1), \\ u(0) = u'(0) = u'(1) = u''(1) = 0. \end{cases} \tag{37}$$

In this example, $K_{\frac{8}{3}} = 5.5637 \times 10^{-2}$, $M_{1,\frac{8}{3}} = 2.1030 \times 10^{-2}$ *and* $f(x, u, v) = -3u^2v^2 + 3u + 4v + \sin(\pi x)$.

As in Example 1, we verify that all conditions of Theorem 1 are satisfied with $M = 3$, $L_1 = 4$ *and* $L_2 = 5$. *Hence problem (37) has a unique solution, and the iterative method converges. Moreover, since in* $\mathcal{D}_3$ *we have* $f'_u \ge 0$ *and* $f'_v \ge 0$, *the function* $f(x, u, v)$ *is nondecreasing in both u and v. Take the initial approximation* $\varphi_0 = f(x, 0, 0) = \sin(\pi x) \ge 0, 0 \le x \le 1$. *By the positivity of the Green's function and Lemma 3, we have*

$$0 \le v_0 := G_{\frac{8}{3}}\varphi_0 \le K_{\frac{8}{3}} \text{ and } 0 \le u_0 := I^1(G_{\frac{8}{3}}\varphi_0) \le M_{1,\frac{8}{3}}.$$

Therefore form the iterative process (29), we obtain

$$\begin{aligned} \varphi_1(x) &= f(x, u_0(x), v_0(x)) \\ &= -3u_0^2v_0^2 + 3u_0 + 4v_0 + \sin(\pi x) \\ &= 3u_0(1 - u_0v_0^2) + 4v_0 + \sin(\pi x) \\ &\ge \sin(\pi x) = \varphi_0. \end{aligned}$$

By Proposition 2, $(u_k := I^\alpha(G_\beta \varphi_k))_{k \ge 0}$ *is a nonnegative increasing sequence which converges to the unique nonnegative solution u. Some iterations are depicted in Figure 5.*

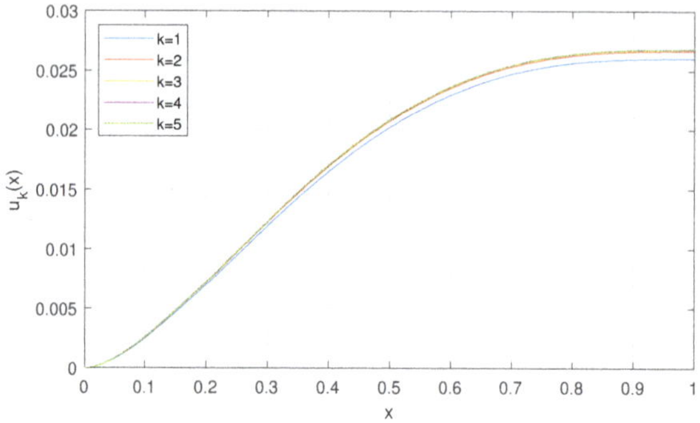

Figure 5. The approximation of the solution of problem (37).

Author Contributions: Investigation, I.B., H.M. and H.E.; Writing–review and editing, I.B. and H.M. All authors contributed equally to the writing of this paper. All authors read and approved the final manuscript.

Funding: Deanship of Scientific Research at King Saud University. Research group NO (RG-1435-043).

Acknowledgments: The authors would like to extend their sincere appreciation to the Deanship of Scientific Research at King Saud University for its funding this Research group NO (RG-1435-043). We also would like to thank the referees for their careful reading of the manuscript and for helpful suggestions which improved the quality of the paper.

Conflicts of Interest: The authors declare no conflict of interest.

References

1. Aftabizadeh, A.R. Existence and uniqueness theorems for fourth-order boundary value problems. *J. Math. Anal. Appl.* **1986**, *116*, 415–426. [CrossRef]
2. Li, Y. Existence of positive solutions for the cantilever beam equations with fully nonlinear terms. *Nonlinear Anal.* **2016**, *27*, 221–237. [CrossRef]
3. Alsaedi, R.S. Existence and global behavior of positive solutions for some fourth order boundary value problems. *Abstr. Appl. Anal.* **2014**, *2014*, 657926. [CrossRef]
4. Bai, Z. The method of lower and upper solutions for a bending of an elastic beam equation. *J. Math. Anal. Appl.* **2010**, *248*, 195–202. [CrossRef]
5. Bai, Z. Positive solutions of some nonlocal fourth-order boundary value problem. *Appl. Math. Comput.* **2010**, *215*, 4191–4197. [CrossRef]
6. Dang, Q.A. Iterative method for solving the Neumann boundary value problem for biharmonic type equation. *J. Comput. Appl. Math.* **2006**, *196*, 634–643. [CrossRef]
7. Dang, Q.A.; Dang, Q.L.; Ngo, T.K.Q. A novel efficient method for nonlinear boundary value problems. *Numer. Algorithms* **2017**, *76*, 427–439.
8. Li, Y. A monotone iterative technique for solving the bending elastic beam equations. *Appl. Math. Comput.* **2010**, *2017*, 2200–2208. [CrossRef]
9. Ma, T.F.; da Silva, J. Iterative solutions for a beam equation with nonlinear boundary conditions of third order. *Appl. Math. Comput.* **2004**, *159*, 11–18. [CrossRef]
10. Ma, R.; Tisdell, C.C. Positive solutions of singular sublinear fourth order boundary value problems. *Appl. Anal.* **2005**, *12*, 1199–1220. [CrossRef]
11. Wei, Y.; Song, Q.; Bai, Z. Existence and iterative method for some fourth order nonlinear boundary value problems. *Appl. Math. Lett.* **2019**, *87*, 101–107. [CrossRef]
12. Dang, Q.; Quy, N. Existence results and iterative method for solving the cantilever beam equation with fully nonlinear term. *Nonlinear Anal. RWA* **2017**, *36*, 56–68. [CrossRef]
13. Agarwal, R.P.; O'Regan, O.; Staněk, S. Positive solutions for Dirichlet problems of singular nonlinear fractional differential equations. *J. Math. Anal. Appl.* **2010**, *371*, 57–68. [CrossRef]

14. Bachar, I.; Mâagli, M.; Toumi, F.; Zine el Abidine, Z. Existence and Global Asymptotic Behavior of Positive Solutions for Sublinear and Superlinear Fractional Boundary Value Problems. *Chin. Ann. Math. Ser. B* **2016**, *37*, 1–28. [CrossRef]
15. Bai, Z.; Lü, H. Positive solutions for boundary value problem of nonlinear fractional differential equation. *J. Math. Anal. Appl.* **2005**, *311*, 495–505. [CrossRef]
16. Diethelm, K.; Freed, A.D. On the solution of nonlinear fractional order differential equations used in the modeling of viscoplasticity. In *Scientific Computing in Chemical Engineering. II-Computational Fluid Dynamics, Reaction Engineering and Molecular Properties*; Keil, F., Mackens, W., Voss, H., Werther, J., Eds.; Springer: Heidelberg, Germany, 1999; pp. 217–224.
17. Gaul, L.; Klein, P.; Kempfle, S. Damping description involving fractional operators. *Mech. Syst. Signal Process.* **1991**, *5*, 81–88. [CrossRef]
18. Glockle, W.G.; Nonnenmacher, T.F. A fractional calculus approach of self-similar protein dynamics. *Biophys. J.* **1995**, *68*, 46–53. [CrossRef]
19. Graef, J.R.; Kong, L.; Kong, Q.; Wang, M. Existence and uniqueness of solutions for a fractional boundary value problem with Dirichlet boundary condition. *Electron. J. Qual. Theory Differ. Equ.* **2013**, *55*, 1–11. [CrossRef]
20. Hilfer, R. *Applications of Fractional Calculus in Physics*; World Scientific: Singapore, 2000.
21. Kaufmann, E.R.; Mboumi, E. Positive solutions of a boundary value problem for a nonlinear fractional differential equation. *Electron. J. Qual. Theory Differ. Equ.* **2008**, *3*, 1–11. [CrossRef]
22. Liang, S.; Zhang, J. Positive solutions for boundary value problems of nonlinear fractional differential equation. *Nonlinear Anal.* **2009**, *71*, 5545–5550. [CrossRef]
23. Mâagli, H.; Mhadhebi, N.; Zeddini, N. Existence and Estimates of Positive Solutions for Some Singular Fractional Boundary Value Problems. *Abstr. Appl. Anal.* **2009**, *2009*, 120781. [CrossRef]
24. Mâagli, H.; Mhadhebi, N.; Zeddini, N. Existence and Exact Asymptotic Behavior of Positive Solutions for a Fractional Boundary Value Problem. *Abstr. Appl. Anal.* **2013**, *2013*, 420514. [CrossRef]
25. Scher, H.; Montroll, E. Anomalous transit-time dispersion in amorphous solids. *Phys. Rev. B.* **1975**, *12*, 2455–2477. [CrossRef]
26. Timoshenko, S.; Gere, J.M. *Theory of Elastic Stability*; McGraw-Hill: New York, NY, USA, 1961.
27. Xu, X.; Jiang, D.; Yuan, C. Multiple positive solutions for the boundary value problem of a nonlinear fractional differenteial equation. *Nonlinear Anal.* **2009**, *71*, 4676–4688. [CrossRef]
28. Xu, X.; Jiang, D.; Yuan, C. Singular Positone and Semipositone Boundary Value Problems of Nonlinear Fractional Differential Equations. *Math. Probl. Eng.* **2009**, *2009*, 535209.
29. Zhang, Y.; Liu, X.; Belić, M.R.; Zhong, W.; Zhang, Y.; Xiao, M. Propagation Dynamics of a Light Beam in a Fractional Schrödinger Equation. *Phys. Rev. Lett.* **2015**, *115*, 180403. [CrossRef]
30. Zhang, X.; Liu, L.; Wu, Y. Multiple positive solutions of a singular fractional differential equation with negatively perturbed term. *Math. Comput. Model.* **2012**, *55*, 1263–1274. [CrossRef]
31. Zhang, D.; Zhang, Y.; Zhang, Z.; Ahmed, N.; Zhang,Y.; Li, F.; Belić, M.R.; Xiao, M. Unveiling the Link Between Fractional Schrödinger Equation and Light Propagation in Honeycomb Lattice. *Ann. Phys. (Berlin)* **2017**, *529*, 1700149. [CrossRef]
32. Zhang, Y.; Zhong, H.; Belić, M.R.; Zhu, Y.; Zhong, W.; Zhang, Y.; Christodoulides, D.N.; Xiao, M. $\mathcal{PT}$ symmetry in a fractional Schrödinger equation. *Laser Photonics Rev.* **2016**, *10*, 526–531. [CrossRef]
33. Kilbas, A.; Srivastava, H.; Trujillo, J. Theory and Applications of Fractional Differential Equations. In *North-Holland Mathematics Studies Vol. 204*; Elsevier: Amsterdam, The Netherlands, 2006.
34. Miller, K.; Ross, B. *An Introduction to the Fractional Calculus and Fractional Differential Equations*; Wiley and Sons: New York, NY, USA, 1993.
35. Podlubny, I. Fractional Differential Equations. In *Mathematics in Science and Engineering*; Academic Press: New York, NY, USA, 1999.
36. Samko, S.; Kilbas, A.; Marichev, O. *Fractional Integrals and Derivative. Theory and Applications*; Gordon and Breach: Yverdon, Switzerland, 1993.

mathematics

MDPI

Article

Positive Solutions for a System of Fractional Integral Boundary Value Problems of Riemann–Liouville Type Involving Semipositone Nonlinearities

Youzheng Ding [1], Jiafa Xu [2] and Zhengqing Fu [3,*]

[1] School of Science, Shandong Jianzhu University, Jinan 250101, China; dingyz@sdjzu.edu.cn
[2] School of Mathematical Sciences, Chongqing Normal University, Chongqing 401331, China; 20150028@cqnu.edu.cn
[3] College of Mathematics and System Sciences, Shandong University of Science and Technology, Qingdao 266590, China
* Correspondence: skd992050@sdust.edu.cn

Received: 18 September 2019; Accepted: 8 October 2019; Published: 14 October 2019

Abstract: In this work by the index of fixed point and matrix theory, we discuss the positive solutions for the system of Riemann–Liouville type fractional boundary value problems

$$D_{0+}^{\alpha}u(t) + f_1(t, u(t), v(t), w(t)) = 0, t \in (0,1),$$

$$D_{0+}^{\alpha}v(t) + f_2(t, u(t), v(t), w(t)) = 0, t \in (0,1),$$

$$D_{0+}^{\alpha}w(t) + f_3(t, u(t), v(t), w(t)) = 0, t \in (0,1),$$

$$u(0) = u'(0) = \cdots = u^{(n-2)}(0) = 0, D_{0+}^{p}u(t)|_{t=1} = \int_0^1 h(t)D_{0+}^{q}u(t)dt,$$

$$v(0) = v'(0) = \cdots = v^{(n-2)}(0) = 0, D_{0+}^{p}v(t)|_{t=1} = \int_0^1 h(t)D_{0+}^{q}v(t)dt,$$

$$w(0) = w'(0) = \cdots = w^{(n-2)}(0) = 0, D_{0+}^{p}w(t)|_{t=1} = \int_0^1 h(t)D_{0+}^{q}w(t)dt,$$

where $\alpha \in (n-1, n]$ with $n \in \mathbb{N}$, $n \geq 3$, $p, q \in \mathbb{R}$ with $p \in [1, n-2]$, $q \in [0, p]$, D_{0+}^{α} is the α order Riemann–Liouville type fractional derivative, and $f_i (i = 1,2,3) \in C([0,1] \times \mathbb{R}^+ \times \mathbb{R}^+ \times \mathbb{R}^+, \mathbb{R})$ are semipositone nonlinearities.

Keywords: Riemann–Liouville type fractional problem; positive solutions; the index of fixed point; matrix theory

1. Introduction

In this work the positive solutions for the system of fractional boundary value problems involving Riemann–Liouville type are considered:

$$\begin{cases} D_{0+}^{\alpha}u(t) + f_1(t, u(t), v(t), w(t)) = 0, t \in (0,1), \\ D_{0+}^{\alpha}v(t) + f_2(t, u(t), v(t), w(t)) = 0, t \in (0,1), \\ D_{0+}^{\alpha}w(t) + f_3(t, u(t), v(t), w(t)) = 0, t \in (0,1), \\ u(0) = u'(0) = \cdots = u^{(n-2)}(0) = 0, D_{0+}^{p}u(t)|_{t=1} = \int_0^1 h(t)D_{0+}^{q}u(t)dt, \\ v(0) = v'(0) = \cdots = v^{(n-2)}(0) = 0, D_{0+}^{p}v(t)|_{t=1} = \int_0^1 h(t)D_{0+}^{q}v(t)dt, \\ w(0) = w'(0) = \cdots = w^{(n-2)}(0) = 0, D_{0+}^{p}w(t)|_{t=1} = \int_0^1 h(t)D_{0+}^{q}w(t)dt, \end{cases} \tag{1}$$

where D_{0+}^{α} is the α order Riemann–Liouville type fractional derivative, the constants α, p, q, n, and the functions h, $f_i(i = 1,2,3)$ satisfy the assumptions

(C0) $n \in \mathbb{N}, n \geq 3, \alpha \in (n-1, n], p \in [1, n-2], q \in [0, p]$,

(C1) there exists h with $h(t) \geq 0(\not\equiv 0)$ on $[0,1]$ such that $A := \frac{\Gamma(\alpha)}{\Gamma(\alpha-p)} - \frac{\Gamma(\alpha)}{\Gamma(\alpha-q)} \int_0^1 h(t)t^{\alpha-q-1}dt > 0$,

(C2) $f_i(i = 1,2,3) \in C([0,1] \times \mathbb{R}^+ \times \mathbb{R}^+ \times \mathbb{R}^+, \mathbb{R})$, and there is a $M > 0$ such that

$$f_i(t, x_1, x_2, x_3) \geq -M, \text{ for } (t, x_1, x_2, x_3) \in [0,1] \times \mathbb{R}^+ \times \mathbb{R}^+ \times \mathbb{R}^+, i = 1,2,3.$$

Fractional calculus theory shows undoubted advantages in aerodynamics, electrodynamics in complex medium, the theory of control, signal and image processing, rheology, and many other issues, see the books [1–3]. The study of such kind of problems has received considerable attention in the previous studies, see for instance [4–79] and the references therein.

In [4] by the fixed point theorem of Guo–Krasnosel'skii, the authors discussed the positive solutions for the multi-point Riemann–Liouville fractional boundary value problems

$$\begin{cases} D_{0+}^{\alpha}u(t) + \lambda f(t, u(t)) = 0, t \in (0,1), \\ u(0) = u'(0) = \cdots = u^{(n-2)}(0) = 0, \\ D_{0+}^{p}u(t)|_{t=1} = \sum_{i=1}^{m} a_i D_{0+}^{q}u(t)|_{t=\xi_i} \end{cases} \tag{2}$$

where f is a sign-changing nonlinearity. In [5], the authors studied the multiple positive solutions for the problem (2) ($\lambda = 1$), where f is a sign-changing nonlinearity, and permits singularities on t and u. In [6], by means of the index of fixed point, the authors researched the positive solutions for the boundary value problems of Hadamard fractional equations

$$\begin{cases} -{}^{H}D^{\alpha}u(t) = f(t, u(t)), \quad t \in [1, e], \\ u(1) = \delta u(1) = \delta u(e) = 0, \end{cases} \tag{3}$$

where f is a sign-changing nonlinearity, and may grow superlinearly and sublinearly at ∞.

The fractional-order equations in systems have also been widely investigated in the literature, see for example [52–79]. In [52], the authors studied the system of Hadamard fractional integral boundary value problems

$$\begin{cases} {}^{H}D^{\beta}u(t) + f_1(t, u(t), v(t)) = 0, \quad 1 < t < e, \\ {}^{H}D^{\beta}v(t) + f_2(t, u(t), v(t)) = 0, \quad 1 < t < e, \\ u(1) = v(1) = u'(1) = v'(1) = 0, \\ u(e) = \int_1^e h(s)v(s)\frac{ds}{s}, \\ v(e) = \int_1^e g(s)u(s)\frac{ds}{s}, \end{cases} \tag{4}$$

where the nonlinearities $f_i(i = 1,2) \in C([1,e] \times \mathbb{R}^+ \times \mathbb{R}^+, \mathbb{R}^+)$.

In [53], by means of the alternative of Leray–Schauder, the authors obtained the uniqueness and existence of solutions for the system of fractional integral boundary value problems

$$\begin{cases} D^{\alpha}x(t) = f(t, x(t), y(t), D^{\gamma}y(t)), t \in [0, T], \\ D^{\beta}y(t) = g(t, x(t), D^{\delta}x(t), y(t)), t \in [0, T], \end{cases} \tag{5}$$

with the integral boundary conditions

$$\begin{cases} x(0) = h(y), \int_0^T y(s)ds = \mu_1 x(\eta), \\ y(0) = \phi(x), \int_0^T x(s)ds = \mu_2 y(\xi), \end{cases}$$

where $D^{\alpha}, D^{\beta}, D^{\delta}, D^{\gamma}$ are the fractional derivatives of Caputo type.

In [54], the authors studied the positive solutions of the abstract fractional semipositone differential system with integral boundary conditions, which arises from HIV infection models

$$\begin{cases} D_{0+}^{\alpha}u(t) + \lambda f\left(t, u(t), D_{0+}^{\beta}u(t), v(t)\right) = 0, \\ D_{0+}^{\gamma}v(t) + \lambda g(t, u(t)) = 0,\ 0 < t < 1, \\ D_{0+}^{\beta}u(0) = D_{0+}^{\beta+1}u(0) = 0,\ D_{0+}^{\beta}u(1) = \int_0^1 D_{0+}^{\beta}u(s)dA(s), \\ v(0) = v'(0) = 0,\ v(1) = \int_0^1 v(s)dB(s), \end{cases} \tag{6}$$

where f, g are the semipositone nonlinearities (so-called semipositone problems), which originally modeled nonlinear phenomena of chemical reactions by Dutch chemist Aris [80]. For some relevant work, we refer the reader to [4–7,71–75].

Motivated by the works aforementioned, in this work we use the index of fixed point and nonnegative matrix theory to study the positive solutions for the system of Riemann–Liouville type fractional boundary value problems (1). We first transform our problem into the equivalent system of Hammerstein type integral equations, and establish some nonnegative operator equations. Then, using some superlinear and sublinear conditions for our nonlinearities, we obtain two existence theorems. Finally, we offer two examples to explain our main theorems.

2. Preliminaries

Now, we offer the definition of the $\alpha(>0)$ order Riemann–Liouville type fractional derivative, which is given by

$$D_{0+}^{\alpha}f(t) = \frac{1}{\Gamma(n-\alpha)}\left(\frac{\mathrm{d}}{\mathrm{d}t}\right)^n \int_0^t (t-s)^{n-\alpha-1}f(s)ds,$$

where $f : (0, +\infty) \to (-\infty, +\infty)$ is a continuous function, and $n = [\alpha] + 1$. For more materials, we refer to the books [1–3].

Lemma 1. *lSuppose that (C0)–(C1) hold. Let $f \in C[0,1]$, then the problem*

$$\begin{cases} D_{0+}^{\alpha}u(t) + f(t) = 0, t \in (0,1), \\ u(0) = u'(0) = \cdots = u^{(n-2)}(0) = 0, D_{0+}^{p}u(t)|_{t=1} = \int_0^1 h(t)D_{0+}^{q}u(t)dt, \end{cases} \tag{7}$$

has a solution, which can take the form

$$u(t) = \int_0^1 G(t,s)f(s)ds,$$

where

$$G(t,s) = g_1(t,s) + \frac{t^{\alpha-1}}{A}\int_0^1 h(t)g_2(t,s)dt,$$

and

$$g_1(t,s) = \frac{1}{\Gamma(\alpha)}\begin{cases} t^{\alpha-1}(1-s)^{\alpha-p-1} - (t-s)^{\alpha-1}, & 0 \le s \le t \le 1, \\ t^{\alpha-1}(1-s)^{\alpha-p-1}, & 0 \le t \le s \le 1, \end{cases} \tag{8}$$

$$g_2(t,s) = \frac{1}{\Gamma(\alpha-q)}\begin{cases} t^{\alpha-q-1}(1-s)^{\alpha-p-1} - (t-s)^{\alpha-q-1}, & 0 \le s \le t \le 1, \\ t^{\alpha-q-1}(1-s)^{\alpha-p-1}, & 0 \le t \le s \le 1. \end{cases} \tag{9}$$

Proof. Using similar arguments in ([4], [Lemma 1 and 2]), we have

$$u(t) = c_1t^{\alpha-1} + c_2t^{\alpha-2} + \cdots + c_nt^{\alpha-n} - \int_0^t \frac{(t-s)^{\alpha-1}}{\Gamma(\alpha)}f(s)ds,$$

where $c_i \in \mathbb{R}, i = 1, 2, ..., n$. Note that $u(0) = u'(0) = \cdots = u^{(n-2)}(0) = 0$, and thus $c_2 = \cdots = c_n = 0$. Consequently, we get

$$u(t) = c_1 t^{\alpha-1} - \int_0^t \frac{(t-s)^{\alpha-1}}{\Gamma(\alpha)} f(s)ds.$$

Therefore, we find

$$D_{0+}^p u(t) = c_1 \frac{\Gamma(\alpha)}{\Gamma(\alpha-p)} t^{\alpha-p-1} - I_{0+}^{\alpha-p} f(t),\ D_{0+}^q u(t) = c_1 \frac{\Gamma(\alpha)}{\Gamma(\alpha-q)} t^{\alpha-q-1} - I_{0+}^{\alpha-q} f(t).$$

Using the condition $D_{0+}^p u(t)|_{t=1} = \int_0^1 h(t) D_{0+}^q u(t)dt$, we have

$$c_1 \frac{\Gamma(\alpha)}{\Gamma(\alpha-p)} - \frac{1}{\Gamma(\alpha-p)} \int_0^1 (1-s)^{\alpha-p-1} f(s)ds = c_1 \frac{\Gamma(\alpha)}{\Gamma(\alpha-q)} \int_0^1 h(t) t^{\alpha-q-1} dt - \frac{1}{\Gamma(\alpha-q)} \int_0^1 h(t) \int_0^t (t-s)^{\alpha-q-1} f(s)dsdt.$$

Solving this equation, we obtain

$$c_1 = \frac{1}{A\Gamma(\alpha-p)} \int_0^1 (1-s)^{\alpha-p-1} f(s)ds - \frac{1}{A\Gamma(\alpha-q)} \int_0^1 h(t) \int_0^t (t-s)^{\alpha-q-1} f(s)dsdt.$$

As a result, we get

$$\begin{aligned}
u(t) &= \frac{1}{A\Gamma(\alpha-p)} \int_0^1 t^{\alpha-1}(1-s)^{\alpha-p-1} f(s)ds - \frac{t^{\alpha-1}}{A\Gamma(\alpha-q)} \int_0^1 h(t) \int_0^t (t-s)^{\alpha-q-1} f(s)dsdt - \int_0^t \frac{(t-s)^{\alpha-1}}{\Gamma(\alpha)} f(s)ds \\
&= \frac{1}{\Gamma(\alpha)} \int_0^1 t^{\alpha-1}(1-s)^{\alpha-p-1} f(s)ds - \int_0^t \frac{(t-s)^{\alpha-1}}{\Gamma(\alpha)} f(s)ds + \left[\frac{1}{A\Gamma(\alpha-p)} - \frac{1}{\Gamma(\alpha)}\right] \int_0^1 t^{\alpha-1}(1-s)^{\alpha-p-1} f(s)ds \\
&\quad - \frac{t^{\alpha-1}}{A\Gamma(\alpha-q)} \int_0^1 h(t) \int_0^t (t-s)^{\alpha-q-1} f(s)dsdt \\
&= \int_0^1 g_1(t,s) f(s)ds + \frac{t^{\alpha-1}}{A\Gamma(\alpha-q)} \left[\int_0^1 \int_0^1 h(t) t^{\alpha-q-1} (1-s)^{\alpha-p-1} f(s)dsdt - \int_0^1 h(t) \int_0^t (t-s)^{\alpha-q-1} f(s)dsdt\right] \\
&= \int_0^1 g_1(t,s) f(s)ds + \frac{t^{\alpha-1}}{A} \int_0^1 \int_0^1 h(t) g_2(t,s)dt f(s)ds \\
&= \int_0^1 G(t,s) f(s)ds.
\end{aligned}$$

□

Lemma 2. *(see ([4], [Lemma 3])). Suppose that (C0) holds. The functions $g_i (i = 1, 2)$ have the properties*
(i) $g_i \in C([0,1] \times [0,1], \mathbb{R}^+)$, and $g_i(t,s) > 0$ for $t, s \in (0,1)$, $i = 1, 2$,
(ii) $t^{\alpha-1}\widetilde{\varphi}(s) \leq g_1(t,s) \leq \widetilde{\varphi}(s)$ for all $t, s \in [0,1]$, where

$$\widetilde{\varphi}(s) = \frac{(1-s)^{\alpha-p-1}(1-(1-s)^p)}{\Gamma(\alpha)}, s \in [0,1],$$

(iii) $g_1(t,s) \leq \frac{t^{\alpha-1}(1-s)^{\alpha-p-1}}{\Gamma(\alpha)}$, $t, s \in [0,1]$.

Lemma 3. *Suppose that (C0)–(C1) hold. The Green's function G has the properties*
(i) $G \in C([0,1] \times [0,1], \mathbb{R}^+)$, and $G(t,s) > 0$ for $t, s \in (0,1)$,
(ii) $t^{\alpha-1}\varphi(s) \leq G(t,s) \leq \varphi(s), \forall t, s \in [0,1]$, where

$$\varphi(s) = \widetilde{\varphi}(s) + \frac{1}{A} \int_0^1 h(t) g_2(t,s)dt, s \in [0,1],$$

(iii) $G(t,s) \leq t^{\alpha-1} \left[\frac{(1-s)^{\alpha-p-1}}{\Gamma(\alpha)} + \frac{1}{A} \int_0^1 h(t) g_2(t,s)dt\right], \forall t, s \in [0,1]$.
This is a direct result of Lemma 2, so we omit its proof.

Lemma 4. *Let* $\kappa_1 = \int_0^1 t^{\alpha-1}\varphi(t)dt$, $\kappa_2 = \int_0^1 \varphi(t)dt$. *Then we have the following inequalities*

$$\kappa_1\varphi(s) \le \int_0^1 G(t,s)\varphi(t)dt \le \kappa_2\varphi(s), \ \forall s \in [0,1]. \tag{10}$$

From Lemma 3(ii), we easily obtain (10).

Next we will consider the problem

$$\begin{cases} D_{0+}^{\alpha}u(t) + \widetilde{f}(t,u(t)) = 0, t \in (0,1), \\ u(0) = u'(0) = \cdots = u^{(n-2)}(0) = 0, D_{0+}^{p}u(t)|_{t=1} = \int_0^1 h(t)D_{0+}^{q}u(t)dt, \end{cases} \tag{11}$$

where $\widetilde{f}$ satisfies the condition

(C2)′ $\widetilde{f} \in C([0,1] \times \mathbb{R}^+, \mathbb{R})$, and there is a $M > 0$ such that

$$\widetilde{f}(t,x_1) \ge -M, \text{ for } (t,x_1) \in [0,1] \times \mathbb{R}^+.$$

Lemma 5. *Suppose that (C0)–(C1) and (C2)′. Then the problem (11) is equivalent to*

$$u(t) = \int_0^1 G(t,s)\widetilde{f}(s,u(s))ds, \tag{12}$$

where G is defined in Lemma 1.

Now, we take care of the following auxiliary problem associated with (11):

$$\begin{cases} D_{0+}^{\alpha}u(t) + \widetilde{F}(t,u(t)-z(t)) = 0, t \in (0,1), \\ u(0) = u'(0) = \cdots = u^{(n-2)}(0) = 0, D_{0+}^{p}u(t)|_{t=1} = \int_0^1 h(t)D_{0+}^{q}u(t)dt, \end{cases} \tag{13}$$

where $\widetilde{F}(t,x_1) = \begin{cases} \widetilde{f}(t,x_1) + M, & t \in [0,1], x_1 \ge 0, \\ \widetilde{f}(t,0) + M, & t \in [0,1], x_1 < 0, \end{cases}$ *and* $z(t) = M\int_0^1 G(t,s)ds$, *for* $t \in [0,1]$. *Then* $\widetilde{F}$ *is nonnegative continuous on* $[0,1] \times \mathbb{R}^+$, *and from Lemma 5 we have (13) is equivalent to*

$$u(t) = \int_0^1 G(t,s)\widetilde{F}(s,u(s)-z(s))ds, \tag{14}$$

where G is as in Lemma 1.

Lemma 6. *(i) If (11) has a positive solution* u^*, *then (13) has a solution* $u^* + z$.

(ii) If u^* *is a solution for (13), and* $u^*(t) \ge z(t)$ *for* $t \in [0,1]$, *then* $u^* - z$ *is a positive solution for (11).*

Proof. Note that z satisfies the fractional boundary value problem

$$\begin{cases} D_{0+}^{\alpha}z(t) + M = 0, t \in (0,1), \\ z(0) = z'(0) = \cdots = z^{(n-2)}(0) = 0, D_{0+}^{p}z(t)|_{t=1} = \int_0^1 h(t)D_{0+}^{q}z(t)dt. \end{cases} \tag{15}$$

Substituting $u^* + z$ into (13), we have

$$D_{0+}^{\alpha}(u^*+z)(t) + \widetilde{F}(t,u^*(t)+z(t)-z(t)) = 0 \Longrightarrow D_{0+}^{\alpha}u^*(t) + D_{0+}^{\alpha}(z)(t) + \widetilde{f}(t,u^*(t)) + M = 0.$$

Using $D_{0+}^{\alpha}(z)(t) = -M$, we have $D_{0+}^{\alpha}u^*(t) + \widetilde{f}(t,u^*(t)) = 0$, and note that u^*, z satisfy the boundary conditions in (11), (15), we obtain Lemma 6(i) holds.

Next, substituting $u^* - z$ into (11), and using $D_{0+}^{\alpha}(z)(t) = -M$ we have

$$D_{0+}^{\alpha}(u^*-z)(t) + \widetilde{f}(t,u^*(t)-z(t)) = 0 \Longrightarrow D_{0+}^{\alpha}u^*(t) - D_{0+}^{\alpha}z(t) + \widetilde{f}(t,u^*(t)-z(t)) = 0,$$

and

$$D_{0+}^{\alpha}u^*(t) + \widetilde{F}(t, u^*(t) - z(t)) = 0.$$

Note that u^*, z satisfy the boundary conditions in (13), (15), we obtain Lemma 6(ii) holds.

Lemma 6 implies that we only need to seek the solution u^* for (13), which is greater than z, we can obtain the positive solution $u^* - z$ for (11).

Let $E := C[0,1]$, $\|u\| := \max_{t\in[0,1]} |u(t)|$, $P := \{u \in E : u(t) \geq 0, \forall t \in [0,1]\}$, $P_0 = \{u \in P : u(t) \geq t^{\alpha-1}\|u\|, \forall t \in [0,1]\}$. Then $(E, \|\cdot\|)$ is a real Banach space, and P, P_0 are cones on E. Note that the relations between (13) and (14), we let an operator $T : P \to P$ as follows:

$$(Tu)(t) = \int_0^1 G(t,s)\widetilde{F}(s, u(s) - z(s))ds, \text{ for } u \in P, t \in [0,1].$$

From the continuity of $G, \widetilde{F}$ we obtain $T : P \to P$ is a completely continuous operator, and if there exists $\overline{u} \in P\backslash\{0\}$ such that $T\overline{u} = \overline{u}$, then this $\overline{u}$ is a positive solution for (13). □

Lemma 7. $T(P) \subset P_0$.

By Lemma 3(ii) we can easily obtain this conclusion, so we omit its proof.

Note that if $\overline{u}$ is a positive fixed point of T, from Lemma 7 we have $\overline{u} \in P_0$. Moreover, when

$$\|\overline{u}\| \geq \widetilde{M} = M\int_0^1 \left[\frac{(1-s)^{\alpha-p-1}}{\Gamma(\alpha)} + \frac{1}{A}\int_0^1 h(t)g_2(t,s)dt\right] ds > 0,$$

we have

$$\begin{aligned}\overline{u}(t) - z(t) &\geq t^{\alpha-1}\|\overline{u}\| - M\int_0^1 G(t,s)ds \\ &\geq t^{\alpha-1}\|\overline{u}\| - M\int_0^1 t^{\alpha-1}\left[\frac{(1-s)^{\alpha-p-1}}{\Gamma(\alpha)} + \frac{1}{A}\int_0^1 h(t)g_2(t,s)dt\right] ds \\ &\geq 0.\end{aligned}$$

Then from Lemma 6 we have $\overline{u} - z$ is a positive solution for (11). Therefore, we only need to study the positive fixed point u^* for T, which the norm is greater than $\widetilde{M}$, then $u^* - z$ is a positive solution for (11).

In the following two lemmas, we let X be a real Banach space and P a cone on X.

Lemma 8. *(see [81]). Let $\Omega \subset X$ be a bounded open set, and $T : \overline{\Omega} \cap P \to P$ a continuous compact operator. If there exists $\mu_0 \in P\backslash\{0\}$ such that*

$$u - Tu \neq \lambda\mu_0, \forall \lambda \geq 0, u \in \partial\Omega \cap P,$$

then $i(T, \Omega \cap P, P) = 0$, where i is the index of fixed point on P.

Lemma 9. *(see [81]). Let $\Omega \subset X$ be a bounded open set with $0 \in \Omega$, and $T : \overline{\Omega} \cap P \to P$ a continuous compact operator. If*

$$u - \lambda Tu \neq 0, \forall \lambda \in [0,1], u \in \partial\Omega \cap P,$$

then $i(T, \Omega \cap P, P) = 1$.

In what follows, in order to build our main theorems, we need to introduce some basic knowledge for nonnegative matrices, for more details see [82,83].

Definition 1. *Let $\mathcal{M}$ be a real matrix. If all elements of $\mathcal{M}$ are nonnegative, then $\mathcal{M}$ is called to be nonnegative.*

Definition 2. *A real square matrix $\mathcal{M} = (m_{ij})_{n\times n}$ is called $\mathbb{R}^n_+$-monotone, if for every column vector $x \in \mathbb{R}^n, \mathcal{M}x \in \mathbb{R}^n_+ \Longrightarrow x \in \mathbb{R}^n_+$.*

Lemma 10. *A real square matrix $\mathcal{M}$ is $\mathbb{R}^n_+$-monotone $\Longleftrightarrow \det \mathcal{M} \neq 0$, and $\mathcal{M}^{-1}$ is nonnegative.*

Remark 1. *Note that our boundary condition at $t = 1$ is integral and generalizes multi-point fractional boundary conditions. However, our problem (7) can be considered as a perturbation of the two-point boundary value problem*

$$\begin{cases} D^\alpha_{0+}u(t) + f(t) = 0, t \in (0,1), \\ u(0) = u'(0) = \cdots = u^{(n-2)}(0) = D^p_{0+}u(t)|_{t=1} = 0, \end{cases} \tag{16}$$

which is equivalent to

$$u(t) = \int_0^1 g_1(t,s)f(s)ds,$$

where g_1 is defined by (8). Therefore, our method, by making good use of the original Green's function for the problem (16), will dispense with constructing a new Green's function, in contrast to some papers dealing with multi-point boundary value problems. For example, in [50] the author studied the problem

$$\begin{cases} D^\alpha_{0+}u(t) + f(t,u(t)) = 0, 0 < t < 1, \\ u(0) = 0, \beta u(\eta) = u(1), \end{cases} \tag{17}$$

where $\alpha \in (1,2], \beta\eta^{\alpha-1}, \eta \in (0,1)$. The author obtained the Green's function associated with (17) is

$$G_{Bai}(t,s) = \begin{cases} \frac{[t(1-s)]^{\alpha-1} - \beta t^{\alpha-1}(\eta-s)^{\alpha-1} - (t-s)^{\alpha-1}(1-\beta\eta^{\alpha-1})}{(1-\beta\eta^{\alpha-1})\Gamma(\alpha)}, & 0 \le s \le t \le 1, s \le \eta, \\ \frac{[t(1-s)]^{\alpha-1} - (t-s)^{\alpha-1}(1-\beta\eta^{\alpha-1})}{(1-\beta\eta^{\alpha-1})\Gamma(\alpha)}, & 0 < \eta \le s \le t \le 1, \\ \frac{[t(1-s)]^{\alpha-1} - \beta t^{\alpha-1}(\eta-s)^{\alpha-1}}{(1-\beta\eta^{\alpha-1})\Gamma(\alpha)}, & 0 \le t \le s \le \eta < 1, \\ \frac{[t(1-s)]^{\alpha-1}}{(1-\beta\eta^{\alpha-1})\Gamma(\alpha)}, & 0 \le t \le s \le 1, \eta \le s. \end{cases} \tag{18}$$

This function is very complicated. However, we note that this function can be expressed by

$$G_{Bai}(t,s) = g_{Bai}(t,s) + \frac{\beta t^{\alpha-1}}{1-\beta\eta^{\alpha-1}} g_{Bai}(\eta,s),$$

$$g_{Bai}(t,s) = \frac{1}{\Gamma(\alpha)} \begin{cases} [t(1-s)]^{\alpha-1} - (t-s)^{\alpha-1}, & 0 \le s \le t \le 1, \\ [t(1-s)]^{\alpha-1} & 0 \le t \le s \le 1, \end{cases}$$

where g_{Bai} is the Green's function for the problem

$$\begin{cases} D^\alpha_{0+}u(t) + f(t,u(t)) = 0, 0 < t < 1, \\ u(0) = u(1) = 0. \end{cases} \tag{19}$$

Compared with G_{Bai}, g_{Bai} is much simpler.

3. Main Results

From the discission of Section 2, we can define the operators $T_i (i = 1,2,3) : P \times P \times P \to P$ and $T : P \times P \times P \to P \times P \times P$ as follows:

$$T_i(u,v,w)(t) = \int_0^1 G(t,s)F_i(s, u(s) - z(s), v(s) - z(s), w(s) - z(s))ds,$$
$$T(u,v,w)(t) = (T_1, T_2, T_3)(u,v,w)(t), \text{ for } t \in [0,1],$$

where $F_i(t,x_1,x_2,x_3) = \begin{cases} f_i(t,x_1,x_2,x_3)+M, & t\in[0,1], \text{ for } x_i\geq 0, i=1,2,3, \\ f_i(t,0,0,0)+M, & t\in[0,1], \text{ for else cases.} \end{cases}$ Consequently, if there exists $(\overline{u},\overline{v},\overline{w})$ is a positive fixed pint of T with $\|\overline{u}\|,\|\overline{v}\|,\|\overline{w}\|\geq \widetilde{M}$, then we obtain $(\overline{u}-z,\overline{v}-z,\overline{w}-z)$ is a positive solution for (1).

Now, we list our assumptions for $F_i(i=1,2,3)$:

(C3) There exist $a_{ji},b_{ji}\geq 0$ and $l_j>0(i,j=1,2,3)$ such that

$$\begin{pmatrix} F_1(t,x_1,x_2,x_3) \\ F_2(t,x_1,x_2,x_3) \\ F_3(t,x_1,x_2,x_3) \end{pmatrix} \geq \begin{pmatrix} a_{11}x_1+a_{12}x_2+a_{13}x_3-l_1 \\ a_{21}x_1+a_{22}x_2+a_{23}x_3-l_2 \\ a_{31}x_1+a_{32}x_2+a_{33}x_3-l_3 \end{pmatrix}, \forall(t,x_1,x_2,x_3)\in[0,1]\times\mathbb{R}^+\times\mathbb{R}^+\times\mathbb{R}^+,$$

and the matrix $\mathcal{M}_1$ is a $\mathbb{R}^3_+$-monotone matrix, where

$$\mathcal{M}_1 = \begin{pmatrix} \kappa_1 a_{11}-1 & \kappa_1 a_{12} & \kappa_1 a_{13} \\ \kappa_1 a_{21} & \kappa_1 a_{22}-1 & \kappa_1 a_{23} \\ \kappa_1 a_{31} & \kappa_1 a_{32} & \kappa_1 a_{33}-1 \end{pmatrix}.$$

(C4) There exists $Q_i(t)$ in $[0,1]$ such that

$$\int_0^1 \varphi(t)Q_i(t)dt<\widetilde{M}, \text{ and } F_i(t,x_1,x_2,x_3)\leq Q_i(t), \forall(t,x_1,x_2,x_3)\in[0,1]\times\left[0,\widetilde{M}\right]^3, i=1,2,3.$$

(C5) There exist $\widetilde{a}_{ji},\widetilde{b}_{ji}\geq 0$ and $\widetilde{l}_j>0(i,j=1,2,3)$ such that

$$\begin{pmatrix} F_1(t,x_1,x_2,x_3) \\ F_2(t,x_1,x_2,x_3) \\ F_3(t,x_1,x_2,x_3) \end{pmatrix} \leq \begin{pmatrix} \widetilde{a}_{11}x_1+\widetilde{a}_{12}x_2+\widetilde{a}_{13}x_3+\widetilde{l}_1 \\ \widetilde{a}_{21}x_1+\widetilde{a}_{22}x_2+\widetilde{a}_{23}x_3+\widetilde{l}_2 \\ \widetilde{a}_{31}x_1+\widetilde{a}_{32}x_2+\widetilde{a}_{33}x_3+\widetilde{l}_3 \end{pmatrix}, \forall(t,x_1,x_2,x_3)\in[0,1]\times\mathbb{R}^+\times\mathbb{R}^+\times\mathbb{R}^+,$$

and the matrix $\mathcal{M}_2$ is a $\mathbb{R}^3_+$-monotone matrix, where

$$\mathcal{M}_2 = \begin{pmatrix} 1-\kappa_2\widetilde{a}_{11} & -\kappa_2\widetilde{a}_{12} & -\kappa_2\widetilde{a}_{13} \\ -\kappa_2\widetilde{a}_{21} & 1-\kappa_2\widetilde{a}_{22} & -\kappa_2\widetilde{a}_{23} \\ -\kappa_2\widetilde{a}_{31} & -\kappa_2\widetilde{a}_{32} & 1-\kappa_2\widetilde{a}_{33} \end{pmatrix}.$$

(C6) There exists $\widetilde{Q}_i(t)$ in $[0,1]$, and $t_0\in(0,1)$ such that

$$\int_0^1 \varphi(t)\widetilde{Q}_i(t)dt>\widetilde{M}t_0^{1-\alpha}, \text{ and } F_i(t,x_1,x_2,x_3)\geq \widetilde{Q}_i(t), \forall(t,x_1,x_2,x_3)\in[1,e]\times\left[0,\widetilde{M}\right]^3, i=1,2,3.$$

Let $B_\rho=\{u\in P:\|u\|<\rho\}$ for $\rho>0$ in the sequel. Then we easily have $\partial B_\rho=\{u\in P:\|u\|=\rho\}$, $\overline{B}_\rho=\{u\in P:\|u\|\leq\rho\}$.

Theorem 1. *Suppose that (C0)–(C4) hold. Then* (1) *has a positive solution.*

Proof. We first show that:

$$(u,v,w)\neq T(u,v,w)+\lambda(\phi_1,\phi_2,\phi_3), \text{ for } u,v,w\in\partial B_{R_1}\cap P,\ \lambda\geq 0, \tag{20}$$

where $\phi_i(i=1,2,3)$ are given elements in cone P_0, and $R_1>\widetilde{M}$. Argument by contrary, there exists $u,v,w\in\partial B_{R_1}\cap P$ and $\lambda_0\geq 0$ such that

$$(u,v,w)=T(u,v,w)+\lambda_0(\phi_1,\phi_2,\phi_3), \text{ for } u,v,w\in\partial B_{R_1}\cap P,\ \lambda\geq 0. \tag{21}$$

This implies that

$$\begin{pmatrix} u(t) \\ v(t) \\ w(t) \end{pmatrix} = \begin{pmatrix} T_1(u,v,w)(t) + \lambda_0\phi_1(t) \\ T_2(u,v,w)(t) + \lambda_0\phi_2(t) \\ T_3(u,v,w)(t) + \lambda_0\phi_3(t) \end{pmatrix} \geq \begin{pmatrix} \int_0^1 G(t,s)F_1(s,u(s)-z(s),v(s)-z(s),w(s)-z(s))ds \\ \int_0^1 G(t,s)F_2(s,u(s)-z(s),v(s)-z(s),w(s)-z(s))ds \\ \int_0^1 G(t,s)F_3(s,u(s)-z(s),v(s)-z(s),w(s)-z(s))ds \end{pmatrix}.$$

Note that Lemma 7 we have

$$u, v, w \in P_0. \tag{22}$$

From (C3) we have

$$\begin{pmatrix} u(t) \\ v(t) \\ w(t) \end{pmatrix} \geq \begin{pmatrix} \int_0^1 G(t,s)(a_{11}(u(s)-z(s)) + a_{12}(v(s)-z(s)) + a_{13}(w(s)-z(s)) - l_1)ds \\ \int_0^1 G(t,s)(a_{21}(u(s)-z(s)) + a_{22}(v(s)-z(s)) + a_{23}(w(s)-z(s)) - l_2)ds \\ \int_0^1 G(t,s)(a_{31}(u(s)-z(s)) + a_{32}(v(s)-z(s)) + a_{33}(w(s)-z(s)) - l_3)ds \end{pmatrix}.$$

Multiplying by $\varphi(t)$ for the above both sides, and integrating on $[0,1]$, by Lemma 4 we get

$$\begin{pmatrix} \int_0^1 u(t)\varphi(t)dt \\ \int_0^1 v(t)\varphi(t)dt \\ \int_0^1 w(t)\varphi(t)dt \end{pmatrix} \geq \begin{pmatrix} \int_0^1 \kappa_1\varphi(t)(a_{11}(u(t)-z(t)) + a_{12}(v(t)-z(t)) + a_{13}(w(t)-z(t)))dt - l_1\kappa_2^2 \\ \int_0^1 \kappa_1\varphi(t)(a_{21}(u(t)-z(t)) + a_{22}(v(t)-z(t)) + a_{23}(w(t)-z(t)))dt - l_2\kappa_2^2 \\ \int_0^1 \kappa_1\varphi(t)(a_{31}(u(t)-z(t)) + a_{32}(v(t)-z(t)) + a_{33}(w(t)-z(t)))dt - l_3\kappa_2^2 \end{pmatrix}.$$

Consequently, we find

$$\begin{aligned} \begin{pmatrix} \kappa_1 a_{11}-1 & \kappa_1 a_{12} & \kappa_1 a_{13} \\ \kappa_1 a_{21} & \kappa_1 a_{22}-1 & \kappa_1 a_{23} \\ \kappa_1 a_{31} & \kappa_1 a_{32} & \kappa_1 a_{33}-1 \end{pmatrix} \begin{pmatrix} \int_0^1 u(t)\varphi(t)dt \\ \int_0^1 v(t)\varphi(t)dt \\ \int_0^1 w(t)\varphi(t)dt \end{pmatrix} &\leq \begin{pmatrix} \kappa_1(a_{11}+a_{12}+a_{13})\int_0^1 \varphi(t)z(t)dt + l_1\kappa_2^2 \\ \kappa_1(a_{21}+a_{22}+a_{23})\int_0^1 \varphi(t)z(t)dt + l_2\kappa_2^2 \\ \kappa_1(a_{31}+a_{32}+a_{33})\int_0^1 \varphi(t)z(t)dt + l_3\kappa_2^2 \end{pmatrix} \\ &\leq \begin{pmatrix} \kappa_1(a_{11}+a_{12}+a_{13})M\kappa_2^2 + l_1\kappa_2^2 \\ \kappa_1(a_{21}+a_{22}+a_{23})M\kappa_2^2 + l_2\kappa_2^2 \\ \kappa_1(a_{31}+a_{32}+a_{33})M\kappa_2^2 + l_3\kappa_2^2 \end{pmatrix}. \end{aligned}$$

Therefore, we obtain

$$\begin{pmatrix} \int_0^1 u(t)\varphi(t)dt \\ \int_0^1 v(t)\varphi(t)dt \\ \int_0^1 w(t)\varphi(t)dt \end{pmatrix} \leq \begin{pmatrix} \kappa_1 a_{11}-1 & \kappa_1 a_{12} & \kappa_1 a_{13} \\ \kappa_1 a_{21} & \kappa_1 a_{22}-1 & \kappa_1 a_{23} \\ \kappa_1 a_{31} & \kappa_1 a_{32} & \kappa_1 a_{33}-1 \end{pmatrix}^{-1} \begin{pmatrix} \kappa_1(a_{11}+a_{12}+a_{13})M\kappa_2^2+l_1\kappa_2^2 \\ \kappa_1(a_{21}+a_{22}+a_{23})M\kappa_2^2+l_2\kappa_2^2 \\ \kappa_1(a_{31}+a_{32}+a_{33})M\kappa_2^2+l_3\kappa_2^2 \end{pmatrix}$$

$$= \frac{1}{\Delta_1} \begin{pmatrix} (\kappa_1 a_{22}-1)(\kappa_1 a_{33}-1)-\kappa_1^2 a_{23}a_{32} & \kappa_1^2 a_{13}a_{32}-\kappa_1 a_{12}(\kappa_1 a_{33}-1) & \kappa_1^2 a_{12}a_{23}-\kappa_1 a_{13}(\kappa_1 a_{22}-1) \\ \kappa_1^2 a_{23}a_{31}-\kappa_1 a_{21}(\kappa_1 a_{33}-1) & (\kappa_1 a_{11}-1)(\kappa_1 a_{33}-1)-\kappa_1^2 a_{13}a_{31} & \kappa_1^2 a_{13}a_{21}-\kappa_1 a_{23}(\kappa_1 a_{11}-1) \\ \kappa_1^2 a_{21}a_{32}-\kappa_1 a_{31}(\kappa_1 a_{22}-1) & \kappa_1^2 a_{12}a_{31}-\kappa_1 a_{32}(\kappa_1 a_{11}-1) & (\kappa_1 a_{11}-1)(\kappa_1 a_{22}-1)-\kappa_1^2 a_{12}a_{21} \end{pmatrix}$$

$$\cdot \begin{pmatrix} \kappa_1(a_{11}+a_{12}+a_{13})M\kappa_2^2+l_1\kappa_2^2 \\ \kappa_1(a_{21}+a_{22}+a_{23})M\kappa_2^2+l_2\kappa_2^2 \\ \kappa_1(a_{31}+a_{32}+a_{33})M\kappa_2^2+l_3\kappa_2^2 \end{pmatrix},$$

where

$$\Delta_1 = \det \begin{pmatrix} \kappa_1 a_{11}-1 & \kappa_1 a_{12} & \kappa_1 a_{13} \\ \kappa_1 a_{21} & \kappa_1 a_{22}-1 & \kappa_1 a_{23} \\ \kappa_1 a_{31} & \kappa_1 a_{32} & \kappa_1 a_{33}-1 \end{pmatrix}.$$

As a result of this, there exist $\mathcal{N}_i > 0 (i = 1,2,3)$ such that

$$\begin{pmatrix} \int_0^1 u(t)\varphi(t)dt \\ \int_0^1 v(t)\varphi(t)dt \\ \int_0^1 w(t)\varphi(t)dt \end{pmatrix} \leq \begin{pmatrix} \mathcal{N}_1 \\ \mathcal{N}_2 \\ \mathcal{N}_3 \end{pmatrix},$$

where $\mathcal{N}_1 = \frac{1}{\Delta_1}[((\kappa_1 a_{22}-1)(\kappa_1 a_{33}-1)-\kappa_1^2 a_{23}a_{32})(\kappa_1(a_{11}+a_{12}+a_{13})M\kappa_2^2+l_1\kappa_2^2)+(\kappa_1^2 a_{13}a_{32}-\kappa_1 a_{12}(\kappa_1 a_{33}-1))(\kappa_1(a_{21}+a_{22}+a_{23})M\kappa_2^2+l_2\kappa_2^2)+(\kappa_1^2 a_{12}a_{23}-\kappa_1 a_{13}(\kappa_1 a_{22}-1))(\kappa_1(a_{31}+a_{32}+a_{33})M\kappa_2^2+l_3\kappa_2^2)]$, $\mathcal{N}_2 = \frac{1}{\Delta_1}[(\kappa_1^2 a_{23}a_{31}-\kappa_1 a_{21}(\kappa_1 a_{33}-1))(\kappa_1(a_{11}+a_{12}+a_{13})M\kappa_2^2+l_1\kappa_2^2)+((\kappa_1 a_{11}-1)(\kappa_1 a_{33}-1)-\kappa_1^2 a_{13}a_{31})(\kappa_1(a_{21}+a_{22}+a_{23})M\kappa_2^2+l_2\kappa_2^2)+(\kappa_1^2 a_{13}a_{21}-\kappa_1 a_{23}(\kappa_1 a_{11}-1))(\kappa_1(a_{31}+a_{32}+a_{33})M\kappa_2^2+l_3\kappa_2^2)]$, $\mathcal{N}_3 = \frac{1}{\Delta_1}[(\kappa_1^2 a_{21}a_{32}-\kappa_1 a_{31}(\kappa_1 a_{22}-1))(\kappa_1(a_{11}+a_{12}+a_{13})M\kappa_2^2+l_1\kappa_2^2)+(\kappa_1^2 a_{12}a_{31}-\kappa_1 a_{32}(\kappa_1 a_{11}-1))(\kappa_1(a_{21}+a_{22}+a_{23})M\kappa_2^2+l_2\kappa_2^2)+((\kappa_1 a_{11}-1)(\kappa_1 a_{22}-1)-\kappa_1^2 a_{21}a_{21})(\kappa_1(a_{31}+a_{32}+a_{33})M\kappa_2^2+l_3\kappa_2^2)]$.

Note that (22), we have

$$\begin{pmatrix} \|u\| \\ \|v\| \\ \|w\| \end{pmatrix} \leq \begin{pmatrix} \mathcal{N}_1\kappa_1^{-1} \\ \mathcal{N}_2\kappa_1^{-1} \\ \mathcal{N}_3\kappa_1^{-1} \end{pmatrix}.$$

Therefore, we can choose $R_1 > \max\{\widetilde{M}, \mathcal{N}_1\kappa_1^{-1}, \mathcal{N}_2\kappa_1^{-1}, \mathcal{N}_3\kappa_1^{-1}\}$ such that when $u, v, w \in \partial B_{R_1} \cap P$, (21) is not satisfied. This also indicates that (20) holds for $u, v, w \in \partial B_{R_1} \cap P$, and Lemma 8 indicates that

$$i(T, B_{R_1} \cap (P \times P \times P), P \times P \times P) = 0. \tag{23}$$

On the other hand, we prove that

$$(u, v, w) \neq \lambda T(u, v, w), \text{ for } u, v, w \in \partial B_{\widetilde{M}} \cap P, \lambda \in [0, 1]. \tag{24}$$

If this claim is not true, there exist $u, v, w \in \partial B_{\widetilde{M}} \cap P, \lambda_1 \in [0,1]$ such that

$$(u, v, w) = \lambda_1 T(u, v, w).$$

This implies that

$$\|u\| \leq \|T_1(u, v, w)\|, \ \|v\| \leq \|T_2(u, v, w)\|, \text{ and } \|w\| \leq \|T_3(u, v, w)\|.$$

However, from (C4) we have

$$\begin{aligned} T_1(u, v, w)(t) &= \int_0^1 G(t, s) F_1(s, u(s) - z(s), v(s) - z(s), w(s) - z(s)) ds \\ &\leq \int_0^1 \varphi(s) Q_1(s) \frac{ds}{s} \\ &< \widetilde{M}. \end{aligned}$$

Note that by (C4), $\|u\| = \widetilde{M}$. Hence, we obtain $\|T_1(u, v, w)\| < \|u\|$. Similarly, $\|T_2(u, v, w)\| < \|v\|$ and $\|T_3(u, v, w)\| < \|w\|$. This has a contradiction. Hence (24) holds. By Lemma 9 we get

$$i(T, B_{\widetilde{M}} \cap (P \times P \times P), P \times P \times P) = 1. \tag{25}$$

By use of (23) and (25) we can calculate

$$\begin{aligned} & i(T, (B_{R_1} \backslash \overline{B}_{\widetilde{M}}) \cap (P \times P \times P), P \times P \times P) \\ &= i(T, B_{R_1} \cap (P \times P \times P), P \times P \times P) - i(T, B_{\widetilde{M}} \cap (P \times P \times P), P \times P \times P) \\ &= -1. \end{aligned}$$

Therefore, T has a fixed point (u^*, v^*, w^*) on $(B_{R_1} \backslash \overline{B}_{\widetilde{M}}) \cap (P \times P \times P)$. Consequently, $(u^* - z, v^* - z, w^* - z)$ is a positive solution for (1), i.e., (1) has a positive solution. □

Theorem 2. *Suppose that (C0)–(C2), (C5)–(C6) hold. Then* (1) *has a positive solution.*

Proof. We first claim that:

$$(u, v, w) \neq \lambda T(u, v, w), \text{ for } u, v, w \in \partial B_{R_2} \cap P, \lambda \in [0,1], \tag{26}$$

where $R_2 > \widetilde{M}$. If this claim does not hold, there exist $u, v, w \in \partial B_{R_2} \cap P, \lambda_2 \in [0,1]$ such that

$$(u, v, w) = \lambda_2 T(u, v, w). \tag{27}$$

This indicates that

$$\begin{pmatrix} u(t) \\ v(t) \\ w(t) \end{pmatrix} = \begin{pmatrix} \lambda_2 T_1(u, v, w)(t) \\ \lambda_2 T_2(u, v, w)(t) \\ \lambda_2 T_3(u, v, w)(t) \end{pmatrix} \leq \begin{pmatrix} \int_0^1 G(t, s) F_1(s, u(s) - z(s), v(s) - z(s), w(s) - z(s)) ds \\ \int_0^1 G(t, s) F_2(s, u(s) - z(s), v(s) - z(s), w(s) - z(s)) ds \\ \int_0^1 G(t, s) F_3(s, u(s) - z(s), v(s) - z(s), w(s) - z(s)) ds \end{pmatrix}.$$

Using Lemma 7, we know $u, v, w \in P_0$. By virtue of (C5), we obtain

$$\begin{pmatrix} u(t) \\ v(t) \\ w(t) \end{pmatrix} \leq \begin{pmatrix} \int_0^1 G(t,s)(\widetilde{a}_{11}(u(s)-z(s)) + \widetilde{a}_{12}(v(s)-z(s)) + \widetilde{a}_{13}(w(s)-z(s)) + \widetilde{l}_1)ds \\ \int_0^1 G(t,s)(\widetilde{a}_{21}(u(s)-z(s)) + \widetilde{a}_{22}(v(s)-z(s)) + \widetilde{a}_{23}(w(s)-z(s)) + \widetilde{l}_2)ds \\ \int_0^1 G(t,s)(\widetilde{a}_{31}(u(s)-z(s)) + \widetilde{a}_{32}(v(s)-z(s)) + \widetilde{a}_{33}(w(s)-z(s)) + \widetilde{l}_3)ds \end{pmatrix}.$$

Multiplying by $\varphi(t)$, and integrating over $[0,1]$, Lemma 4 enables us to get

$$\begin{pmatrix} \int_0^1 u(t)\varphi(t)dt \\ \int_0^1 v(t)\varphi(t)dt \\ \int_0^1 w(t)\varphi(t)dt \end{pmatrix} \leq \begin{pmatrix} \int_0^1 \kappa_2\varphi(t)(\widetilde{a}_{11}(u(t)-z(t)) + \widetilde{a}_{12}(v(t)-z(t)) + \widetilde{a}_{13}(w(t)-z(t)) + \widetilde{l}_1)dt \\ \int_0^1 \kappa_2\varphi(t)(\widetilde{a}_{21}(u(t)-z(t)) + \widetilde{a}_{22}(v(t)-z(t)) + \widetilde{a}_{23}(w(t)-z(t)) + \widetilde{l}_2)dt \\ \int_0^1 \kappa_2\varphi(t)(\widetilde{a}_{31}(u(t)-z(t)) + \widetilde{a}_{32}(v(t)-z(t)) + \widetilde{a}_{33}(w(t)-z(t)) + \widetilde{l}_3)dt \end{pmatrix}$$
$$\leq \begin{pmatrix} \int_0^1 \kappa_2\varphi(t)(\widetilde{a}_{11}u(t) + \widetilde{a}_{12}v(t) + \widetilde{a}_{13}w(t))dt + \widetilde{l}_1\kappa_2^2 \\ \int_0^1 \kappa_2\varphi(t)(\widetilde{a}_{21}u(t) + \widetilde{a}_{22}v(t) + \widetilde{a}_{23}w(t))dt + \widetilde{l}_2\kappa_2^2 \\ \int_0^1 \kappa_2\varphi(t)(\widetilde{a}_{31}u(t) + \widetilde{a}_{32}v(t) + \widetilde{a}_{33}w(t))dt + \widetilde{l}_3\kappa_2^2 \end{pmatrix}.$$

Therefore, we find

$$\begin{pmatrix} 1-\kappa_2\widetilde{a}_{11} & -\kappa_2\widetilde{a}_{12} & -\kappa_2\widetilde{a}_{13} \\ -\kappa_2\widetilde{a}_{21} & 1-\kappa_2\widetilde{a}_{22} & -\kappa_2\widetilde{a}_{23} \\ -\kappa_2\widetilde{a}_{31} & -\kappa_2\widetilde{a}_{32} & 1-\kappa_2\widetilde{a}_{33} \end{pmatrix} \begin{pmatrix} \int_0^1 u(t)\varphi(t)dt \\ \int_0^1 v(t)\varphi(t)dt \\ \int_0^1 w(t)\varphi(t)dt \end{pmatrix} \leq \begin{pmatrix} \widetilde{l}_1\kappa_2^2 \\ \widetilde{l}_2\kappa_2^2 \\ \widetilde{l}_3\kappa_2^2 \end{pmatrix},$$

and

$$\begin{pmatrix} \int_0^1 u(t)\varphi(t)dt \\ \int_0^1 v(t)\varphi(t)dt \\ \int_0^1 w(t)\varphi(t)dt \end{pmatrix} \leq \begin{pmatrix} 1-\kappa_2\widetilde{a}_{11} & -\kappa_2\widetilde{a}_{12} & -\kappa_2\widetilde{a}_{13} \\ -\kappa_2\widetilde{a}_{21} & 1-\kappa_2\widetilde{a}_{22} & -\kappa_2\widetilde{a}_{23} \\ -\kappa_2\widetilde{a}_{31} & -\kappa_2\widetilde{a}_{32} & 1-\kappa_2\widetilde{a}_{33} \end{pmatrix}^{-1} \begin{pmatrix} \widetilde{l}_1\kappa_2^2 \\ \widetilde{l}_2\kappa_2^2 \\ \widetilde{l}_3\kappa_2^2 \end{pmatrix}$$
$$= \frac{1}{\Delta_2} \begin{pmatrix} (1-\kappa_2\widetilde{a}_{22})(1-\kappa_2\widetilde{a}_{33}) - \kappa_2^2\widetilde{a}_{23}\widetilde{a}_{32} & \kappa_2^2\widetilde{a}_{13}\widetilde{a}_{32} + \kappa_2\widetilde{a}_{12}(1-\kappa_2\widetilde{a}_{33}) & \kappa_2^2\widetilde{a}_{12}\widetilde{a}_{23} + \kappa_2\widetilde{a}_{13}(1-\kappa_2\widetilde{a}_{22}) \\ \kappa_2^2\widetilde{a}_{23}\widetilde{a}_{31} + \kappa_2\widetilde{a}_{21}(1-\kappa_2\widetilde{a}_{33}) & (1-\kappa_2\widetilde{a}_{11})(1-\kappa_2\widetilde{a}_{33}) - \kappa_2^2\widetilde{a}_{13}\widetilde{a}_{31} & \kappa_2^2\widetilde{a}_{13}\widetilde{a}_{21} + \kappa_2\widetilde{a}_{23}(1-\kappa_2\widetilde{a}_{11}) \\ \kappa_2^2\widetilde{a}_{21}\widetilde{a}_{32} + \kappa_2\widetilde{a}_{31}(1-\kappa_2\widetilde{a}_{22}) & \kappa_2^2\widetilde{a}_{12}\widetilde{a}_{31} + \kappa_2\widetilde{a}_{32}(1-\kappa_2\widetilde{a}_{11}) & (1-\kappa_2\widetilde{a}_{11})(1-\kappa_2\widetilde{a}_{22}) - \kappa_2^2\widetilde{a}_{12}\widetilde{a}_{21} \end{pmatrix}$$
$$\cdot \begin{pmatrix} \widetilde{l}_1\kappa_2^2 \\ \widetilde{l}_2\kappa_2^2 \\ \widetilde{l}_3\kappa_2^2 \end{pmatrix},$$

where

$$\Delta_2 = \det \begin{pmatrix} 1-\kappa_2\widetilde{a}_{11} & -\kappa_2\widetilde{a}_{12} & -\kappa_2\widetilde{a}_{13} \\ -\kappa_2\widetilde{a}_{21} & 1-\kappa_2\widetilde{a}_{22} & -\kappa_2\widetilde{a}_{23} \\ -\kappa_2\widetilde{a}_{31} & -\kappa_2\widetilde{a}_{32} & 1-\kappa_2\widetilde{a}_{33} \end{pmatrix}.$$

Hence, there exist $\mathcal{N}_i > 0 (i = 4, 5, 6)$ such that

$$\begin{pmatrix} \int_0^1 u(t)\varphi(t)dt \\ \int_0^1 v(t)\varphi(t)dt \\ \int_0^1 w(t)\varphi(t)dt \end{pmatrix} \leq \begin{pmatrix} \mathcal{N}_4 \\ \mathcal{N}_5 \\ \mathcal{N}_6 \end{pmatrix},$$

where $\mathcal{N}_4 = \frac{\kappa_2^2}{\Delta_2}[\widetilde{l}_1((1-\kappa_2\widetilde{a}_{22})(1-\kappa_2\widetilde{a}_{33}) - \kappa_2^2\widetilde{a}_{23}\widetilde{a}_{32}) + \widetilde{l}_2(\kappa_2^2\widetilde{a}_{13}\widetilde{a}_{32} + \kappa_2\widetilde{a}_{12}(1-\kappa_2\widetilde{a}_{33})) + \widetilde{l}_3(\kappa_2^2\widetilde{a}_{12}\widetilde{a}_{23} + \kappa_2\widetilde{a}_{13}(1-\kappa_2\widetilde{a}_{22}))]$, $\mathcal{N}_5 = \frac{\kappa_2^2}{\Delta_2}[\widetilde{l}_1(\kappa_2^2\widetilde{a}_{23}\widetilde{a}_{31} + \kappa_2\widetilde{a}_{21}(1-\kappa_2\widetilde{a}_{33})) + \widetilde{l}_2((1-\kappa_2\widetilde{a}_{11})(1-\kappa_2\widetilde{a}_{33}) - \kappa_2^2\widetilde{a}_{13}\widetilde{a}_{31}) + \widetilde{l}_3(\kappa_2^2\widetilde{a}_{13}\widetilde{a}_{21} + \kappa_2\widetilde{a}_{23}(1-\kappa_2\widetilde{a}_{11}))]$, $\mathcal{N}_6 = \frac{\kappa_2^2}{\Delta_2}[\widetilde{l}_1(\kappa_2^2\widetilde{a}_{21}\widetilde{a}_{32} + \kappa_2\widetilde{a}_{31}(1-\kappa_2\widetilde{a}_{22})) + \widetilde{l}_2(\kappa_2^2\widetilde{a}_{12}\widetilde{a}_{31} + \kappa_2\widetilde{a}_{32}(1-\kappa_2\widetilde{a}_{11})) + \widetilde{l}_3((1-\kappa_2\widetilde{a}_{11})(1-\kappa_2\widetilde{a}_{22}) - \kappa_2^2\widetilde{a}_{12}\widetilde{a}_{21})]$. Note that $u, v, w \in P_0$, we have

$$\begin{pmatrix} \|u\| \\ \|v\| \\ \|w\| \end{pmatrix} \leq \begin{pmatrix} \mathcal{N}_4\kappa_1^{-1} \\ \mathcal{N}_5\kappa_1^{-1} \\ \mathcal{N}_6\kappa_1^{-1} \end{pmatrix}.$$

Therefore, we can choose $R_2 > \max\{\widetilde{M}, \mathcal{N}_4\kappa_1^{-1}, \mathcal{N}_5\kappa_1^{-1}, \mathcal{N}_6\kappa_1^{-1}\}$ such that when $u, v, w \in \partial B_{R_2} \cap P$, (27) is not satisfied. This also indicates that (26) holds for $u, v, w \in \partial B_{R_2} \cap P$, and by Lemma 9 we get

$$i(T, B_{R_2} \cap (P \times P \times P), P \times P \times P) = 1. \tag{28}$$

On the other hand, we prove that

$$(u, v, w) \neq T(u, v, w) + \lambda(\widetilde{\phi}_1, \widetilde{\phi}_2, \widetilde{\phi}_3), \text{ for } u, v, w \in \partial B_{\widetilde{M}} \cap P, \ \forall \lambda \geq 0, \tag{29}$$

where $\widetilde{\phi}_i \in P (i = 1, 2, 3)$ are fixed elements. Otherwise, there exist $u, v, w \in \partial B_{\widetilde{M}} \cap P, \lambda_3 \geq 0$ such that

$$(u, v, w) = T(u, v, w) + \lambda_3(\widetilde{\phi}_1, \widetilde{\phi}_2, \widetilde{\phi}_3).$$

This implies that

$$\begin{pmatrix} \|u\| \\ \|v\| \\ \|w\| \end{pmatrix} \geq \begin{pmatrix} \|T_1(u, v, w)\| \\ \|T_2(u, v, w)\| \\ \|T_3(u, v, w)\| \end{pmatrix}. \tag{30}$$

However, from (C6) we have

$$\begin{aligned} T_i(u, v, w)(t_0) &= \int_0^1 G(t_0, s)F_i(s, u(s) - z(s), v(s) - z(s), w(s) - z(s))ds \\ &\geq t_0^{\alpha-1} \int_0^1 \varphi(s)\widetilde{Q}_i(s)ds \\ &> \widetilde{M}, i = 1, 2, 3. \end{aligned}$$

Note that from (C6), we have $\|u\| = \widetilde{M}$. Hence, we obtain

$$\begin{pmatrix} \|T_1(u, v, w)\| \\ \|T_2(u, v, w)\| \\ \|T_3(u, v, w)\| \end{pmatrix} \geq \begin{pmatrix} T_1(u, v, w)(t_0) \\ T_2(u, v, w)(t_0) \\ T_3(u, v, w)(t_0) \end{pmatrix} > \begin{pmatrix} \|u\| \\ \|v\| \\ \|w\| \end{pmatrix}.$$

This has a contradiction with (30), and thus (29) holds. By Lemma 8 we find

$$i(T, B_{\widetilde{M}} \cap (P \times P \times P), P \times P \times P) = 0. \tag{31}$$

From (28) and (31) we can calculate

$$\begin{aligned} & i(T,(B_{R_2} \backslash \overline{B}_{\widetilde{M}}) \cap (P \times P \times P), P \times P \times P) \\ &= i(T, B_{R_2} \cap (P \times P \times P), P \times P \times P) - i(T, B_{\widetilde{M}} \cap (P \times P \times P), P \times P \times P) \\ &= 1. \end{aligned}$$

Therefore T has a fixed point (u^*, v^*, w^*) on $(B_{R_2} \backslash \overline{B}_{\widetilde{M}}) \cap (P \times P \times P)$. Therefore, $(u^* - z, v^* - z, w^* - z)$ is a positive solution for (1), i.e., (1) has a positive solution.

Let $n = 4, \alpha = 3.5, p = 1.5, q = 0.5$, and $h(t) = t, t \in [0,1]$. Then we have $A = 2.91$, and $\int_0^1 h(t) g_2(t,s) dt = \frac{5}{24}s - \frac{1}{4}s^2 + \frac{1}{24}s^4, s \in [0,1]$. This implies that (C0)–(C1) hold. Moreover, we can calculate

$$\kappa_1 = 0.017, \kappa_2 = 0.075, \widetilde{M} = 0.16M.$$

□

Example 1. *Let $\kappa_1 a_{11} - 1 = \kappa_1 a_{22} - 1 = \kappa_1 a_{33} - 1 = \kappa_1$, and we have $a_{11} = a_{22} = a_{33} = \frac{\kappa_1 + 1}{\kappa_1} = 59.82$. Moreover, we take the matrix*

$$\begin{pmatrix} a_{11} & a_{12} & a_{13} \\ a_{21} & a_{22} & a_{23} \\ a_{31} & a_{32} & a_{33} \end{pmatrix} = \begin{pmatrix} 59.82 & 0 & 0 \\ 0 & 59.82 & 0 \\ 0 & 0 & 59.82 \end{pmatrix},$$

and

$$\begin{pmatrix} F_1(t, x_1, x_2, x_3) \\ F_2(t, x_1, x_2, x_3) \\ F_3(t, x_1, x_2, x_3) \end{pmatrix} = \begin{pmatrix} 2M(9.57M)^{-\gamma_1}(a_{11}x_1 + a_{12}x_2 + a_{13}x_3)^{\gamma_1} \\ 1.8M(9.57M)^{-\gamma_2}(a_{21}x_1 + a_{22}x_2 + a_{23}x_3)^{\gamma_2} \\ 1.5M(9.57M)^{-\gamma_3}(a_{31}x_1 + a_{32}x_2 + a_{33}x_3)^{\gamma_3} \end{pmatrix}, \forall (t, x_1, x_2, x_3) \in [0,1] \times \mathbb{R}^+ \times \mathbb{R}^+ \times \mathbb{R}^+,$$

where $\gamma_i > 1 (i = 1, 2, 3)$. Note that

$$\mathcal{M}_1 = \begin{pmatrix} \kappa_1 a_{11} - 1 & \kappa_1 a_{12} & \kappa_1 a_{13} \\ \kappa_1 a_{21} & \kappa_1 a_{22} - 1 & \kappa_1 a_{23} \\ \kappa_1 a_{31} & \kappa_1 a_{32} & \kappa_1 a_{33} - 1 \end{pmatrix} = \begin{pmatrix} \kappa_1 & 0 & 0 \\ 0 & \kappa_1 & 0 \\ 0 & 0 & \kappa_1 \end{pmatrix}.$$

Hence, $\mathcal{M}_1$ is a $\mathbb{R}^3_+$-monotone matrix. Furthermore, for all $t \in [0,1]$ we have

$$\begin{aligned} \liminf_{a_{11}x_1 + a_{12}x_2 + a_{13}x_3 \to +\infty} \frac{F_1(x_1, x_2, x_3)}{a_{11}x_1 + a_{12}x_2 + a_{13}x_3} &= \liminf_{a_{11}x_1 + a_{12}x_2 + a_{13}x_3 \to +\infty} \frac{2M(9.57M)^{-\gamma_1}(a_{11}x_1 + a_{12}x_2 + a_{13}x_3)^{\gamma_1}}{a_{11}x_1 + a_{12}x_2 + a_{13}x_3} = +\infty, \\ \liminf_{a_{21}x_1 + a_{22}x_2 + a_{23}x_3 \to +\infty} \frac{F_2(x_1, x_2, x_3)}{a_{21}x_1 + a_{22}x_2 + a_{23}x_3} &= \liminf_{a_{21}x_1 + a_{22}x_2 + a_{23}x_3 \to +\infty} \frac{1.8M(9.57M)^{-\gamma_2}(a_{21}x_1 + a_{22}x_2 + a_{23}x_3)^{\gamma_2}}{a_{21}x_1 + a_{22}x_2 + a_{23}x_3} = +\infty, \\ \liminf_{a_{31}x_1 + a_{32}x_2 + a_{33}x_3 \to +\infty} \frac{F_3(x_1, x_2, x_3)}{a_{31}x_1 + a_{32}x_2 + a_{33}x_3} &= \liminf_{a_{31}x_1 + a_{32}x_2 + a_{33}x_3 \to +\infty} \frac{1.5M(9.57M)^{-\gamma_3}(a_{31}x_1 + a_{32}x_2 + a_{33}x_3)^{\gamma_3}}{a_{31}x_1 + a_{32}x_2 + a_{33}x_3} = +\infty. \end{aligned}$$

On the other hand, if $(t, x_1, x_2, x_3) \in [0,1] \times [0, 0.16M]^3$, we have

$$F_1 \leq 2M,\ F_2 \leq 1.8M, F_3 \leq 1.5M.$$

If we choose $Q_1(t) \equiv 2M, Q_2(t) \equiv 1.8M, Q_3(t) \equiv 1.5M$ *for* $t \in [0,1]$, *and we have*

$$\int_0^1 \varphi(t)Q_i(t)dt \le \int_0^1 \varphi(t)Q_1(t)dt = 2\kappa_2 M = 0.15M < \widetilde{M}, i = 1,2,3.$$

Therefore, (C3)–(C4) hold.

Example 2. *Let* $t_0 = 0.5$, $\widetilde{Q}_1(t) = 13M, \widetilde{Q}_2(t) = 14M, \widetilde{Q}_3(t) = 15M$ *for* $t \in [0,1]$, *and*

$$\begin{pmatrix} \widetilde{a}_{11} & \widetilde{a}_{12} & \widetilde{a}_{13} \\ \widetilde{a}_{21} & \widetilde{a}_{22} & \widetilde{a}_{23} \\ \widetilde{a}_{31} & \widetilde{a}_{32} & \widetilde{a}_{33} \end{pmatrix} = \begin{pmatrix} 2 & 5 & 3 \\ 8 & 3 & 4 \\ 6 & 3 & 4 \end{pmatrix},$$

and

$$\begin{pmatrix} F_1(t,x_1,x_2,x_3) \\ F_2(t,x_1,x_2,x_3) \\ F_3(t,x_1,x_2,x_3) \end{pmatrix} = \begin{pmatrix} 13Me^{1.6M}e^{-2x_1-5x_2-3x_3} \\ 14Me^{2.4M}e^{-8x_1-3x_2-4x_3} \\ 15Me^{2.08M}e^{-6x_1-3x_2-4x_3} \end{pmatrix}, \forall (t,x_1,x_2,x_3) \in [0,1] \times \mathbb{R}^+ \times \mathbb{R}^+ \times \mathbb{R}^+.$$

Then if $(t,x_1,x_2,x_3) \in [0,1] \times [0,0.16M]^3$, *we have* $F_1 \ge 13M, F_2 \ge 14M, F_3 \ge 15M$, *and* $\int_0^1 \varphi(t)\widetilde{Q}_i(t)dt \ge \int_0^1 \varphi(t)\widetilde{Q}_1(t)dt = 13\kappa_2 M > 0.16 \times 5.6569M$.

On the other hand, we can calculate

$$\det \mathcal{M}_2 = \det \begin{pmatrix} 1-0.075\times 2 & -0.075\times 5 & -0.075\times 3 \\ -0.075\times 8 & 1-0.075\times 3 & -0.075\times 4 \\ -0.075\times 6 & -0.075\times 3 & 1-0.075\times 4 \end{pmatrix} = 0.0868,$$

and

$$\mathcal{M}_2^{-1} = \frac{1}{0.0868}\begin{pmatrix} 0.475 & 0.313 & 0.287 \\ 0.555 & 0.494 & 0.39 \\ 0.484 & 0.36 & 0.434 \end{pmatrix}.$$

Consequently, $\mathcal{M}_2$ *is a* $\mathbb{R}^3_+$*-monotone matrix. Furthermore, for all* $t \in [0,1]$ *we have*

$$\limsup_{\widetilde{a}_{11}x_1+\widetilde{a}_{12}x_2+\widetilde{a}_{13}x_3\to+\infty} \frac{F_1(t,x_1,x_2,x_3)}{\widetilde{a}_{11}x_1+\widetilde{a}_{12}x_2+\widetilde{a}_{13}x_3} = \limsup_{\widetilde{a}_{11}x_1+\widetilde{a}_{12}x_2+\widetilde{a}_{13}x_3\to+\infty} \frac{13Me^{1.6M}e^{-\widetilde{a}_{11}x_1-\widetilde{a}_{12}x_2-\widetilde{a}_{13}x_3}}{\widetilde{a}_{11}x_1+\widetilde{a}_{12}x_2+\widetilde{a}_{13}x_3} = 0,$$

$$\limsup_{\widetilde{a}_{21}x_1+\widetilde{a}_{22}x_2+\widetilde{a}_{23}x_3\to+\infty} \frac{F_2(t,x_1,x_2,x_3)}{\widetilde{a}_{21}x_1+\widetilde{a}_{22}x_2+\widetilde{a}_{23}x_3} = \limsup_{\widetilde{a}_{21}x_1+\widetilde{a}_{22}x_2+\widetilde{a}_{23}x_3\to+\infty} \frac{14Me^{2.4M}e^{-\widetilde{a}_{21}x_1-\widetilde{a}_{22}x_2-\widetilde{a}_{23}x_3}}{\widetilde{a}_{21}x_1+\widetilde{a}_{22}x_2+\widetilde{a}_{23}x_3} = 0,$$

$$\limsup_{\widetilde{a}_{31}x_1+\widetilde{a}_{32}x_2+\widetilde{a}_{33}x_3\to+\infty} \frac{F_3(t,x_1,x_2,x_3)}{\widetilde{a}_{31}x_1+\widetilde{a}_{32}x_2+\widetilde{a}_{33}x_3} = \limsup_{\widetilde{a}_{31}x_1+\widetilde{a}_{32}x_2+\widetilde{a}_{33}x_3\to+\infty} \frac{15Me^{2.08M}e^{-\widetilde{a}_{31}x_1-\widetilde{a}_{32}x_2-\widetilde{a}_{33}x_3}}{\widetilde{a}_{31}x_1+\widetilde{a}_{32}x_2+\widetilde{a}_{33}x_3} = 0.$$

As a result, (C5)–(C6) hold.

4. Conclusions

In this paper, we utilize the index of fixed point to research the positive solutions for the system of Riemann–Liouville type fractional boundary value problems (1). We first investigate corresponding operator equations for (1), and then establish some coupling behaviors for our nonlinearities $f_i (i = 1,2,3)$ by virtue of nonnegative matrix theory, which ensure that our nonlinearities can grow superlinearly and sublinearly at ∞.

Author Contributions: Conceptualization, Y.D. and J.X.; methodology, J.X.; software, Y.D.; validation, Y.D., J.X. and Z.F.; formal analysis, J.X.; investigation, Y.D.; resources, J.X.; data curation, J.X.; writing—original draft preparation, Y.D.; writing—review and editing, J.X.; visualization, J.X.; supervision, J.X. and Z.F.; project administration, J.X.; funding acquisition, J.X.

Funding: This work is supported by the National Natural Science Foundation of China (Grant No. 11601048), and the Natural Science Foundation of Chongqing Normal University (Grant No. 16XYY24).

Conflicts of Interest: The authors declare no conflict of interest.

References

1. Kilbas, A.A.; Srivastava, H.M.; Trujillo, J.J. *Theory and Applications of Fractional Differential Equations, Volume 204 of North-Holland Mathematics Studies*; Elsevier: Amsterdam, The Netherlands, 2006.
2. Podlubny, I. *Fractional Differential Equations, Volume 198 of Mathematics in Science and Engineering*; Academic Press: San Diego, CA, USA, 1999.
3. Samko, S.G.; Kilbas, A.A.; Marichev, O.I. *Fractional Integrals and Derivatives, Theory and Applications*; Gordon and Breach: Zurich, Switzerland, 1993.
4. Henderson, J.; Luca, R. Existence of positive solutions for a singular fractional boundary value problem. *Nonlinear Anal. Model. Control* **2017**, *22*, 99–114. [CrossRef]
5. Zhang, X.; Shao, Z.; Zhong, Q.; Zhao, Z. Triple positive solutions for semipositone fractional differential equations m-point boundary value problems with singularities and p-q-order derivatives. *Nonlinear Anal. Model. Control* **2018**, *23*, 889–903. [CrossRef]
6. Zhang, K.; Fu, Z. Solutions for a class of Hadamard fractional boundary value problems with sign-changing nonlinearity. *J. Funct. Spaces* **2019**, *2019*, 9046472. [CrossRef]
7. Pu, R.; Zhang, X.; Cui, Y.; Li, P.; Wang, W. Positive solutions for singular semipositone fractional differential equation subject to multipoint boundary conditions. *J. Funct. Spaces* **2017**, *2017*, 5892616. [CrossRef]
8. Wei, Z.; Li, Q.; Che, J. Initial value problems for fractional differential equations involving Riemann–Liouville sequential fractional derivative. *J. Math. Anal. Appl.* **2010**, *367*, 260–272. [CrossRef]
9. Wei, Z.; Dong, W.; Che, J. Periodic boundary value problems for fractional differential equations involving a Riemann–Liouville fractional derivative. *Nonlinear Anal. Theory Methods Appl.* **2010**, *73*, 3232–3238. [CrossRef]
10. Shu, X.; Lai, Y.; Chen, Y. The existence of mild solutions for impulsive fractional partial differential equations. *Nonlinear Anal. Theory Methods Appl.* **2011**, *74*, 2003–2011. [CrossRef]
11. Baleanu, D.; Rezapour, S.; Mohammadi, M. Some existence results on nonlinear fractional differential equations. *Philos. Trans. A* **2013**, *371*, 20120144. [CrossRef]
12. Samet, B.; Aydi, H. On some inequalities involving Caputo fractional derivatives and applications to special means of real numbers. *Mathematics* **2018**, *6*, 193. [CrossRef]
13. Yue, Z.; Zou, Y. New uniqueness results for fractional differential equation with dependence on the first order derivative. *Adv. Differ. Equ.* **2019**, *2019*, 38. [CrossRef]
14. Zou, Y.; He, G. The existence of solutions to integral boundary value problems of fractional differential equations at resonance. *J. Funct. Spaces* **2017**, *2017*, 2785937. [CrossRef]
15. Zou, Y.; He, G. On the uniqueness of solutions for a class of fractional differential equations. *Appl. Math. Lett.* **2017**, *74*, 68–73. [CrossRef]
16. Zou, Y. Positive solutions for a fractional boundary value problem with a perturbation term. *J. Funct. Spaces* **2018**, *2018*, 9070247. [CrossRef]
17. Cui, Y. Uniqueness of solution for boundary value problems for fractional differential equations. *Appl. Math. Lett.* **2016**, *51*, 48–54. [CrossRef]
18. Cui, Y.; Ma, W.; Sun, Q.; Su, X. New uniqueness results for boundary value problem of fractional differential equation. *Nonlinear Anal. Model. Control* **2018**, *23*, 31–39. [CrossRef]
19. Meng, S.; Cui, Y. The extremal solution to conformable fractional differential equations involving integral boundary condition. *Mathematics* **2019**, *7*, 186. [CrossRef]
20. Meng, S.; Cui, Y. Multiplicity results to a conformable fractional differential equations involving integral boundary condition. *Complexity* **2019**, *2019*, 8402347. [CrossRef]

21. Cui, Y.; Sun, Q.; Su, X. Monotone iterative technique for nonlinear boundary value problems of fractional order $p \in (2,3]$. *Adv. Differ. Equ.* **2017**, *2017*, 248. [CrossRef]
22. Zhang, X.; Wu, J.; Liu, L.; Wu, Y.; Cui, Y. Convergence analysis of iterative scheme and error estimation of positive solution for a fractional differential equation. *Math. Model. Anal.* **2018**, *23*, 611–626. [CrossRef]
23. Wu, J.; Zhang, X.; Liu, L.; Wu, Y.; Cui, Y. The convergence analysis and error estimation for unique solution of a p-Laplacian fractional differential equation with singular decreasing nonlinearity. *Bound. Value Probl.* **2018**, *2018*, 82. [CrossRef]
24. He, J.; Zhang, X.; Liu, L.; Wu, Y.; Cui, Y. Existence and asymptotic analysis of positive solutions for a singular fractional differential equation with nonlocal boundary conditions. *Bound. Value Probl.* **2018**, *2018*, 189. [CrossRef]
25. He, J.; Zhang, X.; Liu, L.; Wu, Y.; Cui, Y. A singular fractional Kelvin-Voigt model involving a nonlinear operator and their convergence properties. *Bound. Value Probl.* **2019**, *2019*, 112. [CrossRef]
26. Zhong, Q.; Zhang, X.; Lu, X.; Fu, Z. Uniqueness of successive positive solution for nonlocal singular higher-order fractional differential equations involving arbitrary derivatives. *J. Funct. Spaces* **2018**, *2018*, 6207682. [CrossRef]
27. Wang, F.; Cui, Y. Unbounded solutions to abstract boundary value problems of fractional differential equations on a half line. *Math. Methods Appl. Sci.* **2019**. [CrossRef]
28. Wang, Y.; Liu, Y.; Cui, Y. Infinitely many solutions for impulsive fractional boundary value problem with p-Laplacian. *Bound. Value Probl.* **2018**, *2018*, 94. [CrossRef]
29. Zuo, M.; Hao, X.; Liu, L.; Cui, Y. Existence results for impulsive fractional integro-differential equation of mixed type with constant coefficient and antiperiodic boundary conditions. *Bound. Value Probl.* **2017**, *2017*, 161. [CrossRef]
30. Sun, Q.; Ji, H.; Cui, Y. Positive solutions for boundary value problems of fractional differential equation with integral boundary conditions. *J. Funct. Spaces* **2018**, *2018*, 6461930. [CrossRef]
31. Sun, Q.; Meng, S.; Cui, Y. Existence results for fractional order differential equation with nonlocal Erdélyi-Kober and generalized Riemann–Liouville type integral boundary conditions at resonance. *Adv. Differ. Equ.* **2018**, *2018*, 243. [CrossRef]
32. Ma, W.; Cui, Y. The eigenvalue problem for Caputo type fractional differential equation with Riemann–Stieltjes integral boundary conditions. *J. Funct. Spaces* **2018**, *2018*, 2176809. [CrossRef]
33. Zhang, K.; Wang, J.; Ma, W. Solutions for integral boundary value problems of nonlinear Hadamard fractional differential equations. *J. Funct. Spaces* **2018**, *2018*, 2193234. [CrossRef]
34. Ma, W.; Meng, S.; Cui, Y. Resonant integral boundary value problems for Caputo fractional differential equations. *Math. Probl. Eng.* **2018**, *2018*, 5438592. [CrossRef]
35. Wang, G.; Bai, Z.; Zhang, L. Successive iterations for unique positive solution of a nonlinear fractional q-integral boundary value problem. *J. Appl. Anal. Comput.* **2019**, *9*, 1204–1215.
36. Jia, M.; Li, L.; Liu, X.; Song, J.; Bai, Z. A class of nonlocal problems of fractional differential equations with composition of derivative and parameters. *Adv. Differ. Equ.* **2019**, *2019*, 280. [CrossRef]
37. Zhang, W.; Bai, Z.; Sun, S. Extremal solutions for some periodic fractional differential equations. *Adv. Differ. Equ.* **2016**, *2016*, 179. [CrossRef]
38. Sheng, K.; Zhang, W.; Bai, Z. Positive solutions to fractional boundary-value problems with p-Laplacian on time scales. *Bound. Value Probl.* **2018**, *2018*, 70. [CrossRef]
39. Dong, X.; Bai, Z.; Zhang, S. Positive solutions to boundary value problems of p-Laplacian with fractional derivative. *Bound. Value Probl.* **2017**, *2017*, 5. [CrossRef]
40. Tian, Y.; Sun, S.; Bai, Z. Positive solutions of fractional differential equations with p-Laplacian. *J. Funct. Spaces* **2017**, *2017*, 3187492. [CrossRef]
41. Song, Q.; Bai, Z. Positive solutions of fractional differential equations involving the Riemann–Stieltjes integral boundary condition. *Adv. Differ. Equ.* **2018**, *2018*, 183. [CrossRef]
42. He, L.; Dong, X.; Bai, Z.; Chen, B. Solvability of some two-point fractional boundary value problems under barrier strip conditions. *J. Funct. Spaces* **2017**, *2017*, 1465623. [CrossRef]
43. Song, Q.; Dong, X.; Bai, Z.; Chen, B. Existence for fractional Dirichlet boundary value problem under barrier strip conditions. *J. Nonlinear Sci. Appl.* **2017**, *10*, 3592–3598. [CrossRef]
44. Zhao, Y.; Hou, X.; Sun, Y.; Bai, Z. Solvability for some class of multi-order nonlinear fractional systems. *Adv. Differ. Equ.* **2019**, *2019*, 23. [CrossRef]

45. Bai, Z.; Dong, X.; Yin, C. Existence results for impulsive nonlinear fractional differential equation with mixed boundary conditions. *Bound. Value Probl.* **2016**, *2016*, 63. [CrossRef]
46. Zhai, C.; Li, P.; Li, H. Single upper-solution or lower-solution method for Langevin equations with two fractional orders. *Adv. Differ. Equ.* **2018**, *2018*, 360. [CrossRef]
47. Fu, Z.; Bai, S.; O'Regan, D.; Xu, J. Nontrivial solutions for an integral boundary value problem involving Riemann–Liouville fractional derivatives. *J. Inequal. Appl.* **2019**, *2019*, 104. [CrossRef]
48. Zhang, K.; O'Regan, D.; Xu, J.; Fu, Z. Nontrivial solutions for a higher order nonlinear fractional boundary value problem involving Riemann–Liouville fractional derivatives. *J. Funct. Spaces* **2019**, *2019*, 2381530. [CrossRef]
49. Fazli, H.; Nieto, J.J.; Bahrami, F. On the existence and uniqueness results for nonlinear sequential fractional differential equations. *Appl. Comput. Math.* **2018**, *17*, 36–47.
50. Bai, Z. On positive solutions of a nonlocal fractional boundary value problem. *Nonlinear Anal.* **2010**, *72*, 916–924. [CrossRef]
51. Jiang, J.; O'Regan, D.; Xu, J.; Cui, Y. Positive solutions for a Hadamard fractional p-Laplacian three-point boundary value problem. *Mathematics* **2019**, *7*, 439. [CrossRef]
52. Jiang, J.; O'Regan, D.; Xu, J.; Fu, Z. Positive solutions for a system of nonlinear Hadamard fractional differential equations involving coupled integral boundary conditions. *J. Inequal. Appl.* **2019**, *2019*, 204. [CrossRef]
53. Ahmad, B.; Ntouyas, S.K.; Alsaedi, A. On a coupled system of fractional differential equations with coupled nonlocal and integral boundary conditions. *Chaos Solitons Fractals* **2016**, *83*, 234–241. [CrossRef]
54. Wang, Y.; Liu, L.; Zhang, X.; Wu, Y. Positive solutions of an abstract fractional semipositone differential system model for bioprocesses of HIV infection. *Appl. Math. Comput.* **2015**, *258*, 312–324. [CrossRef]
55. Aljoudi, S.; Ahmad, B.; Nieto, J.J.; Alsaedi, A. A coupled system of Hadamard type sequential fractional differential equations with coupled strip conditions. *Chaos Solitons Fractals* **2016**, *91*, 39–46. [CrossRef]
56. Henderson, J.; Luca, R.; Tudorache, A. On a system of fractional differential equations with coupled integral boundary conditions. *Fract. Calc. Appl. Anal.* **2015**, *18*, 361–386. [CrossRef]
57. Ahmad, B.; Luca, R. Existence of solutions for a system of fractional differential equations with coupled nonlocal boundary conditions. *Fract. Calc. Appl. Anal.* **2018**, *21*, 423–441. [CrossRef]
58. Henderson, J.; Luca, R. Positive solutions for a system of coupled fractional boundary value problems. *Lith. Math. J.* **2018**, *58*, 15–32. [CrossRef]
59. Ahmad, B.; Nieto, J.J. Existence results for a coupled system of nonlinear fractional differential equations with three-point boundary conditions. *Comput. Math. Appl.* **2009**, *58*, 1838–1843. [CrossRef]
60. Ali, A.; Shah, K.; Jarad, F.; Gupta, V.; Abdeljawad, T. Existence and stability analysis to a coupled system of implicit type impulsive boundary value problems of fractional-order differential equations. *Adv. Differ. Equ.* **2019**, *2019*, 101. [CrossRef]
61. Mahmudov, N.I.; Bawaneh, S.; Al-Khateeb, A. On a coupled system of fractional differential equations with four point integral boundary conditions. *Mathematics* **2019**, *7*, 279. [CrossRef]
62. Zhao, Y.; Chen, H.; Xu, C. Nontrivial solutions for impulsive fractional differential equations via Morse theory. *Appl. Math. Comput.* **2017**, *307*, 170–179. [CrossRef]
63. Zhang, X.; Liu, L.; Wu, Y.; Zou, Y. Existence and uniqueness of solutions for systems of fractional differential equations with Riemann–Stieltjes integral boundary condition. *Adv. Differ. Equ.* **2018**, *2018*, 204. [CrossRef]
64. Zhang, X.; Liu, L.; Zou, Y. Fixed-point theorems for systems of operator equations and their applications to the fractional differential equations. *J. Funct. Spaces* **2018**, *2018*, 7469868. [CrossRef]
65. Hao, X.; Wang, H.; Liu, L.; Cui, Y. Positive solutions for a system of nonlinear fractional nonlocal boundary value problems with parameters and p-Laplacian operator. *Bound. Value Probl.* **2017**, *2017*, 182. [CrossRef]
66. Qi, T.; Liu, Y.; Zou, Y. Existence result for a class of coupled fractional differential systems with integral boundary value conditions. *J. Nonlinear Sci. Appl.* **2017**, *10*, 4034–4045. [CrossRef]
67. Qi, T.; Liu, Y.; Cui, Y. Existence of solutions for a class of coupled fractional differential systems with nonlocal boundary conditions. *J. Funct. Spaces* **2017**, *2017*, 6703860. [CrossRef]
68. Zhang, Y. Existence results for a coupled system of nonlinear fractional multi-point boundary value problems at resonance. *J. Inequal. Appl.* **2018**, *2018*, 198. [CrossRef] [PubMed]
69. Li, H.; Zhang, J. Positive solutions for a system of fractional differential equations with two parameters. *J. Funct. Spaces* **2018**, *2018*, 1462505. [CrossRef]

70. Zhai, C.; Wang, W.; Li, H. A uniqueness method to a new Hadamard fractional differential system with four-point boundary conditions. *J. Inequal. Appl.* **2018**, *2018*, 207. [CrossRef]
71. Cheng, W.; Xu, J.; Cui, Y.; Ge, Q. Positive solutions for a class of fractional difference systems with coupled boundary conditions. *Adv. Differ. Equ.* **2019**, *2019*, 249. [CrossRef]
72. Cheng, W.; Xu, J.; Cui, Y. Positive solutions for a system of nonlinear semipositone fractional q-difference equations with q-integral boundary conditions. *J. Nonlinear Sci. Appl.* **2017**, *10*, 4430–4440. [CrossRef]
73. Xu, J.; Goodrich, C.S.; Cui, Y. Positive solutions for a system of first-order discrete fractional boundary value problems with semipositone nonlinearities. *Rev. R. Acad. Cienc. Exactas Fís. Nat. Ser. A Mat.* **2019**, *113*, 1343–1358. [CrossRef]
74. Qiu, X.; Xu, J.; O'Regan, D.; Cui, Y. Positive solutions for a system of nonlinear semipositone boundary value problems with Riemann–Liouville fractional derivatives. *J. Funct. Spaces* **2018**, *2018*, 7351653. [CrossRef]
75. Chen, C.; Xu, J.; O'Regan, D.; Fu, Z. Positive solutions for a system of semipositone fractional difference boundary value problems. *J. Funct. Spaces* **2018**, *2018*, 6835028. [CrossRef]
76. Wang, F.; Cui, Y.; Zhou, H. Solvability for an infinite system of fractional order boundary value problems. *Ann. Funct. Anal.* **2019**, *10*, 395–411. [CrossRef]
77. Wang, F.; Cui, Y. Positive solutions for an infinite system of fractional order boundary value problems. *Adv. Differ. Equ.* **2019**, *2019*, 169. [CrossRef]
78. Riaz, U.; Zada, A.; Ali, Z.; Ahmad, M.; Xu, J.; Fu, Z. Analysis of nonlinear coupled systems of impulsive fractional differential equations with Hadamard derivatives. *Math. Probl. Eng.* **2019**, *2019*, 5093572. [CrossRef]
79. Riaz, U.; Zada, A.; Ali, Z.; Cui, Y.; Xu, J. Analysis of coupled systems of implicit impulsive fractional differential equations involving Hadamard derivatives. *Adv. Differ. Equ.* **2019**, *2019*, 226. [CrossRef]
80. Aris, R. *Introduction to the Analysis of Chemical Reactors*; Prentice Hall: Englewood Cliffs, NJ, USA, 1965.
81. Guo, D.; Lakshmikantham, V. *Nonlinear Problems in Abstract Cones*; Academic Press: Orlando, FL, USA, 1988.
82. Yang, Z.; Zhang, Z. Positive solutions for a system of nonlinear singular Hammerstein integral equations via nonnegative matrices and applications. *Positivity* **2012**, *16*, 783–800. [CrossRef]
83. Abraham, B.; Plemmons, R.J. *Nonnegative Matrices in the Mathematical Sciences*; Academic Press: New York, NY, USA, 1979.

mathematics

MDPI

Article

On a Generalized Langevin Type Nonlocal Fractional Integral Multivalued Problem

Ahmed Alsaedi [1], Bashir Ahmad [1,*], Madeaha Alghanmi [1] and Sotiris K. Ntouyas [1,2]

1 Nonlinear Analysis and Applied Mathematics (NAAM)-Research Group, Department of Mathematics, Faculty of Science, King Abdulaziz University, P.O. Box 80203, Jeddah 21589, Saudi Arabia; aalsaedi@hotmail.com (A.A.); madeaha@hotmail.com (M.A.); sntouyas@uoi.gr (S.K.N.)

2 Department of Mathematics, University of Ioannina, 451 10 Ioannina, Greece

* Correspondence: bmuhammed@kau.edu.sa

Received: 29 August 2019; Accepted: 21 October 2019; Published: 25 October 2019

Abstract: We establish sufficient criteria for the existence of solutions for a nonlinear generalized Langevin-type nonlocal fractional-order integral multivalued problem. The convex and non-convex cases for the multivalued map involved in the given problem are considered. Our results rely on Leray–Schauder nonlinear alternative for multivalued maps and Covitz and Nadler's fixed point theorem. Illustrative examples for the main results are included.

Keywords: differential inclusions; Caputo-type fractional derivative; fractional integral; existence; fixed point

1. Introduction

Fractional calculus is the extension of classical calculus which deals with differential and integral operators of fractional order. It has evolved into a significant and popular branch of mathematical analysis owing to its extensive applications in the mathematical modeling of applied and technical problems. The literature on fractional calculus is now much enriched and covers a wide range of interesting results, for instance [1–6]. For a comprehensive treatment of Hadamard-type fractional differential equations and inclusions, we refer the reader to the text [7].

The Langevin equation is found to be an effective tool to describe stochastic problems in fluctuating situations. A modified type of this equation is used in various functional approaches for fractal media. A variety of boundary value problems involving the Langevin equation have been investigated by several authors. In [8], existence and uniqueness results for a nonlinear Langevin equation involving two fractional orders supplemented with three-point boundary conditions were obtained. An impulsive boundary value problem for a nonlinear Langevin equation involving two different fractional derivatives was investigated in [9]. Some existing results for Langevin fractional differential inclusions with two indices were derived in [10]. In [11], the authors proved the existence of and uniqueness results for an anti-periodic boundary value problem of a system of Langevin fractional differential equations. In [12], the authors investigated a nonlinear fractional Langevin equation with anti-periodic boundary conditions by applying coupled fixed point theorems. In a recent work [13], the authors obtained some existence results for a fractional Langevin equation with nonlinearity depending on Riemann–Liouville fractional integral, and complemented with nonlocal multi-point and multi-strip boundary conditions.

In the present paper, we study the existence of solutions for a nonlinear generalized Langevin type nonlocal fractional-order integral multivalued problem given by

$$\begin{cases} {}^{\rho}_{c}D^{\alpha}_{a+}({}^{\rho}_{c}D^{\beta}_{a+}+\lambda)x(t) \in F(t,x(t)), \;\; t \in J := [a,T], \;\; \lambda \in \mathbb{R}, \\ x(a)=0, \;\; x(\eta)=0, \;\; x(T)=\mu\, {}^{\rho}I^{\gamma}_{a+}x(\xi), \;\; a<\eta<\xi<T, \mu \in \mathbb{R}, \end{cases} \tag{1}$$

where ${}^{\rho}_{c}D^{\alpha}_{a+}$, ${}^{\rho}_{c}D^{\beta}_{a+}$ denote the Caputo-type generalized fractional differential operators of order $1 < \alpha \le 2$, $0 < \beta < 1$, $\rho > 0$, respectively, $F : [a,T] \times \mathbb{R} \to \mathcal{P}(\mathbb{R})$ is a multi-valued map ($\mathcal{P}(\mathbb{R})$ is the family of all nonempty subsets of $\mathbb{R}$), ${}^{\rho}I^{\gamma}_{a+}$ is the generalized fractional integral operator of order $\gamma > 0$ and $\rho > 0$. Here we emphasize that the single-valued analogue of the problem (1) was discussed in [14].

The rest of the paper is arranged as follows. The background material related to our work is outlined in Section 3. The existence results for the problem (1) are presented in Section 3. The first result for the problem (1), associated with the convex valued mutivalued map, is derived with the aid of Leray–Schauder nonlinear alternative for multivalued maps, while the result for non-convex valued map for the problem (1) is proved by applying a fixed point theorem due to Covitz and Nadler. Section 4 contains the illustrative examples for the main results. We summarize the work established in this paper, and its implications, in the last section.

2. Preliminaries

Define by $X^p_c(a,b)$ the space of all complex-valued Lebesgue measurable functions ϕ on (a,b) equipped with the norm:

$$\|\phi\|_{X^p_c} = \Big(\int_a^b |x^c\phi(x)|^p \frac{dx}{x}\Big)^{1/p} < \infty, \ \ c \in \mathbb{R}, 1 \le p \le \infty.$$

Let $AC^n_\delta[a,b]$ denote the class of all absolutely continuous functions g possessing δ^{n-1}-derivative ($\delta^{n-1}g \in AC([a,b],\mathbb{R})$), endowed with the norm $\|g\|_{AC^n_\delta} = \sum_{k=0}^{n-1} \|\delta^k g\|_C$.

Definition 1. *The left-sided and right-sided generalized fractional integrals for $g \in X^p_c(a,b)$ of order $\beta > 0$ and $\rho > 0$, denoted by ${}^{\rho}I^{\beta}_{a+}g$ and ${}^{\rho}I^{\beta}_{b-}g$ respectively, are defined by [15]*

$$({}^{\rho}I^{\beta}_{a+}g)(t) = \frac{\rho^{1-\beta}}{\Gamma(\beta)} \int_a^t \frac{s^{\rho-1}}{(t^\rho - s^\rho)^{1-\beta}} g(s)ds, \ -\infty < a < t < b < \infty, \tag{2}$$

$$({}^{\rho}I^{\beta}_{b-}g)(t) = \frac{\rho^{1-\alpha}}{\Gamma(\beta)} \int_t^b \frac{s^{\rho-1}}{(s^\rho - t^\rho)^{1-\beta}} g(s)ds, \ -\infty < a < t < b < \infty. \tag{3}$$

Definition 2. *Let $\beta > 0$, $n = [\beta] + 1$ and $\rho > 0$. We define the generalized fractional derivatives, associated with the generalized fractional integrals (2) and (3), for $0 \le a < t < b < \infty$, as follows [16]:*

$$({}^{\rho}D^{\beta}_{a+}g)(t) = \Big(t^{1-\rho}\frac{d}{dt}\Big)^n ({}^{\rho}I^{n-\beta}_{a+}g)(t) = \frac{\rho^{\beta-n+1}}{\Gamma(n-\beta)} \Big(t^{1-\rho}\frac{d}{dt}\Big)^n \int_a^t \frac{s^{\rho-1}}{(t^\rho - s^\rho)^{\beta-n+1}} g(s)ds, \tag{4}$$

$$({}^{\rho}D^{\beta}_{b-}g)(t) = \Big(-t^{1-\rho}\frac{d}{dt}\Big)^n ({}^{\rho}I^{n-\beta}_{b-}g)(t) = \frac{\rho^{\beta-n+1}}{\Gamma(n-\beta)} \Big(-t^{1-\rho}\frac{d}{dt}\Big)^n \int_t^b \frac{s^{\rho-1}}{(s^\rho - t^\rho)^{\beta-n+1}} g(s)ds, \tag{5}$$

provided the integrals in the above expressions exist.

Definition 3. *Let $g \in AC^n_\delta[a,b]$ and $\beta > 0, n = [\beta] + 1$. Then the Caputo-type generalized fractional derivatives ${}^{\rho}_{c}D^{\beta}_{a+}g$ and ${}^{\rho}_{c}D^{\beta}_{b-}g$ are respectively defined via (4) and (5) by [17]*

$${}^{\rho}_{c}D^{\beta}_{a+}g(x) = {}^{\rho}D^{\beta}_{a+}\Big[g(t) - \sum_{k=0}^{n-1} \frac{\delta^k g(a)}{k!} \Big(\frac{t^\rho - a^\rho}{\rho}\Big)^k\Big](x), \ \ \delta = x^{1-\rho}\frac{d}{dx}, \tag{6}$$

$${}^{\rho}_{c}D^{\beta}_{b-}g(x) = {}^{\rho}D^{\beta}_{b-}\Big[g(t) - \sum_{k=0}^{n-1} \frac{(-1)^k\delta^k g(b)}{k!} \Big(\frac{b^\rho - t^\rho}{\rho}\Big)^k\Big](x), \ \ \delta = x^{1-\rho}\frac{d}{dx}. \tag{7}$$

Remark 1. *The left and right generalized Caputo derivatives of order β for $g \in AC^n_\delta[a,b]$, are respectively given by [17]*

$$ {}^\rho_c D^\beta_{a+} g(t) = \frac{1}{\Gamma(n-\beta)} \int_a^t \Big(\frac{t^\rho - s^\rho}{\rho}\Big)^{n-\beta-1} \frac{(\delta^n g)(s)ds}{s^{1-\rho}}, \tag{8} $$

$$ {}^\rho_c D^\beta_{b-} g(t) = \frac{1}{\Gamma(n-\beta)} \int_t^b \Big(\frac{s^\rho - t^\rho}{\rho}\Big)^{n-\alpha-1} \frac{(-1)^n(\delta^n g)(s)ds}{s^{1-\rho}}. \tag{9} $$

Lemma 1. *Let $g \in AC^n_\delta[a,b]$ or $C^n_\delta[a,b]$. Then, for $\beta \in \mathbb{R}$, the following results hold [17]:*

$$ {}^\rho I^\beta_{a+} {}^\rho_c D^\beta_{a+} g(x) = g(x) - \sum_{k=0}^{n-1} \frac{(\delta^k g)(a)}{k!} \Big(\frac{x^\rho - a^\rho}{\rho}\Big)^k, $$

$$ {}^\rho I^\beta_{b-} {}^\rho_c D^\beta_{b-} g(x) = g(x) - \sum_{k=0}^{n-1} \frac{(-1)^k(\delta^k g)(a)}{k!} \Big(\frac{b^\rho - x^\rho}{\rho}\Big)^k. $$

In particular, for $0 < \beta \le 1$, we have

$$ {}^\rho I^\beta_{a+} {}^\rho_c D^\beta_{a+} g(x) = g(x) - g(a), \; {}^\rho I^\beta_{b-} {}^\rho_c D^\beta_{b-} g(x) = g(x) - g(b). $$

We need the following known lemma [14] in the sequel.

Lemma 2. *Let $h \in C([a,T],\mathbb{R})$ and $x \in AC^3_\delta(J)$. Then the unique solution of linear problem:*

$$ \begin{cases} {}^\rho_c D^\alpha_{a+}({}^\rho_c D^\beta_{a+} + \lambda)x(t) = h(t), \;\; t \in J := [a,T], \\ x(a) = 0, \;\; x(\eta) = 0, \;\; x(T) = \mu {}^\rho I^\gamma_{a+} x(\xi), \;\; a < \eta < \xi < T, \end{cases} \tag{10} $$

is given by:

$$ \begin{aligned} x(t) \;=\; & {}^\rho I^{\alpha+\beta}_{a+} h(t) - \lambda {}^\rho I^\beta_{a+} x(t) + \frac{(t^\rho - a^\rho)^\beta(\eta^\rho - t^\rho)}{\rho^{\beta+1}\Gamma(\beta+2)\Omega} \Big\{ {}^\rho I^{\alpha+\beta}_{a+} h(T) - \lambda {}^\rho I^\beta_{a+} x(T) \\ & - \mu {}^\rho I^{\alpha+\beta+\gamma}_{a+} h(\xi) + \mu\lambda {}^\rho I^{\beta+\gamma}_{a+} x(\xi) \Big\} - \frac{(t^\rho - a^\rho)^\beta}{\Omega(\eta^\rho - a^\rho)^\beta} \Big(\frac{(T^\rho - a^\rho)^\beta(T^\rho - t^\rho)}{\rho^{\beta+1}\Gamma(\beta+2)} \\ & - \frac{\mu(\xi^\rho - a^\rho)^{\beta+\gamma}[(\beta+1)(\xi^\rho - t^\rho) - \gamma(t^\rho - a^\rho)]}{\rho^{\beta+\gamma+1}\Gamma(\beta+\gamma+2)(\beta+1)} \Big) \Big\{ {}^\rho I^{\alpha+\beta}_{a+} h(\eta) - \lambda {}^\rho I^\beta_{a+} x(\eta) \Big\}, \end{aligned} \tag{11} $$

where it is assumed that

$$ \Omega = \Big[\frac{(T^\rho - a^\rho)^\beta(T^\rho - \eta^\rho)}{\rho^{\beta+1}\Gamma(\beta+2)} - \frac{\mu(\xi^\rho - a^\rho)^{\beta+\gamma}[(\beta+1)(\xi^\rho - \eta^\rho) - \gamma(\eta^\rho - a^\rho)]}{\rho^{\beta+\gamma+1}\Gamma(\beta+\gamma+2)(\beta+1)} \Big] \neq 0. \tag{12} $$

3. Main Results

We begin this section with the definition of a solution for the multi-valued problem (1).

Definition 4. *A function $x \in C(J,\mathbb{R})$ is called a solution of the problem (1) if we can find a function $v \in L^1(J,\mathbb{R})$ with $v(t) \in F(t,x)$ a.e. on J such that $x(a) = 0, x(\eta) = 0, x(T) = \mu {}^\rho I^\gamma_{a+} x(\xi)$ and*

$$ \begin{aligned} x(t) \;=\; & {}^\rho I^{\alpha+\beta}_{a+} v(t) - \lambda {}^\rho I^\beta_{a+} x(t) + \frac{(t^\rho - a^\rho)^\beta(\eta^\rho - t^\rho)}{\rho^{\beta+1}\Gamma(\beta+2)\Omega} \Big\{ {}^\rho I^{\alpha+\beta}_{a+} v(T) - \lambda {}^\rho I^\beta_{a+} x(T) \\ & - \mu {}^\rho I^{\alpha+\beta+\gamma}_{a+} v(\xi) + \mu\lambda {}^\rho I^{\beta+\gamma}_{a+} x(\xi) \Big\} - \frac{(t^\rho - a^\rho)^\beta}{\Omega(\eta^\rho - a^\rho)^\beta} \Big(\frac{(T^\rho - a^\rho)^\beta(T^\rho - t^\rho)}{\rho^{\beta+1}\Gamma(\beta+2)} \\ & - \frac{\mu(\xi^\rho - a^\rho)^{\beta+\gamma}[(\beta+1)(\xi^\rho - t^\rho) - \gamma(t^\rho - a^\rho)]}{\rho^{\beta+\gamma+1}\Gamma(\beta+\gamma+2)(\beta+1)} \Big) \Big\{ {}^\rho I^{\alpha+\beta}_{a+} v(\eta) - \lambda {}^\rho I^\beta_{a+} x(\eta) \Big\}. \end{aligned} \tag{13} $$

For the sake of computational convenience, we set

$$\begin{aligned}\Lambda_1 &= \frac{(T^\rho - a^\rho)^{\alpha+\beta}}{\rho^{\alpha+\beta}\Gamma(\alpha+\beta+1)}\left[1+\frac{\zeta_1}{\rho^{\beta+1}\Gamma(\beta+2)|\Omega|}\right]+\frac{|\mu|(\xi^\rho - a^\rho)^{\alpha+\beta+\gamma}\zeta_1}{\rho^{\alpha+2\beta+\gamma+1}\Gamma(\alpha+\beta+\gamma+1)\Gamma(\beta+2)|\Omega|}\\ &\quad+\frac{(\eta^\rho - a^\rho)^{\alpha}\zeta_2}{\rho^{\alpha+\beta}\Gamma(\alpha+\beta+1)|\Omega|},\end{aligned} \tag{14}$$

$$\begin{aligned}\Lambda_2 &= \frac{|\lambda|(T^\rho - a^\rho)^{\beta}}{\rho^{\beta}\Gamma(\beta+1)}\left[1+\frac{\zeta_1}{\rho^{\beta+1}\Gamma(\beta+2)|\Omega|}\right]+\frac{|\mu||\lambda|(\xi^\rho - a^\rho)^{\beta+\gamma}\zeta_1}{\rho^{2\beta+\gamma+1}\Gamma(\beta+\gamma+1)\Gamma(\beta+2)|\Omega|}\\ &\quad+\frac{|\lambda|\zeta_2}{\rho^{\beta}\Gamma(\beta+1)|\Omega|},\end{aligned} \tag{15}$$

where

$$\zeta_1 := \max_{t\in[a,T]}\left|(t^\rho - a^\rho)^\beta(\eta^\rho - t^\rho)\right|, \tag{16}$$

$$\zeta_2 := \max_{t\in[a,T]}\left|(t^\rho - a^\rho)^\beta\left[\frac{(T^\rho - a^\rho)^\beta(T^\rho - t^\rho)}{\rho^{\beta+1}\Gamma(\beta+2)} - \frac{\mu(\xi^\rho - a^\rho)^{\beta+\gamma}[(\beta+1)(\xi^\rho - t^\rho) - \gamma(t^\rho - a^\rho)]}{\rho^{\beta+\gamma+1}\Gamma(\beta+\gamma+2)(\beta+1)}\right]\right|. \tag{17}$$

We define the set of selections of F by $S_{F,x} := \{y \in L^1(J,\mathbb{R}) : y(t) \in F(t,x(t)) \text{ on } J\}$ for each $x \in C(J,\mathbb{R})$.

3.1. The Upper Semicontinuous Case

In the following result, we assume that the multivalued map F is convex-valued and apply Leray–Schauder nonlinear alternative for multivalued maps [18] to prove the existence of solutions for the problem at hand.

Theorem 1. *Assume that:*

- (A_1) *$F : J\times\mathbb{R} \to \mathcal{P}_{cp,c}(\mathbb{R})$ is L^1-Carathéodory, where $\mathcal{P}_{cp,c}(\mathbb{R}) = \{\mathcal{Y} \in \mathcal{P}(\mathbb{R}) : \mathcal{Y}$ is compact and convex$\}$;*
- (A_2) *there exist a function $P \in C(J,\mathbb{R}^+)$ and a continuous nondecreasing function $Q : [0,\infty) \to (0,\infty)$ such that $\|F(t,x)\|_{\mathcal{P}} := \sup\{|y| : y \in F(t,x)\} \le P(t)Q(|x|)$ for each $(t,x) \in J\times\mathbb{R}$;*
- (A_3) *there exists a constant $M > 0$ such that*

$$\frac{(1-\Lambda_2)M}{\Lambda_1\|P\|Q(M)} > 1, \quad \Lambda_2 < 1,$$

where Λ_1 and Λ_2 are respectively given by (14) and (15).

Then the problem (1) has at least one solution on J.

Proof. Let us first convert the problem (1) into a fixed point problem by introducing a multivalued map: $N : C(J,\mathbb{R}) \to \mathcal{P}(C(J,\mathbb{R}))$ as

$$N(x) = \left\{\begin{array}{l} h \in C(J,\mathbb{R}) : \\ h(t) = \left\{\begin{array}{l} {}^\rho I_{a+}^{\alpha+\beta}v(t) - \lambda^\rho I_{a+}^{\beta}x(t) \\ +\dfrac{(t^\rho - a^\rho)^\beta(\eta^\rho - t^\rho)}{\rho^{\beta+1}\Gamma(\beta+2)\Omega}\Big\{{}^\rho I_{a+}^{\alpha+\beta}v(T) - \lambda^\rho I_{a+}^{\beta}x(T) \\ -\mu^\rho I_{a+}^{\alpha+\beta+\gamma}v(\xi) + \mu\lambda^\rho I_{a+}^{\beta+\gamma}x(\xi)\Big\} - \dfrac{(t^\rho - a^\rho)^\beta}{\Omega(\eta^\rho - a^\rho)^\beta}\Big(\dfrac{(T^\rho - a^\rho)^\beta(T^\rho - t^\rho)}{\rho^{\beta+1}\Gamma(\beta+2)} \\ -\dfrac{\mu(\xi^\rho - a^\rho)^{\beta+\gamma}[(\beta+1)(\xi^\rho - t^\rho) - \gamma(t^\rho - a^\rho)]}{\rho^{\beta+\gamma+1}\Gamma(\beta+\gamma+2)(\beta+1)}\Big)\Big\{{}^\rho I_{a+}^{\alpha+\beta}v(\eta) - \lambda^\rho I_{a+}^{\beta}x(\eta)\Big\}, \end{array}\right. \end{array}\right\}$$

for $v \in S_{F,x}$.

It is clear that fixed points of N are solutions of problem (1). So we need to verify that the operator N satisfies all the conditions of Leray–Schauder nonlinear alternative [18]. This will be done in several steps.

Step 1. *$N(x)$ is convex for each $x \in C(J,\mathbb{R})$.*

Indeed, if h_1, h_2 belongs to $N(x)$, then there exist $v_1, v_2 \in S_{F,x}$ such that, for each $t \in J$, we have

$$\begin{aligned} h_i(t) &= {}^{\rho}I_{a+}^{\alpha+\beta}v_i(t) - \lambda^{\rho}I_{a+}^{\beta}x(t) + \frac{(t^\rho - a^\rho)^\beta(\eta^\rho - t^\rho)}{\rho^{\beta+1}\Gamma(\beta+2)\Omega}\Big\{{}^{\rho}I_{a+}^{\alpha+\beta}v_i(T) - \lambda^{\rho}I_{a+}^{\beta}x(T) \\ &\quad -\mu^{\rho}I_{a+}^{\alpha+\beta+\gamma}v_i(\xi) + \mu\lambda^{\rho}I_{a+}^{\beta+\gamma}x(\xi)\Big\} - \frac{(t^\rho - a^\rho)^\beta}{\Omega(\eta^\rho - a^\rho)^\beta}\Big(\frac{(T^\rho - a^\rho)^\beta(T^\rho - t^\rho)}{\rho^{\beta+1}\Gamma(\beta+2)} \\ &\quad - \frac{\mu(\xi^\rho - a^\rho)^{\beta+\gamma}[(\beta+1)(\xi^\rho - t^\rho) - \gamma(t^\rho - a^\rho)]}{\rho^{\beta+\gamma+1}\Gamma(\beta+\gamma+2)(\beta+1)}\Big)\Big\{{}^{\rho}I_{a+}^{\alpha+\beta}v_i(\eta) - \lambda^{\rho}I_{a+}^{\beta}x(\eta)\Big\}, i = 1,2. \end{aligned}$$

Let $t \in J$ and $\theta \in (0,1)$. Then

$$\begin{aligned} &[\theta h_1 + (1-\theta)h_2](t) \\ =\ & {}^{\rho}I_{a+}^{\alpha+\beta}[\theta v_1(s) + (1-\theta)v_2(s)](t) - \lambda^{\rho}I_{a+}^{\beta}x(t) \\ &+ \frac{(t^\rho - a^\rho)^\beta(\eta^\rho - t^\rho)}{\rho^{\beta+1}\Gamma(\beta+2)\Omega}\Big\{{}^{\rho}I_{a+}^{\alpha+\beta}[\theta v_1(s) + (1-\theta)v_2(s)](T) - \lambda^{\rho}I_{a+}^{\beta}x(T) \\ &-\mu^{\rho}I_{a+}^{\alpha+\beta+\gamma}[\theta v_1(s) + (1-\theta)v_2(s)](\xi) + \mu\lambda^{\rho}I_{a+}^{\beta+\gamma}x(\xi)\Big\} - \frac{(t^\rho - a^\rho)^\beta}{\Omega(\eta^\rho - a^\rho)^\beta}\Big(\frac{(T^\rho - a^\rho)^\beta(T^\rho - t^\rho)}{\rho^{\beta+1}\Gamma(\beta+2)} \\ &- \frac{\mu(\xi^\rho - a^\rho)^{\beta+\gamma}[(\beta+1)(\xi^\rho - t^\rho) - \gamma(t^\rho - a^\rho)]}{\rho^{\beta+\gamma+1}\Gamma(\beta+\gamma+2)(\beta+1)}\Big)\Big\{{}^{\rho}I_{a+}^{\alpha+\beta}[\theta v_1(s) + (1-\theta)v_2(s)](\eta) - \lambda^{\rho}I_{a+}^{\beta}x(\eta)\Big\}. \end{aligned}$$

Since F has convex values ($S_{F,x}$ is convex), therefore, $\theta h_1 + (1-\theta)h_2 \in N(x)$.

Step 2. *$N(x)$ maps bounded sets (balls) into bounded sets in $C(J,\mathbb{R})$.*

Let $B_r = \{x \in C(J,\mathbb{R}) : \|x\| \le r\}$ be a bounded ball in $C(J,\mathbb{R})$, where r is a positive number. Then, for each $h \in N(x), x \in B_r$, there exists $v \in S_{F,x}$ such that

$$\begin{aligned} h(t) &= {}^{\rho}I_{a+}^{\alpha+\beta}v(t) - \lambda^{\rho}I_{a+}^{\beta}x(t) + \frac{(t^\rho - a^\rho)^\beta(\eta^\rho - t^\rho)}{\rho^{\beta+1}\Gamma(\beta+2)\Omega}\Big\{{}^{\rho}I_{a+}^{\alpha+\beta}v(T) - \lambda^{\rho}I_{a+}^{\beta}x(T) \\ &\quad -\mu^{\rho}I_{a+}^{\alpha+\beta+\gamma}v(\xi) + \mu\lambda^{\rho}I_{a+}^{\beta+\gamma}x(\xi)\Big\} - \frac{(t^\rho - a^\rho)^\beta}{\Omega(\eta^\rho - a^\rho)^\beta}\Big(\frac{(T^\rho - a^\rho)^\beta(T^\rho - t^\rho)}{\rho^{\beta+1}\Gamma(\beta+2)} \\ &\quad - \frac{\mu(\xi^\rho - a^\rho)^{\beta+\gamma}[(\beta+1)(\xi^\rho - t^\rho) - \gamma(t^\rho - a^\rho)]}{\rho^{\beta+\gamma+1}\Gamma(\beta+\gamma+2)(\beta+1)}\Big)\Big\{{}^{\rho}I_{a+}^{\alpha+\beta}v(\eta) - \lambda^{\rho}I_{a+}^{\beta}x(\eta)\Big\}. \end{aligned}$$

In view of (H_2), for each $t \in J$, we find that

$$\begin{aligned}
|h(t)| &\leq {}^\rho I_{a+}^{\alpha+\beta}|v(t)| + |\lambda|\, {}^\rho I_{a+}^{\beta}|x(t)| + \frac{|(t^\rho - a^\rho)^\beta(\eta^\rho - t^\rho)|}{\rho^{\beta+1}\Gamma(\beta+2)|\Omega|}\Big\{{}^\rho I_{a+}^{\alpha+\beta}|v(T)| + \lambda^\rho I_{a+}^{\beta}|x(T)| \\
&\quad + |\mu|\, {}^\rho I_{a+}^{\alpha+\beta+\gamma}|v(\xi)| + |\mu\lambda|\, {}^\rho I_{a+}^{\beta+\gamma}|x(\xi)|\Big\} + \Big|\frac{(t^\rho - a^\rho)^\beta}{\Omega(\eta^\rho - a^\rho)^\beta}\Big(\frac{(T^\rho - a^\rho)^\beta(T^\rho - t^\rho)}{\rho^{\beta+1}\Gamma(\beta+2)} \\
&\quad + \frac{\mu(\xi^\rho - a^\rho)^{\beta+\gamma}[(\beta+1)(\xi^\rho - t^\rho) - \gamma(t^\rho - a^\rho)]}{\rho^{\beta+\gamma+1}\Gamma(\beta+\gamma+2)(\beta+1)}\Big)\Big|\Big\{{}^\rho I_{a+}^{\alpha+\beta}|v(\eta)| + |\lambda|\, {}^\rho I_{a+}^{\beta}|x(\eta)|\Big\} \\
&\leq \|P\|Q(\|x\|)\Big(\frac{(T^\rho - a^\rho)^{\alpha+\beta}}{\rho^{\alpha+\beta}\Gamma(\alpha+\beta+1)}\Big[1 + \frac{\zeta_1}{\rho^{\beta+1}\Gamma(\beta+2)|\Omega|}\Big] \\
&\quad + \frac{|\mu|(\xi^\rho - a^\rho)^{\alpha+\beta+\gamma}\zeta_1}{\rho^{\alpha+2\beta+\gamma+1}\Gamma(\alpha+\beta+\gamma+1)\Gamma(\beta+2)|\Omega|} + \frac{(\eta^\rho - a^\rho)^\alpha\zeta_2}{\rho^{\alpha+\beta}\Gamma(\alpha+\beta+1)|\Omega|}\Big) \\
&\quad + \|x\|\Big(\frac{|\lambda|(T^\rho - a^\rho)^\beta}{\rho^\beta\Gamma(\beta+1)}\Big[1 + \frac{\zeta_1}{\rho^{\beta+1}\Gamma(\beta+2)|\Omega|}\Big] \\
&\quad + \frac{|\mu||\lambda|(\xi^\rho - a^\rho)^{\beta+\gamma}\zeta_1}{\rho^{2\beta+\gamma+1}\Gamma(\beta+\gamma+1)\Gamma(\beta+2)|\Omega|} + \frac{|\lambda|\zeta_2}{\rho^\beta\Gamma(\beta+1)|\Omega|}\Big) \\
&= \Lambda_1\|P\|Q(\|x\|) + \Lambda_2\|x\|,
\end{aligned}$$

which leads to $\|h\| \leq \Lambda_1\|P\|Q(r) + \Lambda_2 r$.

Step 3. *$N(x)$ maps bounded sets into equicontinuous sets of $C(J, \mathbb{R})$.*

Let x be any element in B_r and $h \in N(x)$. Then there exists a function $v \in S_{F,x}$ such that, for each $t \in J$ we have

$$\begin{aligned}
h(t) &= {}^\rho I_{a+}^{\alpha+\beta}v(t) - \lambda^\rho I_{a+}^{\beta}x(t) + \frac{(t^\rho - a^\rho)^\beta(\eta^\rho - t^\rho)}{\rho^{\beta+1}\Gamma(\beta+2)\Omega}\Big\{{}^\rho I_{a+}^{\alpha+\beta}v(T) - \lambda^\rho I_{a+}^{\beta}x(T) \\
&\quad - \mu^\rho I_{a+}^{\alpha+\beta+\gamma}v(\xi) + \mu\lambda^\rho I_{a+}^{\beta+\gamma}x(\xi)\Big\} - \frac{(t^\rho - a^\rho)^\beta}{\Omega(\eta^\rho - a^\rho)^\beta}\Big(\frac{(T^\rho - a^\rho)^\beta(T^\rho - t^\rho)}{\rho^{\beta+1}\Gamma(\beta+2)} \\
&\quad - \frac{\mu(\xi^\rho - a^\rho)^{\beta+\gamma}[(\beta+1)(\xi^\rho - t^\rho) - \gamma(t^\rho - a^\rho)]}{\rho^{\beta+\gamma+1}\Gamma(\beta+\gamma+2)(\beta+1)}\Big)\Big\{{}^\rho I_{a+}^{\alpha+\beta}v(\eta) - \lambda^\rho I_{a+}^{\beta}x(\eta)\Big\}.
\end{aligned}$$

Let $\tau_1, \tau_2 \in J$, $\tau_1 < \tau_2$. Then

$$
\begin{aligned}
&|h(t_2) - h(t_1)| \\
\leq\ & \Big| \frac{\rho^{1-(\alpha+\beta)}}{\Gamma(\alpha+\beta)} \Big[\int_0^{t_1} s^{\rho-1}[(t_2^\rho - s^\rho)^{\alpha+\beta-1} - (t_1^\rho - s^\rho)^{\alpha+\beta-1}]v(s)ds + \int_{t_1}^{t_2} s^{\rho-1}(t_2^\rho - s^\rho)^{\alpha+\beta-1}v(s)ds \Big] \\
&+ \Big[\frac{(t_2^\rho - a^\rho)^\beta(\eta^\rho - t_2^\rho)}{\rho^{\beta+1}\Gamma(\beta+2)\Omega} - \frac{(t_1^\rho - a^\rho)^\beta(\eta^\rho - t_1^\rho)}{\rho^{\beta+1}\Gamma(\beta+2)\Omega} \Big] \Big\{ {}^\rho I_{a+}^{\alpha+\beta} v(T) - \mu\, {}^\rho I_{a+}^{\alpha+\beta+\gamma} v(\xi) \Big\} \\
&- \Big[\frac{(t_2^\rho - a^\rho)^\beta}{\Omega(\eta^\rho - a^\rho)^\beta} \Big[\frac{(T^\rho - a^\rho)^\beta(T^\rho - t_2^\rho)}{\rho^{\beta+1}\Gamma(\beta+2)} - \frac{\mu(\xi^\rho - a^\rho)^{\beta+\gamma}[(\beta+1)(\xi^\rho - t_2^\rho) - \gamma(t^\rho - a^\rho)]}{\rho^{\beta+\gamma+1}\Gamma(\beta+\gamma+2)(\beta+1)} \Big] \\
&- \frac{(t_1^\rho - a^\rho)^\beta}{\Omega(\eta^\rho - a^\rho)^\beta} \Big[\frac{(T^\rho - a^\rho)^\beta(T^\rho - t_1^\rho)}{\rho^{\beta+1}\Gamma(\beta+2)} - \frac{\mu(\xi^\rho - a^\rho)^{\beta+\gamma}[(\beta+1)(\xi^\rho - t_1^\rho) - \gamma(t_1^\rho - a^\rho)]}{\rho^{\beta+\gamma+1}\Gamma(\beta+\gamma+2)(\beta+1)} \Big] \Big] \times \\
&\times {}^\rho I_{a+}^{\alpha+\beta} v(\eta) \Big| \\
&+ \Big| \frac{-\lambda\rho^{1-\beta}}{\Gamma(\beta)} \Big[\int_0^{t_1} s^{\rho-1}[(t_2^\rho - s^\rho)^{\beta-1} - (t_1^\rho - s^\rho)^{\beta-1}]x(s)ds + \int_{t_1}^{t_2} s^{\rho-1}(t_2^\rho - s^\rho)^{\beta-1}x(s)ds \Big] \\
&+ \Big[\frac{(t_2^\rho - a^\rho)^\beta(\eta^\rho - t_2^\rho)}{\rho^{\beta+1}\Gamma(\beta+2)\Omega} - \frac{(t_1^\rho - a^\rho)^\beta(\eta^\rho - t_1^\rho)}{\rho^{\beta+1}\Gamma(\beta+2)\Omega} \Big] \Big\{ -\lambda\, {}^\rho I_{a+}^{\beta} x(T) + \mu\lambda\, {}^\rho I_{a+}^{\beta+\gamma} x(\xi) \Big\} \\
&+ \Big[\frac{\lambda(t_2^\rho - a^\rho)^\beta}{\Omega(\eta^\rho - a^\rho)^\beta} \Big[\frac{(T^\rho - a^\rho)^\beta(T^\rho - t_2^\rho)}{\rho^{\beta+1}\Gamma(\beta+2)} - \frac{\mu(\xi^\rho - a^\rho)^{\beta+\gamma}[(\beta+1)(\xi^\rho - t_2^\rho) - \gamma(t^\rho - a^\rho)]}{\rho^{\beta+\gamma+1}\Gamma(\beta+\gamma+2)(\beta+1)} \Big] \\
&- \frac{\lambda(t_1^\rho - a^\rho)^\beta}{\Omega(\eta^\rho - a^\rho)^\beta} \Big[\frac{(T^\rho - a^\rho)^\beta(T^\rho - t_1^\rho)}{\rho^{\beta+1}\Gamma(\beta+2)} - \frac{\mu(\xi^\rho - a^\rho)^{\beta+\gamma}[(\beta+1)(\xi^\rho - t_1^\rho) - \gamma(t_1^\rho - a^\rho)]}{\rho^{\beta+\gamma+1}\Gamma(\beta+\gamma+2)(\beta+1)} \Big] \Big] \times \\
&\times {}^\rho I_{a+}^{\beta} x(\eta) \Big| \\
\leq\ & \frac{\|P\|Q(r)}{\rho^{\alpha+\beta}\Gamma(\alpha+\beta+1)} \Big\{ |t_2^{\rho(\alpha+\beta)} - t_1^{\rho(\alpha+\beta)}| + 2(t_2^\rho - t_1^\rho)^{\alpha+\beta} \Big\} \\
&+ \Big| \frac{(t_2^\rho - a^\rho)^\beta(\eta^\rho - t_2^\rho)}{\rho^{\beta+1}\Gamma(\beta+2)\Omega} - \frac{(t_1^\rho - a^\rho)^\beta(\eta^\rho - t_1^\rho)}{\rho^{\beta+1}\Gamma(\beta+2)\Omega} \Big| \\
&\times \|P\|Q(r) \Big(\frac{(T^\rho - a^\rho)^{\alpha+\beta}}{\rho^{\alpha+\beta}\Gamma(\alpha+\beta+1)} + |\mu| \frac{(\xi^\rho - a^\rho)^{\alpha+\beta+\gamma}}{\rho^{\alpha+\beta+\gamma}\Gamma(\alpha\beta+\gamma+1)} \Big) \\
&+ \Big| \frac{(t_2^\rho - a^\rho)^\beta}{\Omega(\eta^\rho - a^\rho)^\beta} \Big[\frac{(T^\rho - a^\rho)^\beta(T^\rho - t_2^\rho)}{\rho^{\beta+1}\Gamma(\beta+2)} - \frac{\mu(\xi^\rho - a^\rho)^{\beta+\gamma}[(\beta+1)(\xi^\rho - t_2^\rho) - \gamma(t_2^\rho - a^\rho)]}{\rho^{\beta+\gamma+1}\Gamma(\beta+\gamma+2)(\beta+1)} \Big] \\
&- \frac{(t_1^\rho - a^\rho)^\beta}{\Omega(\eta^\rho - a^\rho)^\beta} \Big[\frac{(T^\rho - a^\rho)^\beta(T^\rho - t_1^\rho)}{\rho^{\beta+1}\Gamma(\beta+2)} - \frac{\mu(\xi^\rho - a^\rho)^{\beta+\gamma}[(\beta+1)(\xi^\rho - t_1^\rho) - \gamma(t_1^\rho - a^\rho)]}{\rho^{\beta+\gamma+1}\Gamma(\beta+\gamma+2)(\beta+1)} \Big] \Big| \times \\
&\times \|P\|Q(r) \frac{(\eta^\rho - a^\rho)^{\alpha+\beta}}{\rho^{\alpha+\beta}\Gamma(\alpha+\beta+1)} + \frac{r}{\rho^\beta\Gamma(\beta+1)} \Big\{ |t_2^{\rho\beta} - t_1^{\rho\beta}| + 2(t_2^\rho - t_1^\rho)^\beta \Big\} \\
&+ \Big| \Big[\frac{(t_2^\rho - a^\rho)^\beta(\eta^\rho - t_2^\rho)}{\rho^{\beta+1}\Gamma(\beta+2)\Omega} - \frac{(t_1^\rho - a^\rho)^\beta(\eta^\rho - t_1^\rho)}{\rho^{\beta+1}\Gamma(\beta+2)\Omega} \Big] \Big| |\lambda| r \Big(\frac{(T^\rho - a^\rho)^\beta}{\rho^\beta\Gamma(\beta+1)} + |\mu| \frac{(\xi^\rho - a^\rho)^{\beta+\gamma}}{\rho^{\beta+\gamma}\Gamma(\beta+\gamma+1)} \Big) \\
&+ \Big| \frac{\lambda(t_2^\rho - a^\rho)^\beta}{\Omega(\eta^\rho - a^\rho)^\beta} \Big[\frac{(T^\rho - a^\rho)^\beta(T^\rho - t_2^\rho)}{\rho^{\beta+1}\Gamma(\beta+2)} - \frac{\mu(\xi^\rho - a^\rho)^{\beta+\gamma}[(\beta+1)(\xi^\rho - t_2^\rho) - \gamma(t^\rho - a^\rho)]}{\rho^{\beta+\gamma+1}\Gamma(\beta+\gamma+2)(\beta+1)} \Big] \\
&- \frac{\lambda(t_1^\rho - a^\rho)^\beta}{\Omega(\eta^\rho - a^\rho)^\beta} \Big[\frac{(T^\rho - a^\rho)^\beta(T^\rho - t_1^\rho)}{\rho^{\beta+1}\Gamma(\beta+2)} - \frac{\mu(\xi^\rho - a^\rho)^{\beta+\gamma}[(\beta+1)(\xi^\rho - t_1^\rho) - \gamma(t_1^\rho - a^\rho)]}{\rho^{\beta+\gamma+1}\Gamma(\beta+\gamma+2)(\beta+1)} \Big] \Big| \times \\
&\times \frac{(\eta^\rho - a^\rho)^\beta}{\rho^\beta\Gamma(\beta+1)} \to 0 \text{ when } t_1 \to t_2, \text{ independently of } x \in B_r.
\end{aligned}
$$

Combining the outcome of Steps 1–3 with Arzelá-Ascoli theorem leads to the conclusion that $N : C(J, \mathbb{R}) \to \mathcal{P}(C(J, \mathbb{R}))$ is completely continuous.

Next, we show that N has a closed graph. Then it will follow by Proposition 1.2 in [19] that the operator N is u.s.c.

Step 4. *N has a closed graph.*

Suppose that there exists $x_n \to x_*$, $h_n \in N(x_n)$ and $h_n \to h_*$. Then we have to establish that $h_* \in N(x_*)$. Since $h_n \in N(x_n)$, there exists $v_n \in S_{F,x_n}$. In consequence, for each $t \in J$, we get

$$\begin{aligned} h_n(t) &= {}^\rho I_{a+}^{\alpha+\beta} v_n(t) - \lambda^\rho I_{a+}^{\beta} x_n(t) + \frac{(t^\rho - a^\rho)^\beta(\eta^\rho - t^\rho)}{\rho^{\beta+1}\Gamma(\beta+2)\Omega}\Big\{{}^\rho I_{a+}^{\alpha+\beta} v_n(T) - \lambda^\rho I_{a+}^{\beta} x_n(T) \\ &\quad -\mu^\rho I_{a+}^{\alpha+\beta+\gamma} v_n(\xi) + \mu\lambda^\rho I_{a+}^{\beta+\gamma} x_n(\xi)\Big\} - \frac{(t^\rho - a^\rho)^\beta}{\Omega(\eta^\rho - a^\rho)^\beta}\Big(\frac{(T^\rho - a^\rho)^\beta(T^\rho - t^\rho)}{\rho^{\beta+1}\Gamma(\beta+2)} \\ &\quad - \frac{\mu(\xi^\rho - a^\rho)^{\beta+\gamma}[(\beta+1)(\xi^\rho - t^\rho) - \gamma(t^\rho - a^\rho)]}{\rho^{\beta+\gamma+1}\Gamma(\beta+\gamma+2)(\beta+1)}\Big)\Big\{{}^\rho I_{a+}^{\alpha+\beta} v_n(\eta) - \lambda^\rho I_{a+}^{\beta} x_n(\eta)\Big\}. \end{aligned}$$

Next we show that there exists $v_* \in S_{F,x_*}$ such that, for each $t \in J$,

$$\begin{aligned} h_*(t) &= {}^\rho I_{a+}^{\alpha+\beta} v_*(t) - \lambda^\rho I_{a+}^{\beta} x_*(t) + \frac{(t^\rho - a^\rho)^\beta(\eta^\rho - t^\rho)}{\rho^{\beta+1}\Gamma(\beta+2)\Omega}\Big\{{}^\rho I_{a+}^{\alpha+\beta} v_*(T) - \lambda^\rho I_{a+}^{\beta} x_*(T) \\ &\quad -\mu^\rho I_{a+}^{\alpha+\beta+\gamma} v_*(\xi) + \mu\lambda^\rho I_{a+}^{\beta+\gamma} x_*(\xi)\Big\} - \frac{(t^\rho - a^\rho)^\beta}{\Omega(\eta^\rho - a^\rho)^\beta}\Big(\frac{(T^\rho - a^\rho)^\beta(T^\rho - t^\rho)}{\rho^{\beta+1}\Gamma(\beta+2)} \\ &\quad - \frac{\mu(\xi^\rho - a^\rho)^{\beta+\gamma}[(\beta+1)(\xi^\rho - t^\rho) - \gamma(t^\rho - a^\rho)]}{\rho^{\beta+\gamma+1}\Gamma(\beta+\gamma+2)(\beta+1)}\Big)\Big\{{}^\rho I_{a+}^{\alpha+\beta} v_*(\eta) - \lambda^\rho I_{a+}^{\beta} x_*(\eta)\Big\}. \end{aligned}$$

Consider the continuous linear operator $\Theta : L^1(J, \mathbb{R}) \to C(J, \mathbb{R})$ given by

$$\begin{aligned} v \to \Theta(v)(t) &= {}^\rho I_{a+}^{\alpha+\beta} v(t) - \lambda^\rho I_{a+}^{\beta} x(t) + \frac{(t^\rho - a^\rho)^\beta(\eta^\rho - t^\rho)}{\rho^{\beta+1}\Gamma(\beta+2)\Omega}\Big\{{}^\rho I_{a+}^{\alpha+\beta} v(T) - \lambda^\rho I_{a+}^{\beta} x(T) \\ &\quad -\mu^\rho I_{a+}^{\alpha+\beta+\gamma} v(\xi) + \mu\lambda^\rho I_{a+}^{\beta+\gamma} x(\xi)\Big\} - \frac{(t^\rho - a^\rho)^\beta}{\Omega(\eta^\rho - a^\rho)^\beta}\Big(\frac{(T^\rho - a^\rho)^\beta(T^\rho - t^\rho)}{\rho^{\beta+1}\Gamma(\beta+2)} \\ &\quad - \frac{\mu(\xi^\rho - a^\rho)^{\beta+\gamma}[(\beta+1)(\xi^\rho - t^\rho) - \gamma(t^\rho - a^\rho)]}{\rho^{\beta+\gamma+1}\Gamma(\beta+\gamma+2)(\beta+1)}\Big)\Big\{{}^\rho I_{a+}^{\alpha+\beta} v(\eta) - \lambda^\rho I_{a+}^{\beta} x(\eta)\Big\}. \end{aligned}$$

Notice that $\|h_n(t) - h_*(t)\| \to 0$ as $n \to \infty$. So we deduce by a closed graph result obtained in [20] that $\Theta \circ S_{F,x}$ is a closed graph operator. Furthermore, $h_n \in \Theta(S_{F,x_n})$. Since $x_n \to x_*$, therefore we have

$$\begin{aligned} h_*(t) &= {}^\rho I_{a+}^{\alpha+\beta} v_*(t) - \lambda^\rho I_{a+}^{\beta} x_*(t) + \frac{(t^\rho - a^\rho)^\beta(\eta^\rho - t^\rho)}{\rho^{\beta+1}\Gamma(\beta+2)\Omega}\Big\{{}^\rho I_{a+}^{\alpha+\beta} v_*(T) - \lambda^\rho I_{a+}^{\beta} x_*(T) \\ &\quad -\mu^\rho I_{a+}^{\alpha+\beta+\gamma} v_*(\xi) + \mu\lambda^\rho I_{a+}^{\beta+\gamma} x_*(\xi)\Big\} - \frac{(t^\rho - a^\rho)^\beta}{\Omega(\eta^\rho - a^\rho)^\beta}\Big(\frac{(T^\rho - a^\rho)^\beta(T^\rho - t^\rho)}{\rho^{\beta+1}\Gamma(\beta+2)} \\ &\quad - \frac{\mu(\xi^\rho - a^\rho)^{\beta+\gamma}[(\beta+1)(\xi^\rho - t^\rho) - \gamma(t^\rho - a^\rho)]}{\rho^{\beta+\gamma+1}\Gamma(\beta+\gamma+2)(\beta+1)}\Big)\Big\{{}^\rho I_{a+}^{\alpha+\beta} v_*(\eta) - \lambda^\rho I_{a+}^{\beta} x_*(\eta)\Big\}, \end{aligned}$$

for some $v_* \in S_{F,x_*}$

Step 5. *There exists an open set $\mathcal{V} \subseteq C(J, \mathbb{R})$ with $x \notin \theta N(x)$ for any $\theta \in (0,1)$ and all $x \in \partial\mathcal{V}$.*

Take $\theta \in (0,1)$, $x \in \theta N(x)$ and $t \in J$. Then we show that there exists $v \in L^1(J, \mathbb{R})$ with $v \in S_{F,x}$ such that

$$\begin{aligned} x(t) &= \theta^\rho I_{a+}^{\alpha+\beta} v(t) - \theta\lambda^\rho I_{a+}^{\beta} x(t) + \theta\frac{(t^\rho - a^\rho)^\beta(\eta^\rho - t^\rho)}{\rho^{\beta+1}\Gamma(\beta+2)\Omega}\Big\{{}^\rho I_{a+}^{\alpha+\beta} v_*(T) - \lambda^\rho I_{a+}^{\beta} x(T) \\ &\quad -\mu^\rho I_{a+}^{\alpha+\beta+\gamma} v(\xi) + \mu\lambda^\rho I_{a+}^{\beta+\gamma} x(\xi)\Big\} - \theta\frac{(t^\rho - a^\rho)^\beta}{\Omega(\eta^\rho - a^\rho)^\beta}\Big(\frac{(T^\rho - a^\rho)^\beta(T^\rho - t^\rho)}{\rho^{\beta+1}\Gamma(\beta+2)} \\ &\quad - \frac{\mu(\xi^\rho - a^\rho)^{\beta+\gamma}[(\beta+1)(\xi^\rho - t^\rho) - \gamma(t^\rho - a^\rho)]}{\rho^{\beta+\gamma+1}\Gamma(\beta+\gamma+2)(\beta+1)}\Big)\Big\{{}^\rho I_{a+}^{\alpha+\beta} v(\eta) - \lambda^\rho I_{a+}^{\beta} x(\eta)\Big\}. \end{aligned}$$

Using the computations done in Step 2, for each $t \in J$, we get

$$|x(t)| \leq \Lambda_1 \|P\| Q(\|x\|) + \Lambda_2 \|x\|,$$

which yields

$$\frac{(1-\Lambda_2)\|x\|}{\Lambda_1 \|P\| Q(\|x\|)} \leq 1.$$

By (A_3), there exists M such that $\|x\| \neq M$. Define a set

$$\mathcal{V} = \{x \in C(J, \mathbb{R}) : \|x\| < M\}.$$

Observe that the operator $N : \overline{\mathcal{V}} \to \mathcal{P}(C(J, \mathbb{R}))$ is a compact multivalued map, u.s.c. with convex closed values. With the given choice of $\mathcal{V}$, it is not possible to find $x \in \partial\mathcal{V}$ satisfying $x \in \theta N(x)$ for some $\theta \in (0,1)$. Consequently, by the nonlinear alternative of Leray–Schauder type [18], the operator N has a fixed point $x \in \overline{\mathcal{V}}$, which corresponds to a solution of the problem (1). This finishes the proof. □

3.2. The Lipschitz Case

Let $(\mathcal{X}, d)$ denote a metric space induced from the normed space $(\mathcal{X}; \|\cdot\|)$. Let $H_d : \mathcal{P}(\mathcal{X}) \times \mathcal{P}(\mathcal{X}) \to \mathbb{R} \cup \{\infty\}$ be defined by $H_d(A_1, A_2) = \max\{\sup_{a_1 \in A_1} d(a_1, A_2), \sup_{a_2 \in A_2} d(A_1, a_2)\}$, where $d(A_1, a_2) = \inf_{a_1 \in A_1} d(a_1; a_2)$ and $d(a_1, A_2) = \inf_{a_2 \in A_2} d(a_1; a_2)$. Then $(\mathcal{P}_{b,cl}(\mathcal{X}), H_d)$ is a metric space (see [21]), where $\mathcal{P}_{b,cl}(\mathcal{X}) = \{\mathcal{Y} \in \mathcal{P}(\mathcal{X}) : \mathcal{Y}$ is bounded and closed$\}$,

The following result deals with the non-convex valued case of the problem (1) and is based on Covitz and Nadler's fixed point theorem [22]: "If $N : \mathcal{X} \to \mathcal{P}_{cl}(\mathcal{X})$ is a contraction, then $FixN \neq \emptyset$, where $\mathcal{P}_{cl}(\mathcal{X}) = \{\mathcal{Y} \in \mathcal{P}(\mathcal{X}) : \mathcal{Y}$ is closed$\}$".

Theorem 2. *Assume that*

(A_4) *$F : J \times \mathbb{R} \to \mathcal{P}_{cp}(\mathbb{R})$ is such that $F(\cdot, x) : J \to \mathcal{P}_{cp}(\mathbb{R})$ is measurable for each $x \in \mathbb{R}$, where $\mathcal{P}_{cp}(\mathbb{R}) = \{\mathcal{Y} \in \mathcal{P}(\mathbb{R}) : \mathcal{Y}$ is compact$\}$;*

(A_5) *$H_d(F(t,x), F(t,\widehat{x})) \leq \varpi(t)|x - \widehat{x}|$ for almost all $t \in J$ and $x, \widehat{x} \in \mathbb{R}$ with $\varpi \in C(J, \mathbb{R}^+)$ and $d(0, F(t,0)) \leq \varpi(t)$ for almost all $t \in J$.*

Then the problem (1) has at least one solution on J if

$$\|\varpi\|\Lambda_1 + \Lambda_2 < 1, \tag{18}$$

where Λ_1 and Λ_2 are respectively given by (14) and (15).

Proof. Let us verify that the operator $N : C(J, \mathbb{R}) \to \mathcal{P}(C(J, \mathbb{R}))$, defined in the proof of the last theorem, satisfies the hypothesis of Covitz and Nadler fixed point theorem [22]. We establish it in two steps.

Step I. $N(x)$ is nonempty and closed for every $v \in S_{F,x}$.

Since the set-valued map $F(\cdot, x(\cdot))$ is measurable, it admits a measurable selection $v : J \to \mathbb{R}$ by the measurable selection theorem ([23], Theorem III.6). By (A_5), we have

$$|v(t)| \leq \varpi(t)(1 + |x(t)|),$$

that is, $v \in L^1(J, \mathbb{R})$. So F is integrably bounded. Therefore, $S_{F,x} \neq \emptyset$.

Now we establish that $N(x)$ is closed for each $x \in C(J,\mathbb{R})$. Let $\{u_n\}_{n\geq 0} \in N(x)$ be such that $u_n \to u$ as $n \to \infty$ in $C(J,\mathbb{R})$. Then $u \in C(J,\mathbb{R})$ and we can find $v_n \in S_{F,x_n}$ such that, for each $t \in J$,

$$\begin{aligned} u_n(t) &= {}^{\rho}I_{a+}^{\alpha+\beta}v_n(t) - \lambda^{\rho}I_{a+}^{\beta}x_n(t) + \frac{(t^{\rho}-a^{\rho})^{\beta}(\eta^{\rho}-t^{\rho})}{\rho^{\beta+1}\Gamma(\beta+2)\Omega}\Big\{{}^{\rho}I_{a+}^{\alpha+\beta}v(T) - \lambda^{\rho}I_{a+}^{\beta}x_n(T) \\ &\quad -\mu^{\rho}I_{a+}^{\alpha+\beta+\gamma}v_n(\xi) + \mu\lambda^{\rho}I_{a+}^{\beta+\gamma}x_n(\xi)\Big\} - \frac{(t^{\rho}-a^{\rho})^{\beta}}{\Omega(\eta^{\rho}-a^{\rho})^{\beta}}\Big(\frac{(T^{\rho}-a^{\rho})^{\beta}(T^{\rho}-t^{\rho})}{\rho^{\beta+1}\Gamma(\beta+2)} \\ &\quad -\frac{\mu(\xi^{\rho}-a^{\rho})^{\beta+\gamma}[(\beta+1)(\xi^{\rho}-t^{\rho})-\gamma(t^{\rho}-a^{\rho})]}{\rho^{\beta+\gamma+1}\Gamma(\beta+\gamma+2)(\beta+1)}\Big)\Big\{{}^{\rho}I_{a+}^{\alpha+\beta}v_n(\eta) - \lambda^{\rho}I_{a+}^{\beta}x_n(\eta)\Big\}. \end{aligned}$$

As F has compact values, we can pass onto a subsequence (if necessary) to obtain that v_n converges to v in $L^1(J,\mathbb{R})$. So $v \in S_{F,x}$. Then, for each $t \in J$, we get

$$\begin{aligned} u_n(t) \to v(t) &= {}^{\rho}I_{a+}^{\alpha+\beta}v(t) - \lambda^{\rho}I_{a+}^{\beta}x(t) + \frac{(t^{\rho}-a^{\rho})^{\beta}(\eta^{\rho}-t^{\rho})}{\rho^{\beta+1}\Gamma(\beta+2)\Omega}\Big\{{}^{\rho}I_{a+}^{\alpha+\beta}v(T) - \lambda^{\rho}I_{a+}^{\beta}x(T) \\ &\quad -\mu^{\rho}I_{a+}^{\alpha+\beta+\gamma}v(\xi) + \mu\lambda^{\rho}I_{a+}^{\beta+\gamma}x(\xi)\Big\} - \frac{(t^{\rho}-a^{\rho})^{\beta}}{\Omega(\eta^{\rho}-a^{\rho})^{\beta}}\Big(\frac{(T^{\rho}-a^{\rho})^{\beta}(T^{\rho}-t^{\rho})}{\rho^{\beta+1}\Gamma(\beta+2)} \\ &\quad -\frac{\mu(\xi^{\rho}-a^{\rho})^{\beta+\gamma}[(\beta+1)(\xi^{\rho}-t^{\rho})-\gamma(t^{\rho}-a^{\rho})]}{\rho^{\beta+\gamma+1}\Gamma(\beta+\gamma+2)(\beta+1)}\Big)\Big\{{}^{\rho}I_{a+}^{\alpha+\beta}v(\eta) - \lambda^{\rho}I_{a+}^{\beta}x(\eta)\Big\}, \end{aligned}$$

which implies that $u \in N(x)$.

Step II. We establish that there exists $0 < \hat{\theta} < 1$ $(\hat{\theta} = \Lambda_1\|\varpi\| + \Lambda_2)$ satisfying

$$H_d(N(x), N(\hat{x})) \leq \hat{\theta}\|x - \hat{x}\| \text{ for each } x, \hat{x} \in C(J,\mathbb{R}).$$

Let us take $x, \hat{x} \in C(J,\mathbb{R})$ and $h_1 \in N(x)$. Then there exists $v_1(t) \in F(t, x(t))$ such that, for each $t \in J$,

$$\begin{aligned} h_1(t) &= {}^{\rho}I_{a+}^{\alpha+\beta}v_1(t) - \lambda^{\rho}I_{a+}^{\beta}x(t) + \frac{(t^{\rho}-a^{\rho})^{\beta}(\eta^{\rho}-t^{\rho})}{\rho^{\beta+1}\Gamma(\beta+2)\Omega}\Big\{{}^{\rho}I_{a+}^{\alpha+\beta}v_1(T) - \lambda^{\rho}I_{a+}^{\beta}x(T) \\ &\quad -\mu^{\rho}I_{a+}^{\alpha+\beta+\gamma}v_1(\xi) + \mu\lambda^{\rho}I_{a+}^{\beta+\gamma}x(\xi)\Big\} - \frac{(t^{\rho}-a^{\rho})^{\beta}}{\Omega(\eta^{\rho}-a^{\rho})^{\beta}}\Big(\frac{(T^{\rho}-a^{\rho})^{\beta}(T^{\rho}-t^{\rho})}{\rho^{\beta+1}\Gamma(\beta+2)} \\ &\quad -\frac{\mu(\xi^{\rho}-a^{\rho})^{\beta+\gamma}[(\beta+1)(\xi^{\rho}-t^{\rho})-\gamma(t^{\rho}-a^{\rho})]}{\rho^{\beta+\gamma+1}\Gamma(\beta+\gamma+2)(\beta+1)}\Big)\Big\{{}^{\rho}I_{a+}^{\alpha+\beta}v_1(\eta) - \lambda^{\rho}I_{a+}^{\beta}x(\eta)\Big\}. \end{aligned}$$

By (A_5), we have that $H_d(F(t,x), F(t,\hat{x})) \leq \varpi(t)|x(t) - \hat{x}(t)|$. So, there exists $w(t) \in F(t,\hat{x}(t))$ satisfying $|v_1(t) - w| \leq \varpi(t)|x(t) - \hat{x}(t)|, \ t \in J$.

Define $\mathcal{W} : J \to \mathcal{P}(\mathbb{R})$ by

$$\mathcal{W}(t) = \{w \in \mathbb{R} : |v_1(t) - w| \leq \varpi(t)|x(t) - \hat{x}(t)|\}.$$

As the multivalued operator $\mathcal{W}(t) \cap F(t,\hat{x}(t))$ is measurable by Proposition III.4 [23], we can find a function $v_2(t)$ which is a measurable selection for $\mathcal{W}$. So $v_2(t) \in F(t,\hat{x}(t))$ and for each $t \in J$, we have $|v_1(t) - v_2(t)| \leq \varpi(t)|x(t) - \hat{x}(t)|$. For each $t \in J$, we define

$$\begin{aligned} h_2(t) &= {}^{\rho}I_{a+}^{\alpha+\beta}v_2(t) - \lambda^{\rho}I_{a+}^{\beta}\hat{x}(t) + \frac{(t^{\rho}-a^{\rho})^{\beta}(\eta^{\rho}-t^{\rho})}{\rho^{\beta+1}\Gamma(\beta+2)\Omega}\Big\{{}^{\rho}I_{a+}^{\alpha+\beta}v_2(T) - \lambda^{\rho}I_{a+}^{\beta}\hat{x}(T) \\ &\quad -\mu^{\rho}I_{a+}^{\alpha+\beta+\gamma}v_2(\xi) + \mu\lambda^{\rho}I_{a+}^{\beta+\gamma}\hat{x}(\xi)\Big\} - \frac{(t^{\rho}-a^{\rho})^{\beta}}{\Omega(\eta^{\rho}-a^{\rho})^{\beta}}\Big(\frac{(T^{\rho}-a^{\rho})^{\beta}(T^{\rho}-t^{\rho})}{\rho^{\beta+1}\Gamma(\beta+2)} \\ &\quad -\frac{\mu(\xi^{\rho}-a^{\rho})^{\beta+\gamma}[(\beta+1)(\xi^{\rho}-t^{\rho})-\gamma(t^{\rho}-a^{\rho})]}{\rho^{\beta+\gamma+1}\Gamma(\beta+\gamma+2)(\beta+1)}\Big)\Big\{{}^{\rho}I_{a+}^{\alpha+\beta}v_2(\eta) - \lambda^{\rho}I_{a+}^{\beta}\hat{x}(\eta)\Big\}. \end{aligned}$$

As a result, we get

$$
\begin{aligned}
& |h_1(t) - h_2(t)| \\
= & \Big|{}^{\rho}I_{a+}^{\alpha+\beta}[v_2(s) - v_1(s)](t) - \lambda^{\rho}I_{a+}^{\beta}[x(t) - \hat{x}(t)] \\
& + \frac{(t^{\rho} - a^{\rho})^{\beta}(\eta^{\rho} - t^{\rho})}{\rho^{\beta+1}\Gamma(\beta+2)\Omega}\Big\{{}^{\rho}I_{a+}^{\alpha+\beta}[v_2(s) - v_1(s)](T) - \lambda^{\rho}I_{a+}^{\beta}[x(T) - \hat{x}(T)] \\
& - \mu^{\rho}I_{a+}^{\alpha+\beta+\gamma}[v_2(s) - v_1(s)](\xi) + \mu\lambda^{\rho}I_{a+}^{\beta+\gamma}[x(\xi) - \hat{x}(\xi)]\Big\} - \frac{(t^{\rho} - a^{\rho})^{\beta}}{\Omega(\eta^{\rho} - a^{\rho})^{\beta}}\Big(\frac{(T^{\rho} - a^{\rho})^{\beta}(T^{\rho} - t^{\rho})}{\rho^{\beta+1}\Gamma(\beta+2)} \\
& - \frac{\mu(\xi^{\rho} - a^{\rho})^{\beta+\gamma}[(\beta+1)(\xi^{\rho} - t^{\rho}) - \gamma(t^{\rho} - a^{\rho})]}{\rho^{\beta+\gamma+1}\Gamma(\beta+\gamma+2)(\beta+1)}\Big)\Big\{{}^{\rho}I_{a+}^{\alpha+\beta}[v_2(s) - v_1(s)](\eta) - \lambda^{\rho}I_{a+}^{\beta}[x(\eta) - \hat{x}(\eta)]\Big\}\Big| \\
\leq & \|\omega\|\|x - \hat{x}\|\Big(\frac{(T^{\rho} - a^{\rho})^{\alpha+\beta}}{\rho^{\alpha+\beta}\Gamma(\alpha+\beta+1)}\Big[1 + \frac{\zeta_1}{\rho^{\beta+1}\Gamma(\beta+2)|\Omega|}\Big] \\
& + \frac{|\mu|(\xi^{\rho} - a^{\rho})^{\alpha+\beta+\gamma}\zeta_1}{\rho^{\alpha+2\beta+\gamma+1}\Gamma(\alpha+\beta+\gamma+1)\Gamma(\beta+2)|\Omega|} \\
& + \frac{(\eta^{\rho} - a^{\rho})^{\alpha}\zeta_2}{\rho^{\alpha+\beta}\Gamma(\alpha+\beta+1)|\Omega|}\Big) + \|x - \hat{x}\|\Big(\frac{|\lambda|(T^{\rho} - a^{\rho})^{\beta}}{\rho^{\beta}\Gamma(\beta+1)}\Big[1 + \frac{\zeta_1}{\rho^{\beta+1}\Gamma(\beta+2)|\Omega|}\Big] \\
& + \frac{|\mu||\lambda|(\xi^{\rho} - a^{\rho})^{\beta+\gamma}\zeta_1}{\rho^{2\beta+\gamma+1}\Gamma(\beta+\gamma+1)\Gamma(\beta+2)|\Omega|} + \frac{|\lambda|\zeta_2}{\rho^{\beta}\Gamma(\beta+1)|\Omega|}\Big) \\
= & (\|\omega\|\Lambda_1 + \Lambda_2)\|x - \hat{x}\|.
\end{aligned}
$$

Hence

$$\|h_1 - h_2\| \leq (\|\omega\|\Lambda_1 + \Lambda_2)\|x - \hat{x}\|.$$

Analogously, we can interchange the roles of x and $\hat{x}$ to get

$$H_d(N(x), N(\hat{x})) \leq (\|\omega\|\Lambda_1 + \Lambda_2)\|x - \hat{x}\|,$$

which implies that N is a contraction by the condition (18). Hence, by the conclusion of Covitz and Nadler fixed point theorem [22], N has a fixed point x, which corresponds to a solution of (1). This finishes the proof. □

4. Examples

We illustrate our main results by presenting a numerical example.

Example 1. *Consider the following problem*

$$
\begin{cases}
{}^{1/3}_{\ c}D^{5/4}\big({}^{1/3}_{\ c}D^{1/4} + 1/5\big)x(t) \in F(t, x(t)), \ t \in J := [1, 2], \\
x(1) = 0, \ \ x(3/2) = 0, \ \ x(2) = 2/7\ {}^{1/3}I^{3/4}x(7/4).
\end{cases}
\tag{19}
$$

Here $\rho = 1/3$, $\alpha = 5/4$, $\beta = 1/4$, $\lambda = 1/5$, $a = 1$, $T = 2$ $\eta = 3/2$, $\mu = 2/7$, $\gamma = 3/4$, $\xi = 7/4$. Using the given data, we find that $\zeta_1 \approx 0.082260$, $\zeta_2 \approx 0.232036$, $|\Omega| \approx 0.293634$, $\Lambda_1 \approx 1.336009$ and $\Lambda_2 \approx 0.673563$, where ζ_1, ζ_2, Λ_1 and Λ_2 are given by (16), (17), (14) and (15) respectively.

(i) Let us consider the function

$$F(t, x(t)) = \Big[\frac{1}{\sqrt{t^2 + 63}}\Big(\frac{|x(t)|}{3}\Big(\frac{|x(t)|}{|x(t)| + 1} + 2\Big) + 1\Big), \frac{e^{-t}}{9t + 8}\Big(\sin x(t) + \frac{1}{80}\Big)\Big]. \tag{20}$$

We note that $|F(t, x(t))| \leq P(t)Q(\|x\|)$, where $P(t) = \frac{1}{\sqrt{t^2+63}}$, $Q(\|x\|) = \|x\| + 1$. So the assumption (A_2) holds. Moreover, there exists $M > 1.047447394$ satisfying (A_3). Thus the hypothesis of Theorem 1 holds true and hence there exists at least one solution for the problem (19) with $F(t,x)$ given by (20) on $[1,2]$.

(ii) To illustrate Theorem 2 we consider the function

$$F(t,x) = \left[\frac{e^{-t}}{19+t}, \ \frac{1}{(t+4)^2}\left(x + \tan^{-1}(x) + \frac{1}{15} \right) \right]. \tag{21}$$

Clearly $H_d(F(t,x), F(t,\bar{x})) \leq \varpi(t)|x-\bar{x}|$, where $\varpi(t) = \frac{2}{(t+4)^2}$. Also $d(0, F(t,0)) \leq \varpi(t)$ for almost all $t \in [0,1]$ and $\Lambda_1\|\varpi\| + \Lambda_2 \approx 0.7804433180 < 1$. As the hypothesis of Theorem 2 is satisfied, therefore we conclude that the multivalued problem (19) with $F(t,x)$ given by (21) has at least one solution on $[1,2]$.

5. Conclusions

We have introduced a new class of multivalued (inclusions) boundary value problems on an arbitrary domain containing Caputo-type generalized fractional differential operators of different orders and a generalized integral operator. We have considered convex as well as non-convex valued cases for the multi-valued map involved in the given problem. Leray–Schauder nonlinear alternative for multivalued maps plays a central role in proving the existence of solutions for convex valued case of the given problem, while the existence result for the non-convex valued case is based on Covitz and Nadler fixed point theorem. The work presented in this paper is not only new in the given configuration, but will also lead to some new results as special cases. For example, fixing $\mu = 0$ in the obtained results, we obtain the ones for nonlocal three-point boundary conditions: $x(a) = 0, x(\eta) = 0, x(T) = 0, 0 < \eta < T$. For $\rho = 1$, our results specialize to the ones for Liouville–Caputo type fractional differential inclusions complemented with nonlocal generalized integral boundary conditions on an arbitrary domain.

Author Contributions: Conceptualization, B.A. and M.A.; methodology, A.A. and S.K.N.; validation, A.A., B.A., M.A. and S.K.N.; formal analysis, A.A., B.A., M.A. and S.K.N.; writing—original draft preparation, M.A.; writing—review and editing, A.A., B.A. and S.K.N.; project administration, B.A.; funding acquisition, A.A.

Funding: This project was funded by the Deanship of Scientific Research (DSR), King Abdulaziz University, Jeddah, Saudi Arabia under grant no. (KEP-PhD-70-130-38).

Acknowledgments: This project was funded by the Deanship of Scientific Research (DSR), King Abdulaziz University, Jeddah, under grant no. (KEP-PhD-70-130-38). The authors, therefore, acknowledge with thanks DSR technical and financial support.

Conflicts of Interest: The authors declare no conflict of interest.

References

1. Zaslavsky, G.M. *Hamiltonian Chaos and Fractional Dynamics*; Oxford University Press: Oxford, UK, 2005.
2. Magin, R.L. *Fractional Calculus in Bioengineering*; Begell House Publishers: Danbury, CT, USA, 2006.
3. Kilbas, A.A.; Srivastava, H.M.; Trujillo, J.J. *Theory and Applications of Fractional Differential Equations*; North-Holland Mathematics Studies 204; Elsevier Science B.V.: Amsterdam, The Netherlands, 2006.
4. Diethelm, K. *The Analysis of Fractional Differential Equations. An Application-Oriented Exposition Using Differential Operators of Caputo Type*; Lecture Notes in Mathematics 2004; Springer: Berlin, Germany, 2010.
5. Javidi, M.; Ahmad, B. Dynamic analysis of time fractional order phytoplankton-toxic phytoplankton–zooplankton system. *Ecol. Model.* **2015**, *318*, 8–18. [CrossRef]
6. Fallahgoul, H.A.; Focardi, S.M.; Fabozzi, F.J. *Fractional Calculus and Fractional Processes with Applications to Financial Economics. Theory and Application*; Elsevier/Academic Press: London, UK, 2017.
7. Ahmad, B.; Alsaedi, A.; Ntouyas, S.K.; Tariboon, J. *Hadamard-Type Fractional Differential Equations, Inclusions and Inequalities*; Springer: Cham, Switzerland, 2017.
8. Ahmad, B.; Nieto, J.J.; Alsaedi, A.; El-Shahed, M. A study of nonlinear Langevin equation involving two fractional orders in different intervals. *Nonlinear Anal. Real World Appl.* **2012**, *13*, 599–606. [CrossRef]

9. Wang, G.; Zhang, L.; Song, G. Boundary value problem of a nonlinear Langevin equation with two different fractional orders and impulses. *Fixed Point Theory Appl.* **2012**, *2012*, 200. [CrossRef]
10. Ahmad, B.; Ntouyas, S.K. New existence results for differential inclusions involving Langevin equation with two indices. *J. Nonlinear Convex Anal.* **2013**, *14*, 437–450.
11. Muensawat, T.; Ntouyas, S.K.; Tariboon, J. Systems of generalized Sturm-Liouville and Langevin fractional differential equations. *Adv. Differ. Equ.* **2017**, *2017*, 63. [CrossRef]
12. Fazli, H.; Nieto, J.J. Fractional Langevin equation with anti-periodic boundary conditions. *Chaos Solitons Fractals* **2018**, *114*, 332–337. [CrossRef]
13. Ahmad, B.; Alsaedi, A.; Salem, S. On a nonlocal integral boundary value problem of nonlinear Langevin equation with different fractional orders. *Adv. Differ. Equ.* **2019**, *2019*, 57. [CrossRef]
14. Ahmad, B.; Alghanmi, M.; Alsaedi, A.; Srivastava, H.; Ntouyas, K. The Langevin equation in terms of generalized Liouville-Caputo derivatives with nonlocal boundary conditions involving a generalized fractional integral. *Mathematics* **2019**, *7*, 533. [CrossRef]
15. Katugampola, U.N. New Approach to a generalized fractional integral. *Appl. Math. Comput.* **2015**, *218*, 860–865. [CrossRef]
16. Katugampola, U.N. A new approach to generalized fractional derivatives. *Bull. Math. Anal. Appl.* **2014**, *6*, 1–15.
17. Jarad, F.; Abdeljawad, T.; Baleanu, D. On the generalized fractional derivatives and their caputo modification. *J. Nonlinear Sci. Appl.* **2017**, *10*, 2607–2619. [CrossRef]
18. Granas, A.; Dugundji, J. *Fixed Point Theory*; Springer: New York, NY, USA, 2005.
19. Deimling, K. *Multivalued Differential Equations*; De Gruyter: Berlin, Germany, 1992.
20. Lasota, A.; Opial, Z. An application of the Kakutani-Ky Fan theorem in the theory of ordinary differential equations, *Bull. Acad. Pol. Sci. Ser. Sci. Math. Astronom. Phys.* **1965**, *13*, 781–786.
21. Kisielewicz, M. *Differential Inclusions and Optimal Control*; Kluwer: Dordrecht, The Netherlands, 1991.
22. Covitz, H.; Nadler, S.B., Jr. Multivalued contraction mappings in generalized metric spaces. *ISRAEL J. Math.* **1970**, *8*, 5–11. [CrossRef]
23. Castaing, C.; Valadier, M. *Convex Analysis and Measurable Multifunctions*; Lecture Notes in Mathematics 580; Springer: Berlin/Heidelberg, Germany; New York, NY, USA, 1977.

mathematics

Article

A Mollification Regularization Method for the Inverse Source Problem for a Time Fractional Diffusion Equation

Le Dinh Long [1], Yong Zhou [2,3], Tran Thanh Binh [4] and Nguyen Can [5,*]

1 Faculty of Natural Sciences, Thu Dau Mot University, Thu Dau Mot City 820000, Binh Duong Province, Vietnam; ledinhlong@tdmu.edu.vn
2 Faculty of Information Technology, Macau University of Science and Technology, Macau 999078, China; yzhou@xtu.edu.cn
3 Faculty of Mathematics and Computational Science, Xiangtan University, Xiangtan 411105, China
4 Institute of Research and Development, Duy Tan University, Da Nang 550000, Vietnam; tranthanhbinh6@duytan.edu.vn
5 Applied Analysis Research Group, Faculty of Mathematics and Statistics, Ton Duc Thang University, Ho Chi Minh City 700000, Vietnam
* Correspondence: nguyenhuucan@tdtu.edu.vn

Received: 7 September 2019; Accepted: 29 October 2019; Published: 4 November 2019

Abstract: We consider a time-fractional diffusion equation for an inverse problem to determine an unknown source term, whereby the input data is obtained at a certain time. In general, the inverse problems are ill-posed in the sense of Hadamard. Therefore, in this study, we propose a mollification regularization method to solve this problem. In the theoretical results, the error estimate between the exact and regularized solutions is given by a priori and a posteriori parameter choice rules. Besides, the proposed regularized methods have been verified by a numerical experiment.

Keywords: time-fractional diffusion equation; inverse problem; ill-posed problem; convergence estimates

MSC: 35K05; 35K99; 47J06; 47H10x

1. Introduction

In this work, we study an inverse source problem for the time-fractional diffusion equation in a infinite domain as follows:

$$\begin{cases} \dfrac{\partial^\beta u(x,t)}{\partial t^\beta} = u_{xx}(x,t) + \phi(t)f(x), \quad (x,t) \in \mathbb{R} \times (0,T], \\ u(x,0) = 0, x \in \mathbb{R}, \\ u(x,T) = g(x), \quad x \in \mathbb{R}, \end{cases} \tag{1}$$

where the fractional derivative $\frac{\partial^\beta u}{\partial t^\beta}$ is the Caputo derivative of order β $(0 < \beta < 1)$ as defined by

$$\frac{d^\beta f(t)}{dt^\beta} = \frac{1}{\Gamma(1-\beta)} \int_0^t \frac{df(s)}{ds} \frac{ds}{(t-s)^\beta}, \tag{2}$$

and $\Gamma(\cdot)$ denotes the standard Gamma function.

The biggest motivation for developing the problem (1) is the inverse problems for the heat equation; we recover the unknown source function under different assumptions on the smoothness of input data, which were proposed by Igor Malyshev in Reference [1]. The inverse problems of the restoration of a source function in the heat equation with the classical derivative are studied by many researchers, that is, Geng [2] and Shidfar [3].

The mathematical model (1) arising in control theory, physical, generalized voltage divider, elasticity and the model of neurons in biology is studied in References [4–6].

According to our search, the fractional inverse source problems (1) are the subject of very few works, for example, Sakamoto et al. [7] used the data $u(x_0, t)(x_0 \in \mathbb{R})$ to determine $\phi(t)$ once $f(x)$ was given, where the authors obtained a Lipschitz stability for $\phi(t)$. In Problem (1) for a one-dimensional problem with special coefficients, Wei et al. [8] used the Fourier truncation method to solve an inverse source problem with $\phi(t) = 1$. In Reference [9], using the mollification regularization method, Yang and Fu determined the inverse spatial-dependent heat source problem. In Reference [10], Wei and Wang considered a modified quasi boundary value regularization method for identifying this problem. In Reference [11], using the quasi-reversibility regularization method, Yang and his group identified the unknown source for a time fractional diffusion equation. In Reference [12], with the quasi-reversibility regularization method, Wei and her group investigated a space-dependent source for the time fractional diffusion equation. Actually, to our knowledge, in the case $\phi(t)$, dependent on time, the results of the inverse source problem for the time-fractional diffusion equation still has a limited achievement, if $\phi(t) \neq 0$, we know Huy and his group investigated this problem by way of the Tikhonov regularization method, see Reference [13]. In these regularization methods, the priori parameter choice rule depends on the noise level and the priori bound. But in practice, to know exactly this is very difficult. In the above research, by using Morozov's Discrepancy Principle for choosing the regularization parameter in Tikhonov regularization, the authors show error estimation for both the priori choice rule parameter and the posteriori choice rule parameter.

In this paper, we use the mollification method to solve the inverse source problem. Instead of receiving the correct input data, we only get the approximate input data. We assume that the measured data in functions couple $(g_\varepsilon(x) \in \mathscr{L}_2(\mathbb{R}), \phi_\varepsilon(t) \in C[0,T])$ satisfies

$$\|g - g_\varepsilon\|_{\mathscr{L}^2(\mathbb{R})} \leq \varepsilon, \ \ \|\phi - \phi_\varepsilon\|_{C[0,T]} \leq \varepsilon, \tag{3}$$

where the constant $\varepsilon > 0$ represents a noise level. It is known that the inverse source problem mentioned above is ill-posed in the sense of Hadamard, that is, a solution of this problem (1) does not always exist, if the solution does exist, it is not dependent continuously on the given data, meaning that the error of the initial data is small, the error of the solution will be large. This makes trouble for the numerical solution; here a regularization is required. The Fourier transform of a function $\mathcal{F}$ is defined by

$$\widehat{\mathcal{F}}(\xi) = \frac{1}{\sqrt{2\pi}} \int_{\mathbb{R}} e^{-i\xi x} \mathcal{F}(x) dx. \tag{4}$$

We imposed an a priori bound on the input data, that is,

$$\|\mathcal{F}\|_{\mathbb{H}^k(\mathbb{R})} \leq \mathcal{M}, \ k > 0, \tag{5}$$

where $\mathcal{M} \geq 0$ is a constant, $\|\cdot\|_{\mathbb{H}^k(\mathbb{R})}$ denotes the norm in Sobolev space $\mathbb{H}^k(\mathbb{R})$ is defined

$$\mathbb{H}^k(\mathbb{R}) = \left\{\mathcal{F} \in \mathscr{L}_2(\mathbb{R}), \|\mathcal{F}\|_k \leq \infty\right\}, \ and \ \|\mathcal{F}\|_{\mathbb{H}^k(\mathbb{R})} = \left(\int_{\mathbb{R}} \left|(1+\xi^2)^{k/2} \widehat{\mathcal{F}}(\xi)\right|^2 d\xi \right)^{\frac{1}{2}}. \tag{6}$$

The outline of this paper is divided into the following sections: Section 2 gives some auxiliary results. In Section 3, by the priori bound assumption of the exact solution and the priori parameter choice rule, we present the convergence rate. In Section 4, we show the convergence rate between the exact and regularized solutions under the posteriori parameter choice rule. Next, a numerical example is proposed to show the illustration of the results in theory in Section 5. Finally, a conclusion is given in Section 6.

2. Some Auxiliary Results

Before showing some lemmas, we recall the Mittag-Leffler function which is defined by

$$E_{\beta,\kappa}(z) = \sum_{k=0}^{\infty} \frac{z^k}{\Gamma(\beta k + \kappa)}, \quad z \in \mathbb{C}, \tag{7}$$

where $\beta > 0$ and $\kappa \in \mathbb{R}$ are arbitrary constant. In Reference [14], the properties of the Mittag-Leffler function are discussed. Hereby, we present the following Lemmas of the Mittag-Leffler function which can be found in Reference ([14], Chapter 1).

Lemma 1. *Let $0 < \beta_0 < \beta_1 < 1$. Then there exist the constants $\overline{B}_1, \overline{B}_2, \overline{B}_3$ depending only on β_0, β_1 such that for all $\beta \in [\beta_0, \beta_1]$,*

$$\frac{\overline{B}_1}{\Gamma(1-\beta)}\frac{1}{1-x} \leq E_{\beta,1}(x) \leq \frac{\overline{B}_2}{\Gamma(1-\beta)}\frac{1}{1-x}, \quad E_{\beta,\alpha}(x) \leq \frac{\overline{B}_3}{1-x}, \quad \forall x \leq 0, \forall \alpha \in \mathbb{R}. \tag{8}$$

These estimates are uniform for all $\beta \in [\beta_0, \beta_1]$·

Lemma 2. *(see Reference [7]) For $0 < \beta < 1$, we have:*

$$E_{\beta,\beta}(-\zeta) \geq 0, \quad \zeta \geq 0.$$

Proof. As for the proof, see Miller and Samko [15]. □

Lemma 3. *(see Reference [7]) For $\xi > 0$, $\alpha > 0$ and a positive integer $n \in \mathbb{N}$, we have:*

$$\begin{aligned} &\frac{d^n}{dt^n} E_{\beta,1}(-\xi^2 t^\beta) = -\xi^2 t^{\beta-n} E_{\beta,\beta-n+1}(-\xi^2 t^\beta), \quad t > 0, \\ &\frac{d}{dt}(t E_{\beta,2}(-\xi^2 t^\beta)) = E_{\beta,1}(-\xi^2 t^\beta), \quad t \geq 0. \end{aligned} \tag{9}$$

Lemma 4. *(see Reference [7]) By Lemma 2 and Lemma 3, we have*

$$\begin{aligned} \int_0^\varrho \left| t^{\gamma-1} E_{\beta,\beta}(-\xi^2 t^\gamma) \right| dt &= \int_0^\varrho t^{\gamma-1} E_{\beta,\beta}(-\xi^2 t^\gamma) dt \\ &= -\frac{1}{\xi^2} \int_0^\varrho \frac{d}{dt} E_{\beta,1}(-\xi^2 t^\gamma) dt = \frac{1}{\xi^2}\Big(1 - E_{\alpha,1}(-\xi^2 \varrho^\alpha)\Big), \quad \varrho > 0. \end{aligned} \tag{10}$$

Lemma 5. *(see Reference [16]) For $0 < \alpha < 1$, $\xi \in \mathbb{R}$, the following inequalities hold:*

$$\sup_{\xi \in \mathbb{R}} \left| (1+\xi^2)^{-k} (1 - e^{-\frac{\alpha^2 \xi^2}{4}}) \right| \leq \max\left\{\alpha^{2k}, \alpha^2\right\}. \tag{11}$$

Proof. The proof can be found in Reference [9]. □

Lemma 6. *Let $\beta \in (0,1)$ and $\xi \in \mathbb{R}$, the following estimate holds*

$$\left(\int_0^T s^{\beta-1}E_{\beta,\beta}(-\xi^2 s^\beta)ds\right)^{-1} = \frac{\xi^2}{1-E_{\beta,1}(-\xi^2 T^\beta)} \le \begin{cases} \dfrac{\xi^2}{1-E_{\beta,1}(-T^\beta)}, & \text{if } |\xi| \ge 1, \\ \dfrac{1}{1-E_{\beta,1}(-T^\beta)}, & \text{if } |\xi| < 1. \end{cases} \tag{12}$$

Proof. If $|\xi| \ge 1$ then since $E_{\beta,1}(-y)$ for $0 < \beta < 1$ is a decreasing function for $y > 0$, we get $E_{\beta,1}(-\xi^2 T^\beta) \le E_{\beta,1}(-T^\beta)$. Whereupon

$$\left(\int_0^T s^{\beta-1}E_{\beta,\beta}(-\xi^2 s^\beta)ds\right)^{-1} = \frac{\xi^2}{1-E_{\beta,1}(-\xi^2 T^\beta)} \le \frac{\xi^2}{1-E_{\beta,1}(-T^\beta)}, \quad \text{for } |\xi| \ge 1. \tag{13}$$

If $|\xi| \le 1$ then since $E_{\beta,\beta}(-y)$ with $0 < \beta < 1$ is a decreasing function for $y > 0$, we get $E_{\beta,\beta}(-\xi^2 s^\beta) \ge E_{\beta,\beta}(-s^\beta)$, so

$$\left(\int_0^T s^{\beta-1}E_{\beta,\beta}(-\xi^2 s^\beta)ds\right)^{-1} \le \left(\int_0^T s^{\beta-1}E_{\beta,\beta}(-s^\beta)ds\right)^{-1} = \frac{1}{1-E_{\beta,1}(-T^\beta)}, \quad \text{for } |\xi| \le 1. \tag{14}$$

□

Lemma 7. *For $\alpha \in (0,1)$ and $\xi \in \mathbb{R}$, from Lemma 6, one has:*

$$\frac{1}{\left(\int_0^T s^{\beta-1}E_{\beta,\beta}(-\xi^2 s^\beta)ds\right)e^{\frac{\alpha^2\xi^2}{4}}} = \frac{\xi^2}{\left(1-E_{\beta,1}(-\xi^2 T^\beta)\right)e^{\frac{\alpha^2\xi^2}{4}}}$$
$$\le \begin{cases} \dfrac{\frac{\xi^2\alpha^2}{4}}{\left(1-E_{\beta,1}(-T^\beta)\right)e^{\frac{\alpha^2\xi^2}{4}}} \le \left(\dfrac{4}{\alpha^2}\right)\left(\dfrac{1}{(1-E_{\beta,1}(-T^\beta)}\right), & \text{if } |\xi| \ge 1, \\ \dfrac{1}{\left(1-E_{\beta,1}(-T^\beta)\right)e^{\frac{\alpha^2\xi^2}{4}}} \le \left(\dfrac{4}{\alpha^2}\right)\left(\dfrac{1}{1-E_{\beta,1}(-T^\beta)}\right), & \text{if } |\xi| < 1. \end{cases} \tag{15}$$

This gives

$$\frac{1}{\left(\int_0^T s^{\beta-1}E_{\beta,\beta}(-\xi^2 s^\beta)ds\right)e^{\frac{\alpha^2\xi^2}{4}}} \le \left(\frac{4}{\alpha^2}\right)\left(\frac{1}{1-E_{\beta,1}(-T^\beta)}\right). \tag{16}$$

3. The Priori Parameter Choice

Next, the error estimate of the mollification regularization method will be derived under the priori parameter choice rule in this section. We consider the Gauss function

$$\rho_\alpha(x) := \frac{1}{\alpha\sqrt{\pi}}e^{-\frac{x^2}{\alpha^2}}, \tag{17}$$

as the mollifer kernel, where α is a positive constant.

We define an operator K_α as

$$K_\alpha f(x) := \rho_\alpha f(x) := \int_{\mathbb{R}} \rho_\alpha(t) f(x-t)dt = \int_{\mathbb{R}} \rho_\alpha(x-t) f(t)dt, \tag{18}$$

for $f(x) \in \mathscr{L}_2(\mathbb{R})$. The original ill-posed problem is replaced by a new problem of searching its approximation $f_{\varepsilon,\alpha}(x)$ which is defined by

$$f_{\varepsilon,\alpha}(x) := \frac{1}{\sqrt{2\pi}} \int_{\mathbb{R}} e^{ix\xi} \rho_\alpha \widehat{f_\varepsilon(\xi)} d\xi, \tag{19}$$

The Inverse Source Problem

By using the Fourier transform, the problem (1) is formulated in the following frequency space

$$\begin{cases} \dfrac{\partial^\beta \widehat{u}(\xi,t)}{\partial t^\beta} + \xi^2 \widehat{u}(\xi,t) = \phi(t)\widehat{f}(\xi), & (\xi,t) \in \mathbb{R} \times (0,T], \\ \widehat{u}(\xi,0) = 0, & \xi \in \mathbb{R}, \\ \widehat{u}(\xi,T) = \widehat{g}(\xi), & \xi \in \mathbb{R}. \end{cases} \tag{20}$$

From the equation and the initial value in (20), we obtain

$$\widehat{u}(\xi,t) = \int_0^t (t-s)^{\beta-1} E_{\beta,\beta}(-\xi^2(t-s)^\beta)\phi(s)\widehat{f}(\xi)ds. \tag{21}$$

Or equivalently,

$$u(x,t) = \frac{1}{\sqrt{2\pi}} \int_{\mathbb{R}} e^{i\xi x} \Big(\int_0^t (t-s)^{\beta-1} E_{\beta,\beta}(-\xi^2(t-s)^\beta)\phi(s)ds \Big) \widehat{f}(\xi)d\xi. \tag{22}$$

Set

$$\mathcal{D}_\beta(\xi, t-s) = (t-s)^{\beta-1} E_{\beta,\beta}(-\xi^2(t-s)^\beta).$$

And $\widehat{u}(\xi,T) = \widehat{g}(\xi)$ in (20), one has

$$\widehat{f}(\xi) = \frac{\widehat{g}(\xi)}{\int_0^T \mathcal{D}_\beta(\xi, T-s)\phi(s)ds}. \tag{23}$$

Using the inverse Fourier transform, then we obtain the formula of the source function f

$$f(x) = \frac{1}{\sqrt{2\pi}} \int_{\mathbb{R}} e^{i\xi x} \frac{\widehat{g}(\xi)}{\int_0^T \mathcal{D}_\beta(\xi, T-s)\phi(s)ds} d\xi. \tag{24}$$

On the other hand, if $\phi(t)$ is bounded by $\inf_{t\in[0,T]} |\phi(t)| \le \phi(t) \le \sup_{t\in[0,T]} |\phi(t)| = \|\phi\|_{C[0,T]}$, we have $\Big(\int_0^T \mathcal{D}_\beta(\xi, T-s)\phi(s)ds \Big)^{-1}$ can be written then $\frac{1}{\inf_{t\in[0,T]} |\phi(t)|} \frac{\xi^2}{\left(1-E_{\beta,1}(-\xi^2 T^\beta)\right)}$. The unbounded function $\frac{\xi^2}{\left(1-E_{\beta,1}(-\xi^2 T^\beta)\right)}$ can be seen as an amplification when $\xi \to \infty$. From now on, putting $\inf_{t\in[0,T]} |\phi(t)| = \mathcal{A}_0$, $\inf_{t\in[0,T]} |\phi_\varepsilon(t)| = \mathcal{A}_1$, $\sup_{t\in[0,T]} |\phi(t)| = \|\phi\|_{C[0,T]} = \Phi$. From (19) with α is a regularization parameter and α depends on ε, we found the regularized solution

$$\widehat{f_{\varepsilon,\alpha}}(\xi) = \frac{\widehat{g_\varepsilon}(\xi)}{\left(\int\limits_0^T \mathcal{D}_\beta(\xi, T-s)\phi_\varepsilon(s)ds\right) e^{\frac{\alpha^2\xi^2}{4}}}. \tag{25}$$

Using inverse Fourier transform, we get

$$f_{\varepsilon,\alpha}(x) = \frac{1}{\sqrt{2\pi}} \int\limits_{\mathbb{R}} \frac{\widehat{g_\varepsilon}(\xi)}{\left(\int\limits_0^T \mathcal{D}_\beta(\xi, T-s)\phi_\varepsilon(s)ds\right) e^{\frac{\alpha^2\xi^2}{4}}} e^{i\xi x} d\xi. \tag{26}$$

The main conclusion of this section are given below.

Theorem 1. *Let $f(x)$, given by (24), be the exact solution of (1) with exact data $g \in \mathscr{L}_2(\mathbb{R})$, and $f_{\varepsilon,\alpha}(x)$ is approximation solution of $f(x)$ with measured data $g_\varepsilon \in \mathscr{L}_2(\mathbb{R})$. Then we obtain*

a. If $0 < k < 1$, and choosing $\alpha(\varepsilon) = \left(\frac{\varepsilon}{\mathcal{M}}\right)^{\frac{1}{2(k+1)}}$, we have a convergence estimate

$$\|f(.) - f_{\varepsilon,\alpha}(.)\|_{\mathscr{L}_2(\mathbb{R})} \le \varepsilon^{\frac{k}{k+1}} \mathcal{M}^{\frac{1}{k+1}} \left(\max\left\{1, \left(\frac{\varepsilon}{\mathcal{M}}\right)^{\frac{1-k}{k+1}}\right\} + \mathcal{R}(\mathcal{A}_0, \mathcal{A}_1, \widehat{g})\right). \tag{27}$$

b. If $k > 1$, by choosing $\alpha(\varepsilon) = \left(\frac{\varepsilon}{\mathcal{M}}\right)^{\frac{1}{4}}$, we have a convergence estimate

$$\|f(.) - f_{\varepsilon,\alpha}(.)\|_{\mathscr{L}_2(\mathbb{R})} \le \varepsilon^{\frac{1}{2}} \mathcal{M}^{\frac{1}{2}} \left(1 + \mathcal{R}(\mathcal{A}_0, \mathcal{A}_1, \widehat{g})\right), \tag{28}$$

in which

$$\mathcal{R}(\mathcal{A}_0, \mathcal{A}_1, \widehat{g}) = \frac{4}{(1 - E_{\beta,1}(-T^\beta))} \left(\frac{1}{\mathcal{A}_1} + \frac{\|\widehat{g}\|_{\mathscr{L}_2(\mathbb{R})}}{\mathcal{A}_1 \mathcal{A}_0} \right). \tag{29}$$

Proof. From (24) and (26), by the Parseval formula, the triangle inequality, we obtain

$$\begin{aligned}
\|f(.) - f_{\varepsilon,\alpha}(.)\|_{\mathscr{L}_2(\mathbb{R})} &= \|\widehat{f}(.) - \widehat{f_{\varepsilon,\alpha}}(.)\|_{\mathscr{L}_2(\mathbb{R})} \\
&= \left\| \frac{\widehat{g}(\xi)}{\int\limits_0^T \mathcal{D}_\beta(\xi, T-s)\phi(s)ds} - \frac{\widehat{g_\varepsilon}(\xi)}{\left(\int\limits_0^T \mathcal{D}_\beta(\xi, T-s)\phi_\varepsilon(s)ds\right) e^{\frac{\alpha^2\xi^2}{4}}} \right\|_{\mathscr{L}_2(\mathbb{R})} \\
&= \|\mathcal{I}_1\|_{\mathscr{L}_2(\mathbb{R})} + \|\mathcal{I}_2\|_{\mathscr{L}_2(\mathbb{R})} + \|\mathcal{I}_3\|_{\mathscr{L}_2(\mathbb{R})},
\end{aligned} \tag{30}$$

in which

$$\mathcal{I}_1 = \frac{\widehat{g}(\xi)}{\int\limits_0^T \mathcal{D}_\beta(\xi, T-s)\phi(s)ds} - \frac{\widehat{g}(\xi)}{\left(\int\limits_0^T \mathcal{D}_\beta(\xi, T-s)\phi(s)ds\right)e^{\frac{\alpha^2\xi^2}{4}}},$$

$$\mathcal{I}_2 = \frac{\widehat{g}(\xi)}{\left(\int\limits_0^T \mathcal{D}_\beta(\xi, T-s)\phi_\varepsilon(s)ds\right)e^{\frac{\alpha^2\xi^2}{4}}} - \frac{\widehat{g}_\varepsilon(\xi)}{\left(\int\limits_0^T \mathcal{D}_\beta(\xi, T-s)\phi_\varepsilon(s)ds\right)e^{\frac{\alpha^2\xi^2}{4}}},$$

$$\mathcal{I}_3 = \frac{\widehat{g}(\xi)}{\left(\int\limits_0^T \mathcal{D}_\beta(\xi, T-s)\phi(s)ds\right)e^{\frac{\alpha^2\xi^2}{4}}} - \frac{\widehat{g}(\xi)}{\left(\int\limits_0^T \mathcal{D}_\beta(\xi, T-s)\phi_\varepsilon(s)ds\right)e^{\frac{\alpha^2\xi^2}{4}}}. \tag{31}$$

Next, we estimate the error by diving it into three steps as follows
Step 1: Estimate for $\|\mathcal{I}_1\|^2_{\mathscr{L}_2(\mathbb{R})}$, we have

$$\begin{aligned}
\|\mathcal{I}_1\|^2_{\mathscr{L}_2(\mathbb{R})} &= \left\| \frac{\widehat{g}(\xi)}{\left(\int\limits_0^T \mathcal{D}_\beta(\xi, T-s)\phi(s)ds\right)}(1-e^{-\frac{\alpha^2\xi^2}{4}}) \right\|^2_{\mathscr{L}_2(\mathbb{R})} \\
&= \left\| (1+\xi^2)^{-k}(1-e^{-\frac{\alpha^2\xi^2}{4}})(1+\xi^2)^k \widehat{f}(\xi) \right\|^2_{\mathscr{L}_2(\mathbb{R})} \\
&\le \sup_{\xi\in\mathbb{R}} \left| (1+\xi^2)^{-k}(1-e^{-\frac{\alpha^2\xi^2}{4}}) \right|^2 \|f\|^2_{H^k(\mathbb{R})} \le \mathcal{M}^2 \max\left\{\alpha^{4k}, \alpha^4\right\}.
\end{aligned}$$

Hence,

$$\|\mathcal{I}_1\|_{\mathscr{L}_2(\mathbb{R})} \le \mathcal{M}\max\left\{\alpha^{2k}, \alpha^2\right\}. \tag{32}$$

Step 2: Estimate for $\|\mathcal{I}_2\|^2_{\mathscr{L}_2(\mathbb{R})}$, we get

$$\begin{aligned}
\|\mathcal{I}_2\|^2_{\mathscr{L}_2(\mathbb{R})} &= \left\| \frac{\widehat{g}(\xi)}{\left(\int\limits_0^T \mathcal{D}_\beta(\xi, T-s)\phi_\varepsilon(s)ds\right)e^{\frac{\alpha^2\xi^2}{4}}} - \frac{\widehat{g}_\varepsilon(\xi)}{\left(\int\limits_0^T \mathcal{D}_\beta(\xi, T-s)\phi_\varepsilon(s)ds\right)e^{\frac{\alpha^2\xi^2}{4}}} \right\|^2_{\mathscr{L}_2(\mathbb{R})} \\
&\le \mathcal{A}_1^{-2}\|\widehat{g}(\xi)-\widehat{g}_\varepsilon(\xi)\|^2_{\mathscr{L}_2(\mathbb{R})} \sup_{\xi\in\mathbb{R}} \left| \left(\int\limits_0^T (T-s)^{\beta-1}E_{\beta,\beta}(-\xi^2(T-s)^\beta)ds\right)e^{\frac{\alpha^2\xi^2}{4}} \right|^{-2} \\
&\le \mathcal{A}_1^{-2}\|\widehat{g}(\xi)-\widehat{g}_\varepsilon(\xi)\|^2_{\mathscr{L}_2(\mathbb{R})} \sup_{\xi\in\mathbb{R}} \left| \frac{\xi^2}{(1-E_{\beta,1}(-\xi^2T^\beta)e^{\frac{\alpha^2\xi^2}{4}}} \right|^2 \\
&\le \left(\frac{16\varepsilon^2}{\alpha^4}\right)\Big(\mathcal{A}_1(1-E_{\beta,1}(-T^\beta))\Big)^{-2}.
\end{aligned} \tag{33}$$

Hence, we conclude that

$$\|\mathcal{I}_2\|_{\mathscr{L}_2(\mathbb{R})} \le \left(\frac{4\varepsilon}{\alpha^2}\right)\Big(\mathcal{A}_1(1-E_{\beta,1}(-T^\beta))\Big)^{-1}. \tag{34}$$

Step 3: Estimate for $\|\mathcal{I}_3\|^2_{\mathscr{L}_2(\mathbb{R})}$, we have

$$\|\mathcal{I}_3\|^2_{\mathscr{L}_2(\mathbb{R})} = \left\| \frac{\widehat{g}(\xi)}{\left(\int_0^T \mathcal{D}_\beta(\xi, T-s)\phi(s)ds\right)e^{\frac{\alpha^2\xi^2}{4}}} - \frac{\widehat{g}(\xi)}{\left(\int_0^T \mathcal{D}_\beta(\xi, T-s)\phi_\varepsilon(s)ds\right)e^{\frac{\alpha^2\xi^2}{4}}} \right\|^2_{\mathscr{L}_2(\mathbb{R})}$$

$$= \left\| \frac{\widehat{g}(\xi)}{e^{\frac{\alpha^2\xi^2}{4}}} \frac{\int_0^T \mathcal{D}_\beta(\xi, T-s)\left(\phi_\varepsilon(s) - \phi(s)\right)ds}{\left(\int_0^T \mathcal{D}_\beta(\xi, T-s)\phi_\varepsilon(s)ds\right)\left(\int_0^T \mathcal{D}_\beta(\xi, T-s)\phi(s)ds\right)} \right\|^2_{\mathscr{L}_2(\mathbb{R})}. \tag{35}$$

From (35), we get

$$\begin{aligned}
\|\mathcal{I}_3\|^2_{\mathscr{L}_2(\mathbb{R})} &= \mathcal{A}_1^{-2}\|\phi_\varepsilon - \phi\|^2_{C[0,T]} \left\| \frac{\widehat{g}(\xi)}{\left(\int_0^T \mathcal{D}_\beta(\xi, T-s)\phi(s)ds\right)e^{\frac{\alpha^2\xi^2}{4}}} \right\|^2_{\mathscr{L}_2(\mathbb{R})} \\
&\le \mathcal{A}_1^{-2}\|\phi_\varepsilon - \phi\|^2_{C[0,T]} \left\| \frac{\widehat{g}(\xi)}{\left(\int_0^T \mathcal{D}_\beta(\xi, T-s)\phi(s)ds\right)e^{\frac{\alpha^2\xi^2}{4}}} \right\|^2_{\mathscr{L}_2(\mathbb{R})} \\
&\le (\mathcal{A}_0\mathcal{A}_1)^{-2}\|\phi_\varepsilon - \phi\|^2_{C[0,T]} \left| \frac{\xi^2}{(1 - E_{\beta,1}(-\xi^2 T^\beta))\, e^{\frac{\alpha^2\xi^2}{4}}} \right|^2 \int_{\mathbb{R}} |\widehat{g}(\xi)|^2 d\xi \\
&\le \left(\frac{16}{\alpha^4}\right)\left(\mathcal{A}_0\mathcal{A}_1(1 - E_{\beta,1}(-T^\beta))\right)^{-2}\|\phi_\varepsilon - \phi\|^2_{C[0,T]} \int_{\mathbb{R}} |\widehat{g}(\xi)|^2 d\xi.
\end{aligned} \tag{36}$$

Hence,

$$\|\mathcal{I}_3\|_{\mathscr{L}_2(\mathbb{R})} \le \left(\frac{4\varepsilon}{\alpha^2}\right)\left(\mathcal{A}_0\mathcal{A}_1(1 - E_{\beta,1}(-T^\beta))\right)^{-1}\|\widehat{g}\|_{\mathscr{L}_2(\mathbb{R})}. \tag{37}$$

Combining (32), (34) and (37), we received

(a) If $0 \le k \le 1$ by choosing $\alpha(\varepsilon) = \left(\frac{\varepsilon}{\mathcal{M}}\right)^{\frac{1}{2(k+1)}}$, we have a convergent estimation:

$$\|f(.) - f_{\varepsilon,\alpha}(.)\|_{\mathscr{L}_2(\mathbb{R})} \text{ is of order } \varepsilon^{\frac{k}{k+1}}. \tag{38}$$

(b) If $k > 1$, by choosing $\alpha(\varepsilon) = \left(\frac{\varepsilon}{\mathcal{M}}\right)^{\frac{1}{4}}$, we have a convergent estimation:

$$\|f(.) - f_{\varepsilon,\alpha}(.)\|_{\mathscr{L}_2(\mathbb{R})} \text{ is of order } \varepsilon^{\frac{1}{2}}. \tag{39}$$

□

4. The Discrepancy Principle

Now, we present the posteriori regularization parameter choice rule. The most general of the posteriori rules is the Morozov discrepancy principle [17]. Choosing the regularization parameter α as the solution of the equation

$$l(\alpha) = \left\| e^{-\frac{\alpha^2\xi^2}{4}} \widehat{g}_\varepsilon(\xi) - \widehat{g}_\varepsilon(\xi) \right\|_{\mathscr{L}_2(\mathbb{R})} = \varepsilon + \eta \left(\log \left(\log \left(\frac{T}{\varepsilon} \right) \right) \right)^{-1}, \tag{40}$$

where $\eta > 1$ is a constant.

Remark 1. *To ensure the existence and uniqueness, we can choose η such that*

$$0 < \varepsilon + \eta \left(\log \left(\log \left(\frac{T}{\varepsilon} \right) \right) \right)^{-1} < \|\widehat{g}_\varepsilon\|_{\mathscr{L}_2(\mathbb{R})}.$$

To establish the existence and uniqueness of the solution of Equation (40), we consider the following lemmas

Lemma 8. *If $\varepsilon > 0$ then there holds:*

(a) $l(\alpha)$ *is a continous function.*
(b) $\lim_{\alpha \to 0^+} l(\alpha) = 0$.
(c) $\lim_{\alpha \to +\infty} l(\alpha) = \left\| \widehat{g}_\varepsilon \right\|_{\mathscr{L}_2(\mathbb{R})}$.
(d) $l(\alpha)$ *is a strictly increasing function.*

The proof is very easy and we omit it here.

Lemma 9. *The following inequality holds:*

$$\|e^{-\frac{\alpha^2\xi^2}{4}} \widehat{g}_\varepsilon(\xi) - \widehat{g}(\xi)\|_{\mathscr{L}_2(\mathbb{R})} \leq 2\varepsilon + \eta \left(\log \left(\log \left(\frac{T}{\varepsilon} \right) \right) \right)^{-1}. \tag{41}$$

Proof. Applying the triangle inequality and (40), we have

$$\begin{aligned} \|e^{-\frac{\alpha^2\xi^2}{4}} \widehat{g}_\varepsilon(\xi) - \widehat{g}(\xi)\|_{\mathscr{L}_2(\mathbb{R})} &\leq \|e^{-\frac{\alpha^2\xi^2}{4}} \widehat{g}_\varepsilon(\xi) - \widehat{g}_\varepsilon(\xi)\|_{\mathscr{L}_2(\mathbb{R})} + \|\widehat{g}_\varepsilon(\xi) - \widehat{g}(\xi)\|_{\mathscr{L}_2(\mathbb{R})} \\ &\leq \|e^{-\frac{\alpha^2\xi^2}{4}} \widehat{g}_\varepsilon(\xi) - \widehat{g}_\varepsilon(\xi)\|_{\mathscr{L}_2(\mathbb{R})} + \|\widehat{g}_\varepsilon(\xi) - \widehat{g}(\xi)\|_{\mathscr{L}_2(\mathbb{R})} \\ &\leq 2\varepsilon + \eta \left(\log \left(\log \left(\frac{T}{\varepsilon} \right) \right) \right)^{-1}. \end{aligned} \tag{42}$$

□

Lemma 10. *For any $0 \neq \xi \in \mathbb{R}$, let $s, t \in [0, T]$ such that $0 \leq s \leq t \leq T$, making the substitution ξ^2 and using the inequality: $\frac{\overline{B}_3 s^{\beta-1}}{1+\xi^2 s^\beta} \leq \overline{B}_3 s^{\beta-1}$, we have the following estimate*

$$\int_0^T \mathcal{D}_\beta(\xi, T-s) ds = \int_0^T (T-s)^{\beta-1} E_{\beta,\beta}(-\xi^2 (T-s)^\beta) ds \leq \frac{\overline{B}_3 T^\beta}{\beta}. \tag{43}$$

Lemma 11. *If α is the solution of Equation (40), then the following inequality also holds:*

$$\frac{4}{\alpha^2} \leq \frac{\mathcal{H}_\beta(\overline{B}_3, T, \Phi, \mathcal{M})}{\eta} \left(\log \left(\log \left(\frac{T}{\varepsilon} \right) \right) \right). \tag{44}$$

whereby $\mathcal{M} \geq \|f\|_{\mathbb{H}^k(\mathbb{R})}$.

Proof. Due to (40), we receive

$$\begin{aligned}
&\varepsilon + \eta\Big(\log\Big(\log\Big(\frac{T}{\varepsilon}\Big)\Big)\Big)^{-1} = \Big\|(1-e^{-\frac{\alpha^2\xi^2}{4}})\widehat{g}_\varepsilon(\xi)\Big\|_{\mathscr{L}_2(\mathbb{R})} \\
&\leq \Big\|(1-e^{-\frac{\alpha^2\xi^2}{4}})\widehat{g}_\varepsilon(\xi) - (1-e^{-\frac{\alpha^2\xi^2}{4}})\widehat{g}(\xi) + (1-e^{-\frac{\alpha^2\xi^2}{4}})\widehat{g}(\xi)\Big\|_{\mathscr{L}_2(\mathbb{R})} \\
&\leq \varepsilon + \Big\|(1-e^{-\frac{\alpha^2\xi^2}{4}})(1+\xi^2)^{-k}\Big(\int_0^T \mathcal{D}_\beta(\xi, T-s)\phi(s)ds\Big)(1+\xi^2)^k\widehat{f}(\xi)\Big\|_{\mathscr{L}_2(\mathbb{R})} \\
&\leq \varepsilon + \sup_{\xi\in\mathbb{R}}\Big|(1-e^{-\frac{\alpha^2\xi^2}{4}})(1+\xi^2)^{-k}\Big(\int_0^T \mathcal{D}_\beta(\xi, T-s)\phi(s)ds\Big)\Big|\mathcal{M} \\
&\leq \varepsilon + \frac{\alpha^2}{4}\mathcal{H}_\beta(\overline{B}_3, T, \Phi, \mathcal{M}),
\end{aligned} \tag{45}$$

whereby

$$\mathcal{H}_\beta(\overline{B}_3, T, \Phi, \mathcal{M}) = (\beta)^{-1}\Phi\overline{B}_3 T^\beta\mathcal{M}. \tag{46}$$

So

$$\frac{4}{\alpha^2} \leq \frac{\mathcal{H}_\beta(\overline{B}_3, T, \Phi, \mathcal{M})}{\eta}\log\Big(\log\Big(\frac{T}{\varepsilon}\Big)\Big). \tag{47}$$

□

Lemma 12. *For* $0 < \alpha < 1$, *using the Lemma 7, the following inequality holds:*

$$\sup_{\xi\in\mathbb{R}}\Big|\Big(\frac{\xi^2}{1-E_{\beta,1}(-\xi^2T^\beta)}\Big)^{k+1}e^{-\frac{\alpha^2\xi^2}{4}}\Big| \leq \Big(\frac{k+1}{(1-E_{\beta,1}(-T^\beta))}\Big)^{k+1}\Big(\frac{4}{\alpha^2}\Big)^{k+1}. \tag{48}$$

The proof is similar to Lemma 7 and we omit it here.

Next, the main results of this section are shown under Theorem.

Theorem 2. *Assume the condition* $\|g_\varepsilon - g\| \leq \varepsilon$ *where* $\|.\|$ *denotes the* $\mathscr{L}_2(\mathbb{R})$*-norm with* $\varepsilon > 0$ *is a noise level and the condition (5) holds, then there holds the following error estimate*

$$\begin{aligned}
&\|f(.) - f_{\varepsilon,\alpha}(.)\|_{\mathscr{L}_2(\mathbb{R})} = \|\widehat{f}(.) - \widehat{f_{\varepsilon,\alpha}}(.)\|_{\mathscr{L}_2(\mathbb{R})} \\
&\leq \Bigg(\varepsilon\Big(\log\Big(\log\Big(\frac{T}{\varepsilon}\Big)\Big)\Big)^{k+1}\Big(\frac{\mathcal{L}_\beta(k,T)\,\mathcal{H}_\beta(\overline{B}_3, T, \Phi)}{\mathcal{A}_0\eta}\Big)^{k+1}\mathcal{M}^k \\
&+ \frac{\Phi}{\mathcal{A}_0^{k+1}}\Big(\Big(\frac{1}{1-E_{\beta,1}(-T^\beta)}\Big)^k\Big)^{\frac{1}{k+1}}\mathcal{M}^{\frac{1}{k+1}}.\Big(2\varepsilon + \eta\Big(\log\Big(\log\Big(\frac{T}{\varepsilon}\Big)\Big)\Big)^{-1}\Big)^{\frac{k}{k+1}} \\
&+ \Big(\varepsilon\log\Big(\log\Big(\frac{T}{\varepsilon}\Big)\Big)\Big)\Big(\frac{\mathcal{H}_\beta(\overline{B}_3, T, \Phi, \mathcal{M})}{\eta\mathcal{A}_0(1-E_{\beta,1}(-T^\beta))} + \Big(\frac{\mathcal{H}_\beta(\overline{B}_3, T, \Phi, \mathcal{M})\|\widehat{g}\|_{\mathscr{L}_2(\mathbb{R})}}{\eta(1-E_{\beta,1}(-T^\beta))\mathcal{A}_0\mathcal{A}_1}\Big)\Bigg).
\end{aligned} \tag{49}$$

Proof. By the Parseval formula, we get

$$
\begin{aligned}
\|f(.) - f_{\varepsilon,\alpha}(.)\|_{\mathscr{L}_2(\mathbb{R})} &= \|\widehat{f}(.) - \widehat{f_{\varepsilon,\alpha}}(.)\|_{\mathscr{L}_2(\mathbb{R})} \\
&\leq \left\| \frac{e^{-\frac{\alpha^2\xi^2}{4}}\widehat{g}_\varepsilon(\xi) - \widehat{g}(\xi)}{\int\limits_0^T \mathcal{D}_\beta(\xi, T-s)\phi(s)ds} \right\|_{\mathscr{L}_2(\mathbb{R})} \\
&+ \left\| \frac{\widehat{g}_\varepsilon(\xi)}{\Big(\int\limits_0^T \mathcal{D}_\beta(\xi, T-s)\phi(s)ds\Big)e^{\frac{\alpha^2\xi^2}{4}}} - \frac{\widehat{g}(\xi)}{\Big(\int\limits_0^T \mathcal{D}_\beta(\xi, T-s)\phi(s)ds\Big)e^{\frac{\alpha^2\xi^2}{4}}} \right\|_{\mathscr{L}_2(\mathbb{R})} \\
&+ \left\| \frac{\widehat{g}(\xi)}{\Big(\int\limits_0^T \mathcal{D}_\beta(\xi, T-s)\phi(s)ds\Big)e^{\frac{\alpha^2\xi^2}{4}}} - \frac{\widehat{g}(\xi)}{\Big(\int\limits_0^T \mathcal{D}_\beta(\xi, T-s)\phi_\varepsilon(s)ds\Big)e^{\frac{\alpha^2\xi^2}{4}}} \right\|_{\mathscr{L}_2(\mathbb{R})} \\
&\leq \|\mathcal{J}_1\|_{\mathscr{L}_2(\mathbb{R})} + \|\mathcal{J}_2\|_{\mathscr{L}_2(\mathbb{R})} + \|\mathcal{J}_3\|_{\mathscr{L}_2(\mathbb{R})},
\end{aligned}
\tag{50}
$$

We can divide the proof into three steps as follows:

Step 1: Estimate for $\|\mathcal{J}_1\|^2_{\mathscr{L}_2(\mathbb{R})}$, using the Hölder inequality, we obtain

$$
\begin{aligned}
\|\mathcal{J}_1\|^2_{\mathscr{L}_2(\mathbb{R})} &= \frac{1}{\Big(\int\limits_0^T \mathcal{D}_\beta(\xi, T-s)\phi(s)ds\Big)^2} \left\| e^{-\frac{\alpha^2\xi^2}{4}}\widehat{g}_\varepsilon(\xi) - \widehat{g}(\xi) \right\|^2_{\mathscr{L}_2(\mathbb{R})} \\
&\leq \left\| \frac{\xi^2}{\mathcal{A}_0(1 - E_{\beta,1}(-\xi^2 T^\beta))} \Big(e^{-\frac{\alpha^2\xi^2}{4}}\widehat{g}_\varepsilon(\xi) - \widehat{g}(\xi) \Big) \right\|^2_{\mathscr{L}_2(\mathbb{R})} \\
&\leq (\mathcal{C}_1^2)^{\frac{1}{k+1}} \times (\mathcal{C}_2^2)^{\frac{k}{k+1}},
\end{aligned}
\tag{51}
$$

whereby

$$
\begin{aligned}
\mathcal{C}_1^2 &= \left(\int\limits_{\mathbb{R}} \left(\Big(\frac{\xi^2}{\mathcal{A}_0(1 - E_{\beta,1}(-\xi^2 T^\beta))} \Big)^2 \Big(e^{-\frac{\alpha^2\xi^2}{4}}\widehat{g}_\varepsilon(\xi) - \widehat{g}(\xi) \Big)^{\frac{2}{k+1}} \right)^{k+1} d\xi \right), \\
\mathcal{C}_2^2 &= \left(\int\limits_{\mathbb{R}} \left(\Big(e^{-\frac{\alpha^2\xi^2}{4}}\widehat{g}_\varepsilon(\xi) - \widehat{g}(\xi) \Big)^{\frac{2k}{k+1}} \right)^{k+1} d\xi \right).
\end{aligned}
\tag{52}
$$

From (52), we can check that $(\mathcal{C}_2^2)^{\frac{k}{k+1}}$ as follows

$$
\begin{aligned}
(\mathcal{C}_2^2)^{\frac{k}{k+1}} &= \left(\int\limits_{\mathbb{R}} \Big(e^{-\frac{\alpha^2\xi^2}{4}}\widehat{g}_\varepsilon(\xi) - \widehat{g}(\xi) \Big)^2 d\xi \right)^{\frac{k}{k+1}} \\
&= \left\| e^{-\frac{\alpha^2\xi^2}{4}}\widehat{g}_\varepsilon(\xi) - \widehat{g}(\xi) \right\|^{\frac{2k}{k+1}}_{\mathscr{L}_2(\mathbb{R})} \leq \left(2\varepsilon + \eta\Big(\log\Big(\log\Big(\frac{T}{\varepsilon}\Big)\Big)\Big)^{-1} \right)^{\frac{2k}{k+1}}.
\end{aligned}
\tag{53}
$$

On the other hand, we deduce

$$
\begin{aligned}
(\mathcal{C}_1^2)^{\frac{1}{k+1}} &\le \left(\int_{-\infty}^{+\infty} \Big(\frac{\xi^2}{\mathcal{A}_0(1-E_{\beta,1}(-\xi^2 T^\beta))} \Big)^{2(k+1)} \Big(e^{-\frac{\alpha^2\xi^2}{4}} \widehat{g}_\varepsilon(\xi) - \widehat{g}(\xi) \Big)^2 d\xi \right)^{\frac{1}{k+1}} \\
&= \left\| \Big(\frac{\xi^2}{\mathcal{A}_0(1-E_{\beta,1}(-\xi^2 T^\beta))} \Big)^{k+1} \Big(e^{-\frac{\alpha^2\xi^2}{4}} \widehat{g}_\varepsilon(\xi) - \widehat{g}(\xi) \Big) \right\|_{\mathscr{L}_2(\mathbb{R})}^{\frac{2}{k+1}} \\
&\le \left(\left\| \Big(\frac{\xi^2}{\mathcal{A}_0(1-E_{\beta,1}(-\xi^2 T^\beta))} \Big)^{k+1} e^{-\frac{\alpha^2\xi^2}{4}} \Big(\widehat{g}_\varepsilon(\xi) - \widehat{g}(\xi) \Big) \right\|_{\mathscr{L}_2(\mathbb{R})} \right. \\
&\quad \left. + \left\| \Big(\frac{\xi^2}{\mathcal{A}_0(1-E_{\beta,1}(-\xi^2 T^\beta))} \Big)^{k+1} \Big(e^{-\frac{\alpha^2\xi^2}{4}} \widehat{g}(\xi) - \widehat{g}(\xi) \Big) \right\|_{\mathscr{L}_2(\mathbb{R})} \right)^{\frac{2}{k+1}}. \qquad (54)
\end{aligned}
$$

To estimate $\mathcal{C}_1$, we give two Lemmas as follows:

Lemma 13. *Assume that the condition $\|\widehat{g}_\varepsilon(\xi) - \widehat{g}(\xi)\|_{\mathscr{L}_2(\mathbb{R})} \le \varepsilon$ holds. Then we have the following estimate*

$$
\begin{aligned}
&\left\| \Big(\frac{\xi^2}{\mathcal{A}_0(1-E_{\beta,1}(-\xi^2 T^\beta))} \Big)^{k+1} e^{-\frac{\alpha^2\xi^2}{4}} \Big(\widehat{g}_\varepsilon(\xi) - \widehat{g}(\xi) \Big) \right\|_{\mathscr{L}_2(\mathbb{R})} \\
&\le \varepsilon \left(\log\left(\log\left(\frac{T}{\varepsilon} \right) \right) \right)^{k+1} \left(\frac{\mathcal{L}_\beta(k,T)\, \mathcal{H}_\beta(\overline{B}_3, T, \Phi, \mathcal{M})}{\mathcal{A}_0 \eta} \right)^{k+1}. \qquad (55)
\end{aligned}
$$

Proof. Using the Lemma 12 and setting $\mathcal{L}_\beta(k,T) = \left(\frac{k+1}{(1-E_{\beta,1}(-T^\beta))} \right)$, we get

$$
\begin{aligned}
&\left\| \Big(\frac{\xi^2}{\mathcal{A}_0(1-E_{\beta,1}(-\xi^2 T^\beta))} \Big)^{k+1} e^{-\frac{\alpha^2\xi^2}{4}} \Big(\widehat{g}_\varepsilon(\xi) - \widehat{g}(\xi) \Big) \right\|_{\mathscr{L}_2(\mathbb{R})} \\
&\le \varepsilon \Big(\frac{4}{\alpha^2} \Big)^{k+1} \left(\frac{k+1}{\mathcal{A}_0(1-E_{\beta,1}(-T^\beta))} \right)^{k+1} \\
&\le \varepsilon \left(\log\left(\log\left(\frac{T}{\varepsilon} \right) \right) \right)^{k+1} \left(\frac{\mathcal{L}_\beta(k,T)\, \mathcal{H}_\beta(\overline{B}_3, T, \Phi, \mathcal{M})}{\mathcal{A}_0 \eta} \right)^{k+1} \qquad (56)
\end{aligned}
$$

in which $\mathcal{H}_\beta(\overline{B}_3, T, \Phi, M)$ is defined in Lemma 11. □

Lemma 14. *Let $\xi \in \mathbb{R}$ and exist $\mathcal{M}$ is a positive constant such that $\mathcal{M} \ge \|f\|_{H^k(\mathbb{R})}$, we get*

$$
\left\| \Big(\frac{\xi^2}{\mathcal{A}_0(1-E_{\beta,1}(-\xi^2 T^\beta))} \Big)^{k+1} \Big(e^{-\frac{\alpha^2\xi^2}{4}} \widehat{g}(\xi) - \widehat{g}(\xi) \Big) \right\|_{\mathscr{L}_2(\mathbb{R})} \le \frac{\Phi}{\mathcal{A}_0^{k+1}} \left(\frac{1}{1-E_{\beta,1}(-T^\beta)} \right)^k \mathcal{M}. \qquad (57)
$$

Proof. Applying the Lemma 4, we receive

$$
\begin{aligned}
&\Big\|\Big(\frac{\xi^2}{\mathcal{A}_0(1-E_{\beta,1}(-\xi^2T^\beta))}\Big)^{k+1}\Big(e^{-\frac{\alpha^2\xi^2}{4}}\widehat{g}(\xi)-\widehat{g}(\xi)\Big)\Big\|_{\mathscr{L}_2(\mathbb{R})}\\
&=\Bigg\|\Big(\frac{\xi^2(1-e^{-\frac{\alpha^2\xi^2}{4}})^{\frac{1}{k+1}}}{\mathcal{A}_0(1-E_{\beta,1}(-\xi^2T^\beta))}\Big)^{k+1}(1+\xi^2)^{-k}(1+\xi^2)^k\widehat{f}(\xi)\int_0^T\mathcal{D}_\beta(\xi,T-s)\phi(s)ds\Bigg\|_{\mathscr{L}_2(\mathbb{R})}\\
&\le\frac{\Phi}{\mathcal{A}_0}\sup_{\xi\in\mathbb{R}}\Big|\Big(\frac{\xi^2}{\mathcal{A}_0(1-E_{\beta,1}(-\xi^2T^\beta))}\Big)^k\frac{(1-e^{-\frac{\alpha^2\xi^2}{4}})}{(1+\xi^2)^k}\Big|\mathcal{M}\\
&\le\frac{\Phi}{\mathcal{A}_0^{k+1}}\sup_{\xi\in\mathbb{R}}\Big|\Big(\frac{\xi^2}{(1+\xi^2)(1-E_{\beta,1}(-\xi^2T^\beta))}\Big)^k(1-e^{-\frac{\alpha^2\xi^2}{4}})\Big|\mathcal{M}\\
&\le\frac{\Phi}{\mathcal{A}_0^{k+1}}\left(\frac{1}{1-E_{\beta,1}(-T^\beta)}\right)^k\mathcal{M}. \qquad (58)
\end{aligned}
$$

□

Combining (54), (56) and (58), we have estimate $(\mathcal{C}_1^2)^{\frac{1}{k+1}}$ as follows

$$
\begin{aligned}
(\mathcal{C}_1^2)^{\frac{1}{k+1}}\le&\Bigg(\varepsilon\bigg(\log\Big(\log\Big(\frac{T}{\varepsilon}\Big)\Big)\bigg)^{k+1}\Big(\frac{\mathcal{L}_\beta(k,T)\,\mathcal{H}_\beta(\overline{B}_3,T,\Phi)}{\mathcal{A}_0\eta}\Big)^{k+1}\mathcal{M}^k\\
&+\frac{\Phi}{\mathcal{A}_0^{k+1}}\left(\frac{1}{1-E_{\beta,1}(-T^\beta)}\right)^k\Bigg)^{\frac{2}{k+1}}\mathcal{M}^{\frac{2}{k+1}}. \qquad (59)
\end{aligned}
$$

From (51) to (59), so

$$
\begin{aligned}
\|\mathcal{J}_1\|_{\mathscr{L}_2(\mathbb{R})}\le&\Bigg(\varepsilon\bigg(\log\Big(\log\Big(\frac{T}{\varepsilon}\Big)\Big)\bigg)^{k+1}\Big(\frac{\mathcal{L}_\beta(k,T)\,\mathcal{H}_\beta(\overline{B}_3,T,\Phi)}{\mathcal{A}_0\eta}\Big)^{k+1}\mathcal{M}^k\\
&+\frac{\Phi}{\mathcal{A}_0^{k+1}}\left(\frac{1}{1-E_{\beta,1}(-T^\beta)}\right)^k\Bigg)^{\frac{1}{k+1}}\mathcal{M}^{\frac{1}{k+1}}.\Bigg(2\varepsilon+\eta\bigg(\log\Big(\log\Big(\frac{T}{\varepsilon}\Big)\Big)\bigg)^{-1}\Bigg)^{\frac{k}{k+1}}. \qquad (60)
\end{aligned}
$$

Step 2: Estimate for $\|\mathcal{J}_2\|^2_{\mathscr{L}_2(\mathbb{R})}$, we have

$$
\begin{aligned}
\|\mathcal{J}_2\|^2_{\mathscr{L}_2(\mathbb{R})}&\le\Bigg\|\frac{(\widehat{g}_\varepsilon(\xi)-\widehat{g}(\xi))}{\Big(\int_0^T\mathcal{D}_\beta(\xi,T-s)\phi(s)ds\Big)e^{\frac{\alpha^2\xi^2}{4}}}\Bigg\|^2_{\mathscr{L}_2(\mathbb{R})}\\
&=\Bigg\|\frac{\xi^2}{\mathcal{A}_0\Big(1-E_{\beta,1}(-\xi^2T^\beta)\Big)}e^{-\frac{\alpha^2\xi^2}{4}}\Big(\widehat{g}_\varepsilon(\xi)-\widehat{g}(\xi)\Big)\Bigg\|^2_{\mathscr{L}_2(\mathbb{R})}\\
&\le\Big(\frac{16}{\alpha^4}\Big)\frac{\|\widehat{g}_\varepsilon-\widehat{g}\|^2_{\mathscr{L}_2(\mathbb{R})}}{\mathcal{A}_0^2}\Big(\frac{1}{1-E_{\beta,1}(-T^\beta)}\Big)^2. \qquad (61)
\end{aligned}
$$

Applying the Lemmas 11 and 12 in case $k=0$, we know that

$$
\|\mathcal{J}_2\|^2_{\mathscr{L}_2(\mathbb{R})}\le\left(\varepsilon\log\Big(\log\Big(\frac{T}{\varepsilon}\Big)\Big)\right)^2\Big(\frac{\mathcal{H}_\beta(\overline{B}_3,T,\Phi,\mathcal{M})}{\eta\mathcal{A}_0(1-E_{\beta,1}(-T^\beta))}\Big)^2. \qquad (62)
$$

Hence, we conclude that

$$\|\mathcal{J}_2\|_{\mathscr{L}_2(\mathbb{R})} \leq \left(\varepsilon \log\left(\log\left(\frac{T}{\varepsilon}\right)\right)\right)\left(\frac{\mathcal{H}_\beta(\overline{B}_3, T, \Phi, \mathcal{M})}{\eta \mathcal{A}_0(1 - E_{\beta,1}(-T^\beta))}\right). \tag{63}$$

Step 3: Estimate for $\|\mathcal{J}_3\|^2_{\mathscr{L}_2(\mathbb{R})}$, we have :

$$\begin{aligned}
\|\mathcal{J}_3\|^2_{\mathscr{L}_2(\mathbb{R})} &\leq \left\| \frac{\widehat{g}(\xi)}{\left(\int_0^T \mathcal{D}_\beta(\xi, T-s)\phi(s)ds\right)e^{\frac{\alpha^2\xi^2}{4}}} - \frac{\widehat{g}(\xi)}{\left(\int_0^T \mathcal{D}_\beta(\xi, T-s)\phi_\varepsilon(s)ds\right)e^{\frac{\alpha^2\xi^2}{4}}} \right\|^2_{\mathscr{L}_2(\mathbb{R})} \\
&= \left\| \frac{\left(e^{-\frac{\alpha^2\xi^2}{4}}\widehat{g}(\xi)\right)\int_0^T \mathcal{D}_\beta(\xi, T-s)(\phi_\varepsilon(s) - \phi(s))ds}{\left(\int_0^T \mathcal{D}_\beta(\xi, T-s)\phi(s)ds\right)\left(\int_0^T \mathcal{D}_\beta(\xi, T-s)\phi_\varepsilon(s)ds\right)} \right\|^2_{\mathscr{L}_2(\mathbb{R})}.
\end{aligned} \tag{64}$$

From (64), it gives

$$\begin{aligned}
\|\mathcal{J}_3\|^2_{\mathscr{L}_2(\mathbb{R})} &\leq \left\| \frac{\|\phi_\varepsilon - \phi\|_{C[0,T]} \int_0^T \mathcal{D}_\beta(\xi, T-s)ds}{\mathcal{A}_0\mathcal{A}_1\left(\int_0^T \mathcal{D}_\beta(\xi, T-s)ds\right)^2}\left(e^{-\frac{\alpha^2\xi^2}{4}}\widehat{g}(\xi)\right) \right\|^2_{\mathscr{L}_2(\mathbb{R})} \\
&\leq \frac{\|\phi - \phi_\varepsilon\|^2_{C[0,T]}}{\mathcal{A}_0^2\mathcal{A}_1^2} \left\| \frac{\xi^2}{(1 - E_{\beta,1}(-\xi^2 T^\beta))} e^{-\frac{\alpha^2\xi^2}{4}}\widehat{g}(\xi) \right\|^2_{\mathscr{L}_2(\mathbb{R})}.
\end{aligned} \tag{65}$$

Applying Lemma 12 with $k = 0$ and Lemma 11 , we know that

$$\begin{aligned}
\|\mathcal{J}_3\|^2_{\mathscr{L}_2(\mathbb{R})} &\leq \left(\frac{16}{\alpha^4}\right)\left(\frac{1}{1 - E_{\beta,1}(-T^\beta)}\right)^2 \frac{\|\phi - \phi_\varepsilon\|^2_{C[0,T]}}{\mathcal{A}_0^2\mathcal{A}_1^2}\|\widehat{g}\|^2_{\mathscr{L}_2(\mathbb{R})} \\
&\leq \left(\varepsilon \log\left(\log\left(\frac{T}{\varepsilon}\right)\right)\right)^2\left(\frac{\mathcal{H}_\beta(\overline{B}_3, T, \Phi, \mathcal{M})\|\widehat{g}\|_{\mathscr{L}_2(\mathbb{R})}}{\eta(1 - E_{\beta,1}(-T^\beta))\mathcal{A}_0\mathcal{A}_1}\right)^2.
\end{aligned} \tag{66}$$

Therefore,

$$\|\mathcal{J}_3\|_{\mathscr{L}_2(\mathbb{R})} \leq \left(\varepsilon \log\left(\log\left(\frac{T}{\varepsilon}\right)\right)\right)\left(\frac{\mathcal{H}_\beta(\overline{B}_3, T, \Phi, \mathcal{M})\|\widehat{g}\|_{\mathscr{L}_2(\mathbb{R})}}{\eta(1 - E_{\beta,1}(-T^\beta))\mathcal{A}_0\mathcal{A}_1}\right). \tag{67}$$

Combining (60), (63) and (67), we get:

$$
\begin{aligned}
&\|f(.)-f_{\varepsilon,\alpha}(.)\|_{\mathscr{L}_2(\mathbb{R})} = \|\widehat{f}(.)-\widehat{f_{\varepsilon,\alpha}}(.)\|_{\mathscr{L}_2(\mathbb{R})} \\
&\le \Bigg(\varepsilon\Big(\log\Big(\log\Big(\frac{T}{\varepsilon}\Big)\Big)\Big)^{k+1}\Big(\frac{\mathcal{L}_\beta(k,T)\,\mathcal{H}_\beta(\overline{B}_3,T,\Phi)}{\mathcal{A}_0\eta}\Big)^{k+1}\mathcal{M}^k \\
&+\frac{\Phi}{\mathcal{A}_0^{k+1}}\Big(\frac{1}{1-E_{\beta,1}(-T^\beta)}\Big)^k\Bigg)^{\frac{1}{k+1}}\mathcal{M}^{\frac{1}{k+1}}.\Bigg(2\varepsilon+\eta\Big(\log\Big(\log\Big(\frac{T}{\varepsilon}\Big)\Big)\Big)^{-1}\Bigg)^{\frac{k}{k+1}} \\
&+\Big(\varepsilon\log\Big(\log\Big(\frac{T}{\varepsilon}\Big)\Big)\Big)\Bigg(\frac{\mathcal{H}_\beta(\overline{B}_3,T,\Phi,\mathcal{M})}{\eta\mathcal{A}_0(1-E_{\beta,1}(-T^\beta))}+\Big(\frac{\mathcal{H}_\beta(\overline{B}_3,T,\Phi,\mathcal{M})\|\widehat{g}\|_{\mathscr{L}_2(\mathbb{R})}}{\eta(1-E_{\beta,1}(-T^\beta))\mathcal{A}_0\mathcal{A}_1}\Big)\Bigg).
\end{aligned}
\tag{68}
$$

Nohing that

$$
\lim_{\varepsilon\to 0}\varepsilon\Big(\log\Big(\log\Big(\frac{T}{\varepsilon}\Big)\Big)\Big)=0,\quad \lim_{\varepsilon\to 0}\varepsilon\Big(\log\Big(\log\Big(\frac{T}{\varepsilon}\Big)\Big)\Big)^{k+1}=0. \tag{69}
$$

Combining (68) and (69), we conclude that

$$
\|f(.)-f_{\varepsilon,\alpha}(.)\|_{\mathscr{L}_2(\mathbb{R})}=\|\widehat{f}(.)-\widehat{f_{\varepsilon,\alpha}}(.)\|_{\mathscr{L}_2(\mathbb{R})}\to 0,\text{as }\varepsilon\to 0. \tag{70}
$$

The proof of Theorem 2 is completed. □

5. Numerical Experiments

In this section, in order to illustrate the usefulness of the proposed methods, we consider the numerical examples intended. We carry out numerically above regularization methods to verify our proposed methods. The numerical examples with $T=1$, and $\beta=0.4$, $\beta=0.95$ are shown in this section, respectively. In the following, we give an example which had the exact expression of the solutions $(u(x,t),f(x))$. We use the computations in Matlab codes which are given by Podlubny [18] for computing the generalized Mittag-Leffler function and the accuracy control in computing is 10^{-10}. We will do the numerical tests with $x\in[-7,7]$ and $\eta=1.1$. The couple of $(\phi_\varepsilon,g_\varepsilon)$, which are determined below, play as measured data with a random noise as follows:

$$
\phi_\varepsilon(\cdot)=\phi(\cdot)+\varepsilon\,(2\text{rand}(.)-1),\;g_\varepsilon(\cdot)=g(\cdot)+\varepsilon\,(2\text{rand}(.)-1). \tag{71}
$$

Following Reference [9], we know the function $rand(\cdot)$ generates arrays of random numbers whose elements are normally distributed with mean 0, variance $\sigma^2=1$ and standard deviation $\sigma=1$, and it gives rand(size(.)) and rand(size(.)) returns an array of random entries that is the same size as g and ϕ, respectively. We can easily verify the validity of the inequality:

$$
\|\phi_\varepsilon-\phi\|_{C[0,T]}\le\varepsilon,\;\|g_\varepsilon-g\|_{\mathscr{L}_2(\mathbb{R})}\le\varepsilon. \tag{72}
$$

In this example, we consider particularly a one-dimensional case of the problem (1) for f is an exact data function.

$$
\begin{cases}
\dfrac{\partial^\beta u(x,t)}{\partial t^\beta}=u_{xx}(x,t)+\phi(t)f(x),\;(x,t)\in\mathbb{R}\times(0,T],\\
u(x,0)=0,x\in\mathbb{R},\\
u(x,1)=g(x),\;x\in\mathbb{R}\cdot
\end{cases}
\tag{73}
$$

In this example, we choose the following solution

$$
u(x,t)=\Big(E_{\beta,1}(t^\beta)-E_{\beta,1}(-t^\beta)\Big)\sin\Big(\frac{x}{2}\Big). \tag{74}
$$

Then a simple computation yields

$$\phi(t) = \frac{5}{4} E_{\beta,1}(t^\beta) + \frac{3}{4} E_{\beta,1}(-t^\beta). \tag{75}$$

and $f(x) = \sin\left(\frac{x}{2}\right)$. Moreover, we have $u(x,0) = u_0(x) = 0$ and

$$u(x,1) = u_1(x) = g(x) = \Big(E_{\beta,1}(1) - E_{\beta,1}(-1)\Big) \sin\left(\frac{x}{2}\right). \tag{76}$$

Next, for computing the integral in the latter equality, see Reference [19], we use the fact that

$$\int_0^x u^{\kappa-1} E_{\beta,\kappa}(yu^\beta)(x-u)^{\beta-1} E_{\beta,\beta}(z(x-u)^\beta) du = \frac{yE_{\beta,\kappa+\beta}(yx^\beta) - zE_{\beta,\beta+\kappa}(zx^\beta)}{y-z} x^{\beta+\kappa+1}. \tag{77}$$

From $\phi_\varepsilon(.) = \phi(.) + \varepsilon\Big(2\mathrm{rand}(.) - 1\Big)$, we have

$$\begin{aligned} \int_0^1 s^{\beta-1} E_{\beta,\beta}(-\xi^2 s^\beta) \phi_\varepsilon(1-s) ds &= \int_0^1 s^{\beta-1} E_{\beta,\beta}(-\xi^2 s^\beta) \phi(1-s) ds \\ &+ \varepsilon\left(2\mathrm{rand}(.) - 1\right) \int_0^1 s^{\beta-1} E_{\beta,\beta}(-\xi^2 s^\beta) ds. \end{aligned} \tag{78}$$

Combining (72), (75) and (78), we have

$$\begin{aligned} \int_0^1 s^{\beta-1} E_{\beta,\beta}(-\xi^2 s^{\beta-1}) \phi_\varepsilon(1-s) ds &= \frac{5}{4}\Big(\frac{E_{\beta,\beta+1}(1) + \xi^2 E_{\beta,\beta+1}(-\xi^2)}{1+\xi^2}\Big) \\ &- \frac{3}{4}\Big(\frac{E_{\beta,\beta+1}(-1) - \xi^2 E_{\beta,\beta+1}(-\xi^2)}{-1+\xi^2}\Big) \\ &+ \frac{\varepsilon\Big(2\mathrm{rand}(.) - 1\Big)}{\xi^2}\Big(1 - E_{\beta,1}(-\xi^2)\Big). \end{aligned} \tag{79}$$

In general, the numerical methods referenced by References [20,21] are summarized in three steps as follows.

Step 1: Choose N to generate the spatial and temporal discretization in such a manner as:

$$x_i = i\Delta x,\ \Delta x = \frac{\pi}{N},\ i = \overline{0, N}. \tag{80}$$

Obviously, the higher value of N will provide numerical results that are more accurate and stable. Here we choose $N = 100$ is satisfied.

Step 2: Setting $f_{\varepsilon,\alpha}(x_i) = f^i_{\varepsilon,\alpha}$ and $f(x_i) = f^i$, constructing two vectors contained all discrete values of $f_{\varepsilon,\alpha}$ and f denoted by $\Lambda_{\varepsilon,\alpha}$ and Ψ, respectively.

$$\begin{aligned} \Lambda_{\varepsilon,\alpha} &= [f^0_{\varepsilon,\alpha}\ f^1_{\varepsilon,\alpha} \ldots f^Q_{\varepsilon,\alpha}] \in \mathbb{R}^{Q+1}, \\ \Psi &= [f^0\ f^1 \ldots f^{Q-1}\ f^Q] \in \mathbb{R}^{Q+1}. \end{aligned} \tag{81}$$

Step 3: The estimation is defined:

Relative error estimation:

$$E_1 = \frac{\sqrt{\sum\limits_{i=1}^{N} |f_{\varepsilon,\alpha}(x_i) - f(x_i)|^2_{\mathscr{L}_2(-7,7)}}}{\sqrt{\sum\limits_{i=1}^{N} |f(x_i)|^2_{\mathscr{L}_2(-7,7)}}}. \tag{82}$$

Absolute error estimation:

$$E_2 = \sqrt{\frac{1}{N}\sum_{i=1}^{N} |f_{\varepsilon,\alpha}(x_i) - f(x_i)|^2_{\mathscr{L}_2(-7,7)}}. \tag{83}$$

Figure 1 shows a 2D figures between the sought and its regularized solutions for $N = 100$ and $\beta = 0.95$. All figures are presented with $\varepsilon = 0.1$, $\varepsilon = 0.01$ and $\varepsilon = 0.001$, respectively.

In Tables 1 and 2 of this example, we show the error estimation both the priori and the posteriori within case $N = 100$, that is, in Table 1 we show the error estimation for both the priori and the posteriori at $\beta = 0.95$ with $\varepsilon \in \{0.1, 0.01, 0.001\}$, respectively. In Table 2, we show the relative error estimation and absolute error estimation both the priori and the posteriori with $\varepsilon = 0.01$ with the different values of $\beta \in \{0.2, 0.4, 0.6, 0.8\}$, when ε is fixed and the mesh resolutions are increased, the regularized solution convergence is better than that of the exact solution. From observing the results from the tables and figures above, we conclude that when ε tends to zero, the regularized solution approaches the exact solution.

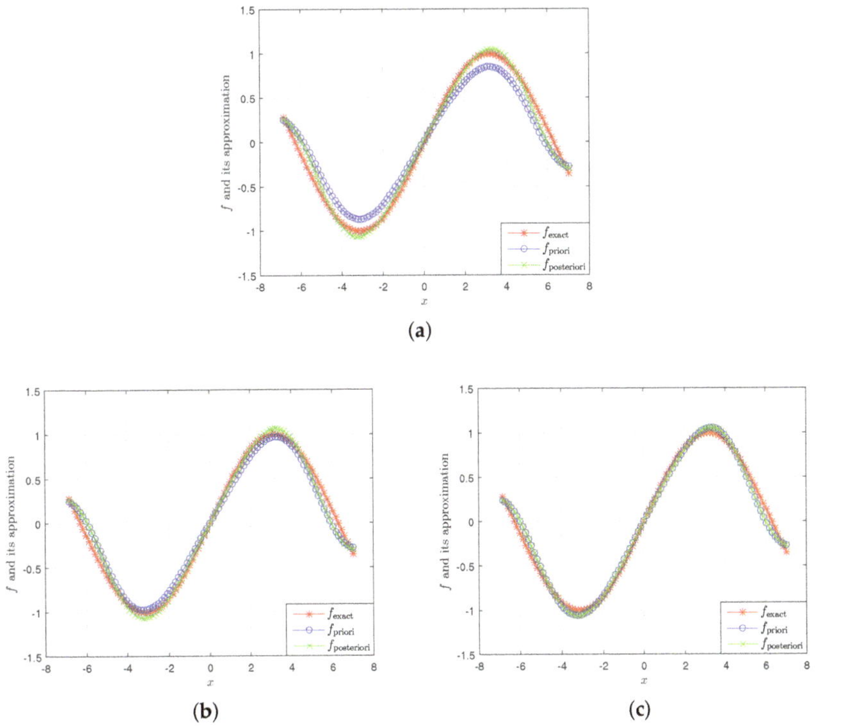

Figure 1. A comparison between the exact and regularized solutions for $k = 1$, $\beta = 0.95$ with $N = 100$. **(a)** $\varepsilon = 0.1$. **(b)** $\varepsilon = 0.01$. **(c)** $\varepsilon = 0.001$.

Table 1. The error estimation between the exact and regularized solutions of this example at $\beta = 0.95$ with $N = 100$.

ε	$E_1^{\beta_{pri}}$	$E_1^{\beta_{pos}}$	$E_2^{\beta_{pri}}$	$E_2^{\beta_{pos}}$
0.1	0.279660141830880	0.163452531664322	0.188256991900635	0.110030273632189
0.01	0.167130513450332	0.146077554813055	0.112506156619184	0.098334073898654
0.001	0.144054212078375	0.144599158066180	0.096972033479447	0.097338871212350

Table 2. The error estimation between the exact and regularized solutions with the different values of β, $\varepsilon = 0.01$ and $N = 100$.

β	$E_1^{\beta_{pri}}$	$E_1^{\beta_{pos}}$	$E_2^{\beta_{pri}}$	$E_2^{\beta_{pos}}$
0.2	0.156401672575436	0.176079016470940	0.078962919638416	0.092189970426402
0.4	0.146364358305196	0.165153671589525	0.073895354649786	0.086469770247512
0.6	0.136338164832119	0.153413164488168	0.068833404246973	0.080322774289912
0.8	0.124692172130227	0.140316883268202	0.062953661590221	0.073465933522836

6. Conclusions

In this study, by using the mollification regularization method, we solved the inverse problem and recovered the source term for time fractional diffusion equation with the time dependent coefficient. In the theoretical results, which we have shown, we obtained the error estimates of both a priori and a posteriori parameter choice rule methods based on a priori condition. In addition, in the numerical results, it shows that the regularized solutions are converged to the exact solution. Furthermore, it also shows that the smaller error of the input data, the better the convergence results.

Author Contributions: Project administration, Y.Z.; Resources, T.T.B.; Methodology, L.D.L.; Writing-review, editing and software, N.C.

Funding: The work was supported by the Fundo para o Desenvolvimento das Ciências e da Tecnologia (FDCT) of Macau under Grant 0074/2019/A2 and NNSF of China (11671339).

Conflicts of Interest: The authors declare no conflict of interest.

References

1. Malyshev, I. An inverse source problem for heat equation. *J. Math. Anal. Appl.* **1989**, *142*, 206–218. [CrossRef]
2. Geng, F.; Lin, Y. Application of the variational iteration method to inverse heat source problems. *Comput. Math. Appl.* **2009**, *58*, 2098–2102. [CrossRef]
3. Shidfar, A.; Babaei, A.; Molabahrami, A. Solving the inverse problem of identifying an unknown source term in a parabolic equation. *Comput. Math. Appl.* **2010**, *60*, 1209–1213. [CrossRef]
4. Barkai, E.; Metzler, R.; Klafter, J. From continuous time random walks to the fractional Fokker-Planck equation. *Phys. Rev. E* **2000**, *61*, 132–138. [CrossRef] [PubMed]
5. Chaves, A. Fractional diffusion equation to describe Levy flights. *Phys. Lett. A* **1998**, *239*, 13–16. [CrossRef]
6. Gorenflo, R.; Mainardi, F.; Scalas, E.; Raberto, M. Fractional calculus and continuoustime finance, the diffusion limit. In *Mathematical Finance. Trends in Mathematics*; Birkhuser: Basel, Switzerland, 2001; pp. 171–180.
7. Sakamoto, K.; Yamamoto, M. Initial value/boundary value problems for fractional diffusion-wave equations and applications to some inverse problems. *J. Math. Anal. Appl.* **2011**, *382*, 426–447. [CrossRef]
8. Zhang, Z.Q.; Wei, T. Identifying an unknown source in time-fractional diffusion equation by a truncation method. *Appl. Math. Comput.* **2013**, *219*, 5972–5983. [CrossRef]
9. Yang, F.; Fu, C.L. The quasi-reversibility regularization method for identifying the unknown source for time fractional diffusion equation. *Appl. Math. Modell.* **2015**, *39*, 1500–1512. [CrossRef]
10. Wei, T.; Wang, J.G. A modified Quasi-Boundary Value method for an inverse source problem of the time fractional diffusion equation. *Appl. Numer. Math.* **2014**, *78*, 95–111. [CrossRef]

11. Yang, F.; Fu, C.L.; Li, X.X. A mollification regularization method for identifying the time-dependent heat source problem. *J. Engine Math.* **2016**, *100*, 67–80. [CrossRef]
12. Wei, T.; Wang, J.G. Quasi-reversibility method to identify a space-dependent source for the time-fractional diffusion equation. *Appl. Math. Modell.* **2015**, *39*, 6139–6149. [CrossRef]
13. Nguyen, H.T.; Le, D.L.; Nguyen, V.T. Regularized solution of an inverse source problem for the time fractional diffusion equation. *Appl. Math. Modell.* **2016**, *40*, 8244–8264. [CrossRef]
14. Podlubny, I. *Fractional Differential Equations*; Academic Press: San Diego, CA, USA, 1999.
15. Samko, S.G.; Kilbas, A.A.; Marichev, O.I. *Fractional Integrals and Derivatives: Theory and Application*; Gordon and Breach: New York, NY, USA, 1993.
16. Yang, F.; Fu, C.L. A mollification regularization method for the inverse spatial-dependent heat source problem. *J. Comput. Appl. Math.* **2014**, *255*, 555–567. [CrossRef]
17. Kirsch, A. *An Introduction to the Mathematical Theory of Inverse Problem*; Applied Mathematical Sciences; Springer Science Business Media: Berlin/Heidelberg, Germany, 2011.
18. Podlubny, I.; Kacenak, M. Mittag-Leffler Function. The MATLAB Routine, 2006. Available online: http://www.mathworks.com/matlabcentral/fileexchange (accessed on 19 September 2019).
19. Mathai, A.M.; Haubold, H.J. Mittag Leffler function and Fractional Calculus. In *Special Functions for Applied Scientists*; Springer: New York, NY, USA, 2008; Chapter 2.
20. Meerschaert, M.; Tadjeran, C. Finite difference approximations for two-sided spacefractional partial differential equations. *Appl. Numer. Math.* **2006**, *56*, 80–90. [CrossRef]
21. Bu, W.; Liu, X.; Tang, Y.; Jiang, Y. Finite element multigrid method for multiterm time fractional advection-diffusion equations. *Int. J. Model. Simul. Sci. Comput.* **2015**, *6*, 1540001. [CrossRef]

mathematics

MDPI

Article

Generalized Integral Inequalities for Hermite–Hadamard-Type Inequalities via *s*-Convexity on Fractal Sets

Ohud Almutairi [1,†] and Adem Kılıçman [2,*,†]

1 Department of Mathematics, University of Hafr Al-Batin, Hafr Al-Batin 31991, Saudi Arabia; AhoudbAlmutairi@gmail.com
2 Department of Mathematics and Institute for Mathematical Research, University Putra Malaysia, Serdang 43400, Malaysia
* Correspondence: akilic@upm.edu.my; Tel.: +603-8946-6813
† These authors contributed equally to this work.

Received: 23 September 2019; Accepted: 1 November 2019; Published: 6 November 2019

Abstract: In this article, we establish new Hermite–Hadamard-type inequalities via Riemann–Liouville integrals of a function ψ taking its value in a fractal subset of $\mathbb{R}$ and possessing an appropriate generalized s-convexity property. It is shown that these fractal inequalities give rise to a generalized s-convexity property of ψ. We also prove certain inequalities involving Riemann–Liouville integrals of a function ψ provided that the absolute value of the first or second order derivative of ψ possesses an appropriate fractal s-convexity property.

Keywords: s-convex function; Hermite–Hadamard inequalities; Riemann–Liouville fractional integrals; fractal space

1. Introduction

Convexity is considered to be an important property in mathematical analysis. The applications of convex functions can be found in many fields of studies including economics, engineering and optimization (see for example [1,2]). A well-known result which was identified as Hermite–Hadamard inequalities is the reformulation through convexity. These inequalities, widely reported in the literature, can be defined as follows:

Theorem 1. *Suppose that $\psi : [u,v] \subset \mathbb{R} \to \mathbb{R}$ is a convex function on $[u,v]$ with $u < v$, then*

$$\psi\left(\frac{u+v}{2}\right) \leq \frac{1}{v-u}\int_u^v \psi(x)dx \leq \frac{\psi(u)+\psi(v)}{2}. \tag{1}$$

These two inequalities, which are refinement of convexity, can be held in reverse order as concave. Following this, many refinements of convex functions using Hermite–Hadamard inequalities have been continuously studied [3–6]. Given the variation of Hermite–Hadamard inequalities, Dragomir and Fitzpatrick [7] established a new generalization of s-convex functions in the second sense.

Theorem 2. *Suppose that $\psi : \mathbb{R}_+ \to \mathbb{R}_+$ is a s-convex function in the second sense, where $0 < s \leq 1$, $u, v \in \mathbb{R}_+$ and $u < v$. If $\psi \in L^1([u,v])$, then*

$$2^{s-1}\psi\left(\frac{u+v}{2}\right) \leq \frac{1}{v-u}\int_u^v \psi(x)dx \leq \frac{\psi(u)+\psi(v)}{s+1}. \tag{2}$$

Though the Hermite–Hadamard inequalities were established for classical integrals [8], the inequalities can also hold for fractional calculus, such as Riemann–Liouville [9–11], Katugampola [12] and local fractional integrals [13]. Some of these were studied through Mittag–Leffler function [14,15]. Other important generalizations include the work of Sarikaya et al. [16], who proved the Hermite–Hadamard inequalities through fractional integrals as follows:

Theorem 3. *Suppose that $\psi : [u,v] \to \mathbb{R}$ is a non-negative function with $0 \le u < v$ and $\psi \in L_1[u,v]$. If ψ is convex function on $[u,v]$, we have:*

$$\psi\left(\frac{u+v}{2}\right) \le \frac{\Gamma(\alpha+1)}{2(v-u)^{\alpha}}\left[J^{\alpha}_{u+}\psi(v) + J^{\alpha}_{v-}\psi(u)\right] \le \frac{\psi(u)+\psi(v)}{2},$$

where $0 < \alpha \le 1$.

The s-convexity mentioned in Hudzik and Maligranda [2] was also given as the generalization on fractal sets.

Definition 1 ([17])**.** *A function $\psi : \mathbb{R}_+ \to \mathbb{R}^{\alpha}$ is called generalized s-convex in the second sense if*

$$\psi(\gamma_1 u + \gamma_2 v) \le \gamma_1^{\alpha s}\psi(u) + \gamma_2^{\alpha s}\psi(v), \tag{3}$$

holds for all $u, v \in \mathbb{R}_+$, $\gamma_1, \gamma_2 \ge 0$, with $\gamma_1 + \gamma_2 = 1$ and for some fixed $s \in (0,1]$. The symbol GK_s^2 denotes the class of this functions.

The Riemann–Liouville fractional integral is introduced here due to its importance.

Definition 2 ([18])**.** *Suppose that $\psi \in L_1[u,v]$. The Riemann–Liouville integrals $J^{\alpha}_{u+}\psi$ and $J^{\alpha}_{v-}\psi$ of order $\alpha \in \mathbb{R}_+$ are defined by*

$$J^{\alpha}_{u+}\psi(x) = \frac{1}{\Gamma(\alpha)}\int_u^x (x-\gamma)^{\alpha-1}\psi(\gamma)d\gamma, \quad x > u,$$

and

$$J^{\alpha}_{v-}\psi(x) = \frac{1}{\Gamma(\alpha)}\int_x^v (\gamma-x)^{\alpha-1}\psi(\gamma)d\gamma, \quad x < v,$$

respectively.

The following lemma for differentiable function is given by Sarikaya et al. [16].

Lemma 1. *Let $\psi : [u,v] \to \mathbb{R}$ be a differentiable function on (u,v) with $u < v$. If $\psi' \in L^1[u,v]$, then we have:*

$$\frac{\psi(u)+\psi(v)}{2} - \frac{\Gamma(\alpha+1)}{2(v-u)^{\alpha}}\left[J^{\alpha}_{u+}\psi(v) + J^{\alpha}_{v-}\psi(u)\right] = \frac{v-u}{2}\int_0^1 \left[(1-\gamma)^{\alpha} - \gamma^{\alpha}\right]\psi'(\gamma u + (1-\gamma)v)d\gamma.$$

Wang et al. [9] extended Lemma 1 to include two cases, one of which involves the second derivative of Riemann–Liouville fractional integrals.

Lemma 2. *Let $\psi : [u,v] \to \mathbb{R}$ be a twice-differentiable function on (u,v) with $u < v$. If $\psi'' \in L^1[u,v]$, then*

$$\begin{aligned}&\frac{\psi(u)+\psi(v)}{2} - \frac{\Gamma(\alpha+1)}{2(v-u)^{\alpha}}\left[J^{\alpha}_{u+}\psi(v) + J^{\alpha}_{v-}\psi(u)\right]\\ =&\frac{(v-u)^2}{2}\int_0^1 \frac{1-(1-\gamma)^{\alpha+1}-\gamma^{\alpha+1}}{\alpha+1}\psi''(\gamma u + (1-\gamma)v)d\gamma,\end{aligned}$$

holds.

Even though studies were conducted on generalized Hermite–Hadamard inequality via Riemann–Liouville fractional integrals for s-convexity [16,19–21], inequalities of this type for generalized s-convexity are lacking. Therefore, this paper is aimed at establishing some new integral inequalities via generalized s-convexity on fractal sets. We show that the newly established inequalities are generalizations of Theorem 2. The new Hermite–Hadamard-type inequalities in the class of functions with derivatives in absolute values are shown to be s-convex function on fractal sets. This was achieved using Riemann–Liouville fractional integrals inequalities.

2. Main Results

Our first main result is obtained in the following theorem.

Theorem 4. *Suppose that $\psi : [u,v] \subseteq \mathbb{R}_+ \to \mathbb{R}^{\alpha}_+$ is a generalized s-convex on $[u,v]$, where $0 < s < 1$, $u, v \in \mathbb{R}_+$ and $u \leq v$. If $\psi \in L^1[u,v]$, then we obtain*

$$2^{\alpha(s-1)}\psi\left(\frac{u+v}{2}\right) \leq \frac{\Gamma(\alpha+1)}{2(v-u)^{\alpha}}\left[J^{\alpha}_{u^+}\psi(v) + J^{\alpha}_{v^-}\psi(u)\right] \leq \left[\frac{1}{s+1} + \frac{\Gamma(\alpha s+1)\Gamma(\alpha+1)}{\Gamma(\alpha(s+1)+1)}\right]\frac{[\psi(u)+\psi(v)]}{2}. \tag{4}$$

Proof. Since $\psi \in GK^2_s$, we get

$$\psi\left(\frac{x+y}{2}\right) \leq \frac{\psi(x)+\psi(y)}{2^{\alpha s}}. \tag{5}$$

Substituting $x = \gamma u + (1-\gamma)v$ and $y = (1-\gamma)u + \gamma v$ with $\gamma \in [0,1]$ in inequality (5), we obtain

$$2^{\alpha s}\psi\left(\frac{x+y}{2}\right) \leq \psi(\gamma u + (1-\gamma)v) + \psi(\gamma v + (1-\gamma)u). \tag{6}$$

Multiplying both sides of (6) by $\gamma^{\alpha-1}$ and integrating the resulting inequality with respect to γ over $[0,1]$ yields

$$\begin{aligned}\frac{2^{\alpha s}}{\alpha}\psi\left(\frac{x+y}{2}\right) &\leq \int_0^1 \gamma^{\alpha-1}\psi(\gamma u + (1-\gamma)v) + \int_0^1 \gamma^{\alpha-1}\psi(\gamma v + (1-\gamma)u)\\ &= \frac{\Gamma(\alpha)}{(v-u)^{\alpha}}\left[J^{\alpha}_{u^+}\psi(v) + J^{\alpha}_{v^-}\psi(u)\right].\end{aligned} \tag{7}$$

Then the first inequality in (4) is proved.

To prove the second inequality in (4), since $\psi \in GK^2_s$, we get

$$\psi(\gamma u + (1-\gamma)v) \leq \gamma^{\alpha s}\psi(u) + (1-\gamma)^{\alpha s}\psi(v), \tag{8}$$

and

$$\psi(\gamma v + (1-\gamma)u) \leq \gamma^{\alpha s}\psi(v) + (1-\gamma)^{\alpha s}\psi(u). \tag{9}$$

Combining the inequalities (8) and (9), we obtain

$$\begin{aligned}\psi(\gamma u+(1-\gamma)v)+\psi((1-\gamma)u+\gamma v) &\leq \gamma^{\alpha s}\psi(u)+(1-\gamma)^{\alpha s}\psi(v)+\gamma^{\alpha s}\psi(u)+(1-\gamma)^{\alpha s}\psi(u) \\ &= [\gamma^{\alpha s}+(1-\gamma)^{\alpha s}][\psi(u)+\psi(v)].\end{aligned} \tag{10}$$

A similar technique used in (6) is applied to inequality (10) to get the following:

$$\begin{aligned}\frac{\Gamma(\alpha)}{(v-u)^{\alpha}}[J^{\alpha}_{u^+}\psi(v)+J^{\alpha}_{v^-}\psi(u)] &\leq \int_0^1 \gamma^{\alpha-1}[\gamma^{\alpha s}+(1-\gamma)^{\alpha s}][\psi(u)+\psi(v)]d\gamma \\ &\leq \left[\frac{1}{\alpha(s+1)}+\frac{\Gamma(\alpha s+1)\Gamma(\alpha)}{\Gamma(\alpha(s+1)+1)}\right][\psi(u)+\psi(v),\end{aligned} \tag{11}$$

where

$$\int_0^1 \gamma^{\alpha s+s-1}d\gamma = \frac{1}{\alpha s+\alpha},$$

and

$$\int_0^1 \gamma^{\alpha-1}(1-\gamma)^{\alpha s}d\gamma = \frac{\Gamma(\alpha s+1)\Gamma(\alpha)}{\Gamma(\alpha s+\alpha+1)}.$$

Using inequalities (7) and (11), we prove Theorem 4. □

Remark 1. *In the second inequality of Theorem 4, the expression* $\frac{1}{2}\left[\frac{1}{s+1}+\frac{\Gamma(\alpha s+1)\Gamma(\alpha+1)}{\Gamma(\alpha(s+1)+1)}\right]$ *for* $0<s\leq 1$ *is the best possible. The map* $\psi:[0,1]\to[0^{\alpha},1^{\alpha}]$ *given by* $\psi(z)=z^{s\alpha}$ *is generalized s-convex in the second sense, and it satisfies the following equalities:*

$$\begin{aligned}\frac{\Gamma(\alpha+1)}{2}[J^{\alpha}_{u^+}\psi(1)+J^{\alpha}_{v^-}\psi(0)] &= \frac{\Gamma(\alpha+1)}{2}\left[\frac{1}{\Gamma(\alpha)}\left(\frac{1}{\alpha(s+1)}+\frac{\Gamma(\alpha s+1)\Gamma(\alpha)}{\Gamma(\alpha s+\alpha+1)}\right)\right] \\ &= \frac{\Gamma(\alpha+1)}{2}\left[\frac{1}{\alpha\Gamma(\alpha)(s+1)}+\frac{\Gamma(\alpha s+1)\Gamma(\alpha)}{\Gamma(\alpha)\Gamma(\alpha s+\alpha+1)}\right] \\ &= \frac{1}{2}\left[\frac{1}{s+1}+\frac{\Gamma(\alpha s+1)\Gamma(\alpha+1)}{\Gamma(\alpha(s+1)+1)}\right],\end{aligned}$$

and

$$\left[\frac{1}{s+1}+\frac{\Gamma(\alpha s+1)\Gamma(\alpha+1)}{\Gamma(\alpha(s+1)+1)}\right]\frac{[\psi(0)+\psi(1)]}{2} = \frac{1}{2}\left[\frac{1}{s+1}+\frac{\Gamma(\alpha s+1)\Gamma(\alpha+1)}{\Gamma(\alpha(s+1)+1)}\right].$$

Corollary 1. *By taking* $\alpha=1$ *in Theorem 4, the inequalities in (2) of Theorem 2 are recovered.*

This result is the same as Theorem 2.1 in Dragomir and Fitzpatrick [7].

Remark 2. *The equality*

$$\beta(u,v)=\frac{\Gamma(u)\Gamma(v)}{\Gamma(u+v)}$$

implies

$$2^{\alpha s-1}\psi(\frac{u+v}{2}) \leq \frac{\Gamma(\alpha+1)}{2(v-u)^{\alpha}}[J^{\alpha}_{u^+}\psi(v)+J^{\alpha}_{v^-}\psi(u)] \leq \left[\frac{1}{s+1}+\frac{\Gamma(\alpha s+1)\Gamma(\alpha+1)}{\Gamma(\alpha(s+1)+1)}\right]\frac{[\psi(u)+\psi(v)]}{2}$$
$$\leq \left[\frac{1}{s+1}+\beta(\alpha s+1,\alpha+1)(\alpha(s+1)+1)\right] \times \frac{[\psi(u)+\psi(v)]}{2}.$$

Theorem 5. *Suppose that* $M:[0,1]\to\mathbb{R}^{\alpha}$ *is the mapping given by*

$$M(\gamma)=\frac{\Gamma(\alpha+1)}{(v-u)^{\alpha}}\left[J^{\alpha}_{u^+}\psi\left(\gamma v+(1-\gamma)\frac{u+v}{2}\right)+J^{\alpha}_{v^-}\psi\left(\gamma u+(1-\gamma)\frac{u+v}{2}\right)\right],\gamma(0,1),$$

where $\psi:[u,v]\to\mathbb{R}^{\alpha}$ *belongs to* GK^2_s, $s\in(0,1]$, $u,v\in\mathbb{R}_+,u<v$ *and* $\psi\in L^1([u,v])$. *Then*

(i) $M\in GK^2_s$ *on* $[0,1]$.
(ii) We have the following inequality:

$$M(\gamma)\geq 2^{\alpha s}\psi\left(\frac{u+v}{2}\right). \tag{12}$$

(iii) We have the following inequality:

$$M(\gamma)\leq \min\{M_1(\gamma),M_2(\gamma)\},\gamma\in[0,1], \tag{13}$$

where

$$M_1(\gamma)=\gamma^{\alpha s}\frac{\Gamma(\alpha+1)}{(v-u)^{\alpha}}[J^{\alpha}_{u^+}\psi(v)+J^{\alpha}_{v^-}\psi(u)]+(1-\gamma)^{\alpha s}\psi\left(\frac{u+v}{2}\right),$$

and

$$M_2(\gamma)=\left[\frac{1}{s+1}+\frac{\Gamma(\alpha s+1)\Gamma(\alpha+1)}{\Gamma(\alpha(s+1)+2)}\right]\left[\psi\left(\gamma u+(1-\gamma)\frac{u+v}{2}\right)+\psi\left(\gamma v+(1-\gamma)\frac{u+v}{2}\right)\right].$$

(iv) If $\tilde{M}=\max\{M_1(\gamma),M_2(\gamma)\},\gamma\in[0,1]$, *then we have*

$$\tilde{M}\leq\left[\frac{1}{s+1}+\frac{\Gamma(\alpha s+1)\Gamma(\alpha+1)}{\Gamma(\alpha(s+1)+1)}\right]\left[\gamma^{\alpha s}[\psi(u)+\psi(v)]+2^{\alpha s}(1-\gamma)^{\alpha s}\psi\left(\frac{u+v}{2}\right)\right].$$

Proof.

(i) Let $\gamma_1,\gamma_2\in[0,1]$ and $\mu_1,\mu_2\geq 0$ with $\mu_1+\mu_2=1$, then

$$\begin{aligned} M(\mu_1\gamma_1+\mu_2\gamma_2) &= \tfrac{\Gamma(\alpha+1)}{(v-u)^{\alpha}}\Big[J^{\alpha}_{u^+}\psi((\mu_1\gamma_1+\mu_2\gamma_2)\tfrac{u+v}{2}+(1-(\mu_1\gamma_1+\mu_2\gamma_2))v) \\ &\quad +J^{\alpha}_{v^-}\psi\Big((\mu_1\gamma_1+\mu_2\gamma_2)u+(1-(\mu_1\gamma_1+\mu_2\gamma_2))\tfrac{u+v}{2}\Big)\Big] \\ &\leq \tfrac{\Gamma(\alpha+1)}{(v-u)^{\alpha}}\Big[\mu_1^{\alpha s}\Big(J^{\alpha}_{u^+}\psi\Big(\gamma_1\tfrac{u+v}{2}+(1-\gamma_1)v\Big)+J^{\alpha}_{v^-}\psi\Big(\gamma_1 u+(1-\gamma_1)\tfrac{u+v}{2}\Big)\Big) \\ &\quad +\mu_2^{\alpha s}\Big(J^{\alpha}_{u^+}\psi\Big(\gamma_2\tfrac{u+v}{2}+(1-\gamma_2)v\Big)+J^{\alpha}_{v^-}\psi\Big(\gamma_2 u+(1-\gamma_2)\tfrac{u+v}{2}\Big)\Big)\Big] \\ &= \mu_1^{\alpha s}M(\gamma_1)+\mu_2^{\alpha s}M(\gamma_2). \end{aligned}$$

(ii) Assume that $\gamma \in (0,1]$. Then by the change of variables $q = \gamma v + (1-\gamma)\frac{u+v}{2}$ and $p = \gamma u + (1-\gamma)\frac{u+v}{2}$, we have

$$\begin{aligned} M(\gamma) &= \frac{\Gamma(\alpha+1)}{\gamma^{\alpha}(v-u)^{\alpha}}\left[J^{\alpha}_{(\gamma v+(1-\gamma)\frac{u+v}{2})+}\psi\left(\gamma u+(1-\gamma)\frac{u+v}{2}\right)\right. \\ &\left.+ J^{\alpha}_{(\gamma u+(1-\gamma)\frac{u+v}{2})-}\psi\left(\gamma v+(1-\gamma)\frac{u+v}{2}\right)\right] \\ &= \frac{\Gamma(\alpha+1)}{(p-q)^{\alpha}}[J^{\alpha}_{q+}\psi(p) + J^{\alpha}_{p-}\psi(q)]. \end{aligned}$$

Applying the first generalized Hermite–Hadamard inequality, we obtain

$$\frac{\Gamma(\alpha+1)}{(p-q)^{\alpha}}[J^{\alpha}_{q+}\psi(p) + J^{\alpha}_{p-}\psi(q)] \geq 2^{\alpha s}\psi\left(\frac{q+p}{2}\right) = 2^{\alpha s}\psi\left(\frac{u+v}{2}\right),$$

and inequality (12) is obtained.
If $\gamma = 0$, the inequality

$$\psi\left(\frac{u+v}{2}\right) \geq 2^{\alpha s-1}\psi\left(\frac{u+v}{2}\right),$$

also holds.

(iii) Applying the second generalized Hermite–Hadamard inequality, we obtain

$$\begin{aligned} \frac{\Gamma(\alpha+1)}{(v-u)^{\alpha}}[J^{\alpha}_{u+}\psi(v) + J^{\alpha}_{v-}\psi(u)] &\leq \left[\frac{1}{s+1} + \frac{\Gamma(\alpha s+1)\Gamma(\alpha+1)}{\Gamma(\alpha(s+1)+1)}\right][\psi(u)+\psi(v)] \\ &= \left[\frac{1}{s+1} + \frac{\Gamma(\alpha s+1)\Gamma(\alpha+1)}{\Gamma(\alpha(s+1)+1)}\right]\left[\psi\left(\gamma u+(1-\gamma)\frac{u+v}{2}\right)\right. \\ &\left.+\psi\left(\gamma v+(1-\gamma)\frac{u+v}{2}\right)\right] \\ &= A_2(\gamma), \forall\gamma \in [0,1]. \end{aligned}$$

Please note that if $\gamma = 0$, then the inequality

$$\psi\left(\frac{u+v}{2}\right) = M(0) \leq M_2(0) = 2^{\alpha}\left[\frac{1}{s+1} + \frac{\Gamma(\alpha s+1)\Gamma(\alpha+1)}{\Gamma(\alpha(s+1)+1)}\right]\psi\left(\frac{u+v}{2}\right),$$

holds as it is equivalent to

$$\left(\left[\frac{\Gamma(\alpha s+\alpha+1)(s+1)}{\Gamma(\alpha s+1)\Gamma(\alpha+1)}\right] - 2^{\alpha}\right)\psi\left(\frac{u+v}{2}\right) \leq 0^{\alpha},$$

which is known to hold for $s \in (0,1]$.

Since for all $\gamma \in [0,1]$ and $x \in [u,v]$ the inequalities

$$\psi\left(\gamma u+(1-\gamma)\frac{u+v}{2}\right) \leq \gamma^{\alpha s}\psi(u) + (1-\gamma)^{\alpha s}\psi\left(\frac{u+v}{2}\right),$$

and

$$\psi\left(\gamma v+(1-\gamma)\frac{u+v}{2}\right) \leq \gamma^{\alpha s}\psi(v) + (1-\gamma)^{\alpha s}\psi\left(\frac{u+v}{2}\right).$$

are true, we obtain

$$\begin{aligned}M(\gamma) =&\frac{\Gamma(\alpha+1)}{(v-u)^{\alpha}}\left[J_{u^+}^{\alpha}\psi\left(\gamma v+(1-\gamma)\frac{u+v}{2}\right)\right.\\&\left.+J_{v^-}^{\alpha}\psi\left(\gamma u+(1-\gamma)\frac{u+v}{2}\right)\right]\\&\leqslant \gamma^{\alpha s}\frac{\Gamma(\alpha+1)}{(v-u)^{\alpha}}\left[J_{u^+}^{\alpha}\psi(v)+J_{v^-}^{\alpha}\psi(u)\right]+(1-\gamma)^{\alpha s}\psi\left(\frac{u+v}{2}\right)\\=&M_1(\gamma)\end{aligned}$$

and the inequality (13) is proved.

(iv) We have

$$\begin{aligned}M_2(\gamma) =& \left[\frac{1}{s+1}+\frac{\Gamma(\alpha s+1)\Gamma(\alpha+1)}{\Gamma(\alpha(s+1)+2)}\right]\left[\psi\left(\gamma u+(1-\gamma)\frac{u+v}{2}\right)\right.\\&\left.+\psi\left(\gamma v+(1-\gamma)\frac{u+v}{2}\right)\right]\\\leq& \left[\frac{1}{s+1}+\frac{\Gamma(\alpha s+1)\Gamma(\alpha+1)}{\Gamma(\alpha(s+1)+1)}\right]\left[\gamma^{\alpha s}\psi(u)+(1-\gamma)^{\alpha s}\psi\left(\frac{u+v}{2}\right)+\gamma^{\alpha s}\psi(v)\right.\\&\left.+(1-\gamma)^{\alpha s}\psi\left(\frac{u+v}{2}\right)\right]\\=&\left[\frac{1}{s+1}+\frac{\Gamma(\alpha s+1)\Gamma(\alpha+1)}{\Gamma(\alpha(s+1)+1)}\right]\left[\gamma^{\alpha s}[\psi(u)+\psi(v)]+2^{\alpha s}(1-\gamma)^{\alpha s}\psi\left(\frac{u+v}{2}\right)\right].\end{aligned}$$

Since

$$\frac{\Gamma(\alpha+1)}{2(v-u)^{\alpha}}\left[J_{u^+}^{\alpha}\psi(v)+J_{v^-}^{\alpha}\psi(u)\right]\leq\left[\frac{1}{s+1}+\frac{\Gamma(\alpha s+1)\Gamma(\alpha+1)}{\Gamma(\alpha(s+1)+1)}\right]\frac{[\psi(u)+\psi(v)]}{2},$$

and

$$(1-\gamma)^{\alpha s}\psi\left(\frac{u+v}{2}\right)\leq 2^{\alpha}(1-\gamma)^{\alpha s}\left[\frac{1}{s+1}+\frac{\Gamma(\alpha s+1)\Gamma(\alpha+1)}{\Gamma(\alpha(s+1)+1)}\right]\psi\left(\frac{u+v}{2}\right),$$

then

$$\begin{aligned}M_1(\gamma)\leq&\gamma^{\alpha s}\left[\frac{1}{s+1}+\frac{\Gamma(\alpha s+1)\Gamma(\alpha+1)}{\Gamma(\alpha(s+1)+1)}\right](\psi(u)+\psi(v))\\&+2^{\alpha}(1-\gamma)^{\alpha s}\left[\frac{1}{s+1}+\frac{\Gamma(\alpha s+1)\Gamma(\alpha+1)}{\Gamma(\alpha(s+1)+1)}\right]\psi\left(\frac{u+v}{2}\right),\end{aligned}$$

and the proof of Theorem 5 is complete.

□

Corollary 2. *Choosing $s=1$ in Theorem 5, we have*

(i)

$$\frac{\Gamma(\alpha+1)}{(v-u)^{\alpha}}\left[J_{u^+}^{\alpha}\psi\left(\gamma v+(1-\gamma)\frac{u+v}{2}\right)+J_{v^-}^{\alpha}\psi\left(\gamma u+(1-\gamma)\frac{u+v}{2}\right)\right]$$
$$\leq \min\left\{\gamma^{\alpha}\frac{\Gamma(\alpha+1)}{(v-u)^{\alpha}}[J_{u^+}^{\alpha}\psi(v)+J_{v^-}^{\alpha}\psi(u)]\right.$$
$$+(1-\gamma)^{\alpha}\psi\left(\frac{u+v}{2}\right),\left[\frac{1}{2}+\frac{(\Gamma(\alpha+1))^2}{\Gamma(2(\alpha+1))}\right]$$
$$\left.\times\left[\psi\left(\gamma u+(1-\gamma)\frac{u+v}{2}\right)+\psi\left(\gamma v+(1-\gamma)\frac{u+v}{2}\right)\right]\right\}.$$

(ii) Since

$$\tilde{M}=\max\left\{\gamma^{\alpha}\frac{\Gamma(\alpha+1)}{(v-u)^{\alpha}}[J_{u^+}^{\alpha}\psi(v)+J_{v^-}^{\alpha}\psi(u)]+(1-\gamma)^{\alpha}\psi(\frac{u+v}{2}),\right.$$
$$\left.\left[\frac{1}{2}+\frac{(\Gamma(\alpha+1))^2}{\Gamma(2(\alpha+1))}\right]\left[\psi(\gamma u+(1-\gamma)\frac{u+v}{2})+\psi\left(\gamma v+(1-\gamma)\frac{u+v}{2}\right)\right]\right\},$$

we get

$$\tilde{M}=\left[\frac{1}{2}+\frac{\Gamma(\alpha+1)\Gamma(\alpha+1)}{\Gamma(2(\alpha+1))}\right]\left(\gamma^{\alpha}(\psi(u)+\psi(v))+2^{\alpha}(1-\gamma)^{\alpha}\psi\left(\frac{u+v}{2}\right)\right).$$

Theorem 6. *Let $\psi:[u,v]\subset\mathbb{R}_+\to\mathbb{R}^{\alpha}$ be a differentiable function on (u,v) where $0\leq u<v$. For some fixed $q\geq 1$, if $|\psi'|^q$ is generalized s-convex on (u,v), we obtain*

$$\left|\frac{\psi(u)+\psi(v)}{2}-\frac{\Gamma(\alpha+1)}{2(v-u)^{\alpha}}\left[J_{u^+}^{\alpha}\psi(v)+J_{v^-}^{\alpha}\psi(u)\right]\right|\leq\frac{v-u}{2}\left(\frac{2}{\alpha+1}\right)^{1-\frac{1}{q}}$$
$$\times\left(\left[\beta(\alpha+1,\alpha s+1)+\frac{1}{\alpha(s+1)+1}\right]\right.$$
$$\left.\times\left[|\psi'(u)|^q+|\psi'(v)|^q\right]\right)^{\frac{1}{q}}.$$

Proof. Applying Lemma 1, we obtain

$$\left|\frac{\psi(u)+\psi(v)}{2}-\frac{\Gamma(\alpha+1)}{2(v-u)^{\alpha}}\left[J_{u^+}^{\alpha}\psi(v)+J_{v^-}^{\alpha}\psi(u)\right]\right|=\frac{v-u}{2}\left|\int_0^1[(1-\gamma)^{\alpha}-\gamma^{\alpha}]\right.$$
$$\left.\times\psi'(\gamma u+(1-\gamma)v)d\gamma\right|. \tag{14}$$

First, suppose $q=1$. Since the function $|\psi'|$ is generalized s-convex on (u,v), we obtain

$$|\psi'(\gamma u+(1-\gamma)v\gamma|\leq\gamma^{\alpha s}|\psi'(u)|+(1-\gamma)^{\alpha s}|\psi'(v)|. \tag{15}$$

Therefore,

$$\left|\int_0^1 [(1-\gamma)^\alpha - \gamma^\alpha]\,\psi'(\gamma u + (1-\gamma)v)d\gamma\right| \le |\psi'(u)| \int_0^1 [(1-\gamma)^\alpha \gamma^{\alpha s} + \gamma^{\alpha s+\alpha}]d\gamma$$
$$+ |\psi'(v)| \int_0^1 [(1-\gamma)^{\alpha s+\alpha} + (1-\gamma^{\alpha s})\gamma^\alpha]d\gamma$$
$$= \left(\beta(\alpha+1, \alpha s+1) + \frac{1}{\alpha s+\alpha+1}\right)[\psi'(u) + \psi'(v)]. \quad (16)$$

Next suppose that $q > 1$. From the power mean inequality and the generalized s-convexity of the function $|\psi'|^q$ we obtain

$$\begin{aligned}
\left|\int_0^1 [(1-\gamma)^\alpha - \gamma^\alpha]\,\psi'(\gamma u + (1-\gamma)v)d\gamma\right| &\le \left|\int_0^1 [(1-\gamma)^\alpha - \gamma^\alpha]^{1-\frac{1}{q}} [(1-\gamma)^\alpha - \gamma^\alpha]^{\frac{1}{q}}\right.\\
&\quad \left.\times \psi'(\gamma u + (1-\gamma)v)d\gamma\right|\\
&\le \left(\int_0^1 |(1-\gamma)^\alpha - \gamma^\alpha| d\gamma\right)^{1-\frac{1}{q}}\\
&\quad \times \left(\int_0^1 |(1-\gamma)^\alpha - \gamma^\alpha||\psi'(\gamma u + (1-\gamma)v)|^q d\gamma\right)^{\frac{1}{q}}\\
&\le \left(\int_0^1 (1-\gamma)^\alpha + \gamma^\alpha d\gamma\right)^{1-\frac{1}{q}}\\
&\quad \times \left(\int_0^1 [(1-\gamma)^\alpha + \gamma^\alpha][\gamma^{\alpha s}|\psi'(u)|^q + (1-\gamma)^{\alpha s}|\psi'(v)|^q]d\gamma\right)^{\frac{1}{q}}\\
&= \left(\frac{2}{\alpha+1}\right)^{1-\frac{1}{q}} \left(\left[\beta(\alpha+1, \alpha s+1) + \frac{1}{\alpha s+\alpha+1}\right]\right.\\
&\quad \left.\times [|\psi'(u)|^q + |\psi'(v)|^q]\right)^{\frac{1}{q}}.
\end{aligned} \quad (17)$$

In view of inequalities (14), (16) and (17) the proof of Theorem 6 is complete now. □

Corollary 3. *Under the conditions of Theorem 6, we get*

(i) If $q = s = 1$, then

$$\left|\frac{\psi(u)+\psi(v)}{2} - \frac{\Gamma(\alpha+1)}{2(v-u)^\alpha}\left[J^\alpha_{u^+}\psi(v) + J^\alpha_{v^-}\psi(u)\right]\right| \le \frac{v-u}{2}\left[\frac{1}{2\alpha+1} + \beta(\alpha+1, \alpha+1)\right](|\psi'(u)| + |\psi'(v)|).$$

(ii) If $q = \alpha = s = 1$, then

$$\left|\frac{\psi(u)+\psi(v)}{2} - \frac{1}{v-u}\int_u^v \psi(x)dx\right| \le \frac{v-u}{4}(|\psi'(u)| + |\psi'(v)|).$$

(iii) If $q > 1$ and $s = 1$

$$\begin{aligned}
\left|\frac{\psi(u)+\psi(v)}{2} - \frac{\Gamma(\alpha+1)}{2(v-u)^\alpha}\left[J^\alpha_{u^+}\psi(v) + J^\alpha_{v^-}\psi(u)\right]\right| &\le \frac{v-u}{2}\left(\frac{2}{\alpha+1}\right)^{1-\frac{1}{q}} \left(\left[\beta(\alpha+1, \alpha s+1) + \frac{1}{\alpha s+\alpha+1}\right]\right.\\
&\quad \left.\times [|\psi'(u)|^q + |\psi'(v)|^q]\right)^{\frac{1}{q}}.
\end{aligned}$$

(iv) If $q > 1$ and $\alpha = s = 1$ then

$$\left|\frac{\psi(u)+\psi(v)}{2} - \frac{1}{v-u}\int_u^v \psi(x)dx\right| \le \frac{v-u}{2}\left(\frac{|\psi'(u)|^q + |\psi'(v)|^q}{2}\right)^{\frac{1}{q}}.$$

Theorem 7. *Let $\psi : [u, v] \subset \mathbb{R}_+ \to \mathbb{R}^\alpha$ be a differentiable function on (u, v) where $0 \le u < v$. If $|\psi'|^q$ is generalized s-convex on (u, v) for $q > 1$, we get*

$$\left|\frac{\psi(u)+\psi(v)}{2} - \frac{\Gamma(\alpha+1)}{2(v-u)^\alpha}\left[J^\alpha_{u^+}\psi(v) + J^\alpha_{v^-}\psi(u)\right]\right| \le \frac{v-u}{2}\left(\frac{2}{\alpha+1}\left[1-\frac{1}{2^\alpha}\right]\right)^{1-\frac{1}{q}}$$
$$\times\left(\left[\frac{\Gamma(\alpha s+1)\Gamma(\alpha+1)}{\Gamma(\alpha s+\alpha+2)} + \frac{1}{\alpha s+\alpha+1}\right]\right)^{\frac{1}{q}}$$
$$\times\left[|\psi'(u)|^q + |\psi'(v)|^q\right]^{\frac{1}{q}}.$$

Proof. Since $|\psi'|$ is generalized s-convex on (u, v), we obtain

$$|\psi'(\gamma u + (1-\gamma)v\gamma| \le \gamma^{\alpha s}|\psi'(u)| + (1-\gamma)^{\alpha s}|\psi'(v)|.$$

From this fact and applying the Hölder's inequality, we have

$$\begin{aligned}
\left|\int_0^1 [(1-\gamma)^\alpha - \gamma^\alpha]\,\psi'(\gamma u+(1-\gamma)v)d\gamma\right| &\le \left|\int_0^1 [(1-\gamma)^\alpha - \gamma^\alpha]^{1-\frac{1}{q}}[(1-\gamma)^\alpha - \gamma^\alpha]^{\frac{1}{q}}\,\psi'(\gamma u+(1-\gamma)v)d\gamma\right| \\
&\le \left(\int_0^{\frac{1}{2}}[(1-\gamma)^\alpha - \gamma^\alpha]d\gamma + \int_{\frac{1}{2}}^1[\gamma^\alpha - (1-\gamma)^\alpha]d\gamma\right)^{1-\frac{1}{q}} \\
&\times\left(\int_0^1 |(1-\gamma)^\alpha - \gamma^\alpha||\psi'(\gamma u+(1-\gamma)v)|^q d\gamma\right)^{\frac{1}{q}} \\
&\le \left(\int_0^{\frac{1}{2}}[(1-\gamma)^\alpha - \gamma^\alpha]d\gamma + \int_{\frac{1}{2}}^1[\gamma^\alpha - (1-\gamma)^\alpha]d\gamma\right)^{1-\frac{1}{q}} \\
&\times\left(\int_0^1[(1-\gamma)^\alpha + \gamma^\alpha][\gamma^{\alpha s}|\psi'(u)|^q + (1-\gamma)^{\alpha s}|\psi'(v)|^q]d\gamma\right)^{\frac{1}{q}} \\
&= \left(\frac{2}{\alpha+1}\left[1-\frac{1}{2^\alpha}\right]\right)^{1-\frac{1}{q}}\left(\left[\frac{\Gamma(\alpha s+1)\Gamma(\alpha+1)}{\Gamma(\alpha s+\alpha+2)} + \frac{1}{\alpha s+\alpha+1}\right]\right)^{\frac{1}{q}} \\
&\times[|\psi'(u)|^q + |\psi'(v)|^q]^{\frac{1}{q}}.
\end{aligned} \tag{18}$$

Thus, the inequalities (14) and (18) complete the proof of Theorem 7. □

Theorem 8. *Let $\psi : [u, v] \subset \mathbb{R}_+ \to \mathbb{R}^\alpha$ be a differentiable mapping on (u, v) with $0 \le u < v$. For some fixed $q > 1$, if $|\psi'|^q$ is generalized s-convex on $[u, v]$, then we have:*

$$\left|\frac{\psi(u)+\psi(v)}{2} - \frac{\Gamma(\alpha+1)}{2(v-u)^\alpha}\left[J^\alpha_{u^+}\psi(v) + J^\alpha_{v^+}\psi(u)\right]\right| \le \frac{v-u}{2}\left(\frac{2}{\alpha p+1}\left[1-\frac{1}{2^{\alpha p}}\right]\right)^{\frac{1}{p}}\left(\frac{1}{\alpha s+1}\right)^{\frac{1}{q}}$$
$$\times\left(|\psi'(u)|^q + |\psi'(v)|^q\right)^{\frac{1}{q}}.$$

Proof. By applying Hölder's inequality and (15), we obtain

$$\begin{aligned}
\left|\int_0^1 [(1-\gamma)^\alpha - \gamma^\alpha]\,\psi'(\gamma u+(1-\gamma)v)d\gamma\right| &\le \left(\int_0^1 |(1-\gamma)^\alpha - \gamma^\alpha|^p d\gamma\right)^{\frac{1}{p}} \\
&\times\left(\int_0^1 |\psi'(\gamma u+(1-\gamma)v)|^q d\gamma\right)^{\frac{1}{q}} \\
&\le \left(\int_0^{\frac{1}{2}}[(1-\gamma)^\alpha - \gamma^\alpha]^p d\gamma + \int_{\frac{1}{2}}^1[\gamma^\alpha - (1-\gamma)^\alpha]^p d\gamma\right)^{\frac{1}{p}} \\
&\times\left(\int_0^1[\gamma^{\alpha s}|\psi'(u)|^q + (1-\gamma)^{\alpha s}|\psi'(v)|^q]d\gamma\right)^{\frac{1}{q}} \\
&\le \left(\frac{2}{\alpha p+1}\left[1-\frac{1}{2^{\alpha p}}\right]\right)^{\frac{1}{p}}\left(\frac{|\psi'(u)|^q+|\psi'(v)|^q}{\alpha s+1}\right)^{\frac{1}{q}}.
\end{aligned} \tag{19}$$

Finally, from (14) and (19) we get the desired result. □

Remark 3. *From Theorems 6–8, we obtain the following inequality for $q > 1$*

$$\left|\frac{\psi(u)+\psi(v)}{2}-\frac{\Gamma(\alpha+1)}{2(v-u)^{\alpha}}\left[J_{u^+}^{\alpha}\psi(v)+J_{v^-}^{\alpha}\psi(u)\right]\right| \leq \min\{S_1,S_2,S_3\}\frac{(v-u)}{2}[|\psi'(u)|^q+|\psi'(v)|^q]^{\frac{1}{q}},$$

where

$$S_1=\left(\frac{2}{\alpha+1}\right)^{1-\frac{1}{q}}\left(\left[\beta(\alpha+1,\alpha s+1)+\frac{1}{\alpha(s+1)+1}\right]\right)^{\frac{1}{q}},$$

$$S_2=\left(\frac{2}{\alpha+1}\left[1-\frac{1}{2^{\alpha}}\right]\right)^{1-\frac{1}{q}}\left(\left[\frac{\Gamma(\alpha s+1)\Gamma(\alpha+1)}{\Gamma(\alpha(s+1)+2)}+\frac{1}{\alpha(s+1)+1}\right]\right)^{\frac{1}{q}},$$

$$S_3=\left(\frac{2}{\alpha p+1}\left[1-\frac{1}{2^{\alpha p}}\right]\right)^{\frac{1}{p}}\left(\frac{1}{\alpha s+1}\right)^{\frac{1}{q}}.$$

Theorem 9. *Let $\psi : [u,v] \subset \mathbb{R}_+ \to \mathbb{R}^{\alpha}$ be a twice-differentiable function on (u,v) with $0 \leq u < v$. If, for some fixed $q \geq 1$, the function $|\psi''|^q$ is generalized s-convex on the interval $[u,v]$, then we have*

$$\begin{aligned}\left|\frac{\psi(u)+\psi(v)}{2}-\frac{\Gamma(\alpha+1)}{2(v-u)^{\alpha}}\left[J_{u^+}^{\alpha}\psi(v)+J_{v^-}^{\alpha}\psi(u)\right]\right| &\leq \frac{(v-u)^2}{2(\alpha+1)}\left(\frac{\alpha}{\alpha+2}\right)^{1-\frac{1}{q}}\\ &\times\left(\frac{1}{\alpha s+1}-\beta(\alpha s+1,\alpha+2)-\frac{1}{\alpha s+\alpha+2}\right)^{\frac{1}{q}}\\ &\times\left(|\psi''(u)|^q+|\psi''(v)|^q\right)^{\frac{1}{q}}.\end{aligned}$$

Proof. Applying Lemma 2, we have

$$\begin{aligned}&\left|\frac{\psi(u)+\psi(v)}{2}-\frac{\Gamma(\alpha+1)}{2(v-u)^{\alpha}}\left[J_{u^+}^{\alpha}\psi(v)+J_{v^-}^{\alpha}\psi(u)\right]\right|\\ =&\frac{(v-u)^2}{2}\int_0^1\frac{1-(1-\gamma)^{\alpha+1}-\gamma^{\alpha+1}}{\alpha+1}|\psi''(\gamma u+(1-\gamma)v)|d\gamma.\end{aligned} \tag{20}$$

First, suppose that $q = 1$. Since the mapping $|\psi''|$ is generalized s-convex on (u,v), we obtain

$$|\psi''(\gamma u+(1-\gamma)v)| \leq \gamma^{\alpha s}\,|\psi''(u)|+(1-\gamma)^{\alpha s}\,|\psi''(v)|. \tag{21}$$

Therefore,

$$\begin{aligned}&\left|\frac{\psi(u)+\psi(v)}{2}-\frac{\Gamma(\alpha+1)}{2(v-u)^{\alpha}}\left[J_{u^+}^{\alpha}\psi(v)+J_{v^-}^{\alpha}\psi(u)\right]\right|\\ \leq&\frac{(v-u)^2}{2}\int_0^1\frac{1-(1-\gamma)^{\alpha+1}-\gamma^{\alpha+1}}{\alpha+1}\left(\gamma^{\alpha s}|\psi''(u)|+(1-\gamma)^{\alpha s}|\psi''(v)|\right)d\gamma\\ =&\frac{(v-u)^2}{2(\alpha+1)}\left(\frac{1}{\alpha s+1}-\beta(\alpha s+1,\alpha+2)-\frac{1}{\alpha s+\alpha+2}\right)(|\psi''(u)|+|\psi''(v)|,\end{aligned}$$

where

$$\beta(\alpha s+1,\alpha+2)=\beta(\alpha+2,\alpha s+1).$$

Secondly, for $q > 1$. From Lemma 2 and the power mean inequality, we have

$$\begin{aligned}&\left|\frac{\psi(u)+\psi(v)}{2}-\frac{\Gamma(\alpha+1)}{2(v-u)^{\alpha}}\left[J_{u^+}^{\alpha}\psi(v)+J_{v^-}^{\alpha}\psi(u)\right]\right|\\&\leq\frac{(v-u)^2}{2(\alpha+1)}\left(\int_0^1(1-(1-\gamma)^{\alpha+1}-\gamma^{\alpha+1})d\gamma\right)^{1-\frac{1}{q}}\\&\times\left(\int_0^1\left(1-(1-\gamma)^{\alpha+1}-\gamma^{\alpha+1}|\psi''(\gamma u+(1-\gamma)v)|^q\right)d\gamma\right)^{\frac{1}{q}}.\end{aligned}\tag{22}$$

Hence, from inequalities (21) and (22), we obtain

$$\begin{aligned}&\left|\frac{\psi(u)+\psi(v)}{2}-\frac{\Gamma(\alpha+1)}{2(v-u)^{\alpha}}\left[J_{u^+}^{\alpha}\psi(v)+J_{v^-}^{\alpha}\psi(u)\right]\right|\\&\leq\frac{(v-u)^2}{2(\alpha+1)}\left(\int_0^1 1-(1-\gamma)^{\alpha+1}-\gamma^{\alpha+1}\right)^{1-\frac{1}{q}}\left(\int_0^1 1-(1-\gamma)^{\alpha+1}-\gamma^{\alpha+1}|\psi''(\gamma u+(1-\gamma)v)|^q d\gamma\right)^{\frac{1}{q}}\\&\leq\frac{(v-u)^2}{2(\alpha+1)}\left(\int_0^1 1-(1-\gamma)^{\alpha+1}-\gamma^{\alpha+1}\right)^{1-\frac{1}{q}}\\&\times\left(\int_0^1 1-(1-\gamma)^{\alpha+1}-\gamma^{\alpha+1}[\gamma^{\alpha s}|\psi''(u)|^q+(1-\gamma)^{\alpha s}|\psi''(v)|^q]d\gamma\right)^{\frac{1}{q}}\\&\leq\frac{(v-u)^2}{2(\alpha+1)}\left(\frac{\alpha}{\alpha+2}\right)^{1-\frac{1}{q}}\left(\frac{1}{\alpha s+1}-\beta(\alpha s+1,\alpha+2)-\frac{1}{\alpha s+\alpha+2}\right)^{\frac{1}{q}}\left(|\psi''(u)|^q+|\psi''(v)|^q\right)^{\frac{1}{q}}.\end{aligned}$$

This completes the proof of Theorem 9. □

Theorem 10. *Let $0 \leqslant u < v < \infty$ and let the function $\psi : [u,v] \to \mathbb{R}^{\alpha}$ be twice-differentiable on the open interval (u,v), and fix $s \in (0,1]$ and fix $q > 1$. If, in addition, the function $|\psi''|^q$ is generalized s-convex on $[u,v]$, then*

$$\begin{aligned}\left|\frac{\psi(u)+\psi(v)}{2}-\frac{\Gamma(\alpha+1)}{2(v-u)^{\alpha}}\left[J_{u^+}^{\alpha}\psi(v)+J_{v^-}^{\alpha}\psi(u)\right]\right|&\leq\frac{(v-u)^2}{2(\alpha+1)}\left(1-\frac{2}{p(\alpha+1)+}\right)^{\frac{1}{p}}\\&\times\left(\frac{|\psi''(u)|^q+|\psi''(v)|^q}{\alpha s+1}\right)^{\frac{1}{q}},\end{aligned}$$

where $\frac{1}{p}+\frac{1}{q}=1$.

Proof. From (20), (21) and the Hölder's inequality, we have

$$\begin{aligned}&\left|\frac{\psi(u)+\psi(v)}{2}-\frac{\Gamma(\alpha+1)}{2(v-u)^{\alpha}}\left[J_{u^+}^{\alpha}\psi(v)+J_{b^-}^{\alpha}\psi(u)\right]\right|\\&\leq\frac{(v-u)^2}{2(\alpha+1)}\left(\int_0^1(1-(1-\gamma)^{\alpha+1}-\gamma^{\alpha+1})^p d\gamma\right)^{\frac{1}{p}}\left(|\psi''(\gamma u+(1-\gamma)v)|^q d\gamma\right)^{\frac{1}{q}}\\&\leq\frac{(v-u)^2}{2(\alpha+1)}\left(\int_0^1(1-(1-\gamma)^{p(\alpha+1)}-\gamma^{p(\alpha+1)})d\gamma\right)^{\frac{1}{p}}\left(|\psi''(u)|^q\int_0^1\gamma^{\alpha s}d\gamma+|\psi''(v)|^q\int_0^1(1-\gamma)^{\alpha s}d\gamma\right)^{\frac{1}{q}}\\&\leq\frac{(v-u)^2}{2(\alpha+1)}\left(1-\frac{2}{p(\alpha+1)+}\right)^{\frac{1}{p}}\left(\frac{|\psi''(u)|^q+|\psi''(v)|^q}{\alpha s+1}\right)^{\frac{1}{q}}.\end{aligned}$$

We use
$$\left(1-(1-\gamma)^{\alpha+1}-\gamma^{\alpha+1}\right)^{q}\leq 1-(1-\gamma)^{q(\alpha+1)}-t^{q(\alpha+1)},$$
for any $\gamma\in[0,1]$, which follows from
$$(V-N)^{q}\leq V^{q}-N^{q},$$
where
$$V>N\geq 0 \text{ and } q\geq 1.$$

The proof of Theorem 10 is complete now. □

The following result exhibits another Hermite–Hadamard type inequality in terms of the second derivative of a function.

Theorem 11. *Under the same assumptions of Theorem 10, we have*
$$\begin{aligned}\left|\frac{\psi(u)+\psi(v)}{2}-\frac{\Gamma(\alpha+1)}{2(v-u)^{\alpha}}\left[J_{u^{+}}^{\alpha}\psi(v)+J_{v^{-}}^{\alpha}\psi(u)\right]\right| \leq & \frac{(v-u)^{2}}{2(\alpha+1)}\\ &\times\left(\frac{1}{\alpha s+1}-\beta(\alpha s+1,q(\alpha+s)+1)\right.\\ &\left.-\frac{1}{(\alpha+1)q+\alpha s+1}\right)^{\frac{1}{q}}\\ &\times\left(|\psi''(u)|^{q}+|\psi''(v)|^{q}\right)^{\frac{1}{q}}.\end{aligned}$$

Proof. By applying Lemma 2 and the Hölder's inequality, we obtain
$$\begin{aligned}&\left|\frac{\psi(u)+\psi(v)}{2}-\frac{\Gamma(\alpha+1)}{2(v-b)^{\alpha}}\left[J_{u^{+}}^{\alpha}\psi(v)+J_{v^{-}}^{\alpha}\psi(u)\right]\right|\\ \leq\;& \frac{(v-u)^{2}}{2(\alpha+1)}\left(\int_{0}^{1}1d\gamma\right)^{\frac{1}{p}}\left(\int_{0}^{1}(1-(1-\gamma)^{\alpha+1}-\gamma^{\alpha+1})^{q}|\psi''(\gamma u+(1-\gamma)v)|^{q}d\gamma\right)^{\frac{1}{q}}\\ \leq\;& \frac{(v-u)^{2}}{2(\alpha+1)}\left(|\psi''(u)|^{q}\int_{0}^{1}(\gamma^{\alpha s}-(1-\gamma)^{q(\alpha+s)}\gamma^{\alpha s}-\gamma^{q(\alpha+1)+\alpha s})d\gamma\right.\\ &\left.+|\psi''(v)|^{q}\int_{0}^{1}((1-\gamma)^{\alpha s}-(1-\gamma)^{q(\alpha+1)+\alpha s}-\gamma^{q(\alpha+1)}(1-\gamma)^{\alpha s})d\gamma\right)^{\frac{1}{q}}\\ \leq\;& \frac{(v-u)^{2}}{2(\alpha+1)}\times\left(\frac{1}{\alpha s+1}-\beta(\alpha s+1,1+q(\alpha+s))-\frac{1}{(\alpha+1)q+\alpha s+1}\right)^{\frac{1}{q}}\\ &\times\left(|\psi''(u)|^{q}+|\psi''(v)|^{q}\right)^{\frac{1}{q}}.\end{aligned}$$

This completes the proof of Theorem 11. □

Remark 4. *From Theorems 9, 10 and 11, we have*
$$\left|\frac{\psi(u)+\psi(v)}{2}-\frac{\Gamma(\alpha+1)}{2(v-u)^{\alpha}}\left[J_{u^{+}}^{\alpha}\psi(v)+J_{v^{-}}^{\alpha}\psi(u)\right]\right|\leq\min\{K_{1},K_{2},K_{3}\},$$

where

$$K_1 = \frac{(v-u)^2}{2(\alpha+1)}\left(\frac{\alpha}{\alpha+2}\right)^{1-\frac{1}{q}}\left(\frac{1}{\alpha s+1} - \beta(\alpha s+1,\alpha+2) - \frac{1}{\alpha s+\alpha+2}\right)^{\frac{1}{q}}\left(|\psi''(u)|^q + |\psi''(v)|^q\right)^{\frac{1}{q}},$$

$$K_2 = \frac{(v-u)^2}{2(\alpha+1)}\left(1 - \frac{2}{p(\alpha+1)+}\right)^{\frac{1}{p}}\left(\frac{|\psi''(u)|^q + |\psi''(v)|^q}{\alpha s+1}\right)^{\frac{1}{q}},$$

$$K_3 = \frac{(v-u)^2}{2(\alpha+1)}\left(\frac{1}{\alpha s+1} - \beta(\alpha s+1,q(\alpha+s)+1) - \frac{1}{q(\alpha+1)+\alpha s+1}\right)^{\frac{1}{q}}\left(|\psi''(u)|^q + |\psi''(v)|^q\right)^{\frac{1}{q}}.$$

3. Applications to Special Means

Using the obtained results, we examine some applications to special means of non-negative numbers u and v.

1. The arithmetic mean:
   $A = A(u,v) = \frac{u+v}{2}$; $u,v \in \mathbb{R}$, with $u,v > 0$.
2. The logarithmic mean:
   $L(u,v) = \frac{v-u}{\log v - \log u}$; $u,v \in \mathbb{R}$, with $u,v > 0$.
3. The generalized logarithmic mean:
   $L_r(u,v) = \left[\frac{v^{r+1} - u^{r+1}}{(v-u)(r+1)}\right]^{\frac{1}{r}}$; $r \in \mathbb{Z}\setminus\{-1,0\}$ $u,v \in \mathbb{R}$, with $u,v > 0$.

Using the results obtained in Section 2, and the above applications of means, we get the following proposition.

Proposition 1. *Suppose that* $r \in \mathbb{Z}$, $|r| \geq 2$ *and* $u,v \in \mathbb{R}$ *such that* $0 < u < v$. *Then we get the following inequality:*

$$\left|A(u^r,v^r) - L_r^r(u,v)\right| \leq \frac{(v-u)|r|}{2}A(|u|^{r-1},|v|^{r-1}).$$

Proof. This result follows Corollary 3 (ii) applied to the function $\psi(x) = x^r$. □

Proposition 2. *Suppose that* $n \in \mathbb{Z}$, $|r| \geq 2$ *and* $u,v \in \mathbb{R}$ *such that* $0 < u < v$. *Then for* $q \geq 1$, *we get the following:*

$$\left|A(u^r,v^r) - L_r^r(u,v)\right| \leq \frac{(v-u)|r|}{2}A^{\frac{1}{q}}(|u|^{q(r-1)},|v|^{q(r-1)}).$$

Proof. This result follows from Corollary 3 (iv) applied to the function $\psi(x) = x^r$. □

Proposition 3. *Suppose that* $u,v \in \mathbb{R}$ *such that* $0 < u < v$, *then*

$$\left|A(u^{-1},v^{-1}) - L(u,v)\right| \leq \frac{(v-u)|}{2}A(|u|^{-2},|v|^{-2}).$$

Proof. This result follows from Corollary 3 (ii) applied to the function $\psi(x) = x^{-1}$. □

Proposition 4. *Suppose that $u, v \in \mathbb{R}$ such that $0 < u < v$, then*

$$\left|A(u^{-1}, v^{-1}) - L(u, v)\right| \leq \frac{(v-u)|}{2} A^{\frac{1}{q}}(|u|^{-2q}, |v|^{-2q}).$$

Proof. This result follows from Corollary 3 (iv) applied to the function $\psi(x) = x^{-1}$. □

Author Contributions: O.A.: writing—original draft preparation, visualization, A.K.: writing–review and editing, supervision.

Funding: This research received no external funding.

Acknowledgments: The authors would like to thank to referees and editors for their very useful and constructive comments and remarks that improved the present manuscript substantially.

Conflicts of Interest: The authors declare no conflict of interest.

References

1. Yang, X.M. On E-convex sets, E-convex functions, and E-convex programming. *J. Optim. Theory Appl.* **2001**, *109*, 699–704. [CrossRef]
2. Hudzik, H.; Maligranda, L. Some remarks ons-convex functions. *Aequ. Math.* **1994**, *48*, 100–111. [CrossRef]
3. Almutairi, O.; Kılıçman, A. New refinements of the Hadamard inequality on coordinated convex function. *J. Inequal. Appl.* **2019**, *1*, 192. [CrossRef]
4. Dragomir, S.S.; Pearce, C. Selected topics on Hermite-Hadamard inequalities and applications. *Math. Prepr. Arch.* **2003**, *3*, 463–817.
5. Agarwal, P.; Jleli, M.; Tomar, M. Certain Hermite-Hadamard type inequalities via generalized k-fractional integrals. *J. Inequal. Appl.* **2017**, *1*, 55. [CrossRef] [PubMed]
6. Jhanthanam, S.; Tariboon, J.; Ntouyas, S.K.; Nonlaopon, K. On *q*-Hermite-Hadamard Inequalities for Differentiable Convex Functions. *Mathematics* **2019**, *7*, 632. [CrossRef]
7. Dragomir, S.S.; Fitzpatrick, S. The Hadamard inequalities for s-convex functions in the second sense. *Demonstr. Math.* **1999**, *32*, 687–696. [CrossRef]
8. Almutairi, O.; Kılıçman, A. New fractional inequalities of midpoint type via s-convexity and their application. *J. Inequal. Appl.* **2019**, *1*, 1–19. [CrossRef]
9. Chen, F. On Hermite-Hadamard type inequalities for Riemann-Liouville fractional integrals via two kinds of convexity. *Chin. J. Math.* **2014**, *92*, 2241–2253. [CrossRef]
10. Sarıkaya, M.Z. On the Hermite–Hadamard-type inequalities for co-ordinated convex function via fractional integrals. *Integr. Transform. Spec. Funct.* **2014**, *25*, 134–147. [CrossRef]
11. Qaisar, S.; Nasir, J.; Butt, S.I.; Hussain, S. On Some Fractional Integral Inequalities of Hermite-Hadamard's Type through Convexity. *Symmetry* **2019**, *11*, 137. [CrossRef]
12. Set, E.; Karaoglan, A. Hermite-Hadamard and Hermite-Hadamard-Fejer Type Inequalities for (*k*, *h*)-Convex Function via Katugampola Fractional Integrals. *Konuralp J. Math.* **2017**, *5*, 181–191.
13. Sun, W. On generalization of some inequalities for generalized harmonically convex functions via local fractional integrals. *Quaest. Math.* **2018**, 1–25. [CrossRef]
14. Abbas, G.; Farid, G. Some integral inequalities for *m*-convex functions via generalized fractional integral operator containing generalized Mittag-Leffler function. *Cogent Math. Stat.* **2016**, *3*, 1269589. [CrossRef]
15. Kılınc, S.; Akkurt, A.; Yildirim, H. Generalization Mittag-Leffer Function Associated with of the Hadamard and Fejer Hadamard Inequalities for (h-m)-Stronly convex Functions via Fractional Integrals. *J. Univ. Math.* **2019**, *2*, 8–15. [CrossRef]
16. Sarikaya, M.Z.; Set, E.; Yaldiz, H.; Başak, N. Hermite–Hadamard's inequalities for fractional integrals and related fractional inequalities. *Math. Comput. Model.* **2013**, *57*, 2403–2407. [CrossRef]
17. Mo, H.; Sui, X. Generalized-convex functions on fractal sets. *Abstr. Appl. Anal.* **2014**, *2014*, 254737. [CrossRef]
18. Kilbas, A.; Srivastava, H.M.; Trujillo, J.J. *Theory and Applications of Fractional Differential Equations*; Elsevier: Amsterdam, The Netherlands, 2014; Volume 204.
19. Wang, J.; Li, X.; Zhou, Y. Hermite-Hadamard Inequalities Involving Riemann-Liouville Fractional Integrals via s-convex Functions and Applications to Special Means. *Filomat* **2016**, *30*, 1143–1150. [CrossRef]

20. Özdemir, M.E.; Merve, A.A.; Kavurmaci-Önalan, H. Hermite-Hadamard type inequalities for s-convex and s-concave functions via fractional integrals. *Turk. J. Sci.* **2016**, *1*, 28–40.
21. Işcan, I. Generalization of different type integral inequalities for s-convex functions via fractional integrals. *Appl. Anal.* **2014**, *93*, 1846–1862. [CrossRef]

mathematics

MDPI

Article

On Neutral Functional Differential Inclusions involving Hadamard Fractional Derivatives

Bashir Ahmad [1,*], Ahmed Alsaedi [1], Sotiris K. Ntouyas [1,2] and Hamed H. Al-Sulami [1]

1 Nonlinear Analysis and Applied Mathematics (NAAM)-Research Group, Department of Mathematics, Faculty of Science, King Abdulaziz University, P.O. Box 80203, Jeddah 21589, Saudi Arabia; aalsaedi@kau.edu.sa (A.A.); sntouyas@uoi.gr (S.K.N.); hhaalsalmi@kau.edu.sa (H.H.A.-S.)

2 Department of Mathematics, University of Ioannina, 451 10 Ioannina, Greece

* Correspondence: bahmad@kau.edu.sa or bashirahmad_qau@yahoo.com

Received: 29 September 2019; Accepted: 6 November 2019; Published: 10 November 2019

Abstract: We prove the existence of solutions for neutral functional differential inclusions involving Hadamard fractional derivatives by applying several fixed point theorems for multivalued maps. We also construct examples for illustrating the obtained results.

Keywords: functional fractional differential inclusions; Hadamard fractional derivative; existence; fixed point

---

1. Introduction

Fractional calculus has emerged as an important area of investigation in view of the application of its tools in scientific and engineering disciplines. Examples include bio-medical sciences, ecology, finance, reaction-diffusion systems, wave propagation, electromagnetics, viscoelasticity, material sciences, and so forth. Fractional-order operators give rise to more informative and realistic mathematical models in contrast to their integer-order counterparts. It has been due to the non-local nature of fractional-order operators, which enables us to gain insight into the hereditary behavior (past history) of the associated phenomena. For examples and recent development of the topic, see References [1,2] and the references cited therein.

Differential inclusions—known as generalization of differential equations and inequalities—are found to be of great utility in the study of dynamical systems, stochastic processes, optimal control theory, and so forth. One can find a detailed account of the topic in Reference [3]. In recent years, an overwhelming interest in the subject of fractional-order differential equations and inclusions has been shown, for instance, see References [4–14] and the references cited therein.

In Reference [15], the authors obtained some existence results for sequential neutral differential equations involving Hadamard derivatives:

$$\begin{cases} \mathcal{D}^{\alpha}[\mathcal{D}^{\beta}y(t) - g(t,y_t)] = f(t,y(t)), \; t \in J := [1,b], \\ y(t) = \phi(t), \; t \in [1-r,1], \qquad \mathcal{D}^{\beta}y(1) = \eta \in \mathbb{R}, \end{cases} \tag{1}$$

where $\mathcal{D}^{\alpha}, \mathcal{D}^{\beta}$ are the Hadamard fractional derivatives of order $0 < \alpha, \beta < 1$, respectively and $f, g : J \times \mathbb{R} \to \mathbb{R}$ are continuous functions, $J \subseteq \mathbb{R}$ and $\phi \in C([1-r,1],\mathbb{R})$.

In this paper, we cover the multivalued case of problem (1) and investigate the Hadamard type neutral fractional differential inclusions given by

$$\begin{cases} \mathcal{D}^{\alpha}[\mathcal{D}^{\beta}y(t) - g(t,y_t)] \in F(t,y(t)), \; t \in J := [1,b], \\ y(t) = \phi(t), \; t \in [1-r,1], \qquad \mathcal{D}^{\beta}y(1) = \eta \in \mathbb{R}, \end{cases} \tag{2}$$

where $F : J \times \mathbb{R} \to \mathcal{P}(\mathbb{R})$ is a multivalued map, $\mathcal{P}(\mathbb{R})$ represents the family of all nonempty subsets of $\mathbb{R}$, and the other quantities in (2) are the same as taken in (1). Here y_t is an element of the Banach space $C_r := C([-r,0],\mathbb{R})$ equipped with norm $\|\phi\|_C := \sup\{|\phi(\theta)| : -r \le \theta \le 0\}$, and is defined by $y_t(\theta) = y(t+\theta),\ \theta \in [-r,0]$, where y is a function defined on $[1-r,b]$ and $t \in J$. The standard fixed point theorems for multivalued maps are applied to establish the existence results for the problem (2).

The remaining content of the paper is composed as follows. In Section 2, we describe the necessary background material needed for our work. Section 3 deals with the main theorems. In Section 4, we construct illustrative examples for the obtained results.

2. Preliminaries

Let us begin this section with some necessary definitions of fractional calculus [1].

Definition 1. *For a function $h : [1,\infty) \to \mathbb{R}$, the Hadamard derivative of fractional order χ is defined by*

$$D^{\chi}h(t) = \frac{1}{\Gamma(n-\chi)}\left(t\frac{d}{dt}\right)^n \int_1^t \left(\log\frac{t}{s}\right)^{n-\chi-1}\frac{h(s)}{s}ds,\ \ n = [\chi]+1,$$

where $[\chi]$ denotes the integer part of the real number χ and $\log(\cdot) = \log_e(\cdot)$.

Definition 2. *The Hadamard fractional integral of order χ for a function h is defined as*

$$I^{\chi}h(t) = \frac{1}{\Gamma(\chi)}\int_1^t \left(\log\frac{t}{s}\right)^{\chi-1}\frac{h(s)}{s}ds,\ \ \chi > 0,$$

provided the integral exists.

Now we state a known result [15], which plays a key role in the forthcoming analysis.

Lemma 1 (Lemma 2.3 in [15]). *The function y is a solution of the problem*

$$\begin{cases} D^{\alpha}[D^{\beta}y(t) - g(t,y_t)] = f(t,y_t),\ t \in J := [1,b], \\ y(t) = \phi(t),\ t \in [1-r,1], \\ D^{\beta}y(1) = \eta \in \mathbb{R}, \end{cases} \tag{3}$$

if and only if

$$y(t) = \begin{cases} \phi(t), & \text{if } t \in [1-r,1], \\ \Big\{\phi(1) + (\eta - g(1,\phi))\frac{(\log t)^{\beta}}{\Gamma(\beta+1)} + \frac{1}{\Gamma(\alpha)}\int_1^t \left(\log\frac{t}{s}\right)^{\alpha-1}\frac{g(s,y_s)}{s}ds & \\ +\frac{1}{\Gamma(\alpha+\beta)}\int_1^t \left(\log\frac{t}{s}\right)^{\alpha+\beta-1}\frac{f(s,y_s)}{s}ds\Big\}, & \text{if } t \in J. \end{cases} \tag{4}$$

3. Existence Results

For a normed space $(X, \|\cdot\|)$, we define $\mathcal{P}_{cl}(X) = \{Y \in \mathcal{P}(X) : Y \text{ is closed}\}$, $\mathcal{P}_{cp}(X) = \{Y \in \mathcal{P}(X) : Y \text{ is compact}\}$, $\mathcal{P}_{cl,b}(X) = \{Y \in \mathcal{P}(X) : Y \text{ is closed and bounded}\}$, $\mathcal{P}_{cp,c}(X) = \{Y \in \mathcal{P}(X) : Y \text{ is compact and convex}\}$ and $\mathcal{P}_{b,cl,c}(X) = \{Y \in \mathcal{P}(X) : Y \text{ is bounded, closed and convex}\}$. In passing, we remark that a closed and bounded set in a metric space is not necessarily compact in general; however, it is true that a set in a metric space of real or complex numbers is compact if and only if it is closed and bounded.

For each $y \in C(J, \mathbb{R})$, define the set of selections of F by

$$S_{F,y} := \{\xi \in L^1(J, \mathbb{R}) : \xi(t) \in F(t, y(t)) \text{ on } J\}.$$

Denote by $C(J, \mathbb{R})$ the Banach space of all continuous functions from J into $\mathbb{R}$ endowed with the norm $\|y\| := \sup\{|y(t)| : t \in J\}$. $L^1(J, \mathbb{R})$ represents the space of functions $y : J \to \mathbb{R}$ such that $\|y\|_{L^1} = \int_1^b |y(t)| dt$.

Our first existence result deals with the case when F has convex values and is based on nonlinear alternative for Kakutani maps [16] with the assumption that the multivalued map F is Carathéodory.

Definition 3 (Granas, Dugundji [16]). *A multivalued map $F : J \times \mathbb{R} \to \mathcal{P}(\mathbb{R})$ is said to be Carathéodory if*

(i) $t \longmapsto F(t, x)$ is measurable for each $x \in \mathbb{R}$;
(ii) $x \longmapsto F(t, x)$ is upper semicontinuous for almost all $t \in J$.

Further a Carathéodory function F is called L^1-Carathéodory if

(iii) for each $\rho > 0$, there exists $\varphi_\rho \in L^1(J, \mathbb{R}^+)$ such that

$$\|F(t, x)\| = \sup\{|v| : v \in F(t, x)\} \leq \varphi_\rho(t)$$

for all $x \in \mathbb{R}$ with $\|x\| \leq \rho$ and for almost everywhere $t \in J$.

Theorem 1. *Assume that:*

(H_0) there exists a non-negative constant $k < \Gamma(\alpha + 1)(\log b)^{-\alpha}$ such that

$$|g(t, u_1) - g(t, u_2)| \leq k\|u_1 - u_2\|_C, \quad \text{for } t \in J \text{ and every } u_1, u_2 \in C_r.$$

(H_1) $F : J \times \mathbb{R} \to \mathcal{P}_{cp,c}(\mathbb{R})$ is L^1-Carathéodory;
(H_2) there exists a continuous non-decreasing function $\Phi : [0, \infty) \to (0, \infty)$ and a function $p \in C(J, \mathbb{R}^+)$ such that

$$\|F(t, x)\|_{\mathcal{P}} := \sup\{|y| : y \in F(t, x)\} \leq p(t)\Phi(\|x\|) \text{ for each } (t, x) \in J \times \mathbb{R};$$

(H_3) there exists a constant $\omega > 0$ such that

$$\frac{\left(1 - \dfrac{k(\log b)^\alpha}{\Gamma(\alpha+1)}\right)\omega}{\|\phi\|_C + (|\eta| + k\|\phi\|_C + g_0)\dfrac{(\log b)^\beta}{\Gamma(\beta+1)} + \dfrac{g_0(\log b)^\alpha}{\Gamma(\alpha+1)} + \dfrac{\Phi(\omega)\|p\|}{\Gamma(\alpha+\beta+1)}(\log b)^{\alpha+\beta}} > 1,$$

where $g_0 = |g(1, 0)|$.

Then the problem (2) has at least one solution on $[1 - r, b]$.

Proof. Let us first transform the problem (2) into a fixed point problem by introducing an operator $\mathcal{V}: C([1-r,b],\mathbb{R}) \longrightarrow \mathcal{P}(C([1-r,b],\mathbb{R}))$ by

$$\mathcal{V}(y) = \left\{ \begin{array}{l} h \in C([1-r,b],\mathbb{R}): \\ h(t) = \left\{ \begin{array}{ll} \phi(t), & \text{if } t \in [1-r,1], \\ \Big\{\phi(1) + (\eta - g(1,\phi))\dfrac{(\log t)^{\beta}}{\Gamma(\beta+1)} & \\ +\dfrac{1}{\Gamma(\alpha)}\displaystyle\int_1^t \left(\log\frac{t}{s}\right)^{\alpha-1}\frac{g(s,y_s)}{s}ds & \\ +\dfrac{1}{\Gamma(\alpha+\beta)}\displaystyle\int_1^t \left(\log\frac{t}{s}\right)^{\alpha+\beta-1}\frac{\xi(s)}{s}ds\Big\}, & \text{if } t \in J, \end{array}\right. \end{array} \right\} \tag{5}$$

for $\xi \in S_{F,x}$. It is obvious by Lemma 1 that the fixed points of the operator $\mathcal{V}$ are solutions of the problem (2).

We verify the hypothesis of nonlinear alternative for Kakutani maps [16] in several steps.

Step 1. $\mathcal{V}(y)$ is convex for each $y \in C([1-r,b],\mathbb{R})$. It directly follows from the fact that $S_{F,y}$ is convex (F has convex values).

Step 2. $\mathcal{V}$ maps bounded sets (balls) into bounded sets in $C([1-r,b],\mathbb{R})$. Let $B_\zeta = \{y \in C([1-r,b],\mathbb{R}) : \|y\|_{[1-r,b]} \leq \zeta\}$ be a bounded set in $C([1-r,b],\mathbb{R})$. Then, for each $h \in \mathcal{B}(y), y \in B_\zeta$, there exists $\xi \in S_{F,y}$ such that

$$\begin{aligned} h(t) &= \phi(1) + (\eta - g(1,\phi))\frac{(\log t)^{\beta}}{\Gamma(\beta+1)} + \frac{1}{\Gamma(\alpha)}\int_1^t \left(\log\frac{t}{s}\right)^{\alpha-1}\frac{g(s,y_s)}{s}ds \\ &\quad + \frac{1}{\Gamma(\alpha+\beta)}\int_1^t \left(\log\frac{t}{s}\right)^{\alpha+\beta-1}\frac{\xi(s)}{s}ds. \end{aligned}$$

Then, for $t \in J$, we have

$$\begin{aligned} |h(t)| &\leq \|\phi\|_C + (|\eta| + k\|\phi\|_C + g_0)\frac{(\log b)^{\beta}}{\Gamma(\beta+1)} + \frac{k\|y\|_{[1-r,b]} + g_0}{\Gamma(\alpha+1)}(\log b)^{\alpha} \\ &\quad + \frac{\Phi(\|y\|_{[1-r,b]})\|p\|}{\Gamma(\alpha+\beta+1)}(\log b)^{\alpha+\beta}. \end{aligned}$$

Thus,

$$\|h\| \leq \|\phi\|_C + (|\eta| + k\|\phi\|_C + g_0)\frac{(\log b)^{\beta}}{\Gamma(\beta+1)} + \frac{k\zeta + g_0}{\Gamma(\alpha+1)}(\log b)^{\alpha} + \frac{\Phi(\zeta)\|p\|}{\Gamma(\alpha+\beta+1)}(\log b)^{\alpha+\beta}.$$

Step 3. $\mathcal{V}$ maps bounded sets into equicontinuous sets of $C([1-r,b],\mathbb{R})$.

Let $t_1, t_2 \in J$ with $t_1 < t_2$ and $y \in B_\zeta$. Then, for each $h \in \mathcal{B}(y)$, we obtain

$$\begin{aligned} |h(t_2) - h(t_1)| &\leq \frac{|\eta| + k\|\phi\|_C + g_0}{\Gamma(\beta+1)}\Big[(\log t_2)^{\beta} - (\log t_1)^{\beta}\Big] \\ &\quad + \frac{k\zeta + g_0}{\Gamma(\alpha+\beta)}\int_1^{t_1}\left|\left(\log\frac{t_2}{s}\right)^{\alpha+\beta-1} - \left(\log\frac{t_1}{s}\right)^{\alpha+\beta-1}\right|\frac{ds}{s} \\ &\quad + \frac{k\zeta + g_0}{\Gamma(\alpha+\beta)}\int_{t_1}^{t_2}\left(\log\frac{t_2}{s}\right)^{\alpha+\beta-1}\frac{ds}{s} \\ &\quad + \frac{\Phi(\zeta)\|p\|}{\Gamma(\alpha+\beta)}\int_1^{t_1}\left|\left(\log\frac{t_2}{s}\right)^{\alpha+\beta-1} - \left(\log\frac{t_1}{s}\right)^{\alpha+\beta-1}\right|\frac{ds}{s} \end{aligned}$$

$$
\begin{aligned}
& +\frac{\Phi(\zeta)\|p\|}{\Gamma(\alpha+\beta)}\int_{t_1}^{t_2}\left(\log\frac{t_2}{s}\right)^{\alpha+\beta-1}\frac{ds}{s}\\
\leq\ & \frac{|\eta|+k\|\phi\|_C+g_0}{\Gamma(\beta+1)}\left[(\log t_2)^{\beta}-(\log t_1)^{\beta}\right]\\
& +\left\{\frac{k\zeta+g_0}{\Gamma(\alpha+\beta)}+\frac{\Phi(\zeta)\|p\|}{\Gamma(\alpha+\beta+1)}\right\}\left[\left|(\log t_2)^{\alpha+\beta}-(\log t_1)^{\alpha+\beta}\right|+|\log t_2/t_1|^{\alpha+\beta}\right],
\end{aligned}
$$

which tends to zero as $t_2 - t_1 \to 0$ independently of $y \in B_\zeta$. For the cases $t_1 < t_2 \leq 0$ and $t_1 \leq 0 \leq t_2$, the equicontinuity can be established in a similar manner. Thus, by Arzelá-Ascoli theorem [17], we deduce that $\mathcal{V} : C([1-r,b],\mathbb{R}) \to \mathcal{P}(C([1-r,b],\mathbb{R}))$ is completely continuous.

Now we show that $\mathcal{V}$ has a closed graph. Then it will follow by the Proposition 1.2 in Reference [18] that $\mathcal{V}$ is upper semi-continuous, as it is already proved to be completely continuous.

Step 4. $\mathcal{V}$ has a closed graph. We need to show that $h_* \in \mathcal{V}(y_*)$ when $y_n \to x_*, h_n \in \mathcal{V}(y_n)$ and $h_n \to h_*$. Associated with $h_n \in \mathcal{V}(y_n)$, there exists $\xi_n \in S_{F,y_n}$ such that, for each $t \in J$,

$$
\begin{aligned}
h_n(t) &= \phi(1)+(\eta-g(1,\phi))\frac{(\log t)^{\beta}}{\Gamma(\beta+1)}+\frac{1}{\Gamma(\alpha)}\int_1^t\left(\log\frac{t}{s}\right)^{\alpha-1}\frac{g(s,y_s)}{s}ds\\
&\quad +\frac{1}{\Gamma(\alpha+\beta)}\int_1^t\left(\log\frac{t}{s}\right)^{\alpha+\beta-1}\frac{\xi_n(s)}{s}ds.
\end{aligned}
$$

Thus it suffices to show that there exists $\xi_* \in S_{F,y_*}$ such that, for each $t \in J$,

$$
\begin{aligned}
h_*(t) &= \phi(1)+(\eta-g(1,\phi))\frac{(\log t)^{\beta}}{\Gamma(\beta+1)}+\frac{1}{\Gamma(\alpha)}\int_1^t\left(\log\frac{t}{s}\right)^{\alpha-1}\frac{g(s,y_s)}{s}ds\\
&\quad +\frac{1}{\Gamma(\alpha+\beta)}\int_1^t\left(\log\frac{t}{s}\right)^{\alpha+\beta-1}\frac{\xi_*(s)}{s}ds.
\end{aligned}
$$

Let us introduce the linear operator $\Theta : L^1(J,\mathbb{R}) \to C(J,\mathbb{R})$ given by

$$
\begin{aligned}
\xi \mapsto \Theta(\xi)(t) &= \phi(1)+(\eta-g(1,\phi))\frac{(\log t)^{\beta}}{\Gamma(\beta+1)}+\frac{1}{\Gamma(\alpha)}\int_1^t\left(\log\frac{t}{s}\right)^{\alpha-1}\frac{g(s,y_s)}{s}ds\\
&\quad +\frac{1}{\Gamma(\alpha+\beta)}\int_1^t\left(\log\frac{t}{s}\right)^{\alpha+\beta-1}\frac{\xi(s)}{s}ds.
\end{aligned}
$$

Notice that $\|h_n(t) - h_*(t)\| \to 0$, as $n \to \infty$. Therefore, it follows from a result dealing with the closed graph operators derived in Reference [19] that $\Theta \circ S_F$ is a closed graph operator. Further, we have $h_n(t) \in \Theta(S_{F,y_n})$. Since $y_n \to y_*$, we have

$$
\begin{aligned}
h(t) &= \phi(1)+(\eta-g(1,\phi))\frac{(\log t)^{\beta}}{\Gamma(\beta+1)}+\frac{1}{\Gamma(\alpha)}\int_1^t\left(\log\frac{t}{s}\right)^{\alpha-1}\frac{g(s,y_s)}{s}ds\\
&\quad +\frac{1}{\Gamma(\alpha+\beta)}\int_1^t\left(\log\frac{t}{s}\right)^{\alpha+\beta-1}\frac{\xi_*(s)}{s}ds,
\end{aligned}
$$

for some $\xi_* \in S_{F,y_*}$.

Step 5. We can find an open set $U \subseteq C([1-r,b],\mathbb{R})$ with $y \notin \nu\mathcal{V}(y)$ for any $\nu \in (0,1)$ and all $y \in \partial U$.

Let $\nu \in (0,1)$ and $y \in \nu\mathcal{V}(y)$. Then there exists $\xi \in L^1(J,\mathbb{R})$ with $\xi \in S_{F,y}$ such that for $t \in J$,

$$
|y(t)| \leq \|\phi\|_C+(|\eta|+k\|\phi\|_C+g_0)\frac{(\log b)^{\beta}}{\Gamma(\beta+1)}+\frac{k\|y\|_{[1-r,b]}+g_0}{\Gamma(\alpha+1)}(\log b)^{\alpha}
$$

$$+\frac{\Phi(\|y\|_{[1-r,b]})\|p\|}{\Gamma(\alpha+\beta+1)}(\log b)^{\alpha+\beta},\ t\in J,$$

which implies that

$$\begin{aligned}\|y\|_{[1-r,b]}\left\{1-\frac{k(\log b)^{\alpha}}{\Gamma(\alpha+1)}\right\} &\leq \|\phi\|_C+(|\eta|+k\|\phi\|_C+g_0)\frac{(\log b)^{\beta}}{\Gamma(\beta+1)}\\ &\quad+\frac{g_0(\log b)^{\alpha}}{\Gamma(\alpha+1)}+\frac{\Phi(\|y\|_{[1-r,b]})\|p\|}{\Gamma(\alpha+\beta+1)}(\log b)^{\alpha+\beta}.\end{aligned}$$

Consequently

$$\frac{\left(1-\frac{k(\log b)^{\alpha}}{\Gamma(\alpha+1)}\right)\|y\|_{[1-r,b]}}{\|\phi\|_C+(|\eta|+k\|\phi\|_C+g_0)\frac{(\log b)^{\beta}}{\Gamma(\beta+1)}+\frac{g_0(\log b)^{\alpha}}{\Gamma(\alpha+1)}+\frac{\Phi(\|y\|_{[1-r,b]})\|p\|}{\Gamma(\alpha+\beta+1)}(\log b)^{\alpha+\beta}}\leq 1.$$

By (H_3), there exists a real number ω such that $\|y\|_{[1-r,b]}\neq\omega$. Let us consider an open set

$$U=\{y\in C([1-r,b],\mathbb{R})):\|y\|_{[1-r,b]}<\omega\},$$

with $\overline{U}=U\cup\partial U$. Notice that $\mathcal{V}:\overline{U}\to\mathcal{P}(C([1-r,b],\mathbb{R}))$ is compact and upper semi-continuous multivalued map with convex closed values. The choice of U implies that there does not exist any $y\in\partial U$ satisfying $y\in\nu\mathcal{V}(y)$ for some $\nu\in(0,1)$. In consequence, we deduce from the nonlinear alternative for Kakutani maps [16] that $\mathcal{V}$ has a fixed point $y\in\overline{U}$ which corresponds to a solution to the problem (2). This finishes the proof. □

In the following result, we make use of the nonlinear alternative for contractive maps ([20] Corollary 3.8) to show the existence of solutions for the problem (2).

Lemma 2. *(Nonlinear alternative [20]) Let D be a bounded neighborhood of $0\in X$, where X is a Banach space. Let $Z_1:X\to\mathcal{P}_{cp,c}(X)$ and $Z_2:\bar{D}\to\mathcal{P}_{cp,c}(X)$ be multivalued operators such that (a) Z_1 is contraction, and (b) Z_2 is upper semi-continuous and compact. Then, if $G=Z_1+Z_2$, either (i) G has a fixed point in $\overline{D}$ or (ii) there is a point $u\in\partial D$ and $\lambda\in(0,1)$ with $u\in\lambda G(u)$.*

Theorem 2. *If the conditions $(H_0)-(H_3)$ of Theorem 1 hold, then there exists at least one solution for the problem (2) on $[1-r,b]$.*

Proof. In order to verify the hypotheses of Lemma 2, we introduce the operator $\Psi_1:C([1-r,b],\mathbb{R})\longrightarrow C([1-r,b],\mathbb{R})$ by

$$\Psi_1y(t)=\begin{cases}0, & \text{if } t\in[1-r,1],\\ (\eta-g(1,\phi))\frac{(\log t)^{\beta}}{\Gamma(\beta+1)}+\frac{1}{\Gamma(\alpha)}\int_1^t\left(\log\frac{t}{s}\right)^{\alpha-1}\frac{g(s,y_s)}{s}ds, & \text{if } t\in J.\end{cases}\tag{6}$$

and the multivalued operator $\Psi_2:C([1-r,b],\mathbb{R})\longrightarrow\mathcal{P}(C([1-r,b],\mathbb{R}))$ by

$$\Psi_2y(t)=\left\{\begin{array}{l}h\in C([1-r,b],\mathbb{R}):\\ h(t)=\begin{cases}\phi(t), & \text{if } t\in[1-r,1],\\ \phi(1)+\frac{1}{\Gamma(\alpha+\beta)}\int_1^t\left(\log\frac{t}{s}\right)^{\alpha+\beta-1}\frac{\xi(s)}{s}ds, & \text{if } t\in J,\end{cases}\end{array}\right\}\tag{7}$$

for $\xi \in S_{F,y}$. Observe that $\mathcal{V} = \Psi_1 + \Psi_2$, where $\mathcal{V}$ is defined by (5). In the first step, it will be established that the operators Ψ_1 and Ψ_2 define the multivalued operators $\Psi_1, \Psi_2 : B_\theta \to \mathcal{P}_{cp,c}(C([1-r,b],\mathbb{R}))$, where $B_\theta = \{y \in C([1-r,b],\mathbb{R}) : \|y\|_{[1-r,b]} \leq \theta\}$ is a bounded set in $C([1-r,b],\mathbb{R})$. Let us show that Ψ_2 is compact-valued on B_θ. Observe that the operator Ψ_2 is equivalent to the composition $\mathcal{L} \circ S_F$, where $\mathcal{L}$ is the continuous linear operator on $L^1(J,\mathbb{R})$ into $C([1-r,b],\mathbb{R})$, defined by

$$\mathcal{L}(v)(t) = \phi(1) + \frac{1}{\Gamma(\alpha+\beta)} \int_1^t \left(\log\frac{t}{s}\right)^{\alpha+\beta-1} \frac{v(s)}{s} ds.$$

Let $y \in B_\theta$ be arbitrary and let $\{\xi_n\}$ be a sequence in $S_{F,y}$. Then it follows by the definition of $S_{F,y}$ that $\xi_n(t) \in F(t,y(t))$ for almost all $t \in J$. As $F(t,y(t))$ is compact for all $t \in J$, we have a convergent subsequence of $\{\xi_n(t)\}$ (we denote it by $\{\xi_n(t)\}$ again) that converges in measure to some $\xi(t) \in S_{F,y}$ for almost all $t \in J$. On the other hand, $\mathcal{L}$ is continuous, so $\mathcal{L}(\xi_n)(t) \to \mathcal{L}(\xi)(t)$ pointwise on J.

The convergence will be uniform once it is shown that $\{\mathcal{L}(\xi_n)\}$ is an equicontinuous sequence. For $t_1, t_2 \in J$ with $t_1 < t_2$, we have

$$\begin{aligned} |\mathcal{L}(\xi_n)(t_2) - \mathcal{L}(\xi_n)(t_1)| &\leq \frac{\Phi(\theta)\|p\|}{\Gamma(\alpha+\beta)} \int_1^{t_1} \left| \left(\log\frac{t_2}{s}\right)^{\alpha+\beta-1} - \left(\log\frac{t_1}{s}\right)^{\alpha+\beta-1} \right| \frac{ds}{s} \\ &\quad + \frac{\Phi(\theta)\|p\|}{\Gamma(\alpha+\beta)} \int_{t_1}^{t_2} \left(\log\frac{t_2}{s}\right)^{\alpha+\beta-1} \frac{ds}{s} \\ &\leq \frac{\Phi(\theta)\|p\|}{\Gamma(\alpha+\beta+1)} \left[\left| (\log t_2)^{\alpha+\beta} - (\log t_1)^{\alpha+\beta} \right| + |\log t_2/t_1|^{\alpha+\beta} \right] \to 0, \end{aligned}$$

as $t_2 \to t_1$, which shows that the sequence $\{\mathcal{L}(\xi_n)\}$ is equicontinuous. As a consequence of the Arzelá-Ascoli theorem, there exists a uniformly convergent subsequence of $\{\xi_n\}$ (we denote it again by $\{\xi_n\}$) such that $\mathcal{L}(\xi_n) \to \mathcal{L}(\xi)$. Noting that $\mathcal{L}(\xi) \in \mathcal{L}(S_{F,y})$, we deduce that $\mathcal{B}(y) = \mathcal{L}(S_{F,y})$ is compact for all $y \in B_\theta$. So $\Psi_2(y)$ is compact.

Now, we show that $\Psi_2(y)$ is convex for all $y \in C([1-r,b],\mathbb{R})$. Let $h_1, h_2 \in \Psi_2(y)$. We select $\xi_1, \xi_2 \in S_{F,y}$ such that

$$h_i(t) = \phi(1) + \frac{1}{\Gamma(\alpha+\beta)} \int_1^t \left(\log\frac{t}{s}\right)^{\alpha+\beta-1} \frac{\xi_i(s)}{s} ds, \ i = 1,2,$$

for almost all $t \in J$. Then

$$[\lambda h_1 + (1-\lambda)h_2](t) = \phi(1) + \frac{1}{\Gamma(\alpha+\beta)} \int_1^t \left(\log\frac{t}{s}\right)^{\alpha+\beta-1} \frac{[\lambda\xi_1(s) + (1-\lambda)\xi_2(s)]}{s} ds,$$

where $0 \leq \lambda \leq 1$. Since $S_{F,y}$ is convex (as F has convex values), $\lambda\xi_1(s) + (1-\lambda)\xi_2(s) \in S_{F,y}$. Thus $\lambda h_1 + (1-\lambda)h_2 \in \Psi_2(y)$, which shows that Ψ_2 is convex-valued.

On the other hand, it is easy to show that Ψ_1 is compact and convex-valued. Next we prove that Ψ_1 is a contraction on $C([1-r,b],\mathbb{R})$. For $y, z \in C([1-r,b],\mathbb{R})$, we have

$$\begin{aligned} |\Psi_1(y)(t) - \Psi_1(z)(t)| &\leq \frac{1}{\Gamma(\alpha)} \int_1^t \left(\log\frac{t}{s}\right)^{\alpha-1} \frac{|g(s,y_s) - g(s,z_s)|}{s} ds \\ &\leq \frac{k}{\Gamma(\alpha)} \int_1^t \left(\log\frac{t}{s}\right)^{\alpha-1} \frac{\|y_s - z_s\|_C}{s} ds \\ &\leq \frac{k(\log t)^\alpha}{\Gamma(\alpha+1)} \|y-z\|_{[1-r,b]}, \end{aligned}$$

which implies that $\|\Psi_1(y) - \Psi_1(z)\|_{[1-r,b]} \leq \frac{k(\log b)^\alpha}{\Gamma(\alpha+1)} \|y-z\|_{[1-r,b]}$. By the assumption (H_0), we conclude that Ψ_1 is a contraction.

As in the proof of Theorem 1, it can easily be shown that the operator Ψ_2 is compact and upper semi-continuous.

In view of the foregoing steps, we deduce that Ψ_1 and Ψ_2 satisfy the hypothesis of Lemma 2. So, from the conclusion of Lemma 2, either condition (i) or condition (ii) holds. We show that conclusion (ii) is not possible. If $y \in \lambda\Psi_1(y) + \lambda\Psi_2(y)$ for $\lambda \in (0,1)$, then there exist $\xi \in S_{F,y}$ such that

$$\begin{aligned} y(t) &= \lambda\Bigg(\phi(1) + (\eta - g(1,\phi))\frac{(\log t)^\beta}{\Gamma(\beta+1)} + \frac{1}{\Gamma(\alpha)}\int_1^t \left(\log\frac{t}{s}\right)^{\alpha-1}\frac{g(s,y_s)}{s}ds \\ &+\frac{1}{\Gamma(\alpha+\beta)}\int_1^t \left(\log\frac{t}{s}\right)^{\alpha+\beta-1}\frac{\xi(s)}{s}ds\Bigg), \quad t \in J. \end{aligned}$$

By our assumptions, we can obtain

$$\begin{aligned} |y(t)| &\le \|\phi\|_C + [|\eta| + k\|\phi\|_C + g_0]\frac{(\log b)^\beta}{\Gamma(\beta+1)} + \frac{k\|y\|_{[1-r,b]} + g_0}{\Gamma(\alpha)}\int_1^t\left(\log\frac{t}{s}\right)^{\alpha-1}\frac{ds}{s} \\ &+\frac{1}{\Gamma(\alpha+\beta)}\int_1^t\left(\log\frac{t}{s}\right)^{\alpha+\beta-1} p(s)\Phi(\|y_s\|_C)\frac{ds}{s} \\ &\le \|\phi\|_C + [|\eta| + k\|\phi\|_C + g_0]\frac{(\log b)^\beta}{\Gamma(\beta+1)} + \frac{k\|y\|_{[1-r,b]} + g_0}{\Gamma(\alpha+1)}(\log b)^\alpha \\ &+\frac{\|p\|\Phi(\|y\|_{[1-r,b]})}{\Gamma(\alpha+\beta+1)}(\log b)^{\alpha+\beta}. \end{aligned}$$

Thus

$$\frac{\left(1 - \frac{k(\log b)^\alpha}{\Gamma(\alpha+1)}\right)\|y\|_{[1-r,b]}}{\|\phi\|_C + [|\eta| + k\|\phi\|_C + g_0]\frac{(\log b)^\beta}{\Gamma(\beta+1)} + \frac{g_0(\log b)^\alpha}{\Gamma(\alpha+1)} + \frac{\Phi(\|y\|_{[1-r,b]})\|p\|}{\Gamma(\alpha+\beta+1)}(\log b)^{\alpha+\beta}} \le 1. \tag{8}$$

If condition (ii) of Lemma 2 is satisfied, then there exists $\lambda \in (0,1)$ and $y \in \partial B_\omega$ with $y = \lambda\mathcal{V}(y)$. Then, y is a solution of (2) with $\|y\|_{[1-r,b]} = \omega$. Now, by the inequality (8), we get

$$\frac{\left(1 - \frac{k(\log b)^\alpha}{\Gamma(\alpha+1)}\right)\omega}{\|\phi\|_C + [|\eta| + k\|\phi\|_C + g_0]\frac{(\log b)^\beta}{\Gamma(\beta+1)} + \frac{g_0(\log b)^\alpha}{\Gamma(\alpha+1)} + \frac{\Phi(\omega)\|p\|}{\Gamma(\alpha+\beta+1)}(\log b)^{\alpha+\beta}} \le 1,$$

which contradicts (H_3). Hence, $\mathcal{V}$ has a fixed point on $[1-r,b]$ by Lemma 2, which implies that the problem (2) has a solution. The proof is complete. □

Our next result deals with the non-convex valued map in the problem (2) and is based on Covitz and Nadler's fixed point theorem [21] (*If $N : X \to \mathcal{P}_{cl}(X)$ is a contraction, then $FixN \neq \emptyset$, where X is a metric space*).

For a metric space (X,d) induced from the normed space $(X;\|\cdot\|)$, it is argued in Reference [22] that $(\mathcal{P}_{cl,b}(X), H_d)$ is a metric space, where $H_d : \mathcal{P}(X)\times\mathcal{P}(X) \to \mathbb{R}\cup\{\infty\}$ is defined by $H_d(A,B) = \max\{\sup_{a\in A} d(a,B), \sup_{b\in B} d(A,b)\}$, $d(A,b) = \inf_{a\in A} d(a;b)$ and $d(a,B) = \inf_{b\in B} d(a;b)$.

Definition 4 (Granas, Dugundji [16]). *A multivalued operator $N : X \to \mathcal{P}_{cl}(X)$ is called*

(a) *$\gamma-$Lipschitz if and only if there exists $\gamma > 0$ such that*

$$H_d(N(x), N(y)) \le \gamma d(x,y) \text{ for each } x,y \in X;$$

(b) *a contraction if and only if it is $\gamma-$Lipschitz with $\gamma < 1$.*

Theorem 3. *Assume that* (H_0) *and the following conditions hold:*

(A_1) $F : J \times \mathbb{R} \to \mathcal{P}_{cp}(\mathbb{R})$ *is such that* $F(\cdot, y) : J \to \mathcal{P}_{cp}(\mathbb{R})$ *is measurable for each* $y \in \mathbb{R}$.

(A_2) $H_d(F(t,y), F(t,\bar{y})) \leq m(t)|y - \bar{y}|$ *for almost all* $t \in J$ *and* $y, \bar{y} \in \mathbb{R}$ *with* $m \in C(J, \mathbb{R}^+)$ *and* $d(0, F(t,0)) \leq m(t)$ *for almost all* $t \in J$.

Then there exists at least one solution for the problem (2) on J, *provided that*

$$\delta := \frac{k}{\Gamma(\alpha+1)}(\log b)^{\alpha} + \frac{\|m\|}{\Gamma(\alpha+\beta+1)}(\log b)^{\alpha+\beta} < 1. \tag{9}$$

Proof. Observe that the set $S_{F,y}$ is nonempty for each $y \in C(J, \mathbb{R})$ by the assumption (A_1). Therefore F has a measurable selection (see Theorem III.6 [23]). Next we consider the operator $\mathcal{V}$ given by (5) and verify that it satisfies the hypothesis of the Covitz and Nadler theorem [21]. We show that $\mathcal{V}(y) \in \mathcal{P}_{cl}(C(J, \mathbb{R}))$ for each $y \in C(J, \mathbb{R})$. Let $\{v_n\}_{n \geq 0} \in \mathcal{F}(y)$ be such that $v_n \to v$ $(n \to \infty)$ in $C(J, \mathbb{R})$. Then $v \in C(J, \mathbb{R})$ and we can find $\xi_n \in S_{F,y_n}$ such that, for each $t \in J$,

$$\begin{aligned} v_n(t) &= \phi(1) + (\eta - g(1,\phi))\frac{(\log t)^{\beta}}{\Gamma(\beta+1)} + \frac{1}{\Gamma(\alpha)}\int_1^t \left(\log\frac{t}{s}\right)^{\alpha-1}\frac{g(s,y_s)}{s}ds \\ &\quad + \frac{1}{\Gamma(\alpha+\beta)}\int_1^t \left(\log\frac{t}{s}\right)^{\alpha+\beta-1}\frac{\xi_n(s)}{s}ds. \end{aligned}$$

Since F has compact values, we pass onto a subsequence (if necessary) such that ξ_n converges to ξ in $L^1(J, \mathbb{R})$. So $\xi \in S_{F,y}$ and for each $t \in J$, we have

$$\begin{aligned} u_n(t) \to \xi(t) &= \phi(1) + (\eta - g(1,\phi))\frac{(\log t)^{\beta}}{\Gamma(\beta+1)} + \frac{1}{\Gamma(\alpha)}\int_1^t \left(\log\frac{t}{s}\right)^{\alpha-1}\frac{g(s,y_s)}{s}ds \\ &\quad + \frac{1}{\Gamma(\alpha+\beta)}\int_1^t \left(\log\frac{t}{s}\right)^{\alpha+\beta-1}\frac{\xi(s)}{s}ds. \end{aligned}$$

Hence, $v \in \mathcal{V}(y)$.

Next we prove that there exists $0 < \delta < 1$ (δ is defined by (9)) such that

$$H_d(\mathcal{V}(y), \mathcal{V}(\bar{y})) \leq \delta\|y - \bar{y}\| \ \text{ for each } \ y, \bar{y} \in C^2(J, \mathbb{R}).$$

Let $y, \bar{y} \in C^2(J, \mathbb{R})$ and $h_1 \in \mathcal{V}(y)$. Then there exists $\xi_1(t) \in F(t, y(t))$ such that, for each $t \in J$,

$$\begin{aligned} h_1(t) &= \phi(1) + (\eta - g(1,\phi))\frac{(\log t)^{\beta}}{\Gamma(\beta+1)} + \frac{1}{\Gamma(\alpha)}\int_1^t \left(\log\frac{t}{s}\right)^{\alpha-1}\frac{g(s,y_s)}{s}ds \\ &\quad + \frac{1}{\Gamma(\alpha+\beta)}\int_1^t \left(\log\frac{t}{s}\right)^{\alpha+\beta-1}\frac{\xi_1(s)}{s}ds. \end{aligned}$$

By (A_2), we have

$$H_d(F(t,y), F(t,\bar{y})) \leq m(t)|y(t) - \bar{y}(t)|.$$

So, there exists $v \in F(t, \bar{y}(t))$ such that

$$|\xi_1(t) - v(t)| \leq m(t)|y(t) - \bar{y}(t)|, \ \ t \in J.$$

Define $V : J \to \mathcal{P}(\mathbb{R})$ by

$$V(t) = \{v \in \mathbb{R} : |\xi_1(t) - v(t)| \leq m(t)|y(t) - \bar{y}(t)|\}.$$

By Proposition III.4 in Reference [23], it follows that the multivalued operator $V(t) \cap F(t, \bar{y}(t))$ is measurable. So we can find a measurable selection $\xi_2(t)$ for V. So $\xi_2(t) \in F(t, \bar{y}(t))$ and satisfying $|\xi_1(t) - \xi_2(t)| \leq m(t)|y(t) - \bar{y}(t)|$ for each $t \in J$.

For each $t \in J$, we define

$$\begin{aligned} h_2(t) &= \phi(1) + (\eta - g(1,\phi))\frac{(\log t)^\beta}{\Gamma(\beta+1)} + \frac{1}{\Gamma(\alpha)}\int_1^t \left(\log\frac{t}{s}\right)^{\alpha-1}\frac{g(s,\bar{y}_s)}{s}ds \\ &+ \frac{1}{\Gamma(\alpha+\beta)}\int_1^t \left(\log\frac{t}{s}\right)^{\alpha+\beta-1}\frac{\xi_2(s)}{s}ds. \end{aligned}$$

Thus,

$$\begin{aligned} |h_1(t) - h_2(t)| &\leq \frac{1}{\Gamma(\alpha)}\int_1^t \left(\log\frac{t}{s}\right)^{\alpha-1}\frac{|g(s,y_s) - g(s,\bar{y}_s)|}{s}ds \\ &+ \frac{1}{\Gamma(\alpha+\beta)}\int_1^t \left(\log\frac{t}{s}\right)^{\alpha+\beta-1}\frac{|\xi_1(s) - \xi_2(s)|}{s}ds \\ &\leq \frac{k\|y - \bar{y}\|_{[1-r,b]}}{\Gamma(\alpha+1)}(\log b)^\alpha + \frac{\|m\|}{\Gamma(\alpha+\beta+1)}(\log b)^{\alpha+\beta}\|y - \bar{y}\|_{[1-r,b]}. \end{aligned}$$

Hence

$$\|h_1 - h_2\| \leq \left\{\frac{k}{\Gamma(\alpha+1)}(\log b)^\alpha + \frac{\|m\|}{\Gamma(\alpha+\beta+1)}(\log b)^{\alpha+\beta}\right\}\|y - \bar{y}\|_{[1-r,b]}.$$

On the other hand, interchanging the roles of y and $\bar{y}$ leads to

$$H_d(\mathcal{F}(y), \mathcal{F}(\bar{y})) \leq \left\{\frac{k}{\Gamma(\alpha+1)}(\log b)^\alpha + \frac{\|m\|}{\Gamma(\alpha+\beta+1)}(\log b)^{\alpha+\beta}\right\}\|y - \bar{y}\|_{[1-r,b]}.$$

So $\mathcal{V}$ is a contraction. Therefore, from the conclusion of Covitz and Nadler theorem [21], the operator $\mathcal{V}$ has a fixed point y which is indeed a solution of the problem (2). This finishes the proof. □

Finally, we prove an existence result by applying the multivalued version of Krasnoselskii's fixed point theorem [24], which is stated below.

Lemma 3 (Krasnoselskii [24])**.** *Let X be a Banach space, $Y \in \mathcal{P}_{b,cl,c}(X)$ and $W_1, W_2 : Y \to \mathcal{P}_{cp,c}(X)$ be multivalued operators satisfying the conditions: (i) $W_1y + W_2y \subset Y$ for all $y \in Y$; (ii) W_1 is contraction; and (iii) W_2 is upper semicontinuous and compact. Then there exists $y \in Y$ such that $y \in W_1y + W_2y$.*

Theorem 4. *Suppose that (H_0), (H_1) and the following assumption are satisfied*

(B_1) *there exists a function $q \in C([1,b], \mathbb{R}^+)$ such that*

$$\|F(t,u)\|_{\mathcal{P}} := \sup\{|y| : y \in F(t,u)\} \leq q(t), \text{ for each } (t,u) \in [1,b] \times C_r.$$

Then there exists at least one solution for the problem (2) on $[1-r, b]$.

Proof. Let us consider the operators Ψ_1 and Ψ_2 defined by (6) and (7) respectively. As in Theorem 2, one can show that $\Psi_1, \Psi_2 : B_\theta \to \mathcal{P}_{cp,c}(C([1-r,b],\mathbb{R}))$ are indeed multivalued operators, where $B_\theta = \{y \in C([1-r,b],\mathbb{R}) : \|y\|_{[1-r,b]} \leq \theta\}$ is a bounded set in $C([1-r,b],\mathbb{R})$. Moreover, Ψ_1 is a contraction on $C([1-r,b],\mathbb{R})$ and Ψ_2 is upper semi-continuous and compact.

Next we show that $\Psi_1(y) + \Psi_2(y) \subset B_\theta$ for all $y \in B_\theta$. Let $y \in B_\theta$ and suppose that

$$\theta\left(1 - \frac{k(\log b)^\alpha}{\Gamma(\alpha+1)}\right) > \|\phi\|_C + \frac{[|\eta| + k\|\phi\|_C + g_0](\log b)^\beta}{\Gamma(\beta+1)} + \frac{g_0(\log b)^\alpha}{\Gamma(\alpha+1)} + \frac{\|q\|(\log b)^{\alpha+\beta}}{\Gamma(\alpha+\beta+1)}.$$

For $h \in \Psi_1, \Psi_2$ and $\xi \in S_{F,y}$, we have

$$\begin{aligned} h(t) &= \phi(1) + (\eta - g(1,\phi))\frac{(\log t)^\beta}{\Gamma(\beta+1)} + \frac{1}{\Gamma(\alpha)}\int_1^t \left(\log\frac{t}{s}\right)^{\alpha-1}\frac{g(s,y_s)}{s}ds \\ &\quad + \frac{1}{\Gamma(\alpha+\beta)}\int_1^t \left(\log\frac{t}{s}\right)^{\alpha+\beta-1}\frac{\xi(s)}{s}ds, \ t \in J. \end{aligned}$$

With the given assumptions, one can obtain

$$\begin{aligned} |h(t)| &\leq \|\phi\|_C + [|\eta| + k\|\phi\|_C + g_0]\frac{(\log b)^\beta}{\Gamma(\beta+1)} + \frac{k\|y\|_{[1-r,b]} + g_0}{\Gamma(\alpha+1)}(\log b)^\alpha \\ &\quad + \frac{\|q\|}{\Gamma(\alpha+\beta+1)}(\log b)^{\alpha+\beta}. \end{aligned}$$

Thus

$$\|h\| \leq \|\phi\|_C + [|\eta| + k\|\phi\|_C + g_0]\frac{(\log b)^\beta}{\Gamma(\beta+1)} + \frac{k\theta + g_0}{\Gamma(\alpha+1)}(\log b)^\alpha + \frac{\|q\|}{\Gamma(\alpha+\beta+1)}(\log b)^{\alpha+\beta} < \theta,$$

which means that $\Psi_1(y) + \Psi_2(y) \subset B_\theta$ for all $y \in B_\theta$.

Thus, the operators Ψ_1 and Ψ_2 satisfy the hypothesis of Lemma 3 and hence its conclusion implies that $y \in \mathcal{A}(y) + \mathcal{B}(y)$ in B_θ. Therefore the problem (2) has a solution in B_θ and the proof is finished. □

4. Examples

In this section, we demonstrate the application of our main results by considering the following Hadamard type neutral fractional differential inclusions:

$$D^{1/4}\left(D^{2/3}y(t) - g(t,y_t)\right) \in F(t,y_t), \ t \in J = [1,e], \tag{10}$$

$$y(t) = \phi(t), \ t \in [1/2,1], \ D^{2/3}y(1) = 1/4. \tag{11}$$

Here $\alpha = 1/4, \beta = 2/3, r = 1/2, b = e$,

$$F(t,y_t) = \left[\frac{\sqrt{3+\ln t}}{4}\sin(y_t), \frac{\sqrt{3}|y_t|^3}{8(1+|y_t|^3)}\sin(\pi t/2e) + \frac{1}{16}\right],$$

$$g(t,y_t) = \frac{1}{4+\ln t}\tan^{-1}(y_t) + \sin(\pi t/2), \ \phi(t) = \frac{1}{16\sqrt{\frac{3}{4}+t^2}}.$$

With the given data, it is easy to see that (H_0) is satisfied with $k < \Gamma(5/4)$, (H_2) is satisfied with $p(t) = \sqrt{3+\ln t}/4, \|p\| = 1/2, \Phi(\|u\|_C) = \|u\|_C$ and (H_3) holds true for $M > 7.05996548$ ($M_0 = 1.46447352, g_0 = 1$) with a particular choice of $k = 1/4$. Thus all the conditions of Theorem 1 hold true. Hence the problem (10) and (11) has at least one solution on $[1/2, e]$ by the conclusion of Theorem 1. In a similar manner, one can check that the hypotheses of Theorem 2 hold with $M > 1.71978641$ and consequently the conclusion of Theorem 2 applies to the problem (10) and (11).

In order to illustrate Theorem 3, let us take

$$F(t, y_t) = \left[0, \frac{\sqrt{15+(\ln t)^2}}{8} \frac{|y_t|}{(1+|y_t|)} + \frac{1}{4}\right] \tag{12}$$

in (10). Then $\|m\| = 1/2$ and from the condition (9), $\delta \approx 0.74950391 < 1$. Clearly the hypothesis of Theorem 3 is satisfied. Therefore, there exists at least one solution for the problem (10) and (11) with $F(t, y_t)$ given by (12) on $[1/2, e]$.

5. Conclusions

In this paper, we have derived several existence results for an initial value problem of neutral functional Hadamard-type fractional differential inclusions. In our first result (Theorem 1), we apply a nonlinear alternative for Kakutani multivalued maps to prove the existence of solutions for the problem at hand when the multivalued map F is assumed to be convex-valued. The nonlinear alternative for contractive maps is applied to prove the existence of solutions for the given problem in Theorem 2. In Theorem 3, we show the existence of solutions for the given problem involving non-convex valued maps with the aid of Covitz and Nadler's fixed point theorem. Our final existence result (Theorem 4) relies on the multivalued version of Krasnoselskii's fixed point theorem. In the nutshell, we have presented a comprehensive study of neutral functional Hadamard-type fractional differential inclusions by making use of different tools of fixed point theory for multivalued maps. In our future work, we plan to investigate the existence of solutions to an initial value problem for neutral functional fractional differential inclusions involving a combination of Caputo and Hadamard fractional derivatives.

Author Contributions: Conceptualization, B.A. and S.K.N.; methodology, A.A. and S.K.N.; validation, B.A., A.A., S.K.N. and H.H.A.-S.; formal analysis, A.A., B.A., and S.K.N.; writing—original draft preparation, S.K.N.; writing—review and editing, A.A., B.A. S.K.N. and H.H.A.-S.; project administration, B.A.; funding acquisition, B.A. and A.A.

Funding: This project was funded by the Deanship of Scientific Research (DSR), King Abdulaziz University, Jeddah, Saudi Arabia under grant no. (RG-39-130-38).

Acknowledgments: This project was funded by the Deanship of Scientific Research (DSR), King Abdulaziz University, Jeddah, under grant no. (RG-39-130-38). The authors, therefore, acknowledge with thanks DSR technical and financial support. The authors also thank the reviewers for their constructive remarks on our work.

Conflicts of Interest: The authors declare no conflict of interest.

References

1. Kilbas, A.A.; Srivastava, H.M.; Trujillo, J.J. *Theory and Applications of Fractional Differential Equations*; North-Holland Mathematics Studies, 204; Elsevier Science B.V.: Amsterdam, The Netherlands, 2006.
2. Sabatier, J.; Agrawal, O.P.; Machado, J.A.T. (Eds.) *Advances in Fractional Calculus: Theoretical Developments and Applications in Physics and Engineering*; Springer: Dordrecht, The Netherlands, 2007.
3. Kisielewicz, M. *Stochastic Differential Inclusions and Applications*; Springer Optimization and Its Applications, 80; Springer: New York, NY, USA, 2013.
4. Agarwal, R.P.; Baleanu, D.; Hedayati, V.; Rezapour, S. Two fractional derivative inclusion problems via integral boundary condition. *Appl. Math. Comput.* **2015**, *257*, 205–212. [CrossRef]
5. Ahmad, B.; Ntouyas, S.K.; Alsaedi, A. New results for boundary value problems of Hadamard-type fractional differential inclusions and integral boundary conditions. *Bound. Value Probl.* **2013**, *2013*, 275. [CrossRef]
6. Ahmad, B.; Ntouyas, S.K.; Alsaedi, A. Existence theorems for nonlocal multivalued Hadamard fractional integro-differential boundary value problems. *J. Inequal. Appl.* **2014**, *2014*, 454. [CrossRef]
7. Ahmad, B.; Ntouyas, S.K. Existence results for a coupled system of Caputo type sequential fractional differential equations with nonlocal integral boundary conditions. *Appl. Math. Comput.* **2015**, *266*, 615–622. [CrossRef]

8. Ahmad, B.; Ntouyas, S.K. Nonlocal fractional boundary value problems with slit-strips boundary conditions. *Fract. Calc. Appl. Anal.* **2015**, *18*, 261–280. [CrossRef]
9. Balochian, S.; Nazari, M. Stability of particular class of fractional differential inclusion systems with input delay. *Control Intell. Syst.* **2014**, *42*, 279–283. [CrossRef]
10. Liu, X.; Liu, Z.; Fu, X. Relaxation in nonconvex optimal control problems described by fractional differential equations. *J. Math. Anal. Appl.* **2014**, *409*, 446–458. [CrossRef]
11. Ntouyas, S.K.; Etemad, S.; Tariboon, J. Existence results for multi-term fractional differential inclusions. *Adv. Difference Equ.* **2015**, *2015*, 140. [CrossRef]
12. Sun, J.; Yin, Q. Robust fault-tolerant full-order and reduced-order observer synchronization for differential inclusion chaotic systems with unknown disturbances and parameters. *J. Vib. Control* **2015**, *21*, 2134–2148. [CrossRef]
13. Wang, X.; Schiavone, P. Harmonic three-phase circular inclusions in finite elasticity. *Contin. Mech. Thermodyn.* **2015**, *27*, 739–747. [CrossRef]
14. Yukunthorn, W.; Ahmad, B.; Ntouyas, S.K.; Tariboon, J. On Caputo-Hadamard type fractional impulsive hybrid systems with nonlinear fractional integral conditions. *Nonlinear Anal. Hybrid Syst.* **2016**, *19*, 77–92. [CrossRef]
15. Ahmad, B.; Ntouyas, S.K. Existence and uniqueness of solutions for Caputo-Hadamard sequential fractional order neutral functional differential equations. *Electron. J. Differ. Equ.* **2017**, *36*, 1–11.
16. Granas, A.; Dugundji, J. *Fixed Point Theory*; Springer: New York, NY, USA, 2005.
17. Rudin, W. *Principles of Mathematical Analysis*, 3rd ed.; McGraw Hill: Singapore, 1976.
18. Deimling, K. *Multivalued Differential Equations*; Walter De Gruyter: Berlin, Germany; New York, NY, USA, 1992.
19. Lasota, A.; Opial, Z. An application of the Kakutani-Ky Fan theorem in the theory of ordinary differential equations. *Bull. Acad. Polon. Sci. Ser. Sci. Math. Astronom. Phys.* **1965**, *13*, 781–786.
20. Petryshyn, W.V.; Fitzpatric, P.M. A degree theory, fixed point theorems, and mapping theorems for multivalued noncompact maps. *Trans. Amer. Math. Soc.* **1974**, *194*, 1–25. [CrossRef]
21. Covitz, H.; Nadler, S.B., Jr. Multivalued contraction mappings in generalized metric spaces. *Israel J. Math.* **1970**, *8*, 5–11. [CrossRef]
22. Kisielewicz, M. *Differential Inclusions and Optimal Control*; Kluwer: Dordrecht, The Netherlands, 1991.
23. Castaing, C.; Valadier, M. *Convex Analysis and Measurable Multifunctions*; Lecture Notes in Mathematics 580; Springer-Verlag: Berlin/Heidelberg, Germany; New York, NY, USA, 1977.
24. Petrusel, A. Fixed points and selections for multivalued operators. *Semin. Fixed Point Theory-Cluj-Napoca* **2001**, *2*, 3–22.

mathematics MDPI

Article

Integral Inequalities for s-Convexity via Generalized Fractional Integrals on Fractal Sets

Ohud Almutairi [1,†] and Adem Kılıçman [2,*,†]

[1] Department of Mathematics, University of Hafr Al-Batin, Hafr Al-Batin 31991, Saudi Arabia; AhoudbAlmutairi@gmail.com
[2] Department of Mathematics, Putra University of Malaysia, Serdang 43400, Malaysia
* Correspondence: akilic@upm.edu.my; Tel.: +60-3-89466813
† These authors contributed equally to this work.

Received: 20 November 2019; Accepted: 29 December 2019; Published: 1 January 2020

Abstract: In this study, we establish new integral inequalities of the Hermite–Hadamard type for s-convexity via the Katugampola fractional integral. This generalizes the Hadamard fractional integrals and Riemann–Liouville into a single form. We show that the new integral inequalities of Hermite–Hadamard type can be obtained via the Riemann–Liouville fractional integral. Finally, we give some applications to special means.

Keywords: Katugampola fractional integrals; s-convex function; Hermite–Hadamard inequality; fractal space

1. Introduction

Fractional calculus, whose applications can be found in many disciplines including economics, life and physical sciences, as well as engineering, can be considered as one of the modern branches of mathematics [1–4]. Many problems of interests from these fields can be analyzed through fractional integrals, which can also be regarded as an interesting sub-discipline of fractional calculus. Some of the applications of integral calculus can be seen in the following papers [5–10], through which problems in physics, chemistry, and population dynamics were studied. The fractional integrals were extended to include the Hermite–Hadamard inequality, which is classically given as follows.

Consider a convex function, $h : E \subseteq \mathbb{R} \to \mathbb{R}$, $w, z \in E$ if, and only if,

$$h\left(\frac{w+z}{2}\right) \le \frac{1}{z-w}\int_w^z h(x)dx \le \frac{h(w)+h(z)}{2}. \tag{1}$$

Following this, many important generalizations of Hermite–Hadamard inequality were studied [11–17], some of which were formulated via generalized s-convexity, which is defined as follows.

Definition 1. *Let $0 < s \le 1$. The function $h : [w,z] \subset \mathbb{R}_+ \to \mathbb{R}^\alpha$ is said to be generalized s-convex on fractal sets $\mathbb{R}^\alpha$ $(0 < \alpha < 1)$ in the second sense if*

$$h\left(tw + (1-t)z\right) \le (t)^{\alpha s} h(w) + (1-t)^{\alpha s} h(z).$$

This class of function is denoted by GK_s^2 (see Mo and Sui [18]).

Hermite–Hadamard-type inequalities have been extended to include fractional integrals. For example, Chen and Katugampola [19] generalized Equation (1) via generalized fractional integrals. Other important extensions of Equation (1) include the work of Mehran and Anwar [20], who studied the Hermite–Hadamard-type inequalities for s-convex functions involving generalized fractional integrals. The definitions of the generalized fractional integrals were given in [21], and we present them as follows.

Definition 2. *Suppose $[w,z] \subset \mathbb{R}$ is a finite interval. For order $\alpha > 0$, the two sides of Katugampola fractional integrals for $h \in X_c^p(w,z)$ are defined by*

$$^{\rho}I_{w+}^{\alpha}h(x) = \frac{\rho^{1-\alpha}}{\Gamma(\alpha)}\int_w^x (x^\rho - t^\rho)^{\alpha-1} t^{\rho-1}h(t)dt,$$

and

$$^{\rho}I_{z-}^{\alpha}h(x) = \frac{\rho^{1-\alpha}}{\Gamma(\alpha)}\int_x^z (t^\rho - x^\rho)^{\alpha-1} t^{\rho-1}h(t)dt,$$

where $w < x < z$, $\rho > 0$, and $X_c^p(w,z)(c \in \mathbb{R}, 1 \le p \le \infty)$ represents the space of complex-valued Lebesgue measurable functions h on $[w,z]$ for $\|h\|_{X_c^p < \infty}$. The norm is given as

$$\|h\|_{X_c^p} = \left(\int_w^z |t^c h(t)|^p \frac{dt}{t}\right)^{1/p} < \infty$$

for $1 \le p < \infty, c \in \mathbb{R}$. For the case $p = \infty$, we get

$$\|h\|_{X_c^\infty} = \operatorname{ess}\sup_{w \le t \le z} \left[t^c |h(t)|\right],$$

whereby ess sup is the essential supremum.

Even though Katugampola fractional integrals have been used to generalize many inequalities, such as Grüss [22,23], Hermite–Hadamard [24], and Lyapunov [25], this work generalizes Hermite–Hadamard inequality involving Katugampola on fractal sets.

When improving the results in Mehran and Anwar [20], we used Definition 2 together with the following lemma.

Lemma 1. [19] *Suppose that $h : [w^\rho, z^\rho] \subset \mathbb{R}_+ \to \mathbb{R}$ is a differentiable function on (w^ρ, z^ρ), where $0 \le w < z$ for $\alpha > 0$ and $\rho > 0$. If the fractional integrals exist, we get*

$$\frac{h(w^\rho) + h(z^\rho)}{2} - \frac{\alpha\rho^\alpha\Gamma(\alpha+1)}{2(z^\rho - w^\rho)^\alpha}\left[{}^{\rho}I_{w+}^{\alpha}h(z^\rho) + {}^{\rho}I_{z-}^{\alpha}h(w^\rho)\right] = \frac{z^\rho - w^\rho}{2}\int_0^1 \left[(1-t^\rho)^\alpha - (t^\rho)^\alpha\right] t^{\rho-1} h'(t^\rho w^\rho + (1-t^\rho)z^\rho)\, dt.$$

This paper is aimed at establishing some new integral inequalities for generalized s-convexity via Katugampola fractional integrals on fractal sets linked with Equation (1). We presented some inequalities for the class of mappings whose derivatives in absolute value are the generalized s-convexity. In addition, we obtained some new inequalities linked with convexity and generalized s-convexity via classical integrals as well as Riemann–Liouville fractional integrals in form of a corollary. As an application, the inequalities for special means are derived.

2. Main Results

Hermite–Hadamard inequality for s-convexity via generalized fractional integral can be written with the aid of the following theorem.

Theorem 1. *Let $h : [w^\rho, z^\rho] \subset \mathbb{R}_+ \to \mathbb{R}^\alpha$ be a positive function for $0 \leq w < z$ and $h \in X_c^p(w^\rho, z^\rho)$ for $\alpha > 0$ and $\rho > 0$. If h is a generalized s-convex function on $[w^\rho, z^\rho]$, then*

$$2^{\alpha(s-1)} h\left(\frac{w^\rho + z^\rho}{2}\right) \leq \frac{\rho^\alpha \Gamma(\alpha+1)}{2(z^\rho - w^\rho)^\alpha}\left[{}^\rho I^\alpha_{w^+} h(z^\rho) + {}^\rho I^\alpha_{z^-} h(w^\rho)\right] \leq \left[\frac{1}{\rho(1+s)} + \alpha\beta(\alpha, \alpha s+1)\right]\frac{h(w^\rho) + h(z^\rho)}{2}. \tag{2}$$

Proof. Since h is generalized s-convex function on $[w^\rho, z^\rho]$, for $t \in [0,1]$, we get

$$h(t^\rho w^\rho + (1-t^\rho) z^\rho) \leq (t^\rho)^{\alpha s} h(w^\rho) + (1-t^\rho)^{\alpha s} h(z^\rho),$$

and

$$h(t^\rho z^\rho + (1-t^\rho) w^\rho) \leq (t^\rho)^{\alpha s} h(z^\rho) + (1-t^\rho)^{\alpha s} h(w^\rho).$$

Combining the above inequalities, we have

$$h(t^\rho w^\rho + (1-t^\rho) z^\rho) + h(t^\rho z^\rho + (1-t^\rho) w^\rho) \leq \left((t^\rho)^{\alpha s} + (1-t^\rho)^{\alpha s}\right)[h(w^\rho) + h(z^\rho)]. \tag{3}$$

Multiplying both sides of Equation (3) by $t^{\alpha\rho-1}$, for $\alpha > 0$ and integrating it over $[0,1]$ with respect to t, we obtain

$$\frac{\rho^{\alpha-1}\Gamma(\alpha)}{(z^\rho - w^\rho)^\alpha}\left[{}^\rho I^\alpha_{w^+} h(z^\rho) + {}^\rho I^\alpha_{z^-} h(w^\rho)\right] \leq \int_0^1 t^{\alpha\rho-1}\left((t^\rho)^{\alpha s} + (1-t^\rho)^{\alpha s}\right)[h(w^\rho) + h(z^\rho)]\,dt. \tag{4}$$

Since

$$\int_0^1 t^{\alpha s\rho + \alpha\rho - 1} dt = \frac{1}{\alpha\rho(s+1)},$$

applying the change of variable $t^\rho = a$ gives the following

$$\int_0^1 t^{\alpha\rho-1}(1-t^\rho)^{\alpha s}\,dt = \frac{\beta(\alpha, \alpha s+1)}{\rho}.$$

Thus, Equation (4) becomes

$$\frac{\rho^\alpha \Gamma(\alpha+1)}{2(z^\rho - w^\rho)^\alpha}\left[{}^\rho I^\alpha_{w^+} h(z^\rho) + {}^\rho I^\alpha_{z^-} h(w^\rho)\right] \leq \left[\frac{1}{\rho(1+s)} + \alpha\beta(\alpha, \alpha s+1)\right]\frac{h(w^\rho) + h(z^\rho)}{2}.$$

In order to prove the first part of Equation (2), since h is generalized s-convex function on $[w^\rho, z^\rho]$, the following inequality is obtained:

$$h\left(\frac{x^\rho + y^\rho}{2}\right) \leq \frac{h(x^\rho) + h(y^\rho)}{2^{\alpha s}}, \tag{5}$$

for $x^\rho, y^\rho \in [w^\rho, z^\rho], \alpha \geq 0$.

Consider $x^\rho = t^\rho w^\rho + (1-t^\rho) z^\rho$ and $y^\rho = t^\rho z^\rho + (1-t^\rho) w^\rho$, where $t \in [0,1]$. Applying Equation (5), we have

$$2^{\alpha s} h\left(\frac{w^\rho + z^\rho}{2}\right) \leq h\left(t^\rho w^\rho + (1-t^\rho) z^\rho\right) + h\left(t^\rho z^\rho + (1-t^\rho) w^\rho\right). \tag{6}$$

Multiplying both sides of the Equation (6) by $t^{\alpha\rho-1}$, for $\alpha > 0$ and integrating over $[0,1]$ with respect to t gives the following:

$$\begin{aligned}\frac{2^s}{\alpha\rho} h\left(\frac{w^\rho + z^\rho}{2}\right) &\leq \int_0^1 t^{\alpha\rho-1} h\left(t^\rho w^\rho + (1-t^\rho) z^\rho\right) dt + \int_0^1 t^{\alpha\rho-1} h\left(t^\rho z^\rho + (1-t^\rho) w^\rho\right) dt \\ &= \int_z^w \left(\frac{z^\rho - x^\rho}{z^\rho - w^\rho}\right)^{\alpha-1} h\left(x^\rho\right) \frac{x^{\rho-1}}{w^\rho - z^\rho} dx \\ &+ \int_w^z \left(\frac{y^\rho - w^\rho}{z^\rho - w^\rho}\right)^{\alpha-1} h\left(y^\rho\right) \frac{y^{\rho-1}}{z^\rho - w^\rho} dy \\ &= \frac{\rho^{\alpha-1}\Gamma(\alpha)}{(z^\rho - w^\rho)^\alpha} \left[I^\alpha_{w^+} h\left(z^\rho\right) + ^\rho I^\alpha_{z^-} h\left(w^\rho\right)\right].\end{aligned} \tag{7}$$

Then, it follows that

$$2^{\alpha(s-1)} h\left(\frac{w^\rho + z^\rho}{2}\right) \leq \frac{\rho^\alpha \Gamma(\alpha+1)}{2\left(z^\rho - w^\rho\right)^\alpha} \left[{}^\rho I^\alpha_{w^+} h\left(z^\rho\right) + ^\rho I^\alpha_{z^-} h\left(w^\rho\right)\right],$$

where $\beta(w,z)$ is the Beta function. □

Remark 1. When substituting $\rho = 1$ and $\alpha = 1$ in Equation (2), we obtained the results reported by Dragomir and Fitzpatrick [11].

Example 1. *Consider a function $h : [w^\rho, z^\rho] \subset \mathbb{R}_+ \to \mathbb{R}^\alpha$, such that $h(x) = x^{s\alpha}$ belongs to GK^2_s, $s \in (0,1]$ with $h \in X^p_c(w^\rho, z^\rho)$, where $\alpha > 0$ and $\rho > 0$. Suppose $w = 0$ and $z = 1$. For $\alpha = 2, s = \frac{1}{2}$ and $\rho = 1$, the first, second, and third parts of Equation (2) give 0.25, 0.33 and 0.50, respectively. Thus, the Equation (2) holds. Similarly, when $\alpha = 1, s = \frac{1}{2}$ and $\rho = 2$, we get 0.35, 0.50 and 0.80, respectively, which satisfies Theorem 1.*

In the next theorem, the new upper bound for the right-hand side of Equation (1) for generalized s-convexity is proposed. Thus, the generalized beta function is defined as

$$\beta_\rho(w,z) = \int_0^1 \rho(1-x^\rho)^{b-1}(x^\rho)^{a-1}x^{\rho-1}dx.$$

Note that, as $\rho \to 1$, $\beta_\rho(w,z) \to \beta(w,z)$.

Theorem 2. *Let $\alpha > 0$ and $\rho > 0$. Let $h : [w^\rho, z^\rho] \subset \mathbb{R}_+ \to \mathbb{R}^\alpha$ be a differentiable function on (w^ρ, z^ρ), and $h' \in L^1[w,z]$ with $0 \leq w < z$. If $|h'|^q$ is generalized s-convex on $[w^\rho, z^\rho]$ for $q \geq 1$, we obtain*

$$\begin{aligned}\left|\frac{h(w^\rho)+h(z^\rho)}{2} - \frac{\alpha\rho^\alpha\Gamma(\alpha+1)}{2(z^\rho - w^\rho)^\alpha}\left[{}^\rho I^\alpha_{w^+} h\left(z^\rho\right) + ^\rho I^\alpha_{z^-} h\left(w^\rho\right)\right]\right| &\leq \frac{z^\rho - w^\rho}{2}\left(\frac{1}{(\alpha+1)\rho}\right)^{\frac{q-1}{q}} \\ &\times \left[\frac{\beta_\rho(\alpha s+1, \alpha+1)}{\rho} + \frac{1}{(\alpha\rho(s+1)+1)}\right]^{\frac{1}{q}} \\ &\times \left(|h'(w^\rho)|^q + |h'(z^\rho)|^q\right)^{\frac{1}{q}}.\end{aligned}$$

Proof. In view of Lemma 1, we have

$$\left|\frac{h(w^{\rho})+h(z^{\rho})}{2}-\frac{\alpha\rho^{\alpha}\Gamma(\alpha+1)}{2(z^{\rho}-w^{\rho})^{\alpha}}\left[{}^{\rho}I^{\alpha}_{w+}h\left(z^{\rho}\right)+{}^{\rho}I^{\alpha}_{z-}h\left(w^{\rho}\right)\right]\right| = \left|\frac{z^{\rho}-w^{\rho}}{2}\int_0^1\left[(1-t^{\rho})^{\alpha}-(t^{\rho})^{\alpha}\right]t^{\rho-1} \times h'\left(t^{\rho}w^{\rho}+(1-t^{\rho})z^{\rho}\right)dt\right|. \tag{8}$$

For the first case, when $q=1$, and $|h'|$ is generalized s-convex on $[w^{\rho},z^{\rho}]$, we have

$$h'\left(t^{\rho}w^{\rho}+(1-t^{\rho})z^{\rho}\right)\leq(t^{\rho})^{\alpha s}h'\left(w^{\rho}\right)+(1-t^{\rho})^{\alpha s}h'\left(z^{\rho}\right).$$

Therefore,

$$\begin{aligned}\left|\int_0^1\left[(1-t^{\rho})^{\alpha}-(t^{\rho})^{\alpha}\right]t^{\rho-1}h'\left(t^{\rho}w^{\rho}+(1-t^{\rho})z^{\rho}\right)dt\right| &\leq \int_0^1\left[(1-t^{\rho})^{\alpha}+(t^{\rho})^{\alpha}\right]t^{\rho-1}[(t^{\rho})^{\alpha s}|h'(w^{\rho}|) \\ &\quad+(1-t^{\rho})^{\alpha s}|h'(z^{\rho}|)]dt \\ &=|h'(w^{\rho})|\int_0^1\left[(t^{\rho-1}(t^{\rho})^{\alpha s})((1-t^{\rho})^{\alpha}+(t^{\rho})^{\alpha})\right]dt \\ &\quad+|h'(z^{\rho})|\int_0^1\left[(t^{\rho-1}(1-t^{\rho})^{\alpha s})((1-t^{\rho})^{\alpha}+(t^{\rho})^{\alpha})\right]dt \\ &=S_1+S_2.\end{aligned} \tag{9}$$

Calculating S_1 and S_2, we get

$$\begin{aligned}S_1 &= |h'(w^{\rho}|)\left[\int_0^1(1-t^{\rho})^{\alpha}t^{\rho-1}(t^{\rho})^{\alpha s}dt+\int_0^1(t^{\rho})^{\alpha(s+1)}t^{\rho-1}dt\right] \\ &= |h'\left(w^{\rho}\right)|\left[\frac{\beta_{\rho}(\alpha s+1,\alpha+1)}{\rho}+\frac{1}{\rho(\alpha s+\alpha+1)}\right],\end{aligned} \tag{10}$$

and

$$\begin{aligned}S_2 &= |h'(z^{\rho}|)\left[\int_0^1(1-t^{\rho})^{\alpha(s+1)}t^{\rho-1}dt+\int_0^1(t^{\rho})^{\alpha}t^{\rho-1}(1-t^{\rho})^{\alpha s}dt\right] \\ &= |h'(z^{\rho})|\left[\frac{1}{\rho(\alpha s+\alpha+1)}+\frac{\beta_{\rho}(\alpha+1,\alpha s+1)}{\rho}\right].\end{aligned} \tag{11}$$

Thus, if we use Equations (10) and (11) in (9), we obtain

$$\begin{aligned}\left|\int_0^1\left[(1-t^{\rho})^{\alpha}-(t^{\rho})^{\alpha}\right]t^{\rho-1}h'\left(t^{\rho}w^{\rho}+(1-t^{\rho})z^{\rho}\right)dt\right| &\leq |h'\left(w^{\rho}\right)|\left[\frac{\beta_{\rho}(\alpha s+1,\alpha+1)}{\rho}+\frac{1}{\rho(\alpha s+\alpha+1)}\right] \\ &\quad+|h'(z^{\rho})|\left[\frac{1}{\rho(\alpha s+\alpha+1)}+\frac{\beta_{\rho}(\alpha+1,\alpha s+1)}{\rho}\right].\end{aligned} \tag{12}$$

Obtaining Equations (8) and (12) completes the proof for this case. Consider the second case, $q > 1$. Using Equation (8) and the power mean inequality, we obtain

$$\begin{aligned}
\left|\int_0^1 \left[(1-t^\rho)^\alpha - (t^\rho)^\alpha\right] t^{\rho-1} h'\left(t^\rho w^\rho + (1-t^\rho) z^\rho\right) dt\right| &\leq \left(\int_0^1 |(1-t^\rho)^\alpha - (t^\rho)^\alpha| t^{\rho-1} dt\right)^{1-\frac{1}{q}} \\
&\times \left(\int_0^1 |(1-t^\rho)^\alpha - (t^\rho)^\alpha| t^{\rho-1} |h'(t^\rho w^\rho + (1-t^\rho) z^\rho)|^q dt\right)^{\frac{1}{q}} \\
&\leq \int_0^1 \left([(1-t^\rho)^\alpha + (t^\rho)^\alpha] t^{\rho-1} dt\right)^{1-\frac{1}{q}} \\
&\times \left(\int_0^1 [(1-t^\rho)^\alpha + (t^\rho)^\alpha] t^{\rho-1} [(t^\rho)^{\alpha s} |h'(w^\rho)|^q \right. \\
&\left. +(1-t^\rho)^{\alpha s} |h'(z^\rho)|^q] dt\right)^{\frac{1}{q}} \\
&= \left(\frac{1}{\rho(\alpha+1)}\right)^{\frac{q-1}{q}} \\
&\times \left(\left(\frac{\beta_\rho(\alpha s+1, \alpha+1)}{\rho} + \frac{1}{\rho(\alpha s+\alpha+1)}\right) |h'(w^\rho)|^q \right. \\
&\left. + \frac{1}{\rho(\alpha s+\alpha+1)} + \frac{\beta_\rho(\alpha+1, \alpha s+1))}{\rho} |h'(z^\rho)|^q\right)^{\frac{1}{q}}.
\end{aligned} \tag{13}$$

The Equations (8) and (13) complete the proof. □

Corollary 1. *Using the similar assumptions given in Theorem 2.*

1. *If $\rho = 1$, we get*

$$\begin{aligned}
\left|\frac{h(w)+h(z)}{2} - \frac{\alpha\Gamma(\alpha+1)}{2(z-w)^\alpha}[I^\alpha_{w^+} h(z) + I^\alpha_{z^-} h(w)]\right| &\leq \frac{z-w}{2}\left(\frac{1}{\alpha+1}\right)^{\frac{q-1}{q}} \\
&\times \left[\beta(\alpha s+1, \alpha+1) + \frac{1}{1+\alpha(s+1)}\right]^{\frac{1}{q}} \\
&\times (|h'(w)| + |h'(z)|).
\end{aligned}$$

2. *If $\rho = 1$ and $s = 1$, then*

$$\begin{aligned}
\left|\frac{h(w)+h(z)}{2} - \frac{\alpha\Gamma(\alpha+1)}{2(z-w)^\alpha}[I^\alpha_{w^+} h(z) + I^\alpha_{z^-} h(w)]\right| &\leq \frac{z-w}{2}\left(\frac{1}{1+\alpha}\right)^{\frac{q-1}{q}} \\
&\times \left(\beta(\alpha+1, \alpha+1) + \frac{1}{1+2\alpha}\right)^{\frac{1}{q}} \\
&\times (|h'(w)|^q + |h'(z)|^q).
\end{aligned}$$

3. *If $\rho = 1$, $s = 1$ and $\alpha = 1$, we obtain*

$$\left|\frac{h(w)+h(z)}{2} - \frac{1}{z-w}\int_w^z h(x)dx\right| \leq \frac{z-w}{2}\left(\frac{1}{2}\right)^{\frac{q-1}{q}}\left(\frac{h'(w)|^q + h'(z)|^q}{2}\right)^{\frac{1}{q}}.$$

Theorem 3. *With the similar assumptions stated in Theorem 2, we get the following inequality:*

$$\begin{aligned}\left|\frac{h(w^\rho)+h(z^\rho)}{2}-\frac{\alpha\rho^\alpha\Gamma(\alpha+1)}{2(z^\rho-w^\rho)^\alpha}\left[{}^\rho I^\alpha_{w^+}h\left(z^\rho\right)+{}^\rho I^\alpha_{z^-}h\left(w^\rho\right)\right]\right| &\le \left(\frac{1}{\rho}\right)^{1-\frac{1}{q}}\frac{z^\rho-w^\rho}{2}\\ &\times\left[\frac{\beta_\rho(\alpha s+1,\alpha+1)}{\rho}+\frac{1}{(\alpha(s+1)+1)\rho}\right]^{\frac{1}{q}}\\ &\times\left(|h'(w^\rho)|^q+|h'(z^\rho)|^q\right)^{\frac{1}{q}}.\end{aligned} \tag{14}$$

Proof. Using the fact $|h'|^q$, a generalized s-convex on $[w^\rho, z^\rho]$ with $q \ge 1$, we get

$$h'\left(t^\rho w^\rho+\left(1-t^\rho\right)z^\rho\right)\le\left(t^\rho\right)^{\alpha s}h'\left(w^\rho\right)+\left(1-t^\rho\right)^{\alpha s}h'\left(z^\rho\right).$$

Applying Equation (8) together with the power mean inequality, we get

$$\begin{aligned}&\left|\int_0^1\left[\left(1-t^\rho\right)^\alpha-\left(t^\rho\right)^\alpha\right]t^{\rho-1}h'\left(t^\rho w^\rho+\left(1-t^\rho\right)z^\rho\right)dt\right| \le\left(\int_0^1 t^{\rho-1}dt\right)^{1-\frac{1}{q}}\\ &\times\left(\int_0^1|(1-t^\rho)^\alpha-(t^\rho)^\alpha|t^{\rho-1}|h'(t^\rho w^\rho+(1-t^\rho)z^\rho)|^q dt\right)^{\frac{1}{q}}\\ &\le\left(\frac{1}{\rho}\right)^{1-\frac{1}{q}}\left(\int_0^1[(1-t^\rho)^\alpha+(t^\rho)^\alpha]t^{\rho-1}[(t^\rho)^{\alpha s}|h'(w^\rho)|^q\right.\\ &\left.+(1-t^\rho)^{\alpha s}|h'(z^\rho)|^q]dt\right)^{\frac{1}{q}}\\ &\le\left(\frac{1}{\rho}\right)^{1-\frac{1}{q}}\left(|h'(w^\rho)|^q\int_0^1[(1-t^\rho)^\alpha(t^\rho)^{\alpha s}t^{\rho-1}+(t^\rho)^\alpha(t^\rho)^{\alpha s}t^{\rho-1}]dt\right.\\ &\left.+|h'(z^\rho)|^q\int_0^1[(1-t^\rho)^\alpha t^{\rho-1}(1-t^\rho)^{\alpha s}+(t^\rho)^\alpha(1-t^\rho)^{\alpha s}t^{\rho-1}]dt\right)^{\frac{1}{q}}\\ &=\left(\frac{1}{\rho}\right)^{1-\frac{1}{q}}\\ &\times\left(|h'(w^\rho)|^q\left[\frac{\beta_\rho(\alpha s+1,\alpha+1)}{\rho}+\frac{1}{\rho(\alpha s+\alpha+1)}\right]\right.\\ &\left.+|h'(z^\rho)|^q\left[\frac{\beta_\rho(\alpha+1,\alpha s+1)}{\rho}+\frac{1}{\rho(\alpha s+\alpha+1)}\right]\right)^{\frac{1}{q}}.\end{aligned}$$

□

Remark 2. *Choosing $\rho = 1$ in Theorem 3, we get the following*

$$\begin{aligned}\left|\frac{h(w)+h(z)}{2}-\frac{\alpha\Gamma(\alpha+1)}{2(z-w)^\alpha}[I^\alpha_{w^+}h(z)+I^\alpha_{z^-}h(w)]\right| &\le\frac{z-w}{2}\\ &\times\left[\beta(\alpha s+1,\alpha+1)+\frac{1}{\alpha(s+1)+1}\right]^{\frac{1}{q}}\\ &\times(|h'(w)|+|h'(z)|).\end{aligned}$$

Remark 3. *When choosing $\rho = 1$ and $s = \frac{1}{2}$ in Theorem 3, we get*

$$\begin{aligned}\left|\frac{h(w)+h(z)}{2}-\frac{\alpha\Gamma(\alpha+1)}{2(z-w)^\alpha}[I^\alpha_{w^+}h(z)+I^\alpha_{z^-}h(w)]\right| &\le\frac{z-w}{2}\left(\beta\left(\frac{\alpha}{2}+1,\alpha+1\right)+\frac{1}{\frac{3}{2}\alpha+1}\right)^{\frac{1}{q}}\\ &\times(|h'(w)|^q+|h'(z)|^q).\end{aligned}$$

Corollary 2. *Choosing* $\rho = 1$, $s = 1$ *and* $\alpha = 1$*in* Theorem 3, we obtain

$$\left|\frac{h(w)+h(z)}{2} - \frac{1}{z-w}\int_w^z h(x)dx\right| \le \frac{z-w}{2}\left(\frac{|h'(w)|^q + |h'(z)|^q}{2}\right)^{\frac{1}{q}}.$$

The other type is given by the next theorem.

Theorem 4. *Let* $\alpha > 0$ *and* $\rho > 0$. *Let* $h : [w^\rho, z^\rho] \subset \mathbb{R}_+ \to \mathbb{R}^\alpha$ *be a differentiable function on* (w^ρ, z^ρ), *where* $h' \in L^1[w,z]$ *with* $0 \le w < z$. *For* $q > 1$, *if* $|h'|^q$ *is generalized s-convex on* $[w^\rho, z^\rho]$, *we get*

$$\begin{aligned}\left|\frac{h(w^\rho)+h(z^\rho)}{2} - \frac{\alpha\rho^\alpha\Gamma(\alpha+1)}{2(z^\rho-w^\rho)^\alpha}\left[{}^\rho I^\alpha_{w+} h(z^\rho) + {}^\rho I^\alpha_{z-} h(w^\rho)\right]\right| &\le \frac{z^\rho - w^\rho}{2}\left(\frac{1}{p(\rho-1)+1}\right)^{\frac{1}{p}} \\ &\times \left[\frac{\beta_\rho(\alpha s+1,\alpha+1)}{\rho} + \frac{1}{\rho(\alpha s+\alpha+1)}\right]^{\frac{1}{q}} \\ &\times \left(|h'(w^\rho)|^q + |h'(z^\rho)|^q\right)^{\frac{1}{q}},\end{aligned}$$

with $\frac{1}{p} + \frac{1}{q} = 1$.

Proof. Using the Hölder's inequality, we obtain the following:

$$\begin{aligned}\left|\int_0^1 \left[(1-t^\rho)^\alpha - (t^\rho)^\alpha\right] t^{\rho-1} h'(t^\rho w^\rho + (1-t^\rho)z^\rho)\,dt\right| &\le \left(\int_0^1 (t^{\rho-1})^p dt\right)^{\frac{1}{p}} \\ &\times \left(\int_0^1 [(1-t^\rho)^\alpha + (t^\rho)^\alpha] t^{\rho-1} |h'(t^\rho w^\rho + (1-t^\rho)z^\rho)|^q dt\right)^{\frac{1}{q}}.\end{aligned}$$

The fact $|h'|$ is generalized s-convex, and it can be used to obtain the following:

$$\begin{aligned}\left|\int_0^1 \left[(1-t^\rho)^\alpha - (t^\rho)^\alpha\right] t^{\rho-1} h'(t^\rho w^\rho + (1-t^\rho)z^\rho)\,dt\right| &\le \left(\frac{1}{p(\rho-1)+1}\right)^{\frac{1}{p}} \\ &\times \left(\int_0^1 [(1-t^\rho)^\alpha + (t^\rho)^\alpha] t^{\rho-1} [(t^\rho)^{\alpha s} |h'(w^\rho)|^q\right. \\ &\left. + (1-t^\rho)^{\alpha s} |h'(z^\rho)|^q] dt\right)^{\frac{1}{q}} \\ &\le \left(\frac{1}{p(\rho-1)+1}\right)^{\frac{1}{p}} \\ &\times \left(|h'(w^\rho)|^q \int_0^1 [t^{\rho-1}(t^\rho)^{\alpha s}(1-t^\rho)^\alpha + t^{\rho-1}(t^\rho)^\alpha (t^\rho)^{\alpha s}] dt\right. \\ &+ |h'(z^\rho)|^q \int_0^1 [t^{\rho-1}(1-t^\rho)^\alpha (1-t^\rho)^{\alpha s} \\ &\left. + t^{\rho-1}(t^\rho)^\alpha (1-t^\rho)^{\alpha s}] dt\right)^{\frac{1}{q}} \\ &= \left(\frac{1}{p(\rho-1)+1}\right)^{\frac{1}{p}} \\ &\times \left(|h'(w^\rho)|^q \left[\frac{\beta_\rho(\alpha s+1,\alpha+1)}{\rho} + \frac{1}{(\alpha(s+1)+1)\rho}\right]\right. \\ &\left. + |h'(z^\rho)|^q \left[\frac{1}{\rho(\alpha(s+1)+1)} + \frac{\beta_\rho(\alpha+1,\alpha s+1)}{\rho}\right]\right)^{\frac{1}{q}}.\end{aligned}$$

□

Corollary 3. From Theorems 2–4, for $q > 1$, we obtain the following inequality:

$$\left|\frac{h(w^\rho)+h(z^\rho)}{2} - \frac{\alpha\rho^\alpha\Gamma(\alpha+1)}{2(z^\rho-w^\rho)^\alpha}\left[{}^\rho I^\alpha_{w+}h(z^\rho) + {}^\rho I^\alpha_{z-}h(w^\rho)\right]\right| \leq \min(M_1, M_2, M_3)\frac{(z^\rho-w^\rho)}{2},$$

where

$$M_1 = \left(\frac{1}{\rho(\alpha+1)}\right)^{\frac{q-1}{q}}\left[\frac{\beta_\rho(\alpha s+1,\alpha+1)}{\rho} + \frac{1}{((s+1)\alpha+1)\rho}\right]^{\frac{1}{q}}(|h'(w^\rho)|^q + |h'(z^\rho)|^q)^{\frac{1}{q}},$$

$$M_2 = \left(\frac{1}{\rho}\right)^{\frac{q-1}{q}}\left[\frac{\beta_\rho(\alpha s+1,\alpha+1)}{\rho} + \frac{1}{\rho(\alpha(s+1)+1)}\right]^{\frac{1}{q}}(|h'(w^\rho)|^q + |h'(z^\rho)|^q)^{\frac{1}{q}},$$

and

$$M_3 = \left(\frac{1}{1+(\rho-1)p}\right)^{\frac{1}{p}}\left[\frac{\beta_\rho(\alpha s+1,\alpha+1)}{\rho} + \frac{1}{(\alpha(s+1)+1)\rho}\right]^{\frac{1}{q}}(|h'(w^\rho)|^q + |h'(z^\rho)|^q)^{\frac{1}{q}}.$$

3. Applications to Special Means

The applications to special means for positive real numbers w and z can be studied through the results obtained.

1. The arithmetic mean:
   $A = A(w,z) = \frac{w+z}{2}$.
2. The logarithmic mean:
   $L(w,z) = \frac{z-w}{\log z - \log w}$.
3. The generalized logarithmic mean:
   $L_i(w,z) = \left[\frac{z^{i+1}-w^{i+1}}{(z-w)(i+1)}\right]^{\frac{1}{i}}; i \in \mathbb{Z}\setminus\{-1,0\}$.

Applying the results in Section 2, together with the applications of means, gives the following propositions.

Proposition 1. *Let $i \in \mathbb{Z}$, $|i| \geq 2$ and $w, z \in \mathbb{R}$ where $0 < w < z$. For $q \geq 1$, we obtain the following:*

$$\left|A(w^i, z^i) - L_i^i(w,z)\right| \leq \frac{(z-w)|i|}{2^{\frac{q-1}{q}+1}}A^{\frac{1}{q}}(|w|^{q(i-1)}, |z|^{q(i-1)}).$$

Proof. This follows from Corollary 1 (iii) when applied on $h(w) = w^i$. □

Proposition 2. *Let $i \in \mathbb{Z}$, $|i| \geq 2$ and $w, z \in \mathbb{R}$, where $0 < x < y$. For $q \geq 1$, we obtain the following:*

$$\left|A(w^i, z^i) - L_i^i(w,z)\right| \leq \frac{(z-w)|i|}{2}A^{\frac{1}{q}}(|w|^{q(i-1)}, |z|^{q(i-1)}).$$

Proof. This follows from Corollary 2 when applied on $h(w) = w^i$. □

Proposition 3. *Let $w, z \in \mathbb{R}$, where $0 < w < z$. For $q \geq 1$, we obtain*

$$\left|A(w^{-1}, z^{-1}) - L(w,z)\right| \leq \frac{(z-w)}{2^{\frac{q-1}{q}}+1} A^{\frac{1}{q}}(|w|^{-2q}, |z|^{-2q}).$$

Proof. This follows from Corollary 1 (iii) when applied on $h(w) = \frac{1}{w}$. □

Proposition 4. *Let $w, z \in \mathbb{R}$, where $0 < w < z$. For $q \geq 1$, we obtain*

$$\left|A(w^{-1}, z^{-1}) - L(w,z)\right| \leq \frac{(z-w)}{2} A^{\frac{1}{q}}(|w|^{-2q}, |z|^{-2q}).$$

Proof. This follows from Corollary 2 when applied for $h(w) = \frac{1}{w}$. □

Author Contributions: O.A., writing—original draft preparation, visualization; A.K., writing—review and editing, supervision. All authors have read and agreed to the published version of the manuscript.

Funding: This research received no external funding.

Conflicts of Interest: The authors declare no conflict of interest.

References

1. He, J.H. Fractal calculus and its geometrical explanation. *Results Phys.* **2018**, *10*, 272–276. [CrossRef]
2. Tarasov, V.E. On history of mathematical economics: Application of fractional calculus. *Mathematics* **2019**, *7*, 509. [CrossRef]
3. Dragomir, S.S.; Pearce, C. Selected topics on Hermite–Hadamard inequalities and applications. *Math. Prepr. Arch.* **2003**, *2003*, 463–817.
4. Tarasov, V. Generalized memory: Fractional calculus approach. *Fractal Fract.* **2018**, *2*, 23. [CrossRef]
5. Dragomir, S.S. Inequalities for the Generalized k-g-Fractional Integrals in Terms of Double Integral Means. In *Advances in Mathematical Inequalities and Applications*; Birkhäuser: Singapore, 2018; pp. 1–27.
6. Mihai, M.V.; Mitroi, F.C. Hermite–Hadamard type inequalities obtained via Riemann–Liouville fractional calculus. *Acta Math. Univ. Comen.* **2014**, *83*, 209–215.
7. Kirmaci, U.S. Inequalities for differentiable mappings and applications to special means of real numbers and to midpoint formula. *Appl. Math. Comput.* **2004**, *147*, 137–146. [CrossRef]
8. Özdemir, M.E. A theorem on mappings with bounded derivatives with applications to quadrature rules and means. *Appl. Math. Comput.* **2003**, *138*, 425–434. [CrossRef]
9. Almutairi, O.; Kılıçman, A. Some Integral Inequalities for h-Godunova-Levin Preinvexity. *Symmetry* **2019**, *11*, 1500. [CrossRef]
10. Agarwal, R.P.; Kılıçman, A.; Parmar, R.K.; Rathie, A.K. Certain generalized fractional calculus formulas and integral transforms involving (p,q)-Mathieu-type series. *Adv. Differ. Equ.* **2019**, *2019*, 221. [CrossRef]
11. Dragomir, S.S.; Fitzpatrick, S. The Hadamard inequalities for s-convex functions in the second sense. *Demonstr. Math.* **1999**, *32*, 687–696. [CrossRef]
12. Dragomir, S.S. Inequalities of Hermite–Hadamard type for HH-convex functions. *Acta Comment. Univ. Tartu. Math.* **2018**, *22*, 179–190. [CrossRef]
13. Özcan, S.; İşcan, İ. Some new Hermite–Hadamard type inequalities for s-convex functions and their applications. *J. Inequal. Appl.* **2019**, *2019*, 201. [CrossRef]
14. Almutairi, O.; Kılıçman, A. New refinements of the Hadamard inequality on coordinated convex function. *J. Inequal. Appl.* **2019**, *2019*, 192. [CrossRef]

15. Dragomir, S.S.; Agarwal, R.P. Two inequalities for differentiable mappings and applications to special means of real numbers and to trapezoidal formula. *Appl. Math. Lett.* **1998**, *11*, 91–95. [CrossRef]
16. Almutairi, O.; Kılıçman, A. Generalized Integral Inequalities for Hermite–Hadamard-Type Inequalities via s-Convexity on Fractal Sets. *Mathematics* **2019**, *7*, 1065. [CrossRef]
17. Almutairi, O.; Kılıçman, A. New fractional inequalities of midpoint type via s-convexity and their application. *J. Inequal. Appl.* **2019**, *2019*, 267. [CrossRef]
18. Mo, H.; Sui, X. Generalized-convex functions on fractal sets. In *Abstract and Applied Analysis*; Hindawi: London, UK, 2014.
19. Chen, H.; Katugampola, U.N. Hermite–Hadamard and Hermite–Hadamard–Fejér type inequalities for generalized fractional integrals. *J. Math. Anal. Appl.***2017**, *446*, 1274–1291. [CrossRef]
20. Mehreen, N.; Anwar, M. Integral inequalities for some convex functions via generalized fractional integrals. *J. Inequal. Appl.* **2018**, *2018*, 208. [CrossRef]
21. Katugampola, U.N. A new approach to generalized fractional derivatives. *Bull. Math. Anal. Appl.* **2014**, *6*, 1–15.
22. Dubey, R.S.; Goswami, P. Some fractional integral inequalities for the Katugampola integral operator. *AIMS Math.***2019**, *4*, 193–198. [CrossRef]
23. Mercer, A.M. An improvement of the Gruss inequality. *J. Inequal. Pure Appl. Math.* **2005**, *6*, 93.
24. Toplu, T.; Set, E.; Iscan, I.; Maden, S. Hermite-Hadamard type inequalities for p-convex functions via katugampola fractional integrals. *Facta Univ. Ser. Math. Inform.***2019**, *34*, 149–164.
25. Lupinska, B.; Odzijewicz, T. A Lyapunov-type inequality with the Katugampola fractional derivative. *Math. Methods Appl. Sci.* **2018**, *41*, 8985–8996. [CrossRef]

mathematics

MDPI

Article

Fractional q-Difference Inclusions in Banach Spaces

Badr Alqahtani [1], Saïd Abbas [2], Mouffak Benchohra [1,3,*] and Sara Salem Alzaid [1]

[1] Department of Mathematics, College of Science, King Saud University, P.O. Box 2455, Riyadh 11451, Saudi Arabia; balqahtani1@ksu.edu.sa (B.A.); sarsalzaid@ksu.edu.sa (S.S.A.)

[2] Department of Mathematics, Tahar Moulay University of Saïda, P.O. Box 138, EN-Nasr, 20000 Saïda, Algeria; said.abbas@univ-saida.dz

[3] Laboratory of Mathematics, Djillali Liabes University of Sidi Bel-Abbès, P.O. Box 89, 22000 Sidi Bel-Abbès, Algeria

* Correspondence: benchohra@univ-sba.dz

Received: 5 December 2019; Accepted: 31 December 2019; Published: 6 January 2020

Abstract: In this paper, we study a class of Caputo fractional q-difference inclusions in Banach spaces. We obtain some existence results by using the set-valued analysis, the measure of noncompactness, and the fixed point theory (Darbo and Mönch's fixed point theorems). Finally we give an illustrative example in the last section. We initiate the study of fractional q-difference inclusions on infinite dimensional Banach spaces.

Keywords: fractional q-difference inclusion; measure of noncompactness; solution; fixed point

1. Introduction

Fractional differential equations and inclusions have attracted much more interest of mathematicians and physicists which provides an efficiency for the description of many practical dynamical arising in engineering, vulnerability of networks (fractional percolation on random graphs), and other applied sciences [1–8]. Recently, Riemann–Liouville and Caputo fractional differential equations with initial and boundary conditions are studied by many authors; [2,9–14]. In [15–18] the authors present some interesting results for classes of fractional differential inclusions.

q-calculus (quantum calculus) has a rich history and the details of its basic notions, results and methods can be found in [19–21]. The subject of q-difference calculus, initiated in the first quarter of 20th century, has been developed over the years. Some interesting results about initial and boundary value problems of ordinary and fractional q-difference equations can be found in [22–27].

Difference inclusions arise in the mathematical modeling of various problems in economics, optimal control, and stochastic analysis, see for instance [28–30]. However q-difference inclusions are studied in few papers; see for example [31,32]. In this article we consider the Caputo fractional q-difference inclusion

$$({}^cD_q^\alpha u)(t) \in F(t, u(t)),\ t \in I := [0, T], \tag{1}$$

with the initial condition

$$u(0) = u_0 \in E, \tag{2}$$

where $(E, \|\cdot\|)$ is a real or complex Banach space, $q \in (0,1)$, $\alpha \in (0,1]$, $T > 0$, $F : I \times E \to \mathcal{P}(E)$ is a multivalued map, $\mathcal{P}(E) = \{Y \subset E : y \neq \emptyset\}$, and ${}^cD_q^\alpha$ is the Caputo fractional q-difference derivative of order α.

This paper initiates the study of fractional q-difference inclusions on Banach spaces.

2. Preliminaries

Consider the Banach space $C(I) := C(I, E)$ of continuous functions from I into E equipped with the supremum (uniform) norm

$$\|u\|_\infty := \sup_{t\in I} \|u(t)\|.$$

As usual, $L^1(I)$ denotes the space of measurable functions $v : I \to E$ which are Bochner integrable with the norm

$$\|v\|_1 = \int_I \|v(t)\|dt.$$

For $a \in \mathbb{R}$, we set

$$[a]_q = \frac{1-q^a}{1-q}.$$

The q-analogue of the power $(a-b)^n$ is

$$(a-b)^{(0)} = 1,\ (a-b)^{(n)} = \Pi_{k=0}^{n-1}(a-bq^k);\ a,b \in \mathbb{R},\ n \in \mathbb{N}.$$

In general,

$$(a-b)^{(\alpha)} = a^\alpha \Pi_{k=0}^{\infty}\left(\frac{a-bq^k}{a-bq^{k+\alpha}}\right);\ a,b,\alpha \in \mathbb{R}.$$

Definition 1 ([21]). *The q-gamma function is defined by*

$$\Gamma_q(\xi) = \frac{(1-q)^{(\xi-1)}}{(1-q)^{\xi-1}};\ \xi \in \mathbb{R} - \{0,-1,-2,\ldots\}$$

Notice that $\Gamma_q(1+\xi) = [\xi]_q\Gamma_q(\xi)$.

Definition 2 ([21]). *The q-derivative of order $n \in \mathbb{N}$ of a function $u : I \to E$ is defined by $(D_q^0 u)(t) = u(t)$,*

$$(D_q u)(t) := (D_q^1 u)(t) = \frac{u(t)-u(qt)}{(1-q)t};\ t \neq 0,\ (D_q u)(0) = \lim_{t\to 0}(D_q u)(t),$$

and

$$(D_q^n u)(t) = (D_q D_q^{n-1} u)(t);\ t \in I,\ n \in \{1,2,\ldots\}.$$

Set $I_t := \{tq^n : n \in \mathbb{N}\} \cup \{0\}$.

Definition 3 ([21]). *The q-integral of a function $u : I_t \to E$ is defined by*

$$(I_q u)(t) = \int_0^t u(s)d_q s = \sum_{n=0}^{\infty} t(1-q)q^n f(tq^n),$$

provided that the series converges.

We note that $(D_q I_q u)(t) = u(t)$, while if u is continuous at 0, then

$$(I_q D_q u)(t) = u(t) - u(0).$$

Definition 4 ([33]). *The Riemann–Liouville fractional q-integral of order $\alpha \in \mathbb{R}_+ := [0, \infty)$ of a function $u : I \to E$ is defined by $(I_q^0 u)(t) = u(t)$, and*

$$(I_q^\alpha u)(t) = \int_0^t \frac{(t-qs)^{(\alpha-1)}}{\Gamma_q(\alpha)} u(s) d_q s; \ t \in I.$$

Lemma 1 ([34]). *For $\alpha \in \mathbb{R}_+$ and $\lambda \in (-1, \infty)$ we have*

$$(I_q^\alpha (t-a)^{(\lambda)})(t) = \frac{\Gamma_q(1+\lambda)}{\Gamma(1+\lambda+\alpha)} (t-a)^{(\lambda+\alpha)}; \ 0 < a < t < T.$$

In particular,

$$(I_q^\alpha 1)(t) = \frac{1}{\Gamma_q(1+\alpha)} t^{(\alpha)}.$$

Definition 5 ([35]). *The Riemann–Liouville fractional q-derivative of order $\alpha \in \mathbb{R}_+$ of a function $u : I \to E$ is defined by $(D_q^0 u)(t) = u(t)$, and*

$$(D_q^\alpha u)(t) = (D_q^{[\alpha]} I_q^{[\alpha]-\alpha} u)(t); \ t \in I,$$

where $[\alpha]$ is the integer part of α.

Definition 6 ([35]). *The Caputo fractional q-derivative of order $\alpha \in \mathbb{R}_+$ of a function $u : I \to E$ is defined by $({}^C D_q^0 u)(t) = u(t)$, and*

$$({}^C D_q^\alpha u)(t) = (I_q^{[\alpha]-\alpha} D_q^{[\alpha]} u)(t); \ t \in I.$$

Lemma 2 ([35]). *Let $\alpha \in \mathbb{R}_+$. Then the following equality holds:*

$$(I_q^\alpha \, {}^C D_q^\alpha u)(t) = u(t) - \sum_{k=0}^{[\alpha]-1} \frac{t^k}{\Gamma_q(1+k)} (D_q^k u)(0).$$

In particular, if $\alpha \in (0,1)$, then

$$(I_q^\alpha \, {}^C D_q^\alpha u)(t) = u(t) - u(0).$$

We define the following subsets of $\mathcal{P}(E)$:

$P_{cl}(E) = \{Y \in \mathcal{P}(E) : Y \text{ is closed}\}$,
$P_{bd}(E) = \{Y \in \mathcal{P}(E) : Y \text{ is bounded}\}$,
$P_{cp}(E) = \{Y \in \mathcal{P}(E) : Y \text{ is compact}\}$,
$P_{cv}(E) = \{Y \in \mathcal{P}(E) : Y \text{ is convex}\}$,
$P_{cp,cv}(E) = P_{cp}(E) \cap P_{cv}(E)$.

Definition 7. *A multivalued map $G : E \to \mathcal{P}(E)$ is said to be convex (closed) valued if $G(x)$ is convex (closed) for all $x \in E$. A multivalued map G is bounded on bounded sets if $G(B) = \cup_{x \in B} G(x)$ is bounded in E for all $B \in P_b(E)$ (i.e. $\sup_{x \in B}\{\sup\{|y| : y \in G(x)\}$ exists).*

Definition 8. *A multivalued map $G : E \to \mathcal{P}(E)$ is called upper semi-continuous (u.s.c.) on E if $G(x_0) \in P_{cl}(E)$; for each $x_0 \in E$, and for each open set $N \subset E$ with $G(x_0) \in N$, there exists an open neighborhood N_0 of x_0 such that $G(N_0) \subset N$. G is said to be completely continuous if $G(B)$ is relatively compact for every $B \in P_{bd}(E)$. An element $x \in E$ is a fixed point of G if $x \in G(x)$.*

We denote by $FixG$ the fixed point set of the multivalued operator G.

Lemma 3 ([28]). *Let $G : X \to \mathcal{P}(E)$ be completely continuous with nonempty compact values. Then G is u.s.c. if and only if G has a closed graph, that is,*

$$x_n \to x_*, y_n \to y_*, y_n \in G(x_n) \Longrightarrow y_* \in G(x_*).$$

Definition 9. *A multivalued map $G : J \to P_{cl}(E)$ is said to be measurable if for every $y \in E$, the function*

$$t \to d(y, G(t)) = \inf\{|y - z| : z \in G(t)\}$$

is measurable.

Definition 10. *A multivalued map $F : I \times \mathbb{R} \to \mathcal{P}(E)$ is said to be Carathéodory if:*

(1) $t \to F(t, u)$ is measurable for each $u \in E$;
(2) $u \to F(t, u)$ is upper semicontinuous for almost all $t \in I$.

F is said to be L^1-Carathéodory if Equations (1) and (2) and the following condition holds:

(3) For each $q > 0$, there exists $\varphi_q \in L^1(I, \mathbb{R}_+)$ such that

$$\|F(t, u)\|_{\mathcal{P}} = \sup\{|v| : v \in F(t, u)\} \leq \varphi_q \text{ for all } |u| \leq q \text{ and for a.e. } t \in I.$$

For each $u \in C(I)$, define the set of selections of F by

$$S_{F\circ u} = \{v \in L^1(I) : v(t) \in F(t, u(t)) \text{ a.e. } t \in I\}.$$

Let (E, d) be a metric space induced from the normed space $(E, |\cdot|)$. The function $H_d : \mathcal{P}(E) \times \mathcal{P}(E) \to \mathbb{R}_+ \cup \{\infty\}$ given by:

$$H_d(A, B) = \max\{\sup_{a \in A} d(a, B), \sup_{b \in B} d(A, b)\}.$$

is known as the Hausdorff-Pompeiu metric. For more details on multivalued maps see the books of Hu and Papageorgiou [28].

Let $\mathcal{M}_X$ be the class of all bounded subsets of a metric space X.

Definition 11. *A function $\mu : \mathcal{M}_X \to [0, \infty)$ is said to be a measure of noncompactness on X if the following conditions are verified for all $B, B_1, B_2 \in \mathcal{M}_X$.*

(a) Regularity, i.e., $\mu(B) = 0$ if and only if B is precompact,
(b) invariance under closure, i.e., $\mu(B) = \mu(\overline{B})$,
(c) semi-additivity, i.e., $\mu(B_1 \cup B_2) = \max\{\mu(B_1), \mu(B_2)\}$.

Definition 12 ([36,37]). *Let E be a Banach space and denote by Ω_E the family of bounded subsets of E. the map $\mu : \Omega_E \to [0, \infty)$ defined by*

$$\mu(M) = \inf\{\epsilon > 0 : M \subset \cup_{j=1}^{m} M_j, \operatorname{diam}(M_j) \leq \epsilon\},\ M \in \Omega_E,$$

is called the Kuratowski measure of noncompactness.

Theorem 1 ([38]). *Let E be a Banach space. Let $C \subset L^1(I)$ be a countable set with $|u(t)| \leq h(t)$ for a.e. $t \in J$ and every $u \in C$, where $h \in L^1(I, \mathbb{R}_+)$. Then $\phi(t) = \mu(C(t)) \in L^1(I, \mathbb{R}_+)$ and verifies*

$$\mu\left(\left\{\int_0^T u(s)\, ds : u \in C\right\}\right) \leq 2\int_0^T \mu(C(s))\, ds,$$

where μ is the Kuratowski measure of noncompactness on the set E.

Lemma 4 ([39]). *Let F be a Carathéodory multivalued map and $\Theta : L^1(I) \to C(I)$; be a linear continuous map. Then the operator*

$$\Theta \circ S_{F\circ u} : C(I) \to \mathcal{P}_{cv,cp}(C(I)), \quad u \mapsto (\Theta \circ S_{F\circ u})(u) = \Theta(S_{F\circ u})$$

is a closed graph operator in $C(I) \times C(I)$.

Definition 13. *Let E be Banach space. A multivalued mapping $T : E \to \mathcal{P}_{cl,b}(E)$ is called $k-$set-Lipschitz if there exists a constant $k > 0$, such that $\mu(T(X)) \leq k\mu(X)$ for all $X \in \mathcal{P}_{cl,b}(E)$ with $T(X) \in \mathcal{P}_{cl,b}(E)$. If $k < 1$, then T is called a $k-$set-contraction on E.*

Now, we recall the set-valued versions of the Darbo and Mönch fixed point theorems.

Theorem 2 ((Darbo fixed point theorem) [40]). *Let X be a bounded, closed, and convex subset of a Banach space E and let $T : X \to \mathcal{P}_{cl,b}(X)$ be a closed and $k-$set-contraction. Then T has a fixed point.*

Theorem 3 ((Mönch fixed point theorem) [41]). *Let E be a Banach space and $K \subset E$ be a closed and convex set. Also, let U be a relatively open subset of K and $N : \overline{U} \to \mathcal{P}_c(K)$. Suppose that N maps compact sets into relatively compact sets,* $\mathrm{graph}(N)$ *is closed and for some $x_0 \in U$, we have*

$$\mathrm{conv}(x_0 \cup N(M)) \supset M \subset \overline{U} \text{ and } \overline{M} = \overline{U} \ (C \subset M \text{ countable}) \text{ imply } \overline{M} \text{ is compact} \tag{3}$$

and

$$x \notin (1-\lambda)x_0 + \lambda N(x) \quad \forall x \in \overline{U} \backslash U, \ \lambda \in (0,1). \tag{4}$$

Then there exists $x \in \overline{U}$ with $x \in N(x)$.

3. Existence Results

First, we state the definition of a solution of the problem found in Equations (1) and (2).

Definition 14. *By a solution of the problem in Equations (1) and (2) we mean a function $u \in C(I)$ that satisfies the initial condition in Equation (2) and the equation $({}^{C}D_q^{\alpha}u)(t) = v(t)$ on I, where $v \in S_{F\circ u}$.*

In the sequel, we need the following hypotheses.

Hypothesis 1. (H_1). *The multivalued map $F : I \times E \to \mathcal{P}_{cp,cv}(E)$ is Carathéodory.*

Hypothesis 2. (H_2). *There exists a function $p \in L^{\infty}(I, \mathbb{R}_+)$ such that*

$$\|F(t,u)\|_{\mathcal{P}} = \sup\{\|v\|_C : v(t) \in F(t,u)\} \leq p(t);$$

for a.e. $t \in I$, and each $u \in E$,

Hypothesis 3. (H_3). *For each bounded set $B \subset C(I)$ and for each $t \in I$, we have*

$$\mu(F(t, B(t)) \leq p(t)\mu(B(t)),$$

where $B(t) = \{u(t) : u \in B\}$,

Hypothesis 4. (H_4) *The function $\phi \equiv 0$ is the unique solution in $C(I)$ of the inequality*

$$\Phi(t) \leq 2p^*(I_q^{\alpha}\Phi)(t),$$

where p is the function defined in (H_3), *and*

$$p^* = esssup_{t\in I} p(t).$$

Remark 1. *In* (H_3), μ *is the Kuratowski measure of noncompactness on the space E.*

Theorem 4. *If the hypotheses* (H_1)–(H_3) *and the condition*

$$L := \frac{p^* T^{(\alpha)}}{\Gamma_q(1+\alpha)} < 1$$

hold, then the problem in Equations (1) and (2) has at least one solution defined on I.

Proof. Consider the multivalued operator $N : C(I) \to \mathcal{P}(C(I))$ defined by:

$$N(u) = \left\{ h \in C(I) : h(t) = \mu_0 + \int_0^t \frac{(t-qs)^{(\alpha-1)}}{\Gamma_q(\alpha)} v(s) d_q s;\ v \in S_{F \circ u} \right\}. \tag{5}$$

From Lemma 2, the fixed points of N are solutions of the problem in Equations (1) and (2). Set

$$R := \|u_0\| + \frac{p^* T^{(\alpha)}}{\Gamma_q(1+\alpha)},$$

and let $B_R := \{u \in C(I) : \|u\|_\infty \leq R\}$ be the bounded, closed and convex ball of $C(I)$. We shall show in three steps that the multivalued operator $N : B_R \to \mathcal{P}_{cl,b}(C(I))$ satisfies all assumptions of Theorem 2.

Step 1. $N(B_R) \in \mathcal{P}(B_R)$.
Let $u \in B_R$, and $h \in N(u)$. Then for each $t \in I$ we have

$$h(t) = u_0 + \int_0^t \frac{(t-qs)^{(\alpha-1)}}{\Gamma_q(\alpha)} v(s) d_q s,$$

for some $v \in S_{F \circ u}$. On the other hand,

$$\begin{aligned} \|h(t)\| &\leq \|u_0\| + \int_0^t \frac{(t-qs)^{(\alpha-1)}}{\Gamma_q(\alpha)} \|v(s)\| d_q s \\ &\leq \|u_0\| + \int_0^t \frac{(t-qs)^{(\alpha-1)}}{\Gamma_q(\alpha)} p(s) d_q s \\ &\leq \|u_0\| + esssup_{t\in I} p(t) \int_0^T \frac{(t-qs)^{(\alpha-1)}}{\Gamma_q(\alpha)} d_q s \\ &= \|u_0\| + \frac{p^* T^{(\alpha)}}{\Gamma_q(1+\alpha)}. \end{aligned}$$

Hence
$\|h\|_\infty \leq R$, and so $N(B_R) \in \mathcal{P}(B_R)$.

Step 2. $N(u) \in \mathcal{P}_{cl}(B_R)$ *for each* $u \in B_R$.
Let $\{u_n\}_{n\geq 0} \in N(u)$ such that $u_n \longrightarrow \tilde{u}$ in $C(I)$.Then, $\tilde{u} \in B_R$ and there exists $f_n(\cdot) \in S_{F \circ u}$ be such that, for each $t \in I$, we have

$$u_n(t) = u_0 + \int_0^t \frac{(t-qs)^{(\alpha-1)}}{\Gamma_q(\alpha)} f_n(s) d_q s.$$

From (H_1), and since F has compact values, then we may pass to a subsequence if necessary to get that $f_n(\cdot)$ converges to f in $L^1(I)$, and then $f \in S_{F \circ u}$. Thus, for each $t \in I$, we get

$$u_n(t) \longrightarrow \tilde{u}(t) = u_0 + \int_0^t \frac{(t-qs)^{(\alpha-1)}}{\Gamma_q(\alpha)} f(s) d_q s.$$

Hence $\tilde{u} \in N(u)$.

Step 3. *N satisfies the Darbo condition.*
Let $U \subset B_R$, then for each $t \in I$, we have

$$\mu((NU)(t)) = \mu(\{(Nu)(t) : u \in U\}).$$

Let $h \in N(u)$. Then, there exists $f \in S_{F \circ u}$ such that for each $t \in I$, we have

$$h(t) = u_0 + \int_0^t \frac{(t-qs)^{(\alpha-1)}}{\Gamma_q(\alpha)} f(s) d_q s.$$

From Theorem 1 and since $U \subset B_R \subset C(I)$, then

$$\mu((NU)(t)) \leq 2 \int_0^t \mu\left(\left\{ \frac{(t-qs)^{(\alpha-1)}}{\Gamma_q(\alpha)} f(s) : u \in U \right\}\right) d_q s.$$

Now, since $f \in S_{F \circ u}$ and $u(s) \in U(s)$, we have

$$\mu(\{(t-qs)^{(\alpha-1)} f(s)\}) = (t-qs)^{(\alpha-1)} p(s) \mu(U(s)).$$

Then

$$\mu((NU)(t)) \leq 2 \int_0^t \mu\left(\left\{ \frac{(t-qs)^{(\alpha-1)}}{\Gamma_q(\alpha)} f(s) \right\}\right) d_q s.$$

Thus

$$\mu((NU)(t)) \leq 2p^* \int_0^t \frac{(t-qs)^{(\alpha-1)}}{\Gamma_q(\alpha)} \mu(U(s)) d_q s.$$

Hence

$$\mu((NU)(t)) \leq \frac{2p^* T^{(\alpha)}}{\Gamma_q(1+\alpha)} \mu(U).$$

Therefore,

$$\mu(N(U)) \leq L\mu(U),$$

which implies the N is a $L-$set-contraction.
As a consequence of Theorem 2, we deduce that N has a fixed point that is a solution of the problem in Equations (1) and (2). □

Now, we prove an other existence result by applying Theorem 3.

Theorem 5. *If the hypotheses $(H_1) - (H_4)$ hold, then there exists at least one solution of our problem in Equations (1) and (2).*

Proof. Consider the multivalued operator $N : C(I) \to \mathcal{P}(C(I))$ defined in Equation (5). We shall show in five steps that the multivalued operator N satisfies all assumptions of Theorem 3.

Step 1. *$N(u)$ is convex for each $u \in C(I)$.*
Let $h_1, h_2 \in N(u)$, then there exist$v_1, v_2 \in S_{F\circ u}$ such that

$$h_i(t) = \mu_0 + \int_0^t \frac{(t-qs)^{(\alpha-1)}}{\Gamma_q(\alpha)} v_i(s) d_q s; \; t \in I, \; i = 1, 2.$$

Let $0 \leq \lambda \leq 1$. Then, for each $t \in I$, we have

$$(\lambda h_1 + (1-\lambda)h_2)(t) = \int_0^t \frac{(t-qs)^{(\alpha-1)}}{\Gamma_q(\alpha)} (\lambda v_1(s) + (1-\lambda)v_2(s)) d_q s.$$

Since $S_{F\circ u}$ is convex (because F has convex values), we have $\lambda h_1 + (1-\lambda)h_2 \in N(u)$.

Step 2. *For each compact $M \subset C(I)$, $N(M)$ is relatively compact.*
Let (h_n) be any sequence in $N(M)$, where $M \subset C(I)$ is compact.We show that (h_n) has a convergent subsequence from Arzéla–Ascoli compactness criterion in $C(I)$. Since $h_n \in N(M)$ there are $u_n \in M$ and $v_n \in S_{F\circ u_n}$ such that

$$h_n(t) = \mu_0 + \int_0^t \frac{(t-qs)^{(\alpha-1)}}{\Gamma_q(\alpha)} v_n(s) d_q s.$$

Using Theorem 1 and the properties of the measure μ, we have

$$\mu(\{h_n(t)\}) \leq 2\int_0^t \mu\left(\left\{\frac{(t-qs)^{(\alpha-1)}}{\Gamma_q(\alpha)} v_n(s)\right\}\right) d_q s. \tag{6}$$

On the other hand, since M is compact, the set $\{v_n(s) : n \geq 1\}$ is compact. Consequently, $\mu(\{v_n(s) : n \geq 1\}) = 0$ for a.e. $s \in I$. Furthermore

$$\mu(\{(t-qs)^{(\alpha-1)} v_n(s)\}) = (t-qs)^{(\alpha-1)} \mu(\{v_n(s) : n \geq 1\}) = 0.$$

for a.e. $t, s \in I$. Now Equation (6) implies that $\{h_n(t) : n \geq 1\}$ is relatively compact for each $t \in I$. In addition, for each $t_1, t_2 \in I$; with $t_1 < t_2$, we have

$$\begin{aligned}
&\|h_n(t_2) - h_n(t_1)\| \\
&\leq \left\| \int_0^{t_2} \frac{(t_2-qs)^{(\alpha-1)}}{\Gamma_q(\alpha)} p(s) d_q s - \int_0^{t_1} \frac{(t_1-qs)^{(\alpha-1)}}{\Gamma_q(\alpha)} p(s) d_q s \right\| \\
&\leq \int_{t_1}^{t_2} \frac{(t_2-qs)^{(\alpha-1)}}{\Gamma_q(\alpha)} p(s) d_q s \\
&\quad + \int_0^{t_1} \frac{|(t_2-qs)^{(\alpha-1)} - (t_1-qs)^{(\alpha-1)}|}{\Gamma_q(\alpha)} p(s) d_q s \\
&\leq \frac{p^* T^\alpha}{\Gamma_q(1+\alpha)} (t_2 - t_1)^\alpha \\
&\quad + p^* \int_0^{t_1} \frac{|(t_2-qs)^{(\alpha-1)} - (t_1-qs)^{(\alpha-1)}|}{\Gamma_q(\alpha)} d_q s \\
&\to 0 \text{ as } t_1 \longrightarrow t_2.
\end{aligned} \tag{7}$$

This shows that $\{h_n : n \geq 1\}$ is equicontinuous. Consequently, $\{h_n : n \geq 1\}$ is relatively compact in $C(I)$.

Step 3. *The graph of N is closed.*
Let $(u_n, h_n) \in \text{graph}(N)$, $n \geq 1$, with$(\|u_n - u\|, \|h_n - h\|) \to (0.0)$, as $n \to \infty$. We have to show

that $(u,h) \in \text{graph}(N)$. $(u_n,h_n) \in \text{graph}(N)$ means that $h_n \in N(u_n)$, which implies that there exists $v_n \in S_{F\circ u_n}$, such that for each $t \in I$,

$$h_n(t) = u_0 + \int_0^t \frac{(t-qs)^{(\alpha-1)}}{\Gamma_q(\alpha)} v_n(s) d_q s.$$

Consider the continuous linear operator $\Theta : L^1(I) \to C(I)$,

$$\Theta(v)(t) \mapsto h_n(t) = u_0 + \int_0^t \frac{(t-qs)^{(\alpha-1)}}{\Gamma_q(\alpha)} v_n(s) d_q s.$$

Clearly, $\|h_n(t) - h(t)\| \to 0$ as as $n \to \infty$. From Lemma 4 it follows that $\Theta \circ S_F$ is a closed graph operator. Moreover, $h_n(t) \in \Theta(S_{F\circ u_n})$. Since $u_n \to u$, Lemma 4 implies

$$h(t) = u_0 + \int_0^t \frac{(t-qs)^{(\alpha-1)}}{\Gamma_q(\alpha)} v(s) d_q s.$$

for some $v \in S_{F\circ u}$.

Step 4. *M is relatively compact in* $C(I)$.
Let $M \subset \overline{U}$; with $M \subset \text{conv}(\{0\} \cup N(M))$, and let $\overline{M} = \overline{C}$; for some countable set $C \subset M$. the set $N(M)$ is equicontinuous from Equation (7). Therefore,

$$M \subset \text{conv}(\{0\} \cup N(M)) \Longrightarrow M \textit{ is equicontinuous.}$$

By applying the Arzéla–Ascoli theorem; the set $M(t)$ is relatively compact for each $t \in I$. Since $C \subset M \subset \text{conv}(\{0\} \cup N(M))$, then there exists a countable set $H = \{h_n : n \geq 1\} \subset N(M)$ such that $C \subset \text{conv}(\{0\} \cup H)$. Thus, there exist $u_n \in M$ and $v_n \in S_{F\circ u_n}$ such that

$$h_n(t) = u_0 + \int_0^t \frac{(t-qs)^{(\alpha-1)}}{\Gamma_q(\alpha)} v_n(s) d_q s.$$

From Theorem 1, we get

$$M \subset \overline{C} \subset \overline{\text{conv}}(\{0\} \cup H)) \Longrightarrow \mu(M(t)) \leq \mu(\overline{C}(t)) \leq \mu(H(t)) = \mu(\{h_n(t) : n \geq 1\}).$$

Using now the inequality Equation (6) in step 2, we obtain

$$\mu(M(t)) \leq 2\int_0^t \mu\left(\left\{\frac{(t-qs)^{(\alpha-1)}}{\Gamma_q(\alpha)} v_n(s)\right\}\right) d_q s.$$

Since $v_n \in S_{F\circ u_n}$ and $u_n(s) \in M(s)$, we have

$$\mu(M(t)) \leq 2\int_0^t \mu\left(\left\{\frac{(t-qs)^{(\alpha-1)}}{\Gamma_q(\alpha)} v_n(s) : n \geq 1\right\}\right) d_q s.$$

Also, since $v_n \in S_{F\circ u_n}$ and $u_n(s) \in M(s)$, then from (H_3) we get

$$\mu(\{(t-qs)^{(\alpha-1)} v_n(s);\, n \geq 1\}) = (t-qs)^{(\alpha-1)} p(s) \mu(M(s)).$$

Hence

$$\mu(M(t)) \leq 2p^* \int_0^t \frac{(t-qs)^{(\alpha-1)}}{\Gamma_q(\alpha)} \mu(M(s)) d_q s.$$

Consequently, from (H_4), the function Φ given by $\Phi(t) = \mu(M(t))$ satisfies $\Phi \equiv 0$; that is, $\mu(M(t)) = 0$ for all $t \in I$. Finally, the Arzéla–Ascoli theorem implies that M is relatively compact in $C(I)$.

Step 5. *The priori estimate.*
Let $u \in C(I)$ such that $u \in \lambda N(u)$ for some $0 < \lambda < 1$. Then

$$u(t) = \lambda u_0 + \lambda \int_0^t \frac{(t-qs)^{(\alpha-1)}}{\Gamma_q(\alpha)} v(s) d_q s,$$

for each $t \in I$, where $v \in S_{F\circ u}$. On the other hand,

$$\begin{aligned} \|u(t)\| &\leq \|u_0\| + \int_0^t \frac{(t-qs)^{(\alpha-1)}}{\Gamma_q(\alpha)} \|v(s)\| d_q s \\ &\leq \|u_0\| + \int_0^t \frac{(t-qs)^{(\alpha-1)}}{\Gamma_q(\alpha)} p(s) d_q s \\ &\leq \|u_0\| + \frac{p^* T^{(\alpha)}}{\Gamma_q(1+\alpha)}. \end{aligned}$$

Then

$$\|u\| \leq \|u_0\| + \frac{p^* T^{(\alpha)}}{\Gamma_q(1+\alpha)} := d.$$

Set

$$U = \{u \in C_\gamma : \|u\| < 1 + d\}.$$

Hence, the condition in Equation (4) is satisfied. Finally, Theorem 3 implies that N has at least one fixed point $u \in C(I)$ which is a solution of our problem in Equations (1) and (2). □

4. An Example

Let

$$E = l^1 = \left\{ u = (u_1, u_2, \ldots, u_n, \ldots), \sum_{n=1}^{\infty} |u_n| < \infty \right\}$$

be the Banach space with the norm

$$\|u\|_E = \sum_{n=1}^{\infty} |u_n|.$$

Consider now the following problem of fractional $\frac{1}{4}$−difference inclusion

$$\begin{cases} ({}^c D_{\frac{1}{4}}^{\frac{1}{2}} u_n)(t) \in F_n(t, u(t)); \ t \in [0, e], \\ u(0) = (1, 0, \ldots, 0, \ldots), \end{cases} \tag{8}$$

where

$$F_n(t, u(t)) = \frac{t^2 e^{-4-t}}{1 + \|u(t)\|_E} [u_n(t) - 1, u_n(t)]; \ t \in [0, e],$$

with $u = (u_1, u_2, \ldots, u_n, \ldots)$. Set $\alpha = \frac{1}{2}$, and $F = (F_1, F_2, \ldots, F_n, \ldots)$.
For each $u \in E$ and $t \in [0, e]$, we have

$$\|F(t, u)\|_{\mathcal{P}} \leq p(t),$$

with $p(t) = t^2 e^{-t-4}$. Hence, the hypothesis (H_2) is satisfied with $p^* = e^{-2}$. A simple computation shows that conditions of Theorem 5 are satisfied. Hence, the problem in Equation (8) has at least one solution defined on $[0, e]$.

5. Conclusions

We have provided some sufficient conditions guaranteeing the existence of solutions for some fractional q-difference inclusions involving the Caputo fractional derivative in Banach spaces. The achieved results are obtained using the fixed point theory and the notion of measure of noncompactness. Such notion requires the use of the set-valued analysis conditions on the right-hand side, like the upper semi-continuity. In the forthcoming paper we shall provide sufficient conditions ensuring the existence of weak solutions by using the concept measure of weak noncompactness, the Pettis integration and an appropriate fixed point theorem.

Author Contributions: Funding acquisition, B.A.; Writing—original draft, S.A.; Writing—review & editing, M.B. and S.S.A. All authors have read and agreed to the published version of the manuscript.

Funding: The authors would like to extend their sincere appreciation to the Deanship of Scientific Research at King Saud University for funding this group No. RG-1437-017.

Acknowledgments: The authors are grateful to the handling editor and reviewers for their careful reviews and useful comments.

Conflicts of Interest: The authors declare no conflict of interest.

References

1. Abbas, S.; Benchohra, M.; Graef, J.R.; Henderson, J. *Implicit Fractional Differential and Integral Equations: Existence and Stability*; De Gruyter: Berlin, Germany, 2018.
2. Abbas, S.; Benchohra, M.; N'Guérékata, G.M. *Topics in Fractional Differential Equations*; Springer: New York, NY, USA, 2012.
3. Abbas, S.; Benchohra, M.; N'Guérékata, G.M. *Advanced Fractional Differential and Integral Equations*; Nova Science Publishers: New York, NY, USA, 2015.
4. Kilbas, A.A.; Srivastava, H.M.; Juan Trujillo, J. *Theory and Applications of Fractional Differential Equations*; North-Holland Mathematics Studies 204; Elsevier Science B.V.: Amsterdam, The Netherlands, 2006.
5. Tenreiro Machado, J.A.; Kiryakova, V. The chronicles of fractional calculus, *Fract. Calc. Appl. Anal.* **2017**, *20*, 307–336.
6. Samko, S.G.; Kilbas, A.A.; Marichev, O.I. *Fractional Integrals and Derivatives. Theory and Applications*; Gordon and Breach: Amsterdam, The Netherlands, 1987. (Translated from the Russian)
7. Tarasov, V.E. *Fractional Dynamics: Application of Fractional Calculus to Dynamics of Particles, Fields and Media*; Springer: Berlin/Heidelberg, Germany; Higher Education Press: Beijing, China, 2010.
8. Zhou, Y. *Basic Theory of Fractional Differential Equations*; World Scientific: Singapore, 2014.
9. Abdeljawad, T.; Alzabut, J. On Riemann–Liouville fractional q-difference equations and their application to retarded logistic type model. *Math. Methods Appl. Sci.* **2018**, *18*, 8953–8962. [CrossRef]
10. Abbas, S.; Benchohra, M.; Hamani, S.; Henderson, J. Upper and lower solutions method for Caputo–Hadamard fractional differential inclusions. *Math. Moravica* **2019**, *23*, 107–118. [CrossRef]
11. Kilbas, A.A. Hadamard-type fractional calculus. *J. Korean Math. Soc.* **2001**, *38*, 1191–1204.
12. Ren, J.; Zhai, C. Characteristic of unique positive solution for a fractional q-difference equation with multistrip boundary conditions. *Math. Commun.* **2019**, *24*, 181–192.
13. Ren, J.; Zhai, C. A fractional q-difference equation with integral boundary conditions and comparison theorem. *Int. J. Nonlinear Sci. Numer. Simul.* **2017**, *18*, 575–583. [CrossRef]
14. Zhang, T.; Tang, Y. A difference method for solving the q-fractional differential equations. *Appl. Math. Lett.* **2019**, *98*, 292–299. [CrossRef]
15. Abbas, S.; Benchohra, M.; Graef, J.R. Coupled systems of Hilfer fractional differential inclusions in Banach spaces. *Commun. Pure Appl. Anal.* **2018**, *17*, 2479–2493. [CrossRef]
16. Ahmad, B.; Ntouyas, S.K.; Alsaedi, A.; Agarwal, R.P. A study of nonlocal integro-multi-point boundary value problems of sequential fractional integro-differential inclusions. *Dyn. Contin. Disc. Impuls. Syst. Ser. A Math. Anal.* **2018**, *25*, 125–140.

17. Ahmad, B.; Ntouyas, S.K.; Zhou, Y.; Alsaedi, A. A study of fractional differential equations and inclusions with nonlocal Erdélyi-Kober type integral boundary conditions. *Bull. Iran. Math. Soc.* **2018**, *44*, 1315–1328. [CrossRef]
18. Wang, J.; Ibrahim, A.G.; O'Regan, D. Global attracting solutions to Hilfer fractional differential inclusions of Sobolev type with noninstantaneous impulses and nonlocal conditions. *Nonlinear Anal. Model. Control* **2019**, *24*, 775–803. [CrossRef]
19. Adams, C. On the linear ordinary q-difference equation. *Ann. Math.* **1928**, *30*, 195–205. [CrossRef]
20. Carmichael, R.D. The general theory of linear q-difference equations. *Am. J. Math.* **1912**, *34*, 147–168. [CrossRef]
21. Kac, V.; Cheung, P. *Quantum Calculus*; Springer: New York, NY, USA, 2002.
22. Abbas, S.; Benchohra, M.; Laledj, N; Zhou, Y. Existence and Ulam stability for implicit fractional q-difference equation, *Adv. Differ. Equ.* **2019**, *2019*, 480. [CrossRef]
23. Ahmad, B. Boundary value problem for nonlinear third order q-difference equations. *Electron. J. Differ. Equ.* **2011**, *94*, 1–7. [CrossRef]
24. El-Shahed, M.; Hassan, H.A. Positive solutions of q-difference equation. *Proc. Am. Math. Soc.* **2010**, *138*, 1733–1738. [CrossRef]
25. Etemad, S.; Ntouyas, S.K.; Ahmad, B. Existence theory for a fractional q-integro-difference equation with q-integral boundary conditions of different orders. *Mathematics* **2019**, *7*, 659. [CrossRef]
26. Ahmad, B.; Ntouyas, S.K.; Purnaras, L.K. Existence results for nonlocal boundary value problems of nonlinear fractional q-difference equations. *Adv. Differ. Equ.* **2012**, *2012*, 140. [CrossRef]
27. Zuo, M.; Hao, X. Existence results for impulsive fractional q-difference equation with antiperiodic boundary conditions. *J. Funct. Spaces* **2018**, *2018*, 3798342. [CrossRef]
28. Hu, S.; Papageorgiou, N. *Handbook of Multivalued Analysis, Volume I: Theory*; Kluwer: Dordrecht, The Netherlands; Boston, MA, USA; London, UK, 1997.
29. Kisielewicz, M. *Differential Inclusions and Optimal Control*; Kluwer: Dordrecht, The Netherlands, 1991.
30. Smirnov, G.V. *Introduction to the Theory of Differential Inclusions*; Graduate Studies in Mathematics, Volumn 41; American Mathematical Society: Providence, RI, USA, 2002.
31. Ahmad, B.; Ntouyas, S.K. Boundary value problems for q-difference inclusions. *Abstr. Appl. Anal.* **2011**, *2011*, 292860. [CrossRef]
32. Ntouyas, S.K. Existence results for q-difference inclusions with three-point boundary conditions involving different numbers of q. *Discuss. Math. Differ. Incl. Control Optim.* **2014**, *34*, 41–59. [CrossRef]
33. Agarwal, R. Certain fractional q-integrals and q-derivatives. *Proc. Camb. Philos. Soc.* **1969**, *66*, 365–370. [CrossRef]
34. Rajkovic, P.M.; Marinkovic, S.D.; Stankovic, M.S. Fractional integrals and derivatives in q-calculus. *Appl. Anal. Discrete Math.* **2007**, **1**, 311–323.
35. Rajkovic, P.M.; Marinkovic, S.D.; Stankovic, M.S. On q-analogues of Caputo derivative and Mittag-Leffler function. *Fract. Calc. Appl. Anal.* **2007**, *10*, 359–373.
36. Banas, J.; Goebel, K. *Measures of Noncompactness in Banach Spaces*; Marcel Dekker: New York, NY, USA, 1980.
37. Ayerbee Toledano, J.M.; Dominguez Benavides, T.; Lopez Acedo, G. *Measures of Noncompactness in Metric Fixed Point Theory*; Operator Theory, Advances and Applications; Birkhäuser: Basel, Switzerland; Boston, MA, USA; Berlin, Germany, 1997; Volumn 99.
38. Heinz, H.P. On the behaviour of meusure of noncompacteness with respect of differentiation and integration of vector-valued function. *Nonlinear Anal.* **1983**, *7*, 1351–1371. [CrossRef]
39. Losta, A.; Opial, Z. An application of the Kakutani-Ky Fan theorem in the theory of ordinary differential equation. *Bull. Acad. Pol. Sci., Ser. Sci. Math. Astronom. Phys.* **1965**, *13*, 781–786.
40. Dhage, B.C. Some generalization of multi-valued version of Schauder's fixed point theorem with applications. *Cubo* **2010**, *12*, 139–151. [CrossRef]
41. O'Regan, D.; Precup, R. Fixed point theorems for set-valued maps and existence principles for integral inclusions. *J. Math. Anal. Appl.* **2000**, *245*, 594–612. [CrossRef]

mathematics

MDPI

Article

Certain Fractional Proportional Integral Inequalities via Convex Functions

Gauhar Rahman [1], Kottakkaran Sooppy Nisar [2,*], Thabet Abdeljawad [3,4,5,*] and Samee Ullah [6]

1 Department of Mathematics, Shaheed Benazir Bhutto University, Sheringal, Upper Dir 18000, Pakistan; gauhar55uom@gmail.com
2 Department of Mathematics, College of Arts and Sciences, Prince Sattam bin Abdulaziz University, Wadi Aldawaser 11991, Saudi Arabia
3 Department of Mathematics and General Sciences, Prince Sultan University, Riyadh 12345, Saudi Arabia
4 Department of Medical Research, China Medical University, Taichung 40402, Taiwan
5 Department of Computer Science and Information Engineering, Asia University, Taichung 40402, Taiwan
6 Department of Mathematics, University of Malakand, Lower Dir 18800, Chakdara, Pakistan; sameeullah413@gmail.com
* Correspondence: n.sooppy@psau.edu.sa or ksnisar1@gmail.com (K.S.N.); tabdeljawad@psu.edu.sa (T.A.)

Received: 27 December 2019; Accepted: 6 February 2020; Published: 9 February 2020

Abstract: The goal of this article is to establish some fractional proportional integral inequalities for convex functions by employing proportional fractional integral operators. In addition, we establish some classical integral inequalities as the special cases of our main findings.

Keywords: convex function; fractional integrals; proportional fractional integrals; inequalities; Qi inequality

MSC: 26A33; 26D10; 26D53; 05A30

1. Introduction

Integral inequalities play a vital role in the field of fractional differential equations. In the past few decades, researchers have paid their valuable consideration to this area. The significant developments in this area have been investigated, for example, [1–3], and [4] (cf. references cited therein). In [5], Ngo et al. established the following inequalities

$$\int_0^1 g^{\sigma+1}(t)dt \geq \int_0^1 t^{\sigma} g(t)dt \tag{1}$$

and

$$\int_0^1 g^{\sigma+1}(t)dt \geq \int_0^1 t g^{\sigma}(t)dt, \tag{2}$$

where $\sigma > 0$ and the positive continuous function g on $[0,1]$ such that

$$\int_x^1 g(t)dt \geq \int_x^1 tdt, x \in [0,1].$$

Later on, Liu et al. [6] established the following inequalities

$$\int_a^b g^{\sigma+\gamma}(t)dt \geq \int_a^b (t-a)^{\sigma} g^{\gamma}(t)dt, \tag{3}$$

where $\sigma > 0, \gamma > 0$, and the positive continuous g on $[a, b]$ is such that

$$\int_a^b g^{\delta}(t)dt \geq \int_a^b (t-a)^{\delta}dt, \delta = \min(1, \gamma), t \in [a, b].$$

Liu et al. [7] derived two theorems for integral inequalities as follows:

Theorem 1. *Suppose that the functions f_1 and g_1 are positive and continuous on $[a, b], (a < b)$ with $f_1 \leq g_1$ on $[a, b]$ such that the function $\frac{f_1}{g_1}, (g_1 \neq 0)$ is decreasing and the function f_1 is increasing. Assume that the function Φ is a convex with $\Phi(0) = 0$. Then, the following inequality holds*

$$\frac{\int_a^b f_1(t)dt}{\int_a^b g_1(t)dt} \geq \frac{\int_a^b \Phi\left(f_1(t)\right) dt}{\int_a^b \Phi\left(g_1(t)\right) dt}.$$

Theorem 2. *Suppose that the functions f_1, f_2, and f_3 be positive and continuous on $[a, b], (a < b)$ with $f_1 \leq f_2$ on $[a, b]$ such that the function $\frac{f_1}{f_2}, (f_2 \neq 0)$ is decreasing and the functions f_1 and f_3 are increasing. Assume that the function Φ is a convex with $\Phi(0) = 0$. Then, the following inequality holds*

$$\frac{\int_a^b f_1(t)dt}{\int_a^b f_2(t)dt} \geq \frac{\int_a^b \Phi\left(f_1(t)\right) f_3(t)dt}{\int_a^b \Phi\left(f_2(t)\right) f_3(t)dt}.$$

The inequalities in Equations (1)–(3) and their various generalizations have gained attention of the researchers [8–12].

Furthermore, the research of fractional integral inequalities is also of prominent importance. In [13,14], the authors presented some weighted Grüss type and new inequalities involving Riemann–Liouville (R-L) fractional integrals. In [15], Nisar et al. introduced many inequalities for extended gamma and confluent hypergeometric k-functions. Certain Gronwall inequalities for R-L and Hadamard k-fractional derivatives with applications are observed in [16]. The inequalities concerning the generalized (k, ρ)-fractional integral operators can be seen in [17].

The generalized fractional integral and Grüss type inequalities via generalized fractional integrals can be found in [18,19]. In [20], the authors examined the (k, s)-R-L fractional integral and its applications. In [21], the authors presented generalized Hermite–Hadamard type inequalities through fractional integral operators. Dahmani [22] introduced some classes of fractional integral inequalities by employing a family of n positive functions. Further the applications of fractional integral inequalities can be found [23,24].

In the last few decades, the researchers have paid their valuable consideration to the field of fractional calculus. This field has received more attention from various researchers due to its wide applications in various fields. In the growth of fractional calculus, researchers concentrate to develop several fractional integral operators and their applications in distinct fields (see, e.g., [25–33]). Zaher et al. [34] presented a new fractional nonlocal model.

Such types of these new fractional integral operators promote the future study to develop certain new approaches to unify the fractional operators and secure fractional integral inequalities. Especially, several striking inequalities, properties, and applicability for the fractional conformable integrals and derivatives are recently studied by various researchers. We refer the interesting readers to the works by [35–44], and [45]. The applications of conformable derivative can be found in [46–49] (cf. references cited therein).

2. Preliminaries

Jarad et al. [50] proposed the following left and right generalized proportional integral operators, which are sequentially defined by

$$\left({}_a\mathcal{J}^{\xi,\delta}f\right)(\tau)=\frac{1}{\delta^{\xi}\Gamma(\xi)}\int_a^{\tau}\exp[\frac{\delta-1}{\delta}(\tau-t)](\tau-t)^{\xi-1}f(t)dt, a<\tau \tag{4}$$

and

$$\left(\mathcal{J}_b^{\xi,\delta}f\right)(\tau)=\frac{1}{\delta^{\xi}\Gamma(\xi)}\int_{\tau}^{b}\exp[\frac{\delta-1}{\delta}(t-\tau)](t-\tau)^{\xi-1}f(t)dt, \tau<b, \tag{5}$$

where the proportional index $\delta\in(0,1]$ and $\xi\in\mathbb{C}$ with $Re(\xi)>0$ and $\Gamma(\tau)$ is the well-know gamma function defined by $\Gamma(\tau)=\int_0^{\infty}t^{\tau-1}e^{-t}dt$ [51–53].

Remark 1. *Setting $\delta=1$ in Equations (4) and (5), we obtain the following left and right R-L:*

$$\left({}_a\mathcal{J}^{\xi}f\right)(\tau)=\frac{1}{\Gamma(\xi)}\int_a^{\tau}(\tau-t)^{\xi-1}f(t)dt, a<\tau,$$

and

$$\left(\mathcal{J}_b^{\xi}f\right)(\tau)=\frac{1}{\Gamma(\xi)}\int_{\tau}^{b}(t-\tau)^{\xi-1}f(t)dt, \tau<b,$$

where $\xi\in\mathbb{C}$ with $Re(\xi)>0$.

Recently, the generalized proportional derivative, and integral operators are established and studied in [54,55]. Certain new classes of integral inequalities for a class of n $(n\in\mathbb{N})$ positive continuous and decreasing functions on $[a,b]$ via generalized proportional fractional integrals can be found in the work of Rahman et al. [56]. The generalized Hadamard proportional fractional integrals and certain inequalities for convex functions by employing were recently proposed by Rahman et al. [57]. The bounds of proportional integrals in the sense of another function can be found in the work of Rahman et al. [58].

3. Main Results

In this section, we establish proportional fractional integral inequalities for convex functions by employing proportional fractional integral operators.

Theorem 3. *Suppose that the functions f and g are positive and continuous on the interval $[a,b]$, $(a<b)$ and $f\leq g$ on $[a,b]$. If the function $\frac{f}{g}$, $(g\neq 0)$ is decreasing and the function f is increasing on $[a,b]$, then, for any convex function Φ with $\Phi(0)=0$, the following inequality satisfies the proportional fractional integral operator given by Equation (4)*

$$\frac{{}_a\mathcal{J}^{\xi,\delta}\left[f(\tau)\right]}{{}_a\mathcal{J}^{\xi,\delta}\left[g(\tau)\right]}\geq\frac{{}_a\mathcal{J}^{\xi,\delta}\left[\Phi(f(\tau))\right]}{{}_a\mathcal{J}^{\xi,\delta}\left[\Phi(g(\tau))\right]}, \tag{6}$$

where $\delta\in(0,1]$, $\xi\in\mathbb{C}$ with $Re(\xi)>0$.

Proof. Since Φ is convex function with $\Phi(0) = 0$, the function $\frac{f(\tau)}{\tau}$ is increasing. As f is increasing, the function $\frac{\Phi(f(\tau))}{f(\tau)}$ is also increasing. Obviously, $\frac{f(\tau)}{g(\tau)}$ is decreasing function. Thus, for all $\rho, \theta \in [a, b]$, we have

$$\left(\frac{\Phi(f(\rho))}{f(\rho)} - \frac{\Phi(f(\theta))}{f(\theta)}\right)\left(\frac{f(\theta)}{g(\theta)} - \frac{f(\rho)}{g(\rho)}\right) \geq 0.$$

It follows that

$$\frac{\Phi(f(\rho))}{f(\rho)}\frac{f(\theta)}{g(\theta)} + \frac{\Phi(f(\theta))}{f(\theta)}\frac{f(\rho)}{g(\rho)} - \frac{\Phi(f(\theta))}{f(\theta)}\frac{f(\theta)}{g(\theta)} - \frac{\Phi(f(\rho))}{f(\rho)}\frac{f(\rho)}{g(\rho)} \geq 0. \tag{7}$$

Multiplying Equation (7) by $g(\rho)g(\theta)$, we have

$$\frac{\Phi(f(\rho))}{f(\rho)}f(\theta)g(\rho) + \frac{\Phi(f(\theta))}{f(\theta)}f(\rho)g(\theta) - \frac{\Phi(f(\theta))}{f(\theta)}f(\theta)g(\rho) - \frac{\Phi(f(\rho))}{f(\rho)}f(\rho)g(\theta) \geq 0. \tag{8}$$

Multiplying Equation (8) by $\frac{1}{\delta^{\xi}\Gamma(\xi)}\exp[\frac{\delta-1}{\delta}(\tau-\rho)](\tau-\rho)^{\xi-1}$, and integrating with respect to ρ over $[a, \tau]$, $a < \tau \leq b$, we have

$$\begin{aligned}
&\frac{1}{\delta^{\xi}\Gamma(\xi)}\int_a^{\tau}\exp[\frac{\delta-1}{\delta}(\tau-\rho)](\tau-\rho)^{\xi-1}\frac{\Phi(f(\rho))}{f(\rho)}f(\theta)g(\rho)d\rho \\
&+\frac{1}{\delta^{\xi}\Gamma(\xi)}\int_a^{\tau}\exp[\frac{\delta-1}{\delta}(\tau-\rho)](\tau-\rho)^{\xi-1}\frac{\Phi(f(\theta))}{f(\theta)}f(\rho)g(\theta)d\rho \\
&-\frac{1}{\delta^{\xi}\Gamma(\xi)}\int_a^{\tau}\exp[\frac{\delta-1}{\delta}(\tau-\rho)](\tau-\rho)^{\xi-1}\frac{\Phi(f(\theta))}{f(\theta)}f(\theta)g(\rho)d\rho \\
&-\frac{1}{\delta^{\xi}\Gamma(\xi)}\int_a^{\tau}\exp[\frac{\delta-1}{\delta}(\tau-\rho)](\tau-\rho)^{\xi-1}\frac{\Phi(f(\rho))}{f(\rho)}f(\rho)g(\theta)d\rho \geq 0.
\end{aligned}$$

Then, it follows that

$$\begin{aligned}
&f(\theta)\,{}_a\mathcal{J}^{\xi,\delta}\left(\frac{\Phi(f(\tau))}{f(\tau)}g(\tau)\right) + \left(\frac{\Phi(f(\theta))}{f(\theta)}g(\theta)\right){}_a\mathcal{J}^{\xi,\delta}(f(\tau)) \\
&-\left(\frac{\Phi(f(\theta))}{f(\theta)}f(\theta)\right){}_a\mathcal{J}^{\xi,\delta}(g(\tau)) - g(\theta)\,{}_a\mathcal{J}^{\xi,\delta}\left(\frac{\Phi(f(\tau))}{f(\tau)}f(\tau)\right) \geq 0.
\end{aligned} \tag{9}$$

Again, multiplying both sides of Equation (9) by $\frac{1}{\delta^{\xi}\Gamma(\xi)}\exp[\frac{\delta-1}{\delta}(\tau-\theta)](\tau-\theta)^{\xi-1}$, and integrating the resultant inequality with respect to θ over $[a, \tau]$, $a < \tau \leq b$, we get

$$\begin{aligned}
&{}_a\mathcal{J}^{\xi,\delta}(f(\tau))\,{}_a\mathcal{J}^{\xi,\delta}\left(\frac{\Phi(f(\tau))}{f(\tau)}g(\tau)\right) + {}_a\mathcal{J}^{\xi,\delta}\left(\frac{\Phi(f(\tau))}{f(\tau)}g(\tau)\right){}_a\mathcal{J}^{\xi,\delta}(f(\tau)) \\
&\geq {}_a\mathcal{J}^{\xi,\delta}(g(\tau))\,{}_a\mathcal{J}^{\xi,\delta}(\Phi(f(\tau))) + {}_a\mathcal{J}^{\xi,\delta}(\Phi(f(\tau)))\,{}_a\mathcal{J}^{\xi,\delta}(g(\tau)).
\end{aligned}$$

It follows that

$$\frac{{}_a\mathcal{J}^{\xi,\delta}(f(\tau))}{{}_a\mathcal{J}^{\xi,\delta}(g(\tau))} \geq \frac{{}_a\mathcal{J}^{\xi,\delta}(\Phi(f(\tau)))}{{}_a\mathcal{J}^{\xi,\delta}\left(\frac{\Phi(f(\tau))}{f(\tau)}g(\tau)\right)}. \tag{10}$$

Now, since $f \leq g$ on $[a, b]$ and $\frac{\Phi(\tau)}{\tau}$ is an increasing function, for $\rho \in [a, \tau]$, $a < \tau \leq b$, we have

$$\frac{\Phi(f(\rho))}{f(\rho)} \leq \frac{\Phi(g(\rho))}{g(\rho)}. \tag{11}$$

Multiplying both sides of Equation (11) by $\frac{1}{\delta^{\xi}\Gamma(\xi)}\exp[\frac{\delta-1}{\delta}(\tau-\rho)](\tau-\rho)^{\xi-1}g(\rho)$ and integrating the resultant inequality with respect to ρ over $[a,\tau]$, $a<\tau\leq b$, we get

$$\begin{aligned}&\frac{1}{\delta^{\xi}\Gamma(\xi)}\int_a^{\tau}\exp[\frac{\delta-1}{\delta}(\tau-\rho)](\tau-\rho)^{\xi-1}\frac{\Phi(f(\rho))}{f(\rho)}g(\rho)d\rho\\ \leq&\frac{1}{\delta^{\xi}\Gamma(\xi)}\int_a^{\tau}\exp[\frac{\delta-1}{\delta}(\tau-\rho)](\tau-\rho)^{\xi-1}\frac{\Phi(g(\rho))}{g(\rho)}g(\rho)d\rho,\end{aligned}$$

which, in view of Equation (4), can be written as

$$_a\mathcal{J}^{\xi,\delta}\left(\frac{\Phi(f(\tau))}{f(\tau)}g(\tau)\right)\leq{}_a\mathcal{J}^{\xi,\delta}\left(\Phi(g(\tau))\right). \tag{12}$$

Hence, from Equations (10) and (12), we get Equation (6). □

Remark 2. *Applying Theorem 3 for $\delta=1$, we get Theorem 3.1 proved by [59].*

Remark 3. *Applying Theorem 3 for $\xi=\delta=1$ and $x=b$, we get Theorem 1.*

Theorem 4. *Suppose that the functions f and g are positive and continuous on $[a,b]$, $(a<b)$ and $f\leq g$ on $[a,b]$. If the function $\frac{f}{g}$, $(g\neq 0)$ is decreasing and the function f is increasing on $[a,b]$, then, for any convex function Φ with $\Phi(0)=0$, the following inequality satisfies the proportional fractional integral operator given by Equation (4)*

$$\frac{_a\mathcal{J}^{\xi,\delta}\left[f(\tau)\right]\,{}_a\mathcal{J}^{\lambda,\delta}\left[\Phi(g(\tau))\right]+{}_a\mathcal{J}^{\lambda,\delta}\left[f(\tau)\right]\,{}_a\mathcal{J}^{\xi,\delta}\left[\Phi(g(\tau))\right]}{_a\mathcal{J}^{\xi,\delta}\left[g(\tau)\right]\,{}_a\mathcal{J}^{\lambda,\delta}\left[\Phi(f(\tau))\right]+{}_a\mathcal{J}^{\lambda,\delta}\left[g(\tau)\right]\,{}_a\mathcal{J}^{\xi,\delta}\left[\Phi(f(\tau))\right]}\geq 1,$$

where $\delta\in(0,1]$, $\xi,\lambda\in\mathbb{C}$ with $Re(\xi)>0$ and $Re(\lambda)>0$.

Proof. Since Φ is convex function with $\Phi(0)=0$, the function $\frac{f(\tau)}{\tau}$ is increasing. As f is increasing, the function $\frac{\Phi(f(\tau))}{f(\tau)}$ is also increasing. Clearly, the function $\frac{f(\tau)}{g(\tau)}$ is decreasing for all $\rho,\theta\in[a,\tau]$, $a<\tau\leq b$. Multiplying Equation (9) by $\frac{1}{\delta^{\lambda}\Gamma(\lambda)}\exp[\frac{\delta-1}{\delta}(\tau-\theta)](\tau-\theta)^{\lambda-1}$ and integrating the resultant inequality with respect to θ over $[a,\tau]$, $a<\tau\leq b$, we get

$$\begin{aligned}&{}_a\mathcal{J}^{\lambda,\delta}(f(\tau))\,{}_a\mathcal{J}^{\xi,\delta}\left(\frac{\Phi(f(\tau))}{f(\tau)}g(\tau)\right)+{}_a\mathcal{J}^{\lambda,\delta}\left(\frac{\Phi(f(\tau))}{f(\tau)}g(\tau)\right){}_a\mathcal{J}^{\xi,\delta}\left(f(\tau)\right)\\ \geq&{}_a\mathcal{J}^{\xi,\delta}\left(g(\tau)\right){}_a\mathcal{J}^{\lambda,\delta}\left(\frac{\Phi(f(\tau)}{f(\tau)}f(\tau)\right)+{}_a\mathcal{J}^{\xi,\delta}\left(\frac{\Phi(f(\tau)}{f(\tau)}f(\tau)\right){}_a\mathcal{J}^{\lambda,\delta}\left(g(\tau)\right).\end{aligned} \tag{13}$$

Now, since $f\leq g$ on $[a,b]$ and $\frac{\Phi(\tau)}{\tau}$ is an increasing function, for $\rho\in[a,\tau]$, $a<\tau\leq b$, we have

$$\frac{\Phi(f(\rho))}{f(\rho)}\leq\frac{\Phi(g(\rho))}{g(\rho)}. \tag{14}$$

Multiplying both sides of Equation (14) by $\frac{1}{\delta^{\xi}\Gamma(\xi)}\exp[\frac{\delta-1}{\delta}(\tau-\rho)](\tau-\rho)^{\xi-1}g(\rho)$ and integrating the resultant inequality with respect to ρ over $[a,\tau]$, $a<\tau\leq b$, we get

$$\begin{aligned}&\frac{1}{\delta^{\xi}\Gamma(\xi)}\int_a^{\tau}\exp[\frac{\delta-1}{\delta}(\tau-\rho)](\tau-\rho)^{\xi-1}\frac{\Phi(f(\rho))}{f(\rho)}g(\rho)d\rho\\ \leq&\frac{1}{\delta^{\xi}\Gamma(\xi)}\int_a^{\tau}\exp[\frac{\delta-1}{\delta}(\tau-\rho)](\tau-\rho)^{\xi-1}\frac{\Phi(g(\rho))}{g(\rho)}g(\rho)d\rho,\end{aligned}$$

which, in view of Equation (4), can be written as

$$ {}_a\mathcal{J}^{\xi,\delta}\left(\frac{\Phi(f(\tau))}{f(\tau)}g(\tau)\right) \leq {}_a\mathcal{J}^{\xi,\delta}\left(\Phi(g(\tau))\right). \tag{15} $$

Similarly, one can obtain

$$ {}_a\mathcal{J}^{\lambda,\delta}\left(\frac{\Phi(f(\tau))}{f(\tau)}g(\tau)\right) \leq {}_a\mathcal{J}^{\lambda,\delta}\left(\Phi(g(\tau))\right). \tag{16} $$

Hence, from Equations (12), (13), (15), and (16), we get the desired result. □

Remark 4. *Setting $\xi = \lambda$, Theorem 4 leads to Theorem 3.*

Remark 5. *Applying Theorem 4 for $\delta = 1$, we get Theorem 3.3 proved by Dahmani [59].*

Theorem 5. *Suppose that the functions f, h, and g are positive and continuous on $[a,b]$, $(a < b)$ and $f \leq h$ on $[a,b]$. If the function $\frac{f}{g}$ is decreasing and the functions f and h are increasing on $[a,b]$, then, for any convex function Φ with $\Phi(0) = 0$, the following inequality satisfies the proportional fractional integral operator given by Equation (4)*

$$ \frac{{}_a\mathcal{J}^{\xi,\delta}\left[f(\tau)\right]}{{}_a\mathcal{J}^{\xi,\delta}\left[g(\tau)\right]} \geq \frac{{}_a\mathcal{J}^{\xi,\delta}\left[\Phi(f(\tau))h(\tau)\right]}{{}_a\mathcal{J}^{\xi,\delta}\left[\Phi(g(\tau))h(\tau)\right]}, $$

where $\delta \in (0,1]$, $\xi \in \mathbb{C}$ with $Re(\xi) > 0$.

Proof. Since Φ is convex function such that $\Phi(0) = 0$, the function $\frac{\Phi(\tau)}{\tau}$ is increasing. As the function f is increasing, $\frac{\Phi(f(\tau))}{f(\tau)}$ is also increasing. Clearly, the function $\frac{f(\tau)}{g(\tau)}$ is decreasing for all $\rho, \theta \in [a,\tau], a < \tau \leq b$.

$$ \left(\frac{\Phi(f(\rho))}{f(\rho)}h(\rho) - \frac{\Phi(f(\theta))}{f(\theta)}h(\theta)\right)\left(f(\theta)g(\rho) - f(\rho)g(\theta)\right) \geq 0. $$

It follows that

$$ \frac{\Phi(f(\rho))h(\rho)}{f(\rho)}f(\theta)g(\rho) + \frac{\Phi(f(\theta))h(\theta)}{f(\theta)}f(\rho)g(\theta) - \frac{\Phi(f(\theta))h(\theta)}{f(\theta)}f(\theta)g(\rho) - \frac{\Phi(f(\rho))h(\rho)}{f(\rho)}f(\rho)g(\theta) \geq 0. \tag{17} $$

Multiplying Equation (17) by $\frac{1}{\delta^{\xi}\Gamma(\xi)}\exp[\frac{\delta-1}{\delta}(\tau-\rho)](\tau-\rho)^{\xi-1}$ and integrating the resultant inequality with respect to ρ over $[a,\tau], a < \tau \leq b$, we have

$$ \begin{aligned} &\frac{1}{\delta^{\xi}\Gamma(\xi)}\int_a^{\tau}\exp[\frac{\delta-1}{\delta}(\tau-\rho)](\tau-\rho)^{\xi-1}\frac{\Phi(f(\rho))}{f(\rho)}f(\theta)g(\rho)h(\rho)d\rho \\ &+\frac{1}{\delta^{\xi}\Gamma(\xi)}\int_a^{\tau}\exp[\frac{\delta-1}{\delta}(\tau-\rho)](\tau-\rho)^{\xi-1}\frac{\Phi(f(\theta))}{f(\theta)}f(\rho)g(\theta)h(\theta)d\rho \\ &-\frac{1}{\delta^{\xi}\Gamma(\xi)}\int_a^{\tau}\exp[\frac{\delta-1}{\delta}(\tau-\rho)](\tau-\rho)^{\xi-1}\frac{\Phi(f(\theta))}{f(\theta)}f(\theta)h(\theta)g(\rho)d\rho \\ &-\frac{1}{\delta^{\xi}\Gamma(\xi)}\int_a^{\tau}\exp[\frac{\delta-1}{\delta}(\tau-\rho)](\tau-\rho)^{\xi-1}\frac{\Phi(f(\rho))}{f(\rho)}f(\rho)h(\rho)g(\theta)d\rho \geq 0. \end{aligned} $$

It follows that

$$f(\theta)\,{}_a\mathcal{J}^{\xi,\delta}\left(\frac{\Phi(f(\tau))}{f(\tau)}g(\tau)h(\tau)\right)+\left(\frac{\Phi(f(\theta))}{f(\theta)}g(\theta)h(\theta)\right){}_a\mathcal{J}^{\xi,\delta}(f(\tau))$$
$$-\left(\frac{\Phi(f(\theta))}{f(\theta)}f(\theta)h(\theta)\right){}_a\mathcal{J}^{\xi,\delta}(g(\tau))-g(\theta)\,{}_a\mathcal{J}^{\xi,\delta}\left(\frac{\Phi(f(\tau))}{f(\tau)}f(\tau)h(\tau)\right)\geq 0. \tag{18}$$

Again, multiplying both sides of Equation (18) by $\frac{1}{\delta^{\xi}\Gamma(\xi)}\exp[\frac{\delta-1}{\delta}(\tau-\theta)](\tau-\theta)^{\xi-1}$ and integrating the resultant inequality with respect to θ over $[a,\tau]$, $a<\tau\leq b$, we get

$$ {}_a\mathcal{J}^{\xi,\delta}(f(\tau))\,{}_a\mathcal{J}^{\xi,\delta}\left(\frac{\Phi(f(\tau))}{f(\tau)}g(\tau)h(\tau)\right)+{}_a\mathcal{J}^{\xi,\delta}\left(\frac{\Phi(f(\tau))}{f(\tau)}g(\tau)h(\tau)\right){}_a\mathcal{J}^{\xi,\delta}\left(f(\tau)\right)$$
$$\geq {}_a\mathcal{J}^{\xi,\delta}\left(g(\tau)\right)\,{}_a\mathcal{J}^{\xi,\delta}\left(\Phi(f(\tau))h(\tau)\right)+{}_a\mathcal{J}^{\xi,\delta}\left(\Phi(f(\tau))h(\tau)\right)\,{}_a\mathcal{J}^{\xi,\delta}\left(g(\tau)\right).$$

It follows that

$$\frac{{}_a\mathcal{J}^{\xi,\delta}(f(\tau))}{{}_a\mathcal{J}^{\xi,\delta}\left(g(\tau)\right)}\geq\frac{{}_a\mathcal{J}^{\xi,\delta}\left(\Phi(f(\tau))h(\tau)\right)}{{}_a\mathcal{J}^{\xi,\delta}\left(\frac{\Phi(f(\tau))}{f(\tau)}g(\tau)h(\tau)\right)}. \tag{19}$$

In addition, since $f\leq g$ on $[a,b]$ and $\frac{\Phi(\tau)}{\tau}$ is an increasing function, for $\eta,\theta\in[a,b]$, we have

$$\frac{\Phi(f(\eta))}{f(\eta)}\leq\frac{\Phi(g(\eta))}{g(\eta)}. \tag{20}$$

Multiplying both sides of Equation (20) by $\frac{1}{\delta^{\xi}\Gamma(\xi)}\exp[\frac{\delta-1}{\delta}(\tau-\eta)](\tau-\eta)^{\xi-1}g(\eta)h(\eta)$ and integrating the resultant inequality with respect to η over $[a,\tau]$, $a<\tau\leq b$, we get

$$\frac{1}{\delta^{\xi}\Gamma(\xi)}\int_a^{\tau}\exp[\frac{\delta-1}{\delta}(\tau-\eta)](\tau-\eta)^{\xi-1}\frac{\Phi(f(\eta))}{f(\eta)}g(\eta)h(\eta)d\eta$$
$$\leq\frac{1}{\delta^{\xi}\Gamma(\xi)}\int_a^{\tau}\exp[\frac{\delta-1}{\delta}(\tau-\eta)](\tau-\eta)^{\xi-1}\frac{\Phi(g(\eta))}{g(\eta)}g(\eta)h(\eta)d\eta,$$

which, in view of Equation (4), can be written as

$${}_a\mathcal{J}^{\xi,\delta}\left(\frac{\Phi(f(\tau))}{f(\tau)}g(\tau)h(\tau)\right)\leq{}_a\mathcal{J}^{\xi,\delta}\left(\Phi(g(\tau))h(\tau)\right). \tag{21}$$

Hence, from Equations (21) and (19), we obtain the required result. □

Remark 6. *Applying Theorem 5 for $\delta=1$, we get Theorem 3.5 proved by Dahmani [59].*

Remark 7. *Applying Theorem 5 for $\delta=\xi=1$ and $x=b$, we get Theorem 2.*

Theorem 6. *Suppose that the functions f, h, and g are positive and continuous on $[a,b]$, $(a<b)$ and $f\leq g$ on $[a,b]$. If the function $\frac{f}{g}$ is decreasing and the functions f and h are increasing on $[a,b]$, then, for any convex function Φ with $\Phi(0)=0$, the following inequality satisfies the proportional fractional integral operator given by Equation (4)*

$$\frac{{}_a\mathcal{J}^{\xi,\delta}\left[f(\tau)\right]{}_a\mathcal{J}^{\lambda,\delta}\left[\Phi(g(\tau))h(\tau)\right]+{}_a\mathcal{J}^{\lambda,\delta}\left[f(\tau)\right]{}_a\mathcal{J}^{\xi,\delta}\left[\Phi(g(\tau))h(\tau)\right]}{{}_a\mathcal{J}^{\xi,\delta}\left[g(\tau)\right]{}_a\mathcal{J}^{\lambda,\delta}\left[\Phi(f(\tau))h(\tau)\right]+{}_a\mathcal{J}^{\lambda,\delta}\left[g(\tau)\right]{}_a\mathcal{J}^{\xi,\delta}\left[\Phi(f(\tau))h(\tau)\right]}\geq 1, \tag{22}$$

where $\delta\in(0,1]$, $\xi,\lambda\in\mathbb{C}$ with $Re(\xi)>0$ and $Re(\lambda)>0$.

Proof. Multiplying both sides of Equation (18) by $\frac{1}{\delta^{\lambda}\Gamma(\lambda)}\exp[\frac{\delta-1}{\delta}(\tau-\theta)](\tau-\theta)^{\lambda-1}$ and integrating the resultant inequality with respect to θ over $[a,\tau], a<\tau\leq b$, we get

$$\begin{aligned}&{}_a\mathcal{J}^{\lambda,\delta}(f(\tau))\,{}_a\mathcal{J}^{\xi,\delta}\left(\frac{\Phi(f(\tau))}{f(\tau)}g(\tau)h(\tau)\right)+{}_a\mathcal{J}^{\lambda,\delta}\left(\frac{\Phi(f(\tau))}{f(\tau)}g(\tau)h(\tau)\right){}_a\mathcal{J}^{\xi,\delta}(f(\tau))\\ &\geq {}_a\mathcal{J}^{\xi,\delta}(g(\tau))\,{}_a\mathcal{J}^{\lambda,\delta}\left(\frac{\Phi(f(\tau))}{f(\tau)}f(\tau)h(\tau)\right)+{}_a\mathcal{J}^{\xi,\delta}\left(\frac{\Phi(f(\tau))}{f(\tau)}f(\tau)h(\tau)\right){}_a\mathcal{J}^{\lambda,\delta}(g(\tau)).\end{aligned} \tag{23}$$

Since $f\leq g$ on $[a,b]$ and $\frac{\Phi(\tau)}{\tau}$ is an increasing function, for $\eta,\theta\in[1,x], a<\tau\leq b$, we have

$$\frac{\Phi(f(\eta))}{f(\eta)}\leq\frac{\Phi(g(\eta))}{g(\eta)}. \tag{24}$$

Multiplying both sides of Equation (24) by $\frac{1}{\delta^{\xi}\Gamma(\xi)}\exp[\frac{\delta-1}{\delta}(\tau-\eta)](\tau-\eta)^{\xi-1}g(\eta)h(\eta)$, $\eta\in[a,x], a<\tau\leq b$ and integrating the resultant inequality with respect to η over $[a,\tau], a<\tau\leq b$, we get

$${}_a\mathcal{J}^{\xi,\delta}\left(\frac{\Phi(f(\tau))}{f(\tau)}g(\tau)h(\tau)\right)\leq {}_a\mathcal{J}^{\xi,\delta}\left(\Phi(g(\tau))h(\tau)\right)). \tag{25}$$

Similarly, one can obtain

$${}_a\mathcal{J}^{\eta,\delta}\left(\frac{\Phi(f(\tau))}{f(\tau)}g(\tau)h(\tau)\right)\leq {}_a\mathcal{J}^{\eta,\delta}\left(\Phi(g(\tau))h(\tau)\right)). \tag{26}$$

Hence, from Equations (23), (25), and (26), we obtain the required inequality in Equation (22). □

Remark 8. *If we consider $\xi=\lambda$, then Theorem 6 leads to Theorem 5.*

Remark 9. *Applying Theorem 6 for $\delta=1$, we get Theorem 3.7 of Dahmani [59].*

4. Concluding Remarks

Some interesting integral inequalities for convex functions were presented by Liu et al. ([7] Theorems 9 and 10). Later, Dahmani [59] improved these integral inequalities by utilizing the R-L fractional integral operator. Here, we present some new fractional proportional integral inequalities for convex functions by utilizing the proportional fractional integrals. In fact, we established the inequalities presented in Theorem 1 and Theorem 2 using the fractional proportional integrals, which are nonlocal and their orders depend on two indices: δ, which is the proportional index, and ξ, which is the iterated index.

Author Contributions: All the authors contributed equally. All authors have read and agreed to the published version of the manuscript.

Funding: The third author would like to thank Prince Sultan University for the support through the research group "Nonlinear Analysis Methods in Applied Mathematics" (NAMAM), group number RG-DES-2017-01-17.

Conflicts of Interest: The authors declare that they have no competing interests.

References

1. Mitrinovic, D.S.; Pecaric, J.E.; Fink, A.M. *Classical and New Inequalities in Analysis*; Kluwer Academic Publishers: Dordrecht, The Netherlands, 1993.
2. Pachpatte, B.G. *Mathematical Inequalities*, 1st ed.; North-Holland Mathematical Library (Volume 67) (Book 67); Elsevier Science: Amsterdam, The Netherlands, 2005.

3. Qi, F. Several integral inequalities. *JIPAM* **2000**, *1*, 19.
4. Sarikaya, M.Z.; Yildirim, H.; Saglam, A. On Hardy type integral inequality associated with the generalized translation. *Int. J. Contemp. Math. Sci.* **2006**, *1*, 333–340. [CrossRef]
5. Ngo, Q.A.; Thang, D.D.; Dat, T.T.; Tuan, D.A. Notes on an integral inequality. *J. Inequal. Pure Appl. Math.* **2006**, *7*, 120.
6. Liu, W.J.; Cheng, G.S.; Li, C.C. Further development of an open problem concerning an integral inequality. *JIPAM* **2008**, *9*, 14.
7. Liu, W.J.; Ngô, Q.A.; Huy, V.N. Several interesting integral inequalities. *J. Math. Inequal.* **2009**, *3*, 201–212. [CrossRef]
8. Bougoufa, L. An integral inequality similar to Qi inequality. *JIPAM* **2005**, *6*, 27.
9. Boukerrioua, K.; Guezane Lakoud, A. On an open question regarding an integral inequality. *JIPAM* **2007**, *8*, 77.
10. Dahmani, Z.; Bedjaoui, N. Some generalized integral inequalities. *J. Adv. Res. Appl. Math.* **2011**, *3*, 58–66. [CrossRef]
11. Dahmani, Z.; Metakkel Elard, H. Generalizations of some integral inequalities using Riemann-Liouville operator. *Int. J. Open Probl. Compt. Math.* **2011**, *4*, 40–46.
12. Liu, W.J.; Li, C.C.; Dong, J.W. On an open problem concerning an integral inequality. *JIPAM* **2007**, *8*, 74.
13. Dahmani, Z.; Tabharit, L. On weighted Gruss type inequalities via fractional integration. *J. Adv. Res. Pure Math.* **2010**, *2*, 31–38. [CrossRef]
14. Dahmani, Z. New inequalities in fractional integrals. *Int. J. Nonlinear Sci.* **2010**, *9*, 493–497.
15. Nisar, K.S.; Qi, F.; Rahman, G.; Mubeen, S.; Arshad, M. Some inequalities involving the extended gamma function and the Kummer confluent hypergeometric k-function. *J. Inequal. Appl.* **2018**, *2018*, 135. [CrossRef] [PubMed]
16. Nisar, K.S.; Rahman, G.; Choi, J.; Mubeen, S.; Arshad, M. Certain Gronwall type inequalities associated with Riemann-Liouville k- and Hadamard k-fractional derivatives and their applications. *East Asian Math. J.* **2018**, *34*, 249–263.
17. Rahman, G.; Nisar, K.S.; Mubeen, S.; Choi, J. Certain Inequalities involving the (k,η)-fractional integral operator. *Far East J. Math. Sci. (FJMS)* **2018**, *103*, 1879–1888. [CrossRef]
18. Sarikaya, M.Z.; Dahmani, Z.; Kiris, M.E.; Ahmad, F. (k,s)-Riemann-Liouville fractional integral and applications. *Hacet. J. Math. Stat.* **2016**, *45*, 77–89. [CrossRef]
19. Set, E.; Tomar, M.; Sarikaya, M.Z. On generalized Grüss type inequalities for k-fractional integrals. *Appl. Math. Comput.* **2015**, *269*, 29–34. [CrossRef]
20. Sarikaya, M.Z.; Budak, H. Generalized Ostrowski type inequalities for local fractional integrals. *Proc. Am. Math. Soc.* **2017**, *145*, 1527–1538. [CrossRef]
21. Set, E.; Noor, M.A.; Awan, M.U.; Gözpinar, A. Generalized Hermite-Hadamard type inequalities involving fractional integral operators. *J. Inequal. Appl.* **2017**, *169*, 10. [CrossRef]
22. Dahmani, Z. New classes of integral inequalities of fractional order. *LE MATEMATICHE* **2014**, *LXIX*, 237–247. [CrossRef]
23. Podlubny, I. *Fractional Differential Equations*; Academic Press: London, UK, 1999.
24. Samko, S.G.; Kilbas, A.A.; Marichev, O.I. *Fractional Integrals and Derivatives, Theory and Applications*; Taylor & Francis: Abingdon, UK, 1993.
25. Abdeljawad, T. On Conformable Fractional Calculus. *J. Comput. Appl. Math.* **2015**, *279*, 57–66. [CrossRef]
26. Abdeljawad, T.; Baleanu, D. Monotonicity results for fractional difference operators with discrete exponential kernels. *Adv. Differ. Equ.* **2017**, *2017*, 78. [CrossRef]
27. Abdeljawad, T.; Baleanu, D. On Fractional Derivatives with Exponential Kernel and their Discrete Versions. *Rep. Math. Phys.* **2017**, *80*, 11–27. [CrossRef]
28. Anderson, D.R.; Ulness, D.J. Newly defined conformable derivatives. *Adv. Dyn. Syst. Appl.* **2015**, *10*, 109–137.
29. Atangana, A.; Baleanu, D. New fractional derivatives with nonlocal and non-singular kernel. Theory and application to heat transfer model. *Therm. Sci.* **2016**, *20*, 763–769. [CrossRef]
30. Caputo, M.; Fabrizio, M. A new Definition of Fractional Derivative without Singular Kernel. *Progr. Fract. Differ. Appl.* **2015**, *1*, 73–85.
31. Jarad, F.; Ugurlu, E.; Abdeljawad, T.; Baleanu, D. On a new class of fractional operators. *Adv. Differ. Equ.* **2017**, *2017*, 247. [CrossRef]

32. Khalil, R.; Horani, M.A.; Yousef, A.; Sababheh, M. A new definition of fractional derivative. *J. Comput. Appl. Math.* **2014**, *264*, 65–70. [CrossRef]
33. Losada, J.; Nieto, J.J. Properties of a New Fractional Derivative without Singular Kernel. *Progr. Fract. Differ. Appl.* **2015**, *1*, 87–92.
34. Rahimi, Z.; Sumelka, W.; Yang, X.J. A new fractional nonlocal model and its application in free vibration of Timoshenko and Euler-Bernoulli beams. *Eur. Phys. J. Plus* **2017**, *132*, 479. [CrossRef]
35. Khan, M.A.; Khurshid, Y.; Du, T.-S.; Chu, Y.-M. Generalization of Hermite-Hadamard type inequalities via conformable fractional integrals. *J. Funct. Spaces* **2018**, *2018*, 5357463.
36. Khan, M.A.; Iqbal, A.; Suleman, M.; Chu, Y.-M. Hermite-Hadamard type inequalities for fractional integrals via Green's function. *J. Inequal. Appl.* **2018**, *2018*, 161. [CrossRef] [PubMed]
37. Huang, C.J.; Rahman, G.; Nisar, K.S.; Ghaffar, A.; Qi, F. Some Inequalities of Hermite-Hadamard type for *k*-fractional conformable integrals. *Aust. J. Math. Anal. Appl.* **2019**, *16*, 1–9.
38. Khurshid, Y.; Khan, M.A.; Chu, Y.-M. Conformable integral inequalities of the Hermite-Hadamard type in terms of GG- and GA-2 convexities. *J. Funct. Spaces* **2019**, *2019*, 6926107. [CrossRef]
39. Khurshid, Y.; Khan, M.A.; Chu, Y.-M.; Khan, Z.A. Hermite-Hadamard-Fejer inequalities for conformable fractional integrals via preinvex functions. *J. Funct. Spaces* **2019**, *2019*, 3146210. [CrossRef]
40. Mubeen, S.; Habib, S.; Naeem, M.N. The Minkowski inequality involving generalized *k*-fractional conformable integral. *J. Inequal. Appl.* **2019**, *2019*, 81. [CrossRef]
41. Nisar, K.S.; Rahman, G.; Mehrez, K. Chebyshev type inequalities via generalized fractional conformable integrals. *J. Inequal. Appl.* **2019**, *2019*, 245. [CrossRef]
42. Niasr, K.S.; Tassadiq, A.; Rahman, G.; Khan, A. Some inequalities via fractional conformable integral operators. *J. Inequal. Appl.* **2019**, *2019*, 217. [CrossRef]
43. Qi, F.; Rahman, G.; Hussain, S.M.; Du, W.S.; Nisar, K.S. Some inequalities of Čebyšev type for conformable *k*-fractional integral operators. *Symmetry* **2018**, *10*, 614. [CrossRef]
44. Rahman, G.; Nisar, K.S.; Qi, F. Some new inequalities of the Gruss type for conformable fractional integrals. *Aims Math.* **2018**, *3*, 575–583. [CrossRef]
45. Rahman, G.; Ullah, Z.; Khan, A.; Set, E.; Nisar, K.S. Certain Chebyshev type inequalities involving fractional conformable integral operators. *Mathematics* **2019**, *7*, 364. [CrossRef]
46. Ortega, A.; Rosales, J.J. Newton's law of cooling with fractional conformable derivative. *Revista Mexicana de Física* **2018**, *64*, 172–175. [CrossRef]
47. Hammad, M.A.; Khalil, R. Abel's formula and Wronskian for conformable fractional differential equations. *Int. J. Differ. Equ. Appl.* **2014**, *13*, 177–183.
48. Ilie, M.; Biazar, J.; Ayati, Z. General solution of Bernoulli and Riccati fractional differential equations based on conformable fractional derivative. *Int. J. Appl. Math. Res.* **2017**, *6*, 49–51.
49. Meng, S.; Cui, Y. The extremal solution to conformable fractional differential equations involving integral boundary condition. *Mathematics* **2019**, *7*, 186; doi:10.3390/math7020186. [CrossRef]
50. Jarad, F.; Abdeljawad, T.; Alzabut, J. Generalized fractional derivatives generated by a class of local proportional derivatives. *Eur. Phys. J. Spec. Top.* **2017**, *226*, 3457–3471. [CrossRef]
51. Wang, M.-K.; Chu, H.-H.; Chu, Y.-M. Precise Bounds for the Weighted Hölder Mean of the Complete P-Elliptic Integrals. *J. Math. Anal. Appl.* **2019**, *480*. [CrossRef]
52. Yang, Z.-H.; Qian, W.-M.; Chu, Y.-M.; Zhang, W. On rational bounds for the gamma function. *J. Inequal. Appl.* **2017**, *2017*, 210. [CrossRef]
53. Yang, Z.-H.; Qian, W.-M.; Chu, Y.-M.; Zhang, W. Monotonicity rule for the quotient of two functions and its application. *J. Inequal. Appl.* **2017**, *2017*, 106. [CrossRef]
54. Alzabut, J.; Abdeljawad, T.; Jarad, F.; Sudsutad, W. A Gronwall inequality via the generalized proportional fractional derivative with applications. *J. Inequal. Appl.* **2019**, *2019*, 101. [CrossRef]
55. Rahman, G.; Khan, A.; Abdeljawad, T.; Nisar, K.S. The Minkowski inequalities via generalized proportional fractional integral operators. *Adv. Differ. Equ.* **2019**, *2019*, 287. [CrossRef]
56. Rahman, G.; Abdeljawad, T.; Khan, A.; Nisar, K.S. Some fractional proportional integral inequalities. *J. Inequal. Appl.* **2019**, *2019*, 244. [CrossRef]
57. Rahman, G.; Abdeljawad, T.; Jarad, F.; Khan, A.; Nisar, K.S. Certain inequalities via generalized proportional Hadamard fractional integral operators. *Adv. Differ. Equ.* **2019**, *2019*, 454. [CrossRef]

58. Rahman, G.; Abdeljawad, T.; Jarad, F.; Nisar, K.S. Bounds of generalized proportional fractional integrals in general form via convex functions and their applications. *Mathematics* **2020**, *8*, 113. [CrossRef]
59. Dahmani, Z. A note on some new fractional results involving convex functions. *Acta Math. Univ. Comen.* **2012**, *LXXXI*, 241–246.

mathematics

MDPI

Article

Nonlinear Integro-Differential Equations Involving Mixed Right and Left Fractional Derivatives and Integrals with Nonlocal Boundary Data

Bashir Ahmad [1,*], Abrar Broom [1], Ahmed Alsaedi [1] and Sotiris K. Ntouyas [1,2]

1 Nonlinear Analysis and Applied Mathematics (NAAM)-Research Group, Department of Mathematics, Faculty of Science, King Abdulaziz University, P.O. Box 80203, Jeddah 21589, Saudi Arabia; abrarbroom1992@gmail.com (A.B.); aalsaedi@hotmail.com (A.A.); sntouyas@uoi.gr (S.K.N.)

2 Department of Mathematics, University of Ioannina, 451 10 Ioannina, Greece

* Correspondence: bashirahmad_qau@yahoo.com or bahmad@kau.edu.sa

Received: 2 February 2020; Accepted: 25 February 2020; Published: 3 March 2020

Abstract: In this paper, we study the existence of solutions for a new nonlocal boundary value problem of integro-differential equations involving mixed left and right Caputo and Riemann–Liouville fractional derivatives and Riemann–Liouville fractional integrals of different orders. Our results rely on the standard tools of functional analysis. Examples are constructed to demonstrate the application of the derived results.

Keywords: caputo-type fractional derivative; fractional integral; existence; fixed point

MSC: 34A08; 34B10; 34B15

---

1. Introduction

In the last few decades, fractional-order single-valued and multivalued boundary value problems containing different fractional derivatives such as Caputo, Riemann–Liouville, Hadamard, etc., and classical, nonlocal, integral boundary conditions have been extensively studied, for example, see the articles [1–12] and the references cited therein.

In the study of variational principles, fractional differential equations involving both left and right fractional derivatives give rise to a special class of Euler–Lagrange equations, for details, see [13] and the references cited therein. Let us consider some works on mixed fractional-order boundary value problems. In [14], the authors discussed the existence of an extremal solution to a nonlinear system involving the right-handed Riemann–Liouville fractional derivative. In [15], a two-point nonlinear higher order fractional boundary value problem involving left Riemann–Liouville and right Caputo fractional derivatives was investigated, while a problem in terms of left Caputo and right Riemann–Liouville fractional derivatives was studied in [16]. A nonlinear fractional oscillator equation containing left Riemann–Liouville and right Caputo fractional derivatives was investigated in [17]. In a recent paper [18], the authors proved some existence results for nonlocal boundary value problems of differential equations and inclusions containing both left Caputo and right Riemann–Liouville fractional derivatives.

Integro-differential equations appear in the mathematical modeling of several real world problems such as, heat transfer phenomena [19,20], forced-convective flow over a heat-conducting plate [21], etc. In [22], the authors studied the steady heat-transfer in fractal media via the local fractional nonlinear Volterra integro-differential equations. Electromagnetic waves in a variety of dielectric media with susceptibility following a fractional power-law are described by the fractional integro-differential equations [23].

Motivated by aforementioned applications of integro-differential equations and [18], we introduce a new kind of integro-differential equation involving right-Caputo and left-Riemann–Liouville fractional derivatives of different orders and right-left Riemann–Liouville fractional integrals and solve it subject to nonlocal boundary conditions. In precise terms, we prove existence and uniqueness of solutions for the problem given by

$$ {}^{C}D_{1-}^{\alpha}\, {}^{RL}D_{0+}^{\beta} y(t) + \lambda I_{1-}^{p} I_{0+}^{q} h(t, y(t)) = f(t, y(t)), \;\; t \in J := [0,1], \tag{1} $$

$$ y(0) = y(\xi) = 0, \quad y(1) = \delta y(\mu), \;\; 0 < \xi < \mu < 1, \tag{2} $$

where ${}^{C}D_{1-}^{\alpha}$ and ${}^{RL}D_{0+}^{\beta}$ denote the right Caputo fractional derivative of order $\alpha \in (1,2]$ and the left Riemann–Liouville fractional derivative of order $\beta \in (0,1]$, I_{1-}^{p} and I_{0+}^{q} denote the right and left Riemann–Liouville fractional integrals of orders $p, q > 0$ respectively, $f, h : [0,1] \times \mathbb{R} \to \mathbb{R}$ are given continuous functions and $\delta, \lambda \in \mathbb{R}$. It is imperative to notice that the integro-differential equation in (1) and (2) contains mixed type (integral and nonintegral) nonlinearities.

We organize the rest of the paper as follows. Section 2 contains some preliminary concepts related to our work. In Section 3, we prove an auxiliary lemma for the linear variant of the problem (1) and (2). Then we derive the existence results for the problem (1) and (2) by applying a fixed point theorem due to Krasnoselski and Leray–Schauder nonlinear alternative, while the uniqueness result is established via Banach contraction mapping principle. Examples illustrating the main results are also presented.

2. Preliminaries

In this section, we recall some related definitions of fractional calculus [1].

Definition 1. *The left and right Riemann–Liouville fractional integrals of order $\beta > 0$ for an integrable function $g : (0, \infty) \to \mathbb{R}$ are respectively defined by*

$$ I_{0+}^{\beta} g(t) = \int_0^t \frac{(t-s)^{\beta-1}}{\Gamma(\beta)} g(s)ds \;\; \text{and} \;\; I_{1-}^{\beta} g(t) = \int_t^1 \frac{(s-t)^{\beta-1}}{\Gamma(\beta)} g(s)ds. $$

Definition 2. *The left Riemann–Liouville fractional derivative and the right Caputo fractional derivative of order $\beta \in (n-1, n], n \in \mathbb{N}$ for a function $g : (0,\infty) \to \mathbb{R}$ with $g \in \mathbb{C}^n((0,\infty), \mathbb{R})$ are respectively given by*

$$ D_{0+}^{\beta} g(t) = \frac{d^n}{dt^n} \int_0^t \frac{(t-s)^{n-\beta-1}}{\Gamma(n-\beta)} g(s)ds \;\; \text{and} \;\; {}^{c}D_{1-}^{\beta} g(t) = (-1)^n \int_t^1 \frac{(s-t)^{n-\beta-1}}{\Gamma(n-\beta)} g^{(n)}(s)ds. $$

Lemma 1. *If $p > 0$ and $q > 0$, then the following relations hold almost everywhere on $[a,b]$:*

$$ I_{1-}^{p} I_{1-}^{q} f(x) = I_{1-}^{p+q} f(x), \;\; I_{0+}^{p} I_{0+}^{q} f(x) = I_{0+}^{p+q} f(x). $$

3. Main Results

In the following lemma, we solve a linear variant of the problem (1) and (2).

Lemma 2. *Let $H, F \in C[0,1] \cap L(0,1)$. Then the linear problem*

$$ \begin{cases} {}^{C}D_{1-}^{\alpha}\, {}^{RL}D_{0+}^{\beta} y(t) + \lambda I_{1-}^{p} I_{0+}^{q} H(t) = F(t), & t \in J := [0,1], \\ y(0) = y(\xi) = 0, \quad y(1) = \delta y(\mu), \end{cases} \tag{3} $$

is equivalent to the fractional integral equation:

$$\begin{aligned} y(t) = &\int_0^t \frac{(t-s)^{\beta-1}}{\Gamma(\beta)}\Big[I_{1-}^{\alpha}F(s) - \lambda I_{1-}^{\alpha+p}I_{0+}^{q}H(s)\Big]ds \\ &+ a_1(t)\Big\{\delta\int_0^{\mu} \frac{(\mu-s)^{\beta-1}}{\Gamma(\beta)}\Big[I_{1-}^{\alpha}F(s) - \lambda I_{1-}^{\alpha+p}I_{0+}^{q}H(s)\Big]ds \\ &- \int_0^1 \frac{(1-s)^{\beta-1}}{\Gamma(\beta)}\Big[I_{1-}^{\alpha}F(s) - \lambda I_{1-}^{\alpha+p}I_{0+}^{q}H(s)\Big]ds\Big\} \\ &+ a_2(t)\int_0^{\xi} \frac{(\xi-s)^{\beta-1}}{\Gamma(\beta)}\Big[I_{1-}^{\alpha}F(s) - \lambda I_{1-}^{\alpha+p}I_{0+}^{q}H(s)\Big]ds, \end{aligned} \tag{4}$$

where

$$a_1(t) = \frac{1}{\Lambda}\Big[\xi^{\beta+1}t^{\beta} - \xi^{\beta}t^{\beta+1}\Big],\ a_2(t) = \frac{1}{\Lambda}\Big[t^{\beta}(1-\delta\mu^{\beta+1}) - t^{\beta+1}(1-\delta\mu^{\beta})\Big], \tag{5}$$

and it is assumed that

$$\Lambda = \xi^{\beta+1}(1-\delta\mu^{\beta}) - \xi^{\beta}(1-\delta\mu^{\beta+1}) \neq 0. \tag{6}$$

Proof. Applying the left and right fractional integrals I_{1-}^{α} and I_{0+}^{β} successively to the integro-differential equation in (3), and then using Lemma 1, we get

$$y(t) = I_{0+}^{\beta}I_{1-}^{\alpha}F(t) - \lambda I_{0+}^{\beta}I_{1-}^{\alpha+p}I_{0+}^{q}H(t) + c_0\frac{t^{\beta}}{\Gamma(\beta+1)} + c_1\frac{t^{\beta+1}}{\Gamma(\beta+2)} + c_2 t^{\beta-1}, \tag{7}$$

where c_0, c_1 and c_2 are unknown arbitrary constants.

In view of the condition $y(0) = 0$, it follows from (7) that $c_2 = 0$. Inserting $c_2 = 0$ in (7) and then using the nonlocal boundary conditions $y(\xi) = 0, y(1) = \delta y(\mu)$ in the resulting equation, we obtain a system of equations in c_0 and c_1 given by

$$\begin{aligned} c_0\Big(\frac{1-\delta\mu^{\beta}}{\Gamma(\beta+1)}\Big) + c_1\Big(\frac{1-\delta\mu^{\beta+1}}{\Gamma(\beta+2)}\Big) &= \delta A_1 - A_2, \\ c_0\Big(\frac{\xi^{\beta}}{\Gamma(\beta+1)}\Big) + c_1\Big(\frac{\xi^{\beta+1}}{\Gamma(\beta+2)}\Big) &= -A_3, \end{aligned} \tag{8}$$

where

$$A_1 = I_{0+}^{\beta}I_{1-}^{\alpha}F(\mu) - \lambda I_{0+}^{\beta}I_{1-}^{\alpha+p}I_{0+}^{q}H(\mu),\ A_2 = I_{0+}^{\beta}I_{1-}^{\alpha}F(1) - \lambda I_{0+}^{\beta}I_{1-}^{\alpha+p}I_{0+}^{q}H(1),$$
$$A_3 = I_{0+}^{\beta}I_{1-}^{\alpha}F(\xi) - \lambda I_{0+}^{\beta}I_{1-}^{\alpha+p}I_{0+}^{q}H(\xi).$$

Solving the system (8), we find that

$$c_0 = \frac{\Gamma(\beta+1)}{\Lambda}\Big[\xi^{\beta+1}(\delta A_1 - A_2) + (1-\delta\mu^{\beta+1})A_3\Big],$$
$$c_1 = \frac{-\Gamma(\beta+2)}{\Lambda}\Big[\xi^{\beta}(\delta A_1 - A_2) + (1-\delta\mu^{\beta})A_3\Big],$$

where Λ is defined by (6). Substituting the values of c_0 and c_1 together with the notations (5) in (7), we obtain the solution (4). The converse follows by direct computation. This completes the proof. □

Let $\mathcal{X} = C([0,1],\mathbb{R})$ denote the Banach space of all continuous functions from $[0,1] \to \mathbb{R}$ equipped with the norm $\|y\| = \sup\{|y(t)| : t \in [0,1]\}$. By Lemma 2, we define an operator $\mathcal{G} : \mathcal{X} \to \mathcal{X}$ associated with the problem (1) and (2) as

$$\begin{aligned}
\mathcal{G}y(t) &= \int_0^t \frac{(t-s)^{\beta-1}}{\Gamma(\beta)} \Big[I_{1-}^{\alpha} f(s,y(s)) - \lambda I_{1-}^{\alpha+p} I_{0+}^{q} h(s,y(s))\Big] ds \\
&+ a_1(t)\Bigg[\delta \int_0^{\mu} \frac{(\mu-s)^{\beta-1}}{\Gamma(\beta)} \Big[I_{1-}^{\alpha} f(s,y(s)) - \lambda I_{1-}^{\alpha+p} I_{0+}^{q} h(s,y(s))\Big] ds \\
&- \int_0^1 \frac{(1-s)^{\beta-1}}{\Gamma(\beta)} \Big[I_{1-}^{\alpha} f(s,y(s)) - \lambda I_{1-}^{\alpha+p} I_{0+}^{q} h(s,y(s))\Big] ds\Bigg] \\
&+ a_2(t) \int_0^{\xi} \frac{(\xi-s)^{\beta-1}}{\Gamma(\beta)} \Big[I_{1-}^{\alpha} f(s,y(s)) - \lambda I_{1-}^{\alpha+p} I_{0+}^{q} h(s,y(s))\Big] ds.
\end{aligned}$$

Notice that the fixed points of the operator $\mathcal{G}$ are solutions of the problem (1) and (2).

In the forthcoming analysis, we use the following estimates:

$$\begin{aligned}
\int_0^t \frac{(t-s)^{\beta-1}}{\Gamma(\beta)} I_{1-}^{\alpha+p} I_{0+}^{q} ds &= \int_0^t \frac{(t-s)^{\beta-1}}{\Gamma(\beta)} \int_s^1 \frac{(u-s)^{\alpha+p-1}}{\Gamma(\alpha+p)} \int_0^u \frac{(u-r)^{q-1}}{\Gamma(q)} dr\, du\, ds \\
&\leq \frac{t^{\beta}}{\Gamma(\beta+1)\Gamma(\alpha+p+1)\Gamma(q+1)}, \\
\int_0^t \frac{(t-s)^{\beta-1}}{\Gamma(\beta)} I_{1-}^{\alpha} ds &= \int_0^t \frac{(t-s)^{\beta-1}}{\Gamma(\beta)} \int_s^1 \frac{(u-s)^{\alpha-1}}{\Gamma(\alpha)} du\, ds \leq \frac{t^{\beta}}{\Gamma(\alpha+1)\Gamma(\beta+1)},
\end{aligned}$$

where we have used $u^q \leq 1,\ (1-s)^{\alpha+p} < 1;\ p,q > 0, 1 < \alpha \leq 2$.

In the sequel, we set

$$\Omega_1 = \frac{\Delta}{\Gamma(\alpha+1)},\ \Omega_2 = \frac{|\lambda|\Delta}{\Gamma(\alpha+p+1)\Gamma(q+1)}, \tag{9}$$

where

$$\begin{aligned}
\Delta &= \frac{1}{\Gamma(\beta+1)}\Big[1 + \bar{a}_1(|\delta|\mu^{\beta}+1) + \bar{a}_2\xi^{\beta}\Big], \\
\bar{a}_1 &= \max_{t\in[0,1]} |a_1(t)|,\ \bar{a}_2 = \max_{t\in[0,1]} |a_2(t)|.
\end{aligned}$$

3.1. Existence Results

In the following, we prove our first existence result for the problem (1) and (2), which relies on Krasnoselskii's fixed point theorem [24].

Theorem 1. *Assumed that:*

(B_1) *There exist* $L > 0$ *such that* $|f(t,x) - f(t,y)| \leq L|x-y|,\ \forall t \in [0,1],\ x,y \in \mathbb{R}$;
(B_2) *There exist* $K > 0$ *such that* $|h(t,x) - h(t,y)| \leq K|x-y|,\ \forall t \in [0,1],\ x,y \in \mathbb{R}$;
(B_3) $|f(t,y)| \leq \sigma(t)$ *and* $|h(t,y)| \leq \rho(t)$, *where* $\sigma,\rho \in C([0,1],\mathbb{R}^+)$.

Then the problem (1) and (2) has at least one solution on $[0,1]$ *if* $L\gamma_1 + K\gamma_2 < 1$, *where*

$$\gamma_1 = \frac{1}{\Gamma(\beta+1)\Gamma(\alpha+1)},\ \gamma_2 = \frac{|\lambda|}{\Gamma(\beta+1)\Gamma(\alpha+p+1)\Gamma(q+1)}. \tag{10}$$

Proof. Introduce the ball $B_\theta = \{y \in \mathcal{X} : \|y\| \le \theta\}$, where $\|\sigma\| = \sup_{t\in[0,1]} |\sigma(t)|$, $\|\rho\| = \sup_{t\in[0,1]} |\rho(t)|$ and

$$\theta \ge \|\sigma\|\Omega_1 + \|\rho\|\Omega_2. \tag{11}$$

Let us split the operator $\mathcal{G} : \mathcal{X} \to \mathcal{X}$ on B_θ as $\mathcal{G} = \mathcal{G}_1 + \mathcal{G}_2$, where

$$\begin{aligned}
\mathcal{G}_1 y(t) &= \int_0^t \frac{(t-s)^{\beta-1}}{\Gamma(\beta)} I_{1-}^{\alpha} f(s,y(s))ds - \lambda \int_0^t \frac{(t-s)^{\beta-1}}{\Gamma(\beta)} I_{1-}^{\alpha+p} I_{0+}^{q} h(s,y(s))ds, \\
\mathcal{G}_2 y(t) &= a_1(t)\Bigg[\delta\Bigg(\int_0^\mu \frac{(\mu-s)^{\beta-1}}{\Gamma(\beta)} I_{1-}^{\alpha} f(s,y(s))ds - \lambda \int_0^\mu \frac{(\mu-s)^{\beta-1}}{\Gamma(\beta)} I_{1-}^{\alpha+p} I_{0+}^{q} h(s,y(s))ds\Bigg) \\
&\quad -\Bigg(\int_0^1 \frac{(1-s)^{\beta-1}}{\Gamma(\beta)} I_{1-}^{\alpha} f(s,y(s))ds - \lambda \int_0^1 \frac{(1-s)^{\beta-1}}{\Gamma(\beta)} I_{1-}^{\alpha+p} I_{0+}^{q} h(s,y(s))ds\Bigg)\Bigg] \\
&\quad +a_2(t)\Bigg[\int_0^\xi \frac{(\xi-s)^{\beta-1}}{\Gamma(\beta)} I_{1-}^{\alpha} f(s,y(s))ds - \lambda \int_0^\xi \frac{(\xi-s)^{\beta-1}}{\Gamma(\beta)} I_{1-}^{\alpha+p} I_{0+}^{q} h(s,y(s))ds\Bigg].
\end{aligned}$$

Now, we verify that the operators $\mathcal{G}_1$ and $\mathcal{G}_2$ satisfy the hypothesis of Krasnoselskii's theorem [24] in three steps.

(i) For $y, x \in B_\theta$, we have

$$\begin{aligned}
&\|\mathcal{G}_1 y + \mathcal{G}_2 x\| \\
\le\ & \sup_{t\in[0,1]} \Bigg\{ \int_0^t \frac{(t-s)^{\beta-1}}{\Gamma(\beta)} I_{1-}^{\alpha} |f(s,y(s))|ds + |\lambda| \int_0^t \frac{(t-s)^{\beta-1}}{\Gamma(\beta)} I_{1-}^{\alpha+p} I_{0+}^{q} |h(s,y(s))|ds \\
& +|a_1(t)|\Bigg\{|\delta|\Bigg(\int_0^\mu \frac{(\mu-s)^{\beta-1}}{\Gamma(\beta)} I_{1-}^{\alpha} |f(s,x(s))|ds + |\lambda| \int_0^\mu \frac{(\mu-s)^{\beta-1}}{\Gamma(\beta)} I_{1-}^{\alpha+p} I_{0+}^{q} |h(s,x(s))|ds\Bigg) \\
& +\Bigg(\int_0^1 \frac{(1-s)^{\beta-1}}{\Gamma(\beta)} I_{1-}^{\alpha} |f(s,x(s))|ds + |\lambda| \int_0^1 \frac{(1-s)^{\beta-1}}{\Gamma(\beta)} I_{1-}^{\alpha+p} I_{0+}^{q} |h(s,x(s))|ds\Bigg)\Bigg\} \\
& +|a_2(t)|\Bigg\{\int_0^\xi \frac{(\xi-s)^{\beta-1}}{\Gamma(\beta)} I_{1-}^{\alpha} |f(s,x(s))|ds + |\lambda| \int_0^\xi \frac{(\xi-s)^{\beta-1}}{\Gamma(\beta)} I_{1-}^{\alpha+p} I_{0+}^{q} |h(s,x(s))|ds\Bigg\}\Bigg\} \\
\le\ & \|\sigma\| \sup_{t\in[0,1]} \Bigg\{ \int_0^t \frac{(t-s)^{\beta-1}}{\Gamma(\beta)} I_{1-}^{\alpha} ds + |a_1(t)|\Bigg[|\delta| \int_0^\mu \frac{(\mu-s)^{\beta-1}}{\Gamma(\beta)} I_{1-}^{\alpha} ds \\
& + \int_0^1 \frac{(1-s)^{\beta-1}}{\Gamma(\beta)} I_{1-}^{\alpha} ds\Bigg] + |a_2(t)| \int_0^\xi \frac{(\xi-s)^{\beta-1}}{\Gamma(\beta)} I_{1-}^{\alpha} ds\Bigg\} \\
& +\|\rho\||\lambda| \sup_{t\in[0,1]} \Bigg\{ \int_0^t \frac{(t-s)^{\beta-1}}{\Gamma(\beta)} I_{1-}^{\alpha+p} I_{0+}^{q} ds + |a_1(t)|\Bigg[|\delta| \int_0^\mu \frac{(\mu-s)^{\beta-1}}{\Gamma(\beta)} I_{1-}^{\alpha+p} I_{0+}^{q} ds \\
& + \int_0^1 \frac{(1-s)^{\beta-1}}{\Gamma(\beta)} I_{1-}^{\alpha+p} I_{0+}^{q} ds\Bigg] + |a_2(t)| \int_0^\xi \frac{(\xi-s)^{\beta-1}}{\Gamma(\beta)} I_{1-}^{\alpha+p} I_{0+}^{q} ds\Bigg\} \\
\le\ & \left\{ \frac{\|\sigma\|}{\Gamma(\alpha+1)} + \frac{\|\rho\||\lambda|}{\Gamma(\alpha+p+1)\Gamma(q+1)} \right\}\Delta \\
=\ & \|\sigma\|\Omega_1 + \|\rho\|\Omega_2 < \theta,
\end{aligned}$$

where we used (11). Thus $\mathcal{G}_1 y + \mathcal{G}_2 x \in B_\theta$.

(ii) Using (B_1) and (B_2), it is easy to show that

$$\begin{aligned}\|\mathcal{G}_1 y - \mathcal{G}_1 x\| &\leq \sup_{t\in[0,1]} \bigg\{ \int_0^t \frac{(t-s)^{\beta-1}}{\Gamma(\beta)} I_{1-}^{\alpha} |f(s,y(s)) - f(s,x(s))| ds \\ &\quad + |\lambda| \int_0^t \frac{(t-s)^{\beta-1}}{\Gamma(\beta)} I_{1-}^{\alpha+p} I_{0+}^{q} |h(s,y(s)) - h(s,x(s))| ds \bigg\} \\ &\leq (L\gamma_1 + K\gamma_2)\|y - x\|,\end{aligned}$$

which, in view of the condition: $L\gamma_1 + K\gamma_2 < 1$, implies that the operator $\mathcal{G}_1$ is a contraction.

(iii) Continuity of the functions f, h implies that the operator $\mathcal{G}_2$ is continuous. In addition, $\mathcal{G}_2$ is uniformly bounded on B_θ as

$$\begin{aligned}\|\mathcal{G}_2 y\| &\leq \sup_{t\in[0,1]} \bigg\{ |a_1(t)| \bigg\{ |\delta| \int_0^{\mu} \frac{(\mu-s)^{\beta-1}}{\Gamma(\beta)} \Big[I_{1-}^{\alpha} |f(s,y(s))| + |\lambda| I_{1-}^{\alpha+p} I_{0+}^{q} |h(s,y(s))| \Big] ds \\ &\quad + \int_0^1 \frac{(1-s)^{\beta-1}}{\Gamma(\beta)} \Big[I_{1-}^{\alpha} |f(s,y(s))| + |\lambda| I_{1-}^{\alpha+p} I_{0+}^{q} |h(s,y(s))| \Big] ds \bigg\} \\ &\quad + |a_2(t)| \int_0^{\xi} \frac{(\xi-s)^{\beta-1}}{\Gamma(\beta)} \Big[I_{1-}^{\alpha} |f(s,y(s))| + |\lambda| I_{1-}^{\alpha+p} I_{0+}^{q} |h(s,y(s))| \Big] ds \bigg\} \\ &\leq \|\sigma\| \sup_{t\in[0,1]} \bigg\{ |a_1(t)| \bigg[|\delta| \int_0^{\mu} \frac{(\mu-s)^{\beta-1}}{\Gamma(\beta)} I_{1-}^{\alpha} ds + \int_0^1 \frac{(1-s)^{\beta-1}}{\Gamma(\beta)} I_{1-}^{\alpha} ds \bigg] \\ &\quad + |a_2(t)| \int_0^{\xi} \frac{(\xi-s)^{\beta-1}}{\Gamma(\beta)} I_{1-}^{\alpha} ds \bigg\} \\ &\quad + \|\rho\| |\lambda| \sup_{t\in[0,1]} \bigg\{ |a_1(t)| \bigg[|\delta| \int_0^{\mu} \frac{(\mu-s)^{\beta-1}}{\Gamma(\beta)} I_{1-}^{\alpha+p} I_{0+}^{q} ds + \int_0^1 \frac{(1-s)^{\beta-1}}{\Gamma(\beta)} I_{1-}^{\alpha+p} I_{0+}^{q} ds \bigg] \\ &\quad + |a_2(t)| \int_0^{\xi} \frac{(\xi-s)^{\beta-1}}{\Gamma(\beta)} I_{1-}^{\alpha+p} I_{0+}^{q} ds \bigg\} \\ &\leq \|\sigma\|(\Omega_1 - \gamma_1) + \|\rho\|(\Omega_2 - \gamma_2),\end{aligned}$$

where Ω_i, and γ_i $(i = 1,2)$ are defined by (9) and (10) respectively.

To show the compactness of $\mathcal{G}_2$, we fix $\sup_{(t,y)\in[0,1]\times B_\theta} |f(t,y)| = \overline{f}$, $\sup_{(t,y)\in[0,1]\times B_\theta} |h(t,y)| = \overline{h}$. Then, for $0 < t_1 < t_2 < 1$, we have

$$\begin{aligned}&|(\mathcal{G}_2 y)(t_2) - (\mathcal{G}_2 y)(t_1)| \\ &\leq |a_1(t_2) - a_1(t_1)| \bigg\{ |\delta| \int_0^{\mu} \frac{(\mu-s)^{\beta-1}}{\Gamma(\beta)} \Big[I_{1-}^{\alpha} |f(s,y(s))| + |\lambda| I_{1-}^{\alpha+p} I_{0+}^{q} |h(s,y(s))| \Big] ds \\ &\quad + \int_0^1 \frac{(1-s)^{\beta-1}}{\Gamma(\beta)} \Big[I_{1-}^{\alpha} |f(s,y(s))| + |\lambda| I_{1-}^{\alpha+p} I_{0+}^{q} |h(s,y(s))| \Big] ds \bigg\} \\ &\quad + |a_2(t_2) - a_2(t_1)| \bigg\{ \int_0^{\xi} \frac{(\xi-s)^{\beta-1}}{\Gamma(\beta)} \Big[I_{1-}^{\alpha} |f(s,y(s))| + |\lambda| I_{1-}^{\alpha+p} I_{0+}^{q} |h(s,y(s))| \Big] ds \bigg\} \\ &\leq (\gamma_1 \overline{f} + \gamma_2 \overline{h}) \bigg\{ \big(\xi^{\beta+1} |t_2^{\beta} - t_1^{\beta}| + \xi^{\beta} |t_1^{\beta+1} - t_2^{\beta+1}| \big) \frac{(|\delta|\mu^{\beta} + 1)}{|\Lambda|} \\ &\quad + \big(|1 - \delta\mu^{\beta+1}| |t_2^{\beta} - t_1^{\beta}| + |1 - \delta\mu^{\beta}| |t_1^{\beta+1} - t_2^{\beta+1}| \big) \frac{\xi^{\beta}}{|\Lambda|} \bigg\},\end{aligned}$$

which tends to zero independent of y as $t_2 \to t_1$. This shows that $\mathcal{G}_2$ is equicontinuous. It is clear from the foregoing arguments that the operator $\mathcal{G}_2$ is relatively compact on B_θ. Hence, by the Arzelá-Ascoli theorem, $\mathcal{G}_2$ is compact on B_θ.

In view of the foregoing arguments (*i*)-(*iii*), the hypothesis of the Krasnoselskii's fixed point theorem [24] holds true. Thus, the operator $\mathcal{G}_1 + \mathcal{G}_2 = \mathcal{G}$ has a fixed point, which implies that the problem (1) and (2) has at least one solution on $[0,1]$. The proof is finished. □

Remark 1. *If we interchange the roles of the operators $\mathcal{G}_1$ and $\mathcal{G}_2$ in the previous result, the condition $L\gamma_1 + K\gamma_2 < 1$, is replaced with the following one:*

$$L(\Omega_1 - \gamma_1) + K(\Omega_2 - \gamma_2) < 1,$$

where Ω_1, Ω_2 and γ_1, γ_2 are defined by (9), (10) respectively.

The following existence result relies on Leray–Schauder nonlinear alternative [25].

Theorem 2. *Suppose that the following conditions hold:*

(B_4) *There exist continuous nondecreasing functions $\phi, \psi : [0,\infty) \to (0,\infty)$ such that $\forall (t,y) \in [0,1] \times \mathbb{R}$, $|f(t,y)| \le \omega_1(t)\phi(\|y\|)$ and $|h(t,y)| \le \omega_2(t)\psi(\|y\|)$, where $\omega_1, \omega_2 \in C([0,T], \mathbb{R}^+)$.*

(B_5) *There exist a constant $M > 0$ such that*

$$\frac{M}{\|\omega_1\|\phi(M)\Omega_1 + \|\omega_2\|\psi(M)\Omega_2} > 1,$$

Then, the problem (1) and (2) has at least one solution on $[0,1]$.

Proof. First we show that the operator $\mathcal{G}$ is completely continuous. This will be established in several steps.

(*i*) $\mathcal{G}$ maps bounded sets into bounded sets in $\mathcal{X}$.

Let $y \in \mathcal{B}_r = \{y \in \mathcal{X} : \|y\| \le r\}$, where r is a fixed number. Then, using the strategy employed in the proof of Theorem 1, we obtain

$$\begin{aligned} \|\mathcal{G}y(t)\| &\le \left\{ \frac{\|\omega_1\|\phi(r)}{\Gamma(\alpha+1)} + \frac{\|\omega_2\|\psi(r)|\lambda|}{\Gamma(\alpha+p+1)\Gamma(q+1)} \right\}\Delta \\ &= \|\omega_1\|\phi(r)\Omega_1 + \|\omega_2\|\psi(r)\Omega_2 < \infty. \end{aligned}$$

(*ii*) $\mathcal{G}$ maps bounded sets into equicontinuous sets.

Let $0 < t_1 < t_2 < 1$ and $y \in \mathcal{B}_r$, where $\mathcal{B}_r$ is bounded set in $\mathcal{X}$. Then we obtain

$$\begin{aligned}
&|\mathcal{G}y(t_2)-\mathcal{G}y(t_1)| \\
\leq\ & \left|\int_0^{t_1}\frac{(t_2-s)^{\beta-1}-(t_1-s)^{\beta-1}}{\Gamma(\beta)}\left[I_{1-}^{\alpha}|f(s,y(s))|+|\lambda|I_{1-}^{\alpha+p}I_{0+}^{q}|h(s,y(s))|\right]ds\right| \\
&+\left|\int_{t_1}^{t_2}\frac{(t_2-s)^{\beta-1}}{\Gamma(\beta)}\left[I_{1-}^{\alpha}|f(s,y(s))|+|\lambda|I_{1-}^{\alpha+p}I_{0+}^{q}|h(s,y(s))|\right]ds\right| \\
&+|a_1(t_2)-a_1(t_1)|\left\{|\delta|\left(\int_0^{\mu}\frac{(\mu-s)^{\beta-1}}{\Gamma(\beta)}\left[I_{1-}^{\alpha}|f(s,y(s))|+|\lambda|I_{1-}^{\alpha+p}I_{0+}^{q}|h(s,y(s))|\right]ds\right)\right. \\
&\left.+\left(\int_0^{1}\frac{(1-s)^{\beta-1}}{\Gamma(\beta)}\left[I_{1-}^{\alpha}|f(s,y(s))|+|\lambda|I_{1-}^{\alpha+p}I_{0+}^{q}|h(s,y(s))|\right]ds\right)\right\} \\
&+|a_2(t_2)-a_2(t_1)|\left\{\int_0^{\xi}\frac{(\xi-s)^{\beta-1}}{\Gamma(\beta)}\left[I_{1-}^{\alpha}|f(s,y(s))|+|\lambda|I_{1-}^{\alpha+p}I_{0+}^{q}|h(s,y(s))|\right]ds\right\} \\
\leq\ & \left[\frac{\|\omega_1\|\phi(r)}{\Gamma(\beta+1)\Gamma(\alpha+1)}+\frac{\|\omega_2\|\psi(r)|\lambda|}{\Gamma(\beta+1)\Gamma(\alpha+p+1)\Gamma(q+1)}\right] \\
&\times\left\{2(t_2-t_1)^{\beta}+|{t_2}^{\beta}-{t_1}^{\beta}|+\frac{(|\delta|\mu^{\beta}+1)}{|\Lambda|}\left(\xi^{\beta+1}|t_2^{\beta}-t_1^{\beta}|+\xi^{\beta}|t_2^{\beta+1}-t_1^{\beta+1}|\right)\right. \\
&\left.+\frac{\xi^{\beta}}{|\Lambda|}\left(|1-\delta\mu^{\beta+1}||t_2^{\beta}-t_1^{\beta}|+|1-\delta\mu^{\beta}||t_2^{\beta+1}-t_1^{\beta+1}|\right)\right\}.
\end{aligned}$$

Notice that the right-hand side of the above inequality tends to 0 as $t_2 \to t_1$, independent of $y \in \mathcal{B}_r$. In view of the foregoing arguments, it follows by the Arzelá–Ascoli theorem that $\mathcal{G} : \mathcal{X} \to \mathcal{X}$ is completely continuous.

The conclusion of the Leray–Schauder nonlinear alternative [25] will be applicable once it is shown that there exists an open set $U \subset C([0,1],\mathbb{R})$ with $y \neq \nu\mathcal{G}y$ for $\nu \in (0,1)$ and $y \in \partial U$. Let $y \in C([0,1],\mathbb{R})$ such that $y = \nu\mathcal{G}y$ for $\nu \in (0,1)$. As argued in proving that the operator $\mathcal{G}$ is bounded, one can obtain that

$$|y(t)| = |\nu\mathcal{G}y(t)| \ \leq |\omega_1(t)|\phi(\|y\|)\Omega_1 + |\omega_2(t)|\psi(\|y\|)\Omega_2,$$

which can be written as

$$\frac{\|y\|}{\|\omega_1\|\phi(\|y\|)\Omega_1 + \|\omega_2\|\psi(\|y\|)\Omega_2} \leq 1.$$

On the other hand, we can find a positive number M such that $\|y\| \neq M$ by assumption (B_5). Let us set

$$U = \{y \in \mathcal{X} : \|y\| < M\}.$$

Clearly, ∂U contains a solution only when $\|y\| = M$. In other words, there is no solution $y \in \partial U$ such that $y = \nu\mathcal{G}y$ for some $\nu \in (0,1)$. Therefore, $\mathcal{G}$ has a fixed point $y \in \overline{U}$ which is a solution of the problem (1) and (2). The proof is finished. □

3.2. Uniqueness Result

Here we prove a uniqueness result for the problem (1) and (2) with the aid of Banach contraction mapping principle.

Theorem 3. *If the conditions* (B_1) *and* (B_2) *hold, then the problem (1) and (2) has a unique solution on* $[0,1]$ *if*

$$L\Omega_1 + K\Omega_2 < 1, \tag{12}$$

where Ω_1 *and* Ω_2 *are defined by (9).*

Proof. In the first step, we show that $\mathcal{G}\mathcal{B}_r \subset \mathcal{B}_r$, where $\mathcal{B}_r = \{y \in \mathcal{X} : \|y\| \leq r\}$ with

$$r \geq \frac{f_0\Omega_1 + h_0\Omega_2}{1-(L\Omega_1 + K\Omega_2)}, \; f_0 = \sup_{t\in[0,1]} |f(t,0)|, \; h_0 = \sup_{t\in[0,1]} |h(t,0)|.$$

For $y \in \mathcal{B}_r$ and using the condition (B_1), we have

$$\begin{aligned} |f(t,y)| &= |f(t,y) - f(t,0) + f(t,0)| \leq |f(t,y) - f(t,0)| + |f(t,0)| \\ &\leq L\|y\| + f_0 \leq Lr + f_0. \end{aligned} \tag{13}$$

Similarly, using (B_2), we get

$$|h(t,y)| \leq Kr + h_0. \tag{14}$$

In view of (13) and (14), we obtain

$$\begin{aligned} \|\mathcal{G}y\| &= \sup_{t\in[0,1]} |\mathcal{G}y(t)| \\ &\leq \sup_{t\in[0,1]} \bigg\{ \int_0^t \frac{(t-s)^{\beta-1}}{\Gamma(\beta)} \Big[I_{1-}^{\alpha} |f(s,y(s))| + |\lambda| I_{1-}^{\alpha+p} I_{0+}^{q} |h(s,y(s))| \Big] ds \\ &\quad + |a_1(t)| \bigg\{ |\delta| \int_0^{\mu} \frac{(\mu-s)^{\beta-1}}{\Gamma(\beta)} \Big[I_{1-}^{\alpha} |f(s,y(s))| + |\lambda| I_{1-}^{\alpha+p} I_{0+}^{q} |h(s,y(s))| \Big] ds \\ &\quad + \int_0^1 \frac{(1-s)^{\beta-1}}{\Gamma(\beta)} \Big[I_{1-}^{\alpha} |f(s,y(s))| + |\lambda| I_{1-}^{\alpha+p} I_{0+}^{q} |h(s,y(s))| \Big] ds \bigg\} \\ &\quad + |a_2(t)| \int_0^{\xi} \frac{(\xi-s)^{\beta-1}}{\Gamma(\beta)} \Big[I_{1-}^{\alpha} |f(s,y(s))| + |\lambda| I_{1-}^{\alpha+p} I_{0+}^{q} |h(s,y(s))| \Big] ds \bigg\} \\ &\leq (Lr + f_0) \sup_{t\in[0,1]} \bigg\{ \int_0^t \frac{(t-s)^{\beta-1}}{\Gamma(\beta)} I_{1-}^{\alpha} ds + |a_1(t)| \bigg[|\delta| \int_0^{\mu} \frac{(\mu-s)^{\beta-1}}{\Gamma(\beta)} I_{1-}^{\alpha} ds \\ &\quad + \int_0^1 \frac{(1-s)^{\beta-1}}{\Gamma(\beta)} I_{1-}^{\alpha} ds \bigg] + |a_2(t)| \int_0^{\xi} \frac{(\xi-s)^{\beta-1}}{\Gamma(\beta)} I_{1-}^{\alpha} ds \bigg\} \\ &\quad + (Kr + h_0)|\lambda| \sup_{t\in[0,1]} \bigg\{ \int_0^t \frac{(t-s)^{\beta-1}}{\Gamma(\beta)} I_{1-}^{\alpha+p} I_{0+}^{q} ds \\ &\quad + |a_1(t)| \bigg[|\delta| \int_0^{\mu} \frac{(\mu-s)^{\beta-1}}{\Gamma(\beta)} I_{1-}^{\alpha+p} I_{0+}^{q} ds + \int_0^1 \frac{(1-s)^{\beta-1}}{\Gamma(\beta)} I_{1-}^{\alpha+p} I_{0+}^{q} ds \bigg] \\ &\quad + |a_2(t)| \int_0^{\xi} \frac{(\xi-s)^{\beta-1}}{\Gamma(\beta)} I_{1-}^{\alpha+p} I_{0+}^{q} ds \bigg\} \\ &\leq \left\{ \frac{(Lr + f_0)}{\Gamma(\alpha+1)} + \frac{(Kr + h_0)|\lambda|}{\Gamma(\alpha+p+1)\Gamma(q+1)} \right\} \Delta \\ &= (Lr + f_0)\Omega_1 + (Kr + h_0)\Omega_2 < r, \end{aligned}$$

which implies that $\mathcal{G}y \in \mathcal{B}_r$, for any $y \in \mathcal{B}_r$. Therefore, $\mathcal{G}\mathcal{B}_r \subset \mathcal{B}_r$. Next, we prove that $\mathcal{G}$ is a contraction. For that, let $x, y \in \mathcal{X}$ and $t \in [0,1]$. Then, by the conditions (B_1) and (B_2), we obtain

$$
\begin{aligned}
\|\mathcal{G}y-\mathcal{G}x\| &= \sup_{t\in[0,1]}\big|(\mathcal{G}y)(t)-(\mathcal{G}x)(t)\big| \\
&\le \sup_{t\in[0,1]}\Bigg\{\int_0^t \frac{(t-s)^{\beta-1}}{\Gamma(\beta)}I_{1-}^{\alpha}|f(s,y(s))-f(s,x(s))|ds \\
&\quad +|\lambda|\int_0^t \frac{(t-s)^{\beta-1}}{\Gamma(\beta)}I_{1-}^{\alpha+p}I_{0+}^{q}|h(s,y(s))-h(s,x(s))|ds \\
&\quad +|a_1(t)|\Bigg[|\delta|\Bigg(\int_0^{\mu}\frac{(\mu-s)^{\beta-1}}{\Gamma(\beta)}I_{1-}^{\alpha}|f(s,y(s))-f(s,x(s))|ds \\
&\quad +|\lambda|\int_0^{\mu}\frac{(\mu-s)^{\beta-1}}{\Gamma(\beta)}I_{1-}^{\alpha+p}I_{0+}^{q}|h(s,y(s))-h(s,x(s))|ds\Bigg) \\
&\quad +\Bigg(\int_0^1\frac{(1-s)^{\beta-1}}{\Gamma(\beta)}I_{1-}^{\alpha}|f(s,y(s))-f(s,x(s))|ds \\
&\quad +|\lambda|\int_0^1\frac{(1-s)^{\beta-1}}{\Gamma(\beta)}I_{1-}^{\alpha+p}I_{0+}^{q}|h(s,y(s))-h(s,x(s))|ds\Bigg)\Bigg] \\
&\quad +|a_2(t)|\Bigg[\int_0^{\xi}\frac{(\xi-s)^{\beta-1}}{\Gamma(\beta)}I_{1-}^{\alpha}|f(s,y(s))-f(s,x(s))|ds \\
&\quad +|\lambda|\int_0^{\xi}\frac{(\xi-s)^{\beta-1}}{\Gamma(\beta)}I_{1-}^{\alpha+p}I_{0+}^{q}|h(s,y(s))-h(s,x(s))|dr\,du\,ds\Bigg]\Bigg\} \\
&\le L\|y-x\|\sup_{t\in[0,1]}\Bigg\{\int_0^t\frac{(t-s)^{\beta-1}}{\Gamma(\beta)}I_{1-}^{\alpha}ds \\
&\quad +|a_1(t)|\Bigg[|\delta|\int_0^{\mu}\frac{(\mu-s)^{\beta-1}}{\Gamma(\beta)}I_{1-}^{\alpha}ds+\int_0^1\frac{(1-s)^{\beta-1}}{\Gamma(\beta)}I_{1-}^{\alpha}ds\Bigg] \\
&\quad +|a_2(t)|\int_0^{\xi}\frac{(\xi-s)^{\beta-1}}{\Gamma(\beta)}I_{1-}^{\alpha}ds\Bigg\} \\
&\quad +K\|y-x\||\lambda|\sup_{t\in[0,1]}\Bigg\{\int_0^t\frac{(t-s)^{\beta-1}}{\Gamma(\beta)}I_{1-}^{\alpha+p}I_{0+}^{q}ds \\
&\quad +|a_1(t)|\Bigg[|\delta|\int_0^{\mu}\frac{(\mu-s)^{\beta-1}}{\Gamma(\beta)}I_{1-}^{\alpha+p}I_{0+}^{q}ds+\int_0^1\frac{(1-s)^{\beta-1}}{\Gamma(\beta)}I_{1-}^{\alpha+p}I_{0+}^{q}ds\Bigg] \\
&\quad +|a_2(t)|\int_0^{\xi}\frac{(\xi-s)^{\beta-1}}{\Gamma(\beta)}I_{1-}^{\alpha+p}I_{0+}^{q}ds\Bigg\} \\
&\le \left\{\frac{L\Delta}{\Gamma(\alpha+1)}+\frac{K|\lambda|\Delta}{\Gamma(\alpha+p+1)\Gamma(q+1)}\right\}\|y-x\| \\
&= (L\Omega_1+K\Omega_2)\|y-x\|.
\end{aligned}
$$

From the above inequality, it follows by the assumption $(L\Omega_1+K\Omega_2)<1$ that $\mathcal{G}$ is a contraction. Therefore, we deduce by Banach contraction mapping principle that there exists a unique fixed point for the operator $\mathcal{G}$, which corresponds to a unique solution for the problem (1) and (2) on $[0,1]$. The proof is completed. □

3.3. Examples

In this subsection, we construct examples to illustrate the existence and uniqueness results obtained in the last two subsections. Let us consider the following problem:

$$
\begin{cases}
D_{1-}^{3/2}D_{0+}^{1/2}y(t)+2I_{1-}^{4/3}I_{0+}^{5/4}h(t,y(t))=f(t,y(t)),\ t\in J:=[0,1],\\
y(0)=y(2/3)=0,\ \ y(1)=\frac{1}{2}y(3/4).
\end{cases}
\tag{15}
$$

Here $\alpha = 3/2, \beta = 1/2, \lambda = 2, p = 4/3, q = 5/4, \mu = 3/4, \delta = 1/2, \xi = 2/3$, and

$$f(t,y) = \frac{1}{(t^2+8)}\big(\tan^{-1} y + e^{-t}\big), \quad h(t,y) = \frac{1}{2\sqrt{t^2+9}}\big(\frac{|y|}{1+|y|} + e^{-t}\big). \tag{16}$$

Using the given data, it is found that $L = 1/8, K = 1/6$,

$$\bar{a}_1 = \max_{t\in[0,1]} |a_1(t)| = |a_1(t)|_{t=1} \approx 1.121394517474712,$$
$$\bar{a}_2 = \max_{t\in[0,1]} |a_2(t)| = |a_2(t)|_{t=t_{a_2}} \approx 1.168623082364286,$$

where

$$t_{a_2} = \frac{\beta(1-\delta\mu^{\beta+1})}{(1-\delta\mu^{\beta})(\beta+1)} \approx 0.396975661732535.$$

In consequence, we get

$$\Omega_1 \approx 3.022797441671726, \Omega_2 \approx 1.451691300771574, |\Lambda| \approx 0.242702744426469,$$

where Ω_1,Ω_2 are defined by (9) and Λ is given by (6).
(i) For illustrating Theorem 1, we have

$$|f(t,y)| \le \sigma(t) = \frac{e^{-t} + (\pi/2)}{t^2+8}, \quad |h(t,y)| \le \rho(t) = \frac{e^{-t}+1}{2\sqrt{t^2+9}},$$

and that

$$L\gamma_1 + K\gamma_2 \approx 0.174044436618777 < 1,$$

where $\gamma_1 \approx 0.848826363156775$ and $\gamma_2 \approx 0.407646847345084$. Clearly, the hypothesis of Theorem 1 is satisfied and consequently its conclusion applies to the problem (15).
(ii) In order to explain Theorem 2, we take the following values (instead of (16)) in the problem (15):

$$f(t,y) = \frac{1}{\sqrt{t^2+25}}(y\cos y + \pi/2), \quad h(t,y) = \frac{1}{5\sqrt{t^2+4}}\big(\sin y + 1/4\big), \tag{17}$$

and note that $\omega_1(t) = \frac{1}{\sqrt{t^2+25}}, \|\omega_1\| = 1/5, \omega_2(t) = \frac{1}{5\sqrt{t^2+4}}, \|\omega_2\| = 1/10, \phi(\|y\|) = \|y\| + \pi/2$ and $\psi(\|y\|) = \|y\| + 1/4$. By the condition (B_5), we find that $M > 3.939452045479877$. Thus, all the conditions of Theorem 2 are satisfied and, hence the problem (15) with $f(t,y)$ and $h(t,y)$ given by (17) has at least one solution on $[0,1]$.
(iii) It is easy to show that $f(t,y)$ and $h(t,y)$ satisfy the conditions (B_1) and (B_2) respectively with $L = 1/8$ and $K = 1/6$ and that $L\Omega_1 + K\Omega_2 \approx 0.619798230337561 < 1$. Thus, all the assumptions of Theorem 3 hold true and hence the problem (15) has a unique solution on $[0,1]$.

4. Conclusions

We considered a fractional differential equation involving left Caputo and right Riemann–Liouville fractional derivatives of different orders and a pair of nonlinearities: $I_{1-}^{p} I_{0+}^{q} h(t,y(t)) = \int_t^1 \frac{(s-t)^{p-1}}{\Gamma(p)} \int_0^s \frac{(s-v)^{q-1}}{\Gamma(q)} h(v,y(v))dvds$ (integral type) and $f(t,y(t))$, equipped with four-point nonlocal boundary conditions. Different criteria ensuring the existence of solutions for the given problem are presented in Theorems 1 and 2, while the uniqueness of solutions is shown in Theorem 3. An interesting and scientific feature of the fractional differential Equation (1) is that the integral type of nonlinearity can describe composition of a physical quantity (like density) over two different arbitrary subsegments of the given domain. In the case of $p = q = 1$, this composition takes the form

$\int_t^1 \int_0^s h(v, y(v))dvds$. As pointed out in the introduction, fractional differential equations containing mixed (left Caputo and right Riemann–Liouville) fractional derivatives appear as Euler–Lagrange equations in the study of variational principles. So, such equations in the presence of the integral type of nonlinearity of the form introduced in (1) enhances the scope of Euler–Lagrange equations studied in [26]. Moreover, the fractional integro-differential Equation (1) can improve the description of the electromagnetic waves in dielectric media considered in [23]. As a special case, our results correspond to a three-point nonlocal mixed fractional order boundary value problem by letting $\delta = 0$, which is indeed new in the given configuration.

Author Contributions: Conceptualization, B.A.; Formal analysis, B.A., A.B., A.A. and S.K.N.; Funding acquisition, A.A.; Methodology, B.A., A.B., A.A. and S.K.N. All authors have read and agreed to the published version of the manuscript.

Funding: This research was funded by the Deanship of Scientific Research (DSR), King Abdulaziz University, Jeddah, Saudi Arabia under grant number KEP-MSc-23-130-40.

Acknowledgments: This project was funded by the Deanship of Scientific Research (DSR), King Abdulaziz University, Jeddah, Saudi Arabia under grant no. (KEP-MSc-23-130-40). The authors, therefore, acknowledge with thanks DSR technical and financial support. We also thank the reviewers for their useful remarks on our work.

Conflicts of Interest: The authors declare no conflict of interest.

References

1. Kilbas, A.A.; Srivastava, H.M.; Trujillo, J.J. *Theory and Applications of Fractional Differential Equations*; North-Holland Mathematics Studies, 204; Elsevier Science B.V.: Amsterdam, The Netherlands, 2006.
2. Henderson, J.; Luca, R.; Tudorache, A. On a system of fractional differential equations with coupled integral boundary conditions. *Fract. Calc. Appl. Anal.* **2015**, *18*, 361–386. [CrossRef]
3. Peng, L.; Zhou, Y. Bifurcation from interval and positive solutions of the three-point boundary value problem for fractional differential equations. *Appl. Math. Comput.* **2015**, *257*, 458–466. [CrossRef]
4. Ahmad, B.; Alsaedi, A.; Ntouyas, S.K.; Tariboon, J. *Hadamard-Type Fractional Differential Equations, Inclusions and Inequalities*; Springer: Cham, Switzerland, 2017.
5. Ahmad, B.; Ntouyas, S.K. Nonlocal initial value problems for Hadamard-type fractional differential equations and inclusions. *Rocky Mt. J. Math.* **2018**, *48*, 1043–1068. [CrossRef]
6. Cui, Y.; Ma, W.; Sun, Q.; Su, X. New uniqueness results for boundary value problem of fractional differential equation. *Nonlinear Anal. Model. Control* **2018**, *23*, 31–39. [CrossRef]
7. Ahmad, B.; Luca, R. Existence of solutions for sequential fractional integro-differential equations and inclusions with nonlocal boundary conditions. *Appl. Math. Comput.* **2018**, *339*, 516–534. [CrossRef]
8. Alsaedi, A.; Ahmad, B.; Alghanmi, M.; Ntouyas, S.K. On a generalized Langevin type nonlocal fractional integral multivalued problem. *Mathematics* **2019**, *7*, 1015. [CrossRef]
9. Ahmad, B.; Alghamdi, N.; Alsaedi, A.; Ntouyas, S.K. A system of coupled multi-term fractional differential equations with three-point coupled boundary conditions. *Fract. Calc. Appl. Anal.* **2019**, *22*, 601–618. [CrossRef]
10. Alsaedi, A.; Ahmad, B.; Alghanmi, M. Extremal solutions for generalized Caputo fractional differential equations with Steiltjes-type fractional integro-initial conditions. *Appl. Math. Lett.* **2019**, *91*, 113120. [CrossRef]
11. Ahmad, B.; Alsaedi, A.; Alruwaily, Y.; Ntouyas, S.K. Nonlinear multi-term fractional differential equations with Riemann-Stieltjes integro-multipoint boundary conditions. *AIMS Math.* **2020**, *5*, 1446–1461. [CrossRef]
12. Liang, S.; Wang, L.; Yin, G. Fractional differential equation approach for convex optimization with convergence rate analysis. *Optim. Lett.* **2020**, *14*, 145–155. [CrossRef]
13. Atanackovic, T.M.; Stankovic, B. On a differential equation with left and right fractional derivatives. *Fract. Calc. Appl. Anal.* **2007**, *10*, 139–150. [CrossRef]
14. Zhang, L.; Ahmad, B.; Wang, G. The existence of an extremal solution to a nonlinear system with the right-handed Riemann-Liouville fractional derivative. *Appl. Math. Lett.* **2014**, *31*, 1–6. [CrossRef]
15. Khaldi, R.; Guezane-Lakoud, A. Higher order fractional boundary value problems for mixed type derivatives. *J. Nonlinear Funct. Anal.* **2017**, *30*, 9.

16. Lakoud, A.G.; Khaldi, R.; Kilicman, A. Existence of solutions for a mixed fractional boundary value problem. *Adv. Differ. Equ.* **2017**, *2017*, 164. [CrossRef]
17. Guezane-Lakoud, A.; Khaldi, R.; Torres, D.F.M. On a fractional oscillator equation with natural boundary conditions. *Progr. Fract. Differ. Appl.* **2017**, *3*, 191–197. [CrossRef]
18. Ahmad, B.; Ntouyas, S.K.; Alsaedi, A. Existence theory for nonlocal boundary value problems involving mixed fractional derivatives. *Nonlinear Anal. Model. Control* **2019**, *24*, 937–957. [CrossRef]
19. Laitinen, M.; Tiihonen, T. Heat transfer in conducting and radiating bodies. *Appl. Math. Lett.* **1997**, *10*, 5–8. [CrossRef]
20. Laitinen, M.; Tiihonen, T. Integro-differential equation modelling heat transfer in conducting, radiating and semitransparent materials. *Math. Methods Appl. Sci.* **1998**, *21*, 375–392. [CrossRef]
21. Hajmohammadi, M.R.; Nourazar, S.S.; Manesh, A.H. Semi-analytical treatments of conjugate heat transfer. *J. Mech. Eng. Sci.* **2013**, *227*, 492–503. [CrossRef]
22. Yang, A.M.; Han, Y.; Zhang, Y.Z.; Wang, L.T.; Zhang, D.; Yang, X.J. On nonlocal fractional Volterra integro-differential equations in fractional steady heat transfer. *Therm. Sci.* **2016**, *20*, S789–S793. [CrossRef]
23. Tarasov, V.E. Fractional integro-differential equations for electromagnetic waves in dielectric media. *Theor. Math. Phys.* **2009**, *158*, 355–359. [CrossRef]
24. Krasnoselskii, M.A. Two remarks on the method of successive approximations. *Uspekhi Mat. Nauk.* **1955**, *10* , 123–127.
25. Granas, A.; Dugundji, J. *Fixed Point Theory*; Springer: New York, NY, USA, 2005.
26. Agrawal, O.P. Formulation of Euler-Lagrange equations for fractional variational problems. *J. Math. Anal. Appl.* **2002**, *272*, 368–379. [CrossRef]

mathematics

MDPI

Article

Integral Representation for the Solutions of Autonomous Linear Neutral Fractional Systems with Distributed Delay

Ekaterina Madamlieva [1], Mihail Konstantinov [2,*], Marian Milev [3] and Milena Petkova [1]

[1] Faculty of Mathematics and Informatics, University of Plovdiv, 4000 Plovdiv, Bulgaria; ekaterinaa.b.m@gmail.com (E.M.); milenapetkova@uni-plovdiv.bg (M.P.)
[2] Department of Mathematics, University of Architecture, Civil Engineering and Geodesy, 1046 Sofia, Bulgaria
[3] Department of Matematics and Physics, University Of Food Technology, 4000 Plovdiv, Bulgaria; marianmilev2002@gmail.com
* Correspondence: misho.konstantinov@gmail.com

Received: 1 February 2020; Accepted: 2 March 2020; Published: 6 March 2020

Abstract: The aim of this work is to obtain an integral representation formula for the solutions of initial value problems for autonomous linear fractional neutral systems with Caputo type derivatives and distributed delays. The results obtained improve and extend the corresponding results in the particular case of fractional systems with constant delays and will be a useful tool for studying different kinds of stability properties. The proposed results coincide with the corresponding ones for first order neutral linear differential systems with integer order derivatives.

Keywords: fractional derivatives; neutral fractional systems; distributed delay; integral representation

MSC: 34A08; 34A12

---

1. Introduction and Notations

Fractional Calculus has a long history, but it has attracted considerable attention recently as an important tool for modeling of various real problems, such as viscoelastic systems, diffusion processes, signal and control processing, and seismic processes. Detailed information about the fractional calculus theory and its applications can be found in the monographs [1–4]. Some results for fractional linear systems with delays are in given in the book [5]. The monograph [6] is devoted to the impulsive differential and functional differential equations with fractional derivatives, as well as to some of their applications.

It is well known that the study of linear fractional equations (integral representation, several types of stability, etc.) is an evergreen theme for research. Concerning these fields of fundamental and qualitative investigations for linear fractional ordinary differential equations and systems we refer to [2,4,7] and the references therein. Using the Laplace transform method, several interesting results in this direction are obtained in [8,9] as well. Regarding works concerning fractional differential systems with constant delays, we point out [10–13]. Concerning the retarded differential systems with variable or distributed delays—fundamental theory and application (stability properties)—we refer to [11,14–18]. Neutral fractional systems with distributed delays are essentially studied less (see [19–21]). Stability properties of retarded fractional systems with derivatives of distributed order are studied in [22]. One of the existing best applications of fractional order equations with delays is modeling human manual control, in which perceptual and neuromuscular delays introduce a delay term. As interesting studies, we refer to [23,24].

The problem of establishing an integral representation for the solutions for neutral or delayed linear fractional differential equations and/or systems needs a theorem for the existence of a fundamental matrix, i.e., theorem for existence and uniqueness of the solution to the initial value problem (IVP) in the case of discontinuous initial functions. As far as we know, there are only a few results concerning the IVP for delayed and neutral systems with discontinuous initial function, for the delayed case [14,15,25–27] and for the neutral case [28].

The aim of the work is to prove an integral representation formula for the general solution of an autonomous linear fractional neutral system with Caputo type derivatives and distributed delays. Note that our results extend and improve the results obtained in [10,12,15]. The proposed results coincide with the corresponding ones for a first order neutral linear differential system with integer order derivatives.

The paper is organized as follows. In Section 2, we recall some necessary definitions of Riemann-Liouville and Caputo fractional derivatives, as well as part of their properties. In this section, we also present the linear neutral fractional system under consideration together with some conditions. In Section 3, as a main result, integral representations of the solutions of the IVP for autonomous linear fractional neutral system with Caputo type derivatives and distributed delays are obtained for the homogeneous and inhomogeneous case. In Section 4, we present an illustrative example. In Section 5 we explain the practical benefits and application options of the obtained theoretical results.

In what follows, we use the notations: $\mathbb{N}$, $\mathbb{R}$ and $\mathbb{C}$ – the sets of natural, real and complex numbers, respectively; $\langle m, n\rangle$ – the set of integers $m, m+1, \ldots, n$ $(m \leq n)$; $\mathbb{R}^{n\times n}$ – the space of real $n \times n$ matrices A with elements A_{pq}; $\mathbb{R}^n = \mathbb{R}^{n\times 1}$; $A^\top$ – the transposed matrix A with elements $(A^\top)_{pq} = A_{qp}$. The elements of $\mathbb{R}^n$ are the real column n-vectors $x = [x_1; x_2; \ldots; x_n]$ with elements x_k. The row n-vectors are denoted as $\xi = [\xi_1, \xi_2, \ldots, \xi_n]$(note that the elements of a vector column and a vector row are separated by ";" and ",", respectively). The identity and the zero matrices are denoted by E and Θ, respectively.

We also denote $\mathbb{C}_+ = \{p \in \mathbb{C} | \mathrm{Re}(p) > 0\}$, $\overline{\mathbb{C}}_+ = \{p \in \mathbb{C} | \mathrm{Re}\, p \geq 0\}$, $\mathbb{C}_- = \mathbb{C} \setminus \overline{\mathbb{C}}_+$, $\mathbb{R}_+ = (0, \infty)$, and $J_s = [s, \infty)$. For $p \in \mathbb{C}$, $y = [y_1; y_2; \ldots; y_n] \in \mathbb{C}^n$ and $\beta = (\beta_1, \beta_2, \ldots, \beta_n)$, $\beta_k \in [-1, 1]$ we set $I_\beta(p) = \mathrm{diag}\left(p^{\beta_1}, p^{\beta_2}, \ldots, p^{\beta_n}\right)$ and $I_\beta(y) = \mathrm{diag}\left(y_1^{\beta_1}, y_2^{\beta_2}, \ldots, y_n^{\beta_n}\right)$. The linear space of locally Lebesgue integrable functions $f : \mathbb{R} \to \mathbb{R}$ is denoted by $L_1^{\mathrm{loc}}(\mathbb{R}, \mathbb{R})$.

2. Preliminaries and Problem Statement

Below, the definitions of Riemann–Liouville and Caputo fractional derivatives and some of their properties necessary for our exposition are described in order to avoid possible misunderstandings. For more details and other properties, we refer to [2–4].

Let $\alpha \in (0, 1)$ be an arbitrary number. Then for $a \in \mathbb{R}$, each $t > a$ and $f \in L_1^{\mathrm{loc}}(\mathbb{R}, \mathbb{R})$ the left-sided fractional integral operator, the left side Riemann-Liouville and Caputo fractional derivatives of order α are defined by

$$\begin{aligned}
(D_{a+}^{-\alpha} f)(t) &= \frac{1}{\Gamma(\alpha)} \int_a^t (t-s)^{\alpha-1} f(s)\mathrm{d}s, \; {}_{RL}D_{a+}^{\alpha} f(t) = \frac{\mathrm{d}}{\mathrm{d}t}\left(D_{a+}^{-(1-\alpha)} f(t)\right), \\
{}_{C}D_{a+}^{\alpha} f(t) &= {}_{RL}D_{a+}^{\alpha}[f(s) - f(a)](t) = {}_{RL}D_{a+}^{\alpha} f(t) - \frac{f(a)}{\Gamma(1-\alpha)}(t-a)^{-\alpha},
\end{aligned}$$

respectively. The following relations [4] involving fractional derivatives will be used

$$(D_{a+}^{0} f)(t) = f(t), \; {}_{C}D_{a+}^{\alpha} D_{a+}^{-\alpha} f(t) = f(t), \; D_{a+}^{-\alpha}\, {}_{C}D_{a+}^{\alpha} f(t) = f(t) - f(a).$$

Concerning the Laplace transform $\mathfrak{L}$,

$$\mathfrak{L}f(p) = \int_0^\infty \exp(-pt) f(t)\mathrm{d}t, \; p \in \mathbb{C},$$

we shall need the relations

$$\mathfrak{L}D_{0+}^{-\alpha}f(p) = p^{-\alpha}(\mathfrak{L}f)(p),\ \mathfrak{L}_{RL}D_{0+}^{\alpha}f(p) = p^{\alpha}(\mathfrak{L}f)(p) - [_{RL}D_{0+}^{\alpha-1}f(t)]_{t=0},$$
$$\mathfrak{L}_{C}D_{0+}^{\alpha}f(p) = p^{\alpha}(\mathfrak{L}f)(p) - p^{\alpha-1}f(0).$$

In what follows, we consider the autonomous linear neutral fractional system with distributed delay

$$D^{\alpha}\Big(X(t) - \sum_{l=1}^{r}\int_{-\tau}^{0}\Big[\mathrm{d}_{\theta}V^{l}(\theta)\Big]\,X(t+\theta)\Big) = \sum_{i=0}^{m}\int_{-\sigma}^{0}\Big[\mathrm{d}_{\theta}U^{i}(\theta)\Big]\,X(t+\theta) + F(t), \tag{1}$$

as well as the corresponding homogeneous system

$$D^{\alpha}\Big(X(t) - \sum_{l=1}^{r}\int_{-\tau}^{0}\Big[\mathrm{d}_{\theta}V^{l}(\theta)\Big]\,X(t+\theta)\Big) = \sum_{i=0}^{m}\int_{-\sigma}^{0}\Big[\mathrm{d}_{\theta}U^{i}(\theta)\Big]\,X(t+\theta), \tag{2}$$

where

$$X, F : J_0 \to \mathbb{R}^n,\ U^i, V^l : \mathbb{R} \to \mathbb{R}^{n\times n},\ U^i(\theta) = \Big[u^i_{kj}(\theta)\Big],\ V^l(\theta) = \Big[v^l_{kj}(\theta)\Big],$$
$$\tau, \sigma > 0,\ \tau_r \in (0,\tau],\ l \in \langle 1, r\rangle,\ \sigma_i \in (0,\sigma],\ i \in \langle 1, m\rangle,\ h = \max(\sigma,\tau),\ \sigma_0 = 0,$$
$$\alpha = (\alpha_1, \alpha_2, \ldots, \alpha_n),\ \alpha_k \in (0,1),\ k \in \langle 1, n\rangle,\ J_s = [s,\infty).$$

For simplicity, D^{α_k} denotes the left side Caputo fractional derivative $_{C}D_{0+}^{\alpha_k}$ in (1) and (2), and we use the notations

$$D^{\alpha}X(t) = [D^{\alpha_1}x_1(t); D^{\alpha_2}x_2(t); \ldots; D^{\alpha_n}x_n(t)],\ D^{\alpha} = \mathrm{diag}(D^{\alpha_1}, D^{\alpha_2}, \ldots, D^{\alpha_n}),$$
$$X(t) = [x_1(t); x_2(t); \ldots; x_n(t)],\ F(t) = [f_1(t); f_2(t); \ldots; f_n(t)].$$

Denote by $BV[-h,0]$ the linear space of matrix valued functions

$$W : \mathbb{R} \to \mathbb{R}^{n\times n},\ W(\theta) = \Big[\omega_{kj}(\theta)\Big]$$

with bounded variation in θ on $[-h,0]$,

$$\mathrm{Var}_{[-h,0]}W(.) = \sum_{k,j=1}^{n}\mathrm{Var}_{[-h,0]}w_{kj}(.),\ |W(\theta)| = \sum_{k,j=1}^{n}|w_{kj}(\theta)|.$$

As a space of initial functions, we use the Banach space $\widetilde{C} = PC([-h,0],\mathbb{R}^n)$ of the piecewise continuous on $[-h,0]$ vector functions $\Phi = [\phi_1; \phi_2; \ldots; \phi_n] : [-h,0] \to \mathbb{R}^n$ with norm

$$\|\Phi\| = \sum_{k=1}^{n}\sup_{s\in[-h,0]}|\phi_k(s)| < \infty.$$

The initial condition for the system (1) or (2) is

$$X(t) = \Phi(t),\ t \in [-h,0]. \tag{3}$$

Definition 1. *The vector function X is a solution of the IVP (1), (3) in the interval J_{-h} if $X|_{J_0} \in C(J_0,\mathbb{R}^n)$ and if it satisfies the system (1) for $t \in \mathbb{R}_+$ and the initial condition (3) for $t \in [-h,0]$.*

We say that for the kernels $U^i : \mathbb{R} \to \mathbb{R}^{n\times n}$, $V^l : \mathbb{R} \to \mathbb{R}^{n\times n}$ the assumptions (SA) are fulfilled, if for each $i \in \langle 0, m\rangle$ and $l \in \langle 1, r\rangle$ the following conditions hold.

(SA1) The matrix valued functions $\theta \mapsto U^i(\theta)$ and $\theta \mapsto V^l(\theta)$ are measurable in $\theta \in \mathbb{R}$ and normalized so that $U^i(\theta) = 0$ and $V^l(\theta) = 0$ for $\theta \geq 0$, $U^i(\theta) = U^i(-\sigma_i)$ for $\theta \leq -\sigma_i$ and $V^l(\theta) = V^l(-\tau_l)$ for $\theta \leq -\tau_l$.

(SA2) The kernels $U^i(\theta)$ and $V^l(\theta)$ are left continuous for $\theta \in (-\sigma, 0)$ and $\theta \in (-\tau, 0]$ and $U^i(\cdot)$, $V^l(\cdot) \in BV[-h, 0]$.

(SA3) The Lebesgue decomposition of the kernels $U^i(\theta)$ and $V^l(\theta)$ for $\theta \in [-h, 0]$ is

$$\begin{aligned} U^i(\theta) &= \aleph^i(\theta) + \int_{-h}^{\theta} B^i(s)\mathrm{d}s + \Upsilon^i(\theta), \\ V^l(\theta) &= \widetilde{\aleph}^l(\theta) + \int_{-h}^{\theta} \widetilde{B}^l(s)\mathrm{d}s + \widetilde{\Upsilon}^l(\theta), \end{aligned}$$

where $A^i = \left[a^i_{kj}\right]$, $\widetilde{A}^l = \left[\widetilde{a}^l_{kj}\right] \in \mathbb{R}^{n\times n}$ and

$$\begin{aligned} &\aleph^i(\theta) = \left[a^i_{kj}H(\theta+\sigma_i)\right], \ \widetilde{\aleph}^l(\theta) = \left[\widetilde{a}^l_{kj}H(\theta+\tau_l)\right], \\ &\Upsilon^i(\theta) = \left[g^i_{kj}(\theta)\right], \ \widetilde{\Upsilon}^l(\theta) = \left[\widetilde{g}^l_{kj}(\theta)\right] \in C(\mathbb{R}, \mathbb{R}^{n\times n}), \\ &B^i(\theta) = \left[b^i_{kj}(\theta)\right], \ \widetilde{B}^l(\theta) = \left[\widetilde{b}^l_{kj}(\theta)\right] \in L_1^{\mathrm{loc}}(\mathbb{R}^{n\times n}, \mathbb{R}^{n\times n}). \end{aligned}$$

Remark 1. *The conditions (SA) are used essentially in the work [21] to establish an apriory estimate of all solutions of the IP (1), (3), which estimate guaranties that the Laplace transform can be correct applied to System (2) and to System (1) too, when the function F is exponentially bounded.*

Let $s \geq 0$ be an arbitrary number, $J_s = [s, \infty)$ and consider the matrix IVP

$$D^\alpha \left(Q(t,s) - \sum_{l=1}^{r} \int_{-\tau}^{0} \left[\mathrm{d}_\theta V^l(\theta)\right] Q(t+\theta, s) \right) = \sum_{i=0}^{m} \int_{-\sigma}^{0} \left[\mathrm{d}_\theta U^i(t,\theta)\right] Q(t+\theta, s) \tag{4}$$

with initial condition

$$Q(t,t) = I; \ Q(t,s) = 0, \ t < s. \tag{5}$$

Definition 2. *For each $s \geq 0$ the matrix valued function*

$$t \mapsto Q(t,s) = \left[\gamma_{kj}(t,s)\right], \ Q(\cdot, s) : J_s \to \mathbb{R}^{n\times n},$$

is called a solution of the IVP (4), (5) for $t \in J_s$, if $Q(\cdot, s)$ is continuous in t on J_s and satisfies the matrix Equation (4) for $t \in (s, \infty)$ and the initial condition (5).

It is well known that the problem of existence of a fundamental matrix for a linear homogeneous fractional system (delayed or neutral) leads to establishing that the corresponding IVP (4), (5) with discontinuous initial function has a unique solution. In the case when $s = 0$, the matrix $Q(t) = Q(t, 0)$ will be called fundamental (or Cauchy) matrix of system (2).

Following [20,21], we introduce the characteristic matrix of System (2)

$$G(p) = I_\alpha(p) - W(p), \tag{6}$$

where

$$\begin{aligned} W(p) &= \sum_{i=0}^{m} U_i(p) + I_\alpha(p) \sum_{l=1}^{r} V_l(p),\ i \in \langle 0, m\rangle,\ l \in \langle 1, r\rangle, \\ U_i(p) &= \left[\int_{-h}^{0} \exp(p\theta) \mathrm{d}u^i_{kj}(\theta)\right],\ V_l(p) = \left[\int_{-h}^{0} \exp(p\theta) \mathrm{d}v^l_{kj}(\theta)\right]. \end{aligned}$$

3. Main Results

The results in this section are a generalization of the results concerning the autonomous case obtained in [10,15,16,25].

Theorem 1. *Let us assume the conditions (SA) are satisfied. Then the IVP (4), (5) has a unique solution $Q(t,s)$ in J_s for every $s \geq 0$ and the fundamental matrix $Q(t,0) = Q(t)$ of Equation (2) is*

$$Q(t) = \mathfrak{L}^{-1}\left(I_{\alpha-1}(p)G^{-1}(p)\right)(t). \tag{7}$$

Proof. Using the results from [28], we obtain that the IVP (4), (5) has a unique solution $Q(t,s)$ in J_s for every $s \geq 0$, and hence, a fundamental matrix $Q(t,0) = Q(t)$. In virtue of Theorem 3 [21], we can conclude that the Laplace transform can be applied to both sides of Equation (4). Substituting $t + \theta = \eta$ we obtain

$$\begin{aligned} &\int_0^\infty \exp(-pt) \sum_{i=0}^{m} \int_{-h}^{0} \left[\mathrm{d}_\theta U^i(\theta)\right] Q(t+\theta)\mathrm{d}t = \sum_{i=0}^{m} \int_{-h}^{0} \left[\mathrm{d}_\theta U^i(\theta)\right] \left(\exp(p\theta) \int_\theta^0 \exp(-p\eta) Q(\eta)\mathrm{d}\eta\right) \\ &+ \int_0^\infty \exp(-p\eta) Q(\eta)\mathrm{d}\eta \sum_{i=0}^{m} \int_{-h}^{0} \exp(p\theta)\mathrm{d}_\theta U^i(\theta) = \mathfrak{L}Q(p) \sum_{i=0}^{m} \int_{-h}^{0} \exp(p\theta)\mathrm{d}_\theta U^i(\theta). \end{aligned}$$

In a similar way for the left-hand side of Equation (4), we have that

$$\begin{aligned} &\mathfrak{L}D^\alpha \left(Q(t) - \sum_{l=1}^{r} \int_{-\tau}^{0} \left[\mathrm{d}_\theta V^l(\theta)\right] Q(t+\theta)\right)(p) = I_\alpha(p)\mathfrak{L}Q(p) \left[E - \sum_{l=1}^{r} \int_{-h}^{0} \exp(p\theta)\mathrm{d}_\theta V^l(\theta)\right] \\ &- I_{\alpha-1}(p) \left[E - \sum_{l=1}^{r} \int_{-\tau}^{0} \left[\mathrm{d}_\theta V^l(\theta)\right] Q(\theta)\right]. \end{aligned} \tag{8}$$

From Equation (8), it follows that

$$\mathfrak{L}Q(p) \left[I_\alpha - I_\alpha(p) \sum_{l=1}^{r} \int_{-h}^{0} \exp(p\theta)\mathrm{d}_\theta V^l(\theta) - \sum_{l=1}^{m} \int_{-h}^{0} \exp(p\theta)\mathrm{d}_\theta U^i(\theta)\right] = I_{\alpha-1}(p)$$

and hence, $\mathfrak{L}Q(p) = I_{\alpha-1}(p)G^{-1}(p)$, which completes the proof. □

Let us introduce the following functions:

$$\Phi_l(t) = \Phi(t),\ t \in [-\tau_l, 0],\ \ \Phi_l(t) = 0,\ t \in \mathbb{R} \setminus [-\tau_l, 0],\ \ l \in \langle 1, r\rangle,$$

and

$$\Phi_i(t) = \Phi(t),\ t \in [-\sigma_i, 0],\ \ \Phi_i(t) = 0,\ t \in \mathbb{R} \setminus [-\sigma_i, 0],\ \ i \in \langle 1, m\rangle.$$

Then, applying the Laplace transform second shifting theorem, we obtain

$$\begin{aligned}
&\int_{\theta}^{0} \exp(-p(\eta-\theta))\Phi_l(\eta)\mathrm{d}\eta = \exp(p\theta)\mathfrak{L}\Phi_l(t)(p), \\
&\int_{\theta}^{0} \exp(-p(\eta-\theta))\Phi_i(\eta)\mathrm{d}\eta = \exp(p\theta)\mathfrak{L}\Phi_i(t)(p).
\end{aligned} \tag{9}$$

Now we are in position to prove the following theorem.

Theorem 2. *Let us assume the conditions (SA) are satisfied. Then for each $\Phi \in \widetilde{C}$ the IVP (2), (3) has a unique solution $X_\Phi(t)$ with the integral representation:*

$$\begin{aligned}
X_\Phi(t) &= Q(t)\left(\Phi(0) - \sum_{l=1}^{r}\int_{-h}^{0}\left[\mathrm{d}_\theta V^l(\theta)\right]\Phi_l(\theta)\right) + \sum_{l=1}^{r}\int_{-h}^{0}\left[\mathrm{d}_\theta V^l(\theta)\right]\Phi_l(t+\theta) \\
&+ \sum_{l=1}^{r}\int_{-h}^{0}\left[\mathrm{d}_\theta V^l(\theta)\right] D^{\frac{1}{2}}Q(t) * D^{\frac{1}{2}}\Phi_l(t+\theta) + \sum_{i=0}^{m}\int_{-h}^{0}[\mathrm{d}_\theta U^i(\theta)]D^{1-\alpha}Q(t) * \Phi_i(t+\theta) \\
&+ \sum_{i=0}^{m}\int_{-h}^{0}[\mathrm{d}_\theta U^i(\theta)]D^{-\alpha}\Phi_i(t+\theta).
\end{aligned} \tag{10}$$

Proof. Let $\Phi \in \widetilde{C}$. Then using the results from [26], we can conclude that the IVP (2), (3) has a unique solution $X_\Phi(t)$. In virtue of Theorem 3 from [21], we can conclude that the Laplace transform can be applied to both sides of Equation (2). Then, substituting $X_\Phi(t)$ in Equation (2), applying the Laplace transform to Equation (2) and substituting $t+\theta=\eta$, we obtain for the right-hand side of Equation (2)

$$\begin{aligned}
&\mathfrak{L}\left(\sum_{i=0}^{m}\int_{-h}^{0}[\mathrm{d}_\theta U^i(\theta)]X_\Phi(t+\theta)\right)(p) = \int_{0}^{\infty}\exp(-p\eta)X_\Phi(\eta)\mathrm{d}\eta\sum_{i=0}^{m}\int_{-h}^{0}\exp(p\theta)\mathrm{d}_\theta U^i(\theta) \\
&+ \sum_{i=0}^{m}\int_{-h}^{0}[\mathrm{d}_\theta U^i(\theta)](\int_{\theta}^{0}\exp(p(\theta-\eta))X_\Phi(\eta)\mathrm{d}\eta) = \mathfrak{L}X_\Phi(t)(p)\sum_{i=0}^{m}\int_{-h}^{0}\exp(p\theta)\mathrm{d}_\theta U^i(\theta) \\
&+ \sum_{i=0}^{m}\int_{-h}^{0}[\mathrm{d}_\theta U^i(\theta)]\int_{\theta}^{0}\exp(p(\theta-\eta))\Phi_i(\eta)\mathrm{d}\eta.
\end{aligned} \tag{11}$$

Similarly, for the left-hand side of Equation (2), one obtains that

$$\begin{aligned}
&\mathfrak{L}D^\alpha\left(X_\Phi(t) - \sum_{l=1}^{r}\int_{-h}^{0}\left[\mathrm{d}_\theta V^l(\theta)\right]X_\Phi(t+\theta)\right)(p) = -I_{\alpha-1}(p)\left(\Phi(0) - \sum_{l=1}^{r}\int_{-h}^{0}\left[\mathrm{d}_\theta V^l(\theta)\right]\Phi_l(\theta)\right) \\
&+ \mathfrak{L}X_\Phi(t)(p)\left(I_\alpha(p) - I_\alpha(p)\sum_{l=1}^{r}\int_{-h}^{0}\exp(p\theta)\mathrm{d}_\theta V^l(\theta)\right) \\
&- I_\alpha(p)\sum_{l=1}^{r}\int_{-h}^{0}[\mathrm{d}_\theta V^l(\theta)]\int_{\theta}^{0}\exp(p(\theta-\eta))\Phi_l(\eta)\mathrm{d}\eta.
\end{aligned} \tag{12}$$

From Equations (11) and (12), it follows

$$\begin{aligned}
&\mathfrak{L}X_\Phi(p)\left(I_\alpha(p) - I_\alpha(p)\sum_{l=1}^{r}\int_{-h}^{0}\exp(p\theta)d_\theta V^l(\theta) - \sum_{i=0}^{m}\int_{-h}^{0}\exp(p\theta)d_\theta U^i(\theta)\right) \\
&= I_{\alpha-1}(p)\left(\Phi(0) - \sum_{l=1}^{r}\int_{-h}^{0}\left[\mathrm{d}_\theta V^l(\theta)\right]\Phi_l(\theta)\right) + I_\alpha(p)\sum_{l=1}^{r}\int_{-h}^{0}\left[\mathrm{d}_\theta V^l(\theta)\right]\int_{\theta}^{0}\exp(p(\theta-\eta))\Phi_l(\eta)\mathrm{d}\eta \\
&+ \sum_{i=0}^{m}\int_{-h}^{0}\left[\mathrm{d}_\theta U^i(\theta)\right]\int_{\theta}^{0}\exp(p(\theta-\eta))\Phi_i(\eta)\mathrm{d}\eta
\end{aligned}$$

and hence,

$$\begin{aligned}
\mathfrak{L}X_{\Phi}(p) &= G^{-1}(p)I_{\alpha-1}(p)\left(\Phi(0)-\sum_{l=1}^{r}\int_{-h}^{0}\left[\mathrm{d}_{\theta}V^{l}(\theta)\right]\Phi_{l}(\theta)\right)\\
&+G^{-1}(p)I_{\alpha}(p)+\sum_{l=1}^{r}\int_{-h}^{0}\left[\mathrm{d}_{\theta}V^{l}(\theta)\right]\int_{\theta}^{0}\exp(p(\theta-\eta))\Phi_{l}(\eta)\mathrm{d}\eta\\
&+G^{-1}(p)\sum_{i=0}^{m}\int_{-h}^{0}\left[\mathrm{d}_{\theta}U^{i}(\theta)\right]\int_{\theta}^{0}\exp(p(\theta-\eta))\Phi_{i}(\eta)\mathrm{d}\eta.
\end{aligned} \tag{13}$$

The representations of Equations (7) and (13) imply that

$$\begin{aligned}
\mathfrak{L}X_{\Phi}(p) &= \mathfrak{L}Q(t)(p)\left(\Phi(0)-\sum_{l=1}^{r}\int_{-h}^{0}[\mathrm{d}_{\theta}V^{l}(\theta)]\Phi_{l}(\theta)\right)\\
&+I_{1}(p)\mathfrak{L}Q(t)(p)\sum_{l=1}^{r}\int_{-h}^{0}\left[\mathrm{d}_{\theta}V^{l}(\theta)\right]\int_{\theta}^{0}\exp(p(\theta-\eta))\Phi_{l}(\eta)\mathrm{d}\eta\\
&+I_{1-\alpha}(p)\mathfrak{L}Q(t)(p)\sum_{i=0}^{m}\int_{-h}^{0}\left[\mathrm{d}_{\theta}U^{i}(\theta)\right]\int_{\theta}^{0}\exp(p(\theta-\eta))\Phi_{i}(\eta)\mathrm{d}\eta.
\end{aligned} \tag{14}$$

In view of Equation (9), we obtain for the second term in the right-hand side of Equation (14) that

$$\begin{aligned}
&I_{1}(p)\mathfrak{L}Q(t)(p)\sum_{l=1}^{r}\int_{-h}^{0}\left[\mathrm{d}_{\theta}V^{l}(\theta)\right]\int_{\theta}^{0}\exp(p(\theta-\eta))\Phi_{l}(\eta)\mathrm{d}\eta\\
&=I_{\frac{1}{2}}(p)\mathfrak{L}Q(t)(p)I_{\frac{1}{2}}(p)\sum_{l=1}^{r}\int_{-h}^{0}\left[\mathrm{d}_{\theta}V^{l}(\theta)\right]\mathfrak{L}\Phi_{l}(t+\theta)(p)\\
&=\mathfrak{L}D^{\frac{1}{2}}Q(t)(p)+I_{-\frac{1}{2}}(p)I_{\frac{1}{2}}(p)\sum_{l=1}^{r}\int_{-h}^{0}\left[\mathrm{d}_{\theta}V^{l}(\theta)\right]\mathfrak{L}\Phi_{l}(t+\theta)(p)\\
&=\mathfrak{L}D^{\frac{1}{2}}Q(t)(p)\sum_{l=1}^{r}\int_{-h}^{0}\left[\mathrm{d}_{\theta}V^{l}(\theta)\right]I_{\frac{1}{2}}(p)\mathfrak{L}\Phi_{l}(t+\theta)(p)\\
&+\sum_{l=1}^{r}\int_{-h}^{0}\left[\mathrm{d}_{\theta}V^{l}(\theta)\right]\mathfrak{L}\Phi_{l}(t+\theta)(p).
\end{aligned} \tag{15}$$

For the first term in the right-hand side of Equation (15) we have

$$\begin{aligned}
&\mathfrak{L}D^{\frac{1}{2}}Q(t)(p)\sum_{l=1}^{r}\int_{-h}^{0}\left[\mathrm{d}_{\theta}V^{l}(\theta)\right]I_{\frac{1}{2}}(p)\mathfrak{L}\Phi_{l}(t+\theta)(p)\\
&=\sum_{l=1}^{r}\int_{-h}^{0}\left[\mathrm{d}_{\theta}V^{l}(\theta)\right]\mathfrak{L}D^{\frac{1}{2}}Q(t)(p)\mathfrak{L}D^{\frac{1}{2}}\Phi_{l}(t+\theta)(p)
\end{aligned}$$

and hence, from Equation (15) it follows

$$\begin{aligned}
&I_{1}(p)\mathfrak{L}Q(t)(p)\sum_{l=1}^{r}\int_{-h}^{0}\left[\mathrm{d}_{\theta}V^{l}(\theta)\right]\int_{\theta}^{0}\exp(p(\theta-\eta))\Phi_{l}(\eta)\mathrm{d}\eta\\
&=\sum_{l=1}^{r}\int_{-h}^{0}\left[\mathrm{d}_{\theta}V^{l}(\theta)\right](\mathfrak{L}D^{\frac{1}{2}}Q(t)(p)\mathfrak{L}D^{\frac{1}{2}}\Phi_{l}(t+\theta)(p)\\
&+\sum_{l=1}^{r}\int_{-h}^{0}[\mathrm{d}_{\theta}V^{l}(\theta)]\mathfrak{L}\Phi_{l}(t+\theta)(p).
\end{aligned} \tag{16}$$

Analogously, for the third term in the right-hand side of Equation (14), we have

$$\begin{aligned}
&I_{1-\alpha}(p)\mathfrak{L}Q(t)(p)\sum_{i=0}^{m}\int_{-h}^{0}\left[\mathrm{d}_\theta U^i(\theta)\right]\int_{\theta}^{0}\exp(p(\theta-\eta))\Phi_i(\eta)\mathrm{d}\eta\\
&=I_{1-\alpha}(p)\mathfrak{L}Q(t)(p)\sum_{i=0}^{m}\int_{-h}^{0}\left[\mathrm{d}_\theta U^i(\theta)\right]\mathfrak{L}\Phi_i(t+\theta)(p)\\
&=\sum_{i=0}^{m}\int_{-h}^{0}\left[\mathrm{d}_\theta U^i(\theta)\right]\mathfrak{L}D^{1-\alpha}Q(t)(p)\mathfrak{L}\Phi_i(t+\theta)(p)\\
&+\sum_{i=0}^{m}\int_{-h}^{0}[\mathrm{d}_\theta U^i(\theta)]\mathfrak{L}D^{-\alpha}\Phi_i(t+\theta)(p)
\end{aligned} \tag{17}$$

From Equations (14), (16) and (17), it follows that

$$\begin{aligned}
\mathfrak{L}X_\Phi(p)&=\mathfrak{L}Q(t)(p)\left(\Phi(0)-\sum_{l=1}^{r}\int_{-h}^{0}\left[\mathrm{d}_\theta V^l(\theta)\right]\Phi(\theta)\right)+\sum_{l=1}^{r}\int_{-h}^{0}\left[\mathrm{d}_\theta V^l(\theta)\right]\mathfrak{L}\Phi_l(t+\theta)(p)\\
&+\sum_{l=1}^{r}\int_{-h}^{0}\left[\mathrm{d}_\theta V^l(\theta)\right]\mathfrak{L}D^{\frac{1}{2}}Q(t)(p)\mathfrak{L}D^{\frac{1}{2}}\Phi_l(t+\theta)(p)\\
&+\sum_{i=0}^{m}\int_{-h}^{0}\left[\mathrm{d}_\theta U^i(\theta)\right]\mathfrak{L}D^{1-\alpha}Q(t)(p)\mathfrak{L}\Phi_i(t+\theta)(p)+\sum_{i=0}^{m}\int_{-h}^{0}\left[\mathrm{d}_\theta U^i(\theta)\right]\mathfrak{L}D^{-\alpha}\Phi_i(t+\theta)(p)
\end{aligned} \tag{18}$$

Applying the inverse Laplace transform to both sides of Equation (18), we obtain Equation (10). □

Theorem 3. *Let the following conditions be satisfied:*

(i) *The conditions (SA) hold.*
(ii) *The function $F \in L_1^{\mathrm{loc}}(\bar{\mathbb{R}}_+, \mathbb{R}^n)$ is exponentially bounded.*

Then the solution $X^F(t)$ of the IVP (1), (3) with initial function $\Phi(t) \equiv 0$, $t \in [-h, 0]$ has the following representation:

$$X^F(t)=\int_0^t D^{1-\alpha}Q(t-s)F(s)\mathrm{d}s+D^{-\alpha}F(t), \tag{19}$$

where $Q(t)$ is the fundamental matrix of System (2).

Proof. First we substitute $X_\Phi(t)$ in Equation (1) and use the fact that $X^F(t) = 0$, $t \in [-h, 0]$. Since the function F is exponentially bounded, then we can apply to both sides the Laplace transform in order to get

$$\begin{aligned}
&\mathfrak{L}X^F(t)(p)\left(I_\alpha(p)-I_\alpha(p)\sum_{l=1}^{r}\int_{-h}^{0}\exp(p\theta)\mathrm{d}_\theta V^l(\theta)-\sum_{i=0}^{m}\int_{-h}^{0}\exp(p\theta)\mathrm{d}_\theta U^i(\theta)\right)\\
&=\mathfrak{L}X^F(t)(p)G(p)=\mathfrak{L}F(t)(p).
\end{aligned} \tag{20}$$

Now it follows from the equality $G^{-1}(p) = I_{1-\alpha}(p)\mathfrak{L}Q(t)(p)$ that

$$\begin{aligned}
&\mathfrak{L}X^F(t)(p)=I_{1-\alpha}(p)I_{\alpha-1}(p)G^{-1}(p)\mathfrak{L}F(t)(p)\\
&=I_{1-\alpha}(p)\mathfrak{L}Q(t)(p)\mathfrak{L}F(t)(p)=\left(\mathfrak{L}D^{1-\alpha}Q(t)(p)+I_{-\alpha}(p)\right)\mathfrak{L}F(t)(p)\\
&=\mathfrak{L}D^{1-\alpha}Q(t)(p)\mathfrak{L}F(t)(p)+\mathfrak{L}D^{-\alpha}F(t)(p).
\end{aligned} \tag{21}$$

Finally, we apply the inverse Laplace transform to Equation (21) and the representation Equation (19) follows. □

Corollary 1. *Let the conditions of Theorem 3 hold. Then for every initial function $\Phi \in \widetilde{C}$, the corresponding unique solution $X^F_\Phi(t)$ of the IVP (1), (3) has the integral representation*

$$X^F_\phi(t) = \int_0^t D^{1-\alpha}Q(t-s)F(s)\mathrm{d}s + D^{-\alpha}F(t) + Q(t)\left(\Phi(0) - \sum_{l=1}^{r}\int_{-h}^{0}\left[\mathrm{d}_\theta V^l(\theta)\right]\Phi(\theta)\right)$$
$$+\sum_{l=1}^{r}\int_{-h}^{0}\left[\mathrm{d}_\theta V^l(\theta)\right]\Phi(t+\theta) + \sum_{l=1}^{r}\int_{-h}^{0}\left[\mathrm{d}_\theta V^l(\theta)\right]D^{\frac{1}{2}}Q(t) * D^{\frac{1}{2}}\Phi_l(t+\theta)$$
$$+\sum_{i=0}^{m}\int_{-h}^{0}\left[\mathrm{d}_\theta U^i(\theta)\right]D^{1-\alpha}Q(t) * \Phi_i(t+\theta) + \sum_{i=0}^{m}\int_{-h}^{0}\left[\mathrm{d}_\theta U^i(\theta)\right]D^{-\alpha}\Phi_i(t+\theta),$$

where $Q(t)$ is the fundamental matrix of System (2).

Proof. Let $\Phi \in \widetilde{C}$ be an arbitrary initial function and let the functions $X_\Phi(t)$ and $X^F(t)$ be defined by the Equalities (9) and (19), respectively. Then, according to the superposition principle, the function $X_\Phi(t) + X^F(t)$ is the unique solution of the IVP (1), (3). Now the statement of Corollary 1 follows immediately from Theorems 2 and 3. □

4. Example

First, we give some results needed for the illustrative example presented below:

The delayed Mittag-Leffler type matrix function $\mathbf{E}^{B,\tau}_{\alpha,1} : \mathbb{R} \to \mathbb{R}^{n\times n}$ for every matrix $B \in \mathbb{R}^{n\times n}$ and for $\tau \in \mathbb{R}_+$ is defined by

$$\mathbf{E}^{Bt^\alpha}_\tau(t) := I + \sum_{k=1}^{\infty}\frac{B^k(t-(k-1)\tau)^{\alpha k}}{\Gamma(\alpha k+1)}H(k\tau - t), \quad t \geq 0 \tag{22}$$

with $\mathbf{E}^{Bt^\alpha}_\tau(0) := I$, $\mathbf{E}^{Bt^\alpha}_\tau(t) := \Theta$ for $t < 0$ and $H(t)$ is the Heavyside function with $H(0) = 1$. This is a slight modification of the original definition in [29], and note that for each $t \geq 0$, the sum in Equation (22) is finite and for $\tau = 0$ we have

$$\mathbf{E}^{Bt^\alpha}_0(t) := E_\alpha(Bt^\alpha) = \sum_{k=0}^{\infty}\frac{B^k t^{\alpha k}}{\Gamma(\alpha k+1)}, \quad t \geq 0, \tag{23}$$

where the right side is the standard Mittag-Leffler type matrix function.

Example 1. *Consider the nonhomogeneous system for $t > 0$:*

$$\begin{aligned} &D^{0.5}_{0+}x_1(t) = x_1(t-1) + 1 \\ &D^{0.5}_{0+}(x_2(t) + x_1(t-1) + x_2(t-1)) = x_2(t) + x_2(t-1) + x_1(t-2) \end{aligned} \tag{24}$$

with the initial conditions

$$\Phi(t) = (0,2)^T, t \in [-2,0] \quad \textit{i.e.} \quad x_1(t) = 0,\ x_2(t) = 2 \quad \textit{for} \quad t \in [-2,0]. \tag{25}$$

The homogenious system has the form

$$\begin{aligned} &D^{0.5}_{0+}\overline{x}_1(t) = \overline{x}_1(t-1) \\ &D^{0.5}_{0+}(\overline{x}_2(t) + \overline{x}_1(t-1) + \overline{x}_2(t-1)) = \overline{x}_2(t) + \overline{x}_2(t-1) + \overline{x}_1(t-2) \end{aligned} \tag{26}$$

and introduce the following initial conditions necessary for the calculating the fundamental matrix $Q(t)$:

$$1.\ x_1(0) = 1, x_2(0) = 0 \quad \textit{and} \quad x_1(t) = 0, x_2(t) = 0 \quad \textit{for} \quad t \in [-2,0); \tag{27}$$

$$2.\ x_1(0)=0, x_2(0)=1 \quad and \quad x_1(t)=0, x_2(t)=0 \quad for \quad t\in[-2,0). \tag{28}$$

Let consider the IP (26), (27). Then the first Equation of (26) in virtue of Theorem 3.1 in [29] has the solution $\overline{x}_1^{\tau}(t)=\mathbf{E}_1^{t^{0.5}}$ $(\tau=1,\alpha=0.5)$*. Taking into account Equation (27), it is simple to check that* $(D_{0+}^{0.5}\overline{x}_1^1(s-1))(t)=(D_{0+}^{0.5}\overline{x}_1^1(s))(t-1)$*, and then in virtue of Theorem 3.1 in [29] we have that* $(D_{0+}^{0.5}\overline{x}_1^1)(t-1)=\overline{x}_1^1(t-2)$*, and hense, from the second equation and Equation (27), we obtain that* $\overline{x}_2^1(t)\equiv 0$ *for* $t\in[-2,\infty)$*. Thus the IP (26), (27) have the following solution* $\overline{x}_1^1(t)=\mathbf{E}_1^{t^{0.5}}$, $\overline{x}_2^1(t)\equiv 0$ *for* $t\in[-1,\infty)$*.*

Consider the IP (26), (28) . Then obviously $\overline{x}_1^2(t)\equiv 0$ *for* $t\in[-2,\infty)$ *and the second equation become the form:* $D_{0+}^{0.5}(\overline{x}_2(t)+\overline{x}_2(t-1))=\overline{x}_2(t)+\overline{x}_2(t-1)$ *and by making the substitutuon* $y(t)=\overline{x}_2(t)+\overline{x}_2(t-1)$ *we obtain the equations* $D_{0+}^{0.5}y(t)=y(t)$ *with initial codition* $y(0)=1$*, i.e., the following IP*

$$D_{0+}^{0.5}y(t)=y(t), \quad t>0; \quad y(0)=1. \tag{29}$$

Applying Lemma 2.23 in [2] for the case when $\lambda=1$, $\tau=1$, $\alpha=0.5$ *we obtain that the solution of the IP (29) is the fuction* $y(t)=\mathbf{E}_1^{t^{0.5}}(t)=\sum_{k=0}^{\infty}\frac{t^{\alpha k}}{\Gamma(\alpha k+1)}$*. Then, using the step method, we obtain for each* $k\in\mathbb{N}$ *and* $t\in[k-1,k)]$ *that* $\overline{x}_2^2(t)=\sum_{k=1}^{\infty}(-1)^{k-1}\mathbf{E}_1^{t^{0.5}}(t-(k-1))H(k-t)$ *for* $t>0$*. Thus, we obtain that the fundamental matrix have the form:*

$$Q(t)=\begin{pmatrix}\mathbf{E}_1^{t^{0.5}}(t) & 0\\ 0 & \sum_{k=1}^{\infty}(-1)^{k-1}\mathbf{E}_1^{t^{0.5}}(t-(k-1))H(k-t)\end{pmatrix}. \tag{30}$$

In the IP (24), (25) we have that: $\Phi(t)=(0,2)^T, t\in[-2,0]$; $F(t)=(1,0)^T$*. Then from Equation (19), we have*

$$x_1^F(t)=\frac{1}{\Gamma(0.5)}\int_0^t\left(\int_0^{t-s}(t-s-\eta)^{-0.5}\left(\mathbf{E}_1^{t^{0.5}}\right)'(\eta)\mathrm{d}\eta\right)\mathrm{d}s+\frac{\sqrt{t}}{\Gamma(1.5)},$$
$$x_2^F(t)=0.$$

From Equation (10), it follows

$$x_1^{\Phi}(t)=0,$$
$$x_2^{\Phi}(t)=2+2\int_0^t\left(\int_0^s(s-\eta)^{-0.5}\left(\sum_{k=1}^{\infty}(-1)^{k-1}\mathbf{E}_1^{t^{0.5}}(\eta-(k-1))H(k-\eta)\right)'\mathrm{d}\eta\right)\mathrm{d}s+\frac{2\sqrt{t}}{\Gamma(1.5)}.$$

Then, the solution of the IP (24), (25), according Corollary 1, is

$$x_1(t)=x_1^{\Phi}(t)+x_1^F(t)=\frac{1}{\Gamma(0.5)}\int_0^t\left(\int_0^{t-s}(t-s-\eta)^{-0.5}\left(\mathbf{E}_1^{t^{0.5}}\right)'(\eta)\mathrm{d}\eta\right)\mathrm{d}s+\frac{\sqrt{t}}{\Gamma(1.5)},$$

$$x_2(t)=x_2^{\Phi}(t)+x_2^F(t)=2+2\int_0^t\left(\int_0^s(s-\eta)^{-0.5}\left(\sum_{k=1}^{\infty}(-1)^{k-1}\mathbf{E}_1^{t^{0.5}}(\eta-(k-1))H(k-\eta)\right)'\mathrm{d}\eta\right)\mathrm{d}s$$
$$+\frac{2\sqrt{t}}{\Gamma(1.5)}.$$

5. Conclusions

Following the investigations way in the case of functional differential systems with integer order derivatives, we proved a formula for integral representation of the solutions of Cauchy problem for fractional neutral systems, which improves and extends the corresponding former results obtained in the particular case of fractional systems with constant delays. However, the main idea is not only to make a standard generalization of existing results, but as in the case of systems with integer derivatives, the proved formula to be an useful tool for further study of different kinds stability properties of linear neutral fractional systems, which have a lot of practical applications.

As examples in this direction, we refer to the works [29,30], where finite time stability is studied by this approach, i.e., in the partial case of one constant delay. In the mentioned articles, first a formula for integral representation of the solutions of Cauchy problem is proved, and then, using the obtained result, sufficient conditions for finite time stability of the considered fractional delayed system are established. Furthermore, applying the same approach, in [16], the asymptotic stability properties of nonlinear perturbed linear fractional delayed systems are studied .

Author Contributions: The authors contribution in the article are equal. All authors have read and agreed to the published version of the manuscript.

Funding: This research received no external funding.

Acknowledgments: The authors are grateful to the anonymous reviewers for their very helpful comments.

Conflicts of Interest: The authors declare no conflict of interest.

References

1. Diethelm, K. *The Analysis of Fractional Differential Equations, an Application–Oriented Exposition Using Differential Operators of Caputo Type*; Lecture Notes in Mathematics; Springer: Berlin, Germany, 2010; Volume 2004.
2. Kilbas, A.A.; Srivastava, H.M.; Trujillo, J.J. *Theory and Applications of Fractional Differential Equations*; Elsevier Science B.V: Amsterdam, The Netherlands, 2006.
3. Kiryakova, V. *Generalized Fractional Calculus and Applications*; Longman Scientific & Technical: Harlow, UK; John Wiley & Sons, Inc.: New York, NY, USA, 1994.
4. Podlubny, I. *Fractional Differential Equation*; Academic Press: San Diego, CA, USA, 1999.
5. Kaczorek, T. *Selected Problems of Fractional Systems Theory*; Lecture Notes in Control and Information Sciences; Springer: Berlin/Heidelberg, Germany, 2011.
6. Stamova, I.; Stamov, G. *Functional and Impulsive Differential Equations of Fractional Order*; Qualitative analysis and applications; CRC Press: Boca Raton, FL, USA, 2017.
7. Bonilla, B.; Rivero, M.; Trujillo, J. On systems of linear fractional differential equations with constant coefficients. *Appl. Comput. Math.* **2007**, *187*, 68–78. [CrossRef]
8. Li, K.; Peng, J. Laplace transform and fractional differential equations, *Appl. Math. Let.* **2011**, *24*, 2019–2023. [CrossRef]
9. Lin, S.D.; Lu, C.H. Laplace transform for solving some families of fractional differential equations and its applications. *Adv. Differ. Equ.* **2013**, *137*, 1–9. [CrossRef]
10. Golev, A.; Milev, M. Integral representation of the solution of the Cauchy problem for autonomous linear neutral fractional system. *Int. J. Pure Appl. Math.* **2018**, *119*, 235–247.
11. Kiskinov H.; Zahariev A. Asymptotic stability of delayed fractional systems with nonlinear perturbation. *AIP Conf. Proc.* **2018**, *2048*, 050014.
12. Zhang, H.; Cao, J.; Jiang, W. General solution of linear fractional neutral differential difference equations. *Discret. Dyn. Nat. Soc.* **2013**. [CrossRef]
13. Zhang, H.; Wu, D. Variation of constant formulae for time invariant and time varying Caputo fractional delay differential systems. *J. Math. Res. Appl.* **2014**, *34*, 549-560.
14. Kiskinov, H.; Zahariev A. On fractional systems with Riemann-Liouville derivatives and distributed delays—Choice of initial conditions, existence and uniqueness of the solutions. *Eur. Phys. Jo. Spec. Top.* **2017**, *26*, 3473-3487. [CrossRef]
15. Krol, K. Asymptotic properties of fractional delay differential equations, *Appl. Math. Comput.* **2011**, *218*, 1515–1532.
16. Milev, M.; Zlatev, S. A note about stability of fractional retarded linear systems with distributed delays. *Int. J. Pure Appl. Math.* **2017**, *115*, 873–881. [CrossRef]

17. Veselinova, M.; Kiskinov, H.; Zahariev, A. Stability analysis of linear fractional differential system with distributed delays. *AIP Conf. Proc.* **2015**, *1690*, 040013.
18. Veselinova, M.; Kiskinov, H.; Zahariev, A. About stability conditions for retarded fractional differential systems with distributed delays. *Commun. Appl. Anal.* **2016**, *20*, 325–334.
19. Kiskinov H.; Milev N.; Zahariev A. A comparison type theorem for linear neutral fractional systems with distributed delays. *AIP Conf. Proc.* **2017**, *1910*, 050009.
20. Veselinova, M.; Kiskinov, H.; Zahariev, A. Stability analysis of neutral linear fractional system with distributed delays. *Filomat* **2016**, *30*, 841–851. [CrossRef]
21. Veselinova, M.; Kiskinov, H.; Zahariev, A. Explicit conditions for stability of neutral linear fractional system with distributed delays. *AIP Conf. Proc.* **2016**, *1789*, 040005.
22. Boyadzhiev, D.; Kiskinov, H.; Veselinova, M.; Zahariev, A. Stability analysis of linear distributed order fractional systems with distributed delays. *Fract. Calc. Appl. Anal.* **2017**, *20*, 914–935. [CrossRef]
23. Martinez-Garcia, M.; Gordon, T.; Shu, L. Extended crossover model for human-control of fractional order plants. *IEEE Access* **2017**, 5, 27622-27635. [CrossRef]
24. Martinez-Garcia, M.; Zhang, Yu; Gordon T. Memory pattern identification for feedback tracking control in human-machine systems. *Hum. Factors* **2019**. [CrossRef]
25. Boyadzhiev, D.; Kiskinov, H.; Zahariev, A. Integral representation of solutions of fractional system with distributed delays. *Integral Transforms Spec. Funct.* **2018**, *29*, 725–744. [CrossRef]
26. Zahariev, A.; Kiskinov, H.; Angelova E. Linear fractional system of incommensurate type with distributed delay and bounded Lebesgue measurable initial conditions. *Dyn. Syst. Appl.* **2019**, *28*, 491–506.
27. Zahariev, A.; Kiskinov, H.; Angelova, E. Smoothness of the fundamental matrix of linear fractional system with variable delays. *Neural Parallel Sci. Comput.* **2019**, *27*, 71–83.
28. Zahariev, A.; Kiskinov, H. Existence of fundamental matrix for neutral linear fractional system with distributed delays. *Int. J. Pure and Appl. Math.* **2018**, *119*, 31–51.
29. Li, M.; Wang, J. Finite time stability of fractional delay differential equations. *Appl. Math. Lett.* **2017**, *64*, 170–176. [CrossRef]
30. Li, M.; Wang, J. Exploring delayed Mittag-Leffler type matrix functions to study finite time stability of fractional delay differential equations. *Appl. Mathe. Comput.* **2018**, *324*, 254–265. [CrossRef]

mathematics

MDPI

Article

Some Fractional Dynamic Inequalities of Hardy's Type via Conformable Calculus

Samir Saker [1,*], Mohammed Kenawy [2], Ghada AlNemer [3] and Mohammed Zakarya [4,5]

1 Department of Mathematics, Faculty of Science, Mansoura University, Mansoura 35516, Egypt
2 Department of Mathematics, Faculty of Science, Fayoum University, Fayoum 63514, Egypt; mrz00@fayoum.edu.eg
3 Department of Mathematical Science, College of Science, Princess Nourah bint Abdulrahman University, P.O. Box 105862, Riyadh, Saudi 11656, Arabia; gnnemer@pnu.edu.sa
4 Department of Mathematics, College of Science, King Khalid University, P.O. Box 9004, Abha 61413, Saudi Arabia; Mohammed_Zakaria1983@yahoo.com
5 Department of Mathematics, Faculty of Science, Al-Azhar University, Assiut 71524, Egypt
* Correspondence: shsaker@mans.edu.eg

Received: 12 February 2020; Accepted: 10 March 2020; Published: 16 March 2020

Abstract: In this article, we prove some new fractional dynamic inequalities on time scales via conformable calculus. By using chain rule and Hölder's inequality on timescales we establish the main results. When $\alpha = 1$ we obtain some well-known time-scale inequalities due to Hardy, Copson, Bennett and Leindler inequalities.

Keywords: fractional hardy's inequality; fractional bennett's inequality; fractional copson's inequality; fractional leindler's inequality; timescales; conformable fractional calculus; fractional hölder inequality

MSC: 26A15; 26D10; 26D15; 39A13; 34A40; 34N05

In 1920, Hardy [1] established the inequality

$$\sum_{n=1}^{\infty}\left(\frac{1}{n}\sum_{i=1}^{n} w(i)\right)^{k} \leq \left(\frac{k}{k-1}\right)^{k}\sum_{n=1}^{\infty} w^{k}(n), \quad k>1. \tag{1}$$

where $w(n)$ is a positive sequence defined for all $n \geq 1$. After that, Hardy [2], by using the calculus of variations, proved the continuous inequality of (1) which has the form

$$\int_0^{\infty}\left(\frac{1}{x}\int_0^{x} g(s)ds\right)^{k} dx \leq \left(\frac{k}{k-1}\right)^{k}\int_0^{\infty} g^{k}(x)dx, \tag{2}$$

for a given positive function g, which is integrable over $(0,x)$, and g^k is convergent and integrable over $(0,\infty)$ and $k>1$. In (1) and (2), $(k/(k-1))^k$ is a sharp constant. As a generalization of (2), Hardy [3] showed that when $k>1$, then

$$\int_0^{\infty} x^{-h}\left(\int_0^{x} g(s)ds\right)^{k} dx \leq \left(\frac{k}{h-1}\right)^{k}\int_0^{\infty} x^{k-h}g^{k}(x)dx, \text{ for } h>1, \tag{3}$$

and

$$\int_0^{\infty} x^{-h}\left(\int_x^{\infty} g(s)ds\right)^{k} dx \leq \left(\frac{k}{1-h}\right)^{k}\int_0^{\infty} x^{k-h}g^{k}(x)dx, \text{for } h<1. \tag{4}$$

The constants $(k/(h-1))^k$ and $(k/(1-h))^k$ in (3) and (4) are the best possible. Copson [4] demonstrated that if $g(x) > 0, k > 1$ and $g^k(x)$ is integrable on the interval $(0, \infty)$, then

$$\int_x^\infty \left(\frac{g(s)}{s}\right) ds,$$

converges for $x > 0$ and

$$\int_0^\infty \left(\int_x^\infty \frac{g(s)}{s} ds\right)^k dx \le k^k \int_0^\infty g^k(x) dx, \tag{5}$$

where k^k is the best possible constant. Some of the generalizations of the discrete Hardy inequality (1) and the discrete version of (5) and its extensions are due to Leindler, we refer to the papers the papers [5–8]. For example, Leindler in [5] proved that if $p > 1, \lambda(n), g(n) > 0$, then

$$\sum_{n=1}^\infty \lambda(n) \left(\sum_{s=1}^n g(s)\right)^p \le p^p \sum_{n=1}^\infty \lambda^{1-p}(n) \left(\sum_{s=n}^\infty \lambda(s)\right)^p g^p(n), \tag{6}$$

and

$$\sum_{n=1}^\infty \lambda(n) \left(\sum_{k=n}^\infty g(k)\right)^p \le p^p \sum_{n=1}^\infty \lambda^{1-p}(n) \left(\sum_{k=1}^n \lambda(k)\right)^p g^p(n). \tag{7}$$

The converses of (6) and (7) are proved by Leindler in [6]. He proved that if $0 < p \le 1$, then

$$\sum_{n=1}^\infty \lambda(n) \left(\sum_{k=1}^n g(k)\right)^p \ge p^p \sum_{n=1}^\infty \lambda^{1-p}(n) \left(\sum_{k=n}^\infty \lambda(k)\right)^p g^p(n), \tag{8}$$

and

$$\sum_{n=1}^\infty \lambda(n) \left(\sum_{k=n}^\infty g(k)\right)^p \ge p^p \sum_{n=1}^\infty \lambda^{1-p}(n) \left(\sum_{k=1}^n \lambda(p)\right)^p g^p(n). \tag{9}$$

For more generalization Copson in [9] showed that if $k > 1$, $\lambda(j) \ge 0$, $w(j) \ge 0$, $\forall j \ge 1$, $\Omega(m) = \sum_{j=1}^m \lambda(j)$, and $h > 1$, then

$$\sum_{m=1}^\infty \frac{\lambda(m)}{\Omega^h(m)} \left(\sum_{j=1}^m w(j)\lambda(j)\right)^k \le \left(\frac{k}{h-1}\right)^k \sum_{m=1}^\infty \lambda(m)\Omega^{k-h}(m) w^k(m), \tag{10}$$

and if $0 \le h < 1$and $k > 1$, then

$$\sum_{m=1}^\infty \frac{\lambda(m)}{\Omega^h(m)} \left(\sum_{j=m}^\infty w(j)\lambda(j)\right)^k \le \left(\frac{k}{1-h}\right)^k \sum_{m=1}^\infty \lambda(m)\Omega^{k-h}(m) w^k(m). \tag{11}$$

The integral versions of the inequalities (10) and (11) was proved by Copson in [10] (Theorems 1 and 3). In particular, he proved that if $k \ge 1$, $h > 1$, and $\Omega(s) = \int_0^s \lambda(t) dt$, then

$$\int_0^\infty \frac{\lambda(s)}{\Omega^h(s)} \Phi^k(s) ds \le \left(\frac{k}{h-1}\right)^k \int_0^\infty \frac{\lambda(s)}{\Omega^{h-k}(s)} g^k(s) ds, \tag{12}$$

where $\Phi(s) = \int_0^s \lambda(t) g(t) dt$, and if $k > 1$, $0 \le h < 1$, then

$$\int_0^\infty \frac{\lambda(s)}{\Omega^h(s)} \Phi^k(s) ds \le \left(\frac{k}{1-h}\right)^k \int_0^\infty \frac{\lambda(s)}{\Omega^{h-k}(s)} g^k(s) ds, \tag{13}$$

where $\Phi(s) = \int_s^\infty \lambda(t)g(t)dt$. Leindler in [5] and Bennett in [11] presented interesting different inequalities. Leindler established that if $k > 1$, $\Omega^*(m) = \sum_{j=m}^\infty \lambda(j) < \infty$, and $0 \le h < 1$, then

$$\sum_{m=1}^{\infty} \frac{\lambda(m)}{(\Omega^*(m))^h} \left(\sum_{j=1}^{m} w(j)\lambda(j) \right)^k \le \left(\frac{k}{1-h} \right)^k \sum_{m=1}^{\infty} \lambda(m)(\Omega^*(m))^{k-h} w^k(m), \tag{14}$$

and Bennett in [11] showed that if $1 < h \le k$, then

$$\sum_{m=1}^{\infty} \frac{\lambda(m)}{(\Omega^*(m))^h} \left(\sum_{j=m}^{\infty} w(j)\lambda(j) \right)^k \le \left(\frac{k}{h-1} \right)^k \sum_{m=1}^{\infty} \lambda(m)(\Omega^*(m))^{k-h} w^k(m). \tag{15}$$

In last decades, studying the dynamic equations and inequalities on time scales become a main field in applied and pure mathematics, we refer to [12–14] and the references they are cited. In fact, the book [13] includes forms of the above inequalities on time-scale and their extensions. The timescales idea is returned to Stefan Hilger [15], who investigated the research of dynamic equations on timescales. The books by Bohner and Peterson in [16,17] summarized and organized most timescales calculus. The three most common timescales calculuses are difference, differential, and quantum calculus (see [18]), i.e., at $\mathbb{T} = \mathbb{N}$, $\mathbb{T} = \mathbb{R}$, and $\mathbb{T} = q^{\mathbb{N}_0} = \{q^s : s \in \mathbb{N}_0\}$ where $q > 1$.

In recent years, a lot of work has been published for fractional inequalities and the subject becomes an active field of research and several authors were interested in proving inequalities of fractional type by using the Riemann-Liouville and Caputo derivative (see [19–21]).

On the other hand, the authors in [22,23] introduced a new fractional calculus called the conformable calculus and gave a new definition of the derivative with the base properties of the calculus based on the new definition of derivative and integrals. By using conformable calculus, some authors have studied classical inequalities like Chebyshev's inequality [24], Hermite-Hadamard's inequality [25–27], Opial's inequality [28,29] and Steffensen's inequality [30].

The main question that arises now is: Is it possible to prove new fractional inequalities on timescales and give a unified approach of such studies? This in fact needs a new fractional calculus on timescales. Very recently Torres and others, in [31,32], combined a time scale calculus and conformable calculus and obtained the new fractional calculus on timescales. So, it is natural to look on new fractional inequalities on timescales and give an affirmative answer to the above question.

In particular, in this paper, we will prove the fractional forms of the classical Hardy, Bennett, Copson and Leindler inequalities. The paper is divided into two sections. Section 2 is an introduction of basics of fractional calculus on timescales and Section 3 contains the main results.

1. Preliminaries and Basic Lemmas

We present the fundamental results about the fractional timescales calculus. The results are adapted from [16,17,31,32]. A time-scale $\mathbb{T}$ is non-empty closed subset of $\mathbb{R}$ ($\mathbb{R}$ is the real numbers). The operators of backward jump and forward jump express of the closest point $t \in \mathbb{T}$ on the right and left of t is defined by, respectively:

$$\rho(t) := \sup\{s \in \mathbb{T} : s < t\}, \tag{16}$$

$$\sigma(t) := \inf\{s \in \mathbb{T} : s > t\}, \tag{17}$$

where $\sup \phi = \inf \mathbb{T}$ and $\inf \phi = \sup \mathbb{T}$ (ϕ denotes the empty set), for any $t \in \mathbb{T}$ the notation $f^\sigma(t)$ refer to $f(\sigma(t))$, i.e., $f^\sigma = f \circ \sigma$. The graininess function $\mu : \mathbb{T} \to [0, \infty)$, defined by $\mu(t) := \sigma(t) - t$.

Definition 1. *The number $T_\alpha^\Delta(f)(t)$ (provided it exists) of the function $f : \mathbb{T} \to \mathbb{R}$, for $t > 0$ and $\alpha \in (0,1]$ is the number which has the property that for any $\epsilon > 0$, there exists a neighborhood U of t S. T.*

$$\left|[f^\sigma(t) - f(s)]t^{1-\alpha} - T_\alpha^\Delta(f(t))(\sigma(t) - s)\right| \leq \epsilon\,|\sigma(t) - s)|, \quad \textit{for all } t \in U$$

$T_\alpha^\Delta(f(t))$ is called the conformable $\alpha-$fractional derivative of function f of order α at t , for conformable fractional derivative on $\mathbb{T}$ at 0, we define it with $T_\alpha^\Delta(f(0) = \lim_{t\to 0^+} T_\alpha^\Delta(f(t))$.

The conformable fractional derivative has the following properties

Theorem 1. *Let v, $u : \mathbb{T} \to \mathbb{R}$ are conformable fractional derivative from order $\alpha \in (0,1]$, then the following properties are hold:*

(i) *The $v + u : \mathbb{T} \to \mathbb{R}$ is conformable fractional derivative and*

$$T_\alpha^\Delta(v + u) = T_\alpha^\Delta(v) + T_\alpha^\Delta(u).$$

(ii) *Fora all $k \in \mathbb{R}$, then $kv : \mathbb{T} \to \mathbb{R}$ is $\alpha-$fractional differentiable and*

$$T_\alpha^\Delta(kv) = kT_\alpha^\Delta(v).$$

(iii) *If v and u are $\alpha-$fractional differentiable, we have $vu : \mathbb{T} \to \mathbb{R}$ is $\alpha-$fractional differentiable and*

$$T_\alpha^\Delta(vu) = T_\alpha^\Delta(v)\,u + (v \circ \sigma)\,T_\alpha^\Delta(u) = T_\alpha^\Delta(v)\,(u \circ \sigma) + v\,T_\alpha^\Delta(u).$$

(iv) *If v is $\alpha-$fractional differentiable, then $1/v$ is $\alpha-$fractional differentiable with*

$$T\left(\frac{1}{v}\right) = -\frac{T_\alpha^\Delta(v)}{v\,(v \circ \sigma)}.$$

(v) *If v and u are $\alpha-$fractional differentiable, then v/u is $\alpha-$fractional differentiable with*

$$T_\alpha^\Delta(v/u) = \frac{T_\alpha^\Delta(v)u - vT_\alpha^\Delta(u)}{u(u \circ \sigma)},$$

valid $\forall\ t \in \mathbb{T}^k$,where $u(t)(u(\sigma(t)) \neq 0$.

Lemma 1. *Let $v : \mathbb{T} \to \mathbb{R}$ is continuous and $\alpha-$fractional differentiable at $t \in \mathbb{T}$ for $\alpha \in (0,1]$, and $u : \mathbb{R} \to \mathbb{R}$ is continuous and differentiable. Then there exists $d \in [t, \sigma(t)]$ with*

$$T_\alpha^\Delta(u \circ v)(t) = u'(v(d))T_\alpha^\Delta(v(t)). \tag{18}$$

Lemma 2. *Let $u : \mathbb{R} \to \mathbb{R}$ is continuously differentiable, $\alpha \in (0,1]$, and $v : \mathbb{T} \to \mathbb{R}$ be $\alpha-$fractional differentiable. Then $(u \circ v) : \mathbb{T} \to \mathbb{R}$ is $\alpha-$fractional differentiable and we have*

$$T_\alpha^\Delta(u \circ v)(s) = \left(\int_0^1 u'\left(v(s) + h\mu(s)s^{\alpha-1}T_\alpha^\Delta(v(s))\right)dh\right)T_\alpha^\Delta(v(s)). \tag{19}$$

Definition 2. *Let $0 < \alpha \leq 1$, the $\alpha-$fractional integral of f, is defined as*

$$\int f(s)\Delta_\alpha s = \int f(s)s^{\alpha-1}\Delta s.$$

The conformable fractional integral satisfying the next properties

Theorem 2. *Assume $a, b, c \in \mathbb{T}$, $\lambda \in \mathbb{R}$. Let $u, v : \mathbb{T} \to \mathbb{R}$. Then*

(i) $\int_a^b [v(s) + u(s)]\Delta_\alpha s = \int_a^b v(s)\Delta_\alpha s + \int_a^b u(s)\Delta_\alpha s$.
(ii) $\int_a^b \lambda v(s)\Delta_\alpha s = \lambda \int_a^b v(s)\Delta_\alpha s$.
(iii) $\int_a^b v(s)\Delta_\alpha s = -\int_b^a v(s)\Delta_\alpha s$.
(iv) $\int_a^b v(s)\Delta_\alpha s = \int_a^c v(s)\Delta_\alpha s + \int_c^b v(s)\Delta_\alpha s$.
(v) $\int_a^a v(s)\Delta_\alpha s = 0$.

Lemma 3. *Assume $\mathbb{T}$ be a time-scale, $a, b \in \mathbb{T}$ where $b > a$. Let u, v are conformable α—fractional differentiable, $\alpha \in (0, 1]$. Then the formula of integration by parts is given by*

$$\int_a^b v(s)T_\alpha^\Delta u(s)\Delta_\alpha s = [v(s)u(s)]_a^b - \int_a^b u^\sigma(s)T_\alpha^\Delta v(s)\Delta_\alpha s. \tag{20}$$

Lemma 4. *Assume $\mathbb{T}$ be a time-scale, $a, b \in \mathbb{T}$ and $\alpha \in (0, 1]$. Let $u, v : \mathbb{T} \to \mathbb{R}$. Then*

$$\int_a^b |v(s)u(s)|\Delta_\alpha s \leq \left[\int_a^b |v(s)|^k \Delta_\alpha s\right]^{\frac{1}{k}} \left[\int_a^b |u(s)|^l \Delta_\alpha s\right]^{\frac{1}{l}}, \tag{21}$$

where $k > 1$ and $1/k + 1/l = 1$.

2. Main Results

Throughout the paper, we will assume that the functions are nonnegative on $[a, \infty)_{\mathbb{T}}$ and its integrals exist and are finite. We start with the fractional time-scale inequality of Copson's type.

Theorem 3. *Assume $1 < c < k$, define*

$$\Phi(x) := \int_a^x \lambda(s)\Delta_\alpha s \text{ and } \Omega(x) := \int_a^x \lambda(s)g(s)\Delta_\alpha s.$$

If

$$\Omega(\infty) < \infty, \text{and } \int_a^\infty \frac{\lambda(s)}{(\Phi^\sigma(s))^{c-\alpha+1}}\Delta_\alpha s < \infty,$$

then

$$\int_a^\infty \frac{\lambda(x)}{(\Phi^\sigma(x))^{c-\alpha+1}} (\Omega^\sigma(x))^k \Delta_\alpha x \leq \left(\frac{k}{c-\alpha}\right)^k \int_a^\infty \frac{\lambda(x)\Phi^{k(\alpha-c)}(x)}{(\Phi^\sigma(x))^{(c-\alpha+1)(1-k)}} g^k(x)\Delta_\alpha x. \tag{22}$$

Proof. By employing the formula of integration by parts (20) on the term

$$\int_a^\infty \frac{\lambda(x)}{(\Phi^\sigma(x))^{c-\alpha+1}} (\Omega^\sigma(x))^k \Delta_\alpha x,$$

with $u^\sigma(x) = (\Omega^\sigma(x))^k$ and $x_\alpha^\Delta v(x) = \frac{\lambda(x)}{(\Phi^\sigma(x))^{c-\alpha+1}}$, we have that

$$\int_a^\infty \frac{\lambda(x)}{(\Phi^\sigma(x))^{c-\alpha+1}} (\Omega^\sigma(x))^k \Delta_\alpha x = v(x)\Omega^k(x)\Big|_a^\infty + \int_a^\infty -v(x)x_\alpha^\Delta\left(\Omega^k(x)\right)\Delta_\alpha x, \tag{23}$$

where

$$-v(x) = \int_x^\infty \frac{\lambda(s)}{(\Phi^\sigma(s))^{c-\alpha+1}}\Delta_\alpha s = \int_x^\infty x_\alpha^\Delta \Phi(s) (\Phi^\sigma(s))^{\alpha-c-1} \Delta_\alpha s.$$

By using the chain rule (18), we obtain that

$$\begin{aligned} -x_{\alpha}^{\Delta}\left(\Phi^{\alpha-c}(x)\right) &= -(\alpha-c)\Phi^{\alpha-c-1}(d)x_{\alpha}^{\Delta}\Phi(x), \text{ where } d\in[x,\sigma(x)] \\ &= \frac{(c-\alpha)\,x_{\alpha}^{\Delta}\Phi(x)}{\Phi^{c-\alpha+1}(d)} \\ &\geq \frac{(c-\alpha)\,x_{\alpha}^{\Delta}\left(\Phi(x)\right)}{\left(\Phi^{\sigma}(x)\right)^{c-\alpha+1}}. \end{aligned}$$

Then we have

$$x_{\alpha}^{\Delta}\left(\Phi(x)\right)\left(\Phi^{\sigma}(x)\right)^{\alpha-c-1} \leq \frac{-1}{c-\alpha}x_{\alpha}^{\Delta}\Phi^{\alpha-c}(x),$$

and thus

$$-v(x) = \int_x^{\infty} \frac{\lambda(s)}{\left(\Phi^{\sigma}(s)\right)^{c-\alpha+1}}\Delta_{\alpha}s \leq \frac{-1}{c-\alpha}\int_x^{\infty} x_{\alpha}^{\Delta}\Phi^{\alpha-c-1}(s)\Delta_{\alpha}s \leq \frac{\Phi^{\alpha-c}(x)}{c-\alpha}. \tag{24}$$

Again, by using the chain rule (18) to calculate

$$x_{\alpha}^{\Delta}\left(\Omega^{k}(x)\right) = k\Omega^{k-1}(d)x_{\alpha}^{\Delta}\left(\Omega(x)\right), \text{ where } d\in[x,\sigma(x)],$$

at $x_{\alpha}^{\Delta}\left(\Omega(x)\right) = \lambda(x)g(x) \geq 0$ and $d \leq \sigma(x)$, we get that

$$x_{\alpha}^{\Delta}\left(\Omega^{k}(x)\right) \leq k\lambda(x)g(x)\left(\Omega^{\sigma}(x)\right)^{k-1}. \tag{25}$$

Since $\Omega(a) = 0$, $v(\infty) = 0$ and from (24), (25) and (23) we have

$$\int_a^{\infty} \frac{\lambda(x)}{\left(\Phi^{\sigma}(x)\right)^{c-\alpha+1}}\left(\Omega^{\sigma}(x)\right)^{k}\Delta_{\alpha}x \leq \frac{k}{c-\alpha}\int_a^{\infty}\Phi^{\alpha-c}(x)\lambda(x)g(x)\left(\Omega^{\sigma}(x)\right)^{k-1}\Delta_{\alpha}x,$$

which reformulated as

$$\begin{aligned} &\int_a^{\infty} \frac{\lambda(x)}{\left(\Phi^{\sigma}(x)\right)^{c-\alpha+1}}\left(\Omega^{\sigma}(x)\right)^{k}\Delta_{\alpha}x \\ &= \frac{k}{c-\alpha}\int_a^{\infty} \frac{\lambda(x)\Phi^{\alpha-c}(x)g(x)}{\left(\lambda(x)\left(\Phi^{\sigma}(x)\right)^{\alpha-c-1}\right)^{\frac{k-1}{k}}}\left(\frac{\lambda(x)\left(\Omega^{\sigma}(x)\right)^{k}}{\left(\Phi^{\sigma}(x)\right)^{c-\alpha+1}}\right)^{\frac{k-1}{k}}\Delta_{\alpha}x. \end{aligned}$$

Using Hölder's inequality (21) on

$$\int_a^{\infty} \frac{\lambda(x)\Phi^{\alpha-c}(x)g(x)}{\left(\lambda(x)\left(\Phi^{\sigma}(x)\right)^{\alpha-c-1}\right)^{\frac{k-1}{k}}}\left(\frac{\lambda(x)\left(\Omega^{\sigma}(x)\right)^{k}}{\left(\Phi^{\sigma}(x)\right)^{c-\alpha+1}}\right)^{\frac{k-1}{k}}\Delta_{\alpha}x$$

with indices k and $k/(k-1)$, we have

$$\int_a^\infty \frac{\lambda(x)}{(\Phi^\sigma(x))^{c-\alpha+1}} (\Omega^\sigma(x))^k \Delta_\alpha x \leq$$

$$\frac{k}{c-\alpha}\left[\int_a^\infty \left[\frac{\lambda(x)\Phi^{\alpha-c}(x)g(x)}{\left[\lambda(x)(\Phi^\sigma(x))^{\alpha-c-1}\right]^{\frac{k-1}{k}}}\right]^k \Delta_\alpha x\right]^{\frac{1}{k}}$$

$$\times\left[\int_a^\infty \left[\left[\frac{\lambda(x)(\Omega^\sigma(x))^k}{(\Phi^\sigma(x))^{c-\alpha+1}}\right]^{\frac{k-1}{k}}\right]^{\frac{k}{k-1}} \Delta_\alpha x\right]^{\frac{k-1}{k}},$$

then

$$\left[\int_a^\infty \frac{\lambda(x)}{(\Phi^\sigma(x))^{c-\alpha+1}} (\Omega^\sigma(x))^k \Delta_\alpha x\right]^{\frac{1}{k}} \leq \frac{k}{c-\alpha}\left[\int_a^\infty \frac{\lambda(x)\Phi^{k(\alpha-c)}(x)g^k(x)}{(\Phi^\sigma(x))^{(c-\alpha+1)(1-k)}}\Delta_\alpha x\right]^{\frac{1}{k}}.$$

This leads to

$$\int_a^\infty \frac{\lambda(x)}{(\Phi^\sigma(x))^{c-\alpha+1}} (\Omega^\sigma(x))^k \Delta_\alpha x \leq \left(\frac{k}{c-\alpha}\right)^k \int_a^\infty \frac{\lambda(x)\Phi^{k(\alpha-c)}(x)g^k(x)}{(\Phi^\sigma(x))^{(c-\alpha+1)(1-k)}}\Delta_\alpha x,$$

that is the desired inequality (22). The proof is complete. □

Corollary 1. *At $\alpha = 1$ in Theorem 3, we obtain the inequality*

$$\int_a^\infty \frac{\lambda(x)}{(\Phi^\sigma(x))^c} (\Omega^\sigma(x))^k \Delta x \leq \left(\frac{k}{c-1}\right)^k \int_a^\infty \frac{\Phi^{k(1-c)}(x)}{(\Phi^\sigma(x))^{c(1-k)}}\lambda(x)g^k(x)\Delta x.$$

that is the timescales version of inequality (2.8) in [33].

Corollary 2. *At $\alpha = 1$, and $\mathbb{T} = \mathbb{R}$ $(\Phi^\sigma(x) = \Phi(x))$ in Theorem 3, we obtain the integral inequality*

$$\int_a^\infty \frac{\lambda(x)}{\Phi^c(x)}\left(\int_a^x \lambda(s)g(s)ds\right)^k dx \leq \left(\frac{k}{c-1}\right)^k \int_a^\infty \Phi^{k-c}(x)\lambda(x)g^k(x)dx,$$

which is of Copson type.

Corollary 3. *At $\alpha = 1$, $T = \mathbb{R}$, $\lambda(x) = 1$ and $a = 0$, $\left(\Phi(x) = \int_0^x \lambda(s)ds = x\right)$ in Theorem 3, we have Hardy-Littlewood integral inequality (3)*

$$\int_0^\infty \frac{1}{x^c}\left(\int_0^x g(s)ds\right)^k dx \leq \left(\frac{k}{c-1}\right)^k \int_0^\infty \frac{1}{x^{c-k}}g^k(x)dx.$$

Also, if $c = k$, we obtain the standard Hardy inequality (2)

$$\int_0^\infty \frac{1}{x^k}\left(\int_0^x g(s)ds\right)^k dx \leq \left(\frac{k}{k-1}\right)^k \int_0^\infty g^k(x)dx.$$

Theorem 4. *Let $0 \leq c < 1$ and $k > 1$. Define*

$$\Phi(x) := \int_a^x \lambda(s)\Delta_\alpha s \text{ and } \Omega(x) := \int_x^\infty \lambda(s)g(s)\Delta_\alpha s.$$

If

$$\Omega(a) < \infty, \textit{and} \int_a^\infty \frac{\lambda(s)}{(\Phi^\sigma(s))^{c-\alpha+1}} \Delta_\alpha s < \infty,$$

then

$$\int_a^\infty \frac{\lambda(x)}{(\Phi^\sigma(x))^{c-\alpha+1}} \Omega^k(x) \Delta_\alpha x \leq \left(\frac{k}{\alpha - c}\right)^k \int_a^\infty (\Phi^\sigma(x))^{k-c+\alpha-1} \lambda(x) g^k(x) \Delta_\alpha x. \tag{26}$$

Proof. By using the formula of integration by parts (20) on

$$\int_a^\infty \frac{\lambda(x)}{(\Phi^\sigma(x))^{c-\alpha+1}} \Omega^k(x) \Delta_\alpha x,$$

with $v(x) = \Omega^k(x)$ and $x_\alpha^\Delta u(x) = \frac{\lambda(x)}{(\Phi^\sigma(x))^{c-\alpha+1}}$, we have

$$\int_a^\infty \frac{\lambda(x)}{(\Phi^\sigma(x))^{c-\alpha+1}} \Omega^k(x) \Delta_\alpha x = u(x) \Omega^k(x) \Big|_a^\infty + \int_a^\infty u^\sigma(x) x_\alpha^\Delta \left(-\Omega^k(x)\right) \Delta_\alpha x, \tag{27}$$

where

$$u(x) = \int_a^x \frac{\lambda(s)}{(\Phi^\sigma(s))^{c-\alpha+1}} \Delta_\alpha s = \int_a^x x_\alpha^\Delta \Phi(s) \left(\Phi^\sigma(s)\right)^{\alpha-c-1} \Delta_\alpha s.$$

By using chain rule (18), then for $d \in [x, \sigma(x)]$, we get that

$$\begin{aligned} x_\alpha^\Delta \left(\Phi^{\alpha-c}(x)\right) &= (\alpha - c) \Phi^{\alpha-c-1}(d) x_\alpha^\Delta \left(\Phi(x)\right) = \frac{(\alpha - c)\, x_\alpha^\Delta \left(\Phi(x)\right)}{\Phi^{c-\alpha+1}(d)} \\ &\geq \frac{(\alpha - c)\, x_\alpha^\Delta \left(\Phi(x)\right)}{(\Phi^\sigma(x))^{c-\alpha+1}}. \end{aligned}$$

So

$$x_\alpha^\Delta \left(\Phi(x)\right) \left(\Phi^\sigma(x)\right)^{\alpha-c-1} \leq \frac{1}{\alpha - c} x_\alpha^\Delta \left(\Phi^{\alpha-c}(x)\right),$$

and then,

$$\begin{aligned} u^\sigma(x) &= \int_a^{\sigma(x)} x_\alpha^\Delta \left(\Phi(s)\right) \left(\Phi^\sigma(s)\right)^{\alpha-c-1} \Delta_\alpha s \\ &\leq \frac{1}{\alpha - c} \int_a^{\sigma(x)} x_\alpha^\Delta \left(\Phi^{\alpha-c}(s)\right) \Delta_\alpha s \leq \frac{(\Phi^\sigma(x))^{\alpha-c}}{\alpha - c}. \end{aligned} \tag{28}$$

Again, by using chain rule (18), we obtain

$$-x_\alpha^\Delta \left(\Omega^k(x)\right) = -k \Omega^{k-1}(d) x_\alpha^\Delta \left(\Omega(x)\right), \text{where } d \in [x, \sigma(x)],$$

since $x_\alpha^\Delta \left(\Omega(x)\right) = -\lambda(x) g(x) \geq 0$ and $d \geq x$, then

$$-x_\alpha^\Delta \left(\Omega^k(x)\right) \leq k \lambda(x) g(x) \Omega^{k-1}(x). \tag{29}$$

Using $\Phi(a) = 0$, $\Omega(\infty) = 0$ and (28), (29) and (27), we have that

$$\int_a^\infty \frac{\lambda(x)}{(\Phi^\sigma(x))^{c-\alpha+1}} \Omega^k(x) \Delta_\alpha x \leq \frac{k}{\alpha - c} \int_a^\infty \left(\Phi^\sigma(x)\right)^{\alpha-c} \lambda(x) g(x) \Omega^{k-1}(x) \Delta_\alpha x,$$

which reformulated as

$$\int_a^\infty \frac{\lambda(x)}{(\Phi^\sigma(x))^{c-\alpha+1}}\Omega^k(x)\Delta_\alpha x =$$

$$\frac{k}{\alpha-c}\int_a^\infty \frac{(\Phi^\sigma(x))^{\alpha-c}\lambda(x)g(x)}{\left(\lambda(x)\left(\Phi^\sigma(x)\right)^{\alpha-c-1}\right)^{\frac{k-1}{k}}}\left(\frac{\lambda(x)\Omega^k(x)}{(\Phi^\sigma(x))^{c-\alpha+1}}\right)^{\frac{k-1}{k}}\Delta_\alpha x.$$

By employing Hölder's inequality (21) on

$$\int_a^\infty \frac{(\Phi^\sigma(x))^{\alpha-c}\lambda(x)g(x)}{\left(\lambda(x)\left(\Phi^\sigma(x)\right)^{\alpha-c-1}\right)^{\frac{k-1}{k}}}\left(\frac{\lambda(x)\Omega^k(x)}{(\Phi^\sigma(x))^{c-\alpha+1}}\right)^{\frac{k-1}{k}}\Delta_\alpha x$$

with indices k and $k/(k-1)$, we have

$$\int_a^\infty \frac{\lambda(x)}{(\Phi^\sigma(x))^{c-\alpha+1}}\Omega^k(x)\Delta_\alpha x \leq$$

$$\frac{k}{\alpha-c}\left[\int_a^\infty \left[\frac{(\Phi^\sigma(x))^{1-c}\lambda(x)g(x)}{\left[\lambda(x)\left(\Phi^\sigma(x)\right)^{\alpha-c-1}\right]^{\frac{k-1}{k}}}\right]^k \Delta_\alpha x\right]^{\frac{1}{k}}$$

$$\times\left[\int_a^\infty \left[\left[\frac{\lambda(x)\Omega^k(x)}{(\Phi^\sigma(x))^{c-\alpha+1}}\right]^{\frac{k-1}{k}}\right]^{\frac{k}{k-1}}\Delta_\alpha x\right]^{\frac{k-1}{k}},$$

then we have

$$\left[\int_a^\infty \frac{\lambda(x)}{(\Phi^\sigma(x))^{c-\alpha+1}}\Omega^k(x)\Delta_\alpha x\right]^{\frac{1}{k}} \leq \frac{k}{\alpha-c}\left[\int_a^\infty \frac{\lambda(x)g^k(x)}{(\Phi^\sigma(x))^{c-k-\alpha+1}}\Delta_\alpha x\right]^{\frac{1}{k}}.$$

This leads to

$$\int_a^\infty \frac{\lambda(x)}{(\Phi^\sigma(x))^c}\Omega^k(x)\Delta_\alpha x \leq \left(\frac{k}{\alpha-c}\right)^k\int_a^\infty (\Phi^\sigma(x))^{k-c+\alpha-1}\lambda(x)g^k(x)\Delta_\alpha x,$$

that is the desired inequality (26). The proof is complete. □

Corollary 4. *At $\alpha=1$ in Theorem 4, then*

$$\int_a^\infty \frac{\lambda(x)}{(\Phi^\sigma(x))^c}\Omega^k(x)\Delta x \leq \left(\frac{k}{1-c}\right)^k\int_a^\infty (\Phi^\sigma(x))^{k-c}\lambda(x)g^k(x)\Delta x.$$

which is the timescales version inequality (2.22) in [33].

Corollary 5. *At $\alpha=1$, and T= $\mathbb{R}$ in Theorem 4, we obtain the next integral inequality*

$$\int_a^\infty \frac{\lambda(x)}{\Phi^c(x)}\left(\int_x^\infty \lambda(s)g(s)ds\right)^k dx \leq \left(\frac{k}{1-c}\right)^k\int_a^\infty \Phi^{k-c}(x)\lambda(x)g^k(x)dx,$$

which considered an extension of Hardy's inequality (4) as in the following corollary.

Corollary 6. *At $\alpha = 1$, $T = \mathbb{R}$, $\lambda(x) = 1$ and $a = 0$ in Theorem 4, we have Hardy-Littlewood integral inequality (4)*

$$\int_0^\infty \frac{1}{x^c}\left(\int_x^\infty g(s)ds\right)^k dx \leq \left(\frac{k}{1-c}\right)^k \int_0^\infty \frac{1}{x^{c-k}} g^k(x)dx.$$

A generalization of Leindler's inequality (14) on fractional time scales will be proved in the next theorem.

Theorem 5. *Assume $0 \leq c < 1 < k$, define*

$$\Phi(x) := \int_x^\infty \lambda(s)\Delta_\alpha s \text{ and } \Omega(x) := \int_a^x \lambda(s)g(s)\Delta_\alpha s.$$

If

$$\Omega(\infty) < \infty, \text{and } \int_a^\infty \frac{\lambda(s)}{\Phi^{c-\alpha+1}(s)}\Delta_\alpha s < \infty,$$

then

$$\int_a^\infty \frac{\lambda(x)}{\Phi^{c-\alpha+1}(x)}\left(\Omega^\sigma(x)\right)^k \Delta_\alpha x \leq \left(\frac{k}{\alpha-c}\right)^k \int_a^\infty \Phi^{k-c+\alpha-1}(x)\lambda(x)g^k(x)\Delta_\alpha x. \tag{30}$$

Proof. Using the formula of integration by parts (20) on

$$\int_a^\infty \frac{\lambda(x)}{\Phi^{c-\alpha+1}(x)}\Omega^k(x)\Delta_\alpha x,$$

with $u^\sigma(x) = (\Omega^\sigma(x))^k$ and $x_\alpha^\Delta v(x) = \frac{\lambda(x)}{\Phi^{c-\alpha+1}(x)}$, we have

$$\int_a^\infty \frac{\lambda(x)}{\Phi^{c-\alpha+1}(x)}\left(\Omega^\sigma(x)\right)^k \Delta_\alpha x = v(x)\Omega^k(x)\Big|_a^\infty + \int_a^\infty -v(x)x_\alpha^\Delta\left(\Omega^k(x)\right)\Delta_\alpha x, \tag{31}$$

where

$$v(x) = -\int_x^\infty \frac{\lambda(s)}{\Phi^{c-\alpha+1}(s)}\Delta_\alpha s.$$

By chain rule (18), we see for $d \in [x, \sigma(x)]$ that

$$\begin{aligned} -x_\alpha^\Delta \Phi^{\alpha-c}(x) &= -(\alpha-c)\Phi^{\alpha-c-1}(d)x_\alpha^\Delta(\Phi(x)) = \frac{-(\alpha-c)(-\lambda(x))}{\Phi^{c-\alpha+1}(d)} \\ &\geq \frac{(\alpha-c)\lambda(x)}{\Phi^{c-\alpha+1}(x)}. \end{aligned}$$

Hence

$$-v(x) = \int_x^\infty \frac{\lambda(s)}{\Phi^{c-\alpha+1}(s)}\Delta_\alpha s \leq \frac{-1}{\alpha-c}\int_x^\infty x_\alpha^\Delta \Phi^{\alpha-c}(s)\Delta_\alpha s \leq \frac{\Phi^{\alpha-c}(x)}{\alpha-c}, \tag{32}$$

from chain rule (18), we obtain

$$x_\alpha^\Delta\left(\Omega^k(x)\right) = k\Omega^{k-1}(d)x_\alpha^\Delta(\Omega(x)), \text{ where } d \in [x, \sigma(x)],$$

since

$$x_\alpha^\Delta(\Omega(x)) = \lambda(x)g(x) \geq 0 \text{ and } d \leq \sigma(x),$$

we get

$$x_\alpha^\Delta\left(\Omega^k(x)\right) \leq k\lambda(x)g(x)\left(\Omega^\sigma(x)\right)^{k-1}. \tag{33}$$

Using $\Omega(a) = 0$, $v(\infty) = 0$ and (32), (33) and (31), we get that

$$\int_a^\infty \frac{\lambda(x)}{\Phi^{c-\alpha+1}(x)} \left(\Omega^\sigma(x)\right)^k \Delta_\alpha x \leq \frac{k}{\alpha - c} \int_a^\infty \Phi^{\alpha-c}(x)\lambda(x)g(x) \left(\Omega^\sigma(x)\right)^{k-1} \Delta_\alpha x,$$

which reformulated as

$$\int_a^\infty \frac{\lambda(x)}{\Phi^{c-\alpha+1}(x)} \left(\Omega^\sigma(x)\right)^k \Delta_\alpha x \leq$$
$$\frac{k}{\alpha - c} \int_a^\infty \frac{\Phi^{(c-\alpha+1)\left(\frac{k-1}{k}\right)}(x)}{\lambda^{\frac{k-1}{k}}(x)\Phi^{c-\alpha}(x)} \lambda(x)g(x) \frac{\lambda^{\frac{k-1}{k}}(x) \left(\Omega^\sigma(x)\right)^{k-1}}{\Phi^{(c-\alpha+1)\left(\frac{k-1}{k}\right)}(x)} \Delta_\alpha x.$$

Using Hölder's inequality (21) on

$$\int_a^\infty \frac{\Phi^{(c-\alpha+1)\left(\frac{k-1}{k}\right)}(x)}{\lambda^{\frac{k-1}{k}}(x)\Phi^{c-\alpha}(x)} \lambda(x)g(x) \frac{\lambda^{\frac{k-1}{k}}(x) \left(\Omega^\sigma(x)\right)^{k-1}}{\Phi^{(c-\alpha+1)\left(\frac{k-1}{k}\right)}(x)} \Delta_\alpha x$$

with indices k and $k/(k-1)$, we have

$$\int_a^\infty \frac{\lambda(x)}{\Phi^{c-\alpha+1}(x)} \left(\Omega^\sigma(x)\right)^k \Delta_\alpha x \leq$$
$$\frac{k}{\alpha - c} \left[\int_a^\infty \left[\frac{\Phi^{(c-\alpha+1)\left(\frac{k-1}{k}\right)}(x)}{\lambda^{\frac{k-1}{k}}(x)\Phi^{c-\alpha}(x)} \lambda(x)g(x)\right]^k \Delta_\alpha x\right]^{\frac{1}{k}} \times$$
$$\left[\int_a^\infty \left[\lambda^{\frac{k-1}{k}}(x) \frac{\left(\Omega^\sigma(x)\right)^{k-1}}{\Phi^{(c-\alpha+1)\left(\frac{k-1}{k}\right)}(x)}\right]^{\frac{k}{k-1}} \Delta_\alpha x\right]^{\frac{k-1}{k}},$$

then

$$\left[\int_a^\infty \frac{\lambda(x)}{\Phi^{c-\alpha+1}(x)} \left(\Omega^\sigma(x)\right)^k \Delta_\alpha x\right]^{\frac{1}{k}} \leq \frac{k}{\alpha - c} \left[\int_a^\infty \Phi^{k-c+\alpha-1}(x)\lambda(x)g^k(x)\Delta_\alpha x\right]^{\frac{1}{k}}.$$

This leads to

$$\int_a^\infty \frac{\lambda(x)}{\Phi^{c-\alpha+1}(x)} \left(\Omega^\sigma(x)\right)^k \Delta_\alpha x \leq \left(\frac{k}{\alpha - c}\right)^k \int_a^\infty \Phi^{k-c+\alpha-1}(x)\lambda(x)g^k(x)\Delta_\alpha x,$$

that is the desired inequality (30). The proof is complete. □

Corollary 7. *At $\alpha = 1$ in Theorem 5, we get*

$$\int_a^\infty \frac{\lambda(x)}{\Phi^c(x)} \left(\Omega^\sigma(x)\right)^k \Delta x \leq \left(\frac{k}{1 - c}\right)^k \int_a^\infty \Phi^{k-c}(x)\lambda(x)g^k(x)\Delta x,$$

which inequality (2.36) in [33].

A generalization of Bennett's inequality (15) on fractional timescales will be proved in the next theorem.

Theorem 6. *Assume $0 < \alpha \leq 1$, $1 < c \leq k$, and define*

$$\Phi(x) := \int_x^\infty \lambda(s)\Delta_\alpha s \text{ and } \Omega(x) := \int_x^\infty \lambda(s)g(s)\Delta_\alpha s.$$

If

$$\Omega(a) < \infty \text{ and } \int_a^\infty \frac{\lambda(s)}{\Phi^{c-\alpha+1}(s)}\Delta_\alpha s < \infty,$$

then

$$\int_a^\infty \frac{\lambda(x)}{\Phi^{c-\alpha+1}(x)}\Omega^k(x)\Delta_\alpha x \leq \left(\frac{k}{c-\alpha}\right)^k \int_a^\infty \Phi^{k-c+\alpha-1}(x)\lambda(x)g^k(x)\Delta_\alpha x. \tag{34}$$

Proof. Using the formula of integration by parts (20) on

$$\int_a^\infty \frac{\lambda(x)}{\Phi^{c-\alpha+1}(x)}\Omega^k(x)\Delta_\alpha x,$$

with $v(x) = \Omega^k(x)$ and $x_\alpha^\Delta u(x) = \frac{\lambda(x)}{\Phi^{c-\alpha+1}(x)}$, then

$$\int_a^\infty \frac{\lambda(x)}{\Phi^{c-\alpha+1}(x)}\Omega^k(x)\Delta_\alpha x = u(x)\Omega^k(x)\Big|_a^\infty + \int_a^\infty u^\sigma(x)x_\alpha^\Delta\left(-\Omega^k(x)\right)\Delta_\alpha x, \tag{35}$$

where

$$u(x) = \int_a^x \frac{\lambda(s)}{\Phi^{c-\alpha+1}(s)}\Delta_\alpha s.$$

By using chain rule (18), we see for $d \in [x, \sigma(x)]$ that

$$\begin{aligned} x_\alpha^\Delta\left(\Phi^{\alpha-c}(x)\right) &= (\alpha-c)\Phi^{\alpha-c-1}(d)x_\alpha^\Delta\left(\Phi(x)\right) = \frac{(\alpha-c)(-\lambda(x))}{\Phi^{c-\alpha+1}(d)} \\ &\geq \frac{(c-\alpha)\,\lambda(x)}{\Phi^{c-\alpha+1}(x)}. \end{aligned}$$

we get,

$$\begin{aligned} u^\sigma(x) &= \int_a^{\sigma(x)} \frac{\lambda(s)}{\Phi^{c-\alpha+1}(s)}\Delta_\alpha s \leq \frac{1}{c-\alpha}\int_a^{\sigma(x)} x_\alpha^\Delta\Phi^{\alpha-c}(s)\Delta_\alpha s \\ &= \frac{\left(\Phi^\sigma(x)\right)^{\alpha-c}}{c-\alpha} - \frac{\Phi^{\alpha-c}(a)}{c-\alpha} \leq \frac{\Phi^{\alpha-c}(x)}{c-\alpha}, \end{aligned} \tag{36}$$

from chain rule (18), we find that

$$x_\alpha^\Delta\left(\Omega^k(x)\right) = k\Omega^{k-1}(d)x_\alpha^\Delta\Omega(x), \text{ where } d \in [x, \sigma(x)],$$

since

$$x_\alpha^\Delta\left(\Omega(x)\right) = -\lambda(x)g(x) \leq 0 \text{ and } x \leq d,$$

then

$$-x_\alpha^\Delta\left(\Omega^k(x)\right) \leq k\lambda(x)g(x)\Omega^{k-1}(x). \tag{37}$$

Using $v(a) = 0, \Omega(\infty) = 0$ and (36), (37) and (35), we get that

$$\int_a^\infty \frac{\lambda(x)}{\Phi^{c-\alpha+1}(x)}\left(\Omega^\sigma(x)\right)^k\Delta_\alpha x \leq \frac{k}{c-\alpha}\int_a^\infty \Phi^{\alpha-c}(x)\lambda(x)g(x)\Omega^{k-1}(x)\Delta_\alpha x,$$

which reformulated as

$$\int_a^\infty \frac{\lambda(x)}{\Phi^{c-\alpha+1}(x)}\Omega^k(x)\Delta_\alpha x \leq$$

$$\frac{k}{c-\alpha}\int_a^\infty \frac{\Phi^{(c-\alpha+1)\left(\frac{k-1}{k}\right)}(x)}{\lambda^{\frac{k-1}{k}}(x)\Phi^{c-\alpha}(x)}\lambda(x)g(x)\lambda^{\frac{k-1}{k}}(x)\frac{\Omega^{k-1}(x)}{\Phi^{(c-\alpha+1)\left(\frac{k-1}{k}\right)}(x)}\Delta_\alpha x,$$

Using Hölder's inequality (21) on

$$\int_a^\infty \lambda(x)g(x)\frac{\Phi^{(c-\alpha+1)\left(\frac{k-1}{k}\right)}(x)}{\lambda^{\frac{k-1}{k}}(x)\Phi^{c-\alpha}(x)}\frac{\lambda^{\frac{k-1}{k}}(x)\Omega^{k-1}(x)}{\Phi^{(c-\alpha+1)\left(\frac{k-1}{k}\right)}(x)}\Delta_\alpha x$$

with indices k and $k/(k-1)$, we have

$$\int_a^\infty \frac{\lambda(x)}{\Phi^{c-\alpha+1}(x)}\Omega^k(x)\Delta_\alpha x \leq$$

$$\frac{k}{c-\alpha}\left[\int_a^\infty \left[\lambda(x)g(x)\frac{\Phi^{(c-\alpha+1)\left(\frac{k-1}{k}\right)}(x)}{\lambda^{\frac{k-1}{k}}(x)\Phi^{c-\alpha}(x)}\right]^k \Delta_\alpha x\right]^{\frac{1}{k}} \times$$

$$\left[\int_a^\infty \left[\lambda^{\frac{k-1}{k}}(x)\frac{\Omega^{k-1}(x)}{\Phi^{(c-\alpha+1)\left(\frac{k-1}{k}\right)}(x)}\right]^{\frac{k}{k-1}} \Delta_\alpha x\right]^{\frac{k-1}{k}},$$

then

$$\left[\int_a^\infty \frac{\lambda(x)}{\Phi^{c-\alpha+1}(x)}\left(\Omega(x)\right)^k \Delta_\alpha x\right]^{\frac{1}{k}} \leq \frac{k}{c-\alpha}\left[\int_a^\infty \Phi^{k-c+\alpha-1}(x)\lambda(x)g^k(x)\Delta_\alpha x\right]^{\frac{1}{k}}.$$

This leads to

$$\int_a^\infty \frac{\lambda(x)}{\Phi^{c-\alpha+1}(x)}\left(\Omega^\sigma(x)\right)^k \Delta_\alpha x \leq \left(\frac{k}{c-\alpha}\right)^k \int_a^\infty \Phi^{k-c+\alpha-1}(x)\lambda(x)g^k(x)\Delta_\alpha x,$$

that is the desired inequality (34). The proof is complete. □

Corollary 8. *At $\alpha = 1$ in Theorem 5, we have the inequality*

$$\int_a^\infty \frac{\lambda(x)}{\Phi^c(x)}\left(\Omega^\sigma(x)\right)^k \Delta x \leq \left(\frac{k}{c-1}\right)^k \int_a^\infty \Phi^{k-c}(x)\lambda(x)g^k(x)\Delta x,$$

which the inequality (2.49) in [33].

3. Conclusions

The new fractional calculus on timescales is presented with applications to some new fractional inequalities on timescales like Hardy, Bennett, Copson and Leindler types. Inequalities are considered in rather general forms and contain several special integral and discrete inequalities. The technique is based on the applications of well-known inequalities and new tools from fractional calculus.

Author Contributions: S.S. contributed in preparing the introduction, preliminaries and formulate theorem 3, its proof and its corollaries (cor.1, cor. 2 and cor.3). M.K. contributed in preparing the introduction, preliminaries and formulate theorem 4, its proof and its corollaries (cor.4, cor. 5 and cor.6). G.A. contributed in preparing the introduction, preliminaries and formulate theorem 5, its proof and its corollaries (cor.7). M.Z. contributed in preparing the introduction, preliminaries and formulate theorem 6, its proof and its corollaries (cor.8). All authors contributed equally to the writing of this manuscript. All authors have read and agreed to the published version of the manuscript.

Funding: This research was funded by the Deanship of Scientific Research at Princess Nourah bint Abdulrahman University through the Fast-track Research Funding Program.

Acknowledgments: This research was funded by the Deanship of Scientific Research at Princess Nourah bint Abdulrahman University through the Fast-track Research Funding Program. The authors thank the referees for helpful comments that lead to the improvement of the presentation of the results in this paper.

Conflicts of Interest: The authors declare no conflict of interest.

References

1. Hardy, G.H. Notes on a theorem of Hilbert. *Math. Z.* **1920**, *6*, 314–317. [CrossRef]
2. Hardy, G.H. Notes on some points in the integral calculus, LX. An inequality between integrals. *Mess. Math.* **1925**, *54*, 150–156.
3. Hardy, G.H. Notes on some points in the integral calculus, LXIV. Further inequalities between integrals. *Mess. Math.* **1928**, *57*, 12–16.
4. Copson, E.T. Note on series of positive terms. *J. Lond. Math. Soc.* **1927**, *2*, 9–12. [CrossRef]
5. Leindler, L. Generalization of inequalities of Hardy and Littlewood. *Acta Sci. Math.* **1970**, *31*, 297–285.
6. Leindler, L. Further sharpining of inequalities of Hardy and Littlewood. *Acta Sci. Math.* **1990**, *54*, 285–289.
7. Leindler, L. A theorem of Hardy-Bennett-type. *Acta Math. Hungar.* **1998**, *87*, 315–325. [CrossRef]
8. Leindler, L. Two Hardy-Bennett-type theorems. *Acta Math. Hungar.* **1999**, *85*, 265–276. [CrossRef]
9. Copson, E.T. Note on series of positive terms. *J. Lond. Math. Soc.* **1928**, *3*, 49–51. [CrossRef]
10. Copson, E.T. Some integral inequalities. *Prof. R. Soc. Edinburg. Sect. A* **1976**, *75*, 157–164. [CrossRef]
11. Bennett, G. Some elementary inequalities. *Quart. J. Math. Oxford* **1987**, *38*, 401–425. [CrossRef]
12. Agarwal, R.P.; Bohner, M.; O'Regan, D.; Saker, S.H. Some dynamic wirtinger-type inequalities and their applications. *Pacific J. Math.* **2011**, *252*, 1–18. [CrossRef]
13. Agarwal, R.P.; O'Regan, D.; Saker, S.H. *Hardy Type Ineqaualities on Time Scales*; Springer: Basel, Switzerland, 2016.
14. Saker, S.H. Some nonlinear dynamic inequalities on time scales and applications. *J. Math. Inequal.* **2010**, *4*, 561–579. [CrossRef]
15. Hilger, S. Analysis on measure chains—A unified approach to continuous and discrete calculus. *Results Math.* **1990**, *18*, 18–56. [CrossRef]
16. Bohner, M.; Peterson, A. *Dynamic Equations on Time Scales: An Introduction with Applications*; Birkhäuser: Boston, MA, USA, 2001.
17. Bohner, M.; Peterson, A. *Advances in Dynamic Equations on Time Scales*; Birkhäuser: Boston, MA, USA, 2003.
18. Kac, V.; Cheung, P. *Quantum Calculus*; Springer: New York, NY, USA, 2001.
19. Bogdan, K.; Dyda, B. The best constant in a fractional Hardy inequality. *Math. Nach.* **2011**, *284*, 629–638. [CrossRef]
20. Jleli, M.; Samet, B. Lyapunov-type inequalities for a fractional differential equation with mixed boundary conditions. *Math. Inequal. Appl.* **2015**, *18*, 443–451. [CrossRef]
21. Yildiz, C.; Ozdemir, M.E.; Onelan, H.K. Fractional integral inequalities for different functions. *New Trends Math. Sci.* **2015**, *3*, 110–117.
22. Abdeljawad, T. On conformable fractional calculus. *J. Comp. Appl. Math.* **2015**, *279*, 57–66. [CrossRef]
23. Khalil, R.; Horani, M.A.; Yousef, A.; Sababheh, M. A new definition of fractional derivative. *J. Comp. Appl. Math.* **2014**, *264*, 65–70. [CrossRef]
24. Akkurt, A.; Yildirim, M.E.; Yildirim, H. On some integral inequalities for conformable fractional integrals. *RGMIA Res. Rep. Collect.* **2016**, *19*, 107.
25. Chu, Y.M.; Khan, M.A.; Ali, T.; Dragomir, S.S. Inequalities for α-fractional differentiable functions. *J. Inequal. Appl.* **2017**, *1*, 93. [CrossRef]
26. Khan, M.A.; Ali, T.; Dragomir, S.S.; Sarikaya, M.Z. Hermite-Hadamard type inequalities for conformable fractional integrals. *Rev. Real Acad. Cienc. Exactas* **2018**, *112*, 1033–1048. [CrossRef]
27. Set, E.; Gözpınar, A.; Ekinci, A. Hermite-Hadamard type inequalities via conformable fractional integrals. *Acta Math. Univ. Comenian.* **2017**, *86*, 309–320. [CrossRef]
28. Sarikaya, M.Z.; Budak, H. New inequalities of Opial type for conformable fractional integrals. *Turkish J. Math.* **2017**, *41*, 1164–1173. [CrossRef]

29. Sarikaya, M.Z.; Budak, H. Opial type inequalities for conformable fractional integrals. *Rgmia Res. Rep. Collect.* **2016**, *19*, 93.
30. Sarikaya, M.Z.; Yaldiz, H.; Budak, H. Steffensen's integral inequality for conformable fractional integrals. *Int. J. Anal. Appl.* **2017**, *15*, 23–30.
31. Benkhettou, N.; Salima, H.; Torres, D.F.M. A conformable fractional calculus on arbitrary time scales. *J. King Saud Univ. Science* **2016**, *1*, 93–98. [CrossRef]
32. Nwaeze, E.R.; Torres, D.F.M. Chain rules and inequalities for the BHT fractional calculus on arbitrary timescales. *Arab J. Math.* **2017**, *6*, 13–20. [CrossRef]
33. Saker, S.H.; O'Regan, D.; Agarwal, R.P. Generalized Hardy, Copson, Leindler and Bennett inequalities on time scales. *Math. Nachr.* **2014**, *287*, 686–698. [CrossRef]

mathematics

MDPI

Article

On the Nonlocal Fractional Delta-Nabla Sum Boundary Value Problem for Sequential Fractional Delta-Nabla Sum-Difference Equations

Jiraporn Reunsumrit [1,†] and Thanin Sitthiwirattham [2,*,†]

[1] Department of Mathematics, Faculty of Applied Science, King Mongkut's University of Technology North Bangkok, Bangkok 10800, Thailand; jiraporn.r@sci.kmutnb.ac.th

[2] Mathematics Department, Faculty of Science and Technology, Suan Dusit University, Bangkok 10300, Thailand

* Correspondence: thanin_sit@dusit.ac.th

† These authors contributed equally to this work.

Received: 31 January 2020; Accepted: 22 March 2020; Published: 31 March 2020

Abstract: In this paper, we propose sequential fractional delta-nabla sum-difference equations with nonlocal fractional delta-nabla sum boundary conditions. The Banach contraction principle and the Schauder's fixed point theorem are used to prove the existence and uniqueness results of the problem. The different orders in one fractional delta differences, one fractional nabla differences, two fractional delta sum, and two fractional nabla sum are considered. Finally, we present an illustrative example.

Keywords: sequential fractional delta-nabla sum-difference equations; nonlocal fractional delta-nabla sum boundary value problem; existence; uniqueness

JEL Classification: 39A05; 39A12

---

1. Introduction

Nowaday, fractional calculus is attractive knowledge for many reseachers in many fields. In particular, the fractional calculus has been used in many research works related to biological, biomechanics, magnetic fields, echanics of micro/nano structures, and physical problems (see [1–7]). We can find fractional delta difference calculus and fractional nabla difference calculus in [8–24] and [25–36], respectively. Definitions and properties of fractional difference calculus are presented in the book [37].

We note that there are a few papers using the delta-nabla calculus as a tool. For example, Malinowska and Torres [38] presented the delta-nabla calculus of variations. Dryl and Torres [39,40] studied the delta-nabla calculus of variations for composition functionals on time scales, and a general delta-nabla calculus of variations on time scales with application to economics. Ghorbanian and Rezapour [41] proposed a two-dimensional system of delta-nabla fractional difference inclusions. Liu, Jin and Hou [42] investigated existence of positive solutions for discrete delta-nabla fractional boundary value problems with p-Laplacian.

In this paper, we aim to extend the study of delta-nabla calculus that has appeared in discrete fractional boundary value problems. We have found that the research works related to delta-nabla calculus were presented as above. However, the boundary value problem for sequential fractional delta-nabla difference equation has not been studied before. Our problem is sequential fractional delta-nabla sum-difference equations with nonlocal fractional delta-nabla sum boundary conditions as given by

$$\begin{aligned}\Delta^{\alpha}\nabla^{\beta}u(t) &= F\left[t+\alpha-1, u(t+\alpha-1), \left(\mathcal{S}^{\theta}u\right)(t+\alpha-1), \left(\mathcal{T}^{\phi}u\right)(t+\alpha-1)\right] \\ \Delta^{-\omega}u(\alpha+\omega-2) &= \kappa\nabla^{-\gamma}u(\alpha-2) \\ u(T+\alpha) &= \lambda u(\eta), \ \ \eta \in \mathbb{N}_{\alpha-1,T+\alpha-1}\end{aligned} \tag{1}$$

where $t \in \mathbb{N}_{0,T} := \{0,1,\ldots,T\}$; $\alpha \in (1,2]$; $\beta,\theta,\phi,\omega,\gamma \in (0,1]$; $\alpha+\beta \in (2,3]$; $T > 0$; κ,λ are given constants; $F \in C(\mathbb{N}_{\alpha-2,T+\alpha} \times \mathbb{R}^3, \mathbb{R})$; and for $\varphi,\psi \in C(\mathbb{N}_{\alpha-2,T+\alpha} \times \mathbb{N}_{\alpha-2,T+\alpha}, [0,\infty))$, we define

$$\left(\mathcal{S}^{\theta}u\right)(t) := \left[\nabla^{-\theta}\varphi u\right](t) = \frac{1}{\Gamma(\theta)}\sum_{s=\alpha-2}^{t}(t-\rho(s))^{\overline{\theta-1}}\varphi(t,s)\,u(s),$$

$$\left(\mathcal{T}^{\phi}u\right)(t) := \left[\Delta^{-\phi}\psi u\right](t+\phi) = \frac{1}{\Gamma(\phi)}\sum_{s=\alpha-\phi-2}^{t-\phi}(t-\sigma(s))^{\underline{\phi-1}}\psi(t,s+\phi)\,u(s+\phi).$$

The objective of this research is to investigate the solution of the boundary value problem (1). The basic knowledge is disscussed in Section 2, the existence results are presented in Section 3, and an example is provided in Section 4.

2. Preliminaries

We give the notations, definitions, and lemmas as follows. The forward operator and the backward operator are defined as $\sigma(t) := t+1$, and $\rho(t) := t-1$, respectively.

For $t,\alpha \in \mathbb{R}$, we define the generalized falling and rising functions as follows:

- The generalized falling function

$$t^{\underline{\alpha}} := \frac{\Gamma(t+1)}{\Gamma(t+1-\alpha)}$$

  for any $t+1-\alpha$ is not a pole of the Gamma function. If $t+1-\alpha$ is a pole and $t+1$ is not a pole, then $t^{\underline{\alpha}} = 0$.

- The generalized rising function

$$t^{\overline{\alpha}} := \frac{\Gamma(t+\alpha)}{\Gamma(t)}$$

  for any t is not a pole of the Gamma function. If t is a pole and $t+\alpha$ is not a pole, then $t^{\overline{\alpha}} = 0$.

Definition 1. *For $\alpha > 0$ and f defined on $\mathbb{N}_a := \{a, a+1, \ldots\}$, the α-order fractional delta sum of f is defined by*

$$\Delta^{-\alpha}f(t) := \frac{1}{\Gamma(\alpha)}\sum_{s=a}^{t-\alpha}(t-\sigma(s))^{\underline{\alpha-1}}f(s), \ \ t \in \mathbb{N}_{a+\alpha}.$$

The α-order fractional nabla sum of f is defined by

$$\nabla^{-\alpha}f(t) := \frac{1}{\Gamma(\alpha)}\sum_{s=a}^{t}(t-\rho(s))^{\overline{\alpha-1}}f(s), \ \ t \in \mathbb{N}_a.$$

Definition 2. *For $\alpha > 0$, $N \in \mathbb{N}$ where $0 \leq N-1 < \alpha < N$ and f defined on $\mathbb{N}_a$, the α-order Riemann-Liouville fractional delta difference of f is defined by*

$$\Delta^{\alpha}f(t) := \Delta^{N}\Delta^{-(N-\alpha)}f(t) = \frac{1}{\Gamma(-\alpha)}\sum_{s=a}^{t+\alpha}(t-\sigma(s))^{\underline{-\alpha-1}}f(s), \ \ t \in \mathbb{N}_{a+N-\alpha}.$$

The α-order Riemann-Liouville fractional nabla difference of f is defined by

$$\nabla^{\alpha} f(t) := \nabla^{N}\nabla^{-(N-\alpha)} f(t) = \frac{1}{\Gamma(-\alpha)} \sum_{s=a}^{t} (t-\rho(s))^{\overline{-\alpha-1}} f(s), \quad t \in \mathbb{N}_{a+N}.$$

Lemma 1 ([15]). *Let $0 \le N-1 < \alpha \le N$, $N \in \mathbb{N}$ and $y : \mathbb{N}_a \to \mathbb{R}$. Then,*

$$\Delta^{-\alpha}\Delta^{\alpha} y(t) = y(t) + C_1(t-a)^{\underline{\alpha-1}} + C_2(t-a)^{\underline{\alpha-2}} + \ldots + C_N(t-a)^{\underline{\alpha-N}},$$

for some $C_i \in \mathbb{R}$, with $1 \le i \le N$.

Lemma 2 ([28]). *Let $0 \le N-1 < \alpha \le N$, $N \in \mathbb{N}$ and $y : \mathbb{N}_{a+1} \to \mathbb{R}$. Then,*

$$\nabla^{-\alpha}\nabla^{\alpha} y(t) = \begin{cases} y(t), & \alpha \notin \mathbb{N} \\ y(t) - \sum_{k=0}^{N-1} \frac{(t-a)^{\overline{k}}}{k!} \nabla^{k} f(a), & \alpha = N, \end{cases}$$

for some $t \in \mathbb{N}_{a+N}$.

We next provide a linear variant of our problem (1).

Lemma 3. *Let $\Lambda \neq 0$; $\alpha \in (1,2]$; $\beta, \omega, \gamma \in (0,1)$; $\alpha+\beta \in (2,3]$; $T > 0$; κ, λ are given constants; and $h \in C(\mathbb{N}_{\alpha-2,T+\alpha}, \mathbb{R})$. Then the problem*

$$\Delta^{\alpha}\nabla^{\beta} u(t) = h(t+\alpha-1), \quad t \in \mathbb{N}_{0,T} \tag{2}$$

$$\Delta^{-\omega} u(\alpha+\omega-2) = \kappa \nabla^{-\gamma} u(\alpha-2) \tag{3}$$

$$u(T+\alpha) = \lambda u(\eta), \quad \eta \in \mathbb{N}_{\alpha-1,T+\alpha-1} \tag{4}$$

has the unique solution

$$\begin{aligned} u(t) &= \frac{\mathcal{O}[h]}{\Lambda\Gamma(\beta)} \sum_{s=\alpha-1}^{t} (t-\rho(s))^{\overline{\beta-1}} s^{\underline{\alpha-1}} \\ &+ \frac{1}{\Gamma(\beta)\Gamma(\alpha)} \sum_{s=\alpha}^{t}\sum_{r=0}^{s-\alpha} (t-\rho(s))^{\overline{\beta-1}} (s-\sigma(r))^{\underline{\alpha-1}} h(r+\alpha-1) \end{aligned} \tag{5}$$

where the functional $\mathcal{O}[h]$ and the constant Λ are defined by

$$\begin{aligned} \mathcal{O}[h] &= \frac{\lambda}{\Gamma(\beta)\Gamma(\alpha)} \sum_{s=\alpha}^{\eta}\sum_{r=0}^{s-\alpha} (\eta-\rho(s))^{\overline{\beta-1}} (s-\sigma(r))^{\underline{\alpha-1}} h(r+\alpha-1) \\ &- \frac{1}{\Gamma(\beta)\Gamma(\alpha)} \sum_{s=\alpha}^{T+\alpha}\sum_{r=0}^{s-\alpha} (T+\alpha-\rho(s))^{\overline{\beta-1}} (s-\sigma(r))^{\underline{\alpha-1}} h(r+\alpha-1), \end{aligned} \tag{6}$$

$$\Lambda = \frac{1}{\Gamma(\beta)} \sum_{s=\alpha-1}^{T+\alpha} (T+\alpha-\rho(s))^{\overline{\beta-1}} s^{\underline{\alpha-1}} - \frac{\lambda}{\Gamma(\beta)} \sum_{s=\alpha-1}^{\eta} (\eta-\rho(s))^{\overline{\beta-1}} s^{\underline{\alpha-1}}. \tag{7}$$

Proof. Using the fractional delta sum of order α for (2), we obtain

$$\nabla^{\beta} u(t) = C_1 t^{\underline{\alpha-1}} + C_2 t^{\underline{\alpha-2}} + \frac{1}{\Gamma(\alpha)} \sum_{s=0}^{t-\alpha} (t-\sigma(s))^{\underline{\alpha-1}} h(s+\alpha-1), \tag{8}$$

for $t \in \mathbb{N}_{\alpha-2,T+\alpha}$.

Taking the fractional nabla sum of order β for (8), we get

$$u(t) = \frac{1}{\Gamma(\beta)} \sum_{s=\alpha-2}^{t} (t-\rho(s))^{\overline{\beta-1}} \left[C_1 s^{\underline{\alpha-1}} + C_2 s^{\underline{\alpha-2}}\right] + \frac{1}{\Gamma(\beta)\Gamma(\alpha)} \sum_{s=\alpha}^{t} \sum_{r=0}^{s-\alpha} (t-\rho(s))^{\overline{\beta-1}} (s-\sigma(r))^{\underline{\alpha-1}} h(r+\alpha-1), \tag{9}$$

for $t \in \mathbb{N}_{\alpha-2,T+\alpha}$.

Using the fractional delta sum of order ω for (9), we have

$$\Delta^{-\omega}u(t) = \sum_{s=\alpha}^{t-\omega} \sum_{r=\alpha}^{s} \frac{(t-\sigma(s))^{\underline{\omega-1}}(s-\rho(r))^{\overline{\beta-1}}}{\Gamma(\omega)\Gamma(\beta)} \left[C_1 r^{\underline{\alpha-1}} + C_2 r^{\underline{\alpha-2}}\right] + \sum_{s=\alpha}^{t-\omega} \sum_{r=\alpha}^{s} \sum_{\xi=0}^{r-\alpha} \frac{(t-\sigma(s))^{\underline{\omega-1}}(s-\rho(r))^{\overline{\beta-1}}(r-\sigma(\xi))^{\underline{\alpha-1}}}{\Gamma(\omega)\Gamma(\beta)\Gamma(\alpha)} h(\xi+\alpha-1), \tag{10}$$

for $t \in \mathbb{N}_{\alpha+\omega-1,T+\alpha+\omega}$.

Taking the fractional nabla sum of order γ for (9), we have

$$\nabla^{-\gamma}u(t) = \sum_{s=\alpha}^{t} \sum_{r=\alpha-2}^{s} \frac{(t-\rho(s))^{\overline{\gamma-1}}(s-\rho(r))^{\overline{\beta-1}}}{\Gamma(\gamma)\Gamma(\beta)} \left[C_1 r^{\underline{\alpha-1}} + C_2 r^{\underline{\alpha-2}}\right] + \sum_{s=\alpha}^{t} \sum_{r=\alpha}^{s} \sum_{\xi=0}^{r-\alpha} \frac{(t-\rho(s))^{\overline{\gamma-1}}(s-\rho(r))^{\overline{\beta-1}}(r-\sigma(\xi))^{\underline{\alpha-1}}}{\Gamma(\omega)\Gamma(\beta)\Gamma(\alpha)} h(\xi+\alpha-1), \tag{11}$$

for $t \in \mathbb{N}_{\alpha-2,T+\alpha}$.

By substituting $t = \alpha+\omega-2$ and $t = \alpha-2$ into (10) and (11), respectively; and using the condition (3), we obtain

$$C_2 = 0$$

Substitute $C_2 = 0$ and apply the condition (4). Then, we obtain

$$C_1 = \frac{\mathcal{O}[h]}{\Lambda}$$

where $\mathcal{O}[h]$ and Λ are defined by (6)–(7), respectively. Substituting the constants C_1 and C_2 into (9), we obtain (5). □

3. Existence and Uniqueness Result

Define $\mathcal{C} = C(\mathbb{N}_{\alpha-2,T+\alpha}, \mathbb{R})$ is the Banach space of all function u and define the norm as

$$\|u\|_{\mathcal{C}} = \|u\| + \|\nabla^{-\theta}u\| + \|\Delta^{-\phi}u\|$$

where $\|u\| = \max_{t\in\mathbb{N}_{\alpha-2,T+\alpha}} |u(t)|$, $\|\nabla^{-\theta}u\| = \max_{t\in\mathbb{N}_{\alpha-2,T+\alpha}} |\nabla^{-\theta}u(t)|$ and $\|\Delta^{-\phi}u\| = \max_{t\in\mathbb{N}_{\alpha-2,T+\alpha}} |\Delta^{-\phi}u(t+\phi)|$.

In addition, we define the operator $\mathcal{F} : \mathcal{C} \to \mathcal{C}$ by

$$(\mathcal{F}u)(t) = \frac{\mathcal{O}[h]}{\Lambda\Gamma(\beta)} \sum_{s=\alpha-1}^{t} (t-\rho(s))^{\overline{\beta-1}} s^{\underline{\alpha-1}} + \frac{1}{\Gamma(\beta)\Gamma(\alpha)} \sum_{s=\alpha}^{t} \sum_{r=0}^{s-\alpha} (t-\rho(s))^{\overline{\beta-1}} (s-\sigma(r))^{\underline{\alpha-1}} \times F\left[r+\alpha-1, u(r+\alpha-1), \left(\mathcal{S}^{\theta}u\right)(r+\alpha-1), \left(\mathcal{T}^{\phi}u\right)(r+\alpha-1)\right], \tag{12}$$

where $\Lambda \neq 0$ is defined by (7) and the functional $\mathcal{O}[F(u)]$ is defined by

$$\begin{aligned}
\mathcal{O}[F(u)] &= \frac{\lambda}{\Gamma(\beta)\Gamma(\alpha)} \sum_{s=\alpha}^{\eta} \sum_{r=0}^{s-\alpha} (\eta - \rho(s))^{\overline{\beta-1}} (s - \sigma(r))^{\underline{\alpha-1}} \times \\
& F\left[r+\alpha-1, u(r+\alpha-1), \left(\mathcal{S}^{\theta}u\right)(r+\alpha-1), \left(\mathcal{T}^{\phi}u\right)(r+\alpha-1)\right] \\
& - \frac{1}{\Gamma(\beta)\Gamma(\alpha)} \sum_{s=\alpha}^{T+\alpha} \sum_{r=0}^{s-\alpha} (T+\alpha - \rho(s))^{\overline{\beta-1}} (s - \sigma(r))^{\underline{\alpha-1}} \times \\
& F\left[r+\alpha-1, u(r+\alpha-1), \left(\mathcal{S}^{\theta}u\right)(r+\alpha-1), \left(\mathcal{T}^{\phi}u\right)(r+\alpha-1)\right]. \qquad (13)
\end{aligned}$$

Obviously, the operator $\mathcal{F}$ has the fixed points if and only if the boundary value problem (1) has solutions. Firstly, we show the existence and uniqueness result of the boundary value problem (1) by using the Banach contraction principle.

Theorem 1. *Assume that* $F : \mathbb{N}_{\alpha-2,T+\alpha} \times \mathbb{R}^3 \to \mathbb{R}$ *is continuous,* $\varphi, \psi : \mathbb{N}_{\alpha-2,T+\alpha} \times \mathbb{N}_{\alpha-2,T+\alpha} \to [0,\infty)$ *with* $\varphi_0 = \max\{\varphi(t-1,s) : (t,s) \in \mathbb{N}_{\alpha-2,T+\alpha} \times \mathbb{N}_{\alpha-2,T+\alpha}\}$ *and* $\psi_0 = \max\{\psi(t-1,s) : (t,s) \in \mathbb{N}_{\alpha-2,T+\alpha} \times \mathbb{N}_{\alpha-2,T+\alpha}\}$. *Suppose that the following conditions hold:*

(H_1) *there exist constants* $L_1, L_2, L_3 > 0$ *such that for each* $t \in \mathbb{N}_{\alpha-2,T+\alpha}$ *and* $u, v \in \mathbb{R}$

$$\begin{aligned}
& \left|F\left(t,u,\left(\mathcal{S}^{\theta}u\right),\left(\mathcal{T}^{\phi}u\right)\right) - F\left(t,v,\left(\mathcal{S}^{\theta}v\right),\left(\mathcal{T}^{\phi}v\right)\right)\right| \\
\leq\ & L_1|u-v| + L_2\left|\left(\mathcal{S}^{\theta}u\right) - \left(\mathcal{S}^{\theta}v\right)\right| + L_3\left|\left(\mathcal{T}^{\phi}u\right) - \left(\mathcal{T}^{\phi}v\right)\right|.
\end{aligned}$$

Then the problem (1) *has a unique solution on* $\mathbb{N}_{\alpha-2,T+\alpha}$ *provided that*

$$\chi := \left\{L_1 + L_2\varphi_0\frac{(T+3)^{\overline{\theta}}}{\Gamma(\theta+1)} + L_3\psi_0\frac{(T+2)^{\underline{\phi}}}{\Gamma(\phi+1)}\right\}[\Omega_1+\Omega_2+\Omega_3] < 1 \qquad (14)$$

where

$$\Theta = \frac{1}{\Gamma(\alpha+1)\Gamma(\beta+1)}\left[\lambda\eta^{\underline{\alpha}}(\eta-\alpha+1)^{\overline{\beta}} + (T+\alpha)^{\underline{\alpha}}(T+1)^{\overline{\beta}}\right], \qquad (15)$$

$$\Omega_1 = \frac{1}{\Gamma(\beta+1)}\left[\frac{\Theta}{|\Lambda|}(T+\alpha)^{\underline{\alpha-1}}(T+2)^{\overline{\beta-1}} + \frac{(T+\alpha)^{\underline{\alpha}}}{\Gamma(\alpha+1)}(T+1)^{\overline{\beta}}\right], \qquad (16)$$

$$\Omega_2 = \frac{1}{\Gamma(\theta+1)\Gamma(\beta+1)}\left[\frac{\Theta}{|\Lambda|}(T+\alpha)^{\underline{\alpha-1}}(T+2)^{\overline{\beta-1}}(T+2)^{\overline{\theta}} + \frac{(T+\alpha)^{\underline{\alpha}}}{\Gamma(\alpha+1)}(T+1)^{\overline{\beta}}(T+1)^{\overline{\theta}}\right], \qquad (17)$$

$$\Omega_3 = \frac{1}{\Gamma(\phi+1)\Gamma(\beta+1)}\left[\frac{\Theta}{|\Lambda|}(T+\alpha)^{\underline{\alpha-1}}(T+2)^{\overline{\beta-1}}(T+1)^{\underline{\phi}} + \frac{(T+\alpha)^{\underline{\alpha}}}{\Gamma(\alpha+1)}(T+1)^{\overline{\beta}}T^{\underline{\phi}}\right]. \qquad (18)$$

Proof. We shall show that $\mathcal{F}$ is a contraction. For any $u, v \in \mathcal{C}$ and for each $t \in \mathbb{N}_{\alpha-2,T+\alpha}$, we have

$$\begin{aligned}\Big|\mathcal{O}[F(u)] - \mathcal{O}[F(v)]\Big| \leq & \Bigg| \tfrac{\lambda}{\Gamma(\beta)\Gamma(\alpha)} \sum_{s=\alpha}^{\eta}\sum_{r=0}^{s-\alpha} (\eta-\rho(s))^{\overline{\beta-1}} (s-\sigma(r))^{\underline{\alpha-1}} \times \\ & \Big[L_1|u-v| + L_2\Big|(\mathcal{S}^{\theta}u) - (\mathcal{S}^{\theta}v)\Big| + L_3\Big|(\mathcal{T}^{\phi}u) - (\mathcal{T}^{\phi}v)\Big|\Big] \\ & - \tfrac{1}{\Gamma(\beta)\Gamma(\alpha)} \sum_{s=\alpha}^{T+\alpha}\sum_{r=0}^{s-\alpha} (T+\alpha-\rho(s))^{\overline{\beta-1}} (s-\sigma(r))^{\underline{\alpha-1}} \times \\ & \Big[L_1|u-v| + L_2\Big|(\mathcal{S}^{\theta}u) - (\mathcal{S}^{\theta}v)\Big| + L_3\Big|(\mathcal{T}^{\phi}u) - (\mathcal{T}^{\phi}v)\Big|\Big]\Bigg| \\ \leq & \|u-v\|_{\mathcal{C}}\left\{L_1 + L_2\varphi_0\tfrac{(T+3)^{\overline{\theta}}}{\Gamma(\theta+1)} + L_3\psi_0\tfrac{(T+2)^{\underline{\phi}}}{\Gamma(\phi+1)}\right\}\Bigg|\lambda \sum_{s=\alpha}^{\eta}\sum_{r=0}^{s-\alpha} \tfrac{(\eta-\rho(s))^{\overline{\beta-1}}}{\Gamma(\beta)} \times \\ & \tfrac{(s-\sigma(r))^{\underline{\alpha-1}}}{\Gamma(\alpha)} - \sum_{s=\alpha}^{T+\alpha}\sum_{r=0}^{s-\alpha} \tfrac{(T+\alpha-\rho(s))^{\overline{\beta-1}}(s-\sigma(r))^{\underline{\alpha-1}}}{\Gamma(\beta)\Gamma(\alpha)}\Bigg| \\ \leq & \|u-v\|_{\mathcal{C}}\left\{L_1 + L_2\varphi_0\tfrac{(T+3)^{\overline{\theta}}}{\Gamma(\theta+1)} + L_3\psi_0\tfrac{(T+2)^{\underline{\phi}}}{\Gamma(\phi+1)}\right\}\Theta,\end{aligned} \tag{19}$$

and

$$\begin{aligned}& \big|(\mathcal{F}u)(t) - (\mathcal{F}v)(t)\big| \\ \leq & \left|\frac{\mathcal{O}[F(u)] - \mathcal{O}[F(v)]}{\Lambda}\right| \sum_{s=\alpha-1}^{T+\alpha} (T+\alpha-\rho(s))^{\overline{\beta-1}} s^{\underline{\alpha-1}} + \sum_{s=\alpha}^{T+\alpha}\sum_{r=0}^{s-\alpha} \frac{(T+\alpha-\rho(s))^{\overline{\beta-1}}}{\Gamma(\beta)} \times \\ & \frac{(s-\sigma(r))^{\underline{\alpha-1}}}{\Gamma(\alpha)}\Big[L_1|u-v| + L_2\Big|\big(\mathcal{S}^{\theta}u\big) - \big(\mathcal{S}^{\theta}v\big)\Big| + L_3\Big|(\mathcal{T}^{\phi}u) - (\mathcal{T}^{\phi}v)\Big|\Big] \\ \leq & \|u-v\|_{\mathcal{C}}\left\{L_1 + L_2\varphi_0\frac{(T+3)^{\overline{\theta}}}{\Gamma(\theta+1)} + L_3\psi_0\frac{(T+2)^{\underline{\phi}}}{\Gamma(\phi+1)}\right\}\left\{\frac{\Theta(T+\alpha)^{\underline{\alpha-1}}}{|\Lambda|} \times\right. \\ & \left.\sum_{s=\alpha-1}^{T+\alpha} (T+\alpha-\rho(s))^{\overline{\beta-1}} + \sum_{s=\alpha}^{T+\alpha}\sum_{r=0}^{s-\alpha} \frac{(T+\alpha-\rho(s))^{\overline{\beta-1}}(s-\sigma(r))^{\underline{\alpha-1}}}{\Gamma(\beta)\Gamma(\alpha)}\right\} \\ \leq & \|u-v\|_{\mathcal{C}}\left\{L_1 + L_2\varphi_0\frac{(T+3)^{\overline{\theta}}}{\Gamma(\theta+1)} + L_3\psi_0\frac{(T+2)^{\underline{\phi}}}{\Gamma(\phi+1)}\right\}\Omega_1.\end{aligned} \tag{20}$$

Next, we consider the following $(\nabla^{-\theta}\mathcal{F}u)$ and $(\Delta^{-\phi}\mathcal{F}u)$ as

$$\begin{aligned}(\nabla^{-\theta}\mathcal{F}u)(t) = & \frac{\mathcal{O}[F(u)]}{\Lambda}\sum_{s=\alpha}^{t}\sum_{r=\alpha-1}^{s} \frac{(t-\rho(s))^{\overline{\theta-1}}(s-\rho(r))^{\overline{\beta-1}}}{\Gamma(\theta)\Gamma(\beta)} r^{\underline{\alpha-1}} \\ & + \sum_{s=\alpha}^{t}\sum_{r=\alpha}^{s}\sum_{\xi=0}^{r-\alpha} \frac{(t-\rho(s))^{\overline{\theta-1}}(s-\rho(r))^{\overline{\beta-1}}(r-\sigma(\xi))^{\underline{\alpha-1}}}{\Gamma(\theta)\Gamma(\beta)\Gamma(\alpha)} \times \\ & F\Big[\xi+\alpha-1, u(\xi+\alpha-1), \Delta^{\theta}u(\xi+\alpha-\theta+1), \nabla^{\gamma}u(\xi+\alpha+1)\Big],\end{aligned} \tag{21}$$

and

$$\begin{aligned}(\Delta^{-\phi}\mathcal{F}u)(t+\phi) = & \frac{\mathcal{O}[F(u)]}{\Lambda}\sum_{s=\alpha}^{t-\phi}\sum_{r=\alpha-1}^{s} \frac{(t+\phi-\sigma(s))^{\underline{\phi-1}}(s-\rho(r))^{\overline{\beta-1}}}{\Gamma(\phi)\Gamma(\beta)} r^{\underline{\alpha-1}} \\ & + \sum_{s=\alpha}^{t-\phi}\sum_{r=\alpha}^{s}\sum_{\xi=0}^{r-\alpha} \frac{(t+\phi-\sigma(s))^{\underline{\phi-1}}(s-\rho(r))^{\overline{\beta-1}}(r-\sigma(\xi))^{\underline{\alpha-1}}}{\Gamma(\phi)\Gamma(\beta)\Gamma(\alpha)} \times \\ & F\Big[\xi+\alpha-1, u(\xi+\alpha-1), \Delta^{\theta}u(\xi+\alpha-\theta+1), \nabla^{\gamma}u(\xi+\alpha+1)\Big].\end{aligned} \tag{22}$$

Similarly to the above, we have

$$\left|(\nabla^{-\theta}\mathcal{F}u)(t)-(\nabla^{-\theta}\mathcal{F}v)(t)\right|\leq\|u-v\|_C\left\{L_1+L_2\varphi_0\frac{(T+3)^{\overline{\theta}}}{\Gamma(\theta+1)}+L_3\psi_0\frac{(T+2)^{\underline{\phi}}}{\Gamma(\phi+1)}\right\}\Omega_2,\tag{23}$$

$$\left|\Delta^{-\phi}\mathcal{F}u)(t+\phi)-\Delta^{-\phi}\mathcal{F}v)(t+\phi)\right|\leq\|u-v\|_C\left\{L_1+L_2\varphi_0\frac{(T+3)^{\overline{\theta}}}{\Gamma(\theta+1)}+L_3\psi_0\frac{(T+2)^{\underline{\phi}}}{\Gamma(\phi+1)}\right\}\Omega_3.\tag{24}$$

From (20), (23) and (24), we get

$$\begin{aligned}\left\|(\mathcal{F}u)-(\mathcal{F}v)\right\|_C\leq&\ \|u-v\|_C\left\{L_1+L_2\varphi_0\frac{(T+3)^{\overline{\theta}}}{\Gamma(\theta+1)}+L_3\psi_0\frac{(T+2)^{\underline{\phi}}}{\Gamma(\phi+1)}\right\}[\Omega_1+\Omega_2+\Omega_3]\\ =&\ \chi\|u-v\|_C.\end{aligned}\tag{25}$$

By (14), we have $\left\|(\mathcal{F}u)(t)-(\mathcal{F}v)(t)\right\|_C<\|u-v\|_C$.

Consequently, $\mathcal{F}$ is a contraction. Therefore, by the Banach fixed point theorem, we get that $\mathcal{F}$ has a fixed point which is a unique solution of the problem (1) on $t\in\mathbb{N}_{\alpha-2,T+\alpha}$. □

4. Existence of at Least One Solution

Next, we provide Arzelá-Ascoli theorem and Schauder's fixed point theorem that will be used to prove the existence of at least one solution of (1).

Lemma 4 ([43]). *(Arzelá-Ascoli theorem) A set of function in $C[a,b]$ with the sup norm is relatively compact if and only it is uniformly bounded and equicontinuous on $[a,b]$.*

Lemma 5 ([43]). *A set is compact if it is closed and relatively compact.*

Lemma 6 ([44]). *(Schauder's fixed point theorem) Let (D,d) be a complete metric space, U be a closed convex subset of D, and $T: D\to D$ be the map such that the set $Tu: u\in U$ is relatively compact in D. Then the operator T has at least one fixed point $u^*\in U$: $Tu^*=u^*$.*

Theorem 2. *Assuming that (H_1) holds, problem (1) has at least one solution on $\mathbb{N}_{\alpha-3,T+\alpha}$.*

Proof. The proof is organized as follows.
Step I. Verify $\mathcal{F}$ map bounded sets into bounded sets in $B_R=\{u\in C(\mathbb{N}_{\alpha-2,T+\alpha}):\|u\|_C\leq R\}$.

Let $\max\limits_{t\in\mathbb{N}_{\alpha-2,T+\alpha}}|F(t,0,0,0)|=M$ and choose a constant

$$R\geq\frac{M\left[\Omega_1+\Omega_2+\Omega_3\right]}{1-\left[\Omega_1+\Omega_2+\Omega_3\right]\left\{L_1+L_2\varphi_0\frac{(T+3)^{\overline{\theta}}}{\Gamma(\theta+1)}+L_3\psi_0\frac{(T+2)^{\underline{\phi}}}{\Gamma(\phi+1)}\right\}}.\tag{26}$$

For each $u \in B_R$, we obtain

$$\begin{aligned} &\left|\mathcal{O}[F(u)]\right| \\ \leq\ & \left|\lambda \sum_{s=\alpha}^{\eta}\sum_{r=0}^{s-\alpha} \frac{(\eta-\rho(s))^{\overline{\beta-1}}(s-\sigma(r))^{\underline{\alpha-1}}}{\Gamma(\beta)\Gamma(\alpha)} - \sum_{s=\alpha}^{T+\alpha}\sum_{r=0}^{s-\alpha} \frac{(T+\alpha-\rho(s))^{\overline{\beta-1}}(s-\sigma(r))^{\underline{\alpha-1}}}{\Gamma(\beta)\Gamma(\alpha)}\right| \\ & \Big[\Big|F\big[r+\alpha-1, u(r+\alpha-1), \left(\mathcal{S}^{\theta}u\right)(r+\alpha-\theta+1), \left(\mathcal{T}^{\phi}v\right)(r+\alpha+1)\big] \\ & -F(r+\alpha-1,0,0,0)\Big| + \left|F(r+\alpha-1,0,0,0)\right|\Big] \\ \leq\ & \left\{\left[L_1 + L_2\varphi_0\frac{(T+3)^{\overline{\theta}}}{\Gamma(\theta+1)} + L_3\psi_0\frac{(T+2)^{\underline{\phi}}}{\Gamma(\phi+1)}\right]\|u\|_C + M\right\}\Theta \end{aligned} \tag{27}$$

and

$$\begin{aligned} &\left|(\mathcal{F}u)(t)\right| \\ \leq\ & \left|\frac{\mathcal{O}[F(u)]}{\Lambda}\right| \sum_{s=\alpha-1}^{T+\alpha}(T+\alpha-\rho(s))^{\overline{\beta-1}} s^{\underline{\alpha-1}} + \sum_{s=\alpha}^{T+\alpha}\sum_{r=0}^{s-\alpha}\frac{(T+\alpha-\rho(s))^{\overline{\beta-1}}}{\Gamma(\beta)} \times \\ & \frac{(s-\sigma(r))^{\underline{\alpha-1}}}{\Gamma(\alpha)}\Big[\Big|F\big[r+\alpha-1, u(r+\alpha-1), \left(\mathcal{S}^{\theta}u\right)(r+\alpha-\theta+1), \\ & \left(\mathcal{T}^{\phi}v\right)(r+\alpha+1)\big] - F(r+\alpha-1,0,0,0)\Big| + \left|F(r+\alpha-1,0,0,0)\right|\Big] \\ \leq\ & \left\{\left[L_1 + L_2\varphi_0\frac{(T+3)^{\overline{\theta}}}{\Gamma(\theta+1)} + L_3\psi_0\frac{(T+2)^{\underline{\phi}}}{\Gamma(\phi+1)}\right]\|u\|_C + M\right\}\Omega_1. \end{aligned} \tag{28}$$

In addition, we have

$$\left|(\nabla^{-\theta}\mathcal{F}u)(t)\right| \leq \left\{\left[L_1 + L_2\varphi_0\frac{(T+3)^{\overline{\theta}}}{\Gamma(\theta+1)} + L_3\psi_0\frac{(T+2)^{\underline{\phi}}}{\Gamma(\phi+1)}\right]\|u\|_C + M\right\}\Omega_2, \tag{29}$$

$$\left|(\Delta^{-\phi}\mathcal{F}u)(t+\phi)\right| \leq \left\{\left[L_1 + L_2\varphi_0\frac{(T+3)^{\overline{\theta}}}{\Gamma(\theta+1)} + L_3\psi_0\frac{(T+2)^{\underline{\phi}}}{\Gamma(\phi+1)}\right]\|u\|_C + M\right\}\Omega_3. \tag{30}$$

From (28), (29) and (30), we have

$$\begin{aligned} \|(\mathcal{F}u)(t)\|_C \leq\ & \left\{\left[L_1 + L_2\varphi_0\frac{(T+3)^{\overline{\theta}}}{\Gamma(\theta+1)} + L_3\psi_0\frac{(T+2)^{\underline{\phi}}}{\Gamma(\phi+1)}\right]\|u\|_C + M\right\}[\Omega_1+\Omega_2+\Omega_3] \\ \leq\ & R. \end{aligned} \tag{31}$$

So, $\|\mathcal{F}u\|_C \leq R$. Therefore $\mathcal{F}$ is uniformly bounded.

Step II. Since F and H are continuous, the operator $\mathcal{F}$ is continuous on B_R.

Step III. Prove that $\mathcal{F}$ is equicontinuous on B_R. For any $\epsilon > 0$, there exists a positive constant $\rho^* = \max\{\delta_1, \delta_2, \delta_3, \delta_4, \delta_5, , \delta_6\}$ such that for $t_1, t_2 \in \mathbb{N}_{\alpha-2,T+\alpha}$
$\left|(t_2-\alpha+2)^{\overline{\beta}} - (t_1-\alpha+2)^{\overline{\beta}}\right| < \frac{\epsilon\Gamma(\beta+1)|\Lambda|}{6(T+\alpha)^{\underline{\alpha-1}}\Theta\|F\|}$, where $|t_2-t_1| < \delta_1$,
$\left|(t_2-\alpha+1)^{\overline{\beta}} - (t_1-\alpha+1)^{\overline{\beta}}\right| < \frac{\epsilon\Gamma(\alpha+1)\Gamma(\beta+1)}{6(T+\alpha)^{\underline{\alpha}}\|F\|}$, where $|t_2-t_1| < \delta_2$, $\left|(t_2-\alpha+2)^{\overline{\theta}} - (t_1-\alpha+2)^{\overline{\theta}}\right| < \frac{\epsilon\Gamma(\beta+1)\Gamma(\theta+1)|\Lambda|}{6(T+2)^{\overline{\beta}}(T+\alpha)^{\underline{\alpha-1}}\Theta\|F\|}$, where $|t_2-t_1| < \delta_3$,
$\left|(t_2-\alpha+1)^{\overline{\theta}} - (t_1-\alpha+1)^{\overline{\theta}}\right| < \frac{\epsilon\Gamma(\alpha+1)\Gamma(\beta+1)\Gamma(\theta+1)}{6(T+1)^{\overline{\beta}}(T+\alpha)^{\underline{\alpha}}\Theta\|F\|}$, where $|t_2-t_1| < \delta_4$,

$$\left|(t_2-\alpha+1)^{\underline{\phi}}-(t_1-\alpha+1)^{\underline{\phi}}\right| < \frac{\epsilon\Gamma(\beta+1)\Gamma(\phi+1)|\Lambda|}{6(T+2)^{\overline{\beta}}(T+\alpha)^{\underline{\alpha-1}}\Theta\|F\|}, \quad \text{where } |t_2-t_1|<\delta_5,$$
$$\left|(t_2-\alpha)^{\underline{\phi}}-(t_1-\alpha)^{\underline{\phi}}\right| < \frac{\epsilon\Gamma(\alpha+1)\Gamma(\beta+1)\Gamma(\phi+1)}{6(T+1)^{\overline{\beta}}(T+\alpha)^{\underline{\alpha}}\|F\|}, \quad \text{where } |t_2-t_1|<\delta_6.$$

Then we have

$$\begin{aligned}
&|(\mathcal{F}u)(t_2)-(\mathcal{F}u)(t_1)| \\
\leq\; & \left|\frac{\mathcal{O}[F(u)]]}{\Lambda}\right| \left|\frac{1}{\Gamma(\beta)}\sum_{s=\alpha-1}^{t_2}(t_2-\rho(s))^{\overline{\beta-1}}s^{\underline{\alpha-1}}-\frac{1}{\Gamma(\beta)}\sum_{s=\alpha-1}^{t_1}(t_1-\rho(s))^{\overline{\beta-1}}s^{\underline{\alpha-1}}\right| \\
& \left|\sum_{s=\alpha}^{t_2}\sum_{r=0}^{s-\alpha}\frac{(t_2-\rho(s))^{\overline{\beta-1}}(s-\sigma(r))^{\underline{\alpha-1}}}{\Gamma(\beta)\Gamma(\alpha)}\times\right. \\
& F\left[r+\alpha-1,u(r+\alpha-1),\left(\mathcal{S}^{\theta}u\right)(r+\alpha-1),\left(\mathcal{T}^{\phi}u\right)(r+\alpha-1)\right] \\
& -\sum_{s=\alpha}^{t_1}\sum_{r=0}^{s-\alpha}\frac{(t_1-\rho(s))^{\overline{\beta-1}}(s-\sigma(r))^{\underline{\alpha-1}}}{\Gamma(\beta)\Gamma(\alpha)}\times \\
& \left.F\left[r+\alpha-1,u(r+\alpha-1),\left(\mathcal{S}^{\theta}u\right)(r+\alpha-1),\left(\mathcal{T}^{\phi}u\right)(r+\alpha-1)\right]\right| \\
<\; & \frac{\Theta\|F\|(T+\alpha)^{\underline{\alpha-1}}}{|\Lambda|\Gamma(\beta+1)}\left|(t_2-\alpha+2)^{\overline{\beta}}-(t_1-\alpha+2)^{\overline{\beta}}\right|+\frac{\|F\|(T+\alpha)^{\underline{\alpha}}}{\Gamma(\alpha+1)\Gamma(\beta+1)}\times \\
& \left|(t_2-\alpha+1)^{\overline{\beta}}-(t_1-\alpha+1)^{\overline{\beta}}\right| \\
<\; & \frac{\epsilon}{6}+\frac{\epsilon}{6}=\frac{\epsilon}{3}.
\end{aligned} \tag{32}$$

Similarly to the above, we have

$$\begin{aligned}
|(\nabla^{-\theta}\mathcal{F}u)(t_2)-(\nabla^{-\theta}\mathcal{F}u)(t_1)| <\; & \frac{\Theta\|F\|(T+\alpha)^{\underline{\alpha-1}}(T+2)^{\overline{\beta}}}{|\Lambda|\Gamma(\beta+1)\Gamma(\theta+1)}\left|(t_2-\alpha+2)^{\overline{\theta}}-(t_1-\alpha+2)^{\overline{\theta}}\right| \\
& +\frac{\|F\|(T+\alpha)^{\underline{\alpha}}(T+1)^{\overline{\beta}}}{\Gamma(\alpha+1)\Gamma(\beta+1)\Gamma(\theta+1)}\left|(t_2-\alpha+1)^{\overline{\theta}}-(t_1-\alpha+1)^{\overline{\theta}}\right| \\
<\; & \frac{\epsilon}{6}+\frac{\epsilon}{6}=\frac{\epsilon}{3},
\end{aligned} \tag{33}$$

$$\begin{aligned}
|(\Delta^{-\phi}\mathcal{F}u)(t_2+\phi)-(\Delta^{-\phi}\mathcal{F}u)(t_1+\phi)| <\; & \frac{\Theta\|F\|(T+\alpha)^{\underline{\alpha-1}}(T+2)^{\overline{\beta}}}{|\Lambda|\Gamma(\beta+1)\Gamma(\phi+1)}\left|(t_2-\alpha+2)^{\overline{\phi}}-(t_1-\alpha+2)^{\overline{\phi}}\right| \\
& +\frac{\|F\|(T+\alpha)^{\underline{\alpha}}(T+1)^{\overline{\beta}}}{\Gamma(\alpha+1)\Gamma(\beta+1)\Gamma(\phi+1)}\left|(t_2-\alpha)^{\overline{\phi}}-(t_1-\alpha)^{\overline{\phi}}\right| \\
<\; & \frac{\epsilon}{6}+\frac{\epsilon}{6}=\frac{\epsilon}{3}.
\end{aligned} \tag{34}$$

Hence

$$\left\|(\mathcal{F}u)(t_2)-(\mathcal{F}u)(t_1)\right\|_{\mathcal{C}} < \frac{\epsilon}{3}+\frac{\epsilon}{3}+\frac{\epsilon}{3} = \epsilon. \tag{35}$$

It implies that the set $\mathcal{F}(B_R)$ is equicontinuous. By the results of Steps I to III and the Arzelá-Ascoli theorem, $\mathcal{F}:\mathcal{C}\to\mathcal{C}$ is completely continuous. By Schauder fixed point theorem, the boundary value problem (1) has at least one solution. □

5. An Example

Here, we provide a sequential fractional delta-nabla sum-difference equations with nonlocal fractional delta-nabla sum boundary conditions

$$\begin{aligned} \Delta^{\frac{3}{2}}\nabla^{\frac{2}{3}}u(t) &= \frac{e^{-\sin^2\left(t+\frac{1}{2}\right)}}{\left(t+\frac{201}{2}\right)^2}\cdot\left[\frac{2\left|u\left(t+\frac{1}{2}\right)\right|}{\left|u\left(t+\frac{1}{2}\right)\right|+1}+\frac{\left|\left(\mathcal{S}^{\frac{1}{5}}u\right)\left(t+\frac{1}{2}\right)\right|}{\left|\left(\mathcal{S}^{\frac{1}{5}}u\right)\left(t+\frac{1}{2}\right)\right|+1}+\frac{3\left|\left(\mathcal{T}^{\frac{3}{4}}u\right)\left(t+\frac{1}{2}\right)\right|}{\left|\left(\mathcal{T}^{\frac{3}{4}}u\right)\left(t+\frac{1}{2}\right)\right|+1}\right] \\ \Delta^{-\frac{1}{3}}u\left(-\frac{11}{2}\right) &= 2\nabla^{-\frac{2}{5}}u\left(-\frac{1}{2}\right) \\ u\left(\frac{15}{2}\right) &= 3u\left(-\frac{7}{2}\right), \end{aligned} \tag{36}$$

where

$$\left(\mathcal{S}^{\theta}u\right)\left(t+\frac{1}{2}\right)=\frac{1}{\Gamma\left(\frac{1}{5}\right)}\sum_{s=-\frac{1}{2}}^{t+\frac{1}{2}}\left(t+\frac{1}{2}-\rho(s)\right)^{\overline{-\frac{4}{5}}}\frac{e^{-(s+1)}}{\left(t+\frac{21}{2}\right)^4}u(s),$$

$$\left(\mathcal{T}^{\phi}u\right)\left(t+\frac{1}{2}\right)=\frac{1}{\Gamma\left(\frac{3}{4}\right)}\sum_{s=-\frac{5}{4}}^{t-\frac{1}{4}}\left(t+\frac{1}{2}-\sigma(s)\right)^{\underline{-\frac{1}{4}}}\frac{e^{-\left(s+\frac{3}{4}\right)}}{\left(t+\frac{201}{2}\right)^2}u\left(s+\frac{3}{4}\right).$$

Letting $\alpha=\frac{3}{2}$, $\beta=\frac{2}{3}$, $\theta=\frac{1}{5}$, $\phi=\frac{3}{4}$, $\omega=\frac{1}{3}$, $\gamma=\frac{2}{5}$, $\eta=\frac{7}{2}$, $\kappa=2$, $\lambda=3$, $T=6$, $\varphi(t,s)=\frac{e^{-(s+1)}}{(t+10)^4}$, $\psi(t,s+\phi)=\frac{e^{-\left(s+\frac{3}{4}\right)}}{(t+100)^2}$, and $F\left[t,u(t),(\mathcal{S}^{\theta}u)(t),(\mathcal{T}^{\phi}u)(t)\right]=\frac{e^{-\sin^2 t}}{(t+100)^2}\cdot\left[\frac{2|u(t)|}{1+|u(t)|}+\frac{\left|(\mathcal{S}^{\frac{1}{5}}u)(t)\right|}{\left|(\mathcal{S}^{\frac{1}{5}}u)(t)\right|+1}+3\frac{\left|(\mathcal{T}^{\frac{3}{4}}u)(t)\right|}{\left|(\mathcal{T}^{\frac{3}{4}}u)(t)\right|+1}\right]$, we find that

$$|\Lambda|=3.29500,\ \ \Theta=29.34125,\ \ \Omega_1=57.88022,\ \ \Omega_2=246.53142,\ \ \Omega_3=699.50153.$$

where $|\Lambda|,\Theta,\Omega_1,\Omega_2,\Omega_3$ are defined by (7), (15)–(18), respectively. In addition, for $(t,s)\in\mathbb{N}_{-\frac{1}{2},\frac{15}{2}}\times\mathbb{N}_{-\frac{1}{2},\frac{15}{2}}$, we find that

$$\varphi_0=\max\left\{\varphi(t-1,s)\right\}=0.0000745\quad\text{and}\quad\psi_0=\max\left\{\psi(t-1,s)\right\}=0.0000787.$$

For each $t\in\mathbb{N}_{-\frac{1}{2},\frac{15}{2}}$, we obtain

$$\begin{aligned} &\left|F\left[t,u(t),\left(\mathcal{S}^{\theta}u\right)(t),\left(\mathcal{T}^{\phi}u\right)(t)\right]-F\left[t,v(t),\left(\mathcal{S}^{\theta}v\right)(t),\left(\mathcal{T}^{\phi}v\right)(t)\right]\right| \\ &<\frac{1}{\left(-\frac{1}{2}+100\right)^2}\cdot\left[2\frac{|u|-|v|}{[1+|u|]\,[1+|v|]}+\frac{|\left(\mathcal{S}^{\theta}u\right)|-|\left(\mathcal{S}^{\theta}v\right)|}{[|\left(\mathcal{S}^{\theta}u\right)|+1]\,[|\left(\mathcal{S}^{\theta}v\right)|+1]}\right. \\ &\quad\left.+3\frac{|\left(\mathcal{T}^{\phi}u\right)|-|\left(\mathcal{T}^{\phi}v\right)|}{[|\left(\mathcal{T}^{\phi}u\right)|+1]\,[|\left(\mathcal{T}^{\phi}v\right)|+1]}\right] \\ &\le\frac{8}{39601}|u-v|+\frac{4}{39601}\left|\left(\mathcal{S}^{\theta}u\right)-\left(\mathcal{S}^{\theta}v\right)\right|+\frac{12}{39601}\left|\left(\mathcal{T}^{\phi}u\right)-\left(\mathcal{T}^{\phi}v\right)\right|. \end{aligned}$$

Therefore, (H_1) holds with $L_1=\dfrac{8}{39601}$, $L_2=\dfrac{4}{39601}$, and $L_3=\frac{12}{39601}$.

Then, we can show that (25) is true as follows

$$\chi=\left\{L_1+L_2\varphi_0\frac{(9)^{\overline{\frac{1}{5}}}}{\Gamma(\frac{6}{5})}+L_3\psi_0\frac{(8)^{\underline{\frac{3}{4}}}}{\Gamma(\frac{7}{4})}\right\}[\Omega_1+\Omega_2+\Omega_3]\approx 0.20294<1.$$

Hence, by Theorem 1, the problem (36) has a unique solution.

6. Conclusions

We estabish the conditions for the existence and unique results of the solution for a fractional delta-nabla difference equations with fractional delta-nabla sum-difference boundary value conditions by using Banach contraction principle and the conditions for result of at least one solution by using the Schauder's fixed point theorem.

Author Contributions: Conceptualization, J.R. and T.S.; methodology, J.R. and T.S.; formal Analysis, J.R. and T.S.; investigation, J.R. and T.S.; writing—original draft preparation, J.R. and T.S.; writing—review & editing, J.R. and T.S.; funding acquisition, J.R. All authors have read and agreed to the published version of the manuscript.

Funding: This research was funded by King Mongkut's University of Technology North Bangkok. Contract no. KMUTNB-61-KNOW-025.

Acknowledgments: The last author of this research was supported by Suan Dusit University.

Conflicts of Interest: The authors declare no conflicts of interest regarding the publication of this paper.

References

1. Wu, G.C.; Baleanu, D. Discrete fractional logistic map and its chaos. *Nonlinear Dyn.* **2014**, *75*, 283–287. [CrossRef]
2. Wu, G.C.; Baleanu, D. Chaos synchronization of the discrete fractional logistic map. *Signal Process.* **2014**, *102*, 96–99. [CrossRef]
3. Wu, G.C.; Baleanu, D.; Xie, H.P.; Chen, F.L. Chaos synchronization of fractional chaotic maps based on stability results. *Phys. A* **2016**, *460*, 374–383. [CrossRef]
4. Voyiadjis, G.Z.; Sumelka, W. Brain modelling in the framework of anisotropic hyperelasticity with time fractional damage evolution governed by the Caputo-Almeida fractional derivative. *J. Mech. Behav. Biomed.* **2019**, *89*, 209–216. [CrossRef]
5. Caputo, M.; Ciarletta, M.; Fabrizio, M.; Tibullo, V. Melting and solidification of pure metals by a phase-field model. *Rend Lincei-Mat. Appl.* **2017**, *28*, 463–478. [CrossRef]
6. Gómez-Aguilar, J.F. Fractional Meissner–Ochsenfeld effect in superconductors. *Phys Lett. B* **2019**, *33*, 1950316.
7. Rahimi, Z.; Rezazadeh, G.; Sumelka, W.; Yang, X.-J. A study of critical point instability of micro and nano beams under a distributed variable-pressure force in the framework of the inhomogeneous non-linear nonlocal theory. *Arch. Mech.* **2017**, *69*, 413–433.
8. Agarwal, R.P.; Baleanu, D.; Rezapour, S.; Salehi, S. The existence of solutions for some fractional finite difference equations via sum boundary conditions. *Adv. Differ. Equ.* **2014**, *2014*, 282. [CrossRef]
9. Goodrich, C.S. On discrete sequential fractional boundary value problems. *J. Math. Anal. Appl.* **2012**, *385*, 111–124. [CrossRef]
10. Goodrich, C.S. On a discrete fractional three-point boundary value problem. *J. Differ. Equ. Appl.* **2012**, *18*, 397–415. [CrossRef]
11. Lv, W. Existence of solutions for discrete fractional boundary value problems witha p-Laplacian operator. *Adv. Differ. Equ.* **2012**, *2012*, 163. [CrossRef]
12. Ferreira, R. Existence and uniqueness of solution to some discrete fractional boundary value problems of order less than one. *J. Differ. Equ. Appl.* **2013**, *19*, 712–718. [CrossRef]
13. Abdeljawad, T. On Riemann and Caputo fractional differences. *Comput. Math. Appl.* **2011**, *62*, 1602–1611. [CrossRef]
14. Atici, F.M.; Eloe, P.W. Two-point boundary value problems for finite fractional difference equations. *J. Differ. Equ. Appl.* **2011**, *17*, 445–456. [CrossRef]
15. Atici, F.M.; Eloe, P.W. A transform method in discrete fractional calculus. *Int. J. Differ. Equ.* **2007**, *2*, 165–176.
16. Atici, F.M.; Eloe, P.W. Initial value problems in discrete fractional calculus. *Proc. Am. Math. Soc.* **2009**, *137*, 981–989. [CrossRef]
17. Sitthiwirattham, T.; Tariboon, J.; Ntouyas, S.K. Existence results for fractional difference equations with three-point fractional sum boundary conditions. *Discret. Dyn. Nat. Soc.* **2013**, *2013*, 104276. [CrossRef]

18. Sitthiwirattham, T.; Tariboon, J.; Ntouyas, S.K. Boundary value problems for fractional difference equations with three-point fractional sum boundary conditions. *Adv. Differ. Equ.* **2013**, *2013*, 296. [CrossRef]
19. Sitthiwirattham, T. Existence and uniqueness of solutions of sequential nonlinear fractional difference equations with three-point fractional sum boundary conditions. *Math. Methods Appl. Sci.* **2015**, *38*, 2809–2815. [CrossRef]
20. Sitthiwirattham, T. Boundary value problem for p-Laplacian Caputo fractional difference equations with fractional sum boundary conditions. *Math. Methods Appl. Sci.* **2016**, *39*, 1522–1534. [CrossRef]
21. Reunsumrit, J.; Sitthiwirattham, T. Positive solutions of three-point fractional sum boundary value problem for Caputo fractional difference equations via an argument with a shift. *Positivity* **2014**, *20*, 861–876. [CrossRef]
22. Reunsumrit, J.; Sitthiwirattham, T. On positive solutions to fractional sum boundary value problems for nonlinear fractional difference equations. *Math. Methods Appl. Sci.* **2016**, *39*, 2737–2751. [CrossRef]
23. Kaewwisetkul, B.; Sitthiwirattham, T. On nonlocal fractional sum-difference boundary value problems for Caputo fractional functional difference equations with delay. *Adv. Differ. Equ.* **2017**, *2017*, 219. [CrossRef]
24. Chasreechai, S.; Sitthiwirattham, T. On separate fractional sum-difference boundary value problems with n-point fractional sum-difference boundary conditions via arbitrary different fractional orders. *Mathematics* **2019**, *2019*, 471. [CrossRef]
25. Setniker, A. Sequntial Differences in Nabla Fractional Calculus. Ph.D. Thesis, University of Nebraska, Lincoln, NE, USA, 2019.
26. Anastassiou, G.A. Nabla discrete calculus and nabla inequalities. *Math. Comput. Model.* **2010**, *51*, 562–571. [CrossRef]
27. Anastassiou, G.A. Foundations of nabla fractional calculus on time scales and inequalities. *Comput. Math. Appl.* **2010**, *59*, 3750–3762. [CrossRef]
28. Abdeljawad, T.; Atici, F.M. On the definitions of nabla fractional operators. *Abstr. Appl. Anal.* **2012**, *2012*, 406757. [CrossRef]
29. Abdeljawad, T. On delta and nabla Caputo fractional differences and dual identities. *Discret. Dyn. Nat. Soc.* **2013**, *2013*, 406910. [CrossRef]
30. Abdeljawad, T.; Abdall, B. Monotonicity results for delta and nabla Caputo and Riemann fractional differences via dual identities. *Filomat* **2017**, *31*, 3671–3683. [CrossRef]
31. Ahrendt, K.; Castle, L.; Holm, M.; Yochman, K. Laplace transforms for the nabla-difference operator and a fractional variation of parameters formula. *Commun. Appl. Anals.* **2012**, *16*, 317–347.
32. Atici, F.M.; Eloe, P.W. Discrete fractional calculus with the nabla operator. *Electron. Qual. Theory* **2009**. [CrossRef]
33. Atici, F.M.; Eloe, P.W. Linear systems of fractional nabla difference equations. *Rocky Mt. Math.* **2011**, *241*, 353–370. [CrossRef]
34. Baoguoa, J.; Erbe, L.; Peterson, A. Convexity for nabla and delta fractional differences. *J. Differ. Equ. Appl.* **2015**, *21*, 360–373. [CrossRef]
35. Baoguoa, J.; Erbe, L.; Peterson, A. Two monotonicity results for nabla and delta fractional differences. *Arch. Math.* **2015**, *104*, 589–597.
36. Natália, M.; Torres, D.F.M. Calculus of variations on timescales with nabla derivatives. *Nonlinear Anal.* **2018**, *71*, 763–773.
37. Goodrich, C.S.; Peterson, A.C. *Discrete Fractional Calculus*; Springer: New York, NY, USA, 2015.
38. Malinowska, A.B.; Torres, D.F.M. The delta-nabla calculus of variations. *Fasc. Math.* **2009**, *44*, 75–83.
39. Dryl, M.; Torres, D.F.M. The Delta-nabla calculus of variations for composition functionals on time scales. *Int. J. Differ. Equ.* **2003**, *8*, 27–47.
40. Dryl, M.; Torres, D.F.M. A general delta-nabla calculus of variations on time scales with application to economics. *Int. J. Dyn. Syst. Differ. Equ.* **2014**, *5*, 42–71. [CrossRef]
41. Ghorbanian, V.; Rezapour, S. A two-dimensional system of delta-nabla fractional difference inclusions. *Novi. Sad. J. Math.* **2017**, *47*, 143–163.
42. Liu, H.; Jin, Y.; Hou, C. Existence of positive solutions for discrete delta-nabla fractional boundary value problems with p-Laplacian. *Bound. Value Probl.* **2017**, *2017*, 60. [CrossRef]

43. Griffel, D.H. *Applied Functional Analysis*; Ellis Horwood Publishers: Chichester, UK, 1981.
44. Guo, D.; Lakshmikantham, V. *Nonlinear Problems in Abstract Cone*; Academic Press: Orlando, FL, USA, 1988.

mathematics

MDPI

Article

Certain Hadamard Proportional Fractional Integral Inequalities

Gauhar Rahman [1], Kottakkaran Sooppy Nisar [2,*] and Thabet Abdeljawad [3,4,5,*]

1 Department of Mathematics, Shaheed Benazir Bhutto University, Sheringal, Upper Dir 18000, Pakistan; gauhar55uom@gmail.com
2 Department of Mathematics, College of Arts and Sciences, Prince Sattam bin Abdulaziz University, Wadi Aldawaser 11991, Saudi Arabia
3 Department of Mathematics and General Sciences, Prince Sultan University, Riyadh 11586, Saudi Arabia
4 Department of Medical Research, China Medical University, Taichung 40402, Taiwan
5 Department of Computer Science and Information Engineering, Asia University, Taichung 40402, Taiwan
* Correspondence: n.sooppy@psau.edu.sa or ksnisar1@gmail.com (K.S.N.); tabdeljawad@psu.edu.sa (T.A.)

Received: 4 March 2020; Accepted: 29 March 2020; Published: 2 April 2020

Abstract: In this present paper we study the non-local Hadmard proportional integrals recently proposed by Rahman et al. (Advances in Difference Equations, (2019) 2019:454) which containing exponential functions in their kernels. Then we establish certain new weighted fractional integral inequalities involving a family of n ($n \in \mathbb{N}$) positive functions by utilizing Hadamard proportional fractional integral operator. The inequalities presented in this paper are more general than the inequalities existing in the literature.

Keywords: fractional integrals; hadamard proportional fractional integrals; fractional integral inequalities

MSC: 26A33; 26D10; 26D53

1. Introduction

The field of fractional integral inequalities play an important role in the field of differential equations and applied mathematics. These inequalities have many applications in applied sciences such as probability, statistical problems, numerical quadrature and transform theory. In the last few decades, many mathematicians have paid their valuable considerations to this field and obtained a bulk of various fractional integral inequalities and their applications. The interested readers are referred to the work of [1–5] and the references cited therein. A variety of different kinds of certain classical integral inequalities and their extensions have been investigated by considering the classical Riemann–Liouville (RL) fractional integrals, fractional derivatives and their various extensions. In [6], the authors presented integral inequalities via generalized (k,s)-fractional integrals. Dahmani and Tabharit [7] proposed weighted Grüss-type inequalities by utilizing Riemann–Liouville fractional integrals. Dahmani [8] presented several new inequalities in fractional integrals. Nisar et al. [9] investigated several inequalities for extended gamma function and confluent hypergeometric k-function. Gronwall type inequalities associated with Riemann–Liouville k and Hadamard k-fractional derivative with applications can be found in the work of Nisar et al. [10]. Rahman et al. [11] presented certain inequalities for generalized fractional integrals. Sarikaya and Budak [12] investigated Ostrowski type inequalities by employing local fractional integrals. The generalized (k,s)-fractional integrals and their applications can be found in the work of Sarikaya et al. [13]. In [14], Set et al. proposed the generalized Grüss-type inequalities by employing generalized k-fractional integrals. Set et al. [15] have introduced the generalized version of Hermite–Hadamard type inequalities via fractional integrals. Agarwal et al. [16] studied the Hermite–Hadamard type inequalities by considering the Riemann–Liouville k-fractional integrals. Fractional integral inequalities via Hadamard fractional integral, Saigo fractional integral and fractional q-integral operators are found in [17–19]. In [20], Huang et al. investigated Hermite–Hadamard type inequalities for k-fractional conformable integrals. Mubeen et al. [21] proposed the Minkowski's inequalities involving the generalized k-fractional conformable integrals. The chebyshev type inequalities involving generalized k-fractional conformable integrals can be found in the work of Qi et al. [22]. Rahman et al. investigated Chebyshev type inequalities by utilizing fractional conformable integrals [23,24]. In [25,26], Nisar et al. proposed generalized Chebyshev-type inequalities and certain Minkowski's type inequalities by employing generalized conformable integrals. Recently, Tassaddiq et al. [27] investigated certain inequalities for the weighted and the extended Chebyshev functionals by using fractional conformable integrals. Nisar et al. [28] established certain new inequalities for a class of $n(n \in \mathbb{N})$ positive continuous and decreasing functions by employing generalized conformable fractional integrals. Certain generalized fractional integral inequalities for Marichev–Saigo–Maeda (MSM) fractional integral operators were recently established by Nisar et al. [29]. Rahman et al. [30] recently investigated Grüss type inequalities for generalized k-fractional conformable integrals. In [31–33], Rahman et al. recently investigated Minkowski's inequalities, fractional proportional integral inequalities and fractional proportional inequalities for convex functions by employing fractional proportional integrals.

Moreover, the recent research focuses to the development of the theory of fractional calculus and its applications in multiple disciplines of sciences. In last three centuries, the field of fractional calculus has earned more recognition due to its wide applications in diverse domains. Recently, several kinds of various fractional integral and derivative operators have been investigated. The idea of fractional conformable derivative operators with some drawbacks was proposed by Khalil et al. [34]. The properties of the fractional conformable derivative operators was investigated by Abdeljawad [35]. Abdeljawad and Baleanu [36] proposed several monotonicity results for fractional difference operators with discrete exponential kernels. Abdeljawad and Baleanu [37] have presented fractional derivative operator with exponential kernel and their discrete version. A new fractional derivative operator with the non-local and non-singular kernel was proposed by Atangana and Baleanu [38]. Jarad et al. [39] proposed the fractional conformable integral and derivative operators. The idea of conformable derivative by employing local proportional derivatives was proposed by Anderson and Unless [40]. In [41], Caputo and Fabrizio proposed fractional derivative without a singular kernel. Later on, Losada and Nieto [42] gave certain properties of fractional derivative without a singular kernel.

In [43], Liu et al. investigated several integral inequalities. Later on, Dahmani [44] proposed certain classes of weighted fractional integral inequalities for a family of n positive increasing and decreasing functions by utilizing Riemann–Liouville fractional integrals. Houas [45] utilized Hadamard fractional integrals and established several weighted type integral inequalities. Recently, Jarad et al. [46] and Rahman et al. [47] proposed the idea of non-local fractional proportional and Hadamard proportional integrals which concerning exponential functions in their kernels. The aim of this paper is to establish weighted type inequalities by using the non-local Hdamard proportional integrals. The paper is organized as follows. In Section 2, we present some basic definitions and mathematical formulas. In Section 3, we establish certain weighted Hadamard proportional fractional integral inequalities. Section 4 containing concluding remarks.

2. Preliminaries

This section is devoted to some well-known definitions and mathematical preliminaries of fractional calculus which will be used in this article. Jarad et al. [46] presented the left and right proportional integral operators.

Definition 1. *The left and right sided proportional fractional integrals are respectively defined by*

$$\left({}_a\mathcal{J}^{\tau,\nu}\mathcal{U}\right)(\theta) = \frac{1}{\nu^{\tau}\Gamma(\tau)}\int_a^{\theta} \exp[\frac{\nu-1}{\nu}(\theta-t)](\theta-t)^{\tau-1}\mathcal{U}(t)dt, a<\theta \tag{1}$$

and

$$\left(\mathcal{J}_b^{\tau,\nu}\mathcal{U}\right)(x) = \frac{1}{\nu^{\tau}\Gamma(\tau)}\int_{\theta}^{b} \exp[\frac{\nu-1}{\nu}(t-\theta)](t-\theta)^{\tau-1}\mathcal{U}(t)dt, \theta<b. \tag{2}$$

where the proportionality index $\nu \in (0,1]$ *and* $\tau \in \mathbb{C}$ *with* $\Re(\tau) > 0$.

Remark 1. *If we consider* $\nu = 1$ *in* (1) *and* (2)*, then we get the well-known left and right Riemann–Liouville integrals which are respectively defined by*

$$\left({}_a\mathcal{J}^{\tau}\mathcal{U}\right)(x) = \frac{1}{\Gamma(\tau)}\int_a^{\theta}(\theta-t)^{\tau-1}\mathcal{U}(t)dt, a<\theta \tag{3}$$

and

$$\left(\mathcal{J}_b^{\tau}\mathcal{U}\right)(\theta) = \frac{1}{\Gamma(\tau)}\int_{\theta}^{b}(t-\theta)^{\tau-1}\mathcal{U}(t)dt, \theta<b \tag{4}$$

where $\tau \in \mathbb{C}$ *with* $\Re(\tau) > 0$.

Recently, Rahman et al. [47] proposed the following generalized Hadamard proportional fractional integrals.

Definition 2. *The left sided generalized Hadamard proportional fractional integral of order* $\tau > 0$ *and proportional index* $\nu \in (0,1]$ *is defined by*

$$\left({}_a\mathcal{H}^{\tau,\nu}\mathcal{U}\right)(\theta) = \frac{1}{\nu^{\tau}\Gamma(\tau)}\int_a^{\theta} \exp[\frac{\nu-1}{\nu}(\ln\theta-\ln t)](\ln\theta-\ln t)^{\tau-1}\frac{\mathcal{U}(t)}{t}dt, a<\theta. \tag{5}$$

Definition 3. *The right sided generalized Hadamard proportional fractional integral of order* $\tau > 0$ *and proportional index* $\nu \in (0,1]$ *is defined by*

$$\left(\mathcal{H}_b^{\tau,\nu}\mathcal{U}\right)(\theta) = \frac{1}{\nu^{\tau}\Gamma(\tau)}\int_{\theta}^{b} \exp[\frac{\nu-1}{\nu}(\ln t-\ln\theta)](\ln t-\ln\theta)^{\tau-1}\frac{\mathcal{U}(t)}{t}dt, \theta<b. \tag{6}$$

Definition 4. *The one sided generalized Hadamard proportional fractional integral of order* $\tau > 0$ *and proportional index* $\nu \in (0,1]$ *is defined by*

$$\left(\mathcal{H}_{1,\theta}^{\tau,\nu}\mathcal{U}\right)(\theta) = \frac{1}{\nu^{\tau}\Gamma(\tau)}\int_1^{\theta} \exp[\frac{\nu-1}{\nu}(\ln\theta-\ln t)](\ln\theta-\ln t)^{\tau-1}\frac{\mathcal{U}(t)}{t}dt, \theta>1, \tag{7}$$

where $\Gamma(\tau)$ *is the classical well-known gamma function.*

Remark 2. *If we consider* $\nu = 1$*, then* (5)–(7) *will led to the following well-known Hadamard fractional integrals*

$$\left({}_a\mathcal{H}^{\tau}\mathcal{U}\right)(\theta) = \frac{1}{\Gamma(\tau)}\int_a^{\theta}(\ln\theta-\ln t)^{\tau-1}\frac{\mathcal{U}(t)}{t}dt, a<\theta, \tag{8}$$

$$\left(\mathcal{H}_b^{\tau}\mathcal{U}\right)(\theta) = \frac{1}{\Gamma(\tau)}\int_{\theta}^{b}(\ln t-\ln\theta)^{\tau-1}\frac{\mathcal{U}(t)}{t}dt, \theta<b \tag{9}$$

and

$$\left(\mathcal{H}^{\tau}_{1,\theta}\mathcal{U}\right)(\theta) = \frac{1}{\Gamma(\tau)}\int_1^{\theta}(\ln\theta - \ln t)^{\tau-1}\frac{\mathcal{U}(t)}{t}dt, \theta > 1. \tag{10}$$

One can easily prove the following results

Lemma 1.

$$\left(\mathcal{H}^{\tau,\nu}_{1,\theta}\exp[\frac{\nu-1}{\nu}(\ln\theta)](\ln\theta)^{\lambda-1}\right)(\theta) = \frac{\Gamma(\lambda)}{\nu^{\tau}\Gamma(\tau+\lambda)}\exp[\frac{\nu-1}{\nu}(\ln\theta)](\ln\theta)^{\tau+\lambda-1} \tag{11}$$

and the semi group property

$$\left(\mathcal{H}^{\tau,\nu}_{1,\theta}\right)\left(\mathcal{H}^{\lambda,\nu}_{1,\theta}\right)\mathcal{U}(\theta) = \left(\mathcal{H}^{\tau+\lambda,\nu}_{1,\theta}\right)\mathcal{U}(\theta). \tag{12}$$

Remark 3. *If $\nu = 1$, then (11) will reduce to the result of [48] as defined by*

$$\left(\mathcal{H}^{\tau}_{1,\theta}(\ln\theta)^{\lambda-1}\right)(\theta) = \frac{\Gamma(\lambda)}{\Gamma(\tau+\lambda)}(\ln\theta)^{\tau+\lambda-1}. \tag{13}$$

3. Main Results

In this section, we present certain new proportional fractional integral inequalities by utilizing Hadamard proportional fractional integral. Employ the left generalized proportional fractional integral operator to establish the generalization of some classical inequalities.

Theorem 1. *Let the two functions $\mathcal{U}$ and $\mathcal{V}$ be positive and continuous on the interval $[1,\infty)$ and satisfy*

$$\left(\mathcal{V}^{\sigma}(\zeta)\mathcal{U}^{\sigma}(\rho) - \mathcal{V}^{\sigma}(\rho)\mathcal{U}^{\sigma}(\zeta)\right)\left(\mathcal{U}^{\delta-\xi}(\rho) - \mathcal{U}^{\delta-\xi}(\zeta)\right) \geq 0, \tag{14}$$

where $\rho,\zeta \in (1,\theta), \theta > 1$ and for any $\sigma > 0, \delta \geq \xi > 0$. Assume that the function $\mathcal{W} : [1,\infty) \to \mathbb{R}^+$ is positive and continuous. Then for all $\theta > 1$, the following inequality for Hadamard proportional fractional integral operator (7) holds;

$$\begin{aligned}&\mathcal{H}^{\tau,\nu}_{1,\theta}\left[\mathcal{W}(\theta)\mathcal{U}^{\delta+\sigma}(\theta)\right]\mathcal{H}^{\tau,\nu}_{1,\theta}\left[\mathcal{W}(\theta)\mathcal{V}^{\sigma}(\theta)\mathcal{U}^{\xi}(\theta)\right]\\ \geq&\mathcal{H}^{\tau,\nu}_{1,\theta}\left[\mathcal{W}(\theta)\mathcal{U}^{\sigma+\xi}(\theta)\right]\mathcal{H}^{\tau,\nu}_{1,\theta}\left[\mathcal{W}(\theta)\mathcal{V}^{\sigma}(\theta)\mathcal{U}^{\delta}(\theta)\right],\end{aligned} \tag{15}$$

where $\delta \geq \xi > 0$, $\tau,\nu,\sigma > 0$.

Proof. Consider the function

$$\mathcal{F}(\theta,\rho) = \frac{1}{\nu^{\tau}\Gamma(\tau)}e^{\frac{\nu-1}{\nu}(\ln\theta-\ln\rho)}(\ln\theta-\ln\rho)^{\tau-1}\frac{\mathcal{W}(\rho)\mathcal{U}^{\xi}(\rho)}{\rho}, \tag{16}$$

where $\tau > 0, \xi > 0$ and $\rho \in (1,\theta), \theta > 1$.

Since the functions $\mathcal{U}$ and $\mathcal{V}$ satisfy (14) for all $\rho,\zeta \in (1,\theta), \theta > 1$ and for any $\sigma > 0, \delta \geq \xi > 0$. Therefore, from (14), we have

$$\mathcal{V}^{\sigma}(\zeta)\mathcal{U}^{\sigma+\delta-\xi}(\rho) + \mathcal{V}^{\sigma}(\rho)\mathcal{U}^{\sigma+\delta-\xi}(\zeta) \geq \mathcal{V}^{\sigma}(\zeta)\mathcal{U}^{\sigma}(\rho)\mathcal{U}^{\delta-\xi}(\zeta) + \mathcal{V}^{\sigma}(\rho)\mathcal{U}^{\sigma}(\zeta)\mathcal{U}^{\delta-\xi}(\rho). \tag{17}$$

We observe that the function $\mathcal{F}(\theta,\rho)$ remains positive for all $\rho \in (1,\theta)$, $\theta > 1$. Therefore multiplying (17) by $\mathcal{F}(\theta,\rho)$ (where $\mathcal{F}(\theta,\rho)$ is defined in (16)) and integrating the resultant estimates with respect to ρ over $(1,\theta)$, we get

$$\mathcal{V}^{\sigma}(\zeta)\frac{1}{\nu^{\tau}\Gamma(\tau)}\int_{1}^{\theta}e^{\frac{\nu-1}{\nu}(\ln\theta-\ln\rho)}(\ln\theta-\ln\rho)^{\tau-1}\frac{\mathcal{W}(\rho)\mathcal{U}^{\xi}(\rho)}{\rho}\mathcal{U}^{\sigma+\delta-\xi}(\rho)d\rho$$
$$+\mathcal{U}^{\sigma+\delta-\xi}(\zeta)\frac{1}{\nu^{\tau}\Gamma(\tau)}\int_{1}^{\theta}e^{\frac{\nu-1}{\nu}(\ln\theta-\ln\rho)}(\ln\theta-\ln\rho)^{\tau-1}\frac{\mathcal{W}(\rho)\mathcal{U}^{\xi}(\rho)}{\rho}\mathcal{V}^{\sigma}(\rho)d\rho$$
$$\geq\mathcal{V}^{\sigma}(\zeta)\mathcal{U}^{\delta-\xi}(\zeta)\frac{1}{\nu^{\tau}\Gamma(\tau)}\int_{1}^{\theta}e^{\frac{\nu-1}{\nu}(\ln\theta-\ln\rho)}(\ln\theta-\ln\rho)^{\tau-1}\frac{\mathcal{W}(\rho)\mathcal{U}^{\xi}(\rho)}{\rho}\mathcal{U}^{\sigma}(\rho)d\rho$$
$$+\mathcal{U}^{\sigma}(\zeta)\frac{1}{\nu^{\tau}\Gamma(\tau)}\int_{1}^{\theta}e^{\frac{\nu-1}{\nu}(\ln\theta-\ln\rho)}(\ln\theta-\ln\rho)^{\tau-1}\frac{\mathcal{W}(\rho)\mathcal{U}^{\xi}(\rho)}{\rho}\mathcal{V}^{\sigma}(\rho)\mathcal{U}^{\delta-\xi}(\rho)d\rho,$$

which in view of (7) becomes,

$$\begin{aligned}&\mathcal{V}^{\sigma}(\zeta)\mathcal{H}_{1,\theta}^{\tau,\nu}\left[\mathcal{W}(\theta)\mathcal{U}^{\sigma+\delta}(\theta)\right]+\mathcal{U}^{\sigma+\delta-\xi}(\zeta)\mathcal{H}_{1,\theta}^{\tau,\nu}\left[\mathcal{W}(\theta)\mathcal{V}^{\sigma}(\theta)\mathcal{U}^{\xi}(\theta)\right]\\&\geq\mathcal{V}^{\sigma}(\zeta)\mathcal{U}^{\delta-\xi}(\zeta)\mathcal{H}_{1,\theta}^{\tau,\nu}\left[\mathcal{W}(\theta)\mathcal{U}^{\sigma+\xi}(\rho)\right]+\mathcal{U}^{\sigma}(\zeta)\mathcal{H}_{1,\theta}^{\tau,\nu}\left[\mathcal{W}(\theta)\mathcal{V}^{\sigma}(\theta)\mathcal{U}^{\delta}(\theta)\right].\end{aligned}\tag{18}$$

Now, multiplying (18) by $\mathcal{F}(\theta,\zeta)$ (where $\mathcal{F}(\theta,\zeta)$ can be obtain from (16) by replacing ρ by ζ) and integrating the resultant estimates with respect to ζ over $(1,\theta)$ and then by using (7), we obtain

$$\mathcal{H}_{1,\theta}^{\tau,\nu}\left[\mathcal{W}(\theta)\mathcal{V}^{\sigma}(\theta)\mathcal{U}^{\xi}(\theta)\right]\mathcal{H}_{1,\theta}^{\tau,\nu}\left[\mathcal{W}(\theta)\mathcal{U}^{\sigma+\delta}(\theta)\right]+\mathcal{H}_{1,\theta}^{\tau,\nu}\left[\mathcal{W}(\theta)\mathcal{U}^{\sigma+\delta}(\theta)\right]\mathcal{H}_{1,\theta}^{\tau,\nu}\left[\mathcal{W}(\theta)\mathcal{V}^{\sigma}(\theta)\mathcal{U}^{\xi}(\theta)\right]$$
$$\geq\mathcal{H}_{1,\theta}^{\tau,\nu}\left[\mathcal{W}(\theta)\mathcal{V}^{\sigma}(\theta)\mathcal{U}^{\delta}(\theta)\right]\mathcal{H}_{1,\theta}^{\tau,\nu}\left[\mathcal{W}(\theta)\mathcal{U}^{\sigma+\xi}(\rho)\right]+\mathcal{H}_{1,\theta}^{\tau,\nu}\left[\mathcal{W}(\theta)\mathcal{U}^{\sigma+\xi}(\theta)\right]\mathcal{H}_{1,\theta}^{\tau,\nu}\left[\mathcal{W}(\theta)\mathcal{V}^{\sigma}(\theta)\mathcal{U}^{\delta}(\theta)\right],$$

which gives the desired assertion (15). □

Remark 4. *If the inequality (14) reverses, then the inequality (15) will also reverse.*

Theorem 2. *Let the two functions $\mathcal{U}$ and $\mathcal{V}$ be positive and continuous on the interval $[1,\infty)$ and satisfy (14). Assume that the function $\mathcal{W}:[1,\infty)\to\mathbb{R}^{+}$ is positive and continuous. Then for all $\theta>1$, the following inequality for Hadamard proportional fractional integral operator (7) holds;*

$$\begin{aligned}&\mathcal{H}_{1,\theta}^{\tau,\nu}\left[\mathcal{W}(\theta)\mathcal{U}^{\sigma+\delta}(\theta)\right]\mathcal{H}_{1,\theta}^{\lambda,\nu}\left[\mathcal{W}(\theta)\mathcal{V}^{\sigma}(\theta)\mathcal{U}^{\xi}(\theta)\right]+\mathcal{H}_{1,\theta}^{\tau,\nu}\left[\mathcal{W}(\theta)\mathcal{V}^{\sigma}(\theta)\mathcal{U}^{\xi}(\theta)\right]\mathcal{H}_{1,\theta}^{\lambda,\nu}\left[\mathcal{W}(\theta)\mathcal{U}^{\sigma+\delta}(\theta)\right]\\\geq\ &\mathcal{H}_{1,\theta}^{\tau,\nu}\left[\mathcal{W}(\theta)\mathcal{U}^{\sigma+\xi}(\rho)\right]\mathcal{H}_{1,\theta}^{\lambda,\nu}\left[\mathcal{W}(\theta)\mathcal{V}^{\sigma}(\theta)\mathcal{U}^{\delta}(\theta)\right]+\mathcal{H}_{1,\theta}^{\tau,\nu}\left[\mathcal{W}(\theta)\mathcal{V}^{\sigma}(\theta)\mathcal{U}^{\delta}(\theta)\right]\mathcal{H}_{1,\theta}^{\lambda,\nu}\left[\mathcal{W}(\theta)\mathcal{U}^{\sigma+\xi}(\theta)\right],\end{aligned}\tag{19}$$

where $\delta\geq\xi>0$, $\tau,\lambda,\nu,\sigma>0$.

Proof. Taking product on both sides of (18) by $\mathcal{G}(\theta,\zeta)=\frac{1}{\nu^{\lambda}\Gamma(\lambda)}e^{\frac{\nu-1}{\nu}(\ln\theta-\ln\zeta)}(\ln\theta-\ln\zeta)^{\lambda-1}\frac{\mathcal{W}(\zeta)\mathcal{U}^{\xi}(\zeta)}{\zeta}$, where $\lambda>0,\xi>0$ and $\zeta\in(1,\theta),\theta>1$. and integrating the resultant estimates with respect to ζ over $(1,\theta)$, we have

$$\mathcal{H}_{1,\theta}^{\tau,\nu}\left[\mathcal{W}(\theta)\mathcal{U}^{\sigma+\delta}(\theta)\right]\frac{1}{\nu^{\lambda}\Gamma(\lambda)}\int_{1}^{\theta}e^{\frac{\nu-1}{\nu}(\ln\theta-\ln\zeta)}(\ln\theta-\ln\zeta)^{\lambda-1}\frac{\mathcal{W}(\zeta)\mathcal{U}^{\xi}(\zeta)}{\zeta}\mathcal{V}^{\sigma}(\zeta)d\zeta$$
$$+\mathcal{H}_{1,\theta}^{\tau,\nu}\left[\mathcal{W}(\theta)\mathcal{V}^{\sigma}(\theta)\mathcal{U}^{\xi}(\theta)\right]\frac{1}{\nu^{\lambda}\Gamma(\lambda)}\int_{1}^{\theta}e^{\frac{\nu-1}{\nu}(\ln\theta-\ln\zeta)}(\ln\theta-\ln\zeta)^{\lambda-1}\frac{\mathcal{W}(\zeta)\mathcal{U}^{\xi}(\zeta)}{\zeta}\mathcal{U}^{\sigma+\delta-\xi}(\zeta)d\zeta$$
$$\geq\mathcal{H}_{1,\theta}^{\tau,\nu}\left[\mathcal{W}(\theta)\mathcal{U}^{\sigma+\xi}(\rho)\right]\frac{1}{\nu^{\lambda}\Gamma(\lambda)}\int_{1}^{\theta}e^{\frac{\nu-1}{\nu}(\ln\theta-\ln\zeta)}(\ln\theta-\ln\zeta)^{\lambda-1}\frac{\mathcal{W}(\zeta)\mathcal{U}^{\xi}(\zeta)}{\zeta}\mathcal{V}^{\sigma}(\zeta)\mathcal{U}^{\delta-\xi}(\zeta)d\zeta$$
$$+\mathcal{H}_{1,\theta}^{\tau,\nu}\left[\mathcal{W}(\theta)\mathcal{V}^{\sigma}(\theta)\mathcal{U}^{\delta}(\theta)\right]\frac{1}{\nu^{\lambda}\Gamma(\lambda)}e^{\frac{\nu-1}{\nu}(\ln\theta-\ln\zeta)}(\ln\theta-\ln\zeta)^{\lambda-1}\frac{\mathcal{W}(\zeta)\mathcal{U}^{\xi}(\zeta)}{\zeta}\mathcal{U}^{\sigma}(\zeta)d\zeta,$$

which in view of (7) yields,

$$\begin{aligned}&\mathcal{H}_{1,\theta}^{\tau,\nu}\left[\mathcal{W}(\theta)\mathcal{U}^{\sigma+\delta}(\theta)\right]\mathcal{H}_{1,\theta}^{\lambda,\nu}\left[\mathcal{W}(\theta)\mathcal{V}^{\sigma}(\theta)\mathcal{U}^{\xi}(\theta)\right]+\mathcal{H}_{1,\theta}^{\tau,\nu}\left[\mathcal{W}(\theta)\mathcal{V}^{\sigma}(\theta)\mathcal{U}^{\xi}(\theta)\right]\mathcal{H}_{1,\theta}^{\lambda,\nu}\left[\mathcal{W}(\theta)\mathcal{U}^{\sigma+\delta}(\theta)\right]\\ \geq&\mathcal{H}_{1,\theta}^{\tau,\nu}\left[\mathcal{W}(\theta)\mathcal{U}^{\sigma+\xi}(\rho)\right]\mathcal{H}_{1,\theta}^{\lambda,\nu}\left[\mathcal{W}(\theta)\mathcal{V}^{\sigma}(\theta)\mathcal{U}^{\delta}(\theta)\right]+\mathcal{H}_{1,\theta}^{\tau,\nu}\left[\mathcal{W}(\theta)\mathcal{V}^{\sigma}(\theta)\mathcal{U}^{\delta}(\theta)\right]\mathcal{H}_{1,\theta}^{\lambda,\nu}\left[\mathcal{W}(\theta)\mathcal{U}^{\sigma+\xi}(\theta)\right],\end{aligned}$$

which completes the proof of (19). □

Remark 5. *Applying Theorem 2 for $\tau = \lambda$, we get Theorem 1.*

Theorem 3. *Let the two functions $\mathcal{U}$ and $\mathcal{V}$ be positive and continuous on the interval $[1,\infty)$ such that the function $\mathcal{U}$ is decreasing and the function $\mathcal{V}$ is increasing on $[1,\infty)$ and assume that the function $\mathcal{W} : [1,\infty) \to \mathbb{R}^+$ is positive and continuous. Then for all $\theta > 1$, the following inequality for Hadamard proportional fractional integral operator (7) holds;*

$$\begin{aligned}&\mathcal{H}_{1,\theta}^{\tau,\nu}\left[\mathcal{W}(\theta)\mathcal{U}^{\delta}(\theta)\right]\mathcal{H}_{1,\theta}^{\tau,\nu}\left[\mathcal{W}(\theta)\mathcal{V}^{\sigma}(\theta)\mathcal{U}^{\xi}(\theta)\right]\\ \geq&\mathcal{H}_{1,\theta}^{\tau,\nu}\left[\mathcal{W}(\theta)\mathcal{U}^{\xi}(\theta)\right]\mathcal{H}_{1,\theta}^{\tau,\nu}\left[\mathcal{W}(\theta)\mathcal{V}^{\sigma}(\theta)\mathcal{U}^{\delta}(\theta)\right],\end{aligned}\tag{20}$$

where $\delta \geq \xi > 0$, $\tau, \nu, \sigma > 0$.

Proof. Since the two functions $\mathcal{U}$ and $\mathcal{V}$ are positive and continuous on $[1,\infty)$ such that $\mathcal{U}$ is decreasing and $\mathcal{V}$ is increasing on $[1,\infty)$, then for all $\sigma > 0$, $\delta \geq \xi > 0$, $\rho, \zeta \in (1,\theta)$, $\theta > 1$, we have

$$(\mathcal{V}^{\sigma}(\zeta)-\mathcal{V}^{\sigma}(\rho))\left(\mathcal{U}^{\delta-\xi}(\rho)-\mathcal{U}^{\delta-\xi}(\zeta)\right)\geq 0,$$

which follows that

$$\mathcal{V}^{\sigma}(\zeta)\mathcal{U}^{\delta-\xi}(\rho)+\mathcal{V}^{\sigma}(\rho)\mathcal{U}^{\delta-\xi}(\zeta)\geq\mathcal{V}^{\sigma}(\zeta)\mathcal{U}^{\delta-\xi}(\zeta)+\mathcal{V}^{\sigma}(\rho)\mathcal{U}^{\delta-\xi}(\rho).\tag{21}$$

Multiplying (21) by $\mathcal{F}(\theta,\rho)$ (where $\mathcal{F}(\theta,\rho)$ is defined in (16)) and integrating the resultant estimates with respect to ρ over $(1,\theta)$, we get

$$\begin{aligned}&\mathcal{V}^{\sigma}(\zeta)\frac{1}{\nu^{\tau}\Gamma(\tau)}\int_1^{\theta}e^{\frac{\nu-1}{\nu}(\ln\theta-\ln\rho)}(\ln\theta-\ln\rho)^{\tau-1}\frac{\mathcal{W}(\rho)\mathcal{U}^{\xi}(\rho)}{\rho}\mathcal{U}^{\delta-\xi}(\rho)d\rho\\ &+\mathcal{U}^{\delta-\xi}(\zeta)\frac{1}{\nu^{\tau}\Gamma(\tau)}\int_1^{\theta}e^{\frac{\nu-1}{\nu}(\ln\theta-\ln\rho)}(\ln\theta-\ln\rho)^{\tau-1}\frac{\mathcal{W}(\rho)\mathcal{U}^{\xi}(\rho)}{\rho}\mathcal{V}^{\sigma}(\rho)d\rho\\ \geq&\mathcal{V}^{\sigma}(\zeta)\mathcal{U}^{\delta-\xi}(\zeta)\frac{1}{\nu^{\tau}\Gamma(\tau)}\int_1^{\theta}e^{\frac{\nu-1}{\nu}(\ln\theta-\ln\rho)}(\ln\theta-\ln\rho)^{\tau-1}\frac{\mathcal{W}(\rho)\mathcal{U}^{\xi}(\rho)}{\rho}d\rho\\ &+\frac{1}{\nu^{\tau}\Gamma(\tau)}\int_1^{\theta}e^{\frac{\nu-1}{\nu}(\ln\theta-\ln\rho)}(\ln\theta-\ln\rho)^{\tau-1}\frac{\mathcal{W}(\rho)\mathcal{U}^{\xi}(\rho)}{\rho}\mathcal{V}^{\sigma}(\rho)\mathcal{U}^{\delta-\xi}(\rho)d\rho,\end{aligned}$$

which in view of (7) becomes,

$$\begin{aligned}&\mathcal{V}^{\sigma}(\zeta)\mathcal{H}_{1,\theta}^{\tau,\nu}\left[\mathcal{W}(\theta)\mathcal{U}^{\delta}(\theta)\right]+\mathcal{U}^{\delta-\xi}(\zeta)\mathcal{H}_{1,\theta}^{\tau,\nu}\left[\mathcal{W}(\theta)\mathcal{V}^{\sigma}(\theta)\mathcal{U}^{\xi}(\theta)\right]\\ \geq&\mathcal{V}^{\sigma}(\zeta)\mathcal{U}^{\delta-\xi}(\zeta)\mathcal{H}_{1,\theta}^{\tau,\nu}\left[\mathcal{W}(\theta)\mathcal{U}^{\xi}(\theta)\right]+\mathcal{H}_{1,\theta}^{\tau,\nu}\left[\mathcal{W}(\theta)\mathcal{V}^{\sigma}(\theta)\mathcal{U}^{\delta}(\theta)\right].\end{aligned}\tag{22}$$

Again, multiplying (22) by $\mathcal{F}(\theta,\zeta)$ (where $\mathcal{F}(\theta,\zeta)$ can be obtained from (16)) and integrating the resultant estimates with respect to ζ over $(1,\theta)$ and then employing (7), we get

$$\mathcal{H}_{1,\theta}^{\tau,\nu}\left[\mathcal{W}(\theta)\mathcal{V}^{\sigma}(\theta)\mathcal{U}^{\xi}(\theta)\right]\mathcal{H}_{1,\theta}^{\tau,\nu}\left[\mathcal{W}(\theta)\mathcal{U}^{\delta}(\theta)\right]+\mathcal{H}_{1,\theta}^{\tau,\nu}\left[\mathcal{W}(\theta)\mathcal{U}^{\delta}(\theta)\right]\mathcal{H}_{1,\theta}^{\tau,\nu}\left[\mathcal{W}(\theta)\mathcal{V}^{\sigma}(\theta)\mathcal{U}^{\xi}(\theta)\right]$$
$$\geq\mathcal{H}_{1,\theta}^{\tau,\nu}\left[\mathcal{W}(\theta)\mathcal{V}^{\sigma}(\theta)\mathcal{U}^{\delta}(\theta)\right]\mathcal{H}_{1,\theta}^{\tau,\nu}\left[\mathcal{W}(\theta)\mathcal{U}^{\xi}(\theta)\right]+\mathcal{H}_{1,\theta}^{\tau,\nu}\left[\mathcal{W}(\theta)\mathcal{U}^{\xi}(\theta)\right]\mathcal{H}_{1,\theta}^{\tau,\nu}\left[\mathcal{W}(\theta)\mathcal{V}^{\sigma}(\theta)\mathcal{U}^{\delta}(\theta)\right],$$

which gives the desired assertion (20). □

Theorem 4. *Let the two functions $\mathcal{U}$ and $\mathcal{V}$ be positive and continuous on the interval $[1,\infty)$ such that the function $\mathcal{U}$ is decreasing and the function $\mathcal{V}$ is increasing on $[1,\infty)$ and assume that the function $\mathcal{W}:[1,\infty)\to\mathbb{R}^{+}$ is positive and continuous. Then for all $\theta>1$, the following inequality for Hadamard proportional fractional integral operator (7) holds;*

$$\mathcal{H}_{1,\theta}^{\tau,\nu}\left[\mathcal{W}(\theta)\mathcal{U}^{\delta}(\theta)\right]\mathcal{H}_{1,\theta}^{\lambda,\nu}\left[\mathcal{W}(\theta)\mathcal{V}^{\sigma}(\theta)\mathcal{U}^{\xi}(\theta)\right]+\mathcal{H}_{1,\theta}^{\lambda,\nu}\left[\mathcal{W}(\theta)\mathcal{U}^{\delta}(\theta)\right]\mathcal{H}_{1,\theta}^{\tau,\nu}\left[\mathcal{W}(\theta)\mathcal{V}^{\sigma}(\theta)\mathcal{U}^{\xi}(\theta)\right]$$
$$\geq\mathcal{H}_{1,\theta}^{\tau,\nu}\left[\mathcal{W}(\theta)\mathcal{U}^{\xi}(\theta)\right]\mathcal{H}_{1,\theta}^{\lambda,\nu}\left[\mathcal{W}(\theta)\mathcal{V}^{\sigma}(\theta)\mathcal{U}^{\delta}(\theta)\right]+\mathcal{H}_{1,\theta}^{\lambda,\nu}\left[\mathcal{W}(\theta)\mathcal{U}^{\xi}(\theta)\right]\mathcal{H}_{1,\theta}^{\tau,\nu}\left[\mathcal{W}(\theta)\mathcal{V}^{\sigma}(\theta)\mathcal{U}^{\delta}(\theta)\right], \quad (23)$$

where $\delta\geq\xi>0$, $\tau,\lambda,\nu,\sigma>0$.

Proof. Taking product on both sides of (22) by $\mathcal{G}(\theta,\zeta)=\frac{1}{\nu^{\lambda}\Gamma(\lambda)}e^{\frac{\nu-1}{\nu}(\ln\theta-\ln\zeta)}(\ln\theta-\ln\zeta)^{\lambda-1}\frac{\mathcal{W}(\zeta)\mathcal{U}^{\xi}(\zeta)}{\zeta}$, where $\lambda>0,\xi>0$ and $\zeta\in(1,\theta),\theta>1$. and integrating the resultant estimates with respect to ζ over $(1,\theta)$, we have

$$\mathcal{H}_{1,\theta}^{\tau,\nu}\left[\mathcal{W}(\theta)\mathcal{U}^{\delta}(\theta)\right]\frac{1}{\nu^{\lambda}\Gamma(\lambda)}\int_{1}^{\theta}e^{\frac{\nu-1}{\nu}(\ln\theta-\ln\zeta)}(\ln\theta-\ln\zeta)^{\lambda-1}\frac{\mathcal{W}(\zeta)\mathcal{U}^{\xi}(\zeta)}{\zeta}\mathcal{V}^{\sigma}(\zeta)d\zeta$$
$$+\mathcal{H}_{1,\theta}^{\tau,\nu}\left[\mathcal{W}(\theta)\mathcal{V}^{\sigma}(\theta)\mathcal{U}^{\xi}(\theta)\right]\frac{1}{\nu^{\lambda}\Gamma(\lambda)}\int_{1}^{\theta}e^{\frac{\nu-1}{\nu}(\ln\theta-\ln\zeta)}(\ln\theta-\ln\zeta)^{\lambda-1}\frac{\mathcal{W}(\zeta)\mathcal{U}^{\xi}(\zeta)}{\zeta}\mathcal{U}^{\delta-\xi}(\zeta)d\zeta$$
$$\geq\mathcal{H}_{1,\theta}^{\tau,\nu}\left[\mathcal{W}(\theta)\mathcal{U}^{\xi}(\theta)\right]\frac{1}{\nu^{\lambda}\Gamma(\lambda)}\int_{1}^{\theta}e^{\frac{\nu-1}{\nu}(\ln\theta-\ln\zeta)}(\ln\theta-\ln\zeta)^{\lambda-1}\frac{\mathcal{W}(\zeta)\mathcal{U}^{\xi}(\zeta)}{\zeta}\mathcal{V}^{\sigma}(\zeta)\mathcal{U}^{\delta-\xi}(\zeta)d\zeta$$
$$+\mathcal{H}_{1,\theta}^{\tau,\nu}\left[\mathcal{W}(\theta)\mathcal{V}^{\sigma}(\theta)\mathcal{U}^{\delta}(\theta)\right]\frac{1}{\nu^{\lambda}\Gamma(\lambda)}\int_{1}^{\theta}e^{\frac{\nu-1}{\nu}(\ln\theta-\ln\zeta)}(\ln\theta-\ln\zeta)^{\lambda-1}\frac{\mathcal{W}(\zeta)\mathcal{U}^{\xi}(\zeta)}{\zeta}d\zeta.$$

Consequently, in view of (7) it can be written as,

$$\mathcal{H}_{1,\theta}^{\tau,\nu}\left[\mathcal{W}(\theta)\mathcal{U}^{\delta}(\theta)\right]\mathcal{H}_{1,\theta}^{\lambda,\nu}\left[\mathcal{W}(\theta)\mathcal{V}^{\sigma}(\theta)\mathcal{U}^{\xi}(\theta)\right]+\mathcal{H}_{1,\theta}^{\tau,\nu}\left[\mathcal{W}(\theta)\mathcal{V}^{\sigma}(\theta)\mathcal{U}^{\xi}(\theta)\right]\mathcal{H}_{1,\theta}^{\lambda,\nu}\left[\mathcal{W}(\theta)\mathcal{U}^{\delta}(\theta)\right]$$
$$\geq\mathcal{H}_{1,\theta}^{\tau,\nu}\left[\mathcal{W}(\theta)\mathcal{U}^{\xi}(\theta)\right]\mathcal{H}_{1,\theta}^{\lambda,\nu}\left[\mathcal{W}(\theta)\mathcal{V}^{\sigma}(\theta)\mathcal{U}^{\delta}(\theta)\right]+\mathcal{H}_{1,\theta}^{\tau,\nu}\left[\mathcal{W}(\theta)\mathcal{V}^{\sigma}(\theta)\mathcal{U}^{\delta}(\theta)\right]\mathcal{H}_{1,\theta}^{\lambda,\nu}\left[\mathcal{W}(\theta)\mathcal{U}^{\xi}(\theta)\right],$$

which gives the desired inequality (23). □

Remark 6. *Applying Theorem 4 for $\tau=\lambda$, we get Theorem 3.*

Next, we present some fractional proportional inequalities for a family of n positive functions defined on $[1,\infty)$ by utilizing Hadamard proportional fractional integral (7).

Theorem 5. *Let the functions $\mathcal{U}_j$, $(j=1,2,\cdots,n)$ and $\mathcal{V}$ be positive and continuous on the interval $[1,\infty)$. Suppose that for any fixed $k=1,2,\cdots,n$,*

$$(\mathcal{V}^{\sigma}(\zeta)\mathcal{U}_k^{\sigma}(\rho)-\mathcal{V}^{\sigma}(\rho)\mathcal{U}_k^{\sigma}(\zeta))\left(\mathcal{U}_k^{\delta-\xi_k}(\rho)-\mathcal{U}_k^{\delta-\xi_k}(\zeta)\right)\geq 0, \quad (24)$$

where $\rho,\zeta\in(1,\theta)$, $\theta>1$, $\sigma>0$, $\delta>\xi_k>0$. Assume that the function $\mathcal{W}:[1,\infty)\to\mathbb{R}^{+}$ is positive and continuous. Then for all $\theta>1$, the following inequality for Hadamard proportional fractional integral operator (7) holds;

$$\mathcal{H}_{1,\theta}^{\tau,\nu}\left[\mathcal{W}(\theta)\mathcal{U}_k^{\delta+\sigma}(\theta)\prod_{j\neq k}^{n}\mathcal{U}_j^{\xi_j}(\theta)\right]\mathcal{H}_{1,\theta}^{\tau,\nu}\left[\mathcal{W}(\theta)\mathcal{V}^{\sigma}(\theta)\prod_{j=1}^{n}\mathcal{U}_j^{\xi_j}(\theta)\right]$$
$$\geq\mathcal{H}_{1,\theta}^{\tau,\nu}\left[\mathcal{W}(\theta)\mathcal{U}_k^{\sigma}(\theta)\prod_{j=1}^{n}\mathcal{U}_j^{\xi_j}(\theta)\right]\mathcal{H}_{1,\theta}^{\tau,\nu}\left[\mathcal{W}(\theta)\mathcal{V}^{\sigma}(\theta)\mathcal{U}_k^{\delta}(\theta)\prod_{j\neq k}^{n}\mathcal{U}_j^{\xi_j}(\theta)\right], \tag{25}$$

where $\delta \geq \xi_k > 0$, $\tau,\nu,\sigma > 0$ and $k = 1,2,\cdots,n$.

Proof. Consider the function

$$\mathcal{F}_1(\theta,\rho) = \frac{1}{\nu^{\tau}\Gamma(\tau)}e^{\frac{\nu-1}{\nu}(\ln\theta-\ln\rho)}(\ln\theta-\ln\rho)^{\tau-1}\frac{\mathcal{W}(\rho)\prod_{j=1}^{n}\mathcal{U}_j^{\xi_j}(\rho)}{\rho}, \tag{26}$$
$$\xi_j > 0, j = 1,2,\cdots,n,\ \rho\in(1,\theta),\theta > 1.$$

Since the functions $\mathcal{U}_j$, $(j = 1,2,\cdots,n)$ and $\mathcal{V}$ satisfy (24) for any fixed $k = 1,2,\cdots,n$. Therefore, we can write

$$\mathcal{V}^{\sigma}(\zeta)\mathcal{U}_k^{\delta+\sigma-\xi_k}(\rho)+\mathcal{V}^{\sigma}(\rho)\mathcal{U}_k^{\delta+\sigma-\xi_k}(\zeta)\geq\mathcal{V}^{\sigma}(\zeta)\mathcal{U}_k^{\sigma}(\rho)\mathcal{U}_k^{\delta-\xi_k}(\zeta)+\mathcal{V}^{\sigma}(\rho)\mathcal{U}_k^{\sigma}(\zeta)\mathcal{U}_k^{\delta-\xi_k}(\rho). \tag{27}$$

Multiplying (27) by $\mathcal{F}_1(\theta,\rho)$ (where $\mathcal{F}_1(\theta,\rho)$ is defined in (26)) and integrating the resultant estimates with respect to ρ over $(1,\theta),\theta > 1$, we have

$$\mathcal{V}^{\sigma}(\zeta)\frac{1}{\nu^{\tau}\Gamma(\tau)}\int_1^{\infty}e^{\frac{\nu-1}{\nu}(\ln\theta-\ln\rho)}(\ln\theta-\ln\rho)^{\tau-1}\frac{\mathcal{W}(\rho)\prod_{j=1}^{n}\mathcal{U}_j^{\xi_j}(\rho)}{\rho}\mathcal{U}_k^{\delta+\sigma-\xi_k}(\rho)d\rho$$
$$+\mathcal{U}_k^{\delta+\sigma-\xi_k}(\zeta)\frac{1}{\nu^{\tau}\Gamma(\tau)}\int_1^{\infty}e^{\frac{\nu-1}{\nu}(\ln\theta-\ln\rho)}(\ln\theta-\ln\rho)^{\tau-1}\frac{\mathcal{W}(\rho)\prod_{j=1}^{n}\mathcal{U}_j^{\xi_j}(\rho)}{\rho}\mathcal{V}^{\sigma}(\rho)d\rho$$
$$\geq\mathcal{V}^{\sigma}(\zeta)\mathcal{U}_k^{\delta-\xi_k}(\zeta)\frac{1}{\nu^{\tau}\Gamma(\tau)}\int_1^{\infty}e^{\frac{\nu-1}{\nu}(\ln\theta-\ln\rho)}(\ln\theta-\ln\rho)^{\tau-1}\frac{\mathcal{W}(\rho)\prod_{j=1}^{n}\mathcal{U}_j^{\xi_j}(\rho)}{\rho}\mathcal{U}_k^{\sigma}(\rho)d\rho$$
$$+\mathcal{U}_k^{\sigma}(\zeta)\frac{1}{\nu^{\tau}\Gamma(\tau)}\int_1^{\infty}e^{\frac{\nu-1}{\nu}(\ln\theta-\ln\rho)}(\ln\theta-\ln\rho)^{\tau-1}\frac{\mathcal{W}(\rho)\prod_{j=1}^{n}\mathcal{U}_j^{\xi_j}(\rho)}{\rho}\mathcal{V}^{\sigma}(\rho)\mathcal{U}_k^{\delta-\xi_k}(\rho)d\rho,$$

which with aid of (7) gives

$$\mathcal{V}^{\sigma}(\zeta)\mathcal{H}_{1,\theta}^{\tau,\nu}\left[\mathcal{W}(\theta)\mathcal{U}_k^{\delta+\sigma}(\theta)\prod_{j\neq k}^{n}\mathcal{U}_j^{\xi_j}(\theta)\right]+\mathcal{U}_k^{\delta+\sigma-\xi_k}(\zeta)\mathcal{H}_{1,\theta}^{\tau,\nu}\left[\mathcal{W}(\theta)\mathcal{V}^{\sigma}(\theta)\prod_{j=1}^{n}\mathcal{U}_j^{\xi_j}(\theta)\right] \tag{28}$$
$$\geq\mathcal{V}^{\sigma}(\zeta)\mathcal{U}_k^{\delta-\xi_k}(\zeta)\mathcal{H}_{1,\theta}^{\tau,\nu}\left[\mathcal{W}(\theta)\mathcal{U}_k^{\sigma}(\theta)\prod_{j=1}^{n}\mathcal{U}_j^{\xi_j}(\theta)\right]+\mathcal{U}_k^{\sigma}(\zeta)\mathcal{H}_{1,\theta}^{\tau,\nu}\left[\mathcal{W}(\theta)\mathcal{V}^{\sigma}(\theta)\mathcal{U}_k^{\delta+\sigma}(\theta)\prod_{j\neq k}^{n}\mathcal{U}_j^{\xi_j}(\theta)\right]. \tag{29}$$

Now, multiplying (28) by $\mathcal{F}_1(\theta,\zeta)$ (where $\mathcal{F}_1(\theta,\zeta)$ can be obtained from (26) by replacing ρ by ζ) and integrating the resultant estimates with respect to ζ over $(1,\theta),\theta > 1$ and then by applying (7), we obtain

$$\mathcal{H}_{1,\theta}^{\tau,\nu}\left[\mathcal{W}(\theta)\mathcal{U}_k^{\delta+\sigma}(\theta)\prod_{j\neq k}^{n}\mathcal{U}_j^{\xi_j}(\theta)\right]\mathcal{H}_{1,\theta}^{\tau,\nu}\left[\mathcal{W}(\theta)\mathcal{V}^{\sigma}(\theta)\prod_{j=1}^{n}\mathcal{U}_j^{\xi_j}(\theta)\right]$$
$$+\mathcal{H}_{1,\theta}^{\tau,\nu}\left[\mathcal{W}(\theta)\mathcal{V}^{\sigma}(\theta)\prod_{j=1}^{n}\mathcal{U}_j^{\xi_j}(\theta)\right]\mathcal{H}_{1,\theta}^{\tau,\nu}\left[\mathcal{W}(\theta)\mathcal{U}_k^{\delta+\sigma}(\theta)\prod_{j\neq k}^{n}\mathcal{U}_j^{\xi_j}(\theta)\right]$$
$$\geq\mathcal{H}_{1,\theta}^{\tau,\nu}\left[\mathcal{W}(\theta)\mathcal{U}_k^{\sigma}(\theta)\prod_{j=1}^{n}\mathcal{U}_j^{\xi_j}(\theta)\right]\mathcal{H}_{1,\theta}^{\tau,\nu}\left[\mathcal{W}(\theta)\mathcal{V}^{\sigma}(\theta)\mathcal{U}_k^{\delta}(\theta)\prod_{j\neq k}^{n}\mathcal{U}_j^{\xi_j}(\theta)\right]$$
$$+\mathcal{H}_{1,\theta}^{\tau,\nu}\left[\mathcal{W}(\theta)\mathcal{V}^{\sigma}(\theta)\mathcal{U}_k^{\delta}(\theta)\prod_{j\neq k}^{n}\mathcal{U}_j^{\xi_j}(\theta)\right]\mathcal{H}_{1,\theta}^{\tau,\nu}\left[\mathcal{W}(\theta)\mathcal{U}_k^{\sigma}(\theta)\prod_{j=1}^{n}\mathcal{U}_j^{\xi_j}(\theta)\right],$$

which completes the desired assertion (25). □

Theorem 6. *Let the functions $\mathcal{U}_j$, $(j=1,2,\cdots,n)$ and $\mathcal{V}$ be positive and continuous on the interval $[1,\infty)$ and satisfy (24) for any fixed $k=1,2,\cdots,n$. Assume that the function $\mathcal{W}:[1,\infty)\to\mathbb{R}^+$ is positive and continuous. Then for all $\theta>1$, the following inequality for Hadamard proportional fractional integral operator (7) holds;*

$$\mathcal{H}_{1,\theta}^{\tau,\nu}\left[\mathcal{W}(\theta)\mathcal{U}_k^{\delta+\sigma}(\theta)\prod_{j\neq k}^{n}\mathcal{U}_j^{\xi_j}(\theta)\right]\mathcal{H}_{1,\theta}^{\lambda,\nu}\left[\mathcal{W}(\theta)\mathcal{V}^{\sigma}(\theta)\prod_{j=1}^{n}\mathcal{U}_j^{\xi_j}(\theta)\right]$$
$$+\mathcal{H}_{1,\theta}^{\lambda,\nu}\left[\mathcal{W}(\theta)\mathcal{U}_k^{\delta+\sigma}(\theta)\prod_{j\neq k}^{n}\mathcal{U}_j^{\xi_j}(\theta)\right]\mathcal{H}_{1,\theta}^{\tau,\nu}\left[\mathcal{W}(\theta)\mathcal{V}^{\sigma}(\theta)\prod_{j=1}^{n}\mathcal{U}_j^{\xi_j}(\theta)\right]$$
$$\geq\mathcal{H}_{1,\theta}^{\tau,\nu}\left[\mathcal{W}(\theta)\mathcal{U}_k^{\sigma}(\theta)\prod_{j=1}^{n}\mathcal{U}_j^{\xi_j}(\theta)\right]\mathcal{H}_{1,\theta}^{\lambda,\nu}\left[\mathcal{W}(\theta)\mathcal{V}^{\sigma}(\theta)\mathcal{U}_k^{\delta}(\theta)\prod_{j\neq k}^{n}\mathcal{U}_j^{\xi_j}(\theta)\right] \tag{30}$$
$$+\mathcal{H}_{1,\theta}^{\lambda,\nu}\left[\mathcal{W}(\theta)\mathcal{U}_k^{\sigma}(\theta)\prod_{j=1}^{n}\mathcal{U}_j^{\xi_j}(\theta)\right]\mathcal{H}_{1,\theta}^{\tau,\nu}\left[\mathcal{W}(\theta)\mathcal{V}^{\sigma}(\theta)\mathcal{U}_k^{\delta}(\theta)\prod_{j\neq k}^{n}\mathcal{U}_j^{\xi_j}(\theta)\right],$$

where $\delta\geq\xi_k>0$, $\tau,\lambda,\nu,\sigma>0$ and $k=1,2,\cdots,n$.

Proof. Taking product on both sides of (28) by $\mathcal{G}_1(\theta,\zeta)=\frac{1}{\nu^{\lambda}\Gamma(\lambda)}e^{\frac{\nu-1}{\nu}(\ln\theta-\ln\zeta)}(\ln\theta-\ln\zeta)^{\lambda-1}\frac{\mathcal{W}(\zeta)\prod_{j=1}^{n}\mathcal{U}_j^{\xi_j}(\zeta)}{\zeta}$, where $\lambda>0,\xi_j>0$ and $\zeta\in(1,\theta),\theta>1,j=1,2,\cdots,n$ and integrating the resultant estimates with respect to ζ over $(1,\theta)$, we have

$$\mathcal{H}_{1,\theta}^{\tau,\nu}\left[\mathcal{W}(\theta)\mathcal{U}_k^{\delta+\sigma}(\theta)\prod_{j\neq k}^{n}\mathcal{U}_j^{\xi_j}(\theta)\right]\frac{1}{\nu^{\lambda}\Gamma(\lambda)}\int_1^{\theta}e^{\frac{\nu-1}{\nu}(\ln\theta-\ln\zeta)}(\ln\theta-\ln\zeta)^{\lambda-1}\frac{\mathcal{W}(\zeta)\prod_{j=1}^{n}\mathcal{U}_j^{\xi_j}(\zeta)}{\zeta}\mathcal{V}^{\sigma}(\zeta)d\zeta$$
$$+\mathcal{H}_{1,\theta}^{\tau,\nu}\left[\mathcal{W}(\theta)\mathcal{V}^{\sigma}(\theta)\prod_{j=1}^{n}\mathcal{U}_j^{\xi_j}(\theta)\right]\frac{1}{\nu^{\lambda}\Gamma(\lambda)}\int_1^{\theta}e^{\frac{\nu-1}{\nu}(\ln\theta-\ln\zeta)}(\ln\theta-\ln\zeta)^{\lambda-1}\frac{\mathcal{W}(\zeta)\prod_{j=1}^{n}\mathcal{U}_j^{\xi_j}(\zeta)}{\zeta}\mathcal{U}_k^{\delta+\sigma-\xi_k}(\zeta)d\zeta$$
$$\geq\mathcal{H}_{1,\theta}^{\tau,\nu}\left[\mathcal{W}(\theta)\mathcal{U}_k^{\sigma}(\theta)\prod_{j=1}^{n}\mathcal{U}_j^{\xi_j}(\theta)\right]\frac{1}{\nu^{\lambda}\Gamma(\lambda)}\int_1^{\theta}e^{\frac{\nu-1}{\nu}(\ln\theta-\ln\zeta)}(\ln\theta-\ln\zeta)^{\lambda-1}\frac{\mathcal{W}(\zeta)\prod_{j=1}^{n}\mathcal{U}_j^{\xi_j}(\zeta)}{\zeta}\mathcal{V}^{\sigma}(\zeta)\mathcal{U}_k^{\delta-\xi_k}(\zeta)d\zeta$$
$$+\mathcal{H}_{1,\theta}^{\tau,\nu}\left[\mathcal{W}(\theta)\mathcal{V}^{\sigma}(\theta)\mathcal{U}_k^{\delta+\sigma}(\theta)\prod_{j\neq k}^{n}\mathcal{U}_j^{\xi_j}(\theta)\right]\frac{1}{\nu^{\lambda}\Gamma(\lambda)}\int_1^{\theta}e^{\frac{\nu-1}{\nu}(\ln\theta-\ln\zeta)}(\ln\theta-\ln\zeta)^{\lambda-1}\frac{\mathcal{W}(\zeta)\prod_{j=1}^{n}\mathcal{U}_j^{\xi_j}(\zeta)}{\zeta}\mathcal{U}_k^{\sigma}(\zeta)d\zeta,$$

which in view of (7) yields the desired assertion

$$\mathcal{H}_{1,\theta}^{\tau,\nu}\left[\mathcal{W}(\theta)\mathcal{U}_k^{\delta+\sigma}(\theta)\prod_{j\neq k}^{n}\mathcal{U}_j^{\xi_j}(\theta)\right]\mathcal{H}_{1,\theta}^{\lambda,\nu}\left[\mathcal{W}(\theta)\mathcal{V}^{\sigma}(\theta)\prod_{j=1}^{n}\mathcal{U}_j^{\xi_j}(\theta)\right]$$
$$+\mathcal{H}_{1,\theta}^{\lambda,\nu}\left[\mathcal{W}(\theta)\mathcal{U}_k^{\delta+\sigma}(\theta)\prod_{j\neq k}^{n}\mathcal{U}_j^{\xi_j}(\theta)\right]\mathcal{H}_{1,\theta}^{\tau,\nu}\left[\mathcal{W}(\theta)\mathcal{V}^{\sigma}(\theta)\prod_{j=1}^{n}\mathcal{U}_j^{\xi_j}(\theta)\right]$$
$$\geq\mathcal{H}_{1,\theta}^{\tau,\nu}\left[\mathcal{W}(\theta)\mathcal{U}_k^{\sigma}(\theta)\prod_{j=1}^{n}\mathcal{U}_j^{\xi_j}(\theta)\right]\mathcal{H}_{1,\theta}^{\lambda,\nu}\left[\mathcal{W}(\theta)\mathcal{V}^{\sigma}(\theta)\mathcal{U}_k^{\delta}(\theta)\prod_{j\neq k}^{n}\mathcal{U}_j^{\xi_j}(\theta)\right]$$
$$+\mathcal{H}_{1,\theta}^{\lambda,\nu}\left[\mathcal{W}(\theta)\mathcal{U}_k^{\sigma}(\theta)\prod_{j=1}^{n}\mathcal{U}_j^{\xi_j}(\theta)\right]\mathcal{H}_{1,\theta}^{\tau,\nu}\left[\mathcal{W}(\theta)\mathcal{V}^{\sigma}(\theta)\mathcal{U}_k^{\delta}(\theta)\prod_{j\neq k}^{n}\mathcal{U}_j^{\xi_j}(\theta)\right].$$

□

Remark 7. *Applying Theorem 6 for $\tau=\lambda$, we get Theorem 5.*

Theorem 7. *Let the functions $\mathcal{U}_j$, $(j=1,2,\cdots,n)$ and $\mathcal{V}$ be positive and continuous on the interval $[1,\infty)$ such that the function $\mathcal{V}$ is increasing and the function $\mathcal{U}_j$ for $j=1,2,\cdots,n$ are decreasing on $[1,\infty)$. Assume that the function $\mathcal{W}:[1,\infty)\to\mathbb{R}^+$ is positive and continuous. Then for all $\theta>1$, the following inequality for Hadamard proportional fractional integral operator (7) holds;*

$$\mathcal{H}_{1,\theta}^{\tau,\nu}\left[\mathcal{W}(\theta)\mathcal{U}_k^{\delta}(\theta)\prod_{j\neq k}^{n}\mathcal{U}_j^{\xi_j}(\theta)\right]\mathcal{H}_{1,\theta}^{\tau,\nu}\left[\mathcal{W}(\theta)\mathcal{V}^{\sigma}(\theta)\prod_{j=1}^{n}\mathcal{U}_j^{\xi_j}(\theta)\right]$$
$$\geq\mathcal{H}_{1,\theta}^{\tau,\nu}\left[\mathcal{W}(\theta)\mathcal{V}^{\sigma}(\theta)\mathcal{U}_k^{\delta}(\theta)\prod_{j\neq k}^{n}\mathcal{U}_j^{\xi_j}(\theta)\right]\mathcal{H}_{1,\theta}^{\tau,\nu}\left[\mathcal{W}(\theta)\prod_{j=1}^{n}\mathcal{U}_j^{\xi_j}(\theta)\right], \tag{31}$$

where $\delta\geq\xi_k>0$, $\tau,\nu,\sigma>0$ and $k=1,2,\cdots,n$.

Proof. Under the conditions stated in Theorem 7, we can write

$$\left(\mathcal{V}^{\sigma}(\zeta)-\mathcal{V}^{\sigma}(\rho)\right)\left(\mathcal{U}_k^{\delta-\xi_k}(\rho)-\mathcal{U}_k^{\delta-\xi_k}(\zeta)\right)\geq 0 \tag{32}$$

for any $\rho,\zeta\in(1,\theta),\theta>1,\sigma>0,\delta>\xi_k>0,k=1,2,3,\cdots,n$.

From (32), we have

$$\mathcal{V}^{\sigma}(\zeta)\mathcal{U}_k^{\delta-\xi_k}(\rho)+\mathcal{V}^{\sigma}(\rho)\mathcal{U}_k^{\delta-\xi_k}(\zeta)\geq\mathcal{V}^{\sigma}(\zeta)\mathcal{U}_k^{\delta-\xi_k}(\zeta)+\mathcal{V}^{\sigma}(\rho)\mathcal{U}_k^{\delta-\xi_k}(\rho). \tag{33}$$

Multiplying (27) by $\mathcal{F}_1(\theta,\rho)$ (where $\mathcal{F}_1(\theta,\rho)$ is defined in (26)) and integrating the resultant estimates with respect to ρ over $(1,\theta),\theta>1$, we have

$$\mathcal{V}^{\sigma}(\zeta)\frac{1}{\nu^{\tau}\Gamma(\tau)}\int_1^{\theta}e^{\frac{\nu-1}{\nu}(\ln\theta-\ln\rho)}(\ln\theta-\ln\rho)^{\tau-1}\frac{\mathcal{W}(\rho)\prod_{j=1}^{n}\mathcal{U}_j^{\xi_j}(\rho)}{\rho}\mathcal{U}_k^{\delta-\xi_k}(\rho)d\rho$$
$$+\mathcal{U}_k^{\delta-\xi_k}(\zeta)\frac{1}{\nu^{\tau}\Gamma(\tau)}\int_1^{\theta}e^{\frac{\nu-1}{\nu}(\ln\theta-\ln\rho)}(\ln\theta-\ln\rho)^{\tau-1}\frac{\mathcal{W}(\rho)\prod_{j=1}^{n}\mathcal{U}_j^{\xi_j}(\rho)}{\rho}\mathcal{V}^{\sigma}(\rho)d\rho$$
$$\geq\mathcal{V}^{\sigma}(\zeta)\mathcal{U}_k^{\delta-\xi_k}(\zeta)\frac{1}{\nu^{\tau}\Gamma(\tau)}\int_1^{\theta}e^{\frac{\nu-1}{\nu}(\ln\theta-\ln\rho)}(\ln\theta-\ln\rho)^{\tau-1}\frac{\mathcal{W}(\rho)\prod_{j=1}^{n}\mathcal{U}_j^{\xi_j}(\rho)}{\rho}d\rho$$
$$+\frac{1}{\nu^{\tau}\Gamma(\tau)}\int_1^{\theta}e^{\frac{\nu-1}{\nu}(\ln\theta-\ln\rho)}(\ln\theta-\ln\rho)^{\tau-1}\frac{\mathcal{W}(\rho)\prod_{j=1}^{n}\mathcal{U}_j^{\xi_j}(\rho)}{\rho}\mathcal{V}^{\sigma}(\rho)\mathcal{U}_k^{\delta-\xi_k}(\rho)d\rho,$$

which in view of Hadamard proportional fractional integral (7) becomes

$$\begin{aligned}&\mathcal{V}^{\sigma}(\zeta)\mathcal{H}_{1,\theta}^{\tau,\nu}\left[\mathcal{W}(\theta)\mathcal{U}_k^{\delta}(\theta)\prod_{j\neq k}^{n}\mathcal{U}_j^{\xi_j}(\theta)\right]+\mathcal{U}_k^{\delta-\xi_k}(\zeta)\mathcal{H}_{1,\theta}^{\tau,\nu}\left[\mathcal{W}(\theta)\mathcal{V}^{\sigma}(\theta)\prod_{j=1}^{n}\mathcal{U}_j^{\xi_j}(\theta)\right]\\ \geq&\mathcal{V}^{\sigma}(\zeta)\mathcal{U}_k^{\delta-\xi_k}(\zeta)\mathcal{H}_{1,\theta}^{\tau,\nu}\left[\mathcal{W}(\theta)\prod_{j=1}^{n}\mathcal{U}_j^{\xi_j}(\theta)\right]+\mathcal{H}_{1,\theta}^{\tau,\nu}\left[\mathcal{W}(\theta)\mathcal{V}^{\sigma}(\theta)\mathcal{U}_k^{\delta}(\theta)\prod_{j\neq k}^{n}\mathcal{U}_j^{\xi_j}(\theta)\right].\end{aligned}\tag{34}$$

Now, multiplying (34) by $\mathcal{F}_1(\theta,\zeta)$ (where $\mathcal{F}_1(\theta,\zeta)$ can be obtained from (26) by replacing ρ by ζ) and integrating the resultant estimates with respect to ζ over $(1,\theta),\theta>1$ and then by applying (7), we obtain

$$\begin{aligned}&\mathcal{H}_{1,\theta}^{\tau,\nu}\left[\mathcal{W}(\theta)\mathcal{U}_k^{\delta}(\theta)\prod_{j\neq k}^{n}\mathcal{U}_j^{\xi_j}(\theta)\right]\mathcal{H}_{1,\theta}^{\tau,\nu}\left[\mathcal{W}(\theta)\mathcal{V}^{\sigma}(\theta)\prod_{j=1}^{n}\mathcal{U}_j^{\xi_j}(\theta)\right]\\ \geq&\mathcal{H}_{1,\theta}^{\tau,\nu}\left[\mathcal{W}(\theta)\mathcal{V}^{\sigma}(\theta)\mathcal{U}_k^{\delta}(\theta)\prod_{j\neq k}^{n}\mathcal{U}_j^{\xi_j}(\theta)\right]\mathcal{H}_{1,\theta}^{\tau,\nu}\left[\mathcal{W}(\theta)\prod_{j=1}^{n}\mathcal{U}_j^{\xi_j}(\theta)\right],\end{aligned}$$

which is the desired assertion (31). □

Theorem 8. *Let the functions $\mathcal{U}_j$, $(j=1,2,\cdots,n)$ and $\mathcal{V}$ be positive and continuous on the interval $[1,\infty)$ such that the function $\mathcal{V}$ is increasing and the function $\mathcal{U}_j$ for $j=1,2,\cdots,n$ are decreasing on $[1,\infty)$. Assume that the function $\mathcal{W}:[1,\infty)\to\mathbb{R}^+$ is positive and continuous. Then for all $\theta>1$, the following inequality for Hadamard proportional fractional integral operator (7) holds;*

$$\begin{aligned}&\mathcal{H}_{1,\theta}^{\tau,\nu}\left[\mathcal{W}(\theta)\mathcal{U}_k^{\delta}(\theta)\prod_{j\neq k}^{n}\mathcal{U}_j^{\xi_j}(\theta)\right]\mathcal{H}_{1,\theta}^{\lambda,\nu}\left[\mathcal{W}(\theta)\mathcal{V}^{\sigma}(\theta)\prod_{j=1}^{n}\mathcal{U}_j^{\xi_j}(\theta)\right]\\ +&\mathcal{H}_{1,\theta}^{\lambda,\nu}\left[\mathcal{W}(\theta)\mathcal{U}_k^{\delta}(\theta)\prod_{j\neq k}^{n}\mathcal{U}_j^{\xi_j}(\theta)\right]\mathcal{H}_{1,\theta}^{\tau,\nu}\left[\mathcal{W}(\theta)\mathcal{V}^{\sigma}(\theta)\prod_{j=1}^{n}\mathcal{U}_j^{\xi_j}(\theta)\right]\\ \geq&\mathcal{H}_{1,\theta}^{\tau,\nu}\left[\mathcal{W}(\theta)\mathcal{V}^{\sigma}(\theta)\mathcal{U}_k^{\delta}(\theta)\prod_{j\neq k}^{n}\mathcal{U}_j^{\xi_j}(\theta)\right]\mathcal{H}_{1,\theta}^{\lambda,\nu}\left[\mathcal{W}(\theta)\prod_{j=1}^{n}\mathcal{U}_j^{\xi_j}(\theta)\right]\\ +&\mathcal{H}_{1,\theta}^{\tau,\nu}\left[\mathcal{W}(\theta)\mathcal{V}^{\sigma}(\theta)\mathcal{U}_k^{\delta}(\theta)\prod_{j\neq k}^{n}\mathcal{U}_j^{\xi_j}(\theta)\right]\mathcal{H}_{1,\theta}^{\lambda,\nu}\left[\mathcal{W}(\theta)\prod_{j=1}^{n}\mathcal{U}_j^{\xi_j}(\theta)\right],\end{aligned}\tag{35}$$

where $\delta\geq\xi_k>0$, $\tau,\lambda,\nu,\sigma>0$ and $k=1,2,\cdots,n$.

Proof. To obtain the desire assertion (35), we multiply (34) by $\mathcal{G}_1(\theta,\zeta)=\frac{1}{\nu^{\lambda}\Gamma(\lambda)}e^{\frac{\nu-1}{\nu}(\ln\theta-\ln\zeta)}(\ln\theta-\ln\zeta)^{\lambda-1}\frac{\mathcal{W}(\zeta)\prod_{j=1}^{n}\mathcal{U}_j^{\xi_j}(\zeta)}{\zeta}$, where $\lambda>0,\xi_j>0$ and $\zeta\in(1,\theta),\theta>1,j=1,2,\cdots,n$ and integrating the resultant estimates with respect to ζ over $(1,\theta)$, we have

$$\frac{1}{\nu^{\lambda}\Gamma(\lambda)}\int_{1}^{\theta} e^{\frac{\nu-1}{\nu}(\ln\theta-\ln\zeta)}(\ln\theta-\ln\zeta)^{\lambda-1}\frac{\mathcal{W}(\zeta)\prod_{j=1}^{n}\mathcal{U}_j^{\xi_j}(\zeta)}{\zeta}\mathcal{V}^{\sigma}(\zeta)d\zeta\mathcal{H}_{1,\theta}^{\tau,\nu}\left[\mathcal{W}(\theta)\mathcal{U}_k^{\delta}(\theta)\prod_{j\neq k}^{n}\mathcal{U}_j^{\xi_j}(\theta)\right]$$
$$+\frac{1}{\nu^{\lambda}\Gamma(\lambda)}\int_{1}^{\theta} e^{\frac{\nu-1}{\nu}(\ln\theta-\ln\zeta)}(\ln\theta-\ln\zeta)^{\lambda-1}\frac{\mathcal{W}(\zeta)\prod_{j=1}^{n}\mathcal{U}_j^{\xi_j}(\zeta)}{\zeta}\mathcal{U}_k^{\delta-\xi_k}(\zeta)d\zeta\mathcal{H}_{1,\theta}^{\tau,\nu}\left[\mathcal{W}(\theta)\mathcal{V}^{\sigma}(\theta)\prod_{j=1}^{n}\mathcal{U}_j^{\xi_j}(\theta)\right]$$
$$\geq\frac{1}{\nu^{\lambda}\Gamma(\lambda)}\int_{1}^{\theta} e^{\frac{\nu-1}{\nu}(\ln\theta-\ln\zeta)}(\ln\theta-\ln\zeta)^{\lambda-1}\frac{\mathcal{W}(\zeta)\prod_{j=1}^{n}\mathcal{U}_j^{\xi_j}(\zeta)}{\zeta}\mathcal{V}^{\sigma}(\zeta)\mathcal{U}_k^{\delta-\xi_k}(\zeta)d\zeta\mathcal{H}_{1,\theta}^{\tau,\nu}\left[\mathcal{W}(\theta)\prod_{j=1}^{n}\mathcal{U}_j^{\xi_j}(\theta)\right]$$
$$+\mathcal{H}_{1,\theta}^{\tau,\nu}\left[\mathcal{W}(\theta)\mathcal{V}^{\sigma}(\theta)\mathcal{U}_k^{\delta}(\theta)\prod_{j\neq k}^{n}\mathcal{U}_j^{\xi_j}(\theta)\right]\frac{1}{\nu^{\lambda}\Gamma(\lambda)}\int_{1}^{\theta} e^{\frac{\nu-1}{\nu}(\ln\theta-\ln\zeta)}(\ln\theta-\ln\zeta)^{\lambda-1}\frac{\mathcal{W}(\zeta)\prod_{j=1}^{n}\mathcal{U}_j^{\xi_j}(\zeta)}{\zeta}.$$

which in view of Hadamard proportional fractional integral (7) gives

$$\mathcal{H}_{1,\theta}^{\tau,\nu}\left[\mathcal{W}(\theta)\mathcal{U}_k^{\delta}(\theta)\prod_{j\neq k}^{n}\mathcal{U}_j^{\xi_j}(\theta)\right]\mathcal{H}_{1,\theta}^{\lambda,\nu}\left[\mathcal{W}(\theta)\mathcal{V}^{\sigma}(\theta)\prod_{j=1}^{n}\mathcal{U}_j^{\xi_j}(\theta)\right]$$
$$+\mathcal{H}_{1,\theta}^{\lambda,\nu}\left[\mathcal{W}(\theta)\mathcal{U}_k^{\delta}(\theta)\prod_{j\neq k}^{n}\mathcal{U}_j^{\xi_j}(\theta)\right]\mathcal{H}_{1,\theta}^{\tau,\nu}\left[\mathcal{W}(\theta)\mathcal{V}^{\sigma}(\theta)\prod_{j=1}^{n}\mathcal{U}_j^{\xi_j}(\theta)\right]$$
$$\geq\mathcal{H}_{1,\theta}^{\tau,\nu}\left[\mathcal{W}(\theta)\mathcal{V}^{\sigma}(\theta)\mathcal{U}_k^{\delta}(\theta)\prod_{j\neq k}^{n}\mathcal{U}_j^{\xi_j}(\theta)\right]\mathcal{H}_{1,\theta}^{\lambda,\nu}\left[\mathcal{W}(\theta)\prod_{j=1}^{n}\mathcal{U}_j^{\xi_j}(\theta)\right]$$
$$+\mathcal{H}_{1,\theta}^{\tau,\nu}\left[\mathcal{W}(\theta)\mathcal{V}^{\sigma}(\theta)\mathcal{U}_k^{\delta}(\theta)\prod_{j\neq k}^{n}\mathcal{U}_j^{\xi_j}(\theta)\right]\mathcal{H}_{1,\theta}^{\lambda,\nu}\left[\mathcal{W}(\theta)\prod_{j=1}^{n}\mathcal{U}_j^{\xi_j}(\theta)\right],$$

which completes the proof of (35). □

Remark 8. *Applying Theorem 8 for $\tau = \lambda$, we get Theorem 7.*

4. Concluding Remarks

Recently Jarad et al. [46] introduced the idea of generalized proportional fractional integral operators which comprises exponential in their kernels. Later on, Rahman et al. [47] studied these operators and defined Hadamard proportional fractional integrals. They established certain inequalities for convex functions by employing Hadamard proportional fractional integrals. In [49], the authors defined bounds of generalized proportional fractional integral operators for convex functions and their applications. Motivated by the above, here we presented certain inequalities by employing Hadamard proportional fractional integrals. The inequalities obtained in this paper generalized the inequalities presented earlier by Houas [45].

Author Contributions: Conceptualization, K.S.N.; Formal analysis, T.A.; Writing—original draft, G.R. and K.S.N.; Writing—review & editing, T.A. All authors have read and agreed to the published version of the manuscript.

Funding: The third author would like to thank Prince Sultan University for funding this work through research group Nonlinear Analysis Methods in Applied Mathematics (NAMAM) group number RG-DES-2017-01-17.

Conflicts of Interest: The authors declare that they have no competing interests.

References

1. Anber, A.; Dahmani, Z.; Bendoukha, B. New integral inequalities of Feng Qi type via Riemann–Liouville fractional integration. *Facta Univ. Ser. Math. Inform.* **2012**, *27*, 13–22.
2. Belarbi, S.; Dahmani, Z. On some new fractional integral inequalities. *J. Inequal. Pure Appl. Math.* **2009**, *10*, 1–12.

3. Houas, M. Certain weighted integral inequalities involving the fractional hypergeometric operators. *Sci. Ser. A Math. Sci.* **2016**, *27*, 87–97.
4. Kilbas, A.A.; Srivastava, H.M.; Trujillo, J.J. *Theory and Application of Fractional Differential Equations*; Elsevier: Amersterdam, The Netherlands, 2006.
5. Podlubny, I. *Fractional Differential Equations*; Academic Press: London, UK, 1999.
6. Aldhaifallah, M.; Tomar, M.; Nisar, K.S.; Purohit, S.D. Some new inequalities for (k,s)-fractional integrals. *J. Nonlinear Sci. Appl.* **2016**, *9*, 5374–5381. [CrossRef]
7. Dahmani, Z.; Tabharit, L. On weighted Gruss type inequalities via fractional integration. *J. Adv. Res. Pure Math.* **2010**, *2*, 31–38. [CrossRef]
8. Dahmani, Z. New inequalities in fractional integrals. *Int. J. Nonlinear Sci.* **2010**, *9*, 493–497.
9. Nisar, K.S.; Qi, F.; Rahman, G.; Mubeen, S.; Arshad, M. Some inequalities involving the extended gamma function and the Kummer confluent hypergeometric k-function. *J. Inequal. Appl.* **2018**, *2018*, 135. [CrossRef]
10. Nisar, K.S.; Rahman, G.; Choi, J.; Mubeen, S.; Arshad, M. Certain Gronwall type inequalities associated with Riemann–Liouville k- and Hadamard k-fractional derivatives and their applications. *East Asian Math. J.* **2018**, *34*, 249–263.
11. Rahman, G.; Nisar, K.S.; Mubeen, S.; Choi, J. Certain Inequalities involving the (k,ρ)-fractional integral operator. *Far East J. Math. Sci. (FJMS)* **2018**, *103*, 1879–1888. [CrossRef]
12. Sarikaya, M.Z.; Budak, H. Generalized Ostrowski type inequalities for local fractional integrals. *Proc. Am. Math. Soc.* **2017**, *145*, 1527–1538. [CrossRef]
13. Sarikaya, M.Z.; Dahmani, Z.; Kiris, M.E.; Ahmad, F. (k,s)-Riemann–Liouville fractional integral and applications. *Hacet. J. Math. Stat.* **2016**, *45*, 77–89. [CrossRef]
14. Set, E.; Tomar, M.; Sarikaya, M.Z. On generalized Grüss type inequalities for k-fractional integrals. *Appl. Math. Comput.* **2015**, *269*, 29–34. [CrossRef]
15. Set, E.; Noor, M.A.; Awan, M.U.; Gözpinar, A. Generalized Hermite–Hadamard type inequalities involving fractional integral operators. *J. Inequal. Appl.* **2017**, *2017*, 169. [CrossRef]
16. Agarwal, P.; Jleli, M.; Tomar, M. Certain Hermite–Hadamard type inequalities via generalized k-fractional integrals. *J. Inequal. Appl.* **2017**, *2017*, 55. [CrossRef] [PubMed]
17. Chinchane, V.L.; Pachpatte, D.B. Some new integral inequalities using Hadamard fractional integral operator. *Adv. Inequal. Appl.* **2014**, *2014*, 12.
18. Dahmani, Z.; Benzidane, A. New inequalities using Q-fractional theory. *Bull. Math. Anal. Appl.* **2012**, *4*, 190–196.
19. Yang, W. Some new Chebyshev and Gruss-type integral inequalities for Saigo fractional integral operators and Their q-analogues. *Filomat* **2015**, *29*, 1269–1289. [CrossRef]
20. Huang, C.J.; Rahman, G.; Nisar, K.S.; Ghaffar, A.; Qi, F. Some Inequalities of Hermite–Hadamard type for k-fractional conformable integrals. *Aust. J. Math. Anal. Appl.* **2019**, *16*, 1–9.
21. Mubeen, S.; Habib, S.; Naeem, M.N. The Minkowski inequality involving generalized k-fractional conformable integral. *J. Inequalities Appl.* **2019**, *2019*, 81. [CrossRef]
22. Qi, F.; Rahman, G.; Hussain, S.M.; Du, W.S.; Nisar, K.S. Some inequalities of Čebyšev type for conformable k-fractional integral operators. *Symmetry* **2018**, *10*, 614. [CrossRef]
23. Rahman, G.; Nisar, K.S.; Qi, F. Some new inequalities of the Gruss type for conformable fractional integrals. *AIMS Math.* **2018**, *3*, 575–583. [CrossRef]
24. Rahman, G.; Ullah, Z.; Khan, A.; Set, E.; Nisar, K.S. Certain Chebyshev type inequalities involving fractional conformable integral operators. *Mathematics* **2019**, *7*, 364. [CrossRef]
25. Nisar, K.S.; Rahman, G.; Mehrez, K. Chebyshev type inequalities via generalized fractional conformable integrals. *J. Inequal. Appl.* **2019**, *2019*, 245. [CrossRef]
26. Niasr, K.S.; Tassadiq, A.; Rahman, G.; Khan, A. Some inequalities via fractional conformable integral operators. *J. Inequal. Appl.* **2019**, *2019*, 217. [CrossRef]
27. Tassaddiq, A.; Rahman, G.; Nisar, K.S.; Samraiz, M. Certain fractional conformable inequalities for the weighted and the extended Chebyshev functionals. *Adv. Differ. Equ.* **2020**, *2020*, 96. [CrossRef]
28. Nisar, K.S.; Rahman, G.; Khan, A. Some new inequalities for generalized fractional conformable integral operators. *Adv. Differ. Equ.* **2019**, *2019*, 427. [CrossRef]
29. Nisar, K.S.; Rahman, G.; Tassaddiq, A.; Khan, A.; Abouzaid, M.S. Certain generalized fractional integral inequalities. *AIMS Math.* **2020**, *5*, 1588–1602. [CrossRef]

30. Rahman, G.; Nisar, K.S.; Ghaffar, A.; Qi, F. Some inequalities of the Grüss type for conformable k-fractional integral operators. *RACSAM* **2020**, *114*, 9. [CrossRef]
31. Rahman, G.; Abdeljawad, T.; Khan, A.; Nisar, K.S. Some fractional proportional integral inequalities. *J. Inequal. Appl.* **2019**, *2019*, 244. [CrossRef]
32. Rahman, G.; Khan, A.; Abdeljawad, T.; Nisar, K.S. The Minkowski inequalities via generalized proportional fractional integral operators. *Adv. Differ. Equ.* **2019**, *2019*, 287. [CrossRef]
33. Rahman, G.; Nisar, K.S.; Abdeljawad, T.; Ullah, S. Certain Fractional Proportional Integral Inequalities via Convex Functions. *Mathematics* **2020**, *8*, 222. [CrossRef]
34. Khalil, R.; Horani, M.A.; Yousef, A.; Sababheh, M. A new definition of fractional derivative. *J. Comput. Appl. Math.* **2014**, *264*, 65–70. [CrossRef]
35. Abdeljawad, T. On Conformable Fractional Calculus. *J. Comput. Appl. Math.* **2015**, *279*, 57–66. [CrossRef]
36. Abdeljawad, T.; Baleanu, D. Monotonicity results for fractional difference operators with discrete exponential kernels. *Adv. Differ. Equ.* **2017**, *2017*, 78. [CrossRef]
37. Abdeljawad, T.; Baleanu, D. On Fractional Derivatives with Exponential Kernel and their Discrete Versions. *Rep. Math. Phys.* **2017**, *80*, 11–27. [CrossRef]
38. Atangana, A.; Baleanu, D. New fractional derivatives with nonlocal and non-singular kernel: Theory and application to heat transfer model. *Thermal Sci.* **2016**, *20*, 763–769. [CrossRef]
39. Jarad, F.; Ugurlu, E.; Abdeljawad, T.; Baleanu, D. On a new class of fractional operators. *Adv. Differ. Equ.* **2017**, *2017*, 247. [CrossRef]
40. Anderson, D.R.; Ulness, D.J. Newly defined conformable derivatives. *Adv. Dyn. Syst. Appl.* **2015**, *10*, 109–137.
41. Caputo, M.; Fabrizio, M. A new Definition of Fractional Derivative without Singular Kernel. *Progr. Fract. Differ. Appl.* **2015**, *1*, 73–85.
42. Losada, J.; Nieto, J.J. Properties of a New Fractional Derivative without Singular Kernel. *Progr. Fract. Differ. Appl.* **2015**, *1*, 87–92.
43. Liu, W.; Ngô, Q.A.; Huy, V.N. Several interesting integral inequalities. *J. Math. Inequ.* **2009**, *3*, 201–212. [CrossRef]
44. Dahmani, Z. New classes of integral inequalities of fractional order. *Le Mat.* **2014**, *69*, 237–247. [CrossRef]
45. Houas, M. On some generalized integral inequalities for Hadamard fractional integrals. *Med. J. Model. Simul.* **2018**, *9*, 43–52.
46. Jarad, F.; Abdeljawad, T.; Alzabut, J. Generalized fractional derivatives generated by a class of local proportional derivatives. *Eur. Phys. J. Spec. Top.* **2017**, *226*, 3457–3471. [CrossRef]
47. Rahman, G.; Abdeljawad, T.; Jarad, F.; Khan, A.; Nisar, K.S. Certain inequalities via generalized proportional Hadamard fractional integral operators. *Adv. Differ. Equ.* **2019**, *2019*, 454. [CrossRef]
48. Samko, S.G.; Kilbas, A.A.; Marichev, O.I. *Fractional Integrals and Derivatives, Theory and Applications*; Edited and with a foreword by S. M. Nikol'skiĭ, Translated from the 1987 Russian original, Revised by the authors; Gordon and Breach Science Publishers: Yverdon, Switzerland, 1993.
49. Rahmnan, G.; Abdeljawad, T.; Jarad, F.; Nisar, K.S. Bounds of Generalized Proportional Fractional Integrals in General Form via Convex Functions and their Applications. *Mathematics* **2020**, *8*, 113. [CrossRef]

mathematics

Article

Some Fractional Hermite–Hadamard Type Inequalities for Interval-Valued Functions

Fangfang Shi [1], Guoju Ye [1,*], Dafang Zhao [2] and Wei Liu [1]

[1] College of Science, Hohai University, Nanjing 210098, China; 15298387799@163.com (F.S.); liuw626@hhu.edu.cn (W.L.)

[2] School of Mathematics and Statistics, Hubei Normal University, Huangshi 435002, China; dafangzhao@163.com

* Correspondence: yegj@hhu.edu.cn; Tel.: +86-152-9838-7799

Received: 6 February 2020; Accepted: 22 March 2020; Published: 4 April 2020

Abstract: In this paper, firstly we prove the relationship between interval h-convex functions and interval harmonically h-convex functions. Secondly, several new Hermite–Hadamard type inequalities for interval h-convex functions via interval Riemann–Liouville type fractional integrals are established. Finally, we obtain some new fractional Hadamard–Hermite type inequalities for interval harmonically h-convex functions by using the above relationship. Also we discuss the importance of our results and some special cases. Our results extend and improve some previously known results.

Keywords: Hermite–Hadamard type inequalities; interval-valued functions; fractional integrals

1. Introduction

Hermite–Hadamard inequality was firstly discovered by Hermite and Hadamard for convex functions are considerable significant in the literature. Since Hermite–Hadamard inequality has been regarded as one of the most useful inequalities in mathematical analysis and optimization, many papers have provided generalizations, refinements and extensions, see [1–4]. Due to the fractional integral has played a irreplaceable part in various scientific fields and importance of Hermite–Hadamard type inequalities, Sarikaya et al. [5] presented Hermite–Hadamard type inequalities via fractional integrals. Moreover, many papers relating to fractional integral inequalities have been obtained for different classes of functions, see [6–8].

On the other hand, interval analysis was initially developed as an attempt to deal with interval uncertainty that appears in computer graphics [9], automatic error analysis [10], and many others. Recently, several authors have extended their research by combining integral inequalities with interval-valued functions (IVFs), one can see Chalco-Cano et al. [11], Román-Flores et al. [12], Flores-Franulič et al. [13], Zhao et al. [14,15], An et al. [16]. As a further extension, more and more Hermite–Hadamard type inequalities involving interval Riemann–Liouville type fractional integral have been obtained for different classes of IVFs, see for interval convex functions [17], for interval harmonically convex functions [18] and the references therein.

Motivated by the ongoing research, We proved the relationship between interval h-convex functions and interval harmonically h-convex functions, then we establish some new Hermite–Hadamard type inequalities for interval h-convex functions and interval harmonically h-convex functions via interval Riemann–Liouville type fractional integrals. Our results extend and improve some known results. Also we discuss the importance of our results and some special cases. In addition, results obtained in this paper may be extended for other classes of convex functions including interval (h_1, h_2)-convex functions and interval Log-h-convex functions and used as a tool to investigate the research of optimization and probability, among others.

2. Preliminaries and Result

Let us denote by $\mathbb{R}_\mathcal{I}$ the collection of all nonempty closed intervals of the real line $\mathbb{R}$. We call $[z] = [\underline{z}, \overline{z}]$ positive if $\underline{z} > 0$. We denote by $\mathbb{R}_\mathcal{I}^+$ and $\mathbb{R}^+$ the set of all positive intervals and the set of all positive numbers of $\mathbb{R}$, respectively. For $[z] = [\underline{z}, \overline{z}], [s] = [\underline{s}, \overline{s}] \in \mathbb{R}_\mathcal{I}$, the inclusion "$\subseteq$" is defined by

$$[\underline{z}, \overline{z}] \subseteq [\underline{s}, \overline{s}] \Leftrightarrow \underline{s} \leq \underline{z},\ \overline{z} \leq \overline{s}.$$

For $\lambda \in \mathbb{R}$, the Minkowski addition and scalar multiplication are defined by

$$[z] + [s] = [\underline{z}, \overline{z}] + [\underline{s}, \overline{s}] = [\underline{z} + \underline{s}, \overline{z} + \overline{s}];$$

$$\lambda[z] = \lambda[\underline{z}, \overline{z}] = \begin{cases} [\lambda\underline{z}, \lambda\overline{z}], & \lambda > 0, \\ \{0\}, & \lambda = 0, \\ [\lambda\overline{z}, \lambda\underline{z}], & \lambda < 0 \end{cases}$$

respectively. The conception of Riemann integral for interval-valued function is introduced in [19]. Moreover, we have

Definition 1. *[19] Let $f : [a, b] \to \mathbb{R}_\mathcal{I}$ be an interval-valued function such that $f = [\underline{f}, \overline{f}]$. Then the f is Riemann integrable on $[a, b]$ iff $\underline{f}$ and $\overline{f}$ are Riemann integrable on $[a, b]$ and*

$$\int_a^b f(t)dt = \left[\int_a^b \underline{f}(t)dt, \int_a^b \overline{f}(t)dt\right].$$

The set of all Riemann integrable IVFs on $[a, b]$ will be denoted by $\mathcal{IR}_{([a,b])}$. For more basic notations with interval analysis, see [19,20]. Furthermore, we recall the following results in [17].

Definition 2. *Let $f : [a, b] \to \mathbb{R}_\mathcal{I}$ be an interval-valued function and $f \in \mathcal{IR}_{([a,b])}$. Then the interval Riemann–Liouville type fractional integrals of f are defined by*

$$\mathcal{J}_{a^+}^\alpha f(t) = \frac{1}{\Gamma(\alpha)} \int_a^t (t - v)^{\alpha-1} f(v)dv, \quad t > a.$$

and

$$\mathcal{J}_{b^-}^\alpha f(t) = \frac{1}{\Gamma(\alpha)} \int_t^b (v - t)^{\alpha-1} f(v)dv, \quad t < b.$$

where $\alpha > 0$ and Γ is the Gamma function.

Definition 3. *[14] Let $h : [0, 1] \to \mathbb{R}^+$ be a non-negative function. We say that $f : [a, b] \to \mathbb{R}_\mathcal{I}^+$ is interval h-convex function or that $f \in SX(h, [a, b], \mathbb{R}_\mathcal{I}^+)$, if for all $x, y \in [a, b]$ and $v \in [0, 1]$, we have*

$$f(vx + (1 - v)y) \supseteq h(v)f(x) + h(1 - v)f(y).$$

Definition 4. *[15] Let $h : [0, 1] \to \mathbb{R}^+$ be a non-negative function. We say that $f : [a, b] \to \mathbb{R}_\mathcal{I}^+$ is interval harmonically h-convex function or that $f \in SHX(h, [a, b], \mathbb{R}_\mathcal{I}^+)$, if for all $x, y \in [a, b]$ and $v \in [0, 1]$, we have*

$$f\left(\frac{xy}{vx + (1 - v)y}\right) \supseteq h(1 - v)f(x) + h(v)f(y).$$

Next, we will present the relationship between interval h-convex functions and interval harmonically h-convex functions which will be used in Section 4.

Theorem 1. $f(x) \in SHX(h, [a, b], \mathbb{R}_\mathcal{I}^+)$ *iff* $f(\frac{1}{x}) \in SX(h, [a, b], \mathbb{R}_\mathcal{I}^+)$.

Proof. Let $\psi(x) = f(\frac{1}{x})$. Since $f \in SHX(h, [a,b], \mathbb{R}_{\mathcal{I}}^{+})$,

$$f\left(\frac{xy}{vx + (1-v)y}\right) \supseteq h(1-v)f(x) + h(v)f(y). \tag{1}$$

By using $A = \frac{1}{x}$ and $B = \frac{1}{y}$ to replace x and y, respectively, applying (1)

$$\begin{aligned} f\left(\frac{\frac{1}{xy}}{v\frac{1}{x} + (1-v)\frac{1}{y}}\right) &= f\left(\frac{1}{(1-v)x + vy}\right) \\ &= \psi\left((1-v)x + vy\right) \\ &\supseteq h(v)f\left(\frac{1}{y}\right) + h(1-v)f\left(\frac{1}{x}\right) \\ &= h(v)\psi(y) + h(1-v)\psi(x), \end{aligned}$$

which gives that $\psi \in SX(h, [a,b], \mathbb{R}_{\mathcal{I}}^{+})$.

On the other hand, if $\psi \in SX(h, [a,b], \mathbb{R}_{\mathcal{I}}^{+})$, then

$$\psi\left(vx + (1-v)y\right) \supseteq h(v)\psi(x) + h(1-v)\psi(y).$$

In the same way as above, we have

$$\begin{aligned} \psi\left(v\frac{1}{x} + (1-v)\frac{1}{y}\right) = f\left(\frac{1}{v\frac{1}{x} + (1-v)\frac{1}{y}}\right) &= f\left(\frac{xy}{(1-v)x + vy}\right) \\ \supseteq h(v)\psi\left(\frac{1}{x}\right) + h(1-v)\psi\left(\frac{1}{y}\right) \\ = h(v)f(x) + h(1-v)f(y), \end{aligned}$$

which gives that $f \in SHX(h, [a,b], \mathbb{R}_{\mathcal{I}}^{+})$. We have completed the proof. □

Remark 1. *If $h(v) = v$ and $\underline{f} = \overline{f}$, then we get the Lemma 2.1 of [6].*

3. Fractional Hermite–Hadamard Type Inequalities of Interval h-Convex Functions

In this section, we will prove some new Hermite–Hadamard type inequalities for interval h-convex functions via interval Riemann–Liouville type integrals.

Theorem 2. *Let $f : [a,b] \to \mathbb{R}_{\mathcal{I}}^{+}$ be an interval-valued function such that $f = [\underline{f}, \overline{f}]$ and $f \in \mathcal{IR}_{([a,b])}$, $h : [0,1] \to \mathbb{R}^{+}$ be a non-negative function and $h\left(\frac{1}{2}\right) \neq 0$. If $f \in SX(h, [a,b], \mathbb{R}_{\mathcal{I}}^{+})$, then*

$$\begin{aligned} \frac{1}{\alpha h(\frac{1}{2})} f\left(\frac{a+b}{2}\right) &\supseteq \frac{\Gamma(\alpha)}{(b-a)^{\alpha}}\left[\mathcal{J}_{a^{+}}^{\alpha} f(b) + \mathcal{J}_{b^{-}}^{\alpha} f(a)\right] \\ &\supseteq [f(a) + f(b)] \int_0^1 v^{\alpha-1}[h(v) + h(1-v)]dv \end{aligned} \tag{2}$$

with $\alpha > 0$.

Proof. Since $f \in SX(h, [a,b], \mathbb{R}_{\mathcal{I}}^{+})$, we have

$$\frac{1}{h(\frac{1}{2})} f\left(\frac{x+y}{2}\right) \supseteq f(x) + f(y).$$

Let $x = \nu a + (1-\nu)b,\ y = (1-\nu)a + \nu b, \nu \in [0,1]$, then

$$\frac{1}{h(\frac{1}{2})} f\left(\frac{a+b}{2}\right) \supseteq f(\nu a + (1-\nu)b) + f((1-\nu)a + \nu b). \tag{3}$$

Multiplying both sides (3) by $\nu^{\alpha-1}$ and integrating on $[0,1]$, we get

$$\begin{aligned}
&\frac{1}{\alpha h(\frac{1}{2})} f\left(\frac{a+b}{2}\right)\\
=&\frac{1}{h(\frac{1}{2})} f\left(\frac{a+b}{2}\right)\int_0^1 \nu^{\alpha-1} d\nu\\
\supseteq&\left[\int_0^1 \nu^{\alpha-1} f(\nu a + (1-\nu)b) d\nu + \int_0^1 \nu^{\alpha-1} f((1-\nu)a + \nu b) d\nu\right]\\
=&\left[\int_0^1 \nu^{\alpha-1}\left(\underline{f}(\nu a + (1-\nu)b) + \underline{f}((1-\nu)a + \nu b)\right) d\nu,\right.\\
&\left.\int_0^1 \nu^{\alpha-1}\left(\overline{f}(\nu a + (1-\nu)b) + \overline{f}((1-\nu)a + \nu b)\right) d\nu\right]\\
=&\left[\int_b^a (\tau(\mu))^{\alpha-1} \underline{f}(\mu)\frac{d\mu}{a-b} + \int_a^b (1-\tau(\mu))^{\alpha-1} \underline{f}(\mu)\frac{d\mu}{b-a},\right.\\
&\left.\int_b^a (\tau(\mu))^{\alpha-1} \overline{f}(\mu)\frac{d\mu}{a-b} + \int_a^b (1-\tau(\mu))^{\alpha-1} \overline{f}(\mu)\frac{d\mu}{b-a}\right]\\
=&\frac{\Gamma(\alpha)}{(b-a)^\alpha}\left[J_{a^+}^\alpha \underline{f}(b) + J_{b^-}^\alpha \underline{f}(a), J_{a^+}^\alpha \overline{f}(b) + J_{b^-}^\alpha \overline{f}(a)\right]\\
=&\frac{\Gamma(\alpha)}{(b-a)^\alpha}\left[\mathcal{J}_{a^+}^\alpha f(b) + \mathcal{J}_{b^-}^\alpha f(a)\right].
\end{aligned} \tag{4}$$

where $\tau(\mu) = \frac{b-\mu}{b-a}$.
Similarly, since $f \in SX(h, [a,b], \mathbb{R}_{\mathcal{I}}^+)$,

$$f(\nu a + (1-\nu)b) + f((1-\nu)a + \nu b) \supseteq [h(\nu) + h(1-\nu)][f(a) + f(b)]. \tag{5}$$

Multiplying both sides (5) by $\nu^{\alpha-1}$ and integrating on $[0,1]$, we have

$$\frac{\Gamma(\alpha)}{(b-a)^\alpha}\left[\mathcal{J}_{a^+}^\alpha f(b) + \mathcal{J}_{b^-}^\alpha f(a)\right] \supseteq [f(a) + f(b)]\int_0^1 \nu^{\alpha-1}[h(\nu) + h(1-\nu)]d\nu. \tag{6}$$

By combining (4) with (6), and the result follows. □

Example 1. *Suppose that $[a,b] = [1,2]$. Let $h(\nu) = \nu$ for all $\nu \in [0,1]$ and $\alpha = \frac{1}{2}$, $f : [a,b] \to \mathbb{R}_{\mathcal{I}}^+$ be defined by*

$$f(t) = [-\sqrt{t} + 2, \sqrt{t} + 2].$$

We obtain

$$\frac{1}{\alpha h(\frac{1}{2})} f\left(\frac{a+b}{2}\right) = [8 - 2\sqrt{6}, 8 + 2\sqrt{6}],$$

$$\frac{\Gamma(\alpha)}{(b-a)^\alpha}\left[\mathcal{J}_{a^+}^\alpha f(b) + \mathcal{J}_{b^-}^\alpha f(a)\right] = \left[7 - \sqrt{2} - \frac{\pi}{2} - log(\sqrt{2}+1), 9 + \sqrt{2} + \frac{\pi}{2} + log(\sqrt{2}+1)\right],$$

$$[f(a) + f(b)]\int_0^1 \nu^{\alpha-1}\left[h(\nu) + h(1-\nu)\right]d\nu = [6 - 2\sqrt{2}, 6 + 2\sqrt{2}].$$

Then we get

$$[8-2\sqrt{6},8+2\sqrt{6}] \supseteq \left[7-\sqrt{2}-\frac{\pi}{2}-log(\sqrt{2}+1),9+\sqrt{2}+\frac{\pi}{2}+log(\sqrt{2}+1)\right] \supseteq [6-2\sqrt{2},6+2\sqrt{2}].$$

Consequently, Theorem 2 is verified.

Remark 2. *If $\alpha = 1$, then we get Theorem 4.1 of [14]. If $h(\nu) = \nu$, then we get Theorem 2.5 of [17]. If $\underline{f} = \overline{f}$ and $\alpha = 1$, then we get Theorem 6 of [4]. If $\underline{f} = \overline{f}$ and $h(\nu) = \nu$, then we get Theorem 2 of [5].*

Theorem 3. *Let $f, g : [a,b] \to \mathbb{R}^+_{\mathcal{I}}$ be two interval-valued functions such that $f = [\underline{f}, \overline{f}]$, $g = [\underline{g}, \overline{g}]$ and $fg \in \mathcal{IR}_{([a,b])}$, $h_1, h_2 : [0,1] \to \mathbb{R}^+$ be non-negative functions. If $f \in SX(h_1, [a,b], \mathbb{R}^+_{\mathcal{I}})$, $g \in SX(h_2, [a,b], \mathbb{R}^+_{\mathcal{I}})$, then*

$$\begin{aligned}
&\frac{\Gamma(\alpha)}{(b-a)^\alpha}\left[\mathcal{J}^\alpha_{a+} f(b)g(b) + \mathcal{J}^\alpha_{b-} f(a)g(a)\right] \\
\supseteq&\mathcal{M}(a,b)\int_0^1 \nu^{\alpha-1}\left[h_1(\nu)h_2(\nu) + h_1(1-\nu)h_2(1-\nu)\right]d\nu \\
&+\mathcal{N}(a,b)\int_0^1 \nu^{\alpha-1}\left[h_1(\nu)h_2(1-\nu) + h_1(1-\nu)h_2(\nu)\right]d\nu
\end{aligned} \tag{7}$$

where

$$\mathcal{M}(a,b) = f(a)g(a) + f(b)g(b),\ \mathcal{N}(a,b) = f(a)g(b) + f(b)g(a).$$

Proof. By hypothesis, one has

$$f(\nu a + (1-\nu)b) \supseteq h_1(\nu)f(a) + h_1(1-\nu)f(b),$$

$$g(\nu a + (1-\nu)b) \supseteq h_2(\nu)g(a) + h_2(1-\nu)g(b).$$

Since $f, g \in \mathbb{R}^+_{\mathcal{I}}$, we obtain

$$\begin{aligned}
&f(\nu a + (1-\nu)b)g(\nu a + (1-\nu)b) \\
\supseteq&h_1(\nu)h_2(\nu)f(a)g(a) + h_1(1-\nu)h_2(1-\nu)f(b)g(b) \\
&+ h_1(\nu)h_2(1-\nu)f(a)g(b) + h_1(1-\nu)h_2(\nu)f(b)g(a).
\end{aligned} \tag{8}$$

In the same way as above, we have

$$\begin{aligned}
&f((1-\nu)a + \nu b)g((1-\nu)a + \nu b) \\
\supseteq&h_1(1-\nu)h_2(1-\nu)f(a)g(a) + h_1(\nu)h_2(\nu)f(b)g(b) \\
&+ h_1(1-\nu)h_2(\nu)f(a)g(b) + h_1(\nu)h_2(1-\nu)f(b)g(a).
\end{aligned} \tag{9}$$

By adding (8) and (9), we obtain

$$\begin{aligned}
&f(\nu a + (1-\nu)b)g(\nu a + (1-\nu)b) + f((1-\nu)a + \nu b)g((1-\nu)a + \nu b) \\
\supseteq&\left[h_1(\nu)f(a) + h_1(1-\nu)f(b)\right]\left[h_2(\nu)g(a) + h_2(1-\nu)g(b)\right] \\
&+ \left[h_1(1-\nu)f(a) + h_1(\nu)f(b)\right]\left[h_2(1-\nu)g(a) + h_2(\nu)g(b)\right] \\
=&\mathcal{M}(a,b)\left[h_1(\nu)h_2(\nu) + h_1(1-\nu)h_2(1-\nu)\right] \\
&+\mathcal{N}(a,b)\left[h_1(1-\nu)h_2(\nu) + h_1(\nu)h_2(1-\nu)\right].
\end{aligned} \tag{10}$$

Multiplying both sides (10) by $\nu^{\alpha-1}$ and integrating on $[0,1]$, we have

$$\begin{aligned}
&\int_0^1 \nu^{\alpha-1} f(\nu a+(1-\nu)b) g(\nu a+(1-\nu)b) d\nu \\
&+\int_0^1 \nu^{\alpha-1} f((1-\nu)a+\nu b) g((1-\nu)a+\nu b) d\nu \\
\supseteq&\mathcal{M}(a,b)\int_0^1 \nu^{\alpha-1}\left[h_1(\nu)h_2(\nu)+h_1(1-\nu)h_2(1-\nu)\right] d\nu \\
&+\mathcal{N}(a,b)\int_0^1 \nu^{\alpha-1}\left[h_1(1-\nu)h_2(\nu)+h_1(\nu)h_2(1-\nu)\right] d\nu.
\end{aligned} \tag{11}$$

By Definition 2, we obtain

$$\int_0^1 \nu^{\alpha-1} f(\nu a+(1-\nu)b) g(\nu a+(1-\nu)b) d\nu = \frac{\Gamma(\alpha)}{(b-a)^{\alpha}} \mathcal{J}_{a^+}^{\alpha} f(b) g(b), \tag{12}$$

$$\int_0^1 \nu^{\alpha-1} f((1-\nu)a+\nu b) g((1-\nu)a+\nu b) d\nu = \frac{\Gamma(\alpha)}{(b-a)^{\alpha}} \mathcal{J}_{b^-}^{\alpha} f(a) g(a). \tag{13}$$

By substituting the equalities (12) and (13) in (11), then we have inequality (7). □

Remark 3. *If $h(\nu)=\nu$, then we get Theorem 3.5 of [17]. If $\alpha=1$, then we get Theorem 4.5 of [14]. If $\underline{f}=\overline{f}$ and $\alpha=1$, then we get Theorem 7 of [4].*

Theorem 4. *Let $f,g:[a,b]\to\mathbb{R}_{\mathcal{I}}^{+}$ be two interval-valued functions such that $f=[\underline{f},\overline{f}]$, $g=[\underline{g},\overline{g}]$ and $fg\in\mathcal{IR}_{([a,b])}$, $h_1,h_2:[0,1]\to\mathbb{R}^{+}$ be non-negative functions and $h_1(\frac{1}{2})h_2(\frac{1}{2})\neq 0$. If $f\in SX(h_1,[a,b],\mathbb{R}_{\mathcal{I}}^{+})$, $g\in SX(h_2,[a,b],\mathbb{R}_{\mathcal{I}}^{+})$, then*

$$\begin{aligned}
&\frac{1}{\alpha h_1(\frac{1}{2})h_2(\frac{1}{2})} f\Big(\frac{a+b}{2}\Big) g\Big(\frac{a+b}{2}\Big) \\
\supseteq&\frac{\Gamma(\alpha)}{(b-a)^{\alpha}}\left[\mathcal{J}_{a^+}^{\alpha} f(b)g(b)+\mathcal{J}_{b^-}^{\alpha} f(a)g(a)\right] \\
&+\mathcal{M}(a,b)\int_0^1 \left[\nu^{\alpha-1}+(1-\nu)^{\alpha-1}\right] h_1(\nu)h_2(1-\nu) d\nu \\
&+\mathcal{N}(a,b)\int_0^1 \left[\nu^{\alpha-1}+(1-\nu)^{\alpha-1}\right] h_1(1-\nu)h_2(1-\nu) d\nu.
\end{aligned} \tag{14}$$

Proof. Since $f\in SX(h_1,[a,b],\mathbb{R}_{\mathcal{I}}^{+})$, $g\in SX(h_2,[a,b],\mathbb{R}_{\mathcal{I}}^{+})$, we get

$$\begin{aligned}
&f\Big(\frac{a+b}{2}\Big) g\Big(\frac{a+b}{2}\Big) \\
=&f\Big(\frac{\nu a+(1-\nu)b}{2}+\frac{(1-\nu)a+\nu b}{2}\Big) g\Big(\frac{\nu a+(1-\nu)b}{2}+\frac{(1-\nu)a+\nu b}{2}\Big) \\
\supseteq& h_1\big(\frac{1}{2}\big)h_2\big(\frac{1}{2}\big)\Big[f(\nu a+(1-\nu)b)+f((1-\nu)a+\nu b)\Big]\Big[q(\nu a+(1-\nu)b)+q((1-\nu)a+\nu b)\Big] \\
\supseteq& h_1\big(\frac{1}{2}\big)h_2\big(\frac{1}{2}\big)\Big[f(\nu a+(1-\nu)b)g(\nu a+(1-\nu)b)+f((1-\nu)a+\nu b)g((1-\nu)a+\nu b)\Big] \\
&+h_1\big(\frac{1}{2}\big)h_2\big(\frac{1}{2}\big)\Big[\Big(h_1(\nu)h_2(1-\nu)+h_1(1-\nu)h_2(\nu)\Big)\mathcal{M}(a,b) \\
&+\Big(h_1(\nu)h_2(\nu)+h_1(1-\nu)h_2(1-\nu)\Big)\mathcal{N}(a,b)\Big].
\end{aligned} \tag{15}$$

Multiplying both sides (15) by $\nu^{\alpha-1}$ and integrating on $[0,1]$, we have inequality (14). □

Remark 4. *If $h(v) = v$, then we get Theorem 3.6 of [17]. If $\alpha = 1$, then we get Theorem 4.6 of [14]. If $\underline{f} = \overline{f}$ and $\alpha = 1$, then we get Theorem 8 of [4].*

4. Fractional Hermite–Hadamard Type Inequalities of Interval Harmonically h-Convex Functions

In this section, we will use the above results to get some Hermite–Hadamard type inequalities for interval harmonically h-convex functions via interval Riemann–Liouville type integrals and some special cases are also discussed.

Theorem 5. *Let $f : [a,b] \to \mathbb{R}_{\mathcal{I}}^{+}$ be an interval-valued function such that $f = [\underline{f}, \overline{f}]$ and $f \in \mathcal{IR}_{([a,b])}$, $h : [0,1] \to \mathbb{R}^{+}$ be a non-negative function and $h\left(\frac{1}{2}\right) \neq 0$. If $f \in SHX(h, [a,b], \mathbb{R}_{\mathcal{I}}^{+})$, then*

$$\begin{aligned} &\frac{1}{\alpha h\left(\frac{1}{2}\right)} f\left(\frac{2ab}{a+b}\right) \\ &\supseteq \Gamma(\alpha)\left(\frac{ab}{b-a}\right)^{\alpha}\left[\mathcal{J}^{\alpha}_{\left(\frac{1}{a}\right)-}(f\circ\xi)\left(\frac{1}{b}\right) + \mathcal{J}^{\alpha}_{\left(\frac{1}{b}\right)+}(f\circ\xi)\left(\frac{1}{a}\right)\right] \\ &\supseteq [f(a)+f(b)]\int_0^1 v^{\alpha-1}[h(v)+h(1-v)]\,dv \end{aligned} \tag{16}$$

where $\xi(x) = \frac{1}{x}$.

Proof. Let $\psi(x) = f\left(\frac{1}{x}\right)$. By Theorem 1, we have $\psi \in SX\left(h, [a,b], \mathbb{R}_{\mathcal{I}}^{+}\right)$ and

$$\frac{1}{h\left(\frac{1}{2}\right)}\psi\left(\frac{x+y}{2xy}\right) \supseteq \psi\left(\frac{1}{x}\right) + \psi\left(\frac{1}{y}\right).$$

Let $x = \frac{ab}{va+(1-v)b}$, $y = \frac{ab}{(1-v)a+vb}$, $v \in [0,1]$. Then

$$\frac{1}{h\left(\frac{1}{2}\right)}\psi\left(\frac{a+b}{2ab}\right) \supseteq \psi\left(\frac{va+(1-v)b}{ab}\right) + \psi\left(\frac{(1-v)a+vb}{ab}\right), \tag{17}$$

Multiplying both sides (17) by $v^{\alpha-1}$ and integrating on $[0,1]$, we have

$$\begin{aligned} &\frac{1}{\alpha h\left(\frac{1}{2}\right)}\psi\left(\frac{a+b}{2ab}\right) \\ &\supseteq \left[\int_0^1 v^{\alpha-1}\psi\left(\frac{va+(1-v)b}{ab}\right)dv + \int_0^1 v^{\alpha-1}\psi\left(\frac{(1-v)a+vb}{ab}\right)dv\right] \\ &= \Gamma(\alpha)\left(\frac{ab}{b-a}\right)^{\alpha}\left[\mathcal{J}^{\alpha}_{\left(\frac{1}{a}\right)-}\psi\left(\frac{1}{b}\right) + \mathcal{J}^{\alpha}_{\left(\frac{1}{b}\right)+}\psi\left(\frac{1}{a}\right)\right] \end{aligned} \tag{18}$$

Similarly, we have

$$\psi\left(\frac{va+(1-v)b}{ab}\right) \supseteq h(v)\psi\left(\frac{1}{b}\right) + h(1-v)\psi\left(\frac{1}{a}\right),$$

$$\psi\left(\frac{(1-v)a+vb}{ab}\right) \supseteq h(v)\psi\left(\frac{1}{a}\right) + h(1-v)\psi\left(\frac{1}{b}\right).$$

Then,

$$\psi\Big(\frac{\nu a+(1-\nu)b}{ab}\Big)+\psi\Big(\frac{(1-\nu)a+\nu b}{ab}\Big)\supseteq[h(\nu)+h(1-\nu)]\left[\psi\left(\frac{1}{a}\right)+\psi\left(\frac{1}{b}\right)\right]. \tag{19}$$

Multiplying both sides (19) by $\nu^{\alpha-1}$ and integrating on $[0,1]$, we have

$$\begin{aligned}&\Gamma(\alpha)\left(\frac{ab}{b-a}\right)^{\alpha}\left[\mathcal{J}^{\alpha}_{\left(\frac{1}{a}\right)^-}\psi\left(\frac{1}{b}\right)+\mathcal{J}^{\alpha}_{\left(\frac{1}{b}\right)^+}\psi\left(\frac{1}{a}\right)\right]\\&\supseteq\left[\psi\left(\frac{1}{a}\right)+\psi\left(\frac{1}{b}\right)\right]\int_0^1\nu^{\alpha-1}\left[h(\nu)+h(1-\nu)\right]dt.\end{aligned} \tag{20}$$

By (18) and (20), we have inequality (16). □

Remark 5. *If $h(\nu)=\nu$, then we get Theorem 3.6 of [18]. If $\alpha=1$, then we get Theorem 2 of [15]. If $\underline{f}=\overline{f}$ and $\alpha=1$, then we get Theorem 3.2 of [3]. If $\underline{f}=\overline{f}$ and $h(\nu)=\nu$, then we get Theorem 4 of [7].*

Theorem 6. *Let $f,g:[a,b]\to\mathbb{R}^+_{\mathcal{I}}$ be two interval-valued functions such that $f=[\underline{f},\overline{f}]$, $g=[\underline{g},\overline{g}]$ and $fg\in\mathcal{IR}_{([a,b])}$, $h_1,h_2:[0,1]\to\mathbb{R}^+$ be non-negative functions. If $f\in SHX(h_1,[a,b],\mathbb{R}^+_{\mathcal{I}})$, $g\in SHX(h_2,[a,b],\mathbb{R}^+_{\mathcal{I}})$, then*

$$\begin{aligned}&\Gamma(\alpha)\Big(\frac{ab}{b-a}\Big)^{\alpha}\left[\mathcal{J}^{\alpha}_{\left(\frac{1}{b}\right)^+}(f\circ\xi)\left(\frac{1}{a}\right)(g\circ\xi)\left(\frac{1}{a}\right)+\mathcal{J}^{\alpha}_{\left(\frac{1}{a}\right)^-}(f\circ\xi)\left(\frac{1}{b}\right)(g\circ\xi)\left(\frac{1}{b}\right)\right]\\&\supseteq\mathcal{M}(a,b)\int_0^1\left[\nu^{\alpha-1}+(1-\nu)^{\alpha-1}\right]h_1(\nu)h_2(\nu)d\nu\\&\quad+\mathcal{N}(a,b)\int_0^1\left[\nu^{\alpha-1}+(1-\nu)^{\alpha-1}\right]h_1(\nu)h_2(1-\nu)d\nu\end{aligned} \tag{21}$$

where $\xi(x)=\frac{1}{x}$.

Proof. The proof is completed by combining Theorems 1, 3 and 5. □

Remark 6. *If $h_1(\nu)=h_2(\nu)=\nu$, then*

$$\begin{aligned}&\Gamma(\alpha)\Big(\frac{ab}{b-a}\Big)^{\alpha}\left[\mathcal{J}^{\alpha}_{\left(\frac{1}{b}\right)^+}(f\circ\xi)\left(\frac{1}{a}\right)(g\circ\xi)\left(\frac{1}{a}\right)+\mathcal{J}^{\alpha}_{\left(\frac{1}{a}\right)^-}(f\circ\xi)\left(\frac{1}{b}\right)(g\circ\xi)\left(\frac{1}{b}\right)\right]\\&\supseteq\mathcal{M}(a,b)\int_0^1\nu^2\left[\nu^{\alpha-1}+(1-\nu)^{\alpha-1}\right]d\nu+\mathcal{N}(a,b)\int_0^1\nu(1-\nu)\left[\nu^{\alpha-1}+(1-\nu)^{\alpha-1}\right]d\nu.\end{aligned}$$

If $\alpha=1$, then we get Theorem 4 of [15]. If $\underline{f}=\overline{f}$ and $\alpha=1$, then we get Theorem 3.6 of [3].

Theorem 7. *Let $f,g:[a,b]\to\mathbb{R}^+_{\mathcal{I}}$ be two interval-valued functions such that $f=[\underline{f},\overline{f}]$, $g=[\underline{g},\overline{g}]$ and $fg\in\mathcal{IR}_{([a,b])}$, $h_1,h_2:[0,1]\to\mathbb{R}^+$ be non-negative functions and $h_1(\frac{1}{2})h_2(\frac{1}{2})\neq0$. If $f\in SHX(h_1,[a,b],\mathbb{R}^+_{\mathcal{I}})$, $g\in SHX(h_2,[a,b],\mathbb{R}^+_{\mathcal{I}})$, then*

$$\begin{aligned}&\frac{1}{\alpha h_1(\frac{1}{2})h_2\left(\frac{1}{2}\right)}f\Big(\frac{2ab}{a+b}\Big)g\Big(\frac{2ab}{a+b}\Big)\\&\supseteq\Gamma(\alpha)\Big(\frac{ab}{b-a}\Big)^{\alpha}\left[\mathcal{J}^{\alpha}_{\left(\frac{1}{b}\right)^+}(f\circ\xi)\left(\frac{1}{a}\right)(g\circ\xi)\left(\frac{1}{a}\right)+\mathcal{J}^{\alpha}_{\left(\frac{1}{a}\right)^-}(f\circ\xi)\left(\frac{1}{b}\right)(g\circ\xi)\left(\frac{1}{b}\right)\right]\\&\quad+\int_0^1\left[\nu^{\alpha-1}+(1-\nu)^{\alpha-1}\right]\left[h_1(\nu)h_2(1-\nu)\mathcal{M}(a,b)+h_1(\nu)h_2(\nu)\mathcal{N}(a,b)\right]d\nu\end{aligned}$$

where $\xi(x) = \frac{1}{x}$.

Proof. The proof is completed by combining Theorem 1, 4 and 5. □

Remark 7. *If* $h_1(\nu) = h_2(\nu) = \nu$, *then*

$$\begin{aligned}
&4f\Big(\frac{2ab}{a+b}\Big)g\Big(\frac{2ab}{a+b}\Big)\\
&\supseteq\Gamma(\alpha+1)\Big(\frac{ab}{b-a}\Big)^{\alpha}\left[\mathcal{J}^{\alpha}_{(\frac{1}{b})^{+}}(f\circ\xi)\left(\frac{1}{a}\right)(g\circ\xi)\left(\frac{1}{a}\right)+\mathcal{J}^{\alpha}_{(\frac{1}{a})^{-}}(f\circ\xi)\left(\frac{1}{b}\right)(g\circ\xi)\left(\frac{1}{b}\right)\right]\\
&\quad+\int_0^1\left[\nu^{\alpha-1}+(1-\nu)^{\alpha-1}\right]\left[\nu(1-\nu)\mathcal{M}(a,b)+\nu^2\mathcal{N}(a,b)\right]d\nu.
\end{aligned}$$

If $\alpha = 1$, *then we get Theorem 4 of [15].*

5. Conclusions

This paper proved the relationship between interval h-convex functions and interval harmonically h-convex functions. Further, we obtained some Hermite–Hadamard type inequalities for IVFs via interval Riemann–Liouville type fractional integrals. The results obtained in this article are the generalizations and refinements of the earlier works. Moreover, these results may be extended for other kinds of convex functions including interval (h_1, h_2)-convex functions and interval Log-h-convex functions and used as a method to establish the Hermite–Hadamard type inequalities for other types of interval harmonically convex functions. As a future research direction, we intend to investigate Hermite–Hadamard type inequalities for IVFs on time scales and some applications in interval optimization, probability, among others.

Author Contributions: Conceptualization, F.S., G.Y., D.Z., and W.L.; methodology, F.S., G.Y., D.Z., and W.L.; writing—original draft preparation, F.S.; writing—review and editing, F.S.; supervision, D.Z.; and W.L.; project administration, G.Y.; and funding acquisition, G.Y.; All authors have read and agreed to the published version of the manuscript.

Funding: This work was supported in part by the Fundamental Research Funds for Central Universities (2019B44914), Special Soft Science Research Projects of Technological Innovation in Hubei Province (2019ADC146), Key Projects of Educational Commission of Hubei Province of China (D20192501), the Natural Science Foundation of Jiangsu Province (BK20180500) and the National Key Research and Development Program of China (2018YFC1508100).

Acknowledgments: The authors are very grateful to three anonymous referees, for several valuable and helpful comments, suggestions and questions, which helped them to improve the paper into present form.

Conflicts of Interest: The authors declare no conflict of interest.

References

1. Bombardelli, M.; Varošanec, S. Properties of h-convex functions related to the Hermite–Hadamard–Fejér inequalities. *Comput. Math. Appl.* **2009**, *58*, 1869–1877. [CrossRef]
2. Dragomir, S.S. Inequalities of Hermite–Hadamard type for h-convex functions on linear spaces. *Proyecciones* **2015**, *34*, 323–341. [CrossRef]
3. Noor, M.A.; Noor, K.I.; Awan, M.U.; Costache, S. Some integral inequalities for harmonically h-convex functions. *Politehn. Univ. Bucharest Sci. Bull. Ser. A Appl. Math. Phys.* **2015**, *77*, 5–16.
4. Sarikaya, M.Z.; Saglam, A.; Yildirim, H. On some Hadamard-type inequalities for h-convex functions. *J. Math. Inequal* **2008**, *2*, 335–341. [CrossRef]
5. Sarikaya, M.Z.; Set, E.; Yaldiz, H.; Başak, N. Hermite–Hadamard's inequalities for fractional integrals and related fractional inequalities. *Math. Comput. Model.* **2013**, *57*, 2403–2407. [CrossRef]
6. Chen, F. Extensions of the Hermite–Hadamard inequality for harmonically convex functions via fractional integrals. *Appl. Math. Comput.* **2015**, *268*, 121–128. [CrossRef]

7. Işcan, İ.; Wu, S.H. Hermite-Hadamard type inequalities for harmonically convex functions via fractional integrals. *Appl. Math. Comput.* **2014**, *238*, 237–244. [CrossRef]
8. Noor, M.A.; Noor, K.I.; Mihai, M.V.; Awan, M.U. Fractional Hermite–Hadamard inequalities for some classes of differentiable preinvex functions. *Sci. Bull. Politeh. Univ. Buchar. Ser. A Appl. Math. Phys.* **2016**, *78*, 163–174.
9. Snyder, J.M. Interval analysis for computer graphics. *SIGGRAPH Comput. Graph* **1992**, *26*, 121–130. [CrossRef]
10. Rothwell, E.J.; Cloud, M.J. Automatic error analysis using intervals. *IEEE Trans. Ed.* **2011**, *55*, 9–15. [CrossRef]
11. Chalco-Cano, Y.; Lodwick, W.A.; Condori-Equice, W. Ostrowski type inequalities for interval-valued functions using generalized Hukuhara derivative. *Comput. Appl. Math.* **2012**, *31*, 457–472.
12. Flores-Franulič, A.; Chalco-Cano, Y.; Román-Flores, H. An Ostrowski type inequality for interval-valued functions. In Proceedings of the 2013 Joint IFSA World Congress and NAFIPS Annual Meeting (IFSA/NAFIPS), Edmonton, AB, Canada, 24–28 June 2013; Volume 35, pp. 1459–1462.
13. Román-Flores, H.; Chalco-Cano, Y.; Lodwick, W.A. Some integral inequalities for interval-valued functions. *Comput. Appl. Math.* **2016**, *35*, 1–13. [CrossRef]
14. Zhao, D.F.; An, T.Q.; Ye, G.J.; Liu, W. New Jensen and Hermite–Hadamard type inequalities for h-convex interval-valued functions. *J. Inequal. Appl.* **2018**, *2018*, 302. [CrossRef]
15. Zhao, D.F.; An, T.Q.; Ye, G.J.; Torres, D.F.M. On Hermite–Hadamard type inequalities for harmonically h-convex interval-valued functions. *Math. Inequal. Appl.* **2020**, *23*, 95–105.
16. An, Y.R.; Ye, G.J.; Zhao, D.F.; Liu, W. Hermite–Hadamard Type Inequalities for Interval $(h1, h2)$-Convex Functions. *Mathematics* **2019**, *7*, 436. [CrossRef]
17. Büdak, H.; Tunc, T.; Sarikaya, M.Z. Fractional Hermite–Hadamard type inequalities for interval-valued functions. *Proc. Amer. Math. Soc.* **2020**, *148*, 705–718. [CrossRef]
18. Liu, X.L.; Ye, G.J.; Zhao, D.F.; Liu, W. Fractional Hermite–Hadamard type inequalities for interval-valued functions. *J. Inequal. Appl.* **2019**, *2019*, 266. [CrossRef]
19. Markov, S. Calculus for interval functions of a real variable. *Computing* **1979**, *22*, 325–337. [CrossRef]
20. Moore, R.E. *Interval Analysis*; Prentice-Hall: Englewood Cliffs, NJ, USA, 1966.

MDPI
St. Alban-Anlage 66
4052 Basel
Switzerland
Tel. +41 61 683 77 34
Fax +41 61 302 89 18
www.mdpi.com

Mathematics Editorial Office
E-mail: mathematics@mdpi.com
www.mdpi.com/journal/mathematics

www.ingramcontent.com/pod-product-compliance
Lightning Source LLC
LaVergne TN
LVHW070129100526
838202LV00016B/2249
* 9 7 8 3 0 3 9 4 3 2 1 8 9 *